# DIFFERENCE EQUATIONS AND INEQUALITIES

# PURE AND APPLIED MATHEMATICS

**A Program of Monographs, Textbooks, and Lecture Notes**


EXECUTIVE EDITORS

Earl J. Taft
*Rutgers University*
*New Brunswick, New Jersey*

Zuhair Nashed
*University of Delaware*
*Newark, Delaware*

EDITORIAL BOARD

*M. S. Baouendi*
*University of California, San Diego*

*Jane Cronin*
*Rutgers University*

*Jack K. Hale*
*Georgia Institute of Technology*

*S. Kobayashi*
*University of California, Berkeley*

*Marvin Marcus*
*University of California, Santa Barbara*

*W. S. Massey*
*Yale University*

*Anil Nerode*
*Cornell University*

*Donald Passman*
*University of Wisconsin, Madison*

*Fred S. Roberts*
*Rutgers University*

*David L. Russell*
*Virginia Polytechnic Institute and State University*

*Walter Schempp*
*Universität Siegen*

*Mark Teply*
*University of Wisconsin, Milwaukee*

## MONOGRAPHS AND TEXTBOOKS IN PURE AND APPLIED MATHEMATICS

1. *K. Yano*, Integral Formulas in Riemannian Geometry (1970)
2. *S. Kobayashi*, Hyperbolic Manifolds and Holomorphic Mappings (1970)
3. *V. S. Vladimirov*, Equations of Mathematical Physics (A. Jeffrey, ed.; A. Littlewood, trans.) (1970)
4. *B. N. Pshenichnyi*, Necessary Conditions for an Extremum (L. Neustadt, translation ed.; K. Makowski, trans.) (1971)
5. *L. Narici et al.*, Functional Analysis and Valuation Theory (1971)
6. *S. S. Passman*, Infinite Group Rings (1971)
7. *L. Dornhoff*, Group Representation Theory. Part A: Ordinary Representation Theory. Part B: Modular Representation Theory (1971, 1972)
8. *W. Boothby and G. L. Weiss, eds.*, Symmetric Spaces (1972)
9. *Y. Matsushima*, Differentiable Manifolds (E. T. Kobayashi, trans.) (1972)
10. *L. E. Ward, Jr.*, Topology (1972)
11. *A. Babakhanian*, Cohomological Methods in Group Theory (1972)
12. *R. Gilmer*, Multiplicative Ideal Theory (1972)
13. *J. Yeh*, Stochastic Processes and the Wiener Integral (1973)
14. *J. Barros-Neto*, Introduction to the Theory of Distributions (1973)
15. *R. Larsen*, Functional Analysis (1973)
16. *K. Yano and S. Ishihara*, Tangent and Cotangent Bundles (1973)
17. *C. Procesi*, Rings with Polynomial Identities (1973)
18. *R. Hermann*, Geometry, Physics, and Systems (1973)
19. *N. R. Wallach*, Harmonic Analysis on Homogeneous Spaces (1973)
20. *J. Dieudonné*, Introduction to the Theory of Formal Groups (1973)
21. *I. Vaisman*, Cohomology and Differential Forms (1973)
22. *B.-Y. Chen*, Geometry of Submanifolds (1973)
23. *M. Marcus*, Finite Dimensional Multilinear Algebra (in two parts) (1973, 1975)
24. *R. Larsen*, Banach Algebras (1973)
25. *R. O. Kujala and A. L. Vitter, eds.*, Value Distribution Theory: Part A; Part B: Deficit and Bezout Estimates by Wilhelm Stoll (1973)
26. *K. B. Stolarsky*, Algebraic Numbers and Diophantine Approximation (1974)
27. *A. R. Magid*, The Separable Galois Theory of Commutative Rings (1974)
28. *B. R. McDonald*, Finite Rings with Identity (1974)
29. *J. Satake*, Linear Algebra (S. Koh et al., trans.) (1975)
30. *J. S. Golan*, Localization of Noncommutative Rings (1975)
31. *G. Klambauer*, Mathematical Analysis (1975)
32. *M. K. Agoston*, Algebraic Topology (1976)
33. *K. R. Goodearl*, Ring Theory (1976)
34. *L. E. Mansfield*, Linear Algebra with Geometric Applications (1976)
35. *N. J. Pullman*, Matrix Theory and Its Applications (1976)
36. *B. R. McDonald*, Geometric Algebra Over Local Rings (1976)
37. *C. W. Groetsch*, Generalized Inverses of Linear Operators (1977)
38. *J. E. Kuczkowski and J. L. Gersting*, Abstract Algebra (1977)
39. *C. O. Christenson and W. L. Voxman*, Aspects of Topology (1977)
40. *M. Nagata*, Field Theory (1977)
41. *R. L. Long*, Algebraic Number Theory (1977)
42. *W. F. Pfeffer*, Integrals and Measures (1977)
43. *R. L. Wheeden and A. Zygmund*, Measure and Integral (1977)
44. *J. H. Curtiss*, Introduction to Functions of a Complex Variable (1978)
45. *K. Hrbacek and T. Jech*, Introduction to Set Theory (1978)
46. *W. S. Massey*, Homology and Cohomology Theory (1978)
47. *M. Marcus*, Introduction to Modern Algebra (1978)
48. *E. C. Young*, Vector and Tensor Analysis (1978)
49. *S. B. Nadler, Jr.*, Hyperspaces of Sets (1978)
50. *S. K. Segal*, Topics in Group Kings (1978)
51. *A. C. M. van Rooij*, Non-Archimedean Functional Analysis (1978)
52. *L. Corwin and R. Szczarba*, Calculus in Vector Spaces (1979)
53. *C. Sadosky*, Interpolation of Operators and Singular Integrals (1979)
54. *J. Cronin*, Differential Equations (1980)
55. *C. W. Groetsch*, Elements of Applicable Functional Analysis (1980)

56. *I. Vaisman*, Foundations of Three-Dimensional Euclidean Geometry (1980)
57. *H. I. Freedan*, Deterministic Mathematical Models in Population Ecology (1980)
58. *S. B. Chae*, Lebesgue Integration (1980)
59. *C. S. Rees et al.*, Theory and Applications of Fourier Analysis (1981)
60. *L. Nachbin*, Introduction to Functional Analysis (R. M. Aron, trans.) (1981)
61. *G. Orzech and M. Orzech*, Plane Algebraic Curves (1981)
62. *R. Johnsonbaugh and W. E. Pfaffenberger*, Foundations of Mathematical Analysis (1981)
63. *W. L. Voxman and R. H. Goetschel*, Advanced Calculus (1981)
64. *L. J. Corwin and R. H. Szczarba*, Multivariable Calculus (1982)
65. *V. I. Istrățescu*, Introduction to Linear Operator Theory (1981)
66. *R. D. Järvinen*, Finite and Infinite Dimensional Linear Spaces (1981)
67. *J. K. Beem and P. E. Ehrlich*, Global Lorentzian Geometry (1981)
68. *D. L. Armacost*, The Structure of Locally Compact Abelian Groups (1981)
69. *J. W. Brewer and M. K. Smith, eds.*, Emmy Noether: A Tribute (1981)
70. *K. H. Kim*, Boolean Matrix Theory and Applications (1982)
71. *T. W. Wieting*, The Mathematical Theory of Chromatic Plane Ornaments (1982)
72. *D. B. Gauld*, Differential Topology (1982)
73. *R. L. Faber*, Foundations of Euclidean and Non-Euclidean Geometry (1983)
74. *M. Carmeli*, Statistical Theory and Random Matrices (1983)
75. *J. H. Carruth et al.*, The Theory of Topological Semigroups (1983)
76. *R. L. Faber*, Differential Geometry and Relativity Theory (1983)
77. *S. Barnett*, Polynomials and Linear Control Systems (1983)
78. *G. Karpilovsky*, Commutative Group Algebras (1983)
79. *F. Van Oystaeyen and A. Verschoren*, Relative Invariants of Rings (1983)
80. *I. Vaisman*, A First Course in Differential Geometry (1984)
81. *G. W. Swan*, Applications of Optimal Control Theory in Biomedicine (1984)
82. *T. Petrie and J. D. Randall*, Transformation Groups on Manifolds (1984)
83. *K. Goebel and S. Reich*, Uniform Convexity, Hyperbolic Geometry, and Nonexpansive Mappings (1984)
84. *T. Albu and C. Năstăsescu*, Relative Finiteness in Module Theory (1984)
85. *K. Hrbacek and T. Jech*, Introduction to Set Theory: Second Edition (1984)
86. *F. Van Oystaeyen and A. Verschoren*, Relative Invariants of Rings (1984)
87. *B. R. McDonald*, Linear Algebra Over Commutative Rings (1984)
88. *M. Namba*, Geometry of Projective Algebraic Curves (1984)
89. *G. F. Webb*, Theory of Nonlinear Age-Dependent Population Dynamics (1985)
90. *M. R. Bremner et al.*, Tables of Dominant Weight Multiplicities for Representations of Simple Lie Algebras (1985)
91. *A. E. Fekete*, Real Linear Algebra (1985)
92. *S. B. Chae*, Holomorphy and Calculus in Normed Spaces (1985)
93. *A. J. Jerri*, Introduction to Integral Equations with Applications (1985)
94. *G. Karpilovsky*, Projective Representations of Finite Groups (1985)
95. *L. Narici and E. Beckenstein*, Topological Vector Spaces (1985)
96. *J. Weeks*, The Shape of Space (1985)
97. *P. R. Gribik and K. O. Kortanek*, Extremal Methods of Operations Research (1985)
98. *J.-A. Chao and W. A. Woyczynski, eds.*, Probability Theory and Harmonic Analysis (1986)
99. *G. D. Crown et al.*, Abstract Algebra (1986)
100. *J. H. Carruth et al.*, The Theory of Topological Semigroups, Volume 2 (1986)
101. *R. S. Doran and V. A. Belfi*, Characterizations of C*-Algebras (1986)
102. *M. W. Jeter*, Mathematical Programming (1986)
103. *M. Altman*, A Unified Theory of Nonlinear Operator and Evolution Equations with Applications (1986)
104. *A. Verschoren*, Relative Invariants of Sheaves (1987)
105. *R. A. Usmani*, Applied Linear Algebra (1987)
106. *P. Blass and J. Lang*, Zariski Surfaces and Differential Equations in Characteristic $p > 0$ (1987)
107. *J. A. Reneke et al.*, Structured Hereditary Systems (1987)
108. *H. Busemann and B. B. Phadke*, Spaces with Distinguished Geodesics (1987)
109. *R. Harte*, Invertibility and Singularity for Bounded Linear Operators (1988)
110. *G. S. Ladde et al.*, Oscillation Theory of Differential Equations with Deviating Arguments (1987)
111. *L. Dudkin et al.*, Iterative Aggregation Theory (1987)
112. *T. Okubo*, Differential Geometry (1987)

113. *D. L. Stancl and M. L. Stancl*, Real Analysis with Point-Set Topology (1987)
114. *T. C. Gard*, Introduction to Stochastic Differential Equations (1988)
115. *S. S. Abhyankar*, Enumerative Combinatorics of Young Tableaux (1988)
116. *H. Strade and R. Farnsteiner*, Modular Lie Algebras and Their Representations (1988)
117. *J. A. Huckaba*, Commutative Rings with Zero Divisors (1988)
118. *W. D. Wallis*, Combinatorial Designs (1988)
119. *W. Więsław*, Topological Fields (1988)
120. *G. Karpilovsky*, Field Theory (1988)
121. *S. Caenepeel and F. Van Oystaeyen*, Brauer Groups and the Cohomology of Graded Rings (1989)
122. *W. Kozlowski*, Modular Function Spaces (1988)
123. *E. Lowen-Colebunders*, Function Classes of Cauchy Continuous Maps (1989)
124. *M. Pavel*, Fundamentals of Pattern Recognition (1989)
125. *V. Lakshmikantham et al.*, Stability Analysis of Nonlinear Systems (1989)
126. *R. Sivaramakrishnan*, The Classical Theory of Arithmetic Functions (1989)
127. *N. A. Watson*, Parabolic Equations on an Infinite Strip (1989)
128. *K. J. Hastings*, Introduction to the Mathematics of Operations Research (1989)
129. *B. Fine*, Algebraic Theory of the Bianchi Groups (1989)
130. *D. N. Dikranjan et al.*, Topological Groups (1989)
131. *J. C. Morgan II*, Point Set Theory (1990)
132. *P. Biler and A. Witkowski*, Problems in Mathematical Analysis (1990)
133. *H. J. Sussmann*, Nonlinear Controllability and Optimal Control (1990)
134. *J.-P. Florens et al.*, Elements of Bayesian Statistics (1990)
135. *N. Shell*, Topological Fields and Near Valuations (1990)
136. *B. F. Doolin and C. F. Martin*, Introduction to Differential Geometry for Engineers (1990)
137. *S. S. Holland, Jr.*, Applied Analysis by the Hilbert Space Method (1990)
138. *J. Okninski*, Semigroup Algebras (1990)
139. *K. Zhu*, Operator Theory in Function Spaces (1990)
140. *G. B. Price*, An Introduction to Multicomplex Spaces and Functions (1991)
141. *R. B. Darst*, Introduction to Linear Programming (1991)
142. *P. L. Sachdev*, Nonlinear Ordinary Differential Equations and Their Applications (1991)
143. *T. Husain*, Orthogonal Schauder Bases (1991)
144. *J. Foran*, Fundamentals of Real Analysis (1991)
145. *W. C. Brown*, Matrices and Vector Spaces (1991)
146. *M. M. Rao and Z. D. Ren*, Theory of Orlicz Spaces (1991)
147. *J. S. Golan and T. Head*, Modules and the Structures of Rings (1991)
148. *C. Small*, Arithmetic of Finite Fields (1991)
149. *K. Yang*, Complex Algebraic Geometry (1991)
150. *D. G. Hoffman et al.*, Coding Theory (1991)
151. *M. O. González*, Classical Complex Analysis (1992)
152. *M. O. González*, Complex Analysis (1992)
153. *L. W. Baggett*, Functional Analysis (1992)
154. *M. Sniedovich*, Dynamic Programming (1992)
155. *R. P. Agarwal*, Difference Equations and Inequalities (1992)
156. *C. Brezinski*, Biorthogonality and Its Applications to Numerical Analysis (1992)
157. *C. Swartz*, An Introduction to Functional Analysis (1992)
158. *S. B. Nadler, Jr.*, Continuum Theory (1992)
159. *M. A. Al-Gwaiz*, Theory of Distributions (1992)
160. *E. Perry*, Geometry: Axiomatic Developments with Problem Solving (1992)
161. *E. Castillo and M. R. Ruiz-Cobo*, Functional Equations and Modelling in Science and Engineering (1992)
162. *A. J. Jerri*, Integral and Discrete Transforms with Applications and Error Analysis (1992)
163. *A. Charlier et al.*, Tensors and the Clifford Algebra (1992)
164. *P. Biler and T. Nadzieja*, Problems and Examples in Differential Equations (1992)
165. *E. Hansen*, Global Optimization Using Interval Analysis (1992)
166. *S. Guerre-Delabrière*, Classical Sequences in Banach Spaces (1992)
167. *Y. C. Wong*, Introductory Theory of Topological Vector Spaces (1992)
168. *S. H. Kulkarni and B. V. Limaye*, Real Function Algebras (1992)
169. *W. C. Brown*, Matrices Over Commutative Rings (1993)
170. *J. Loustau and M. Dillon*, Linear Geometry with Computer Graphics (1993)
171. *W. V. Petryshyn*, Approximation-Solvability of Nonlinear Functional and Differential Equations (1993)

172. *E. C. Young*, Vector and Tensor Analysis: Second Edition (1993)
173. *T. A. Bick*, Elementary Boundary Value Problems (1993)
174. *M. Pavel*, Fundamentals of Pattern Recognition: Second Edition (1993)
175. *S. A. Albeverio et al.*, Noncommutative Distributions (1993)
176. *W. Fulks*, Complex Variables (1993)
177. *M. M. Rao*, Conditional Measures and Applications (1993)
178. *A. Janicki and A. Weron*, Simulation and Chaotic Behavior of $\alpha$-Stable Stochastic Processes (1994)
179. *P. Neittaanmäki and D. Tiba*, Optimal Control of Nonlinear Parabolic Systems (1994)
180. *J. Cronin*, Differential Equations: Introduction and Qualitative Theory, Second Edition (1994)
181. *S. Heikkilä and V. Lakshmikantham*, Monotone Iterative Techniques for Discontinuous Nonlinear Differential Equations (1994)
182. *X. Mao*, Exponential Stability of Stochastic Differential Equations (1994)
183. *B. S. Thomson*, Symmetric Properties of Real Functions (1994)
184. *J. E. Rubio*, Optimization and Nonstandard Analysis (1994)
185. *J. L. Bueso et al.*, Compatibility, Stability, and Sheaves (1995)
186. *A. N. Michel and K. Wang*, Qualitative Theory of Dynamical Systems (1995)
187. *M. R. Darnel*, Theory of Lattice-Ordered Groups (1995)
188. *Z. Naniewicz and P. D. Panagiotopoulos*, Mathematical Theory of Hemivariational Inequalities and Applications (1995)
189. *L. J. Corwin and R. H. Szczarba*, Calculus in Vector Spaces: Second Edition (1995)
190. *L. H. Erbe et al.*, Oscillation Theory for Functional Differential Equations (1995)
191. *S. Agaian et al.*, Binary Polynomial Transforms and Nonlinear Digital Filters (1995)
192. *M. I. Gil'*, Norm Estimations for Operation-Valued Functions and Applications (1995)
193. *P. A. Grillet*, Semigroups: An Introduction to the Structure Theory (1995)
194. *S. Kichenassamy*, Nonlinear Wave Equations (1996)
195. *V. F. Krotov*, Global Methods in Optimal Control Theory (1996)
196. *K. I. Beidar et al.*, Rings with Generalized Identities (1996)
197. *V. I. Arnautov et al.*, Introduction to the Theory of Topological Rings and Modules (1996)
198. *G. Sierksma*, Linear and Integer Programming (1996)
199. *R. Lasser*, Introduction to Fourier Series (1996)
200. *V. Sima*, Algorithms for Linear-Quadratic Optimization (1996)
201. *D. Redmond*, Number Theory (1996)
202. *J. K. Beem et al.*, Global Lorentzian Geometry: Second Edition (1996)
203. *M. Fontana et al.*, Prüfer Domains (1997)
204. *H. Tanabe*, Functional Analytic Methods for Partial Differential Equations (1997)
205. *C. Q. Zhang*, Integer Flows and Cycle Covers of Graphs (1997)
206. *E. Spiegel and C. J. O'Donnell*, Incidence Algebras (1997)
207. *B. Jakubczyk and W. Respondek*, Geometry of Feedback and Optimal Control (1998)
208. *T. W. Haynes et al.*, Fundamentals of Domination in Graphs (1998)
209. *T. W. Haynes et al.*, Domination in Graphs: Advanced Topics (1998)
210. *L. A. D'Alotto et al.*, A Unified Signal Algebra Approach to Two-Dimensional Parallel Digital Signal Processing (1998)
211. *F. Halter-Koch*, Ideal Systems (1998)
212. *N. K. Govil et al.*, Approximation Theory (1998)
213. *R. Cross*, Multivalued Linear Operators (1998)
214. *A. A. Martynyuk*, Stability by Liapunov's Matrix Function Method with Applications (1998)
215. *A. Favini and A. Yagi*, Degenerate Differential Equations in Banach Spaces (1999)
216. *A. Illanes and S. Nadler, Jr.*, Hyperspaces: Fundamentals and Recent Advances (1999)
217. *G. Kato and D. Struppa*, Fundamentals of Algebraic Microlocal Analysis (1999)
218. *G. X.-Z. Yuan*, KKM Theory and Applications in Nonlinear Analysis (1999)
219. *D. Motreanu and N. H. Pavel*, Tangency, Flow Invariance for Differential Equations, and Optimization Problems (1999)
220. *K. Hrbacek and T. Jech*, Introduction to Set Theory, Third Edition (1999)
221. *G. E. Kolosov*, Optimal Design of Control Systems (1999)
222. *N. L. Johnson*, Subplane Covered Nets (2000)
223. *B. Fine and G. Rosenberger*, Algebraic Generalizations of Discrete Groups (1999)
224. *M. Väth*, Volterra and Integral Equations of Vector Functions (2000)
225. *S. S. Miller and P. T. Mocanu*, Differential Subordinations (2000)

226. *R. Li et al.*, Generalized Difference Methods for Differential Equations: Numerical Analysis of Finite Volume Methods (2000)
227. *H. Li and F. Van Oystaeyen*, A Primer of Algebraic Geometry (2000)
228. *R. P. Agarwal*, Difference Equations and Inequalities: Theory, Methods, and Applications, Second Edition (2000)

Additional Volumes in Preparation

# DIFFERENCE EQUATIONS AND INEQUALITIES

## Theory, Methods, and Applications

### Second Edition, Revised and Expanded

Ravi P. Agarwal
*National University of Singapore*
*Kent Ridge, Singapore*

CRC Press is an imprint of the
Taylor & Francis Group, an **informa** business

CRC Press
Taylor & Francis Group
6000 Broken Sound Parkway NW, Suite 300
Boca Raton, FL 33487-2742

First issued in paperback 2019

© 2000 by Taylor & Francis Group, LLC
CRC Press is an imprint of Taylor & Francis Group, an Informa business

No claim to original U.S. Government works

ISBN-13: 978-0-8247-9007-3 (hbk)
ISBN-13: 978-0-367-39894-1 (pbk)

This book contains information obtained from authentic and highly regarded sources. Reasonable efforts have been made to publish reliable data and information, but the author and publisher cannot assume responsibility for the validity of all materials or the consequences of their use. The authors and publishers have attempted to trace the copyright holders of all material reproduced in this publication and apologize to copyright holders if permission to publish in this form has not been obtained. If any copyright material has not been acknowledged please write and let us know so we may rectify in any future reprint.

Except as permitted under U.S. Copyright Law, no part of this book may be reprinted, reproduced, transmitted, or utilized in any form by any electronic, mechanical, or other means, now known or hereafter invented, including photocopying, microfilming, and recording, or in any information storage or retrieval system, without written permission from the publishers.

For permission to photocopy or use material electronically from this work, please access www.copyright.com (http://www.copyright.com/) or contact the Copyright Clearance Center, Inc. (CCC), 222 Rosewood Drive, Danvers, MA 01923, 978-750-8400. CCC is a not-for-profit organization that provides licenses and registration for a variety of users. For organizations that have been granted a photocopy license by the CCC, a separate system of payment has been arranged.

**Trademark Notice:** Product or corporate names may be trademarks or registered trademarks, and are used only for identification and explanation without intent to infringe.

**Visit the Taylor & Francis Web site at**
**http://www.taylorandfrancis.com**

**and the CRC Press Web site at**
**http://www.crcpress.com**

To
Sadhna

# Preface to the Second Edition

Since its publication in 1992, *Difference Equations and Inequalities* has been received very positively by the international scientific community. Its success prompted a request from the publisher for an updated edition.

In this edition, besides a new chapter on the *qualitative properties of solutions of neutral difference equations*, new material has been added in all the existing chapters of the first edition. This includes a variety of interesting examples from real world applications, new theorems, over 200 additional problems and 400 further references.

The theory of difference equations has grown at an accelerated pace in the past decade. It now occupies a central position in applicable analysis and will no doubt continue to play an important role in mathematics as a whole. It is hoped that this new edition will be a timely and welcomed reference.

It is a pleasure to thank all those who have helped in the preparation of this edition. I would especially like to thank Ms. Maria Allegra of Marcel Dekker, Inc., whose help was instrumental in the successful completion of this project.

Ravi P. Agarwal

# Preface to the First Edition

Examples of discrete phenomena in nature abound and yet somehow the continuous version has commandeered all our attention – perhaps owing to that special mechanism in human nature that permits us to notice only what we have been conditioned to. Although difference equations manifest themselves as mathematical models describing real life situations in probability theory, queuing problems, statistical problems, stochastic time series, combinatorial analysis, number theory, geometry, electrical networks, quanta in radiation, genetics in biology, economics, psychology, sociology, etc., unfortunately, these are only considered as the discrete analogs of differential equations. It is an indisputable fact that difference equations appeared much earlier than differential equations and were instrumental in paving the way for the development of the latter. It is only recently that difference equations have started receiving the attention they deserve. Perhaps this is largely due to the advent of computers, where differential equations are solved by using their approximate difference equation formulations. This self-contained monograph is an in-depth and up-to-date coverage of more than 400 recent publications and may be of interest to practically every user of mathematics in almost every discipline.

It is impossible to acknowledge individually colleagues and friends to whom I am indebted for assistance, inspiration and criticism in writing this monograph. I must, however, express my appreciation and thanks to Ms. Rubiah Tukimin for her excellent and careful typing of the manuscript.

Ravi P. Agarwal

# Contents

## Chapter 3 Miscellaneous Difference Equations



## Chapter 4 Difference Inequalities



## Chapter 5 Qualitative Properties of Solutions of Difference Systems

## Chapter 7 Qualitative Properties of Solutions of Neutral Difference Equations

## Chapter 8 Boundary Value Problems for Linear Systems



## Chapter 9 Boundary Value Problems for Nonlinear Systems

# DIFFERENCE EQUATIONS AND INEQUALITIES

# Chapter 1
# Preliminaries

We begin this chapter with some notations which are used throughout this monograph. This is followed by some classifications namely: linear and nonlinear higher order difference equations, linear and nonlinear first order difference systems, and initial and boundary value problems. We also include several examples of initial and boundary value problems, as well as real world phenomena from diverse fields which are sufficient to convey the importance of the serious qualitative as well as quantitative study of difference equations. The discrete Rolle's theorem, the discrete Mean value theorem, the discrete Taylor's formula, the discrete l'Hospital's rule, the discrete Kneser's theorem are stated and proved by using some simple inequalities.

## 1.1. Notations

Throughout, we shall use some of the following notations: $\mathbb{N} = \{0, 1, \cdots\}$ the set of natural numbers including zero, $\mathbb{N}(a) = \{a, a+1, \cdots\}$ where $a \in \mathbb{N}$, $\mathbb{N}(a, b-1) = \{a, a+1, \cdots, b-1\}$ where $a < b-1 < \infty$ and $a,\ b \in \mathbb{N}$. Any one of these three sets will be denoted by $\overline{\mathbb{N}}$. The scalar valued functions on $\overline{\mathbb{N}}$ will be denoted by the lower case letters $u(k),\ v(k), \cdots$ whereas the vector valued functions by the bold face letters $\mathbf{u}(k),\ \mathbf{v}(k), \cdots$ and the matrix valued functions by the calligraphic letters $\mathcal{U}(k),\ \mathcal{V}(k), \cdots$. Let $f(k)$ be a function defined on $\overline{\mathbb{N}}$, then for all $k_1,\ k_2 \in \overline{\mathbb{N}}$ and $k_1 > k_2$, $\sum_{\ell=k_1}^{k_2} f(\ell) = 0$ and $\prod_{\ell=k_1}^{k_2} f(\ell) = 1$, i.e. empty sums and products are taken to be $0$ and $1$ respectively. If $k$ and $k+1$ are in $\overline{\mathbb{N}}$, then for this function $f(k)$ we define the *shift operator* $E$ as $Ef(k) = f(k+1)$. In general, for a positive integer $m$ if $k$ and $k+m$ are in $\overline{\mathbb{N}}$, then $E^m f(k) = E\left[E^{m-1} f(k)\right] = f(k+m)$. Similarly, the *forward* and *backward difference operators* $\Delta$ and $\nabla$ are defined as $\Delta f(k) = f(k+1) - f(k)$ and $\nabla f(k) = f(k) - f(k-1)$ respectively. The higher order differences for a positive integer $m$ are defined as $\Delta^m f(k) = \Delta\left[\Delta^{m-1} f(k)\right]$. Let $I$ be the *identity operator*, i.e. $If(k) = f(k)$, then obviously $\Delta = E - 1$ and for a positive integer $m$

we may deduce the relations

$$(1.1.1)\quad \Delta^m f(k) = (E-I)^m f(k) = \sum_{i=0}^{m}(-1)^i\binom{m}{i}E^{m-i}f(k), \quad E^0 = I$$

and

$$(1.1.2)\qquad E^m f(k) = (I+\Delta)^m f(k) = \sum_{i=0}^{m}\binom{m}{i}\Delta^i f(k), \quad \Delta^0 = I.$$

As usual $\mathbb{R}$ denotes the real line and $\mathbb{R}_+$ the set of nonnegative reals. For $t \in \mathbb{R}$ and $m$ a nonnegative integer the *factorial expression* $(t)^{(m)}$ is defined as $(t)^{(m)} = \prod_{i=0}^{m-1}(t-i)$. Thus, in particular for each $k \in \mathbb{N}$, $(k)^{(k)} = k!$.

## 1.2. Difference Equations

A *difference equation* in one independent variable $k \in \overline{\mathbb{N}}$ and one unknown $u(k)$ is a functional equation of the form

$$(1.2.1)\qquad f(k, u(k), u(k+1), \cdots, u(k+n)) = 0,$$

where $f$ is a given function of $k$ and the values of $u(k)$ at $k \in \overline{\mathbb{N}}$. If (1.1.2) are substituted in (1.2.1) the latter takes the form

$$(1.2.2)\qquad g\,(k, u(k), \Delta u(k), \cdots, \Delta^n u(k)) = 0.$$

It was this notation which led (1.2.1) to the name difference equation.

The *order* of (1.2.1) is defined to be the difference between the largest and smallest arguments explicitly involved, e.g. the equation $u(k+3) - 3u(k+2) + 7u(k+1) = 0$ is of order two, whereas $u(k+10) = k(k-1)$ is of order zero.

The difference equation (1.2.1) is *linear* if it is of the form

$$(1.2.3)\qquad \sum_{i=0}^{n} a_i(k)u(k+i) = b(k).$$

If $b(k)$ is different from zero for at least one $k \in \overline{\mathbb{N}}$, then (1.2.3) is a *nonhomogeneous* linear difference equation. Corresponding to (1.2.3) the equation

$$(1.2.4)\qquad \sum_{i=0}^{n} a_i(k)u(k+i) = 0$$

is called a *homogeneous* linear difference equation.

Equation (1.2.1) is said to be *normal* if it is of the form

$$u(k+n) = f(k, u(k), u(k+1), \cdots, u(k+n-1)) \tag{1.2.5}$$

or

$$\Delta^n u(k) = f\left(k, u(k), \Delta u(k), \cdots, \Delta^{n-1}u(k)\right) \tag{1.2.6}$$

or

$$\Delta^n u(k) = f(k, u(k), u(k+1), \cdots, u(k+n-1)). \tag{1.2.7}$$

We shall also consider *system* of difference equations

$$\mathbf{u}(k+1) = \mathbf{f}(k, \mathbf{u}(k)), \quad k \in \overline{\mathbb{N}} \tag{1.2.8}$$

where $\mathbf{u}$ and $\mathbf{f}$ are $1 \times n$ vectors with components $u_i$ and $f_i$, $1 \le i \le n$ respectively.

The $n$th order equation (1.2.5) is equivalent to the system

$$\begin{aligned} u_i(k+1) &= u_{i+1}(k), \quad 1 \le i \le n-1 \\ u_n(k+1) &= f(k, u_1(k), u_2(k), \cdots, u_n(k)), \quad k \in \overline{\mathbb{N}} \end{aligned} \tag{1.2.9}$$

in the sense that $u(k)$ is a solution of (1.2.5) if and only if

$$u_i(k) = u(k+i-1), \quad 1 \le i \le n. \tag{1.2.10}$$

A system of linear difference equations has the form

$$\mathbf{u}(k+1) = \mathcal{A}(k)\mathbf{u}(k) + \mathbf{b}(k), \quad k \in \overline{\mathbb{N}} \tag{1.2.11}$$

where $\mathcal{A}(k)$ is a given nonsingular $n \times n$ matrix with elements $a_{ij}(k)$, $1 \le i, j \le n$, $\mathbf{b}(k)$ is a given $n \times 1$ vector with components $b_i(k)$, $1 \le i \le n$, $\mathbf{u}(k)$ is an unknown $n \times 1$ vector with components $u_i(k)$, $1 \le i \le n$.

If $\mathbf{b}(k)$ is different from zero for at least one $k \in \overline{\mathbb{N}}$, then the system (1.2.11) is called nonhomogeneous. Corresponding to (1.2.11) the system

$$\mathbf{u}(k+1) = \mathcal{A}(k)\mathbf{u}(k), \quad k \in \overline{\mathbb{N}} \tag{1.2.12}$$

is said to be homogeneous.

If $a_0(k)a_n(k) \neq 0$ for all $k \in \overline{\mathbb{N}}$, then the $n$th order equation (1.2.3) is equivalent to the system (1.2.11) where

$$(1.2.13) \quad \mathcal{A}(k) = \begin{bmatrix} 0 & 1 & 0 & \cdots & 0 \\ 0 & 0 & 1 & \cdots & 0 \\ \cdots & \cdots & \cdots & \cdots & \cdots \\ 0 & 0 & 0 & \cdots & 1 \\ -\frac{a_0(k)}{a_n(k)} & -\frac{a_1(k)}{a_n(k)} & -\frac{a_2(k)}{a_n(k)} & \cdots & -\frac{a_{n-1}(k)}{a_n(k)} \end{bmatrix}$$

and

$$(1.2.14) \qquad \mathbf{b}(k) = \left[0 \ 0 \ \cdots \ 0 \ \frac{b(k)}{a_n(k)}\right]^T.$$

In the above difference equations (systems) the functions are assumed to be defined in all of their arguments. Therefore, not all the systems can be written as higher order difference equations, e.g.

$$\begin{aligned} u_1(k+1) &= u_1(k) + ku_2(k) \\ u_2(k+1) &= (k-1)u_1(k) + u_2(k), \qquad k \in \mathbb{N}. \end{aligned}$$

## 1.3. Initial Value Problems

A function $u(k)$ defined on $\overline{\mathbb{N}}_n$, where

$$\overline{\mathbb{N}}_n = \begin{cases} \mathbb{N}(a, b-1+n) & \text{if } \overline{\mathbb{N}} = \mathbb{N}(a, b-1) \\ \mathbb{N}(a) & \text{if } \overline{\mathbb{N}} = \mathbb{N}(a) \\ \mathbb{N} & \text{if } \overline{\mathbb{N}} = \mathbb{N} \end{cases}$$

is said to be a *solution* of the given $n$th order difference equation on $\overline{\mathbb{N}}$ if the values of $u(k)$ reduce the difference equation to an identity over $\overline{\mathbb{N}}$. Similarly, a function $\mathbf{u}(k)$ defined on $\overline{\mathbb{N}}_1$ is a solution of the given difference system on $\overline{\mathbb{N}}$ provided the values of $\mathbf{u}(k)$ reduce the difference system to an equality over $\overline{\mathbb{N}}$.

The *general solution* of an $n$th order difference equation is a solution $u(k)$ which depends on $n$ arbitrary constants, i.e. $u(k, c_1, \cdots, c_n)$ where $c_i \in \mathbb{R}$, $1 \leq i \leq n$. We observe that these constants $c_i$ can be taken as *periodic functions* $c_i(k)$ of period one, i.e. $c_i(k+1) = c_i(k)$, $k \in \overline{\mathbb{N}}_{n-1}$. Similarly, for the systems the general solution depends on an arbitrary vector.

For a given $n$th order difference equation on $\overline{\mathbb{N}}$ we are usually interested in a *particular solution* on $\overline{\mathbb{N}}_n$, i.e. the one for which the first $n$ consecutive values termed as *initial conditions*

$$u(a+i-1) = u_i, \quad 1 \le i \le n \tag{1.3.1}$$

or

$$\Delta^{i-1}u(a) = u_i, \quad 1 \le i \le n \quad (a = 0 \text{ if } \overline{\mathbb{N}} = \mathbb{N}) \tag{1.3.2}$$

are prescribed. Each of the difference equations (1.2.1), ..., (1.2.7) together with (1.3.1) or (1.3.2) is called an *initial value problem*. Similarly, the system (1.2.8) together with

$$\mathbf{u}(a) = \mathbf{u}^0 \tag{1.3.3}$$

is called an initial value problem. For the linear systems (1.2.11) and (1.2.12) we shall also consider more general initial condition

$$\mathbf{u}(k_0) = \mathbf{u}^0, \tag{1.3.4}$$

where $k_0 \in \overline{\mathbb{N}}_1$ is fixed.

For $k = a$, equation (1.2.5) becomes

$$u(a+n) = f(a, u(a), u(a+1), \cdots, u(a+n-1)).$$

Using the initial conditions (1.3.1), we find

$$u(a+n) = f(a, u_1, u_2, \cdots, u_n).$$

Hence the value of $u(a+n)$ is uniquely determined in terms of known quantities. Next, setting $k = a+1$ in (1.2.5) and using the values of $u(a+1), \cdots, u(a+n)$ we find that $u(a+1+n)$ is uniquely determined. Now using inductive arguments it is easy to see that the initial value problem (1.2.5), (1.3.1) has a unique solution $u(k)$, $k \in \overline{\mathbb{N}}_n$ and it can be constructed recursively. Because of this reason difference equations are also called *recursive relations*. The *existence* and *uniqueness* of each of the initial value problems (1.2.5), (1.3.2); (1.2.6), (1.3.1) or (1.3.2); (1.2.7), (1.3.1) or (1.3.2); (1.2.8), (1.3.3) follow similarly. For the initial value problem (1.2.11), (1.3.4) the existence and uniqueness of the solution $\mathbf{u}(k)$, $k_0 \le k \in \overline{\mathbb{N}}_1$ is now obvious, whereas for $k_0 \ge k \in \overline{\mathbb{N}}$ we need to write (1.2.11) as

$$\mathbf{u}(k) = \mathcal{A}^{-1}(k)\mathbf{u}(k+1) - \mathcal{A}^{-1}(k)\mathbf{b}(k) \tag{1.3.5}$$

and from this $\mathbf{u}(k_0-1)$ and then $\mathbf{u}(k_0-2)$ and so forth, can be obtained uniquely.

Finally, we note that the initial value problem (1.2.3), (1.3.1) need not have a solution or a unique solution, e.g. the problem $ku(k+2)-u(k)=0,\ k\in\mathbb{N},\ u(0)=1,\ u(1)=0$ has no solution. In fact, for $k=0$ the difference equation gives $u(0)=0$, which violates the initial conditions. Also, the initial value problem $ku(k+2)-u(k)=0,\ k\in\mathbb{N},\ u(0)=u(1)=0$ has infinitely many solutions

$$u(k) = \begin{cases} 0 & \text{for } k=0 \text{ and } k \text{ odd} \\ \dfrac{c}{2^{(k/2)-1}((k/2)-1)!} & \text{for } k \text{ even,} \end{cases}$$

where $c$ is an arbitrary constant. However, if $a_0(k)a_n(k)\neq 0$ for all $k\in\overline{\mathbb{N}}$, then (1.2.3), (1.3.1) has a unique solution.

## 1.4. Some Examples: Initial Value Problems

The following examples provide a variety of situations of occurrence of initial value problems.

**Example 1.4.1.** Let $k\geq 1$ given points in a plane be such that any three of them are noncollinear. We shall find the number of straight lines that can be formed by joining together every pair of points. For this, let $u(k)$ represents the number of such lines. Let a new point be added to the set of $k$ points, which is also noncollinear with any other pair. The number of lines can now be written as $u(k+1)$. This $u(k+1)$ can be found from $u(k)$ by adding the $k$ new possible lines from the new $(k+1)$th point to each of the previous $k$ points. Thus, it follows that

$$u(k+1) = u(k)+k, \quad k\in\mathbb{N}(1). \tag{1.4.1}$$

Since when $k=1$ there is no pair of points, it is obvious that

$$u(1) = 0. \tag{1.4.2}$$

The first order initial value problem (1.4.1), (1.4.2) has a unique solution $u(k)=(1/2)k(k-1),\ k\in\mathbb{N}(1)$.

**Example 1.4.2.** In number theory the following result is fundamental:

**Theorem 1.4.1.** Every positive integer greater than one can be expressed as the product of only a single set of prime numbers.

The classical method of proving that there is no greatest prime number is as follows: Suppose the contrary be true and the finite system of primes is $v(1),\ v(2), \cdots, v(k)$ where $v(1) < v(2) < \cdots < v(k)$. Then, the number $m = v(1)v(2)\cdots v(k) + 1$ is prime to $v(1),\ v(2), \cdots, v(k)$. Hence, from Theorem 1.4.1, $m$ is a prime which is greater than $v(k)$.

Let us write this process of derivation of 'greater primes' from 'lesser primes' thus

$$v(k) \ = \ 1 + v(1)v(2)\cdots v(k-1), \qquad k \in \mathbb{N}(2).$$

Then, we have

$$v(k+1) \ = \ 1 + v(1)v(2)\cdots v(k), \qquad k \in \mathbb{N}(1)$$

and hence

$$v(k+1) \ = \ 1 + (v(k)-1)v(k), \qquad k \in \mathbb{N}(1)$$

which is the same as

$$v(k+1) - \frac{1}{2} \ = \ \left(v(k) - \frac{1}{2}\right)^2 + \frac{1}{4}, \qquad k \in \mathbb{N}(1).$$

Thus, the problem gives rise to a nonlinear difference equation, which by writing $u(k) = v(k) - (1/2)$ takes the compact form

$$u(k+1) \ = \ u^2(k) + \frac{1}{4}, \qquad k \in \mathbb{N}(1). \tag{1.4.3}$$

Further, since $v(1) = 2,$ for the difference equation (1.4.3) we find the initial condition

$$u(1) \ = \ 3/2. \tag{1.4.4}$$

**Example 1.4.3.** Consider the definite integral

$$u(k) \ = \ \int_0^1 t^k e^{t-1} dt, \qquad k \in \mathbb{N}(1). \tag{1.4.5}$$

It can easily be seen that $0 < u(k) < u(k-1)$ and $u(k) \to 0$ as $k \to \infty$, also

$$u(k+1) \ = \ 1 - (k+1)u(k) \tag{1.4.6}$$

$$u(1) \ = \ \frac{1}{e}. \tag{1.4.7}$$

With $1/e$ correct to any number of places, the difference equation (1.4.6) provides unrealistic values. Indeed, rounding all the calculations to six decimal places, we obtain

$$\begin{aligned} u(1) &= 0.367879 & u(7) &= 0.110160 \\ u(2) &= 0.264242 & u(8) &= 0.118720 \\ u(3) &= 0.207274 & u(9) &= -0.068480 \\ u(4) &= 0.170904 & u(10) &= 1.684800 \\ u(5) &= 0.145480 & u(11) &= -17.532800 \\ u(6) &= 0.127120 & u(12) &= 211.393600. \end{aligned}$$

**Example 1.4.4.** Let $P_K(t) = \sum_{k=0}^{K} a(k)t^k$ be a given polynomial of degree $K$. Consider the problem of finding a polynomial $Q_K(t) = \sum_{k=0}^{K} u(k)t^k$ of degree $K$ such that $Q_K(t) - Q'_K(t) = P_K(t)$, $t \in \mathbb{R}$. This leads to the following initial value problem

$$(1.4.8) \qquad u(k) = (k+1)u(k+1) + a(k), \qquad k \in \mathbb{N}(0, K-1)$$

$$(1.4.9) \qquad u(K) = a(K).$$

**Example 1.4.5.** Often we need to compute the value of $P_K(t) = \sum_{k=0}^{K} a(k)t^k$ at some $t_0 \in \mathbb{R}$. The computation of $a(k)t_0^k = a(k) \times t_0 \times \cdots \times t_0$ needs $k$ multiplications, and hence to find $P_K(t_0)$ we require in total $K(K+1)/2$ multiplications and $K$ summations. *Horner's method* is an *algorithm* (a list of instructions specifying a sequence of operations to be used in solving a certain problem) which reduces these multiplications to only $K$ and the same number of summations.

At $t = t_0$, we begin with the representation

$$P_K(t_0) = a(0) + t_0(a(1) + t_0(\cdots + t_0(a(K-2) + t_0(a(K-1) + t_0 a(K)))\cdots)).$$

Thus, if the numbers $u(k)$ are obtained from the scheme

$$(1.4.10) \qquad u(k) = a(k) + t_0 u(k+1), \qquad k \in \mathbb{N}(0, K-1)$$

$$(1.4.11) \qquad u(K) = a(K),$$

then $u(0) = P_K(t_0)$.

It is easy to see that the initial value problem (1.4.10), (1.4.11) is equivalent to

$$(1.4.12) \qquad u(k+1) = a(K-k-1) + t_0 u(k), \qquad k \in \mathbb{N}(0, K-1)$$

$$(1.4.13) \qquad u(0) = a(K)$$

and $u(K) = P_K(t_0)$.

**Example 1.4.6.** Consider the initial value problem

$$(1.4.14) \qquad (t+1)y'' + y' + ty = 0$$

$$(1.4.15) \qquad y(0) = 1, \quad y'(0) = 0.$$

Evidently $t = 0$ is an ordinary point of the differential equation (1.4.14). Insertion of $y(t) = \sum_{k=0}^{\infty} u(k)t^k$ into (1.4.14) yields

$$\sum_{k=0}^{\infty} k(k-1)u(k)t^{k-1} + \sum_{k=0}^{\infty} k(k-1)u(k)t^{k-2} + \sum_{k=0}^{\infty} ku(k)t^{k-1} + \sum_{k=0}^{\infty} u(k)t^{k+1} = 0,$$

which is the same as

$$2u(2) + u(1) + \sum_{k=1}^{\infty} [(k+1)ku(k+1) + (k+2)(k+1)u(k+2)$$
$$+(k+1)u(k+1) + u(k-1)]\, t^k = 0.$$

Thus, on equating the coefficients of $t^k$ to zero, we obtain

$$(1.4.16) \quad 2u(2) + u(1) = 0$$

$$(1.4.17) \quad u(k+2) = -\frac{(k+1)}{(k+2)}u(k+1) - \frac{1}{(k+1)(k+2)}u(k-1), \quad k \in \mathbb{N}(1).$$

From the initial conditions (1.4.15) it is obvious that $u(0) = 1$, $u(1) = 0$ and from (1.4.16) we find $u(2) = 0$. Thus, in turn we have a third order difference equation (1.4.17) together with the initial conditions

$$(1.4.18) \qquad u(0) = 1, \quad u(1) = 0, \quad u(2) = 0.$$

**Example 1.4.7.** A system of polynomials $\{P_k(t)\}$, $k \in \mathbb{N}$ is called an orthonormal system with respect to nonnegative weight function $w(t)$ over the interval $[\alpha, \beta]$, if

1. $P_k(t)$ is a polynomial of degree $k$

2. $\displaystyle\int_\alpha^\beta w(t)P_k(t)P_\ell(t)dt = \begin{cases} 0 & \text{for } k \neq \ell \\ 1 & \text{for } k = \ell. \end{cases}$

We will write

$$(1.4.19) \qquad P_k(t) = a(k)t^k + b(k)t^{k-1} + \cdots.$$

Since $tP_k(t)$ is a polynomial of degree $k+1$, it can be represented as

$$tP_k(t) = \sum_{i=0}^{k+1} c_{k,i}P_i(t),$$

where $c_{k,i}$ are the *Fourier coefficients*

$$c_{k,i} = \int_{\alpha}^{\beta} w(t)tP_k(t)P_i(t)dt.$$

If $i < k-1$, then $tP_i(t)$ is a polynomial of degree $i+1 < k$ and $c_{k,i} = 0$. Thus, we find

$$tP_k(t) = c_{k,k+1}P_{k+1}(t) + c_{k,k}P_k(t) + c_{k,k-1}P_{k-1}(t). \tag{1.4.20}$$

In (1.4.20), we substitute for $P_i(t),\ i = k+1,\ k,\ k-1$ its representation (1.4.19) and compare the coefficients of $t^{k+1}$, to obtain $c_{k,k+1} = a(k)/a(k+1)$. Since for all $k$ and $i$ we have the relation $c_{k,i} = c_{i,k}$, it follows that $c_{k,k-1} = a(k-1)/a(k)$. To obtain $c_{k,k}$ we compare the coefficients of $t^k$, this gives $c_{k,k} = \dfrac{b(k)}{a(k)} - \dfrac{b(k+1)}{a(k+1)}$.

Thus, (1.4.20) takes the form

$$\begin{aligned} tP_k(t) &= \frac{a(k)}{a(k+1)}P_{k+1}(t) + \left(\frac{b(k)}{a(k)} - \frac{b(k+1)}{a(k+1)}\right)P_k(t) \\ &\quad + \frac{a(k-1)}{a(k)}P_{k-1}(t), \quad k \in \mathbb{N}(1). \end{aligned} \tag{1.4.21}$$

Hence, on identifying $P_k(t)$ as $u(k)$ we observe that any three successive orthonormal polynomials satisfy the second order difference equation (1.4.21).

In particular, if $w(t) = (1-t^2)^{-1/2}$ and $\beta = -\alpha = 1$, then (1.4.21) reduces to known recurrence formula for the *Chebyshev polynomials* denoted by $T_k(t)$

$$T_{k+1}(t) = 2tT_k(t) - a(k)T_{k-1}(t), \quad k \in \mathbb{N}(1) \tag{1.4.22}$$

where $a(1) = \sqrt{2}$ and $a(k) = 1$ for all $k \in \mathbb{N}(2)$.

The initial functions for (1.4.22) are defined to be

$$T_0(t) = \frac{1}{\sqrt{\pi}}, \quad T_1(t) = \sqrt{\frac{2}{\pi}}\, t. \tag{1.4.23}$$

**Example 1.4.8.** Let $\mathbf{g}(\mathbf{u}) = 0$ be a system of $n$ nonlinear equations in $n$ unknowns $u_1, \cdots, u_n$. *Newton's method* for solving this system is in fact an initial value problem of the type (1.2.8), (1.3.3) where

$$\mathbf{f}(k, \mathbf{u}(k)) = \mathbf{u}(k) - \mathcal{J}^{-1}(\mathbf{u}(k))\mathbf{g}(\mathbf{u}(k)), \quad k \in \mathbb{N} \tag{1.4.24}$$

$\mathbf{u}(0) = \mathbf{u}^0$ is an initial approximation to the solution, and $\mathcal{J}(\mathbf{u}(k))$ is the $n \times n$ *Jacobian matrix* $(\partial g_i/\partial u_j)\,|_{\mathbf{u}=\mathbf{u}(k)}$.

## 1.5. Boundary Value Problems

Another choice to pin out the solution $\mathbf{u}(k)$ of a given difference system on $\mathbb{N}(a, b-1)$ can be described as follows: Let $B(a,b)$ be the space of all real $n$ vector functions defined on $\mathbb{N}(a,b)$ and $\mathcal{F}$ be an operator mapping $B(a,b)$ into $\mathbb{R}^n$, then our concern is that $\mathbf{u}(k)$ must satisfy the boundary condition

$$\mathcal{F}[\mathbf{u}] \;=\; 0. \tag{1.5.1}$$

System (1.2.8) or (1.2.11) or (1.2.12) together with (1.5.1) is called a *boundary value problem*. Obviously, initial condition (1.3.3) as well as (1.3.4) is a special case of (1.5.1). The term boundary condition comes from the fact that $\mathcal{F}$ allows the possibility of defining conditions at the points $a$ and $b$ of $\mathbb{N}(a,b)$. For example, let $k_1 < \cdots < k_r$ $(r \geq 2)$ be some fixed points in $\mathbb{N}(a,b)$, then we seek a solution $\mathbf{u}(k)$ of the difference system on $\mathbb{N}(k_1, k_r - 1)$ satisfying

$$\phi_i(\mathbf{u}(k_1), \cdots, \mathbf{u}(k_r)) \;=\; 0, \quad 1 \leq i \leq n \tag{1.5.2}$$

where $\phi_i$, $1 \leq i \leq n$ are given functions.

In the case when $\mathcal{F}$ is linear we shall prefer to write the boundary condition (1.5.1) as

$$\mathcal{L}[\mathbf{u}] \;=\; \mathbf{l}, \tag{1.5.3}$$

where the vector $\mathbf{l}$ is known. Similarly, if $\phi_i$, $1 \leq i \leq n$ are linear, then (1.5.2) will be written as

$$\sum_{i=1}^{r} \mathcal{L}^i \mathbf{u}(k_i) \;=\; \mathbf{l}, \tag{1.5.4}$$

where $\mathcal{L}^i$, $1 \leq i \leq r$ are given $n \times n$ matrices. If $\mathcal{L}^i = \left(\alpha^i_{pq}\right)$, then (1.5.4) is the same as

$$\sum_{i=1}^{r} \sum_{q=1}^{n} \alpha^i_{pq} u_q(k_i) \;=\; \ell_p, \quad 1 \leq p \leq n. \tag{1.5.5}$$

It is of interest to note that (1.5.4), or equivalently (1.5.5), include in particular the

(i) **Periodic Conditions:** $r = 2$ and for simplicity we let $k_1 = 0$, $k_2 = K$

$$\mathbf{u}(0) = \mathbf{u}(K). \tag{1.5.6}$$

(ii) **Implicit Separated Conditions:**

$$\sum_{q=1}^{n} \alpha_{i(s_i),q} u_q(k_i) = \ell_{i,i(s_i)}, \quad 1 \le i \le r \ (2 \le r \le n, \text{ but fixed}) \tag{1.5.7}$$

where $s_1 = 1, 2, \cdots, \beta_1; \cdots; s_r = 1, 2, \cdots, \beta_r$ and $\sum_{i=1}^{r} \beta_i = n$.

The subscript $i(s_i)$ allows the possibility that at the same point $k_i$ several boundary conditions are prescribed.

(iii) **Separated Conditions:**

$$u_{i(s_i)}(k_i) = \ell_{i,i(s_i)}, \quad 1 \le i \le r \ (2 \le r \le n, \text{ but fixed}) \tag{1.5.8}$$

where $s_i$, $1 \le i \le r$ are the same as in (1.5.7).

In (1.5.8) the subscript $i(s_i)$ allows the possibility that the set of variables specified at the boundary points may not be disjoint. For instance if $n = 7$, $r = 4$, $u_1(k_1)$, $u_3(k_1)$, $u_2(k_2)$, $u_3(k_3)$, $u_1(k_4)$, $u_6(k_4)$ and $u_7(k_4)$, then $u_1$ is fixed at $k_1$ and $k_4$, and $u_3$ is fixed at $k_1$ and $k_3$, whereas no condition is prescribed for $u_4$ and $u_5$. The indexing for the boundary conditions is specified by $1(1) = 1$, $1(2) = 3$, $2(1) = 2$, $3(1) = 3$, $4(1) = 1, 4(2) = 6$ and $4(3) = 7$.

For a given $n$th order difference equation on $\mathbb{N}(a, b-1)$ we shall consider some of the following conditions.

(i) **Niccoletti Conditions:**

$a = k_1 < k_1 + 1 < k_2 < k_2 + 1 < \cdots < k_{n-1} < k_{n-1} + 1 < k_n = b - 1 + n$,

where each $k_i \in \mathbb{N}(a, b-1+n)$

$$u(k_i) = A_i, \quad 1 \le i \le n. \tag{1.5.9}$$

(ii) **Hermite ($r$ point) Conditions:**

$$a = k_1 < k_1 + p_1 + 1 < k_2 < k_2 + p_2 + 1 < \cdots < k_{r-1} < k_{r-1} + p_{r-1} + 1 < k_r \\ \le k_r + p_r = b - 1 + n,$$

where each $k_i \in \mathbb{N}(a, b-1+n)$, $p_i \in \mathbb{N}$, $\sum_{i=1}^{r} p_i + r = n$

$$\Delta^j u(k_i) = A_{i,j}, \quad 1 \le i \le r, \quad 0 \le j \le p_i. \tag{1.5.10}$$

(iii) **Abel–Gontscharoff Conditions:**

$$k_1 \le k_2 \le \cdots \le k_n \quad (k_n > k_1), \quad \text{where each} \quad k_i \in \mathbb{N}(a,b)$$

$$\Delta^i u(k_{i+1}) \;=\; A_i, \quad 0 \le i \le n-1. \tag{1.5.11}$$

(iv) $(n,p)$ **Conditions:**

$$\begin{aligned} \Delta^i u(a) &= A_i, \quad 0 \le i \le n-2 \\ \Delta^p u(b-1+n-p) &= B, \quad (0 \le p \le n-1, \quad \text{but fixed}). \end{aligned} \tag{1.5.12}$$

(v) $(p,n)$ **Conditions:**

$$\begin{aligned} \Delta^p u(a) &= B, \quad (0 \le p \le n-1, \quad \text{but fixed}) \\ \Delta^i u(b+1) &= A_i, \quad 0 \le i \le n-2. \end{aligned} \tag{1.5.13}$$

(vi) **Lidstone Conditions:** $(n = 2m)$

$$\begin{aligned} \Delta^{2i} u(a) &= A_{2i} \\ \Delta^{2i} u(b-1+2m-2i) &= B_{2i}, \quad 0 \le i \le m-1. \end{aligned} \tag{1.5.14}$$

## 1.6. Some Examples: Boundary Value Problems

The following examples are sufficient to demonstrate how discrete boundary value problems appear.

**Example 1.6.1.** Consider a string of length $K+1$, whose mass may be neglected, which is stretched between two fixed ends A and B with a force $f$ and is loaded at intervals 1 with $K$ equal masses $M$ not under the influence of gravity, and which is slightly disturbed so that the tension in the string is constant along each segment and equal to $f$. Let $v(k)$, $1 \le k \le K$ (Figure 1.6.1) be the ordinates at time $t$ of the $K$ particles. Then, the restoring force in the negative direction is given by $F(k) = f[(v(k-1) - v(k)) + (v(k+1) - v(k))]$. Thus, by *Newton's second law* the equation of motion of the $k$th particle is

$$M\frac{d^2 v(k)}{dt^2} + f(-v(k-1) + 2v(k) - v(k+1)) \;=\; 0.$$

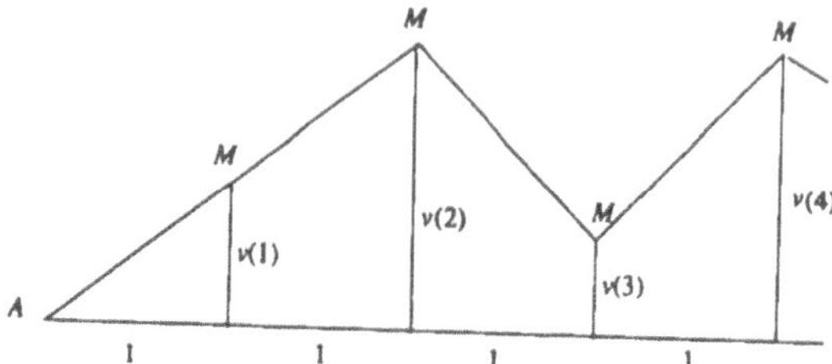


Figure 1.6.1.

Since each particle is vibrating, let $v(k) = u(k)\cos(wt + \phi)$ in the above equation, to obtain

$$-w^2 M u(k) + f(-u(k-1) + 2u(k) - u(k+1)) = 0,$$

which is the same as

$$u(k+1) - cu(k) + u(k-1) = 0, \quad k \in \mathbb{N}(1, K) \tag{1.6.1}$$

where $c = 2 - (w^2 M/f)$.

This second order homogeneous difference equation represents the amplitude of the motion of every particle except the first and last. In order that it may represent these also, it is necessary to suppose that $v(0)$ and $v(K+1)$ are both zero, although there are no particles corresponding to the values of $k$ equal to $0$ and $K+1$. With this understanding, we find that

$$u(0) = u(K+1) = 0. \tag{1.6.2}$$

Equation (1.6.1) together with (1.6.2) is a second order boundary value problem.

**Example 1.6.2.** Consider the electric circuit shown in Figure 1.6.2. Assume that $V_0 = A$ is a given voltage and $V_{K+1} = 0$, and the shaded region indicates the ground where the voltage is zero. Each resistance in the horizontal branch is equal to $R$ and in the vertical branches equal to

$4R$. We want to find the voltage $V_k$ for $1 \le k \le K$. For this, according to *Kirchoff's current law,* the sum of the currents flowing into a junction point is equal to the sum of the currents flowing away from the junction point. Applying this law at the junction point corresponding to the voltage $V_{k+1}$, we have

$$I_{k+1} = I_{k+2} + i_{k+1}.$$

Using *Ohm's law,* $I = V/R$, the above equation can be replaced by

$$\frac{V_k - V_{k+1}}{R} = \frac{V_{k+1} - V_{k+2}}{R} + \frac{V_{k+1} - 0}{4R},$$

which is on identifying $V_k$ as $u(k)$ leads to the second order difference equation

$$(1.6.3) \qquad 4u(k+2) - 9u(k+1) + 4u(k) = 0, \quad k \in \mathbb{N}(0, K-1)$$

and the boundary conditions are

$$(1.6.4) \qquad u(0) = A, \quad u(K+1) = 0.$$

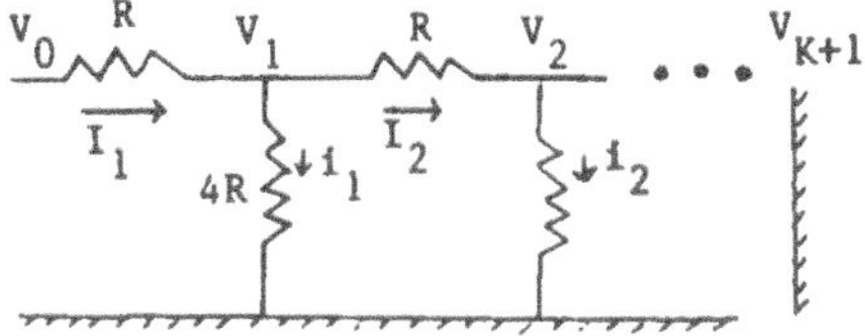


Figure 1.6.2.

**Example 1.6.3.** To test whether a batch of articles is satisfactory, we introduce a scoring system. The score is initially set at $(K/2) + (n-1)$. If a randomly sampled item is found to be defective, we subtract $(n-1)$. If it is acceptable, we add 1. The procedure stops when the score reaches either $K + (n-1)$ or less than $(n-1)$. If $K + (n-1)$, the batch is accepted; if less than $(n-1)$, it is rejected. Suppose that the probability of selecting an acceptable item is $p$, and $q = 1 - p$. Let $P_k$ denote the probability that the batch will be rejected when the score is at $k$. Then after the next choice, the score will be increased by 1 with probability $p$ or decreased by $(n-1)$ with probability $q$. Thus

$$P_k = pP_{k+1} + qP_{k-(n-1)},$$

which on identifying $P_k$ as $u(k)$ can be written as the $n$th order difference equation

$$(1.6.5) \qquad u(k+n) - \frac{1}{p}u(k+n-1) + \frac{q}{p}u(k) = 0, \qquad k \in \mathbb{N}(0, K-1)$$

with the boundary conditions

$$(1.6.6) \qquad u(0) = u(1) = \cdots = u(n-2) = 1, \qquad u(K-1+n) = 0.$$

**Example 1.6.4.** To overcome the difficulty realized in Example 1.4.3, using the known behavior of $u(k)$, Dorn and McCracken [16] used the famous recurrence algorithm proposed by Miller [34]. They took $u(K) = 0$ for sufficiently large $K$ and recursed (1.4.6) backward. To check the accuracy of results, they arbitrarily chose $K_1 > (K)$ and obtained another set of values of the integral (1.4.5). The search for $K$ and $K_1$ continues until the results agree to the desired degree of accuracy. However, this method does not appear to be practicable. To evaluate the integral (1.4.5) we notice that $u(k)$ also satisfies

$$(1.6.7) \qquad u(k+2) = (k+1)(k+2)u(k) - (k+1), \qquad k \in \mathbb{N}(0, K-1)$$

together with

$$(1.6.8) \qquad u(1) = 1 - u(0)$$

and for sufficiently large $K$, $u(K+2) = u(K+1)$ and hence (1.4.6) implies

$$(1.6.9) \qquad u(K+1) = \frac{1}{K+3}.$$

The boundary value problem (1.6.7) - (1.6.9) will be solved satisfactorily later in Example 8.5.2.

**Example 1.6.5.** For the continuous boundary value problem

$$(1.6.10) \qquad y'' = f(t, y)$$

$$(1.6.11) \qquad y(\alpha) = A, \quad y(\beta) = B$$

the following result is well known [30]:

**Theorem 1.6.1.** Let $f(t,y)$ be continuous on $[\alpha,\beta]\times\mathbb{R}$ and

$$\inf_{\substack{\alpha\le t\le\beta\\ -\infty<y<\infty}} \frac{\partial f}{\partial y} = -\eta > -\frac{\pi^2}{(\beta-\alpha)^2}. \tag{1.6.12}$$

Then, the boundary value problem (1.6.10), (1.6.11) has a unique solution.

However, even if $f(t,y)=f(t)y+g(t)$ the analytical solution of (1.6.10), (1.6.11) may not be determined. Faced with this difficulty we find an approximate solution of (1.6.10), (1.6.11) by employing discrete variable methods. One of such well known and widely used discrete methods is due to *Nomerov* which is defined as follows: We introduce the set $\{t_k\}$, where $t_k=\alpha+kh$, $h=(\beta-\alpha)/(K+1)$, $k\in\mathbb{N}(0,K+1)$. Let $u(k)$ be the approximation to the true solution $y(t)$ of (1.6.10), (1.6.11) at $t=t_k$. We assume that $u(k)$ satisfies the following second order difference equation

$$\begin{aligned} u(k+1)-2u(k)+u(k-1) &= \frac{1}{12}h^2(f(\alpha+(k-1)h,u(k-1)) \\ &\quad +10f(\alpha+kh,u(k))+f(\alpha+(k+1)h,u(k+1))),\quad k\in\mathbb{N}(1,K) \end{aligned} \tag{1.6.13}$$

together with the boundary conditions

$$u(0) = A,\quad u(K+1) = B. \tag{1.6.14}$$

The existence and uniqueness of the boundary value problem (1.6.13), (1.6.14) and its usefulness in conjuction with initial value methods will be given in later chapters.

**Example 1.6.6.** Let $[\alpha,\beta]$ be a given interval, and $P:\ \alpha=t_0<t_1<\cdots<t_{K+1}=\beta$ be a fixed partition. We seek a function $S_p(t)\in C^{(2)}[\alpha,\beta]$ which coincides with a cubic polynomial in each subinterval $[t_{k-1},t_k]$, $k\in\mathbb{N}(1,K+1)$ and satisfies $S_p(t_k)=y_k$, $k\in\mathbb{N}(0,K+1)$ where the ordinates $y_k$ are prescribed. The function $S_p(t)$ is called a *cubic spline* with respect to the partition $P$.

Designating $S_P''(t_k)$ by $M_k$, $k\in\mathbb{N}(0,K+1)$ the linearity of $S_P''(t)$ in each subinterval $[t_{k-1},t_k]$, $k\in\mathbb{N}(1,K+1)$ implies that

$$S_P''(t) = M_{k-1}\frac{t_k-t}{h_k}+M_k\frac{t-t_{k-1}}{h_k}, \tag{1.6.15}$$

where $h_k=t_k-t_{k-1}$. If we integrate twice (1.6.15) and evaluate the constants of integration, we obtain the equations

$$(1.6.16)\qquad S_P(t) = M_{k-1}\frac{(t_k-t)^3}{6h_k} + M_k\frac{(t-t_{k-1})^3}{6h_k} + \left(y_{k-1} - \frac{M_{k-1}h_k^2}{6}\right)\frac{t_k-t}{h_k} + \left(y_k - \frac{M_kh_k^2}{6}\right)\frac{t-t_{k-1}}{h_k}$$

and

$$(1.6.17)\quad S'_P(t) = -M_{k-1}\frac{(t_k-t)^2}{2h_k} + M_k\frac{(t-t_{k-1})^2}{2h_k} + \frac{y_k-y_{k-1}}{h_k} - \frac{M_k-M_{k-1}}{6}h_k.$$

From (1.6.17), we have

$$(1.6.18)\qquad S'_P(t_k-) = \frac{h_k}{6}M_{k-1} + \frac{h_k}{3}M_k + \frac{y_k-y_{k-1}}{h_k}$$

and

$$(1.6.19)\qquad S'_P(t_k+) = -\frac{h_{k+1}}{3}M_k - \frac{h_{k+1}}{6}M_{k+1} + \frac{y_{k+1}-y_k}{h_{k+1}}.$$

In virtue of (1.6.15) and (1.6.16), the functions $S''_P(t)$ and $S_P(t)$ are continuous on $[\alpha, \beta]$. The continuity of $S'_P(t)$ at $t = t_k$ yields by means of (1.6.18) and (1.6.19) the following second order difference equation

$$(1.6.20)\quad \frac{h_k}{6}M_{k-1} + \frac{h_k+h_{k+1}}{3}M_k + \frac{h_{k+1}}{6}M_{k+1} = \frac{y_{k+1}-y_k}{h_{k+1}} - \frac{y_k-y_{k-1}}{h_k}, \qquad k \in \mathbb{N}(1, K).$$

Once two appropriate boundary conditions, say, $M_0$ and $M_{K+1}$ are prescribed, the solution of (1.6.20) serves to determine $S_P(t)$ in each subinterval $[t_{k-1}, t_k]$.

Difference equation (1.6.20) can be used to find an approximate solution of the problem (1.6.10), (1.6.11). For this, we observe that $M_k$ are given to be $f(t_k, y_k)$ and the problem is to find $y_k$, $k \in \mathbb{N}(0, K+1)$. Thus, if in particular $h_k = h$, $k \in \mathbb{N}(1, K+1)$ then once again we have $t_k = \alpha + kh$, $h = (\beta-\alpha)/(K+1)$ and on identifying $y_k$ as $u(k)$ we need to solve the following difference equation

$$(1.6.21)\quad u(k+1) - 2u(k) + u(k-1) = \frac{1}{6}h^2(f(\alpha+(k-1)h, u(k-1)) + 4f(\alpha+kh, u(k)) + f(\alpha+(k+1)h, u(k+1))), \quad k \in \mathbb{N}(1, K)$$

together with the boundary conditions (1.6.14).

**Example 1.6.7.** Let in a domain $D \subseteq \mathbb{R}^{K+1}$ the function $\phi(t_1, \cdots, t_{K+1})$ be given for which third order derivatives exist. Then, a necessary condition for a maximum or minimum is that

$$\frac{\partial \phi}{\partial t_k} = 0, \quad k \in \mathbb{N}(1, K+1). \tag{1.6.22}$$

If (1.6.22) is satisfied at $(t_1^1, \cdots, t_{K+1}^1)$, then a sufficient condition for a minimum at this point is that all the determinants

$$\begin{aligned} D(0) &= 1 \\ D(k) &= \begin{vmatrix} a_{11} & a_{12} & & & & \\ a_{21} & a_{22} & a_{23} & & & \\ & \cdots & \cdots & \cdots & & \\ & & & a_{k-1,k-2} & a_{k-1,k-1} & a_{k-1,k} \\ & & & & a_{k,k-1} & a_{k,k} \end{vmatrix}, \\ & \qquad k \in \mathbb{N}(1, K+1) \end{aligned} \tag{1.6.23}$$

where

$$a_{ij} = a_{ji} = \left. \frac{\partial^2 \phi}{\partial t_i \partial t_j} \right|_{(t_1^1, \cdots, t_{K+1}^1)}$$

are positive. Further, a sufficient condition for a maximum is that $D(k)$, $k \in \mathbb{N}(0, K+1)$ alternate in sign.

It is interesting to note that $D(k)$ is the solution of the initial value problem

$$D(k) = a_{kk} D(k-1) - a_{k,k-1}^2 D(k-2), \quad k \in \mathbb{N}(2, K+1) \tag{1.6.24}$$

$$D(0) = 1, \quad D(1) = a_{11}. \tag{1.6.25}$$

Now consider the problem of minimizing the finite sum

$$\phi = \sum_{k=0}^{K} f(k, u, p), \tag{1.6.26}$$

where $u$ is to be determined as a function of $k$, and where $p = \Delta u(k)$.

The necessary condition (1.6.22) for a minimum gives a second order difference equation

$$\frac{\partial \phi}{\partial u(k)} = f_u(k) - f_p(k) + f_p(k-1) = 0, \quad k \in \mathbb{N}(1, K) \tag{1.6.27}$$

where $f(k) = f(k, u, p)$.

In case $u(0)$ and/or $u(K+1)$ are variable as well as $u$'s at intermediate points, we have in addition to (1.6.27) the relations

$$f_u(0) - f_p(0) = 0 \tag{1.6.28}$$

and/or

$$f_p(K) = 0. \tag{1.6.29}$$

In the contrary case, we assume that $u(0)$ and $u(K+1)$ are fixed

$$u(0) = C, \quad u(K+1) = D. \tag{1.6.30}$$

The solution of the boundary value problem (1.6.27), (1.6.30) in the case $u(0)$ and $u(K+1)$ are fixed, or (1.6.27) (1.6.29) in the other case, is called a critical function. This critical function provides a minimum if the solution of (1.6.24), (1.6.25) is positive for all $k$ for which $u(k)$ is variable. If, on the other hand, $D(k)$ alternates in sign then it gives a maximum.

In particular, for the function $f(k,u,p) = 4u^2 + 3p^2$, equation (1.6.27) takes the form

$$3u(k+2) - 10u(k+1) + 3u(k) = 0. \tag{1.6.31}$$

The solution of the boundary value problem (1.6.31), (1.6.30) appears as

$$u(k) = \frac{C - D3^{K+1}}{1 - 3^{2(K+1)}}3^k + \frac{3^{K+1}D - C3^{2(K+1)}}{1 - 3^{2(K+1)}}3^{-k}. \tag{1.6.32}$$

Further, since for this function $\phi_{u(k)u(k)} = 20$ and $\phi_{u(k)u(k-1)} = -6$, the initial value problem (1.6.24), (1.6.25) reduces to

$$D(k+2) - 20D(k+1) + 36D(k) = 0 \tag{1.6.33}$$

$$D(0) = 1, \quad D(1) = 20. \tag{1.6.34}$$

The solution of (1.6.33), (1.6.34) can be written as

$$D(k) = \frac{1}{16}\left(18^{k+1} - 2^{k+1}\right). \tag{1.6.35}$$

From (1.6.35) it is clear that $D(k) > 0$ as long as $k > 0$, and consequently $\phi = \sum_{k=0}^{K}(4u^2 + 3p^2)$ is a minimum when $u(k)$ is given by (1.6.32).

**Example 1.6.8.** Consider the potential equation

$$\frac{\partial^2 u}{\partial x^2} + \frac{\partial^2 u}{\partial y^2} = 0 \tag{1.6.36}$$

over the rectangle $0 \le x \le \alpha$, $0 \le y \le \beta$. We shall assume that the values of $u(x,y)$ are specified on the boundary of the rectangle. If we let $u_{k,\ell}$ denote $u(kh_1, \ell h_2)$ where $(K+1)h_1 = \alpha$, $(L+1)h_2 = \beta$, and replace $\partial^2 u/\partial x^2$ by $(u_{k+1,\ell} - 2u_{k,\ell} + u_{k-1,\ell})/h_1^2$ and $\partial^2 u/\partial y^2$ by $(u_{k,\ell+1} - 2u_{k,\ell} + u_{k,\ell-1})/h_2^2$ in (1.6.36) it takes the form

$$u_{k+1,\ell} + \lambda u_{k,\ell+1} - (2+2\lambda)u_{k,\ell} + \lambda u_{k,\ell-1} + u_{k-1,\ell} = 0, \qquad 1 \le k \le K,\ 1 \le \ell \le L \tag{1.6.37}$$

where $\lambda = h_1^2/h_2^2$.

Let us define the $K+2$ vectors $\mathbf{u}(k)$ of order $L \times 1$ by $\mathbf{u}(k) = (u_{k,\ell})$, $1 \le \ell \le L$. Since we are given the values $(u_{0,\ell})$ and $(u_{K+1,\ell})$, we have, say

$$\mathbf{u}(0) = \mathbf{c}, \quad \mathbf{u}(K+1) = \mathbf{d}. \tag{1.6.38}$$

Next, we define an $L \times L$ matrix $\mathcal{Q} = (q_{ij})$, where

$$q_{ij} = \begin{cases} (2+2\lambda) & \text{if } i = j \\ -\lambda & \text{if } |i-j| = 1 \\ 0 & \text{otherwise.} \end{cases}$$

We also define vectors $\mathbf{r}(k)$ of order $L \times 1$ by $\mathbf{r}(k) = (r_{k,\ell})$, $1 \le \ell \le L$ where

$$r_{k,\ell} = \begin{cases} \lambda u_{k,0} & \text{if } \ell = 1 \\ \lambda u_{k,L+1} & \text{if } \ell = L \\ 0 & \text{otherwise.} \end{cases}$$

Clearly, the vectors $\mathbf{r}(k)$ are given by the boundary conditions on the edges $y = 0$ and $y = \beta$. With these notations, (1.6.37) can be written as the second order difference system

$$\mathbf{u}(k+1) - \mathcal{Q}\mathbf{u}(k) + \mathbf{u}(k-1) + \mathbf{r}(k) = 0, \quad 1 \le k \le K \tag{1.6.39}$$

subject to the boundary conditions (1.6.38).

If we define $\mathbf{v}(k)$ by $\mathbf{u}(k+1) = \mathbf{v}(k)$, then (1.6.39) is equivalent to the first order system

$$\begin{aligned} \mathbf{u}(k+1) &= \mathbf{v}(k) \\ \mathbf{v}(k+1) &= -\mathbf{u}(k) + \mathcal{Q}\mathbf{v}(k) - \mathbf{r}(k+1), \quad 0 \le k \le K-1 \end{aligned} \tag{1.6.40}$$

and the boundary conditions (1.6.38) are the same as

$$\mathbf{u}(0) = \mathbf{c}, \quad \mathbf{v}(K) = \mathbf{d}. \tag{1.6.41}$$

## 1.7. Some Examples: Real World Phenomena

The following examples illustrate the importance of difference equations to some real world problems.

**Example 1.7.1.** (Hardy Weinberg Law). Every characteristic of an individual, like height or color of the hair, is determined by a pair of genes, one obtained from the father and the other from the mother. Every gene occurs in two forms, a dominant denoted by $D$ and a recessive denoted by $R$. Thus with respect to a characteristic, an individual may be a dominant $DD$, a hybrid $DR$ or $RD$ or a recessive $RR$.

In the $k$th generation, let the proportions of dominants, hybrids and recessive be $p(k)$, $q(k)$, $r(k)$ so that

$$p(k) + q(k) + r(k) \;=\; 1, \quad p(k) \geq 0,\; q(k) \geq 0,\; r(k) \geq 0. \tag{1.7.1}$$

We assume that individuals, in this generation mate at random. Then, $p(k+1) =$ the probability that an individual in the $(k+1)$th generation is a dominant $(DD) =$ (probability that this individual gets a $D$ from the father) $\times$ (probability that this individual gets a $D$ from the mother), i.e.

$$p(k+1) \;=\; \left(p(k) + \frac{1}{2}q(k)\right)\left(p(k) + \frac{1}{2}q(k)\right)^2$$

or

$$p(k+1) \;=\; \left(p(k) + \frac{1}{2}q(k)\right)^2. \tag{1.7.2}$$

Similarly,

$$q(k+1) \;=\; 2\left(p(k) + \frac{1}{2}q(k)\right)\left(r(k) + \frac{1}{2}q(k)\right) \tag{1.7.3}$$

and

$$r(k+1) \;=\; \left(r(k) + \frac{1}{2}q(k)\right)^2 \tag{1.7.4}$$

so that

$$\begin{aligned} &p(k+1) + q(k+1) + r(k+1) \\ &\qquad = \left(p(k) + \frac{1}{2}q(k) + \frac{1}{2}q(k) + r(k)\right)^2 \;=\; 1 \end{aligned} \tag{1.7.5}$$

as expected. Similarly,

(1.7.6)

$$\begin{aligned} p(k+2) &= \left(p(k+1) + \frac{1}{2}q(k+1)\right)^2 \\ &= \left(\left(p(k) + \frac{1}{2}q(k)\right)^2 + \left(p(k) + \frac{1}{2}q(k)\right)\left(r(k) + \frac{1}{2}q(k)\right)\right)^2 \\ &= \left(p(k) + \frac{1}{2}q(k)\right)^2 \left(p(k) + \frac{1}{2}q(k) + \frac{1}{2}q(k) + r(k)\right)^2 \\ &= \left(p(k) + \frac{1}{2}q(k)\right)^2 = p(k+1) \end{aligned}$$

and

$$(1.7.7) \qquad q(k+2) = q(k+1), \quad r(k+2) = r(k+1)$$

so that the proportions of dominants, hybrids and recessive in the $(k+2)$th generation are the same as in the $(k+1)$th generation. Thus in any population in which random mating takes place with respect to a characteristic, the proportions of dominants, hybrids and recessive do not change after the first generation. Equations (1.7.2) (1.7.4) is a set of difference equations of the first order.

**Example 1.7.2.** (Improvement Through Elimination of Recessives). Suppose the recessive are undesirable and so we do not allow the recessive in any generation to breed. Let $p(k)$, $q(k)$, $r(k)$ be the proportions of dominants, hybrids and recessives before elimination of recessives and let $p_1(k)$, $q_1(k)$, $0$ be the proportions after the elimination, then

$$(1.7.8) \qquad \frac{p_1(k)}{p(k)} = \frac{q_1(k)}{q(k)} = \frac{p_1(k) + q_1(k)}{p(k) + q(k)} = \frac{1}{1 - r(k)}.$$

Now we assume random mating and let $p(k+1)$, $q(k+1)$, $r(k+1)$ be the proportions in the next generation before elimination of recessives, then using (1.7.2) (1.7.4), we get

$$(1.7.9) \qquad p(k+1) = \left(p_1(k) + \frac{1}{2}q_1(k)\right)^2$$

(1.7.10)

$$q(k+1) = 2\left(p_1(k) + \frac{1}{2}q_1(k)\right)\left(\frac{1}{2}q_1(k)\right) = q_1(k)\left(p_1(k) + \frac{1}{2}q_1(k)\right)$$

$$(1.7.11) \qquad r(k+1) = \left(\frac{1}{2}q_1(k)\right)^2 = \frac{1}{4}q_1^2(k).$$

After elimination of recessives, let the new proportions be $p_1(k+1)$, $q_1(k+1)$, $0$ so that

$$(1.7.12) \quad \frac{p_1(k+1)}{p(k+1)} = \frac{q_1(k+1)}{q(k+1)} = \frac{1}{p(k+1)+q(k+1)} = \frac{1}{1-\frac{1}{4}q_1^2(k)}$$

and

(1.7.13)

$$q_1(k+1) = \frac{q_1(k)\left(p_1(k)+\frac{1}{2}q_1(k)\right)}{1-\frac{1}{4}q_1^2(k)} = \frac{q_1(k)\left(1-\frac{1}{2}q_1(k)\right)}{1-\frac{1}{4}q_1^2(k)} = \frac{q_1(k)}{1+\frac{1}{2}q_1(k)}.$$

Equation (1.7.13) can be solved rather easily, to obtain

$$(1.7.14) \qquad q_1(k) = \frac{1}{c+\frac{1}{2}k},$$

where $c$ is an arbitrary constant. From (1.7.14) it is clear that $q_1(k) \to 0$ as $k \to \infty$, i.e. ultimately we would be left with all dominants. Equation (1.7.14) also determines the rate at which hybrids disappear.

**Example 1.7.3.** (Traffic in Channels). Consider a channel, for example a telephone line, and suppose that two elementary informations $S_1$ and $S_2$ of duration $k_1$ and $k_2$ respectively can be combined in order to obtain a message. Let $k$ be a time interval greater than both $k_1$ and $k_2$. We are interested in the number of messages $u(k)$ of length $k$. These messages can be divided in two groups: those ending with $S_1$ and those ending with $S_2$. The number of messages in the first group is $u(k-k_1)$ while in the second group is $u(k-k_2)$. Then, we have

$$(1.7.15) \qquad u(k) = u(k-k_1) + u(k-k_2).$$

Suppose, for simplicity, that $k_1 = 1$ and $k_2 = 2$. Then, equation (1.7.15) becomes $u(k) = u(k-1) + u(k-2)$. Clearly, for this equation initial conditions are $u(1) = 1$, $u(2) = 2$. This initial value problem defines the Fibonacci numbers (see Problem 1.9.16).

**Example 1.7.4.** (Queuing Theory). Consider a queue of individuals or telephone calls in a channel. Let $P_k(t)$ be the probability of $k$ items arriving at time $t$ and we have $\sum_{k=0}^{\infty} P_k(t) = 1$.

Let $\lambda\Delta t$ be the probability of a single arrival during the small time interval $\Delta t$ and suppose that the probability of more than one arrival in the same interval is negligible. Let $\mu\Delta t$ be the probability of completing the service in the interval $\Delta t$. We shall assume that the service is a Poisson process, i.e. the probability of no arrivals in $\Delta t$ is $1 - \lambda\Delta t$ and the probability of the service not completed in $\Delta t$, no departures, is $1 - \lambda\Delta t$. The following model describes the situation where at the beginning there are no items in the single channel queue and the service is made on a first come, first served basis

$$P_k(t+\Delta t) = P_k(t)(1-\lambda\Delta t)(1-\mu\Delta t) + P_{k-1}(t)\lambda\Delta t + P_{k+1}(t)\mu\Delta t, \; k \geq 1$$
$$P_0(t + \Delta t) = P_0(t)(1 - \lambda\Delta t) + P_1(t)\mu\Delta t.$$

These equations indicate that the probability that at time $t + \Delta t$ there are $k$ items in line is equal to the sum of three terms:

1. The probability of already having $k$ items at time $t$ multiplied by the probability of no arrivals during $\Delta t$ and the probability of no departure in the same interval.
2. The probability of having $k - 1$ items at time $t$ multiplied by the probability of a new arrival in $\Delta t$.
3. The probability of having $k + 1$ items at time $t$ multiplied by the probability of a departure in $\Delta t$.

Taking the limit as $t \to 0$, we obtain

$$\begin{aligned} P_k'(t) &= -(\lambda + \mu)P_k(t) + \lambda P_{k-1}(t) + \mu P_{k+1}(t), \qquad k \geq 1 \\ P_0'(t) &= -\lambda P_0(t) + \mu P_1(t). \end{aligned}$$

It is important to know how the above system behaves for large $t$, i.e. if the limit $P_k = \lim_{t\to\infty} P_k(t)$ exists. The probability of $P_k$ describes the steady state of the problem, and so derivatives are zero. Hence $P_k$ satisfy the difference equations

$$\mu P_{k+1} - (\lambda + \mu)P_k + \lambda P_{k-1} = 0 \tag{1.7.16}$$

$$-\lambda P_0 + \mu P_1 = 0 \tag{1.7.17}$$

and the relation $\sum_{k=0}^{\infty} P_k = 1$, which simply states that in the system we must have either no item, or more items. The solution of this problem can be written as

$$P_k = (1 - \rho)\rho^k, \quad \rho = \frac{\lambda}{\mu} < 1 \text{ (we assume)}$$

which is called the geometric distribution. For this distribution important statistical parameters are:

(i) The expected number in the system, i.e.

$$L = \sum_{k=0}^{\infty} kP_k = \frac{\rho}{1-\rho}.$$

(ii) The variance

$$V = \sum_{k=0}^{\infty} (k-L)^2 P_k = L + L^2.$$

(iii) The expected number in the line

$$L_q = \sum_{k=0}^{\infty} (k-\rho) P_k = \rho L.$$

## 1.8. Finite Difference Calculus

Here we shall develop discrete version of Rolle's theorem, the Mean value theorem, Taylor's formula, l'Hospital's rule, Kneser's theorem etc., which are of independent interest and will be repeatedly used in later chapters.

For the function $u(k) = \{0,1,0\}$ we have $\Delta u(k) = \{1,-1\}$. Thus, Rolle's theorem which plays a key role in the development of continuous calculus is not valid for the discrete functions. However, it can be viewed in terms of the sign changes of the function $u(k)$ and of $\Delta u(k)$.

**Definition 1.8.1.** For a given function $u(k)$ defined on $\mathbb{N}(a,b)$, we say $k = a$ is a *node* for $u(k)$ if $u(a) = 0$, and $a < k \le b$ is a *node* for $u(k)$ if either $u(k) = 0$ or $u(k-1)u(k) < 0$. (This definition can be generalized in several different ways, e.g. see Definition 6.16.1.)

**Theorem 1.8.1.** (Discrete Rolle's Theorem). Suppose that the function $u(k)$ is defined on $\mathbb{N}(1,m)$ and has $P_m$ nodes and that $\Delta u(k)$ on $\mathbb{N}(1,m-1)$ has $Q_m$ nodes. Then, $Q_m \ge P_m - 1$.

**Proof.** The result is obvious if $m = 2$. Assume $m > 2$ and that the result holds if $m$ is replaced by an integer $i < m$. If $P_m = P_{m-1}$, then the result holds. Suppose therefore that $k = m$ is a node, so that $P_m = P_{m-1} + 1$. We can also suppose that $P_{m-1} \ge 1$. Now we have the following cases to consider.

**Case 1.** $u(m) = u(m-1) = 0$. Then, obviously $Q_m = Q_{m-1} + 1$.

**Case 2.** $u(m) = 0$, $u(m-1) \neq 0$, say, $u(m-1) > 0$. Let $k = i$ be the largest node for $u(1), u(2), \cdots$, with $i \leq m-1$. Hence, $P_i = P_{m-1}$ and $Q_i \geq P_{m-1} - 1$. We consider the alternatives $u(i) = 0$ or $u(i)u(i-1) < 0$. In the first alternative $i < m-1$ and $u(i+1) > 0$, so that $\Delta u(i) > 0$, while $\Delta u(m-1) < 0$, so that $Q_m \geq Q_i + 1 \geq P_{m-1} = P_m - 1$. In the second alternative $u(m-1) > 0$ implies that $u(i) > 0$, $u(i-1) < 0$, so that $\Delta u(i-1) > 0$. Again $\Delta u(m-1) < 0$ implies the desired result.

**Case 3.** $u(m)u(m-1) < 0$, say, $u(m) < 0$, $u(m-1) > 0$. The arguments here are as in the last case. ∎

**Theorem 1.8.2.** (Discrete Mean Value Theorem). Suppose that the function $u(k)$ is defined on $\mathbb{N}(a,b)$. Then, there exists a $c \in \mathbb{N}(a+1, b-1)$ such that

$$\Delta u(c) \leq \frac{u(b) - u(a)}{b-a} \leq \nabla u(c) \tag{1.8.1}$$

or

$$\Delta u(c) \geq \frac{u(b) - u(a)}{b-a} \geq \nabla u(c). \tag{1.8.2}$$

**Proof.** Let $f(k)$ be a function defined on $\mathbb{N}(a,b)$, and let it attain its maximum at $c$, where $c \in \mathbb{N}(a+1, b-1)$. Then, $f(c) \geq f(c+k)$, $k \in \mathbb{N}(0, b-c)$ and $f(c) \geq f(c-k)$, $k \in \mathbb{N}(0, c-a)$. Therefore, it follows that $f(c-k) - f(c) \leq 0 \leq f(c) - f(c+k)$, $k \in \mathbb{N}(0, \min\{b-c, c-a\})$. Similarly, if $f(k)$ attains its minimum at $c \in \mathbb{N}(a+1, b-1)$, then $f(c-k) - f(c) \geq 0 \geq f(c) - f(c+k)$, $k \in \mathbb{N}(0, \min\{b-c, c-a\})$.

Let $g(k)$ be a function defined on $\mathbb{N}(a,b)$, such that $g(a) = g(b)$. Then, $g(k)$ will attain its maximum or minimum at some $c \in \mathbb{N}(a+1, b-1)$. (If $g(k)$ is a constant, then we can take any point of $\mathbb{N}(a+1, b-1)$.)

We define an auxiliary function $v(k)$ on $\mathbb{N}(a,b)$ as follows

$$v(k) = u(k) - \frac{u(b) - u(a)}{b-a} k.$$

Obviously, $v(a) = v(b) = (bu(a) - au(b))/(b-a)$. Therefore, there exists some $c \in \mathbb{N}(a+1, b-1)$ such that

$$\left[u(c-k) - \frac{u(b)-u(a)}{b-a}(c-k)\right] - \left[u(c) - \frac{u(b)-u(a)}{b-a}c\right] \leq (\geq)\, 0$$

$$\leq (\geq) \left[u(c) - \frac{u(b)-u(a)}{b-a}c\right] - \left[u(c+k) - \frac{u(b)-u(a)}{b-a}(c+k)\right],$$

$$k \in \mathbb{N}(0, \min\{b-c, c-a\})$$

which is the same as

$$u(c-k)-u(c)+\frac{u(b)-u(a)}{b-a}k \le (\ge)\, 0 \le (\ge)\, u(c)-u(c+k)+\frac{u(b)-u(a)}{b-a}k$$

and hence

$$\frac{u(c-k)-u(c)}{k} \le (\ge)\, -\frac{u(b)-u(a)}{b-a} \le (\ge)\, \frac{u(c)-u(c+k)}{k},$$

$$k \in \mathbb{N}(1, \min\{b-c, c-a\}).$$

Thus, in particular for $k = 1$ it follows that

$$u(c+1)-u(c) \le (\ge)\, \frac{u(b)-u(a)}{b-a} \le (\ge)\, u(c)-u(c-1). \qquad \blacksquare$$

**Corollary 1.8.3.** Suppose that the function $u(k)$ is defined on $\mathbb{N}(a,b)$, and $M = \max\{|\Delta u(k)| : k \in \mathbb{N}(a, b-1)\}$. Then,

$$\left|\frac{u(b)-u(a)}{b-a}\right| \le M. \tag{1.8.3}$$

**Lemma 1.8.4.** (Product Formulae). Let $u(k)$ and $v(k)$ be defined on $\mathbb{N}(a)$. Then, for all $k \in \mathbb{N}(a)$

$$\Delta[u(k)v(k)] = u(k+1)\Delta v(k) + v(k)\Delta u(k) = v(k+1)\Delta u(k) + u(k)\Delta v(k) \tag{1.8.4}$$

and

$$\sum_{\ell=a}^{k-1} u(\ell)\Delta v(\ell) = u(\ell)v(\ell)\Big|_{\ell=a}^{k} - \sum_{\ell=a}^{k-1} \Delta u(\ell)v(\ell+1). \tag{1.8.5}$$

**Theorem 1.8.5.** (Discrete Taylor's Formula). Let $u(k)$ be defined on $\mathbb{N}(a)$. Then, for all $k \in \mathbb{N}(a)$ and $n \ge 1$

$$u(k) = \sum_{i=0}^{n-1} \frac{(k-a)^{(i)}}{i!}\Delta^i u(a) + \frac{1}{(n-1)!}\sum_{\ell=a}^{k-n}(k-\ell-1)^{(n-1)}\Delta^n u(\ell). \tag{1.8.6}$$

**Proof.** The proof is by induction. For $n = 1$, (1.8.6) is the same as

$$u(k) = u(a) + \sum_{\ell=a}^{k-1}\Delta u(\ell) = u(a) + u(k) - u(a) = u(k).$$

Assuming (1.8.6) is true for $n = m$, then since

$$\sum_{\ell=a}^{k-m}(k-\ell-1)^{(m-1)}\Delta^m u(\ell) = -\frac{1}{m}\sum_{\ell=a}^{k-m}\Delta_\ell(k-\ell)^{(m)}\Delta^m u(\ell)$$

identity (1.8.5) provides

$$\begin{aligned}
&\sum_{\ell=a}^{k-m}(k-\ell-1)^{(m-1)}\Delta^m u(\ell)\\
&= -\frac{1}{m}\left\{(k-\ell)^{(m)}\Delta^m u(\ell)\Big|_{\ell=a}^{k-m+1} - \sum_{\ell=a}^{k-m}(k-\ell-1)^{(m)}\Delta^{m+1}u(\ell)\right\}\\
&= \frac{1}{m}(k-a)^{(m)}\Delta^m u(a) + \frac{1}{m}\sum_{\ell=a}^{k-m-1}(k-\ell-1)^{(m)}\Delta^{m+1}u(\ell)
\end{aligned}$$

from which (1.8.6) follows for $n = m+1$. ∎

**Corollary 1.8.6.** Let $u(k)$ be defined on $\mathbb{N}(a)$. Then, for all $k \in \mathbb{N}(a)$ and $0 \le m \le n-1$

$$\begin{aligned}
(1.8.7)\qquad \Delta^m u(k) &= \sum_{i=m}^{n-1}\frac{(k-a)^{(i-m)}}{(i-m)!}\Delta^i u(a)\\
&\quad + \frac{1}{(n-m-1)!}\sum_{\ell=a}^{k-n+m}(k-\ell-1)^{(n-m-1)}\Delta^n u(\ell).
\end{aligned}$$

**Remark 1.8.1.** In the right side of (1.8.6) the first term is *Newton's forward difference interpolating polynomial,* whereas the second term is the *remainder.* Obviously, for the remainder Lagrange's analog is not expected, however since

$$\begin{aligned}
(1.8.8)\qquad &\left|u(k) - \sum_{i=0}^{n-1}\frac{(k-a)^{(i)}}{i!}\Delta^i u(a)\right|\\
&\le \frac{1}{(n-1)!}\sum_{\ell=a}^{k-n}(k-\ell-1)^{(n-1)}\max_{\ell\in\mathbb{N}(a,k-n)}|\Delta^n u(\ell)|\\
&= \frac{1}{n!}(k-a)^{(n)}\max_{\ell\in\mathbb{N}(a,k-n)}|\Delta^n u(\ell)|
\end{aligned}$$

one has an error estimate only in terms of $\Delta^n u(k)$.

**Theorem 1.8.7.** (Discrete l'Hospital's Rule). Let $u(k)$ and $v(k)$ be defined on $\mathbb{N}(a)$ and $v(k) > 0$, $\Delta v(k) < 0$ for all large $k$ in $\mathbb{N}(a)$.

Then, if $\lim_{k\to\infty} u(k) = \lim_{k\to\infty} v(k) = 0$

$$(1.8.9)\quad \liminf \frac{\Delta u(k)}{\Delta v(k)} \le \liminf \frac{u(k)}{v(k)} \le \limsup \frac{u(k)}{v(k)} \le \limsup \frac{\Delta u(k)}{\Delta v(k)}.$$

**Proof.** Let $k_1 \in \mathbb{N}(a)$ be sufficiently large so that for all $k \in \mathbb{N}(k_1)$, $v(k) > 0$ and $\Delta v(k) < 0$. We assume that

$$\frac{\Delta u(k)}{\Delta v(k)} \ge c \quad \text{for all } k \in \mathbb{N}(k_1),$$

where $c \in \mathbb{R}$. Then, $\Delta u(k) \le c\Delta v(k)$ and by summation

$$u(k+p) - u(k) \le c(v(k+p) - v(k)) \quad \text{for all } k \in \mathbb{N}(k_1) \text{ and } 0 < p \in \mathbb{N}.$$

Letting $p \to \infty$, we find $-u(k) \le -cv(k)$, which is the same as

$$\frac{u(k)}{v(k)} \ge c \quad \text{for all } k \in \mathbb{N}(k_1).$$

Since the same holds with the inequalities reversed, (1.8.9) holds. ∎

**Corollary 1.8.8.** Let $u(k)$ and $v(k)$ be as in Theorem 1.8.7. Then,

$$\lim_{k\to\infty} \frac{u(k)}{v(k)} = c \quad \text{provided} \quad \lim_{k\to\infty} \frac{\Delta u(k)}{\Delta v(k)} = c \quad \text{exists.}$$

**Theorem 1.8.9.** (Discrete l'Hospital's Rule). Let $u(k)$ and $v(k)$ be defined on $\mathbb{N}(a)$ and $v(k) > 0$, $\Delta v(k) > 0$ for all large $k$ in $\mathbb{N}(a)$. Then, if $\lim_{k\to\infty} v(k) = \infty$

$$(1.8.10)\qquad \lim_{k\to\infty} \frac{\Delta u(k)}{\Delta v(k)} = c \quad \text{implies} \quad \lim_{k\to\infty} \frac{u(k)}{v(k)} = c.$$

**Proof.** $\lim_{k\to\infty} \frac{\Delta u(k)}{\Delta v(k)} = c$ (finite) implies that for every given $\epsilon > 0$ there exists a large $k_1$ in $\mathbb{N}(a)$ such that

$$(c-\epsilon)\Delta v(k) \le \Delta u(k) \le (c+\epsilon)\Delta v(k) \quad \text{for all } k \in \mathbb{N}(k_1).$$

Summing the above inequality, we find

$$\begin{aligned}(c-\epsilon)[v(k+p) - v(k)] &\le u(k+p) - u(k) \\ &\le (c+\epsilon)[v(k+p) - v(k)] \quad \text{for all } k \in \mathbb{N}(k_1) \text{ and } 0 < p \in \mathbb{N},\end{aligned}$$

which is the same as

$$(c-\epsilon)\left[1-\frac{v(k)}{v(k+p)}\right] \le \frac{u(k+p)}{v(k+p)}-\frac{u(k)}{v(k+p)} \le (c+\epsilon)\left[1-\frac{v(k)}{v(k+p)}\right].$$

Letting $p \to \infty$ in the above inequality leads to (1.8.10).

If $c$ is infinite, say, $\infty$ (the case $-\infty$ can be treated similarly) then for an arbitrary $C > 0$ there exists $k_2 \in \mathbb{N}(a)$ large so that

$$\frac{\Delta u(k)}{\Delta v(k)} \ge C \quad \text{for all} \quad k \in \mathbb{N}(k_2),$$

i.e.

$$\Delta u(k) \ge C\Delta v(k).$$

Summing the above inequality, we find

$$u(k+p)-u(k) \ge C[v(k+p)-v(k)] \quad \text{for all} \ \ k \in \mathbb{N}(k_2) \ \text{and} \ \ 0 < p \in \mathbb{N}.$$

Taking $p \to \infty$ in the above inequality gives $\lim_{k\to\infty} \frac{u(k)}{v(k)} \ge C$. ∎

**Lemma 1.8.10.** Let $1 \le m \le n-1$ and $u(k)$ be defined on $\mathbb{N}(a)$. Then,

(i) $\liminf_{k\to\infty} \Delta^m u(k) > 0$ implies $\lim_{k\to\infty} \Delta^i u(k) = \infty$, $0 \le i \le m-1$

(ii) $\limsup_{k\to\infty} \Delta^m u(k) < 0$ implies $\lim_{k\to\infty} \Delta^i u(k) = -\infty$, $0 \le i \le m-1$.

**Proof.** $\liminf\limits_{k\to\infty} \Delta^m u(k) > 0$ implies that there exists a large $k_1 \in \mathbb{N}(a)$ such that $\Delta^m u(k) \ge c > 0$ for all $k \in \mathbb{N}(k_1)$. Since

$$\Delta^{m-1}u(k) = \Delta^{m-1}u(k_1) + \sum_{\ell=k_1}^{k-1} \Delta^m u(\ell)$$

it follows that $\Delta^{m-1}u(k) \ge \Delta^{m-1}u(k_1)+c(k-k_1)$, and hence $\lim\limits_{k\to\infty} \Delta^{m-1}u(k) = \infty$. The rest of the proof is by induction. The case (ii) can be treated similarly. ∎

**Theorem 1.8.11.** (Discrete Kneser's Theorem). Let $u(k)$ be defined on $\mathbb{N}(a)$, and $u(k) > 0$ with $\Delta^n u(k)$ of constant sign on $\mathbb{N}(a)$ and not identically zero. Then, there exists an integer $m$, $0 \le m \le n$ with $n+m$ odd for $\Delta^n u(k) \le 0$ or $n+m$ even for $\Delta^n u(k) \ge 0$ and such that

$m \le n-1$ implies $(-1)^{m+i}\Delta^i u(k) > 0$ for all $k \in \mathbb{N}(a)$, $m \le i \le n-1$

$m \geq 1$ implies $\Delta^i u(k) > 0$ for all large $k \in \mathbb{N}(a)$, $1 \leq i \leq m-1$.

**Proof.** There are two cases to consider.

**Case 1.** $\Delta^n u(k) \leq 0$ on $\mathbb{N}(a)$. First we shall prove that $\Delta^{n-1}u(k) > 0$ on $\mathbb{N}(a)$. If not, then there exists some $k_1 \geq a$ in $\mathbb{N}(a)$ such that $\Delta^{n-1}u(k_1) \leq 0$. Since $\Delta^{n-1}u(k)$ is decreasing and not identically constant on $\mathbb{N}(a)$, there exists $k_2 \in \mathbb{N}(k_1)$ such that $\Delta^{n-1}u(k) \leq \Delta^{n-1}u(k_2) < \Delta^{n-1}u(k_1) \leq 0$ for all $k \in \mathbb{N}(k_2)$, But, from Lemma 1.8.10 we find $\lim_{k\to\infty} u(k) = -\infty$ which is a contradiction to $u(k) > 0$. Thus, $\Delta^{n-1}u(k) > 0$ on $\mathbb{N}(a)$ and there exists a smallest integer $m$, $0 \leq m \leq n-1$ with $n+m$ odd and

$$(1.8.11) \qquad (-1)^{m+i}\Delta^i u(k) > 0 \quad \text{on} \quad \mathbb{N}(a), \quad m \leq i \leq n-1.$$

Next let $m > 1$ and

$$(1.8.12) \qquad \Delta^{m-1}u(k) < 0 \quad \text{on} \quad \mathbb{N}(a),$$

then once again from Lemma 1.8.10 it follows that

$$(1.8.13) \qquad \Delta^{m-2}u(k) > 0 \quad \text{on} \quad \mathbb{N}(a).$$

Inequalities (1.8.11) (1.8.13) can be unified to

$$(-1)^{(m-2)+i}\Delta^i u(k) > 0 \quad \text{on} \quad \mathbb{N}(a), \quad m-2 \leq i \leq n-1$$

which is a contradiction to the definition of $m$. So, (1.8.12) fails and $\Delta^{m-1}u(k) \geq 0$ on $\mathbb{N}(a)$. From (1.8.11), $\Delta^{m-1}u(k)$ is nondecreasing and hence $\lim_{k\to\infty} \Delta^{m-1}u(k) > 0$. If $m > 2$, we find from Lemma 1.8.10 that $\lim_{k\to\infty} \Delta^i u(k) = \infty$, $1 \leq i \leq m-2$. Thus, $\Delta^i u(k) > 0$ for all large $k \in \mathbb{N}(a)$, $1 \leq i \leq m-1$.

**Case 2.** $\Delta^n u(k) \geq 0$ on $\mathbb{N}(a)$. Let $k_3 \in \mathbb{N}(k_2)$ be such that $\Delta^{n-1}u(k_3) \geq 0$, then since $\Delta^{n-1}u(k)$ is nondecreasing and not identically constant, there exists some $k_4 \in \mathbb{N}(k_3)$ such that $\Delta^{n-1}u(k) > 0$ for all $k \in \mathbb{N}(k_4)$. Thus, $\lim_{k\to\infty} \Delta^{n-1}u(k) > 0$ and from Lemma 1.8.10 $\lim_{k\to\infty} \Delta^i u(k) = \infty$, $1 \leq i \leq n-2$ and so $\Delta^i u(k) > 0$ for all large $k$ in $\mathbb{N}(a)$, $1 \leq i \leq n-1$. This proves the theorem for $m = n$. In case $\Delta^{n-1}u(k) < 0$ for all $k \in \mathbb{N}(a)$, we find from Lemma 1.8.10 that $\Delta^{n-2}u(k) > 0$ for all $k \in \mathbb{N}(a)$. The rest of the proof is the same as in Case 1. ∎

**Corollary 1.8.12.** Let $u(k)$ be defined on $\mathbb{N}(a)$, and $u(k) > 0$ with $\Delta^n u(k) \leq 0$ on $\mathbb{N}(a)$ and not identically zero. Then, there exists a large

$k_1$ in $\mathbb{N}(a)$ such that for all $k \in \mathbb{N}(k_1)$

(1.8.14) $$u(k) \geq \frac{1}{(n-1)!}\Delta^{n-1}u\left(2^{n-m-1}k\right)(k-k_1)^{(n-1)}.$$

**Proof.** From Theorem 1.8.11 it follows that $(-1)^{n+i-1}\Delta^i u(k) > 0$ on $\mathbb{N}(a)$, $m \leq i \leq n-1$, and $\Delta^i u(k) > 0$ for all large $k$ in $\mathbb{N}(a)$, say, for all $k \geq k_1$ in $\mathbb{N}(a)$, $1 \leq i \leq m-1$. Using these inequalities, we obtain

$$\begin{aligned}
-\Delta^{n-2}u(k) &= -\Delta^{n-2}u(\infty) + \sum_{\ell=k}^{\infty}\Delta^{n-1}u(\ell) \\
&\geq \sum_{\ell=k}^{2k}\Delta^{n-1}u(\ell) \;\geq\; \Delta^{n-1}u(2k)(k)^{(1)} \\
\Delta^{n-3}u(k) &= \Delta^{n-3}u(\infty) - \sum_{\ell=k}^{\infty}\Delta^{n-2}u(\ell) \;\geq\; \sum_{\ell=k}^{2k}\ell^{(1)}\Delta^{n-1}u(2\ell) \\
&\geq \sum_{\ell=k}^{2k}(\ell-k)^{(1)}\Delta^{n-1}u(2\ell) \;\geq\; \Delta^{n-1}u\left(2^2k\right)\cdot\frac{1}{2!}(k)^{(2)} \\
&\cdots \quad \cdots \\
\Delta^m u(k) &\geq \Delta^{n-1}u\left(2^{n-m-1}k\right)\frac{1}{(n-m-1)!}(k)^{(n-m-1)}.
\end{aligned}$$

Next, we get

$$\begin{aligned}
\Delta^{m-1}u(k) &= \Delta^{m-1}u(k_1) + \sum_{\ell=k_1}^{k-1}\Delta^m u(\ell) \\
&\geq \sum_{\ell=k_1}^{k-1}\frac{1}{(n-m-1)!}(\ell-k_1)^{(n-m-1)}\Delta^{n-1}u\left(2^{n-m-1}\ell\right) \\
&\geq \frac{1}{(n-m)!}\Delta^{n-1}u\left(2^{n-m-1}k\right)(k-k_1)^{(n-m)}.
\end{aligned}$$

Hence, after $(m-1)$ summations, we obtain (1.8.14). ∎

**Corollary 1.8.13.** Let $u(k)$ be as in Corollary 1.8.12 and bounded. Then,

(i) $\lim_{k\to\infty}\Delta^i u(k) = 0, \quad 1 \leq i \leq n-1$

(ii) $(-1)^{i+1}\Delta^{n-i}u(k) \geq 0$ for all $k \in \mathbb{N}(a)$, $1 \leq i \leq n-1$.

**Proof.** Part (i) follows from Lemma 1.8.10. Also, for Part (ii) we note that in the conclusion of Theorem 1.8.11, $m$ cannot be greater than 1. ∎

**Corollary 1.8.14.** let $u(k)$ be as in Corollary 1.8.12. Then, exactly one of the following is true

(i) $\lim_{k\to\infty} \Delta^i u(k) = 0, \quad 1 \le i \le n-1$

(ii) there is an odd integer $j$, $1 \le j \le n-1$ such that $\lim_{k\to\infty} \Delta^{n-i} u(k) = 0$ for $1 \le i \le j-1$, $\lim_{k\to\infty} \Delta^{n-j} u(k) \ge 0$ (finite), $\lim_{k\to\infty} \Delta^{n-j-1} u(k) > 0$ and $\lim_{k\to\infty} \Delta^i u(k) = \infty$, $0 \le i \le n-j-2$.

**Proof.** The proof is contained in Theorem 1.8.11 and Corollary 1.8.13. ∎

## 1.9. Problems

**1.9.1.** Evaluate the following

(i) $\Delta^3(1-k)(1-2k)(1-3k)$, (ii) $\Delta^n e^{\alpha k+\beta}$, (iii) $\dfrac{\Delta^2}{E}k^3$, (iv) $\dfrac{\Delta^2 k^3}{Ek^3}$.

**1.9.2.** Let $u(k)$ and $v(k)$ be defined on $\overline{\mathbb{N}}_1$. Show that for all $k \in \overline{\mathbb{N}}$

(i) $\Delta[u(k)+v(k)] = \Delta u(k) + \Delta v(k)$

(ii) $\Delta c u(k) = c\Delta u(k)$, $c$ is a constant

(iii) $\Delta\left[\dfrac{u(k)}{v(k)}\right] = \dfrac{v(k)\Delta u(k) - u(k)\Delta v(k)}{v(k+1)v(k)}, \quad v(k+1)v(k) \ne 0.$

**1.9.3.** Show that

(i) $\Delta_t (t)^{(m)} = m(t)^{(m-1)}, \quad t \in \mathbb{R}$

(ii) $\Delta^m \sin(\alpha+\beta k) = \left(2\sin\dfrac{\beta}{2}\right)^m \sin\left(\alpha+\beta k + m\dfrac{(\beta+\pi)}{2}\right),$

$\alpha$ and $\beta$ are constants

(iii) $\Delta^m \cos(\alpha+\beta k) = \left(2\sin\dfrac{\beta}{2}\right)^m \cos\left(\alpha+\beta k + m\dfrac{(\beta+\pi)}{2}\right)$

(iv) $\Delta \tan k = \sec^2 k \dfrac{\tan 1}{1-\tan 1 \tan k}$

(v) $\Delta \sec k = -\dfrac{\Delta \cos k}{\cos(k+1)\cos k}$

(vi) $\Delta \tan^{-1} k = \tan^{-1}\left(\dfrac{1}{1+k+k^2}\right)$

(vii) $\Delta \sinh(\alpha k+\beta) = 2\sinh\dfrac{1}{2}\alpha \cosh\left(\alpha k+\beta+\dfrac{1}{2}\alpha\right)$

(viii) $\Delta \cosh(\alpha k+\beta) = 2\sinh\dfrac{1}{2}\alpha \sinh\left(\alpha k+\beta+\dfrac{1}{2}\alpha\right).$

**1.9.4.** Show that for all $r, k, m \in \mathbb{N}$

(i) $\sum_{i=0}^{r}(-1)^i\binom{r}{i}(k+r-i)^m = \begin{cases} 0 & \text{for } r > m \\ m! & \text{for } r = m \end{cases}$

(ii) $\Delta^r(0^m) = \begin{cases} 0 & \text{for } r > m \\ m! & \text{for } r = m \end{cases}$

(iii) $k^{(m)}(k-m)^{(r)} = (k)^{(m+r)}$.

**1.9.5.** Show that the $n$th forward as well as backward difference of a polynomial of $n$th degree is a constant.

**1.9.6.** The solution of the difference equation $\Delta u(k) = f(k),\ k \in \mathbb{N}$ is $u(k) = c + \sum_{\ell=0}^{k-1} f(\ell)$, where $c$ is an arbitrary constant. We define the operator $\Delta^{-1} = \sum_{\ell=0}^{k-1}$, i.e. it is the inverse process of differencing. The higher order antidifferences for a positive integer $m$ are defined as $\Delta^{-m}u(k) = \Delta^{-1}\left[\Delta^{-m+1}u(k)\right]$. Show that

(i) $\Delta\Delta^{-1} \neq \Delta^{-1}\Delta$, (ii) $\Delta^{-1}$ is linear

(iii) $\Delta^{-1}1 = k$, (iv) $\Delta^{-1}\alpha^k = \dfrac{\alpha^k}{\alpha - 1}$, $\alpha \neq 1$

(v) $\Delta_t^{-1}(\alpha + \beta t)^{(k)} = \dfrac{(\alpha+\beta t)^{(k+1)}}{\beta(k+1)}$

(vi) $\Delta^{-1}\sin(\alpha + \beta k) = -\dfrac{\cos(\alpha + \beta k - \beta/2)}{2\sin(\beta/2)}$.

**1.9.7.** Let the functions $u(k)$ and $v(k)$ be defined on $\mathbb{N}(1)$. Prove that

$$\sum_{\ell=1}^{k} u(\ell)v(\ell) = u(k+1)\sum_{\ell=1}^{k} v(\ell) - \sum_{\ell=1}^{k}\left(\Delta u(\ell)\sum_{\tau=1}^{\ell} v(\tau)\right). \tag{1.9.1}$$

Relation (1.9.1) is called ***Abel's transformation.*** Use it to show that

$$\sum_{\ell=1}^{k} \ell 2^\ell = (k-1)2^{k+1} + 2.$$

**1.9.8.** Let $u(k)$ be defined on $\overline{\mathbb{N}}_n$. Show that for all $k \in \overline{\mathbb{N}}$

$$\Delta^n u(k) = \sum_{i=1}^{n}(-1)^{n-i}\binom{n}{i}(u(k+i) - u(k)).$$

**1.9.9.** Show that

$$u(k) = c_1(k)! + c_2(-1)^k(k)! + (k)!\left(\frac{k^2}{4} + \frac{5k}{4} - \frac{1}{2}\left[\frac{k+1}{2}\right]\right), \quad k \in \mathbb{N}$$

is the general solution of the second order difference equation

$$u(k+2) - (k+2)(k+1)u(k) = (k+3)!, \quad k \in \mathbb{N}.$$

**1.9.10.** Let $u(k)$ be defined on $\mathbb{N}(a)$. Show that for all $k \in \mathbb{N}(a)$

$$1 + \sum_{\ell=a}^{k-1} u(\ell) \prod_{\tau=\ell+1}^{k-1} (1 + u(\tau)) = \prod_{\ell=a}^{k-1}(1 + u(\ell)).$$

**1.9.11.** Let the function $u(k)$ be defined on $\mathbb{N}$ and the series $S(t) = \sum_{k=0}^{\infty} u(k)t^k$ converges absolutely. Show that

$$S(t) = \sum_{k=0}^{\infty} \frac{t^k}{(1-t)^{k+1}} \Delta^k u(0).$$

This result is known as *Montmort's theorem*. In particular deduce that

$$\sum_{k=1}^{\infty} kt^k = \frac{t}{(1-t)^2}.$$

**1.9.12.** Let the function $u(k)$ be defined on $\mathbb{N}$ and the series $S(t) = \sum_{k=0}^{\infty} \frac{u(k)}{k!} t^k$ converges absolutely. Show that

$$S(t) = e^t \sum_{k=0}^{\infty} \frac{1}{k!} t^k \Delta^k u(0).$$

In particular deduce that

$$\sum_{k=0}^{\infty} \frac{(k^2-1)}{k!} t^k = e^t(t^2 + t - 1).$$

**1.9.13.** Show that the continued fraction

$$r_k = \cfrac{a_1}{b_1 + \cfrac{a_2}{b_2 + \cdots + \cfrac{a_k}{b_k}}}$$

can be expressed as $r_k = u(k)/v(k)$, where the numerator $u(k)$ and the denominator $v(k)$ satisfy the linear system

$$\begin{aligned} u(k) &= b_k u(k-1) + a_k u(k-2) \\ v(k) &= b_k v(k-1) + a_k v(k-2), \quad k \in \mathbb{N}(2) \end{aligned}$$

together with the initial conditions

$$\begin{aligned} u(0) &= 0, \quad u(1) = a_1 \\ v(0) &= 1, \quad v(1) = b_1. \end{aligned}$$

**1.9.14.** The *Bernoulli numbers* $B_k$ are defined by the relation

$$\frac{t}{e^t - 1} = \sum_{k=0}^{\infty} \frac{B_k}{k!} t^k, \quad |t| < 2\pi.$$

Show that

(1.9.2) (i) $\dfrac{t}{e^t - 1} + \dfrac{t}{2} = \dfrac{t}{2} \coth \dfrac{t}{2}$

and hence

$$B_1 = -\frac{1}{2}, \quad B_{2k+1} = 0, \quad k \in \mathbb{N}(1)$$

(ii) relation (1.9.2) can be written as

$$\sum_{k=0}^{\infty} \frac{t^{2k}}{2^{2k}(2k)!} = \sum_{i=0}^{\infty} \frac{t^{2i}}{2^{2i}(2i+1)!} \sum_{j=0}^{\infty} \frac{B_{2j}}{(2j)!} t^{2j}$$

and hence deduce the recurrence relation

$$\frac{2^{2k}}{1} \binom{2k}{0} B_{2k} + \frac{2^{2k-2}}{3} \binom{2k}{2} B_{2k-2} + \cdots + \frac{1}{2k+1} \binom{2k}{2k} B_0 = 1, \quad k \in \mathbb{N}(1)$$

with $B_0 = 1$

(iii) use part (ii) to determine $B_2$, $B_4$, $B_6$ and $B_8$.

**1.9.15.** The *Euler numbers* $E_k$ are defined by the relation

(1.9.3) $$\operatorname{sech} t = \sum_{k=0}^{\infty} \frac{E_k}{k!} t^k, \quad |t| < \frac{\pi}{2}.$$

Show that

(i) $E_{2k+1} = 0, \quad k \in \mathbb{N}$

(ii) relation (1.9.3) can be written as

$$1 = \sum_{i=0}^{\infty} \frac{t^{2i}}{(2i)!} \sum_{j=0}^{\infty} \frac{E_{2j}}{(2j)!} t^{2j}$$

and hence deduce the recurrence relation

$$\binom{2k}{0} E_{2k} + \binom{2k}{2} E_{2k-2} + \cdots + \binom{2k}{2k} E_0 = 0, \quad k \in \mathbb{N}(1)$$

with $E_0 = 1$

(iii) use part (ii) to determine $E_2$, $E_4$, $E_6$ and $E_8$.

**1.9.16.** The *Fibonacci numbers* are a sequence of numbers such that each one is the sum of its two predecessors. The first few Fibonacci numbers are $1, 1, 2, 3, 5, 8, 13, \cdots$.

(i) Formulate an initial value problem that will generate the Fibonacci numbers, and verify that its solution can be written as

$$\frac{1}{\sqrt{5}}\left[\left(\frac{1+\sqrt{5}}{2}\right)^k - \left(\frac{1-\sqrt{5}}{2}\right)^k\right].$$

(ii) Show that the ratio of successive Fibonacci numbers tends to $(1 + \sqrt{5})/2$ as $k \to \infty$. This ratio, known as the *golden ratio,* was often used in ancient Greek architecture whenever rectangular structures were constructed. It was believed that when the ratio of the sides of a rectangle was this number, the resulting structure was most pleasing to the eye.

**1.9.17.** Show that for $t \in \mathbb{R}$ the factorial powers $(t)^{(m)}$ and the powers are related by

$$(t)^{(m)} = \sum_{i=1}^{m} s_i^m t^i$$

and

$$t^m = \sum_{i=1}^{m} S_i^m (t)^{(i)},$$

where the coefficients $s_i^m$ are called *Stirling numbers of the first kind,* while the coefficients $S_i^m$ are called *Stirling numbers of the second kind.* These numbers satisfy the relations

$$s_i^{m+1} = s_{i-1}^m - m s_i^m, \quad s_m^m = 1, \quad s_i^m = 0 \text{ for } i \geq m+1$$

and

$$S_i^{m+1} = S_{i-1}^m + i S_i^m, \quad S_m^m = 1, \quad S_i^m = 0 \text{ for } i \geq m+1.$$

In particular, show that

$$\begin{aligned}(t)^{(5)} &= t^5 - 10t^4 + 35t^3 - 50t^2 + 24t\\ t^5 &= (t)^{(5)} + 15(t)^{(4)} + 25(t)^{(3)} + 10(t)^{(2)} + (t)^{(1)}.\end{aligned}$$

**1.9.18.** Let $u(k) = \int_0^1 (\ln t)^k t^\alpha dt$, where $k \in \mathbb{N}$ and $\alpha$ is a positive constant. Show that $u(k)$ is the solution of the initial value problem

$$\begin{aligned}(\alpha+1)u(k+1) + (k+1)u(k) &= 0, \quad k \in \mathbb{N}\\ u(0) &= \frac{1}{\alpha+1}.\end{aligned}$$

**1.9.19.** For $k \in \mathbb{N}(1)$ the *Gamma function* is defined by $\Gamma(k) = \int_0^\infty e^{-t}t^{k-1}dt$. Show that $\Gamma(k)$ is the solution of the initial value problem

$$\Gamma(k+1) = k\Gamma(k), \quad \Gamma(1) = 1$$

and hence $\Gamma(k+1) = (k)!$.

**1.9.20.** Identify $u(k) = I_k(x)$, where

$$I_k(x) = \int_0^\pi \frac{\cos kt - \cos kx}{\cos t - \cos x}dt, \quad k \in \mathbb{N}$$

to show that $u(k)$ satisfies the initial value problem

$$\begin{aligned}u(k+1) - 2\cos x\, u(k) + u(k-1) &= 0, \quad k \in \mathbb{N}(1)\\ u(0) = 0, \quad u(1) &= \pi.\end{aligned}$$

Further, prove that $u(k) = \pi(\sin kx/\sin x)$.

**1.9.21.** Let $u(k) = \int_0^1 \frac{t^k}{5+t}dt$, $k \in \mathbb{N}$. Show that

(i) $u(k)$ is the solution of the initial value problem

$$u(k+1) + 5u(k) = \frac{1}{k+1}, \quad k \in \mathbb{N} \tag{1.9.4}$$

$$u(0) = \ln\frac{6}{5} \tag{1.9.5}$$

(ii) $0 < u(k+1) < u(k)$ and $u(k) \to 0$ as $k \to \infty$

(iii) $u(11)$ obtained from (1.9.4), (1.9.5) by rounding off all the calculations to six decimal places is negative.

**1.9.22.** Let $u(k) = \int_0^2 \frac{t^k}{5+t}dt, \quad k \in \mathbb{N}$. Show that

(i) $u(k)$ is the solution of the initial value problem

$$(1.9.6) \quad u(k+1) + 5u(k) = \frac{2^{k+1}}{k+1}, \quad k \in \mathbb{N}$$

$$(1.9.7) \quad u(0) = \ln\frac{7}{5}$$

(ii) $0 < u(k) < u(k+1), \ k \in \mathbb{N}(1)$ and $u(k) \to \infty$ as $k \to \infty$

(iii) compute $u(50)$ from (1.9.6), (1.9.7) by rounding off all the calculations to eight decimal places and compare it with $u(50) \simeq 0.63425992E+13$.

**1.9.23.** Let $u(k) = \int_0^1 \frac{t^k}{1+t+t^2}dt, \quad k \in \mathbb{N}$. Show that

(i) $u(k)$ satisfies the difference equation

$$(1.9.8) \quad u(k+2) + u(k+1) + u(k) = \frac{1}{k+1}, \quad k \in \mathbb{N}$$

(ii) $0 < u(k+1) < u(k), \ k \in \mathbb{N}$ and $u(k) \to 0$ as $k \to \infty$

(iii) compute $u(0)$ and $u(1)$ directly, and then compute $u(k), \ k \in \mathbb{N}(2,10)$ from (1.9.8).

**1.9.24.** The value of $\pi$ can be calculated by using $\lim_{k\to\infty} k\sin\frac{\pi}{k} = \pi$. Let $v(k) = k\sin\frac{\pi}{k}$ and $u(k) = v(2^k) = 2^k\sin\left(2^{-k}\pi\right)$. Setting $\alpha = 2^{-k}\pi$, from the trigonometric identity $\sin\frac{\alpha}{2} = \sqrt{\frac{1}{2}\left(1-\sqrt{1-\sin^2\alpha}\right)}$ it follows that

$$(1.9.9) \quad u(k+1) = 2^{k+1}\sqrt{\frac{1}{2}\left(1-\sqrt{1-(2^{-k}u(k))^2}\right)}$$

$$(1.9.10) \quad = \sqrt{2}u(k)\Big/\sqrt{1+\sqrt{1-(2^{-k}u(k))^2}}, \quad k \in \mathbb{N}(1).$$

Use (1.9.9) ((1.9.10)) with $u(1) = 2$ to compute $u(2), u(3), \cdots,$ and convince yourself that $u(k) \not\to \pi$ $(u(k) \to \pi)$ as $k \to \infty$. Explain why?

**1.9.25.** We compute the value of $e^t = \sum_{k=0}^{\infty}\frac{t^k}{k!}$ by truncating the infinite series after $K+1$ terms, i.e. $v(K) = \sum_{k=0}^{K} u(k)$, where $u(k) = \frac{t^k}{k!}$, gives an approximate value of $e^t$. Thus, using the computer $e^t$ can be conveniently evaluated by the system of difference equations

$$(1.9.11)\quad \begin{aligned} u(k+1) &= \frac{t}{k+1}u(k), \quad u(0)=1, \quad |t|<1 \\ v(k+1) &= v(k)+u(k+1), \quad v(0)=1, \quad k\in \mathbb{N}(0,K-1). \end{aligned}$$

Also, if $|t|>1$ then $t=[t]+t^*$, where $[t]$ is the integer value of $t$ and $t^*$ its fractional part, and $e^t = e^{[t]}e^{t^*}$. The first factor $e^{[t]}$ is computed by

$$e^{[t]} = \begin{cases} e\cdot e\cdots e, & [t] \text{ times if } t>0 \\ \frac{1}{e}\cdot\frac{1}{e}\cdots\frac{1}{e}, & -[t] \text{ times if } t<0, \end{cases}$$

and the second factor $e^{t^*}$ by the scheme (1.9.11).

Use the above algorithm for $K=10$ to compute the values of $e^{2.3}$ and $e^{-4.7}$.

**1.9.26.** On a machine the sum $S_n = \sum_{i=1}^n a_i$ can be obtained conveniently by using the algorithm: $S_0 = 0$, $S_{i+1} = S_i + a_{i+1}$. However the computation is done on a floating point arithmetic, i.e. instead of the number $a = m\cdot 10^p$ where $0.1 \le m < 1$ one uses $\overline{a} = \overline{m}\cdot 10^p$ where $0.1 \le \overline{m} < 1$ and $\overline{m}$ has only $q$ digits, this means $|a-\overline{a}| < |m-\overline{m}|\cdot 10^p \le 10^{p-q}$. Study the behavior of the errors, considering that $\overline{S}_i = (\overline{S}_{i-1}+\overline{a}_i)(1+10^{-q})$.

**1.9.27.** Show that in system form the boundary value problem (1.6.5), (1.6.6) can be written as

$$\begin{aligned} u_i(k+1) &= u_{i+1}(k), \quad 1\le i\le n-1 \\ u_n(k+1) &= -\frac{q}{p}u_1(k)+\frac{1}{p}u_n(k), \quad k\in \mathbb{N}(0,K-1) \\ u_i(0) &= 1, \quad 1\le i\le n-1, \quad u_n(K) = 0. \end{aligned}$$

**1.9.28.** A drug of amount $u_0$ is administered once every six hours. Let $u(k)$ denote the amount of the drug in the blood system at the $k$th interval, and let body eliminate a certain fraction $q$ of the drug during each time interval. Show that

(i) $u(k+1) = (1-q)u(k)+u_0$

(ii) $u(k) = \left[u_0 - \frac{u_0}{q}\right](1-q)^k + \frac{u_0}{q}$

(iii) $\lim_{k\to\infty} u(k) = \frac{u_0}{q}$.

**1.9.29.** $k$ letters to each of which corresponds an envelope are placed in the envelops at random. If $u(k)$ is the number of ways in which all letters go wrong, show that

(i) $u(k) = (k-1)(u(k-1) + u(k-2))$

(ii) $u(k) - ku(k-1) = (-1)^{k-2}(u(2) - 2u(1)) = (-1)^k$

(iii) $u(k) = k!\left[\frac{1}{2!} - \frac{1}{3!} + \cdots + \frac{(-1)^k}{k!}\right]$

(iv) the probability that all $k$ letters go wrong is given by the first $(k-1)$ terms in the expansion of $1 - e^{-1}$.

**1.9.30.** (Gambler's Ruin). Let a gambler with capital $k$ dollars play against a rich adversary. Let the probabilities of his winning and losing one dollar in any game be $p$ and $q$, respectively, where $p + q = 1$, and let $p(k)$ be the probability of his being ultimately ruined. At the next game, the probability of his winning is $p$ and if he wins, his capital would become $k+1$ and the probability of his ultimate ruin would be $p(k+1)$. On the other hand if he loses at the next game, his capital would become $k-1$ and the probability of his ultimate ruin would be $p(k-1)$. Because these are the only two possibilities, we have

$$p(k) = pp(k+1) + qp(k-1). \tag{1.9.12}$$

Now let the gambler decide to stop this game when his capital becomes $K$ dollars so that the probability of his being ruined when his starting capital is $K$ dollars, is zero, i.e. $p(K) = 0$. Similarly, when his starting capital is zero, he is already ruined, so we have $p(0) = 1$. Thus,

$$p(0) = 1, \quad p(K) = 0. \tag{1.9.13}$$

Verify that the solution of the boundary value problem (1.9.12), (1.9.13) can be written as

$$p(k) = \begin{cases} \dfrac{\lambda^K - \lambda^k}{\lambda^K - 1}, & \lambda = \dfrac{q}{p} \neq 1 \\[2ex] \dfrac{K-k}{K}, & q = p. \end{cases}$$

**1.9.31.** In the previous problem let $N(k)$ denote the expected number of games before the gambler is ruined. If he wins at the next game, his capital becomes $k+1$ and the expected number of games would then be $N(k+1)$ and if he loses, his capital becomes $k-1$ and the expected number of games would be only $N(k-1)$. This leads to the boundary value problem

$$(1.9.14) \qquad N(k) = pN(k+1) + qN(k-1) + 1$$

$$(1.9.15) \qquad N(0) = N(K) = 0.$$

Verify that the solution of the boundary value problem (1.9.14), (1.9.15) can be written as

$$N(k) = \begin{cases} \dfrac{k}{q-p} - \dfrac{K}{q-p}\dfrac{1-\lambda^k}{1-\lambda^K}, & \lambda = \dfrac{q}{p} \neq 1 \\ k(K-k), \quad q = p. \end{cases}$$

**1.9.32.** Suppose a sack contains $r$ red and $w$ white marbles. A marble is drawn at random from the sack, its color is noted and it is replaced. This procedure is repeated $k$ times. Show that $U(k,\ell)$ the number of ways of obtaining $\ell$ red marbles among the $k$ draws satisfies the difference equation

$$(1.9.16) \qquad U(k,\ell) = rU(k-1,\ell-1) + wU(k-1,\ell).$$

Also show that $U(k,\ell) = \binom{k}{\ell} r^\ell w^{k-\ell}$ satisfies the difference equation (1.9.16).

**1.9.33.** Verify the discrete Rolle's theorem for the finite sequence $\{0,1,-1, 2,-2,3,3,-7,0\}$.

**1.9.34.** Verify the discrete mean value theorem and the inequality (1.8.3) for the finite sequence $\{5,7,12,9,36,100\}$.

**1.9.35.** Let $u(k)$ be defined on $\mathbb{N}(a)$. Then, for all $k \in \mathbb{N}(a,p)$, where $p \in \mathbb{N}(a)$, and $0 \le m \le n-1$ show that

$$\Delta^m u(k) = \sum_{i=m}^{n-1} \frac{(p+i-m-1-k)^{(i-m)}}{(i-m)!}(-1)^{i-m}\Delta^i u(p)$$
$$-\frac{(-1)^{n-m-1}}{(n-m-1)!}\sum_{\ell=k}^{p-1}(\ell+n-m-1-k)^{(n-m-1)}\Delta^n u(\ell).$$

**1.9.36.** Let $u(k)$ be defined on $\mathbb{N}(a)$. Show that for all $k \in \mathbb{N}(a)$ and $m \ge 1$

$$\sum_{\ell=a}^{k-1}(\ell)^{(m-1)}\Delta^m u(\ell) = \sum_{i=1}^{m}(-1)^{i+1}\Delta^{i-1}(\ell)^{(m-1)}\Delta^{m-i}u(\ell+i-1)\Big|_{\ell=a}^{k}.$$

**1.9.37.** Let $u(k)$ and $v(k)$ be defined on $\overline{\mathbb{N}}_n$. Show that for all $k \in \overline{\mathbb{N}}$ the following discrete *Leibnitz' formula* holds

$$\Delta^n[u(k)v(k)] = \sum_{i=0}^{n} \binom{n}{i} \Delta^{n-i}u(k)\Delta^i v(k+n-i).$$

**1.9.38.** Construct examples to show that the converse of Corollary 1.8.8 as well as of Theorem 1.8.9 is not necessarily true.

**1.9.39.** Let $u(k)$ be as in Corollary 1.8.12. Show that

$$\lim_{k\to\infty} \frac{\Delta^i u(k)}{u(k)} = 0, \quad 1 \le i \le n-1$$

unless $\lim_{k\to\infty} \Delta^i u(k) = 0,\ 0 \le i \le n-1$. The exceptional case may arise only when $n$ is odd.

## 1.10. Notes

Besides $E$, $\Delta$, $\nabla$ and $\Delta^{-1}$ other operators $\mu$ and $\delta$ and their inter relationships are readily available in several classical books, e.g. Boole [8], Cogan and Norman [13], Fort [19], Gel'fond [20], Jordan [25], Miller [35], Milne Thomson [37], Richardson [39], and Spiegal [41]. Elementary discussions of finite difference equations and various applications are included as a part of difference calculus or differential equations in some of the above books, and also in Brand [9], Brugnano and Trigiante [10], Chorlton [12], Derrick and Grossman [15], Finizio and Ladas [18], Hildebrand [23], Pinney [38]. The books by Agarwal and Wong [3], Ahlbrandt and Peterson [5], Batchelder [7], Elaydi [17], Goldberg [21], Jerri [24], Kelley and Peterson [26], Kocic and Ladas [27], Lakshmikantham and Trigiante [28], Levy and Lessman [31], Mickens [32,33], Miller [36], and Sharkovsky, Maistrenko and Romanenko [40] deal exclusively with difference equations. Example 1.4.2 is taken from Levy and Lessman [31]. Example 1.4.3 and Problem 1.9.21 which show computational difficulties are from Dorn and McCracken [16] and Dahlquist and Björck [14] respectively. Cash [11] and Wimp [42] have devoted their monographs to the solutions of linear as well as nonlinear unstable recurrence relations (also, see our chapter 8). Elementary discussion of boundary value problems for difference equations has been included in several above books, whereas the monograph by Atkinson [6] and the paper by Hartman [22] have attracted many researchers (also, see our Chapters 8 12). For the treatment of spline functions we refer to the most cited monograph by Ahlberg, Nilson and Walsh [4]. Example 1.6.7 is from Fort [19]. The discrete Rolle's theorem is due to Hartman [22], whereas the discrete mean value theorem is essentially new. The rest of the results in Section 1.8 are contained in Agarwal [1], also see [2]. Recently, there has been an attempt to unify the theories of differential and difference equations. For this, the so called *calculus on time scales* has been developed, see Lakshmikantham, Sivasundaram and Kaymakcalan [29].

## 1.11. References


[1]. R.P. Agarwal, Difference calculus with applications to difference equations, in *General Inequalities 4*, ed. W. Walter, ISNM 71, *Birkhäuser Verlag*, Basel, 1984, 95 110.

[2]. R.P. Agarwal, Difference equations and inequalities: A survey, in *Proceedings of the World Congress of Nonlinear Analysts*, Tampa, Florida, August 19 26, 1992, ed. V. Lakshmikantham, *Walter de Gruyter*, Berlin, 1996, 1091 1108.

[3]. R.P. Agarwal and P.J.Y. Wong, *Advanced Topics in Difference Equations*, *Kluwer*, Dordrecht, 1997.

[4]. J.H. Ahlberg, E.N. Nilson and J.L. Walsh, *The Theory of Splines and Their Applications*, *Academic Press*, New York, 1967.

[5]. C.D. Ahlbrandt and A.C. Peterson, *Discrete Hamiltonian Systems: Difference Equations, Continued Fractions and Riccati Equations*, Kluwer, Dordrecht, 1996.

[6]. F.V. Atkinson, *Discrete and Continuous Boundary Value Problems*, *Academic Press*, New York, 1964.

[7]. P.M. Batchelder, *An Introduction to Linear Difference Equations*, *Harvard University Press*, Cambridge, 1927.

[8]. G. Boole, *Calculus of Finite Differences*, 4th ed., *Chelsea*, New York, 1958.

[9]. L. Brand, *Differential and Difference Equations*, *Wiley*, New York, 1966.

[10]. L. Brugnano and D. Trigiante, *Solving Differential Problems to Multistep Initial and Boundary Value Methods*, *Gordon & Breach Sci. Publ.*, Amsterdam, 1998.

[11]. J.R. Cash, *Stable Recursions*, Academic Press, London, 1979.

[12]. F. Chorlton, *Differential and Difference Equations*, *Van Nostrand*, London, 1965.

[13]. E.J. Cogan and R.Z. Norman, *Handbook of Calculus, Differences and Differential Equations, Prentice Hall*, Englewood Cliffs, N.J., 1958.

[14]. G. Dahlquist and A. Björck, *Numerical Methods, Prentice-Hall*, Englewood Cliffs, N.J., 1974.

[15]. W.R. Derrick and S.I. Grossman, *Elementary Differential Equations with Applications, Addison Wesley Publ. Comp.*, Reading, 1981.

[16]. W.S. Dorn and D.D. McCracken, *Numerical Methods with FORTRAN IV Case Studies, Wiley*, New York, 1972.

[17]. S. Elaydi, *An Introduction to Difference Equations, Springer Verlag*, New York, 1996.

[18]. N. Finizio and G. Ladas, *An Introduction to Differential Equations with Difference Equations, Fourier Series, and Partial Differential Equations, Wadsworth Publ. Comp.*, Belmont, 1982.

[19]. T. Fort, *Finite Differences and Difference Equations in the Real Domain, The Clarendon Press*, Oxford, 1948.

[20]. A.O. Gel'fond, *Calculus of Finite Differences, Hindustan*, Delhi, India, 1971.

[21]. S. Goldberg, *Introduction to Difference Equations, Wiley*, New York, 1958.

[22]. P. Hartman, Difference equations: Disconjugacy, principal solutions, Green's functions, complete monotonicity, *Trans. Amer. Math. Soc.* **246**(1978), 1 30.

[23]. F.B. Hildebrand, *Finite Difference Equations and Simulations, Prentice Hall*, Englewood Cliffs, N. J., 1968.

[24]. A.J. Jerri, *Linear Difference Equations with Discrete Transform Methods, Kluwer*, Dordrecht, 1996.

[25]. C. Jordan, *Calculus of Finite Differences, 3rd ed., Chelsa*, New York, 1965.

[26]. W.G. Kelley and A.C. Peterson, *Difference Equations: An Introduction with Applications, Academic Press*, New York, 1991.

[27]. V.L. Kocic and G. Ladas, *Global Behavior of Nonlinear Difference Equations of Higher Order with Applications, Kluwer*, Dordrecht, 1993.

[28]. V. Lakshmikantham and D. Trigiante, *Theory of Difference Equations: Numerical Methods and Applications, Academic Press*, New York, 1988.

[29]. V. Lakshmikantham, S. Sivasundaram and B. Kaymakcalan, *Dynamic Systems on Measure Chains, Kluwer*, Dordrecht, 1996.

[30]. M. Lees, Discrete methods for nonlinear two point boundary value problems, in *Numerical Solutions of Partial Differential Equations*, ed. J.H.

Bramble, *Academic Press*, New York (1966), 59 72.

[31]. H. Levy and F. Lessman, *Finite Difference Equations*, *Sir Issac Pitman and Sons, Ltd.*, London, 1959.

[32]. R.E. Mickens, *Difference Equations: Theory and Applications*, *Van Nostrand Reinhold*, New York, 2nd edition, 1990.

[33]. R.E. Mickens, *Nonstandard Finite Difference Models of Differential Equations*, *World Scientific*, Singapore, 1994.

[34]. J.C.P. Miller, *British Association for the Advancement of Science: Bessel Functions Part II, Mathematical Tables, Vol. 10*, Cambridge University Press, 1952.

[35]. K.S. Miller, *An Introduction to the Calculus of Finite Differences and Difference Equations*, *Holt*, New York, 1960.

[36]. K.S. Miller, *Linear Difference Equations*, *W. A. Benjamin*, New York, 1968.

[37]. L.M. Milne Thomson, *The Calculus of Finite Differences*, *Macmillan*, London, 1960.

[38]. E. Pinney, *Ordinary Difference Differential Equations*, *University of California Press*, Berkeley and Los Angeles, 1959.

[39]. C.H. Richardson, *An Introduction to the Calculus of Finite Differences*, *Van Nostrand*, New York, 1954.

[40]. A.N. Sharkovsky, Y.L. Maistrenko and E.Y. Romanenko, *Difference Equations and Their Applications*, *Kluwer*, Dordrecht, 1993.

[41]. M.R. Spiegal, *Calculus of Finite Differences and Difference Equations*, *McGraw Hill*, New York, 1971.

[42]. J. Wimp, *Computation with Recurrence Relations*, *Pitman Advanced Publishing Program*, Boston, 1984.

# Chapter 2
# Linear Initial Value Problems

This chapter investigates the fundamental theory and the essential techniques employed in the study of linear initial value problems. An inherent property which makes linear systems simple to deal with, is the superposition principle. We begin this chapter with this principle and discuss some of its consequences. Then, we collect several definitions and results from algebra which are used in later chapters also. Next, the concept of linearly independent functions, the Casoratian matrix, the fundamental matrix solution, and its explicit representation along with its properties are discussed in detail. Here an interesting example from Markov chains is also illustrated. This is followed by the method of variation of constants for the solutions of nonhomogeneous difference systems. Next we discuss adjoint systems and develop adjoint identities which are used in Chapter 8. Then, we consider the systems with constant coefficients and provide some constructive methods for their closed form solutions. These methods do not use Jordan form and can easily be mastered. A very important aspect of the qualitative study of the solutions of difference systems is their periodicity. In Sections 2.9 and 2.10 respectively, we provide necessary and sufficient conditions so that the solutions of a given system are periodic and almost periodic. Eventhough, higher order equations are expressible as difference systems, they merit some special attention. In Section 2.11 we incorporate the method of variation of constants, the concepts of exact and adjoint equations, and Lagrange's and Green's identities. This is followed by the method of generating functions, which is a very elegant technique for obtaining the closed form solutions of higher order difference equations. Higher order difference equations with constant coefficients find an application in computing the roots of a given polynomial. This classical method originally due to Bernoulli is presented in Section 2.13. Bernoulli's method also provides the motivation of several important results, e.g. Poincaré's and Perron's theorems, which we discuss in Section 2.14. In Section 2.15 we introduce and illustrate the regular and singular perturbation techniques for the construction of the solutions of difference equations.

## 2.1. Introduction

An important characteristic property of linear systems, which makes them especially simple to treat, is the *superposition principle:* If $\mathbf{u}(k)$ is a solution of the system $\mathbf{u}(k+1) = \mathcal{A}(k)\mathbf{u}(k) + \mathbf{b}^1(k),\ k \in \overline{\mathbb{N}}$ and $\mathbf{v}(k)$ is a solution of $\mathbf{v}(k+1) = \mathcal{A}(k)\mathbf{v}(k) + \mathbf{b}^2(k),\ k \in \overline{\mathbb{N}}$ then $\mathbf{z}(k) = c_1\mathbf{u}(k) + c_2\mathbf{v}(k)$ is a solution of the system $\mathbf{z}(k+1) = \mathcal{A}(k)\mathbf{z}(k) + c_1\mathbf{b}^1(k) + c_2\mathbf{b}^2(k)$. For this, we have

$$\begin{aligned}
\mathbf{z}(k+1) &= c_1\mathbf{u}(k+1) + c_2\mathbf{v}(k+1) \\
&= c_1(\mathcal{A}(k)\mathbf{u}(k) + \mathbf{b}^1(k)) + c_2(\mathcal{A}(k)\mathbf{v}(k) + \mathbf{b}^2(k)) \\
&= \mathcal{A}(k)(c_1\mathbf{u}(k) + c_2\mathbf{v}(k)) + c_1\mathbf{b}^1(k) + c_2\mathbf{b}^2(k) \\
&= \mathcal{A}(k)\mathbf{z}(k) + c_1\mathbf{b}^1(k) + c_2\mathbf{b}^2(k), \qquad k \in \overline{\mathbb{N}}.
\end{aligned}$$

Thus, in particular, if $\mathbf{b}^1(k) = \mathbf{b}^2(k) = 0$ for all $k \in \overline{\mathbb{N}}$, i.e. $\mathbf{u}(k)$ and $\mathbf{v}(k)$ are solutions of the homogeneous system (1.2.12), then $c_1\mathbf{u}(k) + c_2\mathbf{v}(k)$ is also a solution. Hence, solutions of the homogeneous system (1.2.12) form a vector space. Further, if $\mathbf{u}(k)$ is a solution of (1.2.11) on $\overline{\mathbb{N}}_1$, then $\mathbf{v}(k)$ is also a solution of (1.2.11) on $\overline{\mathbb{N}}_1$ if and only if $\mathbf{u}(k) - \mathbf{v}(k)$ is a solution of (1.2.12) on $\overline{\mathbb{N}}_1$. Hence, the general solution of (1.2.11) is obtained by adding to a particular solution of (1.2.11) the general solution of the corresponding homogeneous system (1.2.12).

## 2.2. Preliminary Results from Algebra

For our ready reference we collect here several fundamental concepts and results from algebra.

**Lemma 2.2.1.** Consider the system of $n$ linear equations

$$\mathcal{A}\mathbf{u} = \mathbf{v}, \tag{2.2.1}$$

where $\mathcal{A}$ is an $n \times n$ matrix and $\mathbf{u}$, $\mathbf{v}$ are $n$ dimensional vectors. Then, if

(i) Rank $\mathcal{A} = n$, i.e. *det* $\mathcal{A} \neq 0$, the system (2.2.1) possesses a unique solution. Alternatively, the homogeneous system $\mathcal{A}\mathbf{u} = 0$ possesses only the trivial solution.

(ii) Rank $\mathcal{A} = n - m\ (1 \leq m \leq n)$, the system (2.2.1) possesses a solution if and only if

$$\mathcal{B}\mathbf{v} = 0, \tag{2.2.2}$$

where $\mathcal{B}$ is an $m \times n$ matrix whose row vectors are linearly independent vectors $\mathbf{d}^i,\ 1 \leq i \leq m$ satisfying $\mathbf{d}^i\mathcal{A} = 0$.

In case (2.2.2) holds, any solution of (2.2.1) can be given by

$$\mathbf{u} = \sum_{i=1}^{m} \alpha_i \mathbf{c}^i + \mathcal{S}\mathbf{v},$$

where $\alpha_i$, $1 \le i \le m$ are arbitrary constants and $\mathbf{c}^i$, $1 \le i \le m$ are $m$ linearly independent column vectors satisfying $\mathcal{A}\mathbf{c}^i = 0$ and $\mathcal{S}$ is an $n \times n$ matrix independent of $\mathbf{v}$ such that $\mathcal{A}\mathcal{S}\mathbf{p} = \mathbf{p}$ for any column vector $\mathbf{p}$ satisfying $\mathcal{B}\mathbf{p} = 0$.

The matrix $\mathcal{S}$ in Lemma 2.2.1 is not unique.

The number $\lambda$, real or complex, is called an ***eigenvalue*** of the matrix $\mathcal{A}$ if there exists a nonzero real or complex vector $\mathbf{v}$ such that $\mathcal{A}\mathbf{v} = \lambda\mathbf{v}$. The vector $\mathbf{v}$ is called an ***eigenvector*** corresponding to the eigenvalue $\lambda$. From Lemma 2.2.1, $\lambda$ is an eigenvalue of $\mathcal{A}$ if and only if it is a solution of the *characteristic equation* $p(\lambda) = det\ (\mathcal{A} - \lambda\mathcal{I}) = 0$. Since the matrix $\mathcal{A}$ is of order $n$, $p(\lambda)$ is a polynomial of degree exactly $n$, and is called the *characteristic polynomial* of $\mathcal{A}$. Therefore, the matrix $\mathcal{A}$ has exactly $n$ eigenvalues counting with their multiplicities.

In case the eigenvalues $\lambda_1, \cdots, \lambda_n$ of $\mathcal{A}$ are distinct it is easy to find the corresponding eigenvectors $\mathbf{v}^1, \cdots, \mathbf{v}^n$. For this, first we note that for the fixed eigenvalue $\lambda_j$ of $\mathcal{A}$ at least one of the cofactors of $(a_{ii} - \lambda_j)$ in the matrix $(\mathcal{A} - \lambda_j\mathcal{I})$ is nonzero. If not, then from (2.16.1) it follows that $p'(\lambda) = -$ [cofactor of $(a_{11} - \lambda)$] $- \cdots -$ [cofactor of $(a_{nn} - \lambda)$], and hence $p'(\lambda_j) = 0$, i.e. $\lambda_j$ was a multiple root which is a contradiction to our assumption that $\lambda_j$ is simple. Now let the cofactor of $(a_{kk} - \lambda_j)$ be different from zero, then one of the possible nonzero solution of the system $(\mathcal{A} - \lambda_j\mathcal{I})\mathbf{v}^j = 0$ is $v_i^j =$ cofactor of $a_{ki}$ in $(\mathcal{A} - \lambda_j\mathcal{I})$, $1 \le i \le n$, $i \ne k$, $v_k^j =$ cofactor of $(a_{kk} - \lambda_j)$ in $(\mathcal{A} - \lambda_j\mathcal{I})$. Since for this choice of $\mathbf{v}^j$, it follows from (2.16.3) that every equation, except the $k$th one, of the system $(\mathcal{A} - \lambda_j\mathcal{I})\mathbf{v}^j = 0$ is satisfied, and for the $k$th equation from (2.16.2), we have $\sum_{i=1, i\ne k}^{n} a_{ki} \times$ [cofactor of $a_{ki}$] $+ (a_{kk} - \lambda_j)$[cofactor of $(a_{kk} - \lambda_j)$] $=$ $det\ (\mathcal{A} - \lambda_j\mathcal{I})$, which is also zero. In conclusion this $\mathbf{v}^j$ is the eigenvector corresponding to the eigenvalue $\lambda_j$.

**Example 2.2.1.** The characteristic polynomial for the matrix $\mathcal{A} = \begin{bmatrix} 2 & 1 & 0 \\ 1 & 3 & 1 \\ 0 & 1 & 2 \end{bmatrix}$ is $p(\lambda) = -\lambda^3 + 7\lambda^2 - 14\lambda + 8 = -(\lambda - 1)(\lambda - 2)(\lambda - 4)$. Thus, the eigenvalues are $\lambda_1 = 1$, $\lambda_2 = 2$ and $\lambda_3 = 4$. To find the corresponding eigenvectors we have to consider the systems $(\mathcal{A} - \lambda_i\mathcal{I})\mathbf{v}^i = 0$, $i = 1, 2, 3$.

For $\lambda_1 = 1$, we find $(\mathcal{A} - \lambda_1\mathcal{I}) = \begin{bmatrix} 1 & 1 & 0 \\ 1 & 2 & 1 \\ 0 & 1 & 1 \end{bmatrix}$. Since the cofactor of $(a_{11} - \lambda_1) = 1 \neq 0$, we can take $v_1^1 = 1$, and then $v_2^1 =$ cofactor of $a_{12} = -1$, $v_3^1 =$ cofactor of $a_{13} = 1$, i.e. $\mathbf{v}^1 = \begin{bmatrix} 1 \\ -1 \\ 1 \end{bmatrix}$. Next, for $\lambda_2 = 2$ we have $(\mathcal{A} - \lambda_2\mathcal{I}) = \begin{bmatrix} 0 & 1 & 0 \\ 1 & 1 & 1 \\ 0 & 1 & 0 \end{bmatrix}$. Since the cofactor of $(a_{22} - \lambda_2) = 0$ the choice $v_2^2 =$ cofactor of $(a_{22} - \lambda_2)$ is not correct. However, cofactor of $(a_{11} - \lambda_2) =$ cofactor of $(a_{33} - \lambda_2) = -1 \neq 0$ and we can take $v_1^2 = -1$ $(v_3^2 = -1)$, then $v_2^2 =$ cofactor of $a_{12} = 0$, $v_3^2 =$ cofactor of $a_{13} = 1$ $(v_1^2 =$ cofactor of $a_{31} = 1$, $v_2^2 =$ cofactor of $a_{32} = 0)$, i.e. $\mathbf{v}^2 = \begin{bmatrix} -1 \\ 0 \\ 1 \end{bmatrix} \left( \begin{bmatrix} 1 \\ 0 \\ -1 \end{bmatrix} \right)$. Similarly, we can find $\mathbf{v}^3 = \begin{bmatrix} 1 \\ 2 \\ 1 \end{bmatrix}$.

For the eigenvalues and eigenvectors of an $n \times n$ matrix $\mathcal{A}$ we have the following basic result.

**Theorem 2.2.2.** Let $\lambda_1, \cdots, \lambda_m$ be distinct eigenvalues of an $n \times n$ matrix $\mathcal{A}$ and $\mathbf{v}^1, \cdots, \mathbf{v}^m$ be corresponding eigenvectors. Then, $\mathbf{v}^1, \cdots, \mathbf{v}^m$ are linearly independent.

Since $p(\lambda)$ is a polynomial of degree $n$, and $\mathbf{A}^m$ for all nonnegative integers $m$ is defined, $p(\mathcal{A})$ is a well defined matrix. For this matrix $p(\mathcal{A})$ we state the following well known result.

**Theorem 2.2.3.** (Cayley Hamilton Theorem). Let $\mathcal{A}$ be an $n \times n$ matrix and let $p(\lambda) = det\ (\mathcal{A} - \lambda\mathcal{I})$. Then, $p(\mathcal{A}) = 0$.

If $\mathcal{A}$ is a nonsingular matrix, then for every positive integer $m$, $\mathcal{A}^{1/m}$ is a well defined matrix. This important result is stated in

**Theorem 2.2.4.** Let $\mathcal{A}$ be a nonsingular $n \times n$ matrix. Then, for every positive integer $m$ there exists an $n \times n$ matrix $\mathcal{B}$ such that $\mathcal{B}^m = \mathcal{A}$.

Let $z_1, z_2, \cdots, z_n$ be real or complex numbers. The matrix

$$\mathcal{V}(z_1, z_2, \cdots, z_n) = \begin{bmatrix} 1 & 1 & \cdots & 1 \\ z_1 & z_2 & \cdots & z_n \\ z_1^2 & z_2^2 & \cdots & z_n^2 \\ \cdots & \cdots & \cdots & \cdots \\ z_1^{n-1} & z_2^{n-1} & \cdots & z_n^{n-1} \end{bmatrix}$$

is called *Vandermonde's matrix.* Its determinant is given by

$$det\ \mathcal{V}(z_1, z_2, \cdots, z_n) = \prod_{1 \leq j < i \leq n} (z_i - z_j),$$

which is different from zero if $z_i \neq z_j$ for all $i$ and $j$.

A *real normed vector space* is a real vector space $\mathbf{V}$ in which to each vector $\mathbf{u}$ there corresponds a real number $\|\mathbf{u}\|$, called the norm of $\mathbf{u}$, which satisfies the following conditions

(i) $\|\mathbf{u}\| \geq 0$, and $\|\mathbf{u}\| = 0$ if and only if $\mathbf{u} = 0$

(ii) for each $c \in \mathbb{R}$, $\|c\mathbf{u}\| = |c|\|\mathbf{u}\|$

(iii) the triangle inequality $\|\mathbf{u} + \mathbf{v}\| \leq \|\mathbf{u}\| + \|\mathbf{v}\|$.

In the vector space $\mathbb{R}^n$ the following three norms are in common use: absolute norm $\|\mathbf{u}\|_1 = \sum_{i=1}^n |u_i|$, Euclidean norm $\|\mathbf{u}\|_2 = \left(\sum_{i=1}^n |u_i|^2\right)^{1/2}$, and maximum norm $\|\mathbf{u}\|_\infty = \max_{1 \leq i \leq n} |u_i|$.

The notations $\|\cdot\|_1$, $\|\cdot\|_2$ and $\|\cdot\|_\infty$ are justified because of the fact that all these norms are special cases of a more general norm $\|\mathbf{u}\|_p = \left(\sum_{i=1}^n |u_i|^p\right)^{1/p}$, $p \geq 1$.

The set of all $n \times n$ matrices with real elements can be considered as equivalent to the vector space $\mathbb{R}^{n^2}$, with a special multiplicative operation added into the vector space. Thus, a matrix norm should satisfy the usual three requirements of a vector norm and, in addition, we require

(iv) $\|\mathcal{AB}\| \leq \|\mathcal{A}\|\|\mathcal{B}\|$ for all $n \times n$ matrices $\mathcal{A}$, $\mathcal{B}$

(v) compatibility with the vector norm, i.e. if $\|\cdot\|_*$ is the norm in $\mathbb{R}^n$, then $\|\mathcal{A}\mathbf{u}\|_* \leq \|\mathcal{A}\|\|\mathbf{u}\|_*$ for all $\mathbf{u} \in \mathbb{R}^n$ and any $n \times n$ matrix $\mathcal{A}$.

Once in $\mathbb{R}^n$ a norm $\|\cdot\|_*$ is defined then an associated matrix norm is usually defined by

$$\|\mathcal{A}\| = \sup_{\mathbf{u} \neq 0} \frac{\|\mathcal{A}\mathbf{u}\|_*}{\|\mathbf{u}\|_*}. \tag{2.2.3}$$

From this the condition (v) is immediately satisfied. To show (iv) we use (v) twice, to obtain

$$\|\mathcal{AB}\mathbf{u}\|_* = \|\mathcal{A}(\mathcal{B}\mathbf{u})\|_* \leq \|\mathcal{A}\|\|\mathcal{B}\mathbf{u}\|_* \leq \|\mathcal{A}\|\|\mathcal{B}\|\|\mathbf{u}\|_*$$

and hence for all $\mathbf{u} \neq 0$, we have

$$\frac{\|\mathcal{AB}\mathbf{u}\|_*}{\|\mathbf{u}\|_*} \leq \|\mathcal{A}\|\|\mathcal{B}\|$$

or

$$\|\mathcal{A}\mathcal{B}\| \;=\; \sup_{\mathbf{u}\neq 0} \frac{\|\mathcal{A}\mathcal{B}\mathbf{u}\|_*}{\|\mathbf{u}\|_*} \;\leq\; \|\mathcal{A}\|\|\mathcal{B}\|.$$

The norm of the matrix $\mathcal{A}$ induced by the vector norm $\|\mathbf{u}\|_*$ will be denoted by $\|\mathcal{A}\|_*$. For the three norms $\|\mathbf{u}\|_1$, $\|\mathbf{u}\|_2$ and $\|\mathbf{u}\|_\infty$ the corresponding matrix norms are $\|\mathcal{A}\|_1 = \max_{1\leq j\leq n}\sum_{i=1}^{n}|a_{ij}|$, $\|\mathcal{A}\|_2 = \sqrt{\rho(\mathcal{A}^T\mathcal{A})}$ and $\|\mathcal{A}\|_\infty = \max_{1\leq i\leq n}\sum_{j=1}^{n}|a_{ij}|$, where for a given $n\times n$ matrix $\mathcal{B}$ with eigenvalues $\lambda_1,\cdots,\lambda_n$ not necessarily distinct $\rho(\mathcal{B})$ is called the *spectral radius* of $\mathcal{B}$ and defined as $\rho(\mathcal{B}) = \max\{|\lambda_i|,\ 1\leq i\leq n\}$.

A sequence $\{\mathbf{u}^m\}$ in a normed linear space $\mathbf{V}$ is said to converge to $\mathbf{u}\in\mathbf{V}$ if and only if $\|\mathbf{u}-\mathbf{u}^m\|\to 0$ as $m\to\infty$. In particular, a sequence of $n\times n$ matrices $\{\mathcal{A}_m\}$ is said to converge to a matrix $\mathcal{A}$ if $\|\mathcal{A}-\mathcal{A}_m\|\to 0$ as $m\to\infty$. Further, if $\mathcal{A}_m = \left(a_{ij}^{(m)}\right)$ and $\mathcal{A} = (a_{ij})$, then it is the same as $a_{ij}^{(m)}\to a_{ij}$ for all $1\leq i,j\leq n$. Combining this definition with the Cauchy criterion for sequences of real numbers, we have: the sequence $\{\mathcal{A}_m\}$ converges to a limit if and only if $\|\mathcal{A}_k-\mathcal{A}_\ell\|\to 0$ as $k,\ell\to\infty$. The series $\sum_{n=0}^{\infty}\mathcal{A}_n$ is said to converge if and only if the sequence of its partial sums $\{\sum_{n=0}^{m}\mathcal{A}_n\}$ is convergent. For example, the exponential series $e^{\mathcal{A}} = \mathcal{I}+\displaystyle\sum_{n=1}^{\infty}\frac{\mathcal{A}^n}{n!}$ converges for any matrix $\mathcal{A}$. Indeed, it follows from

$$\left\|\sum_{n=0}^{m+p}\mathcal{A}_n - \sum_{n=0}^{m}\mathcal{A}_n\right\| \;=\; \left\|\sum_{n=m+1}^{m+p}\frac{\mathcal{A}^n}{n!}\right\| \;\leq\; \sum_{n=m+1}^{m+p}\frac{\|\mathcal{A}^n\|}{n!} \;\leq\; e^{\|\mathcal{A}\|}.$$

Hence, for any $n\times n$ matrix $\mathcal{A}$, $e^{\mathcal{A}}$ is an $n\times n$ well defined matrix. Further, since $e^{\mathcal{A}}e^{-\mathcal{A}} = e^{\mathcal{A}-\mathcal{A}} = \mathcal{I}$, it follows that $\left(det\ e^{\mathcal{A}}\right)\left(det\ e^{-\mathcal{A}}\right) = 1$, i.e. the matrix $e^{\mathcal{A}}$ is always nonsingular.

Similarly, for a real number $t$, $e^{\mathcal{A}t}$ is defined as $e^{\mathcal{A}t} = \mathcal{I}+\displaystyle\sum_{n=1}^{\infty}\frac{(\mathcal{A}t)^n}{n!}$. Since each element of $e^{\mathcal{A}t}$ defined as a convergent power series, $e^{\mathcal{A}t}$ is differentiable and it follows that

$$\left(e^{\mathcal{A}t}\right)' \;=\; \sum_{n=1}^{\infty}\frac{\mathcal{A}^n t^{n-1}}{(n-1)!} \;=\; \sum_{n=1}^{\infty}\mathcal{A}\frac{(\mathcal{A}t)^{n-1}}{(n-1)!} \;=\; \mathcal{A}e^{\mathcal{A}t} \;=\; e^{\mathcal{A}t}\mathcal{A}.$$

In a normed linear space $\mathbf{V}$ norms $\|\cdot\|$ and $\|\cdot\|_*$ are said to be *equivalent* if there exist positive constants $m$ and $M$ such that for all

$\mathbf{u} \in \mathbf{V}$, $m\|\mathbf{u}\| \le \|\mathbf{u}\|_* \le M\|\mathbf{u}\|$. It is well known that in $\mathbb{R}^n$ all the norms are equivalent. Hence, unless otherwise stated, in $\mathbb{R}^n$ we shall always consider $\|\cdot\|_1$ norm and the subscript 1 will be dropped.

## 2.3. Linear Dependence and Independence

Let the functions $\mathbf{u}^i(k)$, $1 \le i \le m$ be defined on $\overline{\mathbb{N}}_1$. We say $\mathbf{u}^i(k)$ are *linearly dependent* on $\overline{\mathbb{N}}_1$ if there exist constants $\alpha_i$, $1 \le i \le m$ not all zero, such that

$$\sum_{i=1}^{m} \alpha_i \mathbf{u}^i(k) = 0 \tag{2.3.1}$$

holds for all $k \in \overline{\mathbb{N}}_1$. Conversely, if the relation (2.3.1) implies that $\alpha_i = 0$, $1 \le i \le m$ then $\mathbf{u}^i(k)$ are said to be *linearly independent*.

For the given functions $\mathbf{u}^i(k)$, $1 \le i \le n$ defined on $\overline{\mathbb{N}}_1$ the $n \times n$ matrix $(u_j^i(k))$ is called the *Casoratian matrix* of these functions. We shall denote it by the symbol $C(\mathbf{u}^1, \cdots, \mathbf{u}^n)(k)$ and when there is no ambiguity by $C(k)$. The *det* $C(k)$ is closely related to the question of whether or not $\mathbf{u}^i(k)$ are linearly independent on $\overline{\mathbb{N}}_1$.

**Lemma 2.3.1.** If *det* $C(k)$ of $n$ functions $\mathbf{u}^i(k)$, $1 \le i \le n$ defined on $\overline{\mathbb{N}}_1$ is different from zero for at least one $k_0 \in \overline{\mathbb{N}}_1$, then $\mathbf{u}^i(k)$, $1 \le i \le n$ are linearly independent on $\overline{\mathbb{N}}_1$.

**Proof.** Let $\mathbf{c}$ be a nonzero vector such that $C(k)\mathbf{c} = \sum_{i=1}^{n} c_i \mathbf{u}^i(k) = 0$ for all $k \in \overline{\mathbb{N}}_1$, i.e. $\mathbf{u}^i(k)$ are linearly dependent on $\overline{\mathbb{N}}_1$. Hence, if $k_0 \in \overline{\mathbb{N}}_1$ be such that *det* $C(k_0) \ne 0$, then in particular $C(k_0)\mathbf{c} = 0$. However, from Lemma 2.2.1 this is possible only when $\mathbf{c} = 0$. Thus, $\mathbf{u}^i(k)$ are linearly independent on $\overline{\mathbb{N}}_1$. ∎

The converse of Lemma 2.3.1 is not necessarily true, e.g. $\mathbf{u}^1(k) = \begin{bmatrix} 1 \\ k \end{bmatrix}$, $\mathbf{u}^2(k) = \begin{bmatrix} k \\ k^2 \end{bmatrix}$ are linearly independent on $\mathbb{N}$. But, *det* $C(\mathbf{u}^1, \mathbf{u}^2)(k) = 0$ for all $k \in \mathbb{N}$.

**Lemma 2.3.2.** Let $\mathbf{u}^i(k)$, $1 \le i \le n$ be linearly independent solutions on $\overline{\mathbb{N}}_1$ of the homogeneous system (1.2.12) on $\overline{\mathbb{N}}$. Then, *det* $C(k) \ne 0$ for all $k \in \overline{\mathbb{N}}_1$.

**Proof.** Let $k_0 \in \overline{\mathbb{N}}_1$ be such that *det* $C(k_0) = 0$, then from Lemma 2.2.1 there exists a nonzero vector $\mathbf{c}$ such that $C(k_0)\mathbf{c} = \sum_{i=1}^{n} c_i \mathbf{u}^i(k_0) = 0$. Since $\mathbf{u}(k) = \sum_{i=1}^{n} c_i \mathbf{u}^i(k)$ is a solution of (1.2.12) and $\mathbf{u}(k_0) = 0$, from the uniqueness of the solutions it follows that $\mathbf{u}(k) = 0$ for all $k \in \overline{\mathbb{N}}_1$. From this, the linear independence of $\mathbf{u}^i(k)$ on $\overline{\mathbb{N}}_1$ implies that $\mathbf{c} = 0$. This contradiction completes the proof. ∎

On combining the Lemmas 2.3.1 and 2.3.2 we have the following:

**Theorem 2.3.3.** The solutions $\mathbf{u}^i(k),\ 1 \le i \le n$ of the system (1.2.12) are linearly independent on $\overline{\mathbb{N}}_1$ if and only if there exists at least one $k_0 \in \overline{\mathbb{N}}_1$ such that *det* $C(k_0) \ne 0$.

As a consequence of this result the solutions $\mathbf{u}^i(k),\ 1 \le i \le n$ of the system (1.2.12) satisfying the initial conditions

$$\text{(2.3.2)} \qquad \mathbf{u}^i(k_0) \;=\; \mathbf{e}^i \;=\; (0,\cdots,0,1,0,\cdots,0)^T, \quad 1 \le i \le n, \quad k_0 \in \overline{\mathbb{N}}_1$$

are linearly independent on $\overline{\mathbb{N}}_1$. This proves the existence of $n$ linearly independent solutions of the system (1.2.12) on $\overline{\mathbb{N}}_1$. Now let $\mathbf{u}(k)$ be any solution of the system (1.2.12) on $\overline{\mathbb{N}}_1$, then $\mathbf{u}(k) = \sum_{i=1}^n u_i(k_0)\mathbf{u}^i(k)$, where $\mathbf{u}^i(k)$ are the solutions of the initial value problems (1.2.12), (2.3.2). For this, let $\mathbf{v}(k) = \sum_{i=1}^n u_i(k_0)\mathbf{u}^i(k)$ then $\mathbf{v}(k)$ is a solution of (1.2.12) and $\mathbf{v}(k_0) = \sum_{i=1}^n u_i(k_0)\mathbf{u}^i(k_0) = \sum_{i=1}^n u_i(k_0)\mathbf{e}^i = \mathbf{u}(k_0)$. Thus, from the uniqueness of the solutions it follows that $\mathbf{u}(k) = \mathbf{v}(k)$ for all $k \in \overline{\mathbb{N}}_1$. Thus, every solution of the system (1.2.12) can be expressed as a linear combination of the $n$ linearly independent solutions of (1.2.12), (2.3.2). In conclusion we find that the vector space of all solutions of the system (1.2.12) is of dimension $n$, and any solution $\mathbf{u}(k)$ of (1.2.12) can be written as

$$\text{(2.3.3)} \qquad \mathbf{u}(k) \;=\; \sum_{i=1}^n u_i(k_0)\mathbf{u}^i(k) \;=\; \mathcal{U}(k)\mathbf{u}(k_0),$$

where $\mathcal{U}(k_0) = \mathcal{I}$. Further, if $\mathbf{v}^i(k),\ 1 \le i \le n$ is any set of linearly independent solutions of (1.2.12) then its general solution $\mathbf{u}(k)$ appears as

$$\text{(2.3.4)} \qquad \mathbf{u}(k) \;=\; \sum_{i=1}^n c_i\mathbf{v}^i(k) \;=\; \mathcal{V}(k)\mathbf{c},$$

where $\mathbf{c}$ is an arbitrary vector.

## 2.4. Matrix Linear Systems

Since each column of the matrix $\mathcal{U}(k)$ defined in the previous section is a solution of (1.2.12), it is obvious that it is a solution of the matrix linear system

$$\mathcal{U}(k+1) \;=\; \mathcal{A}(k)\mathcal{U}(k), \quad k \in \overline{\mathbb{N}}. \tag{2.4.1}$$

Further, to emphasize the initial point $k_0$ in (2.3.2) the matrix $\mathcal{U}(k)$ will be denoted as $\mathcal{U}(k,k_0)$. This matrix $\mathcal{U}(k,k_0)$ is called *principal fundamental matrix* and has the property that $\mathcal{U}(k_0,k_0)=\mathcal{I}$.

Any $n \times n$ matrix $\mathcal{V}(k)$ whose columns are linearly independent solutions of the system (1.2.12) is called a *fundamental matrix*. Obviously, $\mathcal{V}(k)$ is a solution of the matrix linear system (2.4.1), however $\mathcal{V}(k_0)$ need not be $\mathcal{I}$. It is straightforward to obtain principal fundamental matrix $\mathcal{U}(k,k_0)$ from a given fundamental matrix $\mathcal{V}(k)$, indeed $\mathcal{U}(k,k_0) = \mathcal{V}(k)\mathcal{V}^{-1}(k_0)$, and conversely $\mathcal{V}(k)=\mathcal{U}(k,k_0)\mathcal{V}(k_0)$ on $\overline{\mathbb{N}}_1$ follows from the uniqueness of the solutions.

If $\mathcal{V}(k)$ is a fundamental matrix of the system (1.2.12) and $\mathcal{C}$ is any nonsingular matrix, then $\mathcal{V}(k)\mathcal{C}$ is also a fundamental matrix, however $\mathcal{C}\mathcal{V}(k)$ need not even be a solution of (2.4.1). Further, if $\mathcal{W}(k)$ is any other fundamental matrix of (1.2.12), then $\mathcal{W}(k)=\mathcal{U}(k,k_0)\mathcal{W}(k_0) = \mathcal{U}(k,k_0)\mathcal{V}(k_0)\mathcal{V}^{-1}(k_0)\mathcal{W}(k_0)=\mathcal{V}(k)\mathcal{V}^{-1}(k_0)\mathcal{W}(k_0)$, i.e. $\mathcal{V}(k)$ and $\mathcal{W}(k)$ are equivalent.

The following result gives an explicit representation of $\mathcal{U}(k,k_0)$.

**Theorem 2.4.1.** The following holds

$$\mathcal{U}(k,k_0) \;=\; \begin{cases} \displaystyle\prod_{\ell=k_0}^{k-1} \mathcal{A}(k_0+k-1-\ell) & \text{for all } k_0 \le k \in \overline{\mathbb{N}}_1 \\ \displaystyle\prod_{\ell=k}^{k_0-1} \mathcal{A}^{-1}(\ell) & \text{for all } k_0 \ge k \in \overline{\mathbb{N}}_1. \end{cases} \tag{2.4.2}$$

**Proof.** Obviously, $\mathcal{U}(k_0,k_0)=\mathcal{I}$ and for all $k_0 \le k \in \overline{\mathbb{N}}$, we have

$$\begin{aligned} \mathcal{U}(k+1,k_0) \;&=\; \prod_{\ell=k_0}^{k} \mathcal{A}(k_0+k-\ell) \;=\; \mathcal{A}(k)\prod_{\ell=k_0+1}^{k} \mathcal{A}(k_0+k-\ell) \\ &=\; \mathcal{A}(k)\prod_{\ell=k_0}^{k-1} \mathcal{A}(k_0+k-1-\ell) \;=\; \mathcal{A}(k)\mathcal{U}(k,k_0). \end{aligned}$$

Similarly, if $k_0 \ge k \in \overline{\mathbb{N}}$, then

$$U(k,k_0) \;=\; \mathcal{A}^{-1}(k)\prod_{\ell=k+1}^{k_0-1} \mathcal{A}^{-1}(\ell) \;=\; \mathcal{A}^{-1}(k)\mathcal{U}(k+1,k_0). \qquad \blacksquare$$

**Corollary 2.4.2.** If $\mathcal{A}(k)$ is a constant matrix $\mathcal{A}$, then

$$(2.4.3) \qquad \mathcal{U}(k,k_0) = \begin{cases} \mathcal{A}^{(k-k_0)} & \text{for all } k_0 \le k \in \overline{\mathbb{N}}_1 \\ \mathcal{A}^{-(k_0-k)} & \text{for all } k_0 \ge k \in \overline{\mathbb{N}}_1. \end{cases}$$

**Corollary 2.4.3.** Let $\mathcal{V}(k)$ be any fundamental matrix of the system (1.2.12). Then,

$$(2.4.4) \quad det\ \mathcal{V}(k) = det\ \mathcal{V}(k_0) \begin{cases} \prod_{\ell=k_0}^{k-1} det\ \mathcal{A}(\ell) & \text{for all } k_0 \le k \in \overline{\mathbb{N}}_1 \\ \prod_{\ell=k}^{k_0-1} (det\ \mathcal{A}(\ell))^{-1} & \text{for all } k_0 \ge k \in \overline{\mathbb{N}}_1. \end{cases}$$

Now as an application of some of the above results we illustrate the following:

**Example 2.4.1.** (Markov Chains). Let a system be capable of being in $n$ possible states $1, \cdots, n$ and let the probability of transition from state $i$ to state $j$ in time interval $k$ to $k+1$ be $p_{ij}$. Let $p_i(k)$ denote the probability that the system is in state $i$ at time $k$, $i = 1, \cdots, n$ then at time $k+1$ it can be in any one of the states $1, \cdots, n$. It can be in the $i$th state at time $k+1$ in $n$ exclusive ways since it could have been in any one of the $n$ states $1, \cdots, n$ at time $k$ and it could have transited from that state to $i$th state in time interval $(k, k+1)$. By using the theorems of total and compound probability, we get

$$(2.4.5) \qquad p_i(k+1) = \sum_{j=1}^{n} p_{ij} p_j(k), \quad i = 1, \cdots, n$$

or

$$(2.4.6) \qquad \mathbf{p}(k+1) = \mathcal{A}\mathbf{p}(k)$$

where $\mathbf{p}(k)$ is the probability vector and $\mathcal{A} = (p_{ij})$ is a matrix, all of whose elements lie between zero and unity. Further the sum of elements of every column is unity, since the sum of elements of the $j$th column $\sum_{i=1}^{n} p_{ij}$ denotes the sum of the probabilities of the system going from the $j$th state to any other state and this sum must be one. It is clear from Corollary 2.4.2 that the solution of (2.4.6) can be written as $\mathbf{p}(k) = \mathcal{A}^k \mathbf{p}(0)$.

For the matrix $\mathcal{A}$ appearing in (2.4.6) it is clear that $|\lambda| \le 1$ for all the eigenvalues, and $\lambda = 1$ is an eigenvalue. Now the probability vector will not change if $\mathbf{p}(k+1) = \mathbf{p}(k)$ so that from (2.4.6), $(\mathcal{I} - \mathcal{A})\mathbf{p}(k) = 0$. Thus, if $\mathbf{p}$ is the eigenvector of the matrix $\mathcal{A}$ corresponding to unit eigenvalue, then $\mathbf{p}$ does not change, i.e. if the system starts with property $\mathbf{p}$ at time 0, it will always remain in this state. We also note that if the system starts from any other probability vector, it will ultimately be described by the probability vector $\mathbf{p}$ as $t \to \infty$.

As a special case, suppose we have a machine which can be in two states working or non working. Let the probability of its transition from working to non working be $\alpha$, of its transition from non working to working be $\beta$, then the system of difference equations is

$$\begin{aligned} p_1(k+1) &= (1-\alpha)p_1(k) + \beta p_2(k) \\ p_2(k+1) &= \alpha p_1(k) + (1-\beta)p_2(k). \end{aligned}$$

The eigenvalues of the matrix $\begin{bmatrix} 1-\alpha & \beta \\ \alpha & 1-\beta \end{bmatrix}$ are $1,\ 1-\alpha-\beta$, and the eigenvector corresponding to the unit eigenvalue is $\dfrac{\beta}{\alpha+\beta},\ \dfrac{\alpha}{\alpha+\beta}$. Thus, ultimately the probability of the machine being found in working order is $\dfrac{\beta}{\alpha+\beta}$ and the probability of its being found in non working state is $\dfrac{\alpha}{\alpha+\beta}$.

## 2.5. Variation of Constants Formula

Let $\mathcal{V}(k)$ be any fundamental matrix of (1.2.12) and $\mathbf{c}(k)$ be a function defined on $\overline{\mathbb{N}}_1$. We define $\mathbf{u}(k) = \mathcal{V}(k)\mathbf{c}(k)$ and demand $\mathbf{u}(k)$ to be the solution of the initial value problem (1.2.11), (1.3.4). For this, it is necessary that $\mathbf{u}(k_0) = \mathcal{V}(k_0)\mathbf{c}(k_0) = \mathbf{u}^0$, i.e. $\mathbf{c}(k_0) = \mathcal{V}^{-1}(k_0)\mathbf{u}^0$. Further, for $k \in \overline{\mathbb{N}}$ we have

$$\begin{aligned} \mathbf{u}(k+1) &= \mathcal{V}(k+1)\mathbf{c}(k+1) = \mathcal{A}(k)\mathbf{u}(k) + \mathbf{b}(k) = \mathcal{A}(k)\mathcal{V}(k)\mathbf{c}(k) + \mathbf{b}(k) \\ &= \mathcal{V}(k+1)\mathbf{c}(k) + \mathbf{b}(k) \end{aligned}$$

and hence

$$\mathcal{V}(k+1)\Delta\mathbf{c}(k) = \mathbf{b}(k),$$

or

$$\Delta\mathbf{c}(k) = \mathcal{V}^{-1}(k+1)\mathbf{b}(k), \quad k \in \overline{\mathbb{N}}.$$

Thus, for all $k_0 \le k \in \overline{\mathbb{N}}_1$

$$\mathbf{c}(k) = \mathbf{c}(k_0) + \sum_{\ell=k_0+1}^{k} \mathcal{V}^{-1}(\ell)\mathbf{b}(\ell-1)$$

and for all $k_0 \geq k \in \overline{\mathbb{N}}_1$

$$\mathbf{c}(k) = \mathbf{c}(k_0) - \sum_{\ell=k+1}^{k_0} \mathcal{V}^{-1}(\ell)\mathbf{b}(\ell-1).$$

Therefore, for all $k_0 \leq k \in \overline{\mathbb{N}}_1$ the solution of the initial value problem (1.2.11), (1.3.4) can be written as

$$\mathbf{u}(k) = \mathcal{V}(k)\left[\mathcal{V}^{-1}(k_0)\mathbf{u}^0 + \sum_{\ell=k_0+1}^{k} \mathcal{V}^{-1}(\ell)\mathbf{b}(\ell-1)\right]$$

$$= \mathcal{U}(k,k_0)\mathbf{u}^0 + \sum_{\ell=k_0+1}^{k} \mathcal{U}(k,k_0)\mathcal{U}^{-1}(\ell,k_0)\mathbf{b}(\ell-1) \tag{2.5.1}$$

and for all $k_0 \geq k \in \overline{\mathbb{N}}_1$

$$\mathbf{u}(k) = \mathcal{U}(k,k_0)\mathbf{u}^0 - \sum_{\ell=k+1}^{k_0} \mathcal{U}(k,k_0)\mathcal{U}^{-1}(\ell,k_0)\mathbf{b}(\ell-1). \tag{2.5.2}$$

Further, the general solution $\mathbf{u}(k)$ of (1.2.11) in terms of $\mathcal{V}(k)$ appears as

$$\mathbf{u}(k) = \mathcal{V}(k)\mathbf{c} + \sum_{\ell=k_0+1}^{k} \mathcal{V}(k)\mathcal{V}^{-1}(\ell)\mathbf{b}(\ell-1) \quad \text{for all } k_0 \leq k \in \overline{\mathbb{N}}_1 \tag{2.5.3}$$

$$= \mathcal{V}(k)\mathbf{c} - \sum_{\ell=k+1}^{k_0} \mathcal{V}(k)\mathcal{V}^{-1}(\ell)\mathbf{b}(\ell-1) \quad \text{for all } k_0 \geq k \in \overline{\mathbb{N}}_1. \tag{2.5.4}$$

## 2.6. Green's Matrix

The kernel $\mathcal{G}(k,\ell) = \mathcal{U}(k,k_0)\mathcal{U}^{-1}(\ell,k_0)$ is called the *Green's matrix* of the system (1.2.12) and it is defined for all $k,\ell \in \overline{\mathbb{N}}_1$ although its use in (2.5.1) is required only for $k \geq \ell \geq k_0+1$, whereas in (2.5.2) only for $k+1 \leq \ell \leq k_0$. The following properties of $\mathcal{G}(k,\ell)$ are immediate

(i) $\mathcal{G}(k,k) = \mathcal{I}$ for all $k \in \overline{\mathbb{N}}_1$

(ii) $\mathcal{G}^{-1}(k,\ell) = \mathcal{G}(\ell,k)$ for all $k,\ell \in \overline{\mathbb{N}}_1$

(iii) $\mathcal{G}(k,\ell) = \mathcal{G}(k,\tau)\mathcal{G}(\tau,\ell)$ for all $k,\tau,\ell \in \overline{\mathbb{N}}_1$

(iv) $\mathcal{G}(k+1,\ell) = \mathcal{A}(k)\mathcal{G}(k,\ell)$ for all $k \in \overline{\mathbb{N}}$, $\ell \in \overline{\mathbb{N}}_1$

(v) $\mathcal{G}(k,\ell+1) = \mathcal{G}(k,\ell)\mathcal{A}^{-1}(\ell)$ for all $k \in \overline{\mathbb{N}}_1$, $\ell \in \overline{\mathbb{N}}$

(vi) $\mathcal{G}(k,\ell) = \prod_{\tau=0}^{k-1-\ell} \mathcal{A}(k-1-\tau)$ for all $k \geq \ell$ in $\overline{\mathbb{N}}_1$

(vii) $\mathcal{G}(k,\ell) = \prod_{\tau=k}^{\ell-1} \mathcal{A}^{-1}(\tau)$ for all $\ell \geq k$ in $\overline{\mathbb{N}}_1$

(viii) if $\mathcal{A}(k)$ is a constant matrix $\mathcal{A}$, then $\mathcal{G}(k,\ell) = \mathcal{G}(k-\ell) = \mathcal{A}^{k-\ell}$ for all $k, \ell \in \overline{\mathbb{N}}_1$.

On combining some of the above results, we have the following:

**Theorem 2.6.1.** The solution of the initial value problem (1.2.11), (1.3.4) can be written as

$$(2.6.1)\quad \mathbf{u}(k) = \prod_{\ell=k_0}^{k-1} \mathcal{A}(k_0+k-1-\ell)\mathbf{u}^0 + \sum_{\ell=k_0+1}^{k} \sum_{\tau=0}^{k-1-\ell} \mathcal{A}(k-1-\tau)\mathbf{b}(\ell-1) \quad \text{for all } k_0 \le k \in \overline{\mathbb{N}}_1$$

$$(2.6.2)\quad = \prod_{\ell=k}^{k_0-1} \mathcal{A}^{-1}(\ell)\mathbf{u}^0 - \sum_{\ell=k+1}^{k_0} \sum_{\tau=k}^{\ell-1} \mathcal{A}^{-1}(\tau)\mathbf{b}(\ell-1) \quad \text{for all } k_0 \ge k \in \overline{\mathbb{N}}_1$$

also, if $\mathcal{A}(k)$ is a constant matrix $\mathcal{A}$, then

$$(2.6.3)\quad \mathbf{u}(k) = \mathcal{A}^{(k-k_0)}\mathbf{u}^0 + \sum_{\ell=k_0+1}^{k} \mathcal{A}^{k-\ell}\,\mathbf{b}(\ell-1) \quad \text{for all } k_0 \le k \in \overline{\mathbb{N}}_1$$

$$(2.6.4)\quad = \mathcal{A}^{-(k_0-k)}\mathbf{u}^0 - \sum_{\ell=k+1}^{k_0} \mathcal{A}^{k-\ell}\,\mathbf{b}(\ell-1) \quad \text{for all } k_0 \ge k \in \overline{\mathbb{N}}_1.$$

## 2.7. Adjoint Systems

Let $\mathbf{u}(k)$ be a nontrivial solution of the homogeneous system (1.2.12). We shall find the function $\mathbf{v}(k)$ so that $\mathbf{v}^T(k)\mathbf{u}(k) = \mathbf{c}$ for all $k \in \overline{\mathbb{N}}_1$. For this, it is necessary that

$$\mathbf{v}^T(k+1)\mathbf{u}(k+1) = \mathbf{v}^T(k)\mathbf{u}(k) = \mathbf{c} \quad \text{for all } k \in \overline{\mathbb{N}},$$

i.e.

$$\mathbf{v}^T(k+1)\mathcal{A}(k)\mathbf{u}(k) = \mathbf{v}^T(k)\mathbf{u}(k) \quad \text{for all } k \in \overline{\mathbb{N}},$$

which implies that $\mathbf{v}(k)$ is a solution of the linear homogeneous system

$$(2.7.1)\quad \mathbf{v}(k) = \mathcal{A}^T(k)\mathbf{v}(k+1), \quad k \in \overline{\mathbb{N}}.$$

The system (2.7.1) is called the *adjoint system* of (1.2.12).

In terms of the principal fundamental matrix $\mathcal{U}(k, k_0)$ of (1.2.12), the function $\mathbf{v}(k)$ can be written as

$$(2.7.2)\quad \mathbf{v}(k) = \left[\mathcal{U}^T(k,k_0)\right]^{-1}\mathbf{v}(k_0), \quad k \in \overline{\mathbb{N}}_1.$$

For this, we have

$$\begin{aligned}\mathbf{v}(k+1) &= \left[\mathcal{U}^T(k+1,k_0)\right]^{-1}\mathbf{v}(k_0) \;=\; \left[(\mathcal{A}(k)\mathcal{U}(k,k_0))^T\right]^{-1}\mathbf{v}(k_0)\\ &= \left[\mathcal{A}^T(k)\right]^{-1}\left[\mathcal{U}^T(k,k_0)\right]^{-1}\mathbf{v}(k_0) \;=\; \left[\mathcal{A}^T(k)\right]^{-1}\mathbf{v}(k),\end{aligned}$$

which is the same as (2.7.1).

Now let $\mathcal{V}(k)$ be any fundamental matrix of the adjoint system (2.7.1), then

$$\mathcal{V}^T(k) \;=\; \mathcal{V}^T(k+1)\mathcal{A}(k), \quad k\in\overline{\mathbb{N}}. \tag{2.7.3}$$

Premultiplying (1.2.11) by $\mathcal{V}^T(k+1)$, to obtain

$$\mathcal{V}^T(k+1)\mathbf{u}(k+1) \;=\; \mathcal{V}^T(k+1)\mathcal{A}(k)\mathbf{u}(k)+\mathcal{V}^T(k+1)\mathbf{b}(k). \tag{2.7.4}$$

Postmultiplying (2.7.3) by $\mathbf{u}(k)$, to get

$$\mathcal{V}^T(k)\mathbf{u}(k) \;=\; \mathcal{V}^T(k+1)\mathcal{A}(k)\mathbf{u}(k). \tag{2.7.5}$$

From (2.7.4) and (2.7.5), we have

$$\Delta\left[\mathcal{V}^T(k)\mathbf{u}(k)\right] \;=\; \mathcal{V}^T(k+1)\mathbf{b}(k). \tag{2.7.6}$$

Thus, it follows that

$$\mathbf{u}(k) \;=\; \left[\mathcal{V}^T(k)\right]^{-1}\mathcal{V}^T(k_0)\mathbf{u}(k_0)+\sum_{\ell=k_0+1}^{k}\left[\mathcal{V}^T(k)\right]^{-1}\mathcal{V}^T(\ell)\mathbf{b}(\ell-1) \quad \text{for all } k_0\le k\in\overline{\mathbb{N}}_1 \tag{2.7.7}$$

$$\;=\; \left[\mathcal{V}^T(k)\right]^{-1}\mathcal{V}^T(k_0)\mathbf{u}(k_0)-\sum_{\ell=k+1}^{k_0}\left[\mathcal{V}^T(k)\right]^{-1}\mathcal{V}^T(\ell)\mathbf{b}(\ell-1) \quad \text{for all } k_0\ge k\in\overline{\mathbb{N}}_1. \tag{2.7.8}$$

On comparing (2.7.7) and (2.7.8) with (2.5.1) and (2.5.2) respectively, it follows that

$$\mathcal{U}(k,k_0) \;=\; \left[\mathcal{V}^T(k)\right]^{-1}\mathcal{V}^T(k_0) \quad \text{for all } k_0,k\in\overline{\mathbb{N}}_1$$

and

$$\mathcal{G}(k,\ell) \;=\; \left[\mathcal{V}^T(k)\right]^{-1}\mathcal{V}^T(\ell) \quad \text{for all } k,\ell\in\overline{\mathbb{N}}_1.$$

Finally, we note that if $\mathbf{v}(k)$ is any column of $\mathcal{V}(k)$, i.e. a solution of (2.7.1), then (2.7.7) and (2.7.8) provide

$$(2.7.9)\quad \sum_{i=1}^{n}[u_i(k)v_i(k) - u_i(k_0)v_i(k_0)] = \sum_{\ell=k_0+1}^{k}\sum_{i=1}^{n} v_i(\ell)b_i(\ell-1) \quad \text{for all } k_0 \le k \in \overline{\mathbf{N}}_1$$

$$(2.7.10)\quad = -\sum_{\ell=k+1}^{k_0}\sum_{i=1}^{n} v_i(\ell)b_i(\ell-1) \quad \text{for all } k_0 \ge k \in \overline{\mathbf{N}}_1.$$

Equations (2.7.9) and (2.7.10) will be referred to as *adjoint identities*.

## 2.8. Systems with Constant Coefficients

For the difference system

$$(2.8.1)\quad \mathbf{u}(k+1) = \mathcal{A}\mathbf{u}(k), \quad k \in \mathbf{Z}$$

where $\mathcal{A}$ is a nonsingular constant matrix and $\mathbf{Z}$ is the set of all integers including zero, the general solution can be written as

$$(2.8.2)\quad \mathbf{u}(k) = \mathcal{A}^k\mathbf{c}, \quad k \in \mathbf{Z}$$

where $\mathbf{c}$ is an arbitrary constant vector. Thus, to find the general solution of (2.8.1) we need to find the general expression for $\mathcal{A}^k$, where $k \in \mathbf{Z}$. This is not an easy task except in few exceptional cases.

**Example 2.8.1.** For the matrix $\mathcal{A} = \begin{bmatrix} 0 & 1 \\ -1 & 0 \end{bmatrix}$ it is easily seen that $\mathcal{A}^k = \begin{bmatrix} \cos(k\pi/2) & \sin(k\pi/2) \\ -\sin(k\pi/2) & \cos(k\pi/2) \end{bmatrix}$ for all $k \in \mathbf{Z}$. Therefore, the general solution of (2.8.1) with this $\mathcal{A}$ can be written as

$$\mathbf{u}(k) = \begin{bmatrix} \cos\frac{k\pi}{2} & \sin\frac{k\pi}{2} \\ -\sin\frac{k\pi}{2} & \cos\frac{k\pi}{2} \end{bmatrix}\mathbf{c} \quad \text{for all } k \in \mathbf{Z}.$$

In the following we shall show that the eigenvalues and eigenfunctions of the matrix $\mathcal{A}$ can be used to find the general solution of (2.8.1).

**Theorem 2.8.1.** Let $\lambda_i$ $(1/\lambda_i)$, $i = 1,\cdots,n$ be the distinct eigenvalues of the matrix $\mathcal{A}(\mathcal{A}^{-1})$ and $\mathbf{v}^i$, $i = 1,\cdots,n$ be the corresponding eigenvectors. Then, the set

$$(2.8.3)\quad \mathbf{u}^i(k) = \mathbf{v}^i\lambda_i^k, \quad i = 1,\cdots,n \quad \text{for all } k \in \mathbf{Z}$$

is a fundamental set of solutions of (2.8.1).

**Proof.** Since $\mathbf{v}^i$ is an eigenvector of $\mathcal{A}(\mathcal{A}^{-1})$ corresponding to the eigenvalue $\lambda_i$ $(1/\lambda_i)$, we find that

$$\mathbf{u}^i(k+1) = \mathbf{v}^i\lambda_i^{k+1} = \lambda_i\mathbf{v}^i\lambda_i^k = \mathcal{A}\mathbf{v}^i\lambda_i^k = \mathcal{A}\mathbf{u}^i(k), \quad k \geq 0$$

and

$$\mathbf{u}^i(k) = \mathbf{v}^i\frac{1}{\lambda_i}\lambda_i^{k+1} = \frac{1}{\lambda_i}\mathbf{v}^i\lambda_i^{k+1} = \mathcal{A}^{-1}\mathbf{v}^i\lambda_i^{k+1} = \mathcal{A}^{-1}\mathbf{u}^i(k+1), \quad k \leq 0.$$

Thus, $\mathbf{u}^i(k)$ is a solution of (2.8.1). To show that (2.8.3) is a fundamental set, we note that $det\ \mathcal{C}(0) = det\ [\mathbf{v}^1, \cdots, \mathbf{v}^n] \neq 0$, since $\mathbf{v}^1, \cdots, \mathbf{v}^n$ are linearly independent from Theorem 2.2.2. The result now follows from Theorem 2.3.3. ∎

Obviously, from Theorem 2.8.1 it follows that

$$\mathcal{A}^k = [\mathbf{v}^1\lambda_1^k, \cdots, \mathbf{v}^n\lambda_n^k]\,[\mathbf{v}^1, \cdots, \mathbf{v}^n]^{-1} \quad \text{for all } k \in \mathbf{Z}.$$

Further, the general solution of (2.8.1) can be written as

$$\mathbf{u}(k) = \sum_{i=1}^{n} c_i\mathbf{v}^i\lambda_i^k \quad \text{for all } k \in \mathbf{Z}. \tag{2.8.4}$$

**Example 2.8.2.** Using the results of Example 2.2.1, Theorem 2.8.1 concludes that the set

$$\mathbf{u}^1(k) = \begin{bmatrix} 1 \\ -1 \\ 1 \end{bmatrix}, \quad \mathbf{u}^2(k) = \begin{bmatrix} -1 \\ 0 \\ 1 \end{bmatrix}(2)^k, \quad \mathbf{u}^3(k) = \begin{bmatrix} 1 \\ 2 \\ 1 \end{bmatrix}(4)^k$$

for all $k \in \mathbf{Z}$

is a fundamental set of solutions of the difference system (2.8.1) with $A = \begin{bmatrix} 2 & 1 & 0 \\ 1 & 3 & 1 \\ 0 & 1 & 2 \end{bmatrix}$.

Unfortunately, when the matrix $\mathcal{A}$ has $m < n$ distinct eigenvalues, then the computation of $\mathcal{A}^k$ is not easy. However, since the solution $\mathcal{Y}(t) = e^{\mathcal{A}t}$ of the matrix differential system

$$\mathcal{Y}'(t) = \mathcal{A}\mathcal{Y}(t), \quad \mathcal{Y}(0) = \mathcal{I} \tag{2.8.5}$$

and $\mathcal{A}^k$ for all nonnegative integers $k$ are related by $\mathcal{A}^k = \mathcal{Y}^{(k)}(0)$, all the known expressions for $e^{\mathcal{A}t}$ can be used to compute $\mathcal{A}^k$.

**Lemma 2.8.2.** Let $\lambda_1, \cdots, \lambda_m$, $m \le n$ be distinct eigenvalues of the matrix $\mathcal{A}$ with multiplicities $r_1, \cdots, r_m$ respectively, so that

$$p(\lambda) = (\lambda - \lambda_1)^{r_1} \cdots (\lambda - \lambda_m)^{r_m}, \tag{2.8.6}$$

then

$$e^{\mathcal{A}t} = \sum_{i=1}^{m} \left[ e^{\lambda_i t} a_i(\mathcal{A}) q_i(\mathcal{A}) \sum_{j=0}^{r_i - 1} \left\{ \frac{1}{j!} (\mathcal{A} - \lambda_i \mathcal{I})^j t^j \right\} \right], \tag{2.8.7}$$

where

$$q_i(\lambda) = p(\lambda)(\lambda - \lambda_i)^{-r_i}, \quad 1 \le i \le m \tag{2.8.8}$$

and $a_i(\lambda)$, $1 \le i \le m$ are the polynomials of degree less than $r_i$ in the expansion

$$\frac{1}{p(\lambda)} = \frac{a_1(\lambda)}{(\lambda - \lambda_1)^{r_1}} + \cdots + \frac{a_m(\lambda)}{(\lambda - \lambda_m)^{r_m}}. \tag{2.8.9}$$

**Proof.** Relations (2.8.8) and (2.8.9) imply that

$$1 = a_1(\lambda) q_1(\lambda) + \cdots + a_m(\lambda) q_m(\lambda).$$

This relation has been derived from the characteristic equation $p(\lambda) = 0$ of $\mathcal{A}$, and therefore, using Cayley Hamilton Theorem 2.2.3, we must have

$$\mathcal{I} = a_1(\mathcal{A}) q_1(\mathcal{A}) + \cdots + a_m(\mathcal{A}) q_m(\mathcal{A}). \tag{2.8.10}$$

Since the matrices $\lambda_i \mathcal{I}$ and $\mathcal{A} - \lambda_i \mathcal{I}$ commute and $e^{\lambda_i \mathcal{I} t} = e^{\lambda_i t} \mathcal{I}$, we have

$$e^{\mathcal{A}t} = e^{\lambda_i \mathcal{I} t} e^{(\mathcal{A} - \lambda_i \mathcal{I})t} = e^{\lambda_i t} \sum_{j=0}^{\infty} \left\{ \frac{1}{j!} (\mathcal{A} - \lambda_i \mathcal{I})^j t^j \right\}.$$

Premultiplying both sides of this equation by $a_i(\mathcal{A}) q_i(\mathcal{A})$, and observing that $q_i(\mathcal{A})(\mathcal{A} - \lambda_i \mathcal{I})^{r_i} = p(\mathcal{A}) = 0$, and consequently, $q_i(\mathcal{A})(\mathcal{A} - \lambda_i \mathcal{I})^j = 0$ for all $j \ge r_i$, it follows that

$$a_i(\mathcal{A}) q_i(\mathcal{A}) e^{\mathcal{A}t} = e^{\lambda_i t} a_i(\mathcal{A}) q_i(\mathcal{A}) \sum_{j=0}^{r_i - 1} \left\{ \frac{1}{j!} (\mathcal{A} - \lambda_i \mathcal{I})^j t^j \right\}.$$

Summing this relation from $i = 1$ to $m$ and using (2.8.10), we obtain (2.8.7). ■

**Theorem 2.8.3.** Let the notations and hypotheses of Lemma 2.8.2 be satisfied. Then, for all nonnegative integers $k$

$$\mathcal{A}^k = \sum_{i=1}^{m}\sum_{j=0}^{r_i-1}\binom{k}{j}\lambda_i^{k-j}a_i(\mathcal{A})q_i(\mathcal{A})(\mathcal{A}-\lambda_i\mathcal{I})^j. \tag{2.8.11}$$

**Proof.** Differentiating (2.8.7), $k$ times and substituting $t = 0$ gives (2.8.11). ■

**Corollary 2.8.4.** If $m = n$, i.e. $\mathcal{A}$ has $n$ distinct eigenvalues, then $a_i(\mathcal{A}) = (1/q_i(\lambda_i))\mathcal{I}$, and (2.8.11) reduces to

$$\begin{aligned}\mathcal{A}^k &= \sum_{i=1}^{n}\frac{q_i(\mathcal{A})}{q_i(\lambda_i)}\lambda_i^k \\ &= \sum_{i=1}^{n}\frac{(\mathcal{A}-\lambda_1\mathcal{I})\cdots(\mathcal{A}-\lambda_{i-1}\mathcal{I})(\mathcal{A}-\lambda_{i+1}\mathcal{I})\cdots(\mathcal{A}-\lambda_n\mathcal{I})}{(\lambda_i-\lambda_1)\cdots(\lambda_i-\lambda_{i-1})(\lambda_i-\lambda_{i+1})\cdots(\lambda_i-\lambda_n)}\lambda_i^k.\end{aligned} \tag{2.8.12}$$

**Corollary 2.8.5.** If $m = 1$, i.e. $\mathcal{A}$ has all the eigenvalues equal to $\lambda_1$, then $a_i(\mathcal{A}) = q_i(\mathcal{A}) = \mathcal{I}$, and (2.8.11) reduces to

$$\mathcal{A}^k = \sum_{j=0}^{n-1}\binom{k}{j}\lambda_1^{k-j}(\mathcal{A}-\lambda_1\mathcal{I})^j. \tag{2.8.13}$$

**Corollary 2.8.6.** If $m = 2$ and $r_1 = (n-1)$, $r_2 = 1$, then we have

$$\begin{aligned} a_1(\mathcal{A}) &= \frac{1}{(\lambda_2-\lambda_1)^{n-1}}\left[(\lambda_2-\lambda_1)^{n-1}\mathcal{I}-(\mathcal{A}-\lambda_1\mathcal{I})^{n-1}\right](\mathcal{A}-\lambda_2\mathcal{I})^{-1}, \\ q_1(\mathcal{A}) &= (\mathcal{A}-\lambda_2\mathcal{I}),\ a_2(\mathcal{A}) = \frac{1}{(\lambda_2-\lambda_1)^{n-1}}\mathcal{I},\ q_2(\mathcal{A}) = (\mathcal{A}-\lambda_1\mathcal{I})^{n-1}\end{aligned}$$

and (2.8.11) reduces to

$$\begin{aligned}\mathcal{A}^k &= \left[\mathcal{I}-\left(\frac{\mathcal{A}-\lambda_1\mathcal{I}}{\lambda_2-\lambda_1}\right)^{n-1}\right]\sum_{j=0}^{n-2}\binom{k}{j}\lambda_1^{k-j}(\mathcal{A}-\lambda_1\mathcal{I})^j+\lambda_2^k\left(\frac{\mathcal{A}-\lambda_1\mathcal{I}}{\lambda_2-\lambda_1}\right)^{n-1} \\ &= \sum_{j=0}^{n-2}\binom{k}{j}\lambda_1^{k-j}(\mathcal{A}-\lambda_1\mathcal{I})^j - \frac{1}{(\lambda_2-\lambda_1)^{n-1}}\sum_{j=0}^{n-2}\binom{k}{j}\lambda_1^{k-j}\times \\ &\qquad (\mathcal{A}-\lambda_1\mathcal{I})^{n-1+j}+\lambda_2^k\left(\frac{\mathcal{A}-\lambda_1\mathcal{I}}{\lambda_2-\lambda_1}\right)^{n-1}.\end{aligned}$$

Now since $(\mathcal{A}-\lambda_2\mathcal{I}) = (\mathcal{A}-\lambda_1\mathcal{I}) - (\lambda_2-\lambda_1)\mathcal{I}$, we find

$$(\mathcal{A}-\lambda_1\mathcal{I})^{n-1}(\mathcal{A}-\lambda_2\mathcal{I}) = (\mathcal{A}-\lambda_1\mathcal{I})^n - (\lambda_2-\lambda_1)(\mathcal{A}-\lambda_1\mathcal{I})^{n-1}.$$

Thus, by Cayley Hamilton Theorem 2.2.3 we get $(\mathcal{A}-\lambda_1\mathcal{I})^n = (\lambda_2-\lambda_1)(\mathcal{A}-\lambda_1\mathcal{I})^{n-1}$. Using this relation repeatedly, we obtain $(\mathcal{A}-\lambda_1\mathcal{I})^{n+j-1} = (\lambda_2-\lambda_1)^j(\mathcal{A}-\lambda_1\mathcal{I})^{n-1}$. It therefore follows that

$$(2.8.14)\quad \mathcal{A}^k = \sum_{j=0}^{n-2}\binom{k}{j}\lambda_1^{k-j}(\mathcal{A}-\lambda_1\mathcal{I})^j + \left[\lambda_2^k - \sum_{j=0}^{n-2}\binom{k}{j}\lambda_1^{k-j}(\lambda_2-\lambda_1)^j\right] \times \left(\frac{\mathcal{A}-\lambda_1\mathcal{I}}{\lambda_2-\lambda_1}\right)^{n-1}.$$

**Lemma 2.8.7.** (Putzer's Algorithm). Let $\lambda_1,\cdots,\lambda_n$ be the eigenvalues of the matrix $\mathcal{A}$ which are arranged in some arbitrary, but specified order. Then,

$$(2.8.15)\qquad e^{\mathcal{A}t} = \sum_{j=0}^{n-1} r_{j+1}(t)\mathcal{P}_j,$$

where $\mathcal{P}_0 = \mathcal{I}$, $\mathcal{P}_j = \prod_{\ell=1}^{j}(\mathcal{A}-\lambda_\ell\mathcal{I})$, $j=1,\cdots,n$, and $r_1(t),\cdots,r_n(t)$ are recursively given by

$$\begin{aligned} r_1'(t) &= \lambda_1 r_1(t), \quad r_1(0)=1 \\ r_j'(t) &= \lambda_j r_j(t) + r_{j-1}(t), \quad r_j(0)=0, \quad j=2,\cdots,n. \end{aligned}$$

(Note that each eigenvalue in the list is repeated according to its multiplicity. Further, since the matrices $(\mathcal{A}-\lambda_i\mathcal{I})$ and $(\mathcal{A}-\lambda_j\mathcal{I})$ commute, we can for convenience adopt the convention that $(\mathcal{A}-\lambda_j\mathcal{I})$ follows $(\mathcal{A}-\lambda_i\mathcal{I})$ if $i>j$.)

**Proof.** It suffices to show that $\mathcal{Y}(t)$ defined by $\mathcal{Y}(t) = \sum_{j=0}^{n-1} r_{j+1}(t)\mathcal{P}_j$ satisfies (2.8.5). For this, we define $r_0(t)=0$. Then, it follows that

$$\begin{aligned} \mathcal{Y}'(t) - \lambda_n\mathcal{Y}(t) &= \sum_{j=0}^{n-1}(\lambda_{j+1}r_{j+1}(t) + r_j(t))\mathcal{P}_j - \lambda_n\sum_{j=0}^{n-1} r_{j+1}(t)\mathcal{P}_j \\ &= \sum_{j=0}^{n-1}(\lambda_{j+1}-\lambda_n)r_{j+1}(t)\mathcal{P}_j + \sum_{j=0}^{n-1} r_j(t)\mathcal{P}_j \end{aligned}$$

$$\begin{aligned}
&= \sum_{j=0}^{n-2}(\lambda_{j+1}-\lambda_n)r_{j+1}(t)\mathcal{P}_j + \sum_{j=0}^{n-2} r_{j+1}(t)\mathcal{P}_{j+1} \\
&= \sum_{j=0}^{n-2}\{(\lambda_{j+1}-\lambda_n)\mathcal{P}_j + (\mathcal{A}-\lambda_{j+1}\mathcal{I})\mathcal{P}_j\}\, r_{j+1}(t) \qquad (2.8.16)\\
&= (\mathcal{A}-\lambda_n\mathcal{I})\sum_{j=0}^{n-2}\mathcal{P}_j r_{j+1}(t) \\
&= (\mathcal{A}-\lambda_n\mathcal{I})\,(\mathcal{Y}(t) - r_n(t)\mathcal{P}_{n-1}) \\
&= (\mathcal{A}-\lambda_n\mathcal{I})\mathcal{Y}(t) - r_n(t)\mathcal{P}_n, \qquad (2.8.17)
\end{aligned}$$

where to obtain (2.8.16) and (2.8.17) we have used $\mathcal{P}_{j+1} = (\mathcal{A}-\lambda_{j+1}\mathcal{I})\mathcal{P}_j$ and $\mathcal{P}_n = (\mathcal{A}-\lambda_n\mathcal{I})\mathcal{P}_{n-1}$ respectively. Now by Cayley Hamilton Theorem 2.2.3, $\mathcal{P}_n = p(\mathcal{A}) = 0$, and therefore (2.8.17) reduces to $\mathcal{Y}'(t) = \mathcal{A}\mathcal{Y}(t)$. Finally, to complete the proof we note that $\mathcal{Y}(0) = \sum_{j=0}^{n-1} r_{j+1}(0)\mathcal{P}_j = r_1(0)\mathcal{I} = \mathcal{I}$. ∎

**Theorem 2.8.8.** (Discrete Putzer's Algorithm). Let the notations and hypotheses of Lemma 2.8.7 be satisfied. Then, for all nonnegative integers $k$

$$\mathcal{A}^k = \sum_{j=0}^{n-1} w_{j+1}(k)\mathcal{P}_j, \qquad (2.8.18)$$

where

$$\begin{aligned}
w_1(k+1) &= \lambda_1 w_1(k), \quad w_1(0) = 1 \\
w_j(k+1) &= \lambda_j w_j(k) + w_{j-1}(k), \quad w_j(0) = 0, \quad j = 2,\cdots,n.
\end{aligned} \qquad (2.8.19)$$

**Proof.** Differentiating (2.8.15), $k$ times and substituting $t = 0$ gives (2.8.18), where $w_j(k) = r_j^{(k)}(0)$, $1 \le j \le n$ (cf. Problem 2.16.7). ∎

**Example 2.8.3.** Consider a $3\times 3$ matrix $\mathcal{A}$ having all the three eigenvalues equal to $\lambda_1$. To use Theorem 2.8.8, we note that $w_1(k) = \lambda_1^k$, $w_2(k) = k\lambda_1^{k-1}$, $w_3(k) = (1/2)k(k-1)\lambda_1^{k-2}$ is the solution set of the system

$$\begin{aligned}
w_1(k+1) &= \lambda_1 w_1(k), \quad w_1(0) = 1 \\
w_2(k+1) &= \lambda_1 w_2(k) + w_1(k), \quad w_2(0) = 0 \\
w_3(k+1) &= \lambda_1 w_3(k) + w_2(k), \quad w_3(0) = 0.
\end{aligned}$$

Thus, it follows that

$$\mathcal{A}^k = \lambda_1^k\mathcal{I} + k\lambda_1^{k-1}(\mathcal{A}-\lambda_1\mathcal{I}) + \frac{1}{2}k(k-1)\lambda_1^{k-2}(\mathcal{A}-\lambda_1\mathcal{I})^2, \qquad (2.8.20)$$

which is exactly the same as (2.8.13) for $n = 3$.

In particular, the matrix $\mathcal{A} = \begin{bmatrix} 2 & 1 & -1 \\ -3 & -1 & 1 \\ 9 & 3 & -4 \end{bmatrix}$ has all its eigenvalues equal to $-1$, and hence from (2.8.20) we obtain

$$\mathcal{A}^k = \frac{1}{2}(-1)^k \begin{bmatrix} 2-3k-3k^2 & -2k & k+k^2 \\ 6k & 2 & -2k \\ -9k-9k^2 & -6k & 2+3k+3k^2 \end{bmatrix}.$$

Similarly, the matrix $\mathcal{A} = \begin{bmatrix} 0 & 1 & 0 \\ 0 & 0 & 1 \\ 1 & -3 & 3 \end{bmatrix}$ has all its eigenvalues equal to 1, and hence from (2.8.20) we obtain

$$\mathcal{A}^k = \frac{1}{2} \begin{bmatrix} (k-1)(k-2) & -2k(k-2) & k(k-1) \\ k(k-1) & -2(k+1)(k-1) & k(k+1) \\ k(k+1) & -2k(k+2) & (k+2)(k+1) \end{bmatrix}.$$

**Example 2.8.4.** Consider a $3\times3$ matrix $\mathcal{A}$ with eigenvalues $\lambda_1, \lambda_1, \lambda_2$. To use Theorem 2.8.8 we note that

$$w_1(k) = \lambda_1^k, \quad w_2(k) = k\lambda_1^{k-1}, \quad w_3(k) = \frac{k\lambda_1^{k-1}}{(\lambda_1-\lambda_2)} + \frac{\lambda_2^k - \lambda_1^k}{(\lambda_1-\lambda_2)^2},$$

and hence

$$(2.8.21)\quad \mathcal{A}^k = \lambda_1^k\mathcal{I} + k\lambda_1^{k-1}(\mathcal{A}-\lambda_1\mathcal{I}) + \left\{\frac{k\lambda_1^{k-1}}{(\lambda_1-\lambda_2)} + \frac{\lambda_2^k-\lambda_1^k}{(\lambda_1-\lambda_2)^2}\right\}(\mathcal{A}-\lambda_1\mathcal{I})^2,$$

which is precisely the same as (2.8.14) for $n = 3$.

In particular, the matrix $\mathcal{A} = \begin{bmatrix} -1 & 0 & 4 \\ 0 & -1 & 2 \\ 0 & 0 & 1 \end{bmatrix}$ has the eigenvalues $-1, -1, 1$ and hence from (2.8.21) we find

$$\mathcal{A}^k = \begin{bmatrix} (-1)^k & 0 & 2\left(1-(-1)^k\right) \\ 0 & (-1)^k & \left(1-(-1)^k\right) \\ 0 & 0 & 1 \end{bmatrix}.$$

**Remark 2.8.1.** From Problem 2.16.8 we note that the explicit representations obtained for $\mathcal{A}^k,\ k \in \mathbb{N}$ in fact hold for all $k \in \mathbf{Z}$.

Once an explicit representation for $\mathcal{A}^k,\ k \in \mathbb{Z}$ is known, the general solution of the nonhomogeneous difference system

$$\text{(2.8.22)} \qquad \mathbf{u}(k+1) = \mathcal{A}\mathbf{u}(k) + \mathbf{b}(k), \quad k \in \mathbb{Z}$$

can be written as

$$\text{(2.8.23)} \quad \mathbf{u}(k) = \mathcal{A}^k \mathbf{c} + \sum_{\ell=1}^{k} \mathcal{A}^{k-\ell}\mathbf{b}(\ell-1) \quad \text{for all } k \in \mathbb{N}$$

$$\text{(2.8.24)} \qquad = \mathcal{A}^k \mathbf{c} - \sum_{\ell=k+1}^{0} \mathcal{A}^{k-\ell}\mathbf{b}(\ell-1) \quad \text{for all nonpositive integers } k.$$

## 2.9. Periodic Linear Systems

A function $u(k)$ defined on $\overline{\mathbb{N}}_K$ is called *periodic* of period $K > 0$ if for all $k \in \overline{\mathbb{N}}$

$$\text{(2.9.1)} \qquad u(k+K) = u(k).$$

Geometrically, this means that the graph of $u(k)$ repeats itself in successive intervals of length $K$. For example, the function $\cos k\pi$ is periodic on $\mathbb{N}$ with the period $K = 2$. For convenience, we shall assume that $K$ is the smallest positive integer for which (2.9.1) holds. If each component $u_i(k),\ 1 \le i \le n$ of $\mathbf{u}(k)$ and each element $a_{ij}(k),\ 1 \le i,j \le n$ of $\mathcal{A}(k)$ are periodic of period $K$, then $\mathbf{u}(k)$ and $\mathcal{A}(k)$ are said to be periodic of period $K$. The system (1.2.11) and in particular (1.2.12) is said to be periodic of period $K$ if $\mathcal{A}(k)$ and $\mathbf{b}(k)$ are periodic of period $K$. Periodicity of solutions of difference systems is an interesting and important aspect of qualitative study. We shall provide certain characterizations for the existence of such solutions of linear difference systems.

**Theorem 2.9.1.** Let the difference system (1.2.11) be periodic of period $K$ on $\mathbb{N}$. Then, it has a periodic solution $\mathbf{u}(k)$ of period $K$ if and only if $\mathbf{u}(0) = \mathbf{u}(K)$.

**Proof.** Let $\mathbf{u}(k)$ be a periodic solution of period $K$, then by definition it is necessary that $\mathbf{u}(0) = \mathbf{u}(K)$. To show sufficiency, let $\mathbf{u}(k)$ be a solution of (1.2.11) satisfying $\mathbf{u}(0) = \mathbf{u}(K)$. If $\mathbf{v}(k) = \mathbf{u}(k+K)$, then it follows that $\mathbf{v}(k+1) = \mathbf{u}(k+1+K) = \mathcal{A}(k+K)\mathbf{u}(k+K) + \mathbf{b}(k+K) = \mathcal{A}(k)\mathbf{v}(k) + \mathbf{b}(k)$, i.e. $\mathbf{v}(k)$ is a solution of (1.2.11). However, since $\mathbf{v}(0) = \mathbf{u}(K) = \mathbf{u}(0)$, the uniqueness of the initial value problems implies that $\mathbf{u}(k) = \mathbf{v}(k) = \mathbf{u}(k+K)$, and hence $\mathbf{u}(k)$ is periodic of period $K$. ∎

**Corollary 2.9.2.** Let the system (1.2.12) be periodic of period $K$ on $\mathbb{N}$. Further, let $\mathcal{V}(k)$ be a fundamental matrix of (1.2.12). Then, the difference system (1.2.12) has a nontrivial periodic solution $\mathbf{u}(k)$ of period $K$ if and only if $det\ (\mathcal{V}(0) - \mathcal{V}(K)) = 0$.

**Proof.** We know that the general solution of (1.2.12) is $\mathbf{u}(k) = \mathcal{V}(k)\mathbf{c}$, where $\mathbf{c}$ is an arbitrary constant vector. This $\mathbf{u}(k)$ is periodic of period $K$ if and only if $\mathcal{V}(0)\mathbf{c} = \mathcal{V}(K)\mathbf{c}$, i.e. the system $(\mathcal{V}(0) - \mathcal{V}(K))\mathbf{c} = 0$ has a nontrivial solution vector $\mathbf{c}$. But, from Lemma 2.2.1 this system has a nontrivial solution if and only if $det\ (\mathcal{V}(0) - \mathcal{V}(K)) = 0$. ∎

**Corollary 2.9.3.** If $\mathcal{A}(k)$ is a constant matrix $\mathcal{A}$ then the difference system (1.2.12) has a nontrivial periodic solution if and only if the matrix $(\mathcal{I} - \mathcal{A}^K)$ is singular.

**Corollary 2.9.4.** Let the difference system (1.2.11) be periodic of period $K$. Then, it has a unique periodic solution of period $K$ if and only if the system (1.2.12) does not have a periodic solution of period $K$ other than the trivial one.

**Proof.** Let $\mathcal{V}(k)$ be a fundamental matrix of (1.2.12). Then, the general solution of (1.2.11) can be written as

$$\mathbf{u}(k) = \mathcal{V}(k)\mathbf{c} + \sum_{\ell=1}^{k} \mathcal{V}(k)\mathcal{V}^{-1}(\ell)\mathbf{b}(\ell - 1),$$

where $\mathbf{c}$ is an arbitrary constant vector. This $\mathbf{u}(k)$ is periodic of period $K$ if and only if

$$\mathcal{V}(0)\mathbf{c} = \mathcal{V}(K)\mathbf{c} + \sum_{\ell=1}^{K} \mathcal{V}(K)\mathcal{V}^{-1}(\ell)\mathbf{b}(\ell - 1),$$

i.e. the system

$$(\mathcal{V}(0) - \mathcal{V}(K))\mathbf{c} = \sum_{\ell=1}^{K} \mathcal{V}(K)\mathcal{V}^{-1}(\ell)\mathbf{b}(\ell - 1)$$

has a unique solution vector $\mathbf{c}$. But, from Lemma 2.2.1 this system has a unique solution if and only if $det\ (\mathcal{V}(0) - \mathcal{V}(K)) \neq 0$. Now the conclusion follows from Corollary 2.9.2. ∎

**Theorem 2.9.5.** Let the difference system (1.2.12) be periodic of period $K$ on $\mathbb{N}$, and $\mathcal{U}(k,0)$ be its principal fundamental matrix. Then, the following hold

(i) $\mathcal{U}(k+K,0) = \mathcal{U}(k,0)\mathcal{U}(K,0)$, and hence $\mathcal{U}(k+K,0)$ is also a fundamental matrix of (1.2.12)

(ii) (Floquet's Theorem) there exists a periodic nonsingular matrix $\mathcal{P}(k)$ of period $K$ and a constant matrix $\mathcal{R}$ such that

$$\mathcal{U}(k,0) = \mathcal{P}(k)\mathcal{R}^k \tag{2.9.2}$$

(iii) the transformation

$$\mathbf{u}(k) = \mathcal{P}(k)\mathbf{v}(k) \tag{2.9.3}$$

reduces the system (1.2.12) to the system

$$\mathbf{v}(k+1) = \mathcal{R}\mathbf{v}(k). \tag{2.9.4}$$

**Proof.** (i) Since the system (1.2.12) is periodic it is clear that $\mathcal{U}(k+K,0)$ is its matrix solution. Further, from the definition of principal fundamental matrix $det\ \mathcal{U}(k+K,0) \neq 0$. Thus, $\mathcal{U}(k+K,0)$ is a fundamental matrix of (1.2.12). Now since both sides of $\mathcal{U}(k+K,0) = \mathcal{U}(k,0)\mathcal{U}(K,0)$ are the same at $k=0$, this identity for all $k \in \mathbb{N}$ follows from the uniqueness of the initial value problems.

(ii) Since the matrix $\mathcal{U}(K,0) = \prod_{\ell=0}^{K-1} \mathcal{A}(K-1-\ell)$ is nonsingular, from Theorem 2.2.4 it is possible to find the matrix $\mathcal{R}$ such that $\mathcal{U}^{1/K}(K,0) = \mathcal{R}$. Thus, from (i) it follows that

$$\mathcal{U}(k+K,0) = \mathcal{U}(k,0)\mathcal{R}^K. \tag{2.9.5}$$

Let $\mathcal{P}(k)$ be a matrix defined by the relation $\mathcal{P}(k) = \mathcal{U}(k,0)\mathcal{R}^{-k}$. Then, using (2.9.5) we have

$$\mathcal{P}(k+K) = \mathcal{U}(k+K,0)\mathcal{R}^{-k-K} = \mathcal{U}(k,0)\mathcal{R}^K\mathcal{R}^{-k-K} = \mathcal{U}(k,0)\mathcal{R}^{-k} = \mathcal{P}(k).$$

Hence, $\mathcal{P}(k)$ is periodic of period $K$. Further, since $\mathcal{U}(k,0)$ and $\mathcal{R}^{-k}$ are nonsingular $det\ \mathcal{P}(k) \neq 0$.

(iii) From the transformation (2.9.3) and the relation (2.9.2), we have

$$\begin{aligned}\mathbf{u}(k+1) &= \mathcal{P}(k+1)\mathbf{v}(k+1) = \mathcal{U}(k+1,0)\mathcal{R}^{-k-1}\mathbf{v}(k+1)\\ &= \mathcal{A}(k)\mathcal{U}(k,0)\mathcal{R}^{-k-1}\mathbf{v}(k+1) = \mathcal{A}(k)\mathcal{P}(k)\mathcal{R}^{-1}\mathbf{v}(k+1),\end{aligned}$$

and hence

$$\mathcal{A}(k)\mathcal{P}(k)\mathcal{R}^{-1}\mathbf{v}(k+1) = \mathcal{A}(k)\mathbf{u}(k) = \mathcal{A}(k)\mathcal{P}(k)\mathbf{v}(k),$$

i.e.

$$\mathcal{A}(k)\mathcal{P}(k)\left[\mathcal{R}^{-1}\mathbf{v}(k+1)-\mathbf{v}(k)\right] \;=\; 0.$$

However, since $\mathcal{A}(k)$ as well as $\mathcal{P}(k)$ is nonsingular it is necessary that $\mathcal{R}^{-1}\mathbf{v}(k+1)-\mathbf{v}(k)=0$. ■

## 2.10. Almost Periodic Linear Systems

A function $\mathbf{u}(k),\ k\in\mathbf{Z}$ is said to be *almost periodic* if for any positive $\epsilon$ there exists a positive integer $N(\epsilon)$ such that any set consisting $N$ consecutive integers contains at least one integer $p$ with the property that

$$\text{(2.10.1)} \qquad \|\mathbf{u}(k+p)-\mathbf{u}(k)\| \;<\; \epsilon, \qquad k\in\mathbf{Z}.$$

In the above definition $p$ is called an $\epsilon$ *almost period* of $\mathbf{u}(k)$, or an $\epsilon$ *translation number*.

A function $\mathbf{u}(k)$ is called *normal* if for any sequence $\{j_\ell\}\subset\mathbf{Z}$ there exists a subsequence $\{j'_\ell\}\subset\{j_\ell\}$ such that $\mathbf{u}(k+j'_\ell)$ converges uniformly with respect to $k\in\mathbf{Z}$ as $\ell\to\infty$, i.e. for any $\epsilon>0$ there exists $L(\epsilon)$ and a function $\overline{\mathbf{u}}(k)$ such that

$$\|\mathbf{u}(k+j'_\ell)-\overline{\mathbf{u}}(k)\| \;<\; \epsilon \quad \text{for} \quad \ell\ge L(\epsilon), \quad k\in\mathbf{Z}.$$

The following properties of almost periodic functions are fundamental:

($P_1$) A function $\mathbf{u}(k),\ k\in\mathbf{Z}$ is almost periodic if and only if it is normal.

($P_2$) A function $\mathbf{u}(k),\ k\in\mathbf{Z}$ is almost periodic if and only if there exists an almost periodic function $\mathbf{f}(t),\ t\in\mathbb{R}$ such that $\mathbf{u}(k)=\mathbf{f}(k),\ k\in\mathbf{Z}$.

($P_3$) If $u(k)$ and $v(k),\ k\in\mathbf{Z}$ are almost periodic functions and $c\in\mathbb{R}$, then the following functions are almost periodic on $\mathbf{Z}$: (i) $cu(k)$, (ii) $u(k)+v(k)$, (iii) $u(k)v(k)$, (iv) $u(k)/v(k)$ provided $|v(k)|\ge m>0$ for every $k\in\mathbf{Z}$, (v) $u(k+\ell)$, where $\ell$ is a fixed integer.

($P_4$) An almost periodic function $\mathbf{u}(k),\ k\in\mathbf{Z}$ is bounded.

($P_5$) Let $v(k),\ k\in\mathbf{Z}$ be a summable function, i.e.

$$V \;=\; \sum_{k\in\mathbf{Z}}|v(k)| \;<\; \infty.$$

Then for any almost periodic function $u(k),\ k\in\mathbf{Z}$ the function $w(k)$ defined by

$$w(k) \;=\; \sum_{\ell\in\mathbf{Z}} v(\ell)u(k-\ell), \qquad k\in\mathbf{Z}$$

is also almost periodic.

($P_6$) Let $u(k),\ k \in \mathbf{Z}$ be an almost periodic function. Then the limit

$$\lim_{m \to \infty} \frac{u(k+1) + \cdots + u(k+m)}{m}$$

exists uniformly with respect to $k \in \mathbf{Z}$, and is independent of $k$. This number is called the *mean value* of $u(k)$ and is usually denoted by $M(u(k))$.

($P_7$) Let $u(k),\ k \in \mathbf{Z}$ be an almost periodic function, and let $U(k)$ be defined as follows: $U(0)$ is chosen arbitrarily, and $U(k+1) - U(k) = u(k),\ k \in \mathbf{Z}$. Then, $U(k),\ k \in \mathbf{Z}$ is almost periodic if and only if it is bounded.

The main result of this section is embodied in the following:

**Theorem 2.10.1.** Let the function $\mathbf{b}(k),\ k \in \mathbf{Z}$ be almost periodic. Then, a solution $\mathbf{u}(k),\ k \in \mathbf{Z}$ of (2.8.22) is almost periodic if and only if it is bounded.

**Proof.** From ($P_4$) an almost periodic solution $\mathbf{u}(k),\ k \in \mathbf{Z}$ of (2.8.22) is bounded, and hence we need only to show that the boundedness of $\mathbf{u}(k),\ k \in \mathbf{Z}$ implies its almost periodicity. It is well known that there exists a nonsingular matrix $\mathcal{T}$ such that $\mathcal{T}^{-1}\mathcal{A}\mathcal{T} = \mathcal{B}$ is an upper triangular matrix, i.e. $\mathcal{B}$ has the form

$$\mathcal{B} = \begin{bmatrix} \lambda_1 & b_{12} & b_{13} & \cdots & b_{1n} \\ 0 & \lambda_2 & b_{23} & \cdots & b_{2n} \\ \cdots & & & \cdots & \cdots \\ 0 & 0 & 0 & \cdots & \lambda_n \end{bmatrix},$$

where $\lambda_i,\ i = 1, \cdots, n$ are the eigenvalues of $\mathcal{A}$ (or $\mathcal{B}$). In (2.8.22) we use the substitution $\mathbf{u}(k) = \mathcal{T}\mathbf{v}(k)$, to obtain

$$\mathbf{v}(k+1) = \mathcal{B}\mathbf{v}(k) + \mathcal{T}^{-1}\mathbf{b}(k), \quad k \in \mathbf{Z}. \tag{2.10.2}$$

Obviously, the system (2.10.2) is of the form as (2.8.22) with $\mathcal{T}^{-1}\mathbf{b}(k)$ an almost periodic function, but reduces considerably the difficulty in discussing the almost periodicity of its solutions. In fact, the general case of an arbitrary matrix $\mathcal{A}$ is now reduced to the scalar case. Indeed, the last equation of the system (2.10.2) is of the form

$$z(k+1) = \lambda z(k) + c(k), \quad k \in \mathbf{Z} \tag{2.10.3}$$

where $\lambda$ is any number, and $c(k),\ k \in \mathbf{Z}$ is an almost periodic function. All we need to show that any bounded solution $z(k),\ k \in \mathbf{Z}$ of (2.10.3) is almost periodic. It will then imply that the $n$th component $v_n(k)$, $k \in \mathbf{Z}$ of the solution $\mathbf{v}(k)$, $k \in \mathbf{Z}$ of (2.10.2) is almost periodic. Then, substituting $v_n(k)$ in the $(n-1)$th equation of (2.10.2) we obtain again an equation of the form (2.10.3) for $v_{n-1}(k)$, and so on. Therefore, we need to discuss only (2.10.3). There are three distinct cases: 1. $|\lambda| < 1$, 2. $|\lambda| > 1$, and 3. $|\lambda| = 1$.

Case 1. From (2.10.3), we have

$$(2.10.4)\quad z(k+p+1)-z(k+1) = \lambda(z(k+p)-z(k))+(c(k+p)-c(k)), \quad k \in \mathbf{Z}$$

and since $z(k)$, $k \in \mathbf{Z}$ is bounded, we obtain

$$(2.10.5)\qquad \sup_{k\in\mathbf{Z}} |z(k+p)-z(k)| \le (1-|\lambda|)^{-1} \sup_{k\in\mathbf{Z}} |c(k+p)-c(k)|.$$

Now (2.10.5) shows that any $(1-|\lambda|)\epsilon$ almost period of $c(k)$ is an $\epsilon$ almost period for $z(k)$. Moreover, if $c(k)$ is periodic then $z(k)$ is periodic of the same period.

Case 2. Once again from (2.10.4), we get

$$(2.10.6)\qquad \sup_{k\in\mathbf{Z}} |z(k+p)-z(k)| \le (|\lambda|-1)^{-1} \sup_{k\in\mathbf{Z}} |c(k+p)-c(k)|.$$

From (2.10.6) the almost periodicity of $z(k)$ is clear.

Case 3. In this case we write $\lambda = e^{-i\alpha}$ for some real $\alpha$, and multiply both sides of (2.10.3) by $e^{i(k+1)\alpha}$, to obtain

$$e^{i(k+1)\alpha} z(k+1) = e^{ik\alpha} z(k) + e^{i\alpha} e^{ik\alpha} c(k), \quad k \in \mathbf{Z}$$

which with obvious notations can be written as

$$(2.10.7)\qquad Z(k+1) - Z(k) = \bar{c}(k), \quad k \in \mathbf{Z}.$$

Now $(P_7)$ implies that $Z(k)$, $k \in \mathbf{Z}$ is almost periodic. Finally, since $z(k) = e^{-ik\alpha} Z(k)$, $k \in \mathbf{Z}$ the function $z(k)$, $k \in \mathbf{Z}$ is almost periodic, as it is the product of two almost periodic functions. ∎

## 2.11. Higher Order Linear Equations

In Section 1.2 we have seen that if $a_0(k)a_n(k) \neq 0$ for all $k \in \overline{\mathbb{N}}$ then the $n$th order difference equation (1.2.3) can be written in the system form (1.2.11) where the matrix $\mathcal{A}(k)$ and the vector $\mathbf{b}(k)$ are defined in (1.2.13) and (1.2.14) respectively, and the relation between the unknown vector $\mathbf{u}(k)$ and the solution $u(k)$ of (1.2.3) is given in (1.2.10). We are interested in finding an explicit representation of the general solution of (1.2.3) in terms of linearly independent solutions of (1.2.4). For this, we note that for the given functions $u_1(k), \cdots, u_n(k)$ on $\overline{\mathbb{N}}_n$ the *Casoratian matrix* $C(u_1, \cdots, u_n)(k)$, or in short $C(k)$, reduces to

$$(2.11.1) \quad C(u_1, \cdots, u_n)(k) = \begin{bmatrix} u_1(k) & \cdots & u_n(k) \\ u_1(k+1) & \cdots & u_n(k+1) \\ \cdots & \cdots & \cdots \\ u_1(k+n-1) & \cdots & u_n(k+n-1) \end{bmatrix},$$

which is defined for all $k \in \overline{\mathbb{N}}_1$. The results analogous to Lemmas 2.3.1, 2.3.2 and Theorem 2.3.3 can be stated as follows:

**Lemma 2.11.1.** If *det* $C(k)$ of $n$ functions $u_i(k)$, $1 \le i \le n$ defined on $\overline{\mathbb{N}}_n$ is different from zero for at least one $k_0 \in \overline{\mathbb{N}}_1$, then $u_i(k)$, $1 \le i \le n$ are linearly independent on $\overline{\mathbb{N}}_n$.

**Lemma 2.11.2.** Let $u_i(k)$, $1 \le i \le n$ be linearly independent solutions of $\overline{\mathbb{N}}_n$ of the homogeneous equation (1.2.4). Then, *det* $C(k) \neq 0$ for all $k \in \overline{\mathbb{N}}_1$.

**Theorem 2.11.3.** The solutions $u_i(k)$, $1 \le i \le n$ of the difference equation (1.2.4) are linearly independent on $\overline{\mathbb{N}}_n$ if and only if there exists at least one $k_0 \in \overline{\mathbb{N}}_1$ such that *det* $C(k_0) \neq 0$.

As a consequence of this result the solutions $u_i(k)$, $1 \le i \le n$ of the equation (1.2.4) satisfying the initial conditions

$$(2.11.2) \qquad u_i(a+j-1) = \delta_{ij}, \quad 1 \le i, j \le n \quad (a = 0 \text{ if } \overline{\mathbb{N}} = \mathbb{N})$$

are linearly independent on $\overline{\mathbb{N}}_n$.

Thus, for the difference equation (1.2.4) there exist exactly $n$ linearly independent solutions $u_i(k)$, $1 \le i \le n$ on $\overline{\mathbb{N}}_n$, and any other solution $u(k)$ of this equation can be written as

$$(2.11.3) \qquad u(k) = \sum_{i=1}^{n} u(a+i-1)u_i(k).$$

Further, if $v_i(k)$, $1 \le i \le n$ is any set of linearly independent solutions of (1.2.4) then its general solution $u(k)$ appears as

$$(2.11.4) \qquad u(k) = \sum_{i=1}^{n} c_i v_i(k),$$

where $c_i$, $1 \le i \le n$ are arbitrary constants.

**Example 2.11.1.** The functions $v_1(k) = 1$, $v_2(k) = k^2$, $k \in \mathbb{N}$ are linearly independent solutions of the difference equation

$$(2.11.5)\quad (2k+1)u(k+2) - 4(k+1)u(k+1) + (2k+3)u(k) = 0, \quad k \in \mathbb{N}$$

whereas its linearly independent solutions satisfying (2.11.2) with $a = 0$ are $u_1(k) = 1 - k^2$, $u_2(k) = k^2$. Similarly, for the difference equation

$$(2.11.6)\quad \left((k+1)^3 - k^3\right)u(k+2) - \left((k+2)^3 - k^3\right)u(k+1) + \left((k+2)^3 - (k+1)^3\right)u(k) = 0, \quad k \in \mathbb{N}$$

linearly independent solutions are $v_1(k) = 1$, $v_2(k) = k^3$, whereas the ones satisfying (2.11.2) are $u_1(k) = 1 - k^3$, $u_2(k) = k^3$.

Now let $v_i(k)$, $1 \le i \le n$ be any fixed set of linearly independent solutions of (1.2.4). We shall compute the first component, say, $z(k,\ell)$ of the vector $\mathcal{V}(k)\mathcal{V}^{-1}(\ell)\mathbf{b}(\ell-1)$. For this, we note that

$$\mathcal{V}(k)\mathcal{V}^{-1}(\ell)\mathbf{b}(\ell-1) = \begin{bmatrix} v_1(k) & \cdots & v_n(k) \\ \cdots & \cdots & \cdots \\ v_1(k+n-1) & \cdots & v_n(k+n-1) \end{bmatrix} \times$$

$$\begin{bmatrix} v_1(\ell) & \cdots & v_n(\ell) \\ \cdots & \cdots & \cdots \\ v_1(\ell+n-1) & \cdots & v_n(\ell+n-1) \end{bmatrix}^{-1} \begin{bmatrix} 0 \\ \vdots \\ 0 \\ \dfrac{b(\ell-1)}{a_n(\ell-1)} \end{bmatrix}$$

$$= \frac{1}{\det \mathcal{V}(\ell)} \begin{bmatrix} v_1(k) & \cdots & v_n(k) \\ \cdots & \cdots & \cdots \\ v_1(k+n-1) & \cdots & v_n(k+n-1) \end{bmatrix} \times$$

$$\begin{bmatrix} \text{cofactor of } v_1(\ell+n-1) \\ \vdots \\ \text{cofactor of } v_n(\ell+n-1) \end{bmatrix} \frac{b(\ell-1)}{a_n(\ell-1)},$$

and hence

$$\begin{aligned} z(k,\ell) &= \frac{1}{\det \mathcal{V}(\ell)} \sum_{i=1}^{n} v_i(k) \text{ cofactor of } v_i(\ell+n-1) \frac{b(\ell-1)}{a_n(\ell-1)} \\ &= G(k,\ell)\frac{b(\ell-1)}{a_n(\ell-1)}, \end{aligned}$$

where

$$G(k,\ell) = \frac{\begin{vmatrix} v_1(\ell) & \cdots & v_n(\ell) \\ \cdots & \cdots & \cdots \\ v_1(\ell+n-2) & \cdots & v_n(\ell+n-2) \\ v_1(k) & \cdots & v_n(k) \end{vmatrix}}{\begin{vmatrix} v_1(\ell) & \cdots & v_n(\ell) \\ \cdots & \cdots & \cdots \\ v_1(\ell+n-1) & \cdots & v_n(\ell+n-1) \end{vmatrix}}. \tag{2.11.7}$$

This function $G(k,\ell)$ is called the *Green's function* of the equation (1.2.4) and it is defined for all $k \in \overline{\mathbb{N}}_n$, $\ell \in \overline{\mathbb{N}}_1$. The following properties of $G(k,\ell)$ are immediate

(i) $G(k,\ell) = 0$ for all $\ell \in \mathbb{N}(k-n+2, k)$ and $k \in \overline{\mathbb{N}}_n$

(ii) $G(k,\ell) = 0$ for all $k \in \mathbb{N}(\ell, \ell+n-2)$ and $\ell \in \overline{\mathbb{N}}_1$, and $G(\ell+n-1,\ell) = 1$

(iii) for a fixed $\ell \in \overline{\mathbb{N}}_1$, $w(k) = G(k,\ell)$ is a solution of (1.2.4)

(iv) $G(k,\ell)$ is independent of the set of linearly independent solutions $v_i(k)$, $1 \le i \le n$ of (1.2.4).

Since the first component in (2.5.3) is the general solution $u(k)$ of (1.2.3), from the above considerations it follows that

$$u(k) = \sum_{i=1}^{n} c_i v_i(k) + \sum_{\ell=a+1}^{k-n+1} G(k,\ell)\frac{b(\ell-1)}{a_n(\ell-1)}, \quad k \in \overline{\mathbb{N}}_n. \tag{2.11.8}$$

**Example 2.11.2.** From Example 2.11.1 and (2.11.8) it is clear that the general solution of the nonhomogeneous difference equation

$$\begin{aligned}(2k+1)u(k+2) - 4(k+1)u(k+1) + (2k+3)u(k) \\ = (2k+1)(2k+3), \quad k \in \mathbb{N}\end{aligned} \tag{2.11.9}$$

can be written as

$$\begin{aligned} u(k) &= c_1 + c_2k^2 + \sum_{\ell=1}^{k-1} \frac{\begin{vmatrix} 1 & \ell^2 \\ 1 & k^2 \end{vmatrix}}{\begin{vmatrix} 1 & \ell^2 \\ 1 & (\ell+1)^2 \end{vmatrix}} \frac{(2\ell-1)(2\ell+1)}{(2\ell-1)} \\ &= c_1 + c_2k^2 + \sum_{\ell=1}^{k-1}(k^2-\ell^2) \\ &= c_1 + c_2k^2 + \frac{1}{6}k(k-1)(4k+1). \end{aligned}$$

Similarly, the general solution of the nonhomogeneous difference equation

$$(2.11.10)\quad \left((k+1)^3-k^3\right)u(k+2)-\left((k+2)^3-k^3\right)u(k+1)+\left((k+2)^3 -(k+1)^3\right)u(k) = \left[(k+1)^3-k^3\right]\left[(k+2)^3-(k+1)^3\right],\ k\in\mathbb{N}$$

appears as

$$u(k) \;=\; c_1+c_2k^3+\frac{1}{4}k^2(k-1)(3k+1).$$

**Definition 2.11.1.** The difference equation (1.2.4) on $\mathbb{N}$ is said to be *exact* if there exist functions $b_i(k)$, $0\le i\le n-1$ defined on $\mathbb{N}$ are such that

$$(2.11.11)\qquad \Delta\left[\sum_{i=0}^{n-1} b_i(k)u(k+i)\right] \;=\; \sum_{i=0}^{n} a_i(k)u(k+i),\quad k\in\mathbb{N}.$$

The above condition holds if and only if

$$(2.11.12)\qquad \begin{aligned} a_0(k) &= -b_0(k)\\ a_i(k) &= b_{i-1}(k+1)-b_i(k),\quad 1\le i\le n-1\\ a_n(k) &= b_{n-1}(k+1). \end{aligned}$$

These equations imply the necessary and sufficient condition for the exactness of the equation (1.2.4)

$$(2.11.13)\qquad \sum_{i=0}^{n} a_{n-i}(k+i) \;=\; 0,\quad k\in\mathbb{N}.$$

**Definition 2.11.2.** A function $v(k)$ defined on $\mathbb{N}$ is said to be a *multiplier* of (1.2.4) if the equation $\sum_{i=0}^{n} v(k)a_i(k)u(k+i)=0$ is exact.

For $v(k)$ to be a multiplier of (1.2.4) it is necessary and sufficient that $v(k)$ is a solution of the equation

$$(2.11.14)\qquad \sum_{i=0}^{n} a_{n-i}(k+i)v(k+i) \;=\; 0,\quad k\in\mathbb{N}.$$

The above equation is called the *adjoint* of (1.2.4). The adjoint of (2.11.14) known as the adjoint of the adjoint is therefore

$$(2.11.15)\qquad \sum_{i=0}^{n} a_i(k+n)w(k+i) \;=\; 0.$$

This equation is the same as (1.2.4). Indeed, if we put $\ell = k+n$, then (2.11.15) becomes

$$\sum_{i=0}^{n} a_i(\ell)w(\ell - n + i) = 0, \quad \ell \in \mathbb{N}(n).$$

Thus, if $u(k)$ is a solution of (1.2.4), then $u(k+n)$ is a solution of (2.11.15).

Let in the equation (2.11.14), $k$ be $k-n$ and $v(k-n) = z(k)$, so that it takes the form

$$\sum_{i=0}^{n} a_{n-i}(k-n+i)z(k+i) = 0, \quad k \in \mathbb{N}. \tag{2.11.16}$$

This equation is called the *transpose* of (1.2.4).

Now we shall develop the discrete Lagrange's identity and the discrete Green's formula. For this, we shall denote the operator

$$L[u(k)] = \sum_{i=0}^{n} a_i(k)u(k+i) \tag{2.11.17}$$

and its adjoint

$$L^*[v(k)] = \sum_{i=0}^{n} a_{n-i}(k+i)v(k+i) = \sum_{i=0}^{n} a_i(k+n-i)v(k+n-i). \tag{2.11.18}$$

From these operators, it is easy to obtain

$$\begin{aligned} v(k)L[u(k)] - u(k)L^*[v(k-n)] \\ &= \sum_{i=1}^{n} [a_i(k)v(k)u(k+i) - a_i(k-i)v(k-i)u(k)] \\ &= \Delta_k \left[\sum_{i=1}^{n}\sum_{\ell=0}^{i-1} a_i(k+\ell-i)v(k+\ell-i)u(k+\ell)\right], \end{aligned} \tag{2.11.19}$$

which is the required *discrete Lagrange's identity*. On summing this identity from $k = k_1$ to $k_2$ where $k_1 < k_2$ and $k_1, k_2 \in \mathbb{N}$, we find the *discrete Green's formula*

$$\begin{aligned} &\sum_{k=k_1}^{k_2} [v(k)L[u(k)] - u(k)L^*[v(k-n)]] \\ &= \left[\sum_{i=1}^{n}\sum_{\ell=0}^{i-1} a_i(k+\ell-i)v(k+\ell-i)u(k+\ell)\right]\Bigg|_{k=k_1}^{k_2+1}. \end{aligned} \tag{2.11.20}$$

This formula can be used to deduce the relation between the solutions of (1.2.4) and its adjoint equation (2.11.14) (cf. Problem 2.16.23).

In particular, in (2.11.20) let $u(k)$ be a solution of the equation (1.2.4), $v(k-n) = z(k)$ be a solution of the nonhomogeneous transpose equation

$$L^*[v(k-n)] = L^*[z(k)] = b(k), \tag{2.11.21}$$

$k_1 = 0,\ k_2 = K$ and $z(K+i) = 0,\ 1 \le i \le n$ to obtain

$$\begin{aligned}\sum_{k=0}^{K} u(k)b(k) &= -\left[\sum_{i=1}^{n}\sum_{\ell=0}^{i-1} a_i(k+\ell-i)v(k+\ell-i)u(k+\ell)\right]\Big|_{k=0}^{K+1} \\ &= -\left[\sum_{\ell=1}^{n}\left[\sum_{i=\ell}^{n} a_i(k-\ell)u(k+i-\ell)\right] z(k+n-\ell)\right]\Big|_{k=0}^{K+1} \\ &= \sum_{\ell=1}^{n}\left[\sum_{i=\ell}^{n} a_i(-\ell)u(i-\ell)\right] z(n-\ell). \qquad (2.11.22)\end{aligned}$$

**Example 2.11.3.** Consider the problem of computing the sum $\sum_{k=0}^{K} b(k) \times u(k)$, where $u(k)$ is a solution of the second order equation

$$a_0(k)u(k) + a_1(k+1)u(k+1) + a_2(k+2)u(k+2) = 0, \tag{2.11.23}$$

and $b(k),\ k \in \mathbb{N}(0, K)$ is a given function. For this, we note that the nonhomogeneous transpose equation (2.11.21) reduces to

$$a_2(k)z(k) + a_1(k)z(k+1) + a_0(k)z(k+2) = b(k), \tag{2.11.24}$$

and thus if $z(K+1) = z(K+2) = 0$, the relation (2.11.22) gives

$$\sum_{k=0}^{K} b(k)u(k) = [a_1(0)u(0) + a_2(1)u(1)]\, z(1) + a_2(0)u(0)z(0). \tag{2.11.25}$$

This ingenious way of computing the sum which avoids the computation of any $u(k)$ except $u(0)$ and $u(1)$ is due to Clenshaw [5].

## 2.12. Method of Generating Functions

To solve the $n$th order nonhomogeneous difference equation

$$P_n(E)u(k) = \sum_{i=0}^{n} a_i u(k+i) = b(k), \quad k \in \mathbb{N}, \quad a_0 a_n \neq 0 \tag{2.12.1}$$

where $a_i$, $0 \le i \le n$ are constants several methods are known. For example, the method of undetermined coefficients and the operational method are given in Problems 2.16.39 and 2.16.41 respectively. Here we shall discuss the method of generating functions whose importance is in its simplicity, and the theory is parallel to the Laplace transform method in ordinary differential equations with constant coefficients.

**Definition 2.12.1.** For a given function $u(k)$, $k \in \mathbb{N}$ the *generating function* is defined by the series

$$U(s) = G(u(k)) = \sum_{k=0}^{\infty} u(k)s^k, \tag{2.12.2}$$

where it is assumed that there exists a constant $c > 0$ so that the above series converges for all $|s| \le c$.

In particular, if $s = 1/z$ then the series (2.12.2) is called *Laurant transformation* or $Z$ transformation of $u(k)$. To resolve the problem of convergence in this case, we note that in most of our applications we will have $|u(k)| \le \rho^k$, where $\rho \ge 0$ is some suitable constant. Therefore, the d'Alembert ratio test for the convergence guarantees that (2.12.2) converges for all $|z| > \rho$.

For example, $G(0) = 0$, $G(\beta^k) = \sum_{k=0}^{\infty} \beta^k s^k = 1/(1-\beta s)$, which converges for all $|s| < 1/|\beta|$, $\beta \ne 0$. However, $G(k!) = \sum_{k=0}^{\infty} k! s^k$ converges only for $s = 0$, i.e. the function $k!$ does not possess a generating function.

In the following Table 2.12.1 the first eight entries provide the general relationships between the generated and generating functions. All of these relations easily follow from the Definition 2.12.1, and find importance in solving the difference equation (2.12.1). In particular, entry 8 can be deduced by first noticing that

$$U(s)V(s) = \left(\sum_{k=0}^{\infty} u(k)s^k\right)\left(\sum_{\ell=0}^{\infty} v(\ell)s^\ell\right) = \sum_{k=0}^{\infty}\sum_{\ell=0}^{\infty} u(k)v(\ell)s^{k+\ell}$$

and then writing $k = i - \ell$ and using the fact that $\ell$ ranges only from 0 to $i$ for a fixed value of $i$, to deduce that

$$U(s)V(s) = \sum_{i=0}^{\infty}\left(\sum_{\ell=0}^{i} u(i-\ell)v(\ell)\right)s^i = G\left(\sum_{\ell=0}^{i} u(i-\ell)v(\ell)\right).$$

Usually, we write

$$u(i) * v(i) = \sum_{\ell=0}^{i} u(i-\ell)v(\ell) = \sum_{\ell=0}^{i} u(\ell)v(i-\ell) = v(i) * u(i),$$

and refer to the result as the *convolution* of $u(i)$ and $v(i)$. Thus, if $U(s)$ generates $u(k)$ and $V(s)$ generates $v(k)$, then $U(s)V(s)$ generates $u(k) * v(k)$.

The remaining entries 9 to 21 in Table 2.12.1 are some of the most frequently used specific relations. Throughout this table $r$ is a nonnegative integer.

Table 2.12.1.

Generating Functions

| | $u(k)$ | $U(s) = G(u(k))$ |
|---|---|---|
| 1. | $u(k)$ | $U(s) = \sum_{k=0}^{\infty} u(k)s^k$ |
| 2. | $\alpha u(k) + \beta v(k)$ | $\alpha U(s) + \beta V(s)$ |
| 3. | $u(k+r),\ r \in \mathbb{N}(1)$ | $\left(U(s) - \sum_{j=0}^{r-1} u(j)s^j\right) / s^r$ |
| 4. | $\begin{cases} u(k-r), & k \in \mathbb{N}(r) \\ 0, & \text{otherwise} \end{cases}$ | $s^r U(s)$ |
| 5. | $k^r u(k),\ r \in \mathbb{N}(1)$ | $\left(s\frac{d}{ds}\right)^r U(s)$ |
| 6. | $(k)^{(r)} u(k),\ r \in \mathbb{N}(1)$ | $s^r \frac{d^r U(s)}{ds^r}$ |
| 7. | $(k+r)^{(r)} u(k+r),\ r \in \mathbb{N}(1)$ | $\frac{d^r U(s)}{ds^r}$ |
| 8. | $\sum_{j=0}^{k} u(k-j)v(j) = \sum_{j=0}^{k} u(j)v(k-j) = u(k) * v(k)$ | $U(s)V(s)$ |
| 9. | $\delta_{k-r,0},\ k \in \mathbb{N}(r)$ | $s^r$ |
| 10. | $e^{\alpha k}$ | $\frac{1}{(1 - e^{\alpha} s)}$ |
| 11. | $\frac{\alpha^k}{k!}$ | $e^{\alpha s}$ |

| | | |
|---|---|---|
| 12. | $(k)^{(r)}$ | $\dfrac{r!s^r}{(1-s)^{r+1}}$ |
| 13. | $k^r\beta^k$ | $\left(s\dfrac{d}{ds}\right)^r \dfrac{1}{1-\beta s}$ |
| 14. | $(k+r)^{(r)}\beta^k$ | $\dfrac{r!}{(1-\beta s)^{r+1}}$ |
| 15. | $\cosh \alpha k$ | $\dfrac{1-s\cosh \alpha}{1-2s\cosh \alpha + s^2}$ |
| 16. | $\sinh \alpha k$ | $\dfrac{s\sinh \alpha}{1-2s\cosh \alpha + s^2}$ |
| 17. | $\beta^k \cos \alpha k$ | $\dfrac{1-\beta s\cos \alpha}{1-2\beta s \cos \alpha + \beta^2 s^2}$ |
| 18. | $\beta^k \sin \alpha k$ | $\dfrac{\beta s \sin \alpha}{1-2\beta s \cos \alpha + \beta^2 s^2}$ |
| 19. | $\binom{r}{k}\alpha^{r-k}\beta^k$ | $(\alpha+\beta s)^r$ |
| 20. | $\dfrac{B_k}{k!}$ (Bernoulli numbers) | $\dfrac{s}{e^s-1}$ |
| 21. | $u(k)$ (Fibonacci numbers) | $\dfrac{s}{1-s-s^2}$ |

For the given function $b(k)$, $k \in \mathbb{N}$ and the solution $u(k)$, $k \in \mathbb{N}$ of the difference equation (2.12.1) let $B(s)$ and $U(s)$ be the corresponding generating functions. Thus, on using the entries 2 and 3 from the Table 2.12.1 in (2.12.1) it follows that

$$\sum_{i=0}^{n} a_i \left[\frac{U(s) - \sum_{j=0}^{i-1} s^j u(j)}{s^i}\right] = B(s),$$

which is on arranging the terms gives

$$U(s) = \frac{\sum_{i=0}^{n-1}\left(\sum_{j=0}^{i} a_{n-j}u(i-j)\right)s^i + s^n B(s)}{a_n s^n p(1/s)} \tag{2.12.3}$$

where $p(\lambda) = \sum_{i=0}^{n}(a_i/a_n)\lambda^i$ is the characteristic equation of the homogeneous difference equation

$$P_n(E)u(k) = \sum_{i=0}^{n} a_i u(k+i) = 0, \quad k \in \mathbb{N}, \quad a_0 a_n \neq 0 \tag{2.12.4}$$

(cf. Problem 2.16.35).

For the right side of (2.12.3) we use the Table 2.12.1 to recover the solution $u(k)$, $k \in \mathbb{N}$ of (2.12.1). If some terms do not appear in the list of special functions given in the table, then entries 1 to 8 may be used to obtain the desired functions. The solution obtained in this way satisfies the given initial conditions $u(0), \cdots, u(n-1)$.

**Example 2.12.1.** For the first order difference equation $u(k+1)-au(k) = b(k)$, $k \in \mathbb{N}$ equation (2.12.3) reduces to

$$U(s) = \frac{u(0)}{1-as} + \frac{sB(s)}{1-as}.$$

Since from the entry 14 of the Table 2.12.1, we have $G(a^k) = 1/(1-as)$, entry 4 gives $G\left\{ \begin{array}{l} a^{k-1}, \ k \in \mathbb{N}(1) \\ 0, \ \text{otherwise} \end{array} \right\} = s/(1-as)$. Now the use of entry 8 provides $G\left[\sum_{j=1}^{k} a^{j-1}b(k-j)\right] = (s/(1-as))B(s)$. Therefore, the solution can be written as

$$u(k) = a^k u(0) + \sum_{j=1}^{k} a^{j-1}b(k-j),$$

which is the same as

$$u(k) = a^k u(0) + \sum_{\ell=1}^{k} a^{k-\ell}b(\ell-1).$$

**Example 2.12.2.** For the second order difference equation

$$u(k+2) + a_1 u(k+1) + a_0 u(k) = b(k), \quad k \in \mathbb{N}$$

equation (2.12.3) reduces to

$$U(s) = \frac{u(0) + (u(1) + a_1 u(0))s + s^2 B(s)}{1 + a_1 s + a_0 s^2}.$$

Thus, if we write $G(v(k)) = 1/(1 + a_1 s + a_0 s^2)$ and $v(-1) = v(-2) = 0$, then it follows that

$$u(k) = u(0)v(k) + (u(1) + a_1 u(0))v(k-1) + \sum_{j=0}^{k} v(j-2)b(k-j).$$

Now let $\lambda_1$ and $\lambda_2$ be the roots of the equation $\lambda^2 + a_1\lambda + a_0 = 0$, so that

$$\frac{1}{1 + a_1 s + a_0 s^2} = \frac{1}{(1-\lambda_1 s)(1-\lambda_2 s)} = \frac{1}{(\lambda_2 - \lambda_1)}\left[\frac{\lambda_2}{1-\lambda_2 s} - \frac{\lambda_1}{1-\lambda_1 s}\right]$$

and hence from the entry 14 of Table 2.12.1 it follows that

$$v(k) = \begin{cases} \left(\lambda_2^{k+1} - \lambda_1^{k+1}\right)/(\lambda_2 - \lambda_1) & \text{if} \quad \lambda_1 \neq \lambda_2 \\ (k+1)\lambda_1^k \quad \text{if} \quad \lambda_1 = \lambda_2, & k \in \mathbb{N}. \end{cases}$$

In particular, if $\lambda_1 = \rho e^{i\theta}$ and $\lambda_2 = \rho e^{-i\theta}$, then we have

$$v(k) = \rho^k \sin(k+1)\theta / \sin\theta, \qquad k \in \mathbb{N}.$$

## 2.13. Bernoulli's Method

Suppose that we are given a polynomial equation

$$\sum_{i=0}^{n} a_i \lambda^i = 0, \quad a_0 \neq 0, \quad a_n = 1 \tag{2.13.1}$$

whose roots $\lambda_1, \cdots, \lambda_n$ are distinct. On the basis of the coefficients $a_i$, $0 \le i \le n$ we consider the difference equation (2.12.4) whose general solution we know can be written as

$$u(k) = c_1 \lambda_1^k + \cdots + c_n \lambda_n^k. \tag{2.13.2}$$

If $|\lambda_i| \le q|\lambda_1|$, for all $i = 2, \cdots, n$ with $q < 1$, then for large $k \in \mathbb{N}$, we have

$$u(k) = c_1 \lambda_1^k + O(|q\lambda_1|^k),$$

which is the same as

$$u(k) = \lambda_1^k \left(c_1 + O(|q|^k)\right).$$

Therefore, we obtain

$$\frac{u(k+1)}{u(k)} = \frac{\lambda_1^{k+1}\left(c_1 + O(|q|^{k+1})\right)}{\lambda_1^k \left(c_1 + O(|q|^k)\right)}$$

and hence for the root $\lambda_1$ we obtain the formula

$$\lim_{k \to \infty} \frac{u(k+1)}{u(k)} = \lambda_1, \quad \text{provided } c_1 \neq 0. \tag{2.13.3}$$

If $c_1 = 0$, then (2.13.2) reduces to $u(k) = c_2 \lambda_2^k + \cdots + c_n \lambda_n^k$ which is a solution of the $(n-1)$th order difference equation

$$\left[\frac{P_n(E)}{E - \lambda_1}\right] u(k) = \sum_{i=0}^{n-1} b_i u(k+i) = 0,$$

where $b_i$, $0 \le i \le n-1$ are suitable constants. Thus, if $c_1 = 0$ then it is necessary that for all $k \in \mathbb{N}$

$$(2.13.4) \qquad D(k) = \begin{vmatrix} u(k) & \cdots & u(k+n-1) \\ u(k+1) & \cdots & u(k+n) \\ \cdots & \cdots & \cdots \\ u(k+n-1) & \cdots & u(k+2n-2) \end{vmatrix} = 0.$$

The above condition we can easily verify for $k = 0$. Of course, if $D(0) \neq 0$ then $c_1 \neq 0$. In particular, we can always choose

$$(2.13.5) \qquad u(i) = 0, \quad 0 \le i \le n-2, \quad u(n-1) = 1$$

for which $D(0) \neq 0$.

**Remark 2.13.1.** By using the substitution $\lambda = 1/\lambda$ in (2.13.1), it is possible to obtain minimum modulus root of (2.13.1).

If the ratio $u(k+n)/u(k+n-1)$ oscillates without tending to limit, then we can suspect that (2.13.1) has complex roots which are largest in modulus. To compute these complex roots, let $\lambda_1 = \alpha + i\beta$ and its conjugate $\lambda_2 = \alpha - i\beta$ have the maximum modulus and are not repeated. If we write $\lambda_1 = \rho e^{i\theta}$ and $\lambda_2 = \rho e^{-i\theta}$ where $\rho^2 = \alpha^2 + \beta^2$ and $\rho\cos\theta = \alpha$, then from Problem 2.16.5 it follows that the terms corresponding to $\lambda_1$ and $\lambda_2$ in (2.13.2) can be written as

$$(2.13.6) \qquad \rho^k(c_1 \cos k\theta + c_2 \sin k\theta).$$

Therefore, as $k \to \infty$ it follows that

$$(2.13.7) \qquad u(k) \simeq \rho^k(c_1 \cos k\theta + c_2 \sin k\theta).$$

If $u(k)$ were given by the right side of (2.13.7), then it would satisfy the difference equation

$$(2.13.8) \qquad u(k+2) - 2\rho\cos\theta u(k+1) + \rho^2 u(k) = 0, \quad k \in \mathbb{N}.$$

The above equation can be considered as a relation involving two unknowns $\rho$ and $\theta$. To determine these unknowns, we replace $k$ by $k-1$ in (2.13.8) to get

$$(2.13.9) \qquad u(k+1) - 2\rho\cos\theta u(k) + \rho^2 u(k-1) = 0, \quad k \in \mathbb{N}(1).$$

Thus, the approximate relations (2.13.8) and (2.13.9) give

$$(2.13.10) \qquad \rho^2 \simeq \frac{v(k)}{v(k-1)} \quad \text{and} \quad 2\rho\cos\theta \simeq \frac{w(k)}{v(k-1)},$$

where

$$(2.13.11) \quad v(k) = u^2(k+1) - u(k)u(k+2) \quad \text{and} \quad w(k) = u(k)u(k+1) - u(k-1)u(k+2).$$

Hence, unless $c_1 = c_2 = 0$ in (2.13.2) the ratios $v(k)/v(k-1)$ and $w(k)/v(k-1)$ will tend to $\rho^2$ and $2\rho\cos\theta$ as $k \to \infty$, from which $\rho$ and $\theta$, and consequently $\alpha$ and $\beta$ can be obtained. In conclusion the desired maximum modulus pair $\lambda_1, \lambda_2$ of complex roots can be computed.

If $\lambda_1$ is a repeated real root of multiplicity 2, i.e. $\lambda_1 = \lambda_2$ and all other roots are of smaller modulus, then the combination of the terms corresponding to $\lambda_1$ and $\lambda_2$ in (2.13.2) is of the form $\lambda_1^k(c_1 + c_2k)$. As $k \to \infty$, $u(k)$ must tend to such a term, and hence $u(k)$ must satisfy the relation

$$(2.13.12) \qquad u(k+2) - 2\lambda_1 u(k+1) + \lambda_1^2 u(k) = 0$$

as $k \to \infty$.

Let $k$ to be $k-1$ in (2.13.12), to get

$$(2.13.13) \qquad u(k+1) - 2\lambda_1 u(k) + \lambda_1^2 u(k-1) = 0.$$

From the approximate relations (2.13.12) and (2.13.13), we easily obtain

$$(2.13.14) \qquad 2\lambda_1 \simeq \frac{w(k)}{v(k-1)},$$

where $v(k)$ and $w(k)$ are defined in (2.13.11).

## 2.14. Poincaré's and Perron's Theorems

The main conclusion from Bernoulli's method which interests here most, can be stated as follows: If $u(k)$ is any arbitrarily chosen solution of the difference equation (2.12.4), then $\lim_{k\to\infty}(u(k+1)/u(k))$ is equal to one of the roots of the characteristic equation (2.13.1) provided all these roots are distinct in modulus. A generalization of this result is embodied in the following:

**Theorem 2.14.1.** (Poincaré's Theorem). Let in the homogeneous difference equation (1.2.4), $a_n(k) = 1$, $a_0(k) \neq 0$ for all $k \in \mathbb{N}$ and $\lim_{k\to\infty} a_i(k) = a_i$, $0 \le i \le n-1$. Further, let the roots $\lambda_i$, $1 \le i \le n$ of the equation (2.13.1) have distinct moduli. Then, for every solution $u(k)$ of (1.2.4)

$$\lim_{k\to\infty} \frac{u(k+1)}{u(k)} = \lambda_i \tag{2.14.1}$$

for some $1 \le i \le n$.

**Proof.** For each $0 \le i \le n-1$ let $a_i(k) = a_i + \alpha_i(k)$, where $\alpha_i(k) = a_i(k) - a_i$. Since $a_i(k) \to a_i$, $\alpha_i(k) \to 0$ as $k \to \infty$. In system form equation (1.2.4) can be written as

$$\mathbf{u}(k+1) = \mathcal{A}\mathbf{u}(k) + \mathcal{B}(k)\mathbf{u}(k), \quad k \in \mathbb{N} \tag{2.14.2}$$

where the matrices $\mathcal{A}$ and $\mathcal{B}(k)$ are

$$\mathcal{A} = \begin{bmatrix} 0 & 1 & 0 & \cdots & 0 \\ 0 & 0 & 1 & \cdots & 0 \\ \cdots & \cdots & \cdots & \cdots & \cdots \\ 0 & 0 & 0 & \cdots & 1 \\ -a_0 & -a_1 & -a_2 & \cdots & -a_{n-1} \end{bmatrix},$$

$$\mathcal{B}(k) = \begin{bmatrix} 0 & \cdots & 0 \\ 0 & \cdots & 0 \\ \cdots & \cdots & \cdots \\ 0 & \cdots & 0 \\ -\alpha_0(k) & \cdots & -\alpha_{n-1}(k) \end{bmatrix}.$$

Since the roots of (2.13.1) have distinct moduli, we can arrange them so that $|\lambda_1| < |\lambda_2| < \cdots < |\lambda_n|$. Now from Problem 2.16.45, $\mathcal{A} = \mathcal{V}\mathcal{D}\mathcal{V}^{-1}$, where $\mathcal{V}$ is the Vandermonde matrix made up of the eigenvalues $\lambda_1, \cdots, \lambda_n$ of $\mathcal{A}$ which are the roots of (2.13.1), and $\mathcal{D}$ is the diagonal matrix $\mathcal{D} = diag\ (\lambda_1, \cdots, \lambda_n)$. Thus, the system (2.14.2) is the same as

$$\mathbf{u}(k+1) = \mathcal{V}\mathcal{D}\mathcal{V}^{-1}\mathbf{u}(k) + \mathcal{B}(k)\mathbf{u}(k), \quad k \in \mathbb{N}. \tag{2.14.3}$$

In the above system, let

$$\mathbf{v}(k) = \mathcal{V}^{-1}\mathbf{u}(k) \tag{2.14.4}$$

to obtain the new system

$$\mathbf{v}(k+1) = \mathcal{D}\mathbf{v}(k) + \mathcal{C}(k)\mathbf{v}(k), \tag{2.14.5}$$

where $\mathcal{C}(k) = \mathcal{V}^{-1}\mathcal{B}(k)\mathcal{V}$.

Since the elements of $\mathcal{B}(k) \to 0$ as $k \to \infty$, in any matrix norm $\|C(k)\| \to 0$ as $k \to \infty$. Suppose now that $\max_{1\le i\le n} |v_i(k)| = |v_\ell(k)|$, where the index $\ell$ is a function of $k$, i.e. $\ell = \ell(k)$, $k \in \mathbb{N}$. We shall show that a sufficiently large $k_0 \in \mathbb{N}$ exists such that for all $k_0 \le k \in \mathbb{N}$, the function $\ell(k)$ is not decreasing. Since, for $1 \le i < j \le n$, $(|\lambda_i|/|\lambda_j|) < 1$, we can take $\epsilon > 0$ small enough so that $[(|\lambda_i| + \epsilon)/(|\lambda_j| - \epsilon)] < 1$, and choose $k_0 \in \mathbb{N}$ large enough such that for all $k_0 \le k \in \mathbb{N}$, $\|C(k)\|_\infty < \epsilon$. Now setting $\ell(k+1) = j$, from (2.14.5) it follows that

$$(2.14.6)\qquad |v_\ell(k+1)| \ \ge\ |\lambda_\ell||v_\ell(k)| - \|C(k)\|_\infty |v_\ell(k)| \ \ge\ (|\lambda_\ell| - \epsilon)|v_\ell(k)|,$$

$$(2.14.7)\qquad |v_j(k+1)| \ \le\ |\lambda_j||v_j(k)| + \epsilon|v_\ell(k)| \ \le\ (|\lambda_j| + \epsilon)|v_\ell(k)|,$$

and

$$(2.14.8)\qquad |v_j(k+1)| \ \ge\ |\lambda_j||v_j(k)| - \epsilon|v_\ell(k)|.$$

Thus, if $\ell(k+1) = j$ were less than $\ell(k)$, then from (2.14.6) and (2.14.7) it follows that

$$\frac{|v_j(k+1)|}{|v_\ell(k+1)|} \le \frac{|\lambda_j| + \epsilon}{|\lambda_\ell| - \epsilon} < 1,$$

but this is a contradiction to the definition of $j$. Hence, there exists a $k_0 \le k_1 \in \mathbb{N}$ sufficiently large so that for all $k_1 \le k \in \mathbb{N}$ the function $\ell(k)$ assumes a fixed value less than or equal to $n$. We shall now show that the ratios

$$(2.14.9)\qquad \frac{|v_j(k)|}{|v_\ell(k)|}, \quad 1 \le j \ne \ell \le n$$

tend to zero. For this, it is clear that $(|v_j(k)|/|v_\ell(k)|) \le c \le 1$ for all $k \in \mathbb{N}(k_1)$. This means that $\alpha$ is an upper limit for (2.14.9), and hence we can extract a subsequence $\{k_i\} \subseteq \mathbb{N}(k_1)$ for which (2.14.9) converges to $c$. If $j > \ell$, then from (2.14.7) (with $j = \ell$) and (2.14.8) we have

$$\frac{|v_j(k_p+1)|}{|v_\ell(k_p+1)|} \ge \frac{|\lambda_j||v_j(k_p)|/|v_\ell(k_p)| - \epsilon}{|\lambda_\ell| + \epsilon}.$$

Thus, on taking the limit, we obtain

$$\liminf_{p\to\infty} \frac{|v_j(k_p+1)|}{|v_\ell(k_p+1)|} \ge \frac{|\lambda_j|c - \epsilon}{|\lambda_\ell| + \epsilon}.$$

This implies that

$$(2.14.10)\qquad \frac{|\lambda_j|c - \epsilon}{|\lambda_\ell| + \epsilon} \le \liminf_{p\to\infty} \frac{|v_j(k_p+1)|}{|v_\ell(k_p+1)|} \le \limsup_{p\to\infty} \frac{|v_j(k_p+1)|}{|v_\ell(k_p+1)|} = c$$

for arbitrary small $\epsilon$. However, since $|\lambda_j|/|\lambda_\ell| > 1$, from (2.14.10) we conclude that $c = 0$. Similarly, if $j < \ell$ then the inequality

$$\frac{|v_j(k_p+1)|}{|v_\ell(k_p+1)|} \le \frac{|\lambda_j||v_j(k_p)|/|v_\ell(k_p)| + \epsilon}{|\lambda_\ell| - \epsilon}$$

leads to the same conclusion that $c = 0$.

Finally, from (2.14.4) we have

$$u(k) = u_1(k) = \sum_{i=1}^{n} v_i(k) = v_\ell(k)\left(1 + \sum_{i=1, i\neq\ell}^{n} \frac{v_i(k)}{v_\ell(k)}\right)$$

and

$$\begin{aligned} u(k+1) = u_1(k+1) = u_2(k) &= \sum_{i=1}^{n} \lambda_i v_i(k) \\ &= \lambda_\ell v_\ell(k)\left(1 + \sum_{i=1, i\neq\ell}^{n} \frac{\lambda_i v_i(k)}{\lambda_\ell v_\ell(k)}\right). \end{aligned}$$

Thus, $\lim_{k\to\infty} \frac{u(k+1)}{u(k)} = \lambda_\ell$ follows immediately from $\lim_{k\to\infty} \frac{v_j(k)}{v_\ell(k)} = 0$, $1 \le j \neq \ell \le n$. ∎

A refinement of Poincaré's theorem is due to Perron which is stated in the following:

**Theorem 2.14.2.** (Perron's Theorem). Let the conditions of Theorem 2.14.1 be satisfied. Then, the difference equation (1.2.4) has a fundamental set of solutions $u_i(k)$, $1 \le i \le n$ with the property

$$\lim_{k\to\infty} \frac{u_i(k+1)}{u_i(k)} = \lambda_i.$$

**Example 2.14.1.** Consider the difference equation

$$u(k+2) - \left(3 + \frac{2k-1}{k^2-2k-1}\right) u(k+1) + 2\left(1 + \frac{2k-1}{k^2-2k-1}\right) u(k) = 0, \quad k \in \mathbb{N}$$

for which $2^k$ and $k^2$ are the solutions. Thus, the general solution of this difference equation can be written as $u(k) = c_1 2^k + c_2 k^2$. Since $k^2/2^k \to 0$ as $k \to \infty$, it is clear that $\lim_{k\to\infty}(u(k+1)/u(k)) = 2$, which is a root of the characteristic equation $\lambda^2 - 3\lambda + 2 = 0$.

**Example 2.14.2.** The condition in Poincaré's theorem that the roots of the characteristic equation have distinct moduli is essential. For this, we consider the difference equation

$$u(k+2) - \left(1 + \frac{(-1)^k}{k+1}\right) u(k) \;=\; 0, \quad k \in \mathbb{N}$$

for which the characteristic equation $\lambda^2 - 1 = 0$ has the roots $\lambda_1 = 1,\ \lambda_2 = -1$ with the same modulus 1. The solution of this difference equation with the initial conditions $u(0) = 0,\ u(1) = 1$ can be written as

$$u(2k+1) \;=\; \prod_{\ell=1}^{k}\left(1 - \frac{1}{2\ell}\right), \quad u(2k) = 0, \quad k \in \mathbb{N}$$

for which obviously $\lim_{k\to\infty}(u(k+1)/u(k))$ does not exist.

## 2.15. Regular and Singular Perturbations

The basic idea of regular perturbation technique relates the unknown solution of the initial value problem (1.2.4), (1.3.1) with the known solutions of an infinite related initial value problems, and can be exhibited as follows: Let the auxiliary difference equation

$$\sum_{i=0}^{n-1} \alpha_i(k)u(k+i) + a_n(k)u(k+n) \;=\; 0 \tag{2.15.1}$$

together with the initial conditions (1.3.1) can be solved explicitly to obtain its solution $u^0(k)$. We write the equation (1.2.4) in the form

$$\sum_{i=0}^{n-1} \left(\alpha_i(k) + a_i(k) - \alpha_i(k)\right) u(k+i) + a_n(k)u(k+n) \;=\; 0,$$

which is the same as

$$\sum_{i=0}^{n-1} \alpha_i(k)u(k+i) + a_n(k)u(k+n) \;=\; \sum_{i=0}^{n-1} c_i(k)u(k+i), \tag{2.15.2}$$

where $c_i(k) = \alpha_i(k) - a_i(k),\ 0 \le i \le n-1$. We introduce a parameter $\epsilon$ and consider the new difference equation

$$\sum_{i=0}^{n-1} \alpha_i(k)u(k+i) + a_n(k)u(k+n) \;=\; \epsilon\left(\sum_{i=0}^{n-1} c_i(k)u(k+i)\right). \tag{2.15.3}$$

Obviously, for $\epsilon = 1$ this new difference equation is the same as (2.15.2). We look for the solution of (2.15.3), (1.3.1) having the form

$$u(k) = \sum_{m=0}^{\infty} \epsilon^m u^m(k). \tag{2.15.4}$$

For this, it is necessary to have

$$\sum_{m=0}^{\infty} \epsilon^m \left[\sum_{i=0}^{n-1} \alpha_i(k) u^m(k+i) + a_n(k) u^m(k+n)\right]$$
$$= \epsilon \sum_{m=0}^{\infty} \epsilon^m \left[\sum_{i=0}^{n-1} c_i(k) u^m(k+i)\right]$$

and

$$\sum_{m=0}^{\infty} \epsilon^m u^m(a+i-1) = u_i, \quad 1 \le i \le n.$$

Thus, on equating the coefficients of $\epsilon^m$, $m = 0, 1, \cdots$ we find the infinite system of initial value problems

$$\text{(2.15.5)} \quad \sum_{i=0}^{n-1} \alpha_i(k) u^0(k+i) + a_n(k) u^0(k+n) = 0, \quad u^0(a+i-1) = u_i, \quad 1 \le i \le n$$

$$\text{(2.15.6)}_m \quad \sum_{i=0}^{n-1} \alpha_i(k) u^m(k+i) + a_n(k) u^m(k+n) = \sum_{i=0}^{n-1} c_i(k) u^{m-1}(k+i),$$
$$u^m(a+i-1) = 0, \quad 1 \le i \le n, \quad m = 1, 2, \cdots.$$

This infinite system can be solved recursively. Indeed, from our initial assumption the solution $u^0(k)$ of (2.15.5) can be obtained explicitly, and thus the term $\sum_{i=0}^{n-1} c_i(k) u^0(k+i)$ in $(2.15.6)_1$ is known; consequently the solution $u^1(k)$ of the nonhomogeneous initial value problem $(2.15.6)_1$ can be obtained by the method of variation of parameters. Continuing in this way the functions $u^2(k)$, $u^3(k)$, $\cdots$ can similarly be obtained. Finally, the solution of the original problem is obtained by summing the series (2.15.4) for $\epsilon = 1$.

The above formal perturbative procedure is not only applicable for the initial value problem (1.2.4), (1.3.1) but also can be employed to a variety of linear as well as nonlinear problems. The implementation of this powerful technique consists in the following three basic steps:

(i) Conversion of the given problem into a perturbation problem by introducing the small parameter $\epsilon$.

(ii) Assumption of the solution in the form of a perturbation series and the computation of the coefficients of that series.

(iii) Finally, obtaining the solution of the original problem by summing the perturbation series for the appropriate value of $\epsilon$.

It is clear that the parameter $\epsilon$ in the original problem can be introduced in an infinite number of ways, however the perturbed problem is meaningful only if the zeroth order solution, i.e. $u^0(k)$ is obtainable explicitly. Further, in a large number of applied problems this parameter occurs naturally.

The perturbation method naturally leads to the question: Under what conditions does the perturbation series converge and actually represent a solution of the original problem? Unfortunately, often perturbation series are divergent, however this is not necessarily bad because a good approximation to the solution when $\epsilon$ is very small can be obtained by summing only first few terms of the series.

**Example 2.15.1.** Consider the initial value problem

$$(2.15.7)\qquad u(k+2)-2u(k+1)+\frac{3}{4}u(k) = 0, \quad u(0) = 1, \quad u(1) = \frac{1}{2}$$

for which $u(k) = 1/2^k$ is the unique solution. We convert (2.15.7) into a perturbation problem

$$(2.15.8)\qquad \Delta^2 u(k) = \epsilon\left(\frac{1}{4}u(k)\right), \quad u(0) = 1, \quad u(1) = \frac{1}{2}$$

and assume that its solution can be written as perturbation series (2.15.4). This leads to an infinite system of initial value problems

$$\Delta^2 u^0(k) = 0, \quad u^0(0) = 1, \quad u^0(1) = \frac{1}{2}$$

$$\Delta^2 u^m(k) = \frac{1}{4}u^{m-1}(k), \quad u^m(0) = u^m(1) = 0, \quad m = 1, 2, \cdots$$

which can be solved recursively, to obtain

$$u^m(k) = \frac{1}{4^m}\left[\binom{k}{2m} - \frac{1}{2}\binom{k}{2m+1}\right], \qquad m = 0, 1, \cdots.$$

Thus, the solution $u(k,\epsilon)$ of the perturbation problem (2.15.8) appears as

$$\begin{aligned} u(k,\epsilon) &= \sum_{m=0}^{\infty} \epsilon^m \frac{1}{4^m}\left[\binom{k}{2m} - \frac{1}{2}\binom{k}{2m+1}\right] \\ &= \sum_{m=0}^{[k/2]} \left(\frac{\epsilon}{4}\right)^m \left[\binom{k}{2m} - \frac{1}{2}\binom{k}{2m+1}\right]. \end{aligned}$$

Hence, the solution $u(k) = u(k,1)$ of the initial value problem (2.15.7) can be written as

$$\begin{aligned} u(k) &= \sum_{m=0}^{[k/2]} \frac{1}{2^{2m}}\left[\binom{k}{2m} - \frac{1}{2}\binom{k}{2m+1}\right] \\ &= \sum_{m=0}^{k} \frac{(-1)^m}{2^m}\binom{k}{m} = \left(1-\frac{1}{2}\right)^k = \frac{1}{2^k}. \end{aligned}$$

**Example 2.15.2.** Consider Airy's differential equation $y'' - ty = 0$, $t \geq 0$ together with the initial conditions $y(0) = 1$, $y'(0) = 0$. The simplest difference equation approximation to this initial value problem is

$$(2.15.9) \qquad \begin{aligned} u(k+2) - 2u(k+1) + u(k) - (k+1)h^3u(k) &= 0, \quad k \in \mathbb{N} \\ u(0) = u(1) &= 1 \end{aligned}$$

where $h > 0$ is an arbitrary constant step size, and $u(k)$ approximates the solution $y(t)$ at $t_k = kh$.

We convert (2.15.9) into a perturbation problem

$$\Delta^2 u(k) = \epsilon((k+1)h^3u(k)), \quad u(0) = u(1) = 1$$

and assume that its solution can be written as perturbation series (2.15.4). This leads to an infinite system of initial value problems

$$\begin{aligned} \Delta^2 u^0(k) &= 0, \quad u^0(0) = u^0(1) = 1 \\ \Delta^2 u^m(k) &= (k+1)h^3u^{m-1}(k), \quad u^m(0) = u^m(1) = 0, \quad m = 1,2,\cdots \end{aligned}$$

which can be solved recursively, to obtain

$$\begin{aligned} u^0(k) &= 1 \\ u^1(k) &= \frac{1}{3!}(k+1)^{(3)}h^3 \\ u^2(k) &= \frac{2}{6!}(k+1)^{(5)}(2k+1)h^6 \\ u^3(k) &= \frac{2}{9!}(k+1)^{(7)}(14k^2+7k-6)h^9 \\ &\cdots. \end{aligned}$$

Thus, a uniform approximation to the solution $u(k)$ of (2.15.9) can be taken as

$$\begin{aligned} u(k) \simeq\ & 1 + \frac{1}{3!}(k+1)^{(3)}h^3 + \frac{2}{6!}(k+1)^{(5)}(2k+1)h^6 \\ & + \frac{2}{9!}(k+1)^{(7)}(14k^2+7k-6)h^9. \end{aligned}$$

This approximation is exact for $k \in \mathbb{N}(0,7)$.

In many practical problems one often meets cases where the parameter $\epsilon$ is involved in the difference equation in such a way that the method of regular perturbation cannot be applied. In the literature such problems are known as singular perturbation problems, and to understand these we consider the following:

**Example 2.15.3.** For the initial value problem

$$(2.15.10) \qquad \begin{aligned} u(k+2) - (1+\epsilon)u(k+1) + \epsilon u(k) &= 0, \quad k \in \mathbb{N} \\ u(0) = \alpha_0, \quad u(1) &= \alpha_1 \end{aligned}$$

explicit solution can be written as

$$(2.15.11) \qquad u(k) = \frac{1}{1-\epsilon}\left[(\alpha_1 - \epsilon\alpha_0) + (\alpha_0 - \alpha_1)\epsilon^k\right],$$

for which it follows that

$$\lim_{k\to 0}\left(\lim_{\epsilon\to 0} u(k)\right) = \alpha_1 \neq \lim_{\epsilon\to 0}\left(\lim_{k\to 0} u(k)\right) = \alpha_0$$

unless $\alpha_0 = \alpha_1$.

Suppressing the small parameter $\epsilon$ in (2.15.10), the resulting degenerate first order equation is

$$(2.15.12) \qquad v^0(k+2) - v^0(k+1) = 0.$$

Obviously, for (2.15.12) the initial conditions $v^0(0) = \alpha_0$, $v^0(1) = \alpha_1$ are inconsistent unless $\alpha_0 = \alpha_1$. Thus, (2.15.10) is said to be in the singularly perturbed form, and a *boundary layer* occurs at $k = 0$.

If we seek the solution of (2.15.10) in the regular perturbation series form (2.15.4), then it leads to the system of first order difference equations

$$\begin{aligned} & u^0(k+2) - u^0(k+1) = 0, \quad u^0(0) = \alpha_0, \quad u^0(1) = \alpha_1 \\ & u^m(k+2) - u^m(k+1) = u^{m-1}(k+1) - u^{m-1}(k), \ u^m(0) = u^m(1) = 0, \\ & \qquad\qquad m = 1, 2, \cdots \end{aligned}$$

which can be solved only if the initial conditions are consistent, i.e. $\alpha_0 = \alpha_1$. Further, in such a case it is easy to obtain $u^0(k) = \alpha_0$, $u^m(k) = 0$, $m = 1, 2, \cdots$, and hence (2.15.4) reduces to just $u(k) = \alpha_0$ which is indeed a solution of (2.15.10).

Now ignoring the terms with coefficients of $\epsilon$ and higher powers of $\epsilon$ in (2.15.11), the zeroth order approximate solution appears as

$$u(k) \simeq \alpha_1 + \epsilon^k(\alpha_0 - \alpha_1). \tag{2.15.13}$$

The first part of this solution, i.e. $\alpha_1$ is called the *outer solution,* as it is valid outside the boundary layer. This satisfies only one of the initial conditions $u(1) = \alpha_1$. The second part of this solution, i.e. $(\alpha_0 - \alpha_1)$ is called the *inner solution* which recovers the lost initial condition $u(0) = \alpha_0$.

The presence of $\epsilon^k$ in (2.15.13) suggests that the inner solution has the transformation $w(k) = u(k)/\epsilon^k$. Using this transformation in (2.15.10) and dividing throughout with $\epsilon^{k+1}$ leads to the difference equation

$$\epsilon w(k+2) - (1+\epsilon)w(k+1) + w(k) = 0. \tag{2.15.14}$$

Putting $\epsilon = 0$ in the above equation gives the degenerate equation

$$-z^0(k+1) + z^0(k) = 0.$$

This equation is solved with the initial condition

$$z^0(0) = u(0) - \alpha_1 = \alpha_0 - \alpha_1$$

to obtain $z^0(k) = \alpha_0 - \alpha_1$, which is the same as the inner solution.

Thus the total zeroth order solution of (2.15.10) is composed of the outer and inner solutions and given by

$$u(k) = v^0(k) + \epsilon^k z^0(k).$$

Utilizing the above ideas we write the solution $u(k)$ of (2.15.10) as the sum of two solutions

$$u(k) = v(k) + \epsilon^k z(k), \tag{2.15.15}$$

where $v(k)$ and $z(k)$ are the outer and inner solutions. Substituting (2.15.15) in (2.15.10) and separating the terms, we obtain two equations

$$v(k+2) - (1+\epsilon)v(k+1) + \epsilon v(k) = 0 \tag{2.15.16}$$

and

$$\epsilon z(k+2) - (1+\epsilon)z(k+1) + z(k) = 0. \tag{2.15.17}$$

For solving these equations, we assume that

$$v(k) = \sum_{m=0}^{\infty} \epsilon^m v^m(k), \qquad z(k) = \sum_{m=0}^{\infty} \epsilon^m z^m(k). \tag{2.15.18}$$

Substituting the above series solutions in (2.15.16) and (2.15.17) respectively leads to the systems

$$\begin{aligned} v^0(k+2) - v^0(k+1) &= 0 \\ v^m(k+2) - v^m(k+1) &= v^{m-1}(k+1) - v^{m-1}(k), \\ &\qquad m = 1, 2, \cdots \end{aligned} \tag{2.15.19}$$

and

$$\begin{aligned} -z^0(k+1) + z^0(k) &= 0 \\ -z^m(k+1) + z^m(k) &= z^{m-1}(k+1) - z^{m-1}(k+2), \\ &\qquad m = 1, 2, \cdots. \end{aligned} \tag{2.15.20}$$

The initial conditions for the above systems are obtained by substituting (2.15.15) in $u(0) = \alpha_0$ and $u(1) = \alpha_1$ and appears as

$$\begin{aligned} v^0(1) &= \alpha_1, \quad z^0(0) = \alpha_0 - v^0(0) \\ v^m(1) &= -z^{m-1}(1), \quad z^m(0) = -v^m(0), \quad m = 1, 2, \cdots. \end{aligned} \tag{2.15.21}$$

Finally, the series solution of (2.15.10) is written as

$$u(k) = \sum_{m=0}^{\infty} \epsilon^m v^m(k) + \epsilon^k \sum_{m=0}^{\infty} \epsilon^m z^m(k). \tag{2.15.22}$$

The above systems (2.15.19) (2.15.21) can easily be solved to obtain approximations of $u(k)$ up to any order. For example, the zeroth order approximation is the same as (2.15.13), as it should be, and the first order approximation appears as

$$u(k) \simeq [\alpha_1 - (\alpha_0 - \alpha_1)\epsilon] + \epsilon^k[(\alpha_0 - \alpha_1) + \epsilon(\alpha_0 - \alpha_1)].$$

## 2.16. Problems

**2.16.1.** Show that the functions $u_1(k) = c(\neq 0)$ and $u_2(k) = 1/(k+1)^{(2)}$

satisfy the nonlinear difference equation

$$\Delta^2 u(k) + \frac{3(k+1)k}{(k+3)} u(k)\Delta u(k) \;=\; 0,$$

but $u_1(k) + u_2(k)$ does not satisfy the given difference equation. (This shows that the principle of superposition holds good only for the linear equations.)

**2.16.2.** Let $\lambda_1, \cdots, \lambda_n$ be the (not necessarily distinct) eigenvalues of an $n \times n$ matrix $\mathcal{A}$. Show that

(i) the eigenvalues of $\mathcal{A}^T$ are $\lambda_1, \cdots, \lambda_n$

(ii) for any constant $\alpha$ the eigenvalues of $\alpha\mathcal{A}$ are $\alpha\lambda_1, \cdots, \alpha\lambda_n$

(iii) $\sum_{i=1}^n \lambda_i = Tr\mathcal{A} = \sum_{i=1}^n a_{ii}$

(iv) $\prod_{i=1}^n \lambda_i = det\ \mathcal{A}$

(v) if $\mathcal{A}^{-1}$ exists then the eigenvalues of $\mathcal{A}^{-1}$ are $1/\lambda_1, \cdots, 1/\lambda_n$

(vi) for any polynomial $P_k(t)$ the eigenvalues of $P_k(\mathcal{A})$ are $P_k(\lambda_1), \cdots, P_k(\lambda_n)$

(vii) if $\mathcal{A}$ is upper (lower) triangular, i.e. $a_{ij} = 0,\ i > j\ (i < j)$, then the eigenvalues of $\mathcal{A}$ are the diagonal elements of $\mathcal{A}$

(viii) if $\mathcal{A}$ is real and $\lambda_1$ is complex with the corresponding eigenvector $\mathbf{v}^1$, then there exists at least one $i$, $2 \le i \le n$ such that $\lambda_i = \overline{\lambda}_1$ and for such an $i$, $\overline{\mathbf{v}}^1$ is the corresponding eigenvector.

**2.16.3.** (i) Let the $n \times n$ matrix $\mathcal{A}(t) = (a_{ij}(t))$ be such that $a_{ij}(t) \in C^{(1)}(\alpha, \beta)$. Show that for all $t \in (\alpha, \beta)$

$$(2.16.1)\quad (det\ \mathcal{A}(t))' = \begin{vmatrix} a'_{11}(t) & \cdots & a'_{1n}(t) \\ a_{21}(t) & \cdots & a_{2n}(t) \\ \cdots & \cdots & \cdots \\ a_{n1}(t) & \cdots & a_{nn}(t) \end{vmatrix} + \cdots + \begin{vmatrix} a_{11}(t) & \cdots & a_{1n}(t) \\ a_{21}(t) & \cdots & a_{2n}(t) \\ \cdots & \cdots & \cdots \\ a'_{n1}(t) & \cdots & a'_{nn}(t) \end{vmatrix}.$$

(ii) An $(n-1) \times (n-1)$ determinant obtained by deleting $i$th row and $j$th column of a given $n \times n$ matrix $\mathcal{A}$ is called the *minor* $\overline{a}_{ij}$ of the element $a_{ij}$. We define the *cofactor* of $a_{ij}$ as $\alpha_{ij} = (-1)^{i+j}\overline{a}_{ij}$. Show that

$$(2.16.2)\qquad det\ \mathcal{A} \;=\; \sum_{j=1}^n a_{ij}\alpha_{ij} \;=\; \sum_{i=1}^n a_{ij}\alpha_{ij}$$

$$(2.16.3)\qquad \sum_{j=1}^n a_{ij}\alpha_{kj} \;=\; 0 \quad \text{if} \quad i \neq k.$$

**2.16.4.** Let the functions $u_i(k),\ 1 \le i \le n$ be defined on $\mathbb{N}$ and $\lim_{k\to\infty}(u_i(k)/u_{i+1}(k)) = 0,\ 1 \le i \le n-1$. Show that these functions are linearly independent.

**2.16.5.** Let $\mathbf{u}(k)$ be a complex solution of the homogeneous system (1.2.12) on $\overline{\mathbb{N}}$. Show that both the real and imaginary parts of $\mathbf{u}(k)$ are solutions of (1.2.12).

**2.16.6.** Let $\mathcal{U}(k, k_0)$ and $\mathcal{V}(k, k_0)$ be the principal fundamental matrix solutions of (1.2.12) and (2.7.1) respectively. Show that $\mathcal{V}^T(k, k_0)\mathcal{U}(k, k_0) = \mathcal{I}$.

**2.16.7.** Let $y(t)$ be the solution of the initial value problem

$$\sum_{i=0}^{n} a_i y^{(i)}(t) \;=\; 0, \qquad y^{(i)}(0) \;=\; \alpha_i, \quad 0 \le i \le n-1$$

where $a_0, \cdots, a_n$ are constants. Show that $u(k) = y^{(k)}(0)$ is the solution of the initial value problem

$$\sum_{i=0}^{n} a_i u(k+i) \;=\; 0, \qquad u(i) \;=\; \alpha_i, \quad 0 \le i \le n-1.$$

**2.16.8.** Let the notations and hypotheses of Lemma 2.8.2 be satisfied. Show that for all $k \in \mathbb{N}$

$$\mathcal{A}^{-k} \;=\; \sum_{i=1}^{m}\sum_{j=0}^{r_i-1}(-1)^j\frac{(k+j-1)^{(j)}}{j!}\lambda_i^{-k-j}a_i(\mathcal{A})q_i(\mathcal{A})(\mathcal{A}-\lambda_i\mathcal{I})^j,$$

which is the same as (2.8.11) with $k$ replaced by $-k$, i.e. (2.8.11) indeed holds for all $k \in \mathbf{Z}$.

**2.16.9.** Let $\mathcal{A}$ and $\mathcal{P}$ be $n \times n$ matrices given by

$$\mathcal{A} = \begin{bmatrix} \lambda & 1 & 0 & \cdots & 0 \\ 0 & \lambda & 1 & \cdots & 0 \\ \cdots & \cdots & \cdots & \cdots & \cdots \\ 0 & 0 & 0 & \cdots & 1 \\ 0 & 0 & 0 & \cdots & \lambda \end{bmatrix}, \qquad \mathcal{P} = \begin{bmatrix} 0 & 1 & 0 & \cdots & 0 \\ 0 & 0 & 1 & \cdots & 0 \\ \cdots & \cdots & \cdots & \cdots & \cdots \\ 0 & 0 & 0 & \cdots & 1 \\ 0 & 0 & 0 & \cdots & 0 \end{bmatrix}.$$

Show that

(i) $\mathcal{P}^n = 0$, (ii) $(\lambda\mathcal{I})\mathcal{P} = \mathcal{P}(\lambda\mathcal{I})$, (iii) $\mathcal{A}^k = \displaystyle\sum_{i=0}^{\min\{n-1,k\}}\binom{k}{i}\lambda^{k-i}\mathcal{P}^i$.

**2.16.10.** Find the general solution of the homogeneous difference system (2.8.1), where the matrix $\mathcal{A}$ is given by

$$\text{(i)}\quad \begin{bmatrix} 4 & -2 \\ 5 & 2 \end{bmatrix}, \quad \text{(ii)}\quad \begin{bmatrix} 7 & 6 \\ 2 & 6 \end{bmatrix}, \quad \text{(iii)}\quad \begin{bmatrix} 0 & 1 & 1 \\ 1 & 0 & 1 \\ 1 & 1 & 0 \end{bmatrix}$$

$$\text{(iv)}\quad \begin{bmatrix} 1 & -1 & 4 \\ 3 & 2 & -1 \\ 2 & 1 & -1 \end{bmatrix}, \quad \text{(v)}\quad \begin{bmatrix} -1 & 1 & 0 \\ 0 & -1 & 0 \\ 0 & 0 & 3 \end{bmatrix}, \quad \text{(vi)}\quad \begin{bmatrix} 5 & -3 & -2 \\ 8 & -5 & -4 \\ -4 & 3 & 3 \end{bmatrix}.$$

**2.16.11.** Find the general solution of the nonhomogeneous difference system (2.8.22) where the matrix $\mathcal{A}$ and the vector $\mathbf{b}(k)$ are given by

$$\text{(i)}\quad \begin{bmatrix} 3 & -1 \\ -3 & 5 \end{bmatrix}, \quad \begin{bmatrix} 2k \\ 4^k \end{bmatrix} \quad \text{(ii)}\quad \begin{bmatrix} -2 & 2 \\ -3 & 4 \end{bmatrix}, \quad \begin{bmatrix} 5^{k+1} \\ 1/2 \end{bmatrix}$$

$$\text{(iii)}\quad \begin{bmatrix} 2 & 1 & -1 \\ -3 & -1 & 1 \\ 9 & 3 & -4 \end{bmatrix}, \quad \begin{bmatrix} 0 \\ k \\ 0 \end{bmatrix} \quad \text{(iv)}\quad \begin{bmatrix} 1 & 0 & 0 \\ 2 & 1 & -2 \\ 3 & 2 & 1 \end{bmatrix}, \quad \begin{bmatrix} 1 \\ k \\ 2^k \end{bmatrix}.$$

**2.16.12.** Two spheres of masses $M$ and $m$ $(M > m)$ and the coefficient of restitution $e$ lie on a smooth horizontal surface with the line of centers right angle to two walls which are perfectly elastic. $M$ is projected at $m$ with velocity $U$ while $m$ is initially at rest. We shall find the velocities just before the $k$th impact between the balls. For this, let $u(k)$ and $v(k)$ be the velocities of $M$ and $m$ just before the $k$th impact. The velocities of $M$ and $m$ just before the $(k+1)$th impact are $u(k+1)$ and $v(k+1)$; thus the velocities just after the $k$th impact are $u(k+1)$ and $v(k+1)$ since the wall is perfectly elastic. The momentum equation is

$$(2.16.4) \qquad Mu(k) + mv(k) \;=\; -mv(k+1) + Mu(k+1).$$

We also have the relation

$$(2.16.5) \qquad -e(u(k) - v(k)) \;=\; u(k+1) + v(k+1).$$

Solve the system (2.16.4), (2.16.5) with the initial condition

$$(2.16.6) \qquad u(1) \;=\; U, \quad v(1) \;=\; 0.$$

In particular, if $M = 4m$ and $e = 1$, show that

$$\begin{aligned} u(k) &= U\cos(k-1)\theta \\ v(k) &= -\frac{3U}{2}\cos(k-1)\theta + \frac{5U}{2}\cos k\theta, \quad \theta = \tan^{-1}\frac{4}{3}. \end{aligned}$$

**2.16.13.** Consider the difference system (2.8.19) to show that

$$\begin{aligned} w_1(k) &= \lambda_1^k \\ w_j(k+1) &= \textstyle\sum_{\ell=0}^{k} \lambda_j^{k-\ell} w_{j-1}(\ell), \quad j = 2, \cdots, n. \end{aligned}$$

**2.16.14.** Let for the given $n \times n$ matrix $\mathcal{A}$ the spectral radius $\rho(\mathcal{A}) = \rho_0 < \beta$. Consider the difference system (2.8.19) to show that

$$|w_j(k)| \leq \beta^k/(\beta - \rho_0)^{j-1}, \quad j = 1, \cdots, n.$$

Further, if $\alpha < r_0 = \min_i |\lambda_i|$, establish a lower bound for $|w_j(k)|$.

**2.16.15.** Consider the system (2.8.1), and let

$$p(\lambda) = \det(\mathcal{A} - \lambda\mathcal{I}) = \lambda^n + c_{n-1}\lambda^{n-1} + \cdots + c_1\lambda + c_0$$

be the characteristic polynomial of the matrix $\mathcal{A}$. Show that

(i) $\mathcal{A}^k$, $k \in \mathbb{N}$ is the unique solution of the $n$th order matrix difference initial value problem

$$\begin{aligned} &\mathcal{U}(n+k) + c_{n-1}\mathcal{U}(n+k-1) + \cdots + c_1\mathcal{U}(k+1) + c_0\mathcal{U}(k) = 0, \quad k \in \mathbb{N} \\ &\mathcal{U}(i) = \mathcal{A}^i, \quad 0 \leq i \leq n-1 \end{aligned}$$

(ii) if $u_i(k)$, $1 \leq i \leq n$ are the solutions of the equation

$$\begin{aligned} u(n+k) + c_{n-1}u(n+k-1) + \cdots + c_1u(k+1) + c_0u(k) &= 0, \quad k \in \mathbb{N} \\ u_i(j-1) &= \delta_{ij}, \quad 1 \leq i, j \leq n \end{aligned}$$

then

$$\mathcal{A}^k = u_1(k)\mathcal{I} + u_2(k)\mathcal{A} + \cdots + u_n(k)\mathcal{A}^{n-1}, \quad k \in \mathbb{N}$$

(iii) use (ii) to find the general solution of the system (2.8.1) with $\mathcal{A} = \begin{bmatrix} 2 & 0 & 1 \\ 0 & 2 & 0 \\ 0 & 0 & 3 \end{bmatrix}$.

**2.16.16.** Let the difference system (1.2.12) be periodic of period $K$ on $\mathbb{N}$. Show that the number of linearly independent periodic solutions of (1.2.12) of period $K$ is the same as of the adjoint system (2.7.1).

**2.16.17.** Let the difference system (1.2.11) be periodic of period $K$ on $\mathbb{N}$. Further, let $\mathbf{v}(k)$ be a periodic solution of (2.7.1) of period $K$. Show that the system (1.2.11) has a periodic solution of period $K$ if and only if $\sum_{\ell=1}^{K} \mathbf{v}^T(\ell)\mathbf{b}(\ell-1) = 0$.

**2.16.18.** A function $u(k)$ defined on $\overline{\mathbb{N}}_K$ is called *periodic of the second kind* of period $K > 0$ and multiplier $\rho$ if for all $k \in \overline{\mathbb{N}}$, $u(k+K) = \rho u(k)$. In particular it is said to be *anti periodic* of period $K$ if $\rho = -1$. Consider the first order difference equation

$$u(k+1) = p(k)u(k), \quad k \in \mathbb{N} \tag{2.16.7}$$

where $p(k)$ is periodic of period $K$ on $\mathbb{N}$. Show that

(i) $\rho_1 = \prod_{i=0}^{K-1} p(k+i)$ is independent of $k$

(ii) every solution of (2.16.7) is periodic of the second kind of period $K$ and multiplier $\rho_1$

(iii) every solution of (2.16.7) is periodic of period $K$ if $\rho_1 = 1$

(iv) every solution of (2.16.7) is anti periodic of period $K$ if $\rho_1 = -1$.

**2.16.19.** Consider the second order difference equation

$$a_0(k)u(k) + a_1(k)u(k+1) + a_2(k)u(k+2) = 0, \quad k \in \mathbb{N}(a) \tag{2.16.8}$$

where the functions $a_0(k)$, $a_1(k)$ and $a_2(k)$ are periodic of period $K$ on $N(a)$. Show that the solution $u(k)$ of (2.16.8) is periodic of period $K$ if and only if $u(a) = u(a+K)$, $\Delta u(a) = \Delta u(a+K)$ and that it is anti periodic of period $K$ if and only if $u(a) = -u(a+K)$, $\Delta u(a) = -\Delta u(a+K)$.

**2.16.20.** In the linear difference system (2.8.22) assume that the function $\mathbf{b}(k)$ is bounded on $\mathbf{Z}$, and moduli of eigenvalues of $\mathcal{A}$ are different from 1. Show that there exists a unique bounded solution $\mathbf{u}(k)$ of (2.8.22). Moreover, the following estimate holds

$$\sup_{k\in\mathbf{Z}} \|\mathbf{u}(k)\| \leq M \sup_{k\in\mathbf{Z}} \|\mathbf{b}(k)\|,$$

where $M > 0$ is a constant which depends on $\mathcal{A}$ only. Further, if $\mathbf{b}(k)$ is almost periodic, then so is $\mathbf{u}(k)$.

**2.16.21.** (Discrete Abel's Formula). Let $u^1(k), \cdots, u^n(k)$ be the solutions on $\overline{\mathbb{N}}_n$ of the homogeneous equation (1.2.4). Show that for all $k \in \overline{\mathbb{N}}_1$

$$det\ \mathcal{C}(k+1) = (-1)^n \frac{a_0(k)}{a_n(k)} det\ \mathcal{C}(k),$$

and hence

$$det\ \mathcal{C}(k) = det\ \mathcal{C}(a)(-1)^{n(k-a)} \prod_{\ell=a}^{k-1} \frac{a_0(\ell)}{a_n(\ell)}.$$

**2.16.22.** Show that the difference equation (1.2.4) has a solution $u(k) \neq 0$ for each $k \in \overline{\mathbb{N}}_n$.

**2.16.23.** Let $u_i(k)$, $k \in \mathbb{N}$, $1 \leq i \leq n$ be a fundamental set of solutions of the equation (1.2.4) on $\mathbb{N}$ and $C(k)$ be its Casoratian matrix. Show that, the functions $v_i(k) = \overline{v}_i(k)/a_0(k) det\, C(k)$ , $1 \leq i \leq n$ where

$$\overline{v}_i(k) = (-1)^{n-1} \begin{vmatrix} u_1(k+1) & u_{i-1}(k+1) & u_{i+1}(k+1) & u_n(k+1) \\ \cdots & \cdots & \cdots & \cdots \\ u_1(k+n-1) & u_{i-1}(k+n-1) & u_{i+1}(k+n-1) & u_n(k+n-1) \end{vmatrix}$$

form a fundamental set of solutions of the adjoint equation (2.11.14). Furthermore,

$$\sum_{i=1}^{n} v_i(k) u_i(k+\ell) \; = \; \begin{cases} 1/a_0(k) & \ell = 0 \\ 0 \quad 1 \leq \ell \leq n-1 \\ -1/a_n(k) & \ell = n. \end{cases}$$

**2.16.24.** Consider the second order homogeneous difference equation (2.16.8) and its adjoint equation

$$(2.16.9) \quad a_2(k)v(k) + a_1(k+1)v(k+1) + a_0(k+2)v(k+2) \; = \; 0, \quad k \in \mathbb{N}(a).$$

Let $v_1(k)$ and $v_2(k)$ be linearly independent solutions of (2.16.9). Show that the functions

$$u_i(k) \; = \; a_0(k-1)v_i(k-1) \prod_{\ell=a}^{k-2} \frac{a_0(\ell)}{a_2(\ell)}, \quad k \in \mathbb{N}(a+1), \quad i = 1, 2$$

are linearly independent solutions of (2.16.8).

**2.16.25.** The adjoint of each of the following difference equations can be solved to obtain two linearly independent solutions. Use Problem 2.16.24 to find their general solutions

(i) $(k+1)(k+2)u(k+2) - u(k) = 0$

(ii) $u(k+2) + (k+1)u(k+1) - ku(k) = 0.$

**2.16.26.** Show that if one solution $u_1(k)$ of the difference equation (1.2.4) is known then its order can be reduced to $n-1$. In particular, if $u_1(k)$ is a solution of (2.16.8) then show that its second solution $u_2(k)$ can be written as

$$u_2(k) \; = \; u_1(k) \sum_{\ell=a}^{k-1} \prod_{\tau=a}^{\ell-1} \frac{a_0(\tau)}{a_2(\tau)} \Big/ u_1(\ell) u_1(\ell+1).$$

**2.16.27.** For each of the following difference equations one solution is known. Use Problem 2.16.26 to find the second solution

(i) $(k+1)^2u(k) - (k^2+3k+1)u(k+1) + ku(k+2) = 0, \quad u_1(k) = 1$

(ii) $(k+1)(k+2)u(k) - 2k(k+2)u(k+1) + k(k+1)u(k+2) = 0, \quad u_1(k) = k$

(iii) $k^2u(k) + u(k+1) - u(k+2) = 0, \quad u_1(k) = (k-1)!$

(iv) $(k+1)u(k) - u(k+1) - (k+4)u(k+2) = 0, \quad u_1(k) = \dfrac{1}{(k+1)(k+2)}.$

**2.16.28.** Let one solution of (2.16.8) be $\phi(k)$ times the other, where $\phi(k),\ k \in \mathbb{N}(a)$ is a known function. Show that the order of (2.16.8) can be reduced to one. In particular for the difference equation

$$(2k+1)u(k+2) - 8(k+1)u(k+1) + 4(2k+3)u(k) = 0, \quad k \in \mathbb{N}$$

it is known that $u(k)$ and $k^2u(k)$ are the solutions. Find $u(k)$.

**2.16.29.** Let one solution of (2.16.8) be the square of the other solution. Show that the order of (2.16.8) can be reduced to one. In particular for the difference equation

$$ku(k+2) - (k+2)(k^2+3k+1)u(k+1) + (k+1)^3(k+2)u(k) = 0, \quad k \in \mathbb{N}(1)$$

it is known that $u(k)$ and $u^2(k)$ are the solutions. Find $u(k)$.

**2.16.30.** Let the product of the two solutions of (2.16.8) be a constant. Show that the order of (2.16.8) can be reduced to one. In particular for the difference equation

$$(k+2)(2k+1)u(k+2) - 4(k+1)^2u(k+1) + k(2k+3)u(k) = 0, \quad k \in \mathbb{N}(1)$$

it is known that $u(k)$ and $1/u(k)$ are the solutions. Find $u(k)$.

**2.16.31.** Let the difference equations

$$u(k+2) + a_1(k)u(k+1) + a_2(k)u(k) = 0, \quad k \in \mathbb{N} \tag{2.16.10}$$

and

$$v(k+2) + b_1(k)v(k+1) + b_2(k)v(k) = 0, \quad k \in \mathbb{N} \tag{2.16.11}$$

have a solution in common. Show that

$$\begin{aligned}(a_1(k) - b_1(k))&(a_1(k-1)b_2(k-1) - b_1(k-1)a_2(k-1)) \\ &= (b_2(k-1) - a_2(k-1))(a_2(k) - b_2(k)), \quad k \in \mathbb{N}(1).\end{aligned}$$

In particular find the general solution of (2.11.5) given that (2.11.5) and (2.11.6) have a solution in common.

**2.16.32.** Show that by means of the transformation $v(k) = r(k)u(k)$ the difference equation (2.16.10) can be reduced to the forms

$$v(k+2) + v(k+1) + b_2(k)v(k) = 0$$

and

$$v(k+2) + b_1(k)v(k+1) + v(k) = 0.$$

**2.16.33.** Let $u_i(k),\ k \in \mathbb{N}(a),\ 1 \le i \le n$ be linearly independent functions. Show that the $n$th order difference equation having these functions as a fundamental system of solutions is

$$\begin{vmatrix} u(k) & u_1(k) & \cdots & u_n(k) \\ u(k+1) & u_1(k+1) & \cdots & u_n(k+1) \\ \cdots & \cdots & \cdots & \cdots \\ u(k+n) & u_1(k+n) & \cdots & u_n(k+n) \end{vmatrix} = 0.$$

**2.16.34.** Show that the function $u(k),\ k \in \mathbb{N}$ is a solution of the $n$th order linear homogeneous difference equation with constant coefficients if and only if for all $k \in \mathbb{N}$

$$D(u(k), \cdots, u(k+n)) = \begin{vmatrix} u(k) & \cdots & u(k+n) \\ u(k+1) & \cdots & u(k+n+1) \\ \cdots & \cdots & \cdots \\ u(k+n) & \cdots & u(k+2n) \end{vmatrix} = 0,$$

and $D(u(k), \cdots, u(k+n-1)) \neq 0$.

**2.16.35.** Consider the difference equation (2.12.4). Show that

(i) its characteristic equation is

$$p(\lambda) = \sum_{i=0}^{n} \frac{a_i}{a_n} \lambda^i = 0 \tag{2.16.12}$$

(ii) if $\lambda_1 \neq \cdots \neq \lambda_n$ are the roots of (2.16.12), then $\lambda_i^k,\ 1 \le i \le n$ are $n$ linearly independent solutions of (2.12.4), and in this case the Green's function defined in (2.11.7) reduces to

$$G(k, \ell) = \sum_{i=1}^{n} \lambda_i^{k-\ell} / p'(\lambda_i)$$

(iii) if $\lambda_1 \neq \cdots \neq \lambda_m$ $(m < n)$ are the roots of (2.16.12) with multiplicities $r_1, \cdots, r_m$ respectively, then $\lambda_i^k, \cdots, (k)^{(r_i-1)}\lambda_i^k$, $1 \le i \le m$ are $n$ linearly independent solutions of (2.12.4)

(iv) if $\lambda_1$ is the root of (2.16.12) with multiplicity $n$, then the Green's function defined in (2.11.7) reduces to

$$G(k,\ell) = \frac{(k-\ell)^{(n-1)}}{(n-1)!}\lambda_1^{k-\ell-n+1}$$

(v) if $\lambda_1, \cdots, \lambda_m$ $(2 \le m < n)$ are the distinct roots of (2.16.12), where $\lambda_1$ has multiplicity $M \ge 1$ and $\lambda_2, \cdots, \lambda_m$ have multiplicity one, so that $n = M + m - 1$, then the Green's function defined in (2.11.7) reduces to

$$G(k,\ell) = \sum_{j=2}^{m} \frac{\lambda_j^{k-\ell}}{p'(\lambda_j)} - \sum_{i=0}^{M-1}\left[\sum_{j=2}^{m}\frac{(\lambda_j-\lambda_1)^i}{p'(\lambda_j)}\right]\frac{(k-\ell)^{(i)}}{i!}\lambda_1^{k-\ell-i}$$

(vi) if $\lambda_1, \lambda_2$ are the distinct roots of (2.16.12) with multiplicities $r_1$ and $r_2$, then the Green's function defined in (2.11.7) reduces to

$$\begin{aligned} G(k,\ell) = {} & \frac{(-1)^{r_2}}{(\lambda_2-\lambda_1)^{n-1}} \sum_{i=0}^{r_1-1} \frac{(n-i-2)^{(r_2-1)}}{(r_2-1)!}(\lambda_2-\lambda_1)^i \frac{(k-\ell)^{(i)}}{i!}\lambda_1^{k-\ell-i} \\ & + \frac{(-1)^{r_1}}{(\lambda_1-\lambda_2)^{n-1}} \sum_{i=0}^{r_2-1} \frac{(n-i-2)^{(r_1-1)}}{(r_1-1)!}(\lambda_1-\lambda_2)^i \frac{(k-\ell)^{(i)}}{i!}\lambda_2^{k-\ell-i}. \end{aligned}$$

**2.16.36.** Find the Green's function for the following difference equations

(i) $(E-2)^3(E-1)(E-3)(E-4)u(k) = 0$

(ii) $(E^2+1)^m u(k) = 0$

(iii) $(E+1)^3(E-2)^2 u(k) = 0$

(iv) $(E-1)^2(E-2)^3(E-3)^2 u(k) = 0$

(v) $(E-1)(E-2)(E+1)^3(E-1/2)^2 u(k) = 0$

(vi) $(E-1)^2(E-2)^2(E-3)(E-4)u(k) = 0.$

**2.16.37.** Show that the $k$th order determinant

$$D_k = \begin{vmatrix} \alpha & 1 & 0 & 0 & & & \\ 1 & \alpha & 1 & 0 & & & \\ 0 & 1 & \alpha & 1 & & & \\ & & \cdot & \cdot & \cdot & & \\ & & & & 1 & \alpha & 1 \\ & & & & 0 & 1 & \alpha \end{vmatrix}$$

satisfies the initial value problem

$$D_{k+2} - \alpha D_{k+1} + D_k = 0, \quad k \in \mathbb{N}(1), \quad D_1 = \alpha, \quad D_2 = \alpha^2 - 1.$$

Hence, deduce that

$$D_k = \begin{cases} \sin(k+1)\theta/\sin\theta, & \alpha = 2\cos\theta, \quad |\alpha| < 2 \\ k+1, & \alpha = 2 \\ \sinh(k+1)\theta/\sinh\theta, & \alpha = 2\cosh\theta, \quad |\alpha| > 2. \end{cases}$$

**2.16.38.** Show that for the $n(\geq 2)$th order homogeneous difference equation

$$u(k+n) = \frac{1}{n}\sum_{i=0}^{n-1} u(k+i), \quad k \in \mathbb{N}$$

(i) the characteristic equation has $\lambda_1 = 1$ as a root, and all other roots are distinct and are less than one in magnitude

(ii) $\lim_{k\to\infty} u(k) = \dfrac{2}{n(n+1)}\sum_{i=0}^{n-1}(i+1)u(i).$

**2.16.39.** (Method of Undetermined Coefficients). This is a very useful method to find the general solution of the nonhomogeneous difference equation (2.12.1). It is assumed that the function $b(k)$ is a solution of the $m$th order difference equation with constant coefficients $P_m(E)b(k) = \sum_{i=0}^m b_i \times b(k+i) = 0$. Thus, all solutions of (2.12.1) are included in the general solution of the $(n+m)$th order equation

$$P_m(E)P_n(E)u(k) = 0. \tag{2.16.13}$$

Since the general solution of (2.16.13) contains $n+m$ arbitrary constants, the $m$ fictitious constants are determined by substituting this general solution in the equation (2.12.1) and simply equating the coefficients of the like functions. Use this method to solve the following difference equations

(i) $u(k+2) - 6u(k+1) + 8u(k) = 3 + 4k^2 - 7\cdot 3^k$

(ii) $u(k+2) - 4u(k+1) + 3u(k) = 27k4^k$

(iii) $u(k+2) - 3u(k+1) + 2u(k) = 3\sin 3k + 2\cos 3k$

(iv) $u(k+2) - 4u(k+1) + 4u(k) = 2 + (-1)^k + 3\cdot 2^k$

(v) $u(k+3) - 7u(k+2) + 16u(k+1) - 12u(k) = 5k2^k + 7\cdot 3^k.$

**2.16.40.** To find a particular solution of (2.12.1) the method of undetermined coefficients requires the difference equation for which $b(k)$ is a solution, and the solution of simultaneous algebraic equations, which is often tedious. Recently, Gupta [10,11] has proposed an easier method to find a particular solution of the difference equation

$$(2.16.14) \qquad \sum_{i=0}^{n} b_i E^{n-i} u(k) \;=\; a^k f_m(k), \quad k \in \mathbb{N}$$

where $a,\ b_i,\ 0 \le i \le n,\ b_0 \neq 0,\ b_n \neq 0$ are constants, and $f_m(k)$ is a polynomial of degree $m$ in $k$ (better written in factorial powers cf. Problem 1.9.17). The first step of the method is to use the substitution $u(k) = a^k v(k)$ which transforms (2.16.14) to

$$(2.16.15) \qquad \sum_{i=0}^{n} b_i (aE)^{n-i} v(k) \;=\; f_m(k), \quad k \in \mathbb{N}$$

which in terms of $\Delta$ appears as

$$(2.16.16) \qquad \sum_{i=0}^{n} c_i \Delta^{n-i} v(k) \;=\; f_m(k), \quad k \in \mathbb{N}$$

where we assume that $c_0 \neq 0,\ c_n \neq 0$. Difference equation (2.16.16) is now operated successively $m$ times by $\Delta$ so that finally the right side reduces to a constant. This results in a set of $m+1$ difference equations, last of which has a particular solution

$$(2.16.17) \qquad \begin{aligned} \Delta^m v(k) &= \Delta^m f_m(k)/c_n = \text{const.} \\ \Delta^{m+s} v(k) &= 0, \quad s = 1, \cdots, n. \end{aligned}$$

Next we substitute (2.16.17) in the last but one equation in the set of $m+1$ difference equations obtained, to get $\Delta^{m-1}v(k)$. We continue this back substitution in this set until from the first equation, i.e. (2.16.16) we find $v(k)$. Finally, a particular solution of (2.16.14) is obtained from the relation $u(k) = a^k v(k)$.

If in (2.16.16), $c_{n-r} = 0,\ 0 \le r \le s-1,\ c_{n-s} \neq 0$, i.e. $\Delta^s v(k)$ is the lowest order term, then the above procedure provides a particular solution for $\Delta^s v(k)$. Now $s$ applications of the process of antidifferencing and the relation $u(k) = a^k v(k)$ lead to a particular solution of (2.16.14).

If the right side of (2.16.14) is of the form $a^k f_p(k) + b^k f_q(k)$ where $f_p(k)$ and $f_q(k)$ are the polynomials of degree $p$ and $q$ in $k$, then a particular solution of (2.16.14) can be obtained by a combination of the principle of superposition and the above method.

Use this method to solve the following difference equations

(i) $u(k+2) - 5u(k+1) + 6u(k) = 4^k(k^2 - k + 5)$

(ii) $(\Delta^3 - 3\Delta^2 + 3\Delta - 1)u(k) = 3k - 2k^{(3)}$

(iii) $(-10\Delta^3 + 10\Delta^2 - 9\Delta + 6)u(k) = 6k^{(4)} + 6k - 250$

(iv) $(\Delta^4 - 3\Delta^3 + 9\Delta^2)u(k) = 540k^{(3)} - 6$

(v) $(\Delta^5 - \Delta^2 + 2\Delta)u(k) = 4k^{(3)} + 2k - 12$

(vi) $u(k+2) + u(k) = \sin k$

(vii) $u(k+2) - 2\cos\frac{1}{2}u(k+1) + u(k) = \cos\frac{k}{2}$

(viii) $u(k+2) - 2u(k+1) + 4u(k) = -2^k\left(6\cos\frac{k\pi}{3} + 2\sqrt{3}\sin\frac{k\pi}{3}\right)$.

**2.16.41.** (Operational Method). The idea of this technique is to determine a particular solution of the nonhomogeneous difference equation (2.12.1) by means of the relation

$$u(k) = P_n^{-1}(E)b(k), \tag{2.16.18}$$

where $P_n^{-1}(E)$ is an operator such that when the right side of (2.16.18) is operated by $P_n(E)$ it gives back $b(k)$. Let $z$ be any complex number, and $u(k)$ be a function defined on $\mathbb{N}$. Show that for all $k, m \in \mathbb{N}$

(i) $P_n(E)(z^k u(k)) = z^k P_n(zE)u(k)$

(ii) $P_n^{-1}(E)(z^k u(k)) = z^k P_n^{-1}(zE)u(k)$

(iii) $(E - zI)^m (z^k u(k)) = z^{k+m}\Delta^m u(k)$

(iv) $(E - zI)^{-m}0 = z^{k-m}Q_{m-1}(k)$, where $Q_{m-1}(k)$ is a polynomial of degree $m-1$

(v) $(E - zI)^{-m} z^k = \dfrac{z^{k-m}(k)^{(m)}}{m!}$

(vi) if $P_n(E) = (E - zI)^m P_{n-m}(E)$, $0 \le m \le n$ and $P_{n-m}(z) \ne 0$, then

$$P_n^{-1}(E)z^k = \frac{z^{k-m}(k)^{(m)}}{P_{n-m}(z)\; m!}$$

(vii) if $Q_r(k)$ is a polynomial of degree $r$, then

$$\begin{aligned} P_n^{-1}(E)Q_r(k) &= P_n^{-1}(I+\Delta)Q_r(k) = \Delta^{-m}P_{n-m}^{-1}(\Delta)Q_r(k) \\ &= \Delta^{-m}\left(\frac{1}{P_{n-m}(0)} + b_1\Delta + \cdots + b_r\Delta^r\right)Q_r(k), \end{aligned}$$

where $0 \le m \le n$ and $P_{n-m}(0) \ne 0$.

Further, use these relations to find the particular solutions of the difference equations given in Problem 2.16.39.

**2.16.42.** Use the method of generating functions to find the solutions of the difference equations given in Problem 2.16.39.

**2.16.43.** The method of generating functions can also be used to solve the difference equation (1.2.3) provided $a_i(k)$, $0 \le i \le n$ are polynomials. Solve the following initial value problems

(i) $(k+1)u(k+1) - u(k) \;=\; 0, \;\; u(0) = 7$

(ii) $(k+1)(k+2)u(k+2) - 2(k+1)u(k+1) - 3u(k) \;=\; 0, \;\; u(0) = u(1) = 2$

(iii) $u(k+2) - u(k+1) - (k+1)u(k) \;=\; 0, \;\; u(0) = u(1) = 1.$

**2.16.44.** Use Bernoulli's method to compute maximum and minimum modulus roots of the following polynomial equations

(i) $3003\lambda^4 - 5660\lambda^3 + 3815\lambda^2 - 1090\lambda + 112 \;=\; 0$

(ii) $\lambda^5 + \lambda^4 - 5 \;=\; 0$

(iii) $49\lambda^4 + 7\lambda^3 + 16\lambda^2 - 33\lambda + 9 \;=\; 0.$

**2.16.45.** Let $\mathcal{A}$ and $\mathcal{B}$ be two $n \times n$ matrices. We say that $\mathcal{A}$ and $\mathcal{B}$ are *similar* if and only if there exists a nonsingular matrix $\mathcal{P}$ such that $\mathcal{P}^{-1}\mathcal{A}\mathcal{P} = \mathcal{B}$. Show that

(i) $\mathbf{v}(k)$ is a solution of the difference system $\mathbf{v}(k+1) = \mathcal{B}\mathbf{v}(k)$ if and only if $\mathbf{u}(k) = \mathcal{P}\mathbf{v}(k)$ is a solution of the difference system (2.8.1)

(ii) if the matrix $\mathcal{A}$ in (2.14.2) has the distinct eigenvalues $\lambda_1, \cdots, \lambda_n$ then $\mathcal{A} = \mathcal{V}\mathcal{D}\mathcal{V}^{-1}$, where $\mathcal{V} = \mathcal{V}(\lambda_1, \cdots, \lambda_n)$ is the Vandermonde's matrix and $\mathcal{D} = diag\ (\lambda_1, \cdots, \lambda_n)$ is the diagonal matrix.

**2.16.46.** Consider the difference equation

$$u(k+2) + pu(k+1) + qu(k) \;=\; 0, \quad k \in \mathbb{N} \tag{2.16.19}$$

and let $\lambda_1$, $\lambda_2$ be the roots of the equation $\lambda^2 + p\lambda + q = 0$. Show that

(i) if $\lambda_1 = \lambda_2 = \lambda$, then for every solution $u(k)$ of (2.16.19), $\displaystyle\lim_{k\to\infty} \frac{u(k+1)}{u(k)} = \lambda$

(ii) if $\lambda_1 = -\lambda_2$, then there exists a solution $u(k)$ of (2.16.19) for which $\displaystyle\lim_{k\to\infty} \frac{u(k+1)}{u(k)}$ does not exist.

**2.16.47.** For each solution $u(k)$ of the either of difference equations

(i) $u(k+2) - \dfrac{(k+2) + 2(-1)^k}{(k+2)^2(k+3)} u(k) = 0, \quad k \in \mathbb{N}$

(ii) $u(k+2) + \dfrac{1}{k+4}u(k+1) - \dfrac{k+1}{k+4}u(k) = 0, \quad k \in \mathbb{N}$

show that $\lim_{k\to\infty} \dfrac{u(k+1)}{u(k)}$ does not exist.

**2.16.48.** Let in the homogeneous difference equation (1.2.4), $a_n(k) = 1$, $a_0(k) \neq 0$ for all $k \in \mathbb{N}$ and $\lim_{k\to\infty} a_i(k) = a_i$, $0 \le i \le n-1$. Further, let $\alpha$ be a number whose modulus is greater than that of every root of (2.13.1). Show that for every solution $u(k)$ of (1.2.4), $\lim_{k\to\infty} \dfrac{u(k)}{\alpha^k} = 0$.

**2.16.49.** Consider the difference equation

$$(2.16.20)\quad u(k+2) - [2 + a_1(k)]u(k+1) + [1 + a_0(k)]u(k) \;=\; 0, \quad k \in \mathbb{N}$$

where $a_0(k)$, $a_1(k) \to 0$ as $n \to \infty$ and for sufficiently large $k$, $a_1(k) \ge 0$, $a_1(k) - a_0(k) \ge 0$. Show that for every solution $u(k)$ of (2.16.20) which is not identically zero for large $k$, $\lim_{k\to\infty} \dfrac{u(k+1)}{u(k)} = 1$.

**2.16.50.** For the difference equation

$$u(k+2) + a_1(k)u(k+1) + a_0(k)u(k) \;=\; 0, \quad k \in \mathbb{N}$$

where $a_0(k)$, $a_1(k) \to 0$ as $k \to \infty$, and $\lim_{k\to\infty} \dfrac{a_0(k)}{a_1(k-1)a_1(k)} = \alpha < \dfrac{1}{4}$, show that for every solution $u(k)$ which is not identically zero for large $k$, $\lim_{k\to\infty} \dfrac{u(k+1)}{u(k)} = 0$.

**2.16.51.** (Poincaré Type Result for Nonlinear Equations). Let the difference equation (1.2.5) be better written as

$$(2.16.21)\qquad u(k+n) \;=\; f_k(u(k), u(k+1), \cdots, u(k+n-1)), \quad k \in \mathbb{N}$$

where $f_k : \mathbb{R}_+^n \to \mathbb{R}_+$ for all $k \in \mathbb{N}$ and $u(i) \ge 0$, $0 \le i \le n-1$. Let $\|\cdot\|$ be a monotonic norm on $\mathbb{R}^n$, i.e. $\|\mathbf{u}\| \le \|\mathbf{v}\|$ for $\mathbf{u}, \mathbf{v} \in \mathbb{R}_+^n$ with $\mathbf{u} \le \mathbf{v}$ and let $X = \{\mathbf{u} \in \mathbb{R}_+^n : \|\mathbf{u}\| = 1\}$. Further, let

(i) $f_k$, $k \in \mathbb{N}$ converges to a continuous function $f : \mathbb{R}_+^n \to \mathbb{R}_+$ with uniform convergence on $X$

(ii) $0 \le \mathbf{u} \le \mathbf{v}$ implies $f_k(\mathbf{u}) \le f_k(\mathbf{v})$ for all $k \in \mathbb{N}$

(iii) $f_k(\lambda\mathbf{u}) = \lambda f_k(\mathbf{u})$ for all $\lambda \ge 0$ and $\mathbf{u} \ge 0$, $k \in \mathbb{N}$

(iv) $0 < f(\mathbf{u})$ for $\mathbf{u} \ge 0$, and if $0 \le \mathbf{u} \le \mathbf{v}$ and $u_1 < v_1$ or $u_n < v_n$ then $f(\mathbf{u}) < f(\mathbf{v})$.

Show that for all nonnegative solutions $u(k)$ of (2.16.21)

$$\lim_{k\to\infty} \frac{u(k+1)}{u(k)} = \lambda^*,$$

where $\lambda^* > 0$ is the unique nonnegative solution of the equation $\lambda^n = f(1, \lambda, \cdots, \lambda^{n-1})$.

**2.16.52.** Consider the initial value problem

$$\begin{aligned} &u(k+2) + p(k,\epsilon)u(k+1) + q(k,\epsilon)u(k) = r(k,\epsilon), \ \ k \in \mathbb{N}(a,b) \\ &u(a) \ = \ A, \quad u(a+1) \ = \ B \end{aligned} \tag{2.16.22}$$

where

$$p(k,\epsilon) = \sum_{i=0}^{\infty} p_i(k)\epsilon^i, \quad q(k,\epsilon) = \sum_{i=0}^{\infty} q_i(k)\epsilon^i, \quad r(k,\epsilon) = \sum_{i=0}^{\infty} r_i(k)\epsilon^i$$

are convergent for $|\epsilon| < \epsilon_0$ and $k \in \mathbb{N}(a,b)$. Show that the solution $u(k)$ of (2.16.22) can be written as

$$u(k) \ = \ \sum_{i=0}^{\infty} u_i(k)\epsilon^i, \tag{2.16.23}$$

which is convergent for $|\epsilon| < \epsilon_0$ and $k \in \mathbb{N}(a, b+2)$. In (2.16.23) functions $u_i(k), \ i = 0, 1, \cdots$ are the solutions of

$$\begin{aligned} u_i(k+2) + p_0(k)u_i(k+1) + q_0(k)u_i(k) \ &= \ \overline{r}_i(k) \\ u_0(a) \ = \ A, \quad u_0(a+1) \ &= \ B \\ u_i(a) \ = \ u_i(a+1) \ = \ 0, \quad i &= 1, 2, \cdots \end{aligned}$$

where

$$\begin{aligned} \overline{r}_0(k) \ &= \ r_0(k) \\ \overline{r}_i(k) \ &= \ r_i(k) - \sum_{j=0}^{i-1} \left[u_j(k+1)p_{i-j}(k) + u_j(k)q_{i-j}(k)\right], \quad i = 1, 2, \cdots. \end{aligned}$$

**2.16.53.** Show that the zeroth order approximation to the singular perturbation problem

$$u(k+2) + au(k+1) + \epsilon u(k) \ = \ 0, \quad u(0) = \alpha_0, \ u(1) = \alpha_1, \quad a \gg \epsilon$$

is $u(k) \simeq \alpha_1(-a)^{k-1} + \epsilon^k \left(\alpha_0 + \frac{1}{a}\alpha_1\right)\left(-\frac{1}{a}\right)^k.$

**2.16.54.** Explain why the singular perturbation series for the difference equation

$$(2.16.24) \qquad \epsilon u(k+2) + au(k+1) + u(k) = 0, \quad a \gg \epsilon$$

satisfying $u(0) = \alpha_0$, $u(1) = \alpha_1$ cannot be developed. However, the terminal point problem (2.16.24), $u(K+1) = \alpha_{K+1}$, $u(K) = \alpha_K$ exhibits a nice boundary layer behavior as $k \to K+1$, and is well behaved for all $k \in \mathbb{N}(0, K)$ as $\epsilon \to 0$.

## 2.17. Notes

For linear difference equations one of the elegant features is the close similarity to the theory of linear differential equations. Therefore, most of our discussion in Sections 2.1 2.8 runs parallel to the theory of differential equations presented in Agarwal and Gupta [1]. In particular some of these results here are in very compact form as compared to what is available in several well known books on difference equations, however there is some similarity and overlapping with Lakshmikantham and Trigiante [18] and Miller [21]. Discrete Putzer's algorithm has appeared in LaSalle [19]. Periodic systems have been treated in Agarwal and Popenda [2], Corduneanu [7], Halanay [12], Pang and Agarwal [25] and Sugiyama [27]. Almost periodicity of a discrete function was first introduced by Walther [28,29]. Theorem 2.10.1 is due to Corduneanu [7]. The content of Section 2.11 is also parallel to the theory of differential equations. The method of generating functions to solve difference equations is one of the classical techniques and can be found in almost every book on difference equations. The classical Bernoulli's method which is presented in Section 2.13 is based on Hildebrand [14] and John [15]. The extension of Bernoulli's method known as the quotient difference algorithm provides simultaneous approximations to all the roots of polynomial equations is available in Henrici [13]. Poincaré's theorem can be found in Gel'fond [9], Milne Thomson [22] and Nörlund [24], however the present compact proof is adapted from Lakshmikantham and Trigiante [18]. The proof of Perron's theorem is available in Evgrafov [8] and Meschkowski [20]. Several refinements of these results are established in Zhao-Hua [30]. Regular and singular perturbation methods are extensively used in obtaining approximate analytic solutions of differential equations. However, for difference equations this powerful technique does not appear to have been explored to its fullest, for instance see Bender and Orszag [4], Comstock and Hsiao [6], and Naidu and Rao [23]. In Section 2.15 we touch upon this technique and look forward to its further developments. Problems 2.16.15, and 2.16.35 and 2.16.36, respectively are based on the work of Kwapisz [17], and Peterson and Schneider [26]. Poincaré's theorem for nonlinear equations given as Problem 2.16.51 is due to Krause [16]. Problem 2.16.52 models a recent contribution of Balla [3].

## 2.18. References

[1]. R.P. Agarwal and R.C. Gupta, *Essentials of Ordinary Differential Equations*, *McGraw Hill*, Singapore, New York, 1991.

[2]. R.P. Agarwal and J. Popenda, On periodic solutions of first order linear difference equations, *Mathl. Comput. Modelling* **22**(1)(1995), 11 19.

[3]. K. Balla, On asymptotic behavior of solutions to some difference equations, in *Proceedings of the 2nd International Conference on Difference Equations and Applications*, Veszprém, Hungary, August 7 11, 1995, eds. S. Elaydi, et. al., *Gordon & Breach*, New York, 1996, 67 80.

[4]. C.M. Bender and S.A. Orszag, *Advanced Mathematical Methods for Scientists and Engineers*, *McGraw Hill*, New York, 1978.

[5]. C.W. Clenshaw, A note on the summation of Chebyshev series, *M.T.A.C.* **9**(1955), 118 120.

[6]. C. Comstock and G.C. Hsiao, Singular perturbations for difference equations, *Rocky Mountain J. Math.* **6**(1976), 561 567.

[7]. C. Corduneanu, Almost periodic discrete processes, *Libertas Math.* **2**(1982), 159 169.

[8]. M.A. Evgrafov, A new proof of a theorem of Perron, *Izv. Akad. Nauk SSSR Ser. Mat.* **17**(1953), 77 82.

[9]. A.O. Gel'fond, *Calculus of Finite Differences*, *Hindustan*, Delhi, India, 1971.

[10]. R.C. Gupta, On linear difference equations with constant coefficients: an alternative to the method of undetermined coefficients, *Mathematics Magazine* **67**(1994), 131 135.

[11]. R.C. Gupta, On particular solutions of linear difference equations with constant coefficients, *SIAM Review* **40**(1998), 680 684.

[12]. A. Halanay, Solution periodiques et presque periodiques des systems d'equationes aux difference finies, *Arch. Rat. Mech.* **12**(1963), 134 149.

[13]. P. Henrici, *Elements of Numerical Analysis*, *Wiley*, New York, 1964.

[14]. F.B. Hildebrand, *Introduction to Numerical Analysis*, *McGraw Hill*, New York, 1956.

[15]. F. John, *Lectures on Advanced Numerical Analysis*, *Thomas and Nelson and Sons*, N.J., 1966.

[16]. U. Krause, A theorem of Poincaré type for non autonomous nonlinear difference equations, in *Proceedings of the 2nd International Conference on Difference Equations and Applications*, Veszprém, Hungary, August 7 11, 1995, eds. S. Elaydi, et. al., *Gordon & Breach*, New York, 1996, 363 369.

[17]. M. Kwapisz, The power of a matrix, *SIAM Rev.* **40**(1998), 703 705.

[18]. V. Lakshmikantham and D. Trigiante, *Theory of Difference Equations: Numerical Methods and Applications*, *Academic Press*, New York, 1988.

[19]. J.P. LaSalle, Stability theory for difference equations, in *Studies in Ordinary Differential Equations*, ed. Jack Hale, *The Mathematical Association of America*, 1977, 1 33.

[20]. H. Meschkowski, *Differenzengleichungen*, *Vandenhoeck and Ruprecht*, Göttingen, 1959.

[21]. K.S. Miller, *Linear Difference Equations*, *W.A. Benjamin*, New York, 1968.

[22]. L.M. Milne Thomson, *The Calculus of Finite Differences*, *Macmillan*, London, 1960.

[23]. D.S. Naidu and A.K. Rao, *Singular Perturbation Analysis of Discrete Control Systems*, Lecture Notes in Math. 1154, *Springer Verlag*, Berlin, 1985.

[24]. N.E. Nörlund, *Vorlesungen Über Differenzenrechnung*, *Chelsea*, New York, 1954.

[25]. P.Y.H. Pang and R.P. Agarwal, On periodicity of difference equations of a general type, *J. Difference Eqns. Applic.* **2**(1996), 271 286.

[26]. A.C. Peterson and J. Schneider, The Cauchy function for $n$th order linear difference equations, *Rocky Mount. J. Math.* **25**(1995), 441 457.

[27]. S. Sugiyama, On periodic solutions of difference equations, *Bull. Sci. Engg. Resh. Lab. Waseda Univ.* **52**(1971), 89 94.

[28]. A. Walther, Fastperiodische Folgen und Potenzreihen mit fastperiodischen Koeffizienten, *Abh. Math. Sem. Hamburg Univ.* **6**(1928), 217 234.

[29]. A. Walther, Fastperiodische Folgen und ihre Fouriersche Analysis, *Atti Congresso Int. Mat. Bologna* **2**(1928), 289 298.

[30]. Zhao Hua Li, The asymptotic estimates of solutions of difference equations, *J. Math. Anal. Appl.* **94**(1983), 181 192.

# Chapter 3
# Miscellaneous Difference Equations

In numerical integration of a differential equation a standard approach is to replace it by a suitable difference equation whose solution can be obtained in a stable manner and without troubles from round off errors. However, often the qualitative properties of the solutions of the difference equation are quite different from the solutions of the corresponding differential equations. In this chapter we shall carefully choose difference equation approximations of several well known ordinary and partial differential equations, and show that the solutions of these difference equations preserve most of the properties of the corresponding differential equations. We begin with Clairaut's, Euler's and Riccati's difference equations which are known for quite sometime. This is followed by Bernoulli's difference equation which can be solved in a closed form. Next we consider the Verhulst difference equation and show that its solutions correctly mimic the true solutions of the Verhulst differential equation. Then, we develop the 'best' discrete approximations of the linear differential equations with constant coefficients. Here, as an example, simple harmonic oscillator differential equation is best discretized. This is followed by Duffing's difference equation which can be solved explicitly and whose solutions have precise agreement with the solutions of Duffing's differential equation. Next we consider van der Pol's difference equation, which like van del Pol's differential equation cannot be solved, however the solutions of both the equations have same qualitative features. Then, we deal with Hill's and in particular Mathieu's difference equations, and provide conditions for basically periodic solutions of period $\pi$ and $2\pi$. This leads to a classification of four different types of periodic solutions, which is a well known result for the solutions of Hill's differential equation. Next we shall show that Weierstrass' elliptic differential equations can be discretized in such a way that the solutions of the resulting difference equations exactly coincide with the corresponding values of the elliptic functions. In Section 3.12 we analyze Volterra's difference equations, their trajectories have the same closed form expression as for the Volterra's differential equations. Then, we provide several methods to solve linear partial difference equations with constant coefficients in two independent variables. This is followed by the best discretizations of Wave equation, FitzHugh Nagumo's equation, Korteweg de Vries' equation and Modified KdV equation. Finally, in Section 3.18 we shall formulate discrete Lagrange's equations of motion.

## 3.1. Clairaut's Equation

The discrete analog of Clairaut's differential equation $y = ty' + f(y')$ appears as

$$u(k) \;=\; k\Delta u(k) + f(\Delta u(k)), \qquad k \in \mathbb{N} \tag{3.1.1}$$

where $f$ is some nonlinear function.

In the above difference equation let $v(k) = \Delta u(k)$, so that

$$u(k) \;=\; kv(k) + f(v(k)) \tag{3.1.2}$$

and

$$v(k) \;=\; (k+1)v(k+1) - kv(k) + f(v(k+1)) - f(v(k)),$$

which is the same as

$$(k+1)\Delta v(k) + f(v(k) + \Delta v(k)) - f(v(k)) \;=\; 0.$$

Therefore, either

$$\Delta v(k) \;=\; 0 \tag{3.1.3}$$

or

$$(k+1) + \frac{f(v(k) + \Delta v(k)) - f(v(k))}{\Delta v(k)} \;=\; 0. \tag{3.1.4}$$

Equation (3.1.3) implies that $v(k) = c$ (constant), and from (3.1.2) we get the solution

$$u(k) \;=\; kc + f(c). \tag{3.1.5}$$

Equation (3.1.4) may lead to a second (singular) solution.

**Example 3.1.1.** Consider the Clairaut difference equation

$$u(k) \;=\; k\Delta u(k) + (\Delta u(k))^2, \qquad k \in \mathbb{N} \tag{3.1.6}$$

for which $u(k) = kc + c^2$ is a solution. Further, the equation (3.1.4) for (3.1.6) reduces to

$$v(k+1) + v(k) + k + 1 = 0. \tag{3.1.7}$$

The solution of (3.1.7) can be written as $v(k) = c(-1)^k - (1/2)k - (1/4)$, and hence the second solution of (3.1.6) is

$$u(k) = \left[c(-1)^k - \frac{1}{4}\right]^2 - \frac{1}{4}k^2.$$

**Example 3.1.2.** Consider the Clairaut difference equation

$$u(k) = k\Delta u(k) + \frac{1}{\Delta u(k)}, \quad k \in \mathbb{N} \tag{3.1.8}$$

for which $u(k) = kc + (1/c)$ is a solution. Further, the equation (3.1.4) for (3.1.8) reduces to

$$v(k)v(k+1) = \frac{1}{k+1}. \tag{3.1.9}$$

The solution of (3.1.9) can be written as

$$v(k) = \begin{cases} \dfrac{k!}{2^k((k/2)!)^2}c & \text{if } k \text{ is even} \\ \dfrac{2^{k-1}(((k-1)/2)!)^2}{k!c} & \text{if } k \text{ is odd.} \end{cases}$$

Therefore, the second solution of (3.1.8) takes the form $u(k) = u(0) + \sum_{\ell=0}^{k-1} v(\ell)$. In particular, for $c = 1$, $u(0) = 1$ the first nine values of $u(k)$ are $1, 2, 3, 7/2, 25/6, 109/24, 203/40, 431/80, 3273/560$.

## 3.2. Euler's Equation

The discrete analog of Euler's differential equation $\sum_{i=0}^{n} a_i t^i y^{(i)} = 0$ appears as

$$\sum_{i=0}^{n} a_i (k+i-1)^{(i)} \Delta^i u(k) = 0, \quad k \in \mathbb{N}(1), \quad a_0 a_n \neq 0 \tag{3.2.1}$$

where $a_i$, $0 \le i \le n$ are constants.

We seek the solution of (3.2.1) in the form

$$u(k) = \frac{\Gamma(k+\lambda)}{\Gamma(k)}, \tag{3.2.2}$$

where $\lambda$ may be a complex number, however $k+\lambda$ is different from a negative integer. Since

$$\Delta^i \frac{\Gamma(k+\lambda)}{\Gamma(k)} = (\lambda)^{(i)} \frac{\Gamma(k+\lambda)}{\Gamma(k+i)}$$

from (3.2.1) it follows that

$$\sum_{i=0}^{n} a_i (k+i-1)^{(i)} (\lambda)^{(i)} \frac{\Gamma(k+\lambda)}{\Gamma(k+i)} = 0. \tag{3.2.3}$$

However, since $(k+i-1)^{(i)} = \Gamma(k+i)/\Gamma(k)$, equation (3.2.3) is the same as

$$\sum_{i=0}^{n} a_i (\lambda)^{(i)} \frac{\Gamma(k+\lambda)}{\Gamma(k)} = 0.$$

But $\Gamma(k+\lambda)/\Gamma(k) \neq 0$, and hence it follows that

$$\sum_{i=0}^{n} a_i (\lambda)^{(i)} = 0. \tag{3.2.4}$$

Thus, (3.2.2) is a solution of (3.2.1) if and only if $\lambda$ is a root of the polynomial (3.2.4).

**Example 3.2.1.** For the Euler difference equation

$$(k+1)k\Delta^2 u(k) - 6k\Delta u(k) + 10u(k) = 0, \quad k \in \mathbb{N}(1) \tag{3.2.5}$$

the polynomial (3.2.4) reduces to $\lambda^2 - 7\lambda + 10 = 0$. Thus,

$$u_1(k) = \frac{\Gamma(k+2)}{\Gamma(k)} = (k+1)^{(2)} \quad \text{and} \quad u_2(k) = \frac{\Gamma(k+5)}{\Gamma(k)} = (k+4)^{(4)}$$

are linearly independent solutions of (3.2.5).

**Example 3.2.2.** For the Euler difference equation

$$(k+1)k\Delta^2 u(k) + 7k\Delta u(k) + 9u(k) = 0, \quad k \in \mathbb{N}(4) \tag{3.2.6}$$

the polynomial (3.2.4) reduces to $\lambda^2 + 6\lambda + 9 = 0$ which has the repeated roots $-3, -3$. Thus,

$$u_1(k) = \frac{\Gamma(k-3)}{\Gamma(k)} = \frac{1}{(k-1)(k-2)(k-3)}$$

is a solution of (3.2.6). To find the second linearly independent solution, we note that (3.2.6) is the same as

$$(k+1)ku(k+2) + k(5-2k)u(k+1) + (k-3)^2 u(k) = 0.$$

Thus, we can use Problem 2.16.26 to find

$$u_2(k) = u_1(k)\sum_{\ell=4}^{k-1}\frac{1}{\ell-3}.$$

**Example 3.2.3.** For the Euler difference equation

$$4(k+1)k\Delta^2 u(k) + 4k\Delta u(k) + 9u(k) = 0 \tag{3.2.7}$$

the polynomial (3.2.4) reduces to $4\lambda^2+9=0$. Thus,

$$u_1(k) = \frac{\Gamma\left(k+\frac{3}{2}i\right)}{\Gamma(k)} \quad \text{and} \quad u_2(k) = \frac{\Gamma\left(k-\frac{3}{2}i\right)}{\Gamma(k)}$$

are the solutions of (3.2.7). However, since

$$\begin{aligned}\Gamma(z) &= \int_0^\infty e^{-t}t^{z-1}dt = \int_0^\infty e^{-t}t^{x-1}t^{iy}dt\\ &= \int_0^\infty e^{-t}t^{x-1}e^{i(y\ln t)}dt\\ &= \int_0^\infty e^{-t}t^{x-1}\cos(y\ln t)dt + i\int_0^\infty e^{-t}t^{x-1}\sin(y\ln t)dt\end{aligned}$$

from Problem 2.16.5 it follows that

$$u_1(k) = \frac{1}{\Gamma(k)}\int_0^\infty e^{-t}t^{k-1}\cos\left(\frac{3}{2}\ln t\right)dt$$

and

$$u_2(k) = \frac{1}{\Gamma(k)}\int_0^\infty e^{-t}t^{k-1}\sin\left(\frac{3}{2}\ln t\right)dt$$

are linearly independent solutions of (3.2.7).

## 3.3. Riccati's Equation

The discrete analog of Riccati's differential equation $y'+\alpha(t)y^2+\beta(t)y+\gamma(t)=0$ appears as

$$u(k)u(k+1) + p(k)u(k+1) + q(k)u(k) + r(k) = 0, \quad k\in\mathbb{N}. \tag{3.3.1}$$

If $r(k)\equiv 0$, then the substitution $u(k)=1/v(k)$ in (3.3.1) gives the first order linear difference equation

$$q(k)v(k+1) + p(k)v(k) + 1 = 0, \tag{3.3.2}$$

which can be solved by using standard methods.

If $r(k) \not\equiv 0$, then the substitution $u(k) = (v(k+1)/v(k)) - p(k)$ in (3.3.1) leads to the second order linear difference equation

$$(3.3.3)\quad v(k+2) + [q(k) - p(k+1)]v(k+1) + [r(k) - p(k)q(k)]v(k) \ = \ 0.$$

Since the equation (3.3.1) is of first order its general solution should depend on only one arbitrary constant, eventhough the solution of the transformed second order equation (3.3.3) contains two arbitrary constants. This fact can easily be seen as follows: Let $v_1(k)$ and $v_2(k)$ be linearly independent solutions of (3.3.3) so that its general solution can be written as $v(k) = c_1v_1(k) + c_2v_2(k)$. Thus, the solution of (3.3.1) takes the form

$$u(k) \ = \ \frac{c_1v_1(k+1) + c_2v_2(k+1)}{c_1v_1(k) + c_2v_2(k)} - p(k),$$

which is the same as

$$u(k) \ = \ \frac{v_1(k+1) + \bar{c}v_2(k+1)}{v_1(k) + \bar{c}v_2(k)} - p(k),$$

where the constant $\bar{c} = c_2/c_1$.

In general it is not possible to find a solution of (3.3.1) but if a particular solution, say, $u_1(k)$ is known then by the substitution $u(k) = u_1(k) + (1/v(k))$ it reduces to a first order linear difference equation

$$(3.3.4)\qquad [q(k) + u_1(k+1)]v(k+1) + [p(k) + u_1(k)]v(k) + 1 \ = \ 0,$$

which can be solved to obtain the solution of the form

$$(3.3.5)\qquad v(k) \ = \ c\phi(k) + \psi(k),$$

and hence the general solution of (3.3.1) can be written as

$$(3.3.6)\qquad u(k) \ = \ u_1(k) + \frac{1}{c\phi(k) + \psi(k)}.$$

Now let $u(k)$, $u_1(k)$, $u_2(k)$ and $u_3(k)$ be any four different solutions of (3.3.1). Then, from the above considerations it is clear that each $v(k) = 1/(u(k) - u_1(k))$, $v_1(k) = 1/(u_2(k) - u_1(k))$ and $v_2(k) = 1/(u_3(k) - u_1(k))$ is a solution of the same first order nonhomogeneous difference equation (3.3.4). Therefore,

$$v(k) - v_1(k) \ = \ \frac{u_2(k) - u(k)}{(u(k) - u_1(k))(u_2(k) - u_1(k))}$$

as well as

$$v_2(k) - v_1(k) = \frac{u_2(k) - u_3(k)}{(u_3(k) - u_1(k))(u_2(k) - u_1(k))}$$

is a solution of the first order homogeneous difference equation

$$[q(k) + u_1(k+1)]v(k+1) + [p(k) + u_1(k)]v(k) = 0. \tag{3.3.7}$$

Hence, it follows that

$$\frac{u_2(k) - u(k)}{(u(k) - u_1(k))(u_2(k) - u_1(k))} = c\,\frac{u_2(k) - u_3(k)}{(u_3(k) - u_1(k))(u_2(k) - u_1(k))},$$

which is the same as

$$\frac{(u(k) - u_2(k))(u_3(k) - u_1(k))}{(u(k) - u_1(k))(u_3(k) - u_2(k))} = c. \tag{3.3.8}$$

In particular, the above relation determines the general solution $u(k)$ of (3.3.1) in terms of the three known solutions $u_1(k)$, $u_2(k)$ and $u_3(k)$.

**Example 3.3.1.** Consider the first order difference equation

$$u(k+1) = \frac{\alpha u(k) + \beta}{\gamma u(k) + \delta}, \quad k \in \mathbb{N} \tag{3.3.9}$$

where the constants $\alpha, \beta, \gamma$ and $\delta$ are such that $\gamma \neq 0$, $D = \begin{vmatrix} \alpha & \beta \\ \gamma & \delta \end{vmatrix} \neq 0$. By the substitution $u(k) = v(k) - (\delta/\gamma)$, equation (3.3.9) reduces to the form

$$v(k+1) = 2\mu - \frac{\nu}{v(k)}, \quad k \in \mathbb{N} \tag{3.3.10}$$

where $2\mu = (\alpha + \delta)/\gamma$ and $\nu = D/\gamma^2$. Equation (3.3.10) is infact a Riccati's equation

$$v(k)v(k+1) - 2\mu v(k) + \nu = 0, \quad k \in \mathbb{N} \tag{3.3.11}$$

and by the substitution $v(k) = w(k+1)/w(k)$ reduces to the linear equation of second order

$$w(k+2) - 2\mu w(k+1) + \nu w(k) = 0, \quad k \in \mathbb{N}. \tag{3.3.12}$$

In particular, if $\alpha = 2$, $\beta = -2$, $\nu = 1$ and $\delta = 0$, then (3.3.9) reduces to

$$u(k+1) = 2 - \frac{2}{u(k)}, \quad k \in \mathbb{N} \tag{3.3.13}$$

which is already in the form (3.3.10). For (3.3.13) equation (3.3.12) becomes

$$w(k+2) - 2w(k+1) + 2w(k) \ = \ 0, \qquad k \in \mathbb{N}$$

whose general solution can be written as

$$w(k) \ = \ 2^{k/2}\left(c_1 \cos\frac{k\pi}{4} + c_2 \sin\frac{k\pi}{4}\right).$$

Therefore, the general solution of (3.3.13) is

$$(3.3.14) \qquad u(k) \ = \ \frac{w(k+1)}{w(k)} \ = \ \sqrt{2}\frac{\left(c_1 \cos\frac{(k+1)\pi}{4} + c_2 \sin\frac{(k+1)\pi}{4}\right)}{\left(c_1 \cos\frac{k\pi}{4} + c_2 \sin\frac{k\pi}{4}\right)}.$$

If we define the constant $\theta$ such that in the interval $-(\pi/2) < \theta \leq (\pi/2)$, $\tan\theta = c_2/c_1$, then (3.3.14) takes the form

$$u(k) \ = \ \sqrt{2}\frac{\cos\left(\frac{(k+1)\pi}{4} - \theta\right)}{\cos\left(\frac{k\pi}{4} - \theta\right)} \ = \ 1 - \tan\left(\frac{k\pi}{4} - \theta\right).$$

Since $u(k+4) = u(k)$, all solutions of (3.3.13) are periodic with period 4.

Riccati difference equations appear in mathematical biology, optics, chemistry and several other branches of mathematics. We present here few such examples.

**Example 3.3.2.** (Distillation of a Binary Liquid). The distillation of a binary ideal mixture of two liquids is realized by a column of $K$ plates, at the top of which there is a condenser and at the bottom there is a heater. At the base of the column there is a feeder of new liquid to still. A stream of vapor, whose composition becomes richer from the more volatile component, proceeds from one plate to the next one until it reaches the condenser from which part of the liquid is removed and part returns to the last plate. On each plate, which is at different temperature, the vapor phase will be in equilibrium with a liquid phase. Thus a liquid steam proceeds from the top to the bottom. We assume that liquids as well as vapors are ideal so that Raoult's and Dalton's laws apply.

Let $v(i)$, $i = 1, 2$ be the mole fraction of the $i$th component in the vapor phase and $u(i)$ the mole fraction of the same component in the liquid phase. Since the sum of the mole fraction in each phase is 1, $v(1) + v(2) = 1$ and $u(1) + u(2) = 1$. In a moderate range of the temperature the relative volatility defined as $\alpha = (v(1)u(2)/v(2)u(1))$ is considered constant. If

$\alpha > 1$ then the first component is more volatile. For simplicity, now we shall consider only the more volatile component. Setting $v(1) = v$ and $u(1) = u$, we have

$$\alpha = \frac{v(1-u)}{u(1-v)}, \tag{3.3.15}$$

which we shall assume holds every time the two phases are in equilibrium.

On the $k$th plate two components are in equilibrium in the two phases. Let $u(k)$ be the mole fraction of the more volatile component in the liquid phase, $v^*(k)$ the mole fraction in vapor phase of the same component, and $v(k)$ the mole fraction of the same component leaving the plate $k$. If we assume that the efficiency of the plate is 100 percent, then $v^*(k) = v(k)$. Moreover, part of the liquid will fall down with mole rate of $d$ and the vapor will go up with mole rate of $V$. Let $D$ be the mole rate of the product, which is withdrawn from the condenser. Consider now the system starting from the $k$th plate (above the point where new liquid enters into the apparatus) and the condenser. We can write the balance equation

$$Vv(k+1) = du(k) + Dz, \tag{3.3.16}$$

where $z$ is the mole fraction of the liquid withdrawn from the condenser. To this equation we need to add the definition of relative volatility that will hold for the equilibrium of the two phases at each plate

$$\alpha = \frac{u(k)(1-u(k))}{u(k)(1-v(k))},$$

which gives

$$v(k) = \frac{\alpha u(k)}{1+(\alpha-1)u(k)}. \tag{3.3.17}$$

From (3.3.16) and (3.3.17), we get

$$u(k)u(k-1) + pu(k) + qu(k-1) + r = 0, \tag{3.3.18}$$

where

$$p = \frac{1}{\alpha-1}, \quad q = \frac{Dz(\alpha-1) - V\alpha}{d(\alpha-1)}, \quad r = \frac{Dz}{d(\alpha-1)}.$$

Equation (3.3.18) is a Riccati equation (3.3.1) with constant coefficients.

**Example 3.3.3.** An object is placed between two concave mirrors $A$ and $B$ of focal lengths $f_1$ and $f_2$ respectively. The mirrors are a distance $2a$

apart. The object is placed at $C$, a distance $d$ from $A$. The object $C$ forms an image in $A$ at distance $v(1)$ from $A$. This image then becomes the object for mirror $B$ and forms an image in it which is in its turn the next object for $A$. This image is now a distance $u(2)$ from $A$. Thus, we have

$$\frac{1}{d} + \frac{1}{v(1)} = \frac{1}{f_1} \quad \text{and} \quad \frac{1}{2a - v(1)} + \frac{1}{2a - u(2)} = \frac{1}{f_2}.$$

Generally,

$$\frac{1}{u(k)} + \frac{1}{v(k)} = \frac{1}{f_1} \quad \text{and} \quad \frac{1}{2a - v(k)} + \frac{1}{2a - u(k+1)} = \frac{1}{f_2}.$$

Eliminating $v(k)$ between these two equations, we obtain

$$u(k+1) = \frac{u(k)\left[4af_2 - f_1f_2 + 2af_1 - 4a^2\right] + 4a\left[af_1 - f_1f_2\right]}{u(k)\left[f_1 + f_2 - 2a\right] + \left[-f_1f_2 + 2af_1\right]},$$

which is exactly of the form (3.3.9).

**Example 3.3.4.** Pielou's logistic equation [25]

$$u(k+1) = \frac{\alpha u(k)}{\gamma u(k) + 1}$$

is a special case of (3.3.9).

## 3.4. Bernoulli's Equation

Bernoulli's differential equation

$$y' + p(t)y = q(t)y^{\alpha}, \quad \alpha \neq 0, 1 \tag{3.4.1}$$

is one of the few nonlinear differential equations which can be solved explicitly. The obvious discretizations of (3.4.1) are

$$u(k+1) - u(k) + hp(k)u(k) = hq(k)u^{\alpha}(k) \tag{3.4.2}$$

and

$$u(k+1) - u(k-1) + 2hp(k)u(k) = 2hq(k)u^{\alpha}(k), \tag{3.4.3}$$

where $u(k)$ approximates the true solution $y(t)$ at the discrete points $t_k = kh,\ k \in \mathbb{N},\ h > 0$ is the step size and $p(k) = p(t_k),\ q(k) = q(t_k)$. However, besides other difficulties (see next section), none of the above discrete equations can be solved in the closed form. Thus, to find a discrete analog of (3.4.1) which can be solved explicitly, we rewrite this differential equation as

$$y^{-\alpha}(t)y'(t) + p(t)y^{1-\alpha}(t) \;=\; q(t). \tag{3.4.4}$$

Since by the mean value theorem of differential calculus

$$y^{1-\alpha}(t_{k+1}) - y^{1-\alpha}(t_k) \;=\; h(1-\alpha)y^{-\alpha}(p)y'(p),$$

where $t_k < p < t_{k+1}$, we find that (3.4.4) can be approximated by

$$\frac{u^{1-\alpha}(k+1) - u^{1-\alpha}(k)}{h(1-\alpha)} + p(k)u^{1-\alpha}(k) \;=\; q(k),$$

which for $k \in \mathbb{N}$ is the same as

$$u^{\alpha-1}(k+1) - u^{\alpha-1}(k) \;=\; h(1-\alpha)(p(k) - q(k)u^{\alpha-1}(k))u^{\alpha-1}(k+1). \tag{3.4.5}$$

This difference equation approximates the solutions of (3.4.1) and is called the *discrete Bernoulli's equation.* Since the substitution $u^{\alpha-1}(k) = v(k)$ in (3.4.5) leads to a homogeneous Riccati's equation in $v(k)$, it can be solved explicitly. Indeed, let $w(k) = 1/u^{\alpha-1}(k)$ to obtain the first order linear difference equation

$$w(k+1) - w(k) \;=\; -h(1-\alpha)(p(k)w(k) - q(k)),$$

which has the closed form solution

$$w(k) \;=\; w(0)\prod_{\ell=0}^{k-1}(1-h(1-\alpha)p(\ell)) + \sum_{\ell=1}^{k}\left[\prod_{\tau=\ell}^{k-1}(1-h(1-\alpha)p(\tau))\right]h(1-\alpha)q(\ell)$$

and hence the solution of (3.4.5) appears as

$$u(k) = \left[u^{1-\alpha}(0)\prod_{\ell=0}^{k-1}(1-h(1-\alpha)p(\ell)) + h(1-\alpha)\sum_{\ell=1}^{k}q(\ell)\prod_{\tau=\ell}^{k-1}(1-h(1-\alpha)p(\tau))\right]^{1/(1-\alpha)} \tag{3.4.6}$$

## 3.5. Verhulst's Equation

The Verhulst differential equation

$$y' \;=\; \beta y - \gamma y^2 \quad (\beta,\ \gamma > 0) \tag{3.5.1}$$

is a particular case of Bernoulli's differential equation (3.4.1). It is used to model a singular population ecological system with a growth term $\beta y$ modified by an inhibiting term $-\gamma y^2$, e.g. Jones and Sleeman [13]. Often (3.5.1) is also called as ***differential equation of logistics*** and its solution known as ***logistic law of growth*** can be written as

$$y(t) = \frac{\beta y(0)}{\gamma y(0) + [\beta - \gamma y(0)]e^{-\beta t}}. \tag{3.5.2}$$

For the differential equation (3.5.1) the discrete approximations (3.4.2) and (3.4.3) reduce to

$$u(k+1) - u(k) = hu(k)(\beta - \gamma u(k)) \tag{3.5.3}$$

and

$$u(k+1) - u(k-1) = 2hu(k)(\beta - \gamma u(k)) \tag{3.5.4}$$

respectively, which cannot be solved explicitly. Besides this difficulty, these difference equations as well as several other nonlinear difference equation approximations of (3.5.1) produce solutions which are qualitatively quite different from the true solutions. This type of solutions in the numerical integration of differential equations have been given several names like ***phantom, ghost*** and ***spurious solutions***. The discrete approximation (3.4.5) for the equation (3.5.1) reduces to

$$u(k+1) - u(k) = h(\beta - \gamma u(k))u(k+1) \tag{3.5.5}$$

for which the solution (3.4.6) becomes

$$u(k) = \frac{\beta u(0)}{\gamma u(0) + [\beta - \gamma u(0)](1 - h\beta)^k}. \tag{3.5.6}$$

For small values of $h$ this solution tends to the solution of the differential equation (3.5.1), i.e. (3.5.2). Further, for $h < 1/\beta$ this solution exhibits the correct qualitative features, namely

$$\begin{array}{lll} 0 < u(0) < \beta/\gamma, & u(k) \to \beta/\gamma & \text{without oscillation from below,} \\ u(0) = \beta/\gamma, & u(k) = \beta/\gamma & \text{for all } k, \\ u(0) > \beta/\gamma, & u(k) \to \beta/\gamma & \text{without oscillation from above.} \end{array}$$

For $h \geq 1/\beta$, phantom solutions arise. In fact for $h = 1/\beta$, $u(k) = \beta/\gamma$ for all $k \in \mathbb{N}$ regardless of the value of $u(0)$, and for $h > 1/\beta$ oscillations occur.

However, phantom solutions can be easily eliminated by considering the following nonlinear difference equation approximation of (3.5.1)

$$(3.5.7) \qquad u(k+1)-u(k) = \frac{(1-e^{-h\beta})}{\beta}(\beta-\gamma u(k))u(k+1),$$

which can also be solved to obtain the closed form solution

$$(3.5.8) \qquad u(k) = \frac{\beta u(0)}{\gamma u(0)+[\beta-\gamma u(0)]e^{-kh\beta}}.$$

Obviously, for all $h>0$ this solution $u(k)=y(t_k)$, and hence (3.5.7) is an approximation of (3.5.1) which has no spurious solutions. In conclusion, although in general the qualitative behavior of the solutions of the discrete approximations is different from the true solutions of differential equations, often it is possible to choose discretizations which reveal true behavior.

## 3.6. Best Discrete Approximations: Harmonic Oscillator Equation

For a given differential equation a difference equation (preferably of the same order) approximation is called *best* if the solution of the difference equation is the same as of the differential equation at the discrete points. Of course, best approximation is not unique, e.g. (3.5.7) as well as (3.19.9) both have the same solution (3.5.8) and $u(k)=y(kh)$.

From the elementary theory of differential equations the general solution of the equation

$$(3.6.1) \qquad (D-\lambda_1)\cdots(D-\lambda_n)y(t) = f(t), \quad D=\frac{d}{dt}$$

where $\lambda_1,\cdots,\lambda_n$ are distinct complex constants can be written as

$$(3.6.2) \qquad y(t) = \sum_{i=1}^{n} c_i e^{\lambda_i t}+w(t),$$

where $c_1,\cdots,c_n$ are arbitrary constants, and $w(t)$ is any particular solution of (3.6.1). We shall find a difference equation

$$(3.6.3) \qquad (E-\mu_1)\cdots(E-\mu_n)u(k) = v(k),$$

where $\mu_1,\cdots,\mu_n$ are complex constants, whose solution $u(k)=y(t_k)$, $t_k=kh$, $k\in\mathbb{N}$, $h>0$. For this, let $w(t_k)=w(k)$. Since

$$Ee^{\lambda_i kh} = e^{\lambda_i kh}e^{\lambda_i h}$$

from (3.6.2) it follows that

$$\left(E-e^{\lambda_1 h}\right)\cdots\left(E-e^{\lambda_n h}\right)u(k) \;=\; \left(E-e^{\lambda_1 h}\right)\cdots\left(E-e^{\lambda_n h}\right)w(k),$$

i.e. in (3.6.3) the constants $\mu_i$ are $e^{\lambda_i h}$, $1 \le i \le n$ and the function $v(k)$ is $\left(E-e^{\lambda_1 h}\right)\cdots\left(E-e^{\lambda_n h}\right)w(k)$.

**Example 3.6.1.** Consider the simple harmonic oscillator equation

$$y'' + \omega^2 y \;=\; \sin\gamma t, \tag{3.6.4}$$

whose general solution can be written as

$$y(t) \;=\; c_1\cos\omega t + c_2\sin\omega t + \begin{cases} \dfrac{1}{\omega^2-\gamma^2}\sin\gamma t & \text{if } \gamma\neq\omega \\ -\dfrac{t}{2\omega}\cos\omega t & \text{if } \gamma=\omega. \end{cases}$$

Therefore, the best difference equation approximation of (3.6.4) is

$$\begin{aligned} &\left(E-e^{i\omega h}\right)\left(E-e^{-i\omega h}\right)u(k) \\ &\quad=\; \left(E-e^{i\omega h}\right)\left(E-e^{-i\omega h}\right)\begin{cases} \dfrac{1}{\omega^2-\gamma^2}\sin\gamma kh & \text{if } \gamma\neq\omega \\ -\dfrac{kh}{2\omega}\cos\omega kh & \text{if } \gamma=\omega, \end{cases} \end{aligned}$$

which is the same as

$$\begin{aligned} &\left(E^2-2\cos\omega hE+1\right)u(k) \\ &\quad=\; \left(E^2-2\cos\omega hE+1\right)\begin{cases} \dfrac{1}{\omega^2-\gamma^2}\sin\gamma kh & \text{if } \gamma\neq\omega \\ -\dfrac{kh}{2\omega}\cos\omega kh & \text{if } \gamma=\omega \end{cases} \end{aligned}$$

or,

$$\begin{aligned} &\frac{\Delta^2 u(k)}{4\omega^{-2}\sin^2\frac{1}{2}\omega h} + \omega^2 u(k+1) \\ &\quad=\; \begin{cases} \dfrac{\omega^2}{\omega^2-\gamma^2}\left[1-\dfrac{\sin^2\frac{1}{2}\gamma h}{\sin^2\frac{1}{2}\omega h}\right]\sin\gamma(k+1)h & \text{if } \gamma\neq\omega \\ \dfrac{\omega h}{2\tan\frac{1}{2}\omega h}\sin\omega(k+1)h & \text{if } \gamma=\omega. \end{cases} \end{aligned} \tag{3.6.5}$$

In particular, if $\gamma=0,\ \omega=1,\ y(0)=0$ then the solution of (3.6.4) is $y(t)=c\sin t$ whose primitive period $T=2\pi$, and the solution of (3.6.5) is $u(k) = c\sin kh$. If the primitive period $T$ is divided into $2p$ equal intervals of length $h$, so that $T = 2ph$, then it follows that $h = \pi/p$, and $u(k) = c\sin(k\pi/p)$ whose period $T = 2\pi$ is the same as for the solution $y(t) = c\sin t$. This is in contrast with the denominator $4\sin^2(h/2)$ in the resulting (3.6.5) replaced by the usual $h^2$, which has a solution with points off the solution curve $y(t) = c\sin t$ and with period $T = 4p\sin(\pi/2p)$. This period tends to the correct value $2\pi$ only in the limit of large $p$, i.e. small $h$.

## 3.7. Duffing's Equation

The classical nonlinear Duffing's equation

$$y'' + ay + by^3 = 0 \tag{3.7.1}$$

describes the undamped unforced vibration of an anharmonic oscillator, of a *hard* or *soft* spring, or of a simple pendulum. The well known [20] analytic solutions of (3.7.1) in terms of the Jacobian elliptic functions [1] $cn$, $dn$, $sn$ and the complete elliptic integral of the first kind $K$, with $m$ as the parameter, $0 \le m < 1$, on which these functions and integral depend, can be written as follows:

| case | $I$ | $II$ | $III$ |
|---|---|---|---|
| boundary conditions | $y(0) = A,$ $y'(0) = 0$ | $y(0) = A,$ $y'(0) = 0$ | $y(0) = 0, y = A,$ $y' = 0$ |
| constant constraints | $b > 0, a > -\frac{1}{2}bA^2$ | $b > 0,$ $-bA^2 < a < \frac{1}{2}bA^2$ | $b < 0, a > -bA^2$ |
| solution $y(t)$ | $Acn\left[(a + bA^2)^{\frac{1}{2}}t\right]$ | $Adn\left[A\left(\frac{b}{2}\right)^{\frac{1}{2}}t\right]$ | $Asn\left[\left(a + \frac{1}{2}bA^2\right)^{\frac{1}{2}}t\right]$ |
| parameter $m$ | $\frac{1}{2}bA^2/\left(a + bA^2\right)$ | $2\left[1 + a/(bA^2)\right]$ | $-\frac{1}{2}bA^2/\left(a + \frac{1}{2}bA^2\right)$ |
| period $T$ | $4K/\left(a + bA^2\right)^{\frac{1}{2}}$ | $2K/A\left(\frac{1}{2}b\right)^{\frac{1}{2}}$ | $4K/\left(a + \frac{1}{2}bA^2\right)^{\frac{1}{2}}$. |

For the differential equation (3.7.1) a variety of discretizations are possible, however

$$\begin{aligned} &h^{-2}(u(k+1) - 2u(k) + u(k-1)) + au(k) \\ &\qquad + \frac{1}{2}bu^2(k)(u(k+1) + u(k-1)) = 0, \quad k \in \mathbb{N} \end{aligned} \tag{3.7.2}$$

or equivalently,

$$(3.7.3)\quad \frac{1}{2}(u(k+1)+u(k-1))(2+bu^2(k)h^2)-(2-ah^2)u(k) \;=\; 0, \quad k\in\mathbb{N}$$

is particularly interesting, because it can be solved in closed form, moreover, gives periodic solutions.

To establish sufficient conditions so that the solutions of (3.7.3) are periodic, we approximate the boundary conditions for the cases I and II as

$$(3.7.4)\qquad u(0) \;=\; A \quad \text{and} \quad u(-1) \;=\; u(1),$$

whereas for the case III as

$$(3.7.5)\quad u(0) \;=\; 0 \quad \text{and} \quad u(p) \;=\; A \quad \text{when} \quad u(p-1) \;=\; u(p+1),$$

where $p$ is a positive integer.

**Theorem 3.7.1.** If for the given constants $a, b$ and $A$ and any positive integer $p$, the step size $h > 0$ can be chosen so that $u(p+1) = u(p-1)$, then the solution $u(k)$ of (3.7.3) satisfying (3.7.4) is periodic of period $T = 2ph$.

**Proof.** Equation (3.7.3) for $k = p-1$ and $k = p+1$ gives

$$\frac{1}{2}(u(p)+u(p-2))(2+bu^2(p-1)h^2)-(2-ah^2)u(p-1) \;=\; 0$$

and

$$\frac{1}{2}(u(p+2)+u(p))(2+bu^2(p+1)h^2)-(2-ah^2)u(p+1) \;=\; 0;$$

the condition $u(p+1) = u(p-1)$, therefore forces $u(p+2) = u(p-2)$. Similarly, now (3.7.3) for $k = p-2$ and $k = p+2$ forces $u(p+3) = u(p-3)$, and so on, until $k = 1$ and $k = 2p-1$ forces $u(2p) = u(0) = A$. Finally, $k = 0$ and $k = 2p$ forces $u(2p+1) = u(-1) = u(1)$.

With $u(2p) = u(0)$ and $u(2p+1) = u(1)$, (3.7.3) gives $u(2p+k) = u(k)$, completing the proof of the theorem. ∎

**Theorem 3.7.2.** If for the given constants $a, b$ and $A$ and any positive integer $p$, the step size $h > 0$ can be chosen so that (3.7.5) is satisfied, then the solution $u(k)$ of (3.7.3) is an odd function which is periodic with period $T = 4ph$.

**Proof.** The proof is similar to that of Theorem 3.7.1. ∎

To solve the difference equation (3.7.3) for the case **I**, we note that the function $cn\left[(a+bA^2)^{1/2}t\right]$ is periodic of period $4K/(a+bA^2)^{1/2}$, and hence $cn(t)$ is periodic of period $4K$, and $cn(2kK/p)$, $k \in \mathbb{N}$ is periodic of period $2p$. For an integer $p > 2$, let

$$u(k) = Acn(2kK/p) \tag{3.7.6}$$

be a trial solution of (3.7.3). Since $cn(0) = 1$, $cn(t) = cn(-t)$, and $u(p+1) = Acn(2(p+1)K/p) = Acn(2K+2K/p) = Acn(2K/p+2K-4K) = Acn(2K/p-2K) = Acn(2K-2K/p)Acn(2(p-1)K/p) = u(p-1)$, this trial solution correctly satisfies the boundary conditions (3.7.4) as well as $u(p+1) = u(p-1)$. Of course, this solution is valid if for the given constraints $b > 0$ and $a > -(1/2)bA^2$, the associated parameter $m$ lies in the allowed interval $(0,1)$ and $h$ is real.

Now since $u(0) = A$ and $u(-1) = u(1)$, equation (3.7.3) gives

$$u(1) = \frac{2-ah^2}{2+bA^2h^2}A. \tag{3.7.7}$$

Thus, the parameter $m$ (and hence the step size $h$ and the period $T$) is determined by first substituting (3.7.6) into (3.7.7), giving

$$cn(2K/p) = \frac{2-ah^2}{2+bA^2h^2} \tag{3.7.8}$$

and then (3.7.6) into (3.7.3), giving

$$\frac{Acn(2kK/p)cn(2K/p)}{1-msn^2(2kK/p)sn^2(2K/p)}\left[2+bA^2h^2cn^2(2kK/p)\right] = (2-ah^2)Acn(2kK/p), \tag{3.7.9}$$

where we have used the formula

$$cn(u+v)+cn(u-v) = \frac{2cnu\ cnv}{1-msn^2u\ sn^2v}.$$

Using (3.7.8) into (3.7.9) simplifies to

$$\left[(2+bA^2h^2)m\ sn^2(2K/p) - bA^2h^2\right]sn^2(2kK/p) = 0,$$

which is satisfied for all $k$ provided

$$msn^2(2K/p) = bA^2h^2/(2+bA^2h^2). \tag{3.7.10}$$

Now eliminating $h^2$ from (3.7.8) and (3.7.10) gives

$$m = \frac{bA^2}{a+bA^2}[1+cn(2K/p)]^{-1}. \tag{3.7.11}$$

This is a transcendental equation for $m$ (noting that $K$ depends on $m$). The constraints $b > 0$ and $a > -(1/2)bA^2$ ensure that the parameter $m$, depending on the ratio $bA^2/(a+bA^2)$ as well as on $p$ (assumed $> 2$), falls in the required interval $0 < m < 1$.

The step size $h$ can now be calculated from (3.7.8)

$$h^2 = \frac{2[1 - cn(2K/p)]}{a + bA^2 cn(2K/p)}, \tag{3.7.12}$$

giving real $h$ for $p > 2$. Alternatively, (3.7.11) gives

$$h^2 = 2\frac{2m(a + bA^2) - bA^2}{m(a^2 - b^2A^4) + b^2A^4}. \tag{3.7.13}$$

The period is finally deduced from $T = 2ph$, completing the details of a valid periodic solution. Thus, if the step size $h$ is determined from (3.7.13) using the solution $m(< 1)$ of (3.7.11), then the solution of (3.7.3) gives, apart from round off errors, an exactly periodic solution.

It is easy to verify that the solution of (3.7.1) is obtained in the limit as $p \to \infty$, for then $K/p \to 0$, $h \to 0$ and $m \to \frac{1}{2}bA^2/(a + bA^2)$. Since $cnu = 1 - (1/2)u^2 + O(u^4)$, equation (3.7.12) gives as $p \to \infty$, $T = 2ph \to 4K/(a+bA^2)^{1/2}$. Finally, $u(k) = Acn(2kK/p) = Acn(2kKh/ph) \to Acn[(a + bA^2)^{1/2}t] = y(t)$.

To solve the difference equation (3.7.3) for the case II, let

$$u(k) = Adn(kK/p), \quad p > 2 \text{ an integer} \tag{3.7.14}$$

be a trial solution. This trial solution correctly satisfies the boundary conditions (3.7.4) as well as $u(p + 1) = u(p - 1)$. Of course, this solution is valid if for the given constraints $b > 0$ and $-bA^2 < a < -(1/2)bA^2$, the associated parameter $m$ lies in the allowed interval $(0, 1)$ and $h$ is real. For this, following exactly as in case I, equations corresponding to (3.7.8), (3.7.11) (3.7.13) for (3.7.14) are obtained and these equations appear as

$$dn(K/p) = \frac{2 - ah^2}{2 + bA^2h^2} \tag{3.7.15}$$

$$m = \frac{a + bA^2}{bA^2}[1 + dn(K/p)] \tag{3.7.16}$$

$$h^2 = \frac{2[1 - dn(K/p)]}{a + bA^2 dn(K/p)} \tag{3.7.17}$$

(3.7.18) $$h^2 = 2\frac{2(a+bA^2)-mbA^2}{(a^2-b^2A^4)+mb^2A^4},$$

where use has been made of the formula

$$dn(u+v)+dn(u-v) = \frac{2dnu\ dnv}{1-msn^2usn^2v}.$$

For $b>0$ and $-bA^2 < a < -(1/2)bA^2$ and for $p>2$, the parameter $m$ in (3.7.16) falls in the required interval $0<m<1$ so that the solution is valid. Further, for $p>2$, $h$ in (3.7.17) or in (3.7.18) is real. The period is finally deduced from $T=2ph$.

Again it is easy to verify that the solution of (3.7.1) is obtained in the limit as $p\to\infty$, for then $K/p\to 0$, $h\to 0$ and $m\to 2[1+a/(bA^2)]$. Since $dnu = 1-(1/2)mu^2+O(u^4)$, equation (3.7.17) gives as $p\to\infty$, $T = 2ph \to 2K/A(b/2)^{1/2}$. Finally, $u(k) = Adn(kK/p) = Adn(khK/ph) \to Adn[a(b/2)^{1/2}t] = y(t)$.

For the case III we try for any positive integer $p$

(3.7.19) $$u(k) = Asn(kK/p),$$

which satisfies (3.7.5). For $k=p$, (3.7.3) gives

(3.7.20) $$u(p-1) = \frac{2-ah^2}{2+bA^2h^2}A$$

and (3.7.19) with $k=p-1$ gives

(3.7.21) $$u(p-1) = Asn[K-(K/p)] = A\frac{cn(K/p)}{dn(K/p)}.$$

Now using the formula

$$sn(u+v)+sn(u-v) = 2\frac{snu\ cnv\ dnv}{1-msn^2usn^2v}$$

and substituting (3.7.19) in (3.7.3) gives

$$\frac{Asn(kK/p)cn(K/p)dn(K/p)}{1-msn^2(kK/p)sn^2(K/p)}[2+bA^2h^2sn^2(kK/p)] = (2-ah^2)Asn(kK/p),$$

which with (3.7.20) and (3.7.21) simplifies to

$$\left[2msn^2(K/p)+bA^2h^2\right]cn^2(kK/p) = 0.$$

The above equation is satisfied for all $k$ provided

$$msn^2(K/p) = -\frac{1}{2}bA^2h^2. \tag{3.7.22}$$

Combining (3.7.20) (3.7.22), we obtain

$$m = -\frac{bA^2}{a}\frac{1-cn(K/p)dn(K/p)}{1-cn^2(K/p)}. \tag{3.7.23}$$

With the constraints $b < 0$ and $a > -bA^2$, we have $0 < m < 1$ as required for a valid solution.

The step size $h$ can be calculated from (3.7.22)

$$h^2 = -\frac{2msn^2(K/p)}{bA^2} \tag{3.7.24}$$

or using (3.7.20) and (3.7.21),

$$h^2 = 2\frac{m(2a+bA^2)+bA^2}{ma^2-b^2A^4}. \tag{3.7.25}$$

The value of $h$ is real and the period is $T = 4ph$.

In the limit as $p \to \infty$, $K/p \to 0$, $h \to 0$ and $m \to -bA^2/(2a + bA^2)$, $T = 4ph \to 4K(a+(b/2)A^2)^{1/2}$, and $u(k) \to Asn[(a+(b/2)A^2)^{1/2}t]$ $= y(t)$.

For the linear problem $b = 0$ and $a > 0$, Duffing's equation (3.7.1) and its approximation (3.7.2) simplifies to

$$y'' + ay = 0 \tag{3.7.26}$$

and

$$h^{-2}(u(k+1) - 2u(k) + u(k-1)) + au(k) = 0 \tag{3.7.27}$$

respectively. For $b = 0$ it follows for the cases I and III that $m = 0$, $K = \pi/2$, $cn = \cos$ and $sn = \sin$. Case II does not arise. Thus, for case I, (3.7.6), (3.7.12) and the time period $T$ reduce to $u(k) = A\cos(k\pi/p)$, $h = 2a^{-1/2}\sin(\pi/2p)$ and $T = 4pa^{-1/2}\sin(\pi/2p)$; whereas for the case III (3.7.19), (3.7.20) and (3.7.21) give $u(k) = A\sin(k\pi/2p)$, $h = (2/a)^{1/2}[1 - \cos(\pi/2p)]^{1/2}$, and finally the time period $T = 4p(2/a)^{1/2}[1-\cos(\pi/2p)]^{1/2}$.

## 3.8. van der Pol's Equation

The following properties of the van der Pol differential equation

$$y'' - \lambda(1-y^2)y' + y = 0, \quad \lambda > 0 \tag{3.8.1}$$

have been studied extensively [14,20]:

(1) When $\lambda = 0$, all solutions are periodic with period $T = 2\pi$ and the trajectories in the $(y, y')$ phase plane are circles of arbitrary radius.

(2) When $\lambda > 0$, periodic solutions exist.

(3) In the phase plane, the limit cycles approach the circle of radius 2 as $\lambda \to 0$.

(4) The period $T$ for periodic solutions increases with $\lambda$, perturbation theory giving the result

$$T = 2\pi\left[1 + \frac{\lambda^2}{16} + \cdots\right]. \tag{3.8.2}$$

For the differential equation (3.8.1) we shall study the discretization

$$\frac{u(k+1) - 2u(k) + u(k-1)}{4\sin^2(h/2)} - \lambda\left\{1 - u(k)\frac{u(k+1)+u(k-1)}{2\cos h}\right\} \times \left[\frac{u(k+1) - u(k-1)}{2\sin h}\right] + u(k) = 0, \quad k \in \mathbb{N}(1) \tag{3.8.3}$$

and show that:

(0) For small $h$, the difference equation (3.8.3) approximates the differential equation (3.8.1).

(I) When $\lambda = 0$, all solutions are periodic with period $2\pi$ and the points in the phase plane $(u(k),\ (u(k+1) - u(k-1))/2\sin h)$ lie on circles of arbitrary radius.

(II) When $\lambda > 0$ periodic solutions exist.

(III) In the phase plane, the points corresponding to periodic solutions in the limit $\lambda \to 0$ lie on the circle of radius 2.

(IV) The period $T$ for periodic solutions increases with $\lambda$, with $T$ as in (3.8.2) as $h \to \infty$.

The property (0) is obvious from the expansions

$$\frac{u(k+1) - 2u(k) + u(k-1)}{4\sin^2(h/2)} = y'' + O(h^2)$$

$$u(k)\frac{u(k+1)+u(k-1)}{2\cos h} = y^2 + O(h^2)$$

$$\frac{u(k+1) - u(k-1)}{2\sin h} = y' + O(h^2).$$

To establish the property (I), (3.8.3) for $\lambda = 0$ reduces to

$$(3.8.4) \qquad \frac{v(k+1) - 2v(k) + v(k-1)}{4\sin^2(h/2)} + v(k) = 0, \qquad k \in \mathbb{N}(1)$$

where $u(k)$ has been replaced by $v(k)$. The solution of (3.8.4) is

$$v(k) = c_1 \cos kh + c_2 \sin kh$$

and since

$$\frac{v(k+1) - v(k-1)}{2\sin h} = -c_1 \sin kh + c_2 \cos kh$$

the points in the phase plane $(v(k),\ (v(k+1) - v(k-1))/2\sin h)$ lie on circles of radius $(c_1^2 + c_2^2)^{1/2}$ fixed by the initial conditions.

For further analysis of (3.8.3) we rewrite it to as

$$(3.8.5) \qquad u(k+1)[c - \mu\{c - u(k)(u(k+1) + u(k-1))\}] - c^2 u(k)$$
$$+ u(k-1)[c + \mu\{c - u(k)(u(k+1) + u(k-1))\}] = 0,$$

where

$$(3.8.6) \qquad \mu = \lambda \tan(h/2) \quad \text{and} \quad c = 2\cos h.$$

To force periodic solutions, $h$ will not be arbitrary but will depend on the parameter $\lambda$, and to emphasize this we shall sometimes write $h = h(\lambda)$. Further, when $\lambda = 0$ (and hence $\mu = 0$) $c$ will be replaced by $d(= 2\cos h(0))$, and as above $u(k)$ by $v(k)$ which satisfies the difference equation

$$(3.8.7) \qquad v(k+1) - dv(k) + v(k-1) = 0.$$

To establish property (II) we shall prove the following:

**Theorem 3.8.1.** If for given $\lambda > 0$ and any integer $p \geq 3$ the step size $h$ is chosen so that

$$(3.8.8) \quad u(0) = 0,\ u(k) > 0,\ 1 \leq k \leq p-1,\ u(p) = 0,\ u(p+1) = -u(1)$$

then the solution of the difference equation (3.8.3) is periodic with primitive period $T = 2ph$.

**Proof.** Equation (3.8.5) for $k = p+1$ and $k = 1$ on using (3.8.8) gives

$$u(p+2)\{c - \mu(c + u(1)u(p+2))\} + c^2 u(1) = 0$$

and

$$u(2)\{c-\mu(c-u(1)u(2))\}-c^2u(1) \;=\; 0.$$

Comparing the above equations, we obtain $u(p+2)=-u(2)$.

Now a similar procedure for $k=p+2$ and $k=2$ gives $u(p+3)=-u(3)$. Continuing in this way until for $k=2p-1$ and $k=p-1$ we deduce $u(2p)=-u(p)=0$. Continuing still further gives $u(2p+k)=u(k)$ for all $k\in\mathbb{N}$, establishing periodicity with period $T=2ph$. ∎

As we shall subsequently be concerned only with periodic solutions we henceforth assume that $u(0)=u(p)=0$ and $u(p+1)=-u(1)$.

For the case $p=3$ with $T=6h$ the periodic solutions of the difference equation (3.8.3) can be obtained in closed form. Indeed with $u(0)=u(3)=0,\ u(4)=-u(1)$, (3.8.5) for $k=1,2$ and $3$ gives

$$\begin{aligned} u(2)[c-\mu(c-u(1)u(2))]-c^2u(1) &= 0\\ -c^2u(2)+u(1)[c+\mu(c-u(1)u(2))] &= 0\\ -u(1)c(1-\mu)+u(2)c(1+\mu) &= 0, \end{aligned}$$

which can be solved, to obtain

$$u^2(1) \;=\; \frac{(3+\mu^2)(1+\mu)^2}{(1+\mu^2)^2},\quad u^2(2) \;=\; \frac{(3+\mu^2)(1-\mu)^2}{(1+\mu^2)^2},\quad c=\frac{1-\mu^2}{1+\mu^2}.$$

Further, from (3.8.6) we find

$$\cos h(\lambda) \;=\; \frac{-(1+\lambda^2)+(9-14\lambda^2+9\lambda^4)^{1/2}}{4(1-\lambda^2)}, \tag{3.8.9}$$

which enables $h$ to be determined, so that exactly periodic solutions with period $T=6h$ are obtained. In particular, for $\lambda=0.5$, equation (3.8.9) determines $h\simeq 1.15$ and consequently $T\simeq 6.93$.

For the case $p=4$ with $T=8h$ the periodic solutions of the difference equation (3.8.3) can again be obtained in closed form. With $u(0)=u(4)=0,\ u(5)=-u(1)$, (3.8.5) and (3.8.6) give

$$u^2(1) \;=\; \frac{2(1+\mu^2)}{(1-\mu)^2},\quad u^2(2) \;=\; \frac{4(1-4\mu^2-\mu^4)}{(1-\mu^2)^2},\quad u^2(3) \;=\; \frac{2(1+\mu^2)}{(1+\mu)^2}$$

and

$$\cos^2 h \;=\; \frac{(1+\mu^2)(1-4\mu^2-\mu^4)}{2(1-\mu^2)^2},\qquad \mu \;=\; \lambda\tan(h/2). \tag{3.8.10}$$

In particular, for $\lambda = 0.5$, equation (3.8.10) determines $h \simeq 0.819$ and hence $T \simeq 6.55$.

In Figure 3.8.1 we illustrate the results for $p = 3$ and 4, the discrete points for $p = 3$ and 4 are marked as $\times$ and o respectively. For the comparison purpose, the solution curve for the differential equation (3.8.1) is also included. Figure 3.8.2 represents the points and the trajectory in the phase plane. In the limiting case $\lambda \to 0$ (recall that we replace $u(k)$ by $v(k)$ and $c$ by $d$) for $p = 3$, we find $h = \pi/3$, $T = 2\pi$, $v(1) = v(2) = \sqrt{3}$, whereas for $p = 4$, $h = \pi/4$, $T = 2\pi$, $v(1) = v(3) = \sqrt{2}$, $v(2) = 2$. As shown in Figure 3.8.3 these points lie exactly on the solution curve $y(t) = 2\sin t$ for the differential equation and the points in the phase plane (Figure 3.8.4) lie on the circle of radius 2.

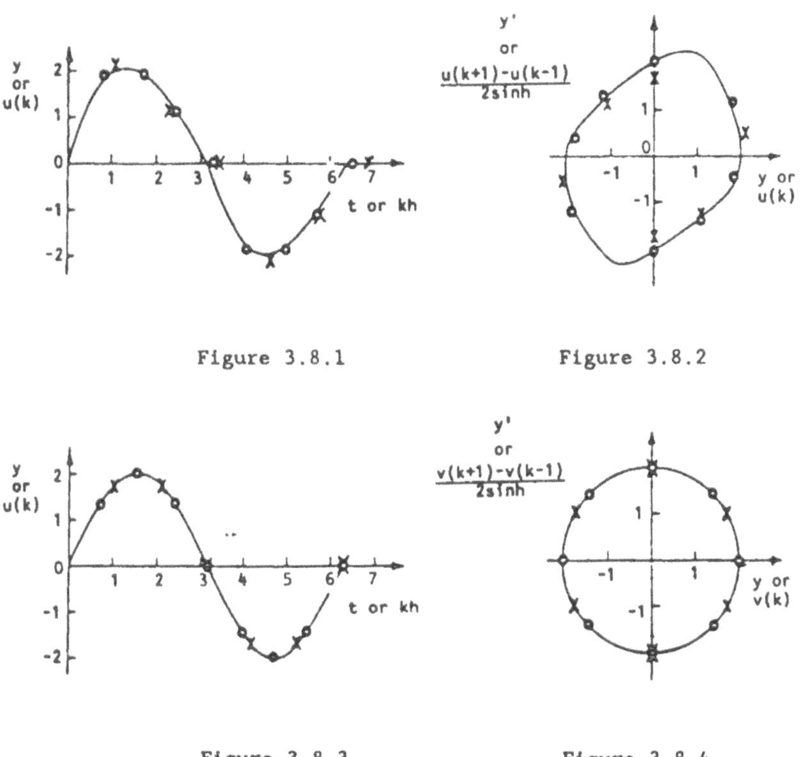


Figure 3.8.1

Figure 3.8.2

Figure 3.8.3

Figure 3.8.4

Now we shall prove the property (III). For this, above we have shown that for $p = 3$ and 4 the periodic solutions of (3.8.3) for $\lambda \to 0$

give discrete points on the solution curve $y(t) = 2\sin t$ of the differential equation and the points in the phase plane lie on the circle of radius 2. To establish this in general, since $T = 2ph$ the solution of (3.8.4) is $v(k) = A\sin(k\pi/p)$, thus, what we need to show is that $A = 2$.

We use (3.8.5) to express $u(k)$ in terms of $u(k+1)$ and $u(k-1)$ as

$$u(k) = \frac{c\{(1+\mu)u(k-1) + (1-\mu)u(k+1)\}}{c^2 - \mu(u^2(k+1) - u^2(k-1))} \tag{3.8.11}$$

and similarly,

$$u(p-k) = \frac{c\{(1+\mu)u(p-k-1) + (1-\mu)u(p-k+1)\}}{c^2 - \mu(u^2(p-k+1) - u^2(p-k-1))} \tag{3.8.12}$$

or

$$u(p-k) = \frac{c\{(1-\mu)u(p-k+1) + (1+\mu)u(p-k+1)\}}{c^2 + \mu(u^2(p-k-1) - u^2(p-k+1))} \tag{3.8.13}$$

so that

$$u_\mu(p-k) = u_{-\mu}(k) \tag{3.8.14}$$

and for $k = p$, (3.8.11) gives

$$u(1) - u(p-1) = \mu(u(1) + u(p-1)). \tag{3.8.15}$$

We shall analyze the difference equation for small $\mu$ (and hence by (3.8.6) small $\lambda$) and consider the limiting case $\mu \to 0$. The cases $p$ even and $p$ odd require slightly different treatment and it is convenient to define $q = [(p-1)/2]$.

From (3.8.11) and (3.8.13) we find the matrix equation

$$\begin{aligned} C^1\left[\mu^{-1}(u(k) - u(p-k))\right] &= c^{-2}[\{(1+\mu)u(k-1) + (1-\mu)u(k+1)\}(u^2(p-k-1) \\ &\quad - u^2(p-k+1))] + c^{-2}[\{(1-\mu)u(p-k+1) \\ &\quad + (1+\mu)u(p-k-1)\}(u^2(k+1) - u^2(k-1))] \\ &\quad - [u(k+1) + u(p-k-1) - (u(k-1) + u(p-k+1))], \end{aligned} \tag{3.8.16}$$

where, for example, $[\mu^{-1}(u(k) - u(p-k))]$ signifies the $q \times 1$ column with $k$th element $\mu^{-1}(u(k) - u(p-k))$ and $C^1$ is the $q \times q$ symmetric tridiagonal matrix

$$C^1 = \begin{bmatrix} c_1^1 & -1 & 0 & 0 & \cdots & 0 & 0 \\ -1 & c_2^1 & -1 & 0 & \cdots & 0 & 0 \\ 0 & -1 & c_3^1 & -1 & \cdots & 0 & 0 \\ \cdots & \cdots & \cdots & \cdots & \cdots & \cdots & \cdots \\ 0 & 0 & 0 & 0 & \cdots & -1 & c_q^1 + \delta \end{bmatrix}$$

with $c_k^1 = c^{-3}\{c^2 - \mu(u^2(k+1) - u^2(k-1))\}\{c^2 + \mu(u^2(p-k-1) - u^2(p-k+1))\}$, $k = 1, 2, \cdots q$ and $\delta = 0$ if $p$ is even and $1$ if $p$ is odd.

If we now let $u(k) = A\sin(k\pi/p)[1+\alpha_k\mu+O(\mu^2)]$, and hence $u(p-k) = A\sin(k\pi/p)[1-\alpha_k\mu+O(\mu^2)]$, then $\mu^{-1}(u(k)-u(p-k)) = 2A\alpha_k \sin(k\pi/p)+O(\mu)$. From (3.8.15) we can deduce $\alpha_1 = 1$ but the remaining $\alpha_k$ are unknown. In the limit $\mu \to 0$, $u(k) \to v(k) = A\sin(k\pi/p)$, $u(p-k) \to v(p-k) = v(k)$, $c_k^1 \to d = 2\cos(\pi/p)$, and the matrix equation (3.8.16) has the limiting form

(3.8.17) $$\mathcal{D}[2A\alpha_k \sin(k\pi/p)] = 4A\sin(\pi/p)\left[\cos(k\pi/p)\{A^2\sin^2(k\pi/p) - 1\}\right],$$

where the matrix $\mathcal{D}$ is the same as $C^1$ with $c_k^1$ replaced by $d$ for all $k = 1, 2, \cdots, q$.

The $q \times 1$ column $\mathbf{z} = [\cos(k\pi/p)]$ is the first column of $\mathcal{D}^{-1}$ and hence $\mathbf{z}^T\mathcal{D} = [1, 0, \cdots, 0]$. Thus, operating on (3.8.17) with $\mathbf{z}^T$ gives

$$2A\sin(\pi/p) = 4A\sin(\pi/p)\mathbf{z}^T[\cos(k\pi/p)\{A^2\sin^2(k\pi/p) - 1\}].$$

This leads to the equation

$$A^2\sum_{k=1}^{q}\sin^2(2k\pi/p) = 2 + 4\sum_{k=1}^{q}\cos^2(k\pi/p).$$

However, since $\sum_{k=1}^{q}\sin^2(2k\pi/p) = p/4$ and $\sum_{k=1}^{q}\cos^2(k\pi/p) = (p-2)/4$, the above equality immediately gives $A = 2$. This completes the proof of property (III). Thus, for all $p \geq 3$ the discrete points in the phase plane lie on the circle of radius 2.

It is convenient to use hereafter $v(k) = 2\sin(k\pi/p)$, and to note that the matrix equation (3.8.17) with $A = 2$ reduces to

(3.8.18) $$\mathcal{D}\,[4\alpha_k \sin(k\pi/p)] = -8\sin(\pi/p)[\cos(3k\pi/p)].$$

The remaining $\alpha_k$ can now be evaluated and appear as

(3.8.19) $$\alpha_k = \frac{\sin(2k\pi/p)}{\sin(2\pi/p)}, \quad k = 1, 2, \cdots, q.$$

To establish property (IV) for the perturbation expansion of the period $T$ in terms of $\lambda$, we need to solve the difference equation $O(\mu^3)$ by letting $u(k) = v(k)\left\{1 + \alpha_k\mu + \beta_k\mu^2 + O(\mu^3)\right\}$, $u(p-k) = v(k)\left\{1 - \alpha_k\mu + \beta_k\mu^2 + O(\mu^3)\right\}$, $c = d\left\{1 - \eta\mu^2 + O(\mu^4)\right\}$, where $v(k) = 2\sin(k\pi/p)$, $d = 2\cos(\pi/p)$, $\alpha_k$ are given by (3.8.19) and use is made of (3.8.14) and the fact that $c$ is an even function of $\mu$. The terms of order $\mu^2$ in (3.8.11) yield the matrix equation

$$\mathcal{D}_+ [\beta_k v(k)] = d\eta[v(k)] + [w(k)] + [y(k)] + [z(k)], \tag{3.8.20}$$

where $w(k) = \alpha_k v^2(k)(v(k+1) - v(k-1))$, $y(k) = 2d^{-1}v(k)\left(\alpha_{k+1}v^2(k+1) - \alpha_{k-1}v^2(k-1)\right)$, $z(k) = \alpha_{k-1}v(k-1) - \alpha_{k+1}v(k+1)$, $k = 1, 2, \cdots, q+1-\delta$ and $\mathcal{D}_+$ is the $(q+1-\delta)\times(q+1-\delta)$ matrix with the same elements as in $\mathcal{D}$ except in the last row the non zero elements are $\delta - 2$, $d - \delta$ instead of $-1$, $d+\delta$. This non symmetric matrix $\mathcal{D}_+$ is of order $(p-1)/2$ when $p$ is odd and order $p/2$ when $p$ is even. The degenerate case $p = 3$ is excluded so that the analysis proceeds on the assumption that $p \geq 4$. To determine $\eta$ without involving the unwanted $\beta_k$ we use the fact that $\mathcal{D}_+$ is a singular matrix with the left eigenvector $\mathbf{r}^T$, corresponding to its eigenvalue zero, given by $\mathbf{r} = [v(k)]$ if $p$ is odd and $\mathbf{r} = \begin{bmatrix} [v(k)] \\ 1 \end{bmatrix}$ if $p$ is even. Since $\mathbf{r}^T\mathcal{D}_+ = [0, \cdots, 0]$, operating on (3.8.20) with $\mathbf{r}^T$ gives

$$0 = d\eta\mathbf{r}^T[v(k)] + \mathbf{r}^T[w(k)] + \mathbf{r}^T[y(k)] + \mathbf{r}^T[z(k)]. \tag{3.8.21}$$

Now an elementary computation gives $d\mathbf{r}^T[v(k)] = 2p\cos(\pi/p)$, $\mathbf{r}^T[w(k)] = p\sec(\pi/p)$, $\mathbf{r}^T[y(k)] = -2p\sec(\pi/p)$ and $\mathbf{r}^T[z(k)] = (p/2)\sec(\pi/p)$. Thus, from (3.8.21) it follows that $\eta = (1/4)\sec^2(\pi/p)$, $p \geq 4$.

The perturbation expansion for the period $T$ now follows. If we let $h = (\pi/p) + \epsilon$, then $c = 2\cosh h \simeq 2\cos(\pi/p)\{1 - \epsilon\tan(\pi/p)\}$, which compared with $c = d\{1 - \eta\mu^2 + O(\mu^4)\}$ and using $\mu = \lambda\tan(h/2)$ gives

$$\epsilon\tan(\pi/p) \simeq \eta\lambda^2\tan^2(\pi/2p) \simeq \frac{\lambda^2}{4}\sec^2(\pi/p)\tan^2(\pi/2p).$$

The period $T$ is now determined from $T = 2ph = 2\pi\{1 + (\epsilon p/\pi)\}$, i.e.

$$T = 2\pi\left\{1 + \frac{p\tan^2(\pi/2p)}{4\pi\cos^2(\pi/p)\tan(\pi/p)}\lambda^2 + \cdots\right\}.$$

This perturbation expansion, validated for $p \geq 4$, gives in the limit $p \to \infty$, or $h \to 0$, (3.8.2).

## 3.9. Hill's Equation

Consider the second order linear homogeneous differential equation

$$y'' + a(t)y = 0, \tag{3.9.1}$$

where the function $a(t)$ is periodic, whose period without loss of generality is taken to be $\pi$. A corresponding difference equation can be obtained by choosing $h = \pi/p$ ($p$ a positive integer) and approximating (3.9.1) by

$$\frac{u(k+1) - 2u(k) + u(k-1)}{H^2} + a(k)u(k) = 0, \quad k \in N(1) \tag{3.9.2}$$

where $H^2 = h^2 + O(h^3)$, $a(k) = a(k+p)$. In the limit $h \to 0$, the difference equation (3.9.2) converges to the differential equation (3.9.1), but the precise form of the function $H = H(h)$ is left undecided.

**Lemma 3.9.1.** If the solution of the difference equation (3.9.2) satisfies

$$u(p) = su(0), \quad u(p+1) = su(1), \tag{3.9.3}$$

where $s$ is a nonzero constant, then for all $k \in N(1)$

$$u(k+p) = su(k). \tag{3.9.4}$$

**Proof.** The proof is similar to that of Theorem 3.7.1. ∎

**Theorem 3.9.2.** The difference equation (3.9.2) has a solution of the form

$$u(k) = \exp(i\nu kh)P_\nu(k), \tag{3.9.5}$$

where $\nu$ is a constant and $P_\nu(k)$ is periodic, i.e. $P_\nu(k+p) = P_\nu(k)$.

**Proof.** Although its proof can be deduced from Theorem 2.9.5, we shall provide an alternative proof which is intrinsic. The difference equation (3.9.2) taken together with the conditions (3.9.3) can be represented by the matrix equation

$$\mathcal{G}(s)\mathbf{u} = 0, \tag{3.9.6}$$

where $\mathcal{G}(s)$ is the $p \times p$ matrix

$$\mathcal{G}(s) = \begin{bmatrix} g(0) & -1 & 0 & \cdots & 0 & 0 & -s^{-1} \\ -1 & g(1) & -1 & \cdots & 0 & 0 & 0 \\ 0 & -1 & g(2) & \cdots & 0 & 0 & 0 \\ \cdots & \cdots & \cdots & \cdots & \cdots & \cdots & \cdots \\ 0 & 0 & 0 & \cdots & g(p-3) & -1 & 0 \\ 0 & 0 & 0 & \cdots & -1 & g(p-2) & -1 \\ -s & 0 & 0 & \cdots & 0 & -1 & g(p-1) \end{bmatrix} \tag{3.9.7}$$

and

$$g(k) = 2 - H^2 a(k), \quad k \in \mathbf{N}(0, p-1), \quad \mathbf{u} = (u(0), \cdots, u(p-1))^T.$$

A nontrivial solution of (3.9.6) requires that $det\, \mathcal{G}(s) = 0$. The Laplace expansion of the determinant using the first and the last rows yields $det\, \mathcal{G}(s) = \phi(\mathbf{g}) - s - s^{-1}$, where $\phi(\mathbf{g})$ is a real valued function of $g(0), \cdots, g(p-1)$ but independent of $s$. Thus, the characteristic equation equivalent to $det\, \mathcal{G}(s) = 0$ is the quadratic $s^2 - \phi(\mathbf{g})s + 1 = 0$, which has at least one root called *characteristic exponent,* and it can be written as $s = \exp(i\nu\pi)$. For this value of $s$, the corresponding solution of (3.9.6) is written as

$$u(k) = \exp(i\nu kh)P_\nu(k). \tag{3.9.8}$$

By Lemma 3.9.1, $u(k+p) = \exp(i\nu hp)u(k)$, so that $P_\nu(k+p) = \exp(-i\nu(k+p)h)u(k+p) = \exp(-i\nu kh)u(k) = P_\nu(k)$, which completes the proof of the theorem. ∎

It is clear that in the characteristic exponent $s = \exp(i\nu\pi)$ the constant $\nu$ can be replaced by $\nu + 2\ell$, where $\ell$ is an arbitrary integer, and the product of the roots of the characteristic equation $s^2 - \phi(\mathbf{g})s + 1 = 0$ is unity.

As for the differential equation (3.9.1), any solution of (3.9.2) or equivalently of (3.9.6) is defined within an arbitrary multiplicative constant. To conform with the usual convention for Mathieu functions we shall adopt the normalization $yy' = p/2$.

As a consequence of Theorem 3.9.2 the unstable and stable solutions of the difference equation (3.9.2) are separated by $\phi(\mathbf{g}) = \pm 2$, or by $det\, \mathcal{G}(\pm 1) = 0$. For $|\phi| < 2$, the imaginary part of $\nu$ is non zero and $u(k) \to \infty$ either as $k \to \infty$ or as $k \to -\infty$, the solution being unstable. On the other hand, if $|\phi| > 2$, $\nu$ is real, $u(k)$ remains finite as $k \to \pm\infty$, and the solution is stable. For $\phi \neq \pm 2$, or $det\, \mathcal{G}(\pm 1) \neq 0$, the characteristic equation has two distinct roots $s$ and $s^{-1}$ and hence yields two linearly independent solutions. But for $\phi = 2$, or $det\, \mathcal{G}(1) = 0$, only one root $s = 1$ is obtained and the corresponding solution has period $ph = \pi$ since $u(k+p) = u(k)$. And for $\phi = -2$, or $det\, \mathcal{G}(-1) = 0$, there is only one root $s = -1$ and the corresponding solution has period $2ph = 2\pi$ since $u(k+p) = -u(k)$. As in the Floquet theory [3], these two periodic solutions, one of period $\pi$ and the other of period $2\pi$ are called *basically periodic solutions.*

For Hill's differential equation the function $a(t)$ in (3.9.1) is not only periodic of period $\pi$ but is also an even function. Thus, in the corresponding difference equation (3.9.2) we take $a(k) = a(-k)$, or equivalently $a(k) = a(p-k)$. Without loss of generality and to simplify the situation, $p$ is hereafter taken to be an even integer, and the integer $r$ is defined by

$r = p/2$. Now, since $g(k) = g(k-p)$ the matrices $\mathcal{G}(\pm 1)$ corresponding to the basically periodic solutions display a symmetry which can be exploited to reduce them to direct sums. For $\mathcal{G}(1)$ and $\mathcal{G}(-1)$ we introduce the symmetric orthogonal transformations

$$\begin{aligned} &u(0) = v(0),\ u(k) = \frac{1}{\sqrt{2}}(v(k) + v(p-k)),\ k = 1, \cdots, r-1 \\ &u(r) = v(r),\ u(k) = \frac{1}{\sqrt{2}}(v(k) - v(p-k)),\ k = r+1, \cdots, p-1 \end{aligned} \tag{3.9.9}$$

and

$$\begin{aligned} &u(0) = w(0),\ u(k) = \frac{1}{\sqrt{2}}(w(k) - w(p-k)),\ k = 1, \cdots, r-1 \\ &u(r) = w(r),\ u(k) = -\frac{1}{\sqrt{2}}(w(k) + w(p-k)),\ k = r, \cdots, p-1. \end{aligned} \tag{3.9.10}$$

These transformations reduce the matrices to the following direct sums

$$\begin{aligned} \mathcal{G}(1) &\sim \mathcal{C}_2 + \mathcal{S}_2 \\ \mathcal{G}(-1) &\sim \mathcal{C}_1 + \mathcal{S}_1, \end{aligned} \tag{3.9.11}$$

where the matrices $\mathcal{C}_2$, $\mathcal{S}_2$, $\mathcal{C}_1$, $\mathcal{S}_1$ are of orders $r+1$, $r-1$, $r$ and $r$, and

$$\mathcal{C}_2 = \begin{bmatrix} g(0) & -\sqrt{2} & 0 & \cdots & 0 & 0 & 0 \\ -\sqrt{2} & g(1) & -1 & \cdots & 0 & 0 & 0 \\ 0 & -1 & g(2) & \cdots & 0 & 0 & 0 \\ \cdots & \cdots & \cdots & \cdots & \cdots & \cdots & \cdots \\ 0 & 0 & 0 & \cdots & g(r-2) & -1 & 0 \\ 0 & 0 & 0 & \cdots & -1 & g(r-1) & -\sqrt{2} \\ 0 & 0 & 0 & \cdots & 0 & -\sqrt{2} & g(r) \end{bmatrix}, \tag{3.9.12}$$

$$\mathcal{S}_2 = \begin{bmatrix} g(r-1) & -1 & 0 & \cdots & 0 & 0 & 0 \\ -1 & g(r-2) & -1 & \cdots & 0 & 0 & 0 \\ 0 & -1 & g(r-3) & \cdots & 0 & 0 & 0 \\ \cdots & \cdots & \cdots & \cdots & \cdots & \cdots & \cdots \\ 0 & 0 & 0 & \cdots & g(3) & -1 & 0 \\ 0 & 0 & 0 & \cdots & -1 & g(2) & -1 \\ 0 & 0 & 0 & \cdots & 0 & -1 & g(1) \end{bmatrix}, \tag{3.9.13}$$

$$\mathcal{C}_1 = \begin{bmatrix} g(0) & -\sqrt{2} & 0 & \cdots & 0 & 0 & 0 \\ -\sqrt{2} & g(1) & -1 & \cdots & 0 & 0 & 0 \\ 0 & -1 & g(2) & \cdots & 0 & 0 & 0 \\ \cdots & \cdots & \cdots & \cdots & \cdots & \cdots & \cdots \\ 0 & 0 & 0 & \cdots & g(r-3) & -1 & 0 \\ 0 & 0 & 0 & \cdots & -1 & g(r-2) & -1 \\ 0 & 0 & 0 & \cdots & 0 & -1 & g(r-1) \end{bmatrix}, \tag{3.9.14}$$

$$(3.9.15)\qquad \mathcal{S}_1 = \begin{bmatrix} g(r) & \sqrt{2} & 0 & \cdots & 0 & 0 & 0 \\ \sqrt{2} & g(r-1) & -1 & \cdots & 0 & 0 & 0 \\ 0 & -1 & g(r-2) & \cdots & 0 & 0 & 0 \\ \cdots & \cdots & \cdots & \cdots & \cdots & \cdots & \cdots \\ 0 & 0 & 0 & \cdots & g(3) & -1 & 0 \\ 0 & 0 & 0 & \cdots & -1 & g(2) & -1 \\ 0 & 0 & 0 & \cdots & 0 & -1 & g(1) \end{bmatrix}.$$

These four matrices are called *basic matrices.* Corresponding to these matrices there are four types of periodic solution vectors, which in anticipation of the notation used in the theory of Mathieu functions, are denoted by *ce* and *se*. The distinguishing properties of these solutions are listed in Table 3.9.1. Thus, for $det\ \mathcal{C}_2 = 0$ or $det\ \mathcal{S}_2 = 0$, $det\ \mathcal{G}(1) = 0$ and the solutions have period $\pi$. For $det\ \mathcal{C}_1 = 0$ or $det\ \mathcal{S}_1 = 0$, $det\ \mathcal{G}(-1) = 0$ so that solutions have period $2\pi$.

Table 3.9.1.

| basic matrix | solution vector | period | symmetry |
|---|---|---|---|
| $\mathcal{C}_2$ | *ce* even | $\pi$ | even, symmetric about $\pi/2$ |
| $\mathcal{S}_2$ | *se* even | $\pi$ | odd, antisymmetric about $\pi/2$ |
| $\mathcal{C}_1$ | *ce* odd | $2\pi$ | even, antisymmetric about $\pi/2$ |
| $\mathcal{S}_1$ | *se* odd | $2\pi$ | odd, symmetric about $\pi/2$ |

For $det\ \mathcal{C}_2 = 0$, the non trivial solution satisfying $\mathcal{C}_2\mathbf{v}^1 = 0$ may be denoted as $\mathbf{v}^1 = (v(0), \cdots, v(r))^T$. Then, the corresponding basically periodic solution of the difference equation is obtained from the transformation (3.9.9): $\mathbf{u}(k) = (v(0), (1/\sqrt{2})v(1), \cdots, (1/\sqrt{2})v(r-1), v(r), (1/\sqrt{2})v(r-1), \cdots, (1/\sqrt{2})v(1))^T$, and since it is of period $ph = \pi$, it is an even function and symmetric about $rh = \pi/2$. It is classified as being of type *ce* even.

For $det\ \mathcal{S}_2 = 0$, if the non trivial solution satisfying $\mathcal{S}_2\mathbf{v}^2 = 0$ is written as $\mathbf{v}^2 = (v(r+1), \cdots, v(p-1))^T$, then the corresponding *se* even solution of period $\pi$ of the difference equation is $\mathbf{u}(k) = (0, (1/\sqrt{2})v(p-1), \cdots, (1/\sqrt{2})v(r+1), 0, -(1/\sqrt{2})v(r+1), \cdots, -(1/\sqrt{2})v(p-1))^T$, which is an odd function and antisymmetric about $\pi/2$.

For $det\ \mathcal{C}_1 = 0$, if the non trivial solution satisfying $\mathcal{C}_1\mathbf{w}^1 = 0$ is $\mathbf{w}^1 = (w(0), \cdots, w(r-1))^T$, then the corresponding *ce* odd solution of period $2\pi$ of the difference equation is $\mathbf{u}(k) = (w(0), (1/\sqrt{2})w(1), \cdots, (1/\sqrt{2})w(r-1), 0, -(1/\sqrt{2})w(r-1), \cdots, -(1/\sqrt{2})w(1))^T$, which is an even function antisymmetric about $\pi/2$.

Finally, if for $det\ S_1 = 0$ the non trivial solution satisfying $S_1\mathbf{w}^2 = 0$ is $\mathbf{w}^2 = (w(r), w(r+1), \cdots, w(p-1))^T$, then the corresponding *se* odd solution of period $2\pi$ of the difference equation is $\mathbf{u}(k) = (0, -(1/\sqrt{2})w(p-1), \cdots, -(1/\sqrt{2})w(r+1), w(r), -(1/\sqrt{2})w(r+1), \cdots, -(1/\sqrt{2})w(p-1))^T$, which is an odd function symmetric about $\pi/2$.

These solutions representing the basic types are mutually orthogonal over the interval $2ph = 2\pi$ is clear from the fact that over this interval the solution vectors are respectively $(\mathbf{u}^T(k),\ \mathbf{u}^T(k))^T$, $(\mathbf{u}^T(k),\ \mathbf{u}^T(k))^T$, $(\mathbf{u}^T(k),\ -\mathbf{u}^T(k))^T$, $(\mathbf{u}^T(k),\ -\mathbf{u}^T(k))^T$.

Thus, the general features of the basically periodic solutions of the Hill's difference equation (3.9.2) are precisely the same as for the Hill's differential equation (3.9.1).

## 3.10. Mathieu's Equation

Mathieu's differential equation

$$y'' + (a - 2q\cos 2t)y = 0, \tag{3.10.1}$$

where $a$ and $q$ are real parameters, occurs in diverse class of applied problems [19]. Infact, it is the most important example of the Hill's differential equation (3.9.1), the function $a(t) = a - 2q\cos 2t$ is even and periodic with period $\pi$. For the corresponding difference equation (3.9.2) we take

$$\frac{u(k+1) - 2u(k) + u(k-1)}{H^2} + (a - 2q\cos 2kh)u(k) = 0 \tag{3.10.2}$$

and so in (3.9.6) the function $g(k) = 2 - H^2(a - 2q\cos 2kh)$.

Thus, the theory developed for the difference equation (3.9.2) applies to the difference equation (3.10.2) and in particular the basically periodic solutions are of the four types designated by *ce* even, *se* even, *ce* odd and *se* odd (Table 3.9.1). In particular, if we choose $H^2 = 4a^{-1}\sin^2\frac{1}{2}h\sqrt{a}$ then (3.10.2) becomes

$$\frac{u(k+1) - 2u(k) + u(k-1)}{4a^{-1}\sin^2\frac{1}{2}h\sqrt{a}} + (a - 2q\cos 2kh)u(k) = 0 \tag{3.10.3}$$

and in (3.9.6) the function $g(k)$ reduces to $g(k) = 2\cos h\sqrt{a} + 8qa^{-1} \times \sin^2\frac{1}{2}h\sqrt{a}\cos 2kh$, with $h = \pi/p = \pi/(2r)$.

A particular advantage of (3.10.3) is that in the case of $q = 0$ solutions of (3.10.1) and (3.10.3) are the same. Further, in this case $g(k) = 2\cos h\sqrt{a}$

which is independent of $k$, and the basic determinants can be evaluated explicitly. Indeed, from Problems 2.16.37 and 3.19.25 it follows that $D_k = \sin kh\sqrt{a}\ \mathrm{cosech}\ h\sqrt{a}$, $k \geq 0$, $det\ \mathcal{C}_2 = -4\sin\pi\sqrt{a}/2\sin h\sqrt{a}$, $det\ \mathcal{S}_2 = \sin\pi\sqrt{a}/2\ \mathrm{cosech}\ h\sqrt{a}$, $det\ \mathcal{C}_1 = det\ \mathcal{S}_1 = 2\cos\pi\sqrt{a}/2$.

## 3.11. Weierstrass' Elliptic Equations

It is well known [1] that the Weierstrass elliptic function $p(z)$ satisfies the first order nonlinear differential equation of the second degree

$$p'(z)^2 = 4p(z)^3 - g_2p(z) - g_3 \tag{3.11.1}$$

and the consequent second order nonlinear differential equation of the first degree

$$p''(z) = 6p(z)^2 - \frac{1}{2}g_2. \tag{3.11.2}$$

The function $p(z)$ is an even function of $z$, $p(z) - z^{-2}$ is analytic at $z = 0$ and equal to $0$ at $z = 0$, and the constants $g_2$ and $g_3$ are the so called invariants.

To obtain the best difference equation approximations of (3.11.1) and (3.11.2) we can use the addition formula for $p(z)$,

$$p(z_1 + z_2) = \frac{1}{4}\left[\frac{p'(z_1) - p'(z_2)}{p(z_1) - p(z_2)}\right]^2 - p(z_1) - p(z_2), \quad z_1 \neq z_2.$$

Indeed, if $p(kh) = u(k)$, $k = 0, 1, \cdots$ then we have

$$u(k+1) = \frac{1}{4}\left[\frac{p'(kh) - p'(h)}{u(k) - u(1)}\right]^2 - u(k) - u(1), \quad k \neq 1.$$

Solving this equation for $p'(kh)$, to obtain

$$p'(kh) = p'(h) \pm 2(u(k) - u(1))(u(k+1) + u(k) + u(1))^{1/2}.$$

Squaring the above equation and using (3.11.1) leads to

$$\begin{aligned} -4(u(k) - u(1))u(k+1) + 4u(1)u(k) + 8u^2(1) - g_2 \\ = \pm\, 4p'(h)(u(k+1) + u(k) + u(1))^{1/2}. \end{aligned}$$

Squaring the above equation and using (3.11.1) now gives the best difference equation approximation of (3.11.1)

$$(u(k) - u(1))^2u^2(k+1) - \left[4u(1)u(k)\frac{1}{2}(u(1) + u(k)) - g_2\frac{1}{2}(u(1)\right. \tag{3.11.3}$$

$$+u(k)) - g_3\Big] u(k+1) + \left[\left(u(1)u(k) + \frac{1}{4}g_2\right)^2 + g_3(u(1) + u(k))\right] = 0.$$

To recognize this difference equation as an approximation for small $h$ to the differential equation (3.11.1) requires rearranging (3.11.3) to the form

$$\text{(3.11.4)} \quad (u(k+1) - u(k))^2 u(1) = 4u(k)u(k+1)\frac{1}{2}(u(k) + u(k+1))$$
$$- g_2\frac{1}{2}(u(k) + u(k+1)) - g_3 - u^{-1}(1)\left[\left(u(k)u(k+1) + \frac{1}{4}g_2\right)^2\right.$$
$$\left. + g_3(u(k) + u(k+1))\right].$$

For small $h$, $u(1) = p(h) = h^{-2} + O(h^2)$ so that $(u(k+1) - u(k))u^{1/2}(1) = p'(z) + O(h)$, and to $O(h)$, (3.11.4) becomes (3.11.1) as required.

The best difference equation approximation to the second order differential equation (3.11.2) can be obtained by differencing (3.11.4), in which for convenience, $k$ is replaced by $k-1$. From the simple identities

$$\Delta(u(k) - u(k-1))^2 = (u(k+1) - u(k-1))(u(k+1) - 2u(k) + u(k-1))$$

$$\Delta[u(k-1)u(k)(u(k-1) + u(k))]$$
$$= (u(k+1) - u(k-1))u(k)(u(k+1) + u(k) + u(k-1))$$

$$\Delta(u(k-1) + u(k)) = u(k+1) - u(k-1)$$

$$\Delta\left(u(k-1)u(k) + \frac{1}{4}g_2\right)^2$$
$$= (u(k+1) - u(k-1))\left[u^2(k)(u(k+1) + u(k-1)) + \frac{1}{2}g_2 u(k)\right]$$

follows the best second order difference equation

$$\text{(3.11.5)} \quad (u(k+1) - 2u(k) + u(k-1))u(1) = 2u(k)(u(k+1) + u(k) + u(k-1))$$
$$-\frac{1}{2}g_2 - u^{-1}(1)\left[u^2(k)(u(k+1) + u(k-1)) + \frac{1}{2}g_2 u(k) + g_3\right].$$

For small $h$, $(u(k+1) - 2u(k) + u(k-1))u(1) = p''(z) + O(h^2)$ and $2u(k)(u(k+1) + u(k) + u(k-1)) = 6p(z) + O(h^2)$, and hence to $O(h)$, (3.11.5) becomes (3.11.2).

## 3.12. Volterra's Equations

Volterra equations [36]

$$(3.12.1)\qquad \begin{aligned} x' &= ax - axy \\ y' &= -cy + cxy, \qquad a > 0, \ c > 0 \end{aligned}$$

describe an ecological system of two competing populations predators $y$ and prey $x$. In (3.12.1) the feasibility constraints $x \geq 0$, $y \geq 0$ are not always explicitly stated but are implicitly understood because the $x$ and $y$ represent the number in the competing populations. The system (3.12.1) admits periodic solutions with closed phase trajectories

$$(3.12.2)\qquad c(x - \ln x) + a(y - \ln y) = \text{const.}$$

A typical trajectory for $a = 2$, $c = 1$ and initial conditions $x_0 = 4.25$, $y_0 = 1$ (originally considered by Volterra [36]) is shown in Figure 3.12.1. As $t$ increases, the point $(x, y)$ traverses the trajectory in the anticlockwise direction ABCDA through the extreme points A, B, C, D where $x = 1$ or $y = 1$.

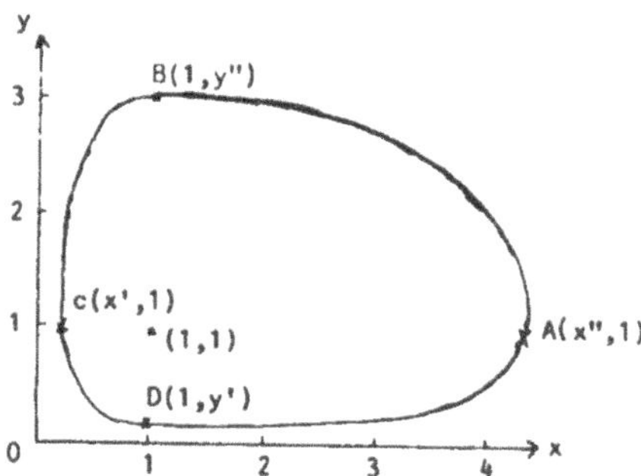


Figure 3.12.1.

An obvious nonlinear difference system approximating (3.12.1) is

$$\begin{aligned} \Delta u(k) &= au(k)(1 - v(k))h \\ \Delta v(k) &= cv(k)(u(k) - 1)h, \end{aligned}$$

where $u(k) = x(t_k)$, $v(k) = y(t_k)$, $t_k = kh$, $k \in \mathbb{N}$, $h > 0$ but numerical experiments for $a \neq c$ have indicated that, if periodic solutions for this system are sought, spurious solutions with negative $u(k)$ or $v(k)$ can arise. To overcome this difficulty, we rewrite the system (3.12.1) as

$$(3.12.3) \qquad \begin{aligned} (x - \ln x)' &= -a(x-1)(y-1) \\ (y - \ln y)' &= c(x-1)(y-1) \end{aligned}$$

and to negotiate with the possible difficulties at the extreme points when $x = 1$ or when $y = 1$ approximate it for $u(k) > 1, \ v(k+1) > 1$ or $u(k) < 1, \ v(k+1) < 1$ by the discrete system

$$(3.12.4) \qquad \begin{aligned} \Delta(u(k) - \ln u(k)) &= -a(u(k)-1)(v(k+1)-1)h \\ \Delta(v(k) - \ln v(k)) &= c(u(k)-1)(v(k+1)-1)h \end{aligned}$$

and for $u(k+1) < 1, \ v(k) > 1$ or $u(k+1) > 1, \ v(k) < 1$ by the discrete system

$$(3.12.5) \qquad \begin{aligned} \Delta(u(k) - \ln u(k)) &= -a(u(k+1)-1)(v(k)-1)h \\ \Delta(v(k) - \ln v(k)) &= c(u(k+1)-1)(v(k)-1)h. \end{aligned}$$

From both the systems (3.12.4), (3.12.5) it trivially follows that

$$c\Delta(u(k) - \ln u(k)) + a\Delta(v(k) - \ln v(k)) = 0,$$

which gives

$$(3.12.6) \qquad c(u(k) - \ln u(k)) + a(v(k) - \ln v(k)) = \text{const}.$$

This is precisely the same as (3.12.2). A particular trajectory for the given values of $a, \ c$ as shown in Figure 3.12.1 can be specified by one initial point, say the extreme point $A(x'', 1)$ with $x'' > 1,$ so that (3.12.6) becomes

$$(3.12.7) \qquad c(u(k) - \ln u(k)) + a(v(k) - \ln v(k)) = c(x'' - \ln x'') + a.$$

The coordinates of the three remaining points $B(1, y''), \ C(x', 1)$ and $D(1, y')$ then can be determined from

$$a(y'' - \ln y'' - 1) = a(y' - \ln y' - 1) = c(x'' - \ln x'' - 1), \quad y' < 1 < y''$$

and

$$x' - \ln x' = x'' - \ln x'', \quad x' < 1 < x''.$$

Finally, we note that once extreme points A, B, C and D of a particular trajectory are known then this trajectory $(u(k), v(k))$ from the systems (3.12.4), (3.12.5) can be realized. For this, it is convenient to consider a trajectory in four segments [AB), [BC), [CD), [DA), where the notation [AB) implies that the point A but not the point B is included. In particular, for the trajectory in Figure 3.12.1 we can begin with the point A so that $u(0) = x''$, $v(0) = 1$ and for the segment [AB) the system (3.12.4) can be executed, similarly over the segments [BC), [CD), [DA) the systems (3.12.5), (3.12.4), (3.12.5) can be used. Thus, in conclusion the difference systems (3.12.4), (3.12.5) correctly approximates the Volterra differential equations (3.12.1).

## 3.13. Elementary Partial Difference Equations: Riccati's Extended Form

For a function of two independent variables $u(k,\ell)$, $(k,\ell) \in \mathbb{N} \times \mathbb{N}$ we introduce four basic difference operators: $E_k u(k,\ell) = u(k+1,\ell)$, $E_\ell u(k,\ell) = u(k,\ell+1)$, $\Delta_k u(k,\ell) = u(k+1,\ell) - u(k,\ell)$, $\Delta_\ell u(k,\ell) = u(k,\ell+1) - u(k,\ell)$. It follows immediately that for all $\tau$, $\eta$ nonnegative integers $E_k^\tau E_\ell^\eta u(k,\ell) = u(k+\tau,\ell+\eta)$, and for $\lambda$, $\mu$ arbitrary constants and $\phi(E_k, E_\ell)$ a polynomial, say, of degree $m$ and $n$ in $E_k$ and $E_\ell$, $\phi(E_k,E_\ell)\lambda^\tau\mu^\eta = \phi(\lambda,\mu)\lambda^\tau\mu^\eta$. A partial difference equation in two independent variables $(k,\ell) \in \mathbb{N} \times \mathbb{N}$ is a functional equation involving $u(k+\tau,\ell+\eta)$, $\tau,\eta = 0,1,2,\cdots$. If a partial difference equation contains $u(k,\ell)$ and $u(k+m,\ell+n)$, where $m$ and $n$ are the largest nonnegative integers, then the equation is said to be of order $(m,n)$, i.e. of order $m$ with respect to $k$, and of order $n$ with respect to $\ell$. The general linear partial difference equation of order $(m,n)$ with constant coefficients has the form

$$\phi(E_k,E_\ell)u(k,\ell) = b(k,\ell), \tag{3.13.1}$$

where $b(k,\ell)$ is a known function of $(k,\ell) \in \mathbb{N} \times \mathbb{N}$. If $b(k,\ell)$ is different from zero for at least one $(k,\ell) \in \mathbb{N} \times \mathbb{N}$, then (3.13.1) is a nonhomogeneous linear partial difference equation. Corresponding to (3.13.1) the equation

$$\phi(E_k,E_\ell)u(k,\ell) = 0 \tag{3.13.2}$$

is called a homogeneous partial difference equation.

The general solution of a given partial difference equation contains certain arbitrary functions of $k$ and $\ell$. In the case of linear partial difference equation of order $(m,n)$ with constant coefficients the general solution contains exactly $m$ arbitrary functions of the variable $k$, or $n$ arbitrary functions of the variable $\ell$. If $v(k,\ell)$ is the general solution of the homogeneous difference equation (3.13.2), and $w(k,\ell)$ is any solution of the nonhomogeneous difference equation (3.13.1), then the general solution of (3.13.1) is given by $u(k,\ell) = v(k,\ell) + w(k,\ell)$.

The main purpose of this section is to discuss several methods to solve linear partial difference equations with constant coefficients.

**Symbolic Method.** In (3.13.2) we can assume that $\phi(E_k, E_\ell) = \prod_{r=1}^{q}(E_k - \psi_r(E_\ell))^{p_r}$, where $p_r$, $1 \le r \le q$ are positive integers, such that $p_1 + \cdots + p_r = m$, and $\psi_r(E_\ell)$, $1 \le r \le q$ may be irrational functions of $E_\ell$. Since the separate factors of $\phi(E_k, E_\ell)$ are commutative, and the equation (3.13.2) is linear, its general solution will be the sum of the general solutions of the separate equations $(E_k - \psi_r(E_\ell))^{p_r} u(k,\ell) = 0$, $1 \le r \le q$.

First, we consider the partial difference equation

$$u(k+1,\ell) - \alpha u(k,\ell+1) - \beta u(k,\ell) \;=\; (E_k - \alpha E_\ell - \beta)u(k,\ell) \;=\; 0, \tag{3.13.3}$$

where $\alpha$ and $\beta$ are constants. Since this equation can be written as $E_k u(k,\ell) = (\alpha E_\ell + \beta)u(k,\ell)$, and $(\alpha E_\ell + \beta)$ has no effect on the variable $k$, it can be solved to obtain its general solution

$$u(k,\ell) \;=\; (\alpha E_\ell + \beta)^k c(\ell) \;=\; \beta^k \sum_{\tau=0}^{k} \binom{k}{\tau} \left(\frac{\alpha}{\beta}\right)^{\tau} c(\ell + \tau), \tag{3.13.4}$$

where $c$ is an arbitrary function of the variable $\ell$.

The solution of (3.13.3) can also be obtained by writing it as $E_\ell u(k,\ell) = \left(-\frac{\beta}{\alpha}\right)\left(1 - \frac{E_k}{\beta}\right) u(k,\ell)$ and proceeding as before. Indeed, we have

$$u(k,\ell) \;=\; \left(-\frac{\beta}{\alpha}\right)^{\ell} \sum_{\eta=0}^{\ell} \binom{\ell}{\eta} \left(-\frac{1}{\beta}\right)^{\eta} d(k+\eta), \tag{3.13.5}$$

where $d$ is an arbitrary function of $k$.

Next we consider the partial difference equation

$$(E_k - \alpha E_\ell - \beta)^r u(k,\ell) \;=\; 0, \tag{3.13.6}$$

where $r$ is a positive integer. Defining $\overline{E}_\ell = \alpha E_\ell + \beta$ and using the fact that $(E_k - \overline{E}_\ell)^r u(k,\ell) = 0$ has the solution $u(k,\ell) = (\overline{E}_\ell)^k \sum_{i=1}^{r} k^{i-1} c_i(\ell)$, where $c_i$, $1 \le i \le r$ are arbitrary functions of $\ell$, we obtain the solution of (3.13.6)

$$u(k,\ell) \;=\; \beta^k \left(1 + \frac{\alpha}{\beta} E_\ell\right)^k \sum_{i=1}^{r} k^{i-1} c_i(\ell). \tag{3.13.7}$$

Similarly, the solution in terms of $r$ arbitrary functions $d_i(k)$, $1 \le i \le r$ can be written as

$$(3.13.8) \qquad u(k,\ell) = \left(-\frac{\beta}{\alpha}\right)^{\ell}\left(1-\frac{E_k}{\beta}\right)^{\ell}\sum_{i=1}^{r}\ell^{i-1}d_i(k).$$

The Symbolic method is not only restricted to linear partial difference equations with constant coefficients, but it can also be used to solve some special partial difference equations with variable coefficients. For example, the equation $u(k+1,\ell+1) - ku(k,\ell) = 0$ can be written as $(E_k - kE_\ell^{-1})u(k,\ell) = 0$, and hence its solution is $u(k,\ell) = \prod_{i=1}^{k-1}(iE_\ell^{-1})c(\ell) = (k-1)!\, E_\ell^{-k+1}c(\ell) = (k-1)!\, c(\ell-k+1)$, where $c$ is an arbitrary function.

**Lagrange's Method.** Let the equation (3.13.2) has a solution of the form $u(k,\ell) = \lambda^k\mu^\ell$, where $\lambda$ and $\mu$ are unspecified constants. For this it is necessary that $\phi(\lambda,\mu) = 0$. Since $\phi(\lambda,\mu)$ is a polynomial of degree $m$ in $\lambda$ it determines the roots $\lambda_i(\mu)$, $1 \le i \le m$. Therefore, for the equation (3.13.2) solutions are $u_i(k,\ell) = (\lambda_i(\mu))^k\mu^\ell$, $1 \le i \le m$. Now since equation (3.12.2) is linear, the sum of all such expressions for all possible values of $\lambda$ will also be a solution. Thus, if we let $c_i(\mu)$ and $d_i(\mu)$ be arbitrary functions of $\mu$, then $u(k,\ell) = \sum_{i=1}^{m}\overline{u}_i(k,\ell)$ is a solution of equation (3.13.2), where for discrete values of $\mu$, $\overline{u}_i(k,\ell) = \sum_\mu c_i(\mu)(\lambda_i(\mu))^k\mu^\ell$, and for continuous values of $\mu$, $\overline{u}_i(k,\ell) = \int_{-\infty}^{\infty} d_i(\mu)(\lambda_i(\mu))^k\mu^\ell d\mu$.

**Example 3.13.1.** For the partial difference equation

$$(3.13.9) \qquad u(k,\ell) = pu(k+1,\ell-1) + qu(k-1,\ell+1), \quad p+q=1$$

Lagrange's method requires that $p\lambda^2 - \lambda\mu + q\mu^2 = 0$. This determines $\lambda_1(\mu) = \mu$ and $\lambda_2(\mu) = \mu q/p$. This gives two particular solutions $u_1(k,\ell) = \mu^{k+\ell}$ and $u_2(k,\ell) = (q/p)^k\mu^{k+\ell}$. Summing these expressions, we find $\overline{u}_1(k,\ell) = \int_{-\infty}^{\infty} d_1(\mu)\mu^{k+\ell}d\mu$ and $\overline{u}_2(k,\ell) = \int_{-\infty}^{\infty} d_2(\mu)(q/p)^k\mu^{k+\ell}d\mu$, where $d_1(\mu)$ and $d_2(\mu)$ are chosen so that the integrals are defined. From this we conclude that the general solution of (3.13.9) is $u(k,\ell) = g(k+\ell) + (q/p)^k h(k+\ell)$, where $g$ and $h$ are arbitrary functions of $k+\ell$.

**Example 3.13.2.** For the partial difference equation

$$(3.13.10) \quad u(k+3,\ell) - 3u(k+2,\ell+1) + 3u(k+1,\ell+2) - u(k,\ell+3) = 0$$

Lagrange's method requires that $\lambda^3 - 3\lambda^2\mu + 3\lambda\mu^2 - \mu^3 = (\lambda-\mu)^3 = 0$. Therefore, $\lambda = \mu$ is a triple root and linearly independent particular solutions are $\mu^{k+\ell}$, $k\mu^{k+\ell}$, and $k^2\mu^{k+\ell}$. Multiplying each of these expressions by an arbitrary function of $\mu$ and summing gives the general solution $u(k,\ell) = f(k+\ell) + kg(k+\ell) + k^2h(k+\ell)$, where $f$, $g$ and $h$ are arbitrary functions of $k+\ell$.

**Separation of Variables Method.** This method is often applicable to linear partial difference equations having the form

$$\phi(E_k, E_\ell, k, \ell) = 0, \tag{3.13.11}$$

where $\phi$ is a polynomial function of $E_k$ and $E_\ell$. We assume that the solution of (3.13.11) can be written as $u(k,\ell) = U(k)V(\ell)$ and its substitution in (3.13.11) leads to

$$\frac{f_1(E_k,k)U(k)}{f_2(E_k,k)U(k)} = \frac{g_1(E_\ell,\ell)V(\ell)}{g_2(E_\ell,\ell)V(\ell)}. \tag{3.13.12}$$

Under these assumptions, $U(k)$ and $V(\ell)$ satisfy the following ordinary difference equations

$$\begin{aligned} f_1(E_k,k)U(k) &= \alpha f_2(E_k,k)U(k) \\ g_1(E_\ell,\ell)V(\ell) &= \alpha g_2(E_\ell,\ell)V(\ell), \end{aligned} \tag{3.13.13}$$

where $\alpha$ is an arbitrary constant.

Once the solution $U(k)$, $V(k)$ of (3.13.13) is known, we can sum over $\alpha$, as in Lagrange's method, to obtain additional solutions.

**Example 3.13.3.** For both the partial difference equations (3.13.9), (3.13.10) the Separation of variables method does not work. For the equation

$$u(k,\ell+1) = u(k-1,\ell) + ku(k,\ell) \tag{3.13.14}$$

Lagrange's method fails, however the Separation of variables method is applicable. Indeed, if we let $u(k,\ell) = U(k)V(\ell)$, then

$$U(k)V(\ell+1) = U(k-1)V(\ell) + kU(k)V(\ell),$$

which is the same as

$$\frac{V(\ell+1)}{V(\ell)} = \frac{U(k-1)+kU(k)}{U(k)} = \alpha,$$

where $\alpha$ is an arbitrary constant. Therefore, $U(k)$ and $V(\ell)$ satisfy the first order ordinary difference equations

$$\begin{aligned} (\alpha - k)U(k) &= U(k-1) \\ V(\ell+1) &= \alpha V(\ell), \end{aligned}$$

which can be solved to obtain

$$U(k) = \frac{c_1(-1)^k}{\Gamma(k-(\alpha-1))}, \qquad V(\ell) = c_2\alpha^\ell.$$

Summing over $\alpha$ now gives

$$u(k,\ell) = (-1)^k \sum_\alpha \frac{c(\alpha)\alpha^\ell}{\Gamma(k-(\alpha-1))},$$

where $c$ is an arbitrary function of $\alpha$.

**Laplace's Method.** This method is applicable when the sum or the difference of the arguments of all $u(k+\tau, \ell+\eta)$ that appear in a partial difference equation is a constant. For example, in the difference equation (3.13.9) the sum of the arguments of all $u(k,\ell)$, $u(k+1,\ell-1)$, $u(k-1,\ell+1)$ is a constant $k+\ell$. If we set $k+\ell = m$ and define $v(k) = u(k, m-k)$, then equation (3.13.9) becomes

$$v(k) = pv(k+1) + qv(k-1),$$

which is a second order ordinary difference equation whose solution is

$$v(k) = c_1 + c_2\left(\frac{q}{p}\right)^k. \tag{3.13.15}$$

However, since an arbitrary constant can be considered a function of another constant, we recover the solution of (3.13.9) by replacing $c_1$ and $c_2$ in (3.13.15) by $c_1(k+\ell)$ and $c_2(k+\ell)$.

There are several nonlinear partial difference equations which can be reduced to linear equations by means of special transformations. As an example, we shall consider *Riccati's extended form,* which is a system of nonlinear partial difference equations

$$\begin{aligned} u(k+1,\ell+1) &= \frac{\alpha u(k,\ell) + \beta v(k,\ell) + \gamma}{pu(k,\ell) + qv(k,\ell) + r} \\ v(k+1,\ell+1) &= \frac{\mu u(k,\ell) + \nu v(k,\ell) + \eta}{pu(k,\ell) + qv(k,\ell) + r}, \end{aligned} \tag{3.13.16}$$

where $\alpha$, $\beta$, $\gamma$, $\mu$, $\nu$, $\eta$, $p$, $q$ and $r$ are constants.

From (3.13.16) it follows that if $\lambda$, $\xi$ and $\zeta$ are undetermined multipliers, then

$$(3.13.17)\qquad \frac{u(k+1,\ell+1)}{\alpha u(k,\ell)+\beta v(k,\ell)+\gamma} = \frac{v(k+1,\ell+1)}{\mu u(k,\ell)+\nu v(k,\ell)+\eta}$$

$$= \frac{1}{pu(k,\ell)+qv(k,\ell)+r}$$

$$= \frac{\lambda u(k+1,\ell+1)+\xi v(k+1,\ell+1)+\zeta}{(\alpha\lambda+\mu\xi+p\zeta)u(k,\ell)+(\beta\lambda+\nu\xi+q\zeta)v(k,\ell)+(\gamma\lambda+\eta\xi+r\zeta)}.$$

Now suppose that $\lambda,\ \xi$ and $\zeta$ are chosen so that

$$(3.13.18)\qquad \begin{aligned} \alpha\lambda+\mu\xi+p\zeta &= h\lambda \\ \beta\lambda+\nu\xi+q\zeta &= h\xi \\ \gamma\lambda+\eta\xi+r\zeta &= h\zeta, \end{aligned}$$

where $h$ is an unknown constant. The condition that $\lambda,\ \xi$ and $\zeta$ are not to be zero demands that

$$(3.13.19)\qquad \begin{vmatrix} \alpha-h & \mu & p \\ \beta & \nu-h & q \\ \gamma & \eta & r-h \end{vmatrix} = 0.$$

This is a cubic equation in $h$, and provides three values of $h$, say, $h_1,\ h_2$ and $h_3$, and for each of these values we can find the corresponding values of $\lambda,\ \xi$ and $\zeta$ from the system (3.13.18), i.e. $h_i \to (\lambda_i,\ \xi_i,\ \zeta_i),\ i=1,2,3.$

This allows us to replace (3.13.17) by the new set of equations

$$(3.13.20)\qquad \frac{U_1(k+1,\ell+1)}{h_1U_1(k,\ell)} = \frac{U_2(k+1,\ell+1)}{h_2U_2(k,\ell)} = \frac{U_3(k+1,\ell+1)}{h_3U_3(k,\ell)},$$

where $U_i(k,\ell)=\lambda_i u(k,\ell)+\xi_i v(k,\ell)+\zeta_i,\ i=1,2,3.$

From the equations (3.13.20) it follows that

$$(3.13.21)\qquad \begin{aligned} \frac{U_1(k+1,\ell+1)}{U_3(k+1,\ell+1)} &= \frac{h_1}{h_3}\frac{U_1(k,\ell)}{U_3(k,\ell)} \\ \frac{U_2(k+1,\ell+1)}{U_3(k+1,\ell+1)} &= \frac{h_2}{h_3}\frac{U_2(k,\ell)}{U_3(k,\ell)}, \end{aligned}$$

which are of the form $u(k+1,\ell+1)=\beta u(k,\ell)$, and hence can be solved to obtain

$$(3.13.22)\qquad \begin{aligned} \frac{U_1(k,\ell)}{U_3(k,\ell)} &= \left(\frac{h_1}{h_3}\right)^k \frac{c_1(k-\ell)}{c_3(k-\ell)} = \phi_1(k,\ell) \\ \frac{U_2(k,\ell)}{U_3(k,\ell)} &= \left(\frac{h_2}{h_3}\right)^k \frac{c_2(k-\ell)}{c_3(k-\ell)} = \phi_2(k,\ell), \end{aligned}$$

where $c_1$, $c_2$ and $c_3$ are arbitrary functions of $k - \ell$.

Thus, on using the expressions for $U_i(k,\ell)$, $i = 1,2,3$ in (3.13.22), we find

$$(3.13.23)\qquad \begin{aligned}(\lambda_1 - \lambda_3\phi_1)u(k,\ell) + (\xi_1 - \xi_3\phi_1)v(k,\ell) &= \zeta_3\phi_1 - \zeta_1\\ (\lambda_2 - \lambda_3\phi_2)u(k,\ell) + (\xi_2 - \xi_3\phi_2)v(k,\ell) &= \zeta_3\phi_2 - \zeta_2.\end{aligned}$$

These equations can be solved to find the solution $u(k,\ell)$, $v(k,\ell)$ of the system (3.13.16).

## 3.14. Wave Equation

From Example 3.6.1 it is clear that the function $u(k) = A\sin\omega kh$, $k \in \mathbb{N}$ is a solution of the second order difference equation

$$\frac{u(k+1) - 2u(k) + u(k-1)}{4\omega^{-2}\sin^2\frac{1}{2}\omega h} + \omega^2 u(k) = 0.$$

Further, as $h \to 0$, $u(k)$ tends to $y(t) = A\sin\omega t$ and the above difference equation converges to the differential equation $y'' + \omega^2 y = 0$.

The function $\phi(x,t) = A\sin\omega(x+ct)$ is a solution of the one dimensional wave equation

$$(3.14.1)\qquad \phi_{tt} = c^2\phi_{xx}.$$

To obtain the partial difference equation corresponding to this solution, we use the discretization $x = kh_1$, $t = \ell h_2$, $k$, $\ell \in \mathbb{N}$ where $h_1 > 0$, $h_2 > 0$ are step sizes, and represent $\phi(x,t) = \phi(kh_1,\ell h_2) = u(k,\ell)$.

Thus, it follows that

$$\frac{u(k+1,\ell) - 2u(k,\ell) + u(k-1,\ell)}{4\omega^{-2}\sin^2\frac{1}{2}\omega h_1} + \omega^2 u(k,\ell) = 0$$

and

$$\frac{u(k,\ell+1) - 2u(k,\ell) + u(k,\ell-1)}{4\omega^{-2}c^{-2}\sin^2\frac{1}{2}\omega c h_2} + \omega^2 c^2 u(k,\ell) = 0,$$

whence

(3.14.2)
$$\frac{u(k,\ell+1) - 2u(k,\ell) + u(k,\ell-1)}{4\omega^{-2}c^{-2}\sin^2\frac{1}{2}\omega c h_2} = c^2\frac{u(k+1,\ell) - 2u(k,\ell) + u(k-1,\ell)}{4\omega^{-2}\sin^2\frac{1}{2}\omega h_1}.$$

As $h_1,\ h_2 \to 0$ this linear second order partial difference equation converges to (3.14.1). The function $\phi(x,t) = A\sin\omega(x-ct)$ is also a solution of both the equations (3.14.1) and (3.14.2).

The analysis can be extended to a solution consisting of a sum of two terms, e.g. $\phi(x,t) = A_1 \sin\omega_1(x+ct) + A_2 \sin\omega_2(x+ct)$. Indeed from the relation

$$\begin{aligned}&\phi(x+h_1,t) - (\cos\omega_1 h_1 + \cos\omega_2 h_1)\phi(x,t) + \phi(x-h_1,t)\\ &\quad = (\cos\omega_1 h_1 - \cos\omega_2 h_1)[A_1 \sin\omega_1(x+ct) - A_2 \sin\omega_2(x+ct)]\end{aligned}$$

follows the required partial difference equation

$$(3.14.3)\qquad \frac{u(k,\ell+1) - (\cos\omega_1 ch_2 + \cos\omega_2 ch_2)u(k,\ell) + u(k,\ell-1)}{(\cos\omega_1 ch_2 - \cos\omega_2 ch_2)}$$

$$= \frac{u(k+1,\ell) - (\cos\omega_1 h_1 + \cos\omega_2 h_1)u(k,\ell) + u(k-1,\ell)}{(\cos\omega_1 h_1 - \cos\omega_2 h_1)}.$$

Since as $h_1 \to 0$ and $h_2 \to 0$, $\cos\omega_1 ch_2 + \cos\omega_2 ch_2 \to 2$, $\cos\omega_1 ch_2 - \cos\omega_2 ch_2 \to c^2h_2^2(\omega_2^2 - \omega_1^2)/2$, $\cos\omega_1 h_1 + \cos\omega_2 h_1 \to 2$, $\cos\omega_1 h_1 - \cos\omega_2 h_1 \to h_1^2(\omega_2^2 - \omega_1^2)/2$ difference equation (3.14.3) converges to (3.14.1).

For the general solution $u(x,t) = \sum_{i=1}^{\infty} A_i \sin\omega_i(x+ct)$ of the Wave equation (3.14.1) it is not possible to find a simple partial difference equation. The appearance of $\omega$ in the denominator $4\omega^{-2}\sin^2\frac{1}{2}\omega h$ loses the advantage enjoyed by the usual approximation term $h^2$.

## 3.15. FitzHugh–Nagumo's Equation

The partial differential equation

$$(3.15.1)\qquad \phi_t = \frac{1}{2}\phi_{xx} + (\omega A - \eta^2\phi)(A^{-2}\phi^2 - 1),$$

was considered by FitzHugh and Nagumo in modelling the propagation of a nerve pulse [6], has the solitary wave solution

$$(3.15.2)\qquad \phi(x,t) = A\tanh(\eta x + \omega t).$$

To obtain the partial difference equation corresponding to this solution, once again we use the discretization $x = kh_1$, $t = \ell h_2$, $k,\ell \in \mathbb{N}$ where $h_1,\ h_2 > 0$ are step sizes, and represent $\phi(x,t) = \phi(kh_1, \ell h_2) = u(k,\ell)$.

From $u(k,\ell) = A\tanh(\eta k h_1 + \omega \ell h_2)$ it follows that

$$u(k,\ell+1) - u(k,\ell) = A\tanh\omega h_2(A^{-2}u(k,\ell)u(k,\ell+1) - 1),$$

which is on combining with an obvious extension of (3.19.25) gives the required partial difference equation

$$\frac{u(k,\ell+1)-u(k,\ell)}{\omega^{-1}\tanh\omega h_2} = \frac{u(k+1,\ell)-2u(k,\ell)+u(k-1,\ell)}{2\eta^{-2}\tanh^2\eta h_1} - \omega A \tag{3.15.3}$$
$$+\omega A^{-1}u(k,\ell)u(k,\ell+1)+\eta^2 u(k,\ell)$$
$$-\eta^2 A^{-2}u^2(k,\ell)(u(k+1,\ell)+u(k-1,\ell))/2.$$

In the limit $h_1,\ h_2 \to 0$, this partial difference equation gives

$$\phi_t = \frac{1}{2}\phi_{xx} - \omega A + \omega A^{-1}\phi^2 + \eta^2\phi - \eta^2 A^{-2}\phi^3,$$

which is the same as (3.15.1). Thus, the partial difference equation (3.15.3) is the best discretization of the partial differential equation (3.15.1).

## 3.16. Korteweg–de Vries' Equation

The partial differential equation

$$\phi_{xxx} - 3\phi_x^2 + \phi_t = 0, \tag{3.16.1}$$

which, when differentiated with respect to $x$ and the substitution $\phi_x = \psi$ is made, gives the Korteweg de Vries equation in the usual form

$$\psi_{xxx} - 6\psi\psi_x + \psi_t = 0. \tag{3.16.2}$$

We shall determine a partial difference equation for the single soliton solution $\phi(x,t) = -2\omega\tanh(\omega x - 4\omega^3 t)$ of the equation (3.16.1) by considering the equations satisfied by $\phi(x,t) = A\tanh(\omega x + \eta t)$ and $u(k,\ell) = A\tanh(\omega k h_1 + \eta \ell h_2)$.

From (3.19.24) follows the relation

$$\frac{u(k+1,\ell)-u(k-1,\ell)}{u(k+1,\ell)+u(k-1,\ell)}\frac{1}{\sinh 2\omega h_1} = \frac{u(k,\ell+1)-u(k,\ell-1)}{u(k,\ell+1)+u(k,\ell-1)}\frac{1}{\sinh 2\eta h_2},$$

which, when used with an obvious extension of (3.19.26) gives the required partial difference equation

$$\frac{u(k+3,\ell)-3u(k+1,\ell)+3u(k-1,\ell)-u(k-3,\ell)}{(\omega^{-1}\sinh 2\omega h_1)^3} \tag{3.16.3}$$
$$+6\omega A^{-1}\cosh 2\omega h_1\left[\frac{u(k+3,\ell)+u(k-3,\ell)}{u(k+1,\ell)+u(k-1,\ell)}\right]\left[\frac{u(k+1,\ell)-u(k-1,\ell)}{\omega^{-1}\sinh 2\omega h_1}\right]^2$$
$$-4\frac{\omega^3}{\eta}\left[\frac{u(k+3,\ell)+u(k-3,\ell)}{u(k,\ell+1)+u(k,\ell-1)}\right]\left[\frac{u(k,\ell+1)-u(k,\ell-1)}{\eta^{-1}\sinh 2\eta h_2}\right] = 0.$$

The limiting partial differential equation is

$$\phi_{xxx} + 6\omega A^{-1}\phi_x^2 - \frac{4\omega^3}{\eta}\phi_t = 0. \tag{3.16.4}$$

When $A = -2\omega$ and $\eta = -4\omega^3$ so that $\phi(x,t) = -2\omega\tanh(\omega x - 4\omega^3 t)$, and the equation (3.16.4) is the same as (3.16.1). Thus, in this case the partial difference equation (3.16.3) is the best discretization of the partial differential equation (3.16.1).

## 3.17. Modified KdV Equation

Consider the modified Korteweg de Vries equation [6,37] in the form

$$\phi_{xxx} + 6\phi^2\phi_x + \phi_t = 0 \tag{3.17.1}$$

for which the solitary wave solution is $\phi(x,t) = \omega\text{sech}\,(\omega x - \omega^3 t)$.

We shall determine differential and difference equations satisfied by $\phi(x,t) = A\text{sech}\,(\omega x + \eta t)$ and $u(k,\ell) = A\text{sech}\,(\omega k h_1 + \eta \ell h_2)$.

From (3.19.27) follows the relation

$$\frac{u(k+1,\ell) - u(k-1,\ell)}{u(k+1,\ell) + u(k-1,\ell)}\frac{1}{\tanh\omega h_1} = \frac{u(k,\ell+1) - u(k,\ell-1)}{u(k,\ell+1) + u(k,\ell-1)}\frac{1}{\tanh\eta h_2},$$

which, when used with an obvious extension of (3.19.29) gives the required partial difference equation

$$\begin{aligned}&\frac{u(k+3,\ell) - 3u(k+1,\ell) + 3u(k-1,\ell) - u(k-3,\ell)}{(2\omega^{-1}\sinh\omega h_1)^3}\cosh 3\omega h_1\\ &+3\omega^2 A^{-2}u(k,\ell)(u(k+3,\ell) + u(k-3,\ell))\times\\ &\cosh\omega h_1\cosh 2\omega h_1\left[\frac{u(k+1,\ell) - u(k-1,\ell)}{2\omega^{-1}\tanh\omega h_1}\right]\\ &-\frac{\omega^3}{\eta}\left[\frac{u(k+3,\ell) + u(k-3,\ell)}{u(k,\ell+1) + u(k,\ell-1)}\right]\left[\frac{u(k,\ell+1) - u(k,\ell-1)}{2\eta^{-1}\tanh\eta h_2}\right] = 0.\end{aligned} \tag{3.17.2}$$

The limiting partial differential equation is

$$\phi_{xxx} + 6\omega^2 A^{-2}\phi^2\phi_x - \frac{\omega^3}{\eta}\phi_t = 0. \tag{3.17.3}$$

When $A = \omega$ and $\eta = -\omega^3$ so that $\phi(x,t) = \omega\tanh(\omega x - \omega^3 t)$, and the equation (3.17.3) is the same as (3.17.1). Thus, in this case the partial difference equation (3.17.2) is the best discretization of the partial differential equation (3.17.1).

## 3.18. Lagrange's Equations

Consider a holonomic mechanical system, with $q = q(t) = q_1(t), \cdots, q_n(t)$ the generalized coordinates, with potential energy $V(q)$, and kinetic energy

$$T(q, \dot{q}) = \frac{1}{2} \sum_{i,j=1}^{n} a_{ij}(q)\dot{q}_i\dot{q}_j \qquad \left(\cdot = \frac{d}{dt}\right) \tag{3.18.1}$$

in which $a_{ij} = a_{ji}$. Lagrange's equations of motion give the nonconservative generalized forces as

$$F_r = \frac{d}{dt}\left(\frac{\partial T}{\partial \dot{q}_r}\right) - \frac{\partial T}{\partial q_r} + \frac{\partial V}{\partial q_r}, \qquad r = 1, \cdots, n. \tag{3.18.2}$$

We shall discretize these equations so that the following two properties are preserved.

$$(P_1) \qquad \frac{d}{dt}(T+V) = \sum_{r=1}^{n} F_r\dot{q}_r, \tag{3.18.3}$$

which leads to the energy integral $T + V =$ constant, when the power of the nonconservative generalized forces, given by the right side of (3.18.3) is zero.

$(P_2)$ If the variable $q_r$ is cyclic, i.e. $T$ and $V$ both are independent of $q_r$, then (3.18.2) reduces to

$$F_r = \frac{d}{dt}\left(\frac{\partial T}{\partial \dot{q}_r}\right), \tag{3.18.4}$$

or in integrated form

$$\int_{t_1}^{t_2} F_r dt = \left.\frac{\partial T}{\partial \dot{q}_r}\right|_{t_2} - \left.\frac{\partial T}{\partial \dot{q}_r}\right|_{t_1}. \tag{3.18.5}$$

Thus, if the impulse of the nonconservative force corresponding to the cyclic variable, given by the left side of (3.18.5) is zero, then the generalized momentum corresponding to this variable, given by the right side of (3.18.5), is also zero.

Since the discretization of (3.18.2) will involve partial differences, first we shall provide a general formula for the partial differences. For this, let

$f = f(q) = f(q_1, \cdots, q_n)$ with

$$df = \sum_{r=1}^{n} \frac{\partial f(q)}{\partial q_r} dq_r. \tag{3.18.6}$$

If $f$ is independent of $q_r$, then

$$\frac{\partial f(q)}{\partial q_r} = 0. \tag{3.18.7}$$

If the time $t$ is discretized at time instants $t(k)$, $k = 0, 1, \cdots$ then the aim is to choose a discrete analog of the partial derivative, which we shall denote by

$$\frac{\Delta_r f(k)}{\Delta q_r(k)} = \frac{\Delta_r f(k)}{q_r(k+1) - q_r(k)}, \tag{3.18.8}$$

so that the discrete analogs of (3.18.6) and (3.18.7) are satisfied. In (3.18.8), $\Delta_r$ will be called the (forward) partial difference operator. The discrete analog of (3.18.6) is

$$\Delta f = f[q(k+1)] - f[q(k)] = \sum_{r=1}^{n} \frac{\Delta_r f(k)}{\Delta q_r(k)} \Delta q_r(k) = \sum_{r=1}^{n} \Delta_r f(k) \tag{3.18.9}$$

and the analog of (3.18.7) is

$$\Delta_r f(k) = 0, \quad f \text{ is independent of } q_r. \tag{3.18.10}$$

In conclusion, we seek an expression for the partial difference $\Delta_r f(k)$ which satisfies (3.18.9) and (3.18.10).

To illustrate the general scheme, consider the case $n = 3$ as demonstrated in Figure 3.18.1. There are 6 different paths from A (for $k$) to D (for $k+1$) along edges of the cube and any such path will give a representation of $\Delta_r f(k)$ which satisfies (3.18.9) and (3.18.10). For example, along the path ABCD, we find

$$\Delta_1 f(k) = f[q_1(k+1), q_2(k+1), q_3(k+1)] - f[q_1(k), q_2(k+1), q_3(k+1)],$$

represented by CD

$$\Delta_2 f(k) = f[q_1(k), q_2(k+1), q_3(k+1)] - f[q_1(k), q_2(k), q_3(k+1)],$$

represented by BC

$$\Delta_3 f(k) = f[q_1(k), q_2(k), q_3(k+1)] - f[q_1(k), q_2(k), q_3(k)],$$

represented by AB.

If all the six paths along the edges of the cube from A to D are given equal weight, then in the representation of $\Delta_1 f(k)$, say, the edges AE and CD occur twice and edges BG and HF once so that the corresponding general representation is

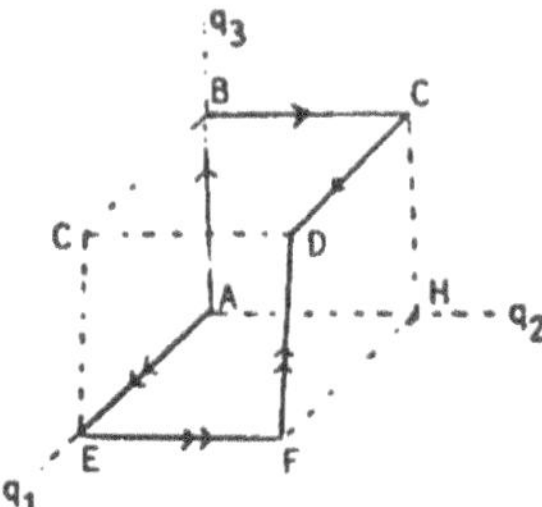


Figure 3.18.1.

$$\text{(3.18.11)} \quad \Delta_1 f(q) = \frac{1}{3}[f[q_1(k+1), q_2(k), q_3(k)] - f[q_1(k), q_2(k), q_3(k)]] + \frac{1}{6}[f[q_1(k+1), q_2(k), q_3(k+1)] - f[q_1(k), q_2(k), q_3(k+1)]] + \frac{1}{6}[f[q_1(k+1), q_2(k+1), q_3(k)] - f[q_1(k), q_2(k+1), q_3(k)]] + \frac{1}{3}[f[q_1(k+1), q_2(k+1), q_3(k+1)] - f[q_1(k), q_2(k+1), q_3(k+1)]].$$

Such a representation uses all eight values of the function at the corners of the cube.

In the case of $n$ variables, the general representation of the partial difference is

$$\text{(3.18.12)} \quad \Delta_r f(k) = \sum_{I,J} \frac{N(I)!\, N(J)!}{n!}[f[q_I(k), q_r(k+1), q_J(k+1)] - f[q_I(k), q_r(k), q_J(k+1)]],$$

where the summation is taken over all disjoint subsets $I$, $J$ of $\{1,2,\cdots,n\}/\{r\}$ such that $I \cup \{r\} \cup J = \{1,2,\cdots,n\}$. For example, the third term in (3.18.11) with $r=1$ corresponds to $I=\{3\}$, $J=\{2\}$, $N(I)=N(J)=1$.

From this construction, it is clear that (3.18.12) satisfies the requirements expressed by (3.18.9) and (3.18.10).

The discrete analog of the Lagrange equations of motion (3.18.2) we shall consider are of the form

$$(3.18.13)\quad F_r(k) = \frac{1}{\Delta t(k)}\Delta\left[\sum_{j=1}^{n} a_{rj}(k)\dot{q}_j(k)\right] - \frac{1}{2}\sum_{i,j=1}^{n}\frac{\Delta_r a_{ij}(k)}{\Delta q_r(k)}\dot{q}_i(k)\dot{q}_j(k+1) + \frac{\Delta_r V(k)}{\Delta q_r(k)},\quad r = 1,2,\cdots,n$$

which are augmented with the trapezoidal smoothing formula

$$(3.18.14)\quad \frac{\Delta q_r(k)}{\Delta t(k)} = \frac{1}{2}[\dot{q}_r(k+1)+\dot{q}_r(k)],\quad r = 1,\cdots,n.$$

These discrete Lagrange's equations (3.18.13), (3.18.14) satisfy the discrete analogs of (3.18.3) and (3.18.4), namely

$$(3.18.15)\quad \Delta(T+V) = \sum_{r=1}^{n} F_r(k)\Delta q_r(k)$$

and for a cyclic variable $q_r(k)$

$$(3.18.16)\quad F_r(k) = \Delta\left(\sum_{j=1}^{n} a_{rj}(k)\dot{q}_j(k)\right)\Big/\Delta t(k).$$

For this, (3.18.16) is immediate from (3.18.13) because the partial difference operator $\Delta_r$ forces $\Delta_r f(k) = 0$ if f is independent of $q$. To show (3.18.15) we multiply (3.18.13) by $\Delta q_r(k)$, use (3.18.14) and sum over from $r = 1$ to $r = n$, to get

$$\sum_{r=1}^{n} F_r(k)\Delta q_r(k) = \sum_{r=1}^{n}\frac{1}{2}[\dot{q}_r(k+1)+\dot{q}_r(k)]\Delta\left[\sum_{j=1}^{n} a_{rj}(k)\dot{q}_j(k)\right] - \frac{1}{2}\sum_{i,j=1}^{n}\left[\sum_{r=1}^{n}\Delta_r a_{ij}(k)\right]\dot{q}_i(k)\dot{q}_j(k+1) + \sum_{r=1}^{n}\Delta_r V(k),$$

which is from (3.18.9) is the same as

$$
\begin{aligned}
\sum_{r=1}^{n} F_r(k)\Delta q_r(k) &= \frac{1}{2}\sum_{r,j=1} [[\dot{q}_r(k+1)+\dot{q}_r(k)]a_{rj}(k+1)\dot{q}_j(k+1) \\
&\quad -[\dot{q}_r(k+1)+\dot{q}_r(k)]a_{rj}(k)\dot{q}_j(k)] \\
&\quad -\frac{1}{2}\sum_{i,j=1}^{n} [a_{ij}(k+1)-a_{ij}(k)]\dot{q}_i(k)\dot{q}_j(k+1)+\Delta V(k) \\
&= \frac{1}{2}\sum_{i,j=1}^{n} [\dot{q}_r(k+1)a_{rj}(k+1)\dot{q}_j(k+1) \\
&\quad -\dot{q}_r(k)a_{rj}(k)\dot{q}_j(k)]+\Delta V(k) \\
&= \frac{1}{2}\Delta\left[\sum_{i,j=1}^{n} a_{ij}(k)\dot{q}_i(k)\dot{q}_j(k)\right]+\Delta V(k) \\
&= \Delta T(k)+\Delta V(k).
\end{aligned}
$$

Now we shall show that the smoothing formula (3.18.14) is not only a sufficient condition for (3.18.15) to be valid but also a necessary condition. For this, we assume Lagrange's equations of the form (multiplying throughout by $\Delta q_r(k)$)

$$
\text{(3.18.17)}\quad F_r(k)\Delta q_r(k) = \frac{\Delta q_r(k)}{\Delta t(k)}\Delta\left[\sum_{j=1}^{n} a_{rj}(k)\dot{q}_j(k)\right] -\frac{1}{2}\sum_{i,j=1} [\Delta_r a_{ij}(k)]Q(\dot{q}_i,\dot{q}_j)+\Delta_r V(k),
$$

where $Q$ is as yet an unspecified discretization of $\dot{q}_i\dot{q}_j$, but symmetric in $i$ and $j$.

Summing (3.18.17) from $r=1$ to $r=n$ and using (3.18.9), to obtain

$$
\sum_{r=1}^{n} F_r(k)\Delta q_r(k) = \sum_{r=1}^{n}\frac{\Delta q_r(k)}{\Delta t(k)}\Delta\left[\sum_{j=1}^{n} a_{rj}(k)\dot{q}_j(k)\right] -\frac{1}{2}\sum_{i,j=1}^{n} [\Delta a_{ij}(k)]Q(\dot{q}_i,\dot{q}_j)+\Delta V(k),
$$

which on comparing with (3.18.15) leads to what is required to prove

$$\begin{aligned}\Delta T &= \frac{1}{2}\Delta\left[\sum_{i,j=1}^{n} a_{ij}(k)\dot{q}_i(k)\dot{q}_r(k)\right] \\ &= \frac{1}{2}\sum_{i,j=1}^{n}\left\{\frac{\Delta q_i(k)}{\Delta t(k)}\Delta[a_{ij}(k)\dot{q}_j(k)] + \frac{\Delta q_j(k)}{\Delta t(k)}\Delta[a_{ij}(k)\dot{q}_i(k)]\right\} \\ &\quad - \frac{1}{2}\sum_{i,j=1}^{n}[\Delta a_{ij}(k)]Q(\dot{q}_i,\dot{q}_j),\end{aligned}$$

where use has been made of the symmetry $a_{ij} = a_{ji}$. Equating the coefficients of $a_{ij}(k+1)$ and $a_{ij}(k)$ in turn leads to

$$\text{(3.18.18)} \qquad \frac{\Delta q_i(k)}{\Delta t(k)}\dot{q}_j(k+1) + \frac{\Delta q_j(k)}{\Delta t(k)}\dot{q}_i(k+1) - Q(\dot{q}_i,\dot{q}_j) = \dot{q}_i(k+1)\dot{q}_j(k+1)$$

$$\text{(3.18.19)} \qquad \frac{\Delta q_i(k)}{\Delta t(k)}\dot{q}_j(k) + \frac{\Delta q_j(k)}{\Delta t(k)}\dot{q}_i(k) - Q(\dot{q}_i,\dot{q}_j) = \dot{q}_i(k)\dot{q}_j(k).$$

Putting $j = i$ and subtracting (3.18.19) from (3.18.18) gives

$$2\frac{\Delta q_i(k)}{\Delta t(k)}[\dot{q}_i(k+1) - \dot{q}_i(k)] = \dot{q}_i^2(k+1) - \dot{q}_i^2(k)$$

from which the smoothing formula (3.18.14) follows immediately. The expression for $Q$ is obtained by substituting (3.18.14) into (3.18.19)

$$\text{(3.18.20)} \qquad Q(\dot{q}_i,\dot{q}_j) = \frac{1}{2}[\dot{q}_i(k)\dot{q}_j(k+1) + \dot{q}_i(k+1)\dot{q}_j(k)].$$

When $Q$ is inserted into (3.18.17) it can be replaced by $\dot{q}_i(k)\dot{q}_j(k+1)$ because of $a_{ij} = a_{ji}$.

## 3.19. Problems

**3.19.1.** Show that $u(k) = ck^2 + c^2$ is a solution of the difference equation

$$u(k) = \frac{\Delta u(k)}{(2k+1)}\left(k^2 + \frac{\Delta u(k)}{(2k+1)}\right), \qquad k \in \mathbb{N}.$$

**3.19.2.** Show that $u(k) = ca^k + c^2$ is a solution of the difference equation

$$\left(\frac{\Delta u(k)}{a-1}\right)^2 + a^{2k}\left(\frac{\Delta u(k)}{a-1}\right) - a^{2k}u(k) = 0, \qquad k \in \mathbb{N}, \quad a \neq 1.$$

**3.19.3.** Show that $u(k) = c2^{2k} - (3/16)c^3$ is a solution of the difference equation

$$\frac{1}{4}u(k) = \frac{1}{3}\Delta u(k-1) - \frac{1}{9}4^{-3k}(\Delta u(k-1))^3.$$

**3.19.4.** Solve the following Euler's difference equations

(i) $(k+1)k\Delta^2 u(k) + k\Delta u(k) - (1/4)u(k) = 0, \quad k \in \mathbb{N}(1)$

(ii) $(k+1)k\Delta^2 u(k) - k\Delta u(k) + u(k) = 0, \quad k \in \mathbb{N}(1)$

(iii) $(k+1)k\Delta^2 u(k) - 3k\Delta u(k) - 9u(k) = 0, \quad k \in \mathbb{N}(1)$.

**3.19.5.** Find the general solution of the Riccati equation

$$u(k)u(k+1) + pu(k+1) + qu(k) + r = 0, \quad k \in \mathbb{N}$$

where $p$, $q$ and $r$ are constants. In particular, solve for

(i) $p = 6,\ q = 1,\ r = 12$

(ii) $p = 5,\ q = 1,\ r = 9$

(iii) $p = 2,\ q = 2,\ r = 8$.

**3.19.6.** Show that

$$u(k) = \begin{cases} \left[c\left(-\frac{p}{q}\right)^k - \frac{1}{p+q}\right]^{-1}, & \text{if } p \neq -q \\ \left[c - \frac{1}{q}k\right]^{-1}, & \text{if } p = -q \end{cases}$$

is the general solution of the Riccati equation $u(k)u(k+1) + pu(k+1) + qu(k) = 0, \quad k \in \mathbb{N}$.

**3.19.7.** Find the general solution $u(k)$ of each of the following difference equations. Further, if $u(k)$ is periodic then find its period, or $\lim_{k\to\infty} u(k)$ if the limit exists.

(i) $u(k+1) = 2 - \dfrac{5}{u(k)}$

(ii) $u(k+1) = 2 + \dfrac{1}{u(k)}$

(iii) $u(k+1) = \dfrac{1}{6}\left(1 + \dfrac{1}{u(k)}\right)$.

**3.19.8.** Use proper transformations to show that the following nonlinear equations can be reduced to linear equations which can be solved easily

(i) $u(k)u(k+2) = u^2(k+1)$

(ii) $(k+1)u^2(k+1) = ku^2(k)$

(iii) $\sqrt{u(k+1)} = k\sqrt{u(k)}$

(iv) $u^2(k+2) - 4u^2(k+1) + 3u^2(k) = 0$

(v) $u(k+1) = (1 + u^{1/3}(k))^3$

(vi) $u(k+1)u(k)u(k-1) = c^2(u(k+1) + u(k) + u(k-1))$, $c$ is a constant

(vii) $u(k+1) = 2u(k)(1 - u(k))$

(viii) $u(k+1) = 2u(k)\sqrt{1 - u^2(k)}$.

**3.19.9.** Show that Newton's method for the computation of the square root of a positive number $A$ reduces to the recurrence relation

$$(3.19.1) \qquad u(k+1) = \frac{1}{2}\left(u(k) + \frac{A}{u(k)}\right),$$

where $u(0)$ is a known initial approximation of $\sqrt{A}$. Further, show that the transformation $u(k) = \sqrt{A}\coth v(k)$ reduces (3.19.1) to the simple equation $v(k+1) = 2v(k)$, and hence the solution of (3.19.1) can be written as $u(k) = \sqrt{A}\coth(c2^k)$, where $c = \coth^{-1}(u(0)/\sqrt{A})$.

**3.19.10.** Let $A$ be a positive number and consider the nonlinear difference equation

$$(3.19.2) \qquad u(k+1) = \frac{1}{2}\left(u(k) - \frac{A}{u(k)}\right).$$

Show that the transformation $u(k) = \sqrt{A}\cot v(k)$ reduces (3.19.2) to the simple equation $v(k+1) = 2v(k)$, and hence the solution of (3.19.2) can be written as $u(k) = \sqrt{A}\cot(c2^k)$, where $c$ is an arbitrary constant.

**3.19.11.** Show that the solution $u(k) = (1/\alpha)2^{-k-1}$ of the recurrence relation

$$(3.19.3) \qquad u(k+1) = \frac{\alpha u^2(k)}{2\left(1 - \alpha\sum_{j=0}^{k} u(j)\right)}$$

$$(3.19.4) \qquad u(0) = \frac{1}{2\alpha}$$

can be obtained as follows:

(i) the transformation $\alpha u(k) = v(k) - v(k-1)$, $v(-1) = 0$ reduces (3.19.3), (3.19.4) to the second order problem

$$(3.19.5) \qquad v(k+1) = v(k) + \frac{1}{2}\frac{(v(k) - v(k-1))^2}{1 - v(k)}$$

$$(3.19.6) \quad v(-1) = 0, \quad v(0) = \frac{1}{2}$$

(ii) the solution of the first order problem

$$(3.19.7) \qquad w(k+1) = \frac{1}{2}(w(k)+1), \quad w(0) = \frac{1}{2}$$

is also the solution of (3.19.5).

**3.19.12.** For an arbitrary constant $c$ show that the solution of the first order equation

$$(3.19.8) \qquad w(k+1) = \frac{\frac{1}{2}w^2(k) - c}{w(k) - 1}$$

is also a solution of the second order equation (3.19.5). The transformation $w(k) = z(k)+1$ reduces (3.19.8) to the form (3.19.1) or (3.19.2), and hence find the general solution of the second order equation (3.19.5).

**3.19.13.** For an arbitrary constant $c$ show that the solution of the first order equation $v(k+1)(1+v^2(k)) + v(k) = c$ is also a solution of the second order equation

$$u(k+2) = u(k)\frac{(1+u(k)u(k+1))}{1+u^2(k+1)}.$$

**3.19.14.** Difference equations which can be expressed in the form

$$f\left(\frac{u(k+1)}{u(k)}, k\right) = 0$$

are called *homogeneous equations*. Solve the following homogeneous difference equations

(i) $u^2(k+1) - 12u(k+1)u(k) + 27u^2(k) = 0$

(ii) $u^2(k+1) + (2k-4)u(k+1)u(k) - 8ku^2(k) = 0.$

**3.19.15.** (*Compound Interest*). Assume that the amount $P$ is deposited in a saving account, and together with the interest, is kept there for $k$ time periods, at the interest rate of $c$ per period. Show that the value $u(k)$ of the account at the end of the $k$th period will be $u(k) = P(1+c)^k$.

**3.19.16.** (*Periodic Payment of Annuities*). Assume that at the beginning of each time period for the next $k$ periods we deposit in a saving account the fixed amount $P$, at the interest rate of $c$ (compounded) per period.

Show that the value of the account immediately after the $k$th payment will be $u(k) = P((1+c)^k - 1)/c$.

**3.19.17.** (*Periodic Payment to Pay off a Loan*). Show that the periodic payment $P$ necessary to pay off a loan $A$ in $K$ periods at an interest rate of $c$ per period is $P = Ac\{1-(1+c)^{-K}\}^{-1}$.

**3.19.18.** Show that for the differential equation (3.5.1) the difference equation approximation

$$u(k+1) - u(k) = \frac{(e^{\beta h}-1)}{\beta}(\beta - \gamma u(k+1))u(k) \tag{3.19.9}$$

has the same solution (3.5.8), and hence it is also a best approximation of (3.5.1).

**3.19.19.** Duffing's equation $y'' + 10y + 90y^3 = 0$ together with the initial conditions $y(0) = 1$, $y'(0) = 0$ has the solution $y(t) = \mathrm{cn}(10t)$, parameter $m = 0.45$ and the time period $T = 0.72555$. For each $p = 2^\ell$, $\ell = 1, 2, \cdots, 5$ use (3.7.3), (3.7.4), (3.7.11), (3.7.13) and $T = 2ph$ with $a = 10$, $b = 90$, $A = 1$ to compute the approximate solution $u(k)$, approximate parameter $m$, step size $h$ and the approximate time period $T$.

**3.19.20.** Duffing's equation $y'' - 24y + 32y^3 = 0$ together with the initial conditions $y(0) = 1$, $y'(0) = 0$ has the solution $y(t) = \mathrm{dn}(4t)$, parameter $m = 0.5$ and the time period $T = 0.92704$. For each $p = 2^\ell$, $\ell = 1, 2, \cdots, 5$ use (3.7.3), (3.7.4), (3.7.16), (3.7.18) and $T = 2ph$ with $a = -24$, $b = 32$, $A = 1$ to compute the approximate solution $u(k)$, approximate parameter $m$, step size $h$ and the approximate time period $T$.

**3.19.21.** Duffing's equation $y''+6y-4y^3 = 0$ together with the boundary conditions $y(0) = 0$, $y(t) = 1$ for the least $t > 0$ for which $y'(t) = 0$ has the solution $y(t) = \mathrm{sn}(2t)$, parameter $m = 0.5$ and the time period $T = 3.70815$. For each $p = 2^\ell$, $\ell = 1, 2, \cdots, 5$ use (3.7.3), (3.7.5), (3.7.23), (3.7.25) and $T = 4ph$ with $a = 6$, $b = -4$, $A = 1$ to compute the approximate solution $u(k)$, approximate parameter $m$, step size $h$ and the approximate time period $T$.

**3.19.22.** Show that in case I for the Duffing equation (3.7.1) the best difference equation approximation is

$$(u(k+1) + u(k-1))\left[\mathrm{dn}^2 + \frac{m}{A^2}u^2(k)\mathrm{sn}^2\right] = 2u(k)\mathrm{cn}$$

satisfying (3.7.4), where $\mathrm{sn} = \mathrm{sn}[(a+bA^2)^{1/2}h]$, and similarly cn and dn, i.e. its solution $u(k) = A\mathrm{cn}[(a+bA^2)^{1/2}kh] = y(kh)$.

**3.19.23.** Show that in case II for the Duffing equation (3.7.1) the best difference equation approximation is

$$(u(k+1)+u(k-1))\left[\mathrm{cn}^2+\frac{1}{A^2}u^2(k)\mathrm{sn}^2\right] = 2u(k)\mathrm{dn}$$

satisfying (3.7.4), where $\mathrm{sn} = \mathrm{sn}[A(b/2)^{1/2}h]$, and similarly cn and dn, i.e. its solution $u(k) = A\mathrm{dn}[A(b/2)^{1/2}kh] = y(kh)$.

**3.19.24.** Show that in case III for the Duffing equation (3.7.1) the best difference equation approximation is

$$(u(k+1)+u(k-1))\left[1-\frac{m}{A^2}u^2(k)\mathrm{sn}^2\right] = 2u(k)\mathrm{cn}\ \mathrm{dn}$$

satisfying (3.7.5), where $\mathrm{sn} = \mathrm{sn}[(a+(bA^2/2))^{1/2}h]$, and similarly cn and dn, i.e. its solution $u(k) = A\mathrm{sn}[(a+(bA^2/2))^{1/2}kh] = y(kh)$.

**3.19.25.** Let $D_1 = 1$, and for $k \ge 2$, $D_k(g(1),\cdots,g(k-1))$

$$= \det\begin{bmatrix} g(k-1) & -1 & 0 & \cdots & 0 & 0 & 0 \\ -1 & g(k-2) & -1 & \cdots & 0 & 0 & 0 \\ 0 & -1 & g(k-3) & \cdots & 0 & 0 & 0 \\ \cdots & \cdots & \cdots & \cdots & \cdots & \cdots & \cdots \\ 0 & 0 & 0 & \cdots & g(3) & -1 & 0 \\ 0 & 0 & 0 & \cdots & -1 & g(2) & -1 \\ 0 & 0 & 0 & \cdots & 0 & -1 & g(1) \end{bmatrix}.$$

Show that for all $k \in \mathbb{N}(1)$

$$D_{k+1}(g(1),\cdots,g(k)) = g(k)D_k(g(1),\cdots,g(k-1)) - D_{k-1}(g(1),\cdots,g(k-2))$$

$$D_0 = 0, \quad D_1 = 1$$

and hence $D_k(g(1),\cdots,g(k-1))$ is a solution of the Hill's difference equation (3.9.2). Further, for the matrices $C_2$, $S_2$, $C_1$ and $S_1$ defined in (3.9.12) (3.9.15) deduce that

$$\begin{aligned}
\det C_2 &= g(0)g(r)D_r(g(1),\cdots,g(r-1)) - 2g(0)D_{r-1}(g(1),\cdots,g(r-2)) \\
&\quad -2g(r)D_{r-1}(g(2),\cdots,g(r-1)) + 4D_{r-2}(g(2),\cdots,g(r-2)), \quad r \ge 3 \\
\det S_2 &= D_r(g(1),\cdots,g(r-1)), \quad r \ge 1 \\
\det C_1 &= g(0)D_r(g(1),\cdots,g(r-1)) - 2D_{r-1}(g(2),\cdots,g(r-1)), \quad r \ge 2 \\
\det S_1 &= g(r)D_r(g(1),\cdots,g(r-1)) - 2D_{r-1}(g(1),\cdots,g(r-2)), \quad r \ge 2.
\end{aligned}$$

**3.19.26.** For the differential equation (3.11.1) show that $O(h)$, $O(h^2)$, and $O(h^4)$ discretizations are

$$(\Delta u(k)/h)^2 = 4u^3(k) - g_2 u(k) - g_3,$$

$$(\Delta u(k)/h)^2 = 4u(k)u(k+1)\frac{1}{2}(u(k)+u(k+1)) - g_2\frac{1}{2}(u(k)+u(k+1)) - g_3,$$

$$(\Delta u(k)/h)^2 = 4u(k)u(k+1)\frac{1}{2}(u(k)+u(k+1)) - g_2\frac{1}{2}(u(k)+u(k+1)) - g_3 - h^2\left[\left(u(k)u(k+1) + \frac{1}{4}g_2\right)^2 + g_3(u(k)+u(k+1))\right].$$

**3.19.27.** For the differential equation (3.11.2) show that $O(h)$, $O(h^2)$, and $O(h^4)$ discretizations are

$$\Delta^2 u(k-1)/h^2 = 6u^2(k) - \frac{1}{2}g_2,$$

$$\Delta^2 u(k-1)/h^2 = 2u(k)(u(k+1) + u(k) + u(k-1)) - \frac{1}{2}g_2,$$

$$\Delta^2 u(k-1)/h^2 = 2u(k)(u(k+1) + u(k) + u(k-1)) - \frac{1}{2}g_2 - h^2\left[u^2(k)(u(k+1) + u(k-1)) + \frac{1}{2}g_2 u(k) + g_3\right].$$

**3.19.28.** Use the Symbolic method to solve the following partial difference equations

(i) $u(k+1,\ell) = 3u(k,\ell+1)$

(ii) $u(k+2,\ell) = 9u(k,\ell+1)$

(iii) $u(k+1,\ell+1) = u(k+1,\ell) + u(k,\ell)$

(iv) $u(k,\ell+2) = u(k+1,\ell) + u(k,\ell).$

**3.19.29.** The Operational method given in Problem 2.16.41 to find a particular solution of the nonhomogeneous ordinary difference equation (2.12.1) can be extended easily to nonhomogeneous partial difference equation (3.13.1). Use this extended method to find particular solutions of the following nonhomogeneous partial difference equations

(i) $u(k+1,\ell) - 2u(k,\ell+1) = 4k^2\ell^2 + 3k^2\ell + 2k\ell^2 + 6k + 7\ell + 8$

(ii) $u(k+1,\ell+1) + 3u(k,\ell) = 4^k(2k^2 + k\ell + \ell^2 + 7k + 3)$

(iii) $u(k+2,\ell+2) - 11u(k+1,\ell+1) + 23u(k,\ell) = k\ell + k + \ell + 2$

(iv) $u(k+1,\ell+1) + 2u(k,\ell) = 3^k(k^2 + \ell^2 + 1).$

**3.19.30.** Consider the first order partial difference equation $(\alpha E_k + \beta E_\ell + \gamma)u(k,\ell) = b(k,\ell)$, where $\alpha$, $\beta$, and $\gamma$ are constants, and $b(k,\ell)$ satisfies $(\alpha E_k + \beta E_\ell + \gamma)b(k,\ell) = 0$. Show that its particular solution can be written as

$$u(k,\ell) = \begin{cases} -\dfrac{1}{\gamma}(k+\ell)b(k,\ell) & \text{if } \gamma \neq 0 \\ \left(-\dfrac{\beta}{\alpha}\right)^k c(k+\ell) & \text{if } \gamma = 0, \end{cases}$$

where $c$ is an arbitrary function of $k+\ell$.

**3.19.31.** Consider the partial difference equations given in Problem 3.19.28 and the corresponding homogeneous partial difference equations given in Problem 3.19.29. If possible use Lagrange's and the method of Separation of variables to solve these homogeneous equations.

**3.19.32.** Use Laplace's method to solve the following partial difference equations

(i) $u(k,\ell) + 2u(k-1,\ell-1) = \ell$

(ii) $u(k,\ell) - ku(k-1,\ell-1) = 0$

(iii) $u(k+4,\ell) - 4u(k+3,\ell+1) + 6u(k+2,\ell+2) - 4u(k+1,\ell+3) + u(k,\ell+4) = 0$

(iv) $u(k+4,\ell) - 16u(k,\ell+4) = 0.$

**3.19.33.** Let $\phi_i(E_k, E_\ell)$, $i = 1,2,3,4$ be polynomial functions of the operators $E_k$ and $E_\ell$, and $f(k,\ell)$, $g(k,\ell)$ be given functions of $k$ and $\ell$. The relationship

$$\begin{aligned} \phi_1(E_k,E_\ell)u(k,\ell) + \phi_2(E_k,E_\ell)v(k,\ell) &= f(k,\ell) \\ \phi_3(E_k,E_\ell)u(k,\ell) + \phi_4(E_k,E_\ell)v(k,\ell) &= g(k,\ell) \end{aligned} \tag{3.19.10}$$

defines a pair of simultaneous linear partial difference equations for the functions $u(k,\ell)$ and $v(k,\ell)$. These equations can be arranged to obtain two partial difference equations involving $u(k,\ell)$ and $v(k,\ell)$ separately. The solutions of these equations contain a number of arbitrary functions, however substitution of these solutions back into (3.19.10) determines proper number of arbitrary functions which should be present in the final solution.

Solve the following simultaneous linear partial difference equations

(i) $$\begin{aligned} E_k u(k,\ell) + E_\ell v(k,\ell) &= 1 \\ E_\ell u(k,\ell) + E_k v(k,\ell) &= k+\ell \end{aligned}$$

(ii) $$\begin{aligned} 2(2E_kE_\ell - 1)u(k,\ell) - (3E_kE_\ell - 1)v(k,\ell) &= 1 \\ 2(E_kE_\ell - 1)u(k,\ell) + (E_kE_\ell - 1)v(k,\ell) &= k. \end{aligned}$$

**3.19.34.** Use proper transformations to show that the following nonlinear equations can be reduced to linear equations which can be solved easily

(i) $u(k+1,\ell) = [u(k,\ell+1)]^p$, $p$ is a constant

(ii) $u(k+1,\ell+1) = [u(k+1,\ell)]^p[u(k,\ell)]^q$, $p$ and $q$ are constants

(iii) $[u(k+1,\ell+1)]^p = a[u(k+1,\ell)]^p + b[u(k,\ell)]^p$, $p$ is a constant.

**3.19.35.** The *Clairaut extended form* is a nonlinear partial difference equation

$$u(k,\ell) = k\Delta_k u(k,\ell) + \ell\Delta_\ell u(k,\ell) + f(\Delta_k u(k,\ell), \Delta_\ell u(k,\ell)).$$

Show that its solution is $u(k,\ell) = c_1k + c_2\ell + f(c_1,c_2)$, where $c_1$ and $c_2$ are arbitrary constants. In particular, solve the following equations

(i) $u(k,\ell) = k\Delta_k u(k,\ell) + \ell\Delta_\ell u(k,\ell) + (\Delta_k u(k,\ell))^3 \sin^2(\Delta_\ell u(k,\ell))^2$

(ii) $$\begin{aligned} &u(k,\ell)[1 + k + \ell + u(k,\ell+1) + u(k+1,\ell) - u(k,\ell)] \\ &\quad = ku(k+1,\ell) + \ell u(k,\ell+1) + u(k+1,\ell)u(k,\ell+1). \end{aligned}$$

**3.19.36.** Show that the partial difference equation

$$u(k+1,\ell+1) - a(k,\ell)u(k,\ell) = b(k,\ell) \tag{3.19.11}$$

can be solved as follows:

(i) the solution of the equation

$$u(k+1,\ell+1) - u(k,\ell) = b(k,\ell) \tag{3.19.12}$$

can be written as $u(k,\ell) = c(k-\ell) + \sum_{\tau=1}^{k-1} b(\tau, \ell - k + \tau)$, where $c$ is an arbitrary function of $k - \ell$

(ii) a particular solution $u_1(k,\ell)$ of the homogeneous difference equation $u(k+1,\ell+1) - a(k,\ell)u(k,\ell) = 0$ can be obtained by using the transformation $v(k,\ell) = \ln u(k,\ell)$.

(iii) elimination of $a(k,\ell)$ between (3.19.11) and the equation $u_1(k+1,\ell+1) - a(k,\ell)u_1(k,\ell) = 0$ leads to a equation of the type (3.19.12).

**3.19.37.** John and Robert engage in transactions with each other. The result of each transaction is that a coin passes from one to the other. The probability of a transaction being to John's advantage is $p$. Each proposes to retire when and if he acquires $K$ coins. We shall find the probability of John's retiring. For this, let $k$ and $\ell$ be the number of coins John and Robert possess at any stage. Then $k$ and $\ell$ are positive integers and $k + \ell = M$, a constant. Let $u(k, \ell)$ be the probability of John finally succeeding, as estimated at the $k$, $\ell$ stage. The probability of John finally retiring is the probability of John gaining a coin at the next step and thereafter finally winning the necessary number $K$, plus the probability of John losing the next coin and thereafter finally winning the necessary number.

Now if John wins a coin at the $(k, \ell)$ stage the probability of his winning $K$ finally is $u(k+1, \ell-1)$, while if John loses a coin at the $(k, \ell)$ stage the probability of his finally acquiring $K$ is $u(k-1, \ell+1)$. Hence

$$u(k, \ell) = pu(k+1, \ell-1) + (1-p)u(k-1, \ell+1). \tag{3.19.13}$$

Next, when John has $K$ coins, Robert has $M - K$, the probability of John retiring is 1. Thus,

$$u(K, M-K) = 1. \tag{3.19.14}$$

Similarly, Robert retires when he has $K$ coins, i.e.

$$u(M-K, K) = 0. \tag{3.19.15}$$

Show that

(i) the solution of (3.19.13) (3.19.15) can be written as

$$u(k, \ell) = p^{K-k} \frac{(1-p)^{K-\ell} - p^{K-\ell}}{(1-p)^{2K-k-\ell} - p^{2K-k-\ell}}$$

(ii) if $k = \ell = K/2$ and $p = 1/2$, then $u = 1/2$

(iii) if $k = \ell = K/2$ and $p = 1/3$, then the probability of Robert retiring is $1/(1 + 2^{-K/2})$.

**3.19.38.** Show that $\dfrac{(k+\ell)!}{k!\ \ell!}$ is the total number of ways to reach the point $(k, \ell) \in \mathbb{N} \times \mathbb{N}$ from $(0,0)$, while always moving parallel to the positive direction of the coordinate axes by one unit.

**3.19.39.** Solve the following Riccati's extended form partial difference systems

(i)

$$\begin{aligned} u(k+1, \ell+1) &= \frac{v(k, \ell) + 2}{u(k, \ell) - v(k, \ell)} \\ v(k+1, \ell+1) &= \frac{u(k, \ell) + 1}{u(k, \ell) - v(k, \ell)} \end{aligned}$$

(ii)
$$u(k+1,\ell+1) = \frac{u(k,\ell)-v(k,\ell)+1}{u(k,\ell)-v(k,\ell)}$$
$$v(k+1,\ell+1) = \frac{u(k,\ell)+v(k,\ell)-2}{u(k,\ell)-v(k,\ell)}.$$

**3.19.40.** Show that both the functions $A\tan\omega kh$, $A\cot\omega kh$ satisfy the same difference equation

$$\frac{\Delta^2 u(k-1)}{\omega^{-2}\tan^2\omega h} - 2\omega^2 u(k) - \omega^2 A^{-2} u^2(k)(u(k+1)+u(k-1)) = 0, \quad (3.19.16)$$

which as $h \to 0$ converges to the differential equation

$$y'' - 2\omega^2 y - 2\omega^2 A^{-2} y^3 = 0 \quad (3.19.17)$$

for which $A\tan\omega t$, $A\cot\omega t$ are solutions, i.e. for the differential equation (3.19.17) the difference equation approximation (3.19.16) is the best.

**3.19.41.** Show that both the functions $A\csc\omega kh$, $A\sec\omega kh$ satisfy the same difference equation

$$\frac{\Delta^2 u(k-1)}{\omega^{-2}\sin^2\omega h} + \frac{\omega^2 u(k)}{\cos^2(\omega h/2)} - \omega^2 A^{-2} u^2(k)(u(k+1)+u(k-1)) = 0, \quad (3.19.18)$$

which as $h \to 0$ converges to the differential equation

$$y'' + \omega^2 y - 2\omega^2 A^{-2} y^3 = 0 \quad (3.19.19)$$

for which $A\csc\omega t$, $A\sec\omega t$ are solutions, i.e. for the differential equation (3.19.19) the difference equation approximation (3.19.18) is the best.

**3.19.42.** Show that both the functions $A\sinh\omega kh$, $A\cosh\omega kh$ satisfy the same difference equation

$$\frac{\Delta^2 u(k-1)}{4\omega^{-2}\sinh^2(\omega h/2)} - \omega^2 u(k) = 0, \quad (3.19.20)$$

which as $h \to 0$ converges to the differential equation

$$y'' - \omega^2 y = 0 \quad (3.19.21)$$

for which $A\sinh\omega t$, $A\cosh\omega t$ are solutions, i.e. for the differential equation (3.19.21) the difference equation approximation (3.19.20) is the best.

**3.19.43.** Show that the function $A\operatorname{csch}\omega kh$ satisfies the difference equation

$$\frac{\Delta^2 u(k-1)}{\omega^{-2}\sinh^2\omega h} - \frac{\omega^2 u(k)}{\cosh^2(\omega h/2)} - \omega^2 A^{-2} u^2(k)(u(k+1)+u(k-1)) = 0, \quad (3.19.22)$$

which as $h \to 0$ converges to the differential equation

$$y'' - \omega^2 y - 2\omega^2 A^{-2} y^3 = 0 \tag{3.19.23}$$

for which $A\text{csch}\,\omega t$ is a solution, i.e. for the differential equation (3.19.23) the difference equation approximation (3.19.22) is the best.

**3.19.44.** Show that the function $A\tanh \omega kh$ satisfies the difference equations

$$\frac{u(k+1) - u(k-1)}{\omega^{-1}\sinh 2\omega h} = \omega \frac{u(k+1) + u(k-1)}{2u(k)} A(1 - A^{-2}u^2(k)), \tag{3.19.24}$$

$$\frac{\Delta^2 u(k-1)}{\omega^{-2}\tanh^2 \omega h} + 2\omega^2 u(k) - \omega^2 A^{-2} u^2(k)(u(k+1) + u(k-1)) = 0, \tag{3.19.25}$$

and

$$\begin{aligned} &\frac{u(k+3) - 3u(k+1) + 3u(k-1) - u(k-3)}{(\omega^{-1}\sinh 2\omega h)^3} \\ &+ 6\omega A^{-1}\cosh 2\omega h \left[\frac{u(k+3) + u(k-3)}{u(k+1) + u(k-1)}\right] \left[\frac{u(k+1) - u(k-1)}{\omega^{-1}\sinh 2\omega h}\right]^2 \\ &- 4\omega^2 \left[\frac{u(k+3) + u(k-3)}{u(k+1) + u(k-1)}\right] \left[\frac{u(k+1) - u(k-1)}{\omega^{-1}\sinh 2\omega h}\right] = 0, \end{aligned} \tag{3.19.26}$$

which as $h \to 0$ converge to the respective differential equations

$$\begin{aligned} y' &= \omega A(1 - A^{-2}y^2), \\ y'' + 2\omega^2 y - 2\omega^2 A^{-2} y^3 &= 0, \end{aligned}$$

and

$$y''' + 6\omega A^{-1} y'^2 - 4\omega^2 y' = 0.$$

**3.19.45.** Show that the function $A\text{sech}\,\omega kh$ satisfies the difference equations

$$\frac{u(k+1) - u(k-1)}{2\omega^{-1}\tanh \omega h} = -\omega(1 - A^{-2}u^2(k))^{1/2} \frac{u(k+1) + u(k-1)}{2}, \tag{3.19.27}$$

$$\frac{\Delta^2 u(k-1)}{\omega^{-2}\sinh^2 \omega h} - \frac{\omega^2 u(k)}{\cosh^2(\omega h/2)} + \omega^2 A^{-2} u^2(k)(u(k+1) + u(k-1)) = 0, \tag{3.19.28}$$

and

$$\frac{u(k+3) - 3u(k+1) + 3u(k-1) - u(k-3)}{(2\omega^{-1}\sinh \omega h)^3} \cosh 3\omega h \tag{3.19.29}$$

$$+ 3\omega^2 A^{-2}u(k)(u(k+3) + u(k-3))\cosh\, \omega h \cosh\, 2\omega h \times$$
$$\frac{u(k+1)-u(k-1)}{2\omega^{-1}\tanh\, \omega h} - \omega^2 \left[\frac{u(k+3)+u(k-3)}{u(k+1)+u(k-1)}\right] \left[\frac{u(k+1)-u(k-1)}{2\omega^{-1}\tanh\, \omega h}\right] = 0,$$

which as $h \to 0$ converges to the respective differential equations

$$y' = -\omega(1 - A^{-2}y^2)^{1/2}y,$$
$$y'' - \omega^2 y + 2\omega^2 A^{-2}y^3 = 0,$$

and

$$y''' + 6\omega^2 A^{-2}y^2 y' - \omega^2 y' = 0.$$

**3.19.46.** The solution $\phi(x,t)$ of the nonlinear *reaction advection equation*

$$\phi_t + \phi_x = \phi(1 - \phi) \tag{3.19.30}$$

with the initial function

$$\phi(x,0) = f(x), \tag{3.19.31}$$

where $f(x)$ is bounded with a bounded derivative, can be written as

$$\phi(x,t) = \frac{f(x-t)}{e^{-t} + (1 - e^{-t})f(x-t)}. \tag{3.19.32}$$

In (3.19.32) use $x = kh,\ t = \ell h$ where $k,\ \ell \in \mathbb{N}$ and $h > 0$ is the step size, and use the representation $\phi(x,t) = \phi(kh, \ell h) = u(k,\ell),\ f(x-t) = f((k-\ell)h) = f_k^\ell,$ and note the property that $f_k^{\ell+1} = f_{k-1}^\ell,$ to obtain the following best possible discretization of (3.19.30)

$$u(k,\ell+1) - u(k-1,\ell) = (e^h - 1)u(k-1,\ell)(1 - u(k,\ell+1)).$$

**3.19.47.** For the nonlinear *diffusion equation*

$$\phi_t = \phi\phi_{xx} \tag{3.19.33}$$

use $x = kh_1,\ t = \ell h_2$ where $k,\ \ell \in \mathbb{N}$ and $h_1,\ h_2 > 0$ are step sizes, and represent $\phi(x,t) = \phi(kh_1, \ell h_2) = u(k,\ell),$ to obtain the following discretizations

(3.19.34)
$$\frac{u(k,\ell+1) - u(k,\ell)}{h_2} = u(k,\ell+1)\left[\frac{u(k+1,\ell) - 2u(k,\ell) + u(k-1,\ell)}{h_1^2}\right]$$

(explicit)

(3.19.35)

$$\frac{u(k,\ell+1)-u(k,\ell)}{h_2} = u(k,\ell)\left[\frac{u(k+1,\ell+1)-2u(k,\ell+1)+u(k-1,\ell+1)}{h_1^2}\right].$$

(implicit)

Show that both (3.19.34) and (3.19.35) are the best discretizations of (3.19. 33) in the sense that

$$\phi(x,t) = \frac{(\alpha/2)x^2+\beta x+\gamma}{\delta-\alpha t}$$

where $\alpha$, $\beta$, $\gamma$, $\delta$ are constants, is a solution of all the three equations (3.19.33) (3.19.35).

## 3.20. Notes

Discrete Clairaut, Euler and Riccati equations are discussed in almost every classical book on difference equations, e.g. Milne Thomson [23]. For several recent results to Riccati equation see Agarwal and Balla [2], and Blandzi and Popenda [4]. The discrete analog of Bernoulli's differential equation considered in Section 3.4 appears to be new. For the Verhulst differential equation several possible difference equation approximations have been analyzed in Hoppensteadt and Hyman [12], May [16 18], Yamaguti et. al. [38 40]. The discretization considered in Section 3.5 is based on Potts [29]. Best possible difference equation approximations of the linear differential equations with constant coefficients are available in Potts [28], and Reid [35]. The nonlinear difference equations approximating Duffing's differential equation considered in Section 3.7 as well as in Problems 3.19.22 3.19.24 are taken from Potts [26,27]. The discrete van der Pol's equation (3.8.3) is studied in Potts [30]. The discussion of Hill's difference equation and in particular of Mathieu's difference equation is borrowed from Potts [31]. The analogy between the continuous and discrete Floquet theories has been discussed from a different point of view by Hochstadt [11]. Best difference equation approximations of Weierstrass' elliptic equations are considered in Potts [33]. The discrete Volterra's systems (3.12.4), (3.12.5) have appeared in the work of Potts [29]. An elementary treatment of partial difference equations is available at several places, e.g. Boole [5], and Levy and Lessman [15]. Best partial difference approximations of Wave equation, FitzHugh Nagumo's equation, Korteweg de Vries' equation and Modified KdV equation are due to Potts [32]. The determination of partial difference equations analogs of nonlinear partial differential equations has also been considered in Hirota [10]. Using the smoothing formula of Greenspan [9], Gotusso [7], and Neumann and Tourasis [24] have considered special discretizations of Lagrange's equations of motion. Our discussion in Section

3.18 is based on Potts [34]. For several other related results to this chapter see Gotusso and Veneziani [8], Meyer Spasche and Düchs [21], and Mickens [22].

## 3.21. References

[1]. M. Abramowitz and I.A. Stegun (ed.), *Handbook of Mathematical Functions, U.S. National Bureau of Standards*, 1964.

[2]. R.P. Agarwal and K. Balla, On Riccati's extended form, *Applicable Analysis* **69**(1998), 133 147.

[3]. F.M. Arscott, *Periodic Differential Equations, Pergamon Press*, Oxford, 1964.

[4]. E.M. Blandzi and J. Popenda, On the behavior of rational first order difference equation with constant coefficients, *Demonstratio Mathematica* **31**(1998), 405 416.

[5]. G. Boole, *Calculus of Finite Differences*, 4th ed., *Chelsea*, New York, 1958.

[6]. G. Eilenberger, *Solitons: Mathematical Methods for Physicists, Springer Verlag*, Berlin, 1983.

[7]. L. Gotusso, On the energy theorem for the Lagrange equations in the discrete case, *Appl. Math. Comp.* **17**(1985), 129 136.

[8]. L. Gotusso and A. Veneziani, Discrete and continuous nonlinear models for the vibrating rod, *Mathl. Comput. Modelling* **24**(4)(1996), 95 115.

[9]. D. Greenspan, *Discrete Models, Addison Wesley*, Mass. USA, 1973.

[10]. R. Hirota, Nonlinear partial difference equations, V, Nonlinear equations reducible to linear equations, *J. Phys. Soc. Japan* **46**(1979), 312 319.

[11]. H. Hochstadt, On the theory of Hill's matrices and related inverse scattering problems, *Linear Alg. Appl.* **11**(1975), 41 52.

[12]. F.C. Hoppensteadt and J.M. Hyman, Periodic solutions of a logistic difference equation, *SIAM J. Appl. Math.* **32**(1977), 73 81.

[13]. D.S. Jones and B.D. Sleeman, *Differential Equations and Mathematical Biology, George Allen and Unwin*, London, 1983.

[14]. D.W. Jordan and P. Smith, *Nonlinear Ordinary Differential Equations*, Clarendon Press, Oxford, 1977.

[15]. H. Levy and F. Lessman, *Finite Difference Equations, Sir Isaac Pitman and Sons, Ltd.*, London, 1959.

[16]. R.M. May, Biological populations with nonoverlapping generations: stable points, stable cycles and chaos, *Science* **186**(1974), 645 647.

[17]. R.M. May, Biological problems obeying difference equations: stable points, stable cycles and chaos, *J. Theor. Biol.* **51**(1975), 511 524.

[18]. R.M. May, Simple mathematical models with very complicated dynamics, *Nature* **261**(1976), 459 467.

[19]. N.W. McLachlan, *Theory and Application of Mathieu Functions, Clarendon Press*, Oxford, 1947.

[20]. N.W. McLachlan, *Ordinary Non linear Differential Equations in Engineering and Physical Sciences, Oxford Univ. Press*, 2nd ed., 1956.

[21]. R. Meyer Spasche and D. Düchs, A general method for obtaining unconventional and nonstandard difference schemes, *Dynamics of Continuous, Discrete and Impulsive Systems* **3**(1997), 453 467.

[22]. R.E. Mickens, *Nonstandard Finite Difference Models of Differential Equations, World Scientific*, Singapore, 1994.

[23]. L.M. Milne Thomson, *The Calculus of Finite Differences, Macmillan*, London, 1960.

[24]. C.P. Neumann and V.D. Tourasis, Discrete dynamic robot models, *IEEE Trans. Systems Man, Cybernet SMC-15*(1985), 193 204.

[25]. E.C. Pielou, *An Introduction to Mathematical Ecology, Wiley*, New York, 1969.

[26]. R.B. Potts, Exact solution of a difference approximation to Duffing's equation, *J. Austral. Math. Soc. (Series B)* **23**(1981), 64 77.

[27]. R. B. Potts, Best difference equation approximation to Duffing's equation, *J. Austral. Math. Soc. (Series B)* **23**(1981), 349 356.

[28]. R.B. Potts, Differential and difference equations, *Amer. Math. Monthly* **89**(1982), 402 407.

[29]. R.B. Potts, Nonlinear difference equations, *Nonlinear Analysis* **6**(1982), 659-665.

[30]. R.B. Potts, van der Pol difference equation, *Nonlinear Analysis* **7**(19 83), 801 812.

[31]. R.B. Potts, Mathieu's difference equation, in *The Wonderful World of Stochastics*, eds. M.F. Shlesinger and G.H. Weiss, *Elsevier Science Publ. B.V.*, 1985, 111 125.

[32]. R.B. Potts, Ordinary and partial difference equations, *J. Austral. Math. Soc. (Series B)* **27**(1986), 488 501.

[33]. R.B. Potts, Weierstrass elliptic difference equations, *Bull. Austral. Math. Soc.* **35**(1987), 43-48.

[34]. R.B. Potts, Discrete Lagrange equations, *Bull. Austral. Math. Soc.* **36**(1988), 227 233.

[35]. J. Gary Reid, *Linear System Fundamentals, Continuous and Discrete, Classic and Modern*, *McGraw Hill*, New York, 1983.

[36]. V. Volterra, *Lecons sur la théorie mathématiques de la lutte pour la vie*, *Paris*, 1931.

[37]. G.B. Whitham, *Linear and Nonlinear Waves*, *Wiley*, Ney York, 1974.

[38]. M. Yamaguti and H. Matano, Euler's finite difference scheme and chaos, *Proc. Japan Acad.* **55**A(1979), 78 80.

[39]. M. Yamaguti and S. Ushiki, Discretization and chaos, *C. R. Acad. Sc. Paris* **290**(1980), 637 640.

[40]. M. Yamaguti and S. Hushiki, Chaos in numerical analysis of ordinary differential equations, *Phisica* **3**D(1981), 618 626.

# Chapter 4
# Difference Inequalities

It is well recognized that the inequalities furnish a very general comparison principle in studying many qualitative as well as quantitative properties of solutions of related equations. The celebrated Gronwall's inequality is but one of the examples for a monotone operator $\mathcal{K}$ in which the exact solution of $\mathbf{w} = \mathbf{p} + \mathcal{K}\mathbf{w}$ provides an upper bound on all solutions of the inequality $\mathbf{u} \leq \mathbf{p} + \mathcal{K}\mathbf{u}$. On the basis of various motivations this inequality has been extended and applied in various contexts. We begin this chapter with Gronwall type inequalities, and include, in particular, the practically important case of weakly singular discrete kernels. This is followed by several nonlinear versions of Gronwall inequality which have been established recently and are of immense value. To deal with inequalities involving higher order differences a usual procedure is to convert them to their equivalent systems and then, either obtain the estimates in terms of maximal solutions of the related difference systems; or use a suitable norm and treat the resulting inequalities as in the scalar case, which provides uniform bounds for all the components of the systems. In Section 4.3 we shall deal with these type of inequalities directly and obtain the estimates in terms of known functions. Then, we shall move to finite linear as well as nonlinear difference inequalities and wherever possible provide upper bounds in terms of known quantities. In Sections 4.5 and 4.6, respectively, we shall consider discrete Opial and Wirtinger type inequalities.

In what follows, unless otherwise stated, all the functions which appear in the inequalities are assumed to be defined and nonnegative in their domains of definition.

## 4.1. Gronwall Inequalities

**Theorem 4.1.1.** Let for all $k \in \mathbb{N}(a)$ the following inequality be satisfied

$$u(k) \leq p(k) + q(k) \sum_{\ell=a}^{k-1} f(\ell) u(\ell). \tag{4.1.1}$$

Then, for all $k \in \mathbb{N}(a)$

$$(4.1.2) \qquad u(k) \leq p(k) + q(k) \sum_{\ell=a}^{k-1} p(\ell)f(\ell) \prod_{\tau=\ell+1}^{k-1} (1 + q(\tau)f(\tau)).$$

**Proof.** Define a function $v(k)$ on $\mathbb{N}(a)$ as follows

$$v(k) = \sum_{\ell=a}^{k-1} f(\ell)u(\ell).$$

For this function, we have

$$(4.1.3) \qquad \Delta v(k) = f(k)u(k), \quad v(a) = 0.$$

Since $u(k) \leq p(k) + q(k)v(k)$, and $f(k) \geq 0$, from (4.1.3) we get

$$(4.1.4) \qquad v(k+1) - (1 + q(k)f(k))v(k) \leq p(k)f(k).$$

Because $1 + q(k)f(k) > 0$ for all $k \in \mathbb{N}(a)$, we can multiply (4.1.4) by $\prod_{\ell=a}^{k}(1 + q(\ell)f(\ell))^{-1}$, to obtain

$$\Delta \left[ \prod_{\ell=a}^{k-1} (1 + q(\ell)f(\ell))^{-1} v(k) \right] \leq p(k)f(k) \prod_{\ell=a}^{k} (1 + q(\ell)f(\ell))^{-1}.$$

Summing the above inequality from $a$ to $k-1$, and using $v(a) = 0$, to get

$$\prod_{\ell=a}^{k-1} (1 + q(\ell)f(\ell))^{-1} v(k) \leq \sum_{\ell=a}^{k-1} p(\ell)f(\ell) \prod_{\tau=a}^{\ell} (1 + q(\tau)f(\tau))^{-1},$$

which is the same as

$$(4.1.5) \qquad v(k) \leq \sum_{\ell=a}^{k-1} p(\ell)f(\ell) \prod_{\tau=\ell+1}^{k-1} (1 + q(\tau)f(\tau)).$$

The result (4.1.2) follows from (4.1.5) and the inequality $u(k) \leq p(k) + q(k)v(k)$. ∎

**Remark 4.1.1.** The above proof obviously holds if $p(k)$ and $u(k)$ in Theorem 4.1.1 change sign on $\mathbb{N}(a)$. Further, the inequality (4.1.2) is the best possible in the sense that equality in (4.1.1) implies equality in (4.1.2).

**Corollary 4.1.2.** Let in Theorem 4.1.1, $p(k) = p$ and $q(k) = q$ for all $k \in \mathbb{N}(a)$. Then, for all $k \in \mathbb{N}(a)$

$$u(k) \le p\prod_{\ell=a}^{k-1}(1 + qf(\ell)).$$

**Proof.** It follows from (4.1.2) and Problem 1.9.10. ∎

**Corollary 4.1.3.** Let in Theorem 4.1.1, $p(k)$ be nondecreasing and $q(k) \ge 1$ for all $k \in \mathbb{N}(a)$. Then, for all $k \in \mathbb{N}(a)$

$$u(k) \le p(k)q(k)\prod_{\ell=a}^{k-1}(1 + q(\ell)f(\ell)).$$

**Proof.** For such $p(k)$ and $q(k)$ the inequality (4.1.2) provides

$$u(k) \le p(k)q(k)\left[1 + \sum_{\ell=a}^{k-1} q(\ell)f(\ell)\prod_{\tau=\ell+1}^{k-1}(1 + q(\tau)f(\tau))\right].$$

Now the result follows from Problem 1.9.10. ∎

**Theorem 4.1.4.** Let for all $k \in \mathbb{N}(a)$ the following inequality be satisfied

$$u(k) \le p(k) + q(k)\sum_{i=1}^{r} E_i(k, u), \tag{4.1.6}$$

where

$$E_i(k, u) = \sum_{\ell_1=a}^{k-1} f_{i1}(\ell_1)\sum_{\ell_2=a}^{\ell_1-1} f_{i2}(\ell_2)\cdots\sum_{\ell_i=a}^{\ell_{i-1}-1} f_{ii}(\ell_i)u(\ell_i). \tag{4.1.7}$$

Then, for all $k \in \mathbb{N}(a)$

$$u(k) \le p(k)+q(k)\sum_{\ell=a}^{k-1}\left[\sum_{i=1}^{r}\Delta E_i(\ell, p)\right]\prod_{\tau=\ell+1}^{k-1}\left[1 + \sum_{i=1}^{r}\Delta E_i(\tau, q)\right]. \tag{4.1.8}$$

**Proof.** Define a function $v(k)$ on $\mathbb{N}(a)$ as follows

$$v(k) = \sum_{i=1}^{r} E_i(k, u).$$

For this function, we have

$$\Delta v(k) = \sum_{i=1}^{r} \Delta E_i(k,u), \quad v(a) = 0. \tag{4.1.9}$$

Since $u(k) \le p(k) + q(k)v(k)$, and $v(k)$ is nondecreasing in $k$, from (4.1.9) we get

$$\begin{aligned}\Delta v(k) &\le \sum_{i=1}^{r} \Delta E_i(k, p+qv) = \sum_{i=1}^{r} \Delta E_i(k,p) + \sum_{i=1}^{r} \Delta E_i(k,qv) \\ &\le \sum_{i=1}^{r} \Delta E_i(k,p) + v(k)\sum_{i=1}^{r} \Delta E_i(k,q).\end{aligned}$$

Rest of the proof is similar to that of Theorem 4.1.1. ∎

**Condition (c).** We say that condition (c) is satisfied if for all $k \in \mathbb{N}(a)$ the inequality (4.1.6) holds, where

$$\begin{aligned} f_{ii}(k) &= f_i(k), \quad 1 \le i \le r, \\ f_{i+1,i}(k) &= f_{i+2,i}(k) = \cdots = f_{r,i}(k) = g_i(k), \quad 1 \le i \le r-1. \end{aligned}$$

In our next result for all $k \in \mathbb{N}(a)$ we shall denote

$$\phi_j(k) = \max\left\{0, \ \sum_{i=1}^{r-j+1} q(k) f_i(k) - g_{r-j+1}(k), \right. \\ \left. g_i(k) - g_{r-j+1}(k), \ 1 \le i \le r-j \right\}, \quad 1 \le j \le r$$

where $g_r(k) = 0$ for all $k \in \mathbb{N}(a)$.

**Theorem 4.1.5.** Let the condition (c) be satisfied. Then, for all $k \in \mathbb{N}(a)$

$$u(k) \le p(k) + q(k)\psi_j(k), \quad 1 \le j \le r \tag{4.1.10$_j$}$$

where

$$\psi_j(k) = \sum_{\ell=a}^{k-1}\left(p(\ell)\sum_{i=1}^{r-j+1} f_i(\ell) + g_{r-j+1}(\ell)\psi_{j-1}(\ell)\right)\prod_{\tau=\ell+1}^{k-1}(1+\phi_j(\tau)), \ 1 \le j \le r.$$

**Proof.** If the condition (c) is satisfied then the inequality (4.1.6) is equivalent to the system

$$u_1(k) \le p(k) + q(k)\sum_{\ell=a}^{k-1}(f_1(\ell)u_1(\ell) + g_1(\ell)u_2(\ell)) \tag{4.1.11}$$

$$(4.1.12)_j \quad u_{j-1}(k) = \sum_{\ell=a}^{k-1}(f_{j-1}(\ell)u_1(\ell) + g_{j-1}(\ell)u_j(\ell)), \quad 3 \le j \le r$$

$$(4.1.13) \quad u_r(k) = \sum_{\ell=a}^{k-1} f_m(\ell)u_1(\ell).$$

We define

$$v_1(k) = \sum_{\ell=a}^{k-1}(f_1(\ell)u_1(\ell) + g_1(\ell)u_2(\ell))$$

and $v_j(k) = u_j(k)$, $2 \le j \le r$, then from (4.1.11) (4.1.13) it follows that

$$(4.1.14) \quad \Delta v_1(k) \le f_1(k)(p(k) + q(k)v_1(k)) + g_1(k)v_2(k)$$

$$(4.1.15)_j \quad \Delta v_{j-1}(k) \le f_{j-1}(k)(p(k)+q(k)v_1(k))+g_{j-1}(k)v_j(k), \quad 3 \le j \le r$$

$$(4.1.16) \quad \Delta v_r(k) \le f_r(k)(p(k) + q(k)v_1(k)).$$

Adding (4.1.14) (4.1.16), to obtain

$$\begin{aligned}\Delta\left(\sum_{i=1}^{r} v_i(k)\right) &\le p(k)\sum_{i=1}^{r} f_i(k) + q(k)\sum_{i=1}^{r} f_i(k)v_1(k) + \sum_{i=1}^{r-1} g_i(k)v_{i+1}(k)\\ &\le p(k)\sum_{i=1}^{r} f_i(k) + \phi_1(k)\left(\sum_{i=1}^{r} v_i(k)\right).\end{aligned}$$

Now as in Theorem 4.1.1, we find

$$(4.1.17) \qquad \sum_{i=1}^{r} v_i(k) \le \psi_1(k).$$

Adding (4.1.14), $(4.1.15)_j$, $3 \le j \le r$ and using (4.1.17), we obtain

$$\begin{aligned}\Delta\left(\sum_{i=1}^{r-1} v_i(k)\right) &\le p(k)\sum_{i=1}^{r-1} f_i(k) + q(k)\sum_{i=1}^{r-1} f_i(k)v_1(k)\\ &\quad + \sum_{i=1}^{r-2} g_i(k)v_{i+1}(k) + g_{r-1}(k)\left(\psi_1(k) - \sum_{i=1}^{r-1} v_i(k)\right)\\ &\le \left(p(k)\sum_{i=1}^{r-1} f_i(k) + g_{r-1}(k)\psi_1(k)\right) + \phi_2(k)\left(\sum_{i=1}^{r-1} v_i(k)\right).\end{aligned}$$

Now once again as in Theorem 4.1.1, we get

$$(4.1.18) \qquad \sum_{i=1}^{r-1} v_i(k) \le \psi_2(k).$$

Continuing in this way, we find

$$(4.1.19)_j \qquad \sum_{i=1}^{r-j+1} v_i(k) \ \le \ \psi_j(k), \quad 3 \le j \le r.$$

Since $u(k) = u_1(k) \le p(k) + q(k)v_1(k)$, the result $(4.1.10)_j$, $1 \le j \le r$ follows from (4.1.17), (4.1.18) and $(4.1.19)_j$, $3 \le j \le r$. ∎

**Theorem 4.1.6.** Let for all $k \in \mathbb{N}(a)$ the following inequality be satisfied

$$(4.1.20) \qquad u(k) \ \le \ p_0(k) + \sum_{i=1}^{r} p_i(k) \sum_{\ell=a}^{k-1} q_i(\ell)u(\ell).$$

Then, for all $k \in \mathbb{N}(a)$

$$(4.1.21)_r \qquad u(k) \ \le \ F_r[p_0(k)],$$

where

$$\begin{aligned} F_i &= D_i D_{i-1} \cdots D_0 \\ D_0[w] &= w \\ D_j[w] &= w + (F_{j-1}[p_j]) \left( \sum_{\ell=a}^{k-1} q_j(\ell)w(\ell) \prod_{\tau=\ell+1}^{k-1} (1 + q_j(\tau)F_{j-1}[p_j(\tau)]) \right), \\ & \qquad\qquad\qquad\qquad\qquad\qquad\qquad\qquad 1 \le j \le r. \end{aligned}$$

**Proof.** The proof is by induction. For $r = 1$, inequality (4.1.20) reduces to (4.1.1) with $p(k) = p_0(k)$, $q(k) = p_1(k)$ and $f(k) = q_1(k)$. Thus, from Theorem 4.1.1, $u(k) \le D_1[p_0(k)] = F_1[p_0(k)]$, i.e. $(4.1.21)_1$ is true. Assume that the result is true for some $j$, where $1 < j \le r-1$. Then, to prove for $j+1$ we have

$$(4.1.22) \quad u(k) \le \left[ p_0(k) + p_{j+1}(k) \sum_{\ell=a}^{k-1} q_{j+1}(\ell)u(\ell) \right] + \sum_{i=1}^{j} p_i(k) \sum_{\ell=a}^{k-1} q_i(\ell)u(\ell)$$

and from $(4.1.21)_j$, we find

$$u(k) \ \le \ F_j \left[ p_0(k) + p_{j+1}(k) \sum_{\ell=a}^{k-1} q_{j+1}(\ell)u(\ell) \right].$$

In the above inequality we use the definition of $F_j$ and the fact that

$\sum_{\ell=a}^{k-1} q_{j+1}(\ell)u(\ell)$ is nondecreasing for all $k \in \mathbb{N}(a)$, to obtain

$$\begin{aligned} u(k) &\le F_j[p_0(k)] + F_j\left[p_{j+1}(k)\sum_{\ell=a}^{k-1} q_{j+1}(\ell)u(\ell)\right] \\ &\le F_j[p_0(k)] + F_j[p_{j+1}(k)]\sum_{\ell=a}^{k-1} q_{j+1}(\ell)u(\ell). \end{aligned}$$

Now an application of Theorem 4.1.1 provides

$$\begin{aligned} u(k) &\le F_j[p_0(k)] + F_j[p_{j+1}(k)]\sum_{\ell=a}^{k-1} q_{j+1}(\ell)F_j[p_0(\ell)] \times \\ &\qquad \prod_{\tau=\ell+1}^{k-1}(1 + q_{j+1}(\tau)F_j[p_{j+1}(\tau)]) \\ &= F_{j+1}[p_0(k)]. \qquad \blacksquare \end{aligned}$$

**Corollary 4.1.7.** Let in addition to hypotheses of Theorem 4.1.6, $p_i(k) \ge 1$ for all $k \in \mathbb{N}(a)$, $1 \le i \le r$. Then, for all $k \in \mathbb{N}(a)$

$$\begin{aligned} u(k) \le \prod_{j=1}^{r} p_j(k)\left[p_0(k) + \sum_{\ell=a}^{k-1}\left(\sum_{i=1}^{r} q_i(\ell)\prod_{j=0}^{r} p_j(\ell)\right) \times \right. \\ \left. \prod_{\tau=\ell+1}^{k-1}\left(1 + \sum_{i=1}^{r} q_i(\tau)\prod_{j=1}^{r} p_j(\tau)\right)\right]. \end{aligned}$$

**Proof.** For such $p_i(k)$, $1 \le i \le r$ inequality (4.1.20) can be written as (4.1.1) with $p(k) = \prod_{i=0}^{r} p_i(k)$, $q(k) = \prod_{i=1}^{r} p_i(k)$ and $f(k) = \sum_{i=1}^{r} q_i(k)$. $\blacksquare$

**Corollary 4.1.8.** Let in addition to hypotheses of Theorem 4.1.6, $p_0(k) > 0$ and nondecreasing, $p_i(k) \ge 1$, $1 \le i \le r$ and nondecreasing when $2 \le i \le r$ for all $k \in \mathbb{N}(a)$. Then, for all $k \in \mathbb{N}(a)$

$$u(k) \le G_r[p_0(k)], \qquad (4.1.23)_r$$

where

$$\begin{aligned} G_0[w] &= w \\ G_j[w] &= wG_{j-1}[p_j]\prod_{\ell=a}^{k-1}(1 + q_j G_{j-1}[p_j]), \quad 1 \le j \le r. \end{aligned}$$

**Proof.** The proof is by induction. For $r = 1$, Corollary 4.1.3 gives that $u(k) \leq G_1[p_0(k)]$. Let the result be true for some $j$, where $1 < j \leq r-1$, then to prove for $j+1$ we have (4.1.22). Since in (4.1.22) the part in brackets is positive and nondecreasing, we find

$$u(k) \leq G_j \left[ p_0(k) + p_{j+1}(k) \sum_{\ell=a}^{k-1} q_{j+1}(\ell) u(\ell) \right].$$

In the above inequality using the definition of $G_j$, we obtain

$$u(k) \leq G_j[p_0(k)] + p_{j+1}(k) G_j[p_0(k)] \sum_{\ell=a}^{k-1} q_{j+1}(\ell) \frac{u(\ell)}{p_0(\ell)},$$

which also gives

$$\begin{aligned} \frac{u(k)}{p_0(k)} &\leq \frac{G_j[p_0(k)] p_{j+1}(k)}{p_0(k)} \left( 1 + \sum_{\ell=a}^{k-1} q_{j+1}(\ell) \frac{u(\ell)}{p_0(\ell)} \right) \\ &= G_j[p_{j+1}(k)] \left( 1 + \sum_{\ell=a}^{k-1} q_{j+1}(\ell) \frac{u(\ell)}{p_0(\ell)} \right). \end{aligned}$$

Now an application of Problem 4.7.1 gives $u(k) \leq G_{j+1}[p_0(k)]$. ∎

**Remark 4.1.2.** In Corollary 4.1.8 the requirement $p_0(k) > 0$ is not essential. Infact, if $p_0(k) = 0$ for some $k$, then we can replace $p_0(k)$ by $p_0(k) + \epsilon$ for any $\epsilon > 0$. The conclusion then follows by letting $\epsilon \to 0$ in the resulting inequalities.

**Theorem 4.1.9.** Let for all $k, r \in \mathbb{N}(a)$ such that $k \leq r$ the following inequality be satisfied

$$u(r) \geq u(k) - q(r) \sum_{\ell=k+1}^{r} f(\ell) u(\ell), \tag{4.1.24}$$

where $u(k)$ is not necessarily nonnegative. Then, for all $k, r \in \mathbb{N}(a)$, $k \leq r$

$$u(r) \geq u(k) \prod_{\ell=k+1}^{r} (1 + q(r) f(\ell))^{-1}, \tag{4.1.25}$$

and (4.1.25) is the best possible.

**Proof.** Inequality (4.1.24) can be written as

$$u(k) \leq u(r) + q(r) \sum_{\ell=k+1}^{r} f(\ell) u(\ell). \tag{4.1.26}$$

Let $v(k)$ be the right side of (4.1.26), then for all $k,\ r \in \mathbb{N}(a),\ k \le r$ it follows that $u(k) \le v(k)$, and

$$\Delta v(k) \;=\; -\,q(r)f(k+1)u(k+1), \qquad v(r) = u(r).$$

Since $q(r)f(k+1) \ge 0$ and $u(k+1) \le v(k+1)$, we find the inequality

$$v(k) \;\le\; (1+q(r)f(k+1))v(k+1), \qquad v(r) = u(r)$$

which easily provides

$$v(k) \;\le\; \prod_{\ell=k+1}^{r} (1+q(r)f(\ell))u(r). \tag{4.1.27}$$

The result (4.1.25) now follows from $u(k) \le v(k)$ and (4.1.27). ∎

**Theorem 4.1.10.** Let for all $k \in \mathbb{N}$ the following inequality be satisfied

$$u(k) \;\le\; c_2 + h^{1/2}c_1\sum_{\ell=0}^{k-1}(k-\ell)^{-1/2}u(\ell), \tag{4.1.28}$$

where $c_1 > 0,\ c_2 > 0$ and $h > 0$. Then, for all $k \in \mathbb{N}$

$$u(k) \;\le\; c_2\left(1+c_1h^{1/2}+2c_1(kh)^{1/2}\right)\left(1+hc_1^2\pi\right)^k. \tag{4.1.29}$$

**Proof.** From (4.1.28), it is immediate that

$$\begin{aligned}
u(k) \;&\le\; c_2 + h^{1/2}c_1\sum_{\ell=0}^{k-1}(k-\ell)^{-1/2}\left[c_2 + h^{1/2}c_1\sum_{\tau=0}^{\ell-1}(\ell-\tau)^{-1/2}u(\tau)\right]\\
&=\; c_2 + h^{1/2}c_1c_2\sum_{\ell=0}^{k-1}(k-\ell)^{-1/2} + hc_1^2\sum_{\ell=0}^{k-1}\sum_{\tau=0}^{\ell-1}(k-\ell)^{-1/2}\times\\
&\qquad\qquad (\ell-\tau)^{-1/2}u(\tau)\\
&=\; c_2 + h^{1/2}c_1c_2k^{-1/2} + h^{1/2}c_1c_2\sum_{\ell=1}^{k-1}(k-\ell)^{-1/2}\\
&\qquad + hc_1^2\sum_{\tau=0}^{k-2}\left[\sum_{\ell=\tau+1}^{k-1}(k-\ell)^{-1/2}(\ell-\tau)^{-1/2}\right]u(\tau)
\end{aligned}$$

$$\le\; c_2 + h^{1/2}c_1c_2 + h^{1/2}c_1c_2\sum_{\ell=1}^{k-1}(k-\ell)^{-1/2} \tag{4.1.30}$$

$$+hc_1^2\sum_{\tau=0}^{k-2}\left[\sum_{\ell=1}^{k-\tau-1}(k-\tau-\ell)^{-1/2}\ell^{-1/2}\right]u(\tau).$$

Now consider the function $\phi(t)=(k-\tau-t)^{-1/2}t^{-1/2}$, $0<t<k-\tau$ ($\geq 2$). This function is strictly convex on the given interval and attains its minimum at $t=(k-\tau)/2$. Thus,

$$\sum_{\ell=1}^{k-\tau-1}(k-\tau-\ell)^{-1/2}\ell^{-1/2} = \sum_{\ell=1}^{k-\tau-1}\phi(\ell) \leq \int_0^{k-\tau}\phi(t)dt;$$

this is an immediate consequence of interpreting the given sum as a lower Riemann sum, with the rectangle for the subinterval $\left[\frac{k-\tau}{2},\frac{k-\tau}{2}+1\right]$ (if $k-\tau$ is even), or $\left[\frac{k-\tau-1}{2},\frac{k-\tau+1}{2}\right]$ (if $k-\tau$ is odd) missing. But

$$\text{(4.1.31)}\qquad \int_0^{k-\tau}\phi(t)dt = \int_0^1(1-t_1)^{-1/2}t_1^{-1/2}dt_1 = B\left(\frac{1}{2},\frac{1}{2}\right) = \pi.$$

In an analogous fashion, we have

$$\text{(4.1.32)}\qquad \sum_{\ell=1}^{k-1}(k-\ell)^{-1/2} \leq \int_0^k(k-t)^{-1/2}dt = 2k^{1/2}.$$

Using (4.1.31) and (4.1.32) in (4.1.30), we obtain

$$u(k) \leq c_2\left(1+c_1h^{1/2}+2c_1(kh)^{1/2}\right)+\sum_{\tau=0}^{k-1}(hc_1^2\pi)u(\tau).$$

Now the result (4.1.29) follows as an application of Corollary 4.1.3. ∎

## 4.2. Nonlinear Inequalities

Our first result for the nonlinear case is connected with the following inequality

$$\text{(4.2.1)}\qquad u(k) \leq p(k)\left[q+\sum_{i=1}^{r}H_i(k,u)\right],$$

where

$$\text{(4.2.2)}\qquad H_i(k,u) = \sum_{\ell_1=a}^{k-1}f_{i1}(\ell_1)u^{\alpha_{i1}}(\ell_1)\cdots\sum_{\ell_i=a}^{\ell_{i-1}-1}f_{ii}(\ell_i)u^{\alpha_{ii}}(\ell_i)$$

and $\alpha_{ij}$, $1 \le j \le i$, $1 \le i \le r$ are nonnegative constants and the constant $q > 0$.

In the following result we shall denote $\alpha_i = \sum_{j=1}^{i} \alpha_{ij}$ and $\alpha = \max_{1 \le i \le r} \alpha_i$.

**Theorem 4.2.1.** Let for all $k \in \mathbb{N}(a)$ the inequality (4.2.1) be satisfied. Then, for all $k \in \mathbb{N}(a)$

$$u(k) \le qp(k) \prod_{\ell=a}^{k-1} (1 + \Delta Q(\ell)), \quad \text{if} \quad \alpha = 1 \tag{4.2.3}$$

$$u(k) \le p(k) \left[q^{1-\alpha} + (1-\alpha) Q(k)\right]^{1/1-\alpha}, \quad \text{if} \quad \alpha \ne 1 \tag{4.2.4}$$

where

$$Q(k) = \sum_{i=1}^{r} H_i(k, p) q^{\alpha_i - \alpha}$$

and when $\alpha > 1$, we assume that $q^{1-\alpha} + (1-\alpha)Q(k) > 0$ for all $k \in \mathbb{N}(a)$.

**Proof.** The inequality (4.2.1) can be written as $u(k) \le p(k)v(k)$, where

$$v(k) = q + \sum_{i=1}^{r} H_i(k, u).$$

Thus, on using the nondecreasing nature of $v(k)$, we find

$$\Delta v(k) \le \sum_{i=1}^{r} \Delta H_i(k, p) v^{\alpha_i}(k).$$

Since $v(k) \ge q$, we get

$$\Delta v(k) \le \sum_{i=1}^{r} \Delta H_i(k, p) q^{\alpha_i - \alpha} v^{\alpha}(k) = \Delta Q(k) v^{\alpha}(k). \tag{4.2.5}$$

If $\alpha = 1$, (4.2.3) immediately follows on using $v(a) = q$, and the fact that $u(k) \le p(k)v(k)$.

If $\alpha \ne 1$, we have

$$\frac{\Delta v^{1-\alpha}(k)}{1-\alpha} = \int_k^{k+1} \frac{dv(t)}{v^{\alpha}(t)} \le \frac{\Delta v(k)}{v^{\alpha}(k)}$$

and from (4.2.5), we obtain

$$(4.2.6) \qquad \frac{\Delta v^{1-\alpha}(k)}{1-\alpha} \;\le\; \Delta Q(k).$$

Summing the inequality (4.2.6), we find

$$v(k) \;\le\; \left[q^{1-\alpha} + (1-\alpha)Q(k)\right]^{1/1-\alpha}$$

and the result (4.2.4) follows from $u(k) \le p(k)v(k)$. ∎

**Theorem 4.2.2.** Let for all $k \in \mathbb{N}(a)$ the following inequality be satisfied

$$(4.2.7) \qquad u(k) \;\le\; p(k) + q(k)\left(\sum_{\ell=a}^{k-1} f(\ell)u^r(\ell)\right)^{1/r},$$

where $1 \le r < \infty$. Then, for all $k \in \mathbb{N}(a)$

$$(4.2.8) \qquad u(k) \;\le\; p(k) + q(k)\frac{\left(\sum_{\ell=a}^{k-1} f(\ell)p^r(\ell)e(\ell+1)\right)^{1/r}}{1-(1-e(k))^{1/r}},$$

where

$$(4.2.9) \qquad e(k) \;=\; \prod_{\ell=a}^{k-1}(1+f(\ell)q^r(\ell))^{-1}.$$

**Proof.** We note that the function $e(k)$ is the solution of the initial value problem

$$(4.2.10) \qquad \Delta e(k) \;=\; -f(k)q^r(k)e(k+1), \qquad e(a) = 1.$$

Define the function $v(k)$ by

$$(4.2.11) \qquad v(k) \;=\; e(k)\sum_{\ell=a}^{k-1} f(\ell)u^r(\ell).$$

For the function $v(k)$, from (4.2.7) and (4.2.10), we obtain

$$(4.2.12) \qquad v(k+1) - v(k) \;\le\; \left(p(k)f^{1/r}(k)e^{1/r}(k+1) + \frac{q(k)f^{1/r}(k)v^{1/r}(k)}{(1+f(k)q^r(k))^{1/r}}\right)^r - \frac{f(k)q^r(k)v(k)}{1+f(k)q^r(k)}.$$

Now we sum (4.2.12) from $a$ to $k-1$, transpose the second sum from the right side to left side, form the $r$th root on both sides, and apply Minkowski's inequality for sums to the right side, to obtain

$$\left(v(k)+\sum_{\ell=a}^{k-1}\frac{f(\ell)q^r(\ell)v(\ell)}{1+f(\ell)q^r(\ell)}\right)^{1/r} \le \left(\sum_{\ell=a}^{k-1} f(\ell)p^r(\ell)e(\ell+1)\right)^{1/r} + \left(\sum_{\ell=a}^{k-1}\frac{f(\ell)q^r(\ell)v(\ell)}{1+f(\ell)q^r(\ell)}\right)^{1/r}. \tag{4.2.13}$$

Transpose the second term of the right side of (4.2.13) to left side to obtain the left side of the form $w(t)=(c+t)^{1/r}-t^{1/r}$ $(c\ge 0,\ r\ge 1)$. Since $w'(t)\le 0$ for all $t\ge 0$, we may replace $t$ by a larger quantity without destroying inequality (4.2.13). In this regard, we note that

$$\begin{aligned}
\sum_{\ell=a}^{k-1}\frac{f(\ell)q^r(\ell)v(\ell)}{1+f(\ell)q^r(\ell)} &= \sum_{\ell=a}^{k-1}\frac{f(\ell)q^r(\ell)e(\ell)}{1+f(\ell)q^r(\ell)}\left(\sum_{\tau=a}^{\ell-1} f(\tau)u^r(\tau)\right)\\
&= \sum_{\ell=a}^{k-1} f(\ell)q^r(\ell)e(\ell+1)\left(\sum_{\tau=a}^{\ell-1} f(\tau)u^r(\tau)\right)\\
&\le \sum_{\ell=a}^{k-1} f(\ell)q^r(\ell)e(\ell+1)\left(\sum_{\ell=a}^{k-1} f(\ell)u^r(\ell)\right)\\
&= (1-e(k))\sum_{\ell=a}^{k-1} f(\ell)u^r(\ell)\\
&= \frac{v(k)}{e(k)}-v(k).
\end{aligned}$$

Hence, (4.2.13) implies that

$$\left(\frac{v(k)}{e(k)}\right)^{1/r}-\left(\frac{v(k)}{e(k)}-v(k)\right)^{1/r} \le \left(\sum_{\ell=a}^{k-1} f(\ell)p^r(\ell)e(\ell+1)\right)^{1/r},$$

i.e.

$$\left(\frac{v(k)}{e(k)}\right)^{1/r} = \left(\sum_{\ell=a}^{k-1} f(\ell)u^r(\ell)\right)^{1/r} \le \frac{\left(\sum_{\ell=a}^{k-1} f(\ell)p^r(\ell)e(\ell+1)\right)^{1/r}}{1-(1-e(k))^{1/r}}. \tag{4.2.14}$$

Using (4.2.14) in (4.2.7) the result (4.2.8) follows. ∎

For the next result we shall need the following class of functions:

**Definition 4.2.1.** A continuous function $W : [0,\infty) \to (0,\infty)$ is said to belong to the class $T$ if (i) $W(u)$ is positive and nondecreasing for all $u \ge 0$, (ii) $\frac{1}{v}W(u) \le W\left(\frac{u}{v}\right)$ for all $u \ge 0$, $v \ge 1$.

**Theorem 4.2.3.** Let for all $k \in \mathbb{N}(a)$ the following inequality be satisfied

$$u(k) \le p(k) + \sum_{i=1}^{r_1} E_i(k,u) + \sum_{i=1}^{r_2} p_i(k) \sum_{\ell=a}^{k-1} q_i(\ell) W_i(u(\ell)), \tag{4.2.15}$$

where (i) $p(k) \ge 1$ and nondecreasing, (ii) $p_i(k) \ge 1$, $1 \le i \le r_2$, (iii) $W_i \in T$, $1 \le i \le r_2$. Then, for all $k \in \mathbb{N}(a)$

$$u(k) \le p(k)v(k)e(k) \prod_{i=1}^{r_2} J_i(k), \tag{4.2.16}$$

where

$$e(k) = \prod_{i=1}^{r_2} p_i(k), \quad v(k) = \prod_{\ell=a}^{k-1} \left(1 + \sum_{i=1}^{r_1} \Delta E_i(\ell, e)\right)$$

$$J_0(k) = 1, \quad J_j(k) = G_j^{-1}\left[G_j(1) + \sum_{\ell=a}^{k-1} q_j(\ell)v(\ell)e(\ell) \prod_{i=1}^{j-1} J_i(\ell)\right], \quad 1 \le j \le r_2$$

and

$$G_j(w) = \int_{w_0}^{w} \frac{dt}{W_j(t)}, \quad w \ge w_0 \ge 1$$

as long as

$$G_j(1) + \sum_{\ell=a}^{k-1} q_j(\ell)v(\ell)e(\ell) \prod_{i=1}^{j-1} J_i(\ell) \in Dom\left(G_j^{-1}\right), \quad 1 \le j \le r_2.$$

**Proof.** From the hypotheses, inequality (4.2.15) provides that

$$\frac{u(k)}{e(k)} \le p^*(k) + \sum_{i=1}^{r_1} E_i\left(k, e\frac{u}{e}\right),$$

where

$$p^*(k) = p(k) + \sum_{i=1}^{r_2} \sum_{\ell=a}^{k-1} q_i(\ell) W_i(u(\ell)).$$

Since $p^*(k)$ is nondecreasing, as in Corollary 4.1.3, we find

$$\frac{u(k)}{e(k)} \le p^*(k)v(k). \tag{4.2.17}$$

Now on using the definition of class $T$, the inequality (4.2.17) implies that

$$w(k) \leq 1+\sum_{i=1}^{r_2}\sum_{\ell=a}^{k-1} q_i(\ell)e(\ell)v(\ell)W_i(w(\ell)),$$

where $w(k) = (u(k)/p(k)v(k)e(k))$.

Thus it is sufficient to show that $w(k) \leq \prod_{i=1}^{r_2} J_i(k)$. We shall prove this by induction. For $r_2 = 1$, we have

$$w(k) \leq 1+\sum_{\ell=a}^{k-1} q_1(\ell)e(\ell)v(\ell)W_1(w(\ell)). \tag{4.2.18}$$

Let $z(k)$ be the right side of (4.2.18), then on using nondecreasing nature of $W_1$, we obtain

$$\Delta z(k) \leq q_1(k)e(k)v(k)W_1(z(k)), \quad z(a) = 1. \tag{4.2.19}$$

Next from the definition of $G_1$ it follows that

$$\Delta G_1(z(k)) = \int_{z(k)}^{z(k+1)} \frac{dt}{W_1(t)} \leq \frac{\Delta z(k)}{W_1(z(k))}. \tag{4.2.20}$$

Using (4.2.20) in (4.2.19) and summing, to obtain

$$z(k) \leq G_1^{-1}\left[G_1(1)+\sum_{\ell=a}^{k-1} q_1(\ell)e(\ell)v(\ell)\right] = J_1(k).$$

This shows that the result is true for $r_2 = 1$. Now assuming that the result is true for some $j$ such that $1 \leq j \leq r_2 - 1$, then to prove for $j+1$ we have

$$w(k) \leq \left[1+\sum_{\ell=a}^{k-1} q_{j+1}(\ell)e(\ell)v(\ell)W_{j+1}(w(\ell))\right]+\sum_{i=1}^{j}\sum_{\ell=a}^{k-1} q_i(\ell)e(\ell)v(\ell)W_i(w(\ell)).$$

Since the part inside the bracket is greater than 1 and nondecreasing, we find

$$w(k) \leq \left[1+\sum_{\ell=a}^{k-1} q_{j+1}(\ell)e(\ell)v(\ell)W_{j+1}(w(\ell))\right]\prod_{i=1}^{j} J_i(k),$$

which also gives

$$\frac{w(k)}{\prod_{i=1}^{j} J_i(k)} \leq 1+\sum_{\ell=a}^{k-1} q_{j+1}(\ell)e(\ell)v(\ell)\prod_{i=1}^{j} J_i(\ell)W_{j+1}\left(\frac{w(\ell)}{\prod_{i=1}^{j} J_i(\ell)}\right)$$

and from this $w(k) \le \prod_{i=1}^{j+1} J_i(k)$ follows on using the same arguments as for the case $r_2 = 1$. This completes the proof. ∎

**Theorem 4.2.4.** In addition to the hypotheses of Theorem 4.2.3 let $p_i(k)$, $1 \le i \le r_2$ be nondecreasing for all $k \in \mathbb{N}(a)$. Then, for all $k \in \mathbb{N}(a)$

$$u(k) \le p(k) v^*(k) \prod_{i=1}^{r_2} J_i^*(k),$$

where $v^*(k)$ is the same as $v(k)$ in Theorem 4.2.3 with $e(k) = 1$,

$$J_0^*(k) = 1, \quad J_j^*(k) = p_j(k) G_j^{-1}\left[G_j(1) + \sum_{\ell=a}^{k-1} q_j(\ell) v^*(\ell) p_j(\ell) \prod_{i=1}^{j-1} J_i^*(\ell)\right], \quad 1 \le j \le r_2$$

as long as

$$G_j(1) + \sum_{\ell=a}^{k-1} q_j(\ell) v^*(\ell) p_j(\ell) \prod_{i=1}^{j-1} J_i^*(\ell) \in Dom\left(G_j^{-1}\right), \quad 1 \le j \le r_2$$

and $G_j$, $1 \le j \le r_2$ are the same as in Theorem 4.2.3.

**Proof.** The proof is similar to that of Theorem 4.2.3. ∎

**Theorem 4.2.5.** Let for all $k \in \mathbb{N}(a)$ the following inequality be satisfied

$$u(k) \le p(k) + \sum_{i=1}^{r_1} E_i(k, u) + \sum_{i=1}^{r_2} E_i(k, W_1(u)), \tag{4.2.21}$$

where (i) $p(k) \ge 1$ and nondecreasing, (ii) $W_1 \in T$. Then, for all $k \in \mathbb{N}(a)$

$$u(k) \le p(k) v^*(k) G_1^{-1}\left[G_1(1) + \sum_{i=1}^{r_2} E_i(k, v^*)\right]$$

as long as

$$G_1(1) + \sum_{i=1}^{r_2} E_i(k, v^*) \in Dom\left(G_1^{-1}\right),$$

where $G_1$ is the same as in Theorem 4.2.3.

**Proof.** The proof is similar to that of Theorem 4.2.3. ∎

**Theorem 4.2.6.** Let in Theorem 4.2.5 hypotheses (i) and (ii) be replaced by (i) $p(k)$ is positive and nondecreasing, (ii) $W_1$ is positive, continuous, nondecreasing and submultiplicative on $[0,\infty)$. Then, for all $k \in \mathbb{N}(a)$

$$u(k) \le p(k)v^*(k)G_1^{-1}\left[G_1(1) + \sum_{i=1}^{r_2} E_i\left(k, \frac{W_1(pv^*)}{p}\right)\right] \tag{4.2.22}$$

as long as

$$G_1(1) + \sum_{i=1}^{r_2} E_i\left(k, \frac{W_1(pv^*)}{p}\right) \in Dom\left(G_1^{-1}\right),$$

where $G_1$ is the same as in Theorem 4.2.3.

**Proof.** We follow as in Corollary 4.1.3 to get

$$u(k) \le \left[p(k) + \sum_{i=1}^{r_2} E_i(k, W_1(u))\right] v^*(k),$$

which provides that

$$\frac{u(k)}{p(k)v^*(k)} \le 1 + \sum_{i=1}^{r_2} E_i\left(k, W_1\left(\frac{u}{pv^*}pv^*\right)/p\right). \tag{4.2.23}$$

Let $w(k)$ be the right side of (4.2.23), then we have

$$\begin{aligned}\Delta w(k) &= \sum_{i=1}^{r_2} \Delta E_i\left(k, W_1\left(\frac{u}{pv^*}pv^*\right)/p\right) \\ &\le \sum_{i=1}^{r_2} \Delta E_i(k, W_1(pv^*)/p)W_1(w(k)).\end{aligned}$$

In the above inequality we use the same arguments as in Theorem 4.2.3, to obtain

$$w(k) \le G_1^{-1}\left[G_1(1) + \sum_{i=1}^{r_2} E_i(k, W_1(pv^*)/p)\right]$$

and from this the inequality (4.2.22) follows. ∎

## 4.3. Inequalities Involving Differences

**Theorem 4.3.1.** Let for all $k \in \mathbb{N}(a)$ the following inequality be satisfied

$$\Delta^n u(k) \le p(k) + q(k)\sum_{i=0}^{n}\sum_{\ell=a}^{k-1} q_i(\ell)\Delta^i u(\ell). \tag{4.3.1}$$

Then, for all $k \in \mathbb{N}(a)$

$$\Delta^n u(k) \le p(k) + q(k)\sum_{\ell=a}^{k-1}\phi_1(\ell)\prod_{\tau=\ell+1}^{k-1}(1+\phi_2(\tau)), \tag{4.3.2}$$

where

$$\begin{aligned}\phi_1(k) &= p(k)q_n(k) + \sum_{i=0}^{n-1}\sum_{j=0}^{i}\Delta^i u(a)q_i(k)\frac{(k-a)^{(i-j)}}{(i-j)!}\\ &\quad + \sum_{i=0}^{n-1} q_{n-i-1}(k)\sum_{\ell=a}^{k-i-1}\frac{(k-\ell-1)^{(i)}}{i!}p(\ell)\end{aligned} \tag{4.3.3}$$

and

$$\phi_2(k) = q(k)q_n(k) + \sum_{i=0}^{n-1} q_{n-i-1}(k)\sum_{\ell=a}^{k-i-1}\frac{(k-\ell-1)^{(i)}}{i!}q(\ell). \tag{4.3.4}$$

**Proof.** Define a function $v(k)$ on $\mathbb{N}(a)$ as follows

$$v(k) = \sum_{i=0}^{n}\sum_{\ell=a}^{k-1} q_i(\ell)\Delta^i u(\ell),$$

then (4.3.1) can be written as

$$\Delta^n u(k) \le p(k) + q(k)v(k). \tag{4.3.5}$$

From the definition of $v(k)$, we have

$$\Delta v(k) = \sum_{i=0}^{n} q_i(k)\Delta^i u(k).$$

Thus, from (1.8.7) and (4.3.5), we obtain

$$\begin{aligned}\Delta v(k) &\le q_n(k)(p(k)+q(k)v(k)) + \sum_{i=0}^{n-1} q_i(k)\left[\sum_{j=i}^{n-1}\frac{(k-a)^{(j-i)}}{(j-i)!}\Delta^j u(a)\right.\\ &\quad \left. + \frac{1}{(n-i-1)!}\sum_{\ell=a}^{k-n+i}(k-\ell-1)^{(n-i-1)}\Delta^n u(\ell)\right]\\ &\le p(k)q_n(k) + \sum_{i=0}^{n-1}\sum_{j=0}^{i}\Delta^i u(a)q_i(k)\frac{(k-a)^{(i-j)}}{(i-j)!} + q(k)q_n(k)v(k)\\ &\quad + \sum_{i=0}^{n-1} q_{n-i-1}(k)\sum_{\ell=a}^{k-i-1}\frac{(k-\ell-1)^{(i)}}{i!}(p(\ell)+q(\ell)v(\ell)).\end{aligned}$$

Now on using the nondecreasing nature of $v(k)$, the above inequality gives

$$\Delta v(k) \le \phi_1(k) + \phi_2(k)v(k).$$

The rest of the proof is similar to that of Theorem 4.1.1. ∎

**Corollary 4.3.2.** Let in Theorem 4.3.1, $\Delta^i u(a) = 0,\ 0 \le i \le n-1$, $p(k)$ be nondecreasing and $q(k) = 1$ for all $k \in \mathbb{N}(a)$. Then, for all $k \in \mathbb{N}(a)$

$$\Delta^n u(k) \le p(k) \prod_{\ell=a}^{k-1} (1 + \phi_3(\ell)),$$

where

$$\phi_3(k) = \sum_{i=0}^{n} \frac{(k-a)^{(i)}}{i!} q_{n-i}(k).$$

**Proof.** The proof is similar to that of Corollary 4.1.3 and uses the equality

$$q_n(k) + \sum_{i=0}^{n-1} q_{n-i-1}(k) \sum_{\ell=a}^{k-i-1} \frac{(k-\ell-1)^{(i)}}{i!} = \sum_{i=0}^{n} \frac{(k-a)^{(i)}}{i!} q_{n-i}(k). \quad ∎$$

**Theorem 4.3.3.** Let in addition to hypotheses of Theorem 4.3.1, $q_i(k) = q^*(k),\ 0 \le i \le n$, and $q(k) \ge 1$ for all $k \in \mathbb{N}(a)$. Then, for all $k \in \mathbb{N}(a)$

$$(4.3.6)_i \qquad \Delta^n u(k) \le p(k) + q(k)B_i(k), \quad 1 \le i \le n+1$$

where

$$B_1(k) = \sum_{\ell=a}^{k-1} q^*(\ell)\phi_4(\ell) \prod_{\tau=\ell+1}^{k-1} (q^*(\tau)q(\tau) + q^*(\tau) + nq(\tau) + n)$$

$$B_i(k) = \sum_{\ell=a}^{k-1} (q^*(\ell)\phi_4(\ell) + B_{i-1}(\ell)) \prod_{\tau=\ell+1}^{k-1} (q^*(\tau)q(\tau) + q^*(\tau) + (n-i-1)q(\tau) + n - i), \quad 2 \le i \le n$$

$$B_{n+1}(k) = \sum_{\ell=a}^{k-1} q^*(\ell)(\phi_4(\ell) + B_n(\ell)) \prod_{\tau=\ell+1}^{k-1} (1 + q^*(\tau)(q(\tau)-1))$$

and $\phi_4(k)$ is the same as $\phi_1(k)$ with $q_i(k) = 1,\ 0 \le i \le n$.

**Proof.** We define

$$v_1(k) = \sum_{i=0}^{n} \sum_{\ell=a}^{k-1} q^*(\ell)\Delta^i u(\ell).$$

Then, as in Theorem 4.3.1, we get

$$(4.3.7)\quad \Delta v_1(k)+q^*(k)v_1(k) \;\le\; q^*(k)\phi_4(k)+q^*(k)q(k)v_1(k)+q^*(k)v_2(k),$$

where

$$v_2(k) \;=\; v_1(k)+\sum_{i=0}^{n-1}\sum_{\ell=a}^{k-i-1}\frac{(k-\ell-1)^{(i)}}{i!}q(\ell)v_1(\ell).$$

Again, as in Theorem 4.3.1 on using $v_1(k)\le v_2(k)$, we find

$$\Delta v_2(k)+v_2(k) \;\le\; q^*(k)\phi_4(k)+(q^*(k)q(k)+q^*(k)+q(k))\,v_2(k)+v_3(k),$$

where

$$v_3(k) \;=\; v_2(k)+\sum_{i=0}^{n-2}\sum_{\ell=a}^{k-i-1}\frac{(k-\ell-1)^{(i)}}{i!}q(\ell)v_2(\ell).$$

Once again, on using $v_2(k)\le v_3(k)$, we obtain

$$\Delta v_3(k)+v_3(k)\le q^*(k)\phi_4(k)+(q^*(k)q(k)+q^*(k)+2q(k)+1)v_3(k)+v_4(k),$$

where

$$v_4(k) \;=\; v_3(k)+\sum_{i=0}^{n-3}\sum_{\ell=a}^{k-i-1}\frac{(k-\ell-1)^{(i)}}{i!}q(\ell)v_3(\ell).$$

Continuing in this way, we get

$$(4.3.8)\quad \Delta v_n(k)+v_n(k) \;\le\; q^*(k)\phi_4(k)+(q^*(k)q(k)+q^*(k) \\ +(n-1)q(k)+(n-2))v_n(k)+v_{n+1}(k),$$

where

$$v_{n+1}(k) \;=\; v_n(k)+\sum_{\ell=a}^{k-1}q(\ell)v_n(\ell)$$

so that from $v_n(k)\le v_{n+1}(k)$, it follows that

$$(4.3.9)\quad \Delta v_{n+1}(k)\le q^*(k)\phi_4(k)+(q^*(k)q(k)+q^*(k)+nq(k) \\ +(n-1))v_{n+1}(k).$$

Obviously, from the above definitions $v_1(k)\le v_2(k)\le\cdots\le v_{n+1}(k)$ and $v_i(a)=0,\ 1\le i\le n+1$. Thus, as in Theorem 4.1.1, (4.3.9) gives $v_{n+1}(k)\le B_1(k)$ and $(4.3.6)_1$ follows from $\Delta^n u(k)\;\le\; p(k)+q(k)v_{n+1}(k)$. Next on using $v_{n+1}(k)\le B_1(k)$ in (4.3.8), we get

$$\Delta v_n(k)+v_n(k) \;\le\; (q^*(k)\phi_4(k)+B_1(k))+(q^*(k)q(k)+q^*(k) \\ +(n-1)q(k)+(n-2))v_n(k),$$

which provides $v_n(k) \le B_2(k)$. Continuing this way, we easily find $v_i(k) \le B_{n-i+2}(k)$, $i = n+1, n, \cdots, 2$. Finally, we use $v_2(k) \le B_n(k)$ in (4.3.7), to obtain

$$\Delta v_1(k) \;\le\; q^*(k)(\phi_4(k) + B_n(k)) + q^*(k)(q(k)-1)v_1(k),$$

which gives $v_1(k) \le B_{n+1}(k)$. ∎

**Remark 4.3.1.** In Theorem 4.3.3 we need $q(k) \ge 1$ only to prove the conclusion $(4.3.6)_{n+1}$. Therefore, instead of $q(k) \ge 1$ it is enough to assume that $1 + q^*(k)(q(k)-1) \ge 0$ for all $k \in \mathbb{N}(a)$. Further, if there is no condition on $q(k)$, then an immediate upper estimate can be obtained from the inequality

$$\Delta v_1(k) \;\le\; q^*(k)(\phi_4(k) + B_n(k)) + q^*(k)q(k)v_1(k).$$

**Theorem 4.3.4.** Let for all $k \in \mathbb{N}(a)$ the following inequality be satisfied

$$\Delta^n u(k) \;\le\; p(k) + \sum_{i=0}^{n}\sum_{\ell=a}^{k-1} q_i(\ell)\Delta^i u(\ell)\Delta^n u(\ell), \tag{4.3.10}$$

where $p(k)$ is positive and nondecreasing. Then, for all $k \in \mathbb{N}(a)$

$$\Delta^n u(k) \;\le\; \frac{p(k)e^{-1}(k)}{1 - \sum_{\ell=a}^{k-1} p(\ell)\phi_3(\ell)e^{-1}(\ell+1)}, \tag{4.3.11}$$

where

$$e(k) \;=\; \prod_{\ell=a}^{k-1}(1 + \phi_5(\ell))^{-1}$$

and $\phi_5(k)$ is the same as $\phi_1(k)$ with $p(k) = 0$, as long as $1 - \sum_{\ell=a}^{k-1} p(\ell) \times \phi_3(\ell)e^{-1}(\ell+1) > 0$.

**Proof.** Since $p(k)$ is positive and nondecreasing, inequality (4.3.10) implies that

$$\frac{\Delta^n u(k)}{p(k)} \;\le\; 1 + \sum_{i=0}^{n}\sum_{\ell=a}^{k-1} q_i(\ell)\Delta^i u(\ell)\frac{\Delta^n u(\ell)}{p(\ell)}. \tag{4.3.12}$$

Let $v(k)$ be the right side of (4.3.12), then

$$\begin{aligned}
\Delta v(k) &= \sum_{i=0}^{n} q_i(k)\Delta^i u(k)\frac{\Delta^n u(k)}{p(k)} \\
&\le q_n(k)p(k)v^2(k) + \sum_{i=0}^{n-1} q_i(k)v(k)\left[\sum_{j=i}^{n-1}\frac{(k-a)^{(j-i)}}{(j-i)!}\Delta^j u(a)\right. \\
&\quad \left. + \frac{1}{(n-i-1)!}\sum_{\ell=a}^{k-n+i}(k-\ell-1)^{(n-i-1)}\Delta^n u(\ell)\right] \\
&\le p(k)\phi_3(k)v^2(k) + \phi_5(k)v(k),
\end{aligned}$$

which is the same as

$$\Delta[e(k)v(k)] \le p(k)\phi_3(k)e^{-1}(k+1)[e(k+1)v(k)]^2. \tag{4.3.13}$$

Now since $v(k)$ is nondecreasing and $e(k)$ is nonincreasing, we have

$$-\Delta[e(k)v(k)]^{-1} = \int_k^{k+1}\frac{d[e(t)v(t)]}{[e(t)v(t)]^2} \le \frac{\Delta[e(k)v(k)]}{[e(k+1)v(k)]^2}.$$

Thus, from (4.3.13) we obtain

$$-\Delta[e(k)v(k)]^{-1} \le p(k)\phi_3(k)e^{-1}(k+1),$$

which is on using $v(a) = 1$ gives

$$v(k) \le \frac{e^{-1}(k)}{1 - \sum_{\ell=a}^{k-1} p(\ell)\phi_3(\ell)e^{-1}(\ell+1)}.$$

Now on substituting this in (4.3.12) the inequality (4.3.11) follows. ∎

## 4.4. Finite Systems of Inequalities

Let the subscript $i$ range over the integers $1,\cdots,n$ and $r$ be some fixed positive integer such that $1 \le r \le n$. The subscripts $p$ and $q$ range over the integers $1,\cdots,r$ and $r+1,\cdots,n$ respectively.

**Definition 4.4.1.** The function $\mathbf{f}(k,\mathbf{u})$ is said to possess *mixed monotone property* if (i) $f_p(k,\mathbf{u})$ is nondecreasing in $u_1,\cdots,u_r$ and nonincreasing in $u_{r+1},\cdots,u_n$ for all fixed $k \in \mathbb{N}(a)$, and (ii) $f_q(k,\mathbf{u})$ is nonincreasing in $u_1,\cdots,u_r$ and nondecreasing in $u_{r+1},\cdots,u_n$. In particular $\mathbf{f}(k,\mathbf{u})$ is said to possess *nondecreasing property* if $f_i(k,\mathbf{u})$ is nondecreasing in $u_1,\cdots,u_n$ for all fixed $k \in \mathbb{N}(a)$.

**Definition 4.4.2.** The function $\mathbf{v}(k)$ defined on $\mathbb{N}(a)$ is said to be a $r$ *under and* $(n-r)$ *over function* with respect to the system $\mathbf{u}(k+1) = \mathbf{f}(k, \mathbf{u}(k))$ if $v_p(k+1) \le f_p(k, \mathbf{v}(k))$ and $v_q(k+1) \ge f_q(k, \mathbf{v}(k))$ for all $k \in \mathbb{N}(a)$. If $\mathbf{v}(k)$ satisfies the reverse inequalities, then it is said to be $r$ *over and* $(n-r)$ *under function.*

**Theorem 4.4.1.** Let the function $\mathbf{f}(k, \mathbf{u})$ possess mixed monotone property. Further, let there exist two functions $\mathbf{v}(k)$ and $\mathbf{w}(k)$ defined on $\mathbb{N}(a)$ such that

$$\begin{aligned} v_p(k+1) \ &\le \ f_p(k, \mathbf{v}(k)), \quad & v_q(k+1) \ &\ge \ f_q(k, \mathbf{v}(k)) \\ w_p(k+1) \ &\ge \ f_p(k, \mathbf{w}(k)), \quad & w_q(k+1) \ &\le \ f_q(k, \mathbf{w}(k)) \end{aligned}$$

$$v_p(a) \ \le \ w_q(a), \quad v_q(a) \ \ge \ w_q(a). \tag{4.4.1}$$

Then, for all $k \in \mathbb{N}(a)$

$$v_p(k) \ \le \ w_p(k), \quad v_q(k) \ \ge \ w_q(k). \tag{4.4.2}$$

**Proof.** Define a function $\mathbf{z}(k)$ as follows: $z_p(k) = w_p(k) - v_p(k)$ and $z_q(k) = v_q(k) - w_q(k)$. By induction we shall show that $z_i(k) \ge 0$ for all $k \in \mathbb{N}(a)$. For this, from (4.4.1), $z_i(a) \ge 0$. Let $z_i(k) \ge 0$ for some fixed $k \in \mathbb{N}(a+1)$, then since $\mathbf{f}(k, \mathbf{u})$ is mixed monotone, we have

$$v_p(k+1) \ \le \ f_p(k, \mathbf{v}(k)) \ \le \ f_p(k, \mathbf{w}(k)) \ \le \ w_p(k+1)$$

and

$$w_q(k+1) \ \le \ f_q(k, \mathbf{w}(k)) \ \le \ f_q(k, \mathbf{v}(k)) \ \le \ v_q(k+1),$$

i.e. $z_i(k+1) \ge 0$. ∎

**Corollary 4.4.2.** Let the function $\mathbf{f}(k, \mathbf{u})$ be nondecreasing. Further, let there exist two functions $\mathbf{v}(k)$ and $\mathbf{w}(k)$ defined on $\mathbb{N}(a)$ such that

$$\mathbf{v}(k+1) \ \le \ \mathbf{f}(k, \mathbf{v}(k)), \quad \mathbf{w}(k+1) \ \ge \ \mathbf{f}(k, \mathbf{w}(k)), \quad \mathbf{v}(a) \le \mathbf{w}(a).$$

Then, for all $k \in \mathbb{N}(a)$, $\mathbf{v}(k) \le \mathbf{w}(k)$.

**Corollary 4.4.3.** Let the functions $\mathbf{v}(k)$, $\mathbf{w}(k)$ be $r$ under and $(n-r)$ over, $r$ over and $(n-r)$ under functions with respect to the system $\mathbf{u}(k+1) = \mathbf{f}(k, \mathbf{u}(k))$ respectively. Further, let the vector valued function $\mathbf{f}(k, \mathbf{u})$ possess mixed monotone property. If $\mathbf{v}(a) = \mathbf{w}(a) = \mathbf{u}(a) = \mathbf{u}^0$, where $\mathbf{u}(k)$ is the solution of the problem $\mathbf{u}(k+1) = \mathbf{f}(k, \mathbf{u}(k))$, $\mathbf{u}(a) = \mathbf{u}^0$, then for all $k \in \mathbb{N}(a)$

$$v_p(k) \ \le \ u_p(k) \ \le \ w_p(k), \quad v_q(k) \ge u_q(k) \ge w_q(k).$$

**Theorem 4.4.4.** Let for all $k \in \mathbb{N}(a)$ the following inequality be satisfied

$$\mathbf{u}(k) \;\leq\; \mathbf{p}(k) + \mathcal{B}(k)\sum_{\ell=a}^{k-1} \mathcal{C}(\ell)\mathbf{u}(\ell), \tag{4.4.3}$$

where $\mathbf{u}(k)$ and $\mathbf{p}(k)$ are not necessarily nonnegative. Then, for all $k \in \mathbb{N}(a)$

$$\mathbf{u}(k) \;\leq\; \mathbf{p}(k) + \mathcal{B}(k)\sum_{\ell=a+1}^{k} \prod_{\tau=0}^{k-1-\ell} (\mathcal{I} + \mathcal{C}(k-1-\tau) \times \mathcal{B}(k-1-\tau))\mathcal{C}(\ell-1)\mathbf{p}(\ell-1). \tag{4.4.4}$$

**Proof.** Define a function $\mathbf{v}(k)$ on $\mathbb{N}(a)$ as follows

$$\mathbf{v}(k) \;=\; \sum_{\ell=a}^{k-1} \mathcal{C}(\ell)\mathbf{u}(\ell).$$

Then, as in Theorem 4.1.1, we have

$$\mathbf{v}(k+1) \;\leq\; (\mathcal{I} + \mathcal{C}(k)\mathcal{B}(k))\mathbf{v}(k) + \mathcal{C}(k)\mathbf{p}(k), \qquad \mathbf{v}(a) = 0.$$

As an application of Corollary 4.4.3, we find that $\mathbf{v}(k) \leq \mathbf{w}(k)$, where $\mathbf{w}(k)$ is the solution of the problem

$$\mathbf{w}(k+1) \;=\; (\mathcal{I} + \mathcal{C}(k)\mathcal{B}(k))\mathbf{w}(k) + \mathcal{C}(k)\mathbf{p}(k), \qquad \mathbf{w}(a) = 0.$$

Thus, from Theorem 2.6.1 it follows that

$$\mathbf{v}(k) \;\leq\; \mathbf{w}(k) \;=\; \sum_{\ell=a+1}^{k} \prod_{\tau=0}^{k-1-\ell} (\mathcal{I} + \mathcal{C}(k-1-\tau)\mathcal{B}(k-1-\tau)) \times \mathcal{C}(\ell-1)\mathbf{p}(\ell-1).$$

The result (4.4.4) now follows from the inequality $\mathbf{u}(k) \;\leq\; \mathbf{p}(k) + \mathcal{B}(k)\mathbf{v}(k)$. ∎

**Remark 4.4.1.** The inequality (4.4.4) is the best possible, however at the cost of several matrix multiplications which may not be feasible. Thus, from a practical point of view it is not of much use. In our next two results we shall provide explicit upper estimates, however these are not the best possible.

**Theorem 4.4.5.** Let for all $k \in \mathbb{N}(a)$ the inequality (4.4.3) be satisfied, and $\mathbf{p}(k)$ is not necessarily nonnegative. Then, for all $k \in \mathbb{N}(a)$

$$u_i(k) \ \le \ p_i(k) + \max_{1 \le j \le n} b_{ij}(k) \sum_{\ell=a}^{k-1} \alpha(\ell) \prod_{\tau=\ell+1}^{k-1} (1 + \beta(\tau)), \tag{4.4.5}$$

where

$$\alpha(k) \ = \ \sum_{j,r=1}^{n} c_{jr}(k) p_r(k), \qquad \beta(k) \ = \ \max_{1 \le s \le n} \sum_{j,r=1}^{n} c_{jr}(k) b_{rs}(k).$$

**Proof.** Taking components of (4.4.3), to obtain

$$u_i(k) \ \le \ p_i(k) + \sum_{j=1}^{n} b_{ij}(k) v_j(k), \tag{4.4.6}$$

where

$$v_j(k) \ = \ \sum_{r=1}^{n} \sum_{\ell=a}^{k-1} c_{jr}(\ell) u_r(\ell). \tag{4.4.7}$$

We define $v(k) = \sum_{j=1}^{n} v_j(k)$, then it follows that

$$\begin{aligned} \Delta v(k) \ &= \ \sum_{j=1}^{n} \sum_{r=1}^{n} c_{jr}(k) u_r(k) \\ &\le \ \sum_{j,r=1}^{n} c_{jr}(k) \left[ p_r(k) + \sum_{s=1}^{n} b_{rs}(k) v_s(k) \right] \\ &\le \ \alpha(k) + \beta(k) v(k) \end{aligned}$$

an hence, as in Theorem 4.1.1 we have

$$v(k) \ \le \ \sum_{\ell=a}^{k-1} \alpha(\ell) \prod_{\tau=\ell+1}^{k-1} (1 + \beta(\tau)).$$

The result (4.4.5) now follows from (4.4.6). ∎

**Theorem 4.4.6.** Let for all $k \in \mathbb{N}(a)$ the inequality (4.4.3) be satisfied, and $\mathbf{p}(k)$ is not necessarily nonnegative. Then, for all $k \in \mathbb{N}(a)$

$$u^*(k) \ \le \ p^*(k) + b^*(k) \sum_{\ell=a}^{k-1} p^*(\ell) c^*(\ell) \prod_{\tau=\ell+1}^{k-1} (1 + b^*(\tau) c^*(\tau)),$$

where

$$u^*(k) = \max_{1\le i\le n} u_i(k), \quad p^*(k) = \max_{1\le i\le n} p_i(k),$$

$$b^*(k) = \sum_{j=1}^{n}\left(\max_{1\le i\le n} b_{ij}(k)\right), \quad \text{and} \quad c^*(k) = \max_{1\le j\le n}\left(\sum_{r=1}^{n} c_{jr}(k)\right).$$

**Proof.** Taking maxima in (4.4.6) over $1 \le i \le n$, we obtain

$$(4.4.8) \qquad u^*(k) \le p^*(k) + \sum_{j=1}^{n} b_j^*(k) v_j(k),$$

where $b_j^*(k) = \max_{1\le j\le n} b_{ij}(k)$. Next from (4.4.7), we find

$$(4.4.9) \qquad v_j(k) \le \sum_{r=1}^{n}\sum_{\ell=a}^{k-1} c_{jr}(\ell)u^*(\ell) = \sum_{\ell=a}^{k-1} c_j(\ell)u^*(\ell),$$

where $c_j(k) = \sum_{r=1}^{n} c_{jr}(k)$.

Using (4.4.9) in (4.4.8), we find

$$(4.4.10) \qquad u^*(k) \le p^*(k) + \sum_{j=1}^{n} b_j^*(k)\sum_{\ell=a}^{k-1} c_j(\ell)u^*(\ell)$$

$$\le p^*(k) + b^*(k)\sum_{\ell=a}^{k-1} c^*(\ell)u^*(\ell).$$

Now the result follows from Theorem 4.1.1. ∎

**Remark 4.4.2.** An explicit upper estimate for $u^*(k)$ can also be provided by the inequality (4.4.10) which has been considered in Theorem 4.1.6. ∎

## 4.5. Opial Type Inequalities

**Theorem 4.5.1.** Let $u(k)$ be nondecreasing for all $k \in \mathbb{N}(a)$ and $u(a) = 0$. Then,

(i) if $p > 0,\ q > 0,\ p + q \ge 1$ or $p < 0,\ q < 0$

$$(4.5.1) \qquad \sum_{\ell=a}^{k-1}(\Delta u(\ell))^q u^p(\ell+1) \le H(k-a)\sum_{\ell=a}^{k-1}(\Delta u(\ell))^{p+q},$$

where $H(0) = q(p+q)^{-1}$, and for $k \in \mathbb{N}(a+1)$

$$H(k-a) = \max\left\{H(k-a-1) + \frac{p(k-a)^{p-1}}{(p+q)}, \frac{q(k-a+1)^p}{(p+q)}\right\}$$

(ii) if $p > 0,\ q < 0,\ p+q \le 1,\ p+q \ne 0$ or $p < 0,\ q > 0,\ p+q \ge 1$

$$\sum_{\ell=a}^{k-1} (\Delta u(\ell))^q u^p(\ell+1) \ \ge\ h(k-a) \sum_{\ell=a}^{k-1} (\Delta u(\ell))^{p+q}, \tag{4.5.2}$$

where $h(0) = q(p+q)^{-1}$, and for $k \in \mathbb{N}(a+1)$

$$h(k-a) \ = \ \min\left\{ h(k-a-1) + \frac{p(k-a)^{p-1}}{(p+q)},\ \frac{q(k-a+1)^p}{(p+q)} \right\}.$$

Further, in particular

(iii) if $p \ge 1,\ q \ge 1$ then (4.5.1) holds with $H(k-a)$ replaced by $q(k-a+1)^p(p+q)^{-1}$

(iv) if $p \le 0,\ q < 0$ then (4.5.1) holds with $H(k-a)$ replaced by $J(k-a)$, where $J(0) = q(p+q)^{-1}$, and for $k \in \mathbb{N}(a+1)$

$$J(k-a) \ = \ 1 + p(p+q)^{-1} \sum_{\ell=a+2}^{k} (\ell-a)^{p-1}$$

(v) if $p \ge 0,\ p+q < 0$ then (4.5.2) holds with $h(k-a)$ replaced by $J(k-a)$.

**Proof.** For all $\ell \in \mathbb{N}(a)$ and $p+q \ne 0$, we define $v(\ell+1) = (\Delta u(\ell))^{p+q}$, so that $(\Delta u(\ell))^q = v^{qr}(\ell+1)$, where $r = (p+q)^{-1}$. Since $u(\ell+1) = \sum_{\tau=a}^{\ell} \Delta u(\tau)$, by Hölder's inequality we have

$$u(\ell+1) \ \le\ (\ell-a+1)^{1-r} \left( \sum_{\tau=a}^{\ell} v(\tau+1) \right)^r \ = \ w(\ell+1), \ \text{ if } p+q \ge 1$$

and

$$u(\ell+1) \ \ge\ w(\ell+1), \quad \text{if} \quad p+q < 0 \quad \text{or} \quad 0 < p+q \le 1.$$

Therefore, if $p \ge 0,\ p+q \ge 1$ or $p \le 0$ and either $p+q < 0$ or $0 < p+q \le 1$, then $u^p(\ell+1) \le w^p(\ell+1)$, and

$$\sum_{\ell=a}^{k-1} (\Delta u(\ell))^q u^p(\ell+1) \ \le\ \sum_{\ell=a}^{k-1} v^{qr}(\ell+1) w^p(\ell+1),$$

while if $p \le 0,\ p+q \ge 1$ or $p \ge 0$ and either $p+q < 0$ or $0 < p+q \le 1$, then $u^p(\ell+1) \ge w^p(\ell+1)$, and

$$\sum_{\ell=a}^{k-1} (\Delta u(\ell))^q u^p(\ell+1) \ \ge\ \sum_{\ell=a}^{k-1} v^{qr}(\ell+1) w^p(\ell+1).$$

Thus, (i) and (ii) will follow if we can prove

$$(4.5.3)\quad \sum_{\ell=a}^{k-1} v^{qr}(\ell+1)w^{p}(\ell+1) \ \le\ H(k-a)\sum_{\ell=a}^{k-1} v(\ell+1) \quad \text{for} \quad pq>0$$

and

$$(4.5.4)\quad \sum_{\ell=a}^{k-1} v^{qr}(\ell+1)w^{p}(\ell+1) \ \ge\ h(k-a)\sum_{\ell=a}^{k-1} v(\ell+1) \quad \text{for} \quad pq<0.$$

We shall prove (4.5.3) by induction on $k \in \mathbb{N}(a)$. Clearly, it holds for $k = a+1$ since $H(1) \ge 1$. Assume that it holds for $k$, and observe that

$$(4.5.5)\ \sum_{\ell=a}^{k} v^{qr}(\ell+1)w^{p}(\ell+1) \ \le\ H(k-a)\sum_{\ell=a}^{k-1} v(\ell+1)+v^{qr}(k+1)w^{p}(k+1).$$

Now since $v(k+1) \ge 0$ for all $k \in \mathbb{N}(a)$, the classical result of arithmetic and geometric means for $pq > 0$ gives

$$\begin{aligned} v^{qr}(k+1)w^{p}(k+1) &= (k-a+1)^{p}\left\{v^{qr}(k+1)\left[(k-a+1)^{-1}\sum_{\ell=a}^{k} v(\ell+1)\right]^{pr}\right\} \\ &\le (k-a+1)^{p}\left\{qrv(k+1)+pr(k-a+1)^{-1}\sum_{\ell=a}^{k} v(\ell+1)\right\} \\ &= Q(k-a+1), \quad \text{say} \end{aligned}$$

since $pr + qr = 1$. Hence, from (4.5.5) we get

$$\begin{aligned} (4.5.6)\quad \sum_{\ell=a}^{k} v^{qr}(\ell+1)w^{p}(\ell+1) \ &\le\ H(k-a)\sum_{\ell=a}^{k-1} v(\ell+1)+qr(k-a+1)^{p}v(k+1) \\ &\quad + pr(k-a+1)^{p-1}\sum_{\ell=a}^{k} v(\ell+1) \\ &\le\ H(k-a+1)\sum_{\ell=a}^{k} v(\ell+1) \end{aligned}$$

since $H(k-a) \ge qr(k-a+1)^{p}$ and $H(k-a+1) \ge H(k-a)+ pr(k-a+1)^{p-1}$, which proves (4.5.3). For $pq < 0$, one can easily see that $v^{qr}(k+1)w^{p}(k+1) \ge Q(k-a+1)$, so that (4.5.4) will follow by proceeding as above.

To prove (iii), consider $H^1(k-a) = qr(k-a+1)^p$ for $p \geq 1$, $q \geq 1$. We have $H^1(1) = qr2^p \geq 1$, and

$$\begin{aligned} \Delta H^1(k-a) &= qr\left[(k-a+2)^p - (k-a+1)^p\right] \\ &\geq qr\left[(k-a+1)^p + p(k-a+1)^{p-1} - (k-a+1)^p\right] \\ &\geq pr(k-a+1)^{p-1}, \end{aligned}$$

where we have used the Bernoulli inequality. Thus, as above up to (4.5.6), we get

$$\begin{aligned} \sum_{\ell=a}^{k} v^{qr}(\ell+1)w^p(\ell+1) &\leq qr(k-a+1)^p \sum_{\ell=a}^{k-1} v(\ell+1) + qr(k-a+1)^p \\ &\quad \times v(k+1) + \Delta H^1(k-a)\sum_{\ell=a}^{k} v(\ell+1) \\ &\leq H^1(k-a+1)\sum_{\ell=a}^{k} v(\ell+1). \end{aligned}$$

This completes the proof of (iii).

Finally, to prove (iv) and (v) we note that for all $k \in \mathbb{N}(a+1)$

$$\Delta J(k-a-1) = pr(k-a)^{p-1}$$

and

$$J(k-a) \geq 1 \geq qr(k-a+1)^p \quad \text{if } p < 0 \text{ and } q < 0$$

but

$$J(k-a) \leq 1 \leq qr(k-a+1)^p \quad \text{if } p \geq 0 \text{ and } p+q < 0.$$

This completes the proof of Theorem 4.5.1. ∎

**Remark 4.5.1.** The conclusion (iii) fails to hold if $p < 1$. For this, consider $p = 1/2$, $q = 1$, $a = 1$, $k = 2$, $u(1) = 1$ and $u(2) = 2$.

**Remark 4.5.2.** Consider $p \geq 1$, $q = 1$, $a = 1$ and $u(k) = k-1$. The conclusion (iii) gives

$$\sum_{\ell=1}^{k-1} \ell^p \leq \frac{k^p(k-1)}{p+1} < \frac{k^{p+1}-1}{p+1} = \int_1^k t^p dt,$$

i.e. (4.5.1) yields a better estimate than that of obtained by simply comparing areas.

**Theorem 4.5.2.** Let $u(k)$ be such that $u(a) = u(b) = 0$ and not necessarily nonnegative on $\mathbb{N}(a, b)$. Then,

$$\sum_{\ell=a+1}^{b-1} |u(\ell)||\Delta u(\ell)| \;\le\; \frac{1}{2}\left\{\left[\frac{a+b+1}{2}\right] - a\right\}\sum_{\ell=a}^{b-1} |\Delta u(\ell)|^2. \tag{4.5.7}$$

If $(a+b)$ is even then the inequality (4.5.7) is the best possible.

**Proof.** Since $u(a) = u(b) = 0$, we have

$$u(k) \;=\; \sum_{\ell=a}^{k-1} \Delta u(\ell), \qquad u(k) \;=\; -\sum_{\ell=k}^{b-1} \Delta u(\ell), \qquad k = a+1, \cdots, b-1.$$

Hence, we find

$$\begin{aligned}
\sum_{\ell=a+1}^{b-1} |u(\ell)||\Delta u(\ell)| \;&=\; \sum_{\ell=a+1}^{k} |u(\ell)||\Delta u(\ell)| + \sum_{\ell=k+1}^{b-1} |u(\ell)||\Delta u(\ell)| \\
&\le\; \sum_{\ell=a+1}^{k} |\Delta u(\ell)| \sum_{\tau=a}^{\ell-1} |\Delta u(\tau)| + \sum_{\ell=k+1}^{b-1} |\Delta u(\ell)| \sum_{\tau=\ell}^{b-1} |\Delta u(\tau)| \\
&\le\; \frac{1}{2} \sum_{\ell=a+1}^{k} \sum_{\tau=a}^{\ell-1} \left(|\Delta u(\ell)|^2 + |\Delta u(\tau)|^2\right) \\
&\qquad + \frac{1}{2} \sum_{\ell=k+1}^{b-1} \sum_{\tau=\ell}^{b-1} \left(|\Delta u(\ell)|^2 + |\Delta u(\tau)|^2\right) \\
&\le\; \frac{1}{2}(k-a) \sum_{\ell=a}^{k} |\Delta u(\ell)|^2 + \frac{1}{2}(b-k) \sum_{\ell=k+1}^{b-1} |\Delta u(\ell)|^2 \\
&\le\; \frac{1}{2}\max\{(k-a),\ (b-k)\} \sum_{\ell=a}^{b-1} |\Delta u(\ell)|^2 \\
&\le\; \frac{1}{2}\left\{\left[\frac{a+b+1}{2}\right] - a\right\} \sum_{\ell=a}^{b-1} |\Delta u(\ell)|^2.
\end{aligned}$$

To complete the proof let $(a+b)$ be even, and $u(k) = \left(\dfrac{a+b}{2} - a\right) - \left|k - \dfrac{a+b}{2}\right|$, $k \in \mathbb{N}(a, b)$ so that $u(a) = u(b) = 0$, $|\Delta u(k)| = 1$ for all $k \in \mathbb{N}(a, b-1)$,

$$\sum_{\ell=a+1}^{b-1} |u(\ell)||\Delta u(\ell)| \;=\; \frac{(b-a)^2}{4}, \qquad \sum_{\ell=a}^{b-1} |\Delta u(\ell)|^2 \;=\; (b-a)$$

and

$$\frac{1}{2}\left\{\left[\frac{a+b+1}{2}\right]-a\right\} = \frac{1}{4}(b-a),$$

i.e. equality holds in (4.5.7). ∎

## 4.6. Wirtinger Type Inequalities

Let $\theta_i \in (0,l)$, $p_i > 0$, $i = 1,\cdots,n$, $P_n = \sum_{i=1}^n p_i$, and $\sigma = (1/P_n)\sum_{i=1}^n p_i\theta_i$.

**Theorem 4.6.1.** Let $f(\theta)$ be a positive $C^{(2)}(0,l)$ function such that $f'(\theta)f''(\theta) \neq 0$ on $(0,l)$, and

$$[f'(\theta)]^2 - f(\theta)f''(\theta) = \mu, \quad 0<\theta<l \tag{4.6.1}$$

where $\mu$ is a constant.

(i) If $f''(\theta)<0$ on $(0,l)$, then

$$\left(\sum_{i=1}^n p_i f(\theta_i)\right)^2 - c_n\sum_{i=1}^n p_i f(\theta_i)f'(\theta_i) \geq \left(P_n f(\sigma) - \sum_{i=1}^n p_i f(\theta_i)\right)^2; \tag{4.6.2}$$

(ii) If $f''(\theta)>0$ on $(0,l)$, then

$$\left(\sum_{i=1}^n p_i f(\theta_i)\right)^2 - c_n\sum_{i=1}^n p_i f(\theta_i)f'(\theta_i) \leq \left(P_n f(\sigma) - \sum_{i=1}^n p_i f(\theta_i)\right)^2, \tag{4.6.3}$$

where $c_n = P_n f(\sigma)/f'(\sigma)$.

In (4.6.2) and (4.6.3) equality holds if and only if $\theta_1 = \cdots = \theta_n = \sigma$.

**Proof.** We shall use the following notation:

$P = (p_1,\cdots,p_n) \in \mathbf{R}_+^n$
$\Theta = (\theta_1,\cdots,\theta_n) \in \mathbf{R}_+^n$, $\theta_i \in (0,l)$
$\Omega = (\sigma,\cdots,\sigma)$, $L_n(\Theta,P) = \sum_{i=1}^n p_i f(\theta_i)$

$$F_p(\Theta) = L_n^2(\Theta,P) - c_n\sum_{i=1}^n p_i f(\theta_i)f'(\theta_i) - [P_n f(\sigma) - L_n(\Theta,P)]^2. \tag{4.6.4}$$

Thus to prove (4.6.2) and (4.6.3) we must establish:

(i) If $f''(\theta)<0$ on $(0,l)$, then $F_p(\Theta) \geq 0$.

(ii) If $f''(\theta)>0$ on $(0,l)$, then $F_p(\Theta) \leq 0$.

Also equality holds in (i) and (ii) if and only if $\theta_i = \sigma,\ i = 1, \cdots, n$.

First, a direct computation in (4.6.4) confirms that $F_p(\Omega) = 0$, and hence in (i) and (ii) equality holds when $\Theta = \Omega$. Next, we consider a line segment which joins $\Theta = (\theta_1, \cdots, \theta_n)$ and $\Omega = (\sigma, \cdots, \sigma)$, i.e.

$$\begin{aligned} \Theta(t) &= (\theta_1(t), \cdots, \theta_n(t)), \\ \theta_i(t) &= t\sigma + (1-t)\theta_i, \quad i = 1, \cdots, n, \quad t \in [0,1]. \end{aligned} \tag{4.6.5}$$

It is clear that if $\sigma < \theta_i$ then $\theta_i(t) > \sigma$, and if $\sigma > \theta_i$ then $\theta_i(t) < \sigma$, $\theta_i'(t) = \sigma - \theta_i$ for $i = 1, \cdots, n$, and

$$\sum_{i=1}^{n} p_i(\sigma - \theta_i) = 0. \tag{4.6.6}$$

Thus, in view of (4.6.1) it follows that

$$\sum_{i=1}^{n} p_i(\sigma - \theta_i) f(\theta_i(t)) f''(\theta_i(t)) = \sum_{i=1}^{n} p_i(\sigma - \theta_i)[f'(\theta_i(t))]^2. \tag{4.6.7}$$

Further, we note that to prove (i) it suffices to show that $F_p(t) = F_p(\Theta(t))$ is decreasing on $[0,1]$, in fact then

$$F_p(\Theta) = F_p(\Theta(0)) = F_p(0) > F_p(1) = F_p(\Theta(1)) = F_p(\Omega) = 0$$

and similarly, to show (ii) it suffices to prove that $F_p(t) = F(\Theta(t))$ is increasing on $[0,1]$, indeed then

$$F_p(\Theta) = F_p(\Theta(0)) = F_p(0) < F_p(1) = F_p(\Theta(1)) = F_p(\Omega) = 0.$$

Case (i) If $f''(\theta) < 0$ on $(0, l)$, then $f'(\theta)$ is decreasing on $(0, l)$ and hence the following inequality holds

$$p_i(\sigma - \theta_i)(f'(\sigma) - f'(\theta_i(t))) < 0, \quad i = 1, \cdots, n, \quad t \in [0,1].$$

Thus, if $f'(\theta) > 0$ then

$$\sum_{i=1}^{n} p_i(\sigma - \theta_i)[f'(\sigma) - f'(\theta_i(t))] f'(\theta_i(t)) < 0, \quad c_n > 0 \tag{4.6.8}$$

and if $f'(\theta) < 0$ then

$$\sum_{i=1}^{n} p_i(\sigma - \theta_i)[f'(\sigma) - f'(\theta_i(t))] f'(\theta_i(t)) > 0, \quad c_n < 0. \tag{4.6.9}$$

Next, differentiating $F_p(\Theta(t)) = F_p(t)$ and using (4.6.7), we find

$$F_p'(t) = 2P_n f(\sigma)\sum_{i=1}^{n} p_i(\sigma-\theta_i)f'(\theta_i(t)) - 2c_n\sum_{i=1}^{n} p_i(\sigma-\theta_i)[f'(\theta_i(t))]^2$$

$$(4.6.10) \qquad = 2c_n\sum_{i=1}^{n} p_i(\sigma-\theta_i)[f'(\sigma) - f'(\theta_i(t))]f'(\theta_i(t)).$$

From this equality for both (4.6.8) and (4.6.9) we have $F_p'(t) = F_p'(\Theta(t)) < 0$, i.e. $F_p(\Theta(t))$ is decreasing on $[0,1]$.

Case (ii) If $f''(\theta) > 0$ on $(0,l)$, then $f'(\theta)$ is increasing on $(0,l)$ and hence the following inequality holds

$$p_i(\sigma-\theta_i)(f'(\sigma) - f'(\theta_i(t))) > 0, \quad i = 1,\cdots,n, \quad t \in [0,1].$$

Thus, if $f'(\theta) > 0$ then

$$(4.6.11) \qquad \sum_{i=1}^{n} p_i(\sigma-\theta_i)[f'(\sigma) - f'(\theta_i(t))]f'(\theta_i(t)) > 0, \quad c_n > 0$$

and if $f'(\theta) < 0$ then

$$(4.6.12) \qquad \sum_{i=1}^{n} p_i(\sigma-\theta_i)[f'(\sigma) - f'(\theta_i(t))]f'(\theta_i(t)) < 0, \quad c_n < 0.$$

Now from the equality (4.6.10) for both (4.6.11) and (4.6.12) we have $F_p'(t) = F_p'(\Theta(t)) > 0$, i.e. $F_p(\Theta(t))$ is increasing on $[0,1]$. ∎

**Corollary 4.6.2.** Let in Theorem 4.6.1, $l = \pi/2$. Then, the following inequalities hold

$$\left(\sum_{i=1}^{n} p_i\sin(\theta_i)\right)^2 - c_n\sum_{i=1}^{n} p_i\sin(\theta_i)\cos(\theta_i) \geq \left(P_n\sin(\sigma) - \sum_{i=1}^{n} p_i\sin(\theta_i)\right)^2,$$

where $c_n = P_n\tan(\sigma)$, and

$$\left(\sum_{i=1}^{n} p_i\cos(\theta_i)\right)^2 - c_n\sum_{i=1}^{n} p_i\sin(\theta_i)\cos(\theta_i) \geq \left(P_n\cos(\sigma) - \sum_{i=1}^{n} p_i\cos(\theta_i)\right)^2,$$

where $c_n = P_n\cot(\sigma)$.

In the above inequalities the equality holds if and only if $\theta_1 = \cdots = \theta_n = \sigma$.

**Proof.** It suffices to note that the functions $\sin(\theta)$ and $\cos(\theta)$, $\theta \in (0, \pi/2)$ satisfy the differential equation (4.6.1) with $\mu = 1$. ∎

In the following result we shall extend the class of functions by relaxing the condition (4.6.1) to

$$(\sigma - \theta)\left([f'(\theta)]^2 - f(\theta)f''(\theta) - \mu\right) \lesseqgtr 0.$$

**Theorem 4.6.3.** Let $f(\theta)$ be a positive $C^{(2)}(0,l)$ function such that $f'(\theta)f''(\theta) \neq 0$ on $(0,l)$.

(I) If

$$(4.6.13) \qquad (\sigma - \theta)\left([f'(\theta)]^2 - f(\theta)f''(\theta) - \mu\right) \leq 0$$

for $\theta \in (0,l)$, where $\mu$ is a constant, then

($I_1$) $f''(\theta) < 0$ and $f'(\theta) > 0$ on $(0,l)$ imply (4.6.2)

($I_2$) $f''(\theta) > 0$ and $f'(\theta) < 0$ on $(0,l)$ imply (4.6.3).

(II) If

$$(4.6.14) \qquad (\sigma - \theta)\left([f'(\theta)]^2 - f(\theta)f''(\theta) - \mu\right) \geq 0$$

for $\theta \in (0,l)$, where $\mu$ is a constant, then

($II_1$) $f''(\theta) < 0$ and $f'(\theta) < 0$ on $(0,l)$ imply (4.6.2)

($II_2$) $f''(\theta) > 0$ and $f'(\theta) > 0$ on $(0,l)$ imply (4.6.3).

In the above inequalities $c_n = P_n f(\sigma)/f'(\sigma)$, and equality holds if and only if $\theta_i = \sigma$, $i = 1, \cdots, n$.

**Proof.** As in Theorem 4.6.1 it suffices to show that

($I_1$) if $f''(\theta) < 0$ and $f'(\theta) > 0$ on $(0,l)$, then $F_p(\Theta) \geq 0$

($I_2$) if $f''(\theta) > 0$ and $f'(\theta) < 0$ on $(0,l)$, then $F_p(\Theta) \leq 0$

($II_1$) if $f''(\theta) < 0$ and $f'(\theta) < 0$ on $(0,l)$, then $F_p(\Theta) \geq 0$

($II_2$) if $f''(\theta) > 0$ and $f'(\theta) > 0$ on $(0,l)$, then $F_p(\Theta) \leq 0$.

(I) From (4.6.6) and (4.6.13), we have

$$(4.6.15) \qquad \sum_{i=1}^{n} p_i(\sigma - \theta_i)f(\theta_i(t))f''(\theta_i(t)) \geq \sum_{i=1}^{n} p_i(\sigma - \theta_i)[f'(\theta_i(t))]^2.$$

Now there are two cases: ($I_1$) $f'' < 0$ and $f' > 0$, and ($I_2$) $f'' > 0$ and $f' < 0$. Let us consider the case (Ii): If $f''(\theta) < 0$ and $f'(\theta) > 0$ on $(0, l)$ then (4.6.8) holds. Differentiating $F_p(\Theta(t))$ and using (4.6.15) and (4.6.8) it follows that

$$\begin{aligned} F_p'(\Theta(t)) & \\ = & -c_n \sum_{i=1}^{n} p_i(\sigma - \theta_i)[f'(\theta_i(t))]^2 - c_n \sum_{i=1}^{n} p_i(\sigma - \theta_i) f(\theta_i(t)) f''(\theta_i(t)) \\ & + 2c_n f'(\sigma) \sum_{i=1}^{n} p_i(\sigma - \theta_i) f'(\theta_i(t)) \\ < & \ 2c_n \sum_{i=1}^{n} p_i(\sigma - \theta_i)[f'(\sigma) - f'(\theta_i(t))] f'(\theta_i(t)) \ < \ 0. \end{aligned}$$

The case ($I_1$) now follows as in Theorem 4.6.1. The other cases can be proved similarly. ∎

**Corollary 4.6.4.** Let $f(\theta)$ be a positive $C^{(3)}(0, l)$ function such that $f'(\theta)f''(\theta) \neq 0$ on $(0, l)$.

(I) If

$$f'(\theta)f''(\theta) - f(\theta)f'''(\theta) \geq 0 \tag{4.6.16}$$

for $\theta \in (0, l)$, then

($I_1$) $f''(\theta) < 0$ and $f'(\theta) > 0$ on $(0, l)$ imply (4.6.2)

($I_2$) $f''(\theta) > 0$ and $f'(\theta) < 0$ on $(0, l)$ imply (4.6.3).

(II) If

$$f'(\theta)f''(\theta) - f(\theta)f'''(\theta) \leq 0 \tag{4.6.17}$$

for $\theta \in (0, l)$, then

($II_1$) $f''(\theta) < 0$ and $f'(\theta) < 0$ on $(0, l)$ imply (4.6.2)

($II_2$) $f''(\theta) > 0$ and $f'(\theta) > 0$ on $(0, l)$ imply (4.6.3).

In the above inequalities $c_n = P_n f(\sigma)/f'(\sigma)$, and equality holds if and only if $\theta_i = \sigma$, $i = 1, \cdots, n$.

**Proof.** Inequality (4.6.16) implies that the function

$$h(\theta) = [f'(\theta)]^2 - f(\theta)f''(\theta)$$

is monotonically increasing on $(0, l)$. Let $h(\sigma) = \mu$, then (4.6.16) leads to the condition (4.6.13). Similarly, (4.6.17) implies condition (4.6.14). ∎

## 4.7. Problems

**4.7.1.** Let in Theorem 4.1.1, $p(k) = q(k)$ for all $k \in \mathbb{N}(a)$. Show that for all $k \in \mathbb{N}(a)$

$$u(k) \leq p(k) \prod_{\ell=a}^{k-1} (1 + p(\ell) f(\ell)).$$

**4.7.2.** Let in Theorem 4.1.4, $r = 2$, $p(k) = u_0$, $q(k) = 1$ and $f_{11}(k) = f_{21}(k)$ for all $k \in \mathbb{N}(a)$. Show that for all $k \in \mathbb{N}(a)$

$$u(k) \leq u_0 \left(1 + \sum_{\ell=a}^{k-1} f_{11}(\ell)(1 - v(\ell)) \prod_{\tau=a}^{\ell-1} (1 + f_{11}(\tau) + f_{22}(\tau))\right),$$

where

$$v(k) = \sum_{\ell=a}^{k-1} f_{22}(\ell) \left(\prod_{\tau=a}^{\ell} (1 + f_{11}(\tau) + f_{22}(\tau))\right)^{-1} \sum_{\tau=a}^{\ell-1} f_{22}(\tau).$$

**4.7.3.** Let for all $k \in \mathbb{N}(a)$ the following inequality be satisfied

$$u(k) \leq p(k) + \sum_{\ell=a}^{k-1} q(k, \ell) u(\ell).$$

Show that for all $k \in \mathbb{N}(a)$

$$u(k) \leq P(k) \prod_{\ell=a}^{k-1} (1 + Q(k, \ell)),$$

where $P(k) = \max\{p(\tau) : \tau \in \mathbb{N}(a, k)\}$, and $Q(k, \ell) = \max\{q(\tau, \ell) : \tau \in \mathbb{N}(a, k)\}$.

**4.7.4.** Let for all $k \in \mathbb{N}(a)$ the following inequality be satisfied

$$u(k+1) \leq p + q u(k),$$

where $p$ and $u(k)$ are not necessarily nennegative. Show that for all $k \in \mathbb{N}(a)$

$$u(k) \leq q^{k-a} u(a) + \begin{cases} \dfrac{q^{k-a} - 1}{q - 1} p & \text{if } q \neq 1 \\ (k - a) p & \text{if } q = 1. \end{cases}$$

**4.7.5.** Let for all $k \in \mathbb{N}$ the following inequality be satisfied

$$u^2(k) \le c^2 + \sum_{\ell=0}^{k-1} (u(\ell+1) + u(\ell))[p(\ell)u(\ell) + q(\ell)].$$

Show that for all $k \in \mathbb{N}$

$$u(k) \le Q(k) \prod_{\ell=0}^{k-1} [1 + p(\ell)]$$

where

$$Q(k) = c + \sum_{\ell=0}^{k-1} q(\ell).$$

**4.7.6.** Let for all $k$, $r - 1 \in \mathbb{N}(a)$ such that $k \le r - 1$ the following inequality be satisfied

$$u(r) \ge u(k) - q(r) \sum_{\ell=k}^{r-1} f(\ell)u(\ell),$$

where $u(k)$ is not necessarily nonnegative. Show that for all $k$, $r - 1 \in \mathbb{N}(a)$, $k \le r - 1$

$$u(r) \ge u(k) \prod_{\ell=k}^{r-1} (1 - q(r)f(\ell)) \tag{4.7.1}$$

as long as $1 - q(r)f(\ell) > 0$. Further, the inequality (4.7.1) is the best possible.

**4.7.7.** Let for all $k \in \mathbb{N}(0, K)$ the following inequality be satisfied

$$u(k) \le c_2 + h^{1-\alpha} c_1 \sum_{\ell=0}^{k-1} (k - \ell)^{-\alpha} u(\ell),$$

where $0 < \alpha < 1$, $c_1 > 0$, $c_2 > 0$ and $h > 0$. Further, let $\nu$ be the smallest positive integer satisfying $\nu(1 - \alpha) \ge 1$. Show that for all $k \in \mathbb{N}(0, K)$

$$u(k) \le c_2 \left(c_2' + hc_1'(Kh)^{\nu(1-\alpha)-1}\right) exp\left(c_1'(Kh)^{\nu(1-\alpha)}\right),$$

where

$$c_1' = (c_1\Gamma(1-\alpha))^\nu \, \Gamma(\nu(1-\alpha)), \quad c_2' = \left(1 + c_1 h^{1-\alpha}\right) \sum_{j=0}^{\nu-2} \lambda^j + \lambda^{\nu-1}$$

$$\lambda = c_1 \frac{(Kh)^{1-\alpha}}{1-\alpha}.$$

**4.7.8.** Let for all $k \in \mathbb{N}(a)$ the following inequality be satisfied

$$u(k+1) \leq q u^p(k).$$

Show that for all $k \in \mathbb{N}(a)$

$$u(k) \leq \begin{cases} q^{\frac{1-p^{k-a}}{1-p}} u^{p^{k-a}}(a) & \text{if} \quad p \neq 1 \\ q^{k-a} u(a) & \text{if} \quad p = 1. \end{cases}$$

**4.7.9.** Let for all $k \in \mathbb{N}(a)$ the following inequality be satisfied

$$u(k) \leq p(k) \left[ q + \sum_{i=1}^{r_1} E_i(k,u) + \sum_{i=1}^{r_2} H_i(k,u) \right],$$

where $E_i(k,u)$ and $H_i(k,u)$ are defined in (4.1.7) and (4.2.2) respectively, and the constant $q > 0$. Then, if $\alpha_i = \sum_{j=1}^{i} \alpha_{ij}$ and $\max_{1 \leq i \leq r_2} \alpha_i = \alpha \neq 1$, show that for all $k \in \mathbb{N}(a)$

$$u(k) \leq p(k) v(k) \left[ q^{1-\alpha} + (1-\alpha) Q(k) \right]^{1/1-\alpha},$$

where

$$v(k) = \prod_{\ell=a}^{k-1} \left( 1 + \sum_{i=1}^{r_1} \Delta E_i(\ell, p) \right), \qquad Q(k) = \sum_{i=1}^{r_2} H_i(k, pv) q^{\alpha_i - \alpha}$$

and when $\alpha > 1$, we assume that $q^{1-\alpha} + (1-\alpha)Q(k) > 0$ for all $k \in \mathbb{N}(a)$.

**4.7.10.** Prove Theorem 4.2.4.

**4.7.11.** Prove Theorem 4.2.5.

**4.7.12.** Let for all $k \in \mathbb{N}(a)$ the following inequality be satisfied

$$u(k) \leq p(k) + q(k) h \left( \sum_{\ell=a}^{k-1} f(\ell) W(u(\ell)) \right),$$

where the functions $h$ and $W$ are continuous, positive and nondecreasing on $[0, \infty)$. Further, in addition $W$ is subadditive and submultiplicative. Show that for all $k \in \mathbb{N}(a)$

$$u(k) \leq p(k) + q(k) h \left( G^{-1} \left[ G \left( \sum_{\ell=a}^{k-1} f(\ell) W(p(\ell)) \right) + \sum_{\ell=a}^{k-1} f(\ell) W(q(\ell)) \right] \right),$$

where

$$G(w) = \int_{w_0}^{w} \frac{dt}{W(h(t))}, \quad w \geq w_0 \geq 0$$

as long as

$$G\left(\sum_{\ell=a}^{k-1} f(\ell)W(p(\ell))\right) + \sum_{\ell=a}^{k-1} f(\ell)W(q(\ell)) \in Dom(G^{-1}).$$

**4.7.13.** Let for all $k \in \mathbb{N}(a)$ the following inequality be satisfied

$$u(k) \leq p(k) + q(k)W^{-1}\left(\sum_{\ell=a}^{k-1} f(\ell)W(u(\ell))\right),$$

where the function $W$ is increasing, convex and submultiplicative on $[0,\infty)$ and $W(0)=0$, $\lim_{u\to\infty} W(u)=\infty$. Show that for all $k \in \mathbb{N}(a)$

$$u(k) \leq p(k) + q(k)W^{-1}\left(\sum_{\ell=a}^{k-1} \alpha(\ell)W(p(\ell)\alpha^{-1}(\ell))f(\ell) \times \prod_{\tau=\ell+1}^{k-1} (1+\beta(\tau)W(q(\tau)\beta^{-1}(\tau))f(\tau))\right),$$

where the functions $\alpha(k)$ and $\beta(k)$ are positive and $\alpha(k)+\beta(k)=1$ for all $k \in \mathbb{N}(a)$.

**4.7.14.** Let for all $k \in \mathbb{N}(a)$ the following inequality be satisfied

$$u(k) \leq p(k) + \sum_{\ell=a}^{k-1} q(k,\ell)W(u(\ell)),$$

where the function $W$ is continuous, positive and nondecreasing on $[0,\infty)$. Show that for all $k \in \mathbb{N}(a)$

$$u(k) \leq G^{-1}\left[G(P(k)) + \sum_{\ell=a}^{k-1} Q(k,\ell)\right],$$

where $P(k)$ and $Q(k,\ell)$ are defined in Problem 4.7.3, and $G(w) = \int_{w_0}^{w} \frac{dt}{W(t)}$, $w \geq w_0 \geq 0$, as long as

$$G(P(k)) + \sum_{\ell=a}^{k-1} Q(k,\ell) \in Dom(G^{-1}).$$

**4.7.15.** Let for all $k,\ r \in \mathbb{N}(a)$ such that $k \le r$ the following inequality be satisfied

$$u(r) \ \ge\ u(k) - q(r) \sum_{\ell=k+1}^{r} f(\ell) W(u(\ell)),$$

where the function $W$ is continuous, positive and nondecresing on $[0,\infty)$. Show that for all $k,\ r \in \mathbb{N}(a),\ k \le r$

$$u(r) \ \ge\ G^{-1}\left( G(u(k)) - q(r) \sum_{\ell=k+1}^{r} f(\ell) \right),$$

where

$$G(w) \ =\ \int_{w_0}^{w} \frac{dt}{W(t)}, \quad w > 0 \ \text{ and arbitrary } \ w_0 \ge 0$$

as long as

$$G(u(k)) - q(r) \sum_{\ell=k+1}^{r} f(\ell) \in Dom(G^{-1}).$$

**4.7.16.** Let for all $k,\ r \in \mathbb{N}(a)$ such that $k \le r$ the following inequality be satisfied

$$u(r) \ \ge\ u(k) - q(r) W^{-1}\left( \sum_{\ell=k+1}^{r} f(\ell) W(u(\ell)) \right),$$

where the function $W$ is positive, increasing, convex and submultiplicative on $(0,\infty)$ and $\lim_{u\to\infty} W(u) = \infty$. Show that for all $k,\ r \in \mathbb{N}(a),\ k \le r$

$$u(r) \ge \alpha(r) W^{-1}\left( \alpha^{-1}(r) W(u(k)) \prod_{\ell=k+1}^{r} \left(1 + \beta(r) W(q(r)\beta^{-1}(r)) f(\ell)\right)^{-1} \right),$$

where the functions $\alpha(k)$ and $\beta(k)$ are positive and $\alpha(k) + \beta(k) = 1$ for all $k \in \mathbb{N}(a)$.

**4.7.17.** Let the conditions of Theorem 4.3.3 be satisfied and $n = 1$. Show that for all $k \in \mathbb{N}(a)$

$$\Delta u(k) \ \le\ p(k) + q(k) C_i(k), \quad i = 0, 1$$

where

$$C_0(k) = \sum_{\ell=a}^{k-1}(\phi(\ell)-\psi(\ell))\prod_{\tau=\ell+1}^{k-1}(1+q(\tau))(1+q^*(\tau))$$

$$C_1(k) = \sum_{\ell=a}^{k-1}(\phi(\ell)+q^*(\ell)C_0(\ell))\prod_{\tau=\ell+1}^{k-1}(1+q^*(\tau)(q(\tau)-1))$$

$$\phi(k) = q^*(k)\left(u(a)+\sum_{\ell=a}^{k}p(\ell)\right)$$

$$\psi(k) = u(a)q(k)(1+q^*(k))\sum_{\ell=a}^{k-1}q(\ell)\sum_{\tau=a}^{\ell-1}q^*(\tau).$$

**4.7.18.** Let for all $k \in \mathbb{N}(a)$ the following inequality be satisfied

$$\Delta^n u(k) \le p(k)+q(k)\left[\sum_{j=1}^{r}E_j\left(k,\sum_{i=0}^{n}\Delta^i u(k)\right)\right],$$

where $E_j(k,*)$, $1 \le j \le r$ are defined in (4.1.7). Show that for all $k \in \mathbb{N}(a)$

$$\Delta^n u(k) \le p(k)+q(k)\sum_{\ell=a}^{k-1}\left(\sum_{j=1}^{r}\Delta E_j(\ell,\phi)\right)\prod_{\tau=\ell+1}^{k-1}\left(1+\sum_{j=1}^{r}\Delta E_j(\tau,\psi)\right),$$

where $\phi(k)$ and $\psi(k)$ are the same as $\phi_1(k)$ and $\phi_2(k)$ with $q_i(k) = 1,\ 0 \le i \le n$ defined in (4.3.3) and (4.3.4) respectively.

**4.7.19.** Let for all $k \in \mathbb{N}(a)$ the following inequality be satisfied

$$\Delta u(k) \le u(a)+\sum_{\ell=a}^{k-1}p(\ell)\left(u(\ell)+\Delta u(\ell)+\sum_{\tau=a}^{\ell-1}q(\tau)\Delta u(\tau)\right).$$

Show that for all $k \in \mathbb{N}(a)$

$$\Delta u(k) \le u(a)\left(1+\sum_{\ell=a}^{k-1}(2-\phi(\ell))p(\ell)\prod_{\tau=a}^{\ell-1}(2+p(\tau)+q(\tau))\right),$$

where

$$\phi(k) = \sum_{\ell=a}^{k-1}(1+q(\ell))\left(1+(\ell-a)+\sum_{\tau=a}^{\ell-1}q(\tau)\right)\prod_{\tau=a}^{\ell}(2+p(\tau)+q(\tau))^{-1}.$$

**4.7.20.** Let for all $k \in \mathbb{N}(a)$ the following inequality be satisfied

$$\Delta^n u(k) \ \le \ p + \sum_{i=0}^{n} \sum_{\ell=a}^{k-1} q_i(\ell)(\Delta^n u(\ell))^{\alpha}(\Delta^i u(\ell))^{\alpha_i},$$

where $\alpha,\ \alpha_i,\ 0 \le i \le n$ are nonnegative constants and the constant $p > 0$. Show that for all $k \in \mathbb{N}(a)$

$$\Delta^n u(k) \ \le \ p\left[1 + (1-\alpha-\beta)\sum_{\ell=a}^{k-1}\phi(\ell)\right]^{1/1-\alpha-\beta},$$

where $\beta = \max_{0\le i\le n}\alpha_i$ such that $1-\alpha-\beta < 0$, and

$$\phi(k) = p^{\alpha-1}\left[p^{\alpha_n}q_n(k) + \sum_{i=0}^{n-1} q_i(k)\left(\sum_{j=i}^{n-1}\frac{(k-a)^{(j-i)}}{(j-i)!}\Delta^j u(a) + p\frac{(k-a)^{(n-i)}}{(n-i)!}\right)^{\alpha_j}\right]$$

as long as $1 + (1-\alpha-\beta)\sum_{\ell=a}^{k-1}\phi(\ell) > 0$.

**4.7.21.** Let for all $k \in \mathbb{N}(a)$ the following inequality be satisfied

$$\Delta^n u(k) \ \le \ p(k) + \sum_{j=1}^{r} p_j(k)\sum_{\ell=a}^{k-1} q_j(\ell) W\left(\sum_{i=0}^{n}\Delta^i u(\ell)\right),$$

where (i) $p(k)$ is positive and nondecreasing, (ii) $p_j(k) \ge 1,\ 1 \le j \le r$, (iii) $W$ is continuous, positive, nondecreasing and submultiplicative on $[0,\infty)$. Show that for all $k \in \mathbb{N}(a)$

$$\Delta^n u(k) \ \le \ p(k)\prod_{j=1}^{r} p_j(k) G^{-1}\left[G(1) + \sum_{j=1}^{r}\sum_{\ell=a}^{k-1}\frac{q_j(\ell)}{p(\ell)}W(\phi(\ell))\right],$$

where $\phi(k)$ is the same as $\phi_1(k)$ defined in (4.3.3) with $p(k)$ replaced by $p(k)\prod_{j=1}^{r} p_j(k)$ and $q_i(k) = 1,\ 0 \le i \le n$; the function $G$ is defined by $G(w) = \int_{w_0}^{w}\frac{dt}{W(t)},\ \ w \ge w_0 \ge 1$, as long as

$$G(1) + \sum_{j=1}^{r}\sum_{\ell=a}^{k-1}\frac{q_j(\ell)}{p(\ell)}W(\phi(\ell)) \in Dom(G^{-1}).$$

**4.7.22.** Let the function $\mathbf{f}(k, \mathbf{u})$ possess mixed monotone property. Further, let $\mathbf{u}(k, \epsilon)$ be the solution of the problem

$$\begin{aligned} u_p(k+1) &= f_p(k, \mathbf{u}(k)) + \epsilon, \quad u_q(k+1) = f_q(k, \mathbf{u}(k)) - \epsilon \\ u_p(a) &= u_p^0 + \epsilon, \quad u_q(a) = u_q^0 - \epsilon \end{aligned}$$

on $\mathbb{N}(a)$. Show that for $0 < \epsilon_1 < \epsilon_2$ and for all $k \in \mathbb{N}(a)$

$$u_p(k, \epsilon_1) < u_p(k, \epsilon_2), \quad u_q(k, \epsilon_1) > u_q(k, \epsilon_2).$$

**4.7.23.** Let the function $\mathbf{f}(k, \mathbf{u})$ possess mixed monotone property. Further, let $\mathbf{v}(k)$ be defined on $\mathbb{N}(a)$ and satisfy the inequalities

$$v_p(k) \le v_p(a) + \sum_{\ell=a}^{k-1} f_p(\ell, \mathbf{v}(\ell)), \quad v_q(k) \ge v_q(a) + \sum_{\ell=a}^{k-1} f_q(\ell, \mathbf{v}(\ell)).$$

Show that for all $k \in \mathbb{N}(a)$

$$v_p(k) \le u_p(k), \quad v_q(k) \ge u_q(k)$$

where $\mathbf{u}(k)$ is the solution of the problem $\Delta \mathbf{u}(k) = \mathbf{f}(k, \mathbf{u}(k))$, $\mathbf{u}(a) = \mathbf{v}(a)$.

**4.7.24.** Let $u(k)$, $k \in \mathbb{N}(0, b)$ be such that $u(0) = 0$. Show that

(i) $\displaystyle\sum_{k=1}^{b-1} |u(k)\Delta u(k)| \le \frac{b-1}{2} \sum_{k=0}^{b-1} |\Delta u(k)|^2$

(ii) $\displaystyle\sum_{k=1}^{b} |u(k)|^\ell |\nabla u(k)| \le \frac{(b+1)^\ell}{\ell+1} \sum_{k=1}^{b} |\Delta u(k)|^{\ell+1}$,

where $\ell \ge 1$.

**4.7.25.** Let $u(k)$, $k \in \mathbb{N}(\tau, b)$ be such that $u(b) = 0$. Show that

$$\sum_{k=\tau}^{b-1} |u(k)\Delta u(k)| \le \frac{b-\tau+1}{2} \sum_{k=\tau}^{b-1} |\Delta u(k)|^2.$$

**4.7.26.** Let $u(k)$, $k \in \mathbb{N}(0, b)$ be such that $u(0) = u(b) = 0$. Show that

$$\sum_{k=1}^{b-1} |u(k)|^\ell |\nabla u(k)|^m \le C \sum_{k=1}^{b-1} |\nabla u(k)|^{\ell+m},$$

where $\ell \geq 1$, $m \geq 1$ and

$$C = \begin{cases} \dfrac{m}{\ell+m}\left(\dfrac{b+1}{2}\right)^{\ell}, & \text{if } b \text{ odd} \\ \dfrac{m}{\ell+m}\left(\dfrac{b+2}{2}\right)^{\ell}, & \text{if } b \text{ even.} \end{cases}$$

**4.7.27.** Let $u(k)$, $v(k)$, $k \in \mathbb{N}(0,b)$ be nondecreasing and $u(0) = v(0) = 0$. Show that

$$\sum_{k=0}^{b-1}[u(k)\Delta v(k) + v(k+1)\Delta u(k)] \leq \frac{b}{2}\sum_{k=0}^{b-1}[(\Delta u(k))^2 + (\Delta v(k))^2].$$

**4.7.28.** Let $u(k)$, $v(k)$, $k \in \mathbb{N}(0,b)$ be such that $u(0) = v(0) = 0$. Show that

(i) $$\sum_{k=0}^{b-1}|u(k)\Delta v(k) + v(k+1)\Delta u(k)| \leq \frac{b}{2}\sum_{k=0}^{b-1}[(\Delta u(k))^2 + (\Delta v(k))^2]$$

(ii) $$\sum_{k=1}^{b}|u(k)\nabla v(k) + v(k)\nabla u(k)| \leq \frac{b+1}{2}\sum_{k=1}^{b}[(\nabla u(k))^2 + (\nabla v(k))^2].$$

**4.7.29.** Let $u(k)$, $k \in \mathbb{N}(0,b)$ be such that $u(0) = 0$. Further, let $f(t)$ be defined for all $t = u(k)$ and for all $t$ of the form $t(j) = \sum_{k=1}^{j}|\nabla u(k)|$; $f(t) \leq f(|t|)$ for all $t$ and that $f(t)$ is nondecreasing for $t \geq 0$. Show that

(i) $$\sum_{k=1}^{b}|f(u(k))\nabla u(k)| \leq F\left(\sum_{k=1}^{b}|\nabla u(k)|\right) + \sum_{k=1}^{b}\left[f\left(\sum_{j=1}^{k}|\nabla u(j)|\right) - f\left(\sum_{j=1}^{k-1}|\nabla u(j)|\right)\right]|\nabla u(k)|,$$

where $F(t) = \int_0^t f(s)ds$, $t \geq 0$

(ii) $$\sum_{k=1}^{b}\left|\frac{\nabla u(k)}{f(u(k))}\right| \geq G\left(\sum_{k=1}^{b}|\nabla u(k)|\right) - \sum_{k=1}^{b}\left[\left(f\left(\sum_{j=1}^{k-1}|\nabla u(j)|\right)\right)^{-1} - \left(f\left(\sum_{j=1}^{k}|\nabla u(j)|\right)\right)^{-1}\right]|\nabla u(k)|,$$

where $G(t) = \int_0^t ds/f(s)$, $t \geq 0$

(iii) $\displaystyle\sum_{k=0}^{b-1}\left|\frac{\Delta u(k)}{f(u(k))}\right| \le G\left(\sum_{k=0}^{b-1}|\Delta u(k)|\right).$

**4.7.30.** Let $u(k)$, $k \in \mathbb{N}(0, b+n-1)$ be such that $\Delta^j u(0) = 0$, $(0 \le)\ i \le j \le n-1$. Further, let the numbers $\ell \ge 0$, $m > 0$ be such that $\ell + m \ge 1$. Show that

$$\sum_{k=0}^{b-1}|\Delta^i u(k)|^\ell |\Delta^n u(k)|^m \le c_{n-i}\left[(b)^{(n-i)}\right]^\ell \sum_{k=0}^{b-1}|\Delta^n u(k)|^{\ell+m},$$

where

$$c_n = \xi m^{m\xi}\left[\frac{n(1-\xi)}{(n-\xi)}\right]^{\ell(1-\xi)}(n!)^{-\ell}, \quad \xi = (\ell+m)^{-1}.$$

**4.7.31.** Let $u(k)$, $k \in \mathbb{N}(0, b+n-1)$ be such that $\Delta^j u(0) = 0$, $(0 \le)\ i \le j \le n-1$. Show that

$$\sum_{k=0}^{b-1}|\Delta^i u(k)||\Delta^{i+1}u(k)| \le C\left[\sum_{k=0}^{b-1}|\Delta^n u(k)|^r\right]^{2/r},$$

where $r > 1$ and

$$C = \frac{\left[(b)^{(n-i)}\right]^2 (b-n+i+1)^{-2/r}}{2((n-i-1)!)^2\left[(n-i-1)r'+1\right]^{2/r'}}, \quad r' = r/(r-1).$$

**4.7.32.** Let $u(k)$, $k \in \mathbb{N}(0, b+n-1)$ be such that $\Delta^j u(0) = 0$, $0 \le j \le n-1$. Show that

$$\sum_{k=0}^{b-1}\prod_{i=0}^{n}|\Delta^i u(k)| \le Cb^{1/2}\prod_{i=0}^{n-1}(b)^{(n-i-1)}(b-n+i+1)^{1/2} \times \left[\sum_{k=0}^{b-1}|\Delta^n u(k)|^2\right]^{(n+1)/2},$$

where

$$C = \frac{1}{(n^2+1)(n+1)\prod_{i=0}^{n-1}(n-i-1)!}\left(\frac{(n^2+1)(n+1)}{\prod_{i=0}^{n-1}(2n-2i-1)}\right)^{1/2}.$$

**4.7.33.** Let in Theorem 4.6.1 condition (4.6.1) is replaced by

$$f(\theta)f''(\theta) = a_0 + f'(\sigma)f'(\theta) + a_2 f'(\theta)[f'(\theta) - f'(\sigma)].$$

Show that

(i) If $a_2 + 1 > 0$, then inequalities (4.6.2) and (4.6.3) hold.

(ii) If $a_2 + 1 < 0$, then inequalities (4.6.2) and (4.6.3) with reverse sign hold.

## 4.8. Notes

In the last few years Gronwall type inequalities have become a subject in its own right, especially because of the required elementary mathematics and their applicability in diverse fields. The notes of Beesack [18] contains an excellent account of these inequalities till 1975, whereas the survey paper of Agarwal and Thandapani [3] gives extensive generalizations of several known results and provides a large number of references. The discrete Gronwall inequality seems to have appeared first in the work of Mikeladze [29], and now it serves as a fundamental tool in proving convergence of the discrete variable methods for ordinary, partial as well as integral equations; and it is therefore, available in almost every book on numerical analysis. Theorem 4.1.1 is due to Pachpatte [31], whereas Theorem 4.1.4 whose several particular cases have been studied by Pachpatte [33,34], Sugiyama [42,43] and several others, is from Agarwal and Thandapani [5]. Theorem 4.1.5 improves a result proved in Agarwal and Thandapani [7]. Theorem 4.1.6 is taken from Agarwal and Thandapani [5]. The several independent variables analog of Theorem 4.1.9 is proved in Chapter 13. Theorem 4.1.10 is essentially due to McKee [28], however its proof is adapted from Brunner and Houwen [20]. More general singular discrete inequalities are available in Beesack [19], Dixon and Mckee [22,23], McKee [28], Popenda [40]. The first nonlinear discrete Gronwall inequality has appeared in Hull and Luxemburg [24]. Theorem 4.2.1 is taken from Agarwal and Thandapani [5], whereas Theorem 4.2.2 is due to Willett and Wong [45]. Rest of the results in Section 4.2 are from Agarwal and Thandapani [4] and in particular include several results of Pachpatte [31 34]. All the results in Section 4.3 are taken from Agarwal and Thandapani [2]. Theorem 4.4.1 is due to Agarwal [1]. Multidimensional analogs of Theorems 4.4.4 and 4.4.5 are provided in Chapter 13. Theorem 4.5.1 is adapted from Lee [27], and it generalizes a result of Wong [46]. An extensive generalization of Theorem 4.5.1 is given by Beesack [17]. Theorem 4.5.2 is taken from the work of Lasota [26]. Wirtinger type inequalities presented in Section 4.6 are due to Agarwal et. al. [15] and extend the results of Tang [44], Zhang [48 51]. Opial and Wirtinger type inequalities once again appear in Chapters 12 and 13. For several other related results refer to Agarwal et. al. [6,8 14], Bainov and Simenov [16], Conlan and Wang [21], Jones [25], Mitrinović, Pečarić and Fink [30], Pachpatte [35], Pang and Agarwal [36], Popenda [37 39], Redheffer and Walter [41], Yang and You [47].

## 4.9. References

[1]. R.P. Agarwal, On finite systems of difference inequalities, *Jour. Math. Phyl. Sci.* **10**(1976), 277 288.

[2]. R.P. Agarwal and E. Thandapani, On some new discrete inequalities, *Appl. Math. Comp.* **7**(1980), 205 224.

[3]. R.P. Agarwal and E. Thandapani, Remarks on generalizations of Gronwall's inequality, *Chinese J. Math.* **9**(1981), 1 22.

[4]. R.P. Agarwal and E. Thandapani, On nonlinear discrete inequalities of Gronwall type, *An. st. Univ. Iasi* **27**(1981), 139 144.

[5]. R.P. Agarwal and E. Thandapani, On discrete generalizations of Gronwall's inequality, *Bull. Inst. Math. Academia Sinica* **9**(1981), 235 248.

[6]. R.P. Agarwal, A note on Grace and Lalli's paper, *J. Math. Anal. Appl.* **86**(1982), 471 475.

[7]. R.P. Agarwal and E. Thandapani, Some inequalities of Gronwall type, *An. st. Univ. Iasi* **28**(1982), 71 75.

[8]. R.P. Agarwal and E. Thandapani, On some new integrodifferential inequalities, *An. st. Univ. Iasi* **28**(1982), 123 126.

[9]. R.P. Agarwal, Opial type inequalities involving $r$ derivatives and their applications, *Tohoku Math. J.* **47**(1995), 567 593.

[10]. R.P. Agarwal and P.Y.H. Pang, Opial type inequalities involving higher order derivatives, *J. Math. Anal. Appl.* **189**(1995), 85 103.

[11]. R.P. Agarwal and P.Y.H. Pang, Remarks on the generalizations of Opial's inequality, *J. Math. Anal. Appl.* **190**(1995), 559 577.

[12]. R.P. Agarwal and P.Y.H. Pang, Opial type inequalities involving higher order differences, *Mathl. Comput. Modelling* **21**(5)(1995), 49 69.

[13]. R.P. Agarwal and P.Y.H. Pang, Sharp Opial type inequalities involving higher order derivatives of two functions, *Mathematische Nachrichten* **174**(1995), 5 20.

[14]. R.P. Agarwal and P.Y.H. Pang, ***Opial Inequalities with Applications in Differential and Difference Equations***, Kluwer, Dordrecht, 1995.

[15]. R.P. Agarwal, V. Čuljak and J. Pečarić, On discrete and continuous Wirtinger inequalities, *Applicable Analysis* **70**(1998), 195 204.

[16]. D. Bainov and P. Simenov, *Integral Inequalities and Applications, Kluwer*, Dordrecht, 1992.

[17]. P.R. Beesack, On certain discrete inequalities involving partial sums, *Cand. J. Math.* **21**(1969), 222–234.

[18]. P.R. Beesack, *Gronwall Inequalities, Carleton Mathematical Lecture Notes*, No. 11, 1975.

[19]. P.R. Beesack, More generalized discrete Gronwall inequalities, *ZAMM* **65**(1985), 589–595.

[20]. H. Brunner and P.J. van der Houwen, *The Numerical Solution of Volterra Equations, North-Holland*, Amsterdam 1986.

[21]. J. Conlan and C.L. Wang, A unified approach to continuous and discrete Gronwall–Bellman inequalities, *Applicable Analysis* **44**(1992), 243–252.

[22]. J. Dixon and S. McKee, Singular Gronwall inequalities, *Numer. Anal. Report* NA/83/44, Hertford College, University of Oxford (1983).

[23]. J. Dixon and S. McKee, Repeated integral inequalities, *IMA Jour. Numer. Anal.* **4**(1984), 99–107.

[24]. T.E. Hull and W.A.J. Luxemburg, Numerical methods and existence theorems for ordinary differential equations, *Numer. Math.* **2**(1960), 30–41.

[25]. G.S. Jones, Fundamental inequalities for discrete and discontinuous functional equations, *J. Soc. Ind. Appl. Math.* **12**(1964), 43–57.

[26]. A. Lasota, A discrete boundary value problem, *Annales Polonici Mathematici* **20**(1968), 183–190.

[27]. C.-M. Lee, On a discrete analogue of inequalities of Opial and Yang, *Cand. Math. Bull.* **11**(1968), 73–77.

[28]. S. McKee, Generalised discrete Gronwall lemmas, *ZAMM* **62** (1982), 429–434.

[29]. Sh.E. Mikeladze, De la résolution numérique des équations intégrales (Russian), *Bull. Acad. Sci. URSS*, **VII**(1935), 255–297.

[30]. D.S. Mitrinović, J. Pečarić and A.M. Fink, *Inequalities Involving Functions and their Integrals and Derivatives, Kluwer*, Dordrecht, 1991.

[31]. B.G. Pachpatte, On the discrete generalizations of Gronwall's inequality, *J. Indian Math. Soc.* **37**(1973), 147–156.

[32]. B.G. Pachpatte, On some nonlinear discrete inequalities of Gronwall type, *Bull. Inst. Math. Academia Sinica* **5**(1977), 305–315.

[33]. B.G. Pachpatte, On some new integral inequalities and their discrete analogues, *Indian Jour. Pure Appl. Math.* **8**(1977), 1093–1107.

[34]. B.G. Pachpatte, On discrete inequalities related to Gronwall's inequality, *Proc. Indian Acad. Sci.* **85A**(1977), 26 40.

[35]. B.G. Pachpatte, Inequalities applicable in the theory of finite difference equations, *J. Math. Anal. Appl.* **222**(1998), 438 459.

[36]. P.Y.H. Pang and R.P. Agarwal, On an Opial type inequality due to Fink, *J. Math. Anal. Appl.* **196**(1995), 748 753.

[37]. J. Popenda, Finite difference inequalities, *Fasciculi Mathematici* **13** (1981), 79 87.

[38]. J. Popenda, On the discrete analogy of Gronwall lemma, *Demonstratio Mathematica* **16**(1983), 11 15.

[39]. J. Popenda, On some discrete Gronwall type inequalities, *Fasciculi Mathematici* **14**(1985), 109 114.

[40]. J. Popenda, Gronwall type inequalities, *ZAMM* **75**(1995), 669 677.

[41]. R. Redheffer and W. Walter, A comparison theorem for difference inequalities, *J. Diff. Eqns.* **44**(1982), 111 117.

[42]. S. Sugiyama, On the stability problems of difference equations, *Bull. Sci. Engg. Res. Lab. Waseda Univ.* **45**(1969). 140 144.

[43]. S. Sugiyama, Difference inequalities and their applications to stability problems, *Lecture Notes in Math.* **243**, Springer Verlag, Berlin (1971), 1 15.

[44]. D. Tang, Discrete Wirtinger and isoperimetric type inequalities, *Bull. Austral. Math. Soc.* **43**(1991), 467 474.

[45]. D. Willett and J.S.W. Wong, On the discrete analogues of some generalizations of Gronwall's inequality, *Monatsch. Math.* **69**(1965), 362 367.

[46]. J.S.W. Wong, A discrete analouge of Opial's inequality, *Cand. Math. Bull.* **10**(1967), 115 118.

[47]. G.S. Yang and C.D. You, A note on discrete Opial's inequality, *Tamkang J. Math.* **23**(1992), 67 78.

[48]. X.M. Zhang, A refinement of the discrete Wirtinger inequality, *J. Math. Anal. Appl.* **200**(1996), 687 697.

[49]. X.M. Zhang, Bonnesen style inequalities and pseudo-perimeters for polygons, *Journal of Geometry* **60**(1997), 188 201.

[50]. X.M. Zhang, Schur-convex functions and isoperimetric inequalities, *Proc. Amer. Math. Soc.* **126**(1998), 461 470.

[51]. X.M. Zhang, A optimization of Schur convex functions, *Mathl. Ineq. Appl.* **1**(1998), 319 330.

# Chapter 5
# Qualitative Properties of Solutions of Difference Systems

This chapter provides methods and suitable criterion that describe the nature and behavior of solutions of difference systems, without actually constructing or approximating them. Since in contrast with differential equations, the existence and uniqueness of solutions of discrete initial value problems is already guaranteed we shall begin with the continuous dependence on the initial conditions and parameters. This is followed by the asymptotic behavior of solutions of linear as well as nonlinear difference systems. In particular, easily verifiable sufficient conditions are obtained so that the solutions of perturbed systems remain bounded or eventually tend to zero, provided the solutions of the unperturbed systems have the same property. Next we introduce various types of stability and give several examples to illustrate these notions. Then, for the stability of linear systems we provide necessary and sufficient conditions in terms of their fundamental matrices. This includes certain concepts which are of computational importance. This is followed by the comparison between the stability and boundedness of the solutions of linear systems with those of perturbed nonlinear systems. Next we develop a nonlinear variation of constants formula and give its application which establishes its importance. Then, for the linear difference systems we define ordinary and exponential dichotomies, provide necessary and sufficient conditions so that these systems have dichotomies, and use these dichotomies to study the behavior of the solutions of perturbed nonlinear difference systems. Then, we introduce Lyapunov functions and emphasize their importance in the study of stability properties of solutions of autonomous as well as non autonomous difference systems. This is followed by the stability of solutions of several discrete models appearing in population dynamics. Next, assuming certain stability properties of the given difference systems, we shall provide the construction of the Lyapunov functions. These results known as converse theorems are then used to study the total stability of the solutions of difference systems. Then, we define the concept of practical stability of the solutions, which goes beyond the classical Lyapunov stability theory and finds some applications in numerical analysis. Finally, we shall introduce the concept of mutual stability of the solutions of two given difference systems, which provides bounds on the solutions in tube like domains.

In what follows, throughout we shall assume that the functions appearing in the nonlinear systems under study are continuous with respect to the dependent variable, although in several results this restriction is not essential.

## 5.1. Dependence on Initial Conditions and Parameters

The initial value problem (1.2.8), (1.3.3) as well as

$$\Delta \mathbf{u}(k) = \mathbf{f}(k, \mathbf{u}(k)), \quad \mathbf{u}(a) = \mathbf{u}^0 \tag{5.1.1}$$

describes a model of a physical problem in which often some parameters such as lengths, masses, temperature etc. are involved. The values of these parameters can be measured only up to a certain degree of accuracy. Thus, in (5.1.1) the initial vector $\mathbf{u}^0$ as well as the function $\mathbf{f}(k, \mathbf{u})$ may be subject to some errors either by necessity or for convenience. Hence, it is important to know how the solution of (5.1.1) changes when $\mathbf{u}^0$ and $\mathbf{f}(k, \mathbf{u})$ are slightly altered. We shall answer this question quantitatively in the following:

**Theorem 5.1.1.** Let the following conditions be satisfied

(i) $\mathbf{f}(k, \mathbf{u})$ is defined on $\mathbb{N}(a) \times \mathbb{R}^n$ and for all $(k, \mathbf{u})$, $(k, \mathbf{v}) \in \mathbb{N}(a) \times \mathbb{R}^n$

$$\|\mathbf{f}(k, \mathbf{u}) - \mathbf{f}(k, \mathbf{v})\| \leq \lambda(k)\|\mathbf{u} - \mathbf{v}\|, \tag{5.1.2}$$

where $\lambda(k)$ is a nonnegative function defined on $\mathbb{N}(a)$

(ii) $\mathbf{g}(k, \mathbf{u})$ is defined on $\mathbb{N}(a) \times \mathbb{R}^n$ and for all $(k, \mathbf{u}) \in \mathbb{N}(a) \times \mathbb{R}^n$

$$\|\mathbf{g}(k, \mathbf{u})\| \leq \mu(k), \tag{5.1.3}$$

where $\mu(k)$ is a nonnegative function defined on $\mathbb{N}(a)$.

Then, for the solutions $\mathbf{u}(k)$ and $\mathbf{v}(k)$ of the initial value problems (5.1.1) and

$$\Delta \mathbf{v}(k) = \mathbf{f}(k, \mathbf{v}(k)) + \mathbf{g}(k, \mathbf{v}(k)), \quad \mathbf{v}(a) = \mathbf{v}^0 \tag{5.1.4}$$

the following inequality holds

$$(5.1.5)\quad \|\mathbf{u}(k)-\mathbf{v}(k)\| \le \left(\|\mathbf{u}^0-\mathbf{v}^0\|+\sum_{\ell=a}^{k-1}\mu(\ell)\right)\prod_{\ell=a}^{k-1}(1+\lambda(\ell)),\quad k\in\mathbb{N}(a).$$

**Proof.** Since the problems (5.1.1) and (5.1.4) are equivalent to

$$\mathbf{u}(k) \ = \ \mathbf{u}^0+\sum_{\ell=a}^{k-1}\mathbf{f}(\ell,\mathbf{u}(\ell))$$

and

$$\mathbf{v}(k) \ = \ \mathbf{v}^0+\sum_{\ell=a}^{k-1}(\mathbf{f}(\ell,\mathbf{v}(\ell))+\mathbf{g}(\ell,\mathbf{v}(\ell)))$$

we find that

$$\mathbf{u}(k)-\mathbf{v}(k) \ = \ \mathbf{u}^0-\mathbf{v}^0+\sum_{\ell=a}^{k-1}(\mathbf{f}(\ell,\mathbf{u}(\ell))-\mathbf{f}(\ell,\mathbf{v}(\ell)))-\sum_{\ell=a}^{k-1}\mathbf{g}(\ell,\mathbf{v}(\ell)).$$

Thus, from (5.1.2) and (5.1.3) it follows that

$$\|\mathbf{u}(k)-\mathbf{v}(k)\| \ \le \ \left(\|\mathbf{u}^0-\mathbf{v}^0\|+\sum_{\ell=a}^{k-1}\mu(\ell)\right)+\sum_{\ell=a}^{k-1}\lambda(\ell)\|\mathbf{u}(\ell)-\mathbf{v}(\ell)\|.$$

Now an application of Corollary 4.1.3 immediately gives (5.1.5). ∎

Hereafter, to emphasize the dependence of the initial point $(a,\mathbf{u}^0)$ we shall denote the solutions of the initial value problems (1.2.8), (1.3.3), and (5.1.1) as $\mathbf{u}(k,a,\mathbf{u}^0)$. In our next result we shall show that $\mathbf{u}(k,a,\mathbf{u}^0)$ is differentiable with respect to $\mathbf{u}^0$.

**Theorem 5.1.2.** Let for all $(k,\mathbf{u})\in\mathbb{N}(a)\times\mathbb{R}^n$ the function $\mathbf{f}(k,\mathbf{u})$ be defined and the partial derivative $\partial\mathbf{f}/\partial\mathbf{u}$ exist. Further, let the solution $\mathbf{u}(k)=\mathbf{u}(k,a,\mathbf{u}^0)$ of the initial value problem (1.2.8), (1.3.3) exist on $\mathbb{N}(a)$, and let $\mathcal{J}(k,a,\mathbf{u}^0)=\partial\mathbf{f}(k,\mathbf{u}(k,a,\mathbf{u}^0))/\partial\mathbf{u}$. Then, the matrix $\mathcal{V}(k,a,\mathbf{u}^0)=\partial\mathbf{u}(k,a,\mathbf{u}^0)/\partial\mathbf{u}^0$ exists and is the solution of the initial value problem

$$(5.1.6)\qquad \mathcal{V}(k+1,a,\mathbf{u}^0) \ = \ \mathcal{J}(k,a,\mathbf{u}^0)\mathcal{V}(k,a,\mathbf{u}^0),\quad \mathcal{V}(a,a,\mathbf{u}^0)=\mathcal{I}.$$

**Proof.** Since $\mathbf{u}(k,a,\mathbf{u}^0)$ is the solution of (1.2.8), (1.3.3) we have

$$\mathbf{u}(k+1,a,\mathbf{u}^0) \ = \ \mathbf{f}(k,\mathbf{u}(k,a,\mathbf{u}^0)),\quad \mathbf{u}(a,a,\mathbf{u}^0) \ = \ \mathbf{u}^0.$$

Thus, differentiation with respect to $\mathbf{u}^0$ gives

$$\frac{\partial \mathbf{u}(k+1)}{\partial \mathbf{u}^0} = \frac{\partial \mathbf{f}(k,\mathbf{u}(k))}{\partial \mathbf{u}} \frac{\partial \mathbf{u}(k)}{\partial \mathbf{u}^0}, \quad \frac{\partial \mathbf{u}(a)}{\partial \mathbf{u}^0} = \mathcal{I}.$$

The result (5.1.6) now follows from the definitions of $\mathcal{J}(k)$ and $\mathcal{V}(k)$. ∎

**Theorem 5.1.3.** Let for all $(k,\mathbf{u}) \in \mathbb{N}(a) \times \mathbb{R}^n$ the function $\mathbf{f}(k,\mathbf{u})$ be defined, and for all $(k,\mathbf{u})$, $(k,\mathbf{v}) \in \mathbb{N}(a) \times \mathbb{R}^n$

$$\|\mathbf{f}(k,\mathbf{u}) - \mathbf{f}(k,\mathbf{v})\| \le g(k, \|\mathbf{u}-\mathbf{v}\|), \tag{5.1.7}$$

where $g(k,r)$ is defined on $\mathbb{N}(a) \times \mathbb{R}_+$ and nondereasing in $r$ for any fixed $k \in \mathbb{N}(a)$. Further, let the solutions $\mathbf{u}(k,a,\mathbf{u}^i)$, $i = 1,2$ of (5.1.1) exist on $\mathbb{N}(a)$. Then, for all $k \in \mathbb{N}(a)$

$$\|\mathbf{u}(k,a,\mathbf{u}^1) - \mathbf{u}(k,a,\mathbf{u}^2)\| \le r(k,a,r^0), \tag{5.1.8}$$

where $r(k) = r(k,a,r^0)$ is the solution of the initial value problem

$$\Delta r(k) = g(k,r(k)), \quad r(a) = r^0 \ (\ge \|\mathbf{u}^1 - \mathbf{u}^2\|). \tag{5.1.9}$$

**Proof.** Since

$$\mathbf{u}(k,a,\mathbf{u}^1) - \mathbf{u}(k,a,\mathbf{u}^2) = \mathbf{u}^1 - \mathbf{u}^2 + \sum_{\ell=a}^{k-1}(\mathbf{f}(\ell,\mathbf{u}(\ell,a,\mathbf{u}^1)) - \mathbf{f}(\ell,\mathbf{u}(\ell,a,\mathbf{u}^2)))$$

it follows that

$$z(k) \le z(a) + \sum_{\ell=a}^{k-1} g(\ell, z(\ell)),$$

where $z(k) = \|\mathbf{u}(k,a,\mathbf{u}^1) - \mathbf{u}(k,a,\mathbf{u}^2)\|$. Further, since

$$r(k) = r(a) + \sum_{\ell=a}^{k-1} g(\ell, r(\ell))$$

and $r(a) \ge z(a)$, the inequality (5.1.8) follows by induction. ∎

**Remark 5.1.1.** If $r(k,a,0) = 0$ for all $k \in \mathbb{N}(a)$, and $r(k,a,r^0) \to 0$ as $r^0 \to 0$, then from (5.1.8) it is clear that the solution $\mathbf{u}(k,a,\mathbf{u}^0)$ of (5.1.1) continuously depends on $\mathbf{u}^0$.

Now we shall consider the following initial value problem

$$\Delta \mathbf{u}(k) = \mathbf{f}(k,\mathbf{u}(k),\mathbf{p}), \quad \mathbf{u}(a) = \mathbf{u}^0 \tag{5.1.10}$$

where $\mathbf{p} \in \mathbb{R}^m$ is a parameter such that $\|\mathbf{p} - \mathbf{p}^0\| \le \delta\ (>0)$ and $\mathbf{p}^0$ is a fixed vector in $\mathbb{R}^m$. For a given $\mathbf{p}$ such that $\|\mathbf{p} - \mathbf{p}^0\| \le \delta$ we shall assume that the solution $\mathbf{u}(k,\mathbf{p}) = \mathbf{u}(k,a,\mathbf{u}^0,\mathbf{p})$ of (5.1.10) exists on $\mathbb{N}(a)$.

**Theorem 5.1.4.** Let for all $k \in \mathbb{N}(a)$, $\mathbf{u} \in \mathbb{R}^n$, $\mathbf{p} \in \mathbb{R}^m$ such that $\|\mathbf{p} - \mathbf{p}^0\| \le \delta$ the function $\mathbf{f}(k,\mathbf{u},\mathbf{p})$ is defined, and the following inequalities hold

$$\|\mathbf{f}(k,\mathbf{u},\mathbf{p}) - \mathbf{f}(k,\mathbf{v},\mathbf{p})\| \ \le\ \lambda(k)\|\mathbf{u} - \mathbf{v}\|$$

and

$$\|\mathbf{f}(k,\mathbf{u},\mathbf{p}^1) - \mathbf{f}(k,\mathbf{u},\mathbf{p}^2)\| \ \le\ \mu(k)\|\mathbf{p}^1 - \mathbf{p}^2\|,$$

where $\lambda(k)$ and $\mu(k)$ are nonnegative functions defined on $\mathbb{N}(a)$. Then, for the solutions $\mathbf{u}(k,a,\mathbf{u}^1,\mathbf{p}^1)$ and $\mathbf{u}(k,a,\mathbf{u}^2,\mathbf{p}^2)$ of (5.1.10) the following inequality holds

$$\begin{aligned}&\|\mathbf{u}(k,a,\mathbf{u}^1,\mathbf{p}^1) - \mathbf{u}(k,a,\mathbf{u}^2,\mathbf{p}^2)\|\\ &\quad \le \left(\|\mathbf{u}^1 - \mathbf{u}^2\| + \|\mathbf{p}^1 - \mathbf{p}^2\|\sum_{\ell=a}^{k-1}\mu(\ell)\right)\prod_{\ell=a}^{k-1}(1+\lambda(\ell)), \quad k \in \mathbb{N}(a).\end{aligned}$$

**Proof.** The proof is similar to that of Theorem 5.1.1. ∎

**Theorem 5.1.5.** Let for all $k \in \mathbb{N}(a)$, $\mathbf{u} \in \mathbb{R}^n$, $\mathbf{p} \in \mathbb{R}^m$ such that $\|\mathbf{p}-\mathbf{p}^0\| \le \delta$ the function $\mathbf{f}(k,\mathbf{u},\mathbf{p})$ be defined and the partial derivatives $\partial\mathbf{f}/\partial\mathbf{u}$ and $\partial\mathbf{f}/\partial\mathbf{p}$ exist. Further, let the solution $\mathbf{u}(k,\mathbf{p}) = \mathbf{u}(k,a,\mathbf{u}^0,\mathbf{p})$ of (5.1.10) exist on $\mathbb{N}(a)$, and let $\mathcal{J}(k,a,\mathbf{u}^0,\mathbf{p}) = \partial\mathbf{f}(k,\mathbf{u}(k,\mathbf{p}),\mathbf{p})/\partial\mathbf{u}$ and $\mathcal{H}(k,a,\mathbf{u}^0,\mathbf{p}) = \partial\mathbf{f}(k,\mathbf{u}(k,\mathbf{p}),\mathbf{p})/\partial\mathbf{p}$. Then, the matrix $\mathcal{V}(k,a,\mathbf{u}^0,\mathbf{p}) = \partial\mathbf{u}(k,a,\mathbf{u}^0,\mathbf{p})/\partial\mathbf{p}$ exists and is the solution of the initial value problem

$$\begin{aligned}\Delta\mathcal{V}(k,a,\mathbf{u}^0,\mathbf{p}) &= \mathcal{J}(k,a,\mathbf{u}^0,\mathbf{p})\mathcal{V}(k,a,\mathbf{u}^0,\mathbf{p}) + \mathcal{H}(k,a,\mathbf{u}^0,\mathbf{p})\\ \mathcal{V}(a,a,\mathbf{u}^0,\mathbf{p}) &= 0.\end{aligned}$$

**Proof.** The proof is similar to that of Theorem 5.1.2. ∎

**Remark 5.1.2.** If $\mathbf{p} \in \mathbb{R}^m$ is such that $\|\mathbf{p} - \mathbf{p}^0\|$ is sufficiently small, then we have a *first order approximation* of the solution $\mathbf{u}(k,\mathbf{p})$ of (5.1.10) which is given by

$$\mathbf{u}(k,\mathbf{p}) \ \simeq\ \mathbf{u}(k,\mathbf{p}^0) + \mathcal{V}(k,a,\mathbf{u}^0,\mathbf{p}^0)(\mathbf{p} - \mathbf{p}^0).$$

As an example, for the problem $\Delta u(k) \ =\ -\lambda u(k)$, $u(0) = 1$, $0 \le \lambda \le 2$ it follows that $u(k,\lambda) = (1-\lambda)^k$, $u(k,0) = 1$, $v(k,0,1,0) = -k$, and hence for small $\lambda > 0$, $u(k,\lambda) \simeq 1 - k\lambda$.

**Theorem 5.1.6.** Let for all $k \in \mathbb{N}(a, b-1)$, $\mathbf{u} \in \mathbb{R}^n$, $\mathbf{p} \in \mathbb{R}^m$ such that $\|\mathbf{p} - \mathbf{p}^0\| \le \delta$ the function $\mathbf{f}(k, \mathbf{u}, \mathbf{p})$ be defined and $\lim_{\mathbf{p} \to \mathbf{p}^0} \mathbf{f}(k, \mathbf{u}, \mathbf{p}) = \mathbf{f}(k, \mathbf{u}, \mathbf{p}^0)$ uniformly in $k$ and $\mathbf{u}$. Then, for a given $\epsilon > 0$ there exists a $\eta(\epsilon) \le \delta$ such that $\|\mathbf{p} - \mathbf{p}^0\| \le \eta$ implies

$$\|\mathbf{u}(k, a, \mathbf{u}^0, \mathbf{p}) - \mathbf{u}(k, a, \mathbf{u}^0, \mathbf{p}^0)\| \le \epsilon, \quad k \in \mathbb{N}(a, b)$$

where $\mathbf{u}(k, a, \mathbf{u}^0, \mathbf{p})$ and $\mathbf{u}(k, a, \mathbf{u}^0, \mathbf{p}^0)$ are the solutions of (5.1.10).

**Proof.** The proof is elementary. ∎

## 5.2. Asymptotic Behavior of Linear Systems

For a given difference system one of the pioneer problems is the study of ultimate behavior of its solutions. In particular, for linear systems we shall provide sufficient conditions on the known quantities so that all their solutions remain bounded or tend to zero as $k \to \infty$. Thus, from the practical point of view the results we shall discuss are very important because an explicit form of the solutions is not needed.

We begin with the difference system (2.8.1) on $\mathbb{N}$. Since every solution of this system is of the form $\mathcal{A}^k \mathbf{c}$, all solutions of (2.8.1) are bounded on $\mathbb{N}$ if and only if $\sup_{k \in \mathbb{N}} \|\mathcal{A}^k\| \le c < \infty$. Further, all solutions of (2.8.1) tend to zero as $k \to \infty$ if and only if $\|\mathcal{A}^k\| \to 0$ as $k \to \infty$.

**Definition 5.2.1.** The eigenvalue $\lambda_i$ of the matrix $\mathcal{A}$ is said to be *semisimple* if $a_i(\mathcal{A}) q_i(\mathcal{A})(\mathcal{A} - \lambda_1 \mathcal{I}) = 0$, where $a_i(\mathcal{A})$ and $q_i(\mathcal{A})$ are the same as in Lemma 2.8.2.

As an example, for the matrix $\mathcal{A} = \mathcal{I}$ the multiple eigenvalue $\lambda = 1$ is semisimple.

From the representation (2.8.11) of $\mathcal{A}^k$ it is clear that $\sup_{k \in \mathbb{N}} \|\mathcal{A}^k\| \le c$ if and only if the eigenvalues of $\mathcal{A}$ have modulus less than or equal to one, and those of modulus one are semisimple. Further, $\|\mathcal{A}^k\| \to 0$ as $k \to \infty$ if and only if the eigenvalues of the matrix $\mathcal{A}$ are inside the unit disc. For the constant matrices of the form (1.2.13), i.e. arising from the difference equation (2.12.4) it is known that the semisimple eigenvalues are only simple. Thus, in particular all solutions of (2.12.4) are bounded on $\mathbb{N}$ if and only if the roots of the characteristic equation (2.16.12) have modulus less than or equal to one, and those of modulus one are simple.

The drawback in the above conclusions is that we must know in advance all the eigenvalues of the matrix $\mathcal{A}$. However, if $n$ is large then solving the characteristic equation $p(\lambda) = 0$ of $\mathcal{A}$ becomes quite complicated.

In such a situation, we assume that $\lambda$ is a complex variable and use the transformation $\lambda = (1+z)/(1-z)$, which maps the circle $|\lambda| = 1$ into the imaginary axis $\Re e\, z = 0$; the interior of the circle $|\lambda| < 1$ into the half plane $\Re e\, z < 0$; and the point $\lambda = 1$ into $z = 0$. Thus, the eigenvalues of $\mathcal{A}$ have modulus less than one if and only if the roots of the polynomial

$$p\left(\frac{1+z}{1-z}\right) = b_0 z^n + b_1 z^{n-1} + \cdots + b_n = 0 \tag{5.2.1}$$

have negative real parts. In (5.2.1) the constants $b_i$, $0 \le i \le n$ are real and we may assume that $b_0 > 0$. For this, the following result is well known.

**Theorem 5.2.1.** (Hurwitz's Theorem). A necessary and sufficient condition for the negativity of the real parts of all the roots of the polynomial (5.2.1) is the positivity of all the principal minors of the *Hurwitz matrix*

$$\mathcal{H}_n = \begin{bmatrix} b_1 & b_0 & 0 & 0 & \cdots & 0 \\ b_3 & b_2 & b_1 & b_0 & \cdots & 0 \\ b_5 & b_4 & b_3 & b_2 & \cdots & 0 \\ \cdots & \cdots & \cdots & \cdots & \cdots & \cdots \\ 0 & 0 & 0 & 0 & \cdots & b_n \end{bmatrix}. \tag{5.2.2}$$

It can be shown that this condition implies $b_i > 0$, $0 \le i \le n$. Thus, positivity of the coefficients in (5.2.1) is a necessary condition but not a sufficient condition for the real parts of all the roots of (5.2.1) to be negative. For $n = 2, 3, 4$ the necessary and sufficient condition reduces to

$$\begin{aligned} n &= 2: \ b_0 > 0,\ b_1 > 0,\ b_2 > 0. \\ n &= 3: \ b_0 > 0,\ b_1 > 0,\ b_2 > 0,\ b_3 > 0,\ b_1 b_2 - b_3 b_0 > 0. \\ n &= 4: \ b_0 > 0,\ b_1 > 0,\ b_2 > 0,\ b_3 > 0,\ b_4 > 0, \\ & \qquad\qquad b_1 b_2 b_3 - b_0 b_3^2 - b_4 b_1^2 > 0. \end{aligned}$$

**Example 5.2.1.** For the matrix $\mathcal{A} = \begin{bmatrix} 0 & 1 & 0 \\ 0 & 0 & 1 \\ -0.2 & -1 & -1 \end{bmatrix}$ the characteristic equation is $\lambda^3 + \lambda^2 + \lambda + 0.2 = 0$. The substitution $\lambda = (1+z)/(1-z)$ transforms this equation to $z^3 + 2z^2 + 3z + 4 = 0$, for which $b_1 b_2 - b_3 b_0 > 0$. Therefore, all solutions of the system (2.8.1) with this $\mathcal{A}$ tend to zero as $k \to \infty$.

Now we shall consider the difference system

$$\mathbf{v}(k+1) = (\mathcal{A} + \mathcal{B}(k))\mathbf{v}(k), \quad k \in \mathbb{N} \tag{5.2.3}$$

where $\mathcal{B}(k)$ is an $n \times n$ matrix with elements $b_{ij}(k)$, $1 \le i,j \le n$. System (5.2.3) can be regarded as a perturbed system of (2.8.1). The following result provides sufficient conditions on the matrix $\mathcal{B}(k)$ so that all solutions of (5.2.3) remain bounded if all solutions of (2.8.1) are bounded.

**Theorem 5.2.2.** Let all solutions of the difference system (2.8.1) be bounded on $\mathbb{N}$. Then, all solutions of (5.2.3) are bounded on $\mathbb{N}$ provided

$$\sum_{\ell=0}^{\infty} \|\mathcal{B}(\ell)\| < \infty. \tag{5.2.4}$$

**Proof.** In (2.6.3) let the nonhomogeneous term $\mathbf{b}(k)$ be $\mathcal{B}(k)\mathbf{v}(k)$, so that each solution $\mathbf{v}(k)$ such that $\mathbf{v}(0) = \mathbf{v}^0$ of (5.2.3) also satisfies

$$\mathbf{v}(k) = \mathcal{A}^k \mathbf{v}^0 + \sum_{\ell=1}^{k} \mathcal{A}^{k-\ell}\mathcal{B}(\ell-1)\mathbf{v}(\ell-1). \tag{5.2.5}$$

Now since all solutions of (2.8.1) are bounded, there exists a constant $c$ such that $\sup_{k\in\mathbb{N}} \|\mathcal{A}^k\| = c$. Hence, for all $k \in \mathbb{N}$ we have

$$\|\mathbf{v}(k)\| \le c_0 + c\sum_{\ell=0}^{k-1} \|\mathcal{B}(\ell)\|\|\mathbf{v}(\ell)\|,$$

where $c_0 = c\|\mathbf{v}^0\|$.

Applying Corollary 4.1.2 to the above inequality, we obtain

$$\|\mathbf{v}(k)\| \le c_0 \prod_{\ell=0}^{k-1}(1 + c\|\mathcal{B}(\ell)\|) \le c_0 \exp\left(c\sum_{\ell=0}^{k-1} \|\mathcal{B}(\ell)\|\right).$$

The result now follows from (5.2.4). ∎

The next result gives sufficient conditions on the matrix $\mathcal{B}(k)$ so that all solutions of (5.2.3) tend to zero as $k \to \infty$ provided all solutions of (2.8.1) tend to zero as $k \to \infty$.

**Theorem 5.2.3.** Let all solutions of the difference system (2.8.1) tend to zero as $k \to \infty$. Then, all solutions of (5.2.3) tend to zero as $k \to \infty$ provided

$$\|\mathcal{B}(k)\| \to 0 \quad \text{as} \quad k \to \infty. \tag{5.2.6}$$

**Proof.** Since all solutions of (2.8.1) tend to zero as $k \to \infty$, all the eigenvalues of $\mathcal{A}$ lie in the unit disc. Thus, there exist constants $c$ and

$0 < \delta < 1$ such that $\|\mathcal{A}^k\| \le c\delta^k$ for all $k \in \mathbb{N}$. Further, because of (5.2.6), for a given constant $c_1 > 0$ there exists a sufficiently large $k_1 \in \mathbb{N}$ such that $\|\mathcal{B}(k)\| \le c_1$ for all $k \in \mathbb{N}(k_1)$. Hence, for all $k \in \mathbb{N}(k_1)$ equation (5.2.5) gives

$$\|\mathbf{v}(k)\| \le c\delta^k\|\mathbf{v}^0\| + \sum_{\ell=1}^{k_1} c\delta^{k-\ell}\|\mathcal{B}(\ell-1)\|\|\mathbf{v}(\ell-1)\| + \sum_{\ell=k_1+1}^{k} c\delta^{k-\ell}c_1\|\mathbf{v}(\ell-1)\|,$$

which can be written as

$$\mathbf{w}(k) \le c_0 + c_2\sum_{\ell=k_1}^{k-1}\mathbf{w}(\ell), \tag{5.2.7}$$

where

$$\mathbf{w}(k) = \|\mathbf{v}(k)\|\delta^{-k}, \quad c_0 = c\|\mathbf{v}^0\| + \frac{c}{\delta}\sum_{\ell=0}^{k_1-1}\delta^{-\ell}\|\mathcal{B}(\ell)\|\|\mathbf{v}(\ell)\|, \quad \text{and } c_2 = \frac{cc_1}{\delta}.$$

Now in view of Corollary 4.1.2 from (5.2.7), we obtain

$$\mathbf{w}(k) \le c_0\prod_{\ell=k_1}^{k-1}(1+c_2) = c_0(1+c_2)^{k-k_1}$$

and hence

$$\|\mathbf{v}(k)\| \le c_0\delta^k(1+c_2)^{k-k_1} \le c_0\left[\delta(1+c_2)\right]^{k-k_1}. \tag{5.2.8}$$

Finally, because of (5.2.6) we can always choose $c_1 < (1-\delta)/c$ so that $\delta(1+c_2) = \delta(1+cc_1/\delta) < 1$, and hence the result follows from (5.2.8). ∎

Conditions (5.2.4) and (5.2.6) are restricted to smallness property on $\mathcal{B}(k)$ as $k \to \infty$. Obviously, condition (5.2.4) is stronger than (5.2.6) and hence in Theorem 5.2.3 condition (5.2.6) can be replaced by (5.2.4), however in Theorem 5.2.2 condition (5.2.4) cannot be replaced by (5.2.6). For this, we have

**Example 5.2.2.** Consider the difference systems

$$\begin{bmatrix} u_1(k+1) \\ u_2(k+1) \end{bmatrix} = \begin{bmatrix} 0 & 1 \\ -1 & 0 \end{bmatrix}\begin{bmatrix} u_1(k) \\ u_2(k) \end{bmatrix} \tag{5.2.9}$$

and

$$\begin{bmatrix} v_1(k+1) \\ v_2(k+1) \end{bmatrix} = \begin{bmatrix} 0 & 1 \\ -1 & 0 \end{bmatrix}\begin{bmatrix} v_1(k) \\ v_2(k) \end{bmatrix} + \begin{bmatrix} 0 & 0 \\ -\dfrac{2}{k+1} & 0 \end{bmatrix}\begin{bmatrix} v_1(k) \\ v_2(k) \end{bmatrix}. \tag{5.2.10}$$

From Example 2.8.1 we know that a fundamental system of solutions of (5.2.9) is $\begin{bmatrix} \cos(k\pi/2) \\ -\sin(k\pi/2) \end{bmatrix}$, $\begin{bmatrix} \sin(k\pi/2) \\ \cos(k\pi/2) \end{bmatrix}$ and hence all solutions of (5.2.9) are bounded. However, a fundamental system of solutions of (5.2.10) is $\begin{bmatrix} (k+1)\cos(k\pi/2) \\ -(k+2)\sin(k\pi/2) \end{bmatrix}$, $\begin{bmatrix} (k+1)\sin(k\pi/2) \\ (k+2)\cos(k\pi/2) \end{bmatrix}$ and hence all nontrivial solutions of (5.2.10) are unbounded as $k \to \infty$. Further, we note that $\|\mathcal{B}(k)\| \to 0$ as $k \to \infty$, while $\sum_{\ell=0}^{\infty} \|\mathcal{B}(\ell)\| = \sum_{\ell=0}^{\infty} 2/(\ell+1) = \infty$.

Next we shall consider the difference system

$$\text{(5.2.11)} \qquad \mathbf{v}(k+1) \;=\; \mathcal{A}\mathbf{v}(k) + \mathbf{b}(k), \quad k \in \mathbb{N}.$$

**Theorem 5.2.4.** Suppose that the function $\mathbf{b}(k)$ is such that

$$\text{(5.2.12)} \qquad \|\mathbf{b}(k)\| \;\le\; c_3\eta^k$$

for all large $k \in \mathbb{N}$, where $c_3$ and $\eta$ are nonnegative constants. Then, every solution $\mathbf{v}(k)$ of the difference system (5.2.11) satisfies

$$\text{(5.2.13)} \qquad \|\mathbf{v}(k)\| \;\le\; c_4\nu^k$$

for all $k \in \mathbb{N}$, where $c_4$ and $\nu$ are nonnegative constants.

**Proof.** From the given hypothesis on $\mathbf{b}(k)$ there exists a $k_1 \in \mathbb{N}$ such that (5.2.12) holds for all $k \ge k_1$. Therefore, for every solution $\mathbf{v}(k)$ such that $\mathbf{v}(0) = \mathbf{v}^0$ of (5.2.11), inequality (5.2.12) in (2.6.3) implies that

$$\begin{aligned} \|\mathbf{v}(k)\| &\le c\delta_1^k\|\mathbf{v}^0\| + \sum_{\ell=1}^{k_1} c\delta_1^{k-\ell}\|\mathbf{b}(\ell-1)\| + \sum_{\ell=k_1+1}^{k} c\delta_1^{k-\ell}c_3\eta^{\ell-1} \\ &= c\delta_1^k\left[\|\mathbf{v}^0\| + \sum_{\ell=0}^{k_1-1}\delta_1^{-\ell-1}\|\mathbf{b}(\ell)\| + \frac{c_3}{(\eta-\delta_1)}\left(\left(\frac{\eta}{\delta_1}\right)^k - \left(\frac{\eta}{\delta_1}\right)^{k_1}\right)\right] \\ &\le c_4\nu^k, \end{aligned}$$

where we have assumed that $\|\mathcal{A}^k\| \le c\delta_1^k$, $\eta \ne \delta_1$, and $\nu = \max\{\eta, \delta_1\}$, and

$$c_4 \;=\; c\left[\|\mathbf{v}^0\| + \sum_{\ell=0}^{k_1-1}\delta_1^{-\ell-1}\|\mathbf{b}(\ell)\| + \frac{c_3}{|\eta-\delta_1|}\right].$$

For the case $\eta = \delta_1$ the above proof needs an obvious modification. ∎

As a consequence of (5.2.13) we find that all solutions of the system (5.2.11) tend to zero as $k \to \infty$ provided $\nu < 1$.

Now we shall study the behavior of solutions of the difference system (1.2.12) on $\mathbb{N}$ as $k \to \infty$. We shall prove two results which involve the eigenvalues of the matrix $\mathcal{A}^T(k)\mathcal{A}(k)$, which obviously are functions of $k$.

**Theorem 5.2.5.** If the largest eigenvalue $M(k)$ of the matrix $\mathcal{A}^T(k)\mathcal{A}(k)$ is such that $\sup_{k\in\mathbb{N}} \prod_{\ell=0}^{k} M(\ell) < \infty$, then all solutions of (1.2.12) are bounded. Further, if $\prod_{\ell=0}^{k} M(\ell) \to 0$ as $k \to \infty$ then all solutions of (1.2.12) tend to zero.

**Proof.** Let $\mathbf{u}(k)$ be a solution of (1.2.12), then $|\mathbf{u}(k)|^2 = \mathbf{u}^T(k)\mathbf{u}(k)$. Thus, it follows that

$$|\mathbf{u}(k+1)|^2 \ = \ \mathbf{u}^T(k+1)\mathbf{u}(k+1) \ = \ \mathbf{u}^T(k)\mathcal{A}^T(k)\mathcal{A}(k)\mathbf{u}(k).$$

Now, since the matrix $\mathcal{A}^T(k)\mathcal{A}(k)$ is symmetric and $M(k)$ is its largest eigenvalue, it is clear that

$$\mathbf{u}^T(k)\left(\mathcal{A}^T(k)\mathcal{A}(k)\right)\mathbf{u}(k) \ \le \ M(k)|\mathbf{u}(k)|^2.$$

Thus, for all $k \in \mathbb{N}$ it follows that

$$0 \ \le \ |\mathbf{u}(k+1)|^2 \ \le \ M(k)|\mathbf{u}(k)|^2$$

and hence

$$|\mathbf{u}(k)|^2 \ \le \ \left(\sum_{\ell=0}^{k-1} M(\ell)\right)|\mathbf{u}(0)|^2$$

from which the conclusions are obvious. ∎

**Theorem 5.2.6.** If the smallest eigenvalue $m(k)$ of the matrix $\mathcal{A}^T(k)\mathcal{A}(k)$ is such that $\limsup_{k\to\infty} \prod_{\ell=0}^{k} m(\ell) = \infty$, then all solutions of (1.2.12) are unbounded.

**Proof.** As in the proof of Theorem 5.2.5 for all $k \in \mathbb{N}$ it is easily seen that

$$|\mathbf{u}(k+1)|^2 \ \ge \ m(k)|\mathbf{u}(k)|^2,$$

which on using the fact that $m(k) > 0$ for all $k \in \mathbb{N}$ gives

$$|\mathbf{u}(k)|^2 \ \ge \ \left(\sum_{\ell=0}^{k-1} m(\ell)\right)|\mathbf{u}(0)|^2$$

from which the conclusion is immediate. ∎

**Example 5.2.3.** Consider the difference system

$$\mathbf{u}(k+1) \ = \ \begin{bmatrix} 0 & 1 \\ \dfrac{1}{k+1} & \dfrac{1}{1+k^2} \end{bmatrix} \mathbf{u}(k). \tag{5.2.14}$$

Since for this system

$$\mathcal{A}^T(k)\mathcal{A}(k) = \begin{bmatrix} \dfrac{1}{(1+k)^2} & \dfrac{1}{(1+k)(1+k^2)} \\ \dfrac{1}{(1+k)(1+k^2)} & 1+\dfrac{1}{(1+k^2)^2} \end{bmatrix}$$

it follows that $M(k) < 1 + \dfrac{1}{(1+k)^2} + \dfrac{1}{(1+k^2)^2}$. Thus, $\prod_{\ell=0}^{k} M(\ell) <$ $\exp\left(\sum_{\ell=0}^{k} \dfrac{1}{(1+\ell)^2} + \dfrac{1}{(1+\ell^2)^2}\right) < \infty$, and hence all solutions of (5.2.14) are bounded.

**Example 5.2.4.** For the difference system

$$\mathbf{u}(k+1) = \begin{bmatrix} \dfrac{1}{1+k} & \dfrac{1}{1+k^2} \\ -\dfrac{1}{1+k^2} & \dfrac{1}{1+k} \end{bmatrix} \mathbf{u}(k) \tag{5.2.15}$$

it is easy to find $M(k) = \dfrac{1}{(1+k)^2} + \dfrac{1}{(1+k^2)^2}$. Since $\prod_{\ell=0}^{k} M(\ell) \to 0$, as $k \to \infty$ all solutions of (5.2.15) tend to zero.

**Example 5.2.5.** Consider the difference system

$$\mathbf{u}(k+1) = \frac{1}{8}\begin{bmatrix} 0 & 9+(-1)^k 7 \\ 9-(-1)^k 7 & 0 \end{bmatrix} \mathbf{u}(k). \tag{5.2.16}$$

Since the system

$$\mathcal{A}^T(k)\mathcal{A}(k) = \frac{1}{64}\begin{bmatrix} 130-(-1)^k 126 & 0 \\ 0 & 130+(-1)^k 126 \end{bmatrix}$$

it follows that for all $k$, $M(k) = 4$ and $m(k) = 1/16$ therefore both the above theorems cannot be applied. However, for this system the principal fundamental matrix is

$$\mathcal{U}(k,0) = \begin{cases} \begin{bmatrix} 2^{-2k} & 0 \\ 0 & 2^k \end{bmatrix} & \text{if } k \text{ is even} \\ \begin{bmatrix} 0 & 2^k \\ 2^{-2k} & 0 \end{bmatrix} & \text{if } k \text{ is odd.} \end{cases}$$

Thus, all solutions of (5.2.16) tend to $\infty$ as $k \to \infty$. It is also interesting to note that the eigenvalues of this matrix $\mathcal{A}(k)$ are $\pm 2^{-1/2}$, which lie inside the unit disc. Thus, for the boundedness of the solutions of (1.2.12) we need stronger conditions compared to those needed for the system (2.8.1).

**Example 5.2.6.** For the difference system

$$\mathbf{u}(k+1) = \begin{bmatrix} 1 & \dfrac{(k+1)^2}{(k+2)} \\ 0 & \left(\dfrac{k+1}{k+2}\right)^3 \end{bmatrix} \mathbf{u}(k) \tag{5.2.17}$$

the principal fundamental matrix $\mathcal{U}(k,0) = \begin{bmatrix} 1 & \dfrac{k}{k+1} \\ 0 & \dfrac{1}{(k+1)^3} \end{bmatrix}$. Hence, all solutions of (5.2.17) are bounded. However,

$$\|\mathcal{U}(2\ell,0)\mathcal{U}^{-1}(\ell,0)\| = \left\| \begin{bmatrix} 1 & \dfrac{\ell(\ell+1)^2}{(2\ell+1)} \\ 0 & \left(\dfrac{\ell+1}{2\ell+1}\right)^3 \end{bmatrix} \right\| \to \infty \quad \text{as} \quad \ell \to \infty.$$

**Definition 5.2.2** Let $\mathcal{V}(k)$ be a fundamental matrix of (1.2.12). The system (1.2.12) is said to be *uniformly bounded* if there exists a constant $c$ such that

$$\sup_{0 \le \ell \le k \in \mathbb{N}} \|\mathcal{V}(k)\mathcal{V}^{-1}(\ell)\| \le c. \tag{5.2.18}$$

Obviously, if (5.2.18) holds then every solution of (1.2.12) is bounded. However, from Example 5.2.6 the converse need not hold. But for the system (2.8.1) condition (5.2.18) is the same as $\sup_{0 \le \ell \le k \in \mathbb{N}} \|\mathcal{A}^{k-\ell}\| = \sup_{k \in \mathbb{N}} \|\mathcal{A}^k\| \le c$, and hence the boundedness of the solutions of (2.8.1) implies (5.2.18).

With respect to the difference system (1.2.12) we shall consider the perturbed system

$$\mathbf{v}(k+1) = (\mathcal{A}(k) + \mathcal{B}(k))\mathbf{v}(k), \quad k \in \mathbb{N} \tag{5.2.19}$$

where $\mathcal{B}(k)$ is an $n \times n$ matrix with elements $b_{ij}(k)$, $1 \le i,j \le n$.

**Theorem 5.2.7.** Let the system (1.2.12) be uniformly bounded, and the condition (5.2.4) be satisfied. Then, all solutions of (5.2.19) are bounded.

**Proof.** Let $\mathcal{V}(k)$ be a fundamental matrix of the difference system (1.2.12). In (2.5.3) let the nonhomogeneous term be $\mathcal{B}(k)\mathbf{v}(k)$ so that each solution $\mathbf{v}(k)$ such that $\mathbf{v}(0) = \mathbf{v}^0$ of (5.2.19) also satisfies

$$\mathbf{v}(k) = \mathcal{V}(k)\mathcal{V}^{-1}(0)\mathbf{v}^0 + \sum_{\ell=1}^{k} \mathcal{V}(k)\mathcal{V}^{-1}(\ell)\mathcal{B}(\ell-1)\mathbf{v}(\ell-1). \tag{5.2.20}$$

Thus, it follows that

$$\|\mathbf{v}(k)\| \leq c\|\mathbf{v}^0\| + c\sum_{\ell=0}^{k-1} \|\mathcal{B}(\ell)\|\|\mathbf{v}(\ell)\|.$$

The rest of the proof is similar to that of Theorem 5.2.2. ∎

**Example 5.2.7.** Consider the difference system

$$\mathbf{u}(k+1) = \begin{bmatrix} \left(\dfrac{k+2}{k+3}\right)^{1/4} & 0 \\ 0 & \dfrac{k+1}{k+3} \end{bmatrix} \mathbf{u}(k) \tag{5.2.21}$$

for which $\mathcal{V}(k) = \begin{bmatrix} \dfrac{1}{(k+2)^{1/4}} & 0 \\ 0 & \dfrac{1}{(k+1)(k+2)} \end{bmatrix}$ is a fundamental matrix. Thus, all solutions of (5.2.21) tend to $0$ as $k \to \infty$. Since $\mathcal{V}(k)\mathcal{V}^{-1}(\ell) = \begin{bmatrix} \left(\dfrac{\ell+2}{k+2}\right)^{1/4} & 0 \\ 0 & \dfrac{(\ell+1)(\ell+2)}{(k+1)(k+2)} \end{bmatrix}$, the system (5.2.21) is uniformly bounded.

For the difference system (5.2.21) we consider the perturbed system

$$\mathbf{v}(k+1) = \begin{bmatrix} \left(\dfrac{k+2}{k+3}\right)^{1/4} & 0 \\ 0 & \dfrac{k+1}{k+3} \end{bmatrix} \mathbf{v}(k) + \begin{bmatrix} 0 & 0 \\ \dfrac{1}{(k+2)^{3/4}} & 0 \end{bmatrix} \mathbf{v}(k), \tag{5.2.22}$$

which can be solved to obtain

$$\mathbf{v}(k) = \begin{bmatrix} \left(\dfrac{2}{k+2}\right)^{1/4} & 0 \\ \dfrac{2^{-3/4}}{(k+1)}\left[k+3-\dfrac{6}{(k+2)}\right] & \dfrac{2}{(k+1)(k+2)} \end{bmatrix} \mathbf{v}(0). \tag{5.2.23}$$

In system (5.2.22), clearly $\|\mathcal{B}(k)\| \to 0$ as $k \to \infty$, however its solution (5.2.23) tends to $\begin{bmatrix} 0 & 0 \\ 2^{-3/4} & 0 \end{bmatrix} \mathbf{v}(0) \neq 0$ (unless $v_1(0) = 0$) as $k \to \infty$. Thus, for the systems (1.2.12) and (5.2.19) Theorem 5.2.3 does not hold.

**Theorem 5.2.8.** Let all solutions of (1.2.12) tend to zero as $k \to \infty$. Then, all solutions of (5.2.19) tend to zero as $k \to \infty$ provided

$$\sum_{\ell=0}^{\infty} \|\mathcal{V}^{-1}(\ell+1)\mathcal{B}(\ell)\| \ < \ \infty, \tag{5.2.24}$$

where $\mathcal{V}(k)$ is a fundamental matrix of (1.2.12).

**Proof.** For each solution $\mathbf{v}(k)$ such that $\mathbf{v}(0) = \mathbf{v}^0$ of (5.2.19) representation (5.2.20) implies that

$$\|\mathbf{v}(k)\| \ \le \ \|\mathcal{V}(k)\| \left[ c_0 + \sum_{\ell=0}^{k-1} \|\mathcal{V}^{-1}(\ell+1)\mathcal{B}(\ell)\| \|\mathbf{v}(\ell)\| \right],$$

where $c_0 = \|\mathcal{V}^{-1}(0)\mathbf{v}^0\|$. Now since all solutions of (1.2.12) tend to zero as $k \to \infty$, there exists a constant $c > 0$ such that $\sup_{k \in \mathbb{N}} \|\mathcal{V}(k)\| \le c$. Therefore, it follows that

$$\|\mathbf{v}(k)\| \ \le \ \|\mathcal{V}(k)\| \left[ c_0 + c\sum_{\ell=0}^{k-1} \|\mathcal{V}^{-1}(\ell+1)\mathcal{B}(\ell)\| \frac{\|\mathbf{v}(\ell)\|}{\|\mathcal{V}(\ell)\|} \right],$$

and hence

$$\frac{\|\mathbf{v}(k)\|}{\|\mathcal{V}(k)\|} \ \le \ c_0 + c\sum_{\ell=0}^{k-1} \|\mathcal{V}^{-1}(\ell+1)\mathcal{B}(\ell)\| \frac{\|\mathbf{v}(\ell)\|}{\|\mathcal{V}(\ell)\|},$$

which gives that

$$\begin{aligned} \|\mathbf{v}(k)\| \ &\le \ \|\mathcal{V}(k)\| c_0 \prod_{\ell=0}^{k-1} (1 + c\|\mathcal{V}^{-1}(\ell+1)\mathcal{B}(\ell)\|) \\ &\le \ \|\mathcal{V}(k)\| c_0 \exp\left( c\sum_{\ell=0}^{k-1} \|\mathcal{V}^{-1}(\ell+1)\mathcal{B}(\ell)\| \right). \end{aligned}$$

Now the conclusion follows from (5.2.24) and the fact that $\|\mathcal{V}(k)\| \to 0$ as $k \to \infty$. ∎

## 5.3. Asymptotic Behavior of Nonlinear Systems

With respect to the difference system (2.8.1), now we shall consider the perturbed system

$$(5.3.1) \qquad \mathbf{v}(k+1) \;=\; \mathcal{A}\mathbf{v}(k) + \mathbf{g}(k, \mathbf{v}(k)), \qquad k \in \mathbb{N}$$

where the function $\mathbf{g}(k, \mathbf{v})$ is defined on $\mathbb{N} \times \mathbb{R}^n$.

**Theorem 5.3.1.** Let for all $(k, \mathbf{v}) \in \mathbb{N} \times \mathbb{R}^n$ the function $\mathbf{g}(k, \mathbf{v})$ satisfy

$$(5.3.2) \qquad \|\mathbf{g}(k, \mathbf{v})\| \;\leq\; h(k)\|\mathbf{v}\|,$$

where $h(k)$ is a nonnegative function defined on $\mathbb{N}$. Then,

(i) all solutions of (5.3.1) are bounded provided all solutions of (2.8.1) are bounded and $\sum_{\ell=0}^{\infty} h(\ell) < \infty$

(ii) all solutions of (5.3.1) tend to zero as $k \to \infty$ provided all solutions of (2.8.1) tend to zero and $h(k) \to 0$ as $k \to \infty$.

**Proof.** In (2.6.3) let the nonhomogeneous term $\mathbf{b}(k)$ be $\mathbf{g}(k, \mathbf{v}(k))$, so that each solution $\mathbf{v}(k)$ such that $\mathbf{v}(0) = \mathbf{v}^0$ of (5.3.1) also satisfies

$$(5.3.3) \qquad \mathbf{v}(k) \;=\; \mathcal{A}^k \mathbf{v}^0 + \sum_{\ell=1}^{k} \mathcal{A}^{k-\ell} \mathbf{g}(\ell - 1, \mathbf{v}(\ell - 1)).$$

Thus, from (5.3.2) it follows that

$$\|\mathbf{v}(k)\| \;\leq\; \|\mathcal{A}^k\| \|\mathbf{v}^0\| + \sum_{\ell=0}^{k-1} \|\mathcal{A}^{k-\ell-1}\| h(\ell) \|\mathbf{v}(\ell)\|.$$

The rest of the proof of part (i) ((ii)) is the same as that of Theorem 5.2.2 (Theorem 5.2.3). ∎

For the difference system (1.2.12) we shall consider the perturbed system

$$(5.3.4) \qquad \mathbf{v}(k+1) \;=\; \mathcal{A}(k)\mathbf{v}(k) + \mathbf{g}(k, \mathbf{v}(k)), \qquad k \in \mathbb{N}$$

where the function $\mathbf{g}(k, \mathbf{v})$ is defined on $\mathbb{N} \times \mathbb{R}^n$.

**Theorem 5.3.2.** Let the function $\mathbf{g}(k, \mathbf{v})$ be as in Theorem 5.3.1. Then,

(i) all solutions of (5.3.4) are bounded provided all solutions of (1.2.12) are uniformly bounded and $\sum_{\ell=0}^{\infty} h(\ell) < \infty$

(ii) all solutions of (5.3.4) tend to zero as $k \to \infty$ provided all solutions of (1.2.12) tend to zero and $\sum_{\ell=0}^{\infty} \|\mathcal{V}^{-1}(\ell+1)\| h(\ell) < \infty$, where $\mathcal{V}(k)$ is a fundamental matrix of (1.2.12).

**Proof.** Let $\mathcal{V}(k)$ be a fundamental matrix of the difference system (1.2.12). In (2.5.3) let the nonhomogeneous term be $\mathbf{g}(k,\mathbf{v}(k))$ so that each solution $\mathbf{v}(k)$ such that $\mathbf{v}(0)=\mathbf{v}^0$ of (5.3.4) also satisfies

$$\mathbf{v}(k) \;=\; \mathcal{V}(k)\mathcal{V}^{-1}(0)\mathbf{v}^0+\sum_{\ell=1}^{k}\mathcal{V}(k)\mathcal{V}^{-1}(\ell)\mathbf{g}(\ell-1,\mathbf{v}(\ell-1)). \tag{5.3.5}$$

The rest of the proof of part (i) ((ii)) is the same as that of Theorem 5.2.7 (Theorem 5.2.8). ∎

**Theorem 5.3.3.** Let for all $(k,\mathbf{u})\in\mathbb{N}\times\mathbb{R}^n$ the function $\mathbf{f}(k,\mathbf{u})$ be defined and

$$\|\mathbf{f}(k,\mathbf{u})\| \;\le\; g(k,\|\mathbf{u}\|), \tag{5.3.6}$$

where $g(k,r)$ is defined for all $(k,r)\in\mathbb{N}\times\mathbb{R}_+$ and monotone nondecreasing in $r$ for any fixed $k\in\mathbb{N}$. Further, let for $r^0\ge 0$ the solution $r(k)=r(k,0,r^0)$ of (5.1.9) be bounded on $\mathbb{N}$. Then, any solution $\mathbf{u}(k)=\mathbf{u}(k,0,\mathbf{u}^0)$ of (5.1.1) such that $\|\mathbf{u}^0\|\le r^0$ is bounded on $\mathbb{N}$ and has a limit as $k\to\infty$.

**Proof.** For $\|\mathbf{u}^0\|\le r^0$ it is easy to deduce that $\|\mathbf{u}(k,0,\mathbf{u}^0)\|\le r(k,0,r^0)$. Since, by assumption, $r(k,0,r^0)$ is bounded on $\mathbb{N}$ it follows that for each $\|\mathbf{u}^0\|\le r^0$ the solution $\mathbf{u}(k,0,\mathbf{u}^0)$ of (5.1.1) is bounded on $\mathbb{N}$. Further, for any $0<k_1<k\in\mathbb{N}$ and $\|\mathbf{u}^0\|\le r^0$, we have

$$\begin{aligned}\|\mathbf{u}(k)-\mathbf{u}(k_1)\| \;&\le\; \sum_{\ell=k_1}^{k-1}\|\mathbf{f}(\ell,\mathbf{u}(\ell))\| \;\le\; \sum_{\ell=k_1}^{k-1} g(\ell,\|\mathbf{u}(\ell)\|)\\ &\le\; \sum_{\ell=k_1}^{k-1} g(\ell,r(\ell)) \;=\; r(k)-r(k_1).\end{aligned}$$

Now since $g(k,r)$ is nonnegative, the solution $r(k,0,r^0)$ of (5.1.9) is nondecreasing in $k$, and hence the boundedness of this solution implies that $r(k,0,r^0)$ tends to a finite limit as $k\to\infty$. Thus, for any $\epsilon>0$ we can choose $k_1>0$ sufficiently large so that $0\le r(k)-r(k_1)\le\epsilon$ for all $k\in\mathbb{N}(k_1)$. But, this implies that $\|\mathbf{u}(k)-\mathbf{u}(k_1)\|\le\epsilon$ for all $k\in\mathbb{N}(k_1)$, which proves that $\mathbf{u}(k)$ tends to a limit as $k\to\infty$. ∎

**Remark 5.3.1.** Let $\lim_{k\to\infty}\mathbf{u}(k,0,\mathbf{u}^0)=\mathbf{u}^\infty$ and $\lim_{k\to\infty}r(k,0,r^0)=r^\infty$. Since $\mathbf{u}^\infty=\mathbf{u}^0+\sum_{\ell=0}^{\infty}\mathbf{f}(\ell,\mathbf{u}(\ell))$, it follows that $\|\mathbf{u}^\infty\|\ge\|\mathbf{u}^0\|-\sum_{\ell=0}^{\infty}g(\ell,r(\ell))=\|\mathbf{u}^0\|-r^\infty+r^0$. Thus, if $0\le r^\infty-r^0<\|\mathbf{u}^0\|\le r^0$, then $\mathbf{u}^\infty\ne 0$.

**Corollary 5.3.4.** Let $\mathcal{U}(k,0)$ be the principal fundamental matrix of the difference system (1.2.12), and the function $\mathbf{g}(k,\mathbf{v})$ be defined for all $(k,\mathbf{v}) \in \mathbb{N}\times\mathbb{R}^n$, and satisfy the inequality $\|\mathcal{U}^{-1}(k+1,0)\mathbf{g}(k,\mathcal{U}(k,0)\mathbf{w})\| \leq h(k)\|\mathbf{w}\|$, where $h(k)$ is a nonnegative function defined on $\mathbb{N}$ such that $\sum_{\ell=0}^{\infty} h(\ell) < \infty$. Then, for any solution $\mathbf{v}(k) = \mathbf{v}(k,0,\mathbf{v}^0)$ of (5.3.4) the function $\mathcal{U}^{-1}(k,0)\mathbf{v}(k)$ has a finite limit as $k \to \infty$.

**Proof.** The transformation $\mathcal{U}(k,0)\mathbf{w}(k) = \mathbf{v}(k)$ reduces (5.3.4) to $\Delta\mathbf{w}(k) = \mathcal{U}^{-1}(k+1,0)\mathbf{g}(k,\mathcal{U}(k,0)\mathbf{w}(k))$. Now the result is a direct consequence of Theorem 5.3.3. ∎

**Remark 5.3.2.** Let $\lim_{k\to\infty}\mathbf{w}(k) = \lim_{k\to\infty}\mathcal{U}^{-1}(k,0)\mathbf{v}(k) = \mathbf{c}$, and $\mathbf{u}(k) = \mathcal{U}(k,0)\mathbf{c}$ be the solution of (1.2.12) satisfying $\mathbf{u}(0) = \mathbf{c}$. Then, since $\mathbf{v}(k) - \mathbf{u}(k) = \mathcal{U}(k,0)(\mathbf{w}(k) - \mathbf{c})$, if all solutions of (1.2.12) are bounded then $\|\mathbf{v}(k) - \mathbf{u}(k)\| \leq \|\mathcal{U}(k,0)\|\|\mathbf{w}(k) - \mathbf{c}\| \to 0$ as $k \to \infty$. Thus, for each solution $\mathbf{v}(k)$ of (5.3.4) there exists a solution of (1.2.12) such that $\lim_{k\to\infty}(\mathbf{v}(k) - \mathbf{u}(k)) = 0$.

## 5.4. Concepts of Stability

Let the solution $\mathbf{u}(k) = \mathbf{u}(k,a,\mathbf{u}^0)$ of (1.2.8) exist for all $k \in \mathbb{N}(a)$. For this solution we shall define various concepts of stability and through examples show that these concepts are not, in general, equivalent.

**Definition 5.4.1.** The solution $\mathbf{u}(k)$ is said to be

(i) *Stable* if, for each $\epsilon > 0$, there exists a $\delta = \delta(\epsilon, a)$ such that, for any solution $\overline{\mathbf{u}}(k) = \mathbf{u}(k,a,\overline{\mathbf{u}}^0)$ of (1.2.8), the inequality $\|\overline{\mathbf{u}}^0 - \mathbf{u}^0\| < \delta$ implies $\|\overline{\mathbf{u}}(k) - \mathbf{u}(k)\| < \epsilon$ for all $k \in \mathbb{N}(a)$.

(ii) *Unstable* if it is not stable.

(iii) *Attractive* if there exists a $\delta = \delta(a)$ such that, for any solution $\overline{\mathbf{u}}(k) = \mathbf{u}(k,a,\overline{\mathbf{u}}^0)$ of (1.2.8), the inequality $\|\overline{\mathbf{u}}^0 - \mathbf{u}^0\| < \delta$ implies $\|\overline{\mathbf{u}}(k) - \mathbf{u}(k)\| \to 0$ as $k \to \infty$.

(iv) *Asymptotically Stable* if it is stable and attractive.

(v) *Uniformly Stable* if it is stable and $\delta$ is independent of $a$, or equivalently, if for each $\epsilon > 0$, there exists a $\delta = \delta(\epsilon) > 0$ such that, for any solution $\overline{\mathbf{u}}(k) = \mathbf{u}(k,a,\overline{\mathbf{u}}^0)$ of (1.2.8), the inequalities $a \leq k_1 \in \mathbb{N}(a)$ and $\|\overline{\mathbf{u}}(k_1) - \mathbf{u}(k_1)\| < \delta$ imply $\|\overline{\mathbf{u}}(k) - \mathbf{u}(k)\| < \epsilon$ for all $k \in \mathbb{N}(k_1)$.

(vi) *Uniformly Attractive* if it is attractive and $\delta$ is independent of $a$.

(vii) *Uniformly Asymptotically Stable* if it is uniformly stable and uniformly attractive.

(viii) *Globally Attractive* if it is attractive for all $\overline{\mathbf{u}}^0 \in \mathbb{R}^n$.

(ix) *Globally Asymptotically Stable* if it is stable and globally attractive.

(x) *Strongly Stable* if, for each $\epsilon > 0$, there exists a $\delta = \delta(\epsilon) > 0$ such that, for any solution $\overline{\mathbf{u}}(k) = \mathbf{u}(k, a, \overline{\mathbf{u}}^0)$ of (1.2.8), the inequalities $a \leq k_1 \in \mathbb{N}(a)$ and $\|\overline{\mathbf{u}}(k_1) - \mathbf{u}(k_1)\| < \delta$ imply $\|\overline{\mathbf{u}}(k) - \mathbf{u}(k)\| < \epsilon$ for all $k \in \mathbb{N}(a)$.

(xi) *Exponentially Asymptotically Stable* if there exists a $\lambda > 0$ and, for any given $\epsilon > 0$, there exists a $\delta = \delta(\epsilon) > 0$ such that, for any solution $\overline{\mathbf{u}}(k) = \mathbf{u}(k, a, \overline{\mathbf{u}}^0)$ of (1.2.8), the inequalities $a \leq k_1 \in \mathbb{N}(a)$ and $\|\overline{\mathbf{u}}(k_1) - \mathbf{u}(k_1)\| < \delta$ imply $\|\overline{\mathbf{u}}(k) - \mathbf{u}(k)\| < \epsilon \exp(-\lambda(k - k_1))$ for all $k \in \mathbb{N}(k_1)$.

(xii) $s_p$ *Stable* if it is stable and for some $p > 0$, $\sum_{\ell=a}^{\infty} \|\overline{\mathbf{u}}(\ell) - \mathbf{u}(\ell)\|^p < \infty$.

**Remark 5.4.1.** Strong stability implies uniform stability which, in turn, leads to stability; and exponential asymptotic stability implies uniform asymptotic stability which, in turn, gives asymptotic stability. However, the converse of these statements is, in general, not true.

**Example 5.4.1.** The solution $u(k) = e^{-k}$ of the difference equation $u(k+1) = \begin{cases} e^{-1}u(k), & u \geq 0 \\ e\, u(k), & u < 0 \end{cases}$ is stable but not uniformly stable, while the trivial solution is unstable.

**Example 5.4.2.** The trivial (in fact any) solution of $u(k+1) = u(k)$ is strongly, but not asymptotically stable.

**Example 5.4.3.** For both the systems (5.2.10) and (5.2.16) the trivial solution is unstable.

**Example 5.4.4.** The trivial solution of $u(k+1) = e^{-1}u(k)$ is exponentially asymptotically stable.

**Example 5.4.5.** The trivial solution of $u(k+1) = u^2(k)$ is uniformly asymptotically stable. For this, we note that for all $a \in \mathbb{N}$ and $c \in \mathbb{R}$ the solution of this difference equation is $\overline{u}(k) = u(k, a, c) = c^{2^{k-a}}$. Thus, for $k_1 \geq a$, $|\overline{u}(k_1) - u(k_1)| = |c|^{2^{k_1 - a}} < \delta = 1$, i.e. $|c| < 1$ implies

$$
\begin{aligned}
|\overline{u}(k) - u(k)| &\leq |c|^{2^{k-k_1} - 2^{k_1 - a}} |c|^{2^{k_1 - a}} \\
&< \delta |c|^{2^{k-k_1} - 2^{k_1 - a}} \to 0 \quad \text{as} \quad k \to \infty,
\end{aligned}
$$

and hence the trivial solution is uniformly attractive. Further, $|\overline{u}(k_1) - u(k_1)| = |c|^{2^{k_1 - a}} < \delta = \min\{1, \epsilon\}$ implies $|\overline{u}(k) - u(k)| = |c|^{2^{k-k_1} - 2^{k_1 - a}} \times$

$|c|^{2^{k_1-a}} < \epsilon$ for all $k \geq k_1$, and hence the trivial solution is uniformly stable. It is also clear that the trivial solution of this equation is not globally attractive.

**Example 5.4.6.** For all $a \in \mathbb{N}$ and $c \in \mathbb{R}$ the solution $\overline{u}(k) = u(k,a,c)$ of the difference equation $u(k+1) = e^{\cos k}u(k)$ is $\overline{u}(k) = c\exp\left(\dfrac{\sin(k-1/2)-\sin(a-1/2)}{2\sin(1/2)}\right)$. Thus, the trivial solution of this difference equation is uniformly stable but not asymptotically stable.

**Example 5.4.7.** For all $a \in \mathbb{N}(1)$ and $c \in \mathbb{R}$ the solution $\overline{u}(k) = u(k,a,c)$ of the difference equation $u(k+1) = e^{-1/k}u(k)$ is $\overline{u}(k) = c\exp\left(-\sum_{\ell=a}^{k-1}\dfrac{1}{\ell}\right)$. Thus, the trivial solution of this difference equation is asymptotically stable. However, it is not uniformly attractive. For this, it suffices to note that $u(2k+1,k+1,c) = c\exp\left(-\sum_{\ell=k+1}^{2k}\dfrac{1}{\ell}\right)$, and $\ln\dfrac{3}{2} \leq \sum_{\ell=k+1}^{2k}\dfrac{1}{\ell} \leq \ln 2$, therefore $u(2k+1,k+1,c) \not\to 0$ as $k \to \infty$. Hence, the trivial solution is not uniformly asymptotically stable.

**Example 5.4.8.** For all $a \in \mathbb{N}(1)$ and $c \in \mathbb{R}$ the solution $\overline{u}(k) = u(k,a,c)$ of the difference equation $u(k+1) = \dfrac{k}{k+1}u(k)$ is $\overline{u}(k) = ca/k$. Thus, the trivial solution of this difference equation is asymptotically stable. However, since $\sum_{\ell=a}^{\infty}|c|a/\ell = \infty$, it is not $s_1$ stable.

**Remark 5.4.2.** Since $\sum_{\ell=a}^{\infty}\|\overline{\mathbf{u}}(\ell)-\mathbf{u}(\ell)\|^p < \infty$ implies $\|\overline{\mathbf{u}}(k)-\mathbf{u}(k)\| \to 0$ as $k \to \infty$, $s_p$ stability implies asymptotic stability. However, from Example 5.4.8 the converse is not necessarily true.

**Remark 5.4.3.** Exponential asymptotic stability implies $s_p$ stability. For this, we note that $\|\overline{\mathbf{u}}(k) - \mathbf{u}(k)\| < \epsilon\exp(-\lambda(k-a))$, $a \in \mathbb{N}$ with $\lambda > 0$ gives $\sum_{\ell=a}^{\infty}\|\overline{\mathbf{u}}(\ell)-\mathbf{u}(\ell)\|^p < \epsilon^p\dfrac{1}{1-e^{-\lambda p}} < \infty$.

**Remark 5.4.4.** Let the difference system (1.2.8) be autonomous, i.e. of the form

$$\mathbf{u}(k+1) = \mathbf{f}(\mathbf{u}(k)) \tag{5.4.1}$$

and $\mathbf{f}(0) = 0$, so that it admits the trivial solution $\mathbf{u}(k) = 0$, $k \in \mathbb{N}(a)$. For the trivial solution of (5.4.1) the uniform stability and stability concepts coincide. For this, if $\mathbf{u}(k) = \mathbf{u}(k,a,\mathbf{u}^0)$ is a solution of (5.4.1) then $\overline{\mathbf{u}}(k) = \mathbf{u}(k-a, 0, \mathbf{u}^0)$ is also a solution of (5.4.1). Further, since $\mathbf{u}(a) = \overline{\mathbf{u}}(a)$ it follows that $\mathbf{u}(k) = \overline{\mathbf{u}}(k)$ for all $k \in \mathbb{N}(a)$. Thus, for (5.4.1) we can always take $a = 0$, and if the trivial solution is stable for $a = 0$, then it is stable for all $a$, which means that stability is uniform. However, for the nontrivial solutions of (5.4.1) Example 5.4.1 shows that stability does not imply uniform stability.

**Example 5.4.9.** Every solution of the difference equation $u(k+1) = u(k) + 1$ is of the form $u(k) = u(a) + k - a$, and hence it is stable but not bounded.

**Example 5.4.10.** The system

$$\begin{aligned} u_1(k+1) &= u_1(k)\cos(u_1^2(k) + u_2^2(k))^{1/2} - u_2(k)\sin(u_1^2(k) + u_2^2(k))^{1/2} \\ u_2(k+1) &= u_1(k)\sin(u_1^2(k) + u_2^2(k))^{1/2} + u_2(k)\cos(u_1^2(k) + u_2^2(k))^{1/2} \end{aligned}$$

has a two parameter family of solutions

$$u_1(k) \;=\; c_1\cos(c_1 k + c_2), \qquad u_2(k) \;=\; c_1\sin(c_1 k + c_2),$$

where $c_1$ and $c_2$ are arbitrary constants. The trivial solution $u_1(k) \equiv 0$, $u_2(k) \equiv 0$ of this system is stable but all its other solutions are unstable. However, every solution of this system is bounded.

From the above two examples it is clear that the concepts of stability and boundedness of solutions are, in general, independent of each other. However, in the case of the homogeneous linear difference system (1.2.12) these concepts are equivalent.

**Theorem 5.4.1.** All solutions of the difference system (1.2.12) are stable if and only if they are bounded on $\mathbb{N}(a)$.

**Proof.** If all solutions of (1.2.12) are bounded, then there exists a positive constant $c$ such that $\|\mathcal{U}(k,a)\| \le c$ for all $k \in \mathbb{N}(a)$, where $\mathcal{U}(k,a)$ is the principal fundamental matrix of (1.2.12). If $\epsilon > 0$, then $\|\overline{\mathbf{u}}^0 - \mathbf{u}^0\| < \epsilon/c = \delta > 0$ implies

$$\|\mathbf{u}(k,a,\overline{\mathbf{u}}^0) - \mathbf{u}(k,a,\mathbf{u}^0)\| \;=\; \|\mathcal{U}(k,a)(\overline{\mathbf{u}}^0 - \mathbf{u}^0)\| \;\le\; c\|\overline{\mathbf{u}}^0 - \mathbf{u}^0\| \;<\; \epsilon,$$

and hence all solutions of (1.2.12) are stable.

Conversely, if all solutions of (1.2.12) are stable, then in particular, the trivial solution, i.e. $\mathbf{u}(k,a,0) \equiv 0$ is stable. Therefore, given any $\epsilon > 0$, there exists a $\delta > 0$ such that $\|\mathbf{u}^0\| < \delta$ implies that $\|\mathbf{u}(k,a,\mathbf{u}^0)\| < \epsilon$ for all $k \in \mathbb{N}(a)$. However, since $\mathbf{u}(k,a,\mathbf{u}^0) = \mathcal{U}(k,a)\mathbf{u}^0$, we find that

$\|\mathbf{u}(k,a,\mathbf{u}^0)\| = \|\mathcal{U}(k,a)\mathbf{u}^0\| < \epsilon$. Now let $\mathbf{u}^0$ be a vector $\frac{\delta}{2}\mathbf{e}^j$, then we have $\|\mathcal{U}(k,a)\mathbf{u}^0\| = \|\mathbf{u}^j(k)\|\frac{\delta}{2} < \epsilon$, where $\mathbf{u}^j(k)$ is the $j$th column of $\mathcal{U}(k,a)$. Therefore, it follows that $\|\mathcal{U}(k,a)\| = \max_{1\le j\le n}\|\mathbf{u}^j(k)\| < \frac{2\epsilon}{\delta}$. Hence, for any solution $\mathbf{u}(k,a,\mathbf{u}^0)$ of the difference system (1.2.12) we have $\|\mathbf{u}(k,0,\mathbf{u}^0)\| = \|\mathcal{U}(k,0)\mathbf{u}^0\| < \frac{2\epsilon}{\delta}\|\mathbf{u}^0\|$, i.e. all solutions of (1.2.12) are bounded. ■

**Corollary 5.4.2.** All solutions of (2.8.1) are stable if and only if the eigenvalues of $\mathcal{A}$ have modulus less than or equal to one, and those of modulus one are semisimple.

**Remark 5.4.5.** In Definition 5.4.1 the existence of the solution $\mathbf{u}(k)$ of (1.2.8) on $\mathbb{N}(a)$ is assumed. In general, we can consider this special solution to be the trivial solution. This assumption would be at once clear if we consider the transformation $\mathbf{w}(k) = \mathbf{v}(k) - \mathbf{u}(k)$, where $\mathbf{v}(k)$ is any solution of (1.2.8). Since $\mathbf{v}(k)$ is a solution of (1.2.8), it follows that $\mathbf{v}(k+1) = \mathbf{w}(k+1) + \mathbf{u}(k+1) = \mathbf{f}(k,\mathbf{w}(k)+\mathbf{u}(k))$ and hence $\mathbf{w}(k+1) = \mathbf{f}(k,\mathbf{w}(k)+\mathbf{u}(k)) - \mathbf{f}(k,\mathbf{u}(k)) = \overline{\mathbf{f}}(k,\mathbf{w}(k))$, say. Obviously, this new system

$$\mathbf{w}(k+1) = \overline{\mathbf{f}}(k,\mathbf{w}(k)) \tag{5.4.2}$$

admits the trivial solution $\mathbf{w}(k) \equiv 0$. Thus, the stability of the solution $\mathbf{u}(k)$ of (1.2.8) is equivalent to the stability of the trivial solution of (5.4.2).

**Definition 5.4.2.** The points $\overline{\mathbf{u}} \in \mathbb{R}^n$ which satisfy the algebraic equation $\mathbf{f}(k,\overline{\mathbf{u}}) = \overline{\mathbf{u}}$ are called *critical points* of (1.2.8). A critical point is also referred to as a *point of equilibrium* or *stationary point* or *rest point* or *singular point* or *fixed point* or *limit point.*

If $\overline{\mathbf{u}} \in \mathbb{R}^n$ is a critical point of (1.2.8), then obviously $\mathbf{u}(k) = \overline{\mathbf{u}}$ is a solution of (1.2.8). From Remark 5.4.5 each nonzero critical point of (1.2.8) can be transformed to the origin.

**Example 5.4.11.** For the difference equation $u(k+1) = u^2(k)$ there are two critical points $\overline{u} = 0$ and $\overline{u} = 1$. From Example 5.4.5 it is clear that the point $\overline{u} = 0$ is uniformly asymptotically stable, whereas $\overline{u} = 1$ is unstable.

**Example 5.4.12.** For the difference equation $u(k+1) = u(k)(2-u(k))$ there are two critical points $\overline{u} = 0$ and $\overline{u} = 1$. From its solution $\overline{u}(k) =$

$u(k,a,c) = 1-(1-c)^{2^{k-a}}$ it is clear that the point $\overline{u}=0$ is unstable, whereas $\overline{u}=1$ is uniformly asymptotically stable.

## 5.5. Stability of Linear Systems

Let the solutions of the nonhomogeneous difference system (1.2.11) exist on $\mathbb{N}(a)$. Since the definition of stability involves only the difference between the neighboring solutions, it follows, from the superposition principle that any solution $\mathbf{u}(k) = \mathbf{u}(k,a,\mathbf{u}^0)$ of (1.2.11) is stable if and only if the trivial solution of the homogeneous system (1.2.12) is stable. This, in turn implies that if the solution $\mathbf{u}(k)$ of (1.2.11) is stable then every other solution of (1.2.11) is stable. This means that the conditions for the stability of linear systems are independent of the particular solution we consider and of the nonhomogeneous term $\mathbf{b}(k)$. Therefore, to say the linear system (1.2.12) is stable (all its solutions are stable) is more appropriate than to say a particular solution of (1.2.12) is stable. The same argument holds for the other types of stability. However, from Examples 5.4.11 and 5.4.12 it is clear that this argument does not hold for nonlinear systems.

**Theorem 5.5.1.** Let $\mathcal{U}(k,a)$ be the principal fundamental matrix of (1.2.12). Then, the difference system (1.2.12) is

(i) stable if and only if there exists a positive constant $c$ such that

$$\|\mathcal{U}(k,a)\| \leq c \quad \text{for all} \quad k \in \mathbb{N}(a) \tag{5.5.1}$$

(ii) uniformly stable if and only if there exists a positive constant $c$ such that

$$\|\mathcal{G}(k,\ell)\| = \|\mathcal{U}(k,a)\mathcal{U}^{-1}(\ell,a)\| \leq c \quad \text{for all} \quad a \leq \ell \leq k \in \mathbb{N}(a), \tag{5.5.2}$$

i.e. uniformly bounded

(iii) strongly stable if and only if there exists a positive constant $c$ such that

$$\|\mathcal{U}(k,a)\| \leq c, \quad \|\mathcal{U}^{-1}(k,a)\| \leq c \quad \text{for all} \quad k \in \mathbb{N}(a) \tag{5.5.3}$$

(iv) asymptotically stable if and only if

$$\|\mathcal{U}(k,a)\| \to 0 \quad \text{as} \quad k \to \infty \tag{5.5.4}$$

(v) uniformly asymptotically stable if and only if there exist positive constants $c$ and $\lambda$ such that

$$\|\mathcal{G}(k,\ell)\| = \|\mathcal{U}(k,a)\mathcal{U}^{-1}(\ell,a)\| \leq c\exp(-\lambda(k-\ell)) \quad \text{for all} \quad a \leq \ell \leq k \in \mathbb{N}(a). \tag{5.5.5}$$

**Proof.** (i) See Theorem 5.4.1.

(ii) Let $\mathbf{u}(k) = \mathbf{u}(k, a, \mathbf{u}^0)$ be a solution of (1.2.12). Then, for any $k_1 \in \mathbb{N}(a)$ we have $\mathbf{u}(k) = \mathcal{U}(k,a)\mathcal{U}^{-1}(k_1,a)\mathbf{u}(k_1)$. If condition (5.5.2) holds, then we have $\|\mathbf{u}(k)\| \le \|\mathcal{G}(k,k_1)\|\|\mathbf{u}(k_1)\| \le c\|\mathbf{u}(k_1)\|$ for all $k \in \mathbb{N}(k_1)$. Therefore, if $\epsilon > 0$ then $a \le k_1$ and $\|\mathbf{u}(k_1)\| < \epsilon/(2c) = \delta(\epsilon) > 0$ imply $\|\mathbf{u}(k)\| < \epsilon$ for all $k \in \mathbb{N}(k_1)$. Conversely, if (1.2.12) is uniformly stable, then for a given $\epsilon > 0$ there exists a $\delta = \delta(\epsilon) > 0$ such that $a \le k_1 \in \mathbb{N}(a)$ and $\|\mathbf{u}(k_1)\| < \delta$ imply $\|\mathbf{u}(k)\| < \epsilon$ for all $k \in \mathbb{N}(k_1)$. Thus, we have $\|\mathcal{U}(k,a)\mathcal{U}^{-1}(k_1,a)\mathbf{u}(k_1)\| < \epsilon$ for all $k \in \mathbb{N}(k_1)$. The rest of the proof is similar to that of Theorem 5.4.1.

(iii) If (5.5.3) holds, then for a given $\epsilon > 0$ we can choose $\delta = (\epsilon/2c^2)$. Thus, if $a \le k_1 \in \mathbb{N}(a)$ and $\|\mathbf{u}(k_1)\| < \delta$, then we have $\|\mathbf{u}(k)\| = \|\mathcal{U}(k,a)\mathcal{U}^{-1}(k_1,a)\mathbf{u}(k_1)\| \le \|\mathcal{U}(k,a)\|\|\mathcal{U}^{-1}(k_1,a)\|\|\mathbf{u}(k_1)\| \le c^2\|\mathbf{u}(k_1)\| < \epsilon$ for all $k \in \mathbb{N}(a)$. Therefore, (1.2.12) is strongly stable. Conversely, if (1.2.12) is strongly stable, then we have $\|\mathcal{U}(k,a)\mathcal{U}^{-1}(k_1,a)\mathbf{u}(k_1)\| < \epsilon$ for all $k \in \mathbb{N}(a)$ whenever $a \le k_1$ and $\|\mathbf{u}(k_1)\| < \delta$ hold. Since $\mathbf{u}(k_1)$, $\|\mathbf{u}(k_1)\| < \delta$ is arbitrary, we can conclude as in the proof of Theorem 5.4.1 that

$$\|\mathcal{U}(k,a)\mathcal{U}^{-1}(k_1,a)\| < c, \tag{5.5.6}$$

where $c = 2\epsilon/\delta$. From the definition of strong stability, it is clear that $\delta$, and hence $c$, is independent of $a$ and $k_1$ as well as of $k$. Putting $k_1 = a$ and $k = a$ in estimate (5.5.6), we obtain respectively, the first and second bound in (5.5.3).

(iv) Every solution of (1.2.12) can be expressed as $\mathbf{u}(k) = \mathbf{u}(k,a,\mathbf{u}^0) = \mathcal{U}(k,a)\mathbf{u}^0$. From (5.5.4) it is clear that there exists a constant $c$ such that $\|\mathcal{U}(k,a)\| \le c$ for all $k \in \mathbb{N}(a)$. Thus, $\|\mathbf{u}(k)\| \le c\|\mathbf{u}^0\|$, and hence every solution of (1.2.12) is bounded. Therefore, from Theorem 5.4.1 the system (1.2.12) is stable. Further, $\|\mathbf{u}(k)\| \to 0$ as $k \to \infty$, and hence the system (1.2.12) is asymptotically stable. Conversely, if (1.2.12) is asymptotically stable, then its trivial solution $\mathbf{u}(k,a,0) \equiv 0$ is asymptotically stable. Hence, $\|\mathbf{u}(k,a,\mathbf{u}^0)\| \to 0$ as $k \to \infty$ provided $\|\mathbf{u}^0\| < \delta$, but this implies that $\|\mathcal{U}(k,a)\| \to 0$ as $k \to \infty$.

(v) Since (5.5.5) implies (5.5.2) the system (1.2.12) is uniformly stable. Further, if $\mathbf{u}(k) = \mathbf{u}(k,a,\mathbf{u}^0)$ is any solution of (1.2.12) then $\mathbf{u}(k) = \mathcal{U}(k,a)\mathbf{u}^0 = \mathcal{U}(k,a)\mathcal{U}^{-1}(k_1,a)\mathbf{u}(k_1)$, and hence for all $k \in \mathbb{N}(k_1)$, $\|\mathbf{u}(k)\| \le \|\mathcal{U}(k,a)\mathcal{U}^{-1}(k_1,a)\|\|\mathbf{u}(k_1)\| \le c\|\mathbf{u}(k_1)\|\exp(-\lambda(k-k_1))$. Thus, it follows that $\|\mathbf{u}(k)\| \to 0$ independent of $k_1$, and hence the system (1.2.12) is uniformly attractive. Therefore, the system (1.2.12) is uniformly asymptotically stable. Conversely, if the system (1.2.12) is uniformly asymptotically stable, then it is uniformly attractive, and hence for a fixed $\epsilon > 0$ and $k_1 \geq a$ there exist $\delta > 0$, and $K(\epsilon) \in \mathbb{N}(1)$ such that $\|\mathbf{u}(k_1)\| < \delta$ implies $\|\mathbf{u}(k)\| = \|\mathcal{G}(k, k_1)\mathbf{u}(k_1)\| < \epsilon$ for all $k_1 + K(\epsilon) \leq k \in \mathbb{N}(k_1)$. Thus, it follows that $\|\mathcal{G}(k, k_1)\| < \eta < 1$ for all $k_1 + K(\epsilon) \leq k \in \mathbb{N}(k_1)$, where $\eta$ can be chosen arbitrary small. Further, since the uniform asymptotic stability implies the uniform stability, we find $\|\mathcal{G}(k, k_1)\| \leq c$ for all $k_1 \leq k \in \mathbb{N}(k_1)$. Thus, for $k \in \mathbb{N}(k_1 + mK(\epsilon), k_1 + (m+1)K(\epsilon))$, where $m \in \mathbb{N}(1)$ we have

$$\begin{aligned} \|\mathcal{G}(k, k_1)\| &\leq \|\mathcal{G}(k, k_1 + mK(\epsilon))\| \|\mathcal{G}(k_1 + mK(\epsilon), k_1 + (m-1)K(\epsilon))\| \\ &\qquad \cdots \|\mathcal{G}(k_1 + K(\epsilon), k_1)\| \\ &\leq c\eta^m = c\eta^{-1}\eta^{\frac{m+1}{K(\epsilon)}K(\epsilon)} \leq c\eta^{-1}\eta^{\frac{(k-k_1)}{K(\epsilon)}} \leq c_1 e^{-\lambda(k-k_1)}, \end{aligned}$$

where $c_1 = c\eta^{-1}$ and $\eta^{1/K(\epsilon)} = e^{-\lambda}$. ∎

**Remark 5.5.1.** The stability of the system (2.8.1) implies the uniform stability of this system. However, from Example 5.2.6 it is clear that this is not true for the system (1.2.12).

**Remark 5.5.2.** As a consequence of Theorem 5.5.1 (v), we see that for linear systems, uniform asymptotic stability implies exponential asymptotic stability. However, this is not true for the nonlinear systems. For example, the general solution of the equation $u(k+1) = \dfrac{u(k)}{\sqrt{1 + 2u^2(k)}}$ is $u(k) = \dfrac{u(a)}{(1 + 2u^2(a)(k-a))^{1/2}}$. Thus, the trivial solution of this equation is uniformly asymptotically stable but not exponentially asymptotically stable.

**Remark 5.5.3.** From Theorem 5.5.1 (iv), it is clear that, the system (2.8.1) is asymptotically stable, and hence uniformly asymptotically stable which, in turn, implies exponentially asymptotically stable, if and only if the eigenvalues of $\mathcal{A}$ lie inside the unit disk.

**Definition 5.5.1.** The difference system (1.2.12) is said to be *restrictively stable* if, together with its adjoint system (2.7.1) it is stable.

If $\mathcal{U}(k, a)$ is the principal fundamental matrix of (1.2.12) then from (2.7.2) it is clear that $[\mathcal{U}^T(k, a)]^{-1}$ is the principal fundamental matrix of (2.7.1). Thus, from Theorem 5.5.1 (i) the following result is immediate.

**Theorem 5.5.2.** Let $\mathcal{U}(k, a)$ be the principal fundamental matrix of (1.2.12). Then, the difference system (1.2.12) is restrictively stable if and only if there exists a positive constant $c$ such that

$$(5.5.7)\qquad \|\mathcal{U}(k,a)\| \le c, \quad \|\mathcal{U}^{-1}(k,a)\| \le c \quad \text{for all} \quad k \in \mathbb{N}(a).$$

**Remark 5.5.4.** From Theorem 5.5.1 (iii) and Theorem 5.5.2 it follows that, for linear homogeneous systems, restrictive stability and strong stability are equivalent. Thus, for such systems restrictive stability implies uniform stability which, in turn, gives stability.

**Definition 5.5.2.** The difference system (1.2.12) is said to be *reducible* (reducible to zero) if there exists an $n \times n$ matrix $\mathcal{L}(k)$ which, together with its inverse $\mathcal{L}^{-1}(k)$, is defined and bounded on $\mathbb{N}(a)$ such that $\mathcal{L}^{-1}(k+1)\mathcal{A}(k)\mathcal{L}(k)$ is a constant (identity) matrix on $\mathbb{N}(a)$.

Since the transformation $\mathbf{u}(k) = \mathcal{L}(k)\mathbf{v}(k)$ converts the system (1.2.12) into $\mathbf{v}(k+1) = \mathcal{L}^{-1}(k+1)\mathcal{A}(k)\mathcal{L}(k)\mathbf{v}(k)$, from the above definition it is clear that $\mathbf{u}(k) = \mathcal{L}(k)\mathbf{v}(k)$ transforms (1.2.12) into a system with constant coefficients (into the system $\mathbf{v}(k+1) = \mathbf{v}(k)$).

**Theorem 5.5.3.** The difference system (1.2.12) is restrictively stable if and only if it is reducible to zero.

**Proof.** Let $\mathcal{U}(k,a)$ be the principal fundamental matrix of (1.2.12). If the system (1.2.12) is restrictively stable then from Theorem 5.5.2 there exists a positive constant $c$ such that (5.5.7) holds. Consider now the transformation $\mathbf{u}(k) = \mathcal{U}(k,a)\mathbf{v}(k)$, which converts the system into $\mathcal{U}(k+1,a)(\mathbf{v}(k+1) - \mathbf{v}(k)) = 0$, which in turn, implies $\mathbf{v}(k+1) = \mathbf{v}(k)$. Hence, (1.2.12) is reducible to zero. Conversely, if (1.2.12) is reducible to zero, then there exists a matrix $\mathcal{L}(k)$ such that $\mathcal{L}^{-1}(k+1)\mathcal{A}(k)\mathcal{L}(k) = \mathcal{I}$, and hence $\mathcal{L}(k+1) = \mathcal{A}(k)\mathcal{L}(k)$, i.e. $\mathcal{L}(k)$ is a fundamental matrix of (1.2.12). Since $\mathcal{L}(k)$, together with its inverse $\mathcal{L}^{-1}(k)$ is bounded on $\mathbb{N}(a)$ from Theorem 5.5.2 the system (1.2.12) is restrictively stable. ∎

**Theorem 5.5.4.** The difference system (1.2.12) is uniformly stable if it is stable, and reducible.

**Proof.** Since (1.2.12) is reducible, $\mathbf{u}(k) = \mathcal{L}(k)\mathbf{v}(k)$ transforms it into $\mathbf{v}(k+1) = \mathcal{B}\mathbf{v}(k)$, where $\mathcal{B} = \mathcal{L}^{-1}(k+1)\mathcal{A}(k)\mathcal{L}(k)$. Let $\mathcal{U}(k,a)$ be the principal fundamental matrix of (1.2.12). By Theorem 5.5.1 (i), the stability of (1.2.12) implies that $\mathcal{U}(k,a)$ is bounded on $\mathbb{N}(a)$. If $\mathcal{V}(k)$ is a fundamental matrix of $\mathbf{v}(k+1) = \mathcal{B}\mathbf{v}(k)$, then it is easy to see that $\mathcal{U}(k,a) = \mathcal{L}(k)\mathcal{V}(k)\mathcal{V}^{-1}(a)\mathcal{L}^{-1}(a)$. Thus, it follows that $\mathcal{V}(k) = \mathcal{L}^{-1}(k)\mathcal{U}(k,a)\mathcal{L}(a)\mathcal{V}(a)$ is bounded on $\mathbb{N}(a)$. Therefore, the system $\mathbf{v}(k+1) = \mathcal{B}\mathbf{v}(k)$ is stable and, in fact, uniformly stable. Hence, from (5.5.2), it is clear that $\|\mathcal{V}(k)\mathcal{V}^{-1}(\ell)\| \le c$ for some positive constant $c$ and $a \le \ell \le k \in \mathbb{N}(a)$. Therefore,

$$\begin{aligned}\|\mathcal{U}(k,a)\mathcal{U}^{-1}(\ell,a)\| &= \|\mathcal{L}(k)\mathcal{V}(k)\mathcal{V}^{-1}(\ell)\mathcal{L}^{-1}(\ell)\| \\ &\le \|\mathcal{L}(k)\|\|\mathcal{V}(k)\mathcal{V}^{-1}(\ell)\|\|\mathcal{L}^{-1}(\ell)\| \le c_1\end{aligned}$$

for some positive constant $c_1$ and $a \le \ell \le k \in \mathbb{N}(a)$. Thus, the system (1.2.12) is uniformly stable. ∎

Let for a fixed $\ell \in \mathbb{N}(a)$ the solution $\mathbf{u}(k) = \mathbf{u}(k,\ell,\mathbf{u}(\ell))$ $(\mathbf{u}(\ell) \ne 0)$ of (1.2.12) exist on $\mathbb{N}(\ell)$ and $\overline{\mathbb{N}}(\ell) = \{k \in \mathbb{N}(\ell) : \mathbf{u}(k) \ne 0\}$ is infinite. Let $\delta(\epsilon,\ell) = \left\{\overline{\mathbf{u}}(\ell) \in \mathbb{R}^n : \frac{\|\overline{\mathbf{u}}(\ell) - \mathbf{u}(\ell)\|}{\|\mathbf{u}(\ell)\|} \le \epsilon\right\}$. Thus, if $\overline{\mathbf{u}}(\ell) \in \delta(\epsilon,\ell)$ then $\overline{\mathbf{u}}(\ell) = \mathbf{u}(\ell) + \|\mathbf{u}(\ell)\|\mathbf{d}$, where $\|\mathbf{d}\| \le \epsilon$, i.e. the vector $\overline{\mathbf{u}}(\ell)$ approximates $\mathbf{u}(\ell)$ with the relative error at most $\epsilon$. Let $\overline{\mathbf{u}}(k) = \mathbf{u}(k,\ell,\overline{\mathbf{u}}(\ell))$ be the solution of (1.2.12) which exists on $\mathbb{N}(\ell)$. Thus, if $\mathcal{V}(k)$ is any fundamental matrix of (1.2.12) then it follows that

$$\overline{\mathbf{u}}(k) - \mathbf{u}(k) = \mathcal{V}(k)\mathcal{V}^{-1}(\ell)(\mathbf{u}(\ell) + \|\mathbf{u}(\ell)\|\mathbf{d} - \mathbf{u}(\ell))$$

and hence

$$\sup_{\substack{\overline{\mathbf{u}}(\ell) \in \delta(\epsilon,\ell) \\ k \in \overline{\mathbb{N}}(\ell)}} \frac{\|\overline{\mathbf{u}}(k) - \mathbf{u}(k)\|}{\|\mathbf{u}(k)\|} = \frac{\|\mathbf{u}(\ell)\|}{\|\mathbf{u}(k)\|} \sup_{\|\mathbf{d}\|=\epsilon} \|\mathcal{V}(k)\mathcal{V}^{-1}(\ell)\mathbf{d}\| = \alpha(\ell,k)\epsilon, \tag{5.5.8}$$

where

$$\alpha(\ell,k) = \frac{\|\mathbf{u}(\ell)\|}{\|\mathbf{u}(k)\|}\|\mathcal{V}(k)\mathcal{V}^{-1}(\ell)\|. \tag{5.5.9}$$

**Definition 5.5.3.** The *index of stability* for the forward computation of $\mathbf{u}(k) = \mathbf{u}(k,\ell,\mathbf{u}(\ell))$ at $\ell$ is $\alpha(k) = \sup_{\ell<k\in\overline{\mathbb{N}}(\ell)} \alpha(\ell,k)$. If $\alpha(k) < \infty$, then (1.2.12) is said to be *stable for* $\mathbf{u}(k)$ *at* $\ell$.

**Definition 5.5.4.** If for each $K \in \overline{\mathbb{N}}(\ell)$ there exists a constant $C = C(K) > 0$ such that $\sup_{\{\ell<k\in\overline{\mathbb{N}}(\ell),\ \ell\le K\}} \alpha(\ell,k) = C < \infty$, then (1.2.12) is said to be *weakly stable for* $\mathbf{u}(k)$.

**Definition 5.5.5.** If $\sup_{\{\ell,k\in\overline{\mathbb{N}}(a),k>\ell\}} \alpha(\ell,k) = C < \infty$, then (1.2.12) is said to be *stable for* $\mathbf{u}(k)$.

**Theorem 5.5.5.** The difference system (1.2.12) is stable for $\mathbf{u}(k)$ at $\ell$ if and only if it is weakly stable for $\mathbf{u}(k)$.

**Proof.** From (5.5.9) for all $k \in \overline{\mathbb{N}}(\ell)$, we have

$$\begin{aligned} \alpha(\ell+1,k) &= \frac{\|\mathcal{A}(\ell)\mathbf{u}(\ell)\|}{\|\mathbf{u}(k)\|}\|\mathcal{V}(k)\mathcal{V}^{-1}(\ell+1)\| \\ &\le \|\mathcal{A}(\ell)\|\frac{\|\mathbf{u}(\ell)\|}{\|\mathbf{u}(k)\|}\|\mathcal{V}(k)\mathcal{V}^{-1}(\ell)\|\|\mathcal{A}^{-1}(\ell)\| \\ &= (\mathit{cond}\ \mathcal{A}(\ell))\alpha(\ell,k), \end{aligned}$$

where $\mathit{cond}\ \mathcal{B} = \|\mathcal{B}\|\|\mathcal{B}^{-1}\|$ is called the *condition number* of $\mathcal{B}$.

Similarly, we have

$$\alpha(\ell,k) = \frac{\|\mathbf{u}(\ell)\|}{\|\mathbf{u}(k)\|}\|\mathcal{V}(k)\mathcal{V}^{-1}(\ell+1)\mathcal{A}(\ell)\|$$

and hence

$$\alpha(\ell+1,k) \ge \frac{\|\mathbf{u}(\ell+1)\|}{\|\mathbf{u}(\ell)\|}\alpha(\ell,k)\frac{1}{\|\mathcal{A}(\ell)\|}.$$

However, since $\dfrac{\|\mathbf{u}(\ell+1)\|}{\|\mathbf{u}(\ell)\|} \ge \dfrac{1}{\|\mathcal{A}^{-1}(\ell)\|}$ we find that

$$\alpha(\ell+1,k) \ge (\mathit{cond}\ \mathcal{A}(\ell))^{-1}\alpha(\ell,k).$$

Now combining the above results, to obtain

$$(\mathit{cond}\ \mathcal{A}(\ell))^{-1}\alpha(\ell,k) \le \alpha(\ell+1,k) \le (\mathit{cond}\ \mathcal{A}(\ell))\alpha(\ell,k),$$

and taking a sup over $k$ leads to

$$(\mathit{cond}\ \mathcal{A}(\ell))^{-1}\alpha(\ell) \le \alpha(\ell+1) \le (\mathit{cond}\ \mathcal{A}(\ell))\alpha(\ell),$$

which proves the desired result. ∎

**Definition 5.5.6.** If any one of the Definitions 5.5.3 5.5.5 is not satisfied then the term stable is replaced by *unstable*.

**Remark 5.5.5.** If (1.2.12) is stable for $\mathbf{u}(k)$ then it is weakly stable for $\mathbf{u}(k)$, however the converse is not true.

**Example 5.5.1.** For the system $\mathbf{u}(k+1) = \begin{bmatrix} 0 & 1 \\ -1 & 2 \end{bmatrix}\mathbf{u}(k)$ a fundamental matrix is $\mathcal{V}(k) = \begin{bmatrix} k & 1 \\ k+1 & 1 \end{bmatrix} = [\mathbf{v}^1(k), \mathbf{v}^2(k)]$. Since $\|\mathbf{v}^1(k)\| = 2k+1$, $\|\mathbf{v}^2(k)\| = 2$, for $\mathbf{u}(k) = \mathbf{v}^1(k)$

$$\alpha(\ell,k) = \frac{(2\ell+1)(2k-2\ell+1)}{(2k+1)}, \quad k > \ell$$

and hence $\alpha(\ell) = 2\ell + 1$. Thus, the system is stable for $\mathbf{v}^1(k)$ at any $\ell$ and also weakly stable, but not stable.

**Example 5.5.2.** For the system $\mathbf{u}(k+1) = \begin{bmatrix} 0 & 1 \\ -2 & 3 \end{bmatrix} \mathbf{u}(k)$ a fundamental matrix is $\mathcal{V}(k) = \begin{bmatrix} 2^k & 1 \\ 2^{k+1} & 1 \end{bmatrix} = [\mathbf{v}^1(k), \mathbf{v}^2(k)]$. Since $\|\mathbf{v}^1(k)\| = 3 \cdot 2^k$, $\|\mathbf{v}^2(k)\| = 2$, for $\mathbf{u}(k) = \mathbf{v}^1(k)$

$$\alpha(\ell, k) = \frac{3 \cdot 2^\ell (3 \cdot 2^k - 2^{\ell+1})}{3 \cdot 2^k 2^\ell}, \quad k > \ell$$

and hence $\alpha(\ell) = 3$. Thus, the system is stable for $\mathbf{v}^1(k)$, which in turn implies weakly stable and stable at any $\ell$.

In the following result for simplicity we shall assume that $\overline{\mathbb{N}}(\ell) = \mathbb{N}(\ell)$.

**Theorem 5.5.6.** Let $n \geq 2$ and let the difference system (1.2.12) have a solution $\mathbf{u}^*(k)$ such that

$$\lim_{k \to \infty} \frac{\|\mathbf{u}(k)\|}{\|\mathbf{u}^*(k)\|} = 0. \tag{5.5.10}$$

Then, (1.2.12) is unstable for $\mathbf{u}(k) = \mathbf{u}(k, \ell, \mathbf{u}(\ell))$ at every $\ell$. Conversely, if (1.2.12) is unstable for $\mathbf{u}(k)$ in the following manner

$$\limsup_{k \to \infty} \alpha(\ell, k) = \infty \quad \text{for some fixed } \ell \in \mathbb{N}(a) \tag{5.5.11}$$

then a solution $\mathbf{u}^*(k)$ of (1.2.12) exists having the property (5.5.10).

**Proof.** Since for any nonsingular constant matrix $\mathcal{B}$ and any matrix $\mathcal{W}(k)$, $\|\mathcal{W}(k)\mathcal{B}\| \leq \|\mathcal{W}(k)\|\|\mathcal{B}\|$ and $\|\mathcal{W}(k)\| = \|\mathcal{W}(k)\mathcal{B}\mathcal{B}^{-1}\| \leq \|\mathcal{W}(k)\mathcal{B}\|\|\mathcal{B}^{-1}\|$ it follows that $\limsup_{k\to\infty} \|\mathcal{W}(k)\mathcal{B}\| = \infty$ if and only if $\limsup_{k\to\infty} \|\mathcal{W}(k)\| = \infty$. If (5.5.10) holds then $\mathbf{u}(k)$ and $\mathbf{u}^*(k)$ are linearly independent solutions of (1.2.12). Therefore, they may be completed to form a fundamental matrix $\mathcal{V}(k) = [\mathbf{u}(k), \mathbf{u}^*(k), \mathbf{u}^3(k), \cdots, \mathbf{u}^n(k)]$. For this fundamental matrix it follows that $\|\mathcal{V}(k)/\|\mathbf{u}(k)\|\| \geq \|\mathbf{u}^*(k)\|/\|\mathbf{u}(k)\|$. Thus, from the previous consideration with $\mathcal{W}(k) = \mathcal{V}(k)/\|\mathbf{u}(k)\|$ and $\mathcal{B} = \|\mathbf{u}(\ell)\|\mathcal{V}^{-1}(\ell)$ it follows that $\limsup_{k\to\infty} \alpha(\ell, k) = \infty$ for every $\ell \in \mathbb{N}(a)$. Thus, (1.2.12) is unstable for $\mathbf{u}(k)$ at every $\ell$.

Conversely, let (5.5.11) holds and $\mathcal{V}(k)$ be a fundamental matrix of (1.2.12). Once again considering $\mathcal{W}(k) = \mathcal{V}(k)/\|\mathbf{u}(k)\|$ and $\mathcal{B} = \|\mathbf{u}(\ell)\|\mathcal{V}^{-1}(\ell)$ in the above relation yields $\limsup_{k\to\infty} \|\mathcal{V}(k)/\|\mathbf{u}(k)\|\| = \infty$. Let $\|\mathcal{V}(k)\| = \|\mathbf{u}^{r(k)}(k)\|$, $1 \leq r(k) \leq n$. Obviously, there exists an integer $r$, $1 \leq r \leq n$ such that $r(k) = r$ for infinitely many $k$. Letting $k \to$

$\infty$ over this subsequence of $\mathbb{N}(\ell)$ shows that $\limsup_{k\to\infty} \|\mathbf{u}^r(k)\|/\|\mathbf{u}(k)\| = \infty$, and taking $\mathbf{u}^*(k) = \mathbf{u}^r(k)$ concludes the proof of the theorem. ∎

Now for a fixed $\ell \in \mathbb{N}(a)$ we shall consider the solution $\mathbf{v}(k) = \mathbf{v}(k, \ell, \mathbf{v}(\ell))$ of (1.2.11) which is assumed to exist on $\mathbb{N}(\ell)$. For this solution we can define the corresponding $\overline{\mathbb{N}}(\ell)$ and $\delta(\epsilon, \ell)$ by replacing $\mathbf{u}(k)$ and $\overline{\mathbf{u}}(\ell)$ by $\mathbf{v}(k)$ and $\overline{\mathbf{v}}(\ell)$. Since the general solution $\mathbf{u}(k)$ of (1.2.11) in terms of the fundamental matrix $\mathcal{V}(k)$ of (1.2.12) can be written as $\mathbf{u}(k) = \mathcal{V}(k)\mathbf{c} + \mathbf{v}(k)$, the solution $\overline{\mathbf{v}}(k) = \mathbf{v}(k, \ell, \overline{\mathbf{v}}(\ell))$ has the representation $\overline{\mathbf{v}}(k) = \mathcal{V}(k)\mathcal{V}^{-1}(\ell)\|\mathbf{v}(\ell)\|\mathbf{d} + \mathbf{v}(k)$. Thus, (5.5.8) and (5.5.9) hold with $\mathbf{u}(k)$ and $\overline{\mathbf{u}}(k)$ replaced by $\mathbf{v}(k)$ and $\overline{\mathbf{v}}(k)$.

**Definition 5.5.7.** The difference system (1.2.11) is said to be stable for $\mathbf{v}(k)$ if there exists a constant $C > 0$ such that $\sup_{\{\ell, k \in \overline{\mathbb{N}}(a),\ k > \ell\}} \overline{\alpha}(\ell, k) = C < \infty$, where $\overline{\alpha}(\ell, k)$ is the same as $\alpha(\ell, k)$ replacing $\mathbf{u}(\ell)$ and $\mathbf{u}(k)$ by $\mathbf{v}(\ell)$ and $\mathbf{v}(k)$. Otherwise, (1.2.11) is said to be unstable for $\mathbf{v}(k)$.

**Theorem 5.5.7.** Let $n \geq 2$ and let the homogeneous system (1.2.12) have a solution $\mathbf{u}^*(k)$ such that $\lim_{k\to\infty} \|\mathbf{v}(k)\|/\|\mathbf{u}^*(k)\| = 0$. Then, the nonhomogeneous difference system (1.2.11) is unstable for $\mathbf{v}(k)$ at any $\ell$.

**Proof.** The proof is clear from Theorem 5.5.6. ∎

## 5.6. Stability of Nonlinear Systems

In Section 5.3 we have considered the difference systems (5.3.1) and (5.3.4) as the perturbed systems of (2.8.1) and (1.2.12) respectively, and provided sufficient conditions on the nonlinear perturbed function $\mathbf{g}(k, \mathbf{v})$ so that the asymptotic properties of the unperturbed systems are maintained for the perturbed systems. Analogously, we expect that under certain conditions on the function $\mathbf{g}(k, \mathbf{v})$ stability properties of the unperturbed systems carry through for the perturbed systems.

**Theorem 5.6.1.** Let for all $(k, \mathbf{v}) \in \mathbb{N}(a) \times \mathbb{R}^n$ the function $\mathbf{g}(k, \mathbf{v})$ satisfy (5.3.2), where $h(k)$ is a nonnegative function defined on $\mathbb{N}(a)$ and $\sum_{\ell=a}^{\infty} h(\ell) < \infty$. Then, the trivial solution $\mathbf{v}(k, a, 0) \equiv 0$ of (5.3.4) is uniformly (and asymptotically; or exponentially asymptotically) stable provided the trivial solution $\mathbf{u}(k, a, 0) \equiv 0$ of (1.2.12) is uniformly (and asymptotically; or uniformly asymptotically) stable.

**Proof.** From the condition (5.3.2) it is clear that the system (5.3.4) admits the trivial solution. Since in terms of the principal fundamental matrix $\mathcal{U}(k, a)$ the solution $\mathbf{v}(k) = \mathbf{v}(k, a, \mathbf{v}^0)$ of (5.3.4) for $k \geq k_1 \in \mathbb{N}(a)$ can be written as

$$(5.6.1)\quad \mathbf{v}(k) = \mathcal{U}(k,a)\mathcal{U}^{-1}(k_1,a)\mathbf{v}(k_1)+\sum_{\ell=k_1}^{k-1}\mathcal{U}(k,a)\mathcal{U}^{-1}(\ell+1,a)\mathbf{g}(\ell,\mathbf{v}(\ell))$$

from Theorem 5.5.1 (ii) it follows that

$$\|\mathbf{v}(k)\| \le c\|\mathbf{v}(k_1)\| + c\sum_{\ell=k_1}^{k-1} h(\ell)\|\mathbf{v}(\ell)\|$$

and therefore for all $k \in \mathbb{N}(k_1)$

$$(5.6.2)\qquad \|\mathbf{v}(k)\| \le c\|\mathbf{v}(k_1)\| \exp\left(c\sum_{\ell=k_1}^{k-1} h(\ell)\right) \le L\|\mathbf{v}(k_1)\|,$$

where $L = c\exp\left(c\sum_{\ell=a}^{\infty} h(\ell)\right)$. From (5.6.2) the uniform stability of the trivial solution of (5.3.4) is obvious.

Now if in addition to the uniform stability of the trivial solution of (1.2.12), it is asymptotically stable also, then from Theorem 5.5.1 (iv), $\|\mathcal{U}(k,a)\| \to 0$ as $k \to \infty$. Therefore, for every $\epsilon > 0$ there exists a $k_1 \in \mathbb{N}(a)$ sufficiently large so that for all $k \ge k_1$

$$\begin{aligned}&\|\mathcal{U}(k,a)\mathcal{U}^{-1}(k_1,a)\mathbf{v}(k_1)\|\\ &\quad = \left\|\mathcal{U}(k,a)\left[\mathbf{v}^0 + \sum_{\ell=a}^{k_1-1}\mathcal{U}^{-1}(\ell+1,a)\mathbf{g}(\ell,\mathbf{v}(\ell))\right]\right\| < \epsilon.\end{aligned}$$

Thus, the inequality (5.6.2) can be replaced by $\|\mathbf{v}(k)\| < \epsilon\exp\left(c\sum_{\ell=k_1}^{k-1} h(\ell)\right)$, from which $\lim_{k\to\infty}\mathbf{v}(k) = 0$ is immediate.

If the trivial solution of (1.2.12) is uniformly asymptotically stable then on using Theorem 5.5.1 (v) in (5.6.1), we obtain

$$\|\mathbf{v}(k)\| \le ce^{-\lambda(k-k_1)}\|\mathbf{v}(k_1)\| + c\sum_{\ell=k_1}^{k-1} e^{-\lambda(k-\ell-1)}h(\ell)\|\mathbf{v}(\ell)\|,$$

which is the same as

$$e^{\lambda k}\|\mathbf{v}(k)\| \le ce^{\lambda k_1}\|\mathbf{v}(k_1)\| + ce^{\lambda}\sum_{\ell=k_1}^{k-1} h(\ell)\left(e^{\lambda\ell}\|\mathbf{v}(\ell)\|\right)$$

and hence

$$\begin{aligned}\|\mathbf{v}(k)\| &\le ce^{-\lambda(k-k_1)}\|\mathbf{v}(k_1)\|\exp\left(ce^{\lambda}\sum_{\ell=k_1}^{k-1}h(\ell)\right)\\ &\le M\|\mathbf{v}(k_1)\|\exp(-\lambda(k-k_1)), \quad k\in\mathbb{N}(k_1)\end{aligned}$$

where $M = c\exp\left(ce^{\lambda}\sum_{\ell=a}^{\infty}h(\ell)\right)$. From the above inequality it is clear that the trivial solution of (5.3.4) is exponentially asymptotically stable. ∎

**Theorem 5.6.2.** Let for all $(k,\mathbf{v})\in\mathbb{N}(a)\times\mathbb{R}^n$ the function $\mathbf{g}(k,\mathbf{v})$ satisfy

$$\|\mathbf{g}(k,\mathbf{v})\| \le \alpha\|\mathbf{v}\|, \tag{5.6.3}$$

where $\alpha>0$ is sufficiently small. Then, the trivial solution $\mathbf{v}(k,a,0)\equiv 0$ of (5.3.4) is exponentially asymptotically stable provided the trivial solution $\mathbf{u}(k,a,0)\equiv 0$ of (1.2.12) is uniformly asymptotically stable.

**Proof.** The proof is similar to that of Theorem 5.6.1. ∎

**Corollary 5.6.3.** Let for all $(k,\mathbf{v})\in\mathbb{N}(a)\times\mathbb{R}^n$ the function $\mathbf{g}(k,\mathbf{v})$ satisfy

$$\|\mathbf{g}(k,\mathbf{v})\| = o(\|\mathbf{v}\|). \tag{5.6.4}$$

Then, the trivial solution $\mathbf{v}(k,a,0)\equiv 0$ of (5.3.1) is exponentially asymptotically stable provided all eigenvalues of the matrix $\mathcal{A}$ are inside the unit disc.

Next we state the following result whose proof differs slightly from Corollary 5.6.3.

**Theorem 5.6.4.** Let $\mathbf{g}(k,\mathbf{v})$ be as in Corollary 5.6.3, and the trivial solution $\mathbf{u}(k,a,0)\equiv 0$ of (2.8.1) is unstable. Then, the trivial solution $\mathbf{v}(k,a,0)\equiv 0$ of (5.3.1) is unstable.

Corollary 5.6.3 and Theorem 5.6.4 fail to embrace the critical case, i.e. when all the eigenvalues of the matrix $\mathcal{A}$ are inside the unit disc, and that of at least one eigenvalue has the modulus one. In this critical case, the nonlinear function $\mathbf{g}(k,\mathbf{v})$ begins to influence the stability of the trivial solution of the system (5.3.1), and generally it is impossible to test for stability on the basis of eigenvalues of $\mathcal{A}$. For example, the trivial solution of the difference equation

$$u(k+1) = u(k)+u(k)\left[\frac{1}{\sqrt{1+cu^2(k)}}-1\right]$$

is stable if $c = 0$, asymptotically stable if $c > 0$, and unstable if $c < 0$.

**Definition 5.6.1.** A matrix $\mathcal{P}$ is said to be a *projection* if $\mathcal{P}^2 = \mathcal{P}$. If $\mathcal{P}$ is a projection, then so is $(\mathcal{I} - \mathcal{P})$. Two such projections, whose sum is $\mathcal{I}$ and hence whose product is $0$, are said to be *supplementary*.

**Lemma 5.6.5.** Let $\mathcal{V}(k)$ be an invertible matrix which is defined on $\mathbb{N}(a)$ and let $\mathcal{P}$ be a projection. If there exists a constant $c > 1$ such that

$$(5.6.5) \qquad \sum_{\ell=a}^{k-1} \|\mathcal{V}(k)\mathcal{P}\mathcal{V}^{-1}(\ell+1)\| \leq c \quad \text{for all } k \in \mathbb{N}(a)$$

then there exists a constant $c_1$ such that

$$(5.6.6) \qquad \|\mathcal{V}(k)\mathcal{P}\| \leq c_1 \left(\frac{c-1}{c}\right)^{k-a} \quad \text{for all } k \in \mathbb{N}(a).$$

**Proof.** We can obviously suppose that $\mathcal{P} \neq 0$. Since for any $\ell \in \mathbb{N}(a)$, we have $\|\mathcal{V}(\ell+1)\mathcal{P}\| > 0$, it follows that

$$\mathcal{V}(k)\mathcal{P}\sum_{\ell=a}^{k-1} \|\mathcal{V}(\ell+1)\mathcal{P}\|^{-1} = \sum_{\ell=a}^{k-1}(\mathcal{V}(k)\mathcal{P}\mathcal{V}^{-1}(\ell+1))\mathcal{V}(\ell+1)\mathcal{P}\|\mathcal{V}(\ell+1)\mathcal{P}\|^{-1}$$

and hence

$$(5.6.7) \qquad \|\mathcal{V}(k)\mathcal{P}\| \sum_{\ell=a}^{k-1} \|\mathcal{V}(\ell+1)\mathcal{P}\|^{-1} \leq c.$$

Setting $r(k) = \sum_{\ell=a}^{k-1} \|\mathcal{V}(\ell+1)\mathcal{P}\|^{-1}$, we obtain

$$r(k) - r(k-1) = \|\mathcal{V}(k)\mathcal{P}\|^{-1}, \quad k \in \mathbb{N}(a+1)$$

and, after substitution in (5.6.7), we get $r(k) \geq (c/c-1)r(k-1)$, which implies $r(k) \geq (c/c-1)^{k-a-1} r(a+1)$, $k \in \mathbb{N}(a+1)$. Using this inequality in (5.6.7), we find

$$(5.6.8) \qquad \|\mathcal{V}(k)\mathcal{P}\| \leq c\left(\frac{c-1}{c}\right)^{k-a-1} \|\mathcal{V}(a+1)\mathcal{P}\|, \quad k \in \mathbb{N}(a+1).$$

Therefore, if we choose $c_1 = \max\left\{\|\mathcal{V}(a)\mathcal{P}\|,\ (c^2/c-1)\|\mathcal{V}(a+1)\mathcal{P}\|\right\}$, the resulting inequality (5.6.6) follows from (5.6.8). ∎

**Lemma 5.6.6.** Let $\mathcal{V}(k)$ be an invertible matrix which is defined on $\mathbb{N}(a)$ and let $\mathcal{P}$ be a projection. If there exists a constant $c > 0$ such that

$$\sum_{\ell=k}^{\infty} \|\mathcal{V}(k)\mathcal{P}\mathcal{V}^{-1}(\ell+1)\| \leq c \quad \text{for all} \quad k \in \mathbb{N}(a) \tag{5.6.9}$$

then for any vector $\mathbf{c}$ such that $\mathcal{P}\mathbf{c} \neq 0$

$$\limsup_{k\to\infty} \|\mathcal{V}(k)\mathcal{P}\mathbf{c}\| = \infty.$$

**Proof.** For any $\ell \in \mathbb{N}(a)$, we have $\|\mathcal{V}(\ell+1)\mathcal{P}\mathbf{c}\| > 0$. Thus, for $k_1 \in \mathbb{N}(k)$ from

$$\sum_{\ell=k}^{k_1} \|\mathcal{V}(\ell+1)\mathcal{P}\mathbf{c}\|^{-1}\mathcal{V}(k)\mathcal{P}\mathbf{c} = \sum_{\ell=k}^{k_1} \|\mathcal{V}(\ell+1)\mathcal{P}\mathbf{c}\|^{-1}\mathcal{V}(k)\mathcal{P}\mathcal{V}^{-1}(\ell+1)\mathcal{V}(\ell+1)\mathcal{P}\mathbf{c},$$

and (5.6.9), we get

$$\|\mathcal{V}(k)\mathcal{P}\mathbf{c}\| \sum_{\ell=k}^{k_1} \|\mathcal{V}(\ell+1)\mathcal{P}\mathbf{c}\|^{-1} \leq c, \qquad k_1 \in \mathbb{N}(k).$$

Therefore, $\sum_{\ell=k}^{\infty} \|\mathcal{V}(\ell+1)\mathcal{P}\mathbf{c}\|^{-1}$ exists and so $\limsup_{k\to\infty} \|\mathcal{V}(k+1)\mathcal{P}\mathbf{c}\|^{-1} = 0$, or $\limsup_{k\to\infty} \|\mathcal{V}(k+1)\mathcal{P}\mathbf{c}\| = \infty$. ∎

**Theorem 5.6.7.** Suppose that there exists a constant $c > 1$ such that for all $k \in \mathbb{N}(a)$

$$\sum_{\ell=a}^{k-1} \|\mathcal{U}(k,a)\mathcal{U}^{-1}(\ell+1,a)\| \leq c, \tag{5.6.10}$$

where $\mathcal{U}(k,a)$ is the principal fundamental matrix of (1.2.12). Further, suppose that for all $(k,\mathbf{v}) \in \mathbb{N}(a) \times \mathbb{R}^n$ the function $\mathbf{g}(k,\mathbf{v})$ satisfies the inequality (5.6.3) with $\alpha < c^{-1}$. Then, the trivial solution $\mathbf{v}(k,a,0) \equiv 0$ of (5.3.4) is asymptotically stable.

**Proof.** By Lemma 5.6.5, $\mathcal{U}(k,a) \to 0$ as $k \to \infty$ and in particular $\mathcal{U}(k,a)$ is bounded, i.e. there exists a constant $\beta > 0$ such that $\|\mathcal{U}(k,a)\| \leq \beta$. Thus, from (5.6.1) with $k_1 = a$, (5.6.3) and (5.6.10) it follows that

$$\|\mathbf{v}(k)\| \leq \beta\|\mathbf{v}(a)\| + \alpha c \sup_{a\leq\ell\leq k} \|\mathbf{v}(\ell)\|$$

and hence

$$\sup_{a\le \ell\le k} \|\mathbf{v}(\ell)\| \;\le\; \beta(1-\alpha c)^{-1}\|\mathbf{v}(a)\|.$$

Therefore, for all $k \in \mathbb{N}(a)$

$$\|\mathbf{v}(k)\| \;\le\; \beta(1-\alpha c)^{-1}\|\mathbf{v}(a)\|.$$

This inequality shows that the trivial solution of (5.3.4) is stable.

Now, let $\mu = \limsup_{k\to\infty} \|\mathbf{v}(k)\|$, and choose $\gamma$ so that $\alpha c < \gamma < 1$. If $\mu > 0$, then $\|\mathbf{v}(k)\| \le \gamma^{-1}\mu$ for $k \ge k_1 \in \mathbb{N}(a)$. Thus, from (5.6.1) we find

$$\|\mathbf{v}(k)\| \;\le\; \|\mathcal{U}(k,a)\|\|\mathcal{U}^{-1}(k_1,a)\mathbf{v}(k_1)\| + c\alpha\gamma^{-1}\mu$$

and hence as $k\to\infty$, we get $\mu \le c\alpha\gamma^{-1}\mu$, which is impossible. Therefore, $\mu = 0$. ■

**Theorem 5.6.8.** Suppose that there exists a constant $c > 1$ such that for all $k \in \mathbb{N}(a)$

$$\sum_{\ell=a}^{k-1} \|\mathcal{U}(k,a)\mathcal{P}\mathcal{U}^{-1}(\ell+1,a)\| + \sum_{\ell=k}^{\infty} \|\mathcal{U}(k,a)(\mathcal{I}-\mathcal{P})\mathcal{U}^{-1}(\ell+1,a)\| \le c, \tag{5.6.11}$$

where $\mathcal{U}(k,a)$ is the principal fundamental matrix of (1.2.12), and $\mathcal{P}$ is a projection. Further, suppose that for all $(k,\mathbf{v}) \in \mathbb{N}(a)\times\mathbb{R}^n$ the function $\mathbf{g}(k,\mathbf{v})$ satisfies the inequality (5.6.3) with $\alpha < c^{-1}$. Then, the following hold

(i) if $\mathbf{v}(k) = \mathbf{v}(k,a,\mathbf{v}^0)$ is a bounded solution of (5.3.4) such that $\|\mathbf{v}(k)\| \le \beta$ for all $k \in \mathbb{N}(a)$, then $\mathbf{v}(k)\to 0$ as $k\to\infty$

(ii) there exists a constant $\gamma > 0$, independent of $\mathbf{g}$, such that for all $k \in \mathbb{N}(a)$

$$\|\mathbf{v}(k)\| \;\le\; (1-\alpha c)^{-1}\gamma\|\mathcal{P}\mathbf{v}(a)\|. \tag{5.6.12}$$

**Proof.** Let the solution $\mathbf{v}(k)$ of (5.3.4) be bounded. Then, from (5.6.3), (5.6.11) and Lemma 5.6.5 the function

$$\begin{aligned}\mathbf{w}(k) \;=\;& \mathbf{v}(k) - \mathcal{U}(k,a)\mathcal{P}\mathbf{v}(a) - \sum_{\ell=a}^{k-1}\mathcal{U}(k,a)\mathcal{P}\mathcal{U}^{-1}(\ell+1,a)\mathbf{g}(\ell,\mathbf{v}(\ell))\\ &+ \sum_{\ell=k}^{\infty}\mathcal{U}(k,a)(\mathcal{I}-\mathcal{P})\mathcal{U}^{-1}(\ell+1,a)\mathbf{g}(\ell,\mathbf{v}(\ell))\end{aligned}$$

exists and is bounded for all $k \in \mathbb{N}(a)$. Moreover, it follows that

$$\begin{aligned}
\mathbf{w}(k+1) &= \mathcal{A}(k)\mathbf{v}(k) + \mathbf{g}(k,\mathbf{v}(k)) - \mathcal{A}(k)\mathcal{U}(k,a)\mathcal{P}\mathbf{v}(a) \\
&\quad - \mathcal{U}(k+1,a)\mathcal{P}\mathcal{U}^{-1}(k+1,a)\mathbf{g}(k,\mathbf{v}(k)) \\
&\quad - \sum_{\ell=a}^{k-1} \mathcal{U}(k+1,a)\mathcal{P}\mathcal{U}^{-1}(\ell+1,a)\mathbf{g}(\ell,\mathbf{v}(\ell)) \\
&\quad + \sum_{\ell=k}^{\infty} \mathcal{U}(k+1,a)(\mathcal{I}-\mathcal{P})\mathcal{U}^{-1}(\ell+1,a)\mathbf{g}(\ell,\mathbf{v}(\ell)) \\
&\quad - \mathcal{U}(k+1,a)(\mathcal{I}-\mathcal{P})\mathcal{U}^{-1}(k+1,a)\mathbf{g}(k,\mathbf{v}(k)) \\
&= \mathcal{A}(k)\Bigg[\mathbf{v}(k) - \mathcal{U}(k,a)\mathcal{P}\mathbf{v}(a) - \sum_{\ell=a}^{k-1} \mathcal{U}(k,a)\mathcal{P}\mathcal{U}^{-1}(\ell+1,a) \times \\
&\qquad \mathbf{g}(\ell,\mathbf{v}(\ell)) + \sum_{\ell=k}^{\infty} \mathcal{U}(k,a)(\mathcal{I}-\mathcal{P})\mathcal{U}^{-1}(\ell+1,a)\mathbf{g}(\ell,\mathbf{v}(\ell))\Bigg] \\
&= \mathcal{A}(k)\mathbf{w}(k),
\end{aligned}$$

i.e. $\mathbf{w}(k)$ is a solution of (1.2.12). Obviously, $\mathcal{P}\mathbf{w}(a) = 0$ and so $\mathbf{w}(k) = \mathcal{U}(k,a)(\mathcal{I}-\mathcal{P})\mathbf{w}(a)$. But, by Lemma 5.6.6 this is possible only if $(\mathcal{I}-\mathcal{P})\mathbf{w}(a) = 0$, i.e. $\mathbf{w}(k) \equiv 0$. Therefore,

$$\begin{aligned}
\mathbf{v}(k) &= \mathcal{U}(k,a)\mathcal{P}\mathbf{v}(a) + \sum_{\ell=a}^{k-1} \mathcal{U}(k,a)\mathcal{P}\mathcal{U}^{-1}(\ell+1,a)\mathbf{g}(\ell,\mathbf{v}(\ell)) \\
&\quad - \sum_{\ell=k}^{\infty} \mathcal{U}(k,a)(\mathcal{I}-\mathcal{P})\mathcal{U}^{-1}(\ell+1,a)\mathbf{g}(\ell,\mathbf{v}(\ell)).
\end{aligned}$$

Since by Lemma 5.6.5 condition (5.6.11) implies that $\|\mathcal{U}(k,a)\mathcal{P}\| \to 0$, there exists a positive constant $\gamma$ such that $\|\mathcal{U}(k,a)\mathcal{P}\| \le \gamma$ for all $k \in \mathbb{N}(a)$. Thus, from (5.6.3) and (5.6.11) it follows that

$$\|\mathbf{v}(k)\| \le \gamma\|\mathcal{P}\mathbf{v}(a)\| + \alpha c \sup_{a \le \ell \le k} \|\mathbf{v}(\ell)\|,$$

and hence (5.6.12) holds. The proof of part (i) is the same as in Theorem 5.6.7. ∎

**Corollary 5.6.9.** If $\mathcal{P} \ne \mathcal{I}$ in (5.6.11), then the trivial solution $\mathbf{v}(k,a,0) \equiv 0$ of (5.3.4) is unstable.

**Proof.** If the trivial solution of (5.3.4) is stable, then for each $\epsilon > 0$ we can find $\delta > 0$ such that $0 < \|\mathbf{v}^0\| < \delta$ implies $\|\mathbf{v}(k,a,\mathbf{v}^0)\| < \epsilon$ for all

$k \in \mathbb{N}(a)$. However, since $\mathcal{P} \neq \mathcal{I}$ we can choose $\mathbf{v}^0$ so that $\mathcal{P}\mathbf{v}^0 = 0$. But, then (5.6.12) gives a contradiction. ∎

**Corollary 5.6.10.** If $\mathcal{A}(k) = \mathcal{A}$ ($\mathcal{A}(k)$ is periodic of period $K$) and no eigenvalue of $\mathcal{A}$ ($\mathcal{U}(K,a)$) lie on the unit circle, and (5.6.3) holds, then the conclusions (i) and (ii) of Theorem 5.6.8 are true.

**Proof.** If $\mathcal{A}(k) = \mathcal{A}$, then $\mathcal{U}(k,a) = \mathcal{A}^{k-a}$ and there exists a projection $\mathcal{P}$ such that

$$\|\mathcal{A}^{k-a}\mathcal{P}\mathcal{A}^{-(\ell+1-a)}\| \leq c\rho^{k-\ell-1} \quad \text{if} \quad k \in \mathbb{N}(\ell+1)$$

and

$$\|\mathcal{A}^{k-a}(\mathcal{I}-\mathcal{P})\mathcal{A}^{-(\ell+1-a)}\| \leq c\rho^{\ell+1-k} \quad \text{if} \quad \ell \in \mathbb{N}(k-1)$$

where $c > 0$ and $0 < \rho < 1$.

If $\mathcal{A}(k)$ is periodic of period $K$, then there exists a projection $\mathcal{P}$ such that

$$\|\mathcal{U}(k,a)\mathcal{P}\mathcal{U}^{-1}(\ell+1,a)\| \leq c_1\rho_1^{k-\ell-1} \quad \text{if} \quad k \in \mathbb{N}(\ell+1)$$

and

$$\|U(k,a)(\mathcal{I}-\mathcal{P})\mathcal{U}^{-1}(\ell+1,a)\| \leq c_1\rho_1^{\ell+1-k} \quad \text{if} \quad \ell \in \mathbb{N}(k-1)$$

where $c_1 > 0$ and $0 < \rho_1 < 1$. ∎

## 5.7. Nonlinear Variation of Constants

In the previous sections the variation of constants formula developed in Section 2.5 for linear difference systems has been repeatedly used to study asymptotic and stability properties of solutions of the perturbed difference systems. The main purpose of this section is to use the same technique to represent the solution $\mathbf{v}(k,a,\mathbf{u}^0)$ of the perturbed system

$$\text{(5.7.1)} \qquad \mathbf{v}(k+1) = \mathbf{f}(k,\mathbf{v}(k)) + \mathbf{g}(k,\mathbf{v}(k)), \quad k \in \mathbb{N}(a)$$

in terms of the solution $\mathbf{u}(k,a,\mathbf{u}^0)$ of the unperturbed system (1.2.8).

**Theorem 5.7.1.** Let for all $k \in \mathbb{N}(a)$, $\mathbf{u} \in \mathbb{R}^n$ the functions $\mathbf{f}(k,\mathbf{u})$ and $\mathbf{g}(k,\mathbf{u})$ be defined, $\partial\mathbf{f}/\partial\mathbf{u}$ exist and continuous and invertible. If for each $\mathbf{u}^0 \in \mathbb{R}^n$ the solution $\mathbf{u}(k,a,\mathbf{u}^0)$ of (1.2.8) exists on $\mathbb{N}(a)$, then any solution $\mathbf{v}(k) = \mathbf{v}(k,a,\mathbf{u}^0)$ of (5.7.1) satisfies the equation

$$\text{(5.7.2)}\ \mathbf{v}(k,a,\mathbf{u}^0) = \mathbf{u}\left(k,a,\mathbf{u}^0 + \sum_{\ell=a}^{k-1} \mathcal{W}^{-1}(\ell+1,a,\mathbf{w}(\ell),\mathbf{w}(\ell+1))\mathbf{g}(\ell,\mathbf{v}(\ell))\right)$$

where

$$(5.7.3)\quad \mathcal{W}(k,a,\mathbf{w}(\ell),\mathbf{w}(\ell+1)) \;=\; \int_0^1 \mathcal{V}(k,a,s\mathbf{w}(\ell+1)+(1-s)\mathbf{w}(\ell))ds,$$

$\mathcal{V}(k,a,\mathbf{u}^0) = \partial\mathbf{u}(k,a,\mathbf{u}^0)/\partial\mathbf{u}^0$ is defined in Theorem 5.1.2, and $\mathbf{w}(k)$ satisfies the implicit equation

$$(5.7.4)\quad \mathbf{w}(k) \;=\; \mathbf{u}^0 + \sum_{\ell=a}^{k-1} \mathcal{W}^{-1}(\ell+1,a,\mathbf{w}(\ell),\mathbf{w}(\ell+1))\mathbf{g}(\ell,\mathbf{u}(\ell,a,\mathbf{w}(\ell))).$$

**Proof.** The method of variation of constants requires determining a function $\mathbf{w}(k)$ so that $\mathbf{v}(k,a,\mathbf{u}^0) = \mathbf{u}(k,a,\mathbf{w}(k))$, $\mathbf{w}(a) = \mathbf{u}^0$. Therefore,

$$\begin{aligned}\mathbf{v}(k+1,a,\mathbf{u}^0) &= \mathbf{u}(k+1,a,\mathbf{w}(k+1)) - \mathbf{u}(k+1,a,\mathbf{w}(k)) + \mathbf{u}(k+1,a,\mathbf{w}(k))\\ &= \mathbf{f}(k,\mathbf{u}(k,a,\mathbf{w}(k))) + \mathbf{g}(k,\mathbf{u}(k,a,\mathbf{w}(k)))\end{aligned}$$

from which, we get

$$\mathbf{u}(k+1,a,\mathbf{w}(k+1)) - \mathbf{u}(k+1,a,\mathbf{w}(k)) \;=\; \mathbf{g}(k,\mathbf{u}(k,a,\mathbf{w}(k))).$$

Thus, the mean value theorem gives

$$\begin{aligned}\int_0^1 \frac{\partial\mathbf{u}(k+1,a,s\mathbf{w}(k+1)+(1-s)\mathbf{w}(k))}{\partial\mathbf{u}^0}ds(\mathbf{w}(k+1)-\mathbf{w}(k))&\\ = \mathbf{g}(k,\mathbf{u}(k,a,\mathbf{w}(k))),&\end{aligned}$$

which is the same as

$$\begin{aligned}\int_0^1 \mathcal{V}(k+1,a,s\mathbf{w}(k+1)+(1-s)\mathbf{w}(k))ds(\mathbf{w}(k+1)-\mathbf{w}(k))&\\ = \mathbf{g}(k,\mathbf{u}(k,a,\mathbf{w}(k)))&\end{aligned}$$

and hence

$$\mathcal{W}(k+1,a,\mathbf{w}(k),\mathbf{w}(k+1))(\mathbf{w}(k+1)-\mathbf{w}(k)) \;=\; \mathbf{g}(k,\mathbf{u}(k,a,\mathbf{w}(k))).$$

The above equation is equivalent to (5.7.4) from which (5.7.2) immediately follows. ∎

**Theorem 5.7.2.** Let the assumptions of Theorem 5.7.1 be satisfied. Then,

$$\begin{aligned}(5.7.5)\quad \mathbf{v}(k,a,\mathbf{u}^0) \;=\; &\mathbf{u}(k,a,\mathbf{u}^0) + \mathcal{W}(k,a,\mathbf{w}(k),\mathbf{u}^0)\,\times\\ &\sum_{\ell=a}^{k-1} \mathcal{W}^{-1}(\ell+1,a,\mathbf{w}(\ell),\mathbf{w}(\ell+1))\mathbf{g}(\ell,\mathbf{v}(\ell)).\end{aligned}$$

**Proof.** Since

$$\frac{d\mathbf{u}}{ds}(k,a,s\mathbf{w}(k)+(1-s)\mathbf{u}^0) = \mathcal{V}(k,a,s\mathbf{w}(k)+(1-s)\mathbf{u}^0)(\mathbf{w}(k)-\mathbf{u}^0)$$

integration from $s=0$ to $1$ yields

$$\mathbf{u}(k,a,\mathbf{w}(k)) = \mathbf{u}(k,a,\mathbf{u}^0)+\int_0^1 \mathcal{V}(k,a,s\mathbf{w}(k)+(1-s)\mathbf{u}^0)ds(\mathbf{w}(k)-\mathbf{u}^0),$$

which from (5.7.3) and (5.7.4) is the same as (5.7.5). ∎

**Corollary 5.7.3.** If $\mathbf{f}(k,\mathbf{u}) = \mathcal{A}(k)\mathbf{u}$, then (5.7.5) reduces to (5.3.5) with $a=0$.

**Proof.** Since in this case $\mathbf{u}(k,0,\mathbf{u}^0) = \mathcal{U}(k,0)\mathbf{u}^0$, where $\mathcal{U}(k,0)$ is the principal fundamental matrix of (1.2.12), it follows that $\mathcal{V}(k,0,\mathbf{u}^0) = \mathcal{U}(k,0) = \mathcal{W}(k,0,\mathbf{w}(\ell),\mathbf{w}(\ell+1))$. From this the result is immediate. ∎

**Example 5.7.1.** For the difference equation $u(k+1) = \dfrac{u(k)}{1+u(k)}$ we have $u(k,a,u^0) = \dfrac{u^0}{1+u^0(k-a)}$. Therefore, $\mathcal{V}(k,a,u^0) = \dfrac{1}{(1+u^0(k-a))^2}$, and

$$W(k,a,w(\ell),w(\ell+1)) = \frac{1}{1+w(\ell)(k-a))(1+w(\ell+1)(k-a))}$$

and $W^{-1}(k,a,w(\ell),w(\ell+1)) = (1+w(\ell)(k-a))(1+w(\ell+1)(k-a))$.

**Theorem 5.7.4.** Let the following conditions be satisfied

(i) for all $k\in\mathbb{N}(a)$ the solution $\mathbf{v}(k,a,\mathbf{u}^0)$ of (5.7.1) admits a representation (5.7.2)

(ii) for each $\mathbf{u}^0\in\mathbb{R}^n$ there exist positive constants $\epsilon$ and $\lambda$ such that $\|\mathbf{u}(k,a,\mathbf{u}^0)\| \le \epsilon\|\mathbf{u}^0\|\exp(-\lambda(k-a))$

(iii) for all $k\in\mathbb{N}(a)$

$$(5.7.6)\quad \|\mathcal{W}^{-1}(k+1,a,\mathbf{w}(k),\mathbf{w}(k+1))\mathbf{g}(k,\mathbf{u}(k,a,\mathbf{w}(k)))\| \le p(k)\|\mathbf{w}(k)\|+q(k)\|\mathbf{w}(k+1)\|,$$

where $p(k)$ and $q(k)$ are nonnegative functions, $q(k)<1$, and $\sum_{\ell=a}^{k-1}\dfrac{p(\ell)+q(\ell)}{1-q(\ell)} \le \mu(k-a)$ with $\mu<\lambda$.

Then, the following inequality holds

$$\|\mathbf{v}(k,a,\mathbf{u}^0)\| \le \epsilon\|\mathbf{u}^0\|\exp(-(\lambda-\mu)(k-a)),\quad k\in\mathbb{N}(a).$$

**Proof.** From (5.7.4) and (5.7.6), we find

$$\|\mathbf{w}(k)\| \le \|\mathbf{u}^0\| + \sum_{\ell=a}^{k-1}(p(\ell)\|\mathbf{w}(\ell)\| + q(\ell)\|\mathbf{w}(\ell+1)\|),$$

which easily determines

$$\begin{aligned} \|\mathbf{w}(k)\| &\le \|\mathbf{u}^0\| \prod_{\ell=a}^{k-1} \frac{1+p(\ell)}{1-q(\ell)} \le \|\mathbf{u}^0\| \exp\left(\sum_{\ell=a}^{k-a} \frac{p(\ell)+q(\ell)}{1-q(\ell)}\right) \\ &\le \|\mathbf{u}^0\| \exp(\mu(k-a)). \end{aligned}$$

Therefore, from (5.7.2) it follows that

$$\begin{aligned} \|\mathbf{v}(k,a,\mathbf{u}^0)\| &= \|\mathbf{u}(k,a,\mathbf{w}(k))\| \le \epsilon\|\mathbf{w}(k)\| \exp(-\lambda(k-a)) \\ &\le \epsilon\|\mathbf{u}^0\| \exp(-(\lambda-\mu)(k-a)), \quad k \in \mathbb{N}(a). \end{aligned}$$ ∎

## 5.8. Dichotomies

Let $\mathcal{U}(k,a)$ be the principal fundamental matrix of the difference system (1.2.12). The system (1.2.12) is said to possess an *exponential dichotomy* if there exists a projection $\mathcal{P}$, and positive constants $\eta$, $\nu$, $\alpha$ and $\beta$ such that for all $k$, $\ell \in \mathbb{N}(a)$

$$(5.8.1) \quad \begin{aligned} &\|\mathcal{U}(k,a)\mathcal{P}\mathcal{U}^{-1}(\ell,a)\| \le \eta\rho_1^{k-\ell} \ (0 < \rho_1 = e^{-\alpha} < 1), \ k \in \mathbb{N}(\ell) \\ &\|\mathcal{U}(k,a)(\mathcal{I}-\mathcal{P})\mathcal{U}^{-1}(\ell,a)\| \le \nu\rho_2^{\ell-k} (0 < \rho_2 = e^{-\beta} < 1), \ \ell \in \mathbb{N}(k). \end{aligned}$$

It is said to possess an *ordinary dichotomy* if the inequalities (5.8.1) hold with $\alpha = \beta = 0$, i.e. $\rho_1 = \rho_2 = 1$. Thus, for the ordinary dichotomy

$$(5.8.2) \quad \begin{aligned} \|U(k,a)\mathcal{P}\mathcal{U}^{-1}(\ell,a)\| &\le \eta, \quad k \in \mathbb{N}(\ell) \\ \|\mathcal{U}(k,a)(\mathcal{I}-\mathcal{P})\mathcal{U}^{-1}(\ell,a)\| &\le \nu, \quad \ell \in \mathbb{N}(k). \end{aligned}$$

In particular the system (2.8.1) has an exponential dichotomy if and only if no eiegnvalues of $\mathcal{A}$ lie on the unit circle. Further, it has an ordinary dichotomy if and only if all eigenvalues of $\mathcal{A}$ which lie on the unit circle are semisimple.

From Theorem 5.5.1 (v) the system (1.2.12) has an exponential dichotomy with $\mathcal{P} = \mathcal{I}$ if and only if it is uniformly asymptotically stable, and from Theorem 5.5.1 (ii) an ordinary dichotomy with $\mathcal{P} = \mathcal{I}$ if and only if it is uniformly stable. In the general case it is convenient to write the inequalities (5.8.1) in the equivalent form

$$
\begin{aligned}
&\|\mathcal{U}(k,a)\mathcal{P}\mathbf{c}\| \;\le\; \eta_1\rho_1^{k-\ell}\|\mathcal{U}(\ell,a)\mathcal{P}\mathbf{c}\|, \quad k\in\mathbb{N}(\ell)\\
(5.8.3)\qquad &\|\mathcal{U}(k,a)(\mathcal{I}-\mathcal{P})\mathbf{c}\| \;\le\; \nu_1\rho_2^{\ell-k}\|\mathcal{U}(\ell,a)(\mathcal{I}-\mathcal{P})\mathbf{c}\|, \quad \ell\in\mathbb{N}(k)\\
&\|\mathcal{U}(k,a)\mathcal{P}\mathcal{U}^{-1}(k,a)\| \;\le\; \xi_1, \quad k\in\mathbb{N}(a)
\end{aligned}
$$

where $\eta_1$, $\nu_1$ and $\xi_1$ are positive constants and $\mathbf{c}$ is an arbitrary vector. Suppose the projection $\mathcal{P}$ has rank $r$, then the first inequality of (5.8.3) says that there is a $r$ dimensional subspace of solutions tending to zero uniformly and exponentially as $k\to\infty$. The second inequality of (5.8.3) says that there is a supplementary $(n-r)$ dimensional subspace of solutions tending to infinity uniformly and exponentially as $k\to\infty$. The third inequality of (5.8.3) says that the angle between two subspaces remains bounded away from zero.

**Remark 5.8.1.** If $\rho_1<1$ or $\rho_2<1$ and $\|\mathcal{A}(k)\|\le M$ for all $k\in\mathbb{N}(a)$, then the third inequality (5.8.3) is implied by the previous two inequalities. For this, it is clear that for all $k\in\mathbb{N}(a)$ and $\tau\in\mathbb{N}$

$$
(5.8.4)\qquad \mathcal{U}(k+\tau,a) \;=\; \prod_{i=1}^{\tau}\mathcal{A}(k+\tau-i)\mathcal{U}(k,a).
$$

Further for any positive integer $\tau$, we have

$$
\begin{aligned}
\|\mathcal{U}(k+\tau,a)\mathcal{P}\mathcal{U}^{-1}(k,a)\| &\;\le\; \eta_1\rho_1^{\tau}\theta\\
\|U(k+\tau,a)(\mathcal{I}-\mathcal{P})\mathcal{U}^{-1}(k,a)\| &\;\ge\; \nu_1^{-1}\rho_2^{-\tau}\phi,
\end{aligned}
$$

where $\theta=\|\mathcal{U}(k,a)\mathcal{P}\mathcal{U}^{-1}(k,a)\|$ and $\phi=\|\mathcal{U}(k,a)(\mathcal{I}-\mathcal{P})\mathcal{U}^{-1}(k,a)\|$. Therefore, it follows that

$$
\|\theta^{-1}\mathcal{U}(k+\tau,a)\mathcal{P}\mathcal{U}^{-1}(k,a)+\phi^{-1}\mathcal{U}(k+\tau,a)(\mathcal{I}-\mathcal{P})\mathcal{U}^{-1}(k,a)\| \;\ge\; \psi,
$$

where $\psi=\nu_1^{-1}\rho_2^{-\tau}-\eta_1\rho_1^{\tau}$. We can always choose $\tau>0$ so large that $\psi>0$. Thus, (5.8.4) gives

$$
\|\theta^{-1}\mathcal{U}(k,a)\mathcal{P}\mathcal{U}^{-1}(k,a)+\phi^{-1}\mathcal{U}(k,a)(\mathcal{I}-\mathcal{P})\mathcal{U}^{-1}(k,a)\| \;\ge\; \psi M^{-\tau}.
$$

The left side of this inequality can be written in the form

$$
\begin{aligned}
\|\phi^{-1}\mathcal{I}+(\theta^{-1}-\phi^{-1})\mathcal{U}(k,a)\mathcal{P}\mathcal{U}^{-1}(k,a)\| &\;\le\; \phi^{-1}+|\theta^{-1}-\phi^{-1}|\theta\\
&\;=\; \phi^{-1}(1+|\phi-\theta|) \;\le\; 2\phi^{-1},
\end{aligned}
$$

where we have used the fact that $|\phi-\theta|\le 1$. Hence, $\phi\le 2\psi^{-1}M^{\tau}$ and, by symmetry, also $\theta\le 2\psi^{-1}M^{\tau}$. Thus, in the third inequality (5.8.3), $\xi_1=2\psi^{-1}M^{\tau}$, where $\tau$ is so large that $\psi=\nu_1^{-1}\rho_2^{-\tau}-\eta_1\rho_1^{\tau}>0$.

**Remark 5.8.2.** If $\rho_1 = \rho_2 = 1$, then $\|\mathcal{A}(k)\| \le M$ for all $k \in \mathbb{N}(a)$ does not imply the third inequality (5.8.3). For this, consider the system $\mathbf{u}(k+1) = \begin{bmatrix} 1 & 1 \\ 0 & 1 \end{bmatrix} \mathbf{u}(k)$ for which $\mathcal{U}(k,0) = \begin{bmatrix} 1 & k \\ 0 & 1 \end{bmatrix}$. If $\mathcal{P} = \begin{bmatrix} 1 & 0 \\ 0 & 0 \end{bmatrix}$ then the first two inequalities (5.8.3) are satisfied with $\rho_1 = \rho_2 = 1$, but the third inequality (5.8.3) is not satisfied.

**Remark 5.8.3.** If $\rho_1 < 1$ and $\rho_2 < 1$ but $\|\mathcal{A}(k)\|$ is not bounded, then the third inequality (5.8.3) need not be satisfied. For this, consider the system $\mathbf{u}(k+1) = \begin{bmatrix} e^{-1} & \frac{1}{4}e^{2k}(e^3 - e^{-1}) \\ 0 & e \end{bmatrix} \mathbf{u}(k)$ for which $\mathcal{U}(k,0) = \begin{bmatrix} e^{-k} & \frac{1}{4}(e^{3k} - e^{-k}) \\ 0 & e^k \end{bmatrix}$. If $\mathcal{P} = \begin{bmatrix} 1 & 0 \\ 0 & 0 \end{bmatrix}$ then the first two inequalities (5.8.3) are satisfied with $\rho_1 = \rho_2 = e^{-1}$, but the third inequality (5.8.3) is not satisfied.

Now we shall provide necessary and sufficient conditions for the exponential dichotomy of the difference system (1.2.12). For simplicity we shall assume that $\|\mathcal{A}(k)\| \le M \ (\ge 1)$ for all $k \in \mathbb{N}(a)$, and $\rho_1 = \rho_2 = \rho < 1$, $\eta_1 = \nu_1 = \zeta$ so that from Remark 5.8.1 the system (1.2.12) has exponential dichotomy if and only if

$$(5.8.5) \qquad \begin{aligned} \|\mathcal{U}(k,a)\mathcal{P}\| &\le \zeta\rho^{k-\ell}\|\mathcal{U}(\ell,a)\mathcal{P}\|, \quad k \in \mathbb{N}(\ell) \\ \|\mathcal{U}(k,a)(\mathcal{I}-\mathcal{P})\| &\le \zeta\rho^{\ell-k}\|\mathcal{U}(\ell,a)(\mathcal{I}-\mathcal{P})\|, \quad \ell \in \mathbb{N}(k). \end{aligned}$$

First we shall prove a lemma which is needed in the main results.

**Lemma 5.8.1.** Suppose that the difference system (1.2.12) has exponential dichotomy for $k \ge K \in \mathbb{N}(a)$. Then, it has exponential dichotomy on $\mathbb{N}(a)$.

**Proof.** Clearly, $\|\mathcal{U}(k,a)(\mathcal{I}-\mathcal{P})\| \le M^{k-a}\|(\mathcal{I}-\mathcal{P})\|$, and from the second inequality (5.8.5), we have $\|\mathcal{U}(K,a)(\mathcal{I}-\mathcal{P})\| \le \zeta\rho^{\ell-K}\|\mathcal{U}(\ell,a)(\mathcal{I}-\mathcal{P})\|$ for all $\ell \ge K$. Let $a \le k \le K \le \ell$. Then, from these inequalities it follows that

$$\begin{aligned} \|\mathcal{U}(k,a)(\mathcal{I}-\mathcal{P})\| &\le \frac{M^{k-a}\|(\mathcal{I}-\mathcal{P})\|}{\|\mathcal{U}(K,a)(\mathcal{I}-\mathcal{P})\|}\|\mathcal{U}(K,a)(\mathcal{I}-\mathcal{P})\| \\ &\le \zeta\frac{(M\rho^{-1})^K\|(\mathcal{I}-\mathcal{P})\|}{\|\mathcal{U}(K,a)(\mathcal{I}-\mathcal{P})\|}\rho^{\ell-k}\|\mathcal{U}(\ell,a)(\mathcal{I}-\mathcal{P})\|. \end{aligned}$$

Let $a \le k \le \ell \le K$ and $\zeta_1 = \min\{\|\mathcal{U}(k,a)(\mathcal{I}-\mathcal{P})\| : a \le k \le K\}$. Since

$\|\mathcal{U}(k,a)(\mathcal{I}-\mathcal{P})\| \neq 0$ for every $k \in \mathbb{N}(a)$, $\zeta_1 \neq 0$. Therefore, it follows that

$$\|\mathcal{U}(k,a)(\mathcal{I}-\mathcal{P})\| \le M^K\|(\mathcal{I}-\mathcal{P})\| \le \zeta_1^{-1}M^K\|(\mathcal{I}-\mathcal{P})\|\|\mathcal{U}(\ell,a)(\mathcal{I}-\mathcal{P})\|.$$

Since $a \le k \le \ell \le K$ we have $1 \le \rho^{-K+\ell-k}$ and hence

$$\|\mathcal{U}(k,a)(\mathcal{I}-\mathcal{P})\| \le \zeta_1^{-1}(M\rho^{-1})^K\|(\mathcal{I}-\mathcal{P})\|\rho^{\ell-k}\|\mathcal{U}(\ell,a)(\mathcal{I}-\mathcal{P})\|.$$

The proof of the first inequality (5.8.5) is similar. ∎

**Theorem 5.8.2.** Suppose that the difference system (1.2.12) has exponential dichotomy on $\mathbb{N}(a)$. Then, there exist constants $0 < \theta < 1$, $K \in \mathbb{N}(1)$, such that every solution $\mathbf{u}(k)$ of (1.2.12) satisfies

$$\|\mathbf{u}(k)\| \le \theta \sup\{\|\mathbf{u}(\tau)\| : |\tau - k| \le K,\ \tau,\ k \in \mathbb{N}(a),\ k \ge K\}. \tag{5.8.6}$$

**Proof.** Let $\mathbf{u}^1(k) = \mathcal{U}(k,a)\mathcal{P}\mathbf{c}$, $\mathbf{u}^2(k) = \mathcal{U}(k,a)(\mathcal{I}-\mathcal{P})\mathbf{c}$. Then, $\mathbf{u}(k) = \mathbf{u}^1(k) + \mathbf{u}^2(k)$. First consider the case $\|\mathbf{u}^2(\ell)\| \ge \|\mathbf{u}^1(\ell)\|$, for some $\ell \in \mathbb{N}(a)$. For $k \ge \ell \ge a$, inequalities (5.8.5) give that

$$\|\mathbf{u}^2(\ell)\| \le \zeta\rho^{k-\ell}\|\mathbf{u}^2(k)\| \quad \text{or} \quad \|\mathbf{u}^2(k)\| \ge \zeta^{-1}\rho^{-(k-\ell)}\|\mathbf{u}^2(\ell)\|$$

$$\|\mathbf{u}^1(k)\| \le \zeta\rho^{k-\ell}\|\mathbf{u}^1(\ell)\| \quad \text{or} \quad -\|\mathbf{u}^1(k)\| \ge -\zeta\rho^{k-\ell}\|\mathbf{u}^1(\ell)\|.$$

Therefore, it follows that

$$\|\mathbf{u}(k)\| \ge \|\mathbf{u}^2(k)\| - \|\mathbf{u}^1(k)\| \ge \left(\zeta^{-1}\rho^{-(k-\ell)} - \zeta\rho^{k-\ell}\right)\|\mathbf{u}^2(\ell)\|$$

and hence

$$\|\mathbf{u}(k)\| \ge \frac{1}{2}\left(\zeta^{-1}\rho^{-(k-\ell)} - \zeta\rho^{k-\ell}\right)\|\mathbf{u}(\ell)\|, \quad k \ge \ell \ge a. \tag{5.8.7}$$

Now consider the case $\|\mathbf{u}^2(\ell)\| \le \|\mathbf{u}^1(\ell)\|$, for some $\ell \in \mathbb{N}(a)$. Similarly, we get

$$\|\mathbf{u}(k)\| \ge \frac{1}{2}\left(\zeta^{-1}\rho^{-(\ell-k)} - \zeta\rho^{\ell-k}\right)\|\mathbf{u}(\ell)\|, \quad \ell \ge k \ge a. \tag{5.8.8}$$

We choose $K \in \mathbb{N}(1)$ sufficiently large and $0 < \theta < 1$ so that $\theta^{-1} = (1/2)(\zeta^{-1}\rho^{-K} - \zeta\rho^K)$. Then, from (5.8.7) and (5.8.8) we obtain

$$\|\mathbf{u}(\ell)\| \le \theta \sup\{\|\mathbf{u}(k)\| : |\ell - k| \le K,\ \ell,\ k \in \mathbb{N}(a),\ \ell \ge K\}. \quad ∎$$

**Theorem 5.8.3.** Suppose that there exist constants $0 < \theta < 1$ and $K \in \mathbb{N}(1)$, such that for every solution $\mathbf{u}(k)$ of (1.2.12) condition (5.8.6) is satisfied. Then, the difference system (1.2.12) has an exponential dichotomy.

**Proof.** Let $\mathbf{V}$ be the set of vectors $\mathbf{u}^0 \in \mathbb{R}^n$ for which the solution $\mathbf{u}(k)$ of (1.2.12) satisfying $\mathbf{u}(a) = \mathbf{u}^0$ is bounded. Obviously, $\mathbf{V}$ is a linear space. Let $\mathbf{u}(k)$ be a solution of (1.2.12) with $\mathbf{u}(a) \in \mathbf{V}$. Since a contradiction with (5.8.6) results from $\limsup_{k\to\infty} \|\mathbf{u}(k)\| > 0$, we have $\lim_{k\to\infty} \|\mathbf{u}(k)\| = 0$. For any $\ell \in \mathbb{N}(a)$ we conclude again by (5.8.6) that $\max\{\|\mathbf{u}(\ell)\|, \cdots, \|\mathbf{u}(\ell + K - 1)\|\} = \max\{\|\mathbf{u}(k)\| : k = \ell, \ell+1, \cdots\}$,

$$\|\mathbf{u}(k)\| \ \le\ \theta \max\{\|\mathbf{u}(\ell)\|, \cdots, \|\mathbf{u}(\ell + K - 1)\|\}$$

for $k = \ell + K, \ell + K + 1, \cdots$ and by induction

$$\|\mathbf{u}(k)\| \ \le\ \theta^m \max\{\|\mathbf{u}(\ell)\|, \cdots, \|\mathbf{u}(\ell + K - 1)\|\} \tag{5.8.9}$$

for $k = \ell + mK,\ \ell + mK + 1, \cdots,\ m = 1, 2, \cdots$. Since $\|\mathcal{A}(k)\| \le M(\ge 1)$ from (5.8.9) it follows that

$$\|\mathbf{u}(k)\| \ \le\ \zeta\rho^{k-\ell}\|\mathbf{u}(\ell)\| \ \text{ for } k \ge \ell \ge a \ \text{ with } \rho = \theta^{1/K} \ \text{ and } \zeta = M^{K-1}\theta^{-1}. \tag{5.8.10}$$

Let $\mathbf{u}(k)$ be a solution of (1.2.12) with $\mathbf{u}(a) \in \mathbb{R}^n \backslash \mathbf{V}$. Since $\mathbf{u}(k)$ is unbounded, there exists such an $s(\mathbf{u}(a)) \in \mathbb{N}(a)$ (we shall write $s$ instead of $s(\mathbf{u}(a))$) that $\|\mathbf{u}(s)\| \ge M^{K-a}\|\mathbf{u}(a)\|$, $\|\mathbf{u}(k)\| < M^{K-a}\|\mathbf{u}(a)\|$ for $k = a, a+1, \cdots, s-1$. Since $\|\mathcal{A}(k)\| \le M(\ge 1)$, $s \ge K$ and hence (5.8.6) implies that a sequence of integers $\{k_i\} \subseteq \mathbb{N}(a)$ exists such that $k_1 = s$, $k_i < k_{i+1} \le k_i + K$, $\|\mathbf{u}(k_{i+1})\| \ge \theta^{-1}\|\mathbf{u}(k_i)\|$, $\|\mathbf{u}(k)\| < \theta^{-1}\|\mathbf{u}(k_i)\|$ for $k_i \le k < k_{i+1}$, $i = 1, 2, \cdots$. Let $s \le k < \ell$, and find $i,\ j \in \mathbb{N}$ so that $k_i \le k < k_{i+1}$, $k_{j-1} < \ell \le k_j$. Then, we have

$$\|\mathbf{u}(\ell)\| \ge M^{-K+1}\|\mathbf{u}(k_j)\| \ge M^{-K+1}\theta^{-j+i}\|\mathbf{u}(t_i)\| \ge M^{-2(K-1)}\theta^{-j+i}\|\mathbf{u}(\ell)\|.$$

Since $(j - i)K \ge \ell - k$, it follows that

$$\|\mathbf{u}(k)\| \ \le\ \zeta\rho^{\ell-k}\|\mathbf{u}(\ell)\| \quad \text{with} \quad \rho = \theta^{1/K} \quad \text{and} \quad \zeta = M^{2(K-1)}. \tag{5.8.11}$$

Let $\mathbf{W}$ be a complementary space to $\mathbf{V}$, (i.e. $\mathbb{R}^n = \mathbf{V} + \mathbf{W}$). Put $S = \sup\{s(\mathbf{w}) : \mathbf{w} \in \mathbf{W}\backslash\{0\}\}$. Since $S = \sup\{s(\mathbf{w}) : \mathbf{w} \in \mathbf{W},\ \|\mathbf{w}\| = 1\}$, we obtain by a compactness argument that $S < \infty$.

Let $\mathcal{P}$ be the projection on $\mathbf{V}$ along $\mathbf{W}$. Then from (5.8.10) and (5.8.11) the dichotomy (5.8.5) holds for $k,\ \ell \ge S$ with $\rho = \theta^{1/K}$ and $\zeta = M^{2(K-1)}\theta^{-1}$. Now Lemma 5.8.1 completes the proof. ∎

As an application of ordinary dichotomy of (1.2.12) on $\mathbb{N}(a)$ we shall prove the following result.

**Theorem 5.8.4.** Suppose that the following conditions are satisfied

(i) the system (1.2.12) has the ordinary dichotomy (5.8.2) with $\eta = \nu$ on $\mathbb{N}(a)$

(ii) for all $(k, \mathbf{u})$, $(k, \mathbf{v}) \in \mathbb{N}(a) \times \mathbb{R}^n$ the inequality (5.1.2) holds, where $\sum_{\ell=a}^{\infty} \|\mathbf{f}(\ell, 0)\| < \infty$, $\sum_{\ell=a}^{\infty} \lambda(\ell) < \infty$.

Then, there exists a homeomorphism between the bounded solutions $\mathbf{v}(k)$ of the difference system (1.2.11) on $\mathbb{N}(a)$ and the bounded solutions $\mathbf{w}(k)$ of the difference system

$$\mathbf{w}(k+1) = \mathcal{A}(k)\mathbf{w}(k) + \mathbf{b}(k) + \mathbf{f}(k, \mathbf{w}(k)), \quad k \in \mathbb{N}(a). \tag{5.8.12}$$

Moreover, the difference between the corresponding solutions of (1.2.11) and (5.8.12) tends to zero as $k \to \infty$ if $\mathcal{U}(k, a)\mathcal{P} \to 0$ as $k \to \infty$.

**Proof.** Let $k_1 \in \mathbb{N}(a)$ be so large that $\theta = \eta \sum_{\ell=k_1}^{\infty} \lambda(\ell) < 1$. Let $B(\mathbb{N}(k_1))$ be the space of all real $n$ vector functions defined and bounded on $\mathbb{N}(k_1)$. On this space we define an operator $\mathcal{T}$ as follows

$$\begin{aligned} \mathcal{T}\mathbf{w}(k) \; = \; & \mathbf{v}(k) + \sum_{\ell=k_1+1}^{k} \mathcal{U}(k, a)\mathcal{P}\mathcal{U}^{-1}(\ell, a)\mathbf{f}(\ell - 1, \mathbf{w}(\ell - 1)) \\ & - \sum_{\ell=k+1}^{\infty} \mathcal{U}(k, a)(\mathcal{I} - \mathcal{P})\mathcal{U}^{-1}(\ell, a)\mathbf{f}(\ell - 1, \mathbf{w}(\ell - 1)). \end{aligned} \tag{5.8.13}$$

The infinite sum is obviously convergent, and since

$$\|\mathcal{T}\mathbf{w}(k)\| \le \|\mathbf{v}(k)\| + \eta \sum_{\ell=k_1}^{\infty} [\lambda(\ell)\|\mathbf{w}(\ell)\| + \|\mathbf{f}(\ell, 0)\|]$$

$\mathcal{T}$ maps $B(\mathbb{N}(k_1))$ into itself. Moreover, for all $\mathbf{w}^1(k)$, $\mathbf{w}^2(k) \in B(\mathbb{N}(k_1))$ it follows that

$$\|\mathcal{T}\mathbf{w}^1(k) - \mathcal{T}\mathbf{w}^2(k)\| \le \theta\|\mathbf{w}^1(k) - \mathbf{w}^2(k)\|.$$

Therefore, by the contraction mapping theorem $\mathcal{T}$ has a unique fixed point $\mathbf{w}(k) \in B(\mathbb{N}(k_1))$, i.e. $\mathbf{w}(k) = \mathcal{T}\mathbf{w}(k)$. Thus, from (5.8.13) it follows that

$$\begin{aligned}\mathbf{w}(k+1)-\mathbf{v}(k+1) &= \mathcal{A}(k)\sum_{\ell=k_1+1}^{k}\mathcal{U}(k,a)\mathcal{P}\mathcal{U}^{-1}(\ell,a)\mathbf{f}(\ell-1,\mathbf{w}(\ell-1))\\ &\quad+\mathcal{U}(k+1,a)\mathcal{P}\mathcal{U}^{-1}(k+1,a)\mathbf{f}(k,\mathbf{w}(k))\\ &\quad-\mathcal{A}(k)\sum_{\ell=k+1}^{\infty}\mathcal{U}(k,a)(\mathcal{I}-\mathcal{P})\mathcal{U}^{-1}(\ell,a)\mathbf{f}(\ell-1,\mathbf{w}(\ell-1))\\ &\quad+\mathcal{U}(k+1,a)(\mathcal{I}-\mathcal{P})\mathcal{U}^{-1}(k+1,a)\mathbf{f}(k,\mathbf{w}(k))\\ &= \mathcal{A}(k)(\mathbf{w}(k)-\mathbf{v}(k))+\mathbf{f}(k,\mathbf{w}(k)).\end{aligned}$$

Hence, if $\mathbf{v}(k)$ is a solution of (1.2.11), then $\mathbf{w}(k)$ is a solution of (5.8.12) with $\mathbf{w}(k)\in\mathcal{B}(\mathbb{N}(k_1))$. Conversely, if $\mathbf{w}(k)$ is a solution of (5.8.12) with $\mathbf{w}(k)\in\mathcal{B}(\mathbb{N}(k_1))$, then $\mathbf{v}(k)$ defined by (5.8.13) with $\mathbf{w}(k)=\mathcal{T}\mathbf{w}(k)$ is a bounded solution of (1.2.11). Therefore, (5.8.13) with $\mathbf{w}(k)=\mathcal{T}\mathbf{w}(k)$ establishes a one to one correspondence between the bounded solutions of (1.2.11) and (5.8.12) for $k\in\mathbb{N}(k_1)$. Consider now, for $k\in\mathbb{N}(k_1)$, $\mathbf{v}^0(k)$ a bounded solution of (1.2.11) and $\mathbf{w}^0(k)$ the corresponding bounded solution of (5.8.12). Then, from (5.8.13) with $\mathbf{w}(k)=\mathcal{T}\mathbf{w}(k)$ and the corresponding equation with the replacement of $\mathbf{v}(k)$ and $\mathbf{w}(k)$ by $\mathbf{v}^0(k)$ and $\mathbf{w}^0(k)$, we obtain

$$\|\mathbf{w}(k)-\mathbf{w}^0(k)\| \;\le\; \|\mathbf{v}(k)-\mathbf{v}^0(k)\|+\theta\|\mathbf{w}(k)-\mathbf{w}^0(k)\|$$

and

$$\|\mathbf{v}(k)-\mathbf{v}^0(k)\|\le\|\mathbf{w}(k)-\mathbf{w}^0(k)\|+\theta\|\mathbf{w}(k)-\mathbf{w}^0(k)\| = (1+\theta)\|\mathbf{w}(k)-\mathbf{w}^0(k)\|.$$

Thus, it follows that

$$(1+\theta)^{-1}\|\mathbf{v}(k)-\mathbf{v}^0(k)\|\le\|\mathbf{w}(k)-\mathbf{w}^0(k)\|\le(1-\theta)^{-1}\|\mathbf{v}(k)-\mathbf{v}^0(k)\|,$$

which shows that the one to one correspondence between the bounded solutions of (1.2.11) and (5.8.12) for $k\in\mathbb{N}(k_1)$ is continuous and its inverse is continuous, so it is a homeomorphism. But the solutions of (1.2.11) and (5.8.12) are defined for all $k\in\mathbb{N}(a)$ and are uniquely determined by the initial data, so we have actually a homeomorphism on $\mathbb{N}(a)$.

Now let $\epsilon>0$ be given and choose $k_2\in\mathbb{N}(k_1)$ so large that

$$\eta\sum_{\ell=k_2+1}^{\infty}\|\mathbf{f}(\ell-1,\mathbf{w}(\ell-1))\| \;\le\; \eta\sum_{\ell=k_2}^{\infty}[\lambda(\ell)\|\mathbf{w}(\ell)\|+\|\mathbf{f}(\ell,0)\|] \;<\; \epsilon.$$

Thus, if $\mathcal{U}(k,a)\mathcal{P} \to 0$ as $k \to \infty$, then we find

$$\|\mathbf{v}(k) - \mathbf{w}(k)\| \le \|\mathcal{U}(k,a)\mathcal{P}\| \sum_{\ell=k_1+1}^{k_2} \|\mathcal{U}^{-1}(\ell,a)\mathbf{f}(\ell-1,\mathbf{w}(\ell-1))\| + \epsilon \le 2\epsilon$$

for all large $k$. Hence, $\mathbf{v}(k) - \mathbf{w}(k) \to 0$ as $k \to \infty$. ∎

**Example 5.8.1.** Let in equation (5.8.12), $\mathcal{A}(k) = \begin{bmatrix} 2^{-k} & 0 \\ 0 & 2^k \end{bmatrix}$, $\mathbf{b}(k) = \begin{bmatrix} 0 \\ 2^{-k} \end{bmatrix}$, $\mathbf{f}(k,\mathbf{u}) = \begin{bmatrix} 2^{-k}\tan^{-1}u_2 \\ 2^{-k}\tan^{-1}u_1 \end{bmatrix}$. Since

$$\mathcal{U}(k,0) = \begin{bmatrix} 2^{-k(k-1)/2} & 0 \\ 0 & 2^{k(k-1)/2} \end{bmatrix},$$

if we take $\mathcal{P} = \begin{bmatrix} 1 & 0 \\ 0 & 0 \end{bmatrix}$, then $\|\mathcal{U}(k,0)\mathcal{P}\mathcal{U}^{-1}(\ell,0)\| = 2^{(\ell-k)(\ell+k-1)/2} \le 1 = \eta$, $k \in \mathbb{N}(\ell)$; $\|\mathcal{U}(k,0)(\mathcal{I}-\mathcal{P})\mathcal{U}^{-1}(\ell,0)\| = 2^{(k-\ell)(\ell+k-1)/2} \le 1 = \eta$, $\ell \in \mathbb{N}(k)$; $\|\mathbf{f}(k,\mathbf{u}) - \mathbf{f}(k,\mathbf{v})\| \le 2^{-k}\|\mathbf{u}-\mathbf{v}\|$, $\sum_{\ell=0}^{\infty} 2^{-\ell} = 2 < \infty$, $\sum_{k=0}^{\infty} \|\mathbf{f}(k,0)\| = 0$. Therefore, all the hypotheses of Theorem 5.8.4 are satisfied.

**Theorem 5.8.5.** Suppose that the following conditions are satisfied

(i) the inequality (5.6.11) holds

(ii) for all $(k,\mathbf{u})$, $(k,\mathbf{v})$ where $k \in \mathbb{N}(a)$ and $\|\mathbf{u}\| \le \delta$, $\|\mathbf{v}\| \le \delta$, $\mathbf{g}(k,0) = 0$, and

$$\|\mathbf{g}(k,\mathbf{u}) - \mathbf{g}(k,\mathbf{v})\| \le \lambda\|\mathbf{u}-\mathbf{v}\|, \tag{5.8.14}$$

where $c\lambda < 1$.

Then, for all $k \in \mathbb{N}(a)$ there exists a unique bounded solution $\mathbf{v}(k)$ of the difference system (5.3.4) on $\mathbb{N}(a)$ such that $\mathcal{P}\mathbf{v}(a) = \mathbf{w}^0$, and $\|\mathbf{w}^0\| \le (1-c\lambda)\delta M^{-1}$, where $M$ is a constant depending only on $\mathcal{A}(k)$. Moreover, $\mathbf{v}(k)$ depends continuously on $\mathbf{w}^0$.

**Proof.** By Lemma 5.6.5 we find that there exists a constant $M$ depending only on $\mathcal{A}(k)$ such that, for all $k \in \mathbb{N}(a)$, $\|\mathcal{U}(k,a)\mathcal{P}\| \le M$. Let $B_\delta(\mathbb{N}(a)) = \{\mathbf{v}(k) \in B(\mathbb{N}(a)) : \|\mathbf{v}(k)\| \le \delta\}$, where $B(\mathbb{N}(a))$ is defined in Theorem 5.8.4. On $B_\delta(\mathbb{N}(a))$ we define an operator $T$ as follows

$$T\mathbf{v}(k) = \mathcal{U}(k,a)\mathcal{P}\mathbf{v}(a) + \sum_{\ell=a}^{k-1} \mathcal{U}(k,a)\mathcal{P}\mathcal{U}^{-1}(\ell+1,a)\mathbf{g}(\ell,\mathbf{v}(\ell)) - \sum_{\ell=k}^{\infty} \mathcal{U}(k,a)(\mathcal{I}-\mathcal{P})\mathcal{U}^{-1}(\ell+1,a)\mathbf{g}(\ell,\mathbf{v}(\ell)). \tag{5.8.15}$$

Choose $0 \le \delta_1 < \delta$ so that $M\|\mathbf{w}^0\| \le (1-c\lambda)\delta_1$. Then, if $\mathbf{v}(k) \in B_{\delta_1}(\mathbb{N}(a))$,

$$\|\mathcal{T}\mathbf{v}(k)\| \le M\|\mathbf{w}^0\| + c\lambda\|\mathbf{v}(k)\| \le (1-c\lambda)\delta_1 + c\lambda\delta_1 = \delta_1.$$

Thus $\mathcal{T}$ maps $B_{\delta_1}(\mathbb{N}(a))$ into itself. Further, for $\mathbf{v}^1(k)$, $\mathbf{v}^2(k) \in B_{\delta_1}(\mathbb{N}(a))$, we obtain

$$\|\mathcal{T}\mathbf{v}^1(k) - \mathcal{T}\mathbf{v}^2(k)\| \le c\lambda\|\mathbf{v}^1(k) - \mathbf{v}^2(k)\|.$$

Therefore, by the contraction mapping theorem $\mathcal{T}$ has a unique fixed point $\mathbf{v}(k) \in B_{\delta_1}(\mathbb{N}(a))$, i.e. $\mathbf{v}(k) = \mathcal{T}\mathbf{v}(k)$. This fixed point is indeed a solution of (5.3.4) on $\mathbb{N}(a)$ follows as in Theorem 5.8.3. Now let $\mathbf{v}^1(k)$ be the solution of (5.3.4) on $\mathbb{N}(a)$ when $\mathbf{w}^0$ is replaced by $\mathbf{w}^1$ such that $\|\mathbf{w}^1\| \le (1-c\lambda)\delta M^{-1}$. Then, we find

$$\|\mathbf{v}(k) - \mathbf{v}^1(k)\| \le M\|\mathbf{w}^0 - \mathbf{w}^1\| + c\lambda\|\mathbf{v}(k) - \mathbf{v}^1(k)\|$$

from which it follows that

$$\|\mathbf{v}(k) - \mathbf{v}^1(k)\| \le (1-c\lambda)^{-1}M\|\mathbf{w}^0 - \mathbf{w}^1\|.$$

Thus, $\mathbf{v}(k)$ continuously depends on $\mathbf{w}^0$. ∎

The last theorem here deals with the situation where the difference system (1.2.12) on $\mathbb{N}(a)$ has an exponential dichotomy and gives an exponential estimate on the bounded solutions of (5.3.4).

**Theorem 5.8.6.** Suppose that the following conditions are satisfied

(i) the system (1.2.12) has the exponential dichotomy (5.8.1) on $\mathbb{N}(a)$

(ii) inequality (5.6.3) holds, where

$$\theta = \alpha\left[\eta(1-\rho_1)^{-1} + \nu\rho_2(1-\rho_2)^{-1}\right] < 1$$

and

$$-\zeta = \ln\rho_1 + \alpha\eta(\rho_1(1-\theta))^{-1} < 0.$$

Then, every bounded solution $\mathbf{v}(k)$ of the difference system (5.3.4) satisfies

$$\|\mathbf{v}(k)\| \le (1-\theta)^{-1}\eta\|\mathbf{v}(k_1)\|\exp(-\zeta(k-k_1)) \tag{5.8.16}$$

for all $a \le k_1 \le k \in \mathbb{N}(a)$.

**Proof.** Clearly, conditions (i) and (ii) of the theorem imply the hypotheses of Theorem 5.6.8. Thus, every bounded solution $\mathbf{v}(k)$ of (5.3.4) on $\mathbb{N}(a)$

tends to 0 as $k \to \infty$. Further, for any $k_1 \in \mathbb{N}(a)$ this solution $\mathbf{v}(k)$ can be written as

$$\mathbf{v}(k) = \mathcal{U}(k,a)\mathcal{P}\mathcal{U}^{-1}(k_1,a)\mathbf{v}(k_1) + \sum_{\ell=k_1+1}^{k} \mathcal{U}(k,a)\mathcal{P}\mathcal{U}^{-1}(\ell,a)\mathbf{g}(\ell-1,\mathbf{v}(\ell-1))$$
$$- \sum_{\ell=k+1}^{\infty} \mathcal{U}(k,a)(\mathcal{I}-\mathcal{P})\mathcal{U}^{-1}(\ell,a)\mathbf{g}(\ell-1,\mathbf{v}(\ell-1)).$$

Therefore, for $k \in \mathbb{N}(k_1)$ we find

$$(5.8.17) \quad \|\mathbf{v}(k)\| \le \eta\rho_1^{k-k_1}\|\mathbf{v}(k_1)\| + \eta \sum_{\ell=k_1+1}^{k} \rho_1^{k-\ell}\alpha\|\mathbf{v}(\ell-1)\|$$
$$+\nu \sum_{\ell=k+1}^{\infty} \rho_2^{\ell-k}\alpha\|\mathbf{v}(\ell-1)\|.$$

Let $v(k) = \sup\{\|\mathbf{v}(\ell)\| : \ell \in \mathbb{N}(k)\}$. Since $\mathbf{v}(k) \to 0$ as $k \to \infty$, for any $k \in \mathbb{N}(a)$ there exists a $k^* \in \mathbb{N}(k)$ such that $v(k) = v(\ell) = \|\mathbf{v}(k^*)\|$ for $k = \ell, \ell+1, \cdots, k^*$ and $v(\ell) < v(k^*)$ for $\ell \in \mathbb{N}(k^*+1)$. Therefore, from (5.8.17) we get

$$v(k) \le \eta\rho_1^{k^*-k_1}\|\mathbf{v}(k_1)\| + \eta \sum_{\ell=k_1+1}^{k^*} \rho_1^{k^*-\ell}\alpha v(\ell-1) + \nu\rho_2(1-\rho_2)^{-1}\alpha v(k)$$
$$\le \eta\rho_1^{k-k_1}\|\mathbf{v}(k_1)\| + \alpha\eta \sum_{\ell=k_1}^{k-1} \rho_1^{k-\ell-1}v(\ell) + \eta\alpha(1-\rho_1)^{-1}v(k)$$
$$+\nu\rho_2(1-\rho_2)^{-1}\alpha v(k), \quad \left(\rho_1^{k^*-k} \le 1\right)$$

and hence

$$v(k) \le (1-\theta)^{-1}\left[\eta\rho_1^{k-k_1}\|\mathbf{v}(k_1)\| + \alpha\eta\rho_1^{-1} \sum_{\ell=k_1}^{k-1} \rho_1^{k-\ell}v(\ell)\right].$$

The above inequality easily gives

$$v(k) \le (1-\theta)^{-1}\eta\rho_1^{k-k_1}\|\mathbf{v}(k_1)\| \exp\left(\alpha\eta(\rho_1(1-\theta))^{-1}(k-k_1)\right),$$

which is the same as (5.8.16). ■

## 5.9. Lyapunov's Direct Method for Autonomous Systems

It is well known that a mechanical system is stable if its total energy

$E$, which is the sum of potential energy $V$ and the kinetic energy $T$, i.e. $E = V + T$ continuously decreases. These two energies are always positive quantities and are zero when the system is completely at rest. Lyapunov's direct method uses a generalized energy function to study the stability of the solutions. For differential systems this method has been used since 1892, while for difference systems its use is recent. The main advantage of this approach is that the stability can be obtained without any prior knowledge of the solutions. Here we shall study this fruitful technique for the autonomous difference system (5.4.1). For this, throughout we shall assume that $\mathbf{f}(0) = 0$ and $\mathbf{f}(\mathbf{u}) \neq 0$ for $\mathbf{u} \neq 0$ in some neighborhood of the origin so that (5.4.1) admits the trivial solution $\mathbf{u}(k) = \mathbf{u}(k, a, 0) \equiv 0$, and the origin is an isolated critical point of the difference system (5.4.1).

Let $\Omega$ be an open set in $\mathbb{R}^n$ containing the origin. Suppose $V(\mathbf{u})$ is a scalar continuous function defined on $\Omega$, i.e. $V \in C[\Omega, \mathbb{R}]$, and $V(0) = 0$. For this function we need the following:

**Definition 5.9.1.** $V(\mathbf{u})$ is said to be *positive definite* on $\Omega$ if and only if $V(\mathbf{u}) > 0$ for $\mathbf{u} \neq 0$, $\mathbf{u} \in \Omega$.

**Definition 5.9.2.** $V(\mathbf{u})$ is said to be *positive semidefinite* on $\Omega$ if $V(\mathbf{u}) \geq 0$ (with equality only at certain points) for all $\mathbf{u} \in \Omega$.

**Definition 5.9.3.** $V(\mathbf{u})$ is said to be *negative definite* (*negative semidefinite*) on $\Omega$ if and only if $-V(\mathbf{u})$ is positive definite (positive semidefinite) on $\Omega$.

**Definition 5.9.4.** A function $\phi(r)$ is said to belong to the class $\mathcal{K}$ if and only if $\phi \in C[[0, \rho), \mathbb{R}_+]$, $\phi(0) = 0$, and $\phi(r)$ is strictly monotonically increasing in $r$.

Since $V(\mathbf{u})$ is continuous, for sufficiently small $r$, $0 < c \leq r \leq d$ we have

$$V(\mathbf{u}) \leq \max_{\|\mathbf{v}\| \leq r} V(\mathbf{v}), \qquad V(\mathbf{u}) \geq \min_{r \leq \|\mathbf{v}\| \leq d} V(\mathbf{v}), \tag{5.9.1}$$

where $\|\mathbf{u}\| = r$. In (5.9.1) the right sides are monotonic functions of $r$ and can be estimated in terms of functions belonging to the class $\mathcal{K}$. Thus, there exist two functions $\phi$, $\psi \in \mathcal{K}$ such that

$$\phi(\|\mathbf{u}\|) \leq V(\mathbf{u}) \leq \psi(\|\mathbf{u}\|). \tag{5.9.2}$$

The left side of (5.9.2) provides an alternative definition for the positive definiteness of $V(\mathbf{u})$ as follows:

**Definition 5.9.5.** The function $V(\mathbf{u})$ is said to be positive definite on $\Omega$ if and only if $V(0) = 0$ and there exists a function $\phi(r) \in \mathcal{K}$ such that $\phi(r) \le V(\mathbf{u})$, $\|\mathbf{u}\| = r$, $\mathbf{u} \in \Omega$.

Let $S_\rho$ be the set $S_\rho = \{\mathbf{u} \in \mathbb{R}^n : \|\mathbf{u}\| \le \rho\}$, and $\mathbf{u}(k) = \mathbf{u}(k, a, \mathbf{u}^0)$ be any solution of (5.4.1) such that $\|\mathbf{u}(k)\| < \rho$ for all $k \in \mathbb{N}(a)$. Since (5.4.1) is autonomous we can always assume that $a = 0$. Along the solution $\mathbf{u}(k) = \mathbf{u}(k, 0, \mathbf{u}^0)$ of (5.4.1) we shall consider the variation of the function $V(\mathbf{u})$ as $\Delta V(\mathbf{u}(k)) = V(\mathbf{u}(k+1)) - V(\mathbf{u}(k)) = V(\mathbf{f}(\mathbf{u}(k))) - V(\mathbf{u}(k))$. The auxiliary function $V(\mathbf{u})$ is called a *Lyapunov function.*

**Theorem 5.9.1.** If there exists a positive definite scalar function $V(\mathbf{u}) \in C[S_\rho, \mathbb{R}_+]$ such that $\Delta V(\mathbf{u}(k, 0, \mathbf{u}^0)) \le 0$ for any solution $\mathbf{u}(k) = \mathbf{u}(k, 0, \mathbf{u}^0)$ of (5.4.1) such that $\|\mathbf{u}(k)\| < \rho$, then the trivial solution $\mathbf{u}(k, 0, 0) \equiv 0$ of the difference system (5.4.1) is stable.

**Proof.** Since $V(\mathbf{u})$ is positive definite, there exists a function $\phi \in \mathcal{K}$ such that $\phi(\|\mathbf{u}\|) \le V(\mathbf{u})$ for all $\mathbf{u} \in S_\rho$. Let $0 < \epsilon < \rho$ be given. Since $V(\mathbf{u})$ is continuous and $V(0) = 0$, we can find a $\delta = \delta(\epsilon) > 0$ such that $\|\mathbf{u}^0\| < \delta$ implies that $V(\mathbf{u}^0) < \phi(\epsilon)$. If the trivial solution of (5.4.1) is unstable, then there exists a solution $\mathbf{u}(k) = \mathbf{u}(k, 0, \mathbf{u}^0)$ of (5.4.1) such that $\|\mathbf{u}^0\| < \delta$ satisfies $\epsilon \le \|\mathbf{u}(k_1)\| < \rho$ for some $k_1 \in \mathbb{N}(1)$. However, since $\Delta V(\mathbf{u}(k)) \le 0$ as long as $\|\mathbf{u}(k)\| < \rho$, we have $V(\mathbf{u}(k_1)) \le V(\mathbf{u}^0)$, and hence

$$\phi(\epsilon) \le \phi(\|\mathbf{u}(k_1)\|) \le V(\mathbf{u}(k_1)) \le V(\mathbf{u}^0) < \phi(\epsilon),$$

which is not true. Thus, if $\|\mathbf{u}^0\| < \delta$ then $\|\mathbf{u}(k)\| < \epsilon$ for all $k \in \mathbb{N}$. This implies that the trivial solution of (5.4.1) is stable. ∎

**Theorem 5.9.2.** If there exists a positive definite scalar function $V(\mathbf{u}) \in C[S_\rho, \mathbb{R}_+]$ such that $\Delta V(\mathbf{u}(k, 0, \mathbf{u}^0)) \le -\alpha(\|\mathbf{u}(k, 0, \mathbf{u}^0)\|)$, where $\alpha \in \mathcal{K}$, for any solution $\mathbf{u}(k) = \mathbf{u}(k, 0, \mathbf{u}^0)$ of (5.4.1) such that $\|\mathbf{u}(k)\| < \rho$, then the trivial solution $\mathbf{u}(k, 0, 0) \equiv 0$ of the difference system (5.4.1) is asymptotically stable.

**Proof.** Since all the conditions of Theorem 5.9.1 are satisfied, the trivial solution of (5.4.1) is stable. Therefore, given $0 < \epsilon < \rho$, suppose that there exist $\delta > 0$, $\lambda > 0$ and a solution $\mathbf{u}(k) = \mathbf{u}(k, 0, \mathbf{u}^0)$ of (5.4.1) such that

$$(5.9.3) \qquad \lambda \le \|\mathbf{u}(k)\| < \epsilon, \quad k \in \mathbb{N}, \quad \|\mathbf{u}^0\| < \delta.$$

Since for this solution $\|\mathbf{u}(k)\| \ge \lambda > 0$ for all $k \in \mathbb{N}$, there exists a constant $d > 0$ such that $\alpha(\|\mathbf{u}(k)\|) \ge d$ for all $k \in \mathbb{N}$. Hence, we have

$\Delta V(\mathbf{u}(k)) \le -d < 0,\ k \in \mathbb{N}$. This implies that

$$V(\mathbf{u}(k)) \;=\; V(\mathbf{u}^0) + \sum_{\ell=0}^{k-1} \Delta V(\mathbf{u}(\ell)) \;\le\; V(\mathbf{u}^0) - kd$$

and for sufficiently large $k$ the right side will become negative, which contradicts $V(\mathbf{u})$ being positive definite. Hence, no such $\lambda$ exists for which (5.9.3) holds. Further, since $V(\mathbf{u}(k))$ is a positive and decreasing function of $k$, it follows that $\lim_{k\to\infty} V(\mathbf{u}(k)) = 0$. Therefore, $\lim_{k\to\infty} \|\mathbf{u}(k)\| = 0$, and this implies that the trivial solution of (5.4.1) is asymptotically stable. ∎

**Theorem 5.9.3.** If there exists a scalar function $V(\mathbf{u}) \in C[S_\rho, \mathbb{R}]$, $V(0) = 0$ such that $\Delta V(\mathbf{u}(k, 0, \mathbf{u}^0)) \ge \alpha(\|\mathbf{u}(k, 0, \mathbf{u}^0)\|)$, where $\alpha \in \mathcal{K}$, for any solution $\mathbf{u}(k) = \mathbf{u}(k, 0, \mathbf{u}^0)$ of (5.4.1) such that $\|\mathbf{u}(k)\| < \rho$, and if in every neighborhood $H$ of the origin $H \subset S_\rho$ there is a point $\mathbf{u}^0$ where $V(\mathbf{u}^0) > 0$, then the trivial solution $\mathbf{u}(k, 0, 0) \equiv 0$ of the difference system (5.4.1) is unstable.

**Proof.** Let $r > 0$ be sufficiently small so that the set $S_r = \{\mathbf{u} \in \mathbb{R}^n : \|\mathbf{u}\| \le r\} \subset S_\rho$. Let $M = \max_{\|\mathbf{u}\| \le r} V(\mathbf{u})$, where $M$ is finite since $V$ is continuous. Let $r_1$ be such that $0 < r_1 < r$, then by the hypotheses there exists a point $\mathbf{u}^0 \in \mathbb{R}^n$ such that $0 < \|\mathbf{u}^0\| < r_1$ and $V(\mathbf{u}^0) > 0$. Along the solution $\mathbf{u}(k) = \mathbf{u}(k, 0, \mathbf{u}^0)$, $k \in \mathbb{N}$, $\Delta V(\mathbf{u}(k)) > 0$, and therefore $V(\mathbf{u}(k))$ is an increasing function and $V(\mathbf{u}(0)) = V(\mathbf{u}^0) > 0$. This implies that this solution $\mathbf{u}(k)$ cannot approach the origin. Thus, it follows that $\inf_{k\in\mathbb{N}} \Delta V(\mathbf{u}(k)) = d > 0$, and therefore, $V(\mathbf{u}(k)) \ge V(\mathbf{u}^0) + kd$ for $k \in \mathbb{N}$. But the right side of this inequality can be made greater than $M$ for $k$ sufficiently large, which implies that $\mathbf{u}(k)$ must leave the set $S_r$. Thus, the trivial solution of (5.4.1) is unstable. ∎

**Example 5.9.1.** For the difference system

$$\begin{aligned} u_1(k+1) &= u_2(k) - cu_1(k)(u_1^2(k) + u_2^2(k)) \\ u_2(k+1) &= u_1(k) + cu_2(k)(u_1^2(k) + u_2^2(k)), \end{aligned} \tag{5.9.4}$$

where $c$ is a constant, we consider the positive definite function $V(u_1, u_2) = u_1^2 + u_2^2$ on $\Omega = \mathbb{R}^2$. A simple computation gives $\Delta V(u_1(k), u_2(k)) = c^2(u_1^2(k) + u_2^2(k))^3$. Thus, if $c = 0$ then $\Delta V(u_1(k), u_2(k)) \equiv 0$, and the trivial solution of the resulting difference system (5.9.4) is stable. However, if $c \neq 0$ then the trivial solution of (5.9.4) is unstable.

**Example 5.9.2.** For the difference system

$$\begin{aligned} u_1(k+1) &= c_1 u_2(k)/(1+u_1^2(k)) \\ u_2(k+1) &= c_2 u_1(k)/(1+u_2^2(k)), \end{aligned} \tag{5.9.5}$$

where $c_1$ and $c_2$ are constants, we consider the positive definite function $V(u_1, u_2) = u_1^2 + u_2^2$ on $\Omega = \mathbb{R}^2$. Then,

$$\Delta V(u_1(k), u_2(k)) = \left(\frac{c_2^2}{(1+u_2^2(k))^2} - 1\right) u_1^2(k) + \left(\frac{c_1^2}{(1+u_1^2(k))^2} - 1\right) u_2^2(k).$$

If $c_1^2 < 1$, $c_2^2 < 1$ then, since

$$\Delta V(u_1(k), u_2(k)) \le (c_2^2 - 1)u_1^2(k) + (c_1^2 - 1)u_2^2(k),$$

the trivial solution of (5.9.5) is asymptotically stable.

If $c_1^2 > 1$, $c_2^2 > 1$ then let $(u_1(k), u_2(k)) \in S_r \subset \mathbb{R}^2$, where $r$ is so small that

$$\Delta V(u_1(k), u_2(k)) \ge \left(\frac{c_2^2}{1+r^2} - 1\right) u_1^2(k) + \left(\frac{c_1^2}{1+r^2} - 1\right) u_2^2(k) > 0.$$

Therefore, the trivial solution of (5.9.5) is unstable.

**Theorem 5.9.4.** If there are positive definite matrices $\mathcal{B}$ and $\mathcal{C}$ such that

$$\mathcal{A}^T \mathcal{B} \mathcal{A} - \mathcal{B} = -\mathcal{C} \tag{5.9.6}$$

then the system (2.8.1) is asymptotically stable. Conversely, if (2.8.1) is asymptotically stable, then given $\mathcal{C}$, (5.9.6) has a unique solution $\mathcal{B}$. Further, if $\mathcal{C}$ is positive definite then $\mathcal{B}$ is positive definite.

**Proof.** For the difference system (2.8.1), we take $V(\mathbf{u}(k)) = \mathbf{u}^T(k)\mathcal{B}\mathbf{u}(k)$, where $\mathcal{B}$ is a symmetric positive definite matrix. The condition $\Delta V(\mathbf{u}(k)) < 0$ forces that $\mathbf{u}^T(k)\mathcal{A}^T\mathcal{B}\mathcal{A}\mathbf{u}(k) - \mathbf{u}^T(k)\mathcal{B}\mathbf{u}(k) < 0$, i.e. we must have $\mathcal{A}^T\mathcal{B}\mathcal{A} - \mathcal{B} = -\mathcal{C}$, where $\mathcal{C}$ is any positive definite matrix.

Conversely, suppose that the system (2.8.1) is asymptotically stable then all the eigenvalues of $\mathcal{A}$ lie in the unit disc. If (5.9.6) has a solution, then

$$-\sum_{k=0}^{m} (\mathcal{A}^T)^k \mathcal{C} \mathcal{A}^k = \sum_{k=0}^{m} (\mathcal{A}^T)^k (\mathcal{A}^T\mathcal{B}\mathcal{A} - \mathcal{B})\mathcal{A}^k = (\mathcal{A}^T)^{m+1}\mathcal{B}\mathcal{A}^{m+1} - \mathcal{B}.$$

Letting $m \to \infty$, we see that the solution must be

$$(5.9.7) \qquad B = \sum_{k=0}^{\infty} (A^T)^k C A^k.$$

This is indeed a solution of (5.9.6), and obviously if $C$ is positive definite then $B$ is positive definite. ∎

**Definition 5.9.6.** The *orbit* $C(\mathbf{u}^0)$ of (5.4.1) through $\mathbf{u}^0$ is defined by $C(\mathbf{u}^0) = \{\mathbf{u} \in \mathbb{R}^n : \mathbf{u} = \mathbf{u}(k, 0, \mathbf{u}^0),\ k \in \mathbb{N}\}$, where $\mathbf{u}(k, 0, \mathbf{u}^0)$ is the solution of (5.4.1) which is assumed to exist for all $k \in \mathbb{N}$.

**Definition 5.9.7.** The *positive limit set* $\Omega(\mathbf{u}^0)$ is the set of all limit points of $C(\mathbf{u}^0)$, i.e. a point $\mathbf{u} \in \Omega(\mathbf{u}^0)$ if there exists a sequence $\{k_\ell\}$ of integers such that $k_\ell \to \infty$ as $\ell \to \infty$ and $\mathbf{u}(k_\ell, 0, \mathbf{u}^0) \to \mathbf{u}$ as $\ell \to \infty$.

**Definition 5.9.8.** A set $H \subset \mathbb{R}^n$ is called an *invariant* set of (5.4.1) if $\mathbf{u}^0 \in H$ implies that $\mathbf{u}(k, 0, \mathbf{u}^0) \in H$ for all $k \in \mathbb{N}$, i.e. $\mathbf{f}(H) = H$.

**Definition 5.9.9.** The *region of attraction of the origin* of (5.4.1) is the set of all points $\mathbf{u}^0 \in \mathbb{R}^n$ such that $\lim_{k\to\infty} \mathbf{u}(k, 0, \mathbf{u}^0) = 0$. If in addition the origin is stable then this set is called the *region of asymptotic stability*.

**Definition 5.9.10.** The distance from a point $\mathbf{u}$ to a set $H$ is defined by

$$d(\mathbf{u}, H) = \inf_{\mathbf{v} \in H} d(\mathbf{u}, \mathbf{v}) \quad \text{where} \quad d(\mathbf{u}, \mathbf{v}) = \|\mathbf{u} - \mathbf{v}\|.$$

**Theorem 5.9.5.** Let $\mathbf{u}(k, 0, \mathbf{u}^0)$ be a bounded solution of (5.4.1) on $\mathbb{N}$. Then, the positive limit set $\Omega(\mathbf{u}^0)$ is a nonempty and invariant set of (5.4.1). Also, $\mathbf{u}(k, 0, \mathbf{u}^0)$ approaches $\Omega(\mathbf{u}^0)$ as $k \to \infty$.

**Proof.** Since $\mathbf{u}(k, 0, \mathbf{u}^0)$ is bounded for all $k \in \mathbb{N}$, its orbit $C(\mathbf{u}^0)$ lies in the interior of some closed sphere $S$ of finite radius. Now, consider an infinite sequence $\{\mathbf{u}(k_\ell, 0, \mathbf{u}^0)\}$, $\ell = 1, 2, \cdots$ in $S$ such that $\|\mathbf{u}(k_\ell, 0, \mathbf{u}^0)\|$ is bounded. Then, there exists a subsequence $\{\mathbf{u}(k_{\ell,m}, 0, \mathbf{u}^0)\}$ which converges to a point $\mathbf{v} \in S$. By Definition 5.9.7, $\mathbf{v} \in \Omega(\mathbf{u}^0)$, and hence $\Omega(\mathbf{u}^0)$ is nonempty. To see that $\Omega(\mathbf{u}^0)$ is invariant, let $\mathbf{w} \in \Omega(\mathbf{u}^0)$, then by Definition 5.9.7 there exists a sequence $\{k_\ell\}$ of integers such that $k_\ell \to \infty$ as $\ell \to \infty$ and $\mathbf{u}(k_\ell, 0, \mathbf{u}^0) \to \mathbf{w}$ as $\ell \to \infty$. However, since the solution $\mathbf{u}(k, 0, \mathbf{u}^0)$ continuously depends on $\mathbf{u}^0$, we have $\lim_{\ell\to\infty} \mathbf{u}(k, 0, \mathbf{u}(k_\ell, 0, \mathbf{u}^0)) = \mathbf{u}(k, 0, \mathbf{w})$. Thus, the relation $\mathbf{u}(k, 0, \mathbf{u}(k_\ell, 0, \mathbf{u}^0)) = \mathbf{u}(k + k_\ell, 0, \mathbf{u}^0)$ gives $\lim_{\ell\to\infty} \mathbf{u}(k + k_\ell, 0, \mathbf{u}^0) = \mathbf{u}(k, 0, \mathbf{w})$. This implies that $\mathbf{u}(k, 0, \mathbf{w}) \in \Omega(\mathbf{u}^0)$, and hence $\Omega(\mathbf{u}^0)$

is invariant with respect to (5.4.1). We now claim that $\mathbf{u}(k,0,\mathbf{u}^0)$ approaches $\Omega(\mathbf{u}^0)$ as $k \to \infty$. Suppose this is not true, then there exists a positive number $\epsilon$ and a sequence $\{k_\ell\}$ of integers such that $k_\ell \to \infty$ as $\ell \to \infty$ and $\|\mathbf{u}(k_\ell,0,\mathbf{u}^0) - \mathbf{w}\| \geq \epsilon$ for all $\mathbf{w} \in \Omega(\mathbf{u}^0)$. This means that the sequence of points $\{\mathbf{u}(k_\ell,0,\mathbf{u}^0)\}$ is away from $\Omega(\mathbf{u}^0)$. From the boundedness property of $\mathbf{u}(k,0,\mathbf{u}^0)$ there is, as we have seen, a subsequence $\{\mathbf{u}(k_{\ell,m},0,\mathbf{u}^0)\}$ which converges to a point $\mathbf{v} \in \Omega(\mathbf{u}^0)$. This is a contradiction. Hence, $\mathbf{u}(k,0,\mathbf{u}^0)$ approaches $\Omega(\mathbf{u}^0)$ as $k \to \infty$. ■

**Theorem 5.9.6.** Assume that there exists a positive definite scalar function $V(\mathbf{u}) \in C[\Omega, \mathbb{R}_+]$ such that $\Delta V(\mathbf{u}(k,0,\mathbf{u}^0)) \leq 0$ for any solution $\mathbf{u}(k) = \mathbf{u}(k,0,\mathbf{u}^0)$ of (5.4.1) which remains in $\Omega$ for all $k \in \mathbb{N}$. Let $\mathbf{u}^0 \in \Omega$ and the solution $\mathbf{u}(k,0,\mathbf{u}^0)$ of (5.4.1) be bounded for all $k \in \mathbb{N}$ and let $C(\mathbf{u}^0) \subset \Omega$. Then, if the positive limit set $\Omega(\mathbf{u}^0)$ of $\mathbf{u}(k,0,\mathbf{u}^0)$ lies in $\Omega$, then $\Delta V(\mathbf{w}) = 0$ for all $\mathbf{w} \in \Omega(\mathbf{u}^0)$.

**Proof.** Let $\mathbf{u}^1$, $\mathbf{u}^2 \in \Omega(\mathbf{u}^0)$. Then, by Definition 5.9.7 there exist two sequences $\{k_\ell\}$, $\{k_m\}$ of integers, each of which approaches infinity, such that $\lim_{\ell\to\infty} \mathbf{u}(k_\ell,0,\mathbf{u}^0) = \mathbf{u}^1$ and $\lim_{m\to\infty} \mathbf{u}(k_m,0,\mathbf{u}^0) = \mathbf{u}^2$. Since $\Delta V(\mathbf{u}(k,0,\mathbf{u}^0)) \leq 0$, $V(\mathbf{u}(k,0,\mathbf{u}^0))$ is a nonincreasing function of $k$. Further, $V(\mathbf{u}(k,0,\mathbf{u}^0))$ is bounded below because $V$ is positive definite in $\Omega$. Therefore, $\lim_{k\to\infty} V(\mathbf{u}(k,0,\mathbf{u}^0)) = \mathbf{v}$, say, exists. Thus, by the continuity of $V$ in $\Omega$, we have $V(\mathbf{u}^1) = V(\mathbf{u}^2) = \mathbf{v}$. This implies that $V(\mathbf{u}) = \mathbf{v}$ in $\Omega(\mathbf{u}^0)$. Moreover, by Theorem 5.9.5, it is clear that $\Omega(\mathbf{u}^0)$ is a positively invariant set, i.e. if $\mathbf{w} \in \Omega(\mathbf{u}^0)$ then $\mathbf{u}(k,0,\mathbf{w}) \in \Omega(\mathbf{u}^0)$ for all $k \in \mathbb{N}$. Therefore, for each $\mathbf{w} \in \Omega(\mathbf{u}^0)$, we obtain

$$\Delta V(\mathbf{w}) = \Delta V(\mathbf{u}(0,0,\mathbf{w})) = \Delta V(\mathbf{u}(k,0,\mathbf{w}))|_{k=0} = \Delta(\mathbf{v}) = 0. \quad ■$$

For any $\mu \geq 0$, let $C_\mu$ be a component of $G_\mu = \{\mathbf{u} \in \mathbb{R}^n : V(\mathbf{u}) \leq \mu\}$ containing the origin. For $\mu = 0$, we get the origin. Assume that $C_\mu$ is a closed and bounded subset of $\Omega$. The following result includes Theorem 5.9.2.

**Theorem 5.9.7.** Assume that there exists a positive definite scalar function $V(\mathbf{u}) \in C[\Omega, \mathbb{R}_+]$ such that $\Delta V(\mathbf{u}(k,0,\mathbf{u}^0)) \leq 0$ for any solution $\mathbf{u}(k) = \mathbf{u}(k,0,\mathbf{u}^0)$ of (5.4.1) which remains in $\Omega$ for all $k \in \mathbb{N}$. Let the set $E = \{\mathbf{u} \in C_\mu : \Delta V(\mathbf{u}) = 0\}$, and $M$ be the largest invariant set of $E \subset C_\mu$. Then, every solution $\mathbf{u}(k,0,\mathbf{u}^0)$ of (5.4.1) starting in $C_\mu$ approaches $M$ as $k \to \infty$.

**Proof.** Let $\mathbf{u}(k,0,\mathbf{u}^0)$ be the solution of (5.4.1) such that $\mathbf{u}^0 \in C_\mu$. From the conditions on $V(\mathbf{u})$ it is clear that this solution $\mathbf{u}(k,0,\mathbf{u}^0)$

must remain in $C_\mu$ for all $k \in \mathbb{N}$. This implies that $\mathbf{u}(k, 0, \mathbf{u}^0)$ is bounded on $\mathbb{N}$. Because $C_\mu$ is closed and bounded, the positive limit set $\Omega(\mathbf{u}^0)$ of $\mathbf{u}(k, 0, \mathbf{u}^0)$ lies in $C_\mu$. Therefore, by Theorem 5.9.6, $\Delta V(\mathbf{w}) = 0$ at all points $\mathbf{w} \in \Omega(\mathbf{u}^0)$, and hence, by the definition of set $E$, $\Omega(\mathbf{u}^0) \subset E$. Moreover, from Theorem 5.9.5 it is clear that $\Omega(\mathbf{u}^0)$ is a nonempty, invariant set and that $\Omega(\mathbf{u}^0) \subset M$. Hence, $\mathbf{u}(k, 0, \mathbf{u}^0)$ approaches $\Omega(\mathbf{u}^0)$ as $k \to \infty$, and consequently approaches $M$ as $k \to \infty$. ∎

**Remark 5.9.1.** From the assumptions of Theorem 5.9.7 it is easy to conclude that the trivial solution $\mathbf{u}(k, 0, 0) \equiv 0$ of (5.4.1) is stable. To obtain its asymptotic stability, we need $M = \{0\}$. For example, if the conditions of Theorem 5.9.2 are satisfied with $S_\rho$ replaced by $C_\mu$, then $E$ is the origin, and therefore $M = \{0\}$.

**Remark 5.9.2.** Theorem 5.9.7 not only gives sufficient conditions for asymptotic stability but also indicates the size of the region of asymptotic stability. Such a region is at least as large as the largest invariant set contained in $\Omega$. In particular, the interior of $C_\mu$ is contained in this region.

**Theorem 5.9.8.** Assume that there exists a positive definite scalar function $V(\mathbf{u}) \in C[\mathbb{R}^n, R_+]$ such that

(i) $V(\mathbf{u}) \to \infty$ as $\|\mathbf{u}\| \to \infty$

(ii) $\Delta V(\mathbf{u}(k, 0, \mathbf{u}^0)) \le 0$ for any solution $\mathbf{u}(k) = \mathbf{u}(k, 0, \mathbf{u}^0)$ of (5.4.1) for all $k \in \mathbb{N}$.

Then, all solutions of (5.4.1) are bounded on $\mathbb{N}$.

**Proof.** For any solution $\mathbf{u}(k) = \mathbf{u}(k, 0, \mathbf{u}^0)$ of (5.4.1) it is easy to see that $\phi(\|\mathbf{u}(k, 0, \mathbf{u}^0)\|) \le V(\mathbf{u}(k, 0, \mathbf{u}^0)) \le V(\mathbf{u}^0)$, where $\phi \in \mathcal{K}$. This shows that $\|\mathbf{u}(k, 0, \mathbf{u}^0)\|$ is bounded by a constant depending only upon $\mathbf{u}^0$. Since $\mathbf{u}^0$ is arbitrary, all solutions of (5.4.1) are bounded on $\mathbb{N}$. ∎

**Corollary 5.9.9.** In addition to the assumptions of Theorem 5.9.8, if the origin is the only invariant subset of $E = \{\mathbf{u} \in \mathbb{R}^n : \Delta V(\mathbf{u}) = 0\}$ then the trivial solution $\mathbf{u}(k, 0, 0) \equiv 0$ of (5.4.1) is globally asymptotically stable.

**Corollary 5.9.10.** In Theorem 5.9.8, if condition (ii) is replaced by $\Delta V(\mathbf{u}(k, 0, \mathbf{u}^0)) \le -\alpha(\|\mathbf{u}(k, 0, \mathbf{u}^0)\|)$, where $\alpha \in \mathcal{K}$, for any solution $\mathbf{u}(k) = \mathbf{u}(k, 0, \mathbf{u}^0)$ of (5.4.1), then the trivial solution $\mathbf{u}(k, 0, 0) \equiv 0$ of (5.4.1) is globally asymptotically stable.

**Remark 5.9.3.** If a system has more than one critical point, then none of these critical points is globally asymptotically stable.

## 5.10. Lyapunov's Direct Method for Non–Autonomous Systems

We shall extend the method of Lyapunov functions to study the stability properties of the solutions of the difference system (1.2.8). For this we shall assume that $\mathbf{f}(k,0) = 0$ for all $k \in \mathbb{N}(a)$ so that (1.2.8) admits the trivial solution. It is clear that a Lyapunov function for the system (1.2.8) must depend on both $k$ and $\mathbf{u}$, i.e. $V = V(k,\mathbf{u})$.

**Definition 5.10.1.** A real valued function $V(k,\mathbf{u})$ defined on $\mathbb{N}(a)\times S_\rho$ is said to be *positive definite* if and only if $V(k,0) = 0$ for all $k \in \mathbb{N}(a)$, and there exists a function $\phi(r) \in \mathcal{K}$ such that $\phi(r) \le V(k,\mathbf{u})$, $\|\mathbf{u}\| = r$, $(k,\mathbf{u}) \in \mathbb{N}(a) \times S_\rho$. It is *negative definite* if $V(k,\mathbf{u}) \le -\phi(r)$.

**Definition 5.10.2.** A real valued function $V(k,\mathbf{u})$ defined on $\mathbb{N}(a)\times S_\rho$ is said to be *decrescent* if and only if $V(k,0) = 0$ for all $k \in \mathbb{N}(a)$, and there exists a function $\psi(r) \in \mathcal{K}$ such that $V(k,\mathbf{u}) \le \psi(r)$, $\|\mathbf{u}\| = r$, $(k,\mathbf{u}) \in \mathbb{N}(a) \times S_\rho$.

Let $\mathbf{u}(k) = \mathbf{u}(k,a,\mathbf{u}^0)$ be any solution of (1.2.8) such that $\|\mathbf{u}(k)\| < \rho$ for all $k \in \mathbb{N}(a)$. Along with this solution we shall consider the variation of the function $V(k,\mathbf{u})$ as $\Delta V(k,\mathbf{u}(k)) = V(k+1,\mathbf{u}(k+1)) - V(k,\mathbf{u}(k)) = V(k+1,\mathbf{f}(k,\mathbf{u}(k))) - V(k,\mathbf{u}(k))$.

The following two theorems regarding the stability and asymptotic stability of the trivial solution of (1.2.8) are parallel to the results in the autonomous case.

**Theorem 5.10.1.** If there exists a positive definite scalar function $V(k,\mathbf{u}) \in C[\mathbb{N}(a) \times S_\rho, \mathbb{R}_+]$ such that $\Delta V(k,\mathbf{u}(k,a,\mathbf{u}^0)) \le 0$ for any solution $\mathbf{u}(k) = \mathbf{u}(k,a,\mathbf{u}^0)$ of (1.2.8) such that $\|\mathbf{u}(k)\| < \rho$, then the trivial solution $\mathbf{u}(k,a,0) \equiv 0$ of the difference system (1.2.8) is stable.

**Theorem 5.10.2.** If there exists a positive definite scalar function $V(k,\mathbf{u}) \in C[\mathbb{N}(a)\times S_\rho, \mathbb{R}_+]$ such that $\Delta V(k,\mathbf{u}(k,a,\mathbf{u}^0)) \le -\alpha(\|\mathbf{u}(k,a,\mathbf{u}^0)\|)$, where $\alpha \in \mathcal{K}$, for any solution $\mathbf{u}(k) = \mathbf{u}(k,a,\mathbf{u}^0)$ of (1.2.8) such that $\|\mathbf{u}(k)\| < \rho$, then the trivial solution $\mathbf{u}(k,a,0) \equiv 0$ of the difference system (1.2.8) is asymptotically stable.

**Theorem 5.10.3.** Let in addition to the hypotheses of Theorem 5.10.1 (Theorem 5.10.2) the function $V(k,\mathbf{u})$ be decrescent also. Then, the trivial solution $\mathbf{u}(k,a,0) \equiv 0$ of the difference system (1.2.8) is uniformly (uniformly asymptotically) stable.

**Proof.** Since $V(k,\mathbf{u})$ is positive definite and decrescent, there exist functions $\phi,\ \psi \in \mathcal{K}$ such that $\phi(\|\mathbf{u}\|) \le V(k,\mathbf{u}) \le \psi(\|\mathbf{u}\|)$ for all $(k,\mathbf{u}) \in \mathbb{N}(a) \times S_\rho$. For each $\epsilon$, $0 < \epsilon < \rho$, we choose a $\delta = \delta(\epsilon) > 0$ such that $\psi(\delta) < \phi(\epsilon)$. We now claim that the trivial solution of (1.2.8) is uniformly stable, i.e. if $k_1 \ge a$ and $\|\mathbf{u}(k_1)\| < \delta$, then $\|\mathbf{u}(k)\| < \epsilon$ for all $k \ge k_1$. If this is not true, then there exists some $k_2 > k_1$ such that $k_1 \ge a$ and $\|\mathbf{u}(k_1)\| < \delta$ imply $\epsilon \le \|\mathbf{u}(k_2)\| < \rho$. However, $\Delta V(k,\mathbf{u}(k)) \le 0$ implies that $V(k,\mathbf{u}(k)) \le V(k_1,\mathbf{u}(k_1))$ for all $k \in \mathbb{N}(k_1)$, thus it follows that

$$\phi(\epsilon) \le \phi(\|\mathbf{u}(k_2)\|) \le V(k_2,\mathbf{u}(k_2)) \le V(k_1,\mathbf{u}(k_1)) \le \psi(\|\mathbf{u}(k_1)\|)$$
$$\le \psi(\delta) < \phi(\epsilon).$$

This contradiction completes the proof. The uniform asymptotic stability of the trivial solution of (1.2.8) can be proved similarly. ∎

We shall now formulate a result which provides sufficient conditions for the trivial solution of the difference system (1.2.8) to be unstable.

**Theorem 5.10.4.** if there exists a scalar function $V(k,\mathbf{u}) \in V[\mathbb{N}(a) \times S_\rho, \mathbb{R}]$ such that

(i) $|V(k,\mathbf{u})| \le \psi(\|\mathbf{u}\|)$ for all $(k,\mathbf{u}) \in \mathbb{N}(a) \times S_\rho$, where $\psi \in \mathcal{K}$

(ii) for every $\delta > 0$ there exists an $\mathbf{u}^0$ with $\|\mathbf{u}^0\| \le \delta$ such that $V(a,\mathbf{u}^0) < 0$

(iii) $\Delta V(k,\mathbf{u}(k,a,\mathbf{u}^0)) \le -\phi(\|\mathbf{u}(k,a,\mathbf{u}^0)\|)$, where $\phi \in \mathcal{K}$, for any solution $\mathbf{u}(k) = \mathbf{u}(k,a,\mathbf{u}^0)$ of (1.2.8) such that $\|\mathbf{u}(k)\| < \rho$,

then the trivial solution $\mathbf{u}(k,a,0) \equiv 0$ of the difference system (1.2.8) is unstable.

**Proof.** Let the trivial solution of (1.2.8) be stable. Then, for every $\epsilon > 0$ such that $\epsilon < \rho$, there exists a $\delta = \delta(\epsilon, a) > 0$ such that $\|\mathbf{u}^0\| < \delta$ implies that $\|\mathbf{u}(k)\| = \|\mathbf{u}(k,a,\mathbf{u}^0)\| < \epsilon$ for all $k \in \mathbb{N}(a)$. Let $\mathbf{u}^0$ be such that $\|\mathbf{u}^0\| < \delta$ and $V(a,\mathbf{u}^0) < 0$. Since $\|\mathbf{u}^0\| < \delta$, we have $\|\mathbf{u}(k)\| < \epsilon$. Hence, condition (i) gives

$$|V(k,\mathbf{u}(k))| \le \psi(\|\mathbf{u}(k)\|) < \psi(\epsilon) \quad \text{for all} \quad k \in \mathbb{N}(a). \tag{5.10.1}$$

Now from condition (iii), it follows that $V(k,\mathbf{u}(k))$ is a decreasing function, and therefore for every $k \in \mathbb{N}(a)$, we obtain $V(k,\mathbf{u}(k)) \le V(a,\mathbf{u}^0) < 0$. This implies that $|V(k,\mathbf{u}(k))| \ge |V(a,\mathbf{u}^0)|$. Hence, from condition (i) we get $\|\mathbf{u}(k)\| \ge \psi^{-1}(|V(a,\mathbf{u}^0)|)$.

From condition (iii) again, we have $\Delta V(k,\mathbf{u}(k)) \le -\phi(\|\mathbf{u}(k)\|)$, and hence on summing this inequality between $a$ and $k-1$, we obtain

$$V(k,\mathbf{u}(k)) \le V(a,\mathbf{u}^0) - \sum_{\ell=a}^{k-1} \phi(\|\mathbf{u}(\ell)\|).$$

However, since $\|\mathbf{u}(k)\| \ge \psi^{-1}(|V(a,\mathbf{u}^0)|)$, it is clear that $\phi(\|\mathbf{u}(k)\|) \ge \phi(\psi^{-1}(|V(a,\mathbf{u}^0)|))$. Thus, we have

$$V(k,\mathbf{u}(k)) \le V(a,\mathbf{u}^0) - (k-a)\phi(\psi^{-1}(|V(a,\mathbf{u}^0)|)).$$

But this shows that $\lim_{k\to\infty} V(k,\mathbf{u}(k)) = -\infty$, which contradicts (5.10.1). Hence, the trivial solution of (1.2.8) is unstable. ∎

**Theorem 5.10.5.** Let $g(k,r)$ be defined on $\mathbb{N}(a)\times\mathbb{R}_+$ and nondecreasing in $r$ for any fixed $k\in\mathbb{N}(a)$. Further, let there exist a positive definite scalar function $V(k,\mathbf{u}) \in C[\mathbb{N}(a)\times S_\rho, \mathbb{R}_+]$ such that for all $k\in\mathbb{N}(a)$

$$\Delta V(k,\mathbf{u}(k,a,\mathbf{u}^0)) \le g(k, V(k,\mathbf{u}(k,a,\mathbf{u}^0)))$$

for any solution $\mathbf{u}(k) = \mathbf{u}(k,a,\mathbf{u}^0)$ of (1.2.8) such that $\|\mathbf{u}(k)\| < \rho$. Then, the trivial solution $\mathbf{u}(k,a,0)\equiv 0$ of (1.2.8) is stable (asymptotically stable) provided the trivial solution $r(k,a,0)\equiv 0$ of (5.1.9) is stable (asymptotically stable).

If in addition the function $V(k,\mathbf{u})$ is decrescent also, then the trivial solution $\mathbf{u}(k,a,0)\equiv 0$ of (1.2.8) is uniformly stable (uniformly asymptotically stable) provided the trivial solution $r(k,a,0)\equiv 0$ of (5.1.9) is uniformly stable (uniformly asymptotically stable).

**Proof.** If $V(a,\mathbf{u}(a)) \le r(a)$, then it is easy to deduce that $V(k,\mathbf{u}(k)) \le r(k)$. From this the conclusions are immediate. ∎

**Theorem 5.10.6.** If there exists a positive definite scalar function $V(k,\mathbf{u}) \in C[\mathbb{N}(a)\times S_\rho,\mathbb{R}_+]$ such that $\Delta V(k,\mathbf{u}(k,a,\mathbf{u}^0)) \le -c\|\mathbf{u}(k,a,\mathbf{u}^0)\|^p$ for any solution $\mathbf{u}(k) = \mathbf{u}(k,a,\mathbf{u}^0)$ of (1.2.8) such that $\|\mathbf{u}(k)\| < \rho$, then the trivial solution $\mathbf{u}(k,a,0)\equiv 0$ of the difference system (1.2.8) is $s_p$ stable.

**Proof.** Since all the conditions of Theorem 5.10.1 are satisfied, the trivial solution of (1.2.8) is stable. Therefore, given $0<\epsilon<\rho$ there exists a $\delta = \delta(\epsilon,a)$ such that $\|\mathbf{u}^0\| < \delta$ implies that $\|\mathbf{u}(k,a,\mathbf{u}^0)\| < \epsilon$. Let $W(k) = V(k,\mathbf{u}(k)) + c\sum_{\ell=a}^{k-1}\|\mathbf{u}(\ell)\|^p$. Then, $\Delta W(k) = \Delta V(k,\mathbf{u}(k)) + c\|\mathbf{u}(k)\|^p \le 0$, and hence $W(k)\le W(a) = V(a,\mathbf{u}^0)$ for all $k\in\mathbb{N}(a)$. Therefore, $\sum_{\ell=a}^{k-1}\|\mathbf{u}(\ell)\|^p \le (1/c)V(a,\mathbf{u}^0)$, and hence $\sum_{\ell=a}^{\infty}\|\mathbf{u}(\ell)\|^p < \infty$. ∎

## 5.11. Stability of Discrete Models in Population Dynamics

The basic model equations which concern us here are of the form (5.4.1) where for each $i$, $1 \le i \le n$, $u_i(k)$ is nonnegative for all $k \in \mathbb{N}$ and the $f_i$ are nonnegative functions of $u_1, \cdots, u_n$. In the context of population biology $u_i(k)$ is related to the magnitude of the population of the $i$th species at time $k$ (although each population varies continuously, $u_i(k)$ can be thought of as representing, say, the maxima, average or total population of the $i$th species at time $k$). To study the stability properties of the critical points of such systems we need to modify slightly some of the results proved in Section 5.9. For this, first we give a variation of the original definition of Lyapunov functions.

**Definition 5.11.1.** Let $\Omega_+$ be any set in $\mathbb{R}^n_+$. The scalar function $V(\mathbf{u})$ defined on $\Omega_+$ is said to be a Lyapunov function of (5.4.1) provided

(i) $V$ is continuous, and

(ii) $\Delta V(\mathbf{u}) = V(\mathbf{f}(\mathbf{u})) - V(\mathbf{u}) \le 0$ for all $\mathbf{u} \in \Omega_+$.

**Theorem 5.11.1.** Let $\mathbf{u} = \overline{\mathbf{u}}$ be a critical point of the difference system (5.4.1), and let there exist a Lyapunov function $V(\mathbf{u})$ of (5.4.1) on $\mathbb{R}^n_+$ with a unique global minimum at $\overline{\mathbf{u}}$, $V(\mathbf{u}) \to \infty$ as $\|\mathbf{u}\| \to \infty$ and $u_i \to 0_+$ for each $i$, $1 \le i \le n$ and $\Delta V(\mathbf{u}) < 0$ for all $\mathbf{u} \in \mathbb{R}^n_+$ with $\mathbf{u} \ne \overline{\mathbf{u}}$ then $\overline{\mathbf{u}}$ is globally asymptotically stable.

**Theorem 5.11.2.** Suppose that there exists a Lyapunov function $V(\mathbf{u})$ of (5.4.1) on $\mathbb{R}^n_+$ and $V(\mathbf{u}) \to \infty$ as $\|\mathbf{u}\| \to \infty$. Let the set $E = \{\mathbf{u} \in \mathbb{R}^n_+ : \Delta V(\mathbf{u}) = 0\}$, and $M$ be the largest invariant set of $E$. If $M$ is compact, then every solution $\mathbf{u}(k, 0, \mathbf{u}^0)$ of (5.4.1) starting in $\mathbb{R}^n_+$ approaches $M$ as $k \to \infty$.

For population models the set $M$ often consists of the origin and the positive critical point $\overline{\mathbf{u}}$ of (5.4.1). To prove that $\overline{\mathbf{u}}$ is globally asymptotically stable by Theorem 5.11.2 we need to establish that no solution starting in $\mathbb{R}^n_+$ can approach the origin as $k \to \infty$. We state as a separate theorem, the case when $M$ consists either of $\overline{\mathbf{u}}$ or the origin as well as $\overline{\mathbf{u}}$.

**Theorem 5.11.3.** Suppose that the conditions of Theorem 5.11.2 are satisfied, and either (i) $M = \{\overline{\mathbf{u}}\}$, or (ii) $M = \{0, \overline{\mathbf{u}}\}$, and no solution starting in $\mathbb{R}^n_+$ can approach $0$ as $k \to \infty$, then $\overline{\mathbf{u}}$ is globally asymptotically stable.

**Remark 5.11.1.** If we can express the system (5.4.1) in the form

$$u_i(k+1) = u_i(k)f_i(\mathbf{u}(k)), \quad 1 \le i \le n$$

where each $f_i$ is positive, then no solution starting in $\mathbb{R}^n_+$ can approach the origin as $k \to \infty$ provided $f_i(0) > 1$ for some $i$, $1 \le i \le n$.

**Example 5.11.1.** Consider the model given by

$$u(k+1) = \lambda u(k)/(1 + \alpha u(k))^\beta, \quad k \in \mathbb{N} \tag{5.11.1}$$

where $u(k)$ and $u(k+1)$ are the populations in successive generations, $\lambda$ is the finite nett rate of increase $(\lambda > 1)$ and $\alpha$ and $\beta$ are constants defining the density dependent feedback term. The positive critical point of this model is at $\overline{u} = (1-\theta)/(\alpha\theta)$, where $\theta = \lambda^{-1/\beta}$ $(0 < \theta < 1)$. If we make the change of variable $v = u/\overline{u}$, then equation (5.11.1) becomes

$$v(k+1) = v(k)/(\theta + (1-\theta)v(k))^\beta. \tag{5.11.2}$$

Let $V(v) = (\ln v)^2$, then

$$\begin{aligned} \Delta V(v) &= [\ln v - \beta \ln(\theta + (1-\theta)v)]^2 - [\ln v]^2 \\ &= -\beta \ln(\theta + (1-\theta)v)[2\ln v - \beta\ln(\theta + (1-\theta)v)]. \end{aligned}$$

The function $\ln(\theta+(1-\theta)v)$ is negative for $v \in (0,1)$ and positive for $v \in (1,\infty)$. It remains to examine the function $h(v) = 2\ln v - \beta\ln(\theta+(1-\theta)v)$. Now, $h(1) = 0$, $h(v) < 0$ as $v \to 0_+$, $h(v) \sim \ln(v^{2-\beta}/(1-\theta)^\beta)$ as $v \to \infty$ and $h'(v) = [2\theta + v(1-\theta)(2-\beta)]/[v(\theta + (1-\theta)v)]$. If we restrict $\beta$ so that $0 < \beta \le 2$, then $h(v) > 0$ as $v \to \infty$ and $h'(v) > 0$ for all $v > 0$. This implies that $h(v) < 0$ for $v \in (0,1)$ and $h(v) > 0$ for $v \in (1,\infty)$, i.e. $\Delta V(v) < 0$ for all $v > 0$ and $v \ne 1$. Hence, from Theorem 5.11.1 the critical point $\overline{v} = 1$ of (5.11.2) (or equivalently, $\overline{u} = (1-\theta)/(\alpha\theta)$ of (5.11.1)) is globally asymptotically stable if $\beta \in (0,2]$.

**Example 5.11.2.** Consider the two species competition model given by

$$\begin{aligned} u_1(k+1) &= \lambda_1 u_1(k)[1 + \alpha_1(u_1(k) + \gamma_1 u_2(k))]^{-\beta_1} \\ u_2(k+1) &= \lambda_2 u_2(k)[1 + \alpha_2(u_2(k) + \gamma_2 u_1(k))]^{-\beta_2}, \quad k \in \mathbb{N} \end{aligned} \tag{5.11.3}$$

where $\lambda_1$ and $\lambda_2$ are the finite rates of increase of the two species, $\gamma_1$ and $\gamma_2$ are the competition coefficients and $\alpha_1$, $\alpha_2$, $\beta_1$ and $\beta_2$ are constants defining the form of the feedback relationships. If we scale the populations with respect to their carrying capacities $r_i = (1/\theta_i - 1)/\alpha_i$, where $\theta_i = \lambda_i^{-1/\beta_i}$, $0 < \theta_i < 1$, $i = 1, 2$, then we can rewrite the equations as

$$\begin{aligned} v_1(k+1) &= v_1(k)[\theta_1 + (1-\theta_1)(v_1(k) + d_1 v_2(k))]^{-\beta_1} \\ v_2(k+1) &= v_2(k)[\theta_2 + (1-\theta_2)(v_2(k) + d_2 v_1(k))]^{-\beta_2}, \end{aligned} \tag{5.11.4}$$

where $d_1 = \gamma_1 r_2/r_1$ and $d_2 = \gamma_2 r_1/r_2$. The positive critical point of (5.11.4) is given by $\overline{v}_1 = (1-d_1)/(1-d_1 d_2)$, $\overline{v}_2 = (1-d_2)/(1-d_1 d_2)$, where $d_i \in (0,1)$. For the critical point we shall show that the system is globally asymptotically stable if $\theta_1 = \theta_2 = \theta$ and $\beta_i \in (0,1]$, $i = 1,2$. For this, we need the following two elementary inequalities

(5.11.5) $\ln(1-t) \le -t$ for all $t \in (-\infty, 1)$, with equality only when $t = 0$,

(5.11.6) $(1-t)^{-p} - 1 \le pt(1-t)^{-1}$ for all $t \in (-\infty, 1)$ and $p \in (0,1]$.

Let $V_i(\mathbf{v}) = v_i/\overline{v}_i - 1 - \ln(v_i/\overline{v}_i)$, $i = 1,2$. Then, it follows that

$$\Delta V_1(\mathbf{v}) = (v_1/\overline{v}_1)\left([\theta + (1-\theta)(v_1 + d_1 v_2)]^{-\beta_1} - 1\right) + \beta_1 \ln[\theta + (1-\theta)(v_1 + d_1 v_2)].$$

We now use inequalities (5.11.5) and (5.11.6) with $t = (1-\theta)(1-v_1-d_1v_2)$ and $p = \beta_1$, to obtain

$$\begin{aligned}\Delta V_1(\mathbf{v}) &\le \frac{\beta_1(v_1/\overline{v}_1)(1-\theta)(1-v_1-d_1v_2)}{\theta + (1-\theta)(v_1+d_1v_2)} - \beta_1(1-\theta)(1-v_1-d_1v_2)\\ &= \frac{\beta_1(1-\theta)(1-v_1-d_1v_2)}{\theta + (1-\theta)(v_1+d_1v_2)}\left\{\frac{d_1}{\overline{v}_1}(v_1\overline{v}_2 - v_2\overline{v_1}) - \theta(1-v_1-d_1v_2)\right\}.\end{aligned}$$

However, since $(1-\overline{v}_1)/d_1 = \overline{v}_2$, we have

$$\Delta V_1(\mathbf{v}) \le -\frac{\beta_1\theta(1-\theta)(1-v_1-d_1v_2)^2}{\theta + (1-\theta)(v_1+d_1v_2)} + \frac{\beta_1 d_1(1-\theta)(1-v_1-d_1v_2)(v_1\overline{v}_2 - v_2\overline{v}_1)}{\overline{v}_1[\theta + (1-\theta)(v_1+d_1v_2)]},$$

for $\beta_1 \in (0,1]$, with equality only when $\mathbf{v} = \overline{\mathbf{v}}$. Similarly, we have for $\beta \in (0,1]$

$$\Delta V_2(\mathbf{v}) \le -\frac{\beta_2\theta(1-\theta)(1-v_2-d_2v_1)^2}{\theta + (1-\theta)(v_2+d_2v_1)} + \frac{\beta_2 d_2(1-\theta)(1-v_2-d_2v_1)(v_2\overline{v}_1 - v_1\overline{v}_2)}{\overline{v}_2[\theta + (1-\theta)(v_2+d_2v_1)]}.$$

Thus, if $V(\mathbf{v}) = c_1V_1(\mathbf{v}) + c_2V_2(\mathbf{v})$, then it follows that

$$\begin{aligned}\Delta V(\mathbf{v}) \le\ & -\frac{c_1\beta_1\theta(1-\theta)(1-v_1-d_1v_2)^2}{\theta + (1-\theta)(v_1+d_1v_2)} - \frac{c_2\beta_2\theta(1-\theta)(1-v_2-d_2v_1)^2}{\theta + (1-\theta)(v_2+d_2v_1)}\\ & - \frac{R(1-\theta)(1-d_1d_2)(v_1\overline{v}_2 - v_2\overline{v}_1)^2}{v_1^* v_2^*[\theta + (1-\theta)(v_1+d_1v_2)][\theta + (1-\theta)(v_2+d_2v_1)]},\end{aligned}$$

where $c_1\beta_1d_1\overline{v}_2 = c_2\beta_2d_2\overline{v}_1 = R$. Hence, $\Delta V(\mathbf{v}) \le 0$ for all $\mathbf{v} > 0$ with equality only when $\mathbf{v} = \overline{\mathbf{v}}$. Therefore, from Theorem 5.11.1 the critical point $\overline{\mathbf{v}}$ of (5.11.4) is globally asymptotically stable, if $\theta_1 = \theta_2$ and $\beta_1, \beta_2 \in (0,1]$.

**Example 5.11.3.** Consider the model given by

$$u(k+1) = u(k)\exp(r(1-u(k)/\lambda)), \quad k \in \mathbb{N} \tag{5.11.7}$$

where $r$ is the growth rate and $\lambda$ is the carrying capacity. We shall show that the critical point $\overline{u} = \lambda$ of (5.11.7) is globally asymptotically stable, if $r \in (0,2]$. For this, let $v = u/\lambda$ so that the equation (5.11.7) becomes

$$v(k+1) = v(k)\exp(r(1-v(k))). \tag{5.11.8}$$

Let $V(v) = (v-1)^2$, then

$$\Delta V(v) = -vh(v)[1-\exp(r(1-v))],$$

where $h(v) = v\exp(r(1-v))+v-2$. Now $h(0) < 0$, $h(1) = 0$ and $h(v) > 0$ for $v \geq 2$. Consider $v \in (0,2)$ with $v \neq 1$. Obviously, $h(v) = 0$ if $r = \left(\frac{1}{1-v}\right)\ln\left(\frac{2-v}{v}\right)$. If $v \in (0,1)$, then let $w = \frac{1}{1-v} > 1$ so that

$$r = w\left\{\ln\left(1+\frac{1}{w}\right) - \ln\left(1-\frac{1}{w}\right)\right\} = w\left\{\sum_{p=1}^{\infty}(-1)^{p+1}\frac{w^{-p}}{p} + \sum_{p=1}^{\infty}\frac{w^{-p}}{p}\right\}$$

$$= 2\sum_{p=0}^{\infty}\frac{w^{-2p}}{2p+1} > 2.$$

Similarly, if $v \in (1,2)$, then let $w = \frac{1}{v-1} > 1$ so that

$$r = w\left\{\ln\left(1+\frac{1}{w}\right) - \ln\left(1-\frac{1}{w}\right)\right\} > 2.$$

Hence, for $r \in (0,2]$ we have that $h(v) < 0$ for $v \in (0,1)$ and $h(v) > 0$ for $v \in (1,\infty)$. Thus, $V(v)$ is a Lyapunov function of (5.11.8) in $\mathbb{R}_+$. The set of points in $\overline{\mathbb{R}}_+$ where $\Delta V(v) = 0$ consists only of $0$ and $1$, and from Remark 5.11.1 no solution starting in $\mathbb{R}_+$ can approach $0$ as $k \to \infty$. Therefore, from Theorem 5.11.3 the critical point $\overline{v} = 1$ of (5.11.8) (or equivalently, $\overline{u} = \lambda$ of (5.11.7)) is globally asymptotically stable if $r \in (0,2]$.

**Example 5.11.4.** A model of two competing species is

$$\begin{aligned} u_1(k+1) &= u_1(k)\exp(r_1[\lambda_1-\alpha_{11}u_1(k)-\alpha_{12}u_2(k)]/\lambda_1) \\ u_2(k+1) &= u_2(k)\exp(r_2[\lambda_2-\alpha_{21}u_1(k)-\alpha_{22}u_2(k)]/\lambda_2), \quad k \in \mathbb{N} \end{aligned} \tag{5.11.9}$$

where $r_i$ and $\lambda_i$ are growth rates and carrying capacities and $\alpha_{ij}$ the competition coefficients. If we scale the populations with respect to their carrying capacities and let $d_{11} = \alpha_{11}$, $d_{12} = \alpha_{12}\lambda_2/\lambda_1$, $d_{21} = \alpha_{21}\lambda_1/\lambda_2$ and $d_{22} = \alpha_{22}$ the system (5.11.9) becomes

$$\begin{aligned} v_1(k+1) &= v_1(k)\exp(r_1[1 - d_{11}v_1(k) - d_{12}v_2(k)]) \\ v_2(k+1) &= v_2(k)\exp(r_2[1 - d_{21}v_1(k) - d_{22}v_2(k)]). \end{aligned} \tag{5.11.10}$$

The positive critical point of this system occurs at $\overline{v}_1 = (d_{22}-d_{12})/D$, $\overline{v}_2 = (d_{11} - d_{21})/D$, where $D = d_{11}d_{22} - d_{12}d_{21}$. Now following similar lines as in Examples 5.11.2 and 5.11.3 it follows that this critical point of (5.11.10) is globally asymptotically stable provided $r_1 = r_2 = r$ where $r \in (0, 1]$.

Now let $u(k)$ denote the adult breeding population in the year $k$ and suppose the recruitment to the breeding population takes place $n - 1$ ($n > 1$) years after birth. A model of this type is provided by the so called *delay difference equation*

$$u(k+1) = Su(k) + (1 - S)g(u(k - n + 1)), \quad k \in \mathbb{N} \tag{5.11.11}$$

where $S \in [0, 1]$ is a survival coefficient and the term $(1-S)g(u(k-n+1))$ represents recruitment. In system form equation (5.11.11) can be written as

$$\begin{aligned} u_i(k+1) &= u_{i+1}(k), \quad i = 1, \cdots, n-1 \\ u_n(k+1) &= Su_n(k) + (1 - S)g(u_1(k)), \end{aligned} \tag{5.11.12}$$

where $\mathbf{u}(k) \in \mathbb{R}^n_+$ for each $k \in \mathbb{N}$ and is defined by

$$u_i(k) = u(k + i - n), \quad i = 1, \cdots, n. \tag{5.11.13}$$

In what follows we assume that no solution of (5.11.11) in which the populations are initially positive can approach the origin as $k \to \infty$. If the origin is a critical point, i.e. $g(0) = 0$, then a sufficient condition for no solution to approach the origin is $g'(0) > 1$.

**Theorem 5.11.4.** Let $u = \overline{u}$ be a positive critical point of the delay difference equation (5.11.11), i.e. $\overline{u} = g(\overline{u})$. If there exists a convex function $V(v)$ which is a Lyapunov function of the scalar equation

$$v(k+1) = g(v(k)) \tag{5.11.14}$$

on $\mathbb{R}_+$ and $V(v) \to \infty$ as $v \to \infty$ then $\overline{u}$ of (5.11.11) is globally asymptotically stable.

**Proof.** Comparing the system (5.11.12) with (5.4.1), we have

$$\begin{aligned} V(f_i(\mathbf{u})) &= V(u_{i+1}), \quad i = 1, \cdots, n-1 \\ V(f_n(\mathbf{u})) &= V(Su_n + (1 - S)g(u_1)) \le SV(u_n) + (1 - S)V(g(u_1)). \end{aligned}$$

Since $V(g(u_1)) \le V(u_1)$ for all $u_1 \in \mathbb{R}_+$, it follows that

$$V(f_n(\mathbf{u})) \ \le\ SV(u_n) + (1-S)V(u_1).$$

Now define $W(\mathbf{u}) = \sum_{i=1}^{n-1} V(u_i) + V(u_n)/(1-S)$. Then, for the system (5.11.12), we find that

$$\begin{aligned} W(\mathbf{f}(\mathbf{u})) - W(\mathbf{u}) &= \sum_{i=1}^{n-1}[V(f_i(\mathbf{u})) - V(u_i)] + [V(f_n(\mathbf{u})) - V(u_n)]/(1-S) \\ &\le \sum_{i=1}^{n-1}[V(u_{i+1}) - V(u_i)] - V(u_n) + V(u_1) \ = \ 0. \end{aligned}$$

Hence, $W(\mathbf{f}(\mathbf{u})) \le W(\mathbf{u})$ for all $\mathbf{u} \in \mathbb{R}_+^n$ and $W(\mathbf{u})$ is a Lyapunov function of the system (5.11.12), with the property that $W(\mathbf{u}) \to \infty$ as $\|\mathbf{u}\| \to \infty$. The only invariant points in $\mathbb{R}_+^n$, are those which satisfy $u_1 = \cdots = u_n = g(u_1)$. Since the only solutions of $g(u) = u$ are $\overline{u}$ and possibly the origin, it follows by Theorem 5.11.3 that $M = \{0, \overline{\mathbf{u}}\}$, where $\overline{\mathbf{u}} = (\overline{u}, \cdots, \overline{u})$. However, since no solution of (5.11.12) beginning in $\mathbb{R}_+^n$ can approach $0$ as $k \to \infty$, the critical point $\overline{\mathbf{u}}$ of (5.11.12) is globally asymptotically stable. Thus, the critical point $\overline{u}$ of (5.11.11) is globally asymptotically stable. ∎

**Example 5.11.5.** Consider the stock recruitment model

$$u(k+1) \ = \ Su(k) + \frac{ru(k-n+1)}{1+wu(k-n+1)}, \tag{5.11.15}$$

where the critical population is given by $w\overline{u} = \dfrac{r}{(1-S)} - 1$, and the parameters $r$ and $S$ satisfy $0 < 1-S < r < 1$. The associated scalar equation corresponding to (5.11.14) is

$$v(k+1) \ = \ \frac{rv(k)}{(1-S)(1+wv(k))}. \tag{5.11.16}$$

Let the convex function $V(v)$ be $V(v) = |v - \overline{u}|$. Since

$$\begin{aligned} V(g(v)) &= \left|\frac{rv}{(1-S)(1+wv)} - \overline{u}\right| = \frac{r}{(1-S)}\left|\frac{v}{1+wv} - \frac{\overline{u}}{1+w\overline{u}}\right| \\ &= |v-\overline{u}|/(1+wv) \end{aligned}$$

we find that $V(g(v)) \le V(v)$ for all $v > 0$. Thus, $V(v)$ is a Lyapunov function of (5.11.16) in $\mathbb{R}_+$. Therefore, by Theorem 5.11.4 the critical point $\overline{u}$ of (5.11.15) is globally asymptotically stable.

**Example 5.11.6.** Consider the stock recruitment model

$$(5.11.17) \qquad u(k+1) = Su(k) + \lambda u(k-n+1)\exp(-qu(k-n+1)),$$

where the critical population is given by $\lambda \exp(-q\overline{u}) = 1 - S$, and the parameters $\lambda$ and $S$ satisfy $0 < \dfrac{\lambda}{(1-S)} \le e^2$. The associated scalar equation corresponding to (5.11.14) is

$$(5.11.18) \qquad v(k+1) = \frac{\lambda}{(1-S)} v(k)\exp(-qv(k)).$$

If we let $r = \ln \dfrac{\lambda}{1-S}$ and $\overline{u} = \lambda = r/q$, then (5.11.18) is the same as (5.11.7). However, in Example 5.11.3 we have seen that for (5.11.7) the function $V(u) = (u-\lambda)^2$ is a Lyapunov function in $\mathbb{R}_+$ provided $r \in (0,2]$. Therefore, by Theorem 5.11.4 the critical point $\overline{u}$ of (5.11.17) is globally asymptotically stable.

## 5.12. Converse Theorems

All the results discussed in Sections 5.9 and 5.10 provide only sufficient conditions, they give no idea on how to construct Lyapunov functions for a given difference system. In fact, there is no general method to construct such functions. But, here assuming certain stability properties of the given difference system we shall provide the construction of the Lyapunov functions. These types of results are called converse theorems and play a very important role in studying the properties of solutions of perturbed systems.

**Theorem 5.12.1.** Suppose that the linear system (1.2.12) is uniformly asymptotically stable. Then, there exist constants $c > 1$, $\lambda > 0$ and a scalar function $V(k, \mathbf{u})$ in $\mathbb{N}(a) \times S_\rho$ such that

(i) $\|\mathbf{u}\| \le V(k,\mathbf{u}) \le c\|\mathbf{u}\|$

(ii) for any solution $\mathbf{u}(k) = \mathbf{u}(k,a,\mathbf{u}^0)$ of (1.2.12) such that $\|\mathbf{u}(k)\| < \rho$

$$(5.12.1) \qquad \Delta V(k,\mathbf{u}(k)) \le -(1-e^{-\lambda})V(k,\mathbf{u}(k))$$

(iii) for $k \in \mathbb{N}(a)$, $\mathbf{u}$, $\mathbf{v} \in S_\rho$

$$(5.12.2) \qquad |V(k,\mathbf{u}) - V(k,\mathbf{v})| \le c\|\mathbf{u}-\mathbf{v}\|.$$

**Proof.** Since (1.2.12) is uniformly asymptotically stable, from Theorem 5.5.1 (v) there exist constants $c > 1$ and $\lambda > 0$ such that the solution $\mathbf{u}(k) = \mathbf{u}(k,a,\mathbf{u}^0)$ of (1.2.12) with $\mathbf{u}(k) \in S_\rho$, $k \in \mathbb{N}(a)$ satisfies

$$(5.12.3) \qquad \|\mathbf{u}(k)\| \le c\exp(-\lambda(k-\ell))\|\mathbf{u}(\ell)\|$$

for $a \le \ell \le k \in \mathbb{N}(a)$ and $\mathbf{u}(\ell) \in S_{\rho(c)}$. Let $V(k, \mathbf{u})$ be defined by

$$V(k, \mathbf{u}) = \sup_{\tau \in \mathbb{N}} \|\mathbf{u}(k+\tau, k, \mathbf{u})\| e^{\lambda \tau}. \tag{5.12.4}$$

Then, clearly $\|\mathbf{u}\| \le V(k, \mathbf{u})$, and from (5.12.3) it follows that $V(k, \mathbf{u}) \le \sup_{\tau \in \mathbb{N}} ce^{-\lambda\tau}\|\mathbf{u}\| e^{\lambda\tau} = c\|\mathbf{u}\|$. Now since the solutions of (1.2.12) are unique, we find that

$$\begin{aligned}
V(k+1, \mathbf{u}(k+1)) &= \sup_{\tau \in \mathbb{N}} \|\mathbf{u}(k+1+\tau, k+1, \mathbf{u}(k+1, a, \mathbf{u}^0))\| e^{\lambda\tau} \\
&= \sup_{\tau \in \mathbb{N}} \|\mathbf{u}(k+1+\tau, a, \mathbf{u}^0)\| e^{\lambda\tau} \\
&= \sup_{\tau \in \mathbb{N}(1)} \|\mathbf{u}(k+\tau, a, \mathbf{u}^0)\| e^{\lambda(\tau-1)} \\
&\le \sup_{\tau \in \mathbb{N}} \|\mathbf{u}(k+\tau, a, \mathbf{u}^0)\| e^{\lambda(\tau-1)} \\
&= \sup_{\tau \in \mathbb{N}} \|\mathbf{u}(k+\tau, k, \mathbf{u}(k, a, \mathbf{u}^0))\| e^{\lambda\tau} \times e^{-\lambda} \\
&= V(k, \mathbf{u}(k)) e^{-\lambda},
\end{aligned}$$

which is the same as (5.12.1).

Finally, from the linearity of (1.2.12), it is obvious that

$$\mathbf{u}(k+\tau, k, \mathbf{u}) - \mathbf{u}(k+\tau, k, \mathbf{v}) = \mathbf{u}(k+\tau, k, \mathbf{u} - \mathbf{v})$$

and hence, from the definition of $V(k, \mathbf{u})$, we have

$$\begin{aligned}
|V(k, \mathbf{u}) - V(k, \mathbf{v})| &\le \sup_{\tau \in \mathbb{N}} \|\mathbf{u}(k+\tau, k, \mathbf{u}) - \mathbf{u}(k+\tau, k, \mathbf{v})\| e^{\lambda\tau} \\
&= \sup_{\tau \in \mathbb{N}} \|\mathbf{u}(k+\tau, k, \mathbf{u} - \mathbf{v})\| e^{\lambda\tau} \\
&= V(k, \mathbf{u} - \mathbf{v}) \le c\|\mathbf{u} - \mathbf{v}\|. \quad \blacksquare
\end{aligned}$$

**Remark 5.12.1.** In view of Remark 5.5.2 and Theorem 5.10.3, Theorem 5.12.1 can also be regarded as a converse theorem for the exponential asymptotic stability of (1.2.12).

**Lemma 5.12.2.** The trivial solution $\mathbf{u}(k, a, 0) \equiv 0$ of (1.2.8) is uniformly stable if and only if there exists a function $\psi \in \mathcal{K}$ such that $\|\mathbf{u}(k)\| \le \psi(\|\mathbf{u}(\ell)\|)$, $k \in \mathbb{N}(\ell)$, whenever $\ell \ge a$ and $\|\mathbf{u}(\ell)\| < \rho$.

**Proof.** The sufficiency of the given condition is obvious. We shall show that this condition is also necessary. Suppose the trivial solution

$\mathbf{u}(k,a,0) \equiv 0$ of (1.2.8) is uniformly stable. Then, for a given $\epsilon$ such that $0 < \epsilon < \rho$, there exists a $\delta = \delta(\epsilon) > 0$ such that the inequalities $\ell \geq a$ and $\|\mathbf{u}(\ell)\| < \delta$ imply $\|\mathbf{u}(k)\| < \epsilon$ for all $k \in \mathbb{N}(\ell)$. Let $\delta_1 = \delta_1(\epsilon)$ be the least upper bound of all numbers $\delta$. Clearly, if $\ell \geq a$ and $\|\mathbf{u}(\ell)\| \leq \delta_1$ then $\|\mathbf{u}(k)\| < \epsilon$ for $k \geq \ell$. Also, for every $\delta_2 > \delta_1$ there exists some $k_1 \geq \ell$ and some $\mathbf{u}(k_1) \in S_\rho$, $\|\mathbf{u}(k_1)\| \leq \delta_2$, such that $\|\mathbf{u}(k)\|$ exceeds the value $\epsilon$ at some value of $k \geq k_1$. Obviously, the function $\delta_1(\epsilon)$ is positive for $\epsilon > 0$, nondecreasing, and tends to zero as $\epsilon \to 0$, but it may be discontinuous. However, we can always choose a positive, continuous, and monotonically increasing function $\hat{\delta} = \hat{\delta}(\epsilon)$ satisfying $\hat{\delta}(\epsilon) \leq \delta_1(\epsilon)$. Let $\psi$ be the inverse function of $\hat{\delta}$. Then, for $\mathbf{u}(k) \in S_\rho$, $\ell \geq a$ and $\|\mathbf{u}(\ell)\| \leq \hat{\delta}$, there exists an $\epsilon_1 > 0$ such that $\ell \geq a$ and $\|\mathbf{u}(\ell)\| \leq \hat{\delta}(\epsilon_1)$ imply $\|\mathbf{u}(k)\| \leq \epsilon_1 \leq \psi(\|\mathbf{u}(\ell)\|)$ for $k \geq \ell$. ∎

**Theorem 5.12.3.** Suppose that the trivial solution $\mathbf{u}(k,a,0) \equiv 0$ of (1.2.8) is uniformly stable. Then, there exists a positive definite and decrescent scalar function $V(k,\mathbf{u})$ in $\mathbb{N}(a) \times S_\rho$ such that $\Delta V(k,\mathbf{u}(k,a,\mathbf{u}^0)) \leq 0$ for any solution $\mathbf{u}(k) = \mathbf{u}(k,a,\mathbf{u}^0)$ of (1.2.8) with $\|\mathbf{u}(k)\| < \rho$.

**Proof.** Define a scalar function

$$V(k,\mathbf{u}) = \sup_{\tau \in \mathbb{N}} \|\mathbf{u}(k+\tau,k,\mathbf{u})\|. \tag{5.12.5}$$

Clearly, $\sup_{\tau \in \mathbb{N}} \|\mathbf{u}(k+\tau,k,\mathbf{u})\| \geq \|\mathbf{u}(k,k,\mathbf{u})\| = \|\mathbf{u}\|$. Moreover, from Lemma 5.12.2 there exists a function $\psi \in \mathcal{K}$ such that $\|\mathbf{u}(k+\tau,k,\mathbf{u})\| \leq \psi(\|\mathbf{u}(k,k,\mathbf{u}\|) = \psi(\|\mathbf{u}\|)$, for all $\tau \in \mathbb{N}$. Thus, $V(k,\mathbf{u}) \leq \psi(\|\mathbf{u}\|)$.

Next, for every solution $\mathbf{u}(k) = \mathbf{u}(k,a,\mathbf{u}^0)$ such that $\|\mathbf{u}(k)\| < \rho$, we have

$$\begin{aligned} V(k,\mathbf{u}(k,a,\mathbf{u}^0)) &= \sup_{\tau \in \mathbb{N}} \|\mathbf{u}(k+\tau,k,\mathbf{u}(k,a,\mathbf{u}^0))\| \\ &= \sup_{\tau \in \mathbb{N}} \|\mathbf{u}(k+\tau,a,\mathbf{u}^0)\|. \end{aligned}$$

Thus, it follows that

$$\begin{aligned} V(k+1,\mathbf{u}(k+1,a,\mathbf{u}^0)) &= \sup_{\tau \in \mathbb{N}} \|\mathbf{u}(k+1+\tau,a,\mathbf{u}^0)\| \\ &\leq \sup_{\tau \in \mathbb{N}} \|\mathbf{u}(k+\tau,a,\mathbf{u}^0)\| \\ &= V(k,\mathbf{u}(k,a,\mathbf{u}^0)). \end{aligned}$$ ∎

**Theorem 5.12.4.** Suppose that the trivial solution $\mathbf{u}(k,a,0) \equiv 0$ of (1.2.8) is $s_p$ stable and $\|\mathbf{u}(k,a,\mathbf{u}^0)\| \leq h(k)\psi(\|\mathbf{u}^0\|)$, where $\psi \in \mathcal{K}$ and the function $h(k)$ is defined and nonnegative on $\mathbb{N}$ with $\sum_{\ell=0}^{\infty} h^p(\ell) = H < \infty$. Then, there exists a positive definite and decrescent scalar function $V(k,\mathbf{u})$ in $\mathbb{N}(a)\times S_\rho$ such that $\Delta V(k,\mathbf{u}(k,a,\mathbf{u}^0)) \le -\|\mathbf{u}(k,a,\mathbf{u}^0)\|^p$ for any solution $\mathbf{u}(k) = \mathbf{u}(k,a,\mathbf{u}^0)$ of (1.2.8) with $\|\mathbf{u}(k)\| < \rho$.

**Proof.** Define a scalar function

$$(5.12.6) \qquad V(k,\mathbf{u}) = \sum_{\tau=0}^{\infty} \|\mathbf{u}(k+\tau,k,\mathbf{u})\|^p.$$

Clearly, $V(k,\mathbf{u}) \ge \|\mathbf{u}\|^p$. Moreover, $V(k,\mathbf{u}) \le \psi^p(\|\mathbf{u}\|)\sum_{\tau=0}^{\infty} h^p(\tau) = H\psi^p(\|\mathbf{u}\|)$. Also,

$$\begin{aligned}\Delta V(k,\mathbf{u}(k,a,\mathbf{u}^0)) &= \sum_{\tau=0}^{\infty}\|\mathbf{u}(k+\tau+1,k+1,\mathbf{u}(k+1,a,\mathbf{u}^0))\|^p \\ &\quad -\sum_{\tau=0}^{\infty}\|\mathbf{u}(k+\tau,k,\mathbf{u}(k,a,\mathbf{u}^0))\|^p \\ &= \sum_{\tau=0}^{\infty}\|\mathbf{u}(k+\tau+1,a,\mathbf{u}^0)\|^p - \sum_{\tau=0}^{\infty}\|\mathbf{u}(k+\tau,a,\mathbf{u}^0)\|^p \\ &= -\|\mathbf{u}(k,a,\mathbf{u}^0)\|^p. \quad \blacksquare\end{aligned}$$

**Theorem 5.12.5.** Suppose that the trivial solution $\mathbf{u}(k,a,0) \equiv 0$ of (1.2.8) is uniformly asymptotically stable, and for all $(k,\mathbf{u})$, $(k,\mathbf{v}) \in \mathbb{N}(a)\times S_\rho$ the function $\mathbf{f}(k,\mathbf{u})$ satisfies the Lipschitz condition

$$(5.12.7) \qquad \|\mathbf{f}(k,\mathbf{u}) - \mathbf{f}(k,\mathbf{v})\| \le L\|\mathbf{u}-\mathbf{v}\|.$$

Then, there exists a positive definite and decrescent scalar function $V(k,\mathbf{u})$ in $\mathbb{N}(a)\times S_\rho$ such that $\Delta V(k,\mathbf{u}(k,a,\mathbf{u}^0)) \le -\psi(\|\mathbf{u}(k+1,a,\mathbf{u}^0)\|)$, where $\psi \in \mathcal{K}$, and $V(k,\mathbf{u})$ satisfies the Lipschitz condition (5.12.2).

**Proof.** Define a scalar function $G(r)$ such that $G(0) = G'(0) = 0$, $G'(r) > 0$, and $G''(r) > 0$. Let $\alpha > 1$. Since $G(r) = \int_0^r dt_1 \int_0^{t_1} G''(t)dt$, and $G(r/\alpha) = \int_0^{r/\alpha} dt_1 \int_0^{t_1} G''(t)dt$, on setting $t_1 = t_2/\alpha$, we obtain $G(r/\alpha) = (1/\alpha)\int_0^r dt_2 \int_0^{t_2/\alpha} G''(t)dt < (1/\alpha)\int_0^r dt_2\int_0^{t_2} G''(t)dt = (1/\alpha) \times G(r)$. Now define a scalar function $V(k,\mathbf{u})$ as

$$(5.12.8) \qquad V(k,\mathbf{u}) = \sup_{\tau\in\mathbb{N}} G(\|\mathbf{u}(k+\tau,k,\mathbf{u})\|)\frac{1+\alpha\tau}{1+\tau}.$$

Then, for $\tau = 0$, we get $G(\|\mathbf{u}\|) \le V(k,\mathbf{u})$. Since the trivial solution $\mathbf{u}(k,a,0) \equiv 0$ of (1.2.8) is uniformly stable, by Lemma 5.12.2, we have

$\|\mathbf{u}(k+\tau,k,\mathbf{u})\| \le \psi(\|\mathbf{u}\|)$, $\psi \in \mathcal{K}$. Therefore, $G(\|\mathbf{u}(k+\tau,k,\mathbf{u})\|) \le G(\psi(\|\mathbf{u}\|))$. From the fact that $\dfrac{1+\alpha\tau}{1+\tau} < \alpha$, it follows that $V(k,\mathbf{u}) \le \alpha G(\psi(\|\mathbf{u}\|))$. The trivial solution of (1.2.8) is asymptotically stable also, and hence for $\tau \ge K(\epsilon)$, $\|\mathbf{u}(k+\tau,k,\mathbf{u})\| < \epsilon$, thus, if $\tau \ge K(\|\mathbf{u}\|/\alpha)$, then $\|\mathbf{u}(k+\tau,k,\mathbf{u})\| < \|\mathbf{u}\|/\alpha$. Therefore, $G(\|\mathbf{u}(k+\tau,k,\mathbf{u})\|) \le G(\|\mathbf{u}\|/\alpha)$, which in turn gives $G(\|\mathbf{u}(k+\tau,k,\mathbf{u})\|)\dfrac{1+\alpha\tau}{1+\tau} \le \alpha G(\|\mathbf{u}\|/\alpha) < G(\|\mathbf{u}\|) \le V(k,\mathbf{u})$. This shows that it is sufficient to consider $\tau \in \mathbb{N}(0,K(\|\mathbf{u}\|/\alpha))$, and thus (5.12.8) becomes

$$\text{(5.12.9)}\qquad \begin{aligned} V(k,\mathbf{u}) &= \sup_{\tau\in\mathbb{N}(0,K(\|\mathbf{u}\|/\alpha))} G(\|\mathbf{u}(k+\tau,k,\mathbf{u})\|)\frac{1+\alpha\tau}{1+\tau} \\ &= G(\|\mathbf{u}(k+\tau_1,k,\mathbf{u})\|)\frac{1+\alpha\tau_1}{1+\tau_1}, \end{aligned}$$

where $\tau_1 \in \mathbb{N}(0,K(\|\mathbf{u}\|/\alpha))$ is the integer where sup is achieved. Now, we have

$$\begin{aligned} V(k+1,\mathbf{u}(k+1,a,\mathbf{u}^0)) &= G(\|\mathbf{u}(k+1+\tau_1,k+1,\mathbf{u}(k+1,a,\mathbf{u}^0))\|) \\ &\quad\times \frac{1+\alpha(\tau_1+1)}{1+(\tau_1+1)}\left[1-\frac{\alpha-1}{(1+\tau_1)(1+\alpha+\alpha\tau_1)}\right] \\ &\le V(k,\mathbf{u}(k,a,\mathbf{u}^0))\left[1-\frac{\alpha-1}{(1+\tau_1)(1+\alpha+\alpha\tau_1)}\right], \end{aligned}$$

from which, we obtain

$$\Delta V(k,\mathbf{u}(k,a,\mathbf{u}^0)) \le -\frac{(\alpha-1)G(\|\mathbf{u}(k)\|)}{[1+K(\|\mathbf{u}(k+1)\|/\alpha)][1+\alpha+\alpha K(\|\mathbf{u}(k+1)\|/\alpha)]}.$$

However, (5.12.7) gives $\|\mathbf{u}(k+1)\| \le \|\mathbf{f}(k,\mathbf{u}(k))\| \le L\|\mathbf{u}(k)\|$, and hence $\|\mathbf{u}(k)\| \ge (1/L)\|\mathbf{u}(k+1)\|$. Therefore, we obtain

$$\begin{aligned} \Delta V(k,\mathbf{u}(k,a,\mathbf{u}^0)) &\le -\frac{(\alpha-1)G(L^{-1}\|\mathbf{u}(k+1)\|)}{[1+K(\|\mathbf{u}(k+1)\|/\alpha)][1+\alpha+\alpha K(\|\mathbf{u}(k+1)\|/\alpha)]} \\ &= -\psi(\|\mathbf{u}(k+1,a,\mathbf{u}^0)\|). \end{aligned}$$

Since $K$ is a decreasing function, the function $\psi$ is strictly increasing; and since $G(0)=0$, it is clear that $\psi \in \mathcal{K}$. Now we shall find a function $G$ so that $V$ satisfies the Lipschitz condition (5.12.2). For this, once again since the trivial solution of (1.2.8) is uniformly asymptotically stable, for $\rho > 0$ there exists a $\delta(\rho)$ such that for $\mathbf{u}^1$, $\mathbf{u}^2 \in S_{\delta(\rho)}$ the solutions $\mathbf{u}(k,a,\mathbf{u}^1)$, $\mathbf{u}(k,a,\mathbf{u}^2)$ of (1.2.8) remain in $S_\rho$. Setting $r_1 = \|\mathbf{u}(k+\tau_1,k,\mathbf{u}^1)\|$ and $r_2 = \|\mathbf{u}(k+\tau_1,k,\mathbf{u}^2)\|$. Since $G(r)$ is monotonically increasing in $r$, for $r_2 \geq r_1$, we have $G(r_2) \geq G(r_1)$, and hence

$$\begin{aligned} V(k, \mathbf{u}^2) &= G(\|\mathbf{u}(k+\tau_1, k, \mathbf{u}^2)\|)\frac{1+\alpha\tau_1}{1+\tau_1} \\ &\geq G(\|\mathbf{u}(k+\tau_1, k, \mathbf{u}^1)\|)\frac{1+\alpha\tau_1}{1+\tau_1} = V(k, \mathbf{u}^1). \end{aligned}$$

Moreover, if $r_2 \leq r_1$, then by mean value theorem, we have $0 \leq G(r_1) - G(r_2) \leq G'(r_1)(r_1 - r_2)$. But, because of (5.12.7) we have $r_1 - r_2 \leq \|\mathbf{u}(k+\tau_1, k, \mathbf{u}^1) - \mathbf{u}(k+\tau_1, k, \mathbf{u}^2)\| \leq L^{\tau_1}\|\mathbf{u}^1 - \mathbf{u}^2\|$, and hence

$$0 \leq G(r_1) - G(r_2) \leq G'(r_1)L^{\tau_1}\|\mathbf{u}^1 - \mathbf{u}^2\|.$$

In particular, let $G(r) = A\int_0^r q^{K(\delta(s)/\alpha)}ds$, where $q \leq \min\{L, L^{-1}\}$, $q < 1$ and $A$ is a positive constant. Since $\delta(0) = 0$ and $K(0) = \infty$, this function $G(r)$ satisfies the required conditions. Thus, we have

$$0 \leq G(r_1) - G(r_2) \leq Aq^{K(\delta(r)/\alpha)}L^{r_1}\|\mathbf{u}^1 - \mathbf{u}^2\| \leq A\|\mathbf{u}^1 - \mathbf{u}^2\|.$$

Multiplying this inequality by $(1+\alpha\tau_1)/(1+\tau_1)$ and using the fact that $(1+\alpha\tau_1)/(1+\tau_1) < \alpha$, we get

$$0 \leq V(k, \mathbf{u}^1) - G(\|\mathbf{u}(k+\tau_1, k, \mathbf{u}^2)\|)\frac{1+\alpha\tau_1}{1+\tau_1} \leq \alpha A\|\mathbf{u}^1 - \mathbf{u}^2\|$$

and hence

$$V(k, \mathbf{u}^2) = G(\|\mathbf{u}(k+\tau_1, k, \mathbf{u}^2)\|)\frac{1+\alpha\tau_1}{1+\tau_1} \geq V(k, \mathbf{u}^1) - \alpha A\|\mathbf{u}^1 - \mathbf{u}^2\|.$$

Thus, in both cases $r_2 \geq r_1$ and $r_1 \geq r_2$, $V(k, \mathbf{u})$ satisfies the inequality

$$V(k, \mathbf{u}^2) - V(k, \mathbf{u}^1) \geq -\alpha A\|\mathbf{u}^1 - \mathbf{u}^2\|.$$

By interchanging the roles of $\mathbf{u}^1$ and $\mathbf{u}^2$ in this inequality, we obtain

$$V(k, \mathbf{u}^1) - V(k, \mathbf{u}^2) \geq -\alpha A\|\mathbf{u}^1 - \mathbf{u}^2\|,$$

and hence, by combining the two foregoing inequalities, we get

$$|V(k, \mathbf{u}^1) - V(k, \mathbf{u}^2)| \leq \alpha A\|\mathbf{u}^1 - \mathbf{u}^2\|,$$

which is the same as (5.12.2) with $c = \alpha A$. ∎

## 5.13. Total Stability

Converse theorems developed in the preceding section can be used to obtain certain stability properties of the solutions of perturbed difference systems. We shall show that the uniform asymptotic stability of the trivial solution of (1.2.8) has also certain stability property under different classes of permanent perturbations.

**Definition 5.13.1.** The trivial solution $\mathbf{u}(k,a,0) \equiv 0$ of (1.2.8) is said to be *totally stable* (or, stable under constantly acting perturbations) if, for every $\epsilon > 0$, there exist two positive numbers $\delta_1 = \delta_1(\epsilon)$ and $\delta_2 = \delta_2(\epsilon)$ such that, for every solution $\mathbf{v}(k) = \mathbf{v}(k,a,\mathbf{v}^0)$ of (5.7.1) the inequalities $a \le k_1 \in \mathbb{N}(a)$ and $\|\mathbf{v}(k_1)\| < \delta_1$ imply that $\|\mathbf{v}(k)\| < \epsilon$ for all $k \in \mathbb{N}(a)$, provided $\|\mathbf{g}(k,\mathbf{v})\| < \delta_2$ for $\|\mathbf{v}\| < \epsilon$, $k \in \mathbb{N}(a)$.

**Theorem 5.13.1.** If the trivial solution $\mathbf{u}(k,a,0) \equiv 0$ of (1.2.8) is uniformly asymptotically stable and for all $(k,\mathbf{u})$, $(k,\mathbf{v}) \in \mathbb{N}(a) \times S_\rho$ the function $\mathbf{f}(k,\mathbf{u})$ satisfies the Lipschitz condition (5.12.7), then it is totally stable.

**Proof.** Let $\mathbf{u}(k) = \mathbf{u}(k,a,\mathbf{u}^0)$ be a solution of (1.2.8). From the hypotheses of uniform asymptotic stability of the trivial solution of (1.2.8) and Theorem 5.12.5 it follows that for $0 < \epsilon < \rho$, there exists a $0 < \delta(\epsilon) < \rho$ such that the inequalities $a \le k_1 \in \mathbb{N}(a)$ and $\|\mathbf{u}(k_1)\| < \delta(\epsilon)$ imply $\|\mathbf{u}(k)\| < \epsilon$ for all $k \in \mathbb{N}(k_1)$. Moreover, there exists a scalar function $V(k,\mathbf{u})$ in $\mathbb{N}(a) \times S_\rho$ such that

(i) $\alpha(\|\mathbf{u}\|) \le V(k,\mathbf{u}) \le \beta(\|\mathbf{u}\|)$, where $\alpha, \beta \in \mathcal{K}$

(ii) $\Delta V(k,\mathbf{u}(k)) \le -\psi(\|\mathbf{u}(k+1)\|)$, $\psi \in \mathcal{K}$

(iii) $V(k,\mathbf{u})$ satisfies the Lipschitz condition (5.12.2).

Let $0 < \epsilon_1 < \delta(\epsilon)$, and choose $\delta_1 > 0$, $\delta_2 > 0$ so that $\beta(2\delta_1) < \alpha(\epsilon_1)$, $\delta_2 \le \delta_1$, and $\delta_2 < \psi(\delta_1)/c$. Suppose that $\epsilon_1$ is so small that $L\epsilon_1 + \delta_2 < \delta(\epsilon)$. Let $\|\mathbf{g}(k,\mathbf{v})\| < \delta_2$ for $\|\mathbf{v}\| < \epsilon_1$. Then, for $\|\mathbf{v}\| < \epsilon_1$, we have $\|\mathbf{u}(k+1,k,\mathbf{v})\| = \|\mathbf{f}(k,\mathbf{u}(k,k,\mathbf{v}))\| = \|\mathbf{f}(k,\mathbf{v})\| \le L\epsilon_1$, and $\|\mathbf{v}(k+1,k,\mathbf{v}) - \mathbf{u}(k+1,k,\mathbf{v})\| = \|\mathbf{g}(k,\mathbf{v})\| < \delta_2$, also $\|\mathbf{v}(k+1,k,\mathbf{v})\| \le L\epsilon_1 + \delta_2 < \delta(\epsilon)$. Therefore, for $\|\mathbf{v}\| < \epsilon$, it follows that

$$
\begin{aligned}
& V(k+1,\mathbf{v}(k+1,k,\mathbf{v})) - V(k,\mathbf{v}(k,k,\mathbf{v})) \\
&\quad = V(k+1,\mathbf{u}(k+1,k,\mathbf{v})) - V(k,\mathbf{v}) + V(k+1,\mathbf{v}(k+1,k,\mathbf{v})) \\
&\qquad - V(k+1,\mathbf{u}(k+1,k,\mathbf{v})) \\
&\quad \le -\psi(\|\mathbf{u}(k+1,k,\mathbf{v})\|) + c\|\mathbf{v}(k+1,k,\mathbf{v}) - \mathbf{u}(k+1,k,\mathbf{v})\| \\
&\quad \le -\psi(\|\mathbf{u}(k+1,k,\mathbf{v})\|) + c\delta_2 \\
&\quad < -\psi(\|\mathbf{u}(k+1,k,\mathbf{v})\|) + \psi(\delta_1).
\end{aligned} \tag{5.13.1}
$$

Now suppose that there is an integer $k_1 > a$ and a $\mathbf{v}^1$, $\|\mathbf{v}^1\| < \delta_1$ such that $\|\mathbf{v}(k,a,\mathbf{v}^1)\| < \epsilon_1$ for all $k \in \mathbb{N}(a,k_1-1)$, and $\|\mathbf{v}(k_1,a,\mathbf{v}^1)\| \geq \epsilon_1$. It then follows that

$$V(k_1,\mathbf{v}(k_1,a,\mathbf{v}^1)) \ \geq \ \alpha(\epsilon_1) \ > \ \beta(2\delta_1) \ \geq \ \beta(\delta_1+\delta_2)$$

and $V(a,\mathbf{v}^1) < \beta(\delta_1)$. Then, there exists an integer $k_2 \in \mathbb{N}(a,k_1-1)$ such that $V(k_2,\mathbf{v}(k_2,a,\mathbf{v}^1)) < \beta(\delta_1+\delta_2)$, and $V(k_2+1,\mathbf{v}(k_2+1,a,\mathbf{v}^1)) \geq \beta(\delta_1+\delta_2)$. Thus, $\|\mathbf{v}(k_2+1,a,\mathbf{v}^1)\| \geq \delta_1+\delta_2$, from which we find

$$\begin{aligned}\|\mathbf{u}(k_2+1,k_2,\mathbf{v}(k_2,a,\mathbf{v}^1))\| \ &\geq \ \|\mathbf{v}(k_2+1,a,\mathbf{v}^1)\| \\ &\quad - \|\mathbf{v}(k_2+1,k_2,\mathbf{v}(k_2,a,\mathbf{v}^1)) - \mathbf{u}(k_2+1,k_2,\mathbf{v}(k_2,a,\mathbf{v}^1))\| \\ &\geq \ \delta_1+\delta_2-\delta_2 \ = \ \delta_1.\end{aligned}$$

Therefore, from (5.13.1) and the fact that $\mathbf{v}(k+1,k,\mathbf{v}(k,a,\mathbf{v}^1)) = \mathbf{v}(k+1,a,\mathbf{v}^1)$, we find

$$\begin{aligned}0 \ &\leq \ V(k_2+1,\mathbf{v}(k_2+1,a,\mathbf{v}^1)) - V(k_2,\mathbf{v}(k_2,a,\mathbf{v}^1)) \\ &< \ -\psi(\|\mathbf{u}(k_2+1,k_2,\mathbf{v}(k_2,a,\mathbf{v}^1))\|) + \psi(\delta_1) \ \leq \ 0,\end{aligned}$$

which is a contradiction. ■

**Corollary 5.13.2.** Suppose that the conditions of Theorem 5.13.1 are satisfied, and for all $(k,\mathbf{v}) \in \mathbb{N}(a) \times S_\rho$, $\|\mathbf{g}(k,\mathbf{v})\| < \mu(k)\|\mathbf{v}\|$, where $\mu(k) \to 0$ monotonically as $k \to \infty$. Then, the trivial solution $\mathbf{v}(k,a,0) \equiv 0$ of (5.7.1) is uniformly asymptotically stable.

**Proof.** As in Theorem 5.13.1, we find

$$\begin{aligned}&V(k+1,\mathbf{v}(k+1,a,\mathbf{v}^0)) - V(k,\mathbf{v}(k,a,\mathbf{v}^0)) \\ &\quad \leq \ -\psi(\|\mathbf{u}(k+1,k,\mathbf{v}(k,a,\mathbf{v}^0))\|) + c\mu(k)\|\mathbf{v}(k,a,\mathbf{v}^0)\|.\end{aligned}$$

Suppose $0 < r < \delta(\epsilon)$ and $r < \|\mathbf{u}(k+1,k,\mathbf{v}(k,a,\mathbf{v}^0))\| < \delta(\epsilon)$. By the hypothesis on $\mu(k)$ we can choose a $k_1 \in \mathbb{N}(a)$ so that for all $k \geq k_1$, $c\mu(k)\|\mathbf{v}(k,a,\mathbf{v}^0)\| < (1/2)\psi(r)$, and then

$$\begin{aligned}\Delta V(k,\mathbf{v}(k,a,\mathbf{v}^0)) \ &\leq \ -\frac{1}{2}\psi(\|\mathbf{u}(k+1,k,\mathbf{v}(k,a,\mathbf{v}^0))\|) \\ &= \ -\frac{1}{2}\psi(\|\mathbf{f}(k,\mathbf{v}(k,a,\mathbf{v}^0))\|).\end{aligned}$$

Now the proof follows from Theorem 5.10.3. ■

## 5.14. Practical Stability

Here we shall introduce another type of stability, which is somewhat connected with total stability, but finds importance in the numerical computation of the solutions of recurrence relations where certain errors cannot be made arbitrarily small.

**Definition 5.14.1.** The trivial solution $\mathbf{u}(k,a,0) \equiv 0$ of (1.2.8) is said to be *practically stable,* if there exists a neighborhood $S$ of the origin and $k_1 \in \mathbb{N}(a)$ such that the solution $\mathbf{v}(k,a,\mathbf{v}^0)$ of (5.7.1) remains in $S$ for all $k \in \mathbb{N}(k_1)$.

**Theorem 5.14.1.** Suppose that for all $(k,\mathbf{u})$, $(k,\mathbf{v}) \in \mathbb{N}(a) \times \Omega$ the function $\mathbf{f}(k,\mathbf{u})$ satisfies the Lipschitz condition (5.12.7) with $L < 1$, and $\|\mathbf{g}(k,\mathbf{v})\| < \delta$. Then, the trivial solution $\mathbf{u}(k,a,0) \equiv 0$ of (1.2.8) is practically stable.

**Proof.** Let $\mathbf{v}(k) = \mathbf{v}(k,a,\mathbf{v}^0)$ be a solution of (5.7.1). For this solution obviously we have $\|\mathbf{v}(k+1)\| \le L\|\mathbf{v}(k)\| + \delta$, and hence $\|\mathbf{v}(k)\| \le L^{k-a}\|\mathbf{v}^0\| + \delta/(1-L)$. Thus, if we choose $k_1 \in \mathbb{N}(a)$ suitably, then for all $k \in \mathbb{N}(k_1)$, $\|\mathbf{v}(k)\| < 1 + \delta(1-L)^{-1} = \rho$, i.e. $\mathbf{v}(k) \in S_\rho$. ∎

**Theorem 5.14.2.** Suppose that there exist two scalar functions $V(\mathbf{u})$, $W(\mathbf{u}) \in C[\Omega, \mathbb{R}]$, such that for all $\mathbf{u} \in \Omega$, $V(\mathbf{u}) \ge 0$, and $\Delta V(\mathbf{u}) = V(\mathbf{f}(k,\mathbf{u}) + \mathbf{g}(k,\mathbf{u})) - V(\mathbf{u}) \le W(\mathbf{u}) \le \alpha$, where $\alpha \ge 0$ is some constant. Let $Z = \{\mathbf{u} \in \overline{\Omega} : W(\mathbf{u}) \ge 0\}$, $\beta = \sup\{V(\mathbf{u}) : \mathbf{u} \in Z\}$ and $S = \{\mathbf{u} \in \overline{\Omega} : V(\mathbf{u}) \le \beta + \alpha\}$. Then, every solution of (5.7.1) which remains in $\Omega$ and enters in $S$ for $k = k_1$ remains in $S$ for all $k \in \mathbb{N}(k_1)$.

**Proof.** Let $\mathbf{v}(k) = \mathbf{v}(k,a,\mathbf{v}^0)$ be a solution of (5.7.1). If $\mathbf{v}(k_1,a,\mathbf{v}^0) \in S$, then $V(\mathbf{v}(k_1,a,\mathbf{v}^0)) \le \beta + \alpha$, and $V(\mathbf{v}(k_1+1,a,\mathbf{v}^0)) \le V(\mathbf{v}(k_1,a,\mathbf{v}^0)) + W(\mathbf{v}(k_1,a,\mathbf{v}^0))$. If $W(\mathbf{v}(k_1,a,\mathbf{v}^0)) \le 0$, then $V(\mathbf{v}(k_1+1,a,\mathbf{v}^0)) \le \beta + \alpha$. Also, if $\mathbf{v}(k_1,a,\mathbf{v}^0) \in Z$, then because $V(\mathbf{v}(k_1,a,\mathbf{v}^0)) \le \beta$, again it follows that $V(\mathbf{v}(k_1+1,a,\mathbf{v}^0)) \le \beta + \alpha$, i.e. $\mathbf{v}(k_1+1,a,\mathbf{v}^0) \in S$. Now an easy induction completes the proof. ∎

**Corollary 5.14.3.** Let in addition to the conditions of Theorem 5.14.2, $d = \sup\{W(\mathbf{u}) : \mathbf{u} \in \Omega - S\} < 0$, then every solution of (5.7.1) which remains in $\Omega$ must enter in $S$ in a finite number of steps.

**Proof.** From $\Delta V(\mathbf{u}) \le W(\mathbf{u})$ for the solution $\mathbf{v}(k) = \mathbf{v}(k,a,\mathbf{v}^0)$, we get $V(\mathbf{v}(k,a,\mathbf{v}^0)) \le V(\mathbf{v}^0) + \sum_{\ell=a}^{k-1} W(\mathbf{v}(\ell,a,\mathbf{v}^0)) \le V(\mathbf{v}^0) + d(k-a)$, from which it follows that $V(\mathbf{v}(k,a,\mathbf{v}^0)) \to -\infty$ as $k \to \infty$. But, $V(\mathbf{v}(k,a,\mathbf{v}^0)) \ge \beta + \alpha$ for $\mathbf{v}(k,a,\mathbf{v}^0) \in \Omega - S$. This contradiction completes the proof. ∎

## 5.15. Mutual Stability

Consider the difference systems (5.1.1) and

$$(5.15.1) \qquad \Delta \mathbf{v}(k) \;=\; \mathbf{g}(k, \mathbf{v}(k)),$$

where the functions $\mathbf{f}(k, \mathbf{u})$ and $\mathbf{g}(k, \mathbf{v})$ are defined on $\mathbb{N}(a) \times \Omega$, and the scalar difference equations (5.1.9) and

$$(5.15.2) \qquad \Delta R(k) \;=\; h(k, R(k)),$$

where the functions $g(k, r)$ and $h(k, R)$ are defined on $\mathbb{N}(a) \times \mathbb{R}_+$, $h(k, r) \leq g(k, r)$, and nondecreasing in the second argument for any fixed $k \in \mathbb{N}(a)$.

**Definition 5.15.1.** Any two solutions $\mathbf{u}(k) = \mathbf{u}(k, a, \mathbf{u}^0)$ and $\mathbf{v}(k) = \mathbf{v}(k, a, \mathbf{v}^0)$ of the difference systems (5.1.1) and (5.15.1) are said to be

(i) *Mutually Stable* if, for each $\epsilon_1 > 0$, there exist $\delta_1 = \delta_1(\epsilon_1, a)$, $\delta_2 = \delta_2(\epsilon_1, a)$ and $\epsilon_2 = \epsilon_2(\epsilon_1, a)$ such that $0 < \epsilon_2 < \delta_2 \leq \delta_1 < \epsilon_1$ and $\delta_2 \leq \|\mathbf{u}^0 - \mathbf{v}^0\| \leq \delta_1$ imply $\epsilon_2 < \|\mathbf{u}(k) - \mathbf{v}(k)\| < \epsilon_1$ for all $k \in \mathbb{N}(a)$.

(ii) *Mutually Attractive* if there exist $\delta_1 = \delta_1(a)$ and $\delta_2 = \delta_2(a)$, such that $0 \leq \delta_2 \leq \delta_1$ and $\delta_2 \leq \|\mathbf{u}^0 - \mathbf{v}^0\| \leq \delta_1$ imply that $\|\mathbf{u}(k) - \mathbf{v}(k)\| \to 0$ as $k \to \infty$.

(iii) *Mutually Asymptotically Stable* if they are mutually stable and mutually attractive.

**Definition 5.15.2.** Any two solutions $r(k) = r(k, a, r^0)$ and $R(k) = R(k, a, R^0)$ of the difference equations (5.1.9) and (5.15.2) are said to be

(i) *Mutually Stable* if, for each $\eta_1 > 0$, there exist $\nu_1 = \nu_1(\eta_1, a)$, $\nu_2 = \nu_2(\eta_1, a)$ and $\eta_2 = \eta_2(\eta_1, a)$ such that $0 < \eta_2 < \nu_2 \leq \nu_1 < \eta_1$ and $\nu_2 \leq R^0 \leq r^0 \leq \nu_1$ imply $\eta_2 < R(k) \leq r(k) < \eta_1$ for all $k \in \mathbb{N}(a)$.

(ii) *Mutually Attractive* if there exist $\delta_1 = \delta_1(a)$ and $\delta_2 = \delta_2(a)$, such that $0 \leq \delta_2 \leq \delta_1$ and $\delta_2 \leq R^0 \leq r^0 \leq \delta_1$ imply that $R(k) \leq r(k)$ for all $k \in \mathbb{N}(a)$ and $r(k) - R(k) \to 0$ as $k \to \infty$.

(iii) *Mutually Asymptotically Stable* if they are mutually stable and mutually attractive.

**Theorem 5.15.1.** In addition to the hypotheses on the functions $\mathbf{f}(k, \mathbf{u})$, $\mathbf{g}(k, \mathbf{v})$, $g(k, r)$ and $h(k, R)$ assume that for all $(k, \mathbf{u})$, $(k, \mathbf{v}) \in \mathbb{N}(a) \times \Omega$

$$h(k, \|\mathbf{u} - \mathbf{v}\|) \;\leq\; \|\mathbf{f}(k, \mathbf{u}) - \mathbf{g}(k, \mathbf{v})\| \;\leq\; g(k, \|\mathbf{u} - \mathbf{v}\|).$$

Then, any two solutions $\mathbf{u}(k) = \mathbf{u}(k, a, \mathbf{u}^0)$ and $\mathbf{v}(k) = \mathbf{v}(k, a, \mathbf{v}^0)$ of (5.1.1) and (5.15.1) are mutually stable; or mutually attractive; or mutually asymptotically stable, provided any two solutions $r(k) = r(k, a, r^0)$ and

$R(k) = R(k, a, R^0)$ of (5.1.9) and (5.15.2) are mutually stable; or mutually attractive; or mutually asymptotically stable.

**Proof.** If $0 \le R^0 \le \|\mathbf{u}^0 - \mathbf{v}^0\| \le r^0$, then it is easy to deduce that

$$R(k, a, R^0) \le \|\mathbf{u}(k, a, \mathbf{u}^0) - \mathbf{v}(k, a, \mathbf{v}^0)\| \le r(k, a, r^0)$$

from which the conclusions are immediate. ∎

Now let the scalar function $V(k, \mathbf{u}, \mathbf{v})$ be defined, nonnegative and continuous in $\mathbb{N}(a) \times \Omega \times \Omega$, and suppose that for each $k \in \mathbb{N}(a)$, $V(k, \mathbf{u}, \mathbf{v}) = 0$ if and only if $\mathbf{u} = \mathbf{v}$. Further, let $\Delta V(k, \mathbf{u}, \mathbf{v})$ denote the expression

$$\Delta V(k, \mathbf{u}, \mathbf{v}) = V(k+1, \mathbf{u} + \mathbf{f}(k, \mathbf{u}), \mathbf{v} + \mathbf{g}(k, \mathbf{v})) - V(k, \mathbf{u}, \mathbf{v}).$$

**Theorem 5.15.2.** In addition to the hypotheses on the functions $g(k, r)$ and $h(k, R)$ assume that there exists a scalar function $V(k, \mathbf{u}, \mathbf{v}) \in C[\mathbb{N}(a) \times \Omega \times \Omega, \mathbb{R}_+]$ such that

$$h(k, V(k, \mathbf{u}, \mathbf{v})) \le \Delta V(k, \mathbf{u}, \mathbf{v}) \le g(k, V(k, \mathbf{u}, \mathbf{v})).$$

Further, assume that

$$\phi(\|\mathbf{u} - \mathbf{v}\|) \le V(k, \mathbf{u}, \mathbf{v}) \le \psi(\|\mathbf{u} - \mathbf{v}\|),$$

where $\phi$ and $\psi \in \mathcal{K}$. Then, any two solutions $\mathbf{u}(k) = \mathbf{u}(k, a, \mathbf{u}^0)$ and $\mathbf{v}(k) = \mathbf{v}(k, a, \mathbf{v}^0)$ of (5.1.1) and (5.15.1) are mutually stable provided any two solutions $r(k) = r(k, a, r^0)$ and $R(k) = R(k, a, R^0)$ of (5.1.9) and (5.15.2) are mutually stable.

**Proof.** If $0 \le R^0 \le V(a, \mathbf{u}^0, \mathbf{v}^0) \le r^0$, then it is easy to deduce that

$$R(k, a, R^0) \le V(k, \mathbf{u}(k, a, \mathbf{u}^0), \mathbf{v}(k, a, \mathbf{v}^0)) \le r(k, a, r^0)$$

from which the conclusion follows. ∎

## 5.16. Problems

**5.16.1.** Let $\mathbf{f}(k, \mathbf{u})$ be defined on $\mathbb{N}(a) \times \mathbb{R}^n$, and let $g(k, r) = \sup_{\|\mathbf{u} - \mathbf{u}^0\| \le r} \|\mathbf{f}(k, \mathbf{u})\|$, where $\mathbf{u}^0 \in \mathbb{R}^n$ is arbitrary. Show that the solution $\mathbf{u}(k, a, \mathbf{u}^0)$ of (5.1.1) existing on $\mathbb{N}(a)$ satisfies the inequality $\|\mathbf{u}(k, a, \mathbf{u}^0) - \mathbf{u}^0\| \le r(k, a, 0)$ where $r(k, a, 0)$ is the solution of (5.1.9).

**5.16.2.** Assume that $g(k, r)$ is nonnegative on $\mathbb{N}(a) \times \mathbb{R}$ and nondecreasing in $r$ for $r > 0$ and fixed $k \in \mathbb{N}(a)$. Show that the following are equivalent

(i) given any $r^\infty > 0$ there exists $k_0 \in \mathbb{N}(a)$ and a solution $r(k, k_0, r^0)$ of the difference equation (5.1.9) such that

$$\lim_{k\to\infty} r(k, k_0, r^0) = r^\infty, \quad r(k_0) = r^0$$

(ii) $\sum_{\ell=k_0}^{\infty} g(k,\gamma) < \infty$ for all $\gamma$ satisfying $0 \le \gamma \le r^0$.

**5.16.3.** Prove Theorem 5.1.4.

**5.16.4.** Prove Theorem 5.1.5.

**5.16.5.** Prove Theorem 5.1.6.

**5.16.6.** Suppose that every solution of (1.2.12) is bounded (tends to zero as $k \to \infty$). Show that every solution of (1.2.11) is bounded (tends to zero as $k \to \infty$) provided at least one of its solutions is bounded (tends to zero as $k \to \infty$).

**5.16.7.** Let the system (1.2.12) be uniformly bounded, and $\sum_{\ell=0}^{\infty} \|\mathbf{b}(\ell)\| < \infty$. Show that every solution of (1.2.11) is bounded.

**5.16.8.** Consider the difference system (1.2.12) where $\mathcal{A}(k)$ is a complex $n \times n$ matrix. Show that

(i) if $\lim_{k\to\infty} \|\mathbf{u}(k)\|$ exists as a finite number for every solution $\mathbf{u}$ of (1.2.12), then it has a nontrivial solution $\mathbf{u}_0$ satisfying $\lim_{k\to\infty} \mathbf{u}_0(k) = 0$, if and only if $\lim_{k\to\infty} \prod_{\ell=0}^{k} \det[\mathcal{A}(\ell)] = 0$

(ii) if $\mathcal{A}^*(k)\mathcal{A}(k) \le \mathcal{I}$ (* denotes the conjugate transpose) on $\mathbb{N}$, then $\lim_{k\to\infty} \|\mathbf{u}(k)\|$ exists as a finite limit for every solution $\mathbf{u}$ of (1.2.12)

(iii) if $\mathcal{A}^*(k)\mathcal{A}(k) \le \mathcal{I}$ on $\mathbb{N}$ and $\lim_{k\to\infty} \prod_{\ell=0}^{k} \det[\mathcal{A}(\ell)] = 0$, then (1.2.12) has a nontrivial solution $\mathbf{u}_0$ with $\lim_{k\to\infty} \mathbf{u}_0(k) = 0$

(iv) if $\lim_{k\to\infty} \|\mathbf{u}(k)\| \le \infty$ exists for all solutions $\mathbf{u}$ of (1.2.12) and $\lim_{k\to\infty} \prod_{\ell=0}^{k} |\det \mathcal{A}(\ell)| = \infty$, then it has a solution $\mathbf{u}_0$ such that $\lim_{k\to\infty} \|\mathbf{u}_0(k)\| = \infty$.

**5.16.9.** Let the difference system (1.2.11) be periodic of period $K$ on $\mathbb{N}$. Show that every solution of (1.2.11) is unbounded if it does not have periodic solutions of period $K$.

**5.16.10.** Let the difference system (1.2.12) be periodic of period $K$ on $N$. Further, let the condition (5.2.4) hold. Show that

(i) all solutions of (5.2.19) are bounded provided all solutions of (1.2.12) are bounded

(ii) all solutions of (5.2.19) tend to zero as $k \to \infty$ provided all solutions of (1.2.12) tend to zero as $k \to \infty$.

**5.16.11.** Let for all $(k, \mathbf{u}) \in \mathbb{N} \times \mathbb{R}^n$ the functions $\mathbf{f}^i(k, \mathbf{u})$, $i = 1, 2$ be defined, and for all $(k, \mathbf{u})$, $(k, \mathbf{v})$

$$\|\mathbf{u} - \mathbf{v} + \mathbf{f}^1(k, \mathbf{u}) - \mathbf{f}^2(k, \mathbf{v})\| \le \|\mathbf{u} - \mathbf{v}\| + g(k, \|\mathbf{u} - \mathbf{v}\|),$$

where $g(k, r)$ is defined for all $(k, r) \in \mathbb{N} \times \mathbb{R}_+$, and $r + g(k, r)$ is nondecreasing in $r$ for any fixed $k \in \mathbb{N}$. Further, let for each $r^0 \ge 0$ the solution $r(k) = r(k, 0, r^0)$ of (5.1.9) tend to zero as $k \to \infty$. Then, for the solutions $\mathbf{u}^i(k) = \mathbf{u}^i(k, 0, \mathbf{u}^i)$, $i = 1, 2$ of $\Delta \mathbf{u}(k) = \mathbf{f}^i(k, \mathbf{u}(k))$, $\mathbf{u}^i(0) = \mathbf{u}^i$, $i = 1, 2$ show that $\lim_{k \to \infty} (\mathbf{u}^1(k) - \mathbf{u}^2(k)) = 0$.

**5.16.12.** Let for all $(k, \mathbf{u}) \in \mathbb{N} \times \mathbb{R}^n$ the function $\mathbf{f}(k, \mathbf{u})$ be defined, and for all $(k, \mathbf{u})$, $(k, \mathbf{v})$

$$\|\mathbf{u} - \mathbf{v} + \mathbf{f}(k, \mathbf{u}) - \mathbf{f}(k, \mathbf{v})\| \le \|\mathbf{u} - \mathbf{v}\| + g(k, \|\mathbf{u} - \mathbf{v}\|),$$

where $g(k, r)$ is as in Problem 5.16.11. Further, let for each $r^0 \ge 0$ the solution $r(k) = r(k, 0, r^0)$ of $\Delta r(k) = g(k, r(k)) + \|\mathbf{f}(k, 0)\|$, $r(0) = r^0$ tend to zero as $k \to \infty$. Show that for each $\mathbf{u}^0 \in \mathbb{R}^n$ the solution $\mathbf{u}(k) = \mathbf{u}(k, 0, \mathbf{u}^0)$ of (5.1.1) tends to zero as $k \to \infty$.

**5.16.13.** Let in addition to the conditions on the functions $\mathbf{f}(k, \mathbf{u})$ and $g(k, r)$ in Problem 5.16.12, $\mathbf{f}(k, \mathbf{u})$ be periodic of period $K$, i.e. $\mathbf{f}(k + K, \mathbf{u}) = \mathbf{f}(k, \mathbf{u})$, and let (5.1.1) have a bounded solution. Show that the system (5.1.1) has a periodic solution of period $K$.

**5.16.14.** Consider the difference system (5.3.1) on $\mathbb{Z}$ and assume that $\mathbf{g}$ is periodic with respect to the first argument. Show that if there exists a bounded solution $\mathbf{v}(k)$ of (5.3.1) which is uniformly stable, then (5.3.1) has an almost periodic solution.

**5.16.15.** Consider the difference system (5.3.1) on $\mathbb{Z}$ and assume that $\mathbf{g}$ is almost periodic from $\mathbb{Z}$ to $\mathbb{R}^n$ uniformly with respect to the second argument in any bounded set. Moreover, let $\mathbf{g}$ satisfy the monotonicity condition

$$< \mathbf{g}(k, \mathbf{v}) - \mathbf{g}(k, \mathbf{w}),\ \mathbf{v} - \mathbf{w} > \ \ge L\|\mathbf{v} - \mathbf{w}\|^2, \quad L > 1 \tag{5.16.1}$$

for any $k \in \mathbb{Z}$ and any $\mathbf{v}$, $\mathbf{w} \in \mathbb{R}^n$. Show that if (5.3.1) has a bounded solution on $\mathbb{Z}$, then this solution is almost periodic.

**5.16.16.** Consider the difference system (1.2.8) on $\mathbb{N}$ and assume that $\lim_{k \to \infty} \mathbf{f}(k, \mathbf{u}) = \mathbf{f}_\infty(\mathbf{u})$ uniformly with respect to the second argument in any bounded set. Moreover, let $\mathbf{f}$ satisfy the monotonicity condition (5.16.1) for any $k \in \mathbb{N}$ and $\mathbf{v}, \mathbf{w} \in \mathbb{R}^n$. Show that if (1.2.8) has a bounded solution $\mathbf{u}(k)$, $k \in \mathbb{N}$ then $\mathbf{u}(k)$ is convergent, i.e. $\mathbf{u}(k) \in c$ (the space of convergent sequences).

**5.16.17.** Consider the second order difference system

$$(5.16.2) \qquad \mathbf{u}(k+2) = \mathbf{f}(k, \mathbf{u}(k), \mathbf{u}(k+1)), \quad k \in \mathbb{Z}$$

and assume that $\mathbf{f}$ is almost periodic from $\mathbb{Z}$ to $\mathbb{R}^n$ uniformly with respect to the second and third arguments in any bounded set. Moreover, let $\mathbf{f}$ satisfy the monotonicity condition

$$< \mathbf{f}(k, \mathbf{u}, \mathbf{v}) - \mathbf{f}(k, \overline{\mathbf{u}}, \overline{\mathbf{v}}),\ \mathbf{u} - \overline{\mathbf{u}} + \mathbf{v} - \overline{\mathbf{v}} > \ \geq L\left(\|\mathbf{u} - \overline{\mathbf{u}}\|^2 + \|\mathbf{v} - \overline{\mathbf{v}}\|^2\right),\ L > 1$$

for any $k \in \mathbb{Z}$ and any $\mathbf{u}, \mathbf{v}, \overline{\mathbf{u}}, \overline{\mathbf{v}} \in \mathbb{R}^n$. Show that if (5.16.2) has a bounded solution on $\mathbb{Z}$, then this solution is almost periodic.

**5.16.18.** Consider the difference system (5.16.2) on $\mathbb{N}$ and assume that $\lim_{k \to \infty} \mathbf{f}(k, \mathbf{u}, \mathbf{v}) = \mathbf{f}_\infty(\mathbf{u}, \mathbf{v})$ uniformly with respect to the second and third arguments in any bounded set. Moreover, let $\mathbf{f}$ satisfy the following monotonicity condition with respect to the second and third arguments

$$< \mathbf{f}(k, \mathbf{u}, \mathbf{v}) - \mathbf{f}(k, \overline{\mathbf{u}}, \mathbf{v}),\ \mathbf{u} - \overline{\mathbf{u}} > \ \geq L\|\mathbf{u} - \overline{\mathbf{u}}\|^2, \quad L > 1$$

and

$$< \mathbf{f}(k, \mathbf{u}, \mathbf{v}) - \mathbf{f}(k, \mathbf{u}, \overline{\mathbf{v}}),\ \mathbf{v} - \overline{\mathbf{v}} > \ \geq L\|\mathbf{v} - \overline{\mathbf{v}}\|^2, \quad L > 1$$

for any $k \in \mathbb{N}$ and any $\mathbf{u}, \mathbf{v}, \overline{\mathbf{u}}, \overline{\mathbf{v}} \in \mathbb{R}^n$. Show that if (5.16.2) has a bounded solution $\mathbf{u}(k)$, $k \in \mathbb{N}$ then $\mathbf{u}(k)$ is convergent.

**5.16.19.** Consider the difference system (5.3.1) on $\mathbb{Z}$ and assume that $\mathbf{g}$ is almost periodic from $\mathbb{Z}$ to $\mathbb{R}^n$ uniformly with respect to the second argument in any bounded set. Moreover, let $\mathbf{g}$ satisfy a Lipschitz condition $\|\mathbf{g}(k, \mathbf{u}) - \mathbf{g}(k, \mathbf{v})\| \leq L\|\mathbf{u} - \mathbf{v}\|$ for all $(k, \mathbf{u}), (k, \mathbf{v}) \in \mathbb{Z} \times \mathbb{R}^n$, and moduli of eigenvalues of $\mathcal{A}$ are different from 1. Show that (5.3.1) admits a unique almost periodic solution provided the Lipschitz constant $L$ is small enough.

**5.16.20.** Consider the difference equation

$$(5.16.3) \qquad u(k+1) = [a(k) + b(k)]u(k), \quad k \in \mathbb{N}.$$

Let $h(k) > 0$, $k \in \mathbb{N}$ be a nonincreasing function, $a(k) \neq 0$, $k \in \mathbb{N}$ and $\sum_{\ell=1}^{\infty} \frac{1}{h(\ell - 1)} \left|\frac{b(\ell)}{a(\ell)}\right| < \infty$. Show that there exists a function $\phi(k)$

so that the change of variable $u(k) = \phi(k)v(k)$ transforms (5.16.3) into $v(k+1) = a(k)v(k)$. Further, show that the solutions of (5.16.3) can be written as

$$u(k) = (1 + o(h(k-1))) \prod_{\ell=k_0}^{k-1} a(\ell)u(k_0).$$

**5.16.21.** Consider the difference equation

$$(5.16.4) \qquad u(k+1) = \lambda(k)u(k) + b(k)u(k-m), \quad k \in \mathbb{N}(1)$$

where $m > 1$ is an integer, and the functions $\lambda$ and $b$ satisfy

$$\left| \left[ \prod_{\ell=k-m}^{k} \lambda(\ell) \right]^{-1} b(k) \right| \in \ell^p, \quad 1 \le p \le 2.$$

Show that (5.16.4) has a solution $u(k)$ such that

$$u(k) = \left[ \prod_{\ell=1}^{k-1} \left( \lambda(\ell) + \left[ \prod_{\tau=\ell-m}^{\ell-1} \lambda(\tau) \right]^{-1} b(\ell) \right) \right] (1 + o(1)) \quad \text{as} \quad k \to \infty.$$

**5.16.22.** Consider the difference equation

$$(5.16.5) \qquad u(k+2) = \alpha^2 u(k) + q(k)u(k-m), \quad k \in \mathbb{N}(1)$$

where $m > 1$ is an integer and $q(k) \in \ell^p$. Show that (5.16.5) has a pair of solutions $u_1(k)$ and $u_2(k)$ such that as $k \to \infty$

$$u_1(k) = \left[ \prod_{\ell=1}^{k-1} \left( -\alpha + (-1)^m \frac{q(\ell)}{2\alpha^{m+1}} \right) \right] (1 + o(1))$$

and

$$u_2(k) = \left[ \prod_{\ell=1}^{k-1} \left( \alpha + \frac{q(\ell)}{2\alpha^{m+1}} \right) \right] (1 + o(1)).$$

**5.16.23.** Consider the $(n+1)$th order difference equation

$$(5.16.6) \qquad u(k+1) = \sum_{i=0}^{n} (c_i + d_i(k))u(k-i), \quad k \in \mathbb{N}$$

where $c_i$, $0 \le i \le n$ are complex constants and $d_i(k)$, $k \in \mathbb{N}$, $0 \le i \le n$ are complex functions. Let $\lambda_0$ be a simple root of the characteristic function $f(\lambda) = \lambda - \sum_{i=0}^{n} c_i \lambda^{-i}$, and suppose that if $\lambda$ is any other root of $f$, then $|\lambda| \neq |\lambda_0|$. Show that, if $\sum_{\ell=0}^{\infty} |d_i(k)|^2 < \infty$, $0 \le i \le n$ then for $k_0$ large enough, (5.16.6) has a solution $u(k)$ such that

$$\lim_{k \to \infty} \frac{u(k)}{\prod_{\ell=k_0}^{k-1} \left( \lambda_0 + \frac{1}{f'(\lambda_0)} \sum_{i=0}^{n} \lambda_0^{-i} d_i(\ell) \right)} = 1.$$

**5.16.24.** (Discrete Levinson's Perturbation Lemma). Consider the discrete matrix system

$$(5.16.7) \qquad \mathcal{U}(k+1) = [\Lambda(k) + \mathcal{B}(k)]\mathcal{U}(k), \quad k \in \mathbb{N}(a)$$

where $\Lambda(k) = \operatorname{diag}\{\lambda_1(k), \cdots, \lambda_n(k)\}$ with $\lambda_i(k) \neq 0$ for all $1 \le i \le n$ and $k \in \mathbb{N}(a)$. The system (5.16.7) is said to have a *Levinson's dichotomy* if there exist constants $M > 0$ and $N > 0$ such that for each index pair $(i,j)$, $i \neq j$, either

$$\prod_{\ell=a}^{k} \left| \frac{\lambda_i(\ell)}{\lambda_j(\ell)} \right| \to 0 \text{ as } k \to \infty \text{ and } \prod_{\ell=k_1}^{k_2} \left| \frac{\lambda_i(\ell)}{\lambda_j(\ell)} \right| \le M \text{ for } a \le k_1 \le k_2$$

or

$$\prod_{\ell=k_1}^{k_2} \left| \frac{\lambda_i(\ell)}{\lambda_j(\ell)} \right| \ge N \quad \text{for} \quad a \le k_1 \le k_2.$$

Use Theorem 5.8.4 to show that if the system (5.16.7) has a Levinson's dichotomy and

$$\sum_{\ell=a}^{\infty} \frac{1}{|\lambda_i(\ell)|} \|\mathcal{B}(\ell)\| < \infty \quad \text{for all} \quad 1 \le i \le n$$

then the fundamental matrix $\mathcal{U}(k)$ of (5.16.7) satisfies

$$\mathcal{U}(k) = (\mathcal{I} + o(1)) \prod_{\ell=a}^{k-1} \Lambda(\ell) \quad \text{as} \quad k \to \infty.$$

**5.16.25.** Consider the difference system

$$(5.16.8) \qquad \mathbf{u}(k+1) = \Lambda(k)\mathbf{u}(k) + \mathcal{B}(k)\mathbf{u}(k-m), \quad k \in \mathbb{N}(1)$$

where $m > 1$ is an integer, $\Lambda(k) = \operatorname{diag}(\lambda_1(k), \cdots, \lambda_n(k))$ with $\lambda_i(k) \neq 0$ for all $1 \le i \le n$ and $k \in \mathbb{N}(1)$. Suppose that there exists a $q \in \{1, \cdots, n\}$ such that

$$\left| \frac{\lambda_i(k)}{\lambda_q(k)} \right| < e^{-\beta} \text{ for } 1 \le i \le q-1 \text{ and } \left| \frac{\lambda_q(k)}{\lambda_i(k)} \right| < e^{-\beta} \text{ for } q+1 \le i \le n$$

where $\beta > 0$, and

$$\left\| \frac{\rho_i(k)}{\lambda_j(k)} \mathcal{B}(k) \right\| \in \ell^p, \quad 1 \le p \le 2, \quad 1 \le i,\ j \le n$$

where

$$\rho_i(k) = \prod_{\ell=k-m}^{k-1} \lambda_i(\ell).$$

Show that for $k_0 \in \mathbb{N}$ sufficiently large, system (5.16.8) has a solution $\mathbf{u}_q(k)$ such that

$$\mathbf{u}_q(k) = \left[ \prod_{\ell=k_0}^{k-1} \overline{\lambda}(\ell) \right] (\mathbf{e}_q + o(1)) \quad \text{as} \quad k \to \infty$$

where $\overline{\lambda}(\ell) = \lambda_q(\ell) + \rho_q(\ell) b_{qq}(\ell)$, $\mathbf{e}_q$ is the $q$th unit vector in $\mathbb{R}^n$, and $\mathcal{B} = (b_{ij})$.

**5.16.26.** (Poincaré Type Result for Linear Systems). Consider the difference system

$$\text{(5.16.9)} \qquad \mathbf{u}(k+1) = (\mathcal{A} + \mathcal{B}(k))\mathbf{u}(k), \quad k \in \mathbb{N}(1)$$

where $\mathcal{A} = (a_{ij})$, $\mathcal{B}(k) = (b_{ij}(k))$, $k \in \mathbb{N}(1)$ are complex matrices and $\mathbf{u}(k)$ is a complex vector function. Assume that

(i) the matrix $\mathcal{A}$ has eigenvalues $\lambda_1, \cdots, \lambda_n$ with $0 < |\lambda_1| \le \cdots \le |\lambda_n|$ and associated linearly independent eigenvectors $\mathbf{v}^1, \cdots, \mathbf{v}^n$

(ii) there exists a nonincreasing function $\phi(k)$ such that $\lim_{k\to\infty} \phi(k) = 0$

(iii) $m$ is the smallest positive integer such that $\left|\frac{\lambda_r}{\lambda_m}\right| \le \alpha$, where $1 \le r \le n$ and the constant $\alpha \ge 1$; if $\left|\frac{\lambda_r}{\lambda_m}\right| = \alpha$, then $s$ is the largest integer such that $|\lambda_s| = |\lambda_m|$; also, if $m > 1$ there is an integer $K$ and a number $\rho$ such that $1 < \rho < \left|\frac{\lambda_r}{\alpha\lambda_{m-1}}\right|$ and $\rho^m \phi(k)$ is nondecreasing for $k \in \mathbb{N}(K)$

(iv) $\sum_{\ell=k}^{\infty} \alpha^\ell b_{ij}(\ell) = O(\phi(k)), \quad 1 \le i,\ j \le n$

(v) $\sum_{\ell=k}^{\infty} |b_{ij}(\ell)| \phi(\ell) = o(\phi(k)), \quad 1 \le i,\ j \le n$

(vi) if $\left|\dfrac{\lambda_r}{\lambda_m}\right| = \alpha$, assume also that

$$\sum_{\ell=k}^{\infty}\left(\frac{\lambda_r}{\lambda_q}\right)^{\ell} b_{ij}(\ell) = O(\phi(k)), \quad m \le q \le s, \ \ 1 \le i, \ j \le n.$$

Then (5.16.9) has a solution $\mathbf{u}(k)$, defined for $k$ sufficiently large, such that

$$\mathbf{u}(k) \ = \ \lambda_r^k\left[\mathbf{v}^r + O\left(\alpha^{-k}\phi(k)\right)\right].$$

**5.16.27.** (Poincaré Type Result for Linear Systems). Consider the difference system

$$\text{(5.16.10)} \qquad \mathbf{u}(k+1) \ = \ (\mathcal{A} + \mathcal{B}(k))\mathbf{u}(k), \quad k \in \mathbb{N}$$

where $\mathcal{A} = (a_{ij})$, $\mathcal{B}(k) = (b_{ij}(k))$, $k \in \mathbb{N}$ are complex matrices and $\mathbf{u}(k)$ is a complex vector function. Suppose that $\lambda_0$ is a simple nonzero eigenvalue of the matrix $\mathcal{A}$ so that $\overline{\lambda}_0$ is a simple (nonzero) eigenvalue of $\mathcal{A}^*$ (conjugate transpose). Suppose also that if $\lambda$ is any other eigenvalue of $\mathcal{A}$, then $|\lambda| \ne |\lambda_0|$. Let $\xi$ and $\eta$ be the eigenvectors of $\mathcal{A}$ and $\mathcal{A}^*$ corresponding to eigenvalues $\lambda_0$ and $\overline{\lambda}_0$, respectively. Show that, if $\sum_{\ell=0}^{\infty}\|\mathcal{B}(\ell)\|^2 < \infty$, then for $k_0$ large enough, (5.16.10) has a solution $\mathbf{u}(k)$ such that

$$\lim_{k\to\infty}\frac{\mathbf{u}(k)}{\prod_{\ell=k_0}^{k-1}(\lambda_0 + \beta(\ell))} \ = \ \xi,$$

where $\beta(k)$, $k \in \mathbb{N}$ is the complex valued function defined by

$$\beta(k) \ = \ (\eta^*\xi)^{-1}\eta^*\mathcal{B}(k)\xi, \quad k \in \mathbb{N}.$$

**5.16.28.** Assume that $\mathcal{U}(k,a)$ is the principal fundamental matrix of the difference system

$$\text{(5.16.11)} \qquad \Delta\mathbf{u}(k) \ = \ \mathcal{A}(k)\mathbf{u}(k), \quad k \in \mathbb{N}(a)$$

such that $\|\mathcal{D}(k)\mathcal{U}(k,a)\| \le \beta(k)$, $k \in \mathbb{N}(a)$ where $\mathcal{D}(k)$ is a nonsingular matrix and $\beta(k)$ is a positive function defined on $\mathbb{N}(a)$. Further assume that the function $\mathbf{g}(k,\mathbf{v})$ is defined and continuous on $\mathbb{N}(a) \times \mathbb{R}^n$ and satisfies $\|\mathcal{U}^{-1}(k+1,a)\mathbf{g}(k,\beta(k)\mathcal{D}^{-1}(k)\mathbf{v})\| \le h(k,\|\mathbf{v}\|)$, where $h(k,r)$ is defined for all $(k,r) \in \mathbb{N}(a) \times \mathbb{R}_+$, and nondecreasing in $r$ for any fixed $k \in \mathbb{N}(a)$. Show that

(i) if $\sum_{\ell=a}^{\infty} h(\ell,\gamma) < \infty$ for all $0 \le \gamma < \infty$, then given any solution $\mathbf{u}(k) = \mathcal{U}(k,a)\xi$ of (5.16.11) with $\|\xi\|$ sufficiently small, there exists a solution $\mathbf{v}(k)$ of

$$\Delta \mathbf{v}(k) = \mathcal{A}(k)\mathbf{v}(k) + \mathbf{g}(k,\mathbf{v}(k)), \quad k \in \mathbb{N}(a) \tag{5.16.12}$$

such that $\|\mathcal{D}(k)(\mathbf{v}(k) - \mathbf{u}(k))\| = o(\beta(k))$ as $k \to \infty$

(ii) if the difference equation (5.1.9) has a positive and bounded solution on $\mathbb{N}(a)$, then corresponding to each solution $\mathbf{v}(k)$ of (5.16.12) with $\|\mathbf{v}(a)\|$ small enough, there is a $\xi \in \mathbb{R}^n$ such that $\|\mathcal{D}(k)(\mathbf{v}(k) - \mathcal{U}(k,a)\xi)\| = o(\beta(k))$ as $k \to \infty$.

(iii) if all solutions of (5.1.9) are bounded on $\mathbb{N}(a)$, then corresponding to each solution $\mathbf{v}(k)$ of (5.16.12) there exists a solution $\mathbf{u}(k)$ of (5.16.11) such that $\|\mathcal{D}(k)(\mathbf{v}(k) - \mathbf{u}(k))\| = o(\beta(k))$ as $k \to \infty$.

**5.16.29.** Let the system (2.8.1) be stable (asymptotically stable) and the condition (5.2.4) be satisfied. Show that the system (5.2.3) is stable (asymptotically stable).

**5.16.30.** Let the difference system (1.2.12) be periodic of period $K$ on $\mathbb{N}$. Show that the stability of (1.2.12) implies its uniform stability.

**5.16.31.** Let $\overline{u}$ be an equilibrium point of the difference equation $u(k+1) = f(u(k))$, where $f$ is continuously differentiable at $\overline{u}$. Show that

(i) if $|f'(\overline{u})| < 1$, then $\overline{u}$ is an asymptotically stable point

(ii) if $|f'(\overline{u})| > 1$, then $\overline{u}$ is an unstable point

(iii) if $f'(\overline{u}) = 1$ and $f''(\overline{u}) \neq 0$, then $\overline{u}$ is an unstable point

(iv) if $f'(\overline{u}) = 1$, $f''(\overline{u}) = 0$ and $f'''(\overline{u}) > 0$, then $\overline{u}$ is an unstable point

(v) if $f'(\overline{u}) = 1$, $f''(\overline{u}) = 0$ and $f'''(\overline{u}) < 0$, then $\overline{u}$ is an asymptotically stable point

(vi) if $f'(\overline{u}) = -1$ and $-2f'''(\overline{u}) - 3[f''(\overline{u})]^2 < 0$, then $\overline{u}$ is an asymptotically stable point

(vii) if $f'(\overline{u}) = -1$ and $-2f'''(\overline{u}) - 3[f''(\overline{u})]^2 > 0$, then $\overline{u}$ is an unstable point.

**5.16.32.** Let $a$ be in the domain of $f$ and $m$ be a positive integer. Then $a$ is called $m$ *periodic* if it is a fixed point of $f^m$, i.e. $f^m(a) = a$. This means $a$ is an equilibrium point of the difference equation $u(k+1) = g(u(k))$, where $g = f^m$. The *periodic orbit* of $a$ denoted as $O^+(a) = \{a, f(a), \cdots, f^{m-1}(a)\}$ is called a $m$ *cycle*.

Euler's method when applied to logistic differential equation

$$y' = \alpha y(1-y), \quad t > 0, \quad y(0) = u(0)$$

leads to the first order difference equation

$$(5.16.13) \qquad u(k+1) = u(k) + \alpha h u(k)(1-u(k)), \quad k \in \mathbb{N}$$

which, after writing $v(k) = \alpha h u(k)/(1+\alpha h)$ and $a = 1 + \alpha h$, becomes the quadratic map

$$(5.16.14) \qquad v(k+1) = av(k)(1-v(k)), \quad k \in \mathbb{N}.$$

This is the *logistic model* of the growth of populations in successive generation. Of particular interest to population biologists is the case $\alpha > 0$ so that $\alpha h > 0$, and $u(0) > 0$. Show that the solution $u(k)$ of (5.16.13)

(i) converges monotonically to the stable equilibrium point $u = 1$ whenever $0 < \alpha h < 1$

(ii) converges oscillating to the stable equilibrium point $u = 1$ whenever $1 < \alpha h < 2$

(iii) when $\alpha h = 2$, the system bifurcates to give 2 cycle which is stable for $2 < \alpha h < \sqrt{6}$

(iv) as $\alpha h$ increases above 2, successive bifurcations give rise to a cascade of period doublings, producing cycles of periods $2, 4, \cdots, 2^s$ where $s$ is a positive integer, for $\alpha h$ in the range $2 < \alpha h < 2.569946\cdots$

(v) as $\alpha h$ increases further to lie in the interval $2.569946\cdots < \alpha h < 3$, the behavior is aperiodic or chaotic with solution curves resembling those of random processes

(vi) the numerical solution diverges for $\alpha h > 3$.

**5.16.33.** An equilibrium point $\overline{u}$ of $u(k+1) = f(u(k))$ is called *semi stable from the right* if given $\epsilon > 0$ there exists $\delta > 0$ such that if $0 < u(0) - \overline{u} < \delta$, then $u(k) - \overline{u} < \epsilon$. *Semi stable from the left* is defined similarly. If in addition $\lim_{k\to\infty} u(k) = \overline{u}$ whenever $0 < u(0) - \overline{u} < \eta$ $(0 < \overline{u} - u(0) < \eta)$, then $\overline{u}$ is said to be *semi asymptotically stable* from the right (left). Suppose that if $f'(\overline{u}) = 1$, $f''(\overline{u}) \neq 0$. Show that $\overline{u}$ is semi asymptotically stable from the right (left) if $f''(\overline{u}) < (>) \ 0$.

**5.16.34.** Show that for the difference equation $u(k+2) + pu(k+1) + qu(k) = 0$ the equilibrium point $\overline{u} = 0$ is asymptotically stable if and only if $1 + p + q > 0$, $1 - p + q > 0$, $1 - q > 0$.

**5.16.35.** (Samuelson's Interaction Model). The economy of a nation is modeled by considering four discrete functions: the national income $u(k)$, the consumer expenditure $C(k)$ used to purchase goods, the investment $I(k)$, and the government expenditure $G(k)$. The definition of $u(k)$ is

$$u(k) = I(k) + C(k) + G(k). \tag{5.16.15}$$

Following Samuelson [88] we make the following assumptions:

(a) Consumer expenditure $C(k)$ is proportional to the national income $u(k-1)$ in the preceding year $k-1$, i.e.

$$C(k) = \alpha u(k-1), \tag{5.16.16}$$

where $0 < \alpha < 1$ is usually called the *marginal propensity to consume.*

(b) Investment $I(k)$ is proportional to the increase in consumption $C(k) - C(k-1)$, i.e.

$$I(k) = \rho(C(k) - C(k-1)), \tag{5.16.17}$$

where $\rho > 0$ is called the *relation.*

(c) Government expenditure $G(k)$ is constant, i.e.

$$G(k) = G. \tag{5.16.18}$$

Combining (5.16.15) (5.16.18), we get

$$u(k+2) - \alpha(1+\rho)u(k+1) + \alpha\rho u(k) - G = 0. \tag{5.16.19}$$

Use Problem 5.16.34 to study the asymptotic stability of the equilibrium state of the national income $\overline{u} = G/(1-\alpha)$.

**5.16.36.** The following difference equation provides how the red blood cells (RBC) count changes in the body

$$\Delta R(k) = -S(k) + M(k), \tag{5.16.20}$$

where $R(k)$, $S(k)$ and $M(k)$, respectively, represent on day $k$ the number of RBC, the number of RBC removed by the spleen, and the number of RBC created by the bone narrow. We assume that

$$S(k) = \alpha R(k), \quad 0 < \alpha < 1 \tag{5.16.21}$$

where $\alpha$ is the fraction of RBC removed by the spleen, and

$$M(k) = \alpha\beta R(k-1), \quad \beta > 0 \tag{5.16.22}$$

where $\beta$ is the production constant, equal to the ratio of the number of new RBC created to those lost on the previous day.

Combination of (5.16.20) (5.16.22) gives

$$R(k+2) \;=\; (1-\alpha)R(k+1)+\alpha\beta R(k). \tag{5.16.23}$$

Find conditions on $\alpha$, $\beta$ so that homeostasis in the RBC count occurs, i.e. $R(k)$ is essentially constant or at least asymptotically constant.

**5.16.37.** (The Cobweb Model). Let $p(k)$ denote the price of a commodity in the year $k$, and let $q(k)$ be the amount of the commodity available in the market in the year $k$. We make the following assumptions:

(a) Amount of the commodity produced this year and available for sale is a linear function of the price of the commodity in the last year, i.e.

$$q(k) \;=\; \alpha+\beta p(k-1), \tag{5.16.24}$$

where $\beta > 0$ since if the last year's price was high, the amount available this year would also be high.

(b) The price of the commodity this year is a linear function of the amount available this year, i.e.

$$p(k) \;=\; \gamma+\delta q(k), \tag{5.16.25}$$

where $\delta < 0$ since if $q(k)$ is large the price would be low.

From (5.16.24) and (5.16.25) show that

(i) $(p(k)-\overline{p}) = (\beta\delta)(p(k-1)-\overline{p})$, $\overline{p} = (\alpha\delta+\gamma)/(1-\beta\delta)$ and hence $p(k)$ oscillates around the equilibrium point $\overline{p}$, i.e. $(p(k)-\overline{p})(p(k+1)-\overline{p}) < 0$. Further, if $|\beta\delta| < 1$ the equilibrium price $\overline{p}$ is asymptotically stable, and if $|\beta\delta| > 1$ it is unstable.

(ii) $(q(k)-\overline{q}) = (\beta\delta)(q(k-1)-\overline{q})$, $\overline{q} = (\alpha+\beta\gamma)/(1-\beta\delta)$ and hence $q(k)$ oscillates around the equilibrium commodity $\overline{q}$. Further, if $|\beta\delta| < 1$ the equilibrium commodity $\overline{q}$ is asymptotically stable, and if $|\beta\delta| > 1$ it is unstable.

**5.16.38.** Discuss the stability of the equilibrium point $\overline{p}$ of the extended Cobweb model for which

$$(p(k)-\overline{p}) \;=\; c(1-\rho)(p(k-1)-\overline{p})+c\rho(p(k-2)-\overline{p}),$$

where $c$ is the ratio of slopes of supply and demand curves and $0 \le \rho \le 1$ represents the expectation of suppliers about price reversal.

**5.16.39.** Discuss the stability of the equilibrium point $\overline{u} = 0$ of the Harrod Domar growth model

$$u(k+2) \;=\; (1+v)u(k+1)-(v+s)u(k),$$

where $s = 1 - c$ is the marginal propensity to save and $v$ is the power of the accelerator.

**5.16.40.** To settle the wage dispute between labor and management the following system occurs (see [59])

$$\begin{aligned} M(k+1) &= M(k) + \alpha(L(k) - M(k)) \\ L(k+1) &= L(k) - \beta(L(k) - M(k)), \end{aligned}$$

where $M(k)$ and $L(k)$, respectively, are management's offer and labor's demand at the $k$th step, $0 < \alpha < 1$, $0 < \beta < 1$ and $M(0) < L(0)$. Show that

(i)
$$\begin{aligned} M(k) &= \left(\frac{\alpha L(0) + \beta M(0)}{\alpha + \beta}\right) - \left(\frac{\alpha(L(0) - M(0))}{\alpha + \beta}\right)(1 - \alpha - \beta)^k \\ L(k) &= \left(\frac{\alpha L(0) + \beta M(0)}{\alpha + \beta}\right) + \left(\frac{\beta(L(0) - M(0))}{\alpha + \beta}\right)(1 - \alpha - \beta)^k \end{aligned}$$

(ii) if $0 < \alpha + \beta < 1$ then $M(k)$ and $L(k)$ converge monotonically to $w = \dfrac{\alpha L(0) + \beta M(0)}{\alpha + \beta}$.

**5.16.41.** Let $u(k)$ and $v(k)$ denote military budgets for the year $k$ for the countries $A$ and $B$, respectively. Richardson [86] proposed that yearly changes in the military budget are related by the following system

$$\begin{aligned} \Delta u(k+1) &= \Gamma v(k) - \alpha u(k) + \gamma \\ \Delta v(k+1) &= \Theta u(k) - \beta v(k) + \delta \end{aligned} \tag{5.16.26}$$

where $\Gamma$ and $\Theta$ are *defense* coefficients and give a measure of how one country reacts to the current military budget of the other country; $\alpha$ and $\beta$ are *fatigue* coefficients and measure the possible negative economic consequences of increasing the military budgets; and $\gamma$ and $\delta$ are called the *grievances* which are independent of the military budgets, but reflect factors such as internal political and economic considerations, dissatisfaction with previous political interactions between the countries $A$ and $B$, and so on. Show that if $\Gamma$, $\Theta$, $\alpha$, $\beta$, $\gamma$, $\delta$ are nonnegative and $\alpha\beta > \Gamma\Theta$ then the equilibrium solution

$$\overline{u} = \frac{\beta\gamma + \Gamma\delta}{\alpha\beta - \Gamma\Theta}, \qquad \overline{v} = \frac{\alpha\delta + \Theta\gamma}{\alpha\beta - \Gamma\Theta}$$

of (5.16.26) is stable.

**5.16.42.** Show that for the difference system

$$\begin{aligned} u_1(k+1) &= u_1(k) + \frac{u_1^2(k)(u_2(k)-u_1(k)) + u_2^5(k)}{r^2(k)+r^6(k)} \\ u_2(k+1) &= u_2(k) + \frac{u_2^2(k)(u_2(k)-2u_1(k))}{r^2(k)+r^6(k)}, \quad r^2(k) = u_1^2(k)+u_2^2(k) \end{aligned}$$

the trivial solution is globally attractive but unstable.

**5.16.43.** Consider the difference system $\mathbf{u}(k+1) = \mathcal{A}\mathbf{u}(k) + \mathbf{b}$, where $\mathcal{A}$ is an $n \times n$ nonnegative matrix and $\mathbf{b}$ is an $n \times 1$ vector. Show that

(i) if $\rho(\mathcal{A}) < 1$ and $\mathbf{b} \geq 0$, then the equation

$$\overline{\mathbf{u}} = \mathcal{A}\overline{\mathbf{u}} + \mathbf{b} \tag{5.16.27}$$

has a nonnegative solution which is asymptotically stable

(ii) if $\mathbf{b} > 0$ and (5.16.27) has a positive solution, then $\rho(\mathcal{A}) < 1$.

**5.16.44.** Let for a fixed $\ell \in \mathbb{N}(a)$ the solution $\mathbf{u}(k) = \mathbf{u}(k, \ell, \mathbf{u}(\ell))$ of (1.2.12) exist on $\mathbb{N}(\ell)$, and $\mathbf{u}(k) \neq 0$ for all $k \in \mathbb{N}(\ell)$. Further, let for some fundamental set of solutions $\{\mathbf{u}^i(k)\}$ of (1.2.12) the quantities $\|\mathbf{u}^i(k)\|/\|\mathbf{u}(k)\|$ be bounded. Show that (1.2.12) is weakly stable for $\mathbf{u}(k)$.

**5.16.45.** Prove Theorem 5.6.2.

**5.16.46.** Prove Corollary 5.6.3.

**5.16.47.** Prove Theorem 5.6.4.

**5.16.48.** Let $\mathbf{f}(\mathbf{u}) \in C^{(1)}[\Omega, \mathbb{R}^n]$, $\mathbf{f}(0) = 0$ and $\mathbf{f}(\mathbf{u}) \neq 0$ for $\mathbf{u} \neq 0$ in $\Omega$, and let $\mathcal{A} = (\partial f_i(0)/\partial u_j)$ be the Jacobian matrix of $\mathbf{f}$ at $\mathbf{u} = 0$. Show that the trivial solution $\mathbf{u}(k, 0, 0) \equiv 0$ of the difference system (5.4.1) is

(i) asymptotically stable if all the eigenvalues of $\mathcal{A}$ are inside the unit disc

(ii) unstable if there is an eigenvalue of $\mathcal{A}$ with magnitude greater than one

(iii) stable or unstable if all the eigenvalues of $\mathcal{A}$ are inside the unit disc, and that of at least one eigenvalue has the modulus one.

**5.16.49.** Let for all $k \in \mathbb{N}(a)$, $\|\mathbf{u}\| < \rho\ (> 0)$ the function $\mathbf{f}(k, \mathbf{u})$ be defined, $\mathbf{f}(k, 0) \equiv 0$, and satisfy the inequality

$$\|\mathbf{u} + \mathbf{f}(k, \mathbf{u})\| \leq \|\mathbf{u}\| + g(k, \|\mathbf{u}\|)$$

where $g(k,r)$ is defined for all $(k,r) \in \mathbb{N}(a) \times \mathbb{R}_+$, $g(k,0) \equiv 0$, and $r + g(k,r)$ is nondecreasing in $r$ for each fixed $k \in \mathbb{N}(a)$. Show that the stability properties of the trivial solution $r(k,a,0) \equiv 0$ of (5.1.9) imply the corresponding stability properties of the trivial solution $\mathbf{u}(k,a,0) \equiv 0$ of (5.1.1).

**5.16.50.** Let $\mathcal{U}(k,a)$ be the principal fundamental matrix of the difference system (1.2.12), and the function $\mathbf{g}(k,\mathbf{v})$ be defined for all $(k,\mathbf{v}) \in \mathbb{N}(a) \times \mathbb{R}^n$, $\mathbf{g}(k,0) \equiv 0$, and satisfy the inequality $\|\mathcal{U}^{-1}(k+1,a)\mathbf{g}(k,\mathcal{U}(k,a)\mathbf{w})\| \le h(k,\|\mathbf{w}\|)$, where $h(k,r)$ is defined for all $(k,r) \in \mathbb{N}(a) \times \mathbb{R}_+$, and nondecreasing in $r$ for any fixed $k \in \mathbb{N}(a)$. Show that the stability properties of (1.2.12) imply the corresponding stability properties of the trivial solution $\mathbf{v}(k,a,0) \equiv 0$ of (5.3.4) provided for each $r^0 \ge 0$ the solution $r(k,a,r^0)$ of $\Delta r(k) = h(k,r(k))$, $r(a) = r^0$ is bounded on $\mathbb{N}(a)$.

**5.16.51.** Let for all $k \in \mathbb{N}(a)$, $\|\mathbf{u}\| < \rho\ (>0)$ the function $\mathbf{f}(k,\mathbf{u})$ be defined, $\mathbf{f}(k,0) \equiv 0$, and satisfy the inequality $\|\mathbf{f}(k,\mathbf{u})\| \le g(k,\|\mathbf{u}\|)$, where $g(k,r)$ is defined for all $(k,r) \in \mathbb{N}(a) \times \mathbb{R}_+$, $g(k,0) \equiv 0$, and $g(k,r)$ is nondecreasing in $r$ for any fixed $k \in \mathbb{N}(a)$. Show that the stability properties of the trivial solution $r(k,a,0) \equiv 0$ of

$$r(k+1) = \|\mathcal{I} + h\mathcal{A}(k)\| r(k) + g(k,r(k)) \tag{5.16.28}$$

imply the corresponding stability properties of the trivial solution $\mathbf{u}(k,a,0) \equiv 0$ of

$$\mathbf{u}(k+1) = (\mathcal{I} + h\mathcal{A}(k))\mathbf{u}(k) + \mathbf{f}(k,\mathbf{u}(k)),$$

where $h$ is a positive constant. (In (5.16.28), $\|\mathcal{I} + h\mathcal{A}\|$ can be less than 1).

**5.16.52.** Suppose that (i) for all $k \in \mathbb{N}(a)$, $\|\mathbf{u}\| < \rho\ (>0)$ the function $\mathbf{f}(k,\mathbf{u})$ is defined, $\mathbf{f}(k,0) \equiv 0$ the Jacobian matrix $\mathbf{f}_{\mathbf{u}}(k,\mathbf{u})$ exists, and for every $\epsilon > 0$ there exists a $\delta = \delta(\epsilon)$ such that

$$\|\mathbf{f}(k,\mathbf{u}) - \mathbf{f}_{\mathbf{u}}(k,0)\mathbf{u}\| \le \epsilon\|\mathbf{u}\|$$

provided $\|\mathbf{u}\| < \delta(\epsilon)$, (ii) the inequality

$$\limsup_{k\to\infty} \frac{1}{k-a} \sum_{\ell=a}^{k-1} \left(\|\mathcal{I} + \mathbf{f}_{\mathbf{u}}(\ell,0)\| - 1\right) < 0$$

holds. Show that the trivial solution $\mathbf{u}(k,a,0) \equiv 0$ of (5.1.1) is asymptotically stable.

**5.16.53.** Suppose that (i) $\mathbf{f}(k,\mathbf{u})$ is as in (i) of Problem 5.16.52, (ii) there exists a positive constant $\sigma$ such that $\|\mathcal{I}+\mathbf{f_u}(k,0)\|-1<-\sigma$ for all $k\in\mathbb{N}(a)$, (iii) $\mathbf{g}(k,\mathbf{u})$ is defined for all $k\in\mathbb{N}(a)$, $\|\mathbf{u}\|<\rho$, $\mathbf{g}(k,0)\equiv 0$, and $\|\mathbf{g}(k,\mathbf{u})\|\leq\mu(k)$, where $\mu(k)\to 0$ as $k\to\infty$. Show that the trivial solution $\mathbf{u}(k,a,0)\equiv 0$ of $\Delta\mathbf{u}(k)=\mathbf{f}(k,\mathbf{u}(k))+\mathbf{g}(k,\mathbf{u}(k))$ is asymptotically stable.

**5.16.54.** Consider the summary difference system of convolution type

$$(5.16.29)\quad \mathbf{u}(k+1) = \mathcal{A}\mathbf{u}(k)+\sum_{\ell=0}^{k-1}\mathcal{B}(k-1-\ell)\mathbf{u}(\ell),\quad k\in\mathbb{N},\quad \mathbf{u}(0)=\mathbf{u}_0$$

and its perturbed system

(5.16.30)

$$\mathbf{v}(k+1) = \mathcal{A}\mathbf{v}(k)+\sum_{\ell=0}^{k-1}\mathcal{B}(k-1-\ell)\mathbf{v}(\ell)+\mathbf{g}(k),\quad k\in\mathbb{N},\quad \mathbf{v}(0)=\mathbf{u}_0$$

where $\mathcal{A}$ and $\mathcal{B}(k)$, $k\in\mathbb{N}$ are $n\times n$ matrices, $g:\mathbb{N}\to\mathbb{R}^n$, and $\mathcal{B}(-k)=0$, $k\in\mathbb{N}(1)$.

(i) The difference resolvent matrix $\mathcal{R}(k)$ associated with the linear system (5.16.29) is the solution of the matrix equation

$$(5.16.31)\qquad \mathcal{R}(k+1) = \mathcal{A}\mathcal{R}(k)+\sum_{\ell=0}^{k-1}\mathcal{B}(k-1-\ell)\mathcal{R}(\ell),\quad \mathcal{R}(0)=\mathcal{I}.$$

Show that in terms of resolvent matrix $\mathcal{R}(k)$ of (5.16.29) the solution $\mathbf{v}(k)$ of (5.16.30) can be written as

$$(5.16.32)\qquad \mathbf{v}(k) = \mathcal{R}(k)\mathbf{u}_0+\sum_{\ell=0}^{k-1}\mathcal{R}(k-1-\ell)\mathbf{g}(\ell).$$

(ii) For the system (5.16.30) with $\mathcal{A}=0$ the resolvent matrix $\overline{\mathcal{R}}(k)$ is the solution of the matrix equation

$$(5.16.33)\qquad \overline{\mathcal{R}}(k) = -\mathcal{B}(k)+\sum_{\ell=0}^{k-1}\mathcal{B}(k-1-\ell)\overline{\mathcal{R}}(\ell),\quad k\in\mathbb{N}.$$

Show that in terms of resolvent matrix $\overline{\mathcal{R}}(k)$ of (5.16.30) with $\mathcal{A}=0$ the solution $\mathbf{v}(k)$ of (5.16.30) with $\mathcal{A}=0$ can be written as

$$(5.16.34)\qquad \mathbf{v}(k) = \mathbf{g}(k)-\sum_{\ell=0}^{k-1}\overline{\mathcal{R}}(k-1-\ell)\mathbf{g}(\ell).$$

**5.16.55.** Let $\mathcal{L}(k,\ell)$, $\mathcal{B}(k,\ell)$, $(k,\ell) \in \mathbb{N}(k_0) \times \mathbb{N}(k_0)$ be $n \times n$ matrices such that $\mathcal{L}$, $\mathcal{B}$ are zero for all $k$, $\ell < k_0$. Show that the following *discrete Fubini's theorem* holds

$$\sum_{\ell=k_0}^{k-1} \mathcal{L}(k,\ell+1) \sum_{\sigma=k_0}^{\ell-1} \mathcal{B}(\ell,\sigma)\mathbf{u}(\sigma) = \sum_{\ell=k_0}^{k-1} \sum_{\sigma=\ell}^{k-1} \mathcal{L}(k,\sigma+1)\mathcal{B}(\sigma,\ell)\mathbf{u}(\ell),$$

where $\mathbf{u} : \mathbb{N}(k_0) \to \mathbb{R}^n$.

**5.16.56.** Consider the summary difference system

(5.16.35)

$$\Delta\mathbf{u}(k) = \mathcal{A}(k)\mathbf{u}(k) + \sum_{\ell=k_0}^{k-1} \mathcal{B}(k,\ell)\mathbf{u}(\ell) + \mathbf{g}(k), \quad k \in \mathbb{N}(k_0), \quad \mathbf{u}(k_0) = \mathbf{u}_0$$

where $\mathcal{A}(k)$, $\mathcal{B}(k,\ell)$, $k,\ell \in \mathbb{N}(k_0)$ are $n \times n$ matrices, and $g : \mathbb{N}(k_0) \to \mathbb{R}^n$. Assume that there exists a $n \times n$ matrix $\mathcal{L}(k,\ell)$ defined on $\mathbb{N}(k_0) \times \mathbb{N}(k_0)$ satisfying

(5.16.36)

$$\mathcal{B}(k,\ell) + \mathcal{L}(k,\ell+1) - \mathcal{L}(k,\ell) + \mathcal{L}(k,\ell+1)\mathcal{A}(\ell) + \sum_{\sigma=\ell}^{k-1} \mathcal{L}(k,\sigma+1)\mathcal{B}(\sigma,\ell) = 0.$$

Use Problem 5.16.55 to show that (5.16.35) is equivalent to the linear difference system

$$(5.16.37) \qquad \Delta\mathbf{v}(k) = \overline{\mathcal{A}}(k)\mathbf{v}(k) + \mathcal{L}(k,k_0)\mathbf{u}_0 + \overline{\mathbf{g}}(k), \quad \mathbf{v}(k_0) = \mathbf{u}_0$$

where

$$\overline{\mathcal{A}}(k) = \mathcal{A}(k) - \mathcal{L}(k,k)$$

$$\overline{\mathbf{g}}(k) = \mathbf{g}(k) + \sum_{\ell=k_0}^{k-1} \mathcal{L}(k,\ell+1)\mathbf{g}(\ell).$$

**5.16.57.** Show that the inequalities (5.8.1) are equivalent to the inequalities (5.8.3).

**5.16.58.** Suppose that in the difference system (1.2.12) the matrix $\mathcal{A}(k)$ is upper triangular and invertible for all $k \in \mathbb{N}(a)$. Show that (1.2.12) has exponential dichotomy if and only if the corresponding diagonal system

$$\mathbf{u}(k+1) = diag\,(a_{11}(k),\cdots,a_{nn}(k))\mathbf{u}(k), \quad k \in \mathbb{N}(a)$$

has an exponential dichotomy.

**5.16.59.** Let for all $\mathbf{u} \in \mathbb{R}^n$ the function $\mathbf{f}(\mathbf{u})$ be defined, $\mathbf{f}(0) = 0$ and $\mathbf{f}(\mathbf{u}) \neq 0$ if $\mathbf{u} \neq 0$. Show that the region of attraction of the origin of (5.4.1) is an open set in $\mathbb{R}^n$.

**5.16.60.** Let $\mathbf{f}(\mathbf{u})$ be as in Problem 5.16.59. Further, let there exist a scalar function $V(\mathbf{u}) \in C[\mathbb{R}^n, \mathbb{R}]$ such that $\Delta^2 V(\mathbf{u}(k, 0, \mathbf{u}^0)) > 0$ if $\mathbf{u}(k, 0, \mathbf{u}^0) \neq 0$, where $\mathbf{u}(k) = \mathbf{u}(k, 0, \mathbf{u}^0)$ is the solution of (5.4.1). Show that for any $\mathbf{u}^0 \in \mathbb{R}^n$, either $\mathbf{u}(k, 0, \mathbf{u}^0)$ is unbounded or it tends to zero as $k \to \infty$. Likewise, if $\Delta^2 V(\mathbf{u}(k, 0, \mathbf{u}^0)) < 0$, $\mathbf{u}(k, 0, \mathbf{u}^0) \neq 0$.

**5.16.61.** Let for all $u \in \mathbb{R}$ the function $f(u)$ be defined, $f(0) = 0$ and $f(u) \neq 0$ if $u \neq 0$. For the system

$$\begin{aligned} u_1(k+1) &= u_2(k) \\ u_2(k+1) &= u_1(k) + f(u_1(k)), \end{aligned}$$

where $\Delta(u_2(k) f(u_1(k))) > 0$ use Problem 5.16.60 with $V(\mathbf{u}(k)) = u_1(k) u_2(k)$ to show that its each solution is either unbounded or tends to zero.

**5.16.62.** Let $\mathbf{f}(\mathbf{u})$ be as in Problem 5.16.59. Further, let there exist two positive definite scalar functions $V(\mathbf{u}) \in C[\mathbb{R}^n, \mathbb{R}_+]$ and $W(\mathbf{u}) \in C[\mathbb{R}^n, \mathbb{R}_+]$ such that

$$V(\mathbf{u}(k+1, 0, \mathbf{u}^0)) = (1 + W(\mathbf{u}(k, 0, \mathbf{u}^0))) V(\mathbf{u}(k, 0, \mathbf{u}^0)) - W(\mathbf{u}(k, 0, \mathbf{u}^0))$$

for any solution $\mathbf{u}(k) = \mathbf{u}(k, 0, \mathbf{u}^0)$ of (5.4.1). Show that the region of asymptotic stability of the origin of (5.4.1) is $D = \{\mathbf{u} \in \mathbb{R}^n : V(\mathbf{u}) < 1\}$.

**5.16.63.** Suppose that (i) $\mathbf{f}(k, \mathbf{u})$ is as in (i) of Problem 5.16.52, (ii) for all $(k, \mathbf{u}) \in \mathbb{N}(a) \times S_\rho$ there exists a scalar function $V(k, \mathbf{u})$ satisfying the Lipschitz condition (5.12.2), and $\|\mathbf{u}\| \leq V(k, \mathbf{u}) \leq c\|\mathbf{u}\|$, (iii) for any solution $\mathbf{v}(k)$ of $\Delta \mathbf{v}(k) = \mathbf{f}_{\mathbf{u}}(k, 0) \mathbf{v}(k)$ the inequality $\Delta V(k, \mathbf{v}(k)) \leq \mu(k) V(k, \mathbf{v}(k))$ is satisfied, where $\limsup_{k \to \infty} \frac{1}{k-a} \sum_{\ell=a}^{k-1} \mu(\ell) < 0$. Show that the trivial solution $\mathbf{u}(k, a, 0) \equiv 0$ of (5.1.1) is asymptotically stable.

**5.16.64.** The trivial solution $\mathbf{u}(k, a, 0) \equiv 0$ of (1.2.8) is said to be *generalized exponentially asymptotically stable* if for any solution $\mathbf{u}(k) = \mathbf{u}(k, a, \mathbf{u}^0)$ of (1.2.8), $\|\mathbf{u}(k, a, \mathbf{u}^0)\| \leq c(k) \|\mathbf{u}^0\| \exp(p(a) - p(k))$, $k \in \mathbb{N}(a)$, where the functions $c(k) > 1$ and $p(k) > 0$ are defined on $\mathbb{N}(a)$, $p(0) = 0$ if $a = 0$, $p(k)$ is strictly monotonically increasing and $p(k) \to \infty$ as $k \to \infty$. Suppose that the difference system (1.2.12) is generalized exponentially asymptotically stable. Show that there exists a scalar function $V(k, \mathbf{u})$ in $\mathbb{N}(a) \times S_\rho$ such that

(i) $\|\mathbf{u}\| \leq V(k, \mathbf{u}) \leq c(k) \|\mathbf{u}\|$

(ii) for any solution $\mathbf{u}(k) = \mathbf{u}(k, a, \mathbf{u}^0)$ of (1.2.12) such that $\|\mathbf{u}(k)\| < \rho$

$$\Delta V(k, \mathbf{u}(k)) \leq -(1 - \exp(-\Delta p(k)))V(k, \mathbf{u}(k))$$

(iii) $|V(k, \mathbf{u}) - V(k, \mathbf{v})| \leq c(k)\|\mathbf{u} - \mathbf{v}\|$ for $k \in \mathbb{N}(a)$, $\mathbf{u}$, $\mathbf{v} \in S_\rho$.

**5.16.65.** Suppose that the difference system (1.2.12) is generalized exponentially asymptotically stable, and the function $\mathbf{g}(k, \mathbf{v})$ is defined on $\mathbb{N}(a) \times S_\rho$, and $\|\mathbf{g}(k, \mathbf{v})\| \leq h(k, \|\mathbf{v}\|)$, where $h(k, r)$ is defined for all $(k, r) \in \mathbb{N}(a) \times \mathbb{R}_+$, $h(k, 0) \equiv 0$, and $r + h(k, r)$ is nondecreasing in $r$ for any fixed $k \in \mathbb{N}(a)$. Show that the stability or asymptotic stability of the trivial solution $r(k, a, 0) \equiv 0$ of $\Delta r(k) = -(1 - \exp(-\Delta p(k)))r(k) + c(k+1)h(k, r(k))$ implies the stability or asymptotic stability of the trivial solution $\mathbf{v}(k, a, 0) \equiv 0$ of (5.3.4).

## 5.17. Notes

The qualitative theory of difference equations is in a process of continuous development, as it is apparent from the large number of research papers dedicated to it. Although several results in the discrete case are similar to those already known in the continuous case, the adaptations from the continuous to the discrete case is not direct but requires some special devices. The dependence on initial conditions and parameters of the solutions has been discussed at several places. Our results in Section 5.1 are based on Agarwal [1] and Sugiyama [96]. Almost all the results discussed on the asymptotic behavior of the solutions of linear and nonlinear systems in Sections 5.2 and 5.3 are parallel to those known in the theory of differential equations, e.g. Agarwal and Gupta [2]. As such the statements of some of these theorems have also appeared in Sugiyama [93,97]. The various concepts of stability defined in Section 5.4 are the same as in the continuous case, e.g. Hahn [38], Lakshmikantham and Leela [50]. Once again Theorems 5.5.1 5.5.4 are the discrete analog of the continuous results, e.g. Agarwal and Gupta [2], Coppel [19]. Some parts of Theorem 5.5.1 as such have appeared in Sugiyama [93]. Theorems 5.5.5 5.5.7 are due to Gautschi [33,34]. Theorems 5.6.1 and 5.6.2 are from Sugiyama [93], whereas Lemmas 5.6.5 and 5.6.6 and Theorems 5.6.7 and 5.6.8 are adapted from Schinas [89]. The nonlinear variation of constants formulae in Theorems 5.7.1 and 5.7.2 are due to Lord [57], whose motivation comes from Alekseev [6], Lord and Mitchell [56]. Further generalizations of these results to the continuous as well as the discrete case discussed by Bernfeld and Lord [11], Brauer [13], Pachpatte [77,78] have an obvious error, which has been corrected by Beesack [9]. For recent results on nonlinear variation of parameters and their applications including in numerical computation of multipoint boundary value problems see Agarwal [3], Lakshmikantham and Deo [53], Sheng and Agarwal [90,91]. Ordinary and exponential dichotomies in the continuous case have been studied extensively by Coppel [20], Massera and Schäffer [63], Palmer [79], and are very useful in the construction of the solutions of boundary value problems on infinite intervals, e.g. Mattheij [65,66]. The exponential dichotomy defined in (5.8.5) is equivalent to that of Henry's [45]. Lemma 5.8.1 and Theorems 5.8.2 and 5.8.3 are borrowed from Papashinopoulos and Schinas [81], whereas Theorems 5.8.4 5.8.6 are from Schinas [89]. The continuous analogs of these results are available in Coppel [19,20]. The theory of Lyapunov functions to study stability properties of differential equations has been extensively exploited since 1892, while its use for the difference equations is recent. Various continuous results established in Hahn [38], Halanay [41], Lakshmikantham and Leela [50], Lassalle and Lefschetz [54], Malkin [60], Yoshizawa [102], Zubov [103] and several others, have been discretize by Diamond [23,24], Freeman [32], Gordon [36,37], Hurt [47], Kalgan and Bertram [48], Lassalle [55], Ortega [75], Pachpatte [76], Sugiyama [93 98], Segö and Kalgan [99]. Applications of Lyapunov functions to study stability properties of several discrete models in population dynamics is mainly due to Fischer [26 31]. Models in Examples 5.11.1 5.11.4 are due to Haskell [42], Haskell and Comines [43], Moran [73] and Bricker [87], and May [67] respectively. Model (5.11.11) has been employed by Allen [7], Clark [17], and the International Whaling Commission [49]. Theorem 5.11.4 is from Fisher and Gogh [30]. Models in Examples 5.11.5 and 5.11.6 are from Beaverton and Holt [12], and Bricker [87]. For the differential equations, total stability has been studied widely in Halanay [41], Lakshmikantham and Leela [50]. Total stability results in the discrete case are due to Ortega [75]. Practical stability discussed in Section 5.14 is defined in Hurt [47] and Ortega [75]. Results on mutual stability are borrowed from Pachpatte [76]. Several related results on the asymptotic behavior and stability properties of solutions of difference systems have also appeared in Agarwal and Pang [4], Agarwal and Pit UK [5], Aulbach [8], Benz aid and Lutz [10], Bykov and Linenko [14], Castillo and Pinto [15], Chen and Wu [16], Coffman [18], Corduneanu [21], Driver [25], Goh [35], Halanay [39,40], Heinen [44], Lakshmikantham, Leela and Martynyuk [52], Luca and Talpalaru [58], Maslovskaya [61,62], Máté and Nevai [64], Medina and Pinto [68], Medina [69], Medina and Pinto [70], Medina [71], Naulin and Pinto [74], Pang and Agarwal [80], Peil and Peterson [82], Petrovanu [83], Pituk [84], Rao [85], Smith [92], Trench [100,101] etc. An alternative treatment of some of the results discussed here is available in Lakshmikantham and Trigiante [51]. Problems 5.16.14 5.16.19 are based on the work of Corduneanu [22] and Moadab [72]. Also for some recent contributions on the almost periodic solutions see Hong and Nunez [46].

## 5.18. References

[1]. R.P. Agarwal, On finite systems of difference inequalities, *Jour. Math. Phyl. Sci.* **10**(1976), 277 288.

[2]. R.P. Agarwal and R.C. Gupta, *Essentials of Ordinary Differential Equations, McGraw Hill*, Singapore, New York, 1991.

[3]. R.P. Agarwal, Variation of parameter methods for difference equations and applications, in *Proceedings of the 2nd International Conference on Difference Equations and Applications,* Veszprém, Hungary, August 7 11, 1995, eds. S. Elaydi, et. al., *Gordon & Breach*, New York, 1996, 19 38.

[4]. R.P. Agarwal and P.Y.H. Pang, On a generalized difference system, *Nonlinear Analysis* **30**(1)(1997), 365 376.

[5]. R.P. Agarwal and M. Pituk, Convergence to equilibria in recurrence equations, *Computers Math. Applic.* **36**(10 12)(1998), 357 368.

[6]. V.M. Alekseev, An estimate for the perturbations of the solutions of ordinary differential equations, *Vestn. Mosk. Univ., Ser. 1, Math. Meh.* **2**(1961), 28 36 (Russian).

[7]. K.R. Allen, Analysis of stock recruitment relations in Antarctic fin whales, *Cons. Int. pour l'Explor. Mer Rapp. et Proc. Verb.* **164**(1963), 132 137.

[8]. B. Aulbach, *Continuous and Discrete Dynamics Near Manifolds of Equilibria, Lecture Notes in Math. Vol.* **1058**, *Springer Verlag*, Berlin, 1984.

[9]. P.R. Beesack, On some variation of parameter methods for integrodifferential, integral, and quasilinear partial integrodifferential equations, *Appl. Math. Comp.* **22**(1987), 189 215.

[10]. Z. Benzaid and D.A. Lutz, Asymptotic representation of solutions of perturbed systems of linear difference equations, *Studies in Appl. Math.* **77**(1987), 195 221.

[11]. S.R. Bernfeld and M.E. Lord, A nonlinear variation of constants method for integrodifferential and integral equations, *Appl. Math. Comp.* **4**(1978), 1 14.

[12]. R.J.H. Beverton and S.J. Holt, On the dynamics of exploited fish populations, *Fish. Invest. Lond.* **19**(1957), 1 533.

[13]. F. Brauer, A nonlinear variation of constants formula for Volterra equations, *Math. Syst. Theory* **6**(1972), 226 234.

[14]. J.A. Bykov and V.G. Linenko, On stability of the solutions of difference equations, *Diff. Urav.* **9**(1973), 349 354 (Russian).

[15]. S. Castillo and M. Pinto, Asymptotic formulae for solutions of delay difference systems, in *Proceedings of the 2nd International Conference on Difference Equations and Applications*, Veszprém, Hungary, August 7 11, 1995, eds. S. Elaydi, et. al., *Gordon & Breach*, New York, 1996, 107 117.

[16]. S. Chen and C. Wu, Riccati techniques and approximation for a second order Poincaré difference equation, *J. Math. Anal. Appl.* **222**(1998), 177 191.

[17]. C.W. Clark, A delayed recruitment model of population dynamics, with an application to baleen whale populations, *J. Math. Biol.* **3**(1976), 381 391.

[18]. C.V. Coffman, Asymptotic behavior of solutions of ordinary difference equations, *Trans. Amer. Math. Soc.* **110**(1964), 22 51.

[19]. W.A. Coppel, *Stability and Asymptotic Behavior of Differential Equations, D.C. Health*, Boston, 1965.

[20]. W.A. Coppel, *Dichotomies in Stability Theory, Lecture Notes in Math. Vol.* **629**, *Springer Verlag*, Berlin, 1978.

[21]. C. Corduneanu, Almost periodic discrete processes, *Libertas Math.* **2**(1982), 159 169.

[22]. C. Corduneanu, Discrete qualitative inequalities and applications, *Nonlinear Analysis* **25**(1995), 933 939.

[23]. P. Diamond, Finite stability domains for difference equations, *J. Austral. Math. Soc. Ser. A* **22**(1976), 177 181.

[24]. P. Diamond, Discrete Liapunov function with $V > 0$, *J. Austral. Math. Soc. Ser. B* **20**(1978), 280 284.

[25]. R.D. Driver, Note on a paper of Halanay on stability of finite difference equations, *Arch. Rat. Mech.* **18**(1965), 241 243.

[26]. M.E. Fisher and B.S. Goh, Stability in a class of discrete time models of interacting populations, *J. Math. Biology* **4**(1977), 265 274.

[27]. M.E. Fisher, Asymptotic behavior of a class of discontinuous difference equations, *J. Austral. Math. Soc. Ser. B* **20**(1978), 370 374.

[28]. M.E. Fisher, B.S. Goh and T.L. Vincent, Some stability conditions for discrete time single species models, *Bull. Math. Biology* **41**(1979), 861 875.

[29]. M.E. Fisher, *Analysis of Difference Equations Models in Population Dynamics, Ph.D. Thesis*, Univ. of Western Australia, 1982.

[30]. M.E. Fisher and B.S. Goh, Stability results for delayed recruitment models in population dynamics, *J. Math. Biology* **19**(1984), 147 156.

[31]. M.E. Fisher, Stability of a class of delay difference equations, *Nonlinear Analysis* **8**(1984), 645 654.

[32]. H. Freeman, *Discrete Time Systems: An Introduction to the Theory, John Wiley*, New York, 1965.

[33]. W. Gautschi, Computational aspects of three term recurrence relations, *SIAM Rev.* **9**(1967), 24 82.

[34]. W. Gautschi, Zur Numerik rekurrenter Relationen, *Computing,* **9**(1972), 107 126.

[35]. B.S. Goh, *Management and Analysis of Biological Populations, Elsevier*, Amsterdam, 1980.

[36]. S.P. Gordon, Stability and summability of solutions of difference equations, *Math. Syst. Theory* **5**(1971), 56 75.

[37]. S.P. Gordon, On converses to the stability theorems for difference equation, *SIAM J. Control* **10**(1972), 76 81.

[38]. W. Hahn, *Stability of Motion, Springer Verlag*, Berlin, 1967.

[39]. A. Halanay, Solution periodiques et presque periodiques des systems d'equationes aux difference finies, *Arch. Rat. Mech.* **12**(1963), 134 149.

[40]. A. Halanay, Quelques questions de la théorie de la stabilité pour les systéms aux différences finies, *Arch. Rat. Mech.* **12** (1963), 150 154.

[41]. A. Halanay, *Differential Equations: Stability, Oscillations, Time Lags, Academic Press*, New York, 1966.

[42]. M.P. Hassell, Density dependence in single species populations, *J. Anim. Ecol.* **44**(1975), 283 295.

[43]. M.P. Hassell and H.N. Comins, Discrete time models for two species competition, *Theor. Pop. Biol.* **9**(1976), 202 221.

[44]. J.A. Heinen, Quantitative stability of discrete systems, *Michigan Math. J.* **17**(1970), 211 216.

[45]. D. Henry, *Geometric Theory of Semilinear Parabolic Equations, Lecture Notes in Math. Vol.* **840**, *Springer Verlag*, Berlin, 1981.

[46]. J. Hong and C. Nunez, The almost periodic type difference equations, *Mathl. Comput. Modelling* **28**(12)(1998), 21 31.

[47]. J. Hurt, Some stability theorems for ordinary difference equations, *SIAM J. Numer. Anal.* **4**(1967), 582-596.

[48]. E. Kalman and J.E. Bertram, Control system analysis and design via the second method of Liapunov. II Discrete time systems, *Trans. ASME,* **D82**(1960), 394 400.

[49]. I.W.C., *Twenty Eighth Annual Report of the International Commission on Whaling*, London, 1978.

[50]. V. Lakshmikantham and S. Leela, *Differential and Integral Inequalities*, *Academic Press*, New York, 1969.

[51]. V. Lakshmikantham and D. Trigiante, *Theory of Difference Equations: Numerical Methods and Applications*, *Academic Press*, New York, 1988.

[52]. V. Lakshmikantham, S. Leela and A.A. Martynyuk, *Practical Stability of Nonlinear Systems*, *World Scientific*, Singapore, 1990.

[53]. V. Lakshmikantham and S.G. Deo, *Method of Variation of Parameters for Dynamic Systems*, *Gordon & Breach*, Amsterdam, 1998.

[54]. J.P. LaSalle and S. Lefschetz, *Stability by Liapunov's Direct Method with Applications*, *Academic Press*, New York, 1961.

[55]. J.P. LaSalle, Stability theory for difference equations, in *Studies in Ordinary Differential Equations*, ed. Jack Hale, *The Mathematical Association of America*, 1977, 1 31.

[56]. M.E. Lord and A.R. Mitchell, A new approach to the method of nonlinear variation of parameters, *Appl. Math. Comp.* **4**(1978), 95 105.

[57]. M.E. Lord, The method of non linear variation of constants for difference equations, *J. Inst. Maths. Applics.* **23**(1979), 285 290.

[58]. N. Luca and P. Talpalaru, Stability and asymptotic behavior of a class of discrete systems, *Ann. Mat. Pure Appl.* **112**(1977), 351 382.

[59]. D.G. Luenberger, *Introduction to Dynamic Systems: Theory, Models and Applications*, *Wiley*, New York, 1979.

[60]. I.G. Malkin, *The Theory of Stability of Motion*, United States Atomic Energy Commission, Tech. Report ABC Tr 3352, 1952; 2nd edn. Nauka, Moscow, 1966 (Russian).

[61]. L.V. Maslovskaya, Stability of difference equations, *Diff. Urav.* **2**(19 66), 1176 1183 (Russian).

[62]. L.V. Maslovskaya, On the problem of stability of difference equations, *Izv. Vys. Ucebn. Zaved. Matematika* **69**(1968), 61 67.

[63]. J.L. Massera and J.J. Schäffer, *Linear Differential Equations and Function Spaces*, *Academic Press*, New York, 1966.

[64]. A. Máté and P. Nevai, A generalization of Poincaré theorem for difference equations, *J. Approx. Theory* **63**(1990), 92 97.

[65]. R.M.M. Mattheij, The conditioning of linear boundary value problems, *SIAM J. Numer. Anal.* **19**(1982), 963 978.

[66]. R.M.M. Mattheij, On the computation of solutions of boundary value problems on infinite intervals, *Math. Comp.* **48**(1987), 533 549.

[67]. R.M. May, Biological populations with nonoverlapping generations: stable points, stable cycles, and chaos, *Science* **186**(1974), 645 647.

[68]. R. Medina and M. Pinto, Stability of nonlinear difference equations, in *Proc. Dynam. Systems Appl.* **2**(1996), 397-404.

[69]. R. Medina, Stability and asymptotic behavior of difference equations, *J. Comput. Appl. Math.* **80**(1997), 17 30.

[70]. R. Medina and M. Pinto, Asymptotic equivalence and asymptotic behavior of difference systems, *Communications in Applied Analysis* **1**(1997), 511 523.

[71]. R. Medina, Asymptotic behavior of nonlinear difference systems, *J. Math. Anal. Appl.* **219**(1998), 294 311.

[72]. M.H. Moadab, *Discrete Dynamical Systems and Applications, Ph.D. Thesis*, The University of Texas at Arlington, U.S.A., 1988.

[73]. P.A.P. Moran, Some remarks on animal population dynamics, *Biometrics* **6**(1950), 250 258.

[74]. R. Naulin and M. Pinto, Reduction of linear difference systems with summable perturbations, in *Proceedings of the Ist International Conference on Difference Equations*, San Antonio, Texas, U.S.A. May 25 28, 1994, eds. S. Elaydi, et. al., *Gordon & Breach*, Luxembourg, 1995, 379 386.

[75]. J.M. Ortega, Stability of difference equations and convergence of iterative processes, *SIAM J. Numer. Anal.* **10**(1973), 268 282.

[76]. B.G. Pachpatte, Finite difference inequalities and an extension of Lyapunov's method, *Michigan Math. J.* **18**(1971), 385 391.

[77]. B.G. Pachpatte, A nonlinear variation of constants method for summary difference equations, *Tamk. J. Math.* **8**(1977), 203 212.

[78]. B.G. Pachpatte, On the behavior of solutions of a certain class of nonlinear integrodifferential equations, *An. st. Univ. Iasi* **24**(1978), 77 86.

[79]. K.J. Palmer, Exponential dichotomy, integral separation and diagonalizability of linear systems of ordinary differential equations, *J. Diff. Eqns.* **43**(1982), 184 203.

[80]. P.Y.H. Pang and R.P. Agarwal, Stability of a general class of difference systems, *Computers Math. Applic.* **36**(10 12)(1998), 423 429.

[81]. G. Papashinopoulos and J. Schinas, Criteria for an exponential dichotomy of difference equations, *Czech. Mathl. J.* **35**(110)(1985), 295 299.

[82]. T. Peil and A. Peterson, A theorem of Milloux for difference equations, *Rocky Mount. J. Math.* **24**(1994), 253 260.

[83]. D. Petrovanu, Equations Hammerstein intégrales et discrétes, *Ann. Mat. Pura Appl.* **70**(1966), 227 254.

[84]. M. Pituk, Asymptotic behavior of a Poincaré recurrence system, *J. Approx. Theory* **91**(1997), 226 243.

[85]. M. Rama Mohana Rao, System of ordinary difference equations and stability with respect to manifolds, *Jour. Math. Phyl. Sci.* **7**(1973), 285 296.

[86]. L.F. Richardson, *Arms and Security, Boxwood Press*, Chicago, 1960.

[87]. W.E. Ricker, Stock and recruitment, *J. Fish. Res. Bd. Can.* **11**(1954), 559 623.

[88]. P.A. Samuelson, Interactions between the multiplier analysis and the principle of acceleration, *Rev. Econom. Stat.* **21**(1939), 75 78.

[89]. J. Schinas, Stability and conditional stability of time dependent difference equations in Banach spaces, *J. Inst. Maths. Applics.* **14**(1974), 335 346.

[90]. Q. Sheng and R.P. Agarwal, On nonlinear variation of parameter methods for summary difference equations, *Dynamic Sys. Appl.* **2**(1993), 227 242.

[91]. Q. Sheng and R.P. Agarwal, Nonlinear variation of parameter methods for summary difference equations in several independent variables, *Appl. Math. Comp.* **61**(1994), 39 60.

[92]. R.A. Smith, Sufficient conditions for stability of a class of difference equations, *Duke Math. J.* **33**(1966), 725 734.

[93]. S. Sugiyama, On the stability problems on difference equations, *Bull. Sci. Eng. Research Lab. Waseda Univ.* **45**(1969), 140 144.

[94]. S. Sugiyama, Stability problems on difference and functional differential equations, *Proc. Japan Acad.* **45**(1969), 526 529.

[95]. S. Sugiyama, Stability and boundedness problems on difference and functional-differential equations, *Memoirs of the School of Sci. Eng. Waseda Univ.* **33**(1969), 79 88.

[96]. S. Sugiyama, Comparison theorems on difference equations, *Bull. Sci. Eng. Research Lab. Waseda Univ.* **47**(1970), 77 82.

[97]. S. Sugiyama, On the asymptotic behaviors of solutions of difference equations I, *Proc. Japan Acad.* **47**(1971), 477 480.

[98]. S. Sugiyama, On the asymptotic behaviors of solutions of difference equations II, *Proc. Japan Acad.* **47**(1971), 481 484.

[99]. G. Szegö and R. Kalman, Sur la stabilité absolue d'une systéme d'équations aux différences finies, *C. R. Acad. Sci. Paris* **257**(1965), 388 390.

[100]. W.F. Trench, Asymptotic behavior of solutions of Poincaré difference equations, *Proc. Amer. Math. Soc.* **119**(1993), 431 438.

[101]. W.F. Trench, Asymptotic behavior of solutions of Poincaré recurrence systems, *Computers Math. Applic.* **28**(1 3)(1994), 317 324.

[102]. T. Yoshizawa, *The Stability Theory by Liapunov's Second Method*, *Mathematical Society of Japan*, Tokyo, 1966.

[103]. V.I. Zubov, *The Methods of Liapunov and Their Applications*, *Noordhoff*, Groningen, 1964.

# Chapter 6

# Qualitative Properties of Solutions of Higher Order Difference Equations

Throughout this chapter by a solution $u(k)$ of a given difference equation we shall mean a nontrivial solution which exists on $\mathbb{N}(a)$ for some $a \in \mathbb{N}$. This solution is called *oscillatory* if for any $k_1 \in \mathbb{N}(a)$ there exists a $k_2 \in \mathbb{N}(k_1)$ such that $u(k_2)u(k_2+1) \leq 0$. The given difference equation itself is called oscillatory if all its solutions are oscillatory. If the solution $u(k)$ is not oscillatory then it is said to be *nonoscillatory*. Equivalently, the solution $u(k)$ is nonoscillatory if it is eventually positive or negative, i.e. there exists a $k_1 \in \mathbb{N}(a)$ such that $u(k)u(k+1) > 0$ for all $k \in \mathbb{N}(k_1)$. The given difference equation is called nonoscillatory if all its solutions are nonoscillatory. A given difference equation can have both oscillatory as well as nonoscillatory solutions, e.g. the equation $\Delta^2 u(k) + \frac{8}{3}\Delta u(k) + \frac{4}{3}u(k) = 0, \quad k \in \mathbb{N}$ has an oscillatory solution $u(k) = (-1)^k$ and a nonoscillatory solution $u(k) = (1/3)^k$. For a nonnegative integer $p$, $F_p$ denotes the class of all functions $u(k)$ defined on $\mathbb{N}(a)$ such that $|u(k)| = 0((k)^{(p)})$ as $k \to \infty$. A solution $u(k)$ which belongs to $F_p$ will be called a $F_p$ solution. For example, for the above difference equation $u(k) = (-1)^k$ is a $F_0$ solution. The solution $u(k)$ is called $T$ type if it changes sign arbitrarily but is ultimately nonnegative or nonpositive. The main objective of this chapter is to offer a systematic treatment of oscillation and nonoscillation theory of difference equations.

## 6.1. General Properties of Solutions of

$$p(k)u(k+1) + p(k-1)u(k-1) = q(k)u(k), \quad k \in \mathbb{N}(1) \tag{6.1.1}$$

where the functions $p$ and $q$ are defined on $\mathbb{N}$ and $\mathbb{N}(1)$ respectively, and $p(k) > 0$ for all $k \in \mathbb{N}$. Equation (6.1.1) equivalently can be written as

$$-\Delta(p(k-1)\Delta u(k-1)) + f(k)u(k) = 0, \quad k \in \mathbb{N}(1) \tag{6.1.2}$$

where $f(k) = q(k) - p(k) - p(k-1)$.

We shall establish the following properties of the solutions of the difference equation (6.1.1) some of which will be needed later.

$(P_1)$ If $v(k)$ and $w(k)$ are two linearly independent solutions of (6.1.1), then $\Delta[p(k-1)(v(k-1)w(k) - v(k)w(k-1))] = 0$, and hence for all $k \in \mathbb{N}$ there exists a constant $c \neq 0$ such that

$$p(k)(v(k)w(k+1) - v(k+1)w(k)) = c. \tag{6.1.3}$$

$(P_2)$ If $w(k)$ is of fixed sign on $\mathbb{N}(k_1)$, where $k_1 \in \mathbb{N}$ then (6.1.3) implies that $\Delta(v(k)/w(k))$ is of fixed sign on $\mathbb{N}(k_1)$, i.e. $v(k)/w(k)$ is monotonic. However, it is not possible if $v(k)$ is oscillatory. Thus, if $v(k)$ is oscillatory then $w(k)$ is also oscillatory, and then every solution of (6.1.1) is oscillatory. In conclusion, if one solution of (6.1.1) is oscillatory (nonoscillatory) then the equation itself is oscillatory (nonoscillatory).

$(P_3)$ For the solution $u(k) = c_1v(k) + c_2w(k)$ of (6.1.1), if the system

$$\begin{aligned} u(k_1) &= c_1v(k_1) + c_2w(k_1) \\ u(k_2) &= c_1v(k_2) + c_2w(k_2), \quad k_1 \neq k_2, \quad k_1,\ k_2 \in \mathbb{N} \end{aligned}$$

has no solution, then there exist values of $c_1$ and $c_2$, not both zero, such that $c_1v(k_1) + c_2w(k_1) = c_1v(k_2) + c_2w(k_2) = 0$, i.e. $u(k)$ vanishes at $k_1$ and $k_2$. Thus, if every solution $u(k)$ of (6.1.1) vanishes at most once on $\mathbb{N}$, then given any two values $u(k_1)$ and $u(k_2)$ uniquely determine $u(k)$.

$(P_4)$ If $|q(k)| \geq p(k-1) + p(k)$, $k \in \mathbb{N}(1)$ and $v(k)$ is a solution of (6.1.1) such that for some $k_1 \in \mathbb{N}$, $|v(k_1+1)| \geq |v(k_1)|$ then by an easy induction it follows that $|v(k+1)| \geq |v(k)|$ for all $k \in \mathbb{N}(k_1)$. Further, if there exists a function $\epsilon(k) \geq 0$, $k \in \mathbb{N}(1)$ such that $\sum^{\infty} \epsilon(k) = \infty$, and $|q(k)| \geq (1+\epsilon(k))p(k) + p(k-1)$, then $|v(k+1)| \geq (1+\epsilon(k))|v(k)|$ for all $k \in \mathbb{N}(k_1)$, and consequently $|v(k)| \to \infty$ as $k \to \infty$.

$(P_5)$ If $q(k) \geq p(k-1) + p(k)$, $k \in \mathbb{N}(1)$ and $v(k)$ is a solution of (6.1.1) such that $v(0) = v(1) = 1$, then $v(k+1) \geq v(k) \geq 1$ for all $k \in \mathbb{N}$, and hence $v(k)$ is nonoscillatory. Therefore, by $(P_2)$ the difference equation (6.1.1) is nonoscillatory.

$(P_6)$ If $|q(k)| \geq p(k-1) + p(k)$, $k \in \mathbb{N}(1)$ then given any two values $u(k_1)$ and $u(k_2)$, $k_1 \neq k_2$, $k_1$, $k_2 \in \mathbb{N}$ uniquely determine the solution $u(k)$ of (6.1.1).

$(P_7)$ If $u(k)$ is a nonoscillatory solution of (6.1.1), say, eventually positive, then $q(k)u(k)$ must be eventually positive. Thus, if the function $q(k)$ is oscillatory or eventually negative, then the equation (6.1.1) is oscillatory.

$(P_8)$ Let $u(k)$ be a nonoscillatory solution of (6.1.2), say, positive for all $k \in \mathbb{N}(k_1)$, and $f(k) \geq 0$ for all $k \in \mathbb{N}(1)$. Since for the function $v(k) = u(k)p(k-1)\Delta u(k-1)$, $\Delta v(k) = p(k)(\Delta u(k))^2 + f(k)u^2(k) \geq 0$, if there exists a $k_2 \in \mathbb{N}(k_1)$ such that $\Delta u(k_2) > 0$ then $\Delta u(k) > 0$ for all $k \in \mathbb{N}(k_2)$. Therefore, either $u(k)$ is eventually increasing or eventually nonincreasing.

$(P_9)$ If in $(P_8)$, in addition $f(k) \neq 0$ for infinitely many $k$, then the solution $u(k) > 0$, $k \in \mathbb{N}(k_1)$ which is eventually nonincreasing is actually decreasing. For this, if $\Delta u(k) \leq 0$ for all $k \in \mathbb{N}(k_2)$, where $k_2 \geq k_1$ then for any $k_3 > k_2$ so that $f(k_3) \neq 0$, from (6.1.2) we have

$$\begin{aligned} 0 \;\geq\; p(k_3)\Delta u(k_3) \;&=\; p(k_3-1)\Delta u(k_3-1) + f(k_3)u(k_3) \\ &>\; p(k_3-1)\Delta u(k_3-1) \\ &=\; p(k_3-2)\Delta u(k_3-2) + f(k_3-1)u(k_3-1) \\ &\geq\; p(k_3-2)\Delta u(k_3-2) \\ &\cdots \\ &\geq\; p(k_2)\Delta u(k_2). \end{aligned}$$

Thus, $\Delta u(k) < 0$ for all $k \in \mathbb{N}(k_2, k_3-1)$. However, since $k_3$ is arbitrary, it follows that $\Delta u(k) < 0$ for all $k \in \mathbb{N}(k_2)$, and hence $u(k)$ is eventually decreasing.

$(P_{10})$ The difference equation

$$\alpha(k)u(k+1) + \beta(k)u(k) + \gamma(k)u(k-1) \;=\; 0, \quad k \in \mathbb{N}(1) \tag{6.1.4}$$

where the functions $\alpha$, $\beta$ and $\gamma$ are defined on $\mathbb{N}(1)$ and $\alpha(k) > 0$, $\gamma(k) > 0$, can be written as (6.1.1) by defining the coefficients $p(k)$ inductively as $p(0) = 1$, $p(k) = p(k-1)\alpha(k)/\gamma(k)$, $k \in \mathbb{N}(1)$ with $q(k) = -p(k)\beta(k)/\alpha(k)$, $k \in \mathbb{N}(1)$.

## 6.2. Boundedness of Solutions of (6.1.1)

The following results provide necessary as well as sufficient conditions so that all solutions of the difference equation (6.1.1) are bounded.

**Theorem 6.2.1.** In the difference equation (6.1.1) assume that $f(k) = q(k) - p(k) - p(k-1) \geq 0$ for all $k \in \mathbb{N}(1)$, and $f(k) \neq 0$ for infinitely many $k$. Then, every solution of (6.1.1) is bounded on $\mathbb{N}$ if and only if

$$\sum_{\ell=1}^{\infty}\sum_{\tau=1}^{\ell}\frac{f(\tau)}{p(\ell)} < \infty. \tag{6.2.1}$$

**Proof.** Let all solutions of (6.1.1) be bounded. For $k_1 \in \mathbb{N}$ we define a solution $u(k)$ of (6.1.1) by setting $u(k_1) = 1$ and $u(k_1+1) = 2$. Then, $\Delta u(k_1) = 1 > 0$, and from (6.1.2) we find

$$\begin{aligned} p(k_1+1)\Delta u(k_1+1) &= p(k_1)\Delta u(k_1) + f(k_1+1)u(k_1+1) \\ &> f(k_1+1)u(k_1+1) \geq 0. \end{aligned}$$

Thus, by induction $u(k) \geq 1$ and $\Delta u(k) > 0$ for all $k \in \mathbb{N}(k_1)$. Now, since from (6.1.2), we have

$$\begin{aligned} u(k+1) &= u(k_1+1) + p(k_1)\Delta u(k_1)\sum_{\ell=k_1+1}^{k}\frac{1}{p(\ell)} \\ &\quad + \sum_{\ell=k_1+1}^{k}\frac{1}{p(\ell)}\sum_{\tau=k_1+1}^{\ell} f(\tau)u(\tau) \end{aligned} \tag{6.2.2}$$

it follows that

$$u(k+1) \geq 2 + p(k_1)\sum_{\ell=k_1+1}^{k}\frac{1}{p(\ell)} + \sum_{\ell=k_1+1}^{k}\sum_{\tau=k_1+1}^{\ell}\frac{f(\tau)}{p(\ell)} \tag{6.2.3}$$

from which it is clear that if $u(k)$ is bounded then (6.2.1) must be satisfied.

Conversely, let $u(k)$ be an unbounded solution of (6.1.1) so that by $(P_5)$ and $(P_8)$ there exists a $k_2 \in \mathbb{N}$ such that $u(k) > 0$ and $\Delta u(k) > 0$ for all $k \in \mathbb{N}(k_2)$. Then, by (6.1.2) we get

$$\begin{aligned} f(k) &= \frac{\Delta(p(k-1)\Delta u(k-1))}{u(k)} \\ &\geq \frac{p(k)\Delta u(k)}{u(k)} - \frac{p(k-1)\Delta u(k-1)}{u(k-1)}, \quad k \in \mathbb{N}(k_2+1) \end{aligned}$$

which yields

$$\frac{1}{p(k)}\sum_{\ell=k_2+1}^{k} f(\ell) + \frac{p(k_2)\Delta u(k_2)}{p(k)u(k_2)} \geq \frac{\Delta u(k)}{u(k)}$$

and hence

$$(6.2.4)\quad \sum_{\ell=k_2+1}^{k}\sum_{\tau=k_2+1}^{\ell}\frac{f(\tau)}{p(\ell)}+\frac{p(k_2)\Delta u(k_2)}{u(k_2)}\sum_{\ell=k_2+1}^{k}\frac{1}{p(\ell)} \ \ge\ \sum_{\ell=k_2+1}^{k}\frac{\Delta u(\ell)}{u(\ell)}.$$

Let $r(t)=u(\ell)+(t-\ell)\Delta u(\ell),\ \ell\le t\le \ell+1$. Then, $r'(t)=\Delta u(\ell)$ and $r(t)\ge u(\ell)$. Hence, we have

$$(6.2.5)\quad \sum_{\ell=k_2+1}^{k}\frac{\Delta u(\ell)}{u(\ell)} \ \ge\ \sum_{\ell=k_2+1}^{k}\int_{\ell}^{\ell+1}\frac{r'(t)}{r(t)}dt \ =\ \ln u(k+1)-\ln u(k_2+1).$$

Next, since $f(k)\ne 0$ for infinitely many $k$, we can choose $k_3\ge k_2+1$ so that $f(k_3)\ne 0$. Then, we find

$$(6.2.6)\qquad \sum_{\ell=1}^{\infty}\frac{1}{p(\ell)}\sum_{\tau=1}^{\ell}f(\tau)\ \ge\ f(k_3)\sum_{\ell=k_3}^{\infty}\frac{1}{p(\ell)}.$$

Thus, if (6.2.1) holds then (6.2.6) implies that $\sum_{\ell=1}^{\infty}(1/p(\ell))<\infty$. This together with (6.2.1) in (6.2.4), in view of (6.2.5) implies that $\ln u(k)$ is bounded. This contradiction completes the proof. ∎

**Corollary 6.2.2.** The difference equation (6.1.1) has unbounded solutions if either of the following holds

(i) $\sum_{\ell=1}^{\infty}(1/p(\ell))=\infty$

(ii) $f(k)\ge \epsilon(k)p(k)$ for all $k\in\mathbb{N}(1)$, where $\epsilon(k)\ge 0$ and $\sum_{\ell=1}^{\infty}\epsilon(\ell)=\infty$

(iii) $\limsup_{k\to\infty}(1/p(k))\sum_{\ell=1}^{k}f(\ell)=c>0$.

**Corollary 6.2.3.** Suppose that $P(k)>0$ for all $k\in\mathbb{N}$, $F(k)\ge 0$ for all $k\in\mathbb{N}(1)$, and $F(k)\ne 0$ for infinitely many $k$. Suppose further that $P(k)\ge p(k)$ and $\sum_{\ell=1}^{k}F(\ell)\le\sum_{\ell=1}^{k}f(\ell)$, for all $k\in\mathbb{N}(1)$. Then, all solutions of the difference equation

$$(6.2.7)\qquad \Delta(P(k-1)\Delta v(k-1)) \ =\ F(k)v(k),\quad k\in\mathbb{N}(1)$$

are bounded provided all solutions of (6.1.2) are bounded.

**Theorem 6.2.4.** Assume that $q(k)-p(k)-p(k-1)\le 0$ for all $k\in\mathbb{N}(1)$. Further, assume that the equation (6.1.1) is nonoscillatory and $\sum^{\infty}(1/p(\ell))<\infty$. Then, all solutions of (6.1.1) are bounded on $\mathbb{N}$.

**Proof.** Let $u(k)$ be any solution of (6.1.1). We can assume that $u(k)>0$ for all $k\in\mathbb{N}(k_1)$, where $k_1\in\mathbb{N}$ is sufficiently large. Since

$\dfrac{q(k)-p(k-1)}{p(k)} \le 1$, from (6.1.1), we have

$$\begin{aligned} u(k+1) &= \frac{q(k)-p(k-1)}{p(k)}u(k) + \frac{p(k-1)}{p(k)}[u(k)-u(k-1)] \\ &\le u(k) + \frac{p(k-1)}{p(k)}[u(k)-u(k-1)] \end{aligned}$$

and hence

$$u(k+1)-u(k) \le \frac{p(k-1)}{p(k)}[u(k)-u(k-1)],$$

which gives

$$u(k+1) \le u(k) + \frac{d}{p(k)}, \quad k \in \mathbb{N}(k_1) \tag{6.2.8}$$

where $d$ is a constant.

Now from (6.2.8) we easily get

$$u(k) \le u(k_1) + d\sum_{\ell=k_1}^{k-1}\frac{1}{p(\ell)}.$$

Thus, from our hypotheses $u(k)$ is bounded on $\mathbb{N}$. ■

**Theorem 6.2.5.** Assume that $f(k) = q(k) - p(k) - p(k-1) \ge 0$ for all $k \in \mathbb{N}(1)$. Then, for every solution $u(k)$ of (6.1.1) the function $\phi(k) = p(k)\Delta u(k)$ is bounded on $\mathbb{N}$ if and only if

$$\sum_{\ell=1}^{\infty}\sum_{\tau=1}^{\ell}\frac{f(\ell+1)}{p(\tau)} < \infty. \tag{6.2.9}$$

**Proof.** Following as in Theorem 6.2.1 the solution $u(k)$ of (6.1.1) satisfying $u(k_1)=1$, $u(k_1+1)=2$ exists and $u(k)>0$ and $\Delta u(k)>0$ for all $k \in \mathbb{N}(k_1)$. For this $u(k)$, since $\Delta(p(k)\Delta u(k)) = f(k+1)u(k+1) \ge 0$ for all $k \in \mathbb{N}(k_1)$, we have $p(k)\Delta u(k) \ge p(k_1)\Delta u(k_1)$, and hence $u(k+1) \ge u(k_1) + p(k_1)\Delta u(k_1)\sum_{\tau=k_1}^{k}\frac{1}{p(\tau)}$. Using this in (6.1.2), we obtain

$$\begin{aligned} \Delta(p(k)\Delta u(k)) &= f(k+1)u(k+1) \\ &\ge f(k+1)u(k_1) + f(k+1)p(k_1)\Delta u(k_1)\sum_{\tau=k_1}^{k}\frac{1}{p(\tau)}, \end{aligned}$$

which gives that

$$\begin{aligned} p(k+1)\Delta u(k+1) \ &\geq \ p(k_1)\Delta u(k_1) + u(k_1)\sum_{\ell=k_1}^{k} f(\ell+1) \\ &\quad + p(k_1)\Delta u(k_1)\sum_{\ell=k_1}^{k}\sum_{\tau=k_1}^{\ell}\frac{f(\ell+1)}{p(\tau)}. \end{aligned}$$

Thus, if $p(k)\Delta u(k)$ is bounded then (6.2.9) must hold.

Conversely, we may assume that $u(k)$ is eventually positive. By $(P_8)$ we may also assume that $u(k)$ is eventually increasing or nonincreasing. If $u(k)$ is eventually nonincreasing, then $p(k)\Delta u(k) \leq 0$ for all large $k$. Further, $\Delta(p(k)\Delta u(k)) = f(k+1)u(k+1) \geq 0$, which means that $p(k)\Delta u(k)$ is nondecreasing as well, and hence $p(k)\Delta u(k)$ must be bounded on $\mathbb{N}$.

Now assume that $u(k)$ is eventually increasing. Then, there exists a $k_2 \in \mathbb{N}$ such that $u(k) > 0$ and $\Delta u(k) > 0$ for all $k \in \mathbb{N}(k_2)$. Thus, it follows that

$$\Delta\left(\frac{u(k+1)}{p(k)\Delta u(k)}\right) = \frac{p(k)\Delta u(k)\Delta u(k+1) - f(k+1)u^2(k+1)}{p(k)p(k+1)\Delta u(k)\Delta u(k+1)} \leq \frac{1}{p(k+1)}$$

and hence

$$\frac{u(k+1)}{p(k)\Delta u(k)} \leq \frac{u(k_2+1)}{p(k_2)\Delta u(k_2)} + \sum_{\tau=k_2+1}^{k}\frac{1}{p(\tau)}.$$

This implies that

$$\frac{\Delta(p(k)\Delta u(k))}{p(k)\Delta u(k)} = \frac{f(k+1)u(k+1)}{p(k)\Delta u(k)} \leq \frac{f(k+1)u(k_2+1)}{p(k_2)\Delta u(k_2)} + \sum_{\tau=k_2+1}^{k}\frac{f(k+1)}{p(\tau)},$$

which on using an argument similar to the one used in the derivation of (6.2.5) leads to

$$\begin{aligned} (6.2.10) \qquad & \sum_{\ell=k_2+1}^{k}\sum_{\tau=k_2+1}^{\ell}\frac{f(\ell+1)}{p(\tau)} + \frac{u(k_2+1)}{p(k_2)\Delta u(k_2)}\sum_{\ell=k_2+1}^{k} f(\ell+1) \\ & \geq \ \sum_{\ell=k_2+1}^{k}\frac{\Delta(p(\ell)\Delta u(\ell))}{p(\ell)\Delta u(\ell)} \\ & \geq \ \ln(p(k+1)\Delta u(k+1)) - \ln(p(k_2+1)\Delta u(k_2+1)). \end{aligned}$$

Furthermore, by reasoning similar to that used in obtaining (6.2.6), we get $\sum_{\ell=k_2+1}^{\infty} f(\ell+1) < \infty$. But, this and (6.2.9) in (6.2.10) then implies that $p(k)\Delta u(k)$ is bounded on $\mathbb{N}$. ∎

## 6.3. Recessive and Dominant Solutions of (6.1.1)

We begin with the following:

**Definition 6.3.1.** If there exist two linearly independent solutions $v(k)$ and $w(k)$ of (6.1.1) such that $v(k)/w(k) \to 0$ as $k \to \infty$, then $v(k)$ is called *recessive* and $w(k)$ is called *dominant solution* of (6.1.1).

Recessive solutions of (6.1.1) are unique up to a constant factor. For this, if $v_1(k)$ and $v_2(k)$ both are linearly independent recessive solutions of (6.1.1), then by the Definition 6.3.1 there exists a solution $w_1(k)$ so that $v_1(k)$ and $w_1(k)$ are linearly independent and $\lim_{k\to\infty} \frac{v_1(k)}{w_1(k)} = 0.$ However, since (6.1.1) is linear and homogeneous, there exist constants $c_1$ and $c_2$ $(\neq 0)$ such that $w_1(k) = c_1 v_1(k) + c_2 v_2(k)$. But, then

$$0 = \lim_{k\to\infty} \frac{v_1(k)}{w_1(k)} = \lim_{k\to\infty} \frac{v_1(k)}{c_1 v_1(k) + c_2 v_2(k)} = \lim_{k\to\infty} \frac{1}{c_1 + c_2(v_2(k)/v_1(k))}$$

implies that $\lim_{k\to\infty} \frac{v_2(k)}{v_1(k)} = \infty$, i.e. $\lim_{k\to\infty} \frac{v_1(k)}{v_2(k)} = 0$, and hence $v_1(k)$ is recessive and $v_2(k)$ is dominant.

**Example 6.3.1.** For the difference equation $u(k+1)+u(k-1) = 2u(k)$, $k \in \mathbb{N}(1)$ the recessive and dominant solutions are $v(k) = 1$, $w(k) = k$.

**Example 6.3.2.** For the difference equation $k^2u(k+1)+(k-1)^2u(k-1) = \frac{k(2k^2-1)}{k+1}u(k)$, the recessive and dominant solutions are $v(k) = 1/k$, $w(k) = \frac{1}{k}\sum_{\ell=1}^{k-1}\left(1+\frac{1}{\ell}\right)$.

**Example 6.3.3.** Since for the difference equation $u(k+1)+u(k-1) = 0$, $k \in \mathbb{N}(1)$ two linearly independent solutions are $\cos(k\pi/2)$ and $\sin(k\pi/2)$, it does not have recessive and dominant solutions.

**Theorem 6.3.1.** If the difference equation (6.1.1) is nonoscillatory, then it has a recessive solution $v(k)$ and a dominant solution $w(k)$ such that

$$\sum^{\infty} \frac{1}{p(\ell)v(\ell)v(\ell+1)} = \infty \tag{6.3.1}$$

and

$$\sum^{\infty} \frac{1}{p(\ell)w(\ell)w(\ell+1)} < \infty. \tag{6.3.2}$$

**Proof.** Let $v(k)$ and $w(k)$ be two linearly independent solutions of (6.1.1) so that the relation (6.1.3) holds. Since (6.1.1) is nonoscillatory there exists a sufficiently large $k_1 \in \mathbb{N}$ so that $v(k) \neq 0$ and $w(k) \neq 0$ for all $k \in \mathbb{N}(k_1)$. Then,

(6.3.3)

$$\Delta\left(\frac{v(k)}{w(k)}\right) = \frac{v(k+1)w(k) - v(k)w(k+1)}{w(k)w(k+1)}\frac{p(k)}{p(k)} = \frac{-c}{p(k)w(k)w(k+1)}.$$

Since $c/p(k)w(k)w(k+1)$ is of one sign for all $k \in \mathbb{N}(k_1)$, we find that $v(k)/w(k)$ is monotone, and hence $\lim_{k\to\infty} \frac{v(k)}{w(k)} = L$ exists. If $L = \pm\infty$, then $w(k)$ is recessive and $v(k)$ is dominant solution. If $L = 0$, then $v(k)$ is recessive and $w(k)$ is dominant solution. If $0 < |L| < \infty$, then we consider the solution $z(k) = v(k) - Lw(k)$. Since $\lim_{k\to\infty} \frac{z(k)}{w(k)} = 0$, from Problem 2.16.4 it is clear that $z(k)$ and $w(k)$ are linearly independent. Thus, renaming if necessary, we can always find a recessive solution $v(k)$ and a dominant solution $w(k)$.

From (6.3.3), we have

$$\frac{v(k)}{w(k)} = \frac{v(k_1)}{w(k_1)} + \sum_{\ell=k_1}^{k-1} \frac{-c}{p(\ell)w(\ell)w(\ell+1)}.$$

Since $\frac{v(k)}{w(k)} \to 0$, as $k \to \infty$ we must have (6.3.2). Starting with $\Delta\left(\frac{w(k)}{v(k)}\right)$ a similar argument proves (6.3.1). ∎

**Corollary 6.3.2.** Suppose that the difference equation (6.1.1) is nonoscillatory. If $w(k)$ is a solution of (6.1.1) such that (6.3.2) holds, then $w(k)$ is dominant and $v(k) = w(k)\sum_{\ell=k}^{\infty} \frac{1}{p(\ell)w(\ell)w(\ell+1)}$ is recessive. Similarly, if $v(k)$ is a solution of (6.1.1) such that (6.3.1) holds, then $v(k)$ is recessive and $w(k) = v(k)\sum_{\ell=k_1}^{k-1} \frac{1}{p(\ell)v(\ell)v(\ell+1)}$ is dominant, where $k_1$ is large enough so that $v(\ell) \neq 0$ for all $\ell \in \mathbb{N}(k_1)$.

**Proof.** If $w(k)$ is a solution of (6.1.1) such that (6.3.2) holds, then $v(k) = w(k)\sum_{\ell=k}^{\infty} \frac{1}{p(\ell)w(\ell)w(\ell+1)}$ is also a solution of (6.1.1). Further, since $\dfrac{v(k)}{w(k)} = \sum_{\ell=k}^{\infty} \dfrac{1}{p(\ell)w(\ell)w(\ell+1)} \to 0$ as $k \to \infty$, $v(k)$ is recessive and $w(k)$ is dominant. The other case can be proved similarly. ∎

**Corollary 6.3.3.** Suppose that the difference equation (6.1.1) is nonoscillatory. Then, $v(k)$ is the recessive solution if and only if (6.3.1) holds. Similarly, $w(k)$ is the dominant solution if and only if (6.3.2) holds.

**Remark 6.3.1.** If all solutions of (6.1.1) are bounded, then the recessive solution must converge to zero.

**Theorem 6.3.4.** If $q(k) \geq p(k-1) + p(k)$, $k \in \mathbb{N}(1)$ then there exists a recessive solution $v(k)$ and a dominant solution $w(k)$ such that $v(k) > 0$, $v(k+1) \leq v(k)$ and $w(k) > 0$, $w(k+1) \geq w(k)$. Suppose there exists a function $\epsilon(k) \geq 0$, $k \in \mathbb{N}(1)$ such that

$$(6.3.4) \qquad q(k) - (1+\epsilon(k))p(k) - p(k-1) \geq 0 \quad \text{and} \quad \sum^{\infty} \epsilon(\ell) = \infty$$

then $w(k) \to \infty$ as $k \to \infty$. If there exists a function $\gamma(k) \geq 0$, $k \in \mathbb{N}(1)$ such that

$$(6.3.5) \qquad q(k) - p(k) - (1+\gamma(k))p(k-1) \geq 0 \quad \text{and} \quad \sum^{\infty} \gamma(\ell) = \infty$$

then $v(k) \to 0$ as $k \to \infty$.

**Proof.** Let $w(k)$ be the solution of the initial value problem (6.1.1), $w(0) = 1$, $w(1) = 1$. Then, by $(P_3)$, $w(k+1) \geq w(k)$ for all $k \in \mathbb{N}$. Now by Problem 6.24.3, there exists a solution $v(k)$ of (6.1.1) such that $v(k) > 0$ and $v(k+1) \leq v(k)$ for all $k \in \mathbb{N}$. Therefore, $v(k)/w(k)$ is positive and monotone decreasing to some limit $L$. If $L = 0$, then $v(k)$ is recessive and $w(k)$ is dominant. If $L > 0$, then $v(k) - Lw(k) > 0$ and $\Delta(v(k) - Lw(k)) \leq 0$ for all $k \in \mathbb{N}$. Indeed, if there is some $k_1 \in \mathbb{N}$ so that $v(k_1) - Lw(k_1) = 0$, then $v(k) - Lw(k) = 0$ for all $k \in \mathbb{N}(k_1)$, but this contradicts the fact that $v(k)$ and $w(k)$ are linearly independent. Since $\lim_{k\to\infty} \dfrac{v(k) - Lw(k)}{w(k)} = 0$, renaming if necessary, we have the existence of a dominant solution $w(k)$ and a recessive solution $v(k)$.

If condition (6.3.4) is satisfied, then $(P_4)$ implies that $w(k) \to \infty$ as $k \to \infty$. Finally, if condition (6.3.5) is satisfied then once again from Problem 6.24.3, there exists a solution $v(k)$ of (6.1.1) such that $v(k) > 0$ and $v(k+1) \leq v(k)$. But, then $v(k-1) = \dfrac{q(k)v(k) - p(k)v(k+1)}{p(k-1)} \geq$

$\dfrac{q(k)-p(k)}{p(k-1)}v(k) \geq (1+\gamma(k))v(k)$, and hence $v(k) \leq v(0)\prod_{\ell=1}^{k}(1+\gamma(\ell))^{-1} \to 0$ as $k \to \infty$, i.e. $v(k) \to 0$ as $k \to \infty$. Clearly, this $v(k)$ must be recessive, because $v(k)/w(k) \to 0$ as $k \to \infty$, where $w(k)$ is the dominant solution defined earlier. ■

**Theorem 6.3.5.** Let $q(k) \geq p(k-1)+p(k)$, $k \in \mathbb{N}(1)$ and let $u(k)$ be an eventually positive and increasing solution of (6.1.1). Then, $u(k)$ is a dominant solution.

**Proof.** By Theorem 6.3.4 there exists a recessive solution $v(k) > 0$ which is nonincreasing, a dominant solution $w(k) > 0$ which is nondecreasing, and constants $c_1$ and $c_2 \neq 0$ such that $u(k) = c_1 v(k) + c_2 w(k)$. Since $\displaystyle\lim_{k\to\infty}\frac{v(k)}{w(k)} = 0$ and $u(k)$ is eventually positive it follows that $\displaystyle\lim_{k\to\infty}\frac{u(k)}{w(k)} = c_2$, and hence $c_2$ must be positive. Further, for all sufficiently large $k \in \mathbb{N}$, $\dfrac{u(k)}{w(k)} \geq \dfrac{1}{2}c_2$, i.e. $\dfrac{1}{u(k)} \leq \dfrac{2}{c_2 w(k)}$, which implies $0 \leq \dfrac{v(k)}{u(k)} \leq \dfrac{2v(k)}{c_2 w(k)}$. Therefore, we have $\dfrac{v(k)}{u(k)} \to 0$ as $k \to \infty$, which means $u(k)$ is a dominant solution. ■

**Theorem 6.3.6.** Assume that $q(k) \geq p(k-1)+p(k)$, $k \in \mathbb{N}(1)$ and the difference equation (6.1.1) has unbounded solutions. Further, assume that $u(k)$ is an eventually positive solution of (6.1.1). Then, $u(k)$ is a dominant solution if and only if $u(k)$ is eventually increasing.

**Proof.** The sufficiency part is Theorem 6.3.5. For the necessity part, suppose $u(k)$ is a dominant solution. Then, by $(P_8)$ either $u(k)$ is eventually increasing or eventually nonincreasing. We shall assume that $u(k)$ is eventually nonincreasing and arrive at a contradiction. Since $u(k)$ is dominant, there exists a recessive solution $v(k)$ of (6.1.1) which is linearly independent from $u(k)$. Further, we can choose $k_1 \in \mathbb{N}$ so large that $u(k) > 0$ and $v(k) > 0$ for all $k \in \mathbb{N}(k_1)$ and by Theorem 6.3.4, $v(k+1) \leq v(k)$. Now, let $w(k)$ be an unbounded solution of (6.1.1). Then, there exists constants $c_1$ and $c_2$ such that $w(k) = c_1 u(k) + c_2 v(k)$. But, this means that an unbounded solution can be written as a linear combination of two eventually positive nonincreasing solutions. This contradiction completes the proof. ■

**Remark 6.3.2.** If the conditions of Theorem 6.3.6 are satisfied, then from $(P_8)$ the solution $u(k)$ is recessive if and only if $u(k)$ is eventually nonincreasing. Further, if $q(k)-p(k-1)-p(k) \neq 0$ for infinitely many

$k$, then by $(P_9)$ nonincreasing can be changed to decreasing.

**Theorem 6.3.7.** Assume that $q(k) \geq p(k-1)+p(k)$, $k \in \mathbb{N}(1)$ and the difference equation (6.1.1) has unbounded solutions. Then, (6.1.1) has a recessive solution which converges to zero provided (6.1.1) has a dominant solution $w(k)$ such that $p(k)\Delta w(k) \to \infty$ as $k \to \infty$. Further, if (6.1.1) has a recessive solution which converges to zero, then all dominant solutions $w(k)$ of (6.1.1) satisfy $p(k)\Delta w(k) \to \infty$ as $k \to \infty$.

**Proof.** Following as in Theorem 6.2.1 the solution $w(k)$ of (6.1.1) satisfying $w(k_1) = 1$, $w(k_1+1) = 2$ exists and $w(k) > 0$ and $\Delta w(k) > 0$ for all $k \in \mathbb{N}(k_1)$. By Theorem 6.3.5, $w(k)$ is dominant and by Corollary 6.3.3, $w(k)$ satisfies (6.3.2). Thus, by Corollary 6.3.2 the function $v(k) = w(k)\sum_{\ell=k}^{\infty}\frac{1}{p(\ell)w(\ell)w(\ell+1)}$ is a recessive solution of (6.1.1). We also note that $w(k) \to \infty$ as $k \to \infty$, otherwise $v(k)$ and $w(k)$ are linearly independent bounded solutions of (6.1.1), and then no solution of (6.1.1) is unbounded. Now since $\Delta(p(k)\Delta u(k)) \geq 0$ for all $k \in \mathbb{N}(k_1)$, we have $p(k)\Delta u(k) \geq p(\ell)\Delta u(\ell)$ for all $k_1 \leq \ell \leq k$, and hence it follows that

$$\begin{aligned} v(k) &= w(k)\sum_{\ell=k}^{\infty}\frac{1}{p(\ell)w(\ell)w(\ell+1)} \\ &= \frac{w(k)}{p(k)\Delta w(k)}\sum_{\ell=k}^{\infty}\frac{p(k)\Delta w(k)}{p(\ell)\Delta w(\ell)}\frac{\Delta w(\ell)}{w(\ell)w(\ell+1)} \\ &\leq \frac{w(k)}{p(k)\Delta w(k)}\sum_{\ell=k}^{\infty}\left(\frac{1}{w(\ell)}-\frac{1}{w(\ell+1)}\right) \\ &= \frac{w(k)}{p(k)\Delta w(k)}\left(\frac{1}{w(k)}-\lim_{k\to\infty}\frac{1}{w(k+1)}\right) \\ &= \frac{1}{p(k)\Delta w(k)}. \qquad (6.3.6) \end{aligned}$$

Next, since $0 \leq p(k)\Delta w(k) \leq p(k+1)\Delta w(k+1)$ and $0 \leq w(k) \leq w(k+1)$, we have $p(k)\Delta w(k)w(k+1) \leq p(k+1)\Delta w(k+1)w(k+2)$, thus $w(k) \to \infty$ as $k \to \infty$ implies that $\lim_{k\to\infty} p(k)\Delta w(k)w(k+1) = \infty$. Therefore, it follows that

$$\begin{aligned} v(k) &= w(k)\sum_{\ell=k}^{\infty}\frac{1}{p(\ell)w(\ell)w(\ell+1)} \\ &= w(k)\sum_{\ell=k}^{\infty}\frac{1}{p(\ell)\Delta w(\ell)}\left(\frac{1}{w(\ell)}-\frac{1}{w(\ell+1)}\right) \end{aligned}$$

$$
\begin{aligned}
&= \frac{1}{p(k)\Delta w(k)} - \sum_{\ell=k}^{\infty} \frac{w(k)}{w(\ell+1)} \left( \frac{1}{p(\ell)\Delta w(\ell)} - \frac{1}{p(\ell+1)\Delta w(\ell+1)} \right) \\
&\qquad\qquad - \lim_{\ell\to\infty} \frac{w(k)}{p(\ell)\Delta w(\ell) w(\ell+1)} \\
&\geq \frac{1}{p(k)\Delta w(k)} - \sum_{\ell=k}^{\infty} \left( \frac{1}{p(\ell)\Delta w(\ell)} - \frac{1}{p(\ell+1)\Delta w(\ell+1)} \right) \\
&= \lim_{\ell\to\infty} \frac{1}{p(\ell)\Delta w(\ell)}, \quad k \in \mathbb{N}(k_1+1).
\end{aligned}
$$

Thus, on combining the above inequality with (6.3.6), we get

$$
\text{(6.3.7)} \qquad \lim_{\ell\to\infty} \frac{1}{p(\ell)\Delta w(\ell)} \leq v(k) \leq \frac{1}{p(k)\Delta w(k)}, \quad k \in \mathbb{N}(k_1+1).
$$

From the above inequality it is clear that $\lim_{k\to\infty} v(k) = 0$ if and only if $\lim_{k\to\infty} p(k)\Delta w(k) = \infty$. Further, since recessive solutions are essentially unique, if $\lim_{k\to\infty} p(k)\Delta w(k) = \infty$, then any recessive solution converges to zero. Thus, only it remains to show that if $\lim_{k\to\infty} p(k)\Delta w(k) = \infty$ then $\lim_{k\to\infty} p(k)\Delta z(k) = \infty$ for any other nonnegative dominant solution $z(k)$. For this, since $v(k)$ and $w(k)$ are linearly independent solutions of (6.1.1), there exist constants $c_1$ and $c_2$ $(> 0)$ such that $z(k) = c_1 v(k) + c_2 w(k)$. Now since $v(k)$ is bounded and $w(k) \to \infty$ as $k \to \infty$, there exists a $k_2 \in \mathbb{N}(k_1)$ such that $|c_1| v(k) \leq c_2 w(k)$ for all $k \in \mathbb{N}(k_2)$, and hence $z(k) \geq (c_2/2) w(k)$ for all $k \in \mathbb{N}(k_2)$. From (6.1.2), $p(k)\Delta w(k) = p(k_2)\Delta w(k_2) + \sum_{\ell=k_2+1}^{k} f(\ell) w(\ell)$, and thus $\sum_{\ell=k_2+1}^{k} f(\ell) w(\ell) \to \infty$ as $k \to \infty$. On the other hand, we have $p(k)\Delta z(k) = p(k_2)\Delta z(k_2) + \sum_{\ell=k_2+1}^{k} f(\ell) z(\ell) \geq p(k_2)\Delta z(k_2) + (c_2/2) \sum_{\ell=k_2+1}^{k} f(\ell) w(\ell)$, and hence $p(k)\Delta z(k) \to \infty$ as $k \to \infty$. ∎

**Theorem 6.3.8.** Assume that $f(k) = q(k) - p(k) - p(k-1) \geq 0$ for all $k \in \mathbb{N}(1)$ and the difference equation (6.1.1) has unbounded solutions. Then, every recessive solution of (6.1.1) converges to zero if and only if

$$
\text{(6.3.8)} \qquad \sum_{\ell=1}^{\infty} \sum_{\tau=1}^{\ell} \frac{f(\ell+1)}{p(\tau)} = \infty.
$$

**Proof.** Following as in Theorem 6.2.1 the solution $u(k)$ of (6.1.1) satisfying $u(k_1) = 1$, $u(k_1+1) = 2$ exists and $u(k) > 0$ and $\Delta u(k) > 0$ for all $k \in \mathbb{N}(k_1)$. Also, $p(k)\Delta u(k)$ is positive and increasing. By Theorem 6.3.6 this $u(k)$ is a dominant solution of (6.1.1). Further, in view of (6.3.8), Theorem 6.2.5 implies that $p(k)\Delta u(k)$ is unbounded. Therefore, as an application of Theorem 6.3.7 we find that the recessive solutions of (6.1.1) tend to zero as $k \to \infty$.

Conversely, if recessive solutions of (6.1.1) converge to zero as $k \to \infty$, then Theorem 6.3.7 implies that there exists a dominant solution $u(k)$ of (6.1.1) such that $p(k)\Delta u(k)$ is unbounded. Now appealing to Theorem 6.2.5 we find that the condition (6.3.8) must be satisfied. ∎

**Corollary 6.3.9.** If $f(k) \geq \epsilon(k)p(k-1)$ for all $k \in \mathbb{N}(1)$ and $\sum^{\infty} \epsilon(\ell) = \infty$, then (6.1.1) has recessive solutions which converge to zero as $k \to \infty$.

**Corollary 6.3.10.** Every recessive solution of

$$\Delta^2 u(k-1) = f(k)u(k), \quad k \in \mathbb{N}(1) \tag{6.3.9}$$

converges to zero if and only if $\sum_{\ell=1}^{\infty} \ell f(\ell+1) = \infty$.

## 6.4. Oscillation and Nonoscillation for (6.1.1)

We shall prove few results which provide sufficient conditions on the functions $p$ and $q$ so that all solutions of (6.1.1) are either oscillatory or nonoscillatory.

**Theorem 6.4.1.** If $q(k) \leq \min(p(k), p(k-1))$ for all sufficiently large $k \in \mathbb{N}$, then (6.1.1) is oscillatory.

**Proof.** Let $u(k)$ be a nonoscillatory solution of (6.1.1), which we can assume to be positive for all large $k \in \mathbb{N}$. Then, from $(P_7)$ we can also assume that $q(k) > 0$ for all large $k \in \mathbb{N}$. However, then equation (6.1.1) simultaneously implies that $u(k+1) < \frac{q(k)}{p(k)}u(k)$ and $u(k-1) < \frac{q(k)}{p(k-1)}u(k)$ for all large $k \in \mathbb{N}$. But, since $\frac{q(k)}{p(k)} \leq 1$ and $\frac{q(k)}{p(k-1)} \leq 1$ we find that $u(k+1) < u(k)$ and $u(k-1) < u(k)$ for all large $k \in \mathbb{N}$. This contradiction completes the proof. ∎

**Corollary 6.4.2.** If $q(k) \leq p(k)$ and if $p(k)$ is eventually nonincreasing, then (6.1.1) is oscillatory.

**Corollary 6.4.3.** If $q(k) \leq p(k-1)$ and if $p(k)$ is eventually nondecreasing, then (6.1.1) is oscillatory.

**Theorem 6.4.4.** If $q(k) \leq p(k-1)$ for all sufficiently large $k \in \mathbb{N}$, and $\sum^{\infty} 1/p(\ell) < \infty$, then (6.1.1) is oscillatory.

**Proof.** Let $u(k)$ be as in Theorem 6.4.1 so that once again we arrive at

$u(k-1) < u(k)$ for all large $k \in \mathbb{N}$. Since we have assumed that (6.1.1) is nonoscillatory by Theorem 6.2.1 there exists a recessive solution $u(k)$ such that $\sum^{\infty} 1/p(\ell)u(\ell)u(\ell+1) = \infty$. But, since for any nonoscillatory solution $u(k)$, we have $u(k) < u(k+1)$ for all large $k$, say, $k \in \mathbb{N}(k_1)$, so it follows that

$$\sum_{\ell=k_1}^{\infty} 1/p(\ell)u(\ell)u(\ell+1) \ \le \ \sum_{\ell=k_1}^{\infty} 1/p(\ell)u^2(\ell) \ \le \ \frac{1}{u^2(k_1)} \sum_{\ell=k_1}^{\infty} \frac{1}{p(\ell)} \ < \ \infty,$$

which is a contradiction. ∎

## 6.5. Riccati Type Transformations for (6.1.1)

In addition to the hypotheses on the functions $p$ and $q$ in (6.1.1), throughout we shall assume that $q(k) > 0$ for all $k \in \mathbb{N}(1)$. Let $u(k)$ be a solution of (6.1.1) such that $u(k) \neq 0$ for all $k \in \mathbb{N}(a)$. For this solution $u(k)$ we use the substitutions $v(k) = u(k+1)/u(k)$, $w(k) = p(k)u(k+1)/u(k)$ and $z(k) = q(k+1)u(k+1)/(p(k)u(k))$, $k \in \mathbb{N}(a)$ in (6.1.1) to obtain the corresponding first order nonlinear difference equations

(6.5.1) $$p(k)v(k) + p(k-1)/v(k-1) \ = \ q(k), \quad k \in \mathbb{N}(a+1)$$

(6.5.2) $$w(k) + p^2(k-1)/w(k-1) \ = \ q(k), \quad k \in \mathbb{N}(a+1)$$

and

(6.5.3) $$h(k)z(k) + \frac{1}{z(k-1)} \ = \ 1, \quad k \in \mathbb{N}(a+1)$$

where $h(k) = p^2(k)/(q(k)q(k+1))$.

Since the above difference equations (6.5.1) (6.5.3) are particular cases of (3.3.1), these will be called *Riccati type difference equations* and the substitutions used to obtain them will be termed as *Riccati type transformations.*

**Theorem 6.5.1.** The following conditions are equivalent:

(i) Equation (6.1.1) is nonoscillatory.

(ii) Equation (6.5.1) has a positive solution $v(k)$, $k \in \mathbb{N}(a)$.

(iii) Equation (6.5.2) has a positive solution $w(k)$, $k \in \mathbb{N}(a)$.

(iv) Equation (6.5.3) has a positive solution $z(k)$, $k \in \mathbb{N}(a)$.

**Proof.** If (6.1.1) is nonoscillatory and $u(k),\ k \in \mathbb{N}$ is its any solution, then there exists an $a \in \mathbb{N}$ such that $u(k)u(k+1) > 0$ for all $k \in \mathbb{N}(a)$. The necessity of the conditions (ii) (iv) then follows immediately from the transformations which lead to equations (6.5.1) (6.5.3).

Conversely, if $v(k),\ k \in \mathbb{N}(a)$ is a positive solution of (6.5.1), then we may let $u(a) = 1,\ u(k+1) = v(k)u(k)$ for all $k \in \mathbb{N}(a)$. This defines a positive solution $u(k)$ of (6.1.1) for all $k \in \mathbb{N}(a)$. Further, given $u(a)$ and $u(a+1),\ u(k)$ for all $k \in \mathbb{N}(0, a-1)$ can be constructed directly from (6.1.1). This $u(k)$ is a nonoscillatory solution of (6.1.1). Similar arguments hold for the equations (6.5.2) and (6.5.3). ∎

**Lemma 6.5.2.** Let $h(k) \geq g(k) > 0,\ k \in \mathbb{N}(1)$ and let $u(k)$ be a solution of

$$h(k)u(k) + 1/u(k-1) = 1, \quad k \in \mathbb{N}(1) \tag{6.5.4}$$

with $u(k) > 0$ for all $k \in \mathbb{N}$. Then, the equation

$$g(k)v(k) + 1/v(k-1) = 1, \quad k \in \mathbb{N}(1) \tag{6.5.5}$$

has a solution $v(k)$ satisfying $v(k) \geq u(k) > 1$ for all $k \in \mathbb{N}$.

**Proof.** Since the solution $u(k) > 0$ for all $k \in \mathbb{N}$, equation (6.5.4) implies that $1/u(k-1) < 1$, i.e. $u(k) > 1$ for all $k \in \mathbb{N}$. Now we define $v(k),\ k \in \mathbb{N}$ by choosing $v(0) \geq u(0)$ and letting $v(k)$ to satisfy (6.5.5) for all $k \in \mathbb{N}(1)$. Since from (6.5.4) and (6.5.5), we have

$$g(k)v(k) = 1 - \frac{1}{v(k-1)} = h(k)u(k) + \frac{1}{u(k-1)} - \frac{1}{v(k-1)}$$

if $v(k-1) \geq u(k-1)$, then it follows that $g(k)v(k) \geq h(k)u(k)$, and hence $v(k) \geq (h(k)/g(k))u(k) \geq u(k)$. Thus, by induction $v(k)$ is well defined and $v(k) \geq u(k) > 1$ for all $k \in \mathbb{N}$. ∎

**Theorem 6.5.3.** If $q(k)q(k+1) \leq (4-\epsilon)p^2(k)$ for some $\epsilon > 0$ and all sufficiently large $k \in \mathbb{N}$, then the difference equation (6.1.1) is oscillatory.

**Proof.** If $\epsilon \geq 4$, then the conclusion is obvious from $(P_7)$. Thus, we can assume that $0 < \epsilon < 4$. If (6.1.1) is nonoscillatory, then (6.5.3) has a positive solution $z(k)$ on $\mathbb{N}(a)$ for sufficiently large $a \in \mathbb{N}$. Since $h(k) = p^2(k)/(q(k)q(k+1)) \geq (4-\epsilon)^{-1}$, say, on $\mathbb{N}(a)$, from Lemma 6.5.2 we conclude that the equation

$$(4-\epsilon)^{-1}\overline{z}(k) + 1/\overline{z}(k-1) = 1, \quad k \in \mathbb{N}(a+1) \tag{6.5.6}$$

has a solution $\overline{z}(k)$ which satisfies $\overline{z}(k) \geq z(k) > 1$ for all $k \in \mathbb{N}(a)$. We now define the positive function $u(k)$ on $\mathbb{N}(a)$ by letting $u(a) = 1$, $u(k+1) = (4-\epsilon)^{-1/2}\overline{z}(k)u(k)$ for all $k \in \mathbb{N}(a)$, i.e. $u(k)$ is a positive solution of the difference equation

$$u(k+1) + u(k-1) \;=\; (4-\epsilon)^{1/2}u(k), \quad k \in \mathbb{N}(a+1). \tag{6.5.7}$$

But this is impossible because (6.5.7) is oscillatory, since it has the solutions $\cos k\theta$ and $\sin k\theta$, $k \in \mathbb{N}(1)$, where $\theta = \tan^{-1}\left(\frac{\epsilon}{4-\epsilon}\right)^{1/2}$. ∎

**Example 6.5.1.** Consider the difference equation (6.1.1) with $p(k) = 1$ and $q(k) = ((k+1)^{1/2} + (k-1)^{1/2})/k^{1/2}$. This equation is nonoscillatory because it has a solution $u(k) = k^{1/2}$, $k \in \mathbb{N}(1)$. Obviously, $q(k) < 2$ and $q(k) \to 2$ as $k \to \infty$, hence $q(k)q(k+1) < 4$ and $\epsilon(k) = 4 - q(k)q(k+1) \to 0$ as $k \to \infty$. Thus, we have $q(k)q(k+1) = 4 - \epsilon(k)$, but the difference equation is nonoscillatory. Therefore, in Theorem 6.5.3 the inequality condition cannot be replaced by the weaker condition

$$q(k)q(k+1) \;\leq\; (4-\epsilon(k))p^2(k),$$

where $\epsilon(k) > 0$ and $\epsilon(k) \to 0$ as $k \to \infty$.

**Corollary 6.5.4.** If $q(k) \leq p(k-1)$ and $p(k)/p(k-1) \geq (4-\epsilon)^{-1}$ for some $\epsilon > 0$ and all sufficiently large $k \in \mathbb{N}$, then (6.1.1) is oscillatory.

**Theorem 6.5.5.** If $q(k)q(k+1) \geq 4p^2(k)$ for all sufficiently large $k \in \mathbb{N}$, then the difference equation (6.1.1) is nonoscillatory.

**Proof.** From the given hypothesis $h(k) = p^2(k)/(q(k)q(k+1)) \leq 1/4$ on $\mathbb{N}(a)$ for sufficiently large $a \in \mathbb{N}$. Construct a solution $z(k)$ of (6.5.3) inductively by defining $z(a) = 2$ and $z(k) = \frac{1}{h(k)}\left(1 - \frac{1}{z(k-1)}\right)$, $k \in \mathbb{N}(a+1)$. We note that if $z(k-1) \geq 2$ for any $k \in \mathbb{N}(a)$, then $h(k)z(k) \geq 1/2$, so $z(k) \geq 4 \cdot \frac{1}{2} = 2$. Therefore, $z(k)$ is well defined. We thus have a positive solution of (6.5.3), and now Theorem 6.5.1 implies that (6.1.1) is nonoscillatory. ∎

**Corollary 6.5.6.** If $q(k) \geq \max(p(k-1),\ 4p(k))$ for all sufficiently large $k \in \mathbb{N}$, then (6.1.1) is nonoscillatory.

**Corollary 6.5.7.** If $q(k) \geq p(k-1)$ and $p(k)/p(k-1) \leq 1/4$ for all sufficiently large $k \in \mathbb{N}$, then (6.1.1) is nonoscillatory.

**Theorem 6.5.8.** If the difference equation (6.1.1) is nonoscillatory, then there exists an $a \in \mathbb{N}(1)$ such that for any $k \in \mathbb{N}(a)$ and any $\ell \geq 0$

$$h(k)h(k+1)\cdots h(k+\ell) \;<\; 4^{-\ell}. \tag{6.5.8}$$

**Proof.** Let $u(k)$ be a solution of (6.1.1) such that $u(k) \neq 0$ for all $k \in \mathbb{N}(a)$. Let $w(k) = p(k)u(k+1)/u(k)$, $k \in \mathbb{N}(a)$. Then, from (6.5.2) we can write

$$q(k)q(k+1) \;=\; p^2(k)(1+1/\alpha(k-1))(1+\alpha(k)),$$

where $\alpha(k) = w(k)w(k+1)/p^2(k) > 0$.

Thus, from a similar expression for $q(k+1)q(k+2)$ it follows that

$$\begin{aligned} &q(k)q^2(k+1)q(k+2) \\ &= p^2(k)p^2(k+1)(1+1/\alpha(k-1))(1+\alpha(k))(1+1/\alpha(k))(1+\alpha(k+1)) \\ &\geq p^2(k)p^2(k+1)(1+1/\alpha(k-1))4(1+\alpha(k+1)). \end{aligned}$$

Proceeding inductively, we obtain

$$\begin{aligned} q(k)q^2(k+1)&\cdots q^2(k+\ell)q(k+\ell+1) \\ &\geq p^2(k)\cdots p^2(k+\ell)(1+1/\alpha(k-1))4^\ell(1+\alpha(k+\ell)) \\ &> 4^\ell p^2(k)\cdots p^2(k+\ell), \end{aligned}$$

which is by the definition of $h(k)$ is the same as (6.5.8). ∎

**Remark 6.5.1.** Theorem 6.5.3 is included in Theorem 6.5.8. Indeed, if $h(k) \geq 1/(4-\epsilon)$ for all $k \in \mathbb{N}(a)$, then $h(k)h(k+1)\cdots h(k+\ell) \geq 1/(4-\epsilon)^{\ell+1} > 4^{-\ell}$, if $k \in \mathbb{N}(a)$ and $\ell$ is large enough. Thus, (6.1.1) is oscillatory.

**Corollary 6.5.9.** If $\liminf p(k)4^{-k} = 0$ and $\prod_{\ell=1}^k q(\ell)/\prod_{\ell=1}^k p(\ell)$ is bounded, say by $M$, as $k \to \infty$ then (6.1.1) is oscillatory.

**Proof.** If (6.1.1) is nonoscillatory, then Theorem 6.5.8 implies that for some $a \in \mathbb{N}(1)$ and all $\ell \geq 0$

$$\frac{q(a)\cdots q(a+\ell)}{p(a)\cdots p(a+\ell)}\,\frac{q(a+1)\cdots q(a+\ell+1)}{p(a)\cdots p(a+\ell)}\,\frac{q(a)}{p(a+\ell+1)}\,\frac{p(a+\ell+1)}{q(a)} \;>\; 4^\ell.$$

However, the left side of the above inequality is bounded above by $M^2p(a+\ell+1)/q(a)$, thus $M^2p(a+\ell+1)/q(a) > 4^\ell$, which implies that $p(a+\ell+1)4^{-(a+\ell+1)} > q(a)/(M^2 4^{a+1})$ for all $\ell \geq 0$. But this contradicts our assumption. ∎

**Example 6.5.2.** In Theorem 6.5.8 inequality (6.5.8) is only a necessary condition for nonoscillation. For this, in (6.1.1) let $p(k) = 1$ for all $k \in \mathbb{N}$, $q(2k) = 2^{-1}4^{2-k}$ and $q(2k-1) = 4^{k-1}$ for all $k \in \mathbb{N}(1)$. For this choice of $p(k)$ and $q(k)$ the inequality (6.5.8) is satisfied. However, (6.1.1) is oscillatory. If not, then (6.5.3) has a positive solution $z(k)$ defined for all $k$ sufficiently large. Further, $1/z(k-1) < 1$, i.e. $z(k-1) > 1$. Since $h(2k) = 1/8$ and $h(2k-1) = 1/2$, (6.5.3) implies that

$$z(2k) = 8(1 - 1/z(2k-1)) \tag{6.5.9}$$

and

$$z(2k-1) = 2(1 - 1/z(2k-2)). \tag{6.5.10}$$

Substitution of (6.5.10) in (6.5.9) yields

$$z(2k) = 4 - 4/(z(2k-2) - 1). \tag{6.5.11}$$

Since $z(2k-2) > 1$ and $z(2k) > 1$, (6.5.11) implies that $z(2k-2) > 2$ and $z(2k) < 4$. Thus,

$$2 < z(\ell) < 4 \quad \text{if } \ell \text{ is even and sufficiently large.} \tag{6.5.12}$$

Now from (6.5.11) and (6.5.12), we find that $4 - 4/(z(\ell) - 1) > 2$ for all even $\ell$ sufficiently large. But this implies that $z(\ell) > 3$, hence from (6.5.11) we get $4 - 4/(z(\ell) - 1) > 3$, i.e. $z(\ell) > 5$ for all even $\ell$ sufficiently large. This contradicts (6.5.12), and thus for this choice of $p(k)$ and $q(k)$ the difference equation (6.1.1) is oscillatory.

**Theorem 6.5.10.** If $p^2(k_\ell) \geq q(k_\ell)q(k_\ell + 1)$ for a sequence $\{k_\ell\} \subseteq \mathbb{N}$ such that $k_\ell \to \infty$ as $\ell \to \infty$, then the difference equation (6.1.1) is oscillatory.

**Proof.** If (6.1.1) is nonoscillatory, then (6.5.3) has a positive solution $z(k)$ for all $k \in \mathbb{N}(a)$. However, then from (6.5.3), $h(k)z(k) < 1$ and $z(k) > 1$ for all $k \in \mathbb{N}(a+1)$, so $h(k) < 1$, i.e. $p^2(k) < q(k)q(k+1)$ for all $k \in \mathbb{N}(a+1)$. This contradiction implies that (6.1.1) must be oscillatory. ∎

**Corollary 6.5.11.** If $\limsup h(k) > 1$, then (6.1.1) is oscillatory.

**Corollary 6.5.12.** If $\limsup (1/k)\sum_{\ell=1}^{k} h(\ell) > 1$, then (6.1.1) is oscillatory.

**Proof.** If (6.1.1) is nonoscillatory, then as in Theorem 6.5.10 we have $p^2(k) < q(k)q(k+1)$ for all $k \in \mathbb{N}(a)$. Thus, it follows that $\sum_{\ell=a}^{k} h(\ell) <$

$k-a+1$, and hence $(1/k)\sum_{\ell=1}^{k} h(\ell) < 1+(L/k)$ for some constant $L$. But this leads to a contradiction to our hypothesis, and (6.1.1) is oscillatory. ∎

**Corollary 6.5.13.** If $\sum_{\ell=1}^{\infty} h^{-\alpha}(\ell) < \infty$ for some $\alpha > 0$, then (6.1.1) is oscillatory.

**Proof.** From Hölder's inequality with indices $\lambda$ and $\mu$, we have

$$k = \sum_{\ell=1}^{k} h^{1/\lambda}(\ell)h^{-1/\lambda}(\ell) \le \left(\sum_{\ell=1}^{k} h(\ell)\right)^{1/\lambda} \left(\sum_{\ell=1}^{k} h^{-\mu/\lambda}(\ell)\right)^{1/\mu}$$

and hence

$$k^{1/\lambda}k^{1/\mu} = k \le \left(\sum_{\ell=1}^{k} h(\ell)\right)^{1/\lambda} \left(\sum_{\ell=1}^{k} h^{1-\mu}(\ell)\right)^{1/\mu}.$$

Thus, it follows that

$$\frac{1}{k}\sum_{\ell=1}^{k} h(\ell) \ge \left[k\Big/\sum_{\ell=1}^{k} h^{1-\mu}(\ell)\right]^{\lambda/\mu}.$$

Hence, if we choose $\lambda = (1+\alpha)/\alpha$ and $\mu = 1+\alpha$, then the above inequality leads to

$$\frac{1}{k}\sum_{\ell=1}^{k} h(\ell) \ge \left[k\Big/\sum_{\ell=1}^{k} h^{-\alpha}(\ell)\right]^{1/\alpha}.$$

Therefore, $\lim_{k\to\infty} \frac{1}{k}\sum_{\ell=1}^{k} h(\ell) = \infty$, and from Corollary 6.5.12 equation (6.1.1) is oscillatory. ∎

**Corollary 6.5.14.** If $\sum_{\ell=1}^{\infty}(q(\ell)/p(\ell-1))^{\alpha} < \infty$ for some $\alpha > 0$, and for sufficiently large $k \in \mathbb{N}$ either of the following holds

(i) $p(k)/p(k-1) \ge \epsilon > 0$

(ii) $q(k) \le p(k)$,

then (6.1.1) is oscillatory.

**Corollary 6.5.15.** If $\sum_{\ell=1}^{\infty}(q(\ell)/p(\ell))^{\alpha} < \infty$ for some $\alpha > 0$, and for sufficiently large $k \in \mathbb{N}$ either of the following holds

(i) $p(k-1)/p(k) \geq \epsilon > 0$

(ii) $q(k+1) \leq p(k)$,

then (6.1.1) is oscillatory.

## 6.6. Riccati Type Transformations for

$$\Delta(p(k)\Delta u(k)) + r(k)u(k+1) \;=\; 0, \quad k \in \mathbb{N} \tag{6.6.1}$$

where the functions $p$ and $r$ are defined on $\mathbb{N}$, and $p(k) > 0$ for all $k \in \mathbb{N}$. Obviously, the difference equation (6.6.1) is equivalent to (6.1.1) with $q(k) = p(k) + p(k-1) - r(k-1)$. If $u(k)$ is a solution of (6.6.1) with $u(k)u(k+1) > 0$ for all $k \in \mathbb{N}(a)$, then the Riccati type transformation we let $v(k) = p(k)\Delta u(k)/u(k)$. Since $v(k) + p(k) = p(k)u(k+1)/u(k) > 0$, this leads to Riccati type difference equation

$$\Delta v(k) + \frac{v(k)v(k+1) + r(k)v(k)}{p(k)} + r(k) \;=\; 0, \quad k \in \mathbb{N}(a) \tag{6.6.2}$$

which is the same as

$$\Delta v(k) + \frac{v^2(k)}{v(k) + p(k)} + r(k) \;=\; 0, \quad k \in \mathbb{N}(a). \tag{6.6.3}$$

**Lemma 6.6.1.** The difference equation (6.6.1) is nonoscillatory if and only if there exists a function $w(k)$ defined on $\mathbb{N}$ with $w(k) > -p(k)$, $k \in \mathbb{N}(a)$ for some $a \in \mathbb{N}$, satisfying

$$\Delta w(k) + \frac{w(k)w(k+1) + r(k)w(k)}{p(k)} + r(k) \;\leq\; 0, \tag{6.6.4}$$

or equivalently

$$\Delta w(k) + \frac{w^2(k)}{w(k) + p(k)} + r(k) \;\leq\; 0. \tag{6.6.5}$$

**Proof.** Since the necessity part is obvious, we need to prove only the sufficiency part. For this, let $z(a) = 1$, $z(k) = \prod_{\ell=a}^{k-1}(1 + w(\ell)/p(\ell))$, $k \in \mathbb{N}(a+1)$ then $z(k) > 0$ for all $k \in \mathbb{N}(a)$ and

$$\Delta(p(k)\Delta z(k)) + r(k)z(k+1) \;\leq\; 0. \tag{6.6.6}$$

Therefore, by Problem 6.24.17 it follows that (6.6.1) is nonoscillatory. ∎

**Theorem 6.6.2.** Assume that

$$\limsup_{k\to\infty} k^{-3/2}\sum_{\ell=0}^{k} p(\ell) \ < \ \infty \tag{6.6.7}$$

and the difference equation (6.6.1) is nonoscillatory. Then, the following are equivalent

$$\text{(i)} \quad \lim_{k\to\infty}\frac{1}{k}\sum_{\ell=0}^{k}\sum_{\tau=0}^{\ell} r(\tau) \ \text{ exists} \tag{6.6.8}$$

$$\text{(ii)} \quad \liminf_{k\to\infty}\frac{1}{k}\sum_{\ell=0}^{k}\sum_{\tau=0}^{\ell} r(\tau) > -\infty \tag{6.6.9}$$

(iii) for any nonoscillatory solution $u(k)$ of (6.6.1) with $u(k) \times u(k+1) > 0$, $k \in \mathbb{N}(a)$, the function $v(k) = p(k)\Delta u(k)/u(k)$, $k \in \mathbb{N}(a)$ satisfies

$$\sum_{\ell=a}^{\infty}\frac{v^2(\ell)}{v(\ell)+p(\ell)} \ < \ \infty. \tag{6.6.10}$$

**Proof.** Clearly (i) implies (ii). To show that (ii) implies (iii) suppose to the contrary that there is a nonoscillatory solution $u(k)$ of (6.6.1) such that $v(k) = p(k)\Delta u(k)/u(k) > -p(k)$ for all $k \in \mathbb{N}(a)$ and

$$\sum_{\ell=a}^{\infty}\frac{v^2(\ell)}{v(\ell)+p(\ell)} \ = \ \infty. \tag{6.6.11}$$

From (6.6.3), we have

$$v(k+1) + \sum_{\ell=a}^{k}\frac{v^2(\ell)}{v(\ell)+p(\ell)} + \sum_{\ell=a}^{k} r(\ell) \ = \ v(a) \tag{6.6.12}$$

and therefore for all $k \in \mathbb{N}(a)$

$$\begin{aligned} \frac{1}{k}\sum_{\ell=a}^{k}(-v(\ell+1)) \ &= \ \frac{1}{k}\sum_{\ell=a}^{k}\sum_{\tau=a}^{\ell}\frac{v^2(\tau)}{v(\tau)+p(\tau)} \\ &= \ \frac{1}{k}\sum_{\ell=a}^{k}\sum_{\tau=a}^{\ell} r(\tau) - \left(\frac{k-a+1}{k}\right)v(a). \end{aligned} \tag{6.6.13}$$

From (6.6.9), (6.6.11) and (6.6.13), we obtain

$$\lim_{k\to\infty}\frac{1}{k}\sum_{\ell=a}^{k}(-v(\ell+1)) \;=\; \infty$$

and hence

$$\lim_{k\to\infty}\frac{1}{k}\sum_{\ell=a}^{k}|v(\ell)| \;=\; \infty. \tag{6.6.14}$$

Let $P(k) = v^2(k)/(v(k)+p(k)),\ k\in\mathbb{N}(a)$. Then, $P(k)\ge 0$ and $P(k)=0$ if and only if $v(k)=0$. Let $A(k)=v^2(k)/P(k)$ if $v(k)\neq 0$ and $A(k)=0$ if $v(k)=0$. Then, we have $p(k)\ge A(k)-v(k)$ and hence

$$k^{-3/2}\sum_{\ell=a}^{k}p(\ell) \;\ge\; k^{-3/2}\sum_{\ell=a}^{k}A(\ell)+k^{-3/2}\sum_{\ell=a}^{k}(-v(\ell)). \tag{6.6.15}$$

Thus, in view of (6.6.7) and $A(k)\ge 0$ it follows that

$$\limsup_{k\to\infty} k^{-3/2}\sum_{\ell=a}^{k}(-v(\ell)) \;<\; \infty. \tag{6.6.16}$$

Therefore, on dividing both sides of (6.6.13) by $k^{1/2}$, and in the resulting equation using (6.6.9) and (6.6.16) leads to

$$\limsup_{k\to\infty} k^{-3/2}\sum_{\ell=a}^{k}\sum_{\tau=a}^{\ell}P(\tau) \;<\; \infty. \tag{6.6.17}$$

Now since

$$\begin{aligned} k^{-1/2}\sum_{\ell=a}^{k}P(\ell) \;&=\; k^{-3/2}k\sum_{\ell=a}^{k}P(\ell) \;\le\; k^{-3/2}\sum_{\ell=a}^{2k}\sum_{\tau=a}^{\ell}P(\tau)\\ &=\; 2^{3/2}(2k)^{-3/2}\sum_{\ell=a}^{2k}\sum_{\tau=a}^{\ell}P(\tau)\end{aligned}$$

from (6.6.17), we have

$$\limsup_{k\to\infty} k^{-1/2}\sum_{\ell=a}^{k}P(\ell) \;<\; \infty. \tag{6.6.18}$$

On the other hand, from (6.6.18) there is an $M>0$ such that

$$\left(\sum_{\ell=a}^{k}|v(\ell)|\right)^2 = \left(\sum_{\ell=a}^{k}(A(\ell)P(\ell))^{1/2}\right)^2 \le \sum_{\ell=a}^{k}A(\ell)\sum_{\ell=a}^{k}P(\ell)\le Mk^{1/2}\sum_{\ell=a}^{k}A(\ell).$$

Therefore, it follows that

$$k^{-3/2}\sum_{\ell=a}^{k} A(\ell) \geq \frac{1}{M}\left(\frac{1}{k}\sum_{\ell=a}^{k} |v(\ell)|\right)^2$$

and hence from (6.6.14) (6.6.16), we have

$$\lim_{k\to\infty} k^{-3/2}\sum_{\ell=a}^{k} p(\ell) = \infty,$$

which contradicts (6.6.7).

Finally, we shall show that (iii) implies (i). Let $v(k)$ be as in (iii) and let $B(k) = \sum_{\ell=a}^{k} |v(\ell)|$. Then, we have

$$\begin{aligned}\left(\sum_{\ell=a}^{k} v(\ell)\right)^2 &\leq B^2(k) = \left(\sum_{\ell=a}^{k}[P(\ell)(v(\ell)+p(\ell))]^{1/2}\right)^2 \\ &\leq \sum_{\ell=a}^{k} P(\ell)\sum_{\ell=a}^{k}(v(\ell)+p(\ell)) \\ &\leq L\left(B(k)+\sum_{\ell=a}^{k} p(\ell)\right) \leq 2L \max\left\{B(k), \sum_{\ell=a}^{k} p(\ell)\right\},\end{aligned}$$

where $L = \sum_{\ell=a}^{\infty} P(\ell)$. Hence, we have

$$B(k) \leq \max\left\{2L, \left(2L\sum_{\ell=a}^{k} p(\ell)\right)^{1/2}\right\}.$$

Thus, from (6.6.7) it follows that $\lim_{k\to\infty}(1/k)B(k) = 0$, so that $\lim_{k\to\infty} (1/k)\sum_{\ell=a}^{k}(-v(\ell+1)) = 0$. The result (i) now follows by letting $k \to \infty$ in (6.6.13). ∎

**Corollary 6.6.3.** Let (6.6.7) hold. Then, (6.6.1) is oscillatory in case either of the following satisfied

$$(6.6.19)\qquad -\infty < \liminf_{k\to\infty}\frac{1}{k}\sum_{\ell=0}^{k}\sum_{\tau=0}^{\ell} r(\tau) < \limsup_{k\to\infty}\frac{1}{k}\sum_{\ell=0}^{k}\sum_{\tau=0}^{\ell} r(\tau)$$

or

$$(6.6.20)\qquad \lim_{k\to\infty}\frac{1}{k}\sum_{\ell=0}^{k}\sum_{\tau=0}^{\ell} r(\tau) = \infty.$$

**Remark 6.6.1.** Suppose that (6.6.7) and (6.6.9) hold. Then, if (6.6.1) is nonoscillatory we can define the constant $c = \lim_{k\to\infty} \frac{1}{k}\sum_{\ell=0}^{k}\sum_{\tau=0}^{\ell} r(\tau)$.

**Theorem 6.6.4.** Let (6.6.7) and (6.6.9) hold.

(i) If (6.6.1) is nonoscillatory, then there exists a function $v(k)$ on $\mathbb{N}$ such that $v(k) > -p(k)$, $k \in \mathbb{N}(a)$ for some $a \in \mathbb{N}$ and

$$v(k) \;=\; c - \sum_{\ell=0}^{k-1} r(\ell) + \sum_{\ell=k}^{\infty} \frac{v^2(\ell)}{v(\ell)+p(\ell)}, \quad k \in \mathbb{N}(a). \tag{6.6.21}$$

(ii) If there exist a function $v(k)$ on $\mathbb{N}$ such that $v(k) > -p(k)$, $k \in \mathbb{N}(a)$, and a constant $c_1$ satisfying

$$v(k) \;\geq\; c_1 - \sum_{\ell=0}^{k-1} r(\ell) + \sum_{\ell=k}^{\infty} \frac{v^2(\ell)}{v(\ell)+p(\ell)} \;\geq\; 0 \tag{6.6.22}$$

or

$$v(k) \;\leq\; c_1 - \sum_{\ell=0}^{k-1} r(\ell) + \sum_{\ell=k}^{\infty} \frac{v^2(\ell)}{v(\ell)+p(\ell)} \;\leq\; 0, \tag{6.6.23}$$

then (6.1.1) is nonoscillatory.

**Proof.** (i) Since

$$\lim_{k\to\infty} \frac{1}{k}\sum_{\ell=a}^{k}\sum_{\tau=a}^{\ell} r(\tau) = \lim_{k\to\infty} \frac{1}{k}\left[\sum_{\ell=0}^{k}\sum_{\tau=0}^{\ell} r(\tau) - \sum_{\ell=0}^{a-1}\sum_{\tau=0}^{\ell} r(\tau) - (k-a)\sum_{\tau=0}^{a-1} r(\tau)\right]$$

$$= c - \sum_{\tau=0}^{a-1} r(\tau)$$

(6.6.21) follows by letting $k \to \infty$ in (6.6.13) and then replacing $a$ by $k$.

(ii) Suppose that (6.6.7) and (6.6.9) hold and there exists a constant $c_1$ such that (6.6.22) or (6.6.23) holds. Let $w(k) = c_1 - \sum_{\ell=0}^{k-1} r(\ell) + \sum_{\ell=k}^{\infty} \frac{v^2(\ell)}{v(\ell)+p(\ell)}$. Then, $\Delta w(k) = -r(k) - \frac{v^2(k)}{v(k)+p(k)}$. But, since $v(k) \geq w(k) \geq 0$ or $v(k) \leq w(k) \leq 0$, we have $\frac{v^2(k)}{v(k)+p(k)} \geq \frac{w^2(k)}{w(k)+p(k)}$, and hence

$$\Delta w(k) + \frac{w^2(k)}{w(k)+p(k)} + r(k) \;\leq\; 0, \quad w(k) \;>\; -p(k), \quad k \in \mathbb{N}(a).$$

Now as an application of Lemma 6.6.1 we find that (6.6.1) is nonoscillatory. ∎

**Theorem 6.6.5.** Assume that

$$\limsup_{k\to\infty}\frac{1}{k}\sum_{\ell=0}^{k}p(\ell) \;<\; \infty \tag{6.6.24}$$

$$\liminf_{k\to\infty}\frac{1}{k}\sum_{\ell=0}^{k}\sum_{\tau=0}^{\ell}r(\tau) \;=\; -\infty \tag{6.6.25}$$

and

$$\limsup_{k\to\infty}\frac{1}{k}\sum_{\ell=0}^{k}\sum_{\tau=0}^{\ell}r(\tau) \;>\; -\infty. \tag{6.6.26}$$

Then, the difference equation (6.6.1) is oscillatory.

**Proof.** Suppose to the contrary that (6.6.1) is nonoscillatory and let $u(k)$ be any nonoscillatory solution. Let $v(k) = p(k)\Delta u(k)/u(k)$ for $k\in\mathbb{N}(a)$. Since condition (6.6.7) follows from (6.6.24), Theorem 6.6.2 and (6.6.25) imply that (6.6.11) holds. But, from (6.6.13) we have

$$\begin{aligned}\limsup_{k\to\infty}\frac{1}{k}\sum_{\ell=a}^{k}(-v(\ell+1)) \;&\geq\; \liminf_{k\to\infty}\frac{1}{k}\sum_{\ell=a}^{k}\sum_{\tau=a}^{\ell}\frac{v^2(\tau)}{v(\tau)+p(\tau)}\\ &\quad+\limsup_{k\to\infty}\frac{1}{k}\sum_{\ell=a}^{k}\sum_{\tau=a}^{\ell}r(\tau)-v(a)\\ &=\; \infty,\end{aligned}$$

which is impossible from $-v(\ell+1) < p(\ell+1)$ and (6.6.24). ∎

**Theorem 6.6.6.** Assume that the difference equation (6.6.1) is nonoscillatory and

(6.6.27) there exists an $M>0$ with $0\leq p(k)\leq M$ for all $k\in\mathbb{N}$.

Then, the following are equivalent

(i) $\sum_{\ell=0}^{\infty}r(\ell)$ exists

(ii) (6.6.8) holds

(iii) (6.6.9) holds

(iv) for any nonoscillatory solution $u(k)$ of (6.6.1) with $u(k)u(k+1)>0$, $k\in\mathbb{N}(a)$, the function $v(k) = p(k)\Delta u(k)/u(k)$, $k\in\mathbb{N}(a)$ satisfies (6.6.10).

**Proof.** Obviously (i) implies (ii), and (ii) implies (iii). Theorem 6.6.2 shows that (iii) and (iv) are equivalent. Therefore, we need only to show that (iv) implies (i). But this is immediate by letting $k \to \infty$ in (6.6.12) and observing that (iv) implies $v(k) \to 0$ as $k \to \infty$. ∎

**Corollary 6.6.7.** If the assumptions of Theorem 6.6.6 hold then the following are equivalent

(i) $\sum_{\ell=0}^{\infty} r(\ell) = -\infty$

(ii) (6.6.25) holds

(iii) there exists a nonoscillatory solution $u(k)$ of (6.6.1) with $u(k)u(k+1) > 0$ on $\mathbb{N}(a)$ for some $a \in \mathbb{N}$ such that the function $v(k) = p(k)\Delta u(k)/u(k) > -p(k)$, $k \in \mathbb{N}(a)$ satisfies (6.6.11).

**Corollary 6.6.8.** Let (6.6.9) and (6.6.27) hold. If $\sum_{\ell=0}^{\infty} r(\ell)$ does not exist then (6.6.1) is oscillatory.

**Corollary 6.6.9.** Let (6.6.27) hold. If

$$-\infty \ = \ \liminf_{k\to\infty} \sum_{\ell=0}^{k} r(\ell) \ < \ \limsup_{k\to\infty} \sum_{\ell=0}^{k} r(\ell)$$

then (6.6.1) is oscillatory.

**Theorem 6.6.10.** If there exist two sequences $\{k_\ell\}$ and $\{m_\ell\}$ of integers with $m_\ell \geq k_\ell + 1$ such that $k_\ell \to \infty$ as $\ell \to \infty$ and

$$\sum_{\tau=k_\ell}^{m_\ell - 1} r(\tau) \ \geq \ p(k_\ell) + p(m_\ell) \tag{6.6.28}$$

then (6.6.1) is oscillatory.

**Proof.** Suppose that (6.6.1) is nonoscillatory. Then, there exists a nonoscillatory solution $u(k)$ such that $u(k)u(k+1) > 0$ for all $k \in \mathbb{N}(a)$ for some $a \in \mathbb{N}$. Let $v(k) = p(k)\Delta u(k)/u(k)$. Then, $v(k)$ satisfies (6.6.3) and $v(k) > -p(k)$ for all $k \in \mathbb{N}(a)$. We will show that

$$\sum_{\ell=a}^{k-1} r(\ell) \ < \ p(a) + p(k) \tag{6.6.29}$$

holds for all $k \in \mathbb{N}(a+1)$ and then this contradiction will prove the theorem.

From (6.6.3), we have

$$
\begin{aligned}
r(a) &= v(a) - v(a+1) - \frac{v^2(a)}{v(a)+p(a)} < p(a+1) + \frac{v(a)p(a)}{v(a)+p(a)} \\
&= p(a+1) + p(a) - \frac{p^2(a)}{v(a)+p(a)} < p(a+1) + p(a).
\end{aligned}
$$

Therefore, (6.6.29) holds for $k = a+1$. For any $k \in \mathbb{N}(a+2)$, from (6.6.3) we have

$$
\sum_{\ell=a+1}^{k-1} r(\ell) = v(a+1) - v(k) - \sum_{\ell=a+1}^{k-1} \frac{v^2(\ell)}{v(\ell)+p(\ell)} < v(a+1) + p(k).
$$

However, since

$$
v(a+1) = p(a)\left(1 - \frac{u(a)}{u(a+1)}\right) - r(a) < p(a) - r(a)
$$

(6.6.29) follows immediately. ∎

## 6.7. Olver Type Comparison Results

We shall develop several comparison theorems which are useful in estimating the growth of the solutions of second order difference equations.

**Theorem 6.7.1.** Let $u(k)$ and $v(k)$, $k \in \mathbb{N}$ be the solutions of the difference equations (2.16.10) and (2.16.11) respectively, where

$$
b_2(k) \geq a_2(k) > 0 \quad \text{and} \quad b_1(k) + b_2(k) \leq a_1(k) + a_2(k) \leq -1.
$$

If $v(1) - u(1) \geq v(0) - u(0) \geq 0$ and $u(1) \geq \max(u(0), 0)$, then $u(k)$ and $v(k)$ are nondecreasing and $u(k) \leq v(k)$ for all $k \in \mathbb{N}$.

**Proof.** Since $u(k+2) - u(k+1) = -(a_1(k) + a_2(k) + 1)u(k+1) + a_2(k)(u(k+1) - u(k))$ from $u(1) \geq u(0)$ and $u(1) \geq 0$ it follows that $u(2) - u(1) \geq 0$, and now by induction we get $u(k+1) - u(k) \geq 0$ for all $k \in \mathbb{N}$. The proof for $v(k+1) - v(k) \geq 0$ for all $k \in \mathbb{N}$ is similar. Next, since

$$
\begin{aligned}
&(v(k+2) - u(k+2)) - (v(k+1) - u(k+1)) \\
&\quad = -(a_1(k) + a_2(k) + 1)(v(k+1) - u(k+1)) \\
&\qquad - [(b_1(k) + b_2(k)) - (a_1(k) + a_2(k))]v(k+1) \\
&\qquad + a_2(k)[(v(k+1) - u(k+1)) - (v(k) - u(k))] \\
&\qquad + (b_2(k) - a_2(k))[v(k+1) - v(k)]
\end{aligned}
$$

if $v(k+1) - u(k+1) \geq v(k) - u(k) \geq 0$, as is the case for $k = 0$, the facts $v(k) \geq 0$ and $v(k+1) - v(k) \geq 0$ imply that $(v(k+2) - u(k+2)) - (v(k+1) - u(k+1)) \geq 0$. The resulting inequality $v(k) \geq u(k)$ now follows by induction. ■

**Corollary 6.7.2.** Let $u(k)$ and $v(k)$, $k \in \mathbb{N}$ be the solutions of (6.1.1) and

$$(6.7.1) \qquad p_1(k)v(k+1) + p_1(k-1)v(k-1) \;=\; q_1(k)v(k), \quad k \in \mathbb{N}(1)$$

where the functions $p_1$ and $q_1$ are defined on $\mathbb{N}$ and $\mathbb{N}(1)$ respectively, and $p_1(k) > 0$ for all $k \in \mathbb{N}$. Further, let

$$\frac{p_1(k-1)}{p_1(k)} \geq \frac{p(k-1)}{p(k)} \quad \text{and} \quad \frac{q_1(k)}{p_1(k)} - \frac{p_1(k-1)}{p_1(k)} \geq \frac{q(k)}{p(k)} - \frac{p(k-1)}{p(k)} \geq 1$$

and $v(1) \geq u(1) \geq 0$, $v(0) \geq u(0) \geq 0$, also $v(1) - v(0) \geq u(1) - u(0) \geq 0$. Then, $v(k+1) - v(k) \geq u(k+1) - u(k)$ and $v(k) \geq u(k)$ for all $k \in \mathbb{N}$.

**Theorem 6.7.3.** Let $u(k)$ be the solution of the difference equation (2.16.10), where $a_1(k) + a_2(k) \leq -1$, $a_2(k) > 0$ and $u(0) > 0$, $\mu = u(1)/u(0) \geq 1$. Then, $u(k)$ is nondecreasing and

$$(6.7.2) \qquad u(0)[\min(\mu, \lambda)]^k \;\leq\; u(k) \;\leq\; u(0)[\max(\mu, \wedge)]^k,$$

where $\lambda$, $\wedge$ are the largest roots of the equations $1 + \alpha\lambda + \beta\lambda^2 = 0$, $1 + A\wedge + B\wedge^2 = 0$, and $\beta = (\inf a_2(k))^{-1}$, $B = (\sup a_2(k))^{-1}$, $\alpha = -1 + \beta \sup(a_1(k) + a_2(k))$ and $A = -1 + B\inf(a_1(k) + a_2(k))$.

**Remark 6.7.1.** $B$ and $\alpha$ always exist. When $\beta = \infty$ the left hand inequality is omitted and when $A = -\infty$ the right hand inequality is omitted. When $A$ and $\beta$ exist, we have $-\alpha \geq 1 + \beta$, $-A \geq 1 + B$, $\beta \geq B$, $\frac{A}{B} \leq \frac{\alpha}{\beta}$, $\wedge \geq \frac{1}{\beta}$, $\wedge \geq \lambda$, and

$$\lambda \;=\; -\frac{\alpha}{2\beta} + \left(\frac{\alpha^2}{4\beta^2} - \frac{1}{\beta}\right)^{1/2} \;\geq\; \frac{\beta + 1 + |\beta - 1|}{2\beta} \;=\; \max\left(\frac{1}{\beta}, 1\right).$$

**Proof.** By Theorem 6.7.1 the solution $u(k)$ of (2.16.10) is a nondecreasing function of $k$. Let $\mu \geq \lambda$, and define $h(k) = u(0)\lambda^k$, to obtain $h(0) = u(0)$, $h(1) = \lambda u(0)$, and $h(k) + \alpha h(k+1) + \beta h(k+2) = 0$. Thus, in Theorem 6.7.1 taking (2.16.10) as $h(k+2) + \frac{\alpha}{\beta}h(k+1) + \frac{1}{\beta}h(k) = 0$ and (2.16.11) as (2.16.10), then $a_2(k) \geq \frac{1}{\beta} > 0$, $a_1(k) + a_2(k) \leq \frac{1+\alpha}{\beta} =$

$\sup(a_1(k)+a_2(k)) \leq -1$, $u(1)-h(1) = (\mu-\lambda)u(0) \geq 0 = u(0)-h(0)$, and $h(1) = \lambda u(0) \geq \max(h(0), 0)$, and in conclusion we have $u(0)\lambda^k \leq u(k)$.

For the case $\mu < \lambda$, let $h(k) = u(0)\mu^k$ so that $h(0) = u(0)$, $h(1) = u(1)$, and $h(k) + \hat{\alpha}h(k+1) + \hat{\beta}h(k+2) = 0$, where $\hat{\alpha} = -\mu\hat{\beta} - \dfrac{1}{\mu}$ and $\hat{\beta} = \max\left(\beta, \dfrac{1}{\mu}\right)$. Thus, in Theorem 6.7.1 taking (2.16.10) as $h(k+2) + \dfrac{\hat{\alpha}}{\hat{\beta}}h(k+1) + \dfrac{1}{\hat{\beta}}h(k) = 0$ and (2.16.11) as (2.16.10), then $a_2(k) \geq \dfrac{1}{\beta} \geq \dfrac{1}{\hat{\beta}} > 0$, $a_1(k) + a_2(k) \leq \dfrac{1+\alpha}{\beta} \leq \dfrac{1+\hat{\alpha}}{\hat{\beta}} \leq -1$, where we have used the identities $\dfrac{1+\hat{\alpha}}{\hat{\beta}} + 1 = (1-\mu)(\hat{\beta} - \dfrac{1}{\mu})/\hat{\beta}$, $\dfrac{1+\hat{\alpha}}{\hat{\beta}} = -1$, if $\hat{\beta} = \dfrac{1}{\mu}$, and if $\hat{\beta} = \beta > \dfrac{1}{\mu}$, then $\dfrac{1+\hat{\alpha}}{\hat{\beta}} - \dfrac{1+\alpha}{\beta} = \dfrac{-\alpha}{\beta} - \mu - \dfrac{1}{\mu\beta} = \lambda + \dfrac{1}{\beta\lambda} - \left(\mu + \dfrac{1}{\mu\beta}\right) = (\lambda-\mu)\left(1 - \dfrac{1}{\beta\mu\lambda}\right)$, and in conclusion we have $u(0)\mu^k \leq u(k)$.

This completes the proof of the left hand inequality (6.7.2). To prove the right hand inequality (6.7.2), for $\mu \leq \Lambda$ we consider the function $h(k) = u(0)\Lambda^k$ to get $h(0) = u(0)$, $h(1) = u(0)\Lambda$ and $h(k+2) + \dfrac{A}{B}h(k+1) + \dfrac{1}{B}h(k) = 0$. Now an application of Theorem 6.7.1 gives $u(k) \leq u(0)\Lambda^k$. For $\mu > \Lambda$ we define $h(k) = u(0)\mu^k$ so that $h(0) = u(0)$, $h(1) = u(1)$ and $h(k+2) + \dfrac{\hat{A}}{\hat{B}}h(k+1) + \dfrac{1}{\hat{B}}h(k) = 0$, where $\hat{B} = B$ and $\hat{A} = -B\mu - \dfrac{1}{\mu}$. Once again Theorem 6.7.1 gives $u(k) \leq u(0)\mu^k$. This completes the proof of the theorem. ∎

**Theorem 6.7.4.** Let $u(k)$ be the solution of the difference equation (2.16.10), where $a_1(k) \leq 0$ and $a_2(k) < 0$ and $u(0) > 0$, $\mu = u(1)/u(0) \geq 1$. Then, (6.7.2) holds, where $\lambda$, $\Lambda$ are the same as in Theorem 6.7.3, but $\beta = (\sup a_2(k))^{-1}$, $B = (\inf a_2(k))^{-1}$, $\alpha = \beta \sup a_1(k)$ and $A = B \inf a_1(k)$.

**Proof.** Let $\mu \geq \lambda$, and assume that $u(k) \geq u(0)\lambda^k$ and $u(k+1) \geq u(0)\lambda^{k+1}$, which is true when $k = 0$. Then, the difference equation (2.16.10) gives

$$\begin{aligned} u(k+2) &= -a_1(k)u(k+1) - a_2(k)u(k) \geq -\frac{\alpha}{\beta}u(0)\lambda^{k+1} - \frac{1}{\beta}u(0)\lambda^k \\ &= \left(-\frac{\alpha}{\beta}\lambda - \frac{1}{\beta}\right)u(0)\lambda^k = u(0)\lambda^{k+2}, \end{aligned}$$

which is the required inequality.

Now let $\mu < \lambda$. Since the roots of the equation $1 + \alpha t + \beta t^2 = 0$ are $\lambda$ and $\frac{1}{\beta\lambda}$, and $\beta < 0$, it follows that $\frac{1}{\beta\lambda} < \mu < \lambda$, and hence $1+\alpha\mu+\beta\mu^2 > 0$. Assume that $u(k) \ge u(0)\mu^k$ and $u(k+1) \ge u(0)\mu^{k+1}$, which is the case when $k = 0$. Then, from (2.16.10) we find

$$\begin{aligned} u(k+2) &= -a_1(k)u(k+1) - a_2(k)u(k) \ \ge \ \left(-\frac{\alpha}{\beta}\mu - \frac{1}{\beta}\right) u(0)\mu^k \\ &> u(0)\mu^{k+2}. \end{aligned}$$

The right hand inequality (6.7.2) can be proved in a similar way. ∎

## 6.8. Sturm Type Comparison Results

In addition to the given hypotheses on the functions $p$, $q$, $p_1$ and $q_1$ in (6.1.1) and (6.7.1), throughout we shall assume that $q(k) > 0$ and $q_1(k) > 0$ for all $k \in \mathbb{N}(1)$.

**Theorem 6.8.1.** If $p(k) \ge p_1(k)$ and $q(k) \le q_1(k)$ for all sufficiently large $k \in \mathbb{N}$ and (6.1.1) is nonoscillatory, then (6.7.1) is also nonoscillatory. Furthermore, if $u(k)$ is a solution of (6.1.1) with $u(k) > 0$ for all $k \in \mathbb{N}(a)$ and if $v(k)$ is a solution of (6.7.1) satisfying $p_1(a)v(a+1)/v(a) \ge p(a)u(a+1)/u(a)$ with $v(a) > 0$, then $v(k+1)/v(k) \ge u(k+1)/u(k)$ for all $k \in \mathbb{N}(a)$. If, in addition $v(a) \ge u(a)$ then $v(k) \ge u(k)$ for all $k \in \mathbb{N}(a)$.

**Proof.** Let $u(k)$ and $v(k)$ be as above, and $w(k) = p(k)u(k+1)/u(k)$ for all $k \in \mathbb{N}(a)$. Then, (6.1.1) implies that

$$w(k+1) = q(k+1) - p^2(k)/w(k), \quad k \in \mathbb{N}(a). \tag{6.8.1}$$

Let $w_1(k) = p_1(k)v(k+1)/v(k)$ for all $k \in \mathbb{N}(a)$ such that $v(k) \ne 0$. Then, for $k \in \mathbb{N}(a)$ such that $v(k)$ and $v(k+1)$ are nonzero, $w_1(k)$ and $w_1(k+1)$ are defined, $w(k) \ne 0$, and (6.7.1) implies

$$w_1(k+1) = q_1(k+1) - p_1^2(k)/w_1(k). \tag{6.8.2}$$

For such values of $k$, we subtract (6.8.1) from (6.8.2), and arrange the terms, to obtain

$$\begin{aligned} w_1(k+1) - w(k+1) &= \left(q_1(k+1) - q(k+1) + \frac{p^2(k) - p_1^2(k)}{w_1(k)}\right) \\ &\quad + \frac{p^2(k)}{w(k)w_1(k)}(w_1(k) - w(k)). \end{aligned} \tag{6.8.3}$$

From the hypotheses, $w(k) > 0$ for all $k \in \mathbb{N}(a)$. If $w_1(k) \geq w(k)$ the right side of (6.8.3) is then nonnegative, hence $w_1(k+1) \geq w(k+1) > 0$. In particular, from the hypotheses, $w_1(a) \geq w(a)$ and $v(a)$ and $v(a+1)$ are positive. Thus, $w_1(a+1)$ is defined and (6.8.3) implies that $w_1(a+1) \geq w(a+1) > 0$. Furthermore, $v(a+2) > 0$ since $v(a+2) = w_1(a+2)v(a+1)/p_1(a+1)$, and hence $w_1(a+2)$ is defined. Proceeding inductively, we conclude that $w_1(k) \geq w(k)$ and $v(k) > 0$ for all $k \in \mathbb{N}(a)$. Hence (6.7.1) is nonoscillatory and that

$$\frac{v(k+1)}{v(k)} \geq \frac{p(k)}{p_1(k)}\frac{u(k+1)}{u(k)} \geq \frac{u(k+1)}{u(k)}, \quad k \in \mathbb{N}(a). \tag{6.8.4}$$

Finally, if $v(a) \geq u(a)$ then (6.8.4) implies that $v(a+1) \geq \dfrac{v(a)}{u(a)}u(a+1) \geq u(a+1)$. Proceeding inductively, we obtain $v(k) \geq u(k)$ for all $k \in \mathbb{N}(a)$. ∎

**Corollary 6.8.2.** Let (6.1.1) be nonoscillatory, and there exist positive functions $v(k)$ and $w(k)$ satisfying

$$p(k)v(k+1) + p(k-1)v(k-1) \leq q(k)v(k), \quad k \in \mathbb{N}(a) \tag{6.8.5}$$

and

$$p(k)w(k+1) + p(k-1)w(k-1) \geq q(k)w(k), \quad k \in \mathbb{N}(a). \tag{6.8.6}$$

If $w(a+1)/w(a) \geq v(a+1)/v(a)$, then (6.1.1) has a solution $u(k)$ satisfying

$$w(k+1)/w(k) \geq u(k+1)/u(k) \geq v(k+1)/v(k), \quad k \in \mathbb{N}(a). \tag{6.8.7}$$

If in addition, $w(a) \geq u(a) \geq v(a)$, then $w(k) \geq u(k) \geq v(k)$, $k \in \mathbb{N}(a)$.

**Proof.** Given $v(k)$ and $w(k)$ as above, we define the functions $q_1(k)$ and $q_2(k)$ by

$$p(k)v(k+1) + p(k-1)v(k-1) = q_1(k)v(k), \quad k \in \mathbb{N}(a)$$

and

$$p(k)w(k+1) + p(k-1)w(k-1) = q_2(k)w(k), \quad k \in \mathbb{N}(a).$$

Then, $q_1(k) \leq q(k) \leq q_2(k)$, $k \in \mathbb{N}(a)$. Let $u(k)$ be the solution of (6.1.1) satisfying $u(a) = v(a)$ and $u(a+1) = v(a+1)$. The conclusion now follows immediately from Theorem 6.8.1. ∎

**Corollary 6.8.3.** The difference equation (6.1.1) is nonoscillatory if and only if there exists a function $v(k)$ satisfying $v(k) > 0$ and $p(k)v(k+1) + p(k-1)v(k-1) \le q(k)v(k)$ for all sufficiently large $k \in \mathbb{N}$.

Now corresponding to (6.5.3) for the equation (6.7.1) we shall consider the difference equation

$$(6.8.8) \qquad H(k)Z(k) + 1/Z(k-1) = 1, \quad k \in \mathbb{N}(a+1)$$

where $H(k) = p_1^2(k)/(q_1(k)q_1(k+1))$.

**Theorem 6.8.4.** If $h(k) \ge H(k)$ for all sufficiently large $k \in \mathbb{N}$ and (6.1.1) is nonoscillatory, then (6.7.1) is also nonoscillatory.

**Proof.** The proof is contained in Lemma 6.5.2. ∎

**Remark 6.8.1.** If $p(k) \ge p_1(k)$ and $q(k) \le q_1(k)$ for all sufficiently large $k \in \mathbb{N}$, then $h(k) \ge H(k)$. Thus, the first part of Theorem 6.8.1 is included in Theorem 6.8.4.

**Remark 6.8.2.** If $p(k) \le p_1(k)$ and $q(k)-p(k)-p(k-1) \le q_1(k)-p_1(k)-p_1(k-1)$ for all sufficiently large $k \in \mathbb{N}$, and (6.1.1) is nonoscillatory, then by Problem 6.24.18 equation (6.7.1) is also nonoscillatory.

## 6.9. Variety of Properties of Solutions of

$$(6.9.1) \quad p(k)z(k+1) + p(k-1)z(k-1) = q(k)z(k) + r(k), \quad k \in \mathbb{N}(1)$$

where the functions $p$, $q$ and $r$ are defined on $\mathbb{N}$, $\mathbb{N}(1)$ and $\mathbb{N}(1)$ respectively, and $p(k) > 0$ for all $k \in \mathbb{N}$.

The following properties of the solutions of (6.9.1) can be deduced rather easily.

$(Q_1)$ Any nontrivial solution $u(k)$ of (6.1.1) can vanish at most once on $\mathbb{N}$ if and only if any two values $z(k_1)$ and $z(k_2)$, $k_1 \ne k_2$, uniquely determine the solution $z(k)$ of (6.9.1).

$(Q_2)$ If (i) $q(k) \ge p(k-1) + p(k)$, $r(k) \ge 0$ for all $k \in \mathbb{N}(a, b)$, $1 \le a < b$, and (ii) $z(k)$ is a solution of (6.9.1) such that $z(a) \ge 0$ and $z(a) \ge z(a-1)$, then $z(k+1) \ge z(k) \ge 0$ for all $k \in \mathbb{N}(a, b)$. If in addition to (i) and (ii) at least one of the following conditions hold, namely (iii) $r(a) > 0$, (iv) $q(a) > p(a-1) + p(a)$ and $z(a) > 0$, or (v) $z(a) > z(a-1)$, then $z(k+1) > z(k)$ for all $k \in \mathbb{N}(a, b)$. Further, if (i) holds for all $k \in \mathbb{N}(a)$, and in addition to (ii) either (vi) $\sum^{\infty} r(\ell)/p(\ell) = \infty$, or (vii) there exists a function $\epsilon(k) \ge 0$ defined on

$\mathbb{N}(a)$, such that $q(k) \geq (1+\epsilon(k))p(k) + p(k-1)$ and $\sum^{\infty}\epsilon(\ell) = \infty$, then $z(k) \to \infty$ as $k \to \infty$.

$(Q_3)$ Suppose that $q(k) \geq p(k-1) + p(k)$ for all $k \in \mathbb{N}(a,b)$, $b > a+1$, and $z(k)$ is a solution of (6.9.1) defined by $z(a) = z(b) = 0$. If $r(k) \geq 0\ (\leq 0)$, $k \in \mathbb{N}(a,b)$ then $z(k) \leq 0\ (\geq 0)$, $k \in \mathbb{N}(a,b)$. Further, if $r(a+1) > 0\ (< 0)$ then $z(k) < 0\ (> 0)$, $k \in \mathbb{N}(a+1, b-1)$.

$(Q_4)$ If $q(k) \geq p(k-1) + p(k)$, $r(k) \geq 0\ (\leq 0)$ for all large $k \in \mathbb{N}$ and $\sum^{\infty} r(\ell)/p(\ell) = \infty\ (-\infty)$, then there exists a solution $z(k)$ of (6.9.1) such that $z(k) \to \infty\ (-\infty)$ as $k \to \infty$.

$(Q_5)$ For the solutions $u(k)$ and $z(k)$ of either (6.1.1) or (6.9.1), we define $W(u,z)(k) = p(k)(u(k+1)z(k) - z(k+1)u(k))$. If $u(k)$ and $z(k)$ are solutions of (6.1.1) and (6.9.1) respectively, then for any $k \in \mathbb{N}(a)$

$$W(u,z)(k) = -\sum_{\ell=a+1}^{k} r(\ell)u(\ell) + W(u,z)(a). \tag{6.9.2}$$

**Definition 6.9.1.** A particular solution $z(k)$ of (6.9.1) is said to be *recessive* if $z(k)/w(k) \to 0$ as $k \to \infty$, where $w(k)$ is a dominant solution of (6.1.1).

**Theorem 6.9.1.** If the difference equation (6.1.1) has a recessive solution $v(k)$ and a dominant solution $w(k)$ such that $\sum^{\infty} r(\ell)v(\ell)$ exists and $(v(k)/w(k))\sum_{\ell=1}^{k} r(\ell)w(\ell) \to 0$ as $k \to \infty$, then (6.9.1) has a recessive solution.

**Proof.** We may assume that the solutions $v(k)$ and $w(k)$ of (6.1.1) are such that (6.1.3) holds for all $k \in \mathbb{N}$ with $c = 1$. Thus, the general solution $z(k)$ of (6.9.1) can be written as

$$\begin{aligned} z(k) &= c_1 v(k) + c_2 w(k) - \sum_{\ell=1}^{k} r(\ell)(v(k)w(\ell) - w(k)v(\ell)) \\ &= v(k)\left(c_1 - \sum_{\ell=1}^{k} r(\ell)w(\ell)\right) + w(k)\left(c_2 + \sum_{\ell=1}^{k} r(\ell)v(\ell)\right). \end{aligned}$$

Therefore, if $c_2 = -\sum_{\ell=1}^{\infty} r(\ell)v(\ell)$, then we find that

$$\frac{z(k)}{w(k)} = \frac{v(k)}{w(k)}c_1 - \frac{v(k)}{w(k)}\sum_{\ell=1}^{k} r(\ell)w(\ell) - \sum_{\ell=k+1}^{\infty} r(\ell)v(\ell)$$

from which it follows that $z(k)/w(k) \to 0$ as $k \to \infty$. ∎

**Corollary 6.9.2.** If for every solution $u(k)$ of (6.1.1), $\sum_{\ell=1}^{\infty} r(\ell)u(\ell)$ exists and if (6.1.1) has recessive and dominant solutions, then (6.9.1) has a recessive solution.

**Example 6.9.1.** For the difference equation $z(k+1)+z(k-1) = 2z(k)+1$, $k \in \mathbb{N}(1)$ the general solution is $z(k) = c_1 + c_2 k + k(k+1)/2$, and hence it has no recessive solution.

Hereafter, we shall assume that in (6.9.1) the function $r(k)$ is not eventually identically equal to zero.

**Theorem 6.9.3.** Let $u(k)$ and $z(k)$ be the solutions of (6.1.1) and (6.9.1) respectively, and $W(u,z)(k)$ is eventually of one sign. Then, (6.1.1) is nonoscillatory if and only if $z(k)$ is a nonoscillatory solution of (6.9.1), which is equivalent to stating that (6.1.1) is oscillatory if and only if $z(k)$ is an oscillatory solution of (6.9.1).

**Proof.** Suppose that (6.1.1) is nonoscillatory and $a \in \mathbb{N}$ is large enough so that $u(k) > 0$ and $p(k)(u(k+1)z(k) - z(k+1)u(k)) \geq 0$ for all $k \in \mathbb{N}(a)$. Then, $u(k+1)z(k)/u(k) \geq z(k+1)$ for all $k \in \mathbb{N}(a)$. Let $k_1 \geq a$ be the first integer so that $z(k_1) \leq 0$, if such an integer exists. Then, $z(k) \leq 0$ for all $k \in \mathbb{N}(k_1)$. If $z(k) \not\equiv 0$ on $\mathbb{N}(k_1)$ then there exists an integer $k_2 \geq k_1$ such that $z(k_2) < 0$, which implies that $z(k) < 0$ for all $k \in \mathbb{N}(k_2)$. If $z(k) \equiv 0$ on $\mathbb{N}(k_1)$, then $r(k) \equiv 0$ on $\mathbb{N}(k_1)$, which we are excluding. If $k_1$ does not exist, then $z(k) > 0$ for all $k \in \mathbb{N}(a)$. Thus, in either case $z(k)$ is nonoscillatory. The arguments are similar if we had assumed $p(k)(u(k+1)z(k) - z(k+1)u(k)) \leq 0$ or $u(k) < 0$.

On the other hand, we assume that $z(k)$ is nonoscillatory, say positive, for all $k \in \mathbb{N}(a)$, and assume that $p(k)(u(k+1)z(k) - z(k+1)u(k)) \geq 0$ on $\mathbb{N}(a)$. Then, $u(k+1) \geq u(k)z(k+1)/z(k)$ for all $k \in \mathbb{N}(a)$. If $u(k)$ is an oscillatory solution of (6.1.1), then there exists a $k_1 \in \mathbb{N}(a)$ such that $u(k_1) > 0$. But then the previous inequality implies that $u(k) > 0$ for all $k \in \mathbb{N}(k_1)$, which is a contradiction. A similar argument holds if $p(k)(u(k+1)z(k) - z(k+1)u(k)) \leq 0$ or if $z(k)$ is eventually negative. ∎

**Corollary 6.9.4.** If the difference equation (6.1.1) is nonoscillatory and $r(k)$ is eventually of one sign, then (6.9.1) is nonoscillatory.

**Proof.** Let $u(k)$ and $z(k)$ be the solutions of (6.1.1) and (6.9.1) respectively. We may choose $a \in \mathbb{N}$ large enough so that $u(k)r(k)$ is of fixed sign for all $k \in \mathbb{N}(a)$. Then, from (6.9.2) it follows that $W(u,z)(k)$ is of one sign. The result now follows from Theorem 6.9.3. ∎

**Corollary 6.9.5.** If the difference equation (6.1.1) is oscillatory (nonoscillatory) and if there exists a solution $u(k)$ of (6.1.1) such that $\sum^{\infty} r(\ell)u(\ell) = \infty$ or $-\infty$, then (6.9.1) is oscillatory (nonoscillatory).

**Corollary 6.9.6.** Suppose $r(k)$ has the form $h(k)u(k)$, where $h(k)$ is of one sign and $u(k)$ is a solution of (6.1.1). If the difference equation (6.1.1) is oscillatory (nonoscillatory), then (6.9.1) is oscillatory (nonoscillatory).

**Corollary 6.9.7.** If $q(k) \le -p(k) - p(k-1)$ for all large $k \in \mathbb{N}$ and $r(k) = (-1)^k h(k)$, where $h(k)$ is of one sign, then (6.9.1) is oscillatory.

**Theorem 6.9.8.** Suppose that $\sum^{\infty} r(\ell)u(\ell)$ exists for every solution $u(k)$ of (6.1.1), and (6.1.1) is oscillatory (nonoscillatory). Then, the difference equation (6.9.1) has at most one nonoscillatory (oscillatory) solution.

**Proof.** Suppose that (6.1.1) is oscillatory and $\overline{z}(k)$ is a nonoscillatory solution of (6.9.1). Consider any other solution $z(k)$ of (6.9.1) of the form $z(k) = \overline{z}(k) + cu(k)$ $(c \ne 0)$, where $u(k)$ is a solution of (6.1.1). Let $v(k)$ be a solution of (6.1.1) which is linearly independent of $u(k)$ such that $W(u,v)(k) = 1$. Then, we have

$$W(z,v)(k) \;=\; W(\overline{z},v)(k) + c. \tag{6.9.3}$$

From (6.9.2) and the hypothesis, $\lim_{k\to\infty} W(\overline{z},v)(k)$ exists. If the limit is nonzero, then the oscillatory behavior of $v(k)$ and Theorem 6.9.3 would imply that $\overline{z}(k)$ oscillates, which is a contradiction. Thus, $\lim_{k\to\infty} W(\overline{z},v)(k) = 0$. However, then (6.9.3) gives $\lim_{k\to\infty} W(z,v)(k) = c \ (\ne 0)$, and now again from Theorem 6.9.3 it follows that $z(k)$ is oscillatory.

Next, assume that (6.1.1) is nonoscillatory and $\overline{z}(k)$ is an oscillatory solution of (6.9.1). Choose $z(k)$, $c$, $u(k)$ and $v(k)$ as above. Based on the previous argument, we must have $\lim_{k\to\infty} W(\overline{z},v)(k) = 0$. Again this implies in (6.9.3) that $\lim_{k\to\infty} W(z,v)(k) = c \ne 0$, which by Theorem 6.9.3 leads to that $z(k)$ is nonoscillatory. ∎

## 6.10. Variety of Properties of Solutions of

$$\Delta^2 u(k-1) + p(k)u^{\gamma}(k) \;=\; 0, \quad k \in \mathbb{N}(1) \tag{6.10.1}$$

where the function $p$ is defined on $\mathbb{N}(1)$ and $\gamma$ is a quotient of odd positive integers.

The following properties of the solutions of (6.10.1) are immediate.

$(R_1)$ If $u(k)$ is a nontrivial solution of (6.10.1) with $u(a)u(a+1) \le 0$ for some $a \in \mathbb{N}$, then either $u(a) \ne 0$ or $u(a+1) \ne 0$. If, in addition,

$a \in \mathbb{N}(1)$ and $u(a) = 0$, then $u(a+1) = -u(a-1)$. Thus, an oscillatory solution of (6.10.1) must change sign infinitely often.

$(R_2)$ Assume that $p(k) \le 0$ for all $k \in \mathbb{N}(1)$, and for every $a \in \mathbb{N}(1)$, $p(k) < 0$ for some $k \in \mathbb{N}(a+1)$. If $u(k)$ is a solution of (6.10.1) with $u(a-1) \le u(a)$ and $u(a) \ge 0$ for some $a \in \mathbb{N}(1)$, then $u(k)$ and $\Delta u(k)$ are nondecreasing and nonnegative for all $k \in \mathbb{N}(a)$. Similarly, if $u(a-1) \ge u(a)$ and $u(a) \le 0$ for some $a \in \mathbb{N}(1)$, then $u(k)$ and $\Delta u(k)$ are nonincreasing and nonpositive for all $k \in \mathbb{N}(a)$.

$(R_3)$ If $p(k)$ is as in $(R_2)$, then all nontrivial solutions of (6.10.1) are nonoscillatory and eventually monotonic.

$(R_4)$ Assume that $p(k) \ge 0$ for all $k \in \mathbb{N}(1)$, and for every $a \in \mathbb{N}(1)$, $p(k) > 0$ for some $k \in \mathbb{N}(a+1)$. If $u(k)$ is a nonoscillatory solution of (6.10.1) such that $u(k) > 0$ for all $k \in \mathbb{N}(a)$, then $u(k+1) > u(k)$ and $0 < \Delta u(k+1) \le \Delta u(k)$ for all $k \in \mathbb{N}(a)$. A similar argument holds if $u(k)$ is eventually negative.

**Theorem 6.10.1.** If $p(k)$ is as in $(R_2)$, and $u(k)$ and $v(k)$ are solutions of (6.10.1) satisfying

$$(6.10.2) \quad u(b) \ \le \ (<)\ v(b) \ \text{ and } \ u(b+1) \ > \ (\ge)\ v(b+1) \ \text{ for some } \ b \in \mathbb{N}$$

then $u(k) > v(k)$ for all $k \in \mathbb{N}(b+2)$, $u(k) < v(k)$ for all $k \in \mathbb{N}(0, b-1)$, and $u(k) - v(k)$ is increasing for all $k \in \mathbb{N}$. Furthermore,

$$(6.10.3) \quad u(k) - v(k) \ \ge \ (k-b)(u(b+1) - v(b+1)) \ \text{ for all } \ k \in \mathbb{N}(b+1)$$

and

$$(6.10.4) \qquad u(k) - v(k) \ \le \ (b-k+1)(u(b) - v(b)) \ \text{ for all } \ k \in \mathbb{N}(0, b).$$

**Proof.** Let $w(k) = u(b+k) - v(b+k)$, then from (6.10.2) it is clear that $w(0) \le (<)\ 0$ and $w(1) > (\ge)\ 0$. By induction we shall show that

$$(6.10.5) \qquad w(k) \ \ge \ \frac{k}{k-1} w(k-1) \ \ge \ 0, \qquad k \in \mathbb{N}(2).$$

For this, since from (6.10.1) we have

$$\Delta^2 u(b) \ = \ -p(b+1)u^{\gamma}(b+1) \ \ge \ -p(b+1)v^{\gamma}(b+1) \ = \ \Delta^2 v(b)$$

it follows that

$$w(2) \ \ge \ 2w(1) - w(0)\ (>0) \ \ge \ 2w(1) \ \ge \ 0,$$

i.e. (6.10.5) is true for $k = 2$. Now let (6.10.5) be true for $k = \ell$, then as before we have $\Delta^2 u(b+\ell-1) \geq \Delta^2 v(b+\ell-1)$, and hence

$$w(\ell+1) \geq 2w(\ell) - w(\ell-1).$$

Thus, from (6.10.5) for $k = \ell$ it follows that

$$w(\ell+1) \geq \left(2 - \frac{\ell-1}{\ell}\right) w(\ell) = \frac{\ell+1}{\ell} w(\ell) > 0,$$

i.e. (6.10.5) holds for $k = \ell + 1$ also.

Now from (6.10.5) and the fact that $w(2) > 0$, it is clear that $u(k) > v(k)$ for all $k \in \mathbb{N}(b+2)$, and $u(k) - v(k)$ is increasing on $N(b+1)$. Further, since $w(k) \geq \frac{k}{k-1} \cdot \frac{k-1}{k-2} \cdots \frac{2}{1} w(1)$ we find that $w(k) \geq kw(1)$, which is the same as (6.10.3).

To prove the conclusions of the theorem for $k \in \mathbb{N}(0, b-1)$, let $u_1(k) = -u(b+1-k)$ and $v_1(k) = -v(b+1-k)$, $k \in \mathbb{N}(0, b+1)$. Then, $u_1(k)$ and $v_1(k)$ are solutions of (6.10.1) with $p(k)$ replaced by $p(b+1-k)$. Now applying the above results to $u_1(k)$ and $v_1(k)$ with $b = 0$ completes the proof. ∎

**Remark 6.10.1.** In $(R_2)$ we assumed that $u(a) \geq u(a-1)$ and $u(a) \geq 0$ and concluded that $u(k)$ was nondecreasing for all $k \in \mathbb{N}(a)$. If we assume that $u(a) > u(a-1) \geq 0$, then Theorem 6.10.1 implies that $u(k)$ is strictly increasing and $u(k) \to \infty$ as $k \to \infty$. For this, let $z(k)$ be a solution of (6.10.1) defined by $z(a) = z(a-1) = u(a-1)$. Now we apply Theorem 6.10.1 with $b = a - 1$. Since $u(b) = z(b)$ and $u(b+1) > z(b+1)$, we have $u(k) - z(k) > u(k-1) - z(k-1)$ for all $k \in \mathbb{N}(a+1)$. Thus, $u(k) - u(k-1) > z(k) - z(k-1) \geq 0$, so $u(k)$ is strictly increasing on $\mathbb{N}(a)$. For all $k \in \mathbb{N}(a)$ we also conclude that $u(k) \geq u(k) - z(k) \geq (k-a+1)(u(a) - z(a)) = (k-a+1)(u(a) - u(a-1))$, where $u(a) - u(a-1) > 0$. Thus, $u(k) \to \infty$ as $k \to \infty$.

**Corollary 6.10.2.** If $p(k)$ is as in $(R_2)$, and $u(k)$ and $v(k)$ are solutions of (6.10.1) satisfying $u(a) = v(a)$ and $u(b) = v(b)$ for some $a < b$, $a$, $b \in \mathbb{N}$, then $u(k) = v(k)$ for all $k \in \mathbb{N}$.

**Lemma 6.10.3.** If $p(k)$ is as in $(R_2)$, then for any $a \geq 1$ there exists a unique solution $u(k)$ of (6.10.1) such that $u(0) = u_0$ and $u(a) = 0$, where $u_0$ is any positive constant.

**Proof.** Let $z(k)$ be a solution of (6.10.1) such that $z(a) = 0$. If $z(a-1) > 0$ and $z(a-2) \leq z(a-1)$, then $(R_2)$ implies that $z(a) \geq$

$z(a-1) > 0$, which is a contradiction. Thus, $z(a-2) > z(a-1) > 0$. Proceeding in this way, we obtain

$$(6.10.6) \qquad z(0) \ > \ z(1) \ > \ \cdots \ > \ z(a-1) \ > \ z(a) \ = \ 0.$$

Since $z(a) = 0$, if $z(a-1)$ is also specified then $z(k)$ is uniquely determined for all $k \in \mathbb{N}(0,a)$ by (6.10.1). Thus, in particular $z(0)$ is determined by $z(a-1)$. Let $f$ be the mapping from $z(a-1)$ to $z(0)$. From (6.10.1), it is clear that each $z(k)$, $k \in \mathbb{N}(0,a-2)$ continuously depends on $z(a-1)$, and so in particular the function $z(0) = f(z(a-1))$ is continuous. If we let $z(a-1) = u_0$, then (6.10.6) implies that $f(u_0) > u_0$, if we let $z(a-1) = 0$, so that $z(a) = z(a-1) = 0$, then $z(k) \equiv 0$ so $f(0) = 0$. Thus, since $f$ is continuous, there exists $\beta$, $0 < \beta < u_0$ such that $f(\beta) = u_0$. Therefore, there exists a solution $u(k)$ of (6.10.1) determined by $u(a) = 0$ and $u(a-1) = \beta$, which must satisfy $u(0) = u_0$. Finally, the uniqueness of this solution follows from Corollary 6.10.2. ∎

**Theorem 6.10.4.** If $p(k)$ is as in $(R_2)$, then (6.10.1) has a positive nonincreasing solution $u(k)$ and a positive strictly increasing solution $v(k)$ such that $v(k) \to \infty$ as $k \to \infty$. In addition, the nonincreasing solution $u(k)$ is uniquely determined once $u(0)$ is specified.

**Proof.** If we choose, say, $v(0) = 1$ and $v(1) > 1$ then the existence of an increasing solution $v(k)$ satisfying the stated properties is an immediate consequence of Remark 6.10.1. To show the existence of a positive nonincreasing solution $u(k)$ of (6.10.1), by Lemma 6.10.3 it is clear that for each $\ell \geq 1$, there is a unique solution $u^\ell(k)$, $k \in \mathbb{N}$ of (6.10.1) such that

$$(6.10.7) \qquad u^\ell(0) \ = \ u_0, \quad u^\ell(\ell) \ = \ 0.$$

Further, in view of (6.10.6) we know that for every $\ell \geq 1$

$$(6.10.8) \qquad u_0 \ \geq \ u^\ell(k) \ > \ u^\ell(k+1) \ \geq \ 0, \quad k \in \mathbb{N}(0,\ell-1).$$

We claim that for every $\ell \geq 1$

$$(6.10.9) \qquad u^{\ell+1}(k) \ > \ u^\ell(k), \quad k \in \mathbb{N}(1).$$

For this, by Theorem 6.10.1 it suffices to show that $u^{\ell+1}(1) > u^\ell(1)$. Suppose to the contrary, that $u^\ell(1) \geq u^{\ell+1}(1)$. If $u^\ell(1) = u^{\ell+1}(1)$, then since $u^\ell(0) = u^{\ell+1}(0)$, the solutions $u^\ell(k)$ and $u^\ell(k+1)$ are identically equal, however, then since $u^\ell(\ell) = u^{\ell+1}(\ell+1) = 0$, both $u^\ell(k)$ and $u^\ell(k+1)$ are identically $0$, which contradicts $u^\ell(0) = u_0 > 0$. If

$u^\ell(1) > u^{\ell+1}(1)$, then from Theorem 6.10.1 we have $u^\ell(k) > u^{\ell+1}(k)$ for all $k \in \mathbb{N}(1)$, but then in particular for $k = \ell$ we find $0 = u^\ell(\ell) > u^{\ell+1}(\ell) > u^{\ell+1}(\ell+1) = 0$, which is also a contradiction. Therefore, (6.10.9) holds.

Combining (6.10.8) and (6.10.9) we find that for each $k \in \mathbb{N}(1)$ the $\ell$ sequence $\{u^\ell(k)\}$ is increasing, bounded above by $u_0$, and is eventually positive. Let $u(k) = \lim_{\ell\to\infty} u^\ell(k)$ for each $k \in \mathbb{N}$. Then, $0 < u(k) \le u_0$, $k \in \mathbb{N}$ and from (6.10.8) we have $u(k) \ge u(k+1)$, $k \in \mathbb{N}$. Now since for each $\ell \in \mathbb{N}(1)$, $u^\ell(k)$ is a solution of (6.1.1), we have $\Delta^2 u^\ell(k-1) = -p(k)(u^\ell(k))^\gamma$. Thus, as $\ell \to \infty$, we find that $u(k)$ is a nonincreasing but positive solution of (6.1.1).

Finally, we shall show that this solution $u(k)$ is unique, once $u_0$ is specified. For this, let $z(k)$ be another positive, nonincreasing solution of (6.1.1) such that $z(0) = u_0$. Then, either $z(1) < u(1)$, $z(1) > u(1)$, or $z(1) = u(1)$.

If $z(1) < u(1)$, then there exists an integer $\ell$ and a solution $u^\ell(k)$ defined by (6.10.7) such that $z(1) < u^\ell(1) < u(1)$. Since $u^\ell(0) = z(0)$ and $u^\ell(1) > z(1)$, Theorem 6.10.1 implies that $u^\ell(k) > z(k)$ for all $k \in \mathbb{N}(1)$. But, then in particular, $0 = u^\ell(\ell) > z(\ell)$, we have a contradiction.

If $z(1) > u(1)$, then Theorem 6.10.1 implies that $z(k) - u(k) \ge k(z(1) - u(1))$, $k \in \mathbb{N}(1)$. This means that $z(k)$ becomes unbounded as $k \to \infty$, which is again a contradiction.

Thus, $z(1) = u(1)$ and hence $z(k) = u(k)$ for all $k \in \mathbb{N}$. ∎

**Theorem 6.10.5.** If $p(k)$ is as in $(R_4)$ and $\gamma > 1$, then the difference equation (6.10.1) is oscillatory if and only if $\sum_{\ell=1}^\infty \ell p(\ell) = \infty$.

**Proof.** Let $u(k)$ be a nonoscillatory solution of (6.10.1), and $u(k) > 0$ for all $k \in \mathbb{N}(a)$. By $(R_4)$, $u(k)$ is increasing and $\Delta u(k)$ is positive and nonincreasing for all $k \in \mathbb{N}(a)$. We multiply both sides of (6.10.1) by $ku^{-\gamma}(k)$ and sum, to obtain

$$\sum_{\ell=a}^{k-1} \ell u^{-\gamma}(\ell)\Delta^2 u(\ell-1) + \sum_{\ell=a}^{k-1} \ell p(\ell) = 0, \qquad k \in \mathbb{N}(a+1)$$

which is from (1.8.5) is the same as

$$ku^{-\gamma}(k)\Delta u(k-1) - au^{-\gamma}(a)\Delta u(a-1) - \sum_{\ell=a}^{k-1} \Delta u(\ell)\Delta(\ell u^{-\gamma}(\ell)) + \sum_{\ell=a}^{k-1} \ell p(\ell) = 0.$$

In view of $(R_4)$ and the hypotheses, the above equality implies that

$$\sum_{\ell=a}^{k-1} \Delta u(\ell)\Delta(\ell u^{-\gamma}(\ell)) \to \infty \quad \text{as} \quad k \to \infty. \tag{6.10.10}$$

We shall show that (6.10.10) is impossible. For this, we note that $\Delta u(k) > 0$ implies that $\Delta(u^{-\gamma}(k)) < 0$, and hence

$$\begin{aligned}\sum_{\ell=a}^{k-1} \Delta u(\ell)\Delta(\ell u^{-\gamma}(\ell)) &= \sum_{\ell=a}^{k-1}[u^{-\gamma}(\ell+1)\Delta u(\ell) + \ell\Delta u(\ell)\Delta(u^{-\gamma}(\ell))] \\ &\le \sum_{\ell=a}^{k-1} u^{-\gamma}(\ell+1)\Delta u(\ell).\end{aligned}$$

Thus, it suffices to show that

$$\sum_{\ell=a}^{\infty} u^{-\gamma}(\ell+1)\Delta u(\ell) < \infty. \tag{6.10.11}$$

Let $r(t) = u(\ell) + (t-\ell)\Delta u(\ell)$, $\ell \le t \le \ell+1$. Then, $r(\ell) = u(\ell)$, $r(\ell+1) = u(\ell+1)$ and $r'(t) = \Delta u(\ell) > 0$, $\ell < t < \ell+1$. Thus, $r(t)$ is continuous and increasing for $t \ge a$. We then have

$$\begin{aligned}u^{-\gamma}(\ell+1)\Delta u(\ell) &= \int_{\ell}^{\ell+1} u^{-\gamma}(\ell+1)\Delta u(\ell)dt = \int_{\ell}^{\ell+1} r^{-\gamma}(\ell+1)r'(t)dt \\ &< \int_{\ell}^{\ell+1} r^{-\gamma}(t)r'(t)dt = \frac{1}{1-\gamma}\left[r^{1-\gamma}(\ell+1) - r^{1-\gamma}(\ell)\right].\end{aligned}$$

This implies that

$$\sum_{\ell=a}^{k} u^{-\gamma}(\ell+1)\Delta u(\ell) \le \frac{1}{1-\gamma}\left[r^{1-\gamma}(k+1) - r^{1-\gamma}(a)\right].$$

However, since $\gamma > 1$ and $r$ is an increasing function, it follows that (6.10.11) holds. This completes the sufficiency proof. The necessity part is contained in the sufficiency part of the next result. ∎

**Theorem 6.10.6.** If $p(k)$ is as in $(R_4)$, then (6.10.1) has a bounded nonoscillatory solution if and only if $\sum^{\infty} \ell p(\ell) < \infty$.

**Proof.** It is easy to verify that any solution $u(k)$ of

$$u(k) = 1 - \sum_{\ell=k+1}^{\infty} (\ell-k)p(\ell)u^{\gamma}(\ell) \tag{6.10.12}$$

is also a solution of (6.10.1). We choose $a \in \mathbb{N}$ so large that

$$\max_{k \in \mathbb{N}(a)} \left\{ \sum_{\ell=k+1}^{\infty} (\ell - k)p(\ell),\ 2\gamma \sum_{\ell=k+1}^{\infty} (\ell - k)p(\ell) \right\} < \frac{1}{2}.$$

Consider the Banach space $\ell_\infty^a$ of all bounded real functions $v(k)$, $k \in \mathbb{N}(a)$ with the norm defined as $\|v\| = \sup |v(k)|$, $k \in \mathbb{N}(a)$. We define a closed, bounded subset $S$ of $\ell_\infty^a$ as, $S = \{v \in \ell_\infty^a : 1/2 \le v(k) \le 1\}$. Let $T$ be an operator $T : S \to S$ such that

$$(Tv)(k) = 1 - \sum_{\ell=k+1}^{\infty} (\ell - k)p(\ell)u^\gamma(\ell), \quad k \in \mathbb{N}(a).$$

To see that the range of $T$ is in $S$, we note that if $v \in S$, then $(Tv)(k) \ge 1 - \sum_{\ell=k+1}^{\infty}(\ell - k)p(\ell) \ge (1/2)$. Clearly, $(Tv)(k) \le 1$. Further, since the mean value theorem applied to the function $r(t) = t^\gamma$ implies that for any $v$ and $w \in S$, $|v^\gamma(k) - w^\gamma(k)| \le 2\gamma|v(k) - w(k)|$ for all $k \in \mathbb{N}(a)$, we have

$$\begin{aligned} |(Tv)(k) - (Tw)(k)| &\le \sum_{\ell=k+1}^{\infty} (\ell - k)p(\ell)|v^\gamma(\ell) - w^\gamma(\ell)| \\ &\le 2\gamma\|v - w\| \sum_{\ell=k+1}^{\infty} (\ell - k)p(\ell) \\ &\le \frac{1}{2}\|v - w\|. \end{aligned}$$

Therefore, $\|Tv - Tw\| \le (1/2)\|v - w\|$, and hence $T$ is contracting on $S$. Thus, $T$ has a unique fixed point in $S$, which is our desired bounded, nonoscillatory solution of (6.10.12).

To prove the converse, let $u(k)$ be a nonoscillatory solution of (6.10.1), and $u(k) > 0$ for all $k \in \mathbb{N}(a)$. By $(R_4)$, $u(k)$ is increasing for all $k \in \mathbb{N}(a)$. Thus, $u(k)$ is bounded above and below by positive constants for all $k \in \mathbb{N}(a)$. Now since any solution $u(k)$ of (6.10.1) also satisfies

$$(6.10.13) \quad k(u(k+1) - u(k)) = a(u(a+1) - u(a)) + u(k) - u(a) - \sum_{\ell=a+1}^{k} \ell p(\ell)u^\gamma(\ell), \quad k \in \mathbb{N}(a)$$

if $\sum_{\ell=a}^{k} \ell p(\ell) \to \infty$ as $k \to \infty$, the right side of (6.10.13) must then approach to $-\infty$. This implies that the left side of (6.10.13) is eventually negative. But this contradicts the fact that $u(k)$ is increasing. This completes the proof. ∎

**Theorem 6.10.7.** If $p(k)$ is as in $(R_4)$ and $0 < \gamma < 1$, then all solutions of (6.10.1) are oscillatory if and only if $\sum^{\infty} \ell^{\gamma} p(\ell) = \infty$.

**Proof.** Let $u(k)$ be a nonoscillatory solution of (6.10.1), and $u(k) > 0$ for all $k \in \mathbb{N}(a)$. By $(R_4)$, $u(k)$ is increasing and $\Delta u(k)$ is positive and nonincreasing for all $k \in \mathbb{N}(a)$. Thus, for all $k \in \mathbb{N}(2a)$ we have

$$u(k) \ = \ u(a) + \sum_{\ell=a}^{k-1} \Delta u(\ell) \ \geq \ (k-a)\Delta u(k-1) \ \geq \ \frac{k}{2}\Delta u(k-1),$$

i.e. $u(k)/\Delta u(k-1) \geq k/2$. Dividing (6.10.1) by $(\Delta u(k-1))^{\gamma}$, using this inequality, and summing from $2a$ to $k$, we obtain

$$\sum_{\ell=2a}^{k} \Delta^2 u(\ell-1)/(\Delta u(\ell-1))^{\gamma} + \sum_{\ell=2a}^{k} p(\ell)\ell^{\gamma}/2^{\gamma} \ \leq \ 0, \quad k \in \mathbb{N}(2a). \tag{6.10.14}$$

By hypothesis, the second sum in (6.10.14) approaches $\infty$ as $k \to \infty$, so the first term approaches $-\infty$. To show this is impossible, let $r(t) = u(\ell) + (t-\ell)\Delta u(\ell)$, $\ell \leq t \leq \ell+1$, $\ell \geq a$ so that $r$ is positive, continuous and increasing. Further, let $s(t) = r(t+1) - r(t) > 0$, $t \geq a$ so that $s$ is continuous, $s'(t) = \Delta u(\ell) - \Delta u(\ell-1) = \Delta^2 u(\ell-1) \leq 0$ for $\ell - 1 < t < \ell$ which implies $s$ is nonincreasing and $s(t) \leq s(\ell-1) = \Delta u(\ell-1)$. Then, for $\ell - 1 < t < \ell$, we have

$$\frac{\Delta^2 u(\ell-1)}{(\Delta u(\ell-1))^{\gamma}} \ = \ \int_{\ell-1}^{\ell} \frac{\Delta^2 u(\ell-1)}{(\Delta u(\ell-1))^{\gamma}} dt \ \geq \ \int_{\ell-1}^{\ell} \frac{s'(t)}{s^{\gamma}(t)} dt.$$

Thus, it follows that

$$\sum_{\ell=2a}^{k} \frac{\Delta^2 u(\ell-1)}{(\Delta u(\ell-1))^{\gamma}} \ \geq \ \int_{2a-1}^{k} \frac{s'(t)}{s^{\gamma}(t)} dt \ = \ \frac{1}{1-\gamma}\left[s^{1-\gamma}(k) - s^{1-\gamma}(2a-1)\right].$$

But $s^{1-\gamma}(k) > 0$ for all $k \in \mathbb{N}(a)$, so the sum on the left in the above inequality is bounded below. This contradiction completes the sufficiency proof. The necessity part is contained in the sufficiency part of the next result. ∎

**Definition 6.10.1.** A solution $u(k)$ of (6.10.1) is said to have *asymptotically positively bounded differences* if there exist positive constants $c_1$ and $c_2$ such that $c_1 \leq \Delta u(k) \leq c_2$ for all $k \in \mathbb{N}(a)$ for some $a \in \mathbb{N}(1)$.

**Theorem 6.10.8.** If $p(k)$ is as in $(R_4)$, then (6.10.1) has a solution with asymptotically positively bounded differences if and only if $\sum^{\infty} \ell^{\gamma} p(\ell) < \infty$.

**Proof.** Assume that $\sum^{\infty} \ell^{\gamma} p(\ell) < \infty$, and choose $a$ large enough so that $\sum_{\ell=a}^{\infty} \ell^{\gamma} p(\ell) < 1/2$. Let $u(k)$ be the solution of (6.10.1) satisfying $u(a) = 0$, $u(a+1) = 1$, so that $\Delta u(a) = 1$. We want to show that $1/2 \le \Delta u(k) \le 1$ for all $k \in \mathbb{N}(a)$. For this, suppose that $1/2 \le \Delta u(k) \le 1$ for all $k \in \mathbb{N}(a, m-1)$. Then, $u(k) > 0$ for all $k \in \mathbb{N}(a, m)$. However, then from (6.10.1), $\Delta^2 u(k-1) \le 0$ for all $k \in \mathbb{N}(a, m)$. Thus, for all $k \in \mathbb{N}(a+1, m)$ it follows that

$$u(k) \ \le \ u(a) + (k-a)\Delta u(a) \ = \ (k-a) \ \le \ k.$$

Now from (6.10.1) and the above inequalities we obtain

$$\Delta u(m) \ = \ \Delta u(a) - \sum_{\ell=a+1}^{m} p(\ell) u^{\gamma}(\ell) \ \ge \ 1 - \sum_{\ell=a+1}^{m} p(\ell)\ell^{\gamma} \ \ge \ \frac{1}{2}.$$

Also, since $\Delta u(k)$ is nonincreasing, we find that $\Delta u(m) \le \Delta u(a) = 1$. Thus, $1/2 \le \Delta u(m) \le 1$, and now by induction $1/2 \le \Delta u(k) \le 1$ holds for all $k \in \mathbb{N}(a)$.

Conversely, let $u(k)$ be a solution of (6.10.1) which has asymptotically positively bounded differences. Then, as in Theorem 6.10.7 we find that $u(k) \ge (k/2)\Delta u(k-1)$ for all $k \in \mathbb{N}(2a)$. Thus, for all $k \in \mathbb{N}(2a+1)$ it follows that

$$\begin{aligned} \Delta u(2a) - \Delta u(k) \ &= \ \sum_{\ell=2a+1}^{k} p(\ell) u^{\gamma}(\ell) \ \ge \ \frac{1}{2^{\gamma}} \sum_{\ell=2a+1}^{k} p(\ell)\ell^{\gamma}(\Delta u(\ell-1))^{\gamma} \\ &\ge \ \left(\frac{c_1}{2}\right)^{\gamma} \sum_{\ell=2a+1}^{k} p(\ell)\ell^{\gamma} \ \ge \ 0. \end{aligned}$$

But this implies that $\sum^{\infty} p(\ell)\ell^{\gamma} < \infty$. ∎

**Example 6.10.1.** Consider the difference equation (6.10.1) with $\gamma = 1/3$ and $p(k) = (k+1)^{4/3}\left\{(k+2)^{-4} + 2(k+1)^{-4} + k^{-4}\right\}$, $k \in \mathbb{N}(1)$. For this difference equation $u(k) = (-1)^k (k+1)^{-4}$ is an oscillatory solution. Further, since $\sum_{\ell=1}^{\infty} \ell p(\ell) < \infty$ and $p(\ell) > 0$, by Theorem 6.10.6 it also has a bounded nonoscillatory solution. Thus, the difference equation (6.10.1) can have both oscillatory as well as nonoscillatory solutions.

## 6.11. Oscillation and Nonoscillation for

$$\Delta(r(k)\Delta u(k)) + f(k)F(u(k)) \ = \ 0, \quad k \in \mathbb{N}(a) \tag{6.11.1}$$

where the functions $r$, $f$ and $F$ are defined in their domain of definition. Further, $uF(u) > 0$ for $u \neq 0$, $r(k) > 0$ for all $k \in \mathbb{N}(a)$, and $R_{a,k} \to \infty$, where $R_{j,k} = \sum_{\ell=j}^{k-1} 1/r(\ell)$, $j \in \mathbb{N}(a)$, $k \in \mathbb{N}(j+1)$.

**Theorem 6.11.1.** Suppose that in (6.11.1) the function $F(u)$ is continuous on $\mathbb{R}$, and

(i) $r(k)$ is nondecreasing on $\mathbb{N}(a)$

(ii) $\displaystyle\lim_{k\to\infty} \frac{1}{r(k)} \sum_{\ell=a}^{k-1} \ell(\mu f^+(\ell) + f^-(\ell)) = \infty$, for every constant $\mu > 0$,

where $f^+(k) = \max(f(k), 0)$ and $f^-(k) = \min(f(k), 0)$.

Then, every bounded solution $u(k)$ of (6.11.1) is either oscillatory or such that $\liminf_{k\to\infty} |u(k)| = 0$.

**Proof.** Suppose that there exists a bounded nonoscillatory solution $u(k)$ of (6.11.1), and let $u(k) > 0$ for all $k \geq k_1 > a$ (a similar argument holds for $u(k) < 0$). If $\liminf_{k\to\infty} u(k) > 0$, then there is a $k_2 \geq k_1$ and constants $c_1$, $c_2$ such that $0 < c_1 \leq u(k) \leq c_2$ for all $k \in \mathbb{N}(k_2)$. Thus, from the given hypotheses there exist constants $M_1$, $M_2 > 0$ such that

$$0 < M_1 < F(u(k)) \leq M_2 \text{ for all } k \in \mathbb{N}(k_2). \tag{6.11.2}$$

From (6.11.1), we have

$$\sum_{\ell=k_2}^{k} \ell\Delta(r(\ell)\Delta u(\ell)) = -\sum_{\ell=k_2}^{k} \ell f(\ell)F(u(\ell)). \tag{6.11.3}$$

Further, since from (1.8.5)

$$\sum_{\ell=k_2}^{k} \ell\Delta(r(\ell)\Delta u(\ell)) = kr(k+1)\Delta u(k+1) - k_2 r(k_2)\Delta u(k_2) - \sum_{\ell=k_2+1}^{k} r(\ell)\Delta u(\ell)$$

and

$$\sum_{\ell=k_2+1}^{k} r(\ell)\Delta u(\ell) = r(k+1)u(k+1) - r(k_2+1)u(k_2+1) - \sum_{\ell=k_2+1}^{k} u(\ell+1)\Delta r(\ell)$$

the hypothesis (i) implies that

(6.11.4)

$$\sum_{\ell=k_2}^{k} \ell\Delta(r(\ell)\Delta u(\ell)) \geq kr(k+1)\Delta u(k+1) - k_2 r(k_2)\Delta u(k_2) - c_2 r(k+1).$$

On the other hand, from (6.11.2) it follows that

$$(6.11.5)\quad \sum_{\ell=k_2}^{k} \ell f(\ell)F(u(\ell)) \geq \sum_{\ell=k_2}^{k} \ell(M_1 f^+(\ell) + M_2 f^-(\ell)) = M_2 \sum_{\ell=k_2}^{k} \ell(\mu f^+(\ell) + f^-(\ell)),$$

where $\mu = M_1/M_2$.

Combining (6.11.3) (6.11.5), we find

$$kr(k+1)\Delta u(k+1) - k_2 r(k_2)\Delta u(k_2) - c_2 r(k+1) \leq -M_2 \sum_{\ell=k_2}^{k} \ell(\mu f^+(\ell) + f^-(\ell)),$$

which implies that

$$\begin{aligned} k\Delta u(k+1) - \lambda k_2 \Delta u(k_2) - c_2 &\leq k\Delta u(k+1) - c_2 - \frac{k_2 r(k_2)\Delta u(k_2)}{r(k+1)} \\ &\leq -\frac{M_2}{r(k+1)} \sum_{\ell=k_2}^{k} \ell(\mu f^+(\ell) + f^-(\ell)), \end{aligned}$$

where $\lambda = \begin{cases} 0 & \text{if } \Delta u(k_2) \leq 0 \\ 1 & \text{if } \Delta u(k_2) > 0. \end{cases}$

Therefore, from the hypothesis (ii) we conclude that $k\Delta u(k+1) \to -\infty$ as $k \to \infty$. But, then there exists a $k_3 \geq k_2$ such that $\Delta u(k+1) \leq -1/k$ for all $k \in \mathbb{N}(k_3)$, and this gives $u(k+1) \leq u(k_3+1) - \sum_{\ell=k_3}^{k-1} 1/k$, which implies that $\lim_{k\to\infty} u(k) = -\infty$. This contradicts our assumption that $u(k) > 0$ for all $k \in \mathbb{N}(k_1)$. ∎

**Theorem 6.11.2.** Suppose that in (6.11.1) the function $F(u)$ is continuous on $\mathbb{R}$, $f(k) \geq 0$ for all $k \in \mathbb{N}(a)$, and there exist functions $\phi(u) \in C^{(1)}[\mathbb{R}, \mathbb{R}]$ and $h(k)$ defined on $\mathbb{N}(a)$ such that

(i) $|F(u)| \geq |\phi(u)|$, $\phi'(u) \geq \epsilon > 0$, $u\phi(u) > 0$ for $u \neq 0$

(ii) $h(k) > 0$ for all $k \in \mathbb{N}(a)$, and

$$\limsup_{k\to\infty} \sum_{\ell=a}^{k} h(\ell)\left[f(\ell) - \frac{r(\ell)}{4\epsilon}\left(\frac{\Delta h(\ell)}{h(\ell)}\right)^2\right] = \infty.$$

Then, the difference equation (6.11.1) is oscillatory.

**Proof.** Suppose that there exists a nonoscillatory solution $u(k)$ of (6.11.1), and let $u(k) > 0$ for all $k \geq k_1 > a$ (a similar argument holds for $u(k) < 0$). It follows from equation (6.11.1) that $\Delta(r(k)\Delta u(k)) \le 0$, and hence $r(k)\Delta u(k)$ is nonincreasing for $k \ge k_1$. We will first show that $r(k)\Delta u(k) \ge 0$ for all $k \in \mathbb{N}(k_1)$. If $r(k_2)\Delta u(k_2) = c < 0$ for some $k_2 \ge k_1$, then $r(k)\Delta u(k) \le c$ for all $k \in \mathbb{N}(k_2)$. But, this implies $u(k) \le u(k_2) + c\sum_{\ell=k_2}^{k-1} 1/r(\ell) \to -\infty$, as $k \to \infty$ which contradicts the fact that $u(k) > 0$ for all $k \in \mathbb{N}(k_1)$. Thus, $r(k)\Delta u(k) \ge 0$ for all $k \in \mathbb{N}(k_1)$, and this implies that $u(k)$ is nondecreasing on $\mathbb{N}(k_1)$. In view of (i), from (6.11.1) we have $\Delta(r(k)\Delta u(k)) + f(k)\phi(u(k)) \le 0$, and so

$$(6.11.6) \qquad \frac{h(k)\Delta(r(k)\Delta u(k))}{\phi(u(k))} \le -f(k)h(k), \quad k \in \mathbb{N}(k_1).$$

For $k \in \mathbb{N}(k_1)$, we define $q(k) = \dfrac{h(k)v(k)}{\phi(u(k))}$, where $v(k) = r(k)\Delta u(k)$. Then,

$$\Delta q(k) = \frac{h(k)\Delta v(k)}{\phi(u(k))} + \frac{v(k+1)\Delta h(k)}{\phi(u(k+1))} - \frac{v(k+1)h(k)\Delta\phi(u(k))}{\phi(u(k))\phi(u(k+1))}$$

and by the mean value theorem and (i), we get

$$\Delta q(k) \le \frac{h(k)\Delta v(k)}{\phi(u(k))} + \frac{v(k+1)\Delta h(k)}{\phi(u(k+1))} - \frac{\epsilon h(k)v(k+1)\Delta u(k)}{\phi(u(k))\phi(u(k+1))}, \quad k \in \mathbb{N}(k_1).$$

Using the inequalities $v(k+1) \le v(k)$, $\phi(u(k)) \le \phi(u(k+1))$ and (6.11.6) in the above inequality, we get

$$\begin{aligned} \Delta q(k) &\le -f(k)h(k) + q(k+1)\frac{\Delta h(k)}{h(k+1)} - q^2(k+1)\frac{\epsilon h(k)}{r(k)h^2(k+1)} \\ &= -\frac{\epsilon h(k)}{r(k)h^2(k+1)}\left[q(k+1) - \frac{\Delta h(k)r(k)h(k+1)}{2\epsilon h(k)}\right]^2 \\ &\quad + \frac{r(k)(\Delta h(k))^2}{4\epsilon h(k)} - f(k)h(k) \\ &\le -h(k)\left[f(k) - \frac{r(k)}{4\epsilon}\left(\frac{\Delta h(k)}{h(k)}\right)^2\right], \quad k \in \mathbb{N}(k_1). \end{aligned}$$

Summing the above inequality from $k_1$ to $k$, we obtain

$$-q(k_1) \le q(k+1) - q(k_1) \le -\sum_{\ell=k_1}^{k} h(\ell)\left[f(\ell) - \frac{r(\ell)}{4\epsilon}\left(\frac{\Delta h(\ell)}{h(\ell)}\right)^2\right],$$

which yields $\displaystyle\sum_{\ell=k_1}^{k} h(\ell)\left[f(\ell) - \frac{r(\ell)}{4\epsilon}\left(\frac{\Delta h(\ell)}{h(\ell)}\right)^2\right] \le c_1$, $k \in \mathbb{N}(k_1)$, where $c_1 > 0$ is a finite constant. But, this contradicts condition (ii) and the proof is complete. ∎

**Corollary 6.11.3.** Suppose that $f(k) \ge 0$ on $\mathbb{N}(a)$ and that there exists a function $h(k) > 0$ on $\mathbb{N}(a)$ such that

$$\limsup_{k\to\infty} \sum_{\ell=a}^{k} h(\ell)\left[f(\ell) - \left(\frac{\Delta h(\ell)}{2h(\ell)}\right)^2\right] = \infty. \tag{6.11.7}$$

Then, the difference equation

$$\Delta^2 u(k) + f(k)u(k) = 0, \quad k \in \mathbb{N}(a) \tag{6.11.8}$$

is oscillatory.

**Remark 6.11.1.** If $f(k) \ge \dfrac{1+\alpha}{4k^2}$ on $\mathbb{N}(1)$, $\alpha > 0$, then if we let $h(k) = k$, the assertion of Corollary 6.11.3 holds.

**Theorem 6.11.4.** Suppose that in (6.11.1) the function $F(u)$ is continuous on $\mathbb{R}$, $f(k) \ge 0$ for all $k \in \mathbb{N}(a)$, and

(i) there exist two nondecreasing functions $\phi \in C[\mathbb{R}, \mathbb{R}]$ and $\psi \in C[(0,\infty),(0,\infty)]$ such that

$$|F(u)| \ge |\phi(u)|, \quad u\phi(u) > 0 \quad \text{for } u \ne 0$$

and

$$\int_{\epsilon}^{\infty} \frac{dt}{\phi(t)\psi(t)} < \infty \text{ and } \int_{-\epsilon}^{-\infty} \frac{dt}{\phi(t)\psi(-t)} < \infty, \text{ for every } \epsilon > 0 \tag{6.11.9}$$

(ii) there exists a nondecreasing function $\rho(k) > 0$ on $\mathbb{N}(a)$ such that $r(k)\Delta\rho(k)$ is nonincreasing on $\mathbb{N}(a)$ and

$$\sum^{\infty} \frac{\rho(\ell)f(\ell)}{\psi(R_{a,\ell})} = \infty. \tag{6.11.10}$$

Then, the difference equation (6.11.1) is oscillatory.

**Proof.** Assume the contrary. Then, as in Theorem 6.11.2 for a nonoscillatory solution $u(k) > 0$, $k \in \mathbb{N}(k_1)$, $k_1 > a$ we have $u(k) \le u(k+1)$, $v(k+1) \le v(k)$ for all $k \in \mathbb{N}(k_1)$, where $v(k) = r(k)\Delta u(k)$. Let $p(k) = \dfrac{\rho(k)v(k)}{\phi(u(k))\psi(R_{k_1},k)}$, $k \in \mathbb{N}(k_1)$, then

$$\begin{aligned}\Delta p(k) &= \frac{\rho(k)\Delta v(k)}{\phi(u(k))\psi(R_{k_1,k})} + \frac{v(k+1)\Delta\rho(k)}{\phi(u(k+1))\psi(R_{k_1,k+1})} \\ &\quad - \frac{v(k+1)\rho(k)\Delta[\phi(u(k))\psi(R_{k_1,k})]}{\phi(u(k))\phi(u(k+1))\psi(R_{k_1,k})\psi(R_{k_1,k+1})}.\end{aligned}$$

Since $\Delta[\phi(u(k))\psi(R_{k_1,k})] \geq 0$, $v(k) \geq 0$ for all $k \in \mathbb{N}(k_1)$, (i) and (ii) in the above equality imply that

$$(6.11.11)\quad \Delta p(k) \leq \frac{\rho(k)\Delta v(k)}{\phi(u(k))\psi(R_{k_1,k})} + \frac{v(k+1)\Delta\rho(k)}{\phi(u(k+1))\psi(R_{k_1,k+1})}, \quad k \in \mathbb{N}(k_1).$$

By the assumptions the equation (6.11.1) gives

$$(6.11.12)\qquad \frac{\rho(k)\Delta v(k)}{\phi(u(k))\psi(R_{k_1,k})} \leq -\frac{\rho(k)f(k)}{\psi(R_{k_1,k})}, \quad k \in \mathbb{N}(k_1).$$

Thus, in view of the monotonicity of $v(k)$ and $r(k)\Delta p(k)$, from (6.11.11) and (6.11.12), we obtain

$$(6.11.13)\quad \Delta p(k) \leq -\frac{\rho(k)f(k)}{\psi(R_{k_1,k})} + r(k_1)\Delta\rho(k_1)\frac{\Delta u(k)}{\phi(u(k+1))\psi(R_{k_1,k+1})}, \quad k \in \mathbb{N}(k_1).$$

Since $\Delta v(k) \leq 0$, $k \in \mathbb{N}(k_1)$ it follows that $u(k) \leq u(k_1) + v(k_1)R_{k_1,k}$, and hence from the assumptions on $r(k)$ there exists a constant $\beta \geq 1$ such that $u(k) \leq \beta R_{k_1,k}$, $k \in \mathbb{N}(k_1)$. Thus, from (6.11.13), we have

$$(6.11.14)\quad \Delta p(k) \leq -\frac{\rho(k)f(k)}{\psi(R_{k_1,k})} + r(k_1)\Delta\rho(k_1)\frac{\Delta u(k)}{\phi(u(k+1)/\beta)\psi(u(k+1)/\beta)}, \quad k \in \mathbb{N}(k_1).$$

Since for $u(k)/\beta \leq t \leq u(k+1)/\beta$, we have $[\phi(t)\psi(t)]^{-1} \geq [\phi(u(k+1)/\beta)\psi(u(k+1)/\beta)]^{-1}$, it follows that

$$\int_{u(k)/\beta}^{u(k+1)/\beta} \frac{dt}{\phi(t)\psi(t)} \geq \frac{1}{\beta}\frac{\Delta u(k)}{\phi(u(k+1)/\beta)\psi(u(k+1)/\beta)}.$$

Using the above inequality in (6.11.14) and summing the resulting inequality from $k_1$ to $k$ leads to

$$p(k+1) - p(k_1) + \sum_{\ell=k_1}^{k} \frac{\rho(\ell)f(\ell)}{\psi(R_{k_1,\ell})} \leq \beta r(k_1)\Delta\rho(k_1)\int_{u(k_1)/\beta}^{u(k+1)/\beta} \frac{dt}{\phi(t)\psi(t)}.$$

The above inequality in view of (6.11.9) and $p(k) \geq 0$, $k \in \mathbb{N}(k_1)$ gives

$$\sum_{\ell=k_1}^{k} \frac{\rho(\ell)f(\ell)}{\psi(R_{k_1,\ell})} \leq c < \infty,$$

which contradicts (6.11.10). ∎

**Corollary 6.11.5.** Suppose that $f(k) \geq 0$ for all $k \in \mathbb{N}(a)$, and there exists a constant $\alpha > 0$ such that

$$(6.11.15) \qquad \sum^{\infty} f(\ell) R_{a,\ell}^{1-\alpha} = \infty.$$

Then, the difference equation

$$(6.11.16) \qquad \Delta(r(k)\Delta u(k)) + f(k)u(k) = 0, \quad k \in \mathbb{N}(a)$$

is oscillatory.

**Proof.** In Theorem 6.11.4 let $\phi(u) = u$, $\psi(u) = u^{\alpha}$ and $\rho(k) = R_{a,k}$. ∎

**Remark 6.11.2.** In (6.11.15) the constant $\alpha$ cannot be zero. For this, we note that the equation $\Delta^2 u(k) + ((2\sqrt{k+2} - \sqrt{k+1} - \sqrt{k+3})/\sqrt{k+1})u(k) = 0$ has a nonoscillatory solution $u(k) = \sqrt{k+1}$ and the condition $\sum^{\infty} \ell f(\ell) = \infty$ holds.

**Corollary 6.11.6.** Suppose that $f(k) \geq 0$ for all $k \in \mathbb{N}(a)$, and $\sum^{\infty} f(\ell) R_{a,\ell} = \infty$. Then, the difference equation

$$(6.11.17) \quad \Delta(r(k)\Delta u(k)) + f(k)|u(k)|^{\alpha}\mathrm{sgn}\, u(k) = 0, \quad \alpha > 1, \quad k \in \mathbb{N}(a)$$

is oscillatory.

**Proof.** In Theorem 6.11.4 let $\phi(u) = |u|^{\alpha}$ sgn $u$, $\alpha > 1$, $\psi(u) = 1$ and $\rho(k) = R_{a,k}$. ∎

## 6.12. Asymptotic Behavior of Solutions of

$$(6.12.1) \qquad \Delta(r(k)\Delta u(k)) + f(k)F(u(k)) = g(k), \quad k \in \mathbb{N}(a)$$

where the functions $r$, $f$, $F$ and $g$ are defined in their domain of definition. Further, $uF(u) > 0$ for $u \neq 0$, $r(k) > 0$ for all $k \in \mathbb{N}(a)$, and $R_{a,k} \to \infty$, where $R_{j,k} = \sum_{\ell=j}^{k-1} 1/r(\ell)$, $j \in \mathbb{N}(a)$, $k \in \mathbb{N}(j+1)$.

**Theorem 6.12.1.** Suppose that the following conditions hold

(i) $f(k) \geq \alpha > 0$ for all $k \in \mathbb{N}(a)$

(ii) $|F(u)|$ is bounded away from zero if $|u|$ is bounded away from zero

(iii) the function $G(k) = \sum_{\ell=a}^{k-1} g(\ell)$ is bounded on $\mathbb{N}(a)$.

Then, for every nonoscillatory solution $u(k)$ of (6.12.1), $\lim_{k\to\infty} u(k) = 0$.

**Proof.** In system form equation (6.12.1) is equivalent to

$$\begin{aligned} \Delta u(k) &= (v(k) + G(k))/r(k) \\ \Delta v(k) &= -f(k)F(u(k)). \end{aligned} \tag{6.12.2}$$

If $u(k)$ is a nonoscillatory solution of (6.12.1), then we can assume that $u(k) > 0$ eventually (the case $u(k) < 0$ can be similarly treated). First we shall show that $\liminf_{k\to\infty} u(k) = 0$. If not, then there exist $k_1 \geq a$ and a positive constant $c_1$ such that $F(u(k)) \geq c_1$ for all $k \in \mathbb{N}(k_1)$. From (6.12.2) it follows that

$$\begin{aligned} v(k+1) - v(k_1) &= -\sum_{\ell=k_1}^{k} f(\ell)F(u(\ell)) \\ &\leq -c_1 \sum_{\ell=k_1}^{k} f(\ell) \to -\infty \text{ as } k \to \infty. \end{aligned}$$

We then have $\Delta u(k) = (v(k)+G(k))/r(k) \leq -1/r(k)$ for all $k \in \mathbb{N}(k_2)$, for some $k_2 \geq k_1$. This implies that $u(k) \leq u(k_2) - \sum_{\ell=k_2}^{k-1} 1/r(\ell) \to -\infty$, as $k \to \infty$. But, this contradicts the fact that $u(k)$ is eventually positive. From the above argument, we also have

$$\sum^{\infty} f(\ell)F(u(\ell)) < \infty. \tag{6.12.3}$$

If $\limsup_{k\to\infty} u(k) = \gamma > 0$, then there exists a sequence $\{k_j\} \subseteq \mathbb{N}$, such that $u(k_j) \to \gamma$ as $j \to \infty$. Hence, there is $j(0)$ $(k_{j(0)} \geq a)$ such that $u(k_j) \geq \gamma/2$ and $F(u(k_j)) \geq c_2$ for all $j \geq j(0)$, where $c_2$ is a positive constant. But, then we have

$$\sum_{\ell=k_{j(0)}}^{k_j} f(\ell)F(u(\ell)) \geq \sum_{\ell=j(0)}^{j} f(k_\ell)F(u(k_\ell)) \geq \alpha c_2(j - j(0) + 1) \to \infty$$

as $j \to \infty$, so that $\sum^{\infty} f(\ell)F(u(\ell)) = \infty$, which contradicts (6.12.3). ∎

**Theorem 6.12.2.** In addition to the condition (ii) let

(iv) $f(k) > 0$ for all $k \in \mathbb{N}(a)$, and $\sum^{\infty} f(\ell) = \infty$

(v) $\lim_{k\to\infty} g(k)/f(k) = 0$.

Then, for every nonoscillatory solution $u(k)$ of (6.12.1), $\liminf\limits_{k\to\infty} |u(k)| = 0$.

**Proof.** Let $u(k)$ be a nonoscillatory solution of (6.12.1), say, $u(k) > 0$ for all $k \in \mathbb{N}(k_1)$, where $k_1 \geq a$. Then, $u(k)$ is also a nonoscillatory solution of

$$\Delta(r(k)\Delta u(k)) + [f(k) - g(k)/F(u(k))]F(u(k)) \ = \ 0, \quad k \in \mathbb{N}(k_1).$$

Suppose that $\liminf_{k\to\infty} u(k) > 0$, then by the hypotheses there exists a positive constant $c$ such that $F(u(k)) \geq c$ for all $k \in \mathbb{N}(k_1)$. Thus, by (v) there exists a $k_2 \geq k_1$ such that $g(k)/(f(k)F(u(k))) < 1/2$ for all $k \in \mathbb{N}(k_2)$. This implies that

$$f(k) - g(k)/F(u(k)) = f(k)\,[1 - g(k)/(f(k)F(u(k)))] \geq \frac{1}{2}f(k), \quad k \in \mathbb{N}(k_2).$$

So from (iv) we get $\sum^{\infty}[f(\ell) - g(\ell)/F(u(\ell))] = \infty$. But, then by Problem 6.24.30, $u(k)$ must be oscillatory. This contradiction completes the proof. ∎

**Theorem 6.12.3.** In addition to the condition (iv) let

(vi) $F(u)$ is continuous at $u = 0$

(vii) $\liminf_{k\to\infty} \sum_{\ell=j}^{k} g(\ell)/\sum_{\ell=j}^{k} f(\ell) \geq c > 0$ for every $j \in \mathbb{N}(a)$.

Then, no solution of (6.12.1) approaches zero.

**Proof.** Let $u(k)$ be a solution of (6.12.1) which approaches zero. Then, by the hypotheses on the function $F$ there exists a $k_1 \geq a$ such that $F(u(k)) < c/4$ for all $k \in \mathbb{N}(k_1)$. Hence, from the equation (6.12.1) we have

$$r(k+1)\Delta u(k+1) - r(k_1)\Delta u(k_1) \ \geq \ -\frac{c}{4}\sum_{\ell=k_1}^{k} f(\ell) + \sum_{\ell=k_1}^{k} g(\ell),$$

which by (vii) yields

$$\begin{aligned} \frac{r(k+1)\Delta u(k+1)}{\sum_{\ell=k_1}^{k} f(\ell)} - \frac{r(k_1)\Delta u(k_1)}{\sum_{\ell=k_1}^{k} f(\ell)} \ &\geq \ -\frac{c}{4} + \sum_{\ell=k_1}^{k} g(\ell)\Big/\sum_{\ell=k_1}^{k} f(\ell) \\ &\geq \ -\frac{c}{4} + \frac{c}{2} \ = \ \frac{c}{4} \ > \ 0, \end{aligned}$$

for all large $k$. But, (iv) in the above inequality implies that $r(k)\Delta u(k) \to \infty$ as $k \to \infty$, which in turn leads to the contradictive conclusion that $u(k) \to \infty$ as $k \to \infty$. ∎

**Remark 6.12.1.** If we replace conditions (iv) and (vii) by

(iv)′ $f(k) < 0$ for all $k \in \mathbb{N}(a)$, and $\sum^{\infty} f(\ell) = -\infty$

(vii)′ $\limsup_{k\to\infty} \sum_{\ell=j}^{k} g(\ell)/\sum_{\ell=j}^{k} f(\ell) \leq c < 0$ for every $j \in \mathbb{N}(a)$, then the assertion of Theorem 6.12.3 holds.

**Theorem 6.12.4.** Suppose that the following conditions hold

(viii) $F(u)$ is locally bounded in $\mathbb{R}$

(ix) $\sum^{\infty} |f(\ell)| < \infty, \quad \sum^{\infty} g(\ell) = \infty.$

Then, every solution of (6.12.1) is unbounded.

**Proof.** Let $u(k)$ be a bounded solution of (6.12.1), i.e. $|u(k)| \leq M$, where $M$ is a positive constant. Then, by (viii) there exist constants $L_1$ and $L_2$ such that $L_1 \leq F(u(k)) \leq L_2$. But then, from (6.12.1) and (ix), we obtain

$$r(k+1)\Delta u(k+1) - r(a)\Delta u(a)$$
$$\geq \sum_{\ell=a}^{k} g(\ell) - L_2 \sum_{\ell=a}^{k} f^{+}(\ell) - L_1 \sum_{\ell=a}^{k} f^{-}(\ell) \rightarrow \infty, \quad \text{as } k \rightarrow \infty.$$

However, this leads to that $u(k) \rightarrow \infty$. This contradiction completes the proof. ∎

## 6.13. $\ell_2$ and $c_0$ Solutions of

$$\Delta^2 u(k) + f(k, u(k)) = 0, \quad k \in \mathbb{N}(a). \tag{6.13.1}$$

**Theorem 6.13.1.** Let for all $(k, u) \in \mathbb{N}(a) \times \mathbb{R}$ the function $f(k, u)$ be defined and

$$|f(k, u)| \leq \frac{1}{2} k^{-2} |u|. \tag{6.13.2}$$

Then, if $u(k) \in \ell_2$ is a solution of (6.13.1), there exists an integer $k_1 \geq a$ $(a \geq 2)$ such that $u(k) = 0$ for all $k \in \mathbb{N}(k_1)$.

**Proof.** Let $u(k)$ be a solution of (6.13.1) such that $\sum_{\ell=a}^{\infty} |u(\ell)|^2 < \infty$. Then, $\lim_{k\rightarrow\infty} u(k) = 0$, and hence $\lim_{k\rightarrow\infty} \Delta u(k) = \lim_{k\rightarrow\infty} \Delta^2 u(k) = 0$. Summing equation (6.13.1) from $k$ to $m$, we obtain $\Delta u(m+1) - \Delta u(k) = -\sum_{\ell=k}^{m} f(\ell, u(\ell))$, and thus as $m \rightarrow \infty$, we find $\Delta u(k) = \sum_{\ell=k}^{\infty} f(\ell, u(\ell))$. Summing this equation from $m$ to $k$, we get

$$u(k+1) - u(m) = \sum_{j=m}^{k} \sum_{\ell=j}^{\infty} f(\ell, u(\ell)) \tag{6.13.3}$$
$$= \sum_{\ell=m}^{\infty} (\ell - m + 1) f(\ell, u(\ell)) - \sum_{\ell=k+1}^{\infty} (\ell - k) f(\ell, u(\ell)).$$

However, since from (6.13.2) and Schwarz's inequality

$$\begin{aligned}\sum_{\ell=m}^{\infty}(\ell-m+1)|f(\ell,u(\ell))| &\le \frac{1}{2}\sum_{\ell=m}^{\infty}(\ell-m+1)\ell^{-2}|u(\ell)| \\ &\le \frac{1}{2}\sum_{\ell=m}^{\infty}\ell^{-1}|u(\ell)| \\ &\le \frac{1}{2}\left(\sum_{\ell=m}^{\infty}\ell^{-2}\right)^{1/2}\left(\sum_{\ell=m}^{\infty}|u(\ell)|^2\right)^{1/2} < \infty, \\ &\qquad m\in\mathbb{N}(a)\end{aligned}$$

from (6.13.3) it follows that

$$u(k) = -\sum_{\ell=k}^{\infty}(\ell-k+1)f(\ell,u(\ell)), \quad k\in\mathbb{N}(a). \tag{6.13.4}$$

Therefore, from (6.13.2), we have $|u(k)| \le (1/2)v(k)$, where $v(k) = \sum_{\ell=k}^{\infty}(\ell-k+1)\ell^{-2}|u(\ell)|$ for all $k\in\mathbb{N}(a)$. Obviously, $v(k)\ge 0$ for all $k\in\mathbb{N}(a)$, and $\lim_{k\to\infty}v(k)=0$. If $v(k)=0$ for some $k=k_1\ge a$, then $(\ell-k+1)\ell^{-2}u(\ell)=0$ for all $\ell\ge k_1$ and this means that $u(\ell)=0$ for all $\ell\in\mathbb{N}(k_1)$. In this case the proof is finished. Now we suppose that $v(k)>0$ for all $k\in\mathbb{N}(a)$. Since $\Delta v(k) = -\sum_{\ell=k}^{\infty}\ell^{-2}|u(\ell)|$, and $\Delta^2 v(k) = k^{-2}|u(k)|$, we have

$$\Delta^2 v(k) \le \frac{1}{2}k^{-2}v(k) \quad \text{for all} \quad k\in\mathbb{N}(a). \tag{6.13.5}$$

From the definition of $v(k)$ and Schwarz's inequality, we obtain

$$\begin{aligned}v(k) \le \sum_{\ell=k}^{\infty}\ell^{-1}|u(\ell)| &\le \left(\sum_{\ell=k}^{\infty}\ell^{-2}\right)^{1/2}\left(\sum_{\ell=k}^{\infty}|u(\ell)|^2\right)^{1/2} \\ &\le \frac{1}{\sqrt{k-1}}\left(\sum_{\ell=k}^{\infty}|u(\ell)|^2\right)^{1/2}.\end{aligned}$$

Thus, it follows that

$$w(k) = \sqrt{k-1}\,v(k) \le \left(\sum_{\ell=k}^{\infty}|u(\ell)|^2\right)^{1/2} \quad \text{for all} \quad k\in\mathbb{N}(a). \tag{6.13.6}$$

Hence, we have

$$w(k)\to 0 \quad \text{and} \quad w(k)>0 \quad \text{for all} \quad k\in\mathbb{N}(a). \tag{6.13.7}$$

From the relations

$$(6.13.8) \qquad \Delta^2 w(k) \;=\; \sqrt{k+1}\Delta^2 v(k) + 2\Delta v(k)\Delta\sqrt{k} + v(k)\Delta^2\sqrt{k-1}$$

and

$$(6.13.9) \qquad \Delta v(k) \;=\; \frac{1}{\sqrt{k}}\Delta w(k) + w(k)\Delta\frac{1}{\sqrt{k-1}}$$

we find that

$$\begin{aligned}\Delta\left(\frac{1}{k-1}\Delta w(k)\right) &= \frac{1}{k}\Delta^2 w(k) - \frac{1}{k(k-1)}\Delta w(k)\\ &= \frac{\sqrt{k+1}}{k}\Delta^2 v(k) + \frac{2}{k}\sqrt{k-1}v(k)\Delta\sqrt{k}\Delta\frac{1}{\sqrt{k-1}}\\ &\quad +\frac{1}{k}v(k)\Delta^2\sqrt{k-1} + \left(\frac{2(k-1)}{k\sqrt{k}}\Delta\sqrt{k} - \frac{1}{k}\right)\frac{1}{k-1}\Delta w(k),\end{aligned}$$

which in view of (6.13.5) (6.13.7) gives

$$(6.13.10) \qquad \Delta z(k) \;\le\; \alpha(k) + \beta(k)z(k),$$

where

$$(6.13.11) \qquad z(k) \;=\; \frac{1}{k-1}\Delta w(k)$$

$$(6.13.12) \quad \alpha(k) = \left(\frac{\sqrt{k+1}}{2k^3} + \frac{2}{k}\sqrt{k-1}\Delta\sqrt{k}\Delta\frac{1}{\sqrt{k-1}} + \frac{1}{k}\Delta^2\sqrt{k-1}\right)v(k)$$

and

$$(6.13.13) \qquad \beta(k) \;=\; \frac{2(k-1)}{k\sqrt{k}}\Delta\sqrt{k} - \frac{1}{k}.$$

It is easy to see that

$$(6.13.14) \qquad -\frac{1}{k} \;<\; \beta(k) \;<\; -\frac{1}{k^2}, \quad k \in \mathbb{N}(a).$$

Further, since

$$\alpha(k) \;=\; \frac{1}{k\sqrt{k}}\left(\frac{\sqrt{k+1}}{2k\sqrt{k}} + \frac{-2+\sqrt{k}(\sqrt{k-1}-\sqrt{k+1})}{(\sqrt{k+1}+\sqrt{k})(\sqrt{k}+\sqrt{k-1})}\right)v(k)$$

from the elementary inequalities

$$\frac{1}{k} < \frac{4}{(\sqrt{k+1}+\sqrt{k})(\sqrt{k}+\sqrt{k-1})} \quad\text{and}\quad \frac{\sqrt{k+1}}{\sqrt{k}} - \frac{\sqrt{k}}{\sqrt{k+1}+\sqrt{k-1}} < 1,$$

$$k \in \mathbb{N}(a)$$

it follows that

$$(6.13.15)\quad \alpha(k) < \frac{1}{k\sqrt{k}}\left(\frac{\sqrt{k+1}}{2\sqrt{k}}\frac{4}{(\sqrt{k+1}+\sqrt{k})(\sqrt{k}+\sqrt{k-1})} + \frac{-2+\sqrt{k}(\sqrt{k-1}-\sqrt{k+1})}{(\sqrt{k+1}+\sqrt{k})(\sqrt{k}+\sqrt{k-1})}\right)v(k)$$

$$= \frac{2}{k\sqrt{k}(\sqrt{k+1}+\sqrt{k})(\sqrt{k}+\sqrt{k-1})}\left(\frac{\sqrt{k+1}}{\sqrt{k}} - \frac{\sqrt{k}}{\sqrt{k+1}+\sqrt{k-1}} - 1\right)v(k)$$

$$= 0 \quad \text{for all} \quad k \in \mathbb{N}(a).$$

Thus, from (6.13.10), we find

$$\Delta\left(z(k)\prod_{\ell=a}^{k-1}(1+\beta(\ell))^{-1}\right) \le \alpha(k)\prod_{\ell=a}^{k-1}(1+\beta(\ell))^{-1} < 0 \quad \text{for all} \quad k \in \mathbb{N}(a),$$

i.e. $z(k)\prod_{\ell=a}^{k-1}(1+\beta(\ell))^{-1}$ is decreasing for all $k \in \mathbb{N}(a)$.

If $z(k)\prod_{\ell=a}^{k-1}(1+\beta(\ell))^{-1} > 0$ for all $k \in \mathbb{N}(a)$, then $z(k) > 0$ for all $k \in \mathbb{N}(a)$, and from (6.13.11) we find $\Delta w(k) > 0$ for all $k \in \mathbb{N}(a)$, and hence $w(k)$ is increasing, but this contradicts (6.13.7). If there exists an integer $K \ge a$ such that $z(K)\prod_{\ell=a}^{K-1}(1+\beta(\ell))^{-1} = p < 0$, then $z(k)\prod_{\ell=a}^{k-1}(1+\beta(\ell))^{-1} < p$ for all $k \in \mathbb{N}(K+1)$, i.e. $z(k) < p\prod_{\ell=a}^{k-1}(1+\beta(\ell))$. However, since $1+\beta(\ell) > (\ell-1)/\ell$ it follows that $z(k) < p(a-1)/(k-1)$, and hence from (6.13.11), we find $\Delta w(k) < p(a-1)$, i.e. $w(k) < w(K+1) + p(a-1)(k-K-1)$ for all $k \in \mathbb{N}(K+2)$. But, this implies that $w(k) \to -\infty$, and again we get a contradiction to (6.13.7).

Combining the above arguments, we find that our assumption $v(k) > 0$ for all $k \in \mathbb{N}(a)$ is not correct, and this completes the proof. ∎

**Theorem 6.13.2.** Let for all $(k,u) \in \mathbb{N}(a) \times \mathbb{R}$ the function $f(k,u)$ be defined and

$$(6.13.16)\qquad |f(k,u)| \le k^{-q}|u|, \quad q > 5/2.$$

Then, if $u(k) \in c_0$ is a solution of (6.13.1) then there exists an integer $k_1 \ge a$ $(a \ge 4)$ such that $u(k) = 0$ for all $k \in \mathbb{N}(k_1)$.

**Proof.** Let $u(k)$ be a solution of (6.13.1) such that $\lim_{k\to\infty}|u(k)| = 0$. Then, $\lim_{k\to\infty}\Delta u(k) = \lim_{k\to\infty}\Delta^2 u(k) = 0$. Thus, for this solution also the relation (6.13.3) holds. Further, since there exists a constant $c > 0$ such that $|u(k)| \le c$ for all $k \in \mathbb{N}(a)$, we find that

$$\sum_{\ell=m}^{\infty}(\ell - m + 1)|f(\ell, u(\ell))| \le \sum_{\ell=m}^{\infty} \ell^{1-q}|u(\ell)| \le c\left[m^{1-q} + \frac{m^{2-q}}{2-q}\right] < \infty$$

for all $m \in \mathbb{N}(a)$. Therefore, this solution also has the representation (6.13.4). Now as in Theorem 6.13.1 we define $\overline{v}(k) = \sum_{\ell=k}^{\infty}(\ell - k + 1)\ell^{-q}|u(\ell)|$, $\overline{w}(k) = \sqrt{k-1}\,\overline{v}(k)$, $\overline{z}(k) = \frac{1}{k-1}\Delta\overline{w}(k)$, $\overline{\alpha}(k) = \left(\frac{\sqrt{k+1}}{k^{1+q}} + \frac{2}{k}\sqrt{k-1}\Delta\sqrt{k}\Delta\frac{1}{\sqrt{k-1}} + \frac{1}{k}\Delta^2\sqrt{k-1}\right)\overline{v}(k)$, $\overline{\beta}(k) = \frac{2(k-1)}{k\sqrt{k}}\Delta\sqrt{k} - \frac{1}{k}$, and apply similar analysis to see that there exists a positive integer $k_1$ such that $u(k) = 0$ for all $k \in \mathbb{N}(k_1)$. ∎

## 6.14. Oscillation and Nonoscillation for

$$\Delta_\alpha^2 u(k) = f(k, u(k), \Delta_\beta u(k)), \quad k \in \mathbb{N} \tag{6.14.1}$$

where $\alpha$ and $\beta$ are real fixed constants, $\Delta_\alpha u(k) = u(k+1) - \alpha u(k)$, $\Delta_\alpha^2 u(k) = \Delta_\alpha[\Delta_\alpha u(k)]$, $\Delta_\beta u(k) = u(k+1) - \beta u(k)$, and $f$ is defined on $\mathbb{N} \times \mathbb{R}^2$.

**Theorem 6.14.1.** Let $\alpha > 0$ and $S = \mathbb{N} \times \{(u, v) \in \mathbb{R}^2 : v + (\beta - \alpha)u = 0\}$. Further, let

(i) $f(k, u, v) = 0$ if $(k, u, v) \in S$

(ii) $f(k, u, v)[v + (\beta - \alpha)u] + \alpha[v + (\beta - \alpha)u]^2 > 0$ if $(k, u, v) \in \mathbb{N} \times \mathbb{R}^2 \backslash S$.

Then, the difference equation (6.14.1) is nonoscillatory.

**Proof.** We observe that $(k, u(k), \Delta_\beta u(k)) \in S$ is equivalent to $u(k+1) - \alpha u(k) = 0$. Therefore, if the solution $u(k)$ of (6.14.1) is such that for a fixed $k_1 \in \mathbb{N}$, $(k_1, u(k_1), \Delta_\beta u(k_1)) \in S$, then from the hypothesis (i), it follows that $\Delta_\alpha^2 u(k_1) = 0$. However, since

$$\Delta_\alpha^2 u(k_1) = \Delta_\alpha u(k_1 + 1) - \alpha\Delta_\alpha u(k_1) = u(k_1 + 2) - \alpha u(k_1 + 1) = 0$$

inductively, we have $u(k_1 + \ell) - \alpha u(k_1 + \ell - 1) = 0$, $\ell \in \mathbb{N}(1)$, and hence $u(\ell) = \alpha^{\ell - k_1}u(k_1)$, $\ell \in \mathbb{N}(k_1)$. This solution is of course nonoscillatory.

Now let $u(k)$ be a solution of (6.14.1) such that for any $k \in \mathbb{N}$, $(k, u(k), \Delta_\beta u(k)) \notin S$, and this solution is oscillatory. Then, there exists a $k_2 \in \mathbb{N}$ such that $u(k_2) > 0$, $u(k_2 + 1) \le 0$ and hence $\Delta_\alpha u(k_2) < 0$.

Setting $k = k_2$ in (6.14.1) and multiplying the resulting equation by $\Delta_\alpha u(k_2)$ gives

$$\begin{aligned}\Delta_\alpha u(k_2)\Delta_\alpha u(k_2+1) &= f\left(k_2, u(k_2), \Delta_\beta u(k_2)\right)[\Delta_\beta u(k_2) \\ &\quad +(\beta-\alpha)u(k_2)] + \alpha[\Delta_\beta u(k_2) + (\beta-\alpha)u(k_2)]^2.\end{aligned}$$

Therefore, from the hypothesis (ii), we find that $\Delta_\alpha u(k_2)\Delta_\alpha u(k_2+1) > 0$, and hence $\Delta_\alpha u(k_2+1) < 0$. Repeating this reasoning we get $\Delta_\alpha u(k) < 0$ for all $k \in \mathbb{N}(k_2)$. This implies that $u(k) < 0$ for all $k \in \mathbb{N}(k_2+2)$, which contradicts our assumption. The proof for the case $u(k_2) \geq 0$, $u(k_2+1) < 0$ is similar. ∎

**Theorem 6.14.2.** Let $\alpha > 0$ and $T = \mathbb{N} \times \left\{(u,v) \in \mathbb{R}^2 : v + \beta u = 0\right\}$. Further, let

$$f(k,u,v)(v+\beta u) + \alpha(v+\beta u)[v + (\beta-\alpha)u] \ > \ 0 \qquad \text{if } \ (k,u,v) \in \mathbb{N} \times \mathbb{R}^2 \backslash T.$$

Then, the difference equation (6.14.1) is nonoscillatory.

**Proof.** We observe that $(k, u(k), \Delta_\beta u(k)) \in T$ is equivalent to $u(k+1) = 0$. Since we consider only nontrivial solution, there exists a $k_1 \in \mathbb{N}$ such that $u(k_1+1) = \Delta_\beta u(k_1) + \beta u(k_1) \neq 0$. Setting $k = k_1$ in (6.14.1) and multiplying the resulting equation by $u(k_1+1)$ gives

$$\begin{aligned}u(k_1+1)\Delta_\alpha u(k_1+1) &= f\left(k_1, u(k_1), \Delta_\beta u(k_1)\right) u(k_1+1) \\ &\qquad\qquad +\alpha u(k_1+1)\Delta_\alpha u(k_1) \\ &= f\left(k_1, u(k_1), \Delta_\beta u(k_1)\right)\left(\Delta_\beta u(k_1) + \beta u(k_1)\right) \\ &\quad +\alpha\left(\Delta_\beta u(k_1) + \beta u(k_1)\right)\left[\Delta_\beta u(k_1) + (\beta-\alpha)u(k_1)\right].\end{aligned}$$

Hence, from the given hypothesis it follows that $u(k_1+1)\Delta_\alpha u(k_1+1) > 0$. If $u(k_1+1) > 0$, then $\Delta_\alpha u(k_1+1) > 0$ implies $u(k_1+2) > \alpha u(k_1+1) > 0$. Repeating the above reasoning we obtain $\Delta_\alpha u(k) > 0$ for all $k \in \mathbb{N}(k_1+1)$, and from this $u(k) > \alpha^{k-k_1-1}u(k_1+1) > 0$ for all $k \in \mathbb{N}(k_1+2)$. This solution is positive and therefore nonoscillatory. A similar proof holds for $u(k_1+1) < 0$. ∎

**Theorem 6.14.3.** Let $\alpha = \beta = 1$ and

$$f(k,u,v)(u+v) \ \geq \ 0 \quad \text{ if } \quad (k,u,v) \in \mathbb{N} \times \mathbb{R}^2.$$

Then, the difference equation (6.14.1) is nonoscillatory.

**Proof.** Suppose there exists an oscillatory solution $u(k)$ of (6.14.1). Then, there exist $k_1,\ k_2 \in \mathbb{N}$ such that $u(k_1) \leq 0$, $u(k_2) \geq 0$, $\mathbb{N}(k_1+1, k_2-1)$

is nonempty and finite, and there exists a $\ell \in \mathbb{N}(k_1+1, k_2-1)$ such that $u(\ell) > 0$ (the case $u(\ell) < 0$ can be considered similarly) and simultaneously $u(\ell) > (\geq) u(\ell+1)$, $u(\ell) \geq (>0) u(\ell-1)$. Thus, $\Delta^2 u(\ell-1) = \Delta u(\ell) - \Delta u(\ell-1) < 0$. But, setting $k = \ell - 1$ in (6.14.1) and multiplying the resulting equation by $u(\ell)$ gives

$$u(\ell)\Delta^2 u(\ell-1) = u(\ell) f(\ell-1, u(\ell-1), \Delta u(\ell-1))$$

and hence, from the given hypothesis we have $u(\ell)\Delta^2 u(\ell-1) > 0$, i.e. $\Delta^2 u(\ell-1) > 0$. This contradiction completes the proof. ∎

**Theorem 6.14.4.** Let in Theorem 6.14.1 the inequality sign $>$ be replaced by $<$ at both the places. Then, the difference equation (6.14.1) is oscillatory.

**Proof.** Suppose $u(k)$ is a nonoscillatory solution of (6.14.1) which is positive for all $k \in \mathbb{N}(a)$, where $a \in \mathbb{N}$. If there is some $k_1 \in \mathbb{N}(a)$ so that $(k_1, u(k_1), \Delta_\beta u(k_1)) \in S$, then as in Theorem 6.14.1, we find that $u(\ell) = \alpha^{\ell - k_1} u(k_1)$, $\ell \in \mathbb{N}(k_1)$. However, since $\alpha < 0$ this solution is oscillatory. Thus, for all $k \in \mathbb{N}(a)$, $(k, u(k), \Delta_\beta u(k)) \in \mathbb{N} \times \mathbb{R}^2 \backslash S$. But, then for all $k \in \mathbb{N}(a)$

$$\Delta_\alpha u(k+1)\Delta_\alpha u(k) = \Delta_\alpha u(k) f(k, u(k), \Delta_\beta u(k)) + \alpha(\Delta_\alpha u(k))^2 < 0,$$

i.e. $\Delta_\alpha u(k+1)\Delta_\alpha u(k) = \alpha^{k+2}\Delta\left(u(k+1)/\alpha^{k+1}\right)\alpha^{k+1}\Delta(u(k)/\alpha^k) < 0$. Thus, if $k$ is even then $\Delta\left(u(k+1)/\alpha^{k+1}\right)\Delta(u(k)/\alpha^k) > 0$. If $\Delta(u(k)/\alpha^k) > 0$, then $u(k+1)/\alpha^{k+1} > u(k)/\alpha^k > 0$. Therefore, $u(k+1) < 0$ and we obtain a contradiction. Hence, it turns out to be that $\Delta(u(k)/\alpha^k) < 0$. Then, $\Delta\left(u(k+1)/\alpha^{k+1}\right) < 0$, i.e. $u(k+2)/\alpha^{k+2} < u(k+1)/\alpha^{k+1} < 0$ which implies that $u(k+2) < 0$. This contradiction completes the proof. ∎

**Theorem 6.14.5.** Let in Theorem 6.14.2 the inequality sign $>$ be replaced by $<$ at both the places. Then, the difference equation (6.14.1) is oscillatory.

**Proof.** Similar reasoning as in the proof of Theorem 6.14.2 gives us $u(k_1+1)\Delta_\alpha u(k_1+1) = u(k_1+1)u(k_1+2) - \alpha u^2(k_1+1) < 0$. But this inequality holds only for an oscillatory solution. ∎

## 6.15. Oscillation and Nonoscillation for

$$\text{(6.15.1)} \qquad \Delta(r(k)\Delta u(k)) + f(k)F(k, u(k), \Delta u(k)) = g(k, u(k), \Delta u(k)), \quad k \in \mathbb{N}(a)$$

where the functions $r$, $f$, $F$ and $g$ are defined in their domain of definition. Further, $r(k) > 0$ for all $k \in \mathbb{N}(a)$, and $R_{a,k} \to \infty$, where $R_{j,k} = \sum_{\ell=j}^{k-1} 1/r(\ell)$, $j \in \mathbb{N}(a)$, $k \in \mathbb{N}(j+1)$.

For the difference equation (6.15.1) each result we shall prove will require some of the following conditions:

($c_1$) $f(k) \geq 0$ for all $k \in \mathbb{N}(a)$

($c_2$) there exists a constant $M_1$ such that $F(k,u,v) \geq M_1$

($c_3$) there exists a constant $M_2$ such that $F(k,u,v) \leq M_2$

($c_4$) there exists a constant $M > 0$ such that $|F(k,u,,v)| \leq M$

($c_5$) there exists a function $\phi(k)$ such that $g(k,u,v) \geq \phi(k)$

($c_6$) there exists a function $\psi(k)$ such that $g(k,u,v) \leq \psi(k)$

($c_7$) $F(k,u,v)$ is bounded from above if $u$ is bounded

($c_8$) $F(k,u,v)$ is bounded from below if $u$ is bounded

($c_9$) $uF(k,u,v) \geq 0$

($c_{10}$) $uF(k,u,v) \leq 0$

($c_{11}$) there exist functions $p(k)$ and $q(k)$ such that $p(k) \leq F(k,u,v) \leq q(k)$.

**Theorem 6.15.1.** Suppose that conditions ($c_1$), ($c_3$) and ($c_5$) hold and for every constant $c > 0$

$$\liminf_{k\to\infty} \left[\sum_{\ell=a}^{k-1} \frac{1}{r(\ell)} \sum_{j=a}^{\ell-1} (\phi(j) - M_2 f(j)) - cR_{a,k}\right] > 0. \tag{6.15.2}$$

Then, all solutions of (6.15.1) are eventually positive.

**Proof.** Let $u(k)$ be a solution of (6.15.1). Applying conditions ($c_1$), ($c_3$) and ($c_5$), we obtain

$$\Delta(r(k)\Delta u(k)) \geq \phi(k) - M_2 f(k), \quad k \in \mathbb{N}(a).$$

Therefore, it follows that

$$u(k) \geq u(a) + r(a)\Delta u(a) \sum_{\ell=a}^{k-1} \frac{1}{r(\ell)} + \sum_{\ell=a}^{k-1} \frac{1}{r(\ell)} \sum_{j=a}^{\ell-1} (\phi(j) - M_2 f(j)).$$

Now in view of the conditions on $r(k)$, there exist constants $c > 0$ and $k_1 \in \mathbb{N}(a)$ such that

$$-cR_{a,k} \leq u(a) + r(a)\Delta u(a) \sum_{\ell=a}^{k-1} \frac{1}{r(\ell)} \leq cR_{a,k}, \quad k \in \mathbb{N}(k_1).$$

Hence, from (6.15.2) we have

$$\liminf_{k\to\infty} u(k) \;\geq\; \liminf_{k\to\infty}\left[\sum_{\ell=a}^{k-1}\frac{1}{r(\ell)}\sum_{j=a}^{\ell-1}(\phi(j)-M_2 f(j)) - cR_{a,k}\right] \;>\; 0.$$

Thus, $u(k)$ is eventually positive. ∎

**Remark 6.15.1** If we replace (6.15.2) in Theorem 6.15.1 by the stronger condition $\sum^{\infty}(\phi(\ell)-M_2 f(\ell))=\infty$, then every solution $u(k)$ of (6.15.1) satisfies $\lim_{k\to\infty} u(k) = \infty$ monotonically. Indeed, then from (6.15.1) we obtain $r(k)\Delta u(k) \geq r(a)\Delta u(a) + \sum_{\ell=a}^{k-1}(\phi(\ell) - M_2 f(\ell)) \to \infty$ as $k\to\infty$. Thus, there exists a $k_1 \in \mathbb{N}(a)$ such that $\Delta u(k) \geq 1/r(k)$ for all $k \in \mathbb{N}(k_1)$, from which the conclusion follows.

The proofs of the following results are similar to that of Theorem 6.15.1 and therefore are omitted.

**Theorem 6.15.2.** Suppose that conditions $(c_1)$, $(c_2)$ and $(c_6)$ hold and for every constant $c>0$

$$\limsup_{k\to\infty}\left[\sum_{\ell=a}^{k-1}\frac{1}{r(\ell)}\sum_{j=a}^{\ell-1}(\psi(j)-M_1 f(j)) + cR_{a,k}\right] \;<\; 0. \tag{6.15.3}$$

Then, all solutions of (6.15.1) are eventually negative.

**Remark 6.15.2.** If in Theorem 6.15.2 we replace (6.15.3) by the condition $\sum^{\infty}(\psi(\ell)-M_1 f(\ell)) = -\infty$, then every solution of (6.15.1) satisfies $\lim_{k\to\infty} u(k) = -\infty$ monotonically.

**Theorem 6.15.3.** Suppose that conditions $(c_1)$, $(c_5)$ and $(c_7)$ hold and for all constants $c_1,\ c_2>0$

$$\liminf_{k\to\infty}\left[\sum_{\ell=a}^{k-1}\frac{1}{r(\ell)}\sum_{j=a}^{\ell-1}(\phi(j)-c_1 f(j)) - c_2 R_{a,k}\right] \;>\; 0. \tag{6.15.4}$$

Then, all bounded solutions of (6.15.1) are eventually positive.

**Theorem 6.15.4.** Suppose that conditions $(c_1)$, $(c_6)$ and $(c_8)$ hold and for all constants $c_1,\ c_2>0$

$$\limsup_{k\to\infty}\left[\sum_{\ell=a}^{k-1}\frac{1}{r(\ell)}\sum_{j=a}^{\ell-1}(\psi(j)+c_1 f(j)) + c_2 R_{a,k}\right] \;<\; 0. \tag{6.15.5}$$

Then, all bounded solutions of (6.15.1) are eventually negative.

**Theorem 6.15.5.** Suppose that conditions $(c_4)$ and $(c_5)$ hold and for every constant $c > 0$

$$\liminf_{k\to\infty}\left[\sum_{\ell=a}^{k-1}\frac{1}{r(\ell)}\sum_{j=a}^{\ell-1}(\phi(j) - M|f(j)|) - cR_{a,k}\right] > 0. \tag{6.15.6}$$

Then, all solution of (6.15.1) are eventually positive.

**Theorem 6.15.6.** Suppose that conditions $(c_4)$ and $(c_6)$ hold and for every constant $c > 0$

$$\limsup_{k\to\infty}\left[\sum_{\ell=a}^{k-1}\frac{1}{r(\ell)}\sum_{j=a}^{\ell-1}(\psi(j) + M|f(j)|) + cR_{a,k}\right] < 0. \tag{6.15.7}$$

Then, all solutions of (6.15.1) are eventually negative.

**Remark 6.15.3.** Replacing (6.15.6) and (6.15.7) by $\sum^{\infty}(\phi(\ell)-M|f(\ell)|) = \infty$ and $\sum^{\infty}(\psi(\ell)+M|f(\ell)|) = -\infty$ respectively, yield analogous results to those in Remarks 6.15.1 and 6.15.2.

**Theorem 6.15.7.** Suppose that conditions $(c_5)$, $(c_7)$ and $(c_8)$ hold and for all constants $c_1,\ c_2 > 0$

$$\liminf_{k\to\infty}\left[\sum_{\ell=a}^{k-1}\frac{1}{r(\ell)}\sum_{j=a}^{\ell-1}(\phi(j) - c_1|f(j)|) - c_2R_{a,k}\right] > 0. \tag{6.15.8}$$

Then, all bounded solutions of (6.15.1) are eventually positive.

**Theorem 6.15.8.** Suppose that conditions $(c_6)$ $(c_8)$ hold and for all constants $c_1,\ c_2 > 0$

$$\limsup_{k\to\infty}\left[\sum_{\ell=a}^{k-1}\frac{1}{r(\ell)}\sum_{j=a}^{\ell-1}(\psi(j) + c_1|f(j)|) + c_2R_{a,k}\right] < 0. \tag{6.15.9}$$

Then, all bounded solutions of (6.15.1) are eventually negative.

**Theorem 6.15.9.** Suppose that $f(k) \equiv 1$ and conditions $(c_5)$, $(c_6)$ and $(c_{11})$ hold. Further, let for every constant $c > 0$ and all large $s \in \mathbb{N}(a)$

$$\liminf_{k\to\infty}\left[\sum_{\ell=s}^{k-1}\frac{1}{r(\ell)}\sum_{j=s}^{\ell-1}(\psi(j) - p(j)) + cR_{s,k}\right] < 0 \tag{6.15.10}$$

and

$$\limsup_{k\to\infty}\left[\sum_{\ell=s}^{k-1}\frac{1}{r(\ell)}\sum_{j=s}^{\ell-1}(\phi(j)-q(j))-cR_{s,k}\right] > 0. \tag{6.15.11}$$

Then, the difference equation (6.15.1) is oscillatory.

**Proof.** Let $u(k)$ be a nonoscillatory solution of (6.15.1), say $u(k) > 0$ for all $s \le k \in \mathbb{N}(a)$. Then, from (6.15.1) we have

$$\phi(k) - q(k) \le \Delta(r(k)\Delta u(k)) \le \psi(k) - p(k), \quad k \in \mathbb{N}(s).$$

Now following as in Theorem 6.15.1, we obtain

$$\sum_{\ell=s}^{k-1}\frac{1}{r(\ell)}\sum_{j=s}^{\ell-1}(\phi(j)-q(j)) - cR_{s,k} \le u(k)$$
$$\le cR_{s,k} + \sum_{\ell=s}^{k-1}\frac{1}{r(\ell)}\sum_{j=s}^{\ell-1}(\psi(j)-p(j)).$$

Condition (6.15.10) then yields a contradiction to the assumption that $u(k) > 0$ for all $k \in \mathbb{N}(s)$. A similar proof holds if $u(k) < 0$ for all $k \in \mathbb{N}(s)$. ∎

**Theorem 6.15.10.** Suppose that $f(k) \equiv 1$ and conditions $(c_5)$, $(c_6)$ and $(c_9)$ hold. Further, let for every constant $c > 0$ and all large $s \in \mathbb{N}(a)$

$$\liminf_{k\to\infty}\left[\sum_{\ell=s}^{k-1}\frac{1}{r(\ell)}\sum_{j=s}^{\ell-1}\psi(j)+cR_{s,k}\right] < 0$$

and

$$\limsup_{k\to\infty}\left[\sum_{\ell=s}^{k-1}\frac{1}{r(\ell)}\sum_{j=s}^{\ell-1}\phi(j)-cR_{s,k}\right] > 0.$$

Then, the difference equation (6.15.1) is oscillatory.

**Theorem 6.15.11.** Suppose that $f(k) \equiv 1$ and conditions $(c_5)$, $(c_6)$ and $(c_{10})$ hold. Further, let for every constant $c > 0$ and large $s \in \mathbb{N}(a)$,

$$\liminf_{k\to\infty}\left[\sum_{\ell=s}^{k-1}\frac{1}{r(\ell)}\sum_{j=s}^{\ell-1}\psi(j)+cR_{s,k}\right] = -\infty$$

and

$$\limsup_{k\to\infty}\left[\sum_{\ell=s}^{k-1}\frac{1}{r(\ell)}\sum_{j=s}^{\ell-1}\phi(j) - cR_{s,k}\right] = \infty.$$

Then, all bounded solutions of the difference equation (6.15.1) are oscillatory.

## 6.16. Variety of Properties of Solutions of

$$\Delta^4 u(k-2) = p(k)u(k), \quad k \in \mathbb{N}(2) \tag{6.16.1}$$

where the function $p$ is defined and positive on $\mathbb{N}(2)$. We begin with the following definition which generalizes the concept of node introduced in Definition 1.8.1 and is due to Hartman [51].

**Definition 6.16.1.** Let $u(k)$ be a function defined on $\mathbb{N}$, we say $k \in \mathbb{N}$ is a *generalized zero* for $u(k)$ if one of the following holds:

(6.16.2) $u(k) = 0$

(6.16.3) $k \in \mathbb{N}(1)$ and $u(k-1)u(k) < 0$, $k \in \mathbb{N}(1)$, and there exists an integer $m$, $1 < m \le k$ such that

(6.16.4) $(-1)^m u(k-m)u(k) > 0$, and $u(j) = 0$ for all $j \in \mathbb{N}(k-m+1, k-1)$.

A generalized zero for $u(k)$ is said to be of order $0$, $1$, or $m > 1$, according to whether condition (6.16.2), (6.16.3), or (6.16.4), respectively, holds. In particular, a generalized zero of order $0$ will simply be called a zero, and a generalized zero of order one will again be called a node.

Obviously, if $u(a) = u(a+1) = u(a+2) = u(a+3) = 0$ for some $a \in \mathbb{N}$, then $u(k) \equiv 0$ is the only solution of (6.16.1). Thus, a nontrivial solution of (6.16.1) can have zeros at no more than three consecutive values of $k$. In Theorem 6.16.1 we shall show that a nontrivial solution of (6.16.1) cannot have a generalized zero of order $m > 3$. However, a solution of (6.16.1) can have arbitrarily many consecutive nodes, as it is clear from $u(k) = (-1)^k$, which is a solution of $\Delta^4 u(k-2) = 16u(k)$.

The following properties of the solutions of (6.16.1) are fundamental and will be used subsequently.

$(S_1)$ If $u(k)$ is a nontrivial solution of (6.16.1) and if

$$\begin{array}{ll}\text{(a)}\ u(k) \ge 0, & \text{(b)}\ \Delta u(k) \ge 0, \\ \text{(c)}\ \Delta^2 u(k-1) \ge 0, & \text{(d)}\ \Delta^3 u(k-2) \ge 0\end{array} \tag{6.16.5}$$

for some $k = a \in \mathbb{N}(2)$, then (6.16.5) holds for all $k \in \mathbb{N}(a)$, with strict inequality in (6.16.5a) for all $k \in \mathbb{N}(a+2)$, strict inequality in (6.16.5b) for all $k \in \mathbb{N}(a+1)$, and strict inequality in (6.16.5c) and (6.16.5d) for all $k \in \mathbb{N}(a+3)$. Furthermore,

$$\Delta^4 u(k-2) \geq 0 \quad \text{for all} \quad k \in \mathbb{N}(a) \tag{6.16.6}$$

with strict inequality for all $k \in \mathbb{N}(a+2)$, and $u(k)$, $\Delta u(k)$, and $\Delta^2 u(k)$ all tend to $\infty$ as $k \to \infty$.

$(S_2)$ If $u(k)$ is a nontrivial solution of (6.16.1) and if

$$\begin{array}{ll} \text{(a)}\ u(k) \geq 0, & \text{(b)}\ \Delta u(k) \geq 0, \\ \text{(c)}\ \Delta^2 u(k) \geq 0, & \text{(d)}\ \Delta^3 u(k) \geq 0 \end{array} \tag{6.16.7}$$

for some $k = a \in \mathbb{N}$, then (6.16.7) holds for all $k \in \mathbb{N}(a)$, with strict inequality in (6.16.7a,b,d) for all $k \in \mathbb{N}(a+3)$, and in (6.16.7c) for all $k \in \mathbb{N}(a+4)$. Furthermore,

$$\Delta^4 u(k) \geq 0 \quad \text{for all} \quad k \in \mathbb{N}(a) \tag{6.16.8}$$

with strict inequality for all $k \in \mathbb{N}(a+2)$, and $u(k)$, $\Delta u(k)$, and $\Delta^2 u(k)$ all tend to $\infty$ as $k \to \infty$.

$(S_3)$ If $u(k)$ is a nontrivial solution of (6.16.1) and if

$$\begin{array}{ll} \text{(a)}\ u(k) \geq 0, & \text{(b)}\ \Delta u(k-1) \leq 0, \\ \text{(c)}\ \Delta^2 u(k-1) \geq 0, & \text{(d)}\ \Delta^3 u(k-1) \leq 0 \end{array} \tag{6.16.9}$$

for some $k = a \in \mathbb{N}(3)$, then (6.16.9) holds for all $k \in \mathbb{N}(2, a)$, and

$$\Delta^4 u(k-2) \geq 0 \quad \text{for all} \quad k \in \mathbb{N}(2, a). \tag{6.16.10}$$

Furthermore, $u(0) > u(1) > 0$, and $\Delta u(0) < 0$. Strict inequality holds in (6.16.9a) and (6.16.10) for all $k \in \mathbb{N}(2, a-2)$ if $a \in \mathbb{N}(4)$, in (6.16.9b) for all $k \in \mathbb{N}(2, a-1)$, and in (6.16.9c,d) for all $k \in \mathbb{N}(2, a-3)$ if $a \in \mathbb{N}(5)$.

$(S_4)$ Let $a \in \mathbb{N}(2)$. If $u(k)$ is a solution of (6.16.1) with $u(a) = 0$, $u(a-1) \geq 0$, $u(a+1) \geq 0$, $u(a-1)$ and $u(a+1)$ not both zero, then at least one of the following conditions must be true: (i) Either $u(k) > 0$ for all $k \in \mathbb{N}(a+2)$, or (ii) $u(k) < 0$ for all $k \in \mathbb{N}(0, a-1)$. In particular, $u(k)$ cannot have generalized zeros of any order at both $\alpha$ and $\beta$, where $\alpha \in \mathbb{N}(0, a-1)$ and $\beta \in \mathbb{N}(a+2)$. An analogous statement holds for the hypotheses $u(a-1) \leq 0$ and $u(a+1) \leq 0$.

**Theorem 6.16.1.** If $u(k)$ is a nontrivial solution of (6.16.1) with zeros at three consecutive values of $k$, say $a$, $a+1$ and $a+2$, then $u(k)$ has no other generalized zeros. If $u(a+3) > 0$ $(< 0)$, then $\Delta u(k) \geq 0$ $(\leq 0)$ for all $k$, and the inequality is strict if $k \in \mathbb{N}(a+2)$ or $k \in \mathbb{N}(0, a-1)$. In particular, if $\alpha \in \mathbb{N}(0, a-1)$ and $\beta \in \mathbb{N}(a+3)$, then $u(\alpha)u(\beta) < 0$.

**Proof.** Clearly $\Delta u(a) = \Delta^2 u(a) = 0$. Since the solution $u(k)$ is nontrivial, we may assume that $u(a+3) > 0$. Thus, $\Delta^3 u(a) > 0$ and by $(S_2)$, $u(k)$ is positive and strictly increasing on $\mathbb{N}(a+3)$. Next, let $v(k) = -u(k)$. Then, $v(a+1) = 0$, $\Delta v(a) = 0$, $\Delta^2 v(a) = 0$ and $\Delta^3 v(a) < 0$. If $a \in \mathbb{N}(2)$, then $(S_3)$ implies that $v(k)$ is positive and strictly decreasing on $\mathbb{N}(0, a)$. Thus, $u(k)$ is negative and strictly increasing on $\mathbb{N}(0, a)$. If $a = 1$, then we again assume that $u(a+3) = u(4) > 0$. Then, by (6.10.1), $\Delta^4 u(0) = p(2)u(2) = 0$. But, $\Delta^4 u(0) = u(4) + u(0)$, so $u(0) = -u(4) < 0$ and $\Delta u(0) = u(1) - u(0) > 0$, as claimed. If $a = 0$, then the part of the conclusion concerning $k \leq a-1$ is empty. This completes the proof. ∎

**Theorem 6.16.2.** Let $a \in \mathbb{N}(1)$, and suppose that $u(k)$ is a solution of (6.16.1) with $u(a) = 0$, $u(a+1) = 0$, $u(a+2) \neq 0$, but $a+2$ is a generalized zero for $u(k)$. Then, $u(k)$ has no other generalized zeros. If $u(a+2) > 0$ $(< 0)$, then $\Delta u(k) \geq 0$ $(\leq 0)$ for all $k \in \mathbb{N}$, with strict inequality for all $k \in \mathbb{N}(a+2)$ or $k \in \mathbb{N}(0, a-1)$. In particular, if $\alpha \in \mathbb{N}(0, a-1)$ and $\beta \in \mathbb{N}(a+2)$, then $u(\alpha)u(\beta) < 0$.

**Proof.** Since $u(a+2) \neq 0$, we can assume that $u(a+2) > 0$. Since $u(a) = u(a+1) = 0$, $a+2$ cannot be a generalized zero of order 1 or 2, and Theorem 6.16.1 implies that the order cannot be greater than 3. Thus, $a+2$ is a generalized zero of order 3, which implies that $u(a-1) < 0$. Now since from (6.16.1), we have $u(a+3) - 4u(a+2) + 6u(a+1) - 4u(a) + u(a-1) = p(a+1)u(a+1)$, or $u(a+3) = 4u(a+2) - u(a-1)$, it follows that $\Delta^3 u(a) = u(a+3) - 3u(a+2) + 3u(a+1) - u(a) = 4u(a+2) - u(a-1) - 3u(a+2) + 3u(a+1) - u(a) = u(a+2) - u(a-1) > 0$. Clearly, $\Delta^2 u(a) > 0$, $\Delta u(a) = 0$ and $u(a) = 0$, thus by $(S_2)$, $u(k)$ is positive and strictly increasing on $\mathbb{N}(a+3)$. For $k \in \mathbb{N}(0, a)$, let $v(k) = -u(k)$. Then, $v(a) = 0$, $\Delta v(a-1) < 0$, $\Delta^2 v(a-1) > 0$ and $\Delta^3 v(a-1) < 0$. If $a \in \mathbb{N}(3)$, then as in Theorem 6.16.1, $(S_3)$ yields the result. If $a = 2$, then $u(2) = u(3) = 0$, $u(1) < 0$, $u(4) > 0$, and $\Delta u(1) > 0$. By (6.16.1), we have $\Delta^4 u(0) = p(2)u(2) = 0$. But, $\Delta^4 u(0) = u(4) - 4u(3) + 6u(2) - 4u(1) + u(0) = u(4) - 4u(1) + u(0)$, and so $4u(1) - u(0) = u(4) > 0$. Hence, $u(0) < 4u(1) < 0$, and $u(0) - u(1) < 3u(1) < 0$. Therefore, $u(0) < 0$ and $\Delta u(0) > 0$, as claimed. If $a = 1$, then $u(1) = u(2) = 0$, $u(3) \neq 0$, and $a+2 = 3$ is a generalized zero. It follows from the definition of a generalized zero that this must be a generalized zero of order 3, so that if $u(3) > 0$ then $u(0) < 0$. Hence, $\Delta u(0) > 0$, which completes the proof. ∎

**Corollary 6.16.3.** If $u(k)$ is a nontrivial solution of (6.16.1) with generalized zeros at $\alpha$ and $\beta$ and a zero at $a$, where $\alpha+1 < a < \beta-1$, then $u(a-1)u(a+1) < 0$. In particular, $u(k)$ does not have a generalized zero at $a+1$.

**Proof.** Since $\alpha + 1 < a < \beta - 1$, from Theorem 6.16.1 it follows that $u(a+1)$ and $u(a-1)$ both cannot be zero. If $u(a+1)u(a-1) \geq 0$, then $(S_4)$ implies that $u(k)$ cannot have generalized zeros at both $\alpha$ and $\beta$, which is a contradiction. Thus, $u(a-1)u(a+1) < 0$. ∎

**Corollary 6.16.4.** If $u(k)$ is a nontrivial solution of (6.16.1) with $u(\alpha) = u(a) = u(\beta) = 0$, where $\alpha < a < \beta - 1$, then $u(a+1) \neq 0$.

**Corollary 6.16.5.** If a nontrivial solution $u(k)$ of (6.16.1) has a zero at $\alpha$ and a generalized zero at $\beta$, where $\alpha < \beta$, then $u(k)$ cannot have consecutive zeros at $a, a+1$ where $\alpha < a < \beta - 1$.

**Remark 6.16.1.** Corollary 6.16.5 says that if a solution $u(k)$ of (6.16.1) has four or more zeros, then no two zeros can occur at consecutive values of $k$, unless they are the first two zeros or the last two zeros. For example, consider the function $u(k) = \{-4, 0, 0, -1, 0, 1, 0, 0, 4, 15, \cdots\}$ which is a solution of $\Delta^4 u(k-2) = 5u(k)$ with $u(1) = -4$. This solution is positive and increasing for all $k \in \mathbb{N}(9)$ follows from $(S_1)$ with $a = 7$. Also, the terms $u(3)$ through $u(7)$ illustrate Corollary 6.16.3.

**Theorem 6.16.6.** If two nontrivial solutions $u(k)$ and $v(k)$ of (6.16.1) have three zeros in common, then $u(k)$ and $v(k)$ are linearly dependent, i.e. specifying any three zeros uniquely determines a nontrivial solution up to a multiplicative constant.

**Proof.** If $u(\alpha) = u(a) = u(a+1) = v(\alpha) = v(a) = v(a+1) = 0$, for some $\alpha$ and $a$, where $0 \leq \alpha < a$, then by Theorem 6.16.1, $u(a+2) \neq 0$ and $v(a+2) \neq 0$. Define $w(k) = v(a+2)u(k) - u(a+2)v(k)$. Since $w(k)$ is a linear combination of $u(k)$ and $v(k)$, it is a solution of (6.16.1). However, $w(\alpha) = w(a) = w(a+1) = w(a+2) = 0$, and so $w(k)$ must be the trivial solution of (6.16.1) by Theorem 6.16.1. Since $u(a+2)$ and $v(a+2)$ are nonzero, $u(k)$ and $v(k)$ must be constant multiples of each other.

Next, if $u(\alpha) = u(a) = u(\beta) = v(\alpha) = v(a) = v(\beta) = 0$, where $\alpha < a < \beta - 1$, then by Corollary 6.16.5, $u(a+1) \neq 0$ and $v(a+1) \neq 0$. Define $w(k) = v(a+1)u(k) - u(a+1)v(k)$. Clearly, $w(\alpha) = w(a) = w(a+1) = w(\beta) = 0$, which contradicts Corollary 6.16.4 unless $w(k) \equiv 0$. But this means $u(k)$ and $v(k)$ are constant multiples of each other. This completes the proof. ∎

**Definition 6.16.2.** A solution $u(k)$ of (6.16.1) is called *recessive* if there exists an $a \in \mathbb{N}$ such that for all $k \in \mathbb{N}(a)$

$$u(k) > 0, \quad \Delta u(k) \le 0, \quad \Delta^2 u(k) \ge 0 \text{ and } \Delta^3 u(k) \le 0. \tag{6.16.11}$$

Let $u^m(k)$ be the solution of (6.16.1) satisfying $u^m(m) = u^m(m+1) = u^m(m+2) = 0$ and $u^m(0) = 1$ where $m \in \mathbb{N}(1)$. For each $m$, $u^m(k)$ exists and is unique. The existence is clear from Theorem 6.16.1 and a normalization, while the uniqueness follows from Theorem 6.16.6. Note that by construction

$$0 \le u^m(k) \le 1 \quad \text{for all } k \in \mathbb{N}(0, m+2). \tag{6.16.12}$$

Also, Theorem 6.16.1 implies that

$$u^m(k) \ge u^m(k+1) \quad \text{for all } k \in \mathbb{N}. \tag{6.16.13}$$

We now consider $m$ sequence $\{u^m(1)\}$. By (6.16.12), $0 \le u^m(1) \le 1$ for all $m \in \mathbb{N}(1)$, thus $\limsup_{m\to\infty}\{u^m(1)\}$ exists, we call it $u(1)$. Then, there exists a subsequence $\{m_{1\ell}\} \subseteq \mathbb{N}(1)$ such that $u^{m_{1\ell}}(1) \to u(1)$ as $\ell \to \infty$. Next, consider $m$ sequence $\{u^m(2)\}$. By (6.16.12) $\limsup_{\ell\to\infty} u^{m_{1\ell}}(2)$ exists, we call it $u(2)$. Also, there exists a subsequence $\{m_{2\ell}\} \subseteq \{m_{1\ell}\}$ such that $u^{m_{2\ell}}(2) \to u(2)$ (and $u^{m_{2\ell}}(1) \to u(1)$) as $\ell \to \infty$. In a similar fashion, by considering $\{u^m(3)\}$, we can arrive at a subsequence $\{m_{3\ell}\}$ and a limit $u(3)$ such that $u^{m_{3\ell}}(k) \to u(k)$ as $\ell \to \infty$, $k \in \mathbb{N}(1,3)$. Clearly, $u^{m_{3\ell}}(0) = 1$, for all $\ell$.

Recall that by definition, for any $k$ and any $m$

$$u^m(k+2) - 4u^m(k+1) + 6u^m(k) - 4u^m(k-1) + u^m(k-2) = p(k)u^m(k). \tag{6.16.14}$$

Consider (6.16.14) with $k = 2$ and $m$ replaced by $m_{3\ell}$. We can conclude that $\lim_{\ell\to\infty} u^{m_{3\ell}}(4)$ exists, we call it $u(4)$. Now replace $k$ by 3 in (6.16.14) and conclude the existence of $\lim_{\ell\to\infty} u^{m_{3\ell}}(5) = u(5)$. Proceeding inductively, we conclude that $\lim_{\ell\to\infty} u^{m_{3\ell}}(k) = u(k)$ exists for any $k \in \mathbb{N}$. Replacing $m$ by $m_{3\ell}$ in (6.16.14) and letting $\ell \to \infty$, we conclude that $u(k)$ is a solution of (6.16.1). Also,

$$u(k) \ge u(k+1) \ge 0. \tag{6.16.15}$$

This follows from (6.16.13) by replacing $m$ by $m_{3\ell}$, fixing $k$, and letting $\ell \to \infty$. From (6.16.15), we conclude that

$$\lim_{k\to\infty} u(k) \text{ exists, and we shall call it } L. \tag{6.16.16}$$

We will now show that this $u(k)$ is a recessive solution of (6.16.1).

**Theorem 6.16.7.** The solution $u(k)$ constructed above is a recessive solution of (6.16.1). In addition $\Delta u(k)$, $\Delta^2 u(k)$ and $\Delta^3 u(k)$ all monotonically approach zero as $k \to \infty$.

**Proof.** We will first show that (6.16.11) is satisfied. By (6.16.13) and Theorem 6.16.1, $u^{m_{3\ell}}(m_{3\ell}+3) < 0$. Choosing $m_{3\ell} \geq 3$ and using $(S_3)$ with $a = m_{3\ell}+1$, we can conclude that for any $k$ such that $2 \leq k \leq m_{3\ell}+1$, $\Delta u^{m_{3\ell}}(k-1) \leq 0$, $\Delta^2 u^{m_{3\ell}}(k-1) \geq 0$ and $\Delta^3 u^{m_{3\ell}}(k-1) \leq 0$. Letting $\ell \to \infty$ implies that $u(k)$ satisfies (6.16.11) for $a = 1$ and is recessive. We note that $u(k)$ also satisfies (6.16.11) for $a = 0$. Concerning the monotonicity, we choose any $k \in \mathbb{N}(2)$ and any $m_{3\ell} \geq k$. Then, $\Delta^2 u^{m_{3\ell}}(k-1) \geq 0$ which means $\Delta u^{m_{3\ell}}(k) \geq \Delta u^{m_{3\ell}}(k-1)$, and hence $0 \leq -\Delta u^{m_{3\ell}}(k) \leq -\Delta u^{m_{3\ell}}(k-1)$. Taking the limit as $\ell \to \infty$ implies that $\Delta u(k)$ is monotonically decreasing in absolute value. By (6.16.16), since $u(k)$ monotonically approaches a finite limit, $\Delta u(k) \to 0$ as $k \to \infty$. The argument that $\Delta^2 u(k)$ and $\Delta^3 u(k)$ monotonically approach zero is similar.

By Theorem 6.16.7 this recessive solution $u(k)$ of (6.16.1) can be written as

(6.16.17) $$u(k-2) = L + \frac{1}{6}\sum_{\ell=k}^{\infty}(\ell-k+1)(\ell-k+2)(\ell-k+3)p(\ell)u(\ell). \quad \blacksquare$$

**Corollary 6.16.8.** If $\sum^{\infty} \ell^3 p(\ell) = \infty$, then the recessive solution $u(k)$ of (6.16.1) constructed above approaches zero as $k \to \infty$.

**Corollary 6.16.9.** Suppose that $u(k)$ and $v(k)$ are two recessive solutions of (6.16.1) such that $u(a) = v(a)$. If $u(k) \geq v(k)$ for all $k \in \mathbb{N}(a)$, then $u(k) \equiv v(k)$.

**Proof.** Let $L = \lim_{k\to\infty} u(k)$ and $M = \lim_{k\to\infty} v(k)$. By hypothesis, $L \geq M$. Thus, if $w(k) = u(k) - v(k)$, then from (6.16.17) with $k = a+2$ we have

$$0 \geq L - M + \frac{1}{6}\sum_{\ell=a+2}^{\infty}(\ell-a-1)(\ell-a)(\ell-a+1)p(\ell)w(\ell) \geq 0.$$

From this we conclude that $u(k) \equiv v(k)$. $\blacksquare$

## 6.17. Asymptotic Behavior of Solutions of

(6.17.1) $$\Delta^n u(k) + f\left(k, u(k), \Delta u(k), \cdots, \Delta^{n-1}u(k)\right) = 0, \quad k \in \mathbb{N}$$

where the function $f$ is defined on $\mathbb{N} \times \mathbb{R}^n$.

**Theorem 6.17.1.** Assume that the function $f(k, u_0, \cdots, u_{n-1})$ for all $(k, u_0, \cdots, u_{n-1}) \in \mathbb{N} \times \mathbb{R}^n$ satisfies

$$|f(k, u_0, \cdots, u_{n-1})| \le \sum_{i=0}^{n-1} p_i(k)|u_i|, \tag{6.17.2}$$

where $p_i(k)$, $0 \le i \le n-1$ are nonnegative functions, defined on $\mathbb{N}$ and

$$\prod^{\infty} \left[1 + \sum_{i=0}^{n-1} (\ell)^{(n-1-i)} p_i(\ell)\right] < \infty. \tag{6.17.3}$$

Then, the difference equation (6.17.1) has solutions which are asymptotic to $\sum_{i=0}^{n-1} a_i(k)^{(i)}$ as $k \to \infty$, where $a_i$, $0 \le i \le n-1$ are constants such that $a_{n-1} \ne 0$.

**Proof.** Let $u(k)$ be a solution of (6.17.1), then from Corollary 1.8.6 for any $a \in \mathbb{N}(1)$, it follows that

(6.17.4)
$$\Delta^m u(k) = \sum_{i=m}^{n-1} \frac{(k-a)^{(i-m)}}{(i-m)!} \Delta^i u(a) - \frac{1}{(n-m-1)!} \sum_{\ell=a}^{k-n+m} (k-\ell-1)^{(n-m-1)}$$
$$\times f\left(\ell, u(\ell), \Delta u(\ell), \cdots, \Delta^{n-1} u(\ell)\right), \quad 0 \le m \le n-1.$$

Thus, from (6.17.2) we find

(6.17.5)
$$|\Delta^m u(k)| \le A_m (k)^{(n-1-m)} + B_m (k)^{(n-m-1)} \sum_{\ell=a}^{k-n+m} \sum_{i=0}^{n-1} p_i(\ell) \left|\Delta^i u(\ell)\right|,$$

where

$$A_m = \left[(k)^{(n-1-m)}\right]^{-1} \sum_{i=m}^{n-1} \frac{(k)^{(i-m)}}{(i-m)!} \left|\Delta^i u(a)\right|,$$

and $B_m = 1/(n-m-1)!$. Define $A = \max_{0 \le m \le n-1,\ k \in \mathbb{N}(a)} A_m$, then since $B_m \le 1$, $0 \le m \le n-1$ from (6.17.5), we get

$$|\Delta^m u(k)| \le (k)^{(n-1-m)} F(k), \quad 0 \le m \le n-1 \tag{6.17.6}$$

where

$$F(k) = A + \sum_{\ell=a}^{k-n+m} \sum_{i=0}^{n-1} p_i(\ell) |\Delta^i u(\ell)|.$$

Using (6.17.6) in the above equality, to obtain

$$F(k) \ \le \ A + \sum_{\ell=a}^{k-1} \sum_{i=0}^{n-1} (\ell)^{(n-1-i)} p_i(\ell) F(\ell).$$

Therefore, as an application of Corollary 4.1.2, we find

$$F(k) \ \le \ A \prod_{\ell=a}^{k-1} \left[ 1 + \sum_{i=0}^{n-1} (\ell)^{(n-1-i)} p_i(\ell) \right]$$

and hence from (6.17.3) there exists a finite constant $c > 0$ such that $F(k) \le c$. Thus, inequality (6.17.6) implies that

$$|\Delta^m u(k)| \ \le \ c(k)^{(n-1-m)}, \quad 0 \le m \le n-1. \tag{6.17.7}$$

Next, from (6.17.4) we have

$$\Delta^{n-1} u(k) \ = \ \Delta^{n-1} u(a) - \sum_{\ell=a}^{k-1} f\left(\ell, u(\ell), \ \Delta u(\ell), \cdots, \Delta^{n-1} u(\ell)\right). \tag{6.17.8}$$

Since condition (6.17.3) implies that $\sum^{\infty} \sum_{i=0}^{n-1} (\ell)^{(n-1-i)} p_i(\ell) \ < \ \infty$, we find from (6.17.2) and (6.17.7) that the sum in (6.17.8) converges as $k \to \infty$ and therefore $\lim_{k\to\infty} \Delta^{n-1} u(k)$ exists and is a finite number. To ensure that this limit is not zero, we choose $a$ so large that $1 - c\sum_{\ell=a}^{\infty} \sum_{i=0}^{n-1} (\ell)^{(n-1-i)} p_i(\ell) > 0$ and impose the condition $\Delta^{n-1} u(a) = 1$ on the solution of (6.17.1). This solution has the desired asymptotic property. ∎

**Corollary 6.17.2.** Under the hypotheses of Theorem 6.17.1 equation (6.17.1) has nonoscillatory solutions.

**Theorem 6.17.3.** If there exists a constant $c > 0$ such that for any function $u(k)$ defined on $\mathbb{N}$, $\liminf_{k\to\infty} u(k) > c$ ($\limsup_{k\to\infty} u(k) < -c$)

$$\sum^{\infty} f\left(\ell, u(\ell), \Delta u(\ell), \cdots, \Delta^{n-1} u(\ell)\right) \ = \ \pm\infty, \tag{6.17.9}$$

then every nonoscillatory solution $u(k)$ of (6.17.1) satisfies $\liminf_{k\to\infty} |u(k)| \le c$.

**Proof.** Let $u(k)$ be a nonoscillatory solution of (6.17.1), say $u(k) > 0$ for all $k \ge a$, and assume that $\liminf_{k\to\infty} u(k) > c$. The case $u(k) < 0$ for all $k \ge a$ can be treated similarly. From (6.17.8) and (6.17.9) it is clear that $\lim_{k\to\infty} \Delta^{n-1} u(k) \ = \ -\infty$, and therefore $\limsup_{k\to\infty} \Delta^{n-1} u(k) \ < \ 0$.

But, then Lemma 1.8.10 implies that $\lim_{k\to\infty} u(k) = -\infty$, which is a contradiction to our assumption that $u(k) > 0$. ∎

**Lemma 6.17.4.** Consider the difference equation

$$\nabla u(k) - \frac{m}{k}u(k) + \frac{f(k)}{k} = 0, \quad k \in \mathbb{N}(m), \quad m \in \mathbb{N}(1) \tag{6.17.10}$$

where the function $f$ is defined on $\mathbb{N}(m)$ and nonoscillatory. If $\lim_{k\to\infty} |f(k)| = \infty$, and $u(k)$ is the solution of (6.17.10) with $u(a) = 0$, where $m < a \in \mathbb{N}(m)$, then $\lim_{k\to\infty} u(k) = \pm\infty$.

**Proof.** By direct substitution, it is easy to verify that

$$u(k) = -(k)^{(m)} \sum_{\ell=a+1}^{k} f(\ell)/(\ell)^{(m+1)} \tag{6.17.11}$$

is the solution of (6.17.10) satisfying $u(a) = 0$. Since $f$ is of constant sign for all large $k$, the summation $\sum_{\ell=a+1}^{\infty} f(\ell)/(\ell)^{(m+1)}$ exists on the extended real line. If the value of this summation is different from zero, then the result follows. If it is zero, then let $p(k) = \sum_{\ell=a+1}^{k} f(\ell)/(\ell)^{(m+1)}$ and $q(k) = 1/(k)^{(m)}$ so that $\Delta p(k) = \dfrac{f(k+1)}{(k+1)^{(m+1)}}$ and $\Delta q(k) = -\dfrac{m}{(k+1)^{(m+1)}}$. Thus, Corollary 1.8.8 is applicable and we find that

$$\lim_{k\to\infty} \frac{p(k)}{q(k)} = \lim_{k\to\infty} \frac{\Delta p(k)}{\Delta q(k)} = \lim_{k\to\infty} \frac{f(k+1)}{-m} = \mp\infty.$$

Therefore, $\lim_{k\to\infty} u(k) = \pm\infty$. ∎

**Theorem 6.17.5.** Assume that there exist integers $p, q, r$ such that $0 \le r \le n-1$, $0 \le q \le p \le n-r-1$, and for every nonoscillatory $u(k) \in F_p$ with $\liminf_{k\to\infty} |u(k)|/(k)^{(q)} \ne 0$

$$\sum^{\infty} (\ell)^{(r)} f\left(\ell, u(\ell), \Delta u(\ell), \cdots, \Delta^{n-1}u(\ell)\right) = \pm\infty. \tag{6.17.12}$$

Then, for all nonoscillatory $F_p$ solutions $u(k)$ of (6.17.1), $\liminf_{k\to\infty} \dfrac{|u(k)|}{(k)^{(q)}} = 0$.

**Proof.** Let $u(k)$ be a nonoscillatory $F_p$ solution with $\liminf_{k\to\infty} \dfrac{|u(k)|}{(k)^{(q)}} \ne 0$. Without loss of generality, we assume that $u(k) > 0$ on $\mathbb{N}(k_1)$, where

$k_1 > \max\{1, r\}$. The case $u(k) < 0$ can be treated similarly. We define

$$R_{ij}^{s}(k) \;=\; \sum_{\ell=k_1}^{k-1} (\ell)^{(i)} \Delta^{j} u(\ell + s)$$

and from (1.8.5) find

$$R_{ij}^{s}(k) \;=\; (\ell)^{(i)} \Delta^{j-1} u(\ell + s)\Big|_{\ell=k_1}^{k} - i \sum_{\ell=k_1}^{k-1} (\ell)^{(i-1)} \Delta^{j-1} u(\ell + s + 1).$$

Since $\nabla R_{i-1,j-1}^{s+1}(k) = (k-1)^{(i-1)} \Delta^{j-1} u(k+s)$, the above equation takes the form

$$R_{ij}^{s}(k) \;=\; k \nabla v(k) - (k_1)^{(i)} \Delta^{j-1} u(k_1 + s) - i v(k),$$

where $v(k) = R_{i-1,j-1}^{s+1}(k)$. Thus, we find that

$$\text{(6.17.13)} \qquad \nabla v(k) - \frac{i}{k} v(k) + \frac{f_{ij}^{s}(k)}{k} \;=\; 0, \quad v(k_1) = 0, \quad k_1 > r$$

where $f_{ij}^{s}(k) = -(k_1)^{(i)} \Delta^{j-1} u(k_1 + s) - R_{ij}^{s}(k)$.

Let $i = r,\ j = n$ and $s = 0$, then from (6.17.1) we have

$$f_{rn}^{0}(k) = -(k_1)^{(r)} \Delta^{n-1} u(k_1) + \sum_{\ell=k_1}^{k-1} (\ell)^{(r)} f\left(\ell, u(\ell), \Delta u(\ell), \cdots, \Delta^{n-1} u(\ell)\right)$$

and from (6.17.12), $f_{rn}^{0}(k)$ is nonoscillatory and $\lim_{k\to\infty} |f_{rn}^{0}(k)| = \infty$. Thus, from Lemma 6.17.4, we get

$$\text{(6.17.14)} \qquad \lim_{k\to\infty} v(k) \;=\; \lim_{k\to\infty} R_{r-1,n-1}^{1}(k) \;=\; \pm\infty.$$

Next, since $f_{r-1,n-1}^{1}(k) = -(k_1)^{(r-1)} \Delta^{n-2} u(k_1+1) - R_{r-1,n-1}^{1}(k)$, from (6.17.14), we find $f_{r-1,n-1}^{1}(k)$ to be nonoscillatory and $\lim_{k\to\infty} |f_{r-1,n-1}^{1}(k)| = \infty$. Thus, Lemma 6.17.4 is again applicable and we obtain $\lim_{k\to\infty} R_{r-2,n-2}^{2}(k) = \pm\infty$. Continuing this way, we find $\lim_{k\to\infty} R_{0,n-r}^{r}(k) = \pm\infty$. However, from the definition $R_{0,n-r}^{r}(k) = \Delta^{n-r-1} u(k+r) - \Delta^{n-r-1} u(k_1 + r)$, and hence we have $\lim_{k\to\infty} \Delta^{n-r-1} u(k) = \pm\infty$. The case $\lim_{k\to\infty} \Delta^{n-r-1} u(k) = -\infty$ is impossible from Lemma 1.8.10, since it contradicts the fact that $u(k)$ is positive, and thus $\lim_{k\to\infty} \Delta^{n-r-1} u(k) = \infty$.

Since $u(k) > 0$ and belongs to $F_p$, there exists a constant $c > 0$ such that $u(k) < c(k)^{(p)}$ for large $k \in \mathbb{N}$. Thus, the function $w(k) =$

$u(k) - c(k)^{(p)}$ is negative for large $k \in \mathbb{N}$, but since $p \leq n - r - 1$, we find $\lim_{k\to\infty} \Delta^{n-r-1} w(k) = \lim_{k\to\infty} \Delta^{n-r-1}\left(u(k) - c(k)^{(p)}\right) = \infty$, which from Lemma 1.8.10 leads to a contradiction that $w(k)$ is negative. This completes the proof. ∎

**Remark 6.17.1.** If in Theorem 6.17.5, $p = 0$ then as conclusion we have that, for all bounded nonoscillatory solutions of (6.17.1), $\liminf_{k\to\infty} |u(k)| = 0$.

## 6.18. Asymptotic Behavior, Oscillation and Nonoscillation for

$$\text{(6.18.1)} \qquad \Delta^n u(k) + h(k) F\left(k, u(k), \Delta u(k), \cdots, \Delta^{n-1} u(k)\right) = g\left(k, u(k), \Delta u(k), \cdots, \Delta^{n-1} u(k)\right), \quad k \in \mathbb{N}$$

where the functions $h$, $F$ and $g$ are defined in their domain of definition.

**Theorem 6.18.1.** Assume that there exist integers $p$, $r$ such that $0 \leq r \leq n-1$, $0 \leq p \leq n-r-1$, and for every nonoscillatory $u(k) \in F_p$ with $\liminf_{k\to\infty} |u(k)| \neq 0$ there exist constants $A$, $B$ (depending on $u(k)$) such that $AB > 0$, and for all large $k \in \mathbb{N}$, $A \leq F(k, u(k), \Delta u(k), \cdots, \Delta^{n-1} u(k)) \leq B$, and for such $u(k)$ there exists a nonnegative function $G(k)$ defined for all large $k$ such that $|g(k, u(k), \Delta u(k), \cdots, \Delta^{n-1} u(k))| \leq G(k)$. Further, assume that for all constants $c_1 > 0$ and $c_2 > 0$

$$\sum^{\infty} (\ell)^{(r)} \left[c_1 h_+(\ell) - h_-(\ell) - c_2 G(\ell)\right] = \infty$$

or

$$\sum^{\infty} (\ell)^{(r)} \left[c_1 h_-(\ell) - h_+(\ell) - c_2 G(\ell)\right] = \infty,$$

where $h_+(k) = \max\{h(k), 0\}$ and $h_-(k) = \max\{-h(k), 0\}$. Then, for all nonoscillatory $F_p$ solutions $u(k)$ of (6.18.1), $\liminf_{k\to\infty} |u(k)| = 0$.

**Proof.** For any nonoscillatory $F_p$ solution with $\liminf_{k\to\infty} |u(k)| \neq 0$, we find for all large $k_1 \leq k \in \mathbb{N}$ that

$$\begin{aligned}
&\sum_{\ell=k_1}^{k} (\ell)^{(r)} \left[A h_+(\ell) - B h_-(\ell) - G(\ell)\right] \\
&\qquad \leq \sum_{\ell=k_1}^{k} (\ell)^{(r)} \left[h(\ell) F(\ell, \cdots) - g(\ell, \cdots)\right] \\
&\qquad \leq \sum_{\ell=k_1}^{k} (\ell)^{(r)} \left[B h_+(\ell) - A h_-(\ell) + G(\ell)\right].
\end{aligned}$$

Thus, for $A$ and $B$ positive

$$
\begin{aligned}
B \sum_{\ell=k_1}^{k} (\ell)^{(r)} \left[ \frac{A}{B} h_+(\ell) - h_-(\ell) - \frac{1}{B} G(\ell) \right] \\
&\le \sum_{\ell=k_1}^{k} (\ell)^{(r)} \left[ h(\ell) F(\ell, \cdots) - g(\ell, \cdots) \right] \\
&\le -B \sum_{\ell=k_1}^{k} (\ell)^{(r)} \left[ \frac{A}{B} h_-(\ell) - h_+(\ell) - \frac{1}{B} G(\ell) \right]
\end{aligned}
$$

and, for $A$ and $B$ negative

$$
\begin{aligned}
-A \sum_{\ell=k_1}^{k} (\ell)^{(r)} \left[ \frac{-B}{-A} h_-(\ell) - h_+(\ell) - \frac{1}{-A} G(\ell) \right] \\
&\le \sum_{\ell=k_1}^{k} (\ell)^{(r)} \left[ h(\ell) F(\ell, \cdots) - g(\ell, \cdots) \right] \\
&\le A \sum_{\ell=k_1}^{k} (\ell)^{(r)} \left[ \frac{-B}{-A} h_+(\ell) - h_-(\ell) - \frac{1}{-A} G(\ell) \right].
\end{aligned}
$$

Now it is obvious that for $f(k, \cdots) = h(k)F(k, \cdots) - g(k, \cdots)$ the conditions of Theorem 6.17.5 with $q = 0$ are satisfied, thus the conclusion follows. ■

Now for the difference equation (6.18.1) each result we shall prove will require some of the following conditions:

($d_1$) $h(k) = 1$ for all $k \in \mathbb{N}$

($d_2$) $h(k) > 0$ for all large $k \in \mathbb{N}$

($d_3$) $h(k) \ge 0$ for all $k \in \mathbb{N}$

($d_4$) $g\left(k, u(k), \Delta u(k), \cdots, \Delta^{n-1} u(k)\right) = g(k)$

($d_5$) there exists a function $G_1(k)$ such that $g(k, u(k), \Delta u(k), \cdots, \Delta^{n-1} u(k)) \ge G_1(k)$

($d_6$) there exists a function $G_2(k)$ such that $g(k, u(k), \Delta u(k), \cdots, \Delta^{n-1} u(k)) \le G_2(k)$

($d_7$) there exists a nonnegative function $G(k)$ such that $|g(k, u(k), \Delta u(k), \cdots, \Delta^{n-1} u(k))| \le G(k)$

($d_8$) if $u_1 \ne 0$, then $u_1 F(k, u_1, u_2, \cdots, u_n) \ge 0$

($d_9$) if $u_1 \ne 0$, then $u_1 F(k, u_1, u_2, \cdots, u_n) \le 0$

$(d_{10})$ condition $(d_8)$ and $F(k, u_1, u_2, \cdots, u_n)$ is bounded away from zero if $u_1$ is bounded away from zero

$(d_{11})$ there exists a constant $L$ such that $F(k, u(k), \Delta u(k), \cdots, \Delta^{n-1} u(k)) \leq L$

$(d_{12})$ there exists a constant $M$ such that $F(k, u(k), \Delta u(k), \cdots, \Delta^{n-1} u(k)) \geq M$.

**Theorem 6.18.2.** Assume that conditions $(d_2)$, $(d_7)$ and $(d_{10})$ hold and

$$\sum^{\infty} h(\ell) = 0, \quad \frac{G(k)}{h(k)} \to 0 \quad \text{as} \quad k \to \infty, \quad \text{or} \quad \sum^{\infty} G(\ell) < \infty.$$

Then, for every nonoscillatory solution $u(k)$ of (6.18.1), $\liminf_{k\to\infty} |u(k)| = 0$.

**Proof.** We shall show that the hypotheses of Theorem 6.17.5 with $p = r = 0$ are satisfied. Let $u(k)$ be some function defined on $\mathbb{N}$ such that $\liminf_{k\to\infty} u(k) > 0$. Then, from $(d_{10})$ there exist $A > 0$ and $k_1 \in \mathbb{N}$ such that $F\left(k, u(k), \Delta u(k), \cdots, \Delta^{n-1} u(k)\right) \geq A$ for all $k \in \mathbb{N}(k_1)$. Let $k_2 \geq k_1$ be large enough so that $h(k) > 0$ for all $k \in \mathbb{N}(k_2)$. If $\sum^{\infty} G(\ell) < \infty$, then

$$\sum_{\ell=k_2}^{k} [h(\ell)F(\ell, \cdots) - g(\ell, \cdots)] \geq \sum_{\ell=k_2}^{k} [Ah(\ell) - G(\ell)] \to \infty$$

as $k \to \infty$ and the conclusion follows. Further, if $G(k)/h(k) \to 0$ as $k \to \infty$, then we can choose $k_3 \geq k_2$ such that $\dfrac{G(k)}{h(k)} \leq \dfrac{A}{2}$ for all $k \in \mathbb{N}(k_3)$, and we find

$$\begin{aligned} \sum_{\ell=k_3}^{k} [h(\ell)F(\ell, \cdots) - g(\ell, \cdots)] &\geq \sum_{\ell=k_3}^{k} h(\ell)\left[A - \frac{G(\ell)}{h(\ell)}\right] \\ &\geq \frac{A}{2} \sum_{\ell=k_3}^{k} h(\ell) \to \infty \end{aligned}$$

as $k \to \infty$. The case $\limsup_{k\to\infty} u(k) < 0$ can be treated similarly. ∎

**Theorem 6.18.3.** Assume that conditions $(d_1)$, $(d_7)$ and $(d_8)$ are satisfied. Then, for any nonoscillatory or $T$ type solution $u(k)$ of (6.18.1)

$$\text{(6.18.2)} \quad |u(k)| = O\left((k)^{(n-1)} + \sum^{k-1}(k - \ell - 1)^{(n-1)} G(\ell)\right) \quad \text{as} \quad k \to \infty.$$

**Proof.** From Theorem 1.8.5 any solution $u(k)$ of (6.18.1) for any $k \in \mathbb{N}$ can be written as

$$(6.18.3) \quad u(k) = \sum_{i=0}^{n-1} \frac{(k-k_1)^{(i)}}{i!} \Delta^i u(k_1) - \frac{1}{(n-1)!} \sum_{\ell=k_1}^{k-n} (k-\ell-1)^{(n-1)} [F(\ell,\cdots) - g(\ell,\cdots)].$$

Thus, if $u(k)$ is nonoscillatory or $T$ type then $u(k) \geq 0$ or $u(k) \leq 0$ for all $k \geq k_2 \geq k_1$. If $u(k) \geq 0$, then from (6.18.3), we find

$$0 \leq u(k) \leq \sum_{i=0}^{n-1} \frac{(k-k_2)^{(i)}}{i!} \Delta^i u(k_2) + \frac{1}{(n-1)!} \sum_{\ell=k_2}^{k-n} (k-\ell-1)^{(n-1)} G(\ell)$$

and, if $u(k) \leq 0$ then

$$0 \geq u(k) \geq \sum_{i=0}^{n-1} \frac{(k-k_2)^{(i)}}{i!} \Delta^i u(k_2) - \frac{1}{(n-1)!} \sum_{\ell=k_2}^{k-n} (k-\ell-1)^{(n-1)} G(\ell)$$

also, in either case

$$|u(k)| \leq \sum_{i=0}^{n-1} \frac{(k-k_2)^{(i)}}{i!} |\Delta^i u(k_2)| + \frac{1}{(n-1)!} \sum_{\ell=k_2}^{k-n} (k-\ell-1)^{(n-1)} G(\ell).$$

This completes the proof. ∎

**Theorem 6.18.4.** Assume that conditions $(d_1)$, $(d_4)$ and $(d_9)$ hold and for every constant $c > 0$

$$\limsup_{k\to\infty} \left[ \sum^{k-n} (k-\ell-1)^{(n-1)} g(\ell) - c(k)^{(n-1)} \right] = \infty$$

and

$$\liminf_{k\to\infty} \left[ \sum^{k-n} (k-\ell-1)^{(n-1)} g(\ell) + c(k)^{(n-1)} \right] = -\infty.$$

Then, every bounded solution of (6.18.1) is oscillatory.

**Proof.** If $u(k)$ is bounded and nonoscillatory or $T$ type, then $u(k) \geq 0$ or $u(k) \leq 0$ for all $k \geq k_2$ in $\mathbb{N}$. If $u(k) \geq 0$, then as in Theorem 6.18.3 it follows that

(6.18.4)

$$0 \le u(k) \ge \sum_{i=0}^{n-1} \frac{(k-k_2)^{(i)}}{i!} \Delta^i u(k_2) + \frac{1}{(n-1)!} \sum_{\ell=k_2}^{k-n} (k-\ell-1)^{(n-1)} g(\ell)$$

and, if $u(k) \le 0$ then

(6.18.5)

$$0 \ge u(k) \le \sum_{i=0}^{n-1} \frac{(k-k_2)^{(i)}}{i!} \Delta^i u(k_2) + \frac{1}{(n-1)!} \sum_{\ell=k_2}^{k-n} (k-\ell-1)^{(n-1)} g(\ell).$$

Next, let $k_3 \ge k_2$ be sufficiently large so that for some $c > 0$

$$-c(k)^{(n-1)} \le \sum_{i=0}^{n-1} \frac{(k-k_2)^{(i)}}{i!} \Delta^i u(k_2) \le c(k)^{(n-1)}$$

for all $k \in \mathbb{N}(k_3)$.

Then, from (6.18.4) and (6.18.5), we find

$$0 \le u(k) \ge \frac{1}{(n-1)!} \sum_{\ell=k_2}^{k-n} (k-\ell-1)^{(n-1)} g(\ell) - c(k)^{(n-1)}$$

and

$$0 \ge u(k) \le \frac{1}{(n-1)!} \sum_{\ell=k_2}^{k-n} (k-\ell-1)^{(n-1)} g(\ell) + c(k)^{(n-1)}$$

and from the hypotheses

$$0 \le \limsup_{k\to\infty} u(k) \ge \limsup_{k\to\infty} \frac{1}{(n-1)!} \left[ \sum_{\ell=k_3}^{k-n} (k-\ell-1)^{(n-1)} g(\ell) - c(n-1)!(k)^{(n-1)} \right] = \infty$$

and

$$0 \ge \liminf_{k\to\infty} u(k) \le \liminf_{k\to\infty} \frac{1}{(n-1)!} \left[ \sum_{\ell=k_3}^{k-n} (k-\ell-1)^{(n-1)} g(\ell) + c(n-1)!(k)^{(n-1)} \right] = -\infty.$$

Thus, in either case we get a contradiction to our assumption that $u(k)$ is bounded. ∎

**Corollary 6.18.5.** Assume that conditions $(d_1)$, $(d_4)$ and $(d_8)$ hold and for every constant $c > 0$

$$\limsup_{k\to\infty}\left[\sum^{k-n}(k-\ell-1)^{(n-1)}g(\ell) - c(k)^{(n-1)}\right] \geq 0$$

and

$$\liminf_{k\to\infty}\left[\sum^{k-n}(k-\ell-1)^{(n-1)}g(\ell) + c(k)^{(n-1)}\right] \leq 0.$$

Then, every solution of (6.18.1) is oscillatory or $T$ type.

**Theorem 6.18.6.** Assume that conditions $(d_3)$, $(d_5)$ and $(d_{11})$ hold and for every constant $c > 0$

$$(6.18.6)\quad \liminf_{k\to\infty}\left[\sum^{k-n}(k-\ell-1)^{(n-1)}(G_1(\ell) - Lh(\ell)) - c(k)^{(n-1)}\right] \geq 0.$$

Then, all solutions of (6.18.1) are nonoscillatory or nonnegative $T$ type. Further, if strict inequality holds in (6.18.6), then equation (6.18.1) is nonoscillatory.

**Proof.** From the given hypotheses, we find that $\Delta^n u(k) = g(k,\cdots)$ $-h(k)F(k,\cdots) \geq G_1(k) - Lh(k)$. Thus, for any $k_1 \in \mathbb{N}$, we have

$$u(k) \geq \sum_{i=0}^{n-1}\frac{(k-k_1)^{(i)}}{i!}\Delta^i u(k_1) + \frac{1}{(n-1)!}\sum_{\ell=k_1}^{k-n}(k-\ell-1)^{(n-1)}(G_1(\ell)-Lh(\ell)).$$

Therefore, there exists a $k_2 \geq k_1$ sufficiently large so that

$$u(k) \geq \frac{1}{(n-1)!}\left[\sum_{\ell=k_2}^{k-n}(k-\ell-1)^{(n-1)}(G_1(\ell) - Lh(\ell)) - c(n-1)!(k)^{(n-1)}\right].$$

Hence, from (6.18.6) we find that $\liminf_{k\to\infty} u(k) \geq 0$, and from this the conclusion follows. ∎

**Theorem 6.18.7.** Assume that conditions $(d_3)$, $(d_6)$ and $(d_{12})$ hold and for every constant $c > 0$

$$(6.18.7)\quad \limsup_{k\to\infty}\left[\sum^{k-n}(k-\ell-1)^{(n-1)}(G_2(\ell) - Mh(\ell)) + c(k)^{(n-1)}\right] \leq 0.$$

Then, all solutions of (6.18.1) are nonoscillatory or nonpositive $T$ type. Further, if strict inequality holds in (6.18.7), then equation (6.18.1) is nonoscillatory.

**Proof.** The proof is similar to that of Theorem 6.18.6. ∎

## 6.19. Oscillation and Nonoscillation for

$$(6.19.1)\quad \Delta^n u(k) + \sum_{i=1}^{m} f_i(k)F_i\left(u(k), \Delta u(k), \cdots, \Delta^{n-1}u(k)\right) = 0, \quad k \in \mathbb{N}$$

where the functions $f_i$, $F_i$, $1 \le i \le m$ are defined in their domain of definition.

**Theorem 6.19.1.** In equation (6.19.1), we assume

(i) $f_i(k) \ge 0$ for all $k \in \mathbb{N}$ and $1 \le i \le m$

(ii) $u_1 F_i(u_1, \cdots, u_n) > 0$ for $u_1 \ne 0$ and $1 \le i \le m$

(iii) there is an index $j \in \mathbb{N}(1, m)$ such that $F_j(u_1, \cdots, u_n)$ is continuous at $(u_1, 0, \cdots, 0)$ with $u_1 \ne 0$, and

(a) $F_j(\lambda u_1, \cdots, \lambda u_n) = \lambda^{2\alpha+1} F_j(u_1, \cdots, u_n)$ for all $(u_1, \cdots, u_n) \in \mathbb{R}^n$ and $\lambda \in \mathbb{R}$, where $\alpha$ is some nonnegative integer

(b) $\sum^{\infty} f_j(\ell) = \infty$.

Then, (1) if $n$ is even, difference equation (6.19.1) is oscillatory

(2) if $n$ is odd, every solution $u(k)$ of (6.19.1) is either oscillatory or tends monotonically to zero together with $\Delta^i u(k)$, $1 \le i \le n-1$.

**Proof.** Let $u(k)$ be a nonoscillatory solution of (6.19.1), which must then eventually be of fixed sign. Let $u(k) > 0$ for all $k \in \mathbb{N}(k_1)$. For this solution our hypothesis implies that $\Delta^n u(k) \le 0$ for all $k \in \mathbb{N}(k_1)$. If $n$ is even, then from Problem 1.9.39, we have $\lim_{k\to\infty} \dfrac{\Delta^i u(k)}{u(k)} = 0$, $1 \le i \le n-1$. Since $F_j(u_1, \cdots, u_n)$ is continuous at $(u_1, 0, \cdots, 0)$ with $u_1 \ne 0$, for any $\epsilon > 0$ there exists $k_2 \ge k_1$ such that for all $k \in \mathbb{N}(k_2)$

$$\left| F_j\left(1, \frac{\Delta u(k)}{u(k)}, \cdots, \frac{\Delta^{n-1}u(k)}{u(k)}\right) - F_j(1, 0, \cdots, 0)\right| < \epsilon.$$

From (ii), $F_j(1, 0, \cdots, 0) > 0$ and we may assume $0 < \epsilon < F_j(1, 0, \cdots, 0)$. From Theorem 1.8.11, we have $\Delta u(k) > 0$, $\Delta^{n-1}u(k) > 0$, also as a consequence $u(k) > c > 0$ for all $k \in \mathbb{N}(k_2)$. Define $v(k) = \Delta^{n-1}u(k)/u(k)$, then we find

$$\begin{aligned}\Delta v(k) &= \frac{\Delta^{n-1}u(k+1)}{u(k+1)} - \frac{\Delta^{n-1}u(k)}{u(k)} \\ &\leq \frac{\Delta^{n-1}u(k+1)}{u(k)} - \frac{\Delta^{n-1}u(k)}{u(k)} = \frac{\Delta^{n}u(k)}{u(k)}.\end{aligned}$$

Thus, from (6.19.1) and the hypotheses (i) and (ii), we get

$$\Delta v(k) \leq -f_j(k)\frac{F_j\left(u(k), \Delta u(k), \cdots, \Delta^{n-1}u(k)\right)}{u(k)}.$$

Summing the above inequality, we obtain from (iii) that

$$\begin{aligned}v(k) - v(k_2) &\leq -\sum_{\ell=k_2}^{k-1} f_j(\ell)u^{2\alpha}(\ell)F_j\left(1, \frac{\Delta u(\ell)}{u(\ell)}, \cdots, \frac{\Delta^{n-1}u(\ell)}{u(\ell)}\right) \\ &\leq -[F_j(1,0,\cdots,0) - \epsilon]c^{2\alpha}\sum_{\ell=k_2}^{k-1} f_j(\ell).\end{aligned}$$

In the above inequality right side tends to $-\infty$ as $k \to \infty$, whereas left side remains bounded. This contradicts our assumption that $u(k) > 0$ for all $k \in \mathbb{N}(k_1)$.

If $n$ is odd, then the case (ii) of Corollary 1.8.14 is impossible because we get a contradiction as in the case $n$ even. Thus, we assume $\lim_{k\to\infty} u(k) = d > 0$, $\lim_{k\to\infty} \Delta^i u(k) = 0$, $1 \leq i \leq n-1$. From the continuity of $F_j$ and hypothesis (ii), we find $\lim_{k\to\infty} F_j\left(u(k), \Delta u(k), \cdots, \Delta^{n-1}u(k)\right) = F_j(d, 0, \cdots, 0) > 0$. Therefore, $F_j > 0$ for all $k \in \mathbb{N}(k_3)$, where $k_3 \geq k_2$. Now, from (6.19.1), we have

$$\Delta^n u(k) \leq -f_j(k)F_j\left(u(k), \Delta u(k), \cdots, \Delta^{n-1}u(k)\right).$$

Summing the above inequality, to obtain

$$-\Delta^{n-1}u(k) + \Delta^{n-1}u(k_3) \geq \sum_{\ell=k_3}^{k-1} f_j(\ell)F_j\left(u(\ell), \Delta u(\ell), \cdots, \Delta^{n-1}u(\ell)\right).$$

If we let $k$ tend to infinity in the above inequality, we have a contradiction that $\Delta^{n-1}u(k_3) \geq \infty$. Thus, $\lim_{k\to\infty} u(k) = 0$.

Finally, we note that with a slight modification in the results Theorem 1.8.11, Corollary 1.8.14 and Problem 1.9.39 the case $u(k) < 0$ eventually can be similarly considered. ∎

**Theorem 6.19.2.** Let in Theorem 6.19.1 the hypotheses (iii) (a) is replaced by

(iii) (a)$'$ for any $r$, $2 \le r \le n$ and any $c \ge 0$, $\liminf F_j(u_1, \cdots, u_n) > 0$ or $\infty$ as $u_1 \to \infty, \cdots, u_{r-1} \to \infty$, $u_r \to c$, $u_{r+1} \to 0, \cdots, u_n \to 0$ and in addition

(iv) $F_i(-u_1, \cdots, -u_n) = -F_i(u_1, \cdots, u_n)$ for all $(u_1, \cdots, u_n) \in \mathbb{R}^n$, $1 \le i \le m$.

Then, the conclusions of Theorem 6.19.1 hold.

**Proof.** Let $u(k)$ be a nonoscillatory solution of (6.19.1). Since condition (iv) implies that $-u(k)$ is again a solution of (6.19.1), without loss of generality we can assume that $u(k) > 0$ for all $k \in \mathbb{N}(k_1)$. Then, as in Theorem 6.19.1, $\Delta^n u(k) \le 0$ and $\Delta^{n-1} u(k) > 0$ for all $k \in \mathbb{N}(k_1)$. Therefore, from (6.19.1), we find

$$\Delta^n u(k) \le -f_j(k) F_j\left(u(k), \Delta u(k), \cdots, \Delta^{n-1} u(k)\right), \quad k \in \mathbb{N}(k_1)$$

and hence

$$(6.19.2) \qquad \Delta^{n-1} u(k_1) > \sum_{\ell=k_1}^{k-1} f_j(\ell) F_j\left(u(\ell), \Delta u(\ell), \cdots, \Delta^{n-1} u(\ell)\right).$$

We distinguish two cases:

**Case 1.** There exists a $\mu$, $1 \le \mu \le n-1$ such that $\lim_{k\to\infty} \Delta^i u(k) = \infty$ for $0 \le i \le \mu - 1$, $\lim_{k\to\infty} \Delta^\mu u(k) = c > 0$ and $\lim_{k\to\infty} \Delta^i u(k) = 0$ for $\mu + 1 \le i \le n-1$. Then, from (iii) (a)$'$, $\liminf_{k\to\infty} F_j\left(u(k), \Delta u(k), \cdots, \Delta^{n-1} u(k)\right) \ge \epsilon > 0$. So, there exists a $k_2 \ge k_1$ such that $F_j(u(k), \Delta u(k), \cdots, \Delta^{n-1} u(k)) \ge \epsilon > 0$ for all $k \in \mathbb{N}(k_2)$. Replacing $k_1$ by $k_2$ in (6.19.2), we find

$$\Delta^{n-1} u(k_2) > \epsilon \sum_{\ell=k_2}^{k-1} f_j(\ell)$$

and this leads to a contradiction.

**Case 2.** $\text{Lim}_{k\to\infty} u(k) = c > 0$ and $\lim_{k\to\infty} \Delta^i u(k) = 0$ for $1 \le i \le n-1$. If $c < \infty$, then since $F_j$ is continuous for every $0 < \epsilon < F_j(c, 0, \cdots, 0)$ there exists a $k_2 \ge k_1$ such that $F_j(c, 0, \cdots, 0) - \epsilon < F_j\left(u(k), \Delta u(k), \cdots, \Delta^{n-1} u(k)\right)$ for all $k \in \mathbb{N}(k_2)$, and from (6.19.2) we get

$$\Delta^{n-1} u(k_2) > [F_j(c, 0, \cdots, 0) - \epsilon] \sum_{\ell=k_2}^{k-1} f_j(\ell),$$

which is again a contradiction. If $c = \infty$, then also from (iii) (a)′ we have a contradiction. This completes the proof. ∎

**Theorem 6.19.3.** Let in addition to the hypotheses (i) and (ii)

(iii)′ there is an index $j \in \mathbb{N}(1,m)$ such that $F_j(u_1,\cdots,u_n)$ is continuous at $(u_1,0,\cdots,0)$ with $u_1 \neq 0$, and

$$\sum^{\infty}(\ell)^{(n-1)}f_j(\ell) = \infty. \tag{6.19.3}$$

Then, (1) if $n$ is even, every bounded solution of (6.19.1) is oscillatory

(2) if $n$ is odd, every bounded solution of (6.19.1) is oscillatory or tends to zero monotonically.

**Proof.** Assuming $u(k) > 0$ (the case $u(k) < 0$ can be similarly treated) is bounded on $\mathbb{N}(k_1)$. Then, $\Delta^n u(k) \le 0$ for all $k \in \mathbb{N}(k_1)$, and from Corollary 1.8.13 we have $\lim_{k\to\infty}\Delta^i u(k) = 0$, $1 \le i \le n-1$. Also, $\Delta u(k) \ge 0$ if $n$ is even, whereas for $n$ odd $\Delta u(k) \le 0$ for all $k \in \mathbb{N}(k_1)$. Since $u(k)$ is bounded, we find for $n$ even $u(\infty) = c > 0$, and for $n$ odd either $u(\infty) = c > 0$ or $u(\infty) = 0$. Thus, to complete the proof we need to consider the case $u(\infty) = c > 0$ whether $n$ is even or odd.

Since $F_j$ is continuous, we find $\lim\limits_{k\to\infty} F_j(u(k),\Delta u(k),\cdots,\Delta^{n-1}u(k)) = F_j(c,0,\cdots,0) > \epsilon > 0$. Hence, there exists a $k_2 \ge k_1$ such that $F_j(c,0,\cdots,0) - \epsilon < F_j\left(u(k),\Delta u(k),\cdots,\Delta^{n-1}u(k)\right)$ for all $k \in \mathbb{N}(k_2)$.

From the equation (6.19.1), we have

$$\Delta^n u(k) + f_j(k)F_j\left(u(k),\Delta u(k),\cdots,\Delta^{n-1}u(k)\right) \le 0, \quad k \in \mathbb{N}(k_2)$$

and hence

$$\Delta^n u(k) + [F_j(c,0,\cdots,0) - \epsilon]f_j(k) < 0, \quad k \in \mathbb{N}(k_2).$$

Multiplying the above inequality by $(k)^{(n-1)}$, and summing from $k_2$ to $k-1$, and using Problem 1.9.36, to obtain

$$\sum_{i=1}^{n}(-1)^{i+1}\Delta^{i-1}(\ell)^{(n-1)}\Delta^{n-i}u(\ell+i-1)\Big|_{\ell=k_2}^{k}$$

$$+ [F_j(c,0,\cdots,0) - \epsilon]\sum_{\ell=k_2}^{k-1}(\ell)^{(n-1)}f_j(\ell) < 0.$$

Thus, from Corollary 1.8.13, we get

$$[F_j(c,0,\cdots,0)-\epsilon]\sum_{\ell=k_2}^{k-1}(\ell)^{(n-1)}f_j(\ell) \;<\; L+(-1)^{n+2}(n-1)!u(k+n-1),$$

where $L$ is some finite constant. But, the above inequality in view of (6.19.3) leads to a contradiction to our assumption that $u(k)$ is bounded. ∎

Our next result is for the even order difference equations, so in (6.19.1) we shall assume that $n=2p$, where $p\geq 1$.

**Theorem 6.19.4.** Let in addition to the hypotheses (i), (ii) and (iv)

(v) $I\neq\phi$, where $I$ denotes the set of all indices $i$ for which the function $F_i(u_1,\cdots,u_{2p})$ is nondecreasing with respect to each variable $u_2,u_4,\cdots,u_{2p}$ and nonincreasing with respect to $u_3,u_5,\cdots,u_{2p-1}$ as well as the function $\frac{1}{u_1}F_i(u_1,0,\cdots,0)$ is nonincreasing on $(0,\infty)$

(vi) there exists an eventually positive function $\phi(k)$, $k\in\mathbb{N}$ such that

$$\sum^{\infty}\left[\phi(\ell)\sum_{i\in I}\frac{f_i\left(2^{2p-2}\ell\right)F_i\left(c(\ell)^{(2p-1)},0,\cdots,0\right)}{c(\ell)^{(2p-1)}}-\frac{1}{4}\frac{(2p-2)!(\Delta\phi(\ell))^2}{(\ell/2)^{(2p-2)}\phi(\ell)}\right]=\infty$$

for every $c\geq 1$.

Then, every bounded solution of (6.19.1) with $n=2p$ is oscillatory.

**Proof.** Assuming $u(k)>0$ is bounded on $\mathbb{N}(k_1)$. Then, $\Delta^{2p}u(k)\leq 0$ for all $k\in\mathbb{N}(k_1)$, and from Corollary 1.8.13 we have $\lim_{k\to\infty}\Delta^i u(k)=0$, $1\leq i\leq 2p-1$, and $(-1)^{i+1}\Delta^{2p-i}u(k)\geq 0$, $1\leq i\leq 2p-1$. We define the transformation $v(k)=-\dfrac{\Delta^{2p-1}u\left(2^{2p-2}k\right)}{u(k)}\phi(k)\leq 0$, $k\in\mathbb{N}(k_1)$, and obtain

$$\Delta v(k) \;=\; -\frac{\Delta^{2p}u\left(2^{2p-2}k\right)}{u(k)}\phi(k)+\frac{\Delta\phi(k)v(k+1)}{\phi(k+1)}-\frac{v(k+1)\phi(k)\Delta u(k)}{u(k)\phi(k+1)}.$$

Therefore, from the equation (6.19.1) and the hypothesis (v) it follows that

$$\text{(6.19.4)}\quad \Delta v(k) \;\geq\; \phi(k)\sum_{i\in I}\frac{f_i\left(2^{2p-2}k\right)F_i(u(k),0,\cdots,0)}{u(k)}+\frac{\Delta\phi(k)v(k+1)}{\phi(k+1)}$$

$$-\frac{v(k+1)\phi(k)\Delta u(k+1)}{u(k)\phi(k+1)},\quad k\in\mathbb{N}(k_1).$$

Since $\Delta u(k+1) > 0$ and $\Delta^{2p-1}(\Delta u(k)) \le 0$, Corollary 1.8.12 is applicable and, we find a $k_2 \ge k_1$ such that

$$\Delta u(k+1) \ \ge \ \frac{1}{(2p-2)!}\Delta^{2p-1}u\left(2^{2p-2}\overline{k+1}\right)(k+1-k_2)^{(2p-2)}, \quad k \in \mathbb{N}(k_2)$$

which is the same as

$$\Delta u(k+1) \ \ge \ -\frac{v(k+1)u(k+1)}{\phi(k+1)}\frac{(k+1-k_2)^{(2p-2)}}{(2p-2)!}, \qquad k \in \mathbb{N}(k_2).$$

Using the above inequality in (6.19.4), we get

$$(6.19.5) \quad \Delta v(k) \ \ge \ \phi(k)\sum_{i\in I}\frac{f_i\left(2^{2p-2}k\right)F_i(u(k),0,\cdots,0)}{u(k)} + \left[\frac{\Delta\phi(k)v(k+1)}{\phi(k+1)}\right.$$
$$\left. + \frac{v^2(k+1)\phi(k)u(k+1)}{\phi^2(k+1)u(k)}\frac{(k+1-k_2)^{(2p-2)}}{(2p-2)!}\right], \quad k \in \mathbb{N}(k_2).$$

Since $\dfrac{u(k+1)}{u(k)} \ge 1$, the terms inside the bracket, say, $A$ are

$$A \ \ge \ \phi(k)\frac{(k+1-k_2)^{(2p-2)}}{(2p-2)!}\left[\frac{v(k+1)}{\phi(k+1)} + \frac{1}{2}\frac{(2p-2)!}{(k+1-k_2)^{(2p-2)}}\frac{\Delta\phi(k)}{\phi(k)}\right]^2$$
$$-\frac{1}{4}\frac{(2p-2)!}{(k+1-k_2)^{(2p-2)}}\frac{(\Delta\phi(k))^2}{\phi(k)}, \quad k \in \mathbb{N}(k_2).$$

Using this in (6.19.5), we obtain

$$(6.19.6) \quad \Delta v(k) \ \ge \ \phi(k)\sum_{i\in I}\frac{f_i\left(2^{2p-2}k\right)F_i(u(k),0,\cdots,0)}{u(k)}$$
$$-\frac{1}{4}\frac{(2p-2)!}{(k+1-k_2)^{(2p-2)}}\frac{(\Delta\phi(k))^2}{\phi(k)}, \quad k \in \mathbb{N}(k_2).$$

Next, from Theorem 1.8.5, we have

$$u(k) \ \le \ \sum_{i=0}^{2p-1}\frac{(k-k_2)^{(i)}}{i!}\Delta^i u(k_2), \quad k \in \mathbb{N}(k_2)$$

and hence there exists some $c \ge 1$ such that

$$u(k) \ \le \ c(k)^{(2p-1)}, \quad k \in \mathbb{N}(k_2).$$

Using this in (6.19.6), we find

$$\Delta v(k) \ \ge \ \phi(k)\sum_{i\in I}\frac{f_i\left(2^{2p-2}k\right)F_i\left(c(k)^{(2p-1)},0,\cdots,0\right)}{c(k)^{(2p-1)}}$$
$$-\frac{1}{4}\frac{(2p-2)!}{(k/2)^{(2p-2)}}\frac{(\Delta\phi(k))^2}{\phi(k)}, \quad k \in \mathbb{N}(2k_2-2).$$

Summing up the above inequality, we find from (vi) that $v(k)$ is eventually positive, which is a contradiction. Hence the result follows. ∎

**Corollary 6.19.5.** Let in addition to the hypotheses (i), (ii) and (iv)

(vii) $I \neq \emptyset$, where $I$ denotes the set of all indices $i$ for which the function $F_i(u_1, \cdots, u_{2p})$ is nondecreasing with respect to each variable $u_2, u_4, \cdots, u_{2p}$ and nonincreasing with respect to $u_3, u_5, \cdots, u_{2p-1}$, and $F_i(\lambda u_1, 0, \cdots, 0) = \lambda F_i(u_1, 0, \cdots, 0)$ for all $u_1 \in \mathbb{R}$ and real $\lambda \neq 0$

(viii) there exists an eventually positive function $\phi(k)$, $k \in \mathbb{N}$ such that

$$\sum^{\infty} \left[ \phi(\ell) \sum_{i \in I} f_i \left(2^{2p-2}\ell\right) F_i(1, 0, \cdots, 0) - \frac{1}{4} \frac{(2p-2)!(\Delta\phi(\ell))^2}{(\ell/2)^{(2p-2)}\phi(\ell)} \right] = \infty.$$

Then, every bounded solution of (6.19.1) with $n = 2p$ is oscillatory.

## 6.20. Oscillation and Nonoscillation for

$$u(k+1) - u(k) + p(k)u(k-m) = 0, \quad k \in \mathbb{N} \tag{6.20.1}$$

where $m \in \mathbb{N}(1)$ is fixed, and the function $p$ is defined on $\mathbb{N}$.

**Theorem 6.20.1.** Let $p(k) \geq 0$ for all $k \in \mathbb{N}$ and $\limsup_{k \to \infty} \sum_{\ell=k-m}^{k} p(\ell) > 1$. Then, the difference equation (6.20.1) is oscillatory.

**Proof.** Let $u(k) > 0$ for all $k \in \mathbb{N}(k_1)$ be a solution of (6.20.1). Since $p(k) \geq 0$, for all $k \in \mathbb{N}(k_1 + m)$ equation (6.20.1) implies that $\Delta u(k) \leq 0$, and hence $u(k)$ is nonincreasing on $\mathbb{N}(k_1 + m)$. Therefore, $\lim_{k \to \infty} u(k) = \alpha \geq 0$ exists. But taking the limit in (6.20.1) ensures that $\alpha = 0$. Now summing (6.20.1) from $k_2 \in \mathbb{N}(k_1 + m)$, to $k_2 + m$, to obtain

$$u(k_2 + m + 1) - u(k_2) + \sum_{\ell=k_2}^{k_2+m} p(\ell)u(\ell - m) = 0,$$

which implies that

$$u(k_2 + m + 1) - u(k_2) \left[ 1 - \sum_{\ell=k_2}^{k_2+m} p(\ell) \right] \leq 0.$$

Therefore, $1 - \sum_{\ell=k_2}^{k_2+m} p(\ell) \geq 0$, and hence $1 \geq \limsup_{k_2 \to \infty} \sum_{\ell=k_2}^{k_2+m} p(\ell)$. This contradiction completes the proof. ∎

**Theorem 6.20.2.** Suppose that

$$\liminf_{k\to\infty} p(k) \;=\; c \;>\; 0 \quad \text{and} \quad \limsup_{k\to\infty} p(k) \;>\; 1-c. \tag{6.20.2}$$

Then, the following hold

$$\text{(i)} \quad v(k+1) - v(k) + p(k)v(k-m) \;\le\; 0, \quad k \in \mathbb{N} \tag{6.20.3}$$

has no eventually positive solution

$$\text{(ii)} \quad w(k+1) - w(k) + p(k)w(k-m) \;\ge\; 0, \quad k \in \mathbb{N} \tag{6.20.4}$$

has no eventually negative solution

(iii) difference equation (6.20.1) is oscillatory.

**Proof.** Assume that $v(k)$ is an eventually positive solution of (6.20.3), i.e. there exists a $k_1 \in \mathbb{N}(1)$ such that $v(k) > 0$ for all $k \in \mathbb{N}(k_1)$. Let $\epsilon > 0,\ 0 < \epsilon < c$ and $k_2 \ge k_1$ be such that $p(k) \ge c-\epsilon > 0$ for all $k \in \mathbb{N}(k_2)$. Let $k_3 = \max\{k_1+m,\ k_2\}$ so that $v(k) \ge p(k)v(k-m) \ge (c-\epsilon)v(k-1)$ for all $k \in \mathbb{N}(k_3)$, since $v(k)$ is nonincreasing for all $k \in \mathbb{N}(k_3)$. On the other hand, we have $0 \ge v(k+1)-v(k)+p(k)v(k-m) \ge v(k+1)+v(k)(p(k)-1)$ for all $k \in \mathbb{N}(k_3)$, so that $v(k)(p(k)-1+c-\epsilon) \le 0$ for all $k \in \mathbb{N}(k_3)$. Thus, it follows that $p(k) \le 1-c+\epsilon$ for all $k \in \mathbb{N}(k_3)$, and hence $\limsup_{k\to\infty} p(k) \le 1-c+\epsilon$. However, since $\epsilon > 0$ is arbitrary we have $\limsup_{k\to\infty} p(k) \le 1-c$. This contradicts (6.20.2) and the proof of (i) is complete. The conclusion (ii) follows from (i) by letting $v(k) = -w(k)$ for an eventually negative solution $w(k)$ of (6.20.4). Finally, (iii) follows from (i) and (ii). ∎

**Theorem 6.20.3.** Suppose that

$$\liminf_{k\to\infty} p(k) \;=\; c > m^m/(m+1)^{m+1}. \tag{6.20.5}$$

Then, the conclusions of Theorem 6.20.2 hold.

**Proof.** Assume the contrary and let $v(k)$ be a solution of (6.20.3) with $v(k) > 0$ for all $k \in \mathbb{N}(k_1)$. Setting $r(k) = v(k)/v(k+1)$ and dividing the inequality (6.20.3) by $v(k)$ and arranging the terms, we obtain

$$[r(k)]^{-1} \;\le\; 1 - p(k)r(k-m)\cdots r(k-1), \quad k \in \mathbb{N}(k_1+m). \tag{6.20.6}$$

From (6.20.5), $p(k) > 0$ for all $k \in \mathbb{N}(k_2)$, where $k_2 \ge k_1$. Setting $k_3 = \max\{k_1+m,\ k_2\}$, it follows that $v(k)$ is nonincreasing on $\mathbb{N}(k_3)$, and so $r(k) \ge 1$ for all $k \in \mathbb{N}(k_3)$. Also, $r(k)$ is bounded above otherwise (6.20.5) and (6.20.6) imply that $r(k) < 0$ for arbitrarily large

$k$. If we set $\liminf_{k\to\infty} r(k) = \beta$, then from (6.20.6), we get

$$\begin{aligned}\limsup_{k\to\infty}\frac{1}{r(k)} &= \frac{1}{\beta} \le 1-\liminf_{k\to\infty}[p(k)r(k-m)\cdots r(k-1)]\\ &\le 1-c\beta^m,\end{aligned}$$

which gives that $c \le \dfrac{\beta-1}{\beta^{m+1}} \le \max\limits_{\beta\ge 1}\dfrac{\beta-1}{\beta^{m+1}} \le \dfrac{m^m}{(m+1)^{m+1}}$. But this contradicts (6.20.5), and the proof of (i) is complete. The conclusions (ii) and (iii) can be proved similarly. ∎

**Remark 6.20.1.** For the difference equation

$$u(k+1)-u(k)+\frac{m^m}{(m+1)^{m+1}}u(k-m) = 0, \quad k\in\mathbb{N}$$

together with $u(i-m) = (m/(m+1))^{i-m}\alpha$, $0\le i\le m$, $\alpha\ne 0$, the solution is $u(k) = (m/(m+1))^k\alpha$, $k\in\mathbb{N}$ which is obviously nonoscillatory. Thus, the inequality (6.20.5) cannot be replaced by an equality.

**Theorem 6.20.4.** Suppose that $p(k)\ge 0$ for all $k\in\mathbb{N}$, and

(6.20.7) $$\sup_{k\in\mathbb{N}} p(k) < m^m/(m+1)^{m+1}.$$

Then, the difference equation (6.20.1) has a nonoscillatory solution.

**Proof.** We shall show that

(6.20.8) $$[\eta(k)]^{-1} = 1-p(k)\eta(k-m)\cdots\eta(k-1), \quad k\in\mathbb{N}$$

has a positive solution. For this, we define

(6.20.9) $$\eta(i-m) = q = \frac{m+1}{m} > 1, \quad 0\le i\le m-1$$

and

(6.20.10) $$\eta(0) = (1-p(0)\eta(-m)\cdots\eta(-1))^{-1} > 1.$$

From (6.20.9) and (6.20.10) it follows that $\eta(0) < q$, so we define

$$\eta(1) = (1-p(1)\eta(-m+1)\cdots\eta(0))^{-1} < q$$

and now by induction $1<\eta(k)<q$ for all $k=2,3,\cdots$ so that $\eta(k)$ is a solution of (6.20.8). Next defining $u(i-m) = \left(\dfrac{m+1}{m}\right)^{m-i}$, $0\le i\le m$,

$u(k) = \dfrac{u(k-1)}{\eta(k-1)}$, $k \in \mathbb{N}(1)$ it follows that this $u(k)$ is a nonoscillatory solution of (6.20.1). ∎

In Theorem 6.20.3 the left side of the inequality (6.20.5) can be improved. This is the content of our next result.

**Theorem 6.20.5.** Assume that $p(k) \geq 0$ for all $k \in \mathbb{N}$ and

$$\liminf_{k \to \infty} \left[ \frac{1}{m} \sum_{\ell=k-m}^{k-1} p(\ell) \right] > \frac{m^m}{(m+1)^{m+1}}. \tag{6.20.11}$$

Then, the conclusions of Theorem 6.20.2 hold.

**Proof.** We shall prove only (iii), whereas (i) and (ii) can be proved analogously. Let $u(k)$ be a nonoscillatory solution of (6.20.1), which we can assume to be positive eventually, and since $p(k) \geq 0$ this solution $u(k)$ is eventually decreasing. Therefore, on using $u(k) \leq u(k-m)$ in (6.20.1), eventually we obtain

$$p(k) \leq 1 - u(k+1)/u(k)$$

and hence on using arithmetic and geometric means inequality, we find

$$\begin{aligned} \frac{1}{m} \sum_{\ell=k-m}^{k-1} p(\ell) &\leq 1 - \frac{1}{m} \sum_{\ell=k-m}^{k-1} u(\ell+1)/u(\ell) \\ &\leq 1 - (u(k)/u(k-m))^{1/m}. \end{aligned} \tag{6.20.12}$$

Setting $\alpha = m^m/(m+1)^{m+1}$, from (6.20.11) we can choose a constant $\beta$ such that for $k$ sufficiently large $\alpha < \beta \leq (1/m)\sum_{\ell=k-m}^{k-1} p(\ell)$. Therefore, from (6.20.12) for all large $k$, $(u(k)/u(k-m))^{1/m} \leq 1 - \beta$, which in particular implies that $0 < \beta < 1$. Now since $\max_{0 \leq \lambda \leq 1} \left[(1-\lambda)\lambda^{1/m}\right] = \alpha^{1/m}$, we have $1 - \lambda \leq \alpha^{1/m}\lambda^{-1/m}$ for $0 < \lambda \leq 1$, and hence it follows that $(u(k)/u(k-m))^{1/m} \leq \alpha^{1/m}\beta^{-1/m}$, which is the same as

$$\frac{\beta}{\alpha} u(k) \leq u(k-m). \tag{6.20.13}$$

Now using (6.20.13) instead of $u(k) \leq u(k-m)$ in (6.20.1) and repeating the arguments, we find $(\beta/\alpha)^2 u(k) \leq u(k-m)$ for all large $k$. Thus, by induction, for every $n \in \mathbb{N}(1)$ there exists an integer $k_n$ such that for all $k \in \mathbb{N}(k_n)$

$$\left(\frac{\beta}{\alpha}\right)^n u(k) \leq u(k-m). \tag{6.20.14}$$

Next, for sufficiently large $k$, $\sum_{\ell=k-m}^{k} p(\ell) \geq \sum_{\ell=k-m}^{k-1} p(\ell) \geq m\beta = M$, say. Since $\beta > \alpha$, we can choose $n$ such that

$$\left(\frac{\beta}{\alpha}\right)^n > \left(\frac{2}{M}\right)^2. \tag{6.20.15}$$

For this specific value of $n$, we consider $k$ sufficiently large, say, $k^*$ so that for all $k \geq k^*$, all the above inequalities are satisfied. Then, for each $k \geq k^* + m$ there exists an integer $\hat{k}$ with $k - m \leq \hat{k} \leq k$ so that $\sum_{\ell=k-m}^{\hat{k}} p(\ell) \geq (M/2)$ and $\sum_{\ell=\hat{k}}^{k} p(\ell) \geq (M/2)$. From (6.20.1) and the nonincreasing nature of $u(k)$, we have

$$\begin{aligned}
-u(k-m) &\leq u(\hat{k}+1) - u(k-m) = \sum_{\ell=k-m}^{\hat{k}} (u(\ell+1) - u(\ell)) \\
&= -\sum_{\ell=k-m}^{\hat{k}} p(\ell)u(\ell-m) \leq -\left(\sum_{\ell=k-m}^{\hat{k}} p(\ell)\right) u(\hat{k}-m) \\
&\leq -\frac{M}{2}u(\hat{k}-m)
\end{aligned}$$

and hence

$$\frac{M}{2}u(\hat{k}-m) \leq u(k-m). \tag{6.20.16}$$

Similarly, we find

$$\begin{aligned}
-u(\hat{k}) &\leq u(k+1) - u(\hat{k}) = \sum_{\ell=\hat{k}}^{k}(u(\ell+1) - u(\ell)) \\
&= -\sum_{\ell=\hat{k}}^{k} p(\ell)u(\ell-m) \leq -\left(\sum_{\ell=\hat{k}}^{k} p(\ell)\right) u(k-m) \\
&\leq -\frac{M}{2}u(k-m)
\end{aligned}$$

and so

$$\frac{M}{2}u(k-m) \leq u(\hat{k}). \tag{6.20.17}$$

Combining (6.20.14), (6.20.16) and (6.20.17), we get

$$\left(\frac{\beta}{\alpha}\right)^n \leq \frac{u(\hat{k}-m)}{u(\hat{k})} \leq \left(\frac{2}{M}\right)^2.$$

But this contradicts (6.20.15) and the proof is complete. ∎

## 6.21. Oscillation and Nonoscillation for

$$\Delta_\alpha u(k) + \delta \sum_{i=1}^{m} f_i(k) F_i(u(g_i(k))) = 0, \quad k \in \mathbb{N}, \quad \delta = \pm 1 \qquad (6.21.1)_\delta$$

where $\alpha > 0$ is a real fixed constant, $\Delta_\alpha u(k) = u(k+1) - \alpha u(k)$, for each $i$, $1 \le i \le m$, $f_i$ is defined on $\mathbb{N}$, $F_i$ is defined on $\mathbb{R}$, and $\{g_i(k)\} \subseteq \mathbb{N}$. Further, we shall assume that

(i) $f_i(k) \ge 0$ for all $k \in \mathbb{N}$, $1 \le i \le m$

(ii) $\lim_{k\to\infty} g_i(k) = \infty$, $1 \le i \le m$

(iii) $uF_i(u) > 0$ for $u \ne 0$, $1 \le i \le m$.

**Theorem 6.21.1.** Let $\alpha \ge 1$ and let there exist an index $j \in \mathbb{N}(1, m)$ such that $|F_j(u)|$ is bounded away from zero if $|u|$ is bounded away from zero, and

$$\sum^{\infty} \alpha^{-\ell} f_j(\ell) = \infty. \qquad (6.21.2)$$

Then, every solution $u(k)$ of $(6.21.1)_1$ is either oscillatory or $u(k) = o(\alpha^k)$.

**Proof.** Let $u(k)$ be a nonoscillatory solution of $(6.21.1)_1$, and suppose that $u(k) > 0$ eventually. Then, there exists a $k_1 \in \mathbb{N}$ such that $u(k) > 0$ and $u(g_i(k)) > 0$, $1 \le i \le m$ for all $k \in \mathbb{N}(k_1)$. Therefore, we have $\Delta_\alpha u(k) = \alpha^{k+1}\Delta(u(k)/\alpha^k) \le 0$ for all $k \in \mathbb{N}(k_1)$. Hence, $\alpha^{-k}u(k)$ is nonincreasing for all $k \in \mathbb{N}(k_1)$, thus $\lim_{k\to\infty} \alpha^{-k}u(k) = \mu \ge 0$ exists. We shall show that $\mu = 0$. Suppose $\mu > 0$, then there exists $k_2 \in \mathbb{N}(k_1)$ such that $u(g_j(k)) \ge \mu\alpha^{g_j(k)} \ge \mu$ for all $k \in \mathbb{N}(k_2)$, and from the given hypotheses there exists a positive constant $c$ such that $F_j(u(g_j(k))) \ge c$ for all $k \in \mathbb{N}(k_2)$.

On the other hand from $(6.21.1)_1$, we have

$$\Delta(u(k)/\alpha^k) + \alpha^{-k-1} f_j(k) F_j(u(g_j(k))) \le 0, \quad k \in \mathbb{N}(k_1) \qquad (6.21.3)$$

and hence

$$\frac{u(k)}{\alpha^k} \le \frac{u(k_2)}{\alpha^{k_2}} - \frac{c}{\alpha} \sum_{\ell=k_2}^{k-1} \alpha^{-\ell} f_j(\ell), \quad k \in \mathbb{N}(k_2).$$

But, in view of (6.21.2) this leads to a contradiction to our assumption that $u(k) > 0$ eventually. The case $u(k) < 0$ eventually can be treated similarly. ∎

**Theorem 6.21.2.** Let there exist an index $j \in \mathbb{N}(1,m)$ and a positive constant $L$ such that

$$|F_j(u)| \ \geq \ L|u| \ \text{ for } \ u \in \mathbb{R},$$
$$\mathbb{N}_j \ = \ \{k \in \mathbb{N} : g_j(k) \leq k\} \ \text{ is an infinite set,}$$
$$Lf_j(k)\alpha^{g_j(k)-k-1} \ \geq \ 1 \ \text{ for all } \ k \in \mathbb{N}_j.$$

Then, the difference equation $(6.21.1)_1$ is oscillatory.

**Proof.** Let $u(k)$ be as in Theorem 6.21.1 so that $u(k)/\alpha^k$ is nonincreasing for all $k \in \mathbb{N}(k_1)$. Thus, for all $k \in \mathbb{N}_j \cap \mathbb{N}(k_1)$ it follows that $u(g_j(k)) \geq u(k)\alpha^{g_j(k)-k}$. On the other hand, we have

$$\begin{aligned} u(k+1) &\leq \alpha u(k) - f_j(k)F_j(u(g_j(k))), \quad k \in \mathbb{N}(k_1) \\ &\leq \alpha u(k) - Lf_j(k)u(g_j(k)), \quad k \in \mathbb{N}(k_1) \\ &\leq \alpha u(k)\left(1 - Lf_j(k)\alpha^{g_j(k)-k-1}\right) \ \leq \ 0, \quad k \in \mathbb{N}_j \cap \mathbb{N}(k_1). \end{aligned}$$

But, this contradicts our assumption that $u(k) > 0$ eventually. A similar contradiction holds for $u(k) < 0$ eventually. ■

**Theorem 6.21.3.** Let $\alpha \geq 1$ and let there exist an index $j \in \mathbb{N}(1,m)$ such that $F_j(u)$ is nondecreasing on $\mathbb{R}\backslash\{0\}$, and

$$\text{(6.21.4)} \quad \int_0^{\beta} \frac{dt}{F_j(t)} \ < \ \infty \quad \text{and} \quad \int_0^{-\beta} \frac{dt}{F_j(t)} \ < \ \infty \quad \text{for every} \quad \beta > 0,$$

$$\text{(6.21.5)} \qquad \sum^{\infty} \chi(\mathbb{N}_j(\ell))\alpha^{-\ell} f_j(\ell) \ = \ \infty,$$

where $\chi(\mathbb{N}_j(\ell))$ is the characteristic function of the set $\mathbb{N}_j(\ell)$ defined in Theorem 6.21.2. Then, the difference equation $(6.21.1)_1$ is oscillatory.

**Proof.** Let $u(k)$ be as in Theorem 6.21.1 so that $u(k)/\alpha^k$ is nonincreasing for all $k \in \mathbb{N}(k_1)$, and the inequality (6.21.3) holds. Let $k_2 \in \mathbb{N}(k_1)$ be so large that $g_j(k) \geq k_1$ for all $k \in \mathbb{N}(k_2)$. Hence, for all $k \in \mathbb{N}_j \cap \mathbb{N}(k_2)$, we have $u(g_j(k)) \geq u(k)/\alpha^k$, and consequently $F_j(u(g_j(k))) \geq F_j(u(k)/\alpha^k)$. Then, from (6.21.3) it follows that

$$-\Delta(u(k)/\alpha^k)/F_j(u(k)/\alpha^k) \ \geq \ \alpha^{-k-1}f_j(k), \quad k \in \mathbb{N}_j \cap \mathbb{N}(k_2).$$

However, since for $\dfrac{u(k+1)}{\alpha^{k+1}} \leq t \leq \dfrac{u(k)}{\alpha^k}$, $[F_j(t)]^{-1} \geq [F_j(u(k)/\alpha^k)]^{-1}$, the above inequality implies

$$\alpha^{-k-1}f_j(k) \ \leq \ \int_{u(k+1)/\alpha^{k+1}}^{u(k)/\alpha^k} \frac{dt}{F_j(t)}, \quad k \in \mathbb{N}_j \cap \mathbb{N}(k_2).$$

This, summing over $k$ leads to the inequality

$$\begin{aligned}\sum_{\ell=k_2}^{k} \chi(\mathbb{N}_j(\ell))\alpha^{-\ell-1} f_j(\ell) &\leq \sum_{\ell=k_2}^{k} \chi(\mathbb{N}_j(\ell)) \int_{u(\ell+1)/\alpha^{\ell+1}}^{u(\ell)/\alpha^{\ell}} \frac{dt}{F_j(t)} \\ &\leq \int_{u(k+1)/\alpha^{k+1}}^{u(k_2)/\alpha^{k_2}} \frac{dt}{F_j(t)}, \quad k \in \mathbb{N}(k_2).\end{aligned}$$

But, from (6.21.4) and (6.21.5) this leads to a contradiction. A similar argument holds for $u(k) < 0$ eventually. ∎

**Theorem 6.21.4.** Let $0 < \alpha \leq 1$ and let there exist an index $j \in \mathbb{N}(1, m)$ such that $F_j(u)$ is nondecreasing on $\mathbb{R}\backslash\{0\}$, and

$$\lim_{k\to\infty} \sum^{k} \alpha^{k-\ell} f_j(\ell) F_j\left(c\alpha^{g_j(\ell)}\right) = \pm\infty \quad \text{for any} \quad c \neq 0. \tag{6.21.6}$$

Then, every solution $u(k)$ of $(6.21.1)_{-1}$ is oscillatory or $|u(k)| \to \infty$ as $k \to \infty$.

**Proof.** Let $u(k)$ be a nonoscillatory solution of $(6.21.1)_{-1}$, and suppose that $u(k) > 0$ eventually. Then, there exists a $k_1 \in \mathbb{N}$ such that $u(k) > 0$ and $u(g_i(k)) > 0$, $1 \leq i \leq m$ for all $k \in \mathbb{N}(k_1)$. Therefore, we have $\Delta(u(k)/\alpha^k) \geq 0$ for all $k \in \mathbb{N}(k_1)$. Hence, $\alpha^{-k}u(k)$ is nondecreasing for all $k \in \mathbb{N}(k_1)$, thus $u(k) \geq \alpha^{k-k_1}u(k_1)$, $k \in \mathbb{N}(k_1)$. Obviously, we can choose $k_2 \in \mathbb{N}(k_1)$ such that $g_j(k) \geq k_1$ for all $k \in \mathbb{N}(k_2)$, and so $u(g_j(k)) \geq c\alpha^{g_j(k)}$, $k \in \mathbb{N}(k_2)$, where $c = u(k_1)/\alpha^{k_1}$.

On the other hand from $(6.21.1)_{-1}$, we have

$$\Delta(u(k)/\alpha^k) \geq \alpha^{-k-1} f_j(k) F_j(u(g_j(k))), \quad k \in \mathbb{N}(k_1)$$

and hence

$$\Delta(u(k)/\alpha^k) \geq \alpha^{-k-1} f_j(k) F_j\left(c\alpha^{g_j(k)}\right), \quad k \in \mathbb{N}(k_2)$$

from which it follows that

$$u(k) \geq \alpha^{k-k_2}u(k_2) + \frac{1}{\alpha}\sum_{\ell=k_2}^{k-1} \alpha^{k-\ell} f_j(\ell) F_j\left(c\alpha^{g_j(\ell)}\right), \quad k \in \mathbb{N}(k_2).$$

The result now follows from (6.21.6). The case $u(k) < 0$ can be treated analogously. ∎

## 6.22. Oscillation and Nonoscillation for

$$\Delta\left(p(k)(\Delta u(k))^{\sigma}\right) + q(k+1)f(u(k+1)) = 0, \quad k \in \mathbb{N} \tag{6.22.1}$$

where $\sigma$ is a quotient of positive odd integers, $p(k)$, $k \in \mathbb{N}$ is a positive function, and $q(k)$ is defined on $\mathbb{N}(1)$. In what follows we shall assume that

(i) $uf(u) > 0$ for $u \neq 0$, and

(ii) $f(u) - f(v) = g(u,v)(u-v)^{\delta}$, $u, v \neq 0$ where $g : \mathbb{R} \times \mathbb{R} \to \mathbb{R}$ is nonnegative and $\delta$ is a quotient of positive odd integers.

We shall need the following:

**Lemma 6.22.1.** Let the function $G(k,\ell,u) : \mathbb{N}(k_0) \times \mathbb{N}(k_0) \times \mathbb{R}_+ \to \mathbb{R}$ be nondecreasing in $u$ for fixed $k$ and $\ell$. Further, let for $k \in \mathbb{N}(k_0)$

$$u(k) \ \geq \ (\leq) \ a(k) + \sum_{\ell=k_0}^{k-1} G(k,\ell,u(\ell)) \tag{6.22.2}$$

and

$$v(k) \ = \ a(k) + \sum_{\ell=k_0}^{k-1} G(k,\ell,v(\ell)), \tag{6.22.3}$$

where $a(k)$, $k \in \mathbb{N}(k_0)$ is a given function. Then, $u(k) \geq (\leq) v(k)$ for all $k \in \mathbb{N}(k_0)$.

**Lemma 6.22.2.** Let $u(k)$, $k \in \mathbb{N}$ be a solution of (6.22.1), and let there exist a positive number $\mu$, and integers $a, b, c$ such that $0 \leq a \leq b < c$. Further, let $\rho(k)$, $k \in \mathbb{N}(a)$ be a positive function, and the following inequality holds

$$\begin{aligned} -\frac{p(a)(\Delta u(a))^{\sigma}\rho(a)}{f(u(a))} &+ \sum_{\ell=a}^{k-1}\left[q(\ell+1)\rho(\ell+1) - \frac{p(\ell)(\Delta u(\ell))^{\sigma}\Delta\rho(\ell)}{f(u(\ell+1))}\right] \\ &+ \sum_{\ell=a}^{b-1}\frac{p(\ell)(\Delta u(\ell))^{\sigma+\delta}\rho(\ell)g(u(\ell+1),u(\ell))}{f(u(\ell))f(u(\ell+1))} \ \geq \ \mu, \quad k \in \mathbb{N}(b,c). \end{aligned} \tag{6.22.4}$$

Then, if $u(k) > 0$, $k \in \mathbb{N}(a,c)$,

$$p(k)\rho(k)(\Delta u(k))^{\sigma} \ \leq \ -\mu f(u(b)), \quad k \in \mathbb{N}(b,c) \tag{6.22.5}$$

and if $u(k) < 0$, $k \in \mathbb{N}(a,c)$,

$$p(k)\rho(k)(\Delta u(k))^{\sigma} \ \geq \ -\mu f(u(b)), \quad k \in \mathbb{N}(b,c). \tag{6.22.6}$$

**Proof.** Let $w(k) = p(k)(\Delta u(k))^{\sigma}\rho(k)$, $k \in \mathbb{N}$. Then, $w(k)\Delta u(k) = p(k)(\Delta u(k))^{\sigma+\delta}\rho(k) > 0$, $k \in \mathbb{N}$ and

$$\Delta w(k) \ = \ \Delta\left(p(k)(\Delta u(k))^{\sigma}\right)\rho(k+1) + p(k)(\Delta u(k))^{\sigma}\Delta\rho(k)$$

so that in view of (6.22.1), we have

$$\frac{\Delta w(k)}{f(u(k+1))} = -q(k+1)\rho(k+1) + \frac{p(k)(\Delta u(k))^{\sigma}\Delta\rho(k)}{f(u(k+1))}.$$

Since

$$\begin{aligned}\Delta\left(\frac{w(k)}{f(u(k))}\right) &= \frac{f(u(k))\Delta w(k) - w(k)(f(u(k+1)) - f(u(k)))}{f(u(k))f(u(k+1))} \\ &= \frac{\Delta w(k)}{f(u(k+1))} - \frac{w(k)g(u(k+1),u(k))(\Delta u(k))^{\delta}}{f(u(k))f(u(k+1))} \\ &= -q(k+1)\rho(k+1) + \frac{p(k)(\Delta u(k))^{\sigma}\Delta\rho(k)}{f(u(k+1))} \\ &\quad -\frac{w(k)g(u(k+1),u(k))(\Delta u(k))^{\delta}}{f(u(k))f(u(k+1))}\end{aligned}$$

and therefore on summing from $a$ to $k-1$, we get

(6.22.7)

$$\begin{aligned}-\frac{w(k)}{f(u(k))} &= -\frac{w(a)}{f(u(a))} + \sum_{\ell=a}^{k-1}\left[q(\ell+1)\rho(\ell+1) - \frac{p(\ell)(\Delta u(\ell))^{\sigma}\Delta\rho(\ell)}{f(u(\ell+1))}\right] \\ &\quad + \sum_{\ell=a}^{k-1}\frac{w(\ell)g(u(\ell+1),u(\ell))(\Delta u(\ell))^{\delta}}{f(u(\ell))f(u(\ell+1))}.\end{aligned}$$

Thus, in view of (6.22.4) it follows that

$$-\frac{w(k)}{f(u(k))} \geq \mu + \sum_{\ell=b}^{k-1}\frac{w(\ell)g(u(\ell+1),u(\ell))(\Delta u(\ell))^{\delta}}{f(u(\ell))f(u(\ell+1))} > 0, \quad k \in \mathbb{N}(b,c). \tag{6.22.8}$$

Now suppose that $u(k)$ is positive. Then, (6.22.8) implies that $-w(k) > 0$, or equivalently $\Delta u(k) < 0$, $k \in \mathbb{N}(b,c)$. Let $z(k) = -w(k) = -p(k)(\Delta u(k))^{\sigma}\rho(k)$. Then, (6.22.8) becomes

$$z(k) \geq \mu f(u(k)) + \sum_{\ell=b}^{k-1}\frac{f(u(k))(-\Delta u(\ell))^{\delta}g(u(\ell+1),u(\ell))}{f(u(\ell))f(u(\ell+1))}z(\ell). \tag{6.22.9}$$

Define

$$G(k,\ell,x) = \frac{f(u(k))(-\Delta u(\ell))^{\delta}g(u(\ell+1),u(\ell))}{f(u(\ell))f(u(\ell+1))}x, \quad k,\ell \in \mathbb{N}(b,c),\ x \in \mathbb{R}_{+}. \tag{6.22.10}$$

Since $\Delta u(k) < 0,\ k \in \mathbb{N}(b,c)$, we observe that for fixed $k, \ell,\ G(k,\ell,x)$ is nondecreasing in $x$. With $a(k) = \mu f(u(k))$, we apply Lemma 6.22.1, to get

$$z(k) \geq v(k), \quad k \in \mathbb{N}(b,c) \tag{6.22.11}$$

where $v(k)$ satisfies

$$v(k) = \mu f(u(k)) + \sum_{\ell=b}^{k-1} \frac{f(u(k))(-\Delta u(\ell))^\delta g(u(\ell+1), u(\ell))}{f(u(\ell))f(u(\ell+1))} v(\ell), \tag{6.22.12}$$

provided $v(k) \in \mathbb{R}_+$ for all $k \in \mathbb{N}(b,c)$. From (6.22.12) we find

$$\begin{aligned} \Delta\left[\frac{v(k)}{f(u(k))}\right] &= \Delta\left[\mu + \sum_{\ell=b}^{k-1} \frac{(-\Delta u(\ell))^\delta g(u(\ell+1), u(\ell))}{f(u(\ell))f(u(\ell+1))} v(\ell)\right] \\ &= \frac{(-\Delta u(k))^\delta g(u(k+1), u(k))}{f(u(k))f(u(k+1))} v(k). \end{aligned} \tag{6.22.13}$$

On the other hand, we have

$$\Delta\left[\frac{v(k)}{f(u(k))}\right] = \frac{\Delta v(k)}{f(u(k+1))} - \frac{v(k)g(u(k+1), u(k))(\Delta u(k))^\delta}{f(u(k))f(u(k+1))}. \tag{6.22.14}$$

Equating (6.22.13) and (6.22.14), we obtain $\Delta v(k) = 0$, and so $v(k) = v(b) = mf(u(b)),\ k \in \mathbb{N}(b,c)$. The inequality (6.22.5) is now immediate from (6.22.11).

If $u(k)$ is negative, then (6.22.8) gives $w(k) > 0$, or equivalently $\Delta u(k) > 0,\ k \in \mathbb{N}(b,c)$. Let $z(k) = w(k) = p(k)(\Delta u(k))^\sigma \rho(k)$. Then, (6.22.8) becomes

$$z(k) \geq -\mu f(u(k)) + \sum_{\ell=b}^{k-1} \frac{(-f(u(k))(\Delta u(\ell))^\delta g(u(\ell+1), u(\ell))}{f(u(\ell))f(u(\ell+1))} z(\ell). \tag{6.22.15}$$

With $G(k,\ell,x)$ defined as in (6.22.10), we note that for fixed $k,\ell,\ G(k,\ell,x)$ is nondecreasing. Applying Lemma 6.22.1 with $a(k) = -\mu f(u(k))$, we get (6.22.11) where $v(k)$ satisfies

$$v(k) = -mf(u(k)) + \sum_{\ell=b}^{k-1} \frac{(-f(u(k))(\Delta u(\ell))^\delta g(u(\ell+1), u(\ell))}{f(u(\ell))f(u(\ell+1))} v(\ell).$$

Now as in the earlier case $\Delta v(k) = 0$, and hence $v(k) = v(b) = -\mu f(u(b))$. Thus, the inequality (6.22.11) is the same as (6.22.6). ∎

**Remark 6.22.1.** If $u(k)$ is a positive solution of (6.22.1) and if there exists a positive function $\rho(k)$ such that $(\Delta u(k))^\sigma \Delta\rho(k) \le 0,\ k \in \mathbb{N}$

$$\liminf_{k\to\infty} \sum_{\ell=0}^{k} q(\ell)\rho(\ell) > -\infty \tag{6.22.16}$$

and

$$\sum_{\ell=0}^{\infty} \frac{1}{(\rho(\ell)p(\ell))^{1/\sigma}} = \infty, \tag{6.22.17}$$

then

$$\sum_{\ell=0}^{\infty} \frac{p(\ell)(\Delta u(\ell))^{\sigma+\delta}\rho(\ell)g(u(\ell+1),u(\ell))}{f(u(\ell))f(u(\ell+1))} < \infty. \tag{6.22.18}$$

For otherwise there would exist nonnegative integers $a, b, c$ and a positive $\mu$ such that (6.22.4) holds. But, then in view of Lemma 6.22.2, (6.22.5) holds for $k \in \mathbb{N}(b)$. This implies that

$$\Delta u(k) \le -\frac{(\mu f(u(b)))^{1/\sigma}}{(\rho(k)p(k))^{1/\sigma}}, \quad k \in \mathbb{N}.$$

However, then by means of (6.22.17), $u(k)$ will tend to $-\infty$, which is a contradiction.

**Lemma 6.22.3.** Suppose that

$$\sum_{\ell=0}^{\infty} \frac{1}{(p(\ell))^{1/\sigma}} = \infty \tag{6.22.19}$$

$$\lim_{|x|\to\infty} |f(x)| = \infty \tag{6.22.20}$$

and

$$-\infty < \sum_{\ell=0}^{\infty} q(\ell+1) < \infty. \tag{6.22.21}$$

If $u(k)$ is a positive solution or a negative solution of (6.22.1), then

$$\sum_{\ell=0}^{\infty} \frac{p(\ell)(\Delta u(\ell))^{\sigma+\delta}g(u(\ell+1),u(\ell))}{f(u(\ell))f(u(\ell+1))} < \infty \tag{6.22.22}$$

$$(6.22.23) \qquad \lim_{k\to\infty} \frac{p(k)(\Delta u(k))^\sigma}{f(u(k))} = 0$$

and

$$(6.22.24) \quad \frac{p(k)(\Delta u(k))^\sigma}{f(u(k))} = \sum_{\ell=k}^{\infty} q(\ell+1) + \sum_{\ell=k}^{\infty} \frac{p(\ell)(\Delta u(\ell))^{\sigma+\delta} g(u(\ell+1), u(\ell))}{f(u(\ell))f(u(\ell+1))}$$

for all large $k$.

**Proof.** Let $u(k)$ be a positive solution of (6.22.1). Taking $\rho(k) = 1$, Remark 6.22.1 shows that (6.22.22) is true. Now by means of the same reasoning used to derive (6.22.7), we have

$$(6.22.25) \quad \frac{p(k)(\Delta u(k))^\sigma}{f(u(k))} = \frac{p(s)(\Delta u(s))^\sigma}{f(u(s))} - \sum_{\ell=s}^{k-1} q(\ell+1)$$
$$- \sum_{\ell=s}^{k-1} \frac{p(\ell)(\Delta u(\ell))^{\sigma+\delta} g(u(\ell+1), u(\ell))}{f(u(\ell))f(u(\ell+1))}$$
$$= \beta + \sum_{\ell=k}^{\infty} q(\ell+1) + \sum_{\ell=k}^{\infty} \frac{p(\ell)(\Delta u(\ell))^{\sigma+\delta} g(u(\ell+1), u(\ell))}{f(u(\ell))f(u(\ell+1))}$$

for any $s$ and $k$ satisfying $0 \le s \le k$, where

$$\beta = \frac{p(s)(\Delta u(s))^\sigma}{f(u(s))} - \sum_{\ell=s}^{\infty} q(\ell+1) - \sum_{\ell=s}^{\infty} \frac{p(\ell)(\Delta u(\ell))^{\sigma+\delta} g(u(\ell+1), u(\ell))}{f(u(\ell))f(u(\ell+1))}.$$

We claim that $\beta = 0$. Indeed, if $\beta < 0$ we choose an integer $k$ so large that

$$\left|\sum_{\ell=k}^{\infty} q(\ell+1)\right| \le -\frac{\beta}{4}$$

and

$$\sum_{\ell=k}^{\infty} \frac{p(\ell)(\Delta u(\ell))^{\sigma+\delta} g(u(\ell+1), u(\ell))}{f(u(\ell))f(u(\ell+1))} < -\frac{\beta}{4}.$$

Then, we have

$$-\frac{p(s)(\Delta u(s))^\sigma}{f(u(s))} + \sum_{\ell=s}^{k-1} q(\ell+1) + \sum_{\ell=s}^{b-1} \frac{p(\ell)(\Delta u(\ell))^{\sigma+\delta} g(u(\ell+1), u(\ell))}{f(u(\ell))f(u(\ell+1))}$$
$$= -\beta - \sum_{\ell=k}^{\infty} q(\ell+1) - \sum_{\ell=b}^{\infty} \frac{p(\ell)(\Delta u(\ell))^{\sigma+\delta} g(u(\ell+1), u(\ell))}{f(u(\ell))f(u(\ell+1))}$$
$$\ge -\frac{\beta}{2} > 0, \quad k \ge b > s.$$

By Lemma 6.22.2, we find

$$\Delta u(k) \leq -\frac{((-\beta/2)f(u(b)))^{1/\sigma}}{(p(s))^{1/\sigma}}, \quad k \in \mathbb{N}(b).$$

But, then (6.22.19) implies that $u(k)$ tends to $-\infty$, which is a contradiction.

If $\beta > 0$, then by (6.22.21), (6.22.22) and (6.22.25), we have

$$\lim_{k\to\infty} \frac{p(k)(\Delta u(k))^{\sigma}}{f(u(k))} = \beta > 0$$

and hence $\Delta u(k) > 0$ for all large $k$. Let $d$ be a nonnegative integer such that

$$\frac{p(k)(\Delta u(k))^{\sigma}}{f(u(k))} \geq \frac{\beta}{2}, \quad k \in \mathbb{N}(d). \tag{6.22.26}$$

From (6.22.26) it follows that

$$\Delta u(k) \geq \left(\frac{\beta f(u(d))}{2p(k)}\right)^{1/\sigma}, \quad k \in \mathbb{N}(d)$$

which on summation, and then employing (6.22.19) leads to

$$\lim_{k\to\infty} u(k) = \infty. \tag{6.22.27}$$

On the other hand, for $k > d$, we have

$$\begin{aligned}
\infty &> \sum_{\ell=d}^{k} \frac{p(\ell)(\Delta u(\ell))^{\sigma+\delta} g(u(\ell+1), u(\ell))}{f(u(\ell))f(u(\ell+1))} \\
&= \sum_{\ell=d}^{k} \frac{p(\ell)(\Delta u(\ell))^{\sigma} \Delta f(u(\ell))}{f(u(\ell))f(u(\ell+1))} \\
&\geq \frac{\beta}{2}\sum_{\ell=d}^{k} \frac{\Delta f(u(\ell))}{f(u(\ell))} \\
&\geq \frac{\beta}{2}\int_{f(u(d))}^{f(u(\ell+1))} \frac{dt}{t} = \frac{\beta}{2}\ln\frac{f(u(k+1))}{f(u(d))},
\end{aligned}$$

which shows that $f(u(k+1)) < \infty$ for all large $k$. But, in view of (6.22.27) and (6.22.20) this is a contradiction.

The proof for the case $u(k)$ is a negative solution of (6.22.1) is similar. ∎

**Remark 6.22.2.** In view of the proof of Lemma 6.22.3, under the conditions (6.22.19) and (6.22.21) equation (6.22.25) holds and $\beta \geq 0$. Thus, if

$$\sum_{\ell=k}^{\infty} q(\ell+1) \geq 0 \quad \text{for all large } k \in \mathbb{N} \tag{6.22.28}$$

then $(\Delta u(k))^{\sigma}/f(u(k)) \geq 0$ for all large $k$.

**Lemma 6.22.4.** Suppose that (6.22.19), (6.22.21) and (6.22.28) hold. If $u(k)$ is a nonoscillatory solution of (6.22.1), then $u(k)(\Delta u(k))^{\sigma} \geq 0$ for all large $k$.

**Theorem 6.22.5.** Suppose in addition to (6.22.19) (6.22.21) the following hold

$$0 < \int_{\epsilon}^{\infty} \frac{dt}{(f(t))^{1/\sigma}} < \infty, \quad \int_{-\epsilon}^{-\infty} \frac{dt}{(f(t))^{1/\sigma}} < \infty \tag{6.22.29}$$

for every $\epsilon > 0$, and

$$\sum_{\ell=0}^{\infty} \frac{1}{(p(\ell))^{1/\sigma}} \left( \sum_{\tau=\ell}^{\infty} q(\tau+1) \right)^{1/\sigma} = \infty. \tag{6.22.30}$$

Then, every solution of (6.22.1) is oscillatory.

**Proof.** Let $u(k)$ be an eventually positive solution of (6.22.1). Then, the equality (6.22.24) holds. Since the second term in the right side of (6.22.24) is nonnegative, we find

$$\frac{(\Delta u(k))^{\sigma}}{f(u(k+1))} \geq \frac{1}{p(k)} \sum_{\ell=k}^{\infty} q(\ell+1).$$

Now taking $\sigma$th root on both sides, summing from a sufficiently large integer $a$ to $k-1$, we get

$$\begin{aligned} \sum_{\ell=a}^{k-1} \frac{1}{(p(\ell))^{1/\sigma}} \left( \sum_{\tau=\ell}^{\infty} q(\tau+1) \right)^{1/\sigma} &\leq \sum_{\ell=a}^{k-1} \frac{\Delta u(k)}{(f(u(k+1)))^{1/\sigma}} \\ &\leq \int_{u(a)}^{u(k+1)} \frac{dt}{(f(t))^{1/\sigma}} \end{aligned}$$

for all $k > a$. But, in view of (6.22.29) and (6.22.30) as $k \to \infty$ the above inequality leads to a contradiction. The case when $u(k)$ is an eventually negative solution is considered similarly. ∎

**Theorem 6.22.6.** Suppose that $\sigma = \tau$ and conditions (6.22.19), (6.22.21) and (6.22.28) hold. Further suppose that $g(u,v) \geq \nu > 0$ for all $u, v \neq 0$, and there exists a positive function $\rho(k)$, $k \in \mathbb{N}$ such that

$$\sum_{\ell=0}^{\infty}\left\{q(\ell+1)\rho(\ell+1) - \frac{p(\ell)(\Delta\rho(\ell))^2}{4\nu\rho(\ell)}\right\} = \infty. \tag{6.22.31}$$

Then, every solution of (6.22.1) is oscillatory.

**Proof.** Let $u(k)$ be an eventually positive solution of (6.22.1). Then, in view of Lemma 6.22.4, $\Delta u(k) \geq 0$ for all large $k$. Further,

$$\begin{aligned}
&\Delta\left(\frac{p(k)\rho(k)\Delta u(k)}{f(u(k))}\right)\\
&= -q(k+1)\rho(k+1) + \frac{p(k)(\Delta u(k))^{\sigma}\Delta\rho(k)}{f(u(k+1))} - \frac{p(k)(\Delta u(k))^{\sigma}\rho(k)\Delta f(u(k))}{f(u(k))f(u(k+1))}\\
&\leq -q(k+1)\rho(k+1) + \frac{p(k)(\Delta u(k))^{\sigma}\Delta\rho(k)}{f(u(k+1))} - \frac{\nu p(k)\rho(k)(\Delta u(k))^{2\sigma}}{f(u(k))f(u(k+1))}\\
&= -q(k+1)\rho(k+1) + \frac{p(k)(\Delta\rho(k))^2 f(u(k))}{4\nu\rho(k)f(u(k+1))}\\
&\quad - \left\{\left(\frac{\nu p(k)\rho(k)}{f(u(k))f(u(k+1))}\right)^{1/2}(\Delta u(k))^{\sigma} - \frac{1}{2}\left(\frac{p(k)f(u(k))}{\nu\rho(k)f(u(k+1))}\right)^{1/2}\Delta\rho(k)\right\}^2\\
&\leq -q(k+1)\rho(k+1) + \frac{p(k)(\Delta\rho(k))^2}{4\nu\rho(k)}
\end{aligned}$$

for all large $k$, where we have used the fact that $f$ is nondecreasing. Summing the above inequality from a sufficiently large integer $a$ to $k-1$, we obtain

$$\begin{aligned}
\sum_{\ell=a}^{k-1}\left\{q(\ell+1)\rho(\ell+1) - \frac{p(\ell)(\Delta\rho(\ell))^2}{4\nu\rho(\ell)}\right\} &\leq \frac{p(a)\rho(a)\Delta u(a)}{f(u(a))} - \frac{p(k)\rho(k)\Delta u(k)}{f(u(k))}\\
&\leq \frac{p(a)\rho(a)\Delta u(a)}{f(u(a))},
\end{aligned}$$

which contradicts (6.22.31). The case when $u(k)$ is negative is similarly proved. ∎

**Example 6.22.1.** Consider the difference equation

$$\Delta\left((k-1)^{-\beta}\Delta u(k)\right) + \frac{1}{k^{\alpha}(k+1)^2}|u(k+1)|^{\gamma}\operatorname{sgn} u(k+1) = 0, \tag{6.22.32}$$

$k \in \mathbb{N}(2)$

where $\gamma > 1$, $\alpha > 0$ and $\beta > 0$. When $\beta \geq \alpha$, we have

$$\sum_{\ell=2}^{\infty} \frac{1}{p(\ell)} \sum_{\tau=\ell}^{\infty} q(\tau+1) \ = \ \sum_{\ell=2}^{\infty} (\ell-1)^{\beta} \sum_{\tau=\ell}^{\infty} \frac{1}{\tau^{\alpha}(\tau+1)^2} \ = \ \infty.$$

Thus, Theorem 6.22.5 implies that every solution of (6.22.32) is oscillatory.

**Example 6.22.2.** Consider the difference equation

$$\Delta^2 u(k) + \frac{\gamma}{(k+1)^2} u(k+1) \ = \ 0, \quad k \in \mathbb{N} \tag{6.22.33}$$

where $\gamma > 1/4$. If we take $\rho(k) = k$, $k \in \mathbb{N}$ then

$$\begin{aligned} &\sum_{\ell=1}^{k} \left\{ q(\ell+1)\rho(\ell+1) - \frac{p(\ell)(\Delta\rho(\ell))^2}{4\nu\rho(\ell)} \right\} \\ &= \sum_{\ell=1}^{k} \left\{ \gamma(\ell+1)(\ell+1)^{-2} - \frac{1}{4\ell} \right\} \\ &= \sum_{\ell=1}^{k} \frac{4\gamma\ell - \ell - 1}{4(\ell+1)\ell} = \sum_{\ell=1}^{k} \frac{4\gamma - 1}{4(\ell+1)} - \sum_{\ell=1}^{k} \frac{1}{4\ell(\ell+1)} \to \infty \end{aligned}$$

as $k \to \infty$. Thus, Theorem 6.22.6 implies that every solution of (6.22.33) is oscillatory.

**Example 6.22.3.** Consider the difference equation

$$\Delta\left(\frac{k}{k+1}\Delta u(k)\right) + \frac{\gamma}{(k+1)^2}(u(k+1) + u^3(k+1)) \ = \ 0, \quad k \in \mathbb{N}(1). \tag{6.22.34}$$

Since the corresponding $g(u,v)$ satisfies

$$g(u,v) \ = \ 1 + u^2 + uv + v^2 \ \geq \ 1 + 2|u||v| - |u||v| \ \geq \ 1$$

if we take $\rho(k) = k$ for $k \geq 1$, then

$$\sum_{\ell=1}^{\infty} \left\{ q(\ell+1)\rho(\ell+1) - \frac{p(\ell)(\Delta u(\ell))^2}{4\nu\rho(\ell)} \right\} \ = \ \sum_{\ell=1}^{\infty} \left\{ \frac{\gamma}{\ell+1} - \frac{1}{4(\ell+1)} \right\} \ = \ \infty,$$

when $\gamma > 1/4$. Thus, Theorem 6.22.6 implies that every solution of (6.22.34) is oscillatory provided $\gamma > 1/4$.

## 6.23. Oscillation and Nonoscillation for

$$u(t) - u(t-\tau) + p(t)u(t-\sigma) \ = \ 0, \tag{6.23.1}$$

where $t \geq 0$ is a continuous variable, $\tau > 0$, $\sigma > 0$ and $p \in C(\mathbb{R}_+, \mathbb{R}_+)$.

By a solution of (6.23.1) we mean a continuous function $u \in C([t_0 - \max\{\tau, \sigma\}, \infty), \mathbb{R})$ which satisfies (6.23.1) for $t \geq t_0$. A solution $u(t)$ of (6.23.1) is said to be ***oscillatory*** if it is neither eventually positive nor eventually negative. Otherwise, the solution is called ***nonoscillatory***.

We shall first present some sufficient conditions under which all solutions of (6.23.1) oscillate. Next, we will establish some oscillation criteria for the nonlinear difference equation

$$u(t) - u(t-\tau) + p(t)H(u(t-\sigma)) = 0. \tag{6.23.2}$$

Finally, we will provide sufficient conditions which guarantee oscillation of all solutions of the forced equation

$$u(t) - u(t-\tau) + p(t)u(t-\sigma) = f(t). \tag{6.23.3}$$

Let $q(t) = \min_{t-\tau \leq s \leq t} p(s)$. Define a set $E$ of real numbers as follows:

$$E = \{\lambda > 0 : 1 - \lambda q(t), \text{ eventually}\}.$$

**Theorem 6.23.1.** Assume that

(i) $\limsup_{t\to\infty} q(t) > 0$

(ii) $\sigma = m\tau + \theta$, $m$ is a positive integer and $\theta \in [0, \tau)$, and there exists a positive number $T$ such that

$$\sup_{\lambda \in E, t \geq T} \lambda \prod_{i=1}^{m-1} (1 - \lambda q(t - i\tau)) < 1. \tag{6.23.4}$$

Then, every solution of (6.23.1) is oscillatory.

**Proof.** Suppose to the contrary, and let $u(t)$ be an eventually positive solution of (6.23.1). Let $z(t) = \int_{t-\tau}^{t} u(s)ds > 0$. Then, $z'(t) = u(t) - u(t-\tau) = -p(t)u(t-\sigma) \leq 0$. Integrating (6.23.1) from $t - \tau$ to $t$, we have

$$z(t) - z(t-\tau) + q(t)z(t-\sigma) \leq 0. \tag{6.23.5}$$

Define a set of real numbers by

$$S(z) = \{\lambda > 0 : z(t) - (1 - \lambda q(t))z(t-\tau) \leq 0, \text{ eventually}\}. \tag{6.23.6}$$

From (6.23.5), $z(t) - (1 - q(t))z(t-\tau) \leq 0$, i.e. $1 \in S(z)$, and so $S(z)$ is nonempty. If $\lambda \in S(z)$, then $0 < z(t) \leq (1 - \lambda q(t))z(t-\tau)$, and hence, $1 - \lambda q(t) > 0$, eventually. Therefore, $S(z) \subset E$. Thus, we reach to a contradiction, if $E$ is empty. Now, condition (i) implies that $E$ is bounded. Hence, $S(z)$ is bounded. Let $\mu \in S(z)$. Then,

$$\begin{aligned} z(t-\tau) &\leq (1-\mu q(t-\tau))z(t-2\tau) \\ &\leq \cdots \leq \prod_{i=1}^{m-1}(1-\mu q(t-i\tau))z(t-m\tau) \\ &\leq \prod_{i=1}^{m-1}(1-\mu q(t-i\tau))z(t-\sigma). \end{aligned}$$

Using the above inequality in (6.23.5), we obtain

$$z(t) - \left(1 - q(t)\left(\prod_{i=1}^{m-1}(1-\mu q(t-i\tau))\right)^{-1}\right) z(t-\tau) \leq 0.$$

By definition (6.23.6) and the above inequality, we find

$$\left(\sup_{t\geq T}\prod_{i=1}^{m-1}(1-\mu q(t-i\tau))\right)^{-1} \in S(z). \tag{6.23.7}$$

On the other hand, (6.23.4) implies that there exists a number $\beta \in (0,1)$ such that

$$\sup_{\lambda\in E, t\geq T} \lambda \prod_{i=1}^{m-1}(1-\lambda q(t-i\tau)) \leq \beta < 1$$

and hence

$$\sup_{t\geq T}\prod_{i=1}^{m-1}(1-\mu q(t-i\tau)) \leq \frac{\beta}{\mu}. \tag{6.23.8}$$

In view of (6.23.7) and (6.23.8), $\mu/\beta \in S(z)$. Repeating this procedure, we obtain $\mu/\beta^r \in S(z)$, $r = 1, 2, \cdots$ which contradicts the boundedness of $S(z)$. ∎

**Corollary 6.23.2.** In addition to the condition (i) assume that for some $m \neq 1$

$$\liminf_{t\to\infty} \frac{1}{m-1}\sum_{i=1}^{m-1} q(t-i\tau) > \frac{(m-1)^{m-1}}{m^m}. \tag{6.23.9}$$

Then, every solution of (6.23.1) is oscillatory.

**Remark 6.23.1.** The above oscillatory criteria for (6.23.1) can be used to the equation

$$u(t) - cu(t-\tau) + p(t)u(t-\sigma) = 0. \tag{6.23.10}$$

In fact, the substitution $v(t) = c^{-t/\tau}u(t)$ reduces (6.23.10) to

$$v(t) - v(t-\tau) + c^{-\sigma/\tau}p(t)v(t-\sigma) = 0.$$

**Theorem 6.23.3.** In addition to conditions (i) and (ii) assume that

(iii) $H \in C(\mathbb{R}, \mathbb{R})$, $uH(u) > 0$ for $0 < |u| \le \alpha$, $H$ is concave and there exists a small positive number $\delta$ such that

$$H(u) \ge u, \quad 0 \le u \le \delta$$

and

$$H(u) \le u, \quad 0 \ge u \ge -\delta.$$

Then, every solution of (6.23.2) is oscillatory.

**Proof.** Suppose to the contrary, and let $u(t)$ be an eventually positive solution of (6.23.2) and $z(t) = \int_{t-\tau}^{t} u(s)ds$. Then, $z'(t) \le 0$ eventually and $\lim_{t\to\infty} z(t) = \beta \ge 0$ exists. Integrating (6.23.2) from $t-\tau$ to $t$ and using Jensen's inequality, we obtain

$$z(t) - z(t-\tau) + q(t)H(z(t-\sigma)) \le 0.$$

If $\beta > 0$, let $t_n \to \infty$ where $t_n$ satisfy $\lim_{n\to\infty} q(t_n) = \limsup_{t\to\infty} q(t)$ in the above inequality, we get $\limsup_{t\to\infty} q(t)H(\beta) \le 0$, which is a contradiction. Therefore, $\lim_{t\to\infty} z(t) = 0$. From (iii) the above inequality leads to (6.23.5). Now as in Theorem 6.23.1 this gives a contradiction. Similarly, we can prove that (6.23.2) has no eventually negative solution. ∎

**Theorem 6.23.4.** Assume that

(iv) $p(t) \equiv p > 0$, and $p = 1$ and $\sigma = \tau$ do not hold simultaneously

(v) $H \in C(\mathbb{R}, \mathbb{R})$, $uH(u) > 0$ for $0 < |u| \le \alpha$, $H$ is concave and

$$\lim_{u\to 0} \frac{H(u)}{u} = 1. \tag{6.23.11}$$

Then, oscillation of (6.23.1) implies oscillation of (6.23.2).

**Proof.** From Problem 6.24.87 it is clear that every solution of (6.23.1) with $p(t) \equiv p$ oscillates if and only if the characteristic equation

$$D(\lambda) = 1 - e^{-\lambda\tau} + pe^{-\lambda\sigma} = 0 \tag{6.23.12}$$

has no real roots.

We claim that there exists $\epsilon_0 > 0$ such that for $|\epsilon| < \epsilon_0$ the equation

$$1 - e^{-\lambda\tau} + (1-\epsilon)pe^{-\lambda\sigma} = 0 \tag{6.23.13}$$

has no real roots. If $\tau > \sigma$, then $D(\infty) = 1$ and $D(-\infty) = -\infty$. It follows that (6.23.12) has real roots. Therefore, $\sigma \geq \tau$. If $\sigma = \tau$, (6.23.1) becomes

$$u(t) + (p-1)u(t-\sigma) = 0.$$

Every solution of the above equation oscillates if and only if $p \geq 1$. In view of (iv) we will not consider the case $p = 1$. If $p > 1$, it is clear that there exists $\epsilon_0 > 0$ such that for $|\epsilon| < \epsilon_0$, $(1-\epsilon)p > 1$. Hence, (6.23.13) has no real roots. Now we consider the case $\tau < \sigma$. In this case, the minimum of $D(\lambda)$ for $\lambda \in \mathbb{R}$ is positive. Let $m = D(\lambda_0) = \min_{\lambda \in \mathbb{R}} D(\lambda)$ and

$$G(\epsilon, \lambda) = 1 - e^{-\lambda\tau} + p(1-\epsilon)e^{-\lambda\sigma},$$

where $|\epsilon| < 1$. $\dfrac{\partial G}{\partial \lambda} = \tau e^{-\lambda\tau} - \sigma p(1-\epsilon)e^{-\lambda\sigma}$. Obviously, $\dfrac{\partial G}{\partial \lambda}(0, \lambda_0) = 0$. We consider the equation $\dfrac{\partial G}{\partial \lambda}(\epsilon, \lambda) = 0$ in the neighborhood of the point $(0, \lambda_0)$. By implicit function theorem, there exists a continuous function $\lambda = \lambda(\epsilon)$ defined on a neighborhood of $\epsilon = 0$, such that $\dfrac{\partial G}{\partial \lambda}(\epsilon, \lambda(\epsilon)) \equiv 0$ and $\lambda_0 = \lambda(0)$. Therefore, the unique minimum point $\lambda(\epsilon)$ of $G(\epsilon, \lambda)$ is continuous with respect to $\epsilon$. Hence, $\lim_{\epsilon \to 0} \lambda(\epsilon) = \lambda_0$, and therefore

$$\lim_{\epsilon \to 0} \left(1 - e^{-\lambda(\epsilon)\tau} + p(1-\epsilon)e^{-\lambda(\epsilon)\sigma}\right) = 1 - e^{\lambda_0\tau} + pe^{-\lambda_0\sigma} = m.$$

Thus, there exists $\epsilon_0 > 0$ such that

$$1 - e^{-\lambda(\epsilon)\tau} + (1-\epsilon)pe^{-\lambda(\epsilon)\sigma} \geq \frac{m}{2}, \quad |\epsilon| < \epsilon_0.$$

Hence, (6.23.13) has no real roots for $|\epsilon| < \epsilon_0$, which implies that every solution of the equation

$$u(t) - u(t-\tau) + p(1-\epsilon)u(t-\sigma) = 0 \tag{6.23.14}$$

is oscillatory.

Now we claim that every solution of (6.23.2) is oscillatory. Suppose to the contrary, and let $u(t)$ be an eventually positive solution of (6.23.2) and $z(t) = \int_{t-\tau}^{t} u(s)ds > 0$. Then, $z'(t) = u(t) - u(t-\tau) = -pH(u(t-\sigma)) \leq 0$. Integrating (6.23.2) from $t-\tau$ to $t$ and using Jensen's inequality, we get

$$z(t) - z(t-\tau) + pH(z(t-\sigma)) \leq 0. \tag{6.23.15}$$

As in the proof of Theorem 6.23.3, $\lim_{t\to\infty} z(t) = 0$. For $\epsilon > 0$, there exists $\delta > 0$ such that

$$(1-\epsilon)u \ < \ H(u) \ < \ (1+\epsilon)u \quad \text{for } 0 < |u| < \delta. \tag{6.23.16}$$

From (6.23.15) and (6.23.16), we obtain

$$z(t) - z(t-\tau) + p(1-\epsilon)z(t-\sigma) \ \le \ 0,$$

which in view of Problem 6.24.88 implies that (6.23.14) has an eventually positive solution. This contradiction completes the proof. ∎

Now we shall consider the difference equation (6.23.3). Let $f$ be continuous for $t \ge T$. We define

$$r(t) \ = \ \begin{cases} f(t), & t \ge T \\ \dfrac{(t-T+\tau)f(T)}{\tau}, & T-\tau \le t < T \\ 0, & t < T-\tau. \end{cases}$$

Then, $r \in C(\mathbb{R}, \mathbb{R})$. Let

$$F(t) \ = \ \sum_{i=0}^{\infty} r(t-i\tau), \quad t \ge T.$$

It is easy to see that $F(t) - F(t-\tau) = f(t)$, $t \ge T$. Finally, let

$$\overline{F}(t) \ = \ \int_{t-\tau}^{t} F(s)ds, \qquad \overline{u}(t) \ = \ \int_{t-\tau}^{t} u(s)ds$$
$$\overline{F}_+(t) \ = \ \max\{\overline{F}(t), 0\}, \qquad \overline{F}_-(t) \ = \ \max\{-\overline{F}(t), 0\}.$$

**Theorem 6.23.5.** Assume that $p \in C(\mathbb{R}_+, \mathbb{R}_+)$ and for any positive number $N$ there exist two sequences $\{t_i\}$, $\{t'_i\}$ such that $t_{i+1} - t_i \ge \tau$, $t'_{i+1} - t'_i \ge \tau$, $i = 1, 2, \cdots$ and

$$\sum_{i=1}^{\infty} q(t_i)\overline{F}_+(t_i - \sigma) \ > \ N \tag{6.23.17}$$

and

$$\sum_{i=1}^{\infty} q(t'_i)\overline{F}_-(t'_i - \sigma) \ > \ N, \tag{6.23.18}$$

where $q$ is defined as before. Then, every solution of (6.23.3) is oscillatory.

**Proof.** We rewrite (6.23.3) in the form

$$(u(t) - F(t)) - (u(t-\tau) - F(t-\tau)) + p(t)u(t-\sigma) = 0. \tag{6.23.19}$$

Suppose to the contrary, and let $u(t)$ be an eventually positive solution of (6.23.19). Let $z(t) = \overline{u}(t) - \overline{F}(t)$. Then (6.23.19) becomes

$$z'(t) + p(t)u(t-\sigma) = 0. \tag{6.23.20}$$

Thus, $z'(t) \leq 0$, $t \geq T$. If $z(t) < 0$ eventually, then $0 < \overline{u}(t) < \overline{F}(t)$ eventually, which contradicts (6.23.18). Therefore, $z(t)$ is positive and $\lim_{t\to\infty} z(t) = \alpha \geq 0$ exists. Integrating (6.23.20) from $T$ to $\infty$, we obtain $\int_T^\infty p(t)u(t-\sigma)dt < \infty$. Since $z(t) > 0$, we have $\overline{u}(t) > \overline{F}(t)$, and hence $\overline{u}(t) \geq \overline{F}_+(t)$, $t \geq T$. There exists $k > 0$ such that $t_k - T \geq T + \sigma$, and hence

$$\begin{aligned}\infty &> \int_{T+\sigma}^\infty p(t)u(t-\sigma)dt \geq \sum_{i=k}^\infty \int_{t_i-\tau}^{t_i} p(s)u(s-\sigma)ds \\ &\geq \sum_{i=k}^\infty q(t_i)\overline{u}(t_i - \sigma) \geq \sum_{i=k}^\infty q(t_i)\overline{F}_+(t_i - \sigma),\end{aligned}$$

which contradicts (6.23.17). Similarly, we can show that (6.23.19) has no eventually negative solutions. ∎

**Theorem 6.23.6.** Assume that

$$\int_T^\infty q(t)\overline{F}_+(t-\sigma)dt = \infty \tag{6.23.21}$$

and

$$\int_T^\infty q(t)\overline{F}_-(t-\sigma)dt = \infty. \tag{6.23.22}$$

Then, every solution of (6.23.3) is oscillatory.

**Proof.** It is sufficient to prove that (6.23.21) and (6.23.22) imply (6.23.17) and (6.23.18). Suppose to the contrary that (6.23.17) is false. Then there exists $N^*$ such that

$$\sum_{i=1}^\infty q(t_i)\overline{F}_+(t_i - \sigma) < N^*$$

for any sequence $\{t_i\}$ with $t_i - t_{i-1} \geq \tau$, $i = 1, 2, \cdots$. In particular, for any $t \in [0, \tau)$,

$$\sum_{i=1}^\infty q(t + i\tau)\overline{F}_+(t + i\tau - \sigma) < N^*.$$

Hence,

$$\int_0^{\tau} \sum_{i=1}^{\infty} q(t+i\tau)\overline{F}_+(t+i\tau-\sigma)dt \; < \; N^*\tau.$$

Now, by the Lebesgue dominated convergence theorem, we have

$$\int_{\tau}^{\infty} q(t)\overline{F}_+(t-\sigma)dt \; = \; \sum_{i=1}^{\infty}\int_0^{\tau} q(t+i\tau)\overline{F}_+(t+i\tau-\sigma)dt \; < \; N^*\sigma,$$

which contradicts (6.23.21). Similarly, we can show that (6.23.18) holds. ∎

**Example 6.23.1.** Consider the difference equation

$$u(t) - u(t-\pi/2) + 2u(t-5\pi/2) \; = \; \sin t + \cos t. \tag{6.23.23}$$

In this case, $f(t) = \sin t + \cos t$, $F(t) = \sin t$, $\overline{F}(t) = \int_{t-\pi/2}^{t} \sin s ds = \sin t - \cos t$, and $q(t) = 2$. Obviously, conditions (6.23.21) and (6.23.22) hold. Therefore, every solution of (6.23.23) oscillates. In fact, $u(t) = \cos t$ is a such solution of (6.23.23).

## 6.24. Problems

**6.24.1.** Consider the difference equation (6.1.1), and assume that in addition to the given hypotheses on the functions $p$ and $q$, $\sum^{\infty} 1/p(\ell) = \infty$ and all its solutions are bounded. Show that (6.1.1) is oscillatory.

**6.24.2.** Consider the difference equation (6.1.1) and assume that in addition to the given hypotheses on the functions $p$ and $q$, $\sum^{\infty} |q(\ell)| < \infty$ and $p(k)$ is eventually either nondecreasing or nonincreasing and bounded below by a positive constant. Show that all solutions of (6.1.1) are bounded.

**6.24.3.** Consider the difference equation (6.1.1) and assume that in addition to the given hypotheses on the functions $p$ and $q$, $q(k) \geq p(k-1) + p(k)$, $k \in \mathbb{N}(1)$. Show that (6.1.1) has a solution $u(k)$ such that $u(k) > 0$ and $u(k+1) \leq u(k)$ for all $k \in \mathbb{N}$.

**6.24.4.** If the difference equation (6.1.1) is nonoscillatory, then show that it has two linearly independent solutions $v(k)$ and $w(k)$ such that $\sum^{\infty}[p(\ell)(v(\ell)v(\ell+1) + w(\ell)w(\ell+1))]^{-1} < \infty$. However, the converse of this is not true.

**6.24.5.** Consider the difference equation (6.1.1) and assume that in addition to the given hypotheses on the functions $p$ and $q$, all its solutions are bounded. Show that all solutions of the perturbed difference equation

$$p(k)u(k+1) + p(k-1)u(k-1) \; = \; (q(k) + r(k))u(k), \quad k \in \mathbb{N}(1)$$

are bounded provided $\sum^{\infty} |r(\ell)| < \infty$.

**6.24.6.** Consider the difference equation (6.1.1) and assume that in addition to the given hypotheses on the functions $p$ and $q$, $q(k) \geq p(k-1)+p(k)$, $k \in \mathbb{N}(1)$. Show that a dominant solution of (6.1.1) cannot converge to zero as $k \to \infty$.

**6.24.7.** Give examples to show that in Theorems 6.3.6 6.3.8 the hypothesis of having "unbounded solutions" cannot be omitted.

**6.24.8.** Consider the difference equation

$$(6.24.1) \;\; p(k)u(k+1)+p(k-1)u(k-1) \; = \; q(k)u(k)+\lambda r(k)u(k), \quad k \in \mathbb{N}(1)$$

where the functions $p$, $q$ and $r$ are defined on $\mathbb{N}$, $\mathbb{N}(1)$ and $\mathbb{N}(1)$ respectively, $p(k) > 0$ and $r(k) > 0$ for all $k$, and $\lambda$ is a real or complex number. Equation (6.24.1) is called *limit point* (LP) if for some $\lambda$ there is a solution $u(k)$ such that $\sum_{\ell=1}^{\infty} r(\ell)|u(\ell)|^2 = \infty$, otherwise (6.24.1) is called *limit circle* (LC). Show that

(i) If $Im\ \lambda \neq 0$, then there is a solution $u(k)$ of (6.24.1) such that $\sum_{\ell=1}^{\infty} r(\ell)|u(\ell)|^2 < \infty$, also if for some $\lambda = \lambda_0$ equation (6.24.1) is LC then it is LC for any value of $\lambda$

(ii) if $\sum^{\infty} p^{-1/2}(\ell) = \infty$ and (6.1.1) is nonoscillatory, then (6.24.1) is LP

(iii) if $\sum^{\infty}(r(\ell)r(\ell+1))^{1/2}/p(\ell) = \infty$, then (6.24.1) is LP

(iv) if $|q(k)| \geq p(k) + p(k-1)$ for all $k \in \mathbb{N}(1)$, then (6.24.1) is LP

(v) if for some real $\lambda$ either

$$q(k) \; \geq \; p(k)(r(k)/r(k+1))^{1/2} + p(k-1)(r(k)/r(k-1))^{1/2} + \lambda r(k),$$

or

$$q(k) \; \leq \; -p(k)(r(k)/r(k+1))^{1/2} - p(k-1)(r(k)/r(k-1))^{1/2} + \lambda r(k)$$

for all $k \in \mathbb{N}(2)$, then equation (6.24.1) is LP.

**6.24.9.** Consider the difference equation (6.1.1) and assume that in addition to the given hypotheses on the functions $p$ and $q$, $q(k) > 0$ for all $k \in \mathbb{N}(1)$, and $\prod_{\ell=1}^{k} q(\ell)/\prod_{\ell=1}^{k} p(\ell)$ is bounded as $k \to \infty$. Show that

(i) if (6.1.1) is nonoscillatory, then all its solutions are bounded

(ii) if $\sum_{\ell=1}^{\infty} p^{-1}(\ell) = \infty$, then (6.1.1) is oscillatory.

**6.24.10.** Consider the difference equation (6.6.1) and assume that in addition to the given hypotheses on the functions $p$ and $r$, (6.6.9) and (6.6.27) hold. Show that

(i) If (6.6.1) is nonoscillatory, then there exists a function $v(k) > -p(k)$, $k \in \mathbb{N}(a)$ for some $a \in \mathbb{N}$, satisfying

$$v(k) = \sum_{\ell=k}^{\infty} \frac{v^2(\ell)}{v(\ell)+p(\ell)} + \sum_{\ell=k}^{\infty} r(\ell), \quad k \in \mathbb{N}(a).$$

(ii) If there exists a function $v(k) > -p(k)$, $k \in \mathbb{N}(a)$ for some $a \in \mathbb{N}$, satisfying

$$v(k) \geq \sum_{\ell=k}^{\infty} \frac{v^2(\ell)}{v(\ell)+p(\ell)} + \sum_{\ell=k}^{\infty} r(\ell) \geq 0$$

or

$$v(k) \leq \sum_{\ell=k}^{\infty} \frac{v^2(\ell)}{v(\ell)+p(\ell)} + \sum_{\ell=k}^{\infty} r(\ell) \leq 0, \quad k \in \mathbb{N}(a)$$

then (6.6.1) is nonoscillatory.

**6.24.11.** Consider the difference equation (6.1.1) and assume that in addition to the given hypotheses on the functions $p$ and $q$, $q(k) > 0$ for all $k \in \mathbb{N}(1)$. Also, let $h(k) = p^2(k)/(q(k)q(k+1))$. Show that equation (6.1.1) is nonoscillatory if and only if there is an eventually positive function $v(k)$, $k \in \mathbb{N}$ such that

$$\left(h(k+1)v(k+1) + \frac{1}{v(k)}\right)\left(h(k)v(k) + \frac{1}{v(k-1)}\right) \leq 1.$$

In particular, equation (6.1.1) is nonoscillatory if there exists an integer $K$ such that the inequality

$$\left(\sqrt{h(k+1)} + \sqrt{h(k)}\right)\left(\sqrt{h(k)} + \sqrt{h(k-1)}\right) \leq 1$$

holds for all $k \in \mathbb{N}(K)$.

**6.24.12.** Let the functions $p$, $q$, $h$ be as in Problem 6.24.11, and let $\overline{h}(k) = \sup_{\ell \in \mathbb{N}} 4^{\ell} h(k)h(k+1)\cdots h(k+\ell)$. Show that equation (6.1.1) is nonoscillatory if there exists an integer $K$ such that the inequality

$$\frac{1}{k}\sum_{\ell=m+1}^{m+k} \sqrt{h(\ell)} < \frac{1}{2}\left[1 + \frac{1-h(m)-\overline{h}(m+k+1)}{k}\right]$$

holds for all $m > K$ and $k \in \mathbb{N}(1)$.

**6.24.13.** Let the functions $p$, $q$ be as in Problem 6.24.11, and let $q(k)$ be periodic of period 2, i.e. $q(2k) = a$ and $q(2k-1) = b$. Show that (6.1.1) is oscillatory if and only if $\sqrt{a} + \sqrt{b} > 1$.

**6.24.14.** Consider the difference equations (6.1.1) and (6.7.1) and assume that in addition to the given hypotheses on the functions $p$, $q$, $p_1$ and $q_1$, $q(k) \geq p(k) + p(k-1)$ and $q_1(k) \geq p_1(k) + p_1(k-1)$ for all $k \in \mathbb{N}(1)$. Further, let $u(k)$ be a positive solution of (6.1.1) such that

$$p_1(k)u(k+1) + p_1(k-1)u(k-1) \leq q_1(k)u(k)$$

for all $k \in \mathbb{N}(1)$. Show that

(i) equation (6.7.1) has a nontrivial solution $v(k)$ such that $0 \leq v(k) \leq cu(k)$ for all $k \in \mathbb{N}$, where $c$ is some constant. Further, if $\Delta u(k) \leq 0$ then $\Delta v(k) \leq 0$ for all $k \in \mathbb{N}$

(ii) if in addition $u(k)$ is recessive and converges to zero, then (6.7.1) has a recessive solution which converges to zero.

**6.24.15.** Consider the difference equations (6.1.1) and (6.7.1) and assume that in addition to the given hypotheses on the functions $p$, $q$, $p_1$ and $q_1$, $q(k) \geq p(k) + p(k-1)$, $q_1(k) \geq p_1(k) + p_1(k-1)$, $p(k) \geq p_1(k)$ and $q(k) \leq q_1(k)$ for all $k \in \mathbb{N}(1)$. Further, let $u(k)$ and $v(k)$, $k \in \mathbb{N}$ be the solutions of (6.1.1) and (6.7.1) satisfying $v(1) - u(1) \geq v(0) - u(0) \geq 0$ and $u(1) \geq u(0) \geq 0$. Show that $v(k) - u(k) \geq v(k-1) - u(k-1) \geq 0$, and in particular $v(k) \geq u(k)$ for all $k \in \mathbb{N}$.

**6.24.16.** Let the functions $p$, $q$, $p_1$ and $q_1$ be as in Problem 6.24.15. Further, let $u(k)$ and $v(k)$, $k \in \mathbb{N}$ be the recessive solutions of (6.1.1) and (6.1.7) satisfying $u(0) \geq v(0) \geq 0$. Show that $u(k) \geq v(k)$ for all $k \in \mathbb{N}$.

**6.24.17.** Show that the difference equation (6.6.1) is nonoscillatory if and only if there exists a function $v(k)$ satisfying $v(k) > 0$ and $\Delta(p(k)\Delta v(k)) + r(k)v(k+1) \leq 0$ for all sufficiently large $k \in \mathbb{N}$.

**6.24.18.** Let in the difference equations (6.1.2) and (6.2.7), $P(k) \leq p(k)$ and $F(k) \leq f(k)$ for all sufficiently large $k \in \mathbb{N}(1)$. Show that if (6.2.7) is nonoscillatory then (6.1.2) is nonoscillatory.

**6.24.19.** Consider the difference equations (6.1.1) and (6.9.1) and assume that in addition to the given hypotheses on the functions $p$, $q$ and $r$, $r(k)$ is not eventually identically equal to zero. If (6.1.1) is oscillatory and if

$r(k)$ is eventually of one sign, then show that any nonoscillatory solution of (6.9.1) must eventually be of the same sign as of $r(k)$.

**6.24.20.** Consider the second order difference equation

$$(6.24.2)\quad \Delta\left(c(k-1)\Delta u(k-1)\right)+p(k-1)\Delta u(k-1)+q(k)u(k) \;=\; 0, \quad k\in\mathbb{N}(1)$$

where $c$, $p$ and $q$ are real functions such that $c(k)-p(k)\geq 0$ and $c(k)>0$ for $k\in\mathbb{N}(k_0)$, $k_0\geq 0$. Let $\rho$ be a positive function defined on $\mathbb{N}$. We define

$$F(k)=\rho(k)q(k)-\left(\sqrt{c(k-1)\rho(k-1)}-\sqrt{\rho(k)(c(k-1)-p(k-1))}\right)^2,\ k\in\mathbb{N}(k_0).$$

Show that

(i) if there exist a positive function $\rho$ on $\mathbb{N}$, and a subsequence $\{k_n\}\subseteq\mathbb{N}$, $k_n\to\infty$ as $n\to\infty$ such that for any $K$

$$(6.24.3)\qquad \rho(k_n)c(k_n)\;\leq\;\sum_{\ell=K}^{k_n}F(\ell)\quad\text{for every}\quad k_n\geq K,$$

then any nonoscillatory solution $u$ of (6.24.2) satisfies $u(k)\Delta u(k)\geq 0$ eventually

(ii) if in addition to (6.24.3), $\limsup_{k\to\infty}\sum_{\ell=k_0}^{k}F(\ell)=\infty$, then (6.24.2) is oscillatory. In particular, let $\rho(k-1)=1/k$, $k\in\mathbb{N}(1)$ to show that the difference equation

$$\Delta(k\Delta u(k-1))+k\Delta u(k-1)+(-1)^k(k+1)^3u(k)\;=\;0,\qquad k\in\mathbb{N}(1)$$

is oscillatory

(iii) if $\displaystyle\limsup_{k\to\infty}\frac{1}{k}\sum_{\ell=K}^{k}\left(\sum_{s=K}^{\ell}F(s)-c(\ell)\rho(\ell)\right)=\infty$, $K\geq k_0$ then (6.24.2) is oscillatory

(iv) if $\displaystyle\limsup_{k\to\infty}\frac{1}{k}\sum_{\ell=K}^{k}\sum_{s=K}^{\ell}\rho(s)q(s)=\infty$, $K\geq k_0$ and

$$\limsup_{k\to\infty}\frac{1}{k}\sum_{\ell=K}^{k}\left(\sum_{s=K}^{\ell}(\sqrt{\rho(s-1)c(s-1)}-\sqrt{\rho(s)(c(s-1)-p(s-1))})\right)^2<\infty,$$

then (6.24.2) is oscillatory

(v) if in addition to (6.24.3) for every $K \ge k_0$ there exists $k \ge K$ such that $c(k-1) - q(k) - p(k-1) < 0$, then (6.24.2) is oscillatory

(vi) if in addition to $\limsup_{k\to\infty} \sum_{\ell=k_0}^{k} F(\ell) = \infty$,

$$\limsup_{k\to\infty} \frac{1}{k} \sum_{\ell=K}^{k} \left( \sum_{s=K}^{\ell} F(s) - c(\ell)\rho(\ell) \right) \;>\; -\infty$$

for every large $K \ge k_0$, then (6.24.2) is oscillatory

(vii) if there exists a positive function $\rho$ defined on $\mathbb{N}$ such that

$$\frac{(c(k) + c(k-1) - p(k-1) - q(k))^2}{4c(k-1)\,(c(k-1) - p(k-1))} \left( \frac{\rho(k)}{\rho(k-1)} \right) \;\le\; d < 1, \quad c(k) > p(k)$$

eventually, and if there exists a subsequence $\{k_n\} \subseteq \mathbb{N}$, $k_n \to \infty$ as $n \to \infty$ and a constant $M \ge 0$ such that

$$c(k_n) + c(k_n - 1) - p(k_n - 1) - q(k_n) - M\rho(k_n - 1)c(k_n - 1)\,(c(k_n - 1) - p(k_n - 1)) \le 0$$

then (6.24.2) is oscillatory.

**6.24.21.** Consider the Riccati equation

(6.24.4) $\quad u(k+1) \;=\; c(k)u(k) + f(k)u(k)u(k+1) + g(k), \quad k \in \mathbb{N}(a)$

where the limits $\lim_{k\to\infty} c(k) = c(\infty)$, $|c(\infty)| < 1$ and $\lim_{k\to\infty} g(k) = 0$ exist, and assume that $f(k)$ is bounded, i.e. $|f(k)| \le F$, say. Show that (6.24.4) has a solution $u(k)$ such that $u(k) \to 0$ as $k \to \infty$. Moreover, such a solution is unique.

**6.24.22.** Let the coefficients of equation (6.24.4) have representations

(6.24.5)

$$c(k) = \sum_{i=0}^{K} \frac{C_i}{k^i} + o\left(\frac{1}{k^K}\right), \quad f(k) = \sum_{i=0}^{K} \frac{F_i}{k^i} + o\left(\frac{1}{k^K}\right), \quad g(k) = \sum_{i=1}^{K} \frac{G_i}{k^i} + o\left(\frac{1}{k^K}\right)$$

as $k \to \infty$ with $|C_0| > 1$ and fixed $K \ge 1$. Show that there exists a unique solution $u(k)$ of (6.24.4) such that $u(k) \to 0$ as $k \to \infty$. Moreover,

(6.24.6) $$u(k) \;=\; \sum_{i=1}^{K} \frac{U_i}{k^i} + o\left(\frac{1}{k^K}\right) \quad \text{as} \quad k \to \infty$$

where $U_i$, $1 \le i \le K$ can be determined by formally substituting (6.24.5) and (6.24.6) into (6.24.4).

**6.24.23.** Consider the difference equations (6.10.1) and

$$\Delta^2 v(k-1) + q(k)v^\gamma(k) = 0, \quad k \in \mathbb{N}(1) \tag{6.24.7}$$

where the functions $p$ and $q$ are defined on $\mathbb{N}(1)$ and for all $k \in \mathbb{N}(1)$, $0 \geq q(k) \geq p(k)$, and $\gamma$ is a quotient of odd positive integers. If $u(k)$ and $v(k)$ are positive solutions of (6.10.1) and (6.24.7), respectively, satisfying $u(1)-v(1) \geq u(0)-v(0) \geq 0$, then show that $u(k+1)-v(k+1) \geq u(k)-v(k) \geq 0$ for all $k \in \mathbb{N}$, and thus $u(k) \geq v(k)$ for all $k \in \mathbb{N}$.

**6.24.24.** Consider the difference equations (6.10.1) and (6.24.7) and assume that $p$, $q$ and $\gamma$ are as in Problem 6.24.23. Further, assume that for every $a \in \mathbb{N}(1)$ there exist $k_1, k_2 \in \mathbb{N}(a+1)$ such that $p(k_1) < 0$, $q(k_2) < 0$. If $u(k)$ and $v(k)$ are unique positive nonincreasing solutions (cf. Theorem 6.10.4) of (6.10.1) and (6.24.7), respectively, satisfying $u(0) = v(0)$, then show that $v(k) \geq u(k)$ for all $k \in \mathbb{N}(1)$.

**6.24.25.** Consider the difference equation

$$u(k+1) + 2u(k) + u(k-1) \ = \ p(k)u^\gamma(k), \quad k \in \mathbb{N}(1), \quad \gamma > 1 \tag{6.24.8}$$

where the function $p$ is defined on $\mathbb{N}(1)$, $p(k) \geq 4$ for all $k \in \mathbb{N}(1)$, and $\sum_{\ell=1}^{\infty} \ln(p(\ell) - 3) = \infty$. If $u(k)$ is a solution of (6.24.8) defined by $u(0) = 1$, $u(1) = 2$, then show that $u(k+1) \geq u(k) \geq 1$ for all $k \in \mathbb{N}(1)$, and $u(k) \to \infty$ as $k \to \infty$.

**6.24.26.** Consider the difference equation (6.10.1) and assume that $p(k)$ is as in $(R_4)$, and $\sum^{\infty} \ell^\gamma p(\ell) < \infty$. Show that

(i) if $\gamma > 1$ and $u(k)$ is an oscillatory solution of (6.10.1), then there exist increasing sequences $\{k_\ell\}$, $\{k_j\} \subset \mathbb{N}$, $k_\ell \to \infty$ and $k_j \to \infty$ such that $\Delta u(k_\ell) \to \infty$ and $\Delta u(k_j) \to -\infty$ as $k \to \infty$ and $j \to \infty$

(ii) if $0 < \gamma < 1$ and $u(k)$ is an oscillatory solution of (6.10.1), then $\Delta u(k) \to 0$ as $k \to \infty$.

**6.24.27.** Consider the $n$th $(n > 1)$ order nonlinear difference equation

$$\Delta\left(p(k-1)\left(\Delta^{n-1}u(k-1)\right)^\delta\right) + q(k)u^\delta(k) \ = \ 0, \quad k \in \mathbb{N}(1) \tag{6.24.9}$$

where $p, q : \mathbb{N}(1) \to \mathbb{R}$ are defined, $\delta$ is a quotient of odd positive integers, and $p(k) > 0$, $k \in \mathbb{N}(1)$, $\sum_{\ell=1}^{\infty}(1/p(\ell))^{1/\delta} = \infty$, $q(k) > 0$, $k \in \mathbb{N}(1)$. Show that

(i) if $u(k)$ is a nonnegative solution of the inequality

$$\Delta\left(p(k-1)\left(\Delta^{n-1}u(k-1)\right)^\delta\right) + q(k)u^\delta(k) \ \leq \ 0, \quad k \in \mathbb{N}(1) \tag{6.24.10}$$

then $\Delta^{n-1}u(k) \ge 0$ eventually

(ii) if $u(k)$ is a nonpositive solution of the inequality

$$\Delta\left(p(k-1)\left(\Delta^{n-1}u(k-1)\right)^{\delta}\right) + q(k)u^{\delta}(k) \ \ge\ 0, \quad k \in \mathbb{N}(1) \tag{6.24.11}$$

then $\Delta^{n-1}u(k) \le 0$ eventually

(iii) if $u(k)$ is a nonoscillatory solution of (6.24.9), then $u(k)\Delta^{n-1}u(k) \ge 0$ eventually

(iv) if (6.24.10) ((6.24.11)) has an eventually nonnegative (nonpositive) solution, then (6.24.9) also has an eventually nonnegative (nonpositive) solution

(v) if $0 < r(k) \le q(k)$, $k \in \mathbb{N}(1)$ and the equation

$$\Delta\left(p(k-1)\left(\Delta^{n-1}u(k-1)\right)^{\delta}\right) + r(k)u^{\delta}(k) \ =\ 0, \quad k \in \mathbb{N}(1)$$

oscillates, then (6.24.9) also oscillates

(vi) if $s(k) \ge p(k)$, $k \in \mathbb{N}(1)$ and the equation

$$\Delta\left(s(k-1)\left(\Delta^{n-1}u(k-1)\right)^{\delta}\right) + q(k)u^{\delta}(k) \ =\ 0, \quad k \in \mathbb{N}(1)$$

oscillates, then (6.24.9) also oscillates

(vii) if $n$ is even all bounded solutions of (6.24.9) oscillate if and only if

$$\sum_{k=k_0}^{\infty} k^{n-2}\left(\frac{1}{p(k)}\sum_{\ell=k+1}^{\infty} q(\ell)\right)^{1/\delta} \ =\ \infty, \quad k_0 \in \mathbb{N}(1) \tag{6.24.12}$$

(viii) if $n$ is odd all bounded solutions of (6.24.9) oscillate or monotonically tend to zero if and only if (6.24.12) holds.

**6.24.28.** Consider the difference equation (6.11.1), and assume that in addition to the given hypotheses on the functions $r$, $f$ and $F$, $F(u)$ is continuous on $\mathbb{R}$, $\sum^{\infty} 1/(r(\ell)R_{a,\ell}) = \infty$, and $\sum^{\infty} R_{a,\ell+1}(cf^{+}(\ell) + f^{-}(\ell)) = \infty$, for every constant $c > 0$. Show that every bounded solution $u(k)$ of (6.11.1) is either oscillatory, or such that $\liminf_{k\to\infty} |u(k)| = 0$.

**6.24.29.** Consider the difference equation (6.11.1), and assume that in addition to the given hypotheses on the functions $r$, $f$ and $F$, $F(u)$ is continuous on $\mathbb{R}$, $f(k) \ge 0$ for all $k \in \mathbb{N}(a)$, and $\sum^{\infty} R_{a,\ell}f(\ell) = \infty$. Show that all bounded solutions of (6.11.1) are oscillatory.

**6.24.30.** Consider the difference equation (6.11.1), and assume that in addition to the given hypotheses on the functions $r$, $f$ and $F$, $|F(u)|$ is bounded away from zero if $|u|$ is bounded away from zero, $f(k) \geq 0$ for all $k \in \mathbb{N}(a)$, and $\sum^{\infty} f(\ell) = \infty$. Show that (6.11.1) is oscillatory.

**6.24.31.** Suppose that the hypotheses of Theorem 6.11.4 are satisfied except the condition (6.11.9). Show that every bounded solution of (6.11.1) is oscillatory.

**6.24.32.** Suppose that in (6.11.1) the function $F(u)$ is continuous on $\mathbb{R}$, $f(k) \geq 0$ for all $k \in \mathbb{N}(a)$, $r(k)$ is nondecreasing on $\mathbb{N}(a)$, there exists a nondecreasing function $\phi \in C[\mathbb{R}, \mathbb{R}]$ such that $|F(u)| \geq |\phi(u)|$, $u\phi(u) > 0$, $u \neq 0$ and

$$\int_{\epsilon}^{\infty} \frac{dt}{\phi(t)} < \infty \quad \text{and} \quad \int_{-\epsilon}^{-\infty} \frac{dt}{\phi(t)} < \infty, \quad \text{for every } \epsilon > 0,$$

and, there exists a nondecreasing function $\mu(k) > 0$ such that $\Delta\mu(k)$ is nonincreasing for all $k \in \mathbb{N}(a)$ and $\sum^{\infty} \mu(\ell) f(\ell)/r(\ell) = \infty$. Show that the difference equation (6.11.1) is oscillatory.

**6.24.33.** Show that the following difference equations are oscillatory

(i) $\Delta^2 u(k) + k^{-2} u(k) = 0, \quad k \in \mathbb{N}(1)$

(ii) $\Delta^2 u(k) + k^{-3/2} (\ln k)^{-1} u(k) = 0, \quad k \in \mathbb{N}(1)$

(iii) $\Delta(k\Delta u(k)) + k^{-1} (\ln k)^{-2} u^3(k) = 0, \quad k \in \mathbb{N}(2)$

(iv) $\Delta(k\Delta u(k)) + 2k^{-1} (\ln k)^{-2} (\ln \ln k)^{-1} |u(k)|^{3/2} \mathrm{sgn}\, u(k) = 0, \ k \in \mathbb{N}(4)$.

**6.24.34.** Let in the difference equation $\Delta(r(k-1)\Delta u(k-1)) + f(k)F(u(k)) = 0$, $k \in \mathbb{N}(1)$ the functions $r$, $f$ and $F$ be defined in their domain of definitions. Further, let $F$ be nondecreasing and $uF(u) > 0$ for $u \neq 0$, $r(k) > 0$ for all $k \in \mathbb{N}$ and $\sum^{\infty} 1/r(\ell) = \infty$, $\sum^{\infty} f(\ell) = \infty$. Show that this difference equation is oscillatory.

**6.24.35.** Let in Theorem 6.12.1 condition (i) be replaced by (i)′ $f(k) \geq 0$ for all $k \in \mathbb{N}(a)$, and $\sum^{\infty} f(\ell) = \infty$. Show that every solution $u(k)$ of (6.12.1) is either oscillatory or $\liminf_{k \to \infty} |u(k)| = 0$.

**6.24.36.** Consider the difference equation (6.12.1), and assume that in addition to the given hypotheses on the functions $r$, $f$, $F$ and $g$, the function $G(k) = \sum_{\ell=a}^{k-1} g(\ell)$ is bounded on $\mathbb{N}(a)$, $\sum^{\infty} f^{+}(\ell) = \infty$, $\sum^{\infty} f^{-}(\ell)$ exists, to every pair of constants $c_1$, $c_2$ with $0 < c_1 < c_2$ there corresponds a pair of constants $L_1$, $L_2$ with $0 < L_1 \leq |F(u)| \leq L_2$ for every $u$ with $c_1 \leq |u| \leq c_2$. Show that every bounded solution $u(k)$ of (6.12.1) is either oscillatory or such that $\liminf_{k \to \infty} |u(k)| = 0$.

**6.24.37.** The classical *secant method* for solving $f(t) = 0$ is given by

$$t_{k+1} = \frac{t_{k-1}f(t_k) - t_k f(t_{k-1})}{f(t_k) - f(t_{k-1})}.$$

For $f(t) = t^2$ it becomes

$$t_{k+1} = \frac{t_{k-1}t_k}{t_{k-1} + t_k}.$$

Show that its solution satisfying $t_0 = 1$, $t_1 = 1/2$ gives the reciprocal of the Fibonacci numbers.

**6.24.38.** Consider the second order difference equation

$$(6.24.13) \qquad \Delta\left(r(k)\Delta u(k)\right) + f\left(k, u(k+1), \Delta u(k)\right) = 0, \quad k \in \mathbb{N}$$

where $r(k)$ is a positive function, and $f : \mathbb{N} \times \mathbb{R}^2 \to \mathbb{R}$. In system from (6.24.13) can be written as

$$(6.24.14) \ \Delta v(k) = w(k)/r(k), \quad \Delta w(k) = -f\left(k, v(k+1), w(k)/r(k)\right).$$

Let the functions $V(k,v,w)$ and $W(k,v,w)$ be defined and continuous for $k \geq K \in \mathbb{N}$, $v > 0$, $|w| < \infty$ and $k \geq K \in \mathbb{N}$, $v < 0$, $|w| < \infty$ respectively. Further, let

(i) $V(k,v,w) \to \infty$ uniformly for $v > 0$ and $|w| < \infty$ as $k \to \infty$, and $W(k,v,w) \to \infty$ uniformly for $v < 0$ and $|w| < \infty$ as $k \to \infty$
(ii) $\Delta V_{(2)}(k,v,w) = V\left(k+1, v(k+1), w(k+1)\right) - V\left(k, v(k), w(k)\right) \leq 0$ for all sufficiently large $k$, where $(v(k), w(k))$ is a solution of (6.24.14) such that $v(k) > 0$ for all large $k$
(iii) $\Delta W_{(2)}(k,v,w) = W\left(k+1, v(k+1), w(k+1)\right) - W\left(k, v(k), w(k)\right) \leq 0$ for all sufficiently large $k$, where $(v(k), w(k))$ is a solution of (6.24.14) such that $v(k) < 0$ for all large $k$.

Show that every solution of (6.24.13) is oscillatory.

**6.24.39.** Let in the difference equation $\Delta_\alpha^2 u(k) = f(k, u(k))$, $k \in \mathbb{N}(a)$, $\alpha \neq 0$ the function $f(k,u)$ be defined for all $(k,u) \in \mathbb{N}(a) \times \mathbb{R}$, and $|f(k,u)| \leq (1/2)\alpha^2 k^{-2}|u|$. If $u(k) \in \ell_2$ is a solution of this difference equation, then show that there exists an integer $k_1 \geq a$ $(a \geq 2)$ such that $u(k) = 0$ for all $k \in \mathbb{N}(k_1)$.

**6.24.40.** Consider the difference equation (6.15.1), and assume that in addition to the given hypotheses on the functions $r$, $F$ and $g$, $f(k) \equiv 1$, and there exist nonnegative functions $\lambda(k)$ and $\mu(k)$ on $\mathbb{N}(a)$ such that $|g(k,u,v)| \leq \lambda(k)$ and $|F(k,u,v)| \leq \mu(k)|u|$. Show that every solution $u(k)$ of (6.15.1) satisfies $|u(k)| = O(R_{a,k})$ as $k \to \infty$ provided $\sum^{\infty} \lambda(\ell) < \infty$ and $\sum^{\infty} \mu(\ell) R_{a,\ell} < \infty$.

**6.24.41.** Consider the difference equation (6.15.1), and assume that in addition to the given hypotheses on the functions $r$, $F$ and $g$, $f(k) \equiv 1$, and $uF(k,u,v) \geq 0$. Further, assume that there exists a nonnegative function $\lambda(k)$ on $\mathbb{N}(a)$ such that $|g(k,u,v)| \leq \lambda(k)$. Show that every nonoscillatory solution $u(k)$ of (6.15.1) satisfies

$$|u(k)| = O\left(R_{a,k} + \sum_{\ell=a}^{k-1} \frac{1}{r(\ell)} \sum_{j=a}^{\ell-1} \lambda(j)\right) \quad \text{as} \quad k \to \infty.$$

**6.24.42.** Let $u(k)$ be defined on $\mathbb{N}(a, b+n)$ and $\Delta^n u(k) \geq 0$ for all $k \in \mathbb{N}(a,b)$, $\Delta^i u(a) > 0$, $0 \leq i \leq n-1$. Show that $\Delta^i u(k) > 0$ for all $k \in \mathbb{N}(a, b+n-i)$, $0 \leq i \leq n-1$.

**6.24.43.** Show that the solution of the initial value problem

$$\begin{aligned} u(k+3) &= \frac{1 + u(k+1)u(k+2)}{u(k)}, \quad k \in \mathbb{N} \\ u(0) &= u(1) = u(2) = 1 \end{aligned}$$

is an integer.

**6.24.44.** Consider the difference equation (6.20.1). Let $C > m^m/(m+1)^{m+1}$, $\nu^2 < \frac{2(m+1)^{m+1}}{m^{2m+1}}\left[C - \frac{m^m}{(m+1)^{m+1}}\right]$, and $G = \bigcup^{\infty} \mathbb{N}(p_k, q_k)$, $q_k - p_k > \frac{\pi}{\nu} + m + 1$, where $\{p_k\} \to \infty$ and $\{q_k\}$ are arbitrary sequences. If $p(k) \geq C$, $k \in G$ then show that all solutions of (6.20.1) are oscillatory.

**6.24.45.** Consider the difference equation

$$(6.24.15) \qquad \Delta^3 u(k) + \sum_{i=1}^{m} f_i(k) F_i(u(k+1), \Delta u(k+1)) = 0, \quad k \in \mathbb{N}$$

where the functions $f_i$ and $F_i$, $1 \leq i \leq m$ are defined in their domain of definition, $f_i(k) > 0$ for all $k \in \mathbb{N}$ and $(1/u)F_i(u,v) \geq c > 0$. Show that

(i) if $u(k)$ is a nonoscillatory solution of (6.24.15), then for all large $k \in \mathbb{N}$ either

$$(6.24.16) \qquad \operatorname{sgn} u(k) = \operatorname{sgn} \Delta^2 u(k) \neq \operatorname{sgn} \Delta u(k),$$

or

$$\text{(6.24.17)} \qquad \operatorname{sgn} u(k) = \operatorname{sgn} \Delta u(k) = \operatorname{sgn} \Delta^2 u(k)$$

(ii) if $u(k)$ is a solution of (6.24.15), then the function $G(k) = 2u(k)\Delta^2 u(k) -(\Delta u(k))^2$ is nonincreasing, and hence either $G(k) \geq 0$ for all $k \in \mathbb{N}$, or there exists a $r \in \mathbb{N}$ such that $G(k) < 0$ for all $k \in \mathbb{N}(r)$

(iii) if for the solution $u(k)$ of (6.24.15), $G(k) \geq 0$ for all $k \in \mathbb{N}$ then

$$\sum_{\ell=0}^{\infty} (\Delta^2 u(\ell))^2 < \infty \quad \text{and}$$

$$\sum_{\ell=0}^{\infty} \left[ u(\ell+1) \sum_{i=1}^{m} f_i(\ell) F_i(u(\ell+1), \Delta u(\ell+1)) \right] < \infty$$

(iv) if for the solution $u(k)$ of (6.24.15), $G(k) \geq 0$ for all $k \in \mathbb{N}$, and there exists an index $j \in \mathbb{N}(1,m)$ such that $f_j(k) \geq d > 0$, then

$$\sum_{\ell=0}^{\infty} u^2(\ell) < \infty \quad \text{and}$$

$$\lim_{k\to\infty} u(k) = \lim_{k\to\infty} \Delta u(k) = \lim_{k\to\infty} \Delta^2 u(k) = 0$$

(v) if there exists an index $j \in \mathbb{N}(1,m)$ such that $f_j(k) \geq d > 0$, then the following are equivalent

(a) for the solution $u(k)$ of (6.24.15), $G(k) \geq 0$ for all $k \in \mathbb{N}$

(b) the solution $u(k) \to 0$ as $k \to \infty$

(c) $G(k) \to 0$ as $k \to \infty$

(vi) if there exists an index $j \in \mathbb{N}(1,m)$ such that $\sum^{\infty} f_j(\ell) = \infty$, then

(a) no nonoscillatory solution $u(k)$ of (6.24.15) can be bounded away from zero

(b) relations (6.24.16) are satisfied

(c) the solution $u(k)$ of (6.24.15) is oscillatory provided for this solution $G(k) < 0$ for all $k \in \mathbb{N}(r)$.

**6.24.46.** Consider the initial value problem

$$\Delta u(k) - f(k)u(k) = g(k)f(k), \quad u(a) = 0$$

where $f(k) \geq 0$ for all $k \in \mathbb{N}(a)$ and $\sum_{\ell=a}^{\infty} f(\ell) = \infty$, and $\lim_{k\to\infty} h(k) = \infty$. Show that $\lim_{k\to\infty} u(k) = \pm\infty$.

**6.24.47.** Consider the difference equation

$$(6.24.18) \qquad (-1)^n \Delta^{2n} u(k) = p(k)u(k+n), \quad k \in \mathbb{N}$$

where $p(k) \geq 0$ and $\sum^\infty p(k) < \infty$. Show that

(i) equation (6.24.18) is nonoscillatory if

$$\lim_{k\to\infty} k^{(2n-1)} \sum_{\ell=k}^{\infty} p(\ell) < \frac{[(2n-1)!!]^2}{4^n}$$

(ii) equation (6.24.18) is oscillatory if

$$\limsup_{k\to\infty} k^{(2n-1)} \sum_{\ell=k}^{\infty} p(\ell) > \frac{(2n-1)!(2n-2)!}{[(n-1)!]^2}.$$

**6.24.48.** Consider the nonlinear difference equations

$$(6.24.19) \qquad \Delta^n u(k) + q(k)f(u(k)) = 0, \quad k \in \mathbb{N}(1)$$

and

$$(6.24.20) \qquad \Delta^n u(k) + Q(k)F(u(k)) = 0, \quad k \in \mathbb{N}(1).$$

We say (6.24.19) or (6.24.20) has Property (A) if for $n$ even, every solution is oscillatory, and for $n$ odd, every solution $u(k)$ is either oscillatory or satisfies $u(k) \to 0$ as $k \to \infty$. Assume that $0 < q(k) \leq Q(k)$, $k \in \mathbb{N}(1)$, $f$, $F : \mathbb{R} \to \mathbb{R}$ are continuous with $uf(u) > 0$ and $uF(u) > 0$ for $u \neq 0$, $f(u)$ is increasing in $u$, and $F(u)\mathrm{sgn}\, u \geq f(u)\mathrm{sgn}\, u$ for all $u$. Show that if (6.24.19) has Property (A), then so does (6.24.20).

**6.24.49.** Consider the difference equation

$$(6.24.21) \qquad u(k+1) = f(u(k), u(k-1), \cdots, u(k-p)), \quad k \in \mathbb{N}$$

with $u(-p), \cdots, u(0)$ given. If there exists a nontrivial function $I_k = I_k(u(k-p), \cdots, u(k))$ such that for every solution $u(k)$, $k \in \mathbb{N}(-p)$ of (6.24.21), $I_k$, $k \in \mathbb{N}$ is a constant, i.e. $I_{k+1} = I_k$, $k \in \mathbb{N}$ then (6.24.21) is said to possess *invariance property*. The function $I_k$ is called the *invariant* of (6.24.21). Show that *Lyness difference equation*

$$(6.24.22) \qquad u(k+1) = \frac{a + u(k)}{u(k-1)}, \quad k \in \mathbb{N}$$

where $u(-1)$, $u(0)$ and $a$ are positive possesses the invariant

$$I_k = (a+u(k-1)+u(k))\left(1+\frac{1}{u(k-1)}\right)\left(1+\frac{1}{u(k)}\right).$$

(Equation (6.24.22) occurs in number theory [76,77], and finds applications in Geometry [72] as well as in frieze patterns [26,27]).

**6.24.50.** Show that the difference equation

$$u(k+1) = \frac{\alpha u(k)+\beta}{(\gamma u(k)+\delta)u(k-1)}, \quad k \in \mathbb{N}$$

where $u(-1)$, $u(0)$, $\alpha$, $\beta$, $\gamma$ and $\delta$ are positive possesses the invariant

$$\begin{aligned} I_k &= (\beta+\alpha u(k-1)+\alpha u(k)+\delta u(k-1)u(k)) \\ &\quad\times \left(\gamma+\frac{\delta}{u(k-1)}+\frac{\delta}{u(k)}+\frac{\alpha}{u(k-1)u(k)}\right). \end{aligned}$$

**6.24.51.** Show that the difference equation

$$u(k+1) = \frac{a+b(k)u(k)+c(k)u^2(k)}{(c(k+1)+e(k)u(k)+fu^2(k))u(k-1)}, \quad k \in \mathbb{N}$$

where $b(k)$, $c(k)$ and $e(k)$ are positive periodic functions of period 1, i.e. constants, or 2; $a$ and $f$ are positive constants, and $u(-1)$, $u(0)$ are positive numbers, possesses the invariant

$$\begin{aligned} I_k &= \frac{a}{u(k-1)u(k)}+\frac{b(k-1)}{u(k)}+c(k-1)\frac{u(k-1)}{u(k)}+c(k)\frac{u(k)}{u(k-1)} \\ &\quad+e(k-1)u(k)+fu(k-1)u(k)+\frac{b(k-2)}{u(k-1)}+e(k-2)u(k-1). \end{aligned}$$

**6.24.52.** Show that the system of difference equations

$$\begin{aligned} u(k+1) &= \frac{a(k)v(k)+A}{u(k-1)} \\ v(k+1) &= \frac{b(k)u(k)+A}{v(k-1)}, \quad k \in \mathbb{N} \end{aligned} \tag{6.24.23}$$

where $a(k)$ and $b(k)$ are positive periodic functions of period 1, i.e. constants, or 2, and $A$, $u(-1)$, $v(-1)$, $u(0)$, $v(0)$ are positive possesses the invariant

$$\begin{aligned} I_k &= \left(1+\frac{b(k-1)}{u(k-1)}\right)\left(1+\frac{a(k-1)}{v(k-1)}\right)\left(1+\frac{b(k)}{u(k)}\right)\left(1+\frac{a(k)}{v(k)}\right)\times \\ &\quad (a(k-1)v(k-1)+A+b(k-2)u(k))\,(b(k-1)u(k-1)+A+a(k-2)v(k)). \end{aligned}$$

**6.24.53.** Show that every positive solution $(u(k), v(k))$ of the system (6.24.23) with $a(k) = b(k) \equiv 1$, $A > 0$ has the property $u(k+5) = v(k)$, $v(k+5) = u(k)$ if and only if $A = 1$. Hence, if $A = 1$ and $u(-1)$, $v(-1)$, $u(0)$, $v(0)$ are positive, all positive solutions are periodic of period 10.

**6.24.54.** Show that the system of difference equations

$$(6.24.24) \qquad \begin{aligned} u(k+1) &= \frac{\max\{a(k)v(k),\ A\}}{u(k-1)} \\ v(k+1) &= \frac{\max\{b(k)u(k),\ A\}}{v(k-1)}, \qquad k \in \mathbb{N} \end{aligned}$$

where $a(k)$ and $b(k)$ are positive periodic functions of period 1, i.e. constants, or 2, and $A$, $u(-1)$, $v(-1)$, $u(0)$, $v(0)$ are positive possesses the invariant

$$I_k = \max\left\{1, \frac{b(k-1)}{u(k-1)}\right\} \max\left\{1, \frac{a(k-1)}{v(k-1)}\right\} \max\left\{1, \frac{b(k)}{u(k)}\right\} \max\left\{1, \frac{a(k)}{v(k)}\right\}$$

$$\times \max\{a(k-1)v(k-1), A, b(k-2)u(k)\} \max\{b(k-1)u(k-1), A, a(k-2)v(k)\}.$$

In particular, for $a(k) = b(k) = A = 1$ show that all positive solutions of (6.24.24) are periodic of period 10.

**6.24.55.** Consider the system

$$(6.24.25) \qquad \begin{aligned} u(k+1) &= a + \frac{v(k)}{u(k-p)} \\ v(k+1) &= a + \frac{u(k)}{v(k-q)}, \qquad k \in \mathbb{N} \end{aligned}$$

where $a \in (0, \infty)$, $p$, $q$ are positive integers, and $u(-p), \cdots, u(0)$, $v(-q)$, $\cdots, v(0)$ are positive numbers. For (6.24.25) it is clear that $(1+a, 1+a)$ is the unique equilibrium point. Show that

(i) every positive solution $(u(k), v(k))$ of (6.24.25) oscillates about $(1+a, 1+a)$, i.e. either $u(k)$ or $v(k)$ oscillates about $1+a$

(ii) if $(u(k), v(k))$ is a positive solution of (6.24.25) and at least one of the integers $p$, $q$ is odd then for $k \geq 2r+3$, $r = \max\{p, q\}$

$$u(k),\ v(k) \ \leq\ a + \sum_{\ell=1}^{r} a^{-r} + a^{-r}$$

(iii) if $(u(k), v(k))$ is a positive solution of (6.24.25), $a > 1$, and both the integers $p$, $q$ are even then the functions $u(k)$, $v(k)$ are bounded above

(iv) if $a > 1$ then the equilibrium point $(1+a, 1+a)$ of (6.24.25) is globally asymptotically stable.

**6.24.56.** Consider the $n$th order nonlinear difference equation

$$u(k) = f(u(k-1), u(k-2), \cdots, u(k-n)), \quad k \in \mathbb{N}(1) \tag{6.24.26}$$

where $f : \mathbb{R}_+^n \to \mathbb{R}_+$ and the initial values $u(1-n), \cdots, u(0)$ are nonnegative. Equation (6.24.26) is said to be ***permanent*** if there exist constants $L,\ M \geq 0$ such that for every solution $u(k)$ of (6.24.26) there is a positive integer $k_0 = k_0(u(1-n), \cdots, u(0))$ such that $u(k) \in [L, M]$ for all $k \in \mathbb{N}(k_0)$. Any compact interval having this property is called ***absorbing interval*** for (6.24.26). It is clear that absorbing interval contains all attracting and limit sets, and thus permanence is a stronger concept than boundedness. As an example, we note that the solution of the first order difference equation

$$u(k) = \frac{1}{u^p(k-1)} \tag{6.24.27}$$

where $p,\ u(0) > 0$ can be written as $u(k) = (u(0))^{(-p)^n}$. Thus, if $0 < p < 1$ the unique equilibrium point of (6.24.27) is globally asymptotically stable, and hence (6.24.27) is permanent. However, for $p = 1$ while every solution of (6.24.27) is bounded and periodic of period 2 (if $u(0) \neq 1$), (6.24.27) is no longer permanent; and if $p > 1$, every solution of (6.24.27) with $u(0) \neq 1$ is unbounded.

(i) Let $g(\mathbf{u}) : \mathbb{R}_+^n \to \mathbb{R}_+$, and $f(\mathbf{u}) \leq g(\mathbf{u})$ for all $\mathbf{u} \in \mathbb{R}_+^n$. Further, let $g$ be nondecreasing in each of its $n$ arguments and the equation

$$v(k) = g(v(k-1), v(k-2), \cdots, v(k-n)), \quad k \in \mathbb{N}(1) \tag{6.24.28}$$

is permanent. Show that (6.24.26) is permanent. Also, if all solutions of (6.24.28) are bounded, then so are all solutions of (6.24.26).

(ii) Let there exist $0 \leq a < 1,\ b \geq 0$ such that for all $\mathbf{u} \in \mathbb{R}_+^n$, $f(\mathbf{u}) \leq a\|\mathbf{u}\|_\infty + b$. Show that (6.24.26) is permanent.

(iii) Let $g(\mathbf{u}) : \mathbb{R}_+^n \to \mathbb{R}_+$, and $\limsup_{\|\mathbf{u}\|_\infty \to \infty} g(\mathbf{u}) < 1$. Show that the equation

$$u(k) = u(k-p)g(u(k-1), u(k-2), \cdots, u(k-n)), \quad k \in \mathbb{N}(1)$$

where $1 \leq p \leq n$ is permanent.

**6.24.57.** Consider the difference equation

$$\Delta^n u(k) + f(k, u(k), u(k+1), \cdots, u(k+n-1)) = g(k), \quad k \in \mathbb{N} \tag{6.24.29}$$

where the functions $f$ and $g$ are defined on $\mathbb{N}\times\mathbb{R}^n$ and $\mathbb{N}$ respectively, and

$$|f(k,u_1,\cdots,u_n)| \le B(k,|u_1|,\cdots,|u_n|),$$

where the function $B(k,v_1,\cdots,v_n)$ is continuous on $\mathbb{R}^n_+$ for each fixed $k\in\mathbb{N}$, and for all $0\le v_i\le w_i$, $1\le i\le n$ and $r(k)\ge\epsilon>0$, $k\in\mathbb{N}$

$$0 \le B(k,v_1,\cdots,v_n) \le B(k,w_1,\cdots,w_n),$$
$$B(k,r(k)v_1,\cdots,r(k)v_n) \le A(r(k))B(k,v_1,\cdots,v_n),$$

where $A\in C[[\epsilon,\infty),\mathbb{R}_+]$ is nondecreasing and $\displaystyle\int_\epsilon^\infty \frac{dt}{A(t)}=\infty$. Further, assume that

$$\sum^\infty B\left(\ell,\ell^{n-1},\cdots,(\ell+n-1)^{n-1}\right) < \infty \quad \text{and} \quad \sum^\infty |g(\ell)| < \infty.$$

Show that every solution $u(k)$ of (6.24.29) has the property

$$\lim_{k\to\infty}\frac{\Delta^{n-p}u(k)}{(k)^{(p-1)}} = \frac{L}{(p-1)!},$$

where $1\le p\le n$, and $L\ne 0$ is a constant.

**6.24.58.** With respect to the difference equation (6.20.1) assume that $m\in\mathbb{N}(1)$, $p(k)\ge 0$, $k\in\mathbb{N}$,

$$a_0 = \liminf_{k\to\infty}\sum_{\ell=k-m}^{k-1}p(\ell) \le \left(\frac{m}{m+1}\right)^{m+1}$$

and

$$\limsup_{k\to\infty}\sum_{\ell=k-m}^{k}p(\ell) > 1-\frac{1-a_0-\sqrt{1-2a_0-a_0^2}}{2}.$$

Show that equation (6.20.1) is oscillatory.

**6.24.59.** Suppose the conditions in Problem 6.24.58 are satisfied. Show that (6.20.3) has no eventually positive solution, and (6.20.4) has no eventually negative solution.

**6.24.60.** With respect to the difference equation (6.20.1) assume that $m\in\mathbb{N}(1)$, $p(k)\ge 0$, $k\in\mathbb{N}$, $\sum_{k=0}^\infty p(k)=\infty$ and $\displaystyle\limsup_{k\to\infty}\sum_{\ell=k-m}^{k-1}p(\ell)<$

1. Show that for every solution $u(k)$, $k \in \mathbb{N} \bigcup \{-m, \cdots, -1, 0\}$ of (6.20.1) there exists a constant $c$ and $k_0 \in \mathbb{N}$ such that

$$\sup_{k \in \mathbb{N}(k_0)} \left[ |u(k)| \prod_{\ell=k_0}^{k-1} \frac{1 - cp(\ell)}{1 - p(\ell)} \right] < \infty.$$

In particular, every solution of (6.20.1) tends to zero as $k \to \infty$.

**6.24.61.** Consider the difference equation

$$(6.24.30) \qquad u(k+1) - u(k) = q(k)u(k-m), \quad k \in \mathbb{N}$$

where $m > 1$ is an integer and $q(k) \in \ell^p$ for some positive integer $p$. Show that for any $k_0 \in \mathbb{N}$ such that

$$q_-(k) \le \frac{m^m}{(m+1)^{m+1}} \quad \text{for} \quad k \in \mathbb{N}(k_0)$$

where $q_-(k) = \max\{0, -q(k)\}$, $k \in \mathbb{N}$ the following hold

(i) the sequence $\{q^{[\nu]}(k)\}_{\nu=1}^{\infty}$ is well defined recursively as follows

$$q^{[1]}(k) = 0 \quad \text{for} \quad k \ge k_0 - m$$

and for $\nu \ge 1$

$$q^{[\nu+1]}(k) = \begin{cases} 0 \quad \text{for} \quad k_0 - m \le k < k_0 \\ -q(k) \prod_{\ell=k-m}^{k-1} \left(1 - q^{[\nu]}(\ell)\right)^{-1} \quad \text{for} \quad k \ge k_0 \end{cases}$$

(ii) for every solution $u(k)$ of (6.24.30) the limit

$$(6.24.31) \qquad \xi = \lim_{k \to \infty} \left[ u(k) \left( \prod_{\ell=k_0}^{k-1} \left(1 - q^{[p]}(\ell)\right) \right)^{-1} \right]$$

exists and is finite

(iii) for every $\xi \in \mathbb{R}$ there exists a solution $u(k)$ of (6.24.30) through $k_0$ which satisfies (6.24.31).

**6.24.62.** Consider the difference equation

$$(6.24.32) \qquad u(k+1) - u(k) + p(k)f(u(k-m)) = 0, \quad k \in \mathbb{N}$$

where $m \in \mathbb{N}(1)$ is fixed, and the functions $p$ and $f$ are defined on $\mathbb{N}$ and $\mathbb{R}$, respectively. Let $uf(u) > 0$, $u \ne 0$ and $\liminf_{u \to 0} \dfrac{f(u)}{u} = M$, $0 < M < \infty$. Show that (6.24.32) is oscillatory if either of the following holds

(i) $cM > m^m/(m+1)^{m+1}$, where $c = \liminf_{k\to\infty} p(k) > 0$,

(ii) $f$ is nondecreasing on $\mathbb{R}$, $p(k) \geq 0$ for all $k \in \mathbb{N}$ and $\limsup_{k\to\infty} \sum_{\ell=k-m}^{k} p(\ell) > \frac{1}{M}$.

**6.24.63.** Consider the difference equation

$$(6.24.33) \qquad u(k+1) - u(k) + p(k)(1+u(k))u(k-m) \ = \ 0, \quad k \in \mathbb{N}$$

where $m \in \mathbb{N}(1)$ is fixed, and the function $p(k) \geq 0$ is defined on $\mathbb{N}$. Show that

(i) if (6.20.5) holds, then every solution $u(k)$ of (6.24.33) such that $1 + u(k) > 0$ for all $k \geq -m$ is oscillatory

(ii) if (6.20.7) holds, then there exists a nonoscillatory solution $u(k)$ of (6.24.33) such that $1 + u(k) > 0$ for all $k \geq -m$.

**6.24.64.** Consider the difference equation

$$(6.24.34) \qquad u(k+1) - u(k) + \sum_{i=1}^{r} p_i(k)u(k-m_i) \ = \ 0, \quad k \in \mathbb{N}$$

where $r$, $m_i \in \mathbb{N}(1)$, $1 \leq i \leq r$ are fixed and the functions $p_i$ are defined on $\mathbb{N}$. Show that (6.24.34) is oscillatory if either of the following holds

(i) $p_i(k) \geq 0$, $k \in \mathbb{N}$, $1 \leq i \leq r$ and $\sum_{i=1}^{r} (\liminf_{k\to\infty} p_i(k)) \frac{(m_i+1)^{m_i+1}}{m_i^{m_i}} > 1$

(ii) $p_i(k) \geq 0$, $k \in \mathbb{N}$, $1 \leq i \leq r$ and $r \left[\prod_{i=1}^{r} \liminf_{k\to\infty} p_i(k)\right]^{1/r} > \frac{\overline{m}^{\overline{m}}}{(\overline{m}+1)^{\overline{m}+1}}$, where $\overline{m} = (1/r)\sum_{i=1}^{r} m_i$

(iii) $p_i(k) \geq 0$, $k \in \mathbb{N}$, $1 \leq i \leq r$ and $\liminf_{k\to\infty} \left[\sum_{i=1}^{r} p_i(k)\right] > \frac{(\hat{m})^{\hat{m}}}{(\hat{m}+1)^{\hat{m}+1}}$, where $\hat{m} = \min\{m_1, \cdots, m_r\}$

(iv) $\liminf_{k\to\infty} \sum_{i=1}^{r} p_i(k) = c > 0$ and $\limsup_{k\to\infty} \sum_{i=1}^{r} p_i(k) = 1 - c$.

**6.24.65.** Consider the difference equation

$$(6.24.35) \qquad u(k+1) - u(k) + p(k)u(k-m) \ = \ \Delta g(k), \quad k \in \mathbb{N}$$

where $m \in \mathbb{N}(1)$ is fixed , and the functions $p$ and $g$ are defined on $\mathbb{N}$, and $p(k) \geq 0$ for all $k \in \mathbb{N}$. Show that (6.24.35) is oscillatory if either of the following holds

(i) for each $k \in \mathbb{N}$ there exists $k_1 \in \mathbb{N}(k+1)$ such that $g(k)g(k_1) < 0$, and

$$\sum^{\infty} p(\ell)g_+(\ell - m) = \infty \quad \text{and} \quad \sum^{\infty} p(\ell)g_-(\ell - m) = \infty$$

(ii) there exist two constants $c_1 < c_2$ and two sequences $\{k_i\}$ and $\{k_j\}$ in $\mathbb{N}$ such that $g(k_i) = c_1$, $g(k_j) = c_2$ and $c_1 \le g(k) \le c_2$ for all $k \in \mathbb{N}$, and the condition (6.20.5) is satisfied.

**6.24.66.** Suppose that $p(k) \ge 0$ and $\sum_{\ell=0}^{m-1} p(k+\ell) > 0$ for all $k \in \mathbb{N}$. Show that

(i) if $v(k)$ is a solution of (6.20.3) such that $v(k) > 0$ for all $k \ge -m$, then (6.20.1) has a solution $u(k)$ such that $0 < u(k) \le v(k)$ for all $k \ge -m$ and $\lim_{k\to\infty} u(k) = 0$

(ii) if there exists a number $\gamma \in (0,1)$ such that $p(k) < \gamma$ and $\prod_{\ell=k-m}^{k-1}\left(1 - \frac{1}{\gamma}\overline{p}(\ell)\right) \ge \gamma$ for all $k \in \mathbb{N}$, where $\overline{p}(k) = \begin{cases} p(k) & \text{for } k \in \mathbb{N} \\ p(0) & \text{for } k < 0, \end{cases}$ then (6.20.1) has a solution $u(k)$ which is positive for $k \ge -m$ and is such that $\lim_{k\to\infty} u(k) = 0$.

**6.24.67.** Consider the Pielou logistic delay equation

$$u(k+1) = \frac{\alpha u(k)}{1 + \beta u(k-m)}, \quad k \in \mathbb{N} \tag{6.24.36}$$

where $\alpha > 1$, $\beta > 0$ and $m \in \mathbb{N}$. Show that

(i) the positive equilibrium $\overline{u} = (\alpha - 1)/\beta$ of (6.24.36) is asymptotically stable if $\alpha < 1 + 2\alpha\cos m\pi/(2m+1)$ and unstable if $\alpha > 1 + 2\alpha\cos m\pi/(2m+1)$

(ii) the positive equilibrium $\overline{u}$ of (6.24.36) is globally asymptotically stable if $(\alpha - 1)(m - 1) \le 1$

(iii) equation (6.24.36) is permanent

(iv) every positive solution of (6.24.36) oscillates about the positive equilibrium $\overline{u}$ if and only if $\alpha > 1 + \alpha m^m/(m+1)^{m+1}$.

**6.24.68.** Consider the difference equation

$$u(k+1) = \frac{a + bu(k)}{A + u(k-1)}, \quad k \in \mathbb{N} \tag{6.24.37}$$

where $a$, $b, A > 0$ and $u(-1)$, $u(0)$ are arbitrary positive numbers. Show that the unique positive equilibrium point of (6.24.37)

$$\overline{u} = \frac{(b - A) + \sqrt{(b-A)^2 + 4a}}{2}$$

is globally asymptotically stable provided one of the following holds:

(i) $A > b$

(ii) $Ab \geq a$ and $b \geq A$

(iii) $Ab < a \leq 2A(b + A)$ and $b \geq A$

(iv) $a > Ab,\ b \geq \left(\frac{1+\sqrt{5}}{2}\right)^{1/2} A$ and $\frac{b^2}{A} \leq \overline{u} \leq 2b.$

**6.24.69.** Let $0 \leq a < 1,\ b \geq 0$. Show that the unique equilibrium point $\overline{u} = b/(1-a)$ of the difference equation

$$u(k) = a\max\{u(k-1), \cdots, u(k-n)\} + b, \quad k \in \mathbb{N}(1)$$

is globally asymptotically stable.

**6.24.70.** Consider the discrete logistic delay equation

$$(6.24.38) \qquad u(k+1) = u(k)\exp\left(r\left(1 - \sum_{i=0}^{m} p_i u(k-i)\right)\right), \quad k \in \mathbb{N}$$

where $r,\ p_m > 0,\ p_i \geq 0,\ 0 \leq i \leq m-1$ and $m \in \mathbb{N}$. Show that

(i) the positive equilibrium $\overline{u} = 1/P$ where $P = \sum_{i=0}^{m} p_i$ of (6.24.38) is a global attractor of all positive solutions of (6.24.38) if $r \leq P[mP - (m-1)p_0 + p_m]^{-1}$

(ii) if $m + r \neq 1$, every positive solution of (6.24.38) oscillates about the positive equilibrium $\overline{u}$ if and only if the equation

$$\lambda - 1 + \frac{r}{P}\sum_{i=0}^{m} p_i \lambda^{-i} = 0$$

has no positive roots.

**6.24.71.** Consider the discrete model of Nicholson's blowflies [85]

$$(6.24.39) \quad u(k+1) - u(k) = -\delta u(k) + pu(k-m)\exp(-au(k-m)), \quad k \in \mathbb{N}$$

where $0 < \delta < 1,\ p,\ a > 0$ and $m \in \mathbb{N}$. Show that

(i) if $p \leq \delta$, then the zero solution of (6.24.39) is uniformly asymptotically stable, and every nonnegative solution of (6.24.39) tends to zero as $k \to \infty$

(ii) equation (6.24.39) is *uniformly persistent* if and only if $p > \delta$, i.e. for the solution $u(k)$ with the given $u(-m), \cdots, u(0) \geq 0$ there exists an $\eta > 0$ such that $\liminf_{k\to\infty} u(k) \geq \eta$

(iii) if $p > \delta$ and $\left[(1-\delta)^{-m-1} - 1\right] \ln(p/\delta) \le 1$, then the positive equilibrium $\overline{u} = (1/a)\ln(p/\delta)$ is globally asymptotically stable

(iv) if $p > \delta$ and $\delta(\ln(p/\delta) - 1)(m+1)^{m+1} > m^m(1-\delta)^{m+1}$, then every positive solution of (6.24.39) oscillates about the positive equilibrium $\overline{u}$.

**6.24.72.** Consider the difference equation

$$\Delta u(k) + p(k)u(\tau(k)) = 0, \quad k \in \mathbb{N} \tag{6.24.40}$$

where $\tau : \mathbb{N} \to \mathbb{Z}$, $\tau(k) < k$, $k \in \mathbb{N}$, $\tau(k) \ge 0$ for $k \in \tau^{-1}(0) = \min\{\ell : \tau(\ell) \ge 0\}$, and $\lim_{k\to\infty} \tau(k) = \infty$. Show that

(i) if there exists a positive function $A(k)$, $k \in \mathbb{N}\bigcup\{-1\}$ such that

$$1 - p^+(k)A(k-1) > 0, \quad k \in [0, \tau^{-1}(0))$$
$$A(k-1)\prod_{\ell=\tau(k)}^{k-1}(1 - p^+(\ell)A(\ell-1)) \ge 1, \quad k \in \mathbb{N}(\tau^{-1}(0))$$

then (6.24.40) has a positive solution. In particular, the equation

$$\Delta u(k) + p(k)u(k-2) = 0, \quad k \in \mathbb{N}$$

where

$$p(k) = \begin{cases} 1/4, & k \text{ is odd} \\ -1/8, & k \text{ is even} \end{cases}$$

has a positive solution

(ii) if $\sum_{\ell=\tau(k)}^{k-1} p^+(\ell) \le 1/4$, $k \in \mathbb{N}$ where $p^+(\ell) \equiv 0$ for $\ell < 0$, then (6.24.40) has a positive solution. In particular, the equation

$$\Delta u(k) + \frac{1}{8(k+1)}u\left(\left[\frac{k}{2}\right] - 1\right) = 0, \quad k \in \mathbb{N}$$

has a positive solution

(iii) if $p(k) \ge 0$ eventually, and

$$\limsup_{k\to\infty} p(k) > 0, \quad \liminf_{k\to\infty} \sum_{\ell=\tau(k)}^{k-1} p(\ell) > \frac{1}{e}$$

then equation (6.24.40) is oscillatory

(iv) if the equation

$$\Delta v(k) + q(k)v(\tau(k)) = 0, \quad k \in \mathbb{N}$$

has a positive solution $v(k)$, $k \in \mathbb{N}(\tau(0))$ where $0 \le p(k) \le q(k)$ and $\sum_{\ell=0}^{k} p(\ell) > 0$ for $k \in \{m : \tau(0) \le \tau(m) \le 0\}$, then (6.24.40) has a positive solution on $\mathbb{N}(\tau(0))$.

**6.24.73.** Consider the difference equation

$$\Delta u(k) + f(k, u(k-m)) = 0, \quad k \in \mathbb{N}(a) \tag{6.24.41}$$

where $m \in \mathbb{N}$, $f : \mathbb{N}(a) \times \mathbb{R} \to \mathbb{R}$ is continuous with $f(k,0) = 0$. Assume that there exists a positive function $p(k)$, $k \in \mathbb{N}(a)$ such that

$$0 < \frac{f(k,z)}{z} \le p(k), \quad k \in \mathbb{N}(a), \quad z \in \mathbb{R}, \quad z \ne 0$$

and

$$\limsup_{k \to \infty} \sum_{\ell=k}^{k+2m} p(\ell) < 2.$$

Further, assume that for every constant $c \ne 0$, $\sum_{\ell=a}^{\infty} |f(k,c)| = \infty$. Show that every solution of (6.24.41) tends to zero as $k \to \infty$.

**6.24.74.** Consider the difference equation $(6.21.1)_1$, and assume that in addition to the given hypotheses on the functions $f_i$, $g_i$ and $F_i$, $0 < \alpha < 1$. Show that every solution $u(k)$ of $(6.21.1)_1$ is either oscillatory or $u(k) = o(1/k)$.

**6.24.75.** Consider the difference equation $(6.21.1)_1$, and assume that in addition to the given hypotheses on the functions $f_i$, $g_i$ and $F_i$, $\alpha \ge 1$, and there exists an index $j \in \mathbb{N}(1,m)$ such that $F_j$ is nonincreasing on $\mathbb{R} \backslash \{0\}$, and

$$\sum^{\infty} \alpha^{-\ell} f_j(\ell) F_j\left(c\alpha^{g_j(\ell)}\right) = \pm\infty \quad \text{for any } c \ne 0.$$

Show that $(6.21.1)_1$ is oscillatory.

**6.24.76.** Consider the difference equation $(6.21.1)_1$, and assume that in addition to the given hypotheses on the functions $f_i$, $g_i$ and $F_i$, $0 < \alpha < 1$, and there exists an index $j \in \mathbb{N}(1,m)$ such that (6.21.2) holds, and

$$\liminf_{u \to 0^+} F_j(u) = \alpha > 0, \qquad \limsup_{u \to 0^-} F_j(u) = \beta < 0.$$

Show that $(6.21.1)_1$ is oscillatory.

**6.24.77.** Consider the difference equation $(6.21.1)_{-1}$, and assume that in addition to the given hypotheses on the functions $f_i$, $g_i$ and $F_i$, $\alpha >$

1. Show that every solution $u(k)$ of $(6.21.1)_{-1}$ is either oscillatory or $|u(k)| \to \infty$ as $k \to \infty$.

**6.24.78.** Consider the difference equation $(6.21.1)_{-1}$, and assume that in addition to the given hypotheses on the functions $f_i$, $g_i$ and $F_i$, $0 < \alpha \leq 1$, and there exists an index $j \in \mathbb{N}(1,m)$ such that $F_j$ is nonincreasing on $\mathbb{R}\backslash\{0\}$, and

$$\lim_{k\to\infty} \sum^{k} \alpha^{k-\ell} f_j(\ell) = \infty.$$

Show that $(6.21.1)_{-1}$ is oscillatory.

**6.24.79.** Consider the difference equation $(6.21.1)_{-1}$, and assume that in addition to the given hypotheses on the functions $f_i$, $g_i$ and $F_i$, $\alpha \geq 1$, and there exists an index $j \in \mathbb{N}(1,m)$ such that $F_j$ is nondecreasing on $\mathbb{R}\backslash\{0\}$, and

$$\int_{\beta}^{\infty} \frac{dt}{F_j(t)} < \infty \quad \text{and} \quad \int_{-\beta}^{-\infty} \frac{dt}{F_j(t)} < \infty \quad \text{for every } \beta > 0,$$

$$\sum^{\infty} \chi(\overline{\mathbb{N}}_j(\ell)) \alpha^{-\ell} f_j(\ell) = \infty,$$

where $\chi(\overline{\mathbb{N}}_j(k))$ is the characteristic function of the set $\overline{\mathbb{N}}_j = \{k \in \mathbb{N} : g_j(k) > k\}$. Show that $(6.21.1)_{-1}$ is oscillatory.

**6.24.80.** Consider the difference equation with constant coefficients

$$u(k+1) - u(k) + \sum_{j=0}^{m} p_j u(k-j) = 0, \quad m \in \mathbb{N}(1), \quad n \in \mathbb{N}.$$

Show that this difference equation is oscillatory if and only if the characteristic equation

$$\lambda - 1 + \sum_{j=0}^{m} p_j \lambda^{-j} = 0$$

has no positive roots. In particular, show that the difference equation $u(k+1) - u(k) + pu(k-m) = 0$ is oscillatory if and only if $p > m^m/(m+1)^{m+1}$.

**6.24.81.** Consider the second order difference equation

$$\Delta^2 u(k) + \sum_{i=1}^{m} p_i(k) u(g_i(k)) = 0, \quad k \in \mathbb{N}(1) \tag{6.24.42}$$

where $p_i : \mathbb{N} \to \mathbb{R}_+$, $\{g_i(k)\} \subseteq \mathbb{N}(1)$ and $\lim_{k\to\infty} g_i(k) = \infty$, $i = 1, \cdots, m$. A function $u$ defined on $\mathbb{N}$ is said to be a *proper solution* of

(6.24.42) if it satisfies (6.24.42) on $\mathbb{N}(1)$ and $\sup\{|u(k)| : k \geq K\} > 0$ for $K \in \mathbb{N}(1)$. Show that if $\liminf_{k\to\infty} \frac{g_i(k)}{k} > 0,\ i = 1,\cdots,m$ and for any $\lambda \in [0,1)$ there exists $\epsilon > 0$ such that

$$\liminf_{k\to\infty}\ k^{1-\lambda}\sum_{\ell=k}^{\infty}\sum_{i=1}^{m} p_i(\ell)\,(g_i(\ell))^{\lambda}\ >\ \lambda+\epsilon,$$

or

$$\liminf_{k\to\infty}\ k\sum_{\ell=k}^{\infty}\sum_{i=1}^{m} p_i(\ell)\left(\frac{g_i(\ell)}{\ell}\right)^{\lambda}\ >\ \lambda(1-\lambda)+\epsilon$$

then every proper solution of (6.24.42) is oscillatory.

**6.24.82.** Consider the second order difference equation

$$\text{(6.24.43)}\qquad \Delta^2 u(k) + p(k)\sum_{i=1}^{m} c_i u\,(g_i(k))\ =\ 0, \qquad k \in \mathbb{N}(1)$$

where $p : \mathbb{N} \to \mathbb{R}_+$, $\{g_i(k)\} \subseteq \mathbb{N}(1)$, $\lim_{k\to\infty} g_i(k) = \infty$ and $c_i \in (0,\infty)$, $i = 1,\cdots,m$. Show that if $0 < \alpha_i = \liminf_{k\to\infty} \frac{g_i(k)}{k}$ and

$$\liminf_{k\to\infty}\ k\sum_{\ell=k}^{\infty} p(\ell)\ >\ \max\left\{\lambda(1-\lambda)\left(\sum_{i=1}^{m} c_i\alpha_i\right)^{-\lambda}\ :\ \lambda\in[0,1]\right\}$$

then every proper solution of (6.24.43) is oscillatory. In particular, show that every solution of the equation $\Delta^2 u(k) + p(k)u(k) = 0,\ k \in \mathbb{N}(1)$ is oscillatory if $\liminf_{k\to\infty} k\sum_{\ell=k}^{\infty} p(\ell) > 1/4$.

**6.24.83.** Consider the difference equation

$$\text{(6.24.44)}\qquad \Delta^2 u(k)\ =\ h(k) + \sum_{i=1}^{m} f_i(k)F_i(u(g_i(k))), \quad k \in \mathbb{N}(a)$$

where for each $i$, $1 \leq i \leq m$, $F_i(u)$ is defined on $\mathbb{R}$, nondecreasing and $uF_i(u) > 0$ for $u \neq 0$, and

$$\sum^{\infty}\ell\sum_{i=1}^{m} f_{+i}(\ell)\ =\ \infty, \qquad \sum^{\infty}\ell\sum_{i=1}^{m} f_{-i}(\ell)\ <\ \infty,$$

also for all $k \in \mathbb{N}(a)$, $\sum^{k} \ell h(\ell)$ is bounded. Show that for every bounded nonoscillatory solution $u(k)$ of (6.24.44), $\liminf_{k\to\infty} |u(k)| = 0$.

**6.24.84.** Consider the difference equation

$$(6.24.45) \qquad \Delta^{2n}u(k) = \sum_{i=1}^{m} f_i(k, u(k), u(g_i(k))), \quad k \in \mathbb{N}(a)$$

where for each $i$, $1 \le i \le m$

(i) $f_i(k,u,v)$ is defined on $\mathbb{N}(a) \times \mathbb{R}^2$, and continuous for all $u$ and $v$

(ii) if $u$, $v > 0$ then $uf_i(k,u,v) > 0$

(iii) $\{g_i(k)\} \subseteq \mathbb{N}(a)$, $g_i(k) \le k$ and $\lim_{k\to\infty} g_i(k) = \infty$.

Further, there exists an index $j$, $1 \le j \le m$ such that $f_j(k,u,v)$ is increasing in $u$ and $v$ for all large $k$, and for every $\alpha \ne 0$, $\sum^{\infty} \ell^{2n-1} f_j(\ell, \alpha, \alpha) = \pm\infty$. Show that for every nonoscillatory solution $u(k)$ of (6.24.45) either $|u(k)| \to 0$ or $|u(k)| \to \infty$ as $k \to \infty$.

**6.24.85.** Consider the second order nonlinear quasilinear difference equation

$$(6.24.46) \qquad \Delta\left(a(k)|\Delta u(k)|^{\alpha-1}\Delta u(k)\right) + q(k+1)f(u(k+1)) = 0, \qquad k \in \mathbb{N}(k_0)$$

where $k_0 \in \mathbb{N}$, $\alpha > 0$ and

(i) $a(k) > 0$ for all $k \in \mathbb{N}(k_0)$ and $\mathcal{R}(k) = \sum_{\ell=k_0}^{k-1} 1/(a(\ell))^{1/\alpha} \to \infty$ as $k \to \infty$

(ii) $q(k)$ is defined on $\mathbb{N}(k_0)$ and is allowed to change sign infinitely often

(iii) $f : \mathbb{R} \to \mathbb{R}$ is continuous and $uf(u) > 0$ for all $u \ne 0$

(iv) $f(u) - f(v) = g(u,v)(u-v)$ for all $u$, $v \ne 0$, where $g$ is a nonnegative function

(v) $q(k)$ is (conditionally) summable on $\mathbb{N}(k_0)$, i.e.

$$\sum_{\ell=k_0}^{\infty} q(\ell+1) = \lim_{k\to\infty} \sum_{\ell=k_0}^{\infty} q(\ell+1)$$

exists and is finite, in which case the function $Q(k) = \sum_{\ell=k}^{\infty} q(\ell+1)$ is well defined on $\mathbb{N}(k_0)$, and

(vi) $\lim_{|u|\to\infty} f(u) = \infty$.

Show that

1. if $u(k)$ is such that $u(k) \neq 0$ on $\mathbb{N}(k_0)$, then for any $\alpha > 0$ and $K \in \mathbb{N}(k_0)$

$$\limsup_{k\to\infty}\left\{\frac{a(k)|\Delta u(k)|^{\alpha-1}\Delta u(k)}{f(u(k))}+\sum_{\ell=K}^{k-1}\frac{a(\ell)|\Delta u(\ell)|^{\alpha+1}g(u(\ell),u(\ell+1))}{f(u(\ell))f(u(\ell+1))}\right\}\geq 0,$$

where $a(k)$ and $f$ satisfy conditions (i), (iii) and (iv)

2. if $u(k)$ is a nonoscillatory solution of (6.24.46) such that $\liminf_{k\to\infty}|u(k)| > 0$, then

$$\frac{a(k)|\Delta u(k)|^{\alpha-1}\Delta u(k)}{f(u(k))} = Q(k)+\sum_{\ell=k}^{\infty}\frac{a(\ell)|\Delta u(\ell)|^{\alpha+1}g(u(\ell),u(\ell+1))}{f(u(\ell))f(u(\ell+1))},$$

$$k \in \mathbb{N}(k_0)$$

3. if $ug(u,v) \geq f(u)$ for $u, v \neq 0$ and $Q(k) \geq 0$ for $k \in \mathbb{N}(k_0)$, then every nonoscillatory solution $u(k)$ of (6.24.46) satisfies eventually 'a priori' estimate $c_1 \leq |u(k)| \leq c_2 R(k)$ for some positive constants $c_1$ and $c_2$ (depending on $u(k)$)

4. if $Q(k) \geq 0$ for all $k \in \mathbb{N}(k_0)$, then a necessary condition for (6.24.46) to have a nonoscillatory solution which tends to a nonzero constant as $k \to \infty$ is that

$$\sum_{k=k_0}^{\infty}\left(\frac{Q(k)}{a(k)}\right)^{1/\alpha} < \infty \quad \text{and} \quad \sum_{k=k_0}^{\infty}\left(\frac{1}{a(k)}\sum_{\ell=k}^{\infty}\frac{Q^{(\alpha+1)/\alpha}(\ell)}{(a(\ell))^{1/\alpha}}\right)^{1/\alpha} < \infty$$

5. if $Q(k) \geq 0$ for all $k \in \mathbb{N}(k_0)$ and $a(k)$ is nondecreasing, then a sufficient condition for (6.24.46) to have a nonoscillatory solution which tends to a nonzero constant as $k \to \infty$ is that

$$\sum_{k=k_0}^{\infty} A^{1/\alpha}(k) < \infty \quad \text{and} \quad \sum_{k=k_0}^{\infty}\left(\sum_{\ell=k}^{\infty} A^{(\alpha+1)/\alpha}(\ell)\right)^{1/\alpha} < \infty$$

where $A(k) = Q(k)/a(k)$

6. if $Q(k) \geq 0$ for all $k \in \mathbb{N}(k_0)$, then a necessary and sufficient condition for (6.24.46) to have a nonoscillatory solution which tends to a nonzero constant as $k \to \infty$ is that

$$\sum_{k=k_0}^{\infty} Q^{1/\alpha}(k) < \infty \quad \text{and} \quad \sum_{k=k_0}^{\infty}\left(\sum_{\ell=k}^{\infty} Q^{(\alpha+1)/\alpha}(\ell)\right)^{1/\alpha} < \infty$$

7. if $Q(k) \geq 0$ for all $k \in \mathbb{N}(k_0)$ and $g(u,v) \geq \lambda > 0$ for $u, v \neq 0$, then a necessary condition for (6.24.46) to have a nonoscillatory solution

$u(k)$ such that $c_1R(k) \le |u(k)| \le c_2R(k),\ k \in \mathbb{N}(k_0)$ for some positive constants $c_1$ and $c_2$ is that

$$\sum_{k=k_0}^{\infty} \frac{Q^{(\alpha+1)/\alpha}(k) f^{1/\alpha}(c_1R(k))}{(a(k))^{1/\alpha}} < \infty$$

8. if

$$\sum_{k=k_0}^{\infty} q(k+1) = \lim_{K\to\infty} \sum_{k=k_0}^{K-1} q(k+1) = \infty$$

then all solutions of (6.24.46) are oscillatory

9. if $Q(k) \ge 0$ for all $k \in \mathbb{N}(k_0)$, $g(u,v) \ge \lambda > 0$ for $u, v \ne 0$,

$$\int_0^c \frac{dt}{f^{1/\alpha}(t^{1/\alpha})} < \infty \quad \text{and} \quad \int_{-c}^0 \frac{dt}{f^{1/\alpha}(t^{1/\alpha})} > -\infty \ \text{ for } c > 0,$$

$$-f(-uv) \ge f(uv) \ge f(u)f(v) \quad \text{for} \quad u, v \ne 0$$

and

$$\sum_{k=k_0}^{\infty} \frac{Q^{(\alpha+1)/\alpha}(k) f^{1/\alpha}(R(k))}{(a(k))^{1/\alpha}} = \infty$$

then all solutions of (6.24.46) are oscillatory

10. the difference equation

$$\Delta\left(k|\Delta u(k)|^{\alpha-1}\Delta u(k)\right) + 2^{\alpha}(2k+1)u^{\beta}(k+1) = 0, \quad k \in \mathbb{N}(1)$$

where $\beta \ge 1$ is the ratio of odd positive integers is oscillatory

11. the difference equation

$$\Delta\left(k^{-\alpha}|\Delta u(k)|^{\alpha-1}\Delta u(k)\right) + \frac{2^{\alpha}\left((k+1)^{\alpha} + k^{\alpha}\right)}{k^{\alpha}(k+1)^{\alpha}} u^{\beta}(k+1) = 0, \quad k \in \mathbb{N}(1)$$

where $\alpha > 1$ and $\beta$ is the ratio of odd positive integers such that $1 \le \beta < \alpha^2 < 2\alpha + 3$ is oscillatory.

**6.24.86.** Consider the Volterra summation equation

$$u(k) = p(k) + \sum_{\ell=0}^{k-1} L(k,\ell) g\left(u(\ell)\right), \quad k \in \mathbb{N} \tag{6.24.47}$$

where

(i) the function $p(k)$ is bounded on $\mathbb{N}$

(ii) $L(k,\ell) > 0$ for $k \ge \ell \in \mathbb{N}$, $L(k,\ell) = 0$ if $k < \ell$, $L(k,\ell)$ is nondecreasing in $k$ for every $\ell \in \mathbb{N}$, $\lim_{k\to\infty} \sum_{\ell=k_0}^{k-1} L(k,\ell) < \infty$ for $k \ge k_0 \in \mathbb{N}$, and $h(k,K) = \sum_{\ell=0}^{K} L(k,\ell) \le \delta(K) < \infty$ for all $k \in \mathbb{N}$
(iii) $g : \mathbb{R} \to \mathbb{R}$ is continuous, $ug(u) > 0$ for $u \ne 0$ and $g(u)/u \le M$ for all $u \ne 0$ and $M > 0$.

Show that all unbounded solutions of (6.24.47) are oscillatory.

**6.24.87.** Consider the difference equation

$$u(t) - u(t-\tau) + \sum_{i=1}^{m} p_i u(t-\sigma_i) = 0, \quad t \ge 0 \tag{6.24.48}$$

where $0 < \tau \le \sigma_1 < \cdots < \sigma_m$ and $p_i > 0$, $i = 1, \cdots, m$. Show that the following are equivalent

(i) every solution of (6.24.48) is oscillatory
(ii) the characteristic equation

$$1 - e^{-\lambda\tau} + \sum_{i=1}^{m} p_i e^{-\lambda\sigma_i} = 0$$

has no real roots.

**6.24.88.** Show that every solution of (6.24.48) oscillates if and only if the inequality

$$u(t) - u(t-\tau) + \sum_{i=1}^{m} p_i u(t-\sigma_i) \le 0, \quad t \ge 0$$

has no eventually positive solutions.

**6.24.89.** Consider the difference equation (6.23.1) with $p(t) \equiv p > 0$. Show that every solution of (6.23.1) oscillates if and only if $p^\tau \sigma^\sigma > \tau^\tau(\sigma - \tau)^{\sigma-\tau}$.

**6.24.90.** Suppose that conditions of Theorem 6.23.4 are satisfied, and $p^\tau\sigma^\sigma > \tau^\tau(\sigma-\tau)^{\sigma-\tau}$. Show that equation (6.23.2) is oscillatory.

**6.24.91.** Consider the difference equation

$$u(t) - u(t-\tau) - p(t)u(t+\sigma) = 0, \quad t \ge 0 \tag{6.24.49}$$

where $\tau > 0$, $\sigma > 0$ and $p \in C(\mathbb{R}_+, \mathbb{R}_+)$. Assume that

$$\liminf_{t\to\infty} \int_t^{t+\sigma} q(s)ds > \frac{\tau}{e},$$

where $q(t) = \min_{t-\tau \le s \le t} p(s)$. Show that equation (6.24.49) is oscillatory. In particular, examine the difference equation

$$u(t) - u(t - \pi/2) = \frac{1}{\sqrt{2}} u(t + 7\pi/4)$$

for which $u(t) = \sin t$ is an oscillatory solution.

**6.24.92.** Consider the nonlinear difference equation

$$u(t) - u(t-\tau) + \sum_{i=1}^{m} p_i f_i(u(t - \sigma_i)) = 0, \quad t \ge 0 \tag{6.24.50}$$

where $0 < \tau \le \sigma_1 < \cdots < \sigma_m$ and $p_i > 0$, $i = 1, \cdots, m$; $f_i \in C(\mathbf{R}, \mathbf{R})$, $uf_i(u) > 0$ for $u \ne 0$, $\lim_{u \to 0} f_i(u)/u = 1$, $f_i(u)$ is convex for $u > 0$, and $f_i(u)$ is concave for $u < 0$, $i = 1, \cdots, m$. Show that

(i) oscillation of the difference equation

$$u(t) - u(t-\tau) + (1-\epsilon) \sum_{i=1}^{m} p_i u(t - \sigma_i) = 0, \quad \epsilon \in (0,1), \quad t \ge 0$$

implies the oscillation of (6.24.50)

(ii) if

$$u(t) - u(t-\tau) + (1+\epsilon) \sum_{i=1}^{m} p_i u(t - \sigma_i) = 0, \quad \epsilon \in (0,1), \quad t \ge 0$$

has positive solutions, so does (6.24.50)

(iii) if

$$G(\lambda) = 1 - e^{-\lambda\tau} + \sum_{i=1}^{m} p_i e^{-\lambda \sigma_i} = 0$$

has no real roots, then there exists $\epsilon_0 > 0$ such that

$$1 - e^{-\lambda\tau} + (1-\epsilon) \sum_{i=1}^{m} p_i e^{-\lambda \sigma_i} = 0, \quad |\epsilon| < \epsilon_0$$

also has no real roots

(iv) equation (6.24.50) is oscillatory if and only if (6.24.49) is oscillatory.

**6.24.93.** Consider the difference equation

$$(-1)^{n+1} \Delta^n u(k) + \sum_{i=0}^{\infty} p(i) u(k - \tau_i) = 0, \quad (n \ge 1) \tag{6.24.51}$$

where $p(k) > 0$, $k \in \mathbb{N}$ and $\{\tau_i\}$ is a sequence of integers such that $0 \le \tau_0 < \tau_1 < \cdots$. A solution $u(k)$, $k \in \mathbf{Z}$ of (6.24.51) is called positive if $u(k) > 0$ for every $k \in \mathbf{Z}$, and bounded at $\infty$ if $\limsup_{k\to\infty} u(k) < \infty$. Show that

(i) for $n$ even, (6.24.51) has a positive solution which is bounded at $\infty$ if and only if the characteristic equation of (6.24.51), i.e.

$$-(1-\lambda)^n + \sum_{i=0}^{\infty} p(i)\lambda^{-\tau_i} = 0 \tag{6.24.52}$$

has a root in $(0,1)$

(ii) for $n$ odd, (6.24.51) has a positive solution if and only if (6.24.52) has a root in $(0,1)$.

## 6.25. Notes

The qualitative properties of solutions of higher order differential equations with and without deviating arguments has been the subject of many investigations, e.g. Graef [39] has cited over 100 publications, while the recent monograph by Ladde, Lakshmikantham and Zhang [70] refers to over 300 papers. But, the similar investigations for the higher order difference equations have gained momentum only recently. The general properties of solutions of (6.1.1) collected in Section 6.1 are from Cheng, Li and Patula [22], Fort [36], and Patula [89]. Theorems 6.2.1 and 6.2.5 and Corollaries 6.2.2 and 6.2.3 are due to Cheng, Li and Patula [22], whereas Theorem 6.2.4 is taken from Patula [89]. Recessive and dominant solutions of (6.1.1) are introduced in Olver and Sookne [87]. Theorems 6.3.1 and 6.3.4 and Corollaries 6.3.2 and 6.3.3 are adapted from Patula [89], whereas all other results in Section 6.3 have appeared in Cheng, Li and Patula [22]. Results in Section 6.4 are borrowed from Patula [89]. Section 6.5 is based on Hooker and Patula [55], Kwong, Hooker and Patula [66]. Several other related results are available in Hooker, Kwong and Patula [59]. For the second order difference systems similar results have been investigated in Ahlbrandt and Hooker [9], Chen and Erbe [19]. Results in Section 6.6 are adapted from Chen and Erbe [18]. Olver's type comparison results in Section 6.7 have appeared in Olver [86]. Section 6.8 contains the work of Hooker and Patula [55], Kwong, Hooker and Patula [66]. More precise Sturmian comparison theorems are available in Cheng [21]. Results in Section 6.9 are based on Patula [90]. All the results in Section 6.10 are due to Hooker and Patula [57]. Results in Section 6.11 are adapted from Szmanda [114]. Theorems 6.12.1 6.12.4 are taken from Szmanda [112], whereas Theorems 6.13.1 and 6.13.2 are by Popenda and Schmeidel [96]. Theorems 6.14.1 6.14.5 are borrowed from Popenda [100]. Results in Section 6.15 consist the work of Szmanda [115]. Results in Section 6.16 are based on Hooker and Patula [58]. All the results in Sections 6.17 6.19 are due to Agarwal [2 4]. Similar results for the differential equations with deviating arguments are also available in Agarwal [1], Thandapani and Agarwal [119]. Theorems 6.20.1 6.20.4 are from Erbe and Zhang [35], whereas Theorem 6.20.5 is by Ladas, Philos and Sficas [67]. Results in Section 6.21 are taken from Popenda and Szmanda [98]. All the results in Section 6.22 are adapted from Li and Cheng [74]. However, more general results which are given in Problem 6.24.85 were published earlier by Thandapani, Manuel and Agarwal [128]. Section 6.23 is based on the recent work of Zhang, Yan and Choi [147]. The terms limit point and limit circle are due to Atkinson [10]. Related properties of solutions of higher order difference equations have also been discussed in Agarwal et. al. [5,6], Ahlbrandt and Hooker [7,8], Balla [11], Bykov et. al. [12 14], Camouzis et. al. [15], Chen et. al. [16,17], Cheng et. al. [20,23 25], Domshlak [28 30], Došlý [31], Driver et. al. [32], Drozdowicz and Popenda [33], Dunkel [34], Gopalsamy [37], Grace et. al. [38], Graef et. al. [40 42], Grove et. al. [43], Gurney et. al. [44], Györi et. al. [45 48], Hartman and Wintner [49,50], He [52,53], Hinton and Lewis [54], Hooker [56], Jaroma [60], Kiventidis [61], Kocic and Ladas [62], Kong and Zettl [63], Korczak and Migda [64], Kulenovic and Budincevic [65], Ladas et. al. [68,69], Lalli and Zhang [71], Li et. al. [73,75], McCarthy [78], Medina et. al. [79 82], Mingarelli [83], Moulton [84], Papaschinopoulos and Schinas [88], Philos et. al. [91 93], Popenda et. al. [94,95,97,99,101], Risha [102], Schinas [103], Sedaghat [104], Smith and Taylor, Jr. [105], So and Yu [106], Stavroulakis [107], Szafranski et. al. [108 110], Szmanda [111,113,116,117], Tang and Yu [118], Thandapani et. al. [120 127], Trench [129], J. Wong et. al. [130], P.J.Y. Wong and Agarwal [131 139], Wouk [140], Yan and Yan [141], Yang and Zhang [142], Yu et. al. [143], Zhang et. al. [144 148], Zhou and Yan [149].

## 6.26. References

[1]. R.P. Agarwal, Oscillation and asymptotic behaviour of solutions of differential equations with nested arguments, *Bollettino U.M.I. Analisi Funzionale e Applicazioni*, Serie VI, Vol. 1 C, N.1(1982), 137 146.

[2]. R.P. Agarwal, Properties of solutions of higher order nonlinear difference equations, *An. st. Univ. Iasi* **29**(1983), 85 96.

[3]. R.P. Agarwal, Difference calculus with applications to difference equations, in *General Inequalities 4*, ed. W. Walter, ISNM 71, *Birkhäuser Ver-*

*lag*, Basel, (1984), 95 110.

[4]. R.P. Agarwal, Properties of solutions of higher order nonlinear difference equations, *An. st. Univ. Iasi* **31**(1985), 165 172.

[5]. R.P. Agarwal, S. Pandian and E. Thandapani, Oscillatory property for second order nonlinear difference equations via Lyapunov's second method, in *Advances in Nonlinear Dynamics, Gordon and Breach Publ.* U.K. (1997), 11 21.

[6]. R.P. Agarwal and P.J.Y. Wong, On the oscillation of second order nonlinear difference equations, *Mathl. Ineq. Appl.* **1**(1998), 349 365.

[7]. C.D. Ahlbrandt and J.W. Hooker, A variational view of nonoscillation theory for linear difference equations, in *Proc. Thirteenth Midwest Differential Equations Conference*, ed. J.L. Henderson, University of Missouri Rolla, Rolla, MO, (1985), 1 21.

[8]. C.D. Ahlbrandt and J.W. Hooker, Recessive solutions of symmetric three term recurrence relations, in *C.M.S. Conference Proc., Vol. 8, Oscillation, Bifurcation and Chaos, AMS*, (1987), 3 42.

[9]. C.D. Ahlbrandt and J.W. Hooker, Riccati matrix difference equations and disconjugacy of discrete linear systems, *SIAM J. Math. Anal.* **19**(1988), 1183 1197.

[10]. F.V. Atkinson, *Discrete and Continuous Boundary Value Problems, Academic Press*, New York, 1964.

[11]. K. Balla, On asymptotic behavior of solutions to some difference equations, in *Proceedings of the 2nd International Conference on Difference Equations and Applications,* Veszprém, Hungary, August 7 11, 1995, eds. S. Elaydi, et. al., *Gordon & Breach*, New York, 1996, 67 79.

[12]. Ya.V. Bykov, L.V. Živogladova and E.I. Ševcov, Sufficient conditions for oscillation of solutions of nonlinear finite difference equations, *Differencial'nye Uravnenija* **9**(1973), 1523 1524 (Russian).

[13]. Ya.V. Bykov and L.V. Živogladova, On the oscillation of solutions of nonlinear finite difference equations, *Differencial'nye Uravnenija* **9**(1973), 2080 2081 (Russian).

[14]. Ya.V. Bykov and E.I. Ševcov, Sufficient conditions for oscillation of solutions of nonlinear finite difference equations, *Differencial'nya Uravnenija* **9**(1973), 2241 2244 (Russian).

[15]. E. Camouzis, E.A. Grove and G. Ladas, Oscillation of a difference equation with periodic coefficients, *Applicable Analysis* **53**(1994), 143 148.

[16]. M. P. Chen and J.S. Yu, Oscillations of delay difference equations with variable coefficients, in *Proceedings of the 1st International Conference on*

*Difference Equations*, San Antonio, Texas, U.S.A. May 25 28, 1994, eds. S. Elaydi, et. al., *Gordon & Breach*, Luxembourg, 1995, 105 114.

[17]. M. P. Chen and B. Liu, Asymptotic behavior of solutions of first order nonlinear delay difference equations, *Computers Math. Applic.* **32**(1996), 9 13.

[18]. S. Chen and L.H. Erbe, Riccati techniques and discrete oscillations, *J. Math. Anal. Appl.* **142**(1989), 468 487.

[19]. S. Chen and L.H. Erbe, Oscillation and nonoscillation for systems of self adjoint second order difference equations, *SIAM J. Math. Anal.* **20**(1989), 939 949.

[20]. S.S. Cheng, Monotone solutions of $\Delta^2\mathbf{x}(k) = Q(k)\mathbf{x}(k+1)$, *Chinese J. Math.* **10**(1982), 71 75.

[21]. S.S. Cheng, Sturmian comparison theorems for three–term recurrence equations, *J. Math. Anal. Appl.* **111**(1985), 465 474.

[22]. S.S. Cheng, H.J. Li and W.T. Patula, Bounded and zero convergent solutions of second order difference equations, *J. Math. Anal. Appl.* **141**(1989), 463 483.

[23]. S.S. Cheng and W.T. Patula, An existence theorem for a nonlinear difference equation, *Nonlinear Analysis* **20**(1993), 193 203.

[24]. S.S. Cheng, Hille Wintner type comparison theorem for nonlinear difference equations, *Funkcialaj Ekvacioj* **37**(1994), 531 535.

[25]. S.S. Cheng and B.G. Zhang, Monotone solutions of a class of nonlinear difference equations, *Computers Math. Applic.* **28**(1 3)(1994), 71 79.

[26]. J.H. Conway and H.S.M. Coxeter, Triangulated polygons and frieze patterns, *Math. Gaz.* **57**(1973), 87 94.

[27]. J.H. Conway and H.S.M. Coxeter, Triangulated polygons and frieze patterns, *Math. Gaz.* **57**(1973), 175 183.

[28]. Y. Domshlak, Oscillatory properties of linear difference equations with continuous time, *Diff. Eqns. Dyn. Sys.* **1**(1993), 311 324.

[29]. Y. Domshlak, Sturmian comparison method in the oscillation study for discrete difference equations I, *Differential and Integral Equations* **7**(1994), 571 582.

[30]. Y. Domshlak, Does there exist a discrete version of Chanturia Koplatadze lemma, preprint.

[31]. D. Došlý, Reciprocity principle for Sturm Liouville difference equations and some of its applications, in *Proceedings of the 2nd International Conference on Difference Equations and Applications,* Veszprém, Hungary, August 7 11, 1995, eds. S. Elaydi, et. al., *Gordon & Breach*, New York,

1996, 145 153.

[32]. R.D. Driver, G. Ladas and P.N. Vlahos, Asymptotic behavior of a linear delay difference equation, *Proc. Amer. Math. Soc.* **115**(1992), 105 112.

[33]. A. Drozdowicz and J. Popenda, Asymptotic behavior of the solutions of the second order difference equations, *Proc. Amer. Math. Soc.* **99**(1987), 135 140.

[34]. O. Dunkel, The alternation of nodes of linearly independent solutions of second order difference equations, *Bull. AMS* **32**(1926), 333 334.

[35]. L.H. Erbe and B.G. Zhang, Oscillation of discrete analogues of delay equations, *Differential and Integral Equations* **2**(1989), 300 309.

[36]. T. Fort, *Finite Differences and Difference Equations in the Real Domain, The Clarendon Press*, Oxford, 1948.

[37]. K. Gopalsamy, *Stability and Oscillation in Delay Differential Equations of Population Dynamics, Kluwer*, Dordrecht, 1992.

[38]. S.R. Grace, A.A. Abadeer and H.A. El Morshedy, On the oscillation of certain second order difference equations, *Communications in Applied Analysis* **2**(1998), 447 455.

[39]. J.R. Graef, Oscillation, nonoscillation, and growth of solutions of nonlinear functional differential equations of arbitrary order, *J. Math. Anal. Appl.* **60**(1977), 398 409.

[40]. J.R. Graef and P.W. Spikes, Asymptotic decay of oscillatory solutions of forced nonlinear difference equations, *Dynamic Systems Appl.* **3**(1994), 95 102.

[41]. J.R. Graef, A. Miciano Cariño and C. Qian, A Sturm type comparison theorem for higher order nonlinear difference equations, in *Proceedings of the 2nd International Conference on Difference Equations and Applications*, Veszprém, Hungary, August 7 11, 1995, eds. S. Elaydi, et. al., *Gordon & Breach*, New York, 1996, 263 270.

[42]. J.R. Graef and E. Thandapani, Oscillatory behavior of solutions of Volterra summation equations, *Appl. Math. Letters*, to appear.

[43]. E.A. Grove, E.J. Janowski, C.M. Kent and G. Ladas, On the rational recursive sequence $x_{n+1} = (\alpha x_n + \beta)/(\gamma x_n + \delta)x_{n-1}$, *Comm. Appl. Nonlinear Anal.* **1**(1994), 61 72.

[44]. W.S. Gurney, S.P. Blythe and R.M. Nisbet, Nicolson's blowflies revisited, Nature **287**(1980), 17-21.

[45]. I. Györi and G. Ladas, Linearized oscillations for equations with piecewise constant arguments, *Differential and Integral Equations* **2**(1989),

123 131.

[46]. I. Györi and G. Ladas, *Oscillation Theory of Delay Differential Equations, Clarendon Press*, Oxford, 1991.

[47]. I Györi and M. Pituk, Asymptotic formulae for the solutions of a linear delay difference equation, *J. Math. Anal. Appl.* **195**(1995), 376 392.

[48]. I. Györi and M. Pituk, Asymptotic stability in a linear delay difference equation, in *Proceedings of the 2nd International Conference on Difference Equations and Applications*, Veszprém, Hungary, August 7 11, 1995, eds. S. Elaydi, et. al., *Gordon & Breach*, New York, 1996, 295 299.

[49]. P. Hartman and A. Wintner, On linear difference equations of the second order, *Amer. J. Math.* **72**(1950), 124 128.

[50]. P. Hartman and A. Wintner, Linear differential and difference equations with monotone solutions, *Amer. J. Math.* **75**(1953), 731 743.

[51]. P. Hartman, Difference equations: Disconjugacy, principal solutions, Green's functions, complete monotonicity, *Trans. Amer. Math. Soc.* **246**(1978), 1 30.

[52]. X.Z. He, Oscillatory property of solutions of a second order difference equation, *Ann. Diff. Eqns.* **7**(1991), 415 422.

[53]. X.Z. He, Oscillatory and asymptotic behavior of second order nonlinear difference equations, *J. Math. Anal. Appl.* **175**(1993), 482 498.

[54]. D. Hinton and R. Lewis, Spectral analysis of second order difference equations, *J. Math. Anal. Appl.* **63**(1978), 421 438.

[55]. J.W. Hooker and W.T. Patula, Riccati type transformations for second order linear difference equations, *J. Math. Anal. Appl.* **82**(1981), 451 462.

[56]. J.W. Hooker, A Hille Winter type comparison theorem for second order difference equations, *Internat. J. Math. & Math. Sci.* **6**(1983), 387 394.

[57]. J.W. Hooker and W.T. Patula, A second order nonlinear difference equation: oscillation and asymptotic behaviour, *J. Math. Anal. Appl.* **91**(1983), 9 29.

[58]. J.W. Hooker and W.T. Patula, Growth and oscillation properties of solutions of a fourth order linear difference equation, *J. Austral. Math. Soc. B* **26**(1985), 310-328.

[59]. J.W. Hooker, M.K. Kwong and W.T. Patula, Oscillatory second order linear difference equations and Riccati equations, *SIAM J. Math. Anal.* **18**(1987), 54 63.

[60]. J.H. Jaroma, On the global asymptotic stability of $x_{n+1} = (a + bx_n)/(A + x_{n-1})$, in *Proceedings of the Ist International Conference on*

*Difference Equations*, San Antonio, Texas, U.S.A. May 25 28, 1994, eds. S. Elaydi, et. al., *Gordon & Breach*, Luxembourg, 1995, 283 295.

[61]. T. Kiventidis, Positive solutions of integrodifferential and difference equations with unbounded delay, *Glasgow Math. J.* **35**(1993), 105 113.

[62]. V.L. Kocic and G. Ladas, *Global Behavior of Nonlinear Difference Equations of Higher Order with Applications*, *Kluwer*, Dordrechet, 1993.

[63]. Q. Kong and A. Zettl, Interval oscillation conditions for difference equations, *SIAM J. Math. Anal.* **26**(1995), 1047 1060.

[64]. J. Korczak and M. Migda, On the asymptotic behaviour of solution of the $m$ th order difference equation, *Demonstratio Mathematica* **21**(1988), 615 630.

[65]. M.R.S. Kulenovic and M. Budincevic, Asymptotic analysis of nonlinear second order difference equations, *An. St. Univ. Iasi* **30**(1984), 39 52.

[66]. M.K. Kwong, J.W. Hooker and W.T. Patula, Riccati type transformations for second order linear difference equations II, *J. Math. Anal. Appl.* **107**(1985), 182 196.

[67]. G. Ladas, Ch.G. Philos and Y.G. Sficas, Sharp conditions for the oscillation of delay difference equations, *J. Appl. Math. Simulation* **2**(1989), 101 111.

[68]. G. Ladas, L. Pakula and Z. Wang, Necessary and sufficient conditions for the oscillation of difference equations, *PanAmerican Math. J.* **2**(1992), 17 26.

[69]. G. Ladas, Invariants for generalized Lyness equation, *J. Difference Eqns. Appl.* **1**(1995), 209 214.

[70]. G.S. Ladde, V. Lakshmikantham and B.G. Zhang, *Oscillation Theory of Differential Equations with Deviating Arguments*, *Marcel Dekker*, New York, 1987.

[71]. B.S. Lalli and B.G. Zhang, Oscillation of difference equations, *Colloquium Math.* **65**(1993), 25 32.

[72]. J. Leech, The rational cuboid revisited, *Amer. Math. Monthly* **84**(1977), 518 533.

[73]. H.J. Li and C.C. Yeh, Existence of positive nondecreasing solutions of nonlinear difference equations, *Nonlinear Analysis* **22**(1994), 1271 1284.

[74]. W.T. Li and S.S. Cheng, Oscillation criteria for a nonlinear difference equation, *Computers Math. Applic.* **36**(8)(1998), 87 94.

[75]. Z. H. Li, A note on the oscillatory property for nonlinear difference equations and differential equations, *J. Math. Anal. Appl.* **103**(1984),

344 352.

[76]. R.C. Lyness, Note 1581, *Math. Gaz.* **26**(1942), 62.

[77]. R.C. Lyness, Note 1847, *Math. Gaz.* **29**(1945), 231.

[78]. P.J. McCarthy, Note on the oscillation of solutions of second order linear difference equations, *Portugaliae Mathematica* **18**(1959), 203 205.

[79]. R. Medina and M. Pinto, Asymptotic behavior of solutions of second order nonlinear difference equations, *Nonlinear Analysis* **19**(1992), 187 195.

[80]. R. Medina and M. Pinto, Asymptotic representation of solutions of linear second order difference equations, *J. Math. Anal. Appl.* **165**(1992), 505 516.

[81]. R. Medina, Asymptotic behavior of the solutions of second order difference equations, *J. Comput. Appl. Math.* **64**(1995), 233 246.

[82]. R. Medina, Asymptotic properties of solutions of nonlinear difference equations, *J. Comput. Appl. Math.* **70**(1996), 57 66.

[83]. A.B. Mingarelli, *Volterra Stieljes Integral Equations and Generalized Ordinary Differential Expressions, Lecture Notes in Math. Vol.* **989**, *Springer Verlag*, Berlin, 1983.

[84]. E.J. Moulton, A theorem in difference equations on the alternation of nodes of linearly independent solutions, *Ann. Math.* **13**(1911 1912), 137 139.

[85]. A.J. Nicholson, An outline of the dynamics of animal populations, *Austral. Jour. Zool.* **2**(1954), 9 25.

[86]. F.W.J. Olver, Bounds for the solutions of second order difference equations, *J. Res. Nat. Bur. Standards Sect.* **B71**(1967), 161 166.

[87]. F.W.J. Olver and D.J. Sookne, Note on backward recurrence algorithms, *Math. Comput.* **26**, No. 120(1972), 941 947.

[88]. G. Papaschinopoulos and C.J. Schinas, On a system of two nonlinear difference equations, *J. Math. Anal. Appl.* **219**(1998), 415 426.

[89]. W.T. Patula, Growth and oscillation properties of second order linear difference equations, *SIAM J. Math. Anal.* **10**(1979), 55 61.

[90]. W.T. Patula, Growth, oscillation and comparison theorems for second order linear difference equations, *SIAM J. Math. Anal.* **10**(1979), 1272 1279.

[91]. Ch.G. Philos and Y.G. Sficas, Positive solutions of difference equations, *Proc. Amer. Math. Soc.* **108**(1990), 107 115.

[92]. Ch.G. Philos, Oscillations in a nonautonomous delay logistic difference equation, *Proc. Edinburgh Math. Soc.* **35**(1992), 121 131.

[93]. Ch.G. Philos, On the existence of positive solutions for certain difference equations and inequalities, *Jour. Ineq. Applic.* **2**(1998), 57 69.

[94]. J. Popenda and E. Schmeidel, On the asymptotic behaviour of nonoscillatory solutions of difference equations, *Fasciculi Mathematici* **12**(1980), 43 53.

[95]. J. Popenda and J. Werbowski, On the asymptotic behaviour of the solutions of difference equations of second order, *Comm. Math.* **22** (1980), 135 142.

[96]. J. Popenda and E. Schmeidel, Some properties of solutions of difference equations, *Fasciculi Mathematici* **13**(1981), 89 98.

[97]. J. Popenda, On the asymptotic behaviour of the solutions of an $n$ th order difference equation, *Annales Polonici Mathematici* **44**(1984), 95 111.

[98]. J. Popenda and B. Szmanda, On the oscillation of solutions of certain difference equations, *Demonstratio Mathematica* **17**(1984), 153 164.

[99]. J. Popenda, On the boundedness of the solutions of difference equations, *Fasciculi Mathematici* **14**(1985), 101 108.

[100]. J. Popenda, Oscillation and nonoscillation theorems for second order difference equations, *J. Math. Anal. Appl.* **123**(1987), 34 38.

[101]. J. Popenda, The oscillation of solutions of difference equations, *Computers Math. Applic.* **28**(1 3)(1994), 271 279.

[102]. M.H.A. Risha, On oscillation of second order linear difference equations, *Appl. Math. Letters*, to appear.

[103]. C.J. Schinas, Invariants for difference equations and systems of difference equations of rational form, *J. Math. Anal. Appl.* **216**(1997), 164 179.

[104]. H. Sedaghat, General permanence conditions for nonlinear difference equations of higher order, *J. Math. Anal. Appl.* **213**(1997), 496 510.

[105]. B. Smith and W.E. Taylor, JR., Nonlinear third order difference equations: oscillatory and asymptotic behaviour, *Tamk. Jour. Math.* **19**(1988), 91 95.

[106]. J.W. H. So and J.S. Yu, On the stability and uniform persistence of a discrete model of Nicholson's blowflies, *J. Math. Anal. Appl.* **193**(1995), 233 244.

[107]. I.P. Stavroulakis, Oscillation of delay difference equations, *Computers Math. Applic.* **29**(7)(1995), 83 88.

[108]. Z. Szafranski, On some oscillation criteria for difference equations of second order, *Fasc. Math.* **11**(1979), 135 142.

[109]. Z. Szafranski and B. Szmanda, Oscillation and asymptotic behavior of certain nonlinear difference equations, *Riv. Math. Univ. Parma*

**4**(5)(1995), 231 240.

[110]. Z. Szafranski and B. Szmanda, Oscillation of solutions of some nonlinear difference equations, *Publ. Math.* **40**(1996), 127 133.

[111]. S. Szmanda, Oscillation of solutions of second order difference equations, *Portugaliac Mathematica* **37**(1978), 251 254.

[112]. B. Szmanda, Note on the behaviour of solutions of a second order nonlinear difference equation, *Atti Accad. Naz. Lincei Rend. Sc. fis. mat. e nat.* **69**(1980), 120 125.

[113]. B. Szmanda, Oscillation theorems for nonlinear second order difference equations, *J. Math. Anal. Appl.* **79**(1981), 90 95.

[114]. B. Szmanda, Oscillation criteria for second order nonlinear difference equations, *Annales Polonici Mathematici* **43**(1983), 225 235.

[115]. B. Szmanda, Nonoscillation, oscillation and growth of solutions of nonlinear difference equations of second order, *J. Math. Anal. Appl.* **109**(1985), 22 30.

[116]. B. Szmanda, Characterization of oscillation of second order nonlinear difference equations, *Bull. Polish. Acad. Sci. Math.* **34**(1986), 133 141.

[117]. B. Szmanda, Note on the oscillation of certain difference equations, *Glas. Math. Ser. III* **31**(1996), 115 121.

[118]. X.H. Tang and J.S. Yu, New oscillation criterion for delay difference equations, *Dynamics of Continuous, Discrete and Impulsive Systems*, to appear.

[119]. E. Thandapani and R.P. Agarwal, Asymptotic behaviour and oscillation of solutions of differential equations with deviating arguments, *Bolletino U.M.I.* **17-B**(1980), 82 93.

[120]. E. Thandapani, Asymptotic and oscillatory behavior of solutions of nonlinear second order difference equations, *Indian J. Pure Appl. Math.* **24**(1993), 365 372.

[121]. E. Thandapani, Oscillation theorems for higher order nonlinear difference equations, *Indian J. Pure Appl. Math.* **25**(1994), 519 524.

[122]. E. Thandapani, Oscillation theorems for perturbed nonlinear second order difference equations, *Computers Math. Applic.* **28**(1 3)(1994), 309 316.

[123]. E. Thandapani, I. Györi and B.S. Lalli, An application of discrete inequality to second order nonlinear oscillation, *J. Math. Anal. Appl.* **186**(1994), 200 208.

[124]. E. Thandapani and B.S. Lalli, Asymptotic behavior and oscillation of difference equations of Volterra type, *Appl. Math. Letters* **7**(1994), 89 93.

[125]. E. Thandapani, Oscillation theorems for second order damped nonlinear difference equations, *Czech. Math. J.* **45**(120)(1995), 327 335.

[126]. E. Thandapani, J.R. Graef and P.W. Spikes, Monotonicity and summability of solutions of a second order nonlinear difference equation, *Bull. Inst. Math. Acad. Sinica* **23**(1995), 343 356.

[127]. E. Thandapani, J.R. Graef and P.W. Spikes, On the oscillation of solutions of second order quasilinear difference equations, *Nonlinear World* **3**(1996), 545 565.

[128]. E. Thandapani, M. Maria Susai Manuel and R.P. Agarwal, Oscillation and nonoscillation theorems for second order quasilinear difference equations, *Facta Universitatis (Niš) Ser. Math. Inform.* **11**(1996), 49 65.

[129]. W.F. Trench, Asymptotic behavior of solutions of Emden Fower difference equations with oscillating coefficients, *J. Math. Anal. Appl.* **179**(1973), 135 153.

[130]. J. Wong, F. Meng and J. Li, Asymptotic behavior of solutions of second order nonlinear difference equations, *Kodai Math. J.* **19**(1996), 200 206.

[131]. P.J.Y. Wong and R.P. Agarwal, Oscillation theorems and existence of positive monotone solutions for second order nonlinear difference equations, *Mathl. Comput. Modelling* **21**(3)(1995), 63 84.

[132]. P.J.Y. Wong and R.P. Agarwal, Oscillation theorems and existence criteria of asymptotically monotone solutions for second order differential equations, *Dynamic Systems Appl.* **4**(1995), 477 496.

[133]. P.J.Y. Wong and R.P. Agarwal, On the oscillation and asymptotically monotone solutions of second order quasilinear differential equations, *Appl. Math. Comput.* **79**(1996), 207 237.

[134]. P.J.Y. Wong and R.P. Agarwal, Oscillatory behaviour of solutions of certain second order nonlinear differential equations, *J. Math. Anal. Appl.* **198**(1996), 337 354.

[135]. P.J.Y. Wong and R.P. Agarwal, On the oscillation of an $m$th order perturbed nonlinear difference equation, *Arch. Math. (Brnö)* **32**(1996), 13 27.

[136]. P.J.Y. Wong and R.P. Agarwal, Summation averages and the oscillation of second order nonlinear difference equations, *Mathl. Comput. Modelling* **24**(9)(1996), 21 35.

[137]. P.J.Y. Wong and R.P. Agarwal, Oscillation theorems for certain second order nonlinear difference equations, *J. Math. Anal. Appl.* **204**(1996), 813 829.

[138]. P.J.Y. Wong and R.P. Agarwal, Comparison theorems for the oscilla-

tion of higher order difference equations with deviating arguments, *Mathl. Comput. Modelling* **24**(12)(1996), 39 48.

[139]. P.J.Y. Wong and R.P. Agarwal, Oscillation and monotone solutions of second order quasilinear difference equations, *Funkcialaj Ekvacioj* **39**(1996), 491 517.

[140]. A. Wouk, Difference equations and $J$ matrices, *Duke Math. J.* **20**(1953), 141 159.

[141]. W.P. Yan and J. Yan, Comparison and oscillation results for delay difference equations with oscillatory coefficients, *Internat. J. Math. Math. Sci.* **19**(1996), 171 176.

[142]. Y. Yang and W. Zhang, Oscillation for second order superlinear difference equations, *J. Math. Anal. Appl.* **189**(1995), 631 639.

[143]. J. Yu, B.G. Zhang and X. Qian, Oscillations of delay difference equations with oscillatory coefficients, *J. Math. Anal. Appl.* **177**(1993), 432 444.

[144]. B.G. Zhang, Oscillation and asymptotic behavior of second order difference equations, *J. Math. Anal. Appl.* **173**(1993), 58 68.

[145]. B.G. Zhang and G.D. Chen, Oscillation of certain second order nonlinear difference equations, *J. Math. Anal. Appl.* **199**(1996), 827 841.

[146]. B.G. Zhang and C.J. Tian, Nonexistence and existence of positive solutions for difference equations with unbounded delay, *Computers Math. Applic.* **36**(1)(1998), 1 8.

[147]. B.G. Zhang, J. Yan and S.K. Choi, Oscillation for difference equations with continuous variable, *Computers Math. Applic.* **36**(9)(1998), 11 18.

[148]. B.G. Zhang and B.M. Liu, Linearized oscillation theorems for certain nonlinear difference equations with continuous arguments, *Mathl. Comput. Modelling*, to appear.

[149]. X. Zhou and J. Yan, Oscillatory and asymptotic properties of higher order nonlinear difference equations, *Nonlinear Analysis* **31**(1998), 493 502.

# Chapter 7
# Qualitative Properties of Solutions of Neutral Difference Equations

A neutral differential equation involves the highest order derivative of the unknown function with and without delays. Analogous to this, in the discrete case, a neutral difference equation involves the highest order difference of the unknown function with and without delays. Although, often neutral difference equations can be written as higher order difference equations, it is convenient to study the behavior of solutions of these equations directly. The main aim of this chapter is to present recently developed theory of oscillation and nonoscillation of neutral difference equations.

## 7.1. Oscillation and Nonoscillation for

$$\Delta(u(k) + pu(k-\tau)) + q(k)u(k-\sigma) = 0, \quad k \in \mathbb{N} \tag{7.1.1}$$

where $p$ is a constant, the function $q(k)$ is defined on $\mathbb{N}$, $\tau$ is a positive integer and $\sigma$ is a nonnegative integer.

Let $\mu = \max\{\tau, \sigma\}$. Then, by a *solution* of (7.1.1) we mean a function $u(k)$ which is defined for $k \geq -\mu$ and satisfies the equation (7.1.1) for $k \in \mathbb{N}$. Clearly, if

$$u(k) = A_k, \quad k = -\mu, \cdots, -1, 0 \tag{7.1.2}$$

are given, then (7.1.1) has a unique solution, and it can be constructed recursively. In what follows we shall assume that the function $q(k)$ is not identically zero. A nontrivial solution $u(k)$ of (7.1.1) is said to be *oscillatory*, and similarly for later equations, if for every $0 < K \in \mathbb{N}$ there exists a $k \geq K$ such that $u(k)u(k+1) \leq 0$. The equation (7.1.1) itself is called oscillatory if all its solutions are oscillatory. Otherwise, it is called *nonoscillatory*. A given difference equation of the type (7.1.1) can have both oscillatory as well as nonoscillatory solutions, e.g. the difference equation

$$\Delta\left(u(k) - \frac{1}{4}u(k-1)\right) + \frac{1}{16}u(k-2) \;=\; 0$$

has an oscillatory solution $\left\{(-1)^k\left((\sqrt{17}-3)/8\right)^k\right\}$ and a nonoscillatory solution $\{2^{-k}\}$.

We begin with the following:

**Lemma 7.1.1.** Let $q(k) \geq 0$ for all $k \in \mathbb{N}$ and let $u(k)$ be an eventually positive solution of (7.1.1). Set $z(k) = u(k) + pu(k-\tau)$.

(a) If $p = -1$, then $z(k) > 0$ and $\Delta z(k) \leq 0$ eventually.

(b) If $-1 < p < 0$, then $z(k) > 0$ and $\Delta z(k) < 0$ eventually.

(c) If $p < -1$ and $\sum_{k=1}^{\infty} q(k) = \infty$, then $z(k) < 0$ and $\Delta z(k) \leq 0$ eventually.

**Proof.** Since $q(k) \not\equiv 0$, from the equation (7.1.1), we have $\Delta z(k) = -q(k)u(k-\sigma) \leq 0$, eventually, so $z(k)$ cannot be eventually identically zero. Thus, it follows that $z(k)$ is eventually positive or eventually negative.

If $z(k) < 0$ eventually, then $z(k) \leq z(K) < 0$ for $k \geq K \in \mathbb{N}$. Hence,

$$u(K+\tau k) \;\leq\; z(K) + u(K+(k-1)\tau) \;\leq\; \cdots \;\leq\; kz(K) + u(K).$$

On letting $k \to \infty$ in the above inequality, we find $u(K+\tau k)$ to be negative, which is a contradiction to $u(k) > 0$. This proves (a).

The proof of (b) is similar to that of (a).

To prove (c), again from (7.1.1), we have $\Delta z(k) = -q(k)u(k-\sigma) \leq 0$, for all large $k$. We shall prove that $z(k) < 0$, eventually. If not, then

$$z(k) \;=\; u(k) + pu(k-\tau) \;\geq\; 0, \quad k \geq K$$

i.e.

$$u(k) \;\geq\; -pu(k-\tau), \quad k \geq K$$

which implies that

$$0 < u(K-\tau) \leq (-1/p)u(K) \;\leq\; \cdots \;\leq\; (-1/p)^j u(K+(j-1)\tau),\; j = 1, 2, \cdots.$$

On letting $j \to \infty$ in the above inequality, we get $u(k) \to \infty$ as $k \to \infty$. But, then since

$$\Delta z(k) = -q(k)u(k-\tau) \le -Lq(k) \text{ for large } k,$$

where $L$ is a positive number. On summing the last inequality, we obtain

$$z(k+1) - z(K) \le -L\sum_{\ell=K}^{k} q(\ell),$$

which implies that $z(k) \to -\infty$ as $k \to \infty$. This contradicts the fact that $z(k) \ge 0$ for $k \ge K$. ∎

Now we shall establish several oscillation criteria for the difference equation (7.1.1). The obtained results depend on the values of the parameter $p$.

**Theorem 7.1.2.** Assume that $p = -1$, $q(k) \ge 0$ for $k \in \mathbb{N}(1)$, and for a positive integer $K$,

$$\sum_{k=K}^{\infty} q(k) = \infty. \tag{7.1.3}$$

Then, the equation (7.1.1) is oscillatory.

**Proof.** Assume the contrary. Without loss of generality let $u(k)$ be an eventually positive solution of (7.1.1). By Lemma 7.1.1(a), $z(k) = u(k) + pu(k-\tau) > 0$ and $\Delta z(k) \le 0$, eventually. This implies that $\lim_{k\to\infty} z(k) = \alpha \ge 0$ exists.

On summing (7.1.1) from $K$ to $k$, we get

$$z(k+1) - z(K) + \sum_{\ell=K}^{k} q(\ell)u(\ell-\sigma) = 0.$$

On letting $k \to \infty$ in the above equation, we obtain

$$z(K) \ge \sum_{\ell=K}^{\infty} q(\ell)u(\ell-\sigma). \tag{7.1.4}$$

Now setting $\min_{K\le\ell\le K+\tau} u(\ell-\tau) = s > 0$, we find from $z(k) = u(k) - u(k-\tau) > 0$ for $k \ge K$ that $u(k) \ge s$ for $k \ge K$. Thus, from (7.1.4) we have

$$\infty > z(K+\sigma) \ge \sum_{\ell=K+\sigma}^{\infty} q(\ell)u(\ell-\sigma) \ge s\sum_{\ell=K+\sigma}^{\infty} q(\ell),$$

which contradicts condition (7.1.3). ∎

**Example 7.1.1.** Consider the difference equation

$$\Delta(u(k) - u(k-\tau)) + q(k)u(k-\sigma) \;=\; 0, \qquad k \in \mathbb{N} \tag{7.1.5}$$

where $\tau$ and $\sigma$ are positive integers, and

$$q(k) \;=\; \frac{\tau}{(k+1)(k-\tau+1)}\left(\sum_{\ell=1}^{k-\sigma}\frac{1}{\ell}\right)^{-1}.$$

It is clear that $\sum_{\ell=K}^{\infty} q(\ell) < \infty$ for $K \ge \tau$. Thus, the equation (7.1.5) does not satisfy the assumption (7.1.3) of Theorem 7.1.2. In fact, (7.1.5) has a nonoscillatory solution $u(k) = \sum_{\ell=1}^{k} 1/\ell$.

**Example 7.1.2.** Consider the difference equation

$$\Delta(u(k) - u(k-\tau)) + 4u(k-\sigma) \;=\; 0, \qquad k \in \mathbb{N} \tag{7.1.6}$$

where $\tau$ and $\sigma$ are odd and even positive integers. Equation (7.1.6) satisfies the assumptions of Theorem 7.1.2, and therefore the equation (7.1.6) is oscillatory. In fact, $u(k) = (-1)^{k+1}$ is an oscillatory solution of (7.1.6).

In Theorem 7.1.2 the condition (7.1.3) can be weakened by

$$\sum_{k=K}^{\infty} kq(k)\sum_{\ell=k}^{\infty} q(\ell) \;=\; \infty. \tag{7.1.7}$$

This we shall prove in the following:

**Theorem 7.1.3.** The conclusion of Theorem 7.1.2 holds even if (7.1.3) is replaced by (7.1.7).

**Proof.** Since (7.1.3) implies that the equation (7.1.1) is oscillatory, it suffices to show that all solutions of (7.1.1) oscillate in the case that

$$\sum_{k=K}^{\infty} q(k) \;<\; \infty. \tag{7.1.8}$$

Assume, for the sake of contradiction, that (7.1.1) has an eventually positive solution $u(k)$. Then, by Lemma 7.1.1(a), $z(k) = u(k) - u(k-\tau) > 0$ and $\Delta z(k) \le 0$, eventually. Thus, eventually $u(k) > u(k-\tau)$, which implies that there exist a constant $L > 0$ and $K \in \mathbb{N}$ sufficiently large such that $u(k-\mu) \ge L$, $k \ge K$. Thus, from $\Delta z(k) = -q(k)u(k-\sigma)$ it follows that $\Delta z(k) \le -Lq(k)$, $k \ge K$ and hence $z(k) \ge L\sum_{\ell=k}^{\infty} q(\ell)$, $k \ge K$, which is the same as

$$u(k) \;\ge\; u(k-\tau) + L\sum_{\ell=k}^{\infty} q(\ell), \qquad k \ge K. \tag{7.1.9}$$

Now let $I(k)$ denote the greatest integer part of $(k-K)/\tau$, then we have

$$u(k) \geq L\left(\sum_{\ell=k}^{\infty} q(\ell) + \sum_{\ell=k-\tau}^{\infty} q(\ell) + \cdots + \sum_{\ell=k-(I(k)-1)\tau}^{\infty} q(\ell)\right) + u(k-I(k)\tau),$$

which together with $\Delta z(k) = -q(k)u(k-\sigma)$ yields

$$\Delta z(k) \leq -H(k), \tag{7.1.10}$$

where

$$H(k) = I(k)Lq(k)\sum_{\ell=k}^{\infty} q(\ell).$$

By noting the fact that $I(k)/k \to 1/\tau$ as $k \to \infty$, we have

$$H(k)\left(kq(k)\sum_{\ell=k}^{\infty} q(\ell)\right)^{-1} = \frac{I(k)L}{k} \to \frac{L}{\tau} \quad \text{as } k \to \infty. \tag{7.1.11}$$

Thus, (7.1.7) and (7.1.11) imply that $\sum_{k=K}^{\infty} H(k) = \infty$, which together with (7.1.10) leads to $z(k) \to -\infty$ as $k \to \infty$. This contradicts the hypothesis that $z(k)$ is eventually positive. ∎

**Example 7.1.3.** For the neutral difference equation

$$\Delta(u(k) - u(k-\tau)) + k^{-\alpha}u(k-\sigma) = 0, \quad \alpha \in (1, 3/2] \tag{7.1.12}$$

condition (7.1.7) is satisfied. Therefore, by Theorem 7.1.3 the equation (7.1.12) is oscillatory. However, the condition (7.1.3) does not satisfy.

**Theorem 7.1.4.** Assume that $-1 < p < 0$, $\sigma > \tau$, $q(k) \geq q(k-\tau)$ for all large $k$, and

$$\left(\frac{1}{1+p}\right)\liminf_{k\to\infty} q(k) > \frac{\sigma^{\sigma}}{(\sigma+1)^{\sigma+1}}. \tag{7.1.13}$$

Then, the equation (7.1.1) is oscillatory.

**Proof.** If not, we can assume that $u(k)$ is an eventually positive solution. Let $z(k) = u(k) + pu(k-\tau)$ and $w(k) = z(k) + pz(k-\tau)$. By Lemma 7.1.1(b), we know that $z(k) > 0$, $\Delta z(k) < 0$ and $w(k) > 0$, $\Delta w(k) < 0$. In fact,

$$\begin{aligned} \Delta w(k) &= \Delta z(k) + p\Delta z(k-\tau) \\ &= -q(k)u(k-\sigma) - pq(k-\tau)u(k-\tau-\sigma) \\ &\leq -q(k)\left(u(k-\sigma) + pu(k-\tau-\sigma)\right) \\ &= -q(k)z(k-\sigma) \leq 0. \end{aligned} \tag{7.1.14}$$

Since $\lim_{k\to\infty} z(k) = b \geq 0$ exists, we get

$$\lim_{k\to\infty} w(k) = b + pb = (1+p)b \geq 0.$$

Therefore, $w(k) > 0$ for all large $k$. On the other hand,

$$w(k) = z(k) + pz(k-\tau) \leq (1+p)z(k),$$

which is the same as

$$z(k) \geq \frac{w(k)}{1+p}. \tag{7.1.15}$$

From (7.1.14) and (7.1.15), we have

$$\Delta w(k) \leq -q(k)z(k-\sigma) \leq -\frac{q(k)}{1+p}w(k-\sigma). \tag{7.1.16}$$

By Theorem 6.20.3, in view of condition (7.1.13), the equation (7.1.16) has no eventually positive solution, which is a contradiction. ∎

**Example 7.1.4.** Consider the difference equation

$$\Delta\left(u(k) - \frac{1}{2}u(k-1)\right) + \frac{(6k^2+k-3)(k-2)}{8k(k+1)(k-1)}u(k-2) = 0, \tag{7.1.17}$$

$k \geq 3$.

It is easy to see that $q(k) > q(k-1)$ for all large $k$ and that the condition (7.1.13) is satisfied. Therefore, by Theorem 7.1.4 the equation (7.1.17) is oscillatory. In fact, $u(k) = (-1)^k 2^{-k}/k$ is such a solution.

**Theorem 7.1.5.** Assume that $-1 < p < 0$, $\tau > \sigma$, $q(k) \geq 0$, $q(k) \geq q(k-\tau)$ for all large $k$, and

$$\liminf_{k\to\infty} q(k) > F(\bar{b}) \equiv \frac{(\bar{b}-1)(1+p\bar{b}^{\tau})}{\bar{b}^{\sigma+1}}, \tag{7.1.18}$$

where $\bar{b} \in (1, (-p)^{-1/\tau})$ is the unique real root of the equation

$$1 + pb^{\tau} = (b-1)(\sigma + \sigma p b^{\tau} - p\tau b^{\tau}). \tag{7.1.19}$$

Then, the equation (7.1.1) is oscillatory.

**Proof.** Suppose the contrary, and let $u(k)$ be an eventually positive solution of (7.1.1). By Lemma 7.1.1(b), $z(k) > 0$ and $\Delta z(k) < 0$ eventually. From (7.1.14), we have

$$\Delta(z(k) + pz(k-\tau)) + q(k)z(k-\sigma) \leq 0. \tag{7.1.20}$$

Setting $\gamma(k) = z(k-1)/z(k)$, then $\gamma(k) \geq 1$ for all large $k$. Dividing (7.1.20) by $z(k)$, we get

$$\text{(7.1.21)} \quad \frac{1}{\gamma(k+1)} \leq 1 + p\left(\frac{z(k-\tau)}{z(k)} - \frac{z(k-\tau+1)}{z(k)}\right) - q(k)\frac{z(k-\sigma)}{z(k)}$$
$$\leq 1 + p\left(\gamma(k-\tau+1)\cdots\gamma(k) - \gamma(k-\tau+2)\cdots\gamma(k)\right) - q(k)\gamma(k-\sigma+1)\cdots\gamma(k).$$

From (7.1.21), $\gamma(k)$ is bounded above. We set, $\liminf_{k\to\infty}\gamma(k) = b \geq 1$. It is clear that $b$ is finite. From (7.1.21), we have

$$\limsup_{k\to\infty}\frac{1}{\gamma(k+1)} = \frac{1}{b} \leq 1 + pb^{\tau-1}(b-1) - b^{\sigma}\liminf_{k\to\infty} q(k).$$

Hence, it follows that

$$\text{(7.1.22)} \qquad \liminf_{k\to\infty} q(k) \leq F(b),$$

where

$$\text{(7.1.23)} \qquad F(b) = \frac{(1+pb^{\tau})(b-1)}{b^{\sigma+1}}.$$

From $F'(b) = 0$, we get the equation

$$\text{(7.1.24)} \qquad 1 + pb^{\tau} + (b-1)\left[p\tau b^{\tau} - \sigma(1+pb^{\tau})\right] = 0.$$

Equation (7.1.24) has a unique real root $\bar{b}$ in $[1,\infty)$. Further, it is easy to see that $F(\bar{b})$ is a maximum value of $F(b)$ on $[1,\infty)$. Thus, we have $\liminf_{k\to\infty} q(k) \leq F(\bar{b})$. However, this contradicts the condition (7.1.18). ∎

**Theorem 7.1.6.** Assume that $p < -1$, $\tau > \sigma$, $\sum_{k=1}^{\infty} q(k) = \infty$, $q(k) \leq q(k-\tau)$ for all large $k$, and

$$\text{(7.1.25)} \qquad -\left(\frac{1}{1+p}\right)\liminf_{k\to\infty} q(k) > \frac{(\tau-\sigma-1)^{\tau-\sigma-1}}{(\tau-\sigma)^{\tau-\sigma}}.$$

Then, the equation (7.1.1) is oscillatory.

**Proof.** Suppose the contrary, and let $u(k)$ be an eventually positive solution of the equation (7.1.1). By Lemma 7.1.1(c), we have $z(k) < 0$ and $\Delta z(k) \leq 0$. Define $w(k) = z(k) + pz(k-\tau)$, then we have $w(k) > 0$, $\Delta w(k) \geq 0$ eventually. Note that $w(k) \leq (1+p)z(k-\tau)$, and hence $z(k-\tau) \leq w(k)/(1+p)$. Thus, it follows that

$$\begin{aligned}
\Delta w(k) &= \Delta z(k) + p\Delta z(k-\tau) \\
&= -q(k)u(k-\sigma) - pq(k-\tau)u(k-\sigma+\tau) \\
&\geq -q(k)z(k-\sigma) \geq 0,
\end{aligned}$$

which provides

$$(7.1.26)\quad 0 \le \Delta w(k) + q(k)z(k-\sigma) \le \Delta w(k) + \frac{q(k)}{1+p} w(k-\sigma+\tau).$$

Set $\gamma(k) = w(k+1)/w(k) \ge 1$. Then, the inequality (7.1.26) leads to

$$(7.1.27)\qquad \gamma(k) \ge 1 - \frac{q(k)}{1+p}\gamma(k)\cdots\gamma(k+(\tau-\sigma)-1).$$

From condition (7.1.25), we know that $\gamma(k)$ is bounded above. Now, taking limit infimum on both sides of (7.1.27), we get

$$(7.1.28)\qquad b \ge 1 - \frac{1}{1+p}\liminf_{k\to\infty} q(k) b^{\tau-\sigma},$$

where $b = \liminf_{k\to\infty}\gamma(k)$. From (7.1.28), we have

$$-\left(\frac{1}{1+p}\right)\liminf_{k\to\infty} q(k) \le \frac{b-1}{b^{\tau-\sigma}} \le \frac{(\tau-\sigma-1)^{\tau-\sigma-1}}{(\tau-\sigma)^{\tau-\sigma}},$$

which contradicts condition (7.1.25). ∎

**Theorem 7.1.7.** Assume that $p < -1$, $\tau > \sigma$, $\sum_{k=1}^{\infty} q(k) = \infty$, and

$$(7.1.29)\qquad -\frac{1}{p}\liminf_{k\to\infty} q(k) > \frac{(\tau-\sigma-1)^{\tau-\sigma-1}}{(\tau-\sigma)^{\tau-\sigma}}.$$

Then, the equation (7.1.1) is oscillatory.

**Proof.** If not, let $u(k)$ be an eventually positive solution. Note that $z(k) = u(k) + pu(k-\tau) > pu(k-\tau)$. From the equation (7.1.1), we have

$$(7.1.30)\qquad \Delta z(k) = -q(k)u(k-\sigma) \le -\frac{q(k)}{p} z(k+(\tau-\sigma)).$$

The rest of the proof is similar to that of Theorem 7.1.6. In fact, the condition (7.1.29) leads to the fact that the inequality (7.1.30) has no eventually negative solution, whereas Lemma 7.1.1(c) implies that $z(k)$ is eventually negative. ∎

**Theorem 7.1.8.** Assume that $p > 0$, $\sigma > \tau$, $\sum_{k=1}^{\infty} q(k) = \infty$, $q(k) \ge q(k-\tau)$, and

$$(7.1.31)\qquad \left(\frac{1}{1+p}\right)\liminf_{k\to\infty} q(k) > \frac{(\sigma-\tau)^{\sigma-\tau}}{(\sigma-\tau+1)^{\sigma-\tau+1}}.$$

Then, the equation (7.1.1) is oscillatory.

**Proof.** The proof is similar to that of Theorem 7.1.6. ∎

**Example 7.1.5.** For the difference equation

$$\Delta\left(u(k)+\frac{4}{3}u(k-1)\right)+\frac{2k-7}{3(k-3)}u(k-3) = 0, \quad k\ge 4 \tag{7.1.32}$$

all the conditions of Theorem 7.1.8 are satisfied. Therefore, the equation (7.1.32) is oscillatory. In fact, $u(k)=(-1)^k k$ is such a solution.

## 7.2. Existence and Asymptotic Behavior of Nonoscillatory Solutions of (7.1.1)

Here, we shall study the existence and asymptotic behavior of nonoscillatory solutions of the equation (7.1.1) when $q(k)$ is either eventually positive or negative. We begin with the following theorem which gives sufficient conditions for the existence of a positive solution of (7.1.1), when $q(k)$ is eventually positive.

**Theorem 7.2.1.** Assume that $-1\le p\le 0$, $q(k)>0$ for all large $k\ge K-\tau$, and there exists a constant $\nu^*>1$ such that

$$\sup_{k\ge K}\left\{\frac{1}{\nu^*}+q(k)(\nu^*)^{\sigma}-p\frac{q(k)}{q(k-\tau)}(\nu^*)^{\tau-1}(\nu^*-1)\right\} \le 1. \tag{7.2.1}$$

Then, the equation (7.1.1) has a positive solution.

**Proof.** Let $r$ be a constant such that $r\in(1,\nu^*)$, and let

$$v(K-\mu) = \cdots = v(K-1) = r. \tag{7.2.2}$$

We define

$$v(k) = 1+q(k)\prod_{\ell=k-\sigma}^{k}v(\ell)-p\frac{q(k)}{q(k-\tau)}\prod_{\ell=k-\tau+1}^{k}v(\ell)[v(k-\tau)-1], \tag{7.2.3}$$

$$k=K,K+1,\cdots.$$

From (7.2.3), it is clear that

$$v(K) = 1+q(K)v(K)r^{\sigma}-p\frac{q(K)}{q(K-\tau)}v(K)r^{\tau-1}(r-1),$$

or

$$v(K)\left(1-q(K)r^{\sigma}+p\frac{q(K)}{q(K-\tau)}r^{\tau-1}(r-1)\right) = 1.$$

From (7.2.1), we have

$$1 - q(K)r^{\sigma} + p\frac{q(K)}{q(K-\tau)}r^{\tau-1}(r-1) \ \geq \ \frac{1}{r}$$

so that $v(K)/r \leq 1$. Hence, it follows that $1 < v(K) \leq r < \nu^*$. By induction, we can show that $v(k)$ is well defined by (7.2.3). Next, we define

$$w(\ell) \ = \ \frac{v(\ell)-1}{v(\ell)}, \qquad \ell = K-\mu, K-\mu+1, \cdots$$

then from (7.2.3), we have

$$\begin{aligned}\frac{1}{1-w(k)} \ &= \ 1 + q(k)\prod_{\ell=k-\sigma}^{k}(1-w(\ell))^{-1} \\ &\quad -p\frac{q(k)}{q(k-\tau)}\prod_{\ell=k-\tau+1}^{k}\frac{1}{(1-w(\ell))}\frac{w(k-\tau)}{(1-w(k-\tau))},\end{aligned}$$

or

$$\begin{aligned}w(k) \ &= \ q(k)\prod_{\ell=k-\sigma}^{k-1}(1-w(\ell))^{-1} \\ &\quad -p\frac{q(k)}{q(k-\tau)}w(k-\tau)\prod_{\ell=k-\tau}^{k-1}(1-w(\ell))^{-1},\end{aligned} \tag{7.2.4}$$

where $k = K, K+1, \cdots$ and $1 > w(k) > 0$. Let $z(K-\mu) = 1$, and

$$z(k+1) \ = \ z(k)(1-w(k)), \qquad k > K-\mu$$

so that

$$z(k) \ = \ \prod_{\ell=K-\mu}^{k-1}(1-w(\ell)), \qquad z(k) > 0$$

and

$$\Delta z(k) \ = \ -w(k)z(k) \ < \ 0, \qquad \text{i.e.} \quad w(k) \ = \ -\frac{\Delta z(k)}{z(k)}. \tag{7.2.5}$$

On substituting (7.2.5) into (7.2.4), we get

$$-\frac{\Delta z(k)}{z(k)} \ = \ q(k)\frac{z(k-\sigma)}{z(k)} - p\frac{q(k)}{q(k-\tau)}\left(1 - \frac{z(k-\tau+1)}{z(k-\tau)}\right)\frac{z(k-\tau)}{z(k)},$$

which is the same as

$$-\frac{\Delta z(k)}{q(k)} \ = \ z(k-\sigma) + \frac{p}{q(k-\tau)}\Delta z(k-\tau). \tag{7.2.6}$$

Finally, we define

$$(7.2.7) \qquad y(k) = -\frac{\Delta z(k+\sigma)}{q(k+\sigma)} > 0, \quad k = K-\mu, K-\mu+1, \cdots.$$

From (7.2.6) and (7.2.7), we have

$$(7.2.8) \qquad z(k-\sigma) = y(k-\sigma) + py(k-\tau-\sigma).$$

On combining (7.2.8) and (7.2.7), we find that $y(k)$ defined in (7.2.7) is a positive solution of (7.1.1). ■

In view of Theorem 7.2.1 and Problem 7.17.9 the following corollary is immediate.

**Corollary 7.2.2.** If $q(k) = q > 0$, then the equation (7.1.1) is oscillatory if and only if

$$(7.2.9) \qquad \inf_{\nu>1}\left\{\frac{1}{\nu} + q\nu^{\sigma} - p\nu^{\tau-1}(\nu-1)\right\} > 1.$$

**Remark 7.2.1.** If $\tau = 0$, $\sigma \ge 1$, then (7.2.9) is equivalent to

$$(7.2.10) \qquad \frac{q}{1+p} > \frac{\sigma^{\sigma}}{(\sigma+1)^{\sigma+1}}$$

and hence, (7.2.10) is a necessary and sufficient condition for (7.1.1) to be oscillatory.

**Remark 7.2.2.** If $\tau > 0$, $\sigma \ge 1$, then (7.2.10) implies (7.2.9) and therefore (7.2.10) is a sufficient condition for (7.1.1) to be oscillatory.

**Remark 7.2.3.** If $\sigma = 0$, $\tau \ge 1$, then (7.2.9) reduces to

$$(7.2.11) \qquad \inf_{\nu>1}\left\{\frac{1}{\nu} + q - p\nu^{\tau-1}(\nu-1)\right\} > 1.$$

Thus, it is sufficient to have

$$(7.2.12) \qquad \inf_{\nu>1}\left\{\frac{1}{\nu} + q - p(\nu-1)\right\} > 1.$$

From (7.2.12) we obtain the sufficient condition

$$q > 1 - p - 2\sqrt{-p}$$

for (7.1.1) to be oscillatory.

Next, we shall establish conditions for the existence of a decaying positive solution of (7.1.1). For this, we shall need the Knaster Tarski fixed point theorem [58]. We recall that a non empty and closed subset $E$ of a Banach space $X$ is called a *cone* if it possesses the following properties:

(1) if $\alpha \in \mathbb{R}_+$ and $u \in E$, then $\alpha u \in E$

(2) if $u, v \in E$, then $u + v \in E$, and

(3) if $u \in E - \{0\}$, then $-u \notin E$.

A Banach space $X$ is ***partially ordered*** if it contains a cone $E$ with nonempty interior. The ordering $\preceq$ in $X$ is then defined as follows:

$$u \preceq v \text{ if and only if } v - u \in E.$$

Let $S$ be a subset of a partially ordered Banach space $X$. Set

$$\overline{S} = \{u \in X : v \preceq u \text{ for every } v \in S\}.$$

We say that the point $u_0 \in X$ is the supremum of $S$ if $u_0 \in \overline{S}$ and for every $u \in \overline{S}$, $u \preceq u_0$. The infimum of $S$ is defined in a similar way.

**Theorem 7.2.3.** (Knaster Tarski Fixed Point Theorem). Let $X$ be a partially ordered Banach space with ordering $\preceq$. Let $S$ be a subset of $X$ with the property that the infimum of $S$ belongs to $S$ and every non empty subset of $S$ has a supremum which belongs to $S$. Let $\mathcal{T} : S \to S$ be an increasing mapping, i.e. $u \preceq v$ implies that $\mathcal{T}u \preceq \mathcal{T}v$. Then, $\mathcal{T}$ has a fixed point in $S$.

**Remark 7.2.4.** In Knaster Tarski fixed point theorem the continuity of $\mathcal{T}$ is not required.

**Theorem 7.2.4.** Assume that $-1 < p \le 0$, and

$$(7.2.13) \qquad -\frac{pk}{k-\tau} + k\sum_{\ell=k}^{\infty} \frac{q(\ell)}{\ell - \sigma} \le 1 \quad \text{for all large } k.$$

Then, equation (7.1.1) has a positive solution $u(k)$ satisfying $\lim_{k\to\infty} u(k) = 0$.

**Proof.** Consider the Banach space $l_\infty^K$ of all real functions $v = v(k)$ where $k \ge K$ with sup norm $\|v\| = \sup_{k\ge K} |v(k)|$. We define a subset $S$ of $l_\infty^K$ as

$$S = \{v \in l_\infty^K : 0 \le v(k) \le 1, \ k \ge K\}.$$

We define a partial ordering on $l_\infty^K$ in the usual way, i.e. for $u, v \in l_\infty^K$, $u \preceq v$ means that $u(k) \le v(k)$ for $k \ge K$. Thus, if for any $u, v \in$

$l_\infty^K$, $u(k) = v(k)$ for all $k \gg 1$ we will consider such functions to be the same. Then, for every subset $A$ of $S$ both $\inf A$ and $\sup A$ exist in $S$. Now, we define an operator $\mathcal{T} : S \to l_\infty^K$ as follows

$$\text{(7.2.14)} \qquad \mathcal{T}v(k) = \begin{cases} -pu(k-\tau)\left(\dfrac{k}{k-\tau}\right) + k\displaystyle\sum_{\ell=k}^{\infty}\frac{q(\ell)v(\ell-\sigma)}{\ell-\sigma}, \\ \qquad\qquad\qquad\qquad k \ge K_1\ (\ge K+\mu) \\ \dfrac{k}{K_1}\mathcal{T}v(K_1) + \left(1 - \dfrac{k}{K_1}\right), \quad K \le k < K_1. \end{cases}$$

By (7.2.13) it is clear that $\mathcal{T}S \subseteq S$, and that $\mathcal{T}$ is an increasing mapping. By Knaster Tarski fixed point Theorem 7.2.3 there is a $v \in S$ such that

$$\text{(7.2.15)} \quad v(k) = -pv(k-\tau)\left(\frac{k}{k-\tau}\right) + k\sum_{\ell=k}^{\infty}\frac{q(\ell)v(\ell-\sigma)}{\ell-\sigma}, \qquad k \ge K_1$$

and $v(k) > 0$ for $k \ge K_1$. If we set $u(k) = v(k)/k$, then from (7.2.15) we have

$$u(k) = -pu(k-\tau) + \sum_{\ell=k}^{\infty} q(\ell)u(\ell-\sigma)$$

and so $\Delta(u(k) + pu(k-\tau)) + q(k)u(k-\sigma) = 0$, $k \ge K_1$, which shows that $u(k)$ is a positive solution of the equation (7.1.1) with $\lim_{k\to\infty} u(k) = 0$. ∎

**Example 7.2.1.** For the difference equation

$$\text{(7.2.16)} \qquad \Delta\left(u(k) - \frac{1}{2}u(k-1)\right) + \frac{k-1}{2(k+2)(k+3)}u(k-1) = 0$$

it is easy to see that conditions of Theorem 7.2.4 are satisfied. Therefore, the equation (7.2.16) has a decaying positive solution. In fact, $u(k) = 1/(k+2)$ is such a solution.

**Theorem 7.2.5.** Assume that $-1 < p < 0$, $q(k) \ge 0$, $k \in \mathbb{N}$, $\sum_{k=1}^{\infty} q(k) = \infty$, and for any subsequence $\{k_\ell\} \subseteq \{k\}$, $\sum_{\ell=1}^{\infty} q(k_\ell) = \infty$. Then, every nonoscillatory solution of the equation (7.1.1) tends to zero as $k \to \infty$.

**Proof.** If not, let $u(k)$ be an eventually positive solution of (7.1.1). By Lemma 7.1.1(b), $z(k) > 0$ and $\Delta z(k) < 0$ eventually, where $z(k) = u(k) + pu(k-\tau)$. Then, $\lim_{k\to\infty} z(k) = r \ge 0$ exists. On summing (7.1.1) from $K$ to $k$, we have

$$z(k+1) - z(K) = -\sum_{\ell=K}^{k} q(\ell)u(\ell-\sigma),$$

which implies that

$$\sum_{\ell=K}^{\infty} q(\ell)u(\ell-\sigma) \; < \; \infty.$$

On the other hand, if $\limsup_{k\to\infty} u(k) > 0$, then there exists a subsequence $\{k_\ell\} \subseteq \{k\}$ such that $\lim_{\ell\to\infty} u(k_\ell) = s > 0$. Then, we have

$$\infty \; > \; \sum_{\ell=K_1}^{\infty} q(k_\ell+\sigma)u(k_\ell) \; \ge \; \frac{s}{2}\sum_{\ell=K_1}^{\infty} q(k_\ell+\sigma), \tag{7.2.17}$$

where $K_1$ is a sufficiently large integer such that $u(k_\ell) \ge s/2$ for $\ell \ge K_1$. The inequality (7.2.17) contradicts the hypotheses. ∎

**Example 7.2.2.** Consider the difference equation

$$\Delta\left(u(k) - \frac{1}{2e}u(k-1)\right) + \frac{1}{2e^2}\left(1-\frac{1}{e}\right)u(k-2) \; = \; 0. \tag{7.2.18}$$

By Theorem 7.2.5 every nonoscillatory solution of (7.2.18) tends to zero as $k \to \infty$. In fact, $u(k) = e^{-k}$ is such a solution.

Our next result improves Theorem 7.2.5 considerably.

**Theorem 7.2.6.** Assume that $|p| < 1$, $q(k) \ge 0$, $k \in \mathbb{N}$ and $\sum_{k=1}^{\infty} q(k) = \infty$. Then, every nonoscillatory solution of the equation (7.1.1) tends to zero as $k \to \infty$.

**Proof.** Assume that $u(k)$ be an eventually positive solution of (7.1.1). To be specific, let $u(k) > 0$ for $k \ge K$. Let $z(k) = u(k) + pu(k-\tau)$, $k \ge K+\mu$, then $\Delta z(k) = -q(k)u(k-\sigma) \le 0$. Thus, $z(k)$ is nonincreasing for $k \ge K+\mu$.

We shall first show that $u(k)$ is bounded. In fact, if $u(k)$ is unbounded, then there exists an increasing sequence of integers $\{k_i\}$ such that $k_i \ge K+\tau$, $k_i \to \infty$ as $i \to \infty$ and that $u(k_i) \to \infty$, where $u(k_i) = \max\{u(k) : K-\mu \le k \le k_i\}$. This implies $z(k_i) \ge u(k_i) - |p|u(k_i-\tau) \ge (1-|p|)u(k_i) \to \infty$ as $i \to \infty$, which is impossible since $z(k)$ is nonincreasing. Thus, $u(k)$ is bounded, and therefore $z(k)$ is bounded. Hence, $\lim_{k\to\infty} z(k) = \alpha \in \mathbb{R}$ exists. This implies that

$$\sum_{k=K+\sigma}^{\infty} q(k)u(k-\sigma) \; = \; \sum_{k=K+\sigma}^{\infty} (-\Delta z(k)) \; = \; z(K+\sigma) - \alpha$$

and hence the condition $\sum_{k=1}^{\infty} q(k) = \infty$ implies that $\liminf_{k\to\infty} u(k) = 0$.

Now we shall prove that $\alpha = 0$. If not, then we shall arrive on a contradiction for the case $\alpha > 0$. The proof for $\alpha < 0$ is similar.

For $\alpha > 0$, we note that $p > 0$. In fact, $p \leq 0$ implies that $u(k) = z(k) - pu(k-\tau) \geq z(k) \geq \alpha$. Thus, $\liminf_{k\to\infty} u(k) \geq \alpha > 0$ which contradicts $\liminf_{k\to\infty} u(k) = 0$. Now in view of $\liminf_{k\to\infty} u(k) = 0$ we can find an increasing sequence of integers $\{k_j\}$ such that

$$\lim_{j\to\infty} k_j = \infty \quad \text{and} \quad \lim_{j\to\infty} u(k_j) = 0. \tag{7.2.19}$$

Thus, from $\lim_{k\to\infty} z(k) = \alpha$ we have

$$z(k_j) - z(k_j - \tau) = u(k_j) - (1-p)u(k_j - \tau) - pu(k_j - 2\tau) \to 0 \tag{7.2.20}$$

as $j \to \infty$. Now (7.2.19), (7.2.20) and the fact that $p > 0$ imply that $\lim_{j\to\infty} u(k_j - \tau) = 0$ and $\lim_{j\to\infty} u(k_j - 2\tau) = 0$, which in turn imply that $z(k_i - \tau) = u(k_j - \tau) + pu(k_j - 2\tau) \to 0$ as $j \to \infty$. But, this contradicts the assumption that $\alpha > 0$. Hence, $\lim_{k\to\infty} z(k) = 0$.

Next, let $\beta = \limsup_{k\to\infty} u(k)$. Then, there exists a sequence of integers $\{k_m\}$ such that $\lim_{m\to\infty} k_m = \infty$ and $\lim_{m\to\infty} u(k_m) = \beta$. We shall show that $\beta = 0$. Since $\lim_{k\to\infty} z(k) = 0$, if $p = 0$ then obviously $\beta = 0$. If $p \neq 0$, then the relation $z(k_m) - u(k_m) = pu(k_m - \tau)$ implies that $\lim_{m\to\infty} u(k_m - \tau) = -\beta/p$, which in view of $\beta > 0$, for $0 < p < 1$ leads to a contradiction that $u(k)$ is eventually positive, and for $-1 < p < 0$ contradicts the fact that $\limsup_{k\to\infty} u(k) = \beta$.

Summing up the above discussion it follows that $\lim_{k\to\infty} u(k) = 0$. ∎

**Theorem 7.2.7.** Assume that the conditions of Lemma 7.1.1(c) hold. Then, every nonoscillatory solution of the equation (7.1.1) tends to $\infty$ or $-\infty$ as $k \to \infty$.

**Proof.** Let $u(k)$ be an eventually positive solution of (7.1.1). By Lemma 7.1.1(c), we have $z(k) < 0$, $\Delta z(k) \leq 0$ eventually. Therefore, $0 > \lim_{k\to\infty} z(k) = r \geq -\infty$. We shall show that $r = -\infty$. Assume that $-\infty < r < 0$. On summing (7.1.1) from $K$ to $k$, we get

$$z(k+1) - z(K) + \sum_{\ell=K}^{k} q(\ell)u(\ell - \sigma) = 0,$$

which implies that

$$\sum_{\ell=K}^{\infty} q(\ell)u(\ell - \sigma) < \infty.$$

Since $\sum_{\ell=K}^{\infty} q(\ell) = \infty$, we have $\liminf_{k\to\infty} u(k) = 0$, i.e. there exists a subsequence $\{k_j\}$ such that $\lim_{j\to\infty} k_j = \infty$ and $\lim_{j\to\infty} u(k_j - \tau) = 0$. On the other hand $z(k_j) > pu(k_j - \tau)$, i.e.

$$(7.2.21) \qquad 0 < \frac{z(k_j)}{p} < u(k_j - \tau),$$

which implies that $\lim_{j\to\infty} z(k_j) = 0$, a contradiction. Therefore, $\lim_{k\to\infty} z(k) = -\infty$. From (7.2.21) we have $\lim_{k\to\infty} u(k) = \infty$. The proof for the eventually negative solution is similar. ∎

**Example 7.2.3.** Consider the difference equation

$$(7.2.22) \quad \Delta\left(u(k) - 8u(k-2)\right) + \frac{8(k^2 - 4k - 6)}{(k-3)^2} u(k-3) = 0, \quad k \ge 6.$$

By Theorem 7.2.7, every nonoscillatory solution of (7.2.22) tends to $\infty$ or $-\infty$. In fact, $u(k) = k^2 2^k$ is such a solution.

Next, we shall discuss the asymptotic behavior of nonoscillatory solutions of the equation (7.1.1) when $q(k)$ is eventually negative. For this it is convenient to write (7.1.1) in the following form

$$(7.2.23) \quad \Delta\left(u(k) + pu(k-\tau)\right) = q(k)u(k-\sigma), \quad q(k) \ge 0, \quad k \in \mathbb{N}.$$

We begin with the remark that the equation (7.2.23) can have oscillatory solutions. For example,

$$\Delta(u(k) - u(k-1)) = \frac{4k^2 - 2}{k(k+1)} u(k-1)$$

has a solution $u(k) = (-1)^k/k$ which is oscillatory. However, the equation (7.2.23) always has a nonoscillatory solution. This we shall prove in our next theorem.

**Theorem 7.2.8.** The equation (7.2.23) always has a positive solution.

**Proof.** For $q(k) \ge 0$ one can find a function $h(k)$ such that $\sum_{\ell=k}^{\infty} q(\ell)h(\ell) = \infty$, and

$$\frac{q(k)}{\sum_{\ell=k_0}^{\infty} q(\ell)h(\ell)} \to 0 \quad \text{as} \quad k \to \infty.$$

Now we define a sequence as

$$z(k) = \prod_{\ell=k_0}^{k} \sum_{s=k_0}^{\ell} \prod_{j=k_0}^{s} \sum_{i=k_0}^{j} q(i)h(i).$$

Then, it follows that

$$\frac{z(k-\tau)}{z(k)} \to 0 \quad \text{as} \quad k \to \infty. \tag{7.2.24}$$

By l'Hopital's rule Theorem 1.8.9, we have

$$\frac{1}{z(k)} \sum_{\ell=k_0}^{k} q(\ell) z(\ell - \sigma) \to 0 \quad \text{as} \quad k \to \infty. \tag{7.2.25}$$

With $l_\infty^K$ and $S$ as defined in Theorem 7.2.4 we define an operator $T$ as follows

(7.2.26)

$$Tv(k) = \begin{cases} \dfrac{1}{2z(k)} - p\dfrac{z(k-\tau)}{z(k)} v(k-\tau) + \dfrac{1}{z(k)} \displaystyle\sum_{\ell=K_1}^{k-1} q(\ell) z(\ell-\sigma) v(\ell-\sigma), \\ \hfill k \geq K_1 + 1 \\ 1, \quad K \leq k \leq K_1 \end{cases}$$

where $K = K_1 - \max\{\sigma, \tau\}$ and $K_1$ is chosen so large that

(7.2.27)

$$z(k) \geq 1, \quad -p\frac{z(k-\tau)}{z(k)} < \frac{1}{2}, \quad -p\frac{z(k-\tau)}{z(k)} + \frac{1}{z(k)} \sum_{\ell=K_1}^{k} q(\ell) z(\ell-\sigma) \leq \frac{1}{2}$$

for $k \geq K_1$. We note that, in view of (7.2.24) and (7.2.25), such an integer $K_1$ does exist. By (7.2.26) and (7.2.27) we have $0 \leq Tv(k) \leq 1$ for $k \geq K$, which implies that $TS \subseteq S$. Thus, by Schauder's fixed point theorem it follows that there exists an element $v \in S$ such that $Tv = v$. Hence, we have

(7.2.28)

$$v(k) = \begin{cases} \dfrac{1}{2z(k)} - p\dfrac{z(k-\tau)}{z(k)} v(k-\tau) + \dfrac{1}{z(k)} \displaystyle\sum_{\ell=K_1}^{k-1} q(\ell) z(\ell-\sigma) v(\ell-\sigma), \\ \hfill k \geq K_1 + 1 \\ 1, \quad K \leq k \leq K_1. \end{cases}$$

It is clear that $v(k) > 0$ for $k \geq K$. Now we set $u(k) = v(k)z(k) > 0$, so that from (7.2.28), we find

$$u(k) = \frac{1}{2} - pu(k-\tau) + \sum_{\ell=K_1}^{k-1} q(\ell) u(\ell-\sigma), \quad k \geq K_1 + 1 \tag{7.2.29}$$

which is the same as (7.2.23). It is clear that this $u(k)$ is a positive solution of (7.2.23). ∎

From the expression (7.2.29) describing the solution of (7.2.23), we can get more information about the solutions of (7.2.23). In fact, we have the following:

**Corollary 7.2.9.** (i) If $p \le 0$, (7.2.23) has a positive solution $u(k)$ with $\lim_{k\to\infty} u(k) = r$, where $r$ is either a positive number or $\infty$.
(ii) If $p = -1$, (7.2.23) has a solution $u(k)$ with $\lim_{k\to\infty} u(k) = \infty$.
(iii) If $p < -1$, (7.2.23) has a solution $u(k)$ with $u(k) \ge (-p)^{(k-k_0)/\tau} \times u(k_0)$, which implies that the solution tends to infinity exponentially.
(iv) If $-1 \le p \le 0$ and $\sum_{\ell=k_0}^{\infty} q(\ell) = \infty$, (7.2.23) has a solution $u(k)$ with $\lim_{k\to\infty} u(k) = \infty$.
(v) If $-1 < p < 0$ and $\sum_{\ell=k_0}^{\infty} q(\ell) = \infty$, then every bounded solution of (7.2.23) is either oscillatory or tends to zero as $k \to \infty$.

**Proof.** From (7.2.29), we have

$$u(k) \ \ge\ \frac{1}{2} - pu(k-\tau) \ \ge\ \frac{1}{2} - p\left(\frac{1}{2} - pu(k-2\tau)\right) \ \ge\ \cdots$$

from which (i) and (ii) follow immediately, and (iii) follows from

$$u(k) \ \ge\ -pu(k-\tau) \ \ge\ \cdots\ \ge\ (-p)^j u(k-j\tau).$$

The assertion (iv) follows from (7.2.29) directly. In order to prove (v), we let $u(k)$ to be a bounded positive solution of (7.2.23), then $\Delta z(k) \ge 0$, $\lim_{k\to\infty} z(k) = r$ exists, where $r$ is finite. If $r > 0$, then

$$r - z(k_0) \ =\ \sum_{\ell=k_0}^{\infty} q(\ell)u(\ell-\sigma). \tag{7.2.30}$$

In this case $u(k) \ge r/2$, and (7.2.30) leads to $\sum_{\ell=k_0}^{\infty} q(\ell) < \infty$ which is a contradiction. If $r < 0$

$$u(k) \ \le\ -pu(k-\tau) \ \le\ \cdots\ \le (-p)^j u(k-j\tau),$$

which implies that $\lim_{k\to\infty} u(k) = 0$. ∎

**Example 7.2.4.** Consider the difference equation

$$\Delta\left(u(k) - u(k-1)\right) \ =\ q(k)u(k-1), \quad k \ge 2 \tag{7.2.31}$$

where $q(k) = k^2 - k + 1$. By Corollary 7.2.9, (7.2.31) has a solution for which $\lim_{k\to\infty} u(k) = \infty$. In fact, $u(k) = k!$ is such a solution of (7.2.31).

When $p < -1$ the difference equation (7.2.23), besides having a solution tending to infinity exponentially, can have a nonoscillatory solution which tends to a nonzero constant as $k \to \infty$. For example, the difference equation

$$\Delta\left(u(k) - 2u(k-1)\right) = \frac{k+3}{k^2(k+1)}u(k-1) \tag{7.2.32}$$

has a bounded nonoscillatory solution $u(k) = (k+1)/k$.

**Theorem 7.2.10.** Assume that $p < -1$, and

$$\sum_{\ell=k_0}^{\infty} q(\ell) = \infty. \tag{7.2.33}$$

Then, every bounded nonoscillatory solution of the equation (7.2.23) satisfies $\liminf_{k\to\infty} |u(k)| = 0$.

**Proof.** Let $u(k)$ be a bounded positive solution of (7.2.23). Then, $\Delta z(k) \geq 0$, where $z(k) = u(k) + pu(k-\tau)$. If $z(k) > 0$, then the boundedness implies that $\lim_{k\to\infty} z(k) = r > 0$. In which case $u(k) \geq z(k) \geq r/2 > 0$ for all sufficiently large $k$. From (7.2.23), we have

$$\frac{r}{2}\sum_{\ell=k_0}^{\infty} q(\ell) \leq \sum_{\ell=k_0}^{\infty} q(\ell)u(\ell-\sigma) < \infty,$$

which is a contradiction. If $z(k) < 0$ then $\lim_{k\to\infty} z(k) = \beta \leq 0$, we also have $\sum_{\ell=k_0}^{\infty} q(\ell)u(\ell-\sigma) < \infty$, which implies that $\liminf_{k\to\infty} u(k) = 0$. We shall prove that $\limsup_{k\to\infty} u(k) = 0$ also. If there exist two subsequences $\{k'\}$ and $\{k''\}$ such that

$$\limsup_{k\to\infty} u(k) = \lim_{k'\to\infty} u(k') = \alpha > 0$$

and

$$\liminf_{k\to\infty} u(k) = \lim_{k''\to\infty} u(k'') = 0,$$

then

$$\lim_{k\to\infty} z(k+\tau) = \lim_{k''\to\infty} u(k''+\tau) \geq 0,$$

which implies that $\beta = 0$. Then, we have

$$\lim_{k'\to\infty} z(k'+\tau) = 0 = \lim_{k'\to\infty} u(k'+\tau) + p\lim_{k'\to\infty} u(k'),$$

i.e. $\lim_{k' \to \infty} u(k' + \tau) = -p\alpha > \alpha$, which contradicts the definition of $\alpha$. Therefore, $\lim_{k \to \infty} u(k) = 0$. ∎

**Example 7.2.5.** Consider the difference equation

$$\Delta\left(u(k) - 2u(k-1)\right) = \frac{k+3}{k(k+1)}u(k-1), \tag{7.2.34}$$

which satisfies the assumptions of Theorem 7.2.10. In fact, (7.2.34) has a solution $u(k) = 1/k$.

Next, we shall study the existence of positive solutions of the equation (7.1.1) defined for $k \geq K \in \mathbb{N}$, subject to the condition

$$\sum_{k=K}^{\infty} |q(k)| < \infty. \tag{7.2.35}$$

**Theorem 7.2.11.** Assume that $p \neq -1$, and that the condition (7.2.35) holds. Then, the equation (7.1.1) has a positive solution.

**Proof.** We need to consider the following five cases:

Case 1. $-1 < p \leq 0$.

Choose a positive integer $K_0 \geq K$ sufficiently large such that $K_0 - \mu \geq K$, and

$$\sum_{k=K_0}^{\infty} |q(k)| \leq \frac{1+p}{4}.$$

Consider the Banach space $l_\infty^K$ of all real functions $u = u(k)$ where $k \geq K$ with the sup norm, i.e. $\|u\| = \sup_{k \geq K} |u(k)|$. We define a subset $S$ of $l_\infty^K$ as

$$S = \left\{u \in l_\infty^K : \frac{2(1+p)}{3} \leq u(k) \leq \frac{4}{3},\ k \geq K\right\}.$$

Clearly, $S$ is a closed and convex subset of $l_\infty^K$. We define an operator $\mathcal{T} : S \to l_\infty^K$ as follows

$$\mathcal{T}u(k) = \begin{cases} 1 + p - pu(k-\tau) + \sum_{\ell=k}^{\infty} q(\ell)u(\ell - \sigma), & k \geq K_0 \\ \mathcal{T}u(K_0), \quad K \leq k \leq K_0. \end{cases}$$

This mapping $\mathcal{T}$ is completely continuous, and for every $u = u(k) \in S$, $k \geq K_0$, we have

$$\mathcal{T}u(k) \leq 1 + p - \frac{4}{3}p + \frac{4}{3}\sum_{\ell=k}^{\infty} |q(\ell)| \leq 1 + p - \frac{4}{3}p + \frac{4}{3}\frac{(1+p)}{4} = \frac{4}{3},$$

and

$$Tu(k) \ \geq \ 1+p+\frac{4}{3}\frac{(-1-p)}{4} \ = \ \frac{2(1+p)}{3}.$$

Hence, $TS \subset S$. Thus, from Schauder's fixed point theorem it follows that there exists an element $u \in S$ such that $Tu = u$. It is easily seen that this $u = u(k)$ is a positive solution of the equation (7.1.1).

Case 2. $p < -1$.

Let $K_0 \geq K$ be such that $K_0 + \tau - \sigma \geq K$, and

$$\sum_{k=K_0+\tau}^{\infty} |q(k)| \ \leq \ -\frac{1+p}{4}.$$

Consider the Banach space $l_\infty^K$ as in the proof of Case 1. Set

$$S \ = \ \left\{u \in l_\infty^K : -\frac{p}{2} \leq u(k) \leq -2p, \ k \geq K\right\}.$$

Clearly, $S$ is a closed and convex subset of $l_\infty^K$. We define an operator $T : S \to l_\infty^K$ as follows

$$Tu(k) \ = \ \begin{cases} -1-p-\dfrac{1}{p}u(k+\tau)+\dfrac{1}{p}\displaystyle\sum_{\ell=k+\tau}^{\infty} q(\ell)u(\ell-\sigma), & k \geq K_0 \\ Tu(K_0), \quad K \leq k \leq K_0. \end{cases}$$

This mapping $T$ is completely continuous, and for every $u = u(k) \in S$, $k \geq K_0$ it follows that $Tu \in S$, i.e. $TS \subset S$. Thus, from Schauder's fixed point theorem there exists an element $u \in S$ such that $Tu = u$. This $u = u(k)$ is a positive solution of (7.1.1).

Case 3. $0 < p < 1$.

Let $K_0 \geq K$ be such that $K_0 - \mu \geq K$, and

$$\sum_{k=K_0}^{\infty} |q(k)| \ \leq \ \frac{1-p}{4}.$$

Let $l_\infty^K$ be as in the proof of Case 1. Set

$$S \ = \ \left\{u \in l_\infty^K : 2(1-p) \leq u(k) \leq 4, \ k \geq K\right\}.$$

Clearly, $S$ is a closed and convex subset of $l_\infty^K$. Define $T : S \to l_\infty^K$ as follows

$$Tu(k) \ = \ \begin{cases} 3+p-pu(k-\tau)+\displaystyle\sum_{\ell=k}^{\infty} q(\ell)u(\ell-\sigma), & k \geq K_0 \\ Tu(K_0), \quad K \leq k \leq K_0. \end{cases}$$

This mapping $\mathcal{T}$ is completely continuous, and $\mathcal{T}S \subset S$. Thus, from Schauder's fixed point theorem there exists an element $u \in S$ such that $\mathcal{T}u = u$. This $u = u(k)$ is a positive solution of (7.1.1).

Case 4. $p = 1$.

Let $K_0 \geq K$ be such that $K_0 + \tau - \sigma \geq K$, and

$$\sum_{k=K_0+\tau}^{\infty} |q(k)| \leq \frac{1}{4}.$$

Let $l_\infty^K$ be as in Case 1. Then,

$$S = \left\{u \in l_\infty^K : 2 \leq u(k) \leq 4,\ k \geq K\right\}$$

is a closed and convex subset of $l_\infty^K$. Define $\mathcal{T} : S \to l_\infty^K$ as follows

$$\mathcal{T}u(k) = \begin{cases} 3 + \displaystyle\sum_{j=1}^{\infty} \sum_{\ell=k+(2j-1)\tau}^{k+2\tau j-1} q(\ell)u(\ell-\sigma), & k \geq K_0 \\ \mathcal{T}u(K_0), \quad K \leq k \leq K_0. \end{cases}$$

Clearly $\mathcal{T}$ is completely continuous, and $\mathcal{T}S \subset S$. Thus, from Schauder's fixed point theorem there exists an element $u \in S$ such that $\mathcal{T}u = u$, i.e.

$$u(k) = \begin{cases} 3 + \displaystyle\sum_{j=1}^{\infty} \sum_{\ell=k+(2j-1)\tau}^{k+2\tau j-1} q(\ell)u(\ell-\sigma), & k \geq K_0 \\ u(K_0), \quad K \leq k \leq K_0. \end{cases}$$

It follows that

$$\begin{aligned} & u(k) + u(k-\tau) \\ &= 6 + \sum_{j=1}^{\infty}\left[\sum_{\ell=k-\tau+(2j-1)\tau}^{k-\tau+2\tau j-1} q(\ell)u(\ell-\sigma) + \sum_{\ell=k+(2j-1)\tau}^{k+2\tau j-1} q(\ell)u(\ell-\sigma)\right] \\ &= 6 + \sum_{\ell=k}^{\infty} q(\ell)u(\ell-\sigma), \quad k \geq K_0 + \tau. \end{aligned}$$

Thus, this $u = u(k)$ is a positive solution of (7.1.1) for $k \geq K_0 + \tau$.

Case 5. $p > 1$.

Let $K_0 \geq K$ be such that $K_0 + \tau - \sigma \geq K$, and

$$\sum_{k=K_0+\tau}^{\infty} |q(k)| \leq \frac{p-1}{4}.$$

Let $l_\infty^K$ be as in Case 1. Then,

$$S = \left\{u \in l_\infty^K : 2(p-1) \le u(k) \le 4p, \ k \ge K\right\}$$

is a closed and convex subset of $l_\infty^K$. Define $\mathcal{T} : S \to l_\infty^K$ as follows

$$\mathcal{T}u(k) = \begin{cases} 3p + 1 - \dfrac{1}{p}u(k+\tau) + \displaystyle\sum_{\ell=k+\tau}^{\infty} q(\ell)u(\ell-\sigma), & k \ge K_0 \\ \mathcal{T}u(K_0), \quad K \le k \le K_0. \end{cases}$$

Clearly $\mathcal{T}$ is completely continuous, and $\mathcal{T}S \subset S$. Thus, from Schauder's fixed point theorem there exists an element $u \in S$ such that $\mathcal{T}u = u$. This $u = u(k)$ is a positive solution of (7.1.1) for $k \ge K_0 + \tau$. ∎

In the following example we shall show that Theorem 7.2.11 does not hold when $p = -1$.

**Example 7.2.6.** Consider the difference equation

$$\Delta\left(u(k) - u(k-1)\right) + \frac{1}{k\ln^2 k}u(k-1) = 0, \quad k \ge 2. \tag{7.2.36}$$

Since

$$\sum_{k=2}^{\infty} q(k) = \sum_{k=2}^{\infty} \frac{1}{k\ln^2 k} < \infty,$$

it is clear that the condition (7.2.35) holds. We shall show that the equation (7.2.36) has no positive solutions. Otherwise, assume that (7.2.36) has a positive solution $u(k)$ satisfying $u(k-1) > 0, \ k \ge K_0 \ge 2$ for some $K_0 \ge 2$. Set $z(k) = u(k) - u(k-1)$ for $k \ge K_0$. Then, by (7.2.36), we have

$$\Delta z(k) = -\frac{1}{k\ln^2 k}u(k-1) < 0, \quad k \ge K_0. \tag{7.2.37}$$

We need to consider the following two possible cases:

(a) If $z(k)$ is eventually negative, then by (7.2.37) we see that there exist an $\alpha > 0$ and a positive integer $K_1 \ge K_0$ such that $z(k) \le -\alpha$, i.e. $u(k) \le -\alpha + u(k-1)$ for $k \ge K_1$. But, then by induction it follows that

$$u(k + K_1) \le -k\alpha + u(K_1) \to -\infty \quad \text{as} \quad k \to \infty$$

which contradicts the positivity of $u(k)$.

(b) If $z(k)$ is eventually positive, then there exists an integer $K_2 \ge K_1$ such that $z(k) > 0$ for all $k \ge K_2$. Now, we let $\lim_{k\to\infty} z(k) = \beta \ge 0$, and sum (7.2.37) from $k \geq K_2$ to $\infty$, to obtain

$$\beta - z(k) + \sum_{s=k}^{\infty} \frac{1}{s \ln^2 s} u(s-1) = 0, \tag{7.2.38}$$

which implies that

$$\sum_{k=K_2}^{\infty} \frac{1}{k \ln^2 k} u(k-1) < \infty.$$

On the other hand, since $u(k) - u(k-1) > 0$ for $k \geq K_2$, it follows that there exists a positive constant $L$ such that $u(k) \geq L$ for $k \geq K_2$. On substituting this into (7.2.38), we get

$$z(k) \geq \beta + L \sum_{s=k}^{\infty} \frac{1}{s \ln^2 s} \geq \beta + L \int_k^{\infty} \frac{1}{t \ln^2 t} dt = \beta + \frac{L}{\ln k}, \quad k \geq K_2 + 1$$

i.e.

$$u(k) \geq u(k-1) + \frac{L}{\ln k}, \qquad k \geq K_2 + 1.$$

From the above inequality it follows that

$$\begin{aligned} u(k) &\geq L \left[ \frac{1}{\ln k} + \frac{1}{\ln(k-1)} + \cdots + \frac{1}{\ln(k+1-(k-K_2))} \right] + u(K_2) \\ &\geq L \frac{k - K_2}{\ln k}, \quad k \geq K_2 + 1. \end{aligned}$$

Hence, we find

$$\frac{1}{k \ln^2 k} u(k-1) \geq L \frac{k - K_2 - 1}{k \ln^2 k \ \ln(k-1)}, \qquad k \geq K_2 + 2.$$

Since

$$\sum_{k=K_2+2}^{\infty} \frac{k - K_2 - 1}{k \ln^2 k \ \ln(k-1)} = \infty,$$

we obtain from the last inequality that

$$\sum_{k=K_2+2}^{\infty} \frac{1}{k \ln^2 k} u(k-1) = \infty.$$

This shows that the case (b) is also not possible.

**Remark 7.2.5.** On combining Theorem 7.2.11 and Example 7.2.6, we conclude that $\sum_{k=K}^{\infty} |q(k)| = \infty$ is a necessary condition for the oscillation of all solutions of (7.1.1) when $p \neq -1$.

## 7.3. Oscillation and Comparison Theorems for (7.1.1)

Here, we shall present some oscillation criteria for the equation (7.1.1) by using comparison results. For this, first we shall establish a comparison theorem which gives a necessary and sufficient condition for the oscillation of all solutions of (7.1.1). This result will be used further in later sections for determining sufficient conditions for the oscillation of difference equations with mixed arguments as well as the equations with nonlinear terms.

**Theorem 7.3.1.** Assume that $-1 < p \le 0$ and $q(k) > 0$ eventually. Then, the equation (7.1.1) is oscillatory if and only if

$$\Delta\left(v(k) + pv(k-\tau)\right) + q(k)v(k-\sigma) \le 0, \quad k \in \mathbb{N} \tag{7.3.1}$$

has no eventually positive solution.

**Proof.** The sufficiency is obvious. Suppose $u(k)$ is an eventually positive solution of (7.3.1). We shall prove that (7.1.1) also has a positive solution. Let $z(k) = u(k) + pu(k-\tau)$, then $\Delta z(k) < 0$, $z(k) > 0$ eventually. We define

$$w(k) = -\frac{\Delta z(k)}{z(k)} > 0, \quad k \ge K. \tag{7.3.2}$$

It is clear that $w(k) < 1$ for $k \ge K$. Inequality (7.3.1) can be written as

$$\Delta z(k) + q(k)z(k-\sigma) - p\frac{q(k)}{q(k-\tau)}q(k-\tau)u(k-\tau-\sigma) \le 0. \tag{7.3.3}$$

Dividing (7.3.3) by $z(k)$ and using (7.3.2), we obtain

(7.3.4)
$$w(k) \ge q(k)\prod_{\ell=k-\sigma}^{k-1}(1-w(\ell))^{-1} - p\frac{q(k)}{q(k-\tau)}w(k-\tau)\prod_{\ell=k-\tau}^{k-1}(1-w(\ell))^{-1}.$$

Further, we define

$$\left\{\lambda^{(0)}(k)\right\} = 0, \quad k = K, K+1, \cdots$$

$$\left\{\lambda^{(r)}(k)\right\} = \begin{cases} 0, \quad k = K, K+1, \cdots, K+\mu-1 \\ q(k)\displaystyle\prod_{\ell=k-\sigma}^{k-1}\left(1-\lambda^{(r-1)}(\ell)\right)^{-1} \\ -p\dfrac{q(k)}{q(k-\tau)}\lambda^{(r-1)}(k-\tau)\displaystyle\prod_{\ell=k-\tau}^{k-1}\left(1-\lambda^{(r-1)}(\ell)\right)^{-1}, \\ \qquad k \ge K+\mu, \quad r = 1, 2, \cdots. \end{cases} \tag{7.3.5}$$

It is not difficult to prove that

$$\lambda^{(0)}(k) \le \lambda^{(1)}(k) \le \cdots \le \lambda^{(r)}(k) \le \cdots \le w(k), \quad k \ge K, \ r \ge 0.$$

Thus, for each $k \ge K$, we have $\lim_{r\to\infty} \lambda^{(r)}(k) = \lambda(k)$. By taking the limit in (7.3.5), we get

$$(7.3.6)\quad \lambda(k) = \begin{cases} 0, & k = K, K+1, \cdots, K+\mu-1 \\ q(k)\displaystyle\prod_{\ell=k-\sigma}^{k-1}(1-\lambda(\ell))^{-1} - p\frac{q(k)}{q(k-\tau)}\lambda(k-\tau) \times \\ \qquad\qquad \displaystyle\prod_{\ell=k-\tau}^{k-1}(1-\lambda(\ell))^{-1}, & k \ge K+\mu. \end{cases}$$

Next, we define

$$z(K) = 1, \quad z(k+1) = z(k)(1-\lambda(k)), \quad k > K$$

so that

$$z(k) = \prod_{\ell=K}^{k-1}(1-\lambda(\ell)) > 0, \quad k \ge K.$$

Thus, $\Delta z(k) = -z(k)\lambda(k) < 0$, or

$$(7.3.7)\qquad \lambda(k) = -\frac{\Delta z(k)}{z(k)} > 0.$$

Substituting (7.3.7) into (7.3.6), we get

$$-\frac{\Delta z(k)}{z(k)} = q(k)\frac{z(k-\sigma)}{z(k)} - p\frac{q(k)}{q(k-\tau)}\left(1 - \frac{z(k-\tau+1)}{z(k-\tau)}\right)\frac{z(k-\tau)}{z(k)},$$

$$k \ge K+\mu$$

which is the same as

$$(7.3.8)\qquad -\frac{\Delta z(k)}{q(k)} = z(k-\sigma) + \frac{p}{q(k-\tau)}\Delta z(k-\tau), \quad k \ge K+\mu.$$

Let

$$(7.3.9)\qquad u(k) = -\frac{\Delta z(k+\sigma)}{q(k+\sigma)} > 0, \quad k \ge K+\mu-\sigma.$$

Combining (7.3.8) and (7.3.9), we obtain

$$(7.3.10)\qquad z(k-\sigma) - u(k-\sigma) - pu(k-\tau-\sigma) = 0, \quad k \ge K+\mu.$$

Now substituting (7.3.10) into (7.3.9), we find

$$\Delta\left(u(k)+pu(k-\tau)\right)+q(k)u(k-\sigma) \;=\; 0, \qquad k \ge K+\mu.$$

Hence, $u(k),\;\; k \ge K+\mu-\sigma$ is a positive solution of (7.1.1). ∎

The following results illustrate some applications of Theorem 7.3.1.

For the difference equation

(7.3.11)
$$\Delta\left(u(k)+pu(k-\tau)\right)+q(k)u(k-\sigma)+f(k,u(k-h_1),\cdots,u(k-h_r))=0$$

we have the following:

**Theorem 7.3.2.** Assume that the assumptions of Theorem 7.3.1 hold, and that

$$f(k,u_1,\cdots,u_r)u_1 \;\ge\; 0 \quad \text{whenever} \quad u_1u_j>0,\;\; j=1,\cdots,r.$$

Then, the oscillation of (7.1.1) implies that of (7.3.11).

**Proof.** If not, without loss of generality let $u(k)$ be an eventually positive solution of (7.3.11). Then, it is clear that (7.3.1) has an eventually positive solution which contradicts the conclusion of Theorem 7.3.1. ∎

**Remark 7.3.1.** From Theorem 7.3.1 the oscillation of (7.1.1) implies the same for the difference equation

$$\Delta\left(u(k)+pu(k-\tau)\right)+q(k)u(k-\sigma)+h(k)u(k+\eta) \;=\; 0, \qquad k\in\mathbb{N}$$

where $-1<p\le 0,\; q(k)\ge 0,\; h(k)\ge 0$ eventually, $\tau\in\mathbb{N}(1)$, and $\sigma,\;\eta\in\mathbb{N}$.

Our next result is for the equations (7.1.1) and

(7.3.12) $$\Delta\left(v(k)+\overline{p}v(k-\tau)\right)+h(k)v(k-\sigma) \;=\; 0, \qquad k\in\mathbb{N}.$$

**Theorem 7.3.3.** Assume that $p,\;\overline{p}\in(-1,0]$, and for large $k\in\mathbb{N}$

(7.3.13) $$h(k) \;\ge\; q(k) \;>\; 0$$

and

(7.3.14) $$\overline{p}\frac{h(k)}{h(k-\tau)} \;\le\; p\frac{q(k)}{q(k-\tau)}.$$

Then, the oscillation of (7.1.1) implies that of (7.3.12).

**Proof.** If not, let $v(k)$ be a positive solution of (7.3.12) for $k \ge K$. Then, as in the proof of Theorem 7.3.1, there exists a function $w(k) \in (0,1)$ such that

(7.3.15)
$$w(k) = h(k) \prod_{\ell=k-\sigma}^{k-1} (1-w(\ell))^{-1} - \overline{p}\frac{h(k)}{h(k-\tau)} w(k-\tau) \prod_{\ell=k-\tau}^{k-1} (1-w(\ell))^{-1}.$$

Thus, in view of (7.3.13) and (7.3.14), the inequality (7.3.4) follows from (7.3.15). This leads to the contradiction that (7.1.1) has a positive solution. ∎

**Corollary 7.3.4.** If $p = \overline{p} = 0$, then for the equations

$$\Delta u(k) + q(k)u(k-\sigma) = 0, \quad k \in \mathbb{N} \tag{7.3.16}$$

and

$$\Delta v(k) + h(k)v(k-\sigma) = 0, \quad k \in \mathbb{N} \tag{7.3.17}$$

with $h(k) \ge q(k) > 0$, for large $k \in \mathbb{N}$ the oscillation of (7.3.16) implies that of (7.3.17).

**Proof.** The proof is immediate from Theorem 7.3.3. ∎

**Remark 7.3.2.** It is easy to extend the above results for the equation

$$\Delta\left(u(k) + pu(k-\tau)\right) + \sum_{i=1}^{r} q_i(k)u(k-\sigma_i) = 0, \quad k \in \mathbb{N} \tag{7.3.18}$$

where $\sigma_i$, $1 \le i \le r$ are nonnegative integers.

## 7.4. Global Asymptotic Stability Criterion for (7.1.1)

The trivial solution of (7.1.1) is said to be *globally asymptotically stable* if every solution $u(k)$ of (7.1.1) tends to zero as $k \to \infty$.

The main result of this section is embodied in the following:

**Theorem 7.4.1.** Suppose $|p| < 1$, $q(k) > 0$, $k \in \mathbb{N}$ and $\sum_{k=0}^{\infty} q(k) = \infty$. Then, the trivial solution of (7.1.1) is globally asymptotically stable provided

$$\lim_{k\to\infty} \left\{ |p|\left[1 + \frac{q(k+\tau+\sigma)}{q(k+\sigma)}\right] + \sum_{\ell=k}^{k+2\sigma} q(\ell) \right\} < 2. \tag{7.4.1}$$

**Proof.** We split the proof into the following steps:

Step 1. Let $u(k)$ be a convergent solution of (7.1.1) which tends to some constant $\alpha$. If $\alpha \neq 0$, we can assume without loss of generality that $\alpha > 0$. Then, there is an integer $K$ such that $u(k-\sigma) > \alpha/2$ for $k \geq K$. Thus, from (7.1.1) it follows that

$$\Delta(u(k) + pu(k-\tau)) \geq -\alpha \frac{q(k)}{2}, \quad k \geq K.$$

Summing the above inequality, we get

$$u(k) + pu(k-\tau) \leq u(K) + pu(K-\tau) - \frac{\alpha}{2} \sum_{\ell=K}^{k} q(\ell).$$

This in view of $\sum_{k=0}^{\infty} q(k) = \infty$ implies that $u(k) + pu(k-\tau)$ tends to $-\infty$ as $k \to \infty$. But, this contradicts our assumption that $u(k)$ is convergent. Hence, every solution of (7.1.1) which is convergent will tend to zero.

Step 2. Equation (7.1.1) can be written as

$$\Delta \left\{ u(k) + pu(k-\tau) - \sum_{\ell=k-\sigma}^{k-1} q(\ell+\sigma)u(\ell) \right\} + q(k+\sigma)u(k) = 0, \quad k \in \mathbb{N}.$$

Let

$$v(k) = u(k) + pu(k-\tau) - \sum_{\ell=k-\sigma}^{k-1} q(\ell+\sigma)u(\ell) \tag{7.4.2}$$

then

$$\Delta v(k) = -q(k+\sigma)u(k), \quad k \in \mathbb{N} \tag{7.4.3}$$

and

$$v(k+1) + v(k) = 2v(k) - q(k+\sigma)u(k).$$

Therefore, we have

$$\begin{aligned} v^2(k+1) - v^2(k) &= (-q(k+\sigma)u(k))(2v(k) - q(k+\sigma)u(k)) \\ &= q(k+\sigma) \Big\{ q(k+\sigma)u^2(k) - 2u^2(k) - 2pu(k)u(k-\tau) \end{aligned} \tag{7.4.4}$$

$$+ \sum_{\ell=k-\sigma}^{k-1} 2q(\ell+\sigma)u(\ell)u(k)\Bigg\}$$

$$\leq q(k+\sigma)\Bigg\{q(k+\sigma)u^2(k) - 2u^2(k) + |p|(u^2(k) + u^2(k-\tau))$$

$$+ \sum_{\ell=k-\sigma}^{k-1} q(\ell+\sigma)\left(u^2(\ell) + u^2(k)\right)\Bigg\}.$$

Step 3. Let $u(k)$ be a solution of (7.1.1) such that $v(k)$ defined in (7.4.2) converges to, say, $\beta$. We shall show that $u(k)$ converges to $\beta/(1+p)$. For this, we note from (7.4.3) that

$$\sum_{\ell=K}^{\infty} q(\ell+\sigma)u(\ell) = v(K) - \beta,$$

which implies that $\lim_{k\to\infty} q(k+\sigma)u(k) = 0$, and $\lim_{k\to\infty} \sum_{\ell=k-\sigma}^{k-1} q(\ell+\sigma)u(\ell) = 0$. Thus,

$$\text{(7.4.5)} \quad \lim_{k\to\infty}(u(k) + pu(k-\tau)) = \lim_{k\to\infty}\left\{v(k) + \sum_{\ell=k-\sigma}^{k-1} q(\ell+\sigma)u(\ell)\right\} = \beta.$$

We claim that $u(k)$ is bounded. Otherwise, there is a sequence $\{k_j\}$ of integers such that $k_j \to \infty$, $|u(k_j)| \to \infty$ as $j \to \infty$, and $|u(k_j)| = \sup_{0\leq k\leq k_j}|u(k)|$. Consequently,

$$\begin{aligned} |u(k_j) + pu(k_j - \tau)| &\geq |u(k_j)| - |p||u(k_j-\tau)| \\ &\geq (1-|p|)|u(k_j)| \to \infty \quad \text{as} \quad j \to \infty, \end{aligned}$$

which contradicts (7.4.5).

Let $\omega^+ = \limsup u(k)$, $\omega^- = \liminf u(k)$ and let $\{k_m\}$ and $\{s_m\}$ be two integer sequences such that $k_m \to \infty$ and $s_m \to \infty$ as $m \to \infty$ and $\omega^+ = \lim_{m\to\infty} u(k_m)$, $\omega^- = \lim_{m\to\infty} u(s_m)$. Consider the following two cases: (i) $0 \leq p < 1$, and (ii) $-1 < p < 0$. In the first case, we have

$$\beta = \lim_{m\to\infty}(u(k_m) + pu(k_m - \tau)) = \omega^+ + p\lim_{m\to\infty} u(k_m - \tau) \geq \omega^+ + p\omega^-$$

and

$$\beta = \lim_{m\to\infty}(u(s_m) + pu(s_m - \tau)) = \omega^- + p\lim_{m\to\infty} u(s_m - \tau) < \omega^- + p\omega^+.$$

Thus, $0 \le \omega^+ - \omega^- \le p(\omega^+ - \omega^-)$ so that $\omega^+ = \omega^- = \beta/(1+p)$.

In the second case, we have

$$\begin{aligned} \omega^+ &= \lim_{m\to\infty} u(k_m) = \lim_{m\to\infty}(u(k_m) + pu(k_m - \tau) - pu(k_m - \tau)) \\ &= \beta - p \lim_{m\to\infty} u(k_m - \tau) \le \beta - p\omega^+ \end{aligned}$$

and

$$\begin{aligned} \omega^- &= \lim_{m\to\infty} u(s_m) = \lim_{m\to\infty}(u(s_m) + pu(s_m - \tau) - pu(s_m - \tau)) \\ &= \beta - p \lim_{m\to\infty} u(s_m - \tau) \ge \beta - p\omega^-. \end{aligned}$$

Thus, $\omega^+ \le \beta/(1+p) \le \omega^-$, which implies $\omega^+ = \omega^- = \beta/(1+p)$. Hence, if $v(k)$ defined in (7.4.2) converges, then so does $u(k)$.

Step 4. We shall construct a nonnegative Lyapunov functional $V$ which is eventually nonincreasing along the solutions of (7.1.1). Let $u(k)$ be a solution of (7.1.1), and let $v(k)$ be defined by (7.4.3). Let

$$\begin{aligned} P(k) &= v^2(k), \quad k \in \mathbb{N} \\ Q(k) &= \sum_{\ell=k-\tau}^{k-1} |p| q(\ell+\tau+\sigma) u^2(\ell), \quad k \in \mathbb{N} \\ R(k) &= \sum_{\ell=k-\sigma}^{k-1} \sum_{j=\ell}^{k-1} q(\ell+2\sigma) q(j+\sigma) u^2(j), \quad k \in \mathbb{N}. \end{aligned}$$

Then, $P(k),\ Q(k),\ R(k) \ge 0,\ k \in \mathbb{N}$ and

$$\begin{aligned} \Delta P(k) &= v^2(k+1) - v^2(k) \\ \Delta Q(k) &= |p| q(k+\tau+\sigma) u^2(k) - |p| q(k+\sigma) u^2(k-\tau) \\ \Delta R(k) &= \sum_{\ell=k+1-\sigma}^{k} q(k+\sigma) q(\ell+2\sigma) u^2(k) - \sum_{\ell=k-\sigma}^{k-1} q(k+\sigma) q(\ell+\sigma) u^2(\ell). \end{aligned}$$

Therefore,

$$V(u(k)) = P(k) + Q(k) + R(k) \ge 0, \quad k \in \mathbb{N}.$$

Further, in view of (7.4.4) it follows that

$$\begin{aligned} \Delta V(u(k)) &= \Delta(P(k) + Q(k) + R(k)) \\ &\le \left\{ -2 + |p| \left[ 1 + \frac{q(k+\tau+\sigma)}{q(k+\sigma)} \right] + \sum_{\ell=k}^{k+2\sigma} q(\ell) \right\} \{ q(k+\sigma) u^2(k) \}. \end{aligned}$$

Thus, (7.4.1) implies

$$\Delta V(u(k)) \le -\epsilon\left\{q(k+\sigma)u^2(k)\right\} \le 0$$

for all large $k$, where $\epsilon$ is some sufficiently small positive number. This shows that

$$\lim_{k\to\infty} V(u(k)) = \lim_{k\to\infty}(P(k)+Q(k)+R(k)) = \rho \ge 0$$

and

$$\epsilon\sum_{k=K}^{\infty} q(k+\sigma)u^2(k) \le V(u(K)) - \lim_{k\to\infty} V(u(k+1)) = V(u(K)) - \rho,$$

where $K$ is some large integer, which implies that

$$\sum_{k=0}^{\infty} q(k+\sigma)u^2(k) < \infty. \tag{7.4.6}$$

Next, we note that condition (7.4.1) implies

$$\limsup_{\ell\to\infty}\left\{|p|\frac{q(\ell+\tau+\sigma)}{q(\ell+\sigma)}\right\} < 2$$

so that

$$Q(k) = \sum_{\ell=k-\tau}^{k-1} |p|q(\ell+\tau+\sigma)u^2(\ell) < 2\sum_{\ell=k-\tau}^{k-1} q(\ell+\sigma)u^2(\ell)$$

for all large $k$. Similarly, the condition (7.4.1) implies

$$\sum_{\ell=k-\sigma}^{k-1} q(\ell+2\sigma) < 2$$

and

$$\sum_{\ell=k-\sigma}^{k-1} q(\ell+2\sigma)q(j+\sigma)u^2(j) < 2q(j+\sigma)u^2(j)$$

for all large $k$. Thus,

$$\begin{aligned} R(k) &< \sum_{\ell=k-\sigma}^{k-1}\sum_{j=k-\sigma}^{k-1} q(\ell+2\sigma)q(j+\sigma)u^2(j) \\ &< 2\sum_{j=k-\sigma}^{k-1} q(j+\sigma)u^2(j) \end{aligned}$$

for all large $k$. Therefore, in view of (7.4.6), $R(k)$ and $Q(k)$ tend to zero. This shows that

$$\lim_{k\to\infty} |v(k)| = \lim_{k\to\infty} |P(k)|^{1/2} = \rho^{1/2}.$$

We now claim that $v(k)$ converges. If $\rho = 0$ it is trivial, and if $\rho > 0$ it suffices to show that $v(k)$ is eventually positive or negative. If not, we pick a number $\epsilon$ such that $0 < \epsilon < \rho^{1/2}$ and let $K$ be a positive integer such that

$$\rho^{1/2} - \epsilon < |v(k)| < \rho^{1/2} + \epsilon, \quad k = K, K+1, \cdots \tag{7.4.7}$$

and let

$$I = \{k \geq K : v(k) > 0\} \quad \text{and} \quad J = \{k \geq K : v(k) < 0\}.$$

Since $v(k)$ is neither eventually positive nor eventually negative, $I$ and $J$ are unbounded, thus we can pick a divergent sequence of integers $k_i$ such that $K \leq k_1 < k_2 < \cdots < k_i$, $k_i \in I$ and $k_i + 1 \in J$. Then, $v(k_i + 1) < 0$ and $v(k_i) > 0$. Furthermore, in view of (7.4.7) we have

$$2\left(-\rho^{1/2} - \epsilon\right) < v(k_i + 1) - v(k_i) < 2\left(-\rho^{1/2} + \epsilon\right) < 0, \quad i \geq 1.$$

But then (7.4.3) leads to

$$0 < 2(\rho^{1/2} - \epsilon) < q(k_i + \sigma)u(k_i) = v(k_i) - v(k_i + 1) < 2(\rho^{1/2} + \epsilon), \quad i \geq 1 \tag{7.4.8}$$

i.e. $q(k_i + \sigma)u(k_i)$ is bounded. But, then (7.4.6) and (7.4.8) imply that $u(k_i)$ converges to zero, for otherwise $0 < \lim_{i\to\infty} q(k_i+1)u(k_i)u(k_i) = 0$. Finally, condition (7.4.1) implies that $q(k)$ is bounded so that $q(k_i + 1)u(k_i)$ converges to zero as $i \to \infty$, but this contradicts (7.4.8). We have thus shown that $v(k)$ converges. ∎

## 7.5. Oscillation and Nonoscillation for

$$\Delta\left(u(k) + pu(k-\tau)\right) + q(k)u(k-\sigma_1) - h(k)u(k-\sigma_2) = 0, \quad k \in \mathbb{N} \tag{7.5.1}$$

where $\tau \in \mathbb{N}(1)$, $\sigma_1$, $\sigma_2 \in \mathbb{N}$ and the functions $q(k)$, $h(k)$ are nonnegative on $\mathbb{N}$.

**Theorem 7.5.1.** Assume that $-1 < p \leq 0$, $q(k) \geq 0$, $h(k) \geq 0$, $\sigma_1 > \sigma_2 + 1$ and $q(k) - h(k - (\sigma_1 - \sigma_2)) \geq 0\ (\not\equiv 0)$,

$$\lim_{k\to\infty} \sum_{\ell=k-(\sigma_1-\sigma_2)}^{k-1} h(\ell) = 0. \tag{7.5.2}$$

Further assume that the equation

$$\Delta z(k) + (q(k) - h(k - (\sigma_1 - \sigma_2)))\, z(k - \sigma_1) \;=\; 0 \tag{7.5.3}$$

is oscillatory. Then, the equation (7.5.1) is oscillatory.

**Proof.** If not, let $u(k)$ be a positive solution of (7.5.1). We define $z(k) = u(k) + pu(k - \tau)$, and

$$w(k) \;=\; z(k) - \sum_{\ell=k-(\sigma_1-\sigma_2)}^{k-1} h(\ell)u(\ell - \sigma_2) \tag{7.5.4}$$

so that

$$\Delta w(k) \;=\; (h(k - (\sigma_1 - \sigma_2)) - q(k))\, u(k - \sigma_1). \tag{7.5.5}$$

Since $h(k - (\sigma_1 - \sigma_2)) - q(k) \le 0$, we have $\Delta w(k) \le 0$. If $w(k) \to -\infty$ as $k \to \infty$, then $u(k)$ must be unbounded, which implies that there exist an integer $K$ such that $w(K) < 0$, $u(K) = \max_{k \le K} u(k) > 0$ and $1 + p - \sum_{\ell=K-(\sigma_1-\sigma_2)}^{K-1} h(\ell) \ge 0$.

On the other hand, we have

$$\begin{aligned} 0 \;>\; w(K) \;&=\; z(K) - \sum_{\ell=K-(\sigma_1-\sigma_2)}^{K-1} h(\ell)u(\ell - \sigma_2) \\ &=\; u(K) + pu(K - \tau) - \sum_{\ell=K-(\sigma_1-\sigma_2)}^{K-1} h(\ell)u(\ell - \sigma_2) \\ &\ge\; u(K)\left(1 + p - \sum_{\ell=K-(\sigma_1-\sigma_2)}^{K-1} h(\ell)\right) \;\ge\; 0, \end{aligned}$$

which is impossible. Thus, we must have $\lim_{k\to\infty} w(k) = r$, where $r$ is finite. Hence, $u(k)$ is bounded and therefore from (7.5.4), we find $\lim_{k\to\infty} z(k) = r$. If $r < 0$, then for $k$ sufficiently large $u(k) + pu(k - \tau) \le r/2 < 0$. It follows that $u(k) \to 0$ and hence $z(k) \to 0$ as $k \to \infty$. Consequently, $r$ cannot be negative. Therefore, $r \ge 0$, and since $\Delta w(k) \le 0$, we have $w(k) \ge 0$, i.e. $u(k) \ge z(k) \ge w(k)$. On substituting this into (7.5.5), we obtain

$$\Delta w(k) + (q(k) - h(k - (\sigma_1 - \sigma_2)))\, w(k - \sigma_1) \;\le\; 0.$$

Since $q(k) - h(k - (\sigma_1 - \sigma_2)) \not\equiv 0$, $\Delta w(k) \not\equiv 0$, it follows that $w(k) > 0$ eventually, which implies that (7.5.3) has a positive solution. This contradicts the assumption that (7.5.3) is oscillatory. ∎

**Remark 7.5.1.** From Theorem 6.20.3 it is clear that the equation (7.5.3) is oscillatory if

$$\liminf_{k\to\infty} \left(q(k) - h(k - (\sigma_1 - \sigma_2))\right) > \frac{\sigma_1^{\sigma_1}}{(\sigma_1 + 1)^{\sigma_1+1}}.$$

**Example 7.5.1.** Consider the difference equation

$$\Delta u(k) + q(k)u(k-4) - h(k)u(k-1) = 0, \quad k \ge 4 \tag{7.5.6}$$

where

$$q(k) = \frac{2(k-3)}{k+1}, \quad h(k) = \frac{k-1}{k(k+1)}.$$

Equation (7.5.6) satisfies all the assumptions of Theorem 7.5.1. Therefore, all solutions of (7.5.6) are oscillatory. In fact, $u(k) = (-1)^k/k$ is such a solution.

## 7.6. Oscillation and Nonoscillation for

$$\Delta\left(u(k) + pu(k-\tau)\right) + q(k)u(k-\sigma) = F(k), \quad k \in \mathbb{N}. \tag{7.6.1}$$

**Theorem 7.6.1.** Assume that $p \ge 0$, $q(k) \le 0$ for all large $k$, and there exists a function $h(k)$ such that

$$\Delta h(k) = F(k) \quad \text{and} \quad \limsup_{k\to\infty} h(k) = \infty, \quad \liminf_{k\to\infty} h(k) = -\infty. \tag{7.6.2}$$

Then, every bounded solution of (7.6.1) is oscillatory.

**Proof.** Let $u(k)$ be a bounded positive solution of (7.6.1). Then, for $z(k) = u(k) + pu(k-\tau)$, we have $\Delta(z(k) - h(k)) = -q(k)u(k-\sigma) \ge 0$, i.e. $z(k) - h(k)$ is nondecreasing eventually. Since $z(k) > 0$ and bounded, and $h(k)$ is oscillatory, we must have $z(k) \ge h(k)$ for all large $k$. But, this contradicts the assumption (7.6.2). ∎

**Example 7.6.1.** Consider the difference equation (7.6.1) with $p = 1$, $\tau = \sigma = 2$, and

$$q(k) = -(2k+1)(k-2) + \frac{4k^3 - 6k^2 - 2k + 2}{k(k^2-1)}$$

$$F(k) = (-1)^{k+1}(2k+1), \quad h(k) = (-1)^k k.$$

This equation satisfies all the conditions of Theorem 7.6.1, and therefore every bounded solution of this equation is oscillatory. One such solution is $u(k) = (-1)^k/k$.

**Theorem 7.6.2.** Assume that $p \geq 0$, $q(k) \geq 0$ for all large $k$, and the assumption (7.6.2) holds. Then, the equation (7.6.1) is oscillatory.

**Proof.** Suppose the contrary, and without loss of generality, let $u(k)$ be a positive solution of (7.6.1). Then, for $z(k) = u(k) + pu(k-\tau)$, we have $\Delta(z(k) - h(k)) = -q(k)u(k-\sigma) \leq 0$, i.e. $z(k) - h(k)$ is nonincreasing eventually. If $z(K) - h(K) \leq 0$ for some $K \in \mathbb{N}$, then $z(k) \leq h(k)$ for all $k \geq K$, which in view of (7.6.2) contradicts the positiveness of $z(k)$. Therefore, $z(k) - h(k) > 0$ for all $k \geq K$, and so $\lim_{k\to\infty}(z(k) - h(k)) = \alpha \geq 0$ exists. From (7.6.2) there exists a sequence $\{k_j\}$ such that $\lim_{j\to\infty} h(k_j) = -\infty$. However, since $\lim_{j\to\infty}(z(k_j) - h(k_j)) = \alpha > 0$, $z(k_j)$ cannot be eventually positive. This contradiction completes the proof. ∎

**Example 7.6.2.** Consider the difference equation

$$(7.6.3) \quad \Delta\left(u(k) + u(k-2)\right) + \frac{2k-3}{k-2}u(k-2) = (-1)^{k+1}(2k+1), \quad k \geq 3.$$

For this equation all the assumptions of Theorem 7.6.2 are satisfied, and therefore this equation is oscillatory. One such solution of (7.6.3) is $u(k) = (-1)^k k$.

**Remark 7.6.1.** For $p \geq 0$, $q(k) \geq 0$, condition (7.6.2) ensures that all solutions of (7.6.1) are oscillatory. But, for $p \geq 0$, $q(k) \leq 0$ condition (7.6.2) guarantees oscillation of only bounded solutions. In fact, the difference equation (7.6.1) for $k \geq 2$ with $p = 1$, $\tau = 2$, $\sigma = 1$,

$$p(k) = -\frac{4k + 2(-1)^k k + (-1)^k - 2}{(k-1)^2}, \quad F(k) = (-1)^{k+1}(2k+1)$$

has a nonoscillatory solution $u(k) = k^2$.

**Remark 7.6.2.** From the proofs it is clear that Theorems 7.6.1 and 7.6.2 hold even for the forced difference equations with variable coefficients of the type

$$(7.6.4) \qquad \Delta\left(u(k) + p(k)u(k-\tau)\right) + q(k)u(k-\sigma) = F(k), \qquad k \in \mathbb{N}$$

provided $p(k) \geq 0$ for all large $k$.

**Theorem 7.6.3.** Assume that $p \geq p(k) \geq 0$, where $p$ is a positive number, $q(k) \geq 0$ and there exists a constant $L > 0$ such that $q(k) \leq Lq(k-\tau)$. Let $\Delta h(k) = F(k)$ and denote $h_+(k) = \max\{h(k), 0\}$, $h_-(k) =$

$\max\{-h(k), 0\}$. If

$$\sum_{k=K}^{\infty} q(k+\sigma)h_+(k) = \infty, \tag{7.6.5}$$

then the equation (7.6.4) is oscillatory.

**Proof.** If not, we can assume that $u(k)$ is an eventually positive solution. Then, for $z(k) = u(k)+p(k)u(k-\tau)$, we have $\Delta(z(k)-h(k)) = -q(k)u(k-\sigma) \leq 0$. Thus, as in Theorem 7.6.2, we have $z(k)-h(k) > 0$ for all $k \geq K$, and hence $z(k) \geq h_+(k)$ for all $k \geq K$, also $\lim_{k\to\infty}(z(k) - h(k)) = \alpha \geq 0$ exists. Consequently, $\sum_{\ell=K}^{\infty} q(\ell)u(\ell - \sigma) < \infty$. Now, in view of $p \geq p(k) \geq 0$ and $q(k) \leq Lq(k - \tau)$ it follows that

$$\begin{aligned}\sum_{k=K}^{\infty} q(k+\sigma)h_+(k) &\leq \sum_{k=K}^{\infty} q(k+\sigma)z(k) \\ &= \sum_{k=K}^{\infty} q(k+\sigma)u(k) + \sum_{k=K}^{\infty} q(k+\sigma)p(k)u(k-\tau) < \infty,\end{aligned}$$

which contradicts (7.6.5). ∎

**Example 7.6.3.** Consider the difference equation (7.6.4) for $k \geq 2$ with $p(k) = 1$, $\tau = \sigma = 2$,

$$q(k) = \frac{4k^3 - 6k^2 - 2k + 2}{k(k^2 - 1)} - \frac{(k-2)(2k+1)}{k(k+1)}, \quad F(k) = (-1)^{k+1}\frac{2k+1}{k(k+1)}.$$

It is easy to see that $q(k) \to 2$ as $k \to \infty$, $h(k) = (-1)^k/k$, and $\sum_{k=1}^{\infty} p(k+\sigma)h_+(k) = \infty$. Therefore, this equation is oscillatory. In fact, $u(k) = (-1)^k/k$ is such a solution.

## 7.7. Oscillation and Nonoscillation for

$$\Delta\left(u(k) - p(k)u(k-\tau)\right) + q(k)u(k-\sigma) = 0, \quad k \in \mathbb{N} \tag{7.7.1}$$

where $\tau$ is a positive and $\sigma$ is a nonnegative integer, and the functions $p(k)$ and $q(k)$ are defined on $\mathbb{N}$, $q(k) \geq 0$ and is not identically zero.

We shall need the following:

**Lemma 7.7.1.** Assume that there exists a positive integer $K$ such that

$$p(K + \ell\tau) \leq 1 \quad \text{for } \ell \in \mathbb{N} \tag{7.7.2}$$

then for any eventually positive solution $v(k)$ of the inequality

$$\Delta(v(k) - p(k)v(k-\tau)) + q(k)v(k-\sigma) \leq 0, \quad k \in \mathbb{N} \tag{7.7.3}$$

the function $z(k)$ defined by $z(k) = v(k) - p(k)v(k-\tau)$ satisfies $z(k) > 0$ and $\Delta z(k) \leq 0$ for all large $k \in \mathbb{N}$.

**Proof.** Let $k_1$ be a positive integer such that $v(k-\mu) > 0$ for all $k \in \mathbb{N}(k_1)$, where again $\mu = \max\{\tau, \sigma\}$. Then, by (7.7.3) we have

$$\Delta z(k) \leq -q(k)v(k-\sigma) \leq 0 \quad \text{for all } k \in \mathbb{N}(k_1),$$

which implies that $z(k)$ is nonincreasing for $k \in \mathbb{N}(k_1)$. Hence, if $z(k) > 0$ does not hold, then we would have eventually $z(k) < 0$. Thus, there exist an integer $k_2 > k_1$ and a constant $\alpha > 0$ such that $z(k) \leq -\alpha$ for all $k \in \mathbb{N}(k_2)$. Therefore, we have

$$v(k) \leq -\alpha + p(k)v(k-\tau) \quad \text{for all } k \in \mathbb{N}(k_2). \tag{7.7.4}$$

Now choose a positive integer $k_3$ such that $K + k_3\tau \geq k_2$. Then, by (7.7.2) and (7.7.4), we have for $j = 0, 1, \cdots$

$$v(K + k_3\tau + j\tau) \leq -(j+1)\alpha + v(K + (k_3-1)\tau) \to -\infty \quad \text{as } j \to \infty.$$

But, this contradicts the assumption that $v(k)$ is eventually positive. ∎

**Lemma 7.7.2.** Assume that there exists a nonnegative integer $s$ such that the functions

$$\begin{aligned} H_0(k) &= \sum_{\ell=k}^{\infty} q(\ell) \\ H_j(k) &= \sum_{\ell=k}^{\infty} \ell q(\ell) H_{j-1}(\ell), \quad j = 1, 2, \cdots, s \end{aligned} \tag{7.7.5}$$

exist, and let

$$\sum_{\ell=1}^{\infty} \ell q(\ell) H_s(\ell) = \infty. \tag{7.7.6}$$

Further, assume that $p(k) \geq 1$, $k \in \mathbb{N}$, and the functions $v(k)$ and $z(k)$ be as in Lemma 7.7.1. Then, $z(k) < 0$ and $\Delta z(k) \leq 0$ for all large $k \in \mathbb{N}$.

**Proof.** Clearly, $z(k)$ is eventually nonincreasing. We assume that $z(k) > 0$ eventually. Since $p(k) \geq 1$, $k \in \mathbb{N}$ it follows that $v(k) > v(k-\tau)$.

Hence, there exists a constant $M > 0$ and a positive integer $k_1$ such that $v(k-\mu) \ge M$ for all $k \in \mathbb{N}(k_1)$. Then, by (7.7.3) we have $\Delta z(k) \le -Mq(k)$ for $k \in \mathbb{N}(k_1)$. Hence, it follows that

$$z(k) \ \ge \ M\sum_{\ell=k}^{\infty} q(\ell) \ = \ MH_0(k) \quad \text{for } \ k \in \mathbb{N}(k_1),$$

which in view of $p(k) \ge 1, \ k \in \mathbb{N}$ yields

$$v(k) \ \ge \ v(k-\tau) + MH_0(k), \quad k \in \mathbb{N}(k_1). \tag{7.7.7}$$

Let $I_1(k)$ be the greatest integer value of $(k-k_1)/\tau$. Then, we have

$$\begin{aligned} v(k) \ &\ge \ M\left[H_0(k) + \cdots + H_0(k-(I_1(k)-1)\tau)\right] + v(k - I_1(k)\tau) \\ &\ge \ M\left[H_0(k) + \cdots + H_0(k-(I_1(k)-1)\tau)\right], \quad k \ge k_1 + \tau. \end{aligned}$$

This in view of the fact that $H_0(k)$ is decreasing, we get

$$v(k) \ > \ MI_1(k)H_0(k), \quad k \ge k_1 + \tau.$$

Using the above inequality in (7.7.3), we obtain for $k \ge k_1 + \tau + \sigma$,

$$\Delta z(k) \ \le \ -Mq(k)I_1(k-\sigma)H_0(k-\sigma) \ \le \ -Mq(k)I_1(k-\sigma)H_0(k). \tag{7.7.8}$$

Since $k/I_1(k-\sigma) \to \tau$ as $k \to \infty$, it follows that

$$\frac{Mq(k)I_1(k-\sigma)H_0(k)}{kq(k)H_0(k)} \ \to \ \frac{M}{\tau} \quad \text{as} \quad k \to \infty.$$

Thus, there exists an integer $k_2 \ge k_1 + \tau + \sigma$ such that

$$Mq(k)I_1(k-\sigma)H_0(k) \ \ge \ \frac{M}{2\tau}kq(k)H_0(k) \quad \text{for} \quad k \in \mathbb{N}(k_2)$$

and consequently (7.7.8) yields

$$\Delta z(k) \ \le \ -\frac{M}{2\tau}kq(k)H_0(k) \quad \text{for} \quad k \in \mathbb{N}(k_2). \tag{7.7.9}$$

In case $s = 0$, a direct summation of (7.7.9) leads to a contradiction. Therefore, suppose that $s \ne 0$. Summing (7.7.9) from $k$ to $\infty$, we have $z(k) \ge (M/2\tau)H_1(k)$ for $k \in \mathbb{N}(k_2)$. This together with (7.7.3) and $p(k) \ge 1, \ k \in \mathbb{N}$ yields

$$v(k) \ \ge \ v(k-\tau) + \frac{M}{2\tau}H_1(k) \quad \text{for } \ k \in \mathbb{N}(k_2). \tag{7.7.10}$$

Now let $I_2(k)$ be the greatest integer value of $(k-k_2)/\tau$. Then, as earlier, on using the fact that $H_1(k)$ is decreasing, we get

$$v(k) > \frac{M}{2\tau} I_2(k) H_1(k), \qquad k \ge k_2 + \tau.$$

Thus, for $k \ge k_2 + \tau + \sigma$,

$$\Delta z(k) \le -\frac{M}{2\tau} q(k) I_2(k-\sigma) H_1(k-\sigma) \le -\frac{M}{2\tau} q(k) I_2(k-\sigma) H_1(k).$$

Since $k/I_2(k-\sigma) \to \tau$ as $k \to \infty$, there exists a positive integer $k_3 \ge k_2 + \tau + \sigma$ such that $I_2(k-\sigma) \ge k/(2\tau)$ for $k \in \mathbb{N}(k_3)$. Hence,

$$\Delta z(k) \le -\frac{M}{(2\tau)^2} k q(k) H_1(k), \qquad k \in \mathbb{N}(k_3)$$

which gives

$$z(k) \ge \frac{M}{(2\tau)^2} H_2(k), \qquad k \in \mathbb{N}(k_3).$$

Thus, we have

$$v(k) \ge v(k-\tau) + \frac{M}{(2\tau)^2} H_2(k), \qquad k \in \mathbb{N}(k_3).$$

By repeating the above procedure, we can choose a positive integer $k_4 \ge k_3 + \tau + \sigma$ such that

$$\Delta z(k) \le -\frac{M}{(2\tau)^3} k q(k) H_2(k), \qquad k \in \mathbb{N}(k_4).$$

In general, there exists a sequence $\{k_i\}$ of positive integers such that

$$\Delta z(k) \le -\frac{M}{(2\tau)^i} k q(k) H_{i-1}(k), \qquad k \in \mathbb{N}(k_{i+1}).$$

In particular, we have

$$\Delta z(k) \le -\frac{M}{(2\tau)^{s+1}} k q(k) H_s(k), \qquad k \in \mathbb{N}(k_{s+2}),$$

which in view of (7.7.6) implies that $z(k) \to -\infty$ as $k \to \infty$. This is a contradiction, and hence $z(k) < 0$ eventually. ∎

**Theorem 7.7.3.** Assume that there exists a nonnegative integer $s$ such that the functions $H_j(k)$, $j = 0, 1, \cdots, s$ defined in (7.7.5) exist and (7.7.6) holds. Further, let $p(k) \equiv 1$. Then, the equation (7.7.1) is oscillatory.

**Proof.** The proof is immediate from Lemmas 7.7.1 and 7.7.2. ■

**Remark 7.7.1.** Theorem 7.7.3 with $s = 0$ reduces to Theorem 7.1.3.

**Example 7.7.1.** Consider the difference equation

$$\Delta(u(k) - u(k-\tau)) + \frac{1}{(k+1)^{1.6}}u(k-\sigma) \;=\; 0, \qquad k \in \mathbb{N}. \tag{7.7.11}$$

Since $(k+1)^{-1.6}\left[k^{-0.6} - (k+1)^{-0.6}\right]^{-1} \to 5/3$ as $k \to \infty$, there exists an integer $k_0$ such that

$$k^{-0.6} - (k+1)^{-0.6} \;<\; \frac{1}{(k+1)^{1.6}} < 2\left[k^{-0.6} - (k+1)^{-0.6}\right], \qquad k \geq k_0$$

and hence

$$k^{-0.6} \;<\; H_0(k) \;=\; \sum_{\ell=k}^{\infty} \frac{1}{(\ell+1)^{1.6}} \;<\; 2k^{-0.6}, \qquad k \geq k_0.$$

Thus, we have

$$\frac{k}{(k+1)^{1.6}}k^{-0.6} \;<\; kq(k)H_0(k) \;<\; \frac{k}{(k+1)^{1.6}}2k^{-0.6}, \qquad k \geq k_0$$

and hence

$$\sum_{k=k_0}^{\infty} kq(k)\sum_{\ell=k}^{\infty} q(\ell) \;<\; \infty,$$

i.e. condition (7.1.7) is not satisfied, and so Theorem 7.1.3 is not applicable.

Next, since $k^{0.4}(k+1)^{-1.6}\left[k^{-0.2} - (k+1)^{-0.2}\right]^{-1} \to 5$ as $k \to \infty$ there exists an integer $k_1 > k_0$ such that

$$k^{-0.2} - (k+1)^{-0.2} \;<\; \frac{k^{0.4}}{(k+1)^{1.6}} \;<\; 6\left[k^{-0.2} - (k+1)^{-0.2}\right], \qquad k \geq k_1$$

which implies

$$H_1(k) \;=\; \sum_{\ell=k}^{\infty} \ell q(\ell)H_0(\ell) \;>\; k^{-0.2}, \qquad k \geq k_1.$$

Finally, since

$$kq(k)H_1(k) \;>\; \frac{k}{(k+1)^{1.6}}k^{-0.2} \;=\; \frac{k^{0.8}}{(k+1)^{1.6}}, \qquad k \geq k_1$$

it follows that

$$\sum_{k=k_1}^{\infty} kq(k)H_1(k) \;=\; \infty.$$

Thus, with $s = 2$ conditions of Theorem 7.7.3 are satisfied, and hence the equation (7.7.11) is oscillatory.

**Theorem 7.7.4.** In Theorem 7.7.3 the condition $p(k) \equiv 1$ can be replaced by $p(k) \geq 1$, $k \in \mathbb{N}$ and that there exists a positive integer $K$ such that $p(K + \ell\tau) = 1$ for $\ell \in \mathbb{N}$.

**Proof.** The proof is immediate from Lemmas 7.7.1 and 7.7.2. ∎

**Example 7.7.2.** The difference equation

$$(7.7.12)\qquad \Delta(u(k) - p(k)u(k-2)) + \frac{1}{(k+1)^{1.8}}u(k-2) \;=\; 0, \qquad k \in \mathbb{N}$$

where $p(0) = 4$, $p(1) = 1$, $p(k+2) = p(k)$, $k \in \mathbb{N}$ satisfies all the conditions of Theorem 7.7.4. Thus, the equation (7.7.12) is oscillatory.

**Theorem 7.7.5.** In Theorem 7.7.3 the condition $p(k) \equiv 1$ can be replaced by $p(k-\sigma)q(k) \geq q(k-\tau)$, $k \geq \mu$ and that there exists a positive integer $K$ such that (7.7.2) holds.

**Proof.** If not, (7.7.1) would have an eventually positive solution $u(k)$. Then, by Lemma 7.7.1 there exists a positive integer $k_0$ such that $z(k) = u(k) - p(k)u(k-\tau) > 0$ for $k \in \mathbb{N}(k_0)$. Now from (7.7.1), we have

$$\begin{aligned}
\Delta z(k) &= -q(k)u(k-\sigma) \\
&= -q(k)\left[z(k-\sigma) + p(k-\sigma)u(k-\sigma-\tau)\right] \\
&= -q(k)z(k-\sigma) - q(k)p(k-\sigma)u(k-\sigma-\tau) \\
&\leq -q(k)z(k-\sigma) - q(k-\tau)u(k-\sigma-\tau) \\
&= -q(k)z(k-\sigma) + \Delta z(k-\tau), \qquad k \geq k_0 + \tau + \sigma
\end{aligned}$$

i.e.

$$(7.7.13)\qquad \Delta(z(k) - z(k-\tau)) + q(k)z(k-\sigma) \;\leq\; 0, \qquad k \geq k_0 + \tau + \sigma.$$

But, Lemma 7.7.2 implies that (7.7.13) cannot have an eventually positive solution, which is a contradiction. ∎

**Theorem 7.7.6.** In Theorem 7.7.3 the condition $p(k) \equiv 1$ can be replaced by $p(k) \geq 1$, $k \in \mathbb{N}$ and that $p(k-\sigma)q(k) \leq q(k-\tau)$ for $k \geq \mu$.

**Proof.** If not, (7.7.1) would have an eventually positive solution $u(k)$. Then, by Lemma 7.7.2 there exists a positive integer $k_1$ such that $z(k) = u(k) - p(k)u(k-\tau) < 0$ for $k \in \mathbb{N}(k_1)$. Now from (7.7.1), we have

$$\begin{aligned}
\Delta z(k) &= -q(k)u(k-\sigma) \\
&= -q(k)\left[z(k-\sigma) + p(k-\sigma)u(k-\sigma-\tau)\right] \\
&\geq -q(k)z(k-\sigma) - q(k-\sigma)u(k-\sigma-\tau) \\
&= -q(k)z(k-\sigma) + \Delta z(k-\tau),
\end{aligned}$$

which shows that $-z(k)$ is an eventually positive solution of the inequality

$$(7.7.14) \qquad \Delta(w(k) - w(k-\tau)) + q(k)w(k-\sigma) \leq 0.$$

But, Lemmas 7.7.1 and 7.7.2 imply that (7.7.14) cannot have an eventually positive solution, which is a contradiction. ∎

**Example 7.7.3.** The difference equation

$$(7.7.15) \quad \Delta\left(u(k) - \frac{k+2}{k+1}u(k-1)\right) + \frac{1}{(k+1)^{5/3}}u(k-1) = 0, \; k \in \mathbb{N}$$

satisfies all the conditions of Theorem 7.7.6. Thus, the equation (7.7.15) is oscillatory.

## 7.8. Oscillation and Nonoscillation for

$$(7.8.1) \quad \Delta\left(r(k)u(k) - p(k)u(k-\tau)\right) + q(k)u(k-\sigma(k)) = 0, \qquad k \in \mathbb{N}$$

where $\tau \in \mathbb{N}(1)$ and the functions $r(k) > 0$, $p(k) \geq 0$, $q(k) \geq 0$, $\sigma(k) \in \mathbb{N}$ for all $k \in \mathbb{N}$, and $\lim_{k\to\infty}(k - \sigma(k)) = \infty$.

It is clear that the solutions of equation (7.8.1) are defined for $k \geq -\mu$, where $\mu = \max\left\{\tau, \max_{k\in\mathbb{N}}(\sigma(k) - k)\right\}$. We will also need the corresponding difference inequality

$$(7.8.2) \quad \Delta\left(r(k)u(k) - p(k)u(k-\tau)\right) + q(k)u(k-\sigma(k)) \leq 0, \qquad k \in \mathbb{N}.$$

We begin with the following:

**Lemma 7.8.1.** Assume that there exists a positive integer $K$ such that either

$$(7.8.3) \qquad \frac{p(K+\ell\tau)}{r(K+(\ell-1)\tau)} \leq 1, \quad \ell \in \mathbb{N}(1)$$

or

$$\text{(7.8.4)} \qquad \frac{p(K+\ell\tau)}{r(K+\ell\tau)} \le 1, \quad \ell \in \mathbb{N}(1) \quad \text{and} \quad \sum_{\ell=1}^{\infty} \frac{1}{r(K+\ell\tau)} = \infty.$$

Further, assume that there does not exist a $k_1 \in \mathbb{N}$ such that $q(k) \equiv 0$ on $\mathbb{N}(k_1)$. Let $u(k)$ be an eventually positive solution of the equation (7.8.1) (or inequality (7.8.2)) and define $z(k) = r(k)u(k) - p(k)u(k-\tau)$. Then, there is a $k_2 \in \mathbb{N}$ such that $z(k) > 0$ and $\Delta z(k) \le 0$ on $\mathbb{N}(k_2)$.

**Proof.** We shall prove the lemma for (7.8.1), the proof for (7.8.2) is similar. From (7.8.1), we have $\Delta z(k) = -q(k)u(k-\sigma(k)) \le 0$ eventually. Now since $q(k) \not\equiv 0$ on $\mathbb{N}(k_1)$ for any $k_1 \in \mathbb{N}$, $z(k)$ must be nonincreasing and cannot be eventually constant. Hence, there exists a $k_2 \in \mathbb{N}(k_1)$ such that either $z(k) > 0$ or $z(k) < 0$, $k \in \mathbb{N}(k_2)$. Let $z(k) < 0$, $k \in \mathbb{N}(k_2)$. Then, $0 < r(k)u(k) < p(k)u(k-\tau)$ which implies that $p(k) > 0$ on $\mathbb{N}(k_2)$. There exists a constant $m < 0$ and $k_3 \ge k_2$ such that $z(k) \le m < 0$ on $\mathbb{N}(k_3)$, i.e.

$$\text{(7.8.5)} \qquad r(k)u(k) \le m + p(k)u(k-\tau).$$

We choose an integer $k^*$ sufficiently large so that $K + k^*\tau \ge k_3$.

If (7.8.3) holds, then from (7.8.5) we have

$$\begin{aligned} & r(K+(k^*+k)\tau)u(K+(k^*+k)\tau) \\ & \quad \le m + p(K+(k^*+k)\tau)u(K+(k^*+k-1)\tau) \\ & \quad = m + \frac{p(K+(k^*+k)\tau)}{r(K+(k^*+k-1)\tau)} r(K+(k^*+k-1)\tau)u(K+(k^*+k-1)\tau) \\ & \quad \le m + r(K+(k^*+k-1)\tau)u(K+(k^*+k-1)\tau). \end{aligned}$$

This inequality by induction leads to

$$r(K+(k^*+k)\tau)u(K+(k^*+k)\tau) \le km + r(K+k^*\tau)u(K+k^*\tau).$$

The right side of the above inequality tends to $-\infty$ as $k \to \infty$, and this implies that $u(k)$ is eventually negative, which is a contradiction. Hence, $z(k) > 0$ on $\mathbb{N}(k_2)$.

If (7.8.4) holds, then from (7.8.5) we have

$$\begin{aligned} u(K+(k^*+k)\tau) & \le \frac{m}{r(K+(k^*+k)\tau)} + \frac{p(K+(k^*+k)\tau)}{r(K+(k^*+k)\tau)} \times \\ & \qquad u(K+(k^*+k-1)\tau) \\ & \le \frac{m}{r(K+(k^*+k)\tau)} + u(K+(k^*+k-1)\tau). \end{aligned}$$

The right side of the above inequality tends to $-\infty$ as $k \to \infty$, which is again a contradiction. Hence, $z(k) > 0$ on $\mathbb{N}(k_2)$. ∎

**Theorem 7.8.2.** with respect to difference equation (7.8.1) assume that (7.8.3) or (7.8.4) hold and $q(k) \not\equiv 0$ on any infinite interval. Further, assume that either

$$p(k) + q(k)\sigma(k) \;>\; 0 \tag{7.8.6}$$

or

$$\sigma(k) \;>\; 0 \text{ and } q(\ell) \;\not\equiv\; 0 \text{ for } \ell \in \mathbb{N}(k, K^*) \tag{7.8.7}$$

where $K^* = \min\{\ell : \ell - \sigma(\ell) < k \le \ell + 1 - \sigma(\ell+1)\}$. Then, the equation (7.8.1) is oscillatory if and only if (7.8.2) does not have an eventually positive solution.

**Proof.** The sufficiency is obvious. Suppose $v(k)$ is an eventually positive solution of (7.8.2). We shall prove that (7.8.1) also has a positive solution. Let $z(k) = r(k)u(k) - p(k)u(k-\tau)$. Then, by Lemma 7.8.1 there exists a $k_2 \in \mathbb{N}$ such that $\Delta z(k) \le 0$ and $z(k) > 0$ on $\mathbb{N}(k_2)$. Thus, $\lim_{k\to\infty} z(k) \ge 0$ exists. Hence, $\Delta z(k) + q(k)u(k-\sigma(k)) \le 0$ on $\mathbb{N}(k_2)$. Summing this inequality from $k$ to $\infty$, we obtain

$$-z(k) + \sum_{\ell=k}^{\infty} q(\ell)u(\ell - \sigma(\ell)) \;\le\; 0 \text{ on } \mathbb{N}(k_2).$$

Therefore, it follows that

$$z(k) \;\ge\; \sum_{\ell=k}^{\infty} q(\ell)u(\ell - \sigma(\ell)) \text{ on } \mathbb{N}(k_2).$$

This implies that

$$r(k)u(k) \;\ge\; p(k)u(k-\tau) + \sum_{\ell=k}^{\infty} q(\ell)u(\ell - \sigma(\ell)) \text{ on } \mathbb{N}(k_2). \tag{7.8.8}$$

Let $k_3 \in \mathbb{N}$ be the least integer such that (7.8.8) holds for all $k \in \mathbb{N}(k_3)$, and set $\beta = \max\{\tau, \max_{k\ge k_3}\{k_3 - (k - \sigma(k))\}\}$. We also consider the set of functions

$$S \;=\; \{v : 0 \;\le\; v(k) \;\le\; 1 \text{ for } k \;\ge\; k_3 - \beta\}$$

and define a mapping $\mathcal{T}$ on $S$ as follows

$$\mathcal{T}v(k) = \begin{cases} \dfrac{1}{r(k)u(k)}\Bigg[p(k)v(k-\tau)u(k-\tau) \\ \qquad\qquad + \displaystyle\sum_{\ell=k}^{\infty} q(\ell)v(\ell-\sigma(\ell))u(\ell-\sigma(\ell))\Bigg] & \text{for } k \in \mathbb{N}(k_3) \\ 1-\dfrac{1}{\beta}(k-k_3+\beta)(1-\mathcal{T}v(k_3)) \text{ for } k_3-\beta \le k \le k_3-1. \end{cases}$$

From (7.8.8) and the definition of the set $S$ it is clear that

$$0 \le \mathcal{T}v(k) \le 1 \tag{7.8.9}$$

and

$$\mathcal{T}v(k) > 0 \ \text{ for } \ k_3-\beta \le k \le k_3-1. \tag{7.8.10}$$

Now we define a sequence of functions $\{v_m(k)\}$ in $S$ as follows

$$\begin{aligned} v_0(k) &= 1 \ \text{ for } \ k \ge k_3-\beta \\ v_{m+1}(k) &= \mathcal{T}v_m(k) \ \text{ for } \ k \ge k_3-\beta \ \text{ and } \ m = 0,1,2,\cdots. \end{aligned}$$

We shall show that $0 \le v_{m+1}(k) \le v_m(k) \le 1$ for $k \ge k_3-\beta$, $m = 0,1,2,\cdots$. Since $v_1(k) = \mathcal{T}v_0(k)$ from (7.8.9) we have $0 \le v_1(k) \le 1 = v_0(k)$. Now assume the induction hypothesis $v_i(k) \le v_{i-1}(k)$ for $i = 1,\cdots,m$. Then,

$$\begin{aligned} v_{m+1}(k) &= \mathcal{T}v_m(k) \\ &= \begin{cases} \dfrac{1}{r(k)u(k)}\Bigg[p(k)v_m(k-\tau)u(k-\tau) \\ \qquad + \displaystyle\sum_{\ell=k}^{\infty} q(\ell)v_m(\ell-\sigma(\ell))u(\ell-\sigma(\ell))\Bigg] & \text{for } k \in \mathbb{N}(k_3) \\ 1-\dfrac{1}{\beta}(k-k_3+\beta)(1-\mathcal{T}v_m(k_3)) \text{ for } k_3-\beta \le k \le k_3-1 \end{cases} \\ &\le \begin{cases} \dfrac{1}{r(k)u(k)}\Bigg[p(k)v_{m-1}(k-\tau)u(k-\tau) \\ \qquad + \displaystyle\sum_{\ell=k}^{\infty} q(\ell)v_{m-1}(\ell-\sigma(\ell))u(\ell-\sigma(\ell))\Bigg] & \text{for } k \in \mathbb{N}(k_3) \\ 1-\dfrac{1}{\beta}(k-k_3+\beta)(1-\mathcal{T}v_{m-1}(k_3)) \text{ for } k_3-\beta \le k \le k_3-1 \end{cases} \\ &= \mathcal{T}v_{m-1}(k) = v_m(k). \end{aligned}$$

Thus, for each $k$, $\{v_m(k)\}$ is a decreasing sequence and $v_m(k) \geq 0$. Therefore, the limit exists. We let $v(k) = \lim_{m\to\infty} v_m(k)$ for $k \geq k_3 - \beta$. Then, it follows that $\mathcal{T}v(k) = v(k)$, i.e.

$$(7.8.11) \quad v(k) = \frac{1}{r(k)u(k)}\left[p(k)v(k-\tau)u(k-\tau) + \sum_{\ell=k}^{\infty} q(\ell)v(\ell-\sigma(\ell))u(\ell-\sigma(\ell))\right] \quad \text{for } k \in \mathbb{N}(k_3).$$

Now define a function $w(k)$ by $w(k) = v(k)u(k)$, $k \geq k_3 - \beta$. It is clear that $w(k) \geq 0$ for $k \geq k_3 - \beta$. Then, by (7.8.10) we have $w(k) > 0$ for $k_3 - \beta \leq k \leq k_3 - 1$, and by (7.8.11) we have

$$(7.8.12) \qquad w(k) = \frac{1}{r(k)}\left[p(k)w(k-\tau) + \sum_{\ell=k}^{\infty} q(\ell)w(\ell-\sigma(\ell))\right].$$

We need to show that $w(k) > 0$ for $k \geq k_3 - \beta$. Suppose there exists a $K \geq k_3 - \beta$ such that $w(K) = 0$ and $w(k) > 0$ for $k_3 - \beta \leq k < K$. Then, $K \geq k_3$ and from (7.8.12), we get

$$\frac{1}{r(K)}\left[p(K)w(K-\tau) + \sum_{\ell=K}^{\infty} q(\ell)w(\ell-\sigma(\ell))\right] = 0.$$

Thus, we must have $p(K) = 0$ and

$$(7.8.13) \qquad q(\ell)w(\ell-\sigma(\ell)) = 0 \quad \text{for all } \ell \in \mathbb{N}(K).$$

Hence, either $q(K) = 0$ or $\sigma(K) = 0$, so condition (7.8.6) cannot hold. Therefore, we assume that condition (7.8.7) holds, i.e. $\sigma(K) > 0$ and $q(\ell) \not\equiv 0$ for $\ell \in \mathbb{N}(K, K^*)$, where $K^* = \min\{\ell : \ell - \sigma(\ell) < K \leq \ell + 1 - \sigma(\ell+1)\}$. But, this contradicts (7.8.13), since $\ell - \sigma(\ell) < K$ implies that $w(\ell - \sigma(\ell)) > 0$ and this in turn implies $q(\ell) \equiv 0$ on $\mathbb{N}(K, K^*)$. Hence, $w(k) > 0$ for $k \geq k_3 - \beta$. Clearly, this $w(k)$ is a solution of (7.8.1) on $\mathbb{N}(k_3)$. ∎

## 7.9. Oscillatory and Asymptotic Behavior for

$$(7.9.1) \qquad \Delta\left(u(k) + p(k)u(k-\tau)\right) + q(k)f(u(k-\sigma)) = 0, \qquad k \in \mathbb{N}$$

where $p(k)$ and $q(k)$ are such that $q(k) \geq 0$ for all $k \in \mathbb{N}$, and are not zero for infinitely many values of $k$, $\tau \in \mathbb{N}(1)$, $\sigma \in \mathbb{N}$, and $f : \mathbb{R} \to \mathbb{R}$ is continuous with $uf(u) > 0$ for $u \neq 0$. We shall need the following condition in several results that follow:

(7.9.2) $f(u)$ is bonded away from zero if $u$ is bounded away from zero.

We begin with the following lemma which will be used frequently.

**Lemma 7.9.1.** Let $u(k)$ be a nonoscillatory solution of (7.9.1). Set $z(k) = u(k) + p(k)u(k-\tau)$.

(a) If $u(k)$ is eventually positive (negative), then $z(k)$ is eventually nonincreasing (nondecreasing).

(b) If $u(k)$ is eventually positive (negative) and there exists a constant $P_1$ such that

$$-1 < P_1 \leq p(k), \tag{7.9.3}$$

then eventually $z(k) > 0$ $(z(k) < 0)$.

(c) If in addition to (7.9.2)

$$\sum^{\infty} q(k) = \infty \tag{7.9.4}$$

and there exists a constant $P_2$ such that

$$-1 < P_2 \leq p(k) \leq 0, \tag{7.9.5}$$

then $z(k) \to 0$ as $k \to \infty$.

(d) Suppose that (7.9.2) and (7.9.4) hold, and there exists a constant $P_3$ such that

$$p(k) \leq P_3 < -1. \tag{7.9.6}$$

If $u(k)$ is eventually positive (negative), then eventually $z(k) < 0$ $(z(k) > 0)$.

(e) Suppose that (7.9.2) and (7.9.4) hold, and there exists a constant $P_4$ such that

$$P_4 \leq p(k) \leq 0. \tag{7.9.7}$$

If $u(k)$ is eventually positive (negative), then $\lim_{k\to\infty} z(k)$ exists and its value is either $0$ or $-\infty$ ($0$ or $\infty$).

(f) Suppose that (7.9.2) and (7.9.4) hold, and there exist constants $P_3$ and $P_4$ such that

$$P_4 \leq p(k) \leq P_3 < -1. \tag{7.9.8}$$

If $u(k)$ is eventually positive (negative), then $\lim_{k\to\infty} z(k) = -\infty$ $(\lim_{k\to\infty} z(k) = \infty)$.

**Proof.** All parts will be proved for solutions that are eventually positive. The arguments for the case of eventually negative solutions are similar.

(a) From (7.9.1) it follows that $\Delta z(k) = -q(k)f(u(k-\sigma)) \leq 0$ for all large $k$. Thus, $z(k)$ is eventually nonincreasing.

(b) If $u(k)$ is eventually positive and the conclusion does not hold, then since by (a) $z(k)$ is nonincreasing, it follows that eventually either $z(k) \equiv 0$ or $z(k) < 0$. Now $z(k) \equiv 0$ implies that $\Delta z(k) = -q(k)f(u(k-\sigma)) \equiv 0$, but this contradicts the fact that $q(k) \not\equiv 0$ for infinitely many $k$. If $z(k) < 0$, then $u(k) < -p(k)u(k-\tau)$ so $p(k) < 0$. From (7.9.3) it follows that $-1 < P_1 < 0$ and $u(k) < -P_1 u(k-\tau)$. Thus, by induction, we obtain $u(k+j\tau) \leq (-P_1)^j u(k)$ for all positive integers $j$. Hence, $u(k) \to 0$ as $k \to \infty$. But this, together with the fact that $p(k)$ is bounded, implies that $z(k)$ decreases to zero as $k \to \infty$. This contradicts the fact that $z(k) < 0$ and completes the proof of (b).

(c) Note first that (7.9.5) implies (7.9.3) with $P_1$ replaced by $P_2$. If $u(k)$ is eventually positive, then (a) and (b) imply that $z(k)$ is eventually positive and nonincreasing. Therefore, $z(k) \to L \geq 0$ as $k \to \infty$. Now suppose that $L > 0$. By (7.9.5), we have $z(k) \leq u(k)$. Thus, there exists an integer $k_1 \geq k_0 \in \mathbb{N}$ such that $L \leq z(k-\sigma) \leq u(k-\sigma)$ for $k \geq k_1$. Hence, from (7.9.1) and (7.9.2), it follows that $\Delta z(k) \leq -L_1 q(k)$ for some positive constant $L_1$. Summing the last inequality, we obtain

$$z(k) \;\leq\; z(k_1) - L_1 \sum_{\ell=k_1}^{k-1} q(\ell),$$

which as $k \to \infty$, in view of (7.9.4), implies that $z(k) \to -\infty$. This is a contradiction.

(d) Suppose $u(k)$ is eventually positive. If the conclusion does not hold, then eventually $z(k) \geq 0$, i.e. $u(k) + p(k)u(k-\tau) \geq 0$. Thus, from (7.9.6), we have $u(k) \geq -P_3 u(k-\tau)$ for sufficiently large $k$. It follows by induction that for each positive integer $j$, $u(k+j\tau) \geq (-P_3)^j u(k)$, which implies that $u(k) \to \infty$ as $k \to \infty$. Thus, (7.9.1) and (7.9.2) imply that there exist a positive integer $k_2$ and a positive constant $L_2$ such that $\Delta z(k) = -q(k)f(u(k-\sigma)) \leq -L_2 q(k)$ for $k \geq k_2$. On summing the last inequality, we get

$$0 \;\leq\; z(k) \;\leq\; z(k_2) - L_2 \sum_{\ell=k_2}^{k-1} q(\ell) \;\to\; -\infty \quad \text{as } k \to \infty,$$

which is a contradiction.

(e) For $u(k)$ eventually positive, we have from (7.9.1) that $\Delta z(k) \leq 0$ for all sufficiently large $k$. Hence, $z(k) \to L_3 < \infty$ as $k \to \infty$. If

$L_3 > -\infty$, then summing (7.9.1), we obtain

$$\sum_{\ell=k}^{\infty} q(\ell)f(u(\ell-\sigma)) = z(k) - L_3 < \infty.$$

The last inequality, in view of (7.9.2) and (7.9.4) implies that $\liminf_{k\to\infty} u(k) = 0$. Let $k_3 \geq k_0$ be such that $u(k-\tau) > 0$ for $k \geq k_3$. If $L_3 < 0$, there exists $k_4 \geq k_3$ so that

$$u(k_4 - \tau) < \frac{L_3}{4P_4} \quad \text{and} \quad \sum_{\ell=k_4}^{\infty} q(\ell)f(u(\ell-\sigma)) < -\frac{L_3}{4}.$$

We then have from (7.9.7) that

$$\begin{aligned} u(k_4) = z(k_4) - p(k_4)u(k_4-\tau) &\leq L_3 + \sum_{\ell=k_4}^{\infty} q(\ell)f(u(\ell-\sigma)) - P_4 u(k_4-\tau) \\ &< L_3 - \frac{L_3}{4} - \frac{L_3}{4} = \frac{L_3}{2}, \end{aligned}$$

which contradicts $u(k) > 0$. If $L_3 > 0$, then by (7.9.7) we find that $u(k) \geq z(k) > L_3$ for $k \geq k_3$. Therefore, it follows from (7.9.2) that there exists a constant $L_4 > 0$ such that

$$z(k) > L_3 + L_4 \sum_{\ell=k}^{\infty} q(\ell),$$

which implies that $z(k) \to \infty$ as $k \to \infty$ contradicting $z(k) \to L_3 < \infty$. Thus, either $L_3 = 0$ or $-\infty$.

(f) Note that (7.9.8) implies that (7.9.6) and (7.9.7) hold. Hence, for $u(k)$ eventually positive from (a), (d) and (e), we have $\Delta z(k) \leq 0$, $z(k) < 0$ and $z(k) \to 0$ or $-\infty$ as $k \to \infty$. Clearly, $z(k) \to -\infty$ as $k \to \infty$. ∎

**Theorem 7.9.2.** Suppose there exist positive constants $q$ and $B$ such that

(7.9.9) $$q(k) \geq q, \quad k \in \mathbb{N}$$

(7.9.10) $$|f(u)| \geq B|u| \quad \text{for all } u,$$

and

(7.9.11) $$0 \leq p(k), \quad k \in \mathbb{N}.$$

Then, every nonoscillatory solution of (7.9.1) tends to 0 as $k \to \infty$.

**Proof.** Let $u(k)$ be an eventually positive solution of (7.9.1), say $u(k) > 0$, $u(k-\tau) > 0$ and $u(k-\sigma) > 0$ for $k \geq k_0 \in \mathbb{N}$. By parts (a) and (b) of Lemma 7.9.1, $z(k)$ is eventually positive and nonincreasing, so $z(k) \to L_5 \geq 0$ as $k \to \infty$. Summing the equation (7.9.1) from $k$ to $\infty$ for $k \geq k_0$, we obtain

$$z(k) \;=\; L_5 + \sum_{\ell=k}^{\infty} q(\ell) f(u(\ell-\sigma)), \qquad k \geq k_0.$$

Now by (7.9.9) and (7.9.10), we get

$$qB \sum_{\ell=k}^{\infty} u(\ell-\sigma) \;\leq\; \sum_{\ell=k}^{\infty} q(\ell) f(u(\ell-\sigma)) \;=\; z(k) - L_5 \;<\; \infty,$$

which implies that $u(k) \to 0$ as $k \to \infty$. The proof is similar when $u(k)$ is eventually negative. ∎

**Example 7.9.1.** The difference equation

$$\Delta\left(u(k) + \frac{1}{k}u(k-2)\right) + \frac{2^{k-3}(k^2+5k+8)}{(1+2^{k-1})k(k+1)}(1+|u(k-1)|)u(k-1) = 0, \quad n \geq 3$$

satisfies all the conditions of Theorem 7.9.2, and hence all nonoscillatory solutions tend to zero as $k \to \infty$. In fact, $u(k) = 1/2^k$ is such a solution.

**Theorem 7.9.3.** If (7.9.2), (7.9.4) and (7.9.5) are satisfied, then every nonoscillatory solution of (7.9.1) tends to zero as $k \to \infty$.

**Proof.** Let $u(k)$ be an eventually positive solution of (7.9.1). First we will prove that $u(k)$ is bounded. By Lemma 7.9.1(a), (b) and (c), we see that $0 < z(k) \leq M$ for some positive constant $M$ and so, by (7.9.5) we have

$$u(k) \;\leq\; -P_2 u(k-\tau) + M. \tag{7.9.12}$$

Assume that $u(k)$ is not bounded. Then, there exists a subsequence $\{k_i\}$ of $\mathbb{N}$, so that $\lim_{i\to\infty} u(k_i) = \infty$ and $u(k_i) = \max_{k_0 \leq j \leq k_i} u(j)$, $i = 1, 2, \cdots$. From (7.9.12), for $i$ sufficiently large, we get

$$u(k_i) \;\leq\; -P_2 u(k_i) + M$$

and so

$$(1+P_2)u(k_i) \;\leq\; M,$$

which as $i \to \infty$ leads to a contradiction.

Now suppose that $\limsup_{k\to\infty} u(k) = \alpha > 0$. Then, there exists a subsequence $\{k_i\}$ of $\mathbb{N}$, with $k_1$ large enough so that $u(k) > 0$ for $k > k_1 - \tau$ and $u(k_i) \to \alpha$ as $i \to \infty$. Then, from (7.9.5), we have

$$z(k_i) \ \geq \ u(k_i) + P_2 u(k_i - \tau)$$

and so

$$u(k_i - \tau) \ \geq \ -\frac{1}{P_2}(u(k_i) - z(k_i)).$$

As $i \to \infty$, we obtain

$$\alpha \ \geq \ \lim_{i\to\infty} u(k_i - \tau) \ \geq \ -\frac{\alpha}{P_2}.$$

Since $-P_2 \in (0,1)$, it follows that $\alpha = 0$, i.e. $u(k) \to 0$ as $k \to \infty$. The arguments when $u(k)$ is eventually negative are similar. ■

**Example 7.9.2.** Theorem 7.9.3 implies that all nonoscillatory solutions of the difference equation

$$\Delta\left(u(k) - \frac{1}{2}u(k-2)\right) + \frac{(k-1)^2(k^2-7k+4)}{2k(k+1)(k-2)}u^3(k-1) \ = \ 0, \quad k \geq 7$$

tend to zero as $k \to \infty$. One such solution is $u(k) = 1/k$.

The above results guarantee that under certain appropriate conditions all nonoscillatory solutions of (7.9.1) tend to zero if $q(k) \geq q$ or $\sum_{k=K}^{\infty} q(k) = \infty$ holds. It is natural to ask if the same conclusion holds provided

$$\sum_{k=K}^{\infty} q(k) \ < \ \infty. \tag{7.9.13}$$

The following result gives a partial answer to this question.

**Theorem 7.9.4.** Let in addition to the condition (7.9.5) the function $f$ be nondecreasing. Further, let there exist a positive nondecreasing function $h(k)$ such that $h(k) \to \infty$ as $k \to \infty$, and for sufficiently large $k$

$$-P_2\frac{h(k)}{h(k-\tau)} + h(k)\sum_{\ell=k}^{\infty} q(\ell) f\left(\frac{1}{h(\ell-\sigma)}\right) \ \leq \ 1. \tag{7.9.14}$$

Then, the equation (7.9.1) has a positive solution which tends to zero as $k \to \infty$.

**Proof.** Consider the Banach space $l_\infty^K$ of all real functions $v = v(k)$, $k \geq K$ with the sup norm $\|v\| = \sup_{k\geq K} |v(k)|$. Let $S$ be a subset of $l_\infty^K$ defined as

$$S \ = \ \{v \in l_\infty^K : 0 \leq v(k) \leq 1, \ \ k \geq K\}.$$

Next, we define a partial ordering $\preceq$ on $l_\infty^K$ as follows: for $u, v \in l_\infty^K$, $u \preceq v$ if and only if $u(k) \le v(k)$ for $k \ge K$. Let $\mathcal{T} : S \to l_\infty^K$ be an operator defined as

$$\mathcal{T}v(k) \;=\; -p(k)v(k-\tau)\frac{h(k)}{h(k-\tau)} + h(k)\sum_{\ell=k}^{\infty} q(\ell) f\left(\frac{v(\ell-\sigma)}{h(\ell-\sigma)}\right), \quad k \ge K.$$

By (7.9.14), it is clear that $\mathcal{T}S \subseteq S$, and the nondecreasing nature of $f$ implies that $\mathcal{T}$ is an increasing mapping. Thus, Theorem 7.2.3 is applicable, and there exists a $v \in S$ such that $\mathcal{T}v = v$, i.e.

$$(7.9.15)\quad v(k) \;=\; -p(k)v(k-\tau)\frac{h(k)}{h(k-\tau)} + h(k)\sum_{\ell=k}^{\infty} q(\ell) f\left(\frac{v(\ell-\sigma)}{h(\ell-\sigma)}\right), \quad k \ge K$$

and $v(k) \ge 0$ for $k \ge K$. If we set $u(k) = v(k)/h(k)$, then from (7.9.15), we have

$$u(k) \;=\; -p(k)u(k-\tau) + \sum_{\ell=k}^{\infty} q(\ell) f(u(\ell-\sigma)), \quad k \ge K$$

i.e. $u(k)$ is a positive solution of the equation (7.9.1) with $\lim_{k\to\infty} u(k) = 0$. ∎

**Example 7.9.3.** Consider the difference equation

$$(7.9.16)\quad \Delta\left(u(k) - \frac{1}{4}u(k-1)\right) + \frac{1}{2^{(4k+13)/5}} u^{1/5}(k-3) \;=\; 0, \quad k \ge 3.$$

It is easy to see that all conditions of Theorem 7.9.4 are satisfied with $h(k) = 2^k$. Therefore, (7.9.16) has a positive solution which tends to zero as $k \to \infty$. In fact, $u(k) = 1/2^k$ is such a solution.

**Theorem 7.9.5.** If (7.9.2), (7.9.4) and (7.9.8) are satisfied, then every nonoscillatory solution of (7.9.1) tends to $\infty$ or $-\infty$ as $k \to \infty$.

**Proof.** The proof is immediate from Lemma 7.9.1(f). ∎

**Example 7.9.4.** The difference equation

$$(7.9.17)\quad \Delta\left(u(k) - \frac{4k+1}{k-1}u(k-1)\right) + \frac{3}{(k-2)^\alpha} u^\alpha(k-2) \;=\; 0, \quad k \ge 3$$

where $0 < \alpha \le 1$ is a ratio of odd positive integers satisfies all conditions of Theorem 7.9.5. Therefore, all nonoscillatory solutions of (7.9.17) tend to $\pm\infty$ as $k \to \infty$. In fact, $u(k) = k$ is such a solution.

**Theorem 7.9.6.** Suppose that (7.9.2), (7.9.4) and (7.9.5) hold. If $f$ is a nondecreasing function such that

$$(7.9.18) \quad \int_0^\alpha \frac{dt}{f(t)} < \infty \quad \text{and} \quad \int_{-\alpha}^0 \frac{dt}{f(t)} > -\infty \quad \text{for all } \alpha > 0,$$

then the equation (7.9.1) is oscillatory.

**Proof.** Suppose that (7.9.1) has a nonoscillatory solution $u(k)$. If $u(k) > 0$ for $k \geq k_0$, then by Lemma 7.9.1(a) and (b) there exists a $k_1 \geq k_0$ such that $u(k-\tau) > 0$, $u(k-\sigma) > 0$, $z(k) > 0$ and $\Delta z(k) \leq 0$ for $k \geq k_1$. We note that (7.9.5) implies that $z(k) \leq u(k)$ and therefore from (7.9.1), we have

$$\Delta z(k) + q(k) f(z(k-\sigma)) \leq 0$$

and so, we obtain

$$\Delta z(k) + q(k) f(z(k)) \leq 0$$

or

$$q(k) \leq -\frac{\Delta z(k)}{f(z(k))} \quad \text{for} \quad k \geq k_2 = k_1 + \sigma.$$

Now for $z(k+1) \leq t \leq z(k)$ we have $f(t) \leq f(z(k))$, and so

$$q(k) \leq \int_{z(k+1)}^{z(k)} \frac{dt}{f(t)} \quad \text{for} \quad k \geq k_2.$$

Summing both sides of the above inequality from $k_2$ to $k$, we get

$$\sum_{\ell=k_2}^{k} q(\ell) \leq \int_{z(k+1)}^{z(k_2)} \frac{dt}{f(t)} < \int_0^{z(k_2)} \frac{dt}{f(t)} < \infty,$$

which contradicts the condition (7.9.4). The proof for the case $u(k)$ eventually negative is similar. ∎

**Theorem 7.9.7.** Suppose that (7.9.2), (7.9.4) and (7.9.8) hold. If $\tau > \sigma$ and $f$ is a nondecreasing function such that

$$(7.9.19) \quad \int_\epsilon^\infty \frac{dt}{f(t)} < \infty \quad \text{and} \quad \int_{-\epsilon}^{-\infty} \frac{dt}{f(t)} < \infty \quad \text{for all } \epsilon > 0,$$

then the equation (7.9.1) is oscillatory.

**Proof.** If not, then there exists a nonoscillatory solution $u(k)$ of (7.9.1), and let $u(k) > 0$ for $k \geq k_0$. From Lemma 7.9.1(a) and (f) there exists a

$k_1 \geq k_0$ such that $u(k-\tau) > 0$, $u(k-\sigma) > 0$, $z(k) < 0$ and $\Delta z(k) \leq 0$ for $k \geq k_1$. Then, from (7.5.8) we have

$$P_4 u(k-\tau) \ \leq \ p(k)u(k-\tau) \ < \ z(k) \ < \ 0$$

and hence

$$0 \ < \ \frac{z(k+\tau)}{P_4} \ < \ u(k) \quad \text{for} \quad k \geq k_1.$$

Thus, it follows that

$$f\left(\frac{z(k+\tau-\sigma)}{P_4}\right) \ \leq \ f(u(k-\sigma)) \quad \text{for} \quad k \geq k_2 \geq k_1 + \sigma.$$

Since $k+\tau-\sigma \geq k+1$ the above inequality gives

$$f\left(\frac{z(k+1)}{P_4}\right) \ \leq \ f\left(\frac{z(k+\tau-\sigma)}{P_4}\right) \ \leq \ f(u(k-\sigma)).$$

Hence, from (7.9.1) we find

$$\Delta z(k) + q(k) f\left(\frac{z(k+1)}{P_4}\right) \ \leq \ 0$$

or

$$q(k) \ \leq \ -\frac{\Delta z(k)}{f\left(\frac{z(k+1)}{P_4}\right)} \quad \text{for} \quad k \geq k_2. \tag{7.9.20}$$

Now for $z(k)/P_4 \leq t \leq z(k+1)/P_4$, we have $f(z(k+1)/P_4) \geq f(t)$, and so

$$\frac{1}{P_4}\frac{\Delta z(k)}{f\left(\frac{z(k+1)}{P_4}\right)} \ \leq \ \int_{z(k)/P_4}^{z(k+1)/P_4} \frac{dt}{f(t)} \quad \text{for} \quad k \geq k_2. \tag{7.9.21}$$

Using (7.9.21) in (7.9.20) and summing both sides from $k_2$ to $k$, we get

$$\sum_{\ell=k_2}^{k} q(\ell) \ \leq \ -P_4 \int_{z(k_2)/P_4}^{z(k+1)/P_4} \frac{dt}{f(t)}, \quad k \geq k_2.$$

But this in view of (7.9.19) contradicts (7.9.4). The proof for the case $u(k)$ eventually negative is similar. ∎

**Example 7.9.5.** Theorem 7.9.6 implies that all solutions of

(7.9.22)
$$\Delta\left(u(k) - \frac{1}{e}u(k-2)\right) + (e^3-1)(e+1)e^{(2k-7)/3}u^{1/3}(k-2) = 0, \quad k \geq 2$$

is oscillatory. In fact, one such solution of (7.9.22) is $u(k) = (-1)^k e^k$.

Similarly, Theorem 7.9.7 implies that all solutions of

(7.9.23)
$$\Delta\left(u(k) - (k-2)u(k-2)\right) + \frac{(2k^2-1)}{k(k+1)}(k-1)^3 u^3(k-1) = 0, \quad k \geq 4$$

are oscillatory. One such solution of (7.9.23) is $u(k) = (-1)^k/k$.

**Theorem 7.9.8.** Suppose that (7.9.11) holds. Then, all unbounded solutions of the equation (7.9.1) are oscillatory.

**Proof.** Suppose the contrary. Without loss of generality, let $u(k)$ be an unbounded and eventually positive solution of (7.9.1). By Lemma 7.9.1(a) and (b), we have $z(k) > 0$ and $\Delta z(k) \leq 0$ eventually. Hence,

(7.9.24)
$$z(k) \to L \in [0, \infty) \quad \text{as} \quad k \to \infty.$$

Now, in view of (7.9.11), $z(k) \geq u(k)$ and (7.9.24) show that $u(k)$ is bounded, which is a contradiction. ∎

**Example 7.9.6.** Theorem 7.9.8 implies that all unbounded solutions of

(7.9.25)
$$\Delta(u(k) + ku(k-1)) + \frac{(2k^2-2k-1)}{(k-3)^\alpha}u^\alpha(k-3) = 0, \quad k \geq 4$$

are oscillatory, where $\alpha \in (0, \infty)$ is the ratio of odd positive integers. One such solution of (7.9.25) is $u(k) = (-1)^k k$.

Now we shall study the difference equation (7.9.1) when $q(k)$ is eventually negative. For this it is convenient to write (7.9.1) in the following form

(7.9.26)
$$\Delta(u(k) + p(k)u(k-\tau)) = q(k)f(u(k-\sigma)), \quad k \in \mathbb{N}.$$

**Theorem 7.9.9.** Suppose that conditions (7.9.11) and (7.9.13) hold, and $f$ is nondecreasing. Further, suppose that

(7.9.27)
$$\int_c^\infty \frac{dt}{f(t)} = \infty \quad \text{and} \quad \int_{-\infty}^{-c} \frac{dt}{f(t)} = -\infty \quad \text{for all } c > 0.$$

Then, all nonoscillatory solutions of (7.9.26) are bounded.

**Proof.** Let $u(k)$ be a nonoscillatory solution of (7.9.26), and let $k_0 \in \mathbb{N}$ be such that $|u(k)| \neq 0$ for all $k \in \mathbb{N}(k_0)$. If $u(k) > 0$ for all $k \geq k_0$, then $u(k-\tau-\sigma) > 0$ for all $k \ge k_0 + \tau + \sigma = k_1$. Thus, from (7.9.11) it follows that $z(k) = u(k) + p(k)u(k-\tau) > 0$ and $\Delta z(k) \ge 0$ for all $k \in \mathbb{N}(k_1)$. Hence, $z(k)$ is nondecreasing and satisfies $z(k) \ge u(k)$ for all $k \ge k_1$. Therefore, in view of the fact that $f$ is nondecreasing, we find

$$\Delta z(k) \;=\; q(k)f(u(k-\sigma)) \;\le\; q(k)f(z(k-\sigma)) \;\le\; q(k)f(z(k))$$

or

$$\text{(7.9.28)} \qquad \frac{\Delta z(k)}{f(z(k))} \;\le\; q(k), \qquad k \in \mathbb{N}(k_1).$$

Since $z(k) \le t \le z(k+1)$ implies $f(t) \ge f(z(k))$, from (7.9.28) it follows that

$$\int_{z(k)}^{z(k+1)} \frac{dt}{f(t)} \;\le\; \frac{\Delta z(k)}{f(z(k))} \;\le\; q(k), \qquad k \in \mathbb{N}(k_1).$$

Summing the last inequality from $k_1$ to $k-1$, we obtain

$$\int_{z(k_1)}^{z(k)} \frac{dt}{f(t)} \;\le\; \sum_{\ell=k_1}^{k-1} q(\ell)$$

and by (7.9.13) and (7.9.27) we find that $z(k)$ is bounded from above. Now $0 < u(k) \le z(k)$ implies that $u(k)$ is also bounded from above. The proof for the case when $u(k)$ is eventually negative is similar. ∎

**Corollary 7.9.10.** Suppose that in addition to the hypothesis of Theorem 7.9.9, $p(k) \to \infty$ as $k \to \infty$. Further, suppose that $u(k)$ is a nonoscillatory solution of (7.9.26). Then, $u(k) \to 0$ as $k \to \infty$.

**Proof.** If $u(k)$ is eventually positive, then from Theorem 7.9.9, $z(k)$ is eventually positive, nondecreasing and bounded above. Thus, there exists a constant $L > 0$ such that

$$p(k)u(k-\tau) \;<\; z(k) \;<\; L$$

for sufficiently large $k$. Thus,

$$u(k-\tau) \;<\; \frac{L}{p(k)} \;\to\; 0 \quad \text{as } k \to \infty.$$

The proof is similar when $u(k)$ is eventually negative. ∎

**Example 7.9.7.** The difference equation

$$\text{(7.9.29)} \quad \Delta(u(k)+u(k-2)) \;=\; \frac{(e^4-1)e^{-2k/3}}{e(e+1)(e^k-e)^{1/3}}u^{1/3}(k-1), \quad k \in \mathbb{N}(1)$$

satisfies all the hypothesis of Theorem 7.9.9 and has a bounded nonoscillatory solution $u(k) = 1 - e^{-k}$. However, Corollary 7.9.10 does not apply to (7.9.29) since $p(k) \equiv 1$.

The difference equation

$$(7.9.30)\quad \Delta(u(k)+u(k-2)) = (e-1)(1+e^{-2})e^{-2k+3}u^3(k-1), \quad k \in \mathbb{N}$$

has an unbounded nonoscillatory solution $u(k) = e^k$. Clearly, (7.9.30) satisfies all the conditions of Theorem 7.9.9 except (7.9.27).

**Theorem 7.9.11.** In addition to (7.9.2) and (7.9.4) assume that $p(k) \le p$ for some constant $p$. Further, assume that $u(k)$ is a bounded nonoscillatory solution of (7.9.26). Then, $\liminf_{k\to\infty} |u(k)| = 0$.

**Proof.** Let $u(k)$ be a bounded nonoscillatory solution of (7.9.26) with $u(k) \ge c > 0$ for $k \in \mathbb{N}(k_0)$. Then, $u(k-\tau-\sigma) \ge c$ for all $k \ge k_0+\tau+\sigma = k_1$. Now (7.9.2) implies that $f(u(k-\sigma)) \ge M$, $k \in \mathbb{N}(k_1)$ for some $M > 0$. Hence, $\Delta z(k) = q(k)f(u(k-\sigma)) \ge Mq(k)$, $k \in \mathbb{N}(k_1)$. Thus,

$$z(k) = z(k_1) + M\sum_{\ell=k_1}^{k-1} q(\ell) \to \infty \quad \text{as } k \to \infty.$$

However, since $u(k)$ is bounded, say $0 < u(k) \le B$, we have $z(k) \le B + pB$. This contradiction completes the proof. ∎

**Corollary 7.9.12.** Suppose that hypothesis of Theorem 7.9.11 hold. Then, every bounded solution of (7.9.26) is either oscillatory or tends to zero as $k \to \infty$.

**Corollary 7.9.13.** If (7.9.2), (7.9.4) hold, and $p(k) \le 0$ then every nonoscillatory solution $u(k)$ satisfies $\liminf_{k\to\infty} |u(k)| = 0$, or $|u(k)| \to \infty$ as $k \to \infty$.

**Example 7.9.8.** The difference equation

$$(7.9.31)\quad \Delta(u(k)-2u(k-1)) = \frac{(k-1)^{\alpha-1}(k+3)}{k(k+1)}|u(k-1)|^\alpha \operatorname{sgn} u(k-1), \quad k \in \mathbb{N}(2)$$

where $\alpha \ge 1$ satisfies all the assumptions of Theorem 7.9.11 and it has a bounded nonoscillatory solution $u(k) = 1/k$ for which $\liminf\limits_{k\to\infty} u(k) = 0$.

The difference equation

$$(7.9.32)\quad \Delta(u(k)-2u(k-1)) = \frac{(k+3)(k-1)^{\alpha-1}}{k^{\alpha+1}(k+1)}|u(k-1)|^\alpha \operatorname{sgn} u(k-1), \quad k \in \mathbb{N}(2)$$

where $\alpha > 0$ has a bounded nonoscillatory solution $u(k) = 1 + 1/k$ for which $\liminf_{k\to\infty} u(k) = 1$. Clearly, the equation (7.9.32) satisfies all the assumptions of Theorem 7.9.11 except condition (7.9.4).

Corollary 7.9.12 implies that every bounded solution of the difference equation

$$\text{(7.9.33)}\quad \Delta(u(k)+u(k-2)) = (k-1)^3\left(\frac{2k+1}{k(k+1)} + \frac{2k-3}{(k-1)(k-2)}\right)u^3(k-1),$$

$$k \in \mathbb{N}(3)$$

is either oscillatory or tends to zero as $k \to \infty$. One such solution of (7.9.33) is $u(k) = (-1)^k/k$.

Corollary 7.9.13 implies that every nonoscillatory solution $u(k)$ of the equation

$$\text{(7.9.34)}\quad \Delta\left(u(k) - \frac{1}{k-2}u(k-2)\right) = \frac{1}{(k-\sigma)^\alpha}|u(k-\sigma)|^\alpha \mathrm{sgn}\ u(k-\sigma),$$

$$k \geq \sigma + 1$$

where $0 < \alpha \leq 1$ and $\sigma \in \mathbb{N}(1)$ satisfies either $\liminf_{k\to\infty} |u(k)| = 0$, or $|u(k)| \to \infty$ as $k \to \infty$. One such solution of (7.9.34) is $u(k) = k$.

**Theorem 7.9.14.** In addition to (7.9.4), (7.9.8) and (7.9.18) assume that $\sigma \geq \tau$ and $f$ is nondecreasing. Then, every nonoscillatory solution $u(k)$ of (7.9.26) satisfies $|u(k)| \to \infty$ as $k \to \infty$.

**Proof.** Let $u(k)$ be a nonoscillatory solution of (7.9.26). If $u(k)$ is eventually positive, then there exists $k_0 \in \mathbb{N}$ such that $u(k-\tau-\sigma) > 0$ for $k \in \mathbb{N}(k_0)$. Again, let $z(k) = u(k) + p(k)u(k-\tau)$ then since $\Delta z(k) \geq 0$ for $k \in \mathbb{N}(k_0)$, $z(k)$ is nondecreasing for $k \in \mathbb{N}(k_0)$. Therefore, $z(k) \to L > -\infty$ as $k \to \infty$. If $L \leq 0$ then $z(k) < 0$ for $k \geq k_0$, and hence

$$0 > z(k) = u(k) + p(k)u(k-\tau) > P_4 u(k-\tau), \quad k \geq k_0.$$

Thus, it follows that $u(k) > z(k+\tau)/P_4$ for $k \geq k_0$. Now since $\sigma \geq \tau$ and $f$ is nondecreasing, we find

$$\Delta z(k) \geq q(k) f\left(\frac{z(k+\tau-\sigma)}{P_4}\right) \geq q(k) f\left(\frac{z(k)}{P_4}\right)$$

so that

$$\frac{\Delta z(k)}{f\left(\frac{z(k)}{P_4}\right)} \geq q(k) \quad \text{for} \quad k \geq k_0.$$

Since for $z(k+1)/P_4 \le t \le z(k)/P_4$ we have $f(z(k)/P_4) \ge f(t)$, it follows that

$$-P_4 \int_{z(k+1)/P_4}^{z(k)/P_4} \frac{dt}{f(t)} \ \ge\ \frac{\Delta z(k)}{f\left(\frac{z(k)}{P_4}\right)} \ \ge\ q(k) \quad \text{for} \quad k \ge k_0.$$

Summing the last inequality from $k_0$ to $k-1$ and taking the limit as $k \to \infty$, we obtain

$$-P_4 \int_{L/P_4}^{z(k_0)/P_4} \frac{dt}{f(t)} \ \ge\ \sum_{\ell=k_0}^{\infty} q(\ell),$$

which in view of (7.9.4) and (7.9.18) is impossible. Thus, $L > 0$. Now let $k_1 \ge k_0$ be such that $0 < z(k) \le u(k) + P_3 u(k-\tau)$ for $k \ge k_1$. Then, $u(k) \ge -P_3 u(k-\tau)$ and by induction, we have $u(k+j\tau) \ge (-P_3)^j u(k)$ for each positive integer $j$. But, this implies that $u(k) \to \infty$ as $k \to \infty$. The proof is similar when $u(k)$ is eventually negative. ■

**Corollary 7.9.15.** If the conditions of Theorem 7.9.14 hold, then every solution $u(k)$ of (7.9.26) is either oscillatory or satisfies $|u(k)| \to \infty$ as $k \to \infty$.

**Example 7.9.9.** The difference equation

$$(7.9.35) \qquad \Delta(u(k) - 2u(k-1)) \ =\ 12(2)^{2k/3} u^{1/3}(k-3), \qquad k \in \mathbb{N}(1)$$

satisfies all the conditions of Corollary 7.9.15. Hence, every solution of (7.9.35) is either oscillatory or tends to $\pm\infty$ as $k \to \infty$. One such solution of (7.9.35) is $u(k) = (-1)^k 2^k$.

## 7.10. Oscillation and Nonoscillation for

$$(7.10.1)_\delta \quad \Delta(u(k) + p(k)u(k+\delta\tau)) - q(k)f(u(\sigma(k))) \ =\ F(k), \quad k \in \mathbb{N}$$

where $\delta = \pm 1$, $\tau \in \mathbb{N}(1)$, $\sigma(k) \in \mathbb{N}$, $k \in \mathbb{N}$ with $\lim_{k\to\infty} \sigma(k) = \infty$, $p(k) \ge 0$ eventually, $q(k) \ge 0$ eventually, unless otherwise stated, and $f : \mathbb{R} \to \mathbb{R}$ is continuous with $uf(u) > 0$ for $u \ne 0$.

**Theorem 7.10.1.** Assume that $q(k) \ge 0$ eventually, and

(i) there exists an oscillatory function $h(k)$ such that $\Delta h(k) = F(k)$

(ii) $\limsup_{k\to\infty} h(k) = \infty$ and $\liminf_{k\to\infty} h(k) = -\infty$.

Then, every bounded solution of $(7.10.1)_\delta$ is oscillatory.

**Proof.** The proof is similar to that of Theorem 7.6.1. ■

**Theorem 7.10.2.** Assume that $q(k) \le 0$ eventually, and the assumptions (i) and (ii) hold. Then, the equation $(7.10.1)_\delta$ is oscillatory.

**Proof.** The proof is similar to that of Theorem 7.6.2. ∎

Now we shall consider the case when $(7.10.1)_\delta$ is superlinear, i.e. when $f$ satisfies the condition

$$(7.10.2)\qquad \begin{array}{l} f(u) \text{ is nondecreasing for } u \neq 0, \ \displaystyle\sum_{\ell=k_0}^{\infty} \frac{\Delta u(\ell)}{f(u(\ell+1))} < \infty \\ \text{and } \displaystyle\sum_{\ell=k_0}^{\infty} \frac{\Delta u(\ell)}{f(-u(\ell+1))} > -\infty \quad \text{where} \quad k_0 \in \mathbb{N}(1). \end{array}$$

For convenience, we set

$$A_\alpha = \{k \in \mathbb{N} : \sigma(k) > k + \alpha + 1\}, \qquad \alpha \text{ is a positive integer.}$$

**Theorem 7.10.3.** In addition to the assumptions (i), (ii) and (7.10.2) assume that

(iii) the function $h(k)$ is periodic with period $\tau$, i.e. $h(k \pm \tau) = h(k)$

(iv) $\sum_{k \in A_\alpha} q(k) = \infty$.

Then, (1) the equation $(7.10.1)_{-1}$ is oscillatory provided $0 \le p(k) < p < 1$ and $\alpha = 0$

(2) the equation $(7.10.1)_1$ is oscillatory provided $1 < \mu \le p(k) < \nu$ and $\alpha = \tau$.

**Proof.** Let $u(k)$ be a nonoscillatory solution of $(7.10.1)_\delta$, which we assume to be eventually positive. Then, there exists an integer $k_0 \in \mathbb{N}$ such that $u(k) > 0$, $u(k + \delta\tau) > 0$ and $u(\sigma(k)) > 0$ for $k \in \mathbb{N}(k_0)$. We define $z(k) = u(k) + p(k)u(k + \delta\tau)$ and $w(k) = z(k) - h(k)$. Then, $\Delta w(k) = q(k)f(u(\sigma(k))) \ge 0$, $k \in \mathbb{N}(k_0)$. We claim that $w(k) > 0$ for $k \ge k_0$. Otherwise, $w(k) \le 0$ for $k \ge k_1$ for some $k_1 \ge k_0$, and hence $z(k) - h(k) \le 0$, i.e. $0 < z(k) \le h(k)$, $k \in \mathbb{N}(k_1)$ which contradicts the fact that $h(k)$ is oscillatory. Now we consider the following two cases:

Case 1. $\delta = -1$ and $0 \le p(k) < p < 1$. Since

$$\begin{aligned} u(k) &= w(k) + h(k) - p(k)u(k - \tau) \\ &= w(k) + h(k) - p(k)(w(k - \tau) + h(k - \tau) - p(k - \tau)u(k - 2\tau)) \end{aligned}$$

in view of (iii) and $\Delta w(k) \ge 0$, $k \in \mathbb{N}(k_0)$ we can choose $k_2$ sufficiently large so that

$$(7.10.3)\qquad u(k) \ \ge \ (1 - p)(w(k) + h(k)) \quad \text{for} \quad k \in \mathbb{N}(k_2).$$

Now we choose an integer $k_3 \geq k_2$ such that

$$(7.10.4) \qquad u(k) \ \geq \ (1-p)(w(k)+h(k_3)) \ = \ \xi(k), \qquad k \in \mathbb{N}(k_3).$$

It is clear that $\Delta w(k) = \Delta\xi(k)/(1-p)$, and $\xi(k) \geq (1-p)(w(k_3)+h(k_3)) > 0$ for $k \geq k_3$.

Case 2. $\delta = 1$ and $1 < \mu \leq p(k) < \nu$. Once again from the definitions of $z(k)$ and $w(k)$, we have

$$\begin{aligned} u(k) &= \frac{1}{p(k-\tau)}(w(k-\tau)+h(k-\tau)-u(k-\tau)) \\ &= \frac{1}{p(k-\tau)}\left(w(k-\tau)+h(k-\tau)-\frac{1}{p(k-2\tau)}(w(k-2\tau)\right. \\ &\qquad \left. +h(k-2\tau)-u(k-2\tau))\right). \end{aligned}$$

Thus, in view of (iii) and $\Delta w(k) \geq 0$, $k \in \mathbb{N}(k_0)$ we can choose $k_4 \geq k_0$ sufficiently large so that

$$(7.10.5) \qquad u(k) \ \geq \ \frac{\mu-1}{\mu\nu}(w(k-\tau)+h(k-\tau)), \quad k \in \mathbb{N}(k_4).$$

Now select an integer $k_5 \geq k_4 + \tau$ such that

$$(7.10.6) \quad u(k) \ \geq \ \frac{\mu-1}{\mu\nu}(w(k-\tau)+h(k_5-\tau)) \ = \ \zeta(k-\tau), \quad k \geq k_5.$$

It is clear that

$$\Delta w(k) \ = \ \frac{\mu\nu}{\mu-1}\Delta\zeta(k), \quad \text{and} \quad \zeta(k) > 0 \quad \text{for} \quad k \geq k_5.$$

Let $k_6 = \max\{k_3, k_5\}$. Then, from $\Delta w(k) = q(k)f(u(\sigma(k))) \geq 0$, $k \geq k_0$, (7.10.4) and (7.10.6), we get

$$(7.10.7) \qquad \Delta\theta(k) \ \geq \ \gamma q(k)f(\theta(\sigma(k)-\alpha)), \quad k \in \mathbb{N}(k_6)$$

where

$$(7.10.8) \qquad \gamma = \begin{cases} 1-p, & \alpha = 0 \quad \text{if} \quad \theta(k) = \xi(k) \\ \dfrac{\mu-1}{\mu\nu}, & \alpha = \tau \quad \text{if} \quad \theta(k) = \zeta(k). \end{cases}$$

Now dividing (7.10.7) by $f(\theta(k+1))$ and then summing over $A_\alpha \cap \mathbb{N}(k_6, k) = D$, and using the fact that $\theta(k+1) \geq \theta(\sigma(k)-\alpha)$ on the set $D$, we obtain

$$\sum_{\ell=k_6}^{k} \frac{\Delta\theta(\ell)}{f(\theta(\ell+1))} \ \geq \ \gamma\sum_{\ell\in D} q(\ell)$$

and hence

$$\gamma \sum_{\ell \in D} q(\ell) \ < \ \sum_{\ell=k_6}^{\infty} \frac{\Delta\theta(\ell)}{f(\theta(\ell+1))} \ < \ \infty$$

which contradicts $\sum_{k \in A_a} q(k) \ = \ \infty$. This completes the proof of the theorem. ∎

In our next result we shall assume that the function $f$ satisfies the condition

(7.10.9) $\quad \frac{f(u)}{u} \ \geq \ M \quad$ for $\ u \neq 0\ $ where $M$ is a positive constant.

**Theorem 7.10.4.** Assume that there exists an integer $\eta \geq 2$ such that $\sigma(k) - \alpha \geq k + \eta$, $k \in \mathbb{N}$ ($\alpha$ is defined below). Further, in addition to (i), (iii) and (7.10.9) suppose that

(7.10.10) $$\liminf_{k \to \infty} \sum_{\ell=k-\eta+1}^{k-1} q(\ell) \ > \ \gamma^* \left(\frac{\eta}{\eta+1}\right)^{\eta+1}.$$

Then, (1) the equation $(7.10.1)_{-1}$ is oscillatory provided $0 \leq p(k) < p < 1$, $\gamma^* = 1/M(1-p)$ and $\alpha = 0$

(2) the equation $(7.10.1)_1$ is oscillatory provided $1 < \mu \leq p(k) < \nu$, $\gamma^* = \mu\nu/M(\mu - 1)$ and $\alpha = \tau$.

**Proof.** Suppose for the sake of contradiction $(7.10.1)_\delta$ has a nonoscillatory solution $u(k)$ which is eventually positive. Then, as in Theorem 7.10.3 in view of (7.10.9) it follows that

(7.10.11) $$\Delta w(k) \ \geq \ Mq(k)u(\sigma(k)), \qquad k \in \mathbb{N}(k_0).$$

Now we consider the two cases 1. $\delta = -1$ and $0 \leq p(k) < p < 1$, and 2. $\delta = 1$ and $1 < \mu \leq p(k) < \nu$, and proceed as in Theorem 7.10.3 to obtain (7.10.4) and (7.10.6), respectively. Next, we use (7.10.4) and (7.10.6) in (7.10.11), to get

(7.10.12) $$\Delta\omega(k) \ \geq \ \beta q(k)\omega(k+\eta-\alpha), \quad k \geq K$$

where $K$ is sufficiently large, and

$$\beta \ = \ \begin{cases} M(1-p), & \alpha = 0 \quad \text{if} \quad \omega(k) = \xi(k) \\ M\dfrac{\mu-1}{\mu\nu}, & \alpha = \tau \quad \text{if} \quad \omega(k) = \zeta(k). \end{cases}$$

But, now (7.10.10) implies that the inequality (7.10.12) has no eventually positive solution, which is a contradiction. ∎

**Example 7.10.1.** Consider the difference equation

$$\Delta(u(k)+u(k-2))-(2k-3)|u(k+4)|^{\lambda}\mathrm{sgn}\ u(k+4) = (-1)^{k+1}(2k+1), \quad k\in\mathbb{N}, \quad \lambda\geq 1. \tag{7.10.13}$$

Here $F(k)=(-1)^{k+1}(2k+1)$ and $h(k)=(-1)^k k$. All the hypotheses of Theorem 7.10.1 are satisfied and hence every bounded solution of (7.10.13) is oscillatory. When $\lambda=1$ one such solution is $u(k)=(-1)^k$.

**Example 7.10.2.** Consider the difference equation

$$\Delta(u(k)+pu(k+2\delta))-3\left(\frac{2k+3}{k+7}\right)|u(\sigma(k))|^{\lambda}\mathrm{sgn}\ u(\sigma(k)) = 2(-1)^{k+1}, \quad k\in\mathbb{N}, \quad \lambda\geq 1 \tag{7.10.14$_\delta$}$$

where $\delta=\pm 1$, $p$ is a nonnegative real number, and $\sigma(k)\in\mathbb{N}$, $k\in\mathbb{N}$ with $\lim_{k\to\infty}\sigma(k)=\infty$. We take $h(k)=(-1)^k$ with period 2. First, let $\lambda>1$ and suppose that $\sigma(k)$ has the forms $k+a$, $a$ is a positive integer; $k+(-1)^k$ or $k^a$, $a>1$, $a$ is a real number. We apply Theorem 7.10.3 to conclude that

(i) $(7.10.14)_{-1}$ is oscillatory when $0\leq p<1$

(ii) $(7.10.14)_1$ is oscillatory when $p>1$.

Now suppose that $\lambda=1$ and that $\sigma(k)-\alpha\geq k+\eta$, where $\eta\geq 2$, $\alpha=0$ when $\delta=-1$ and $\alpha=2$ when $\delta=1$. We use Theorem 7.10.4 to conclude that

(i) $(7.10.14)_{-1}$ is oscillatory if $0\leq p<1$, $\alpha=0$ and

$$\liminf_{k\to\infty}\sum_{\ell=k-\eta+1}^{k-1}3\left(\frac{2\ell+3}{\ell+7}\right) > \frac{1}{1-p}\left(\frac{\eta}{\eta+1}\right)^{\eta+1}$$

(i) $(7.10.14)_1$ is oscillatory if $p>1$, $\alpha=2$ and

$$\liminf_{k\to\infty}\sum_{\ell=k-\eta+1}^{k-1}3\left(\frac{2\ell+3}{\ell+7}\right) > \frac{p^2}{p-1}\left(\frac{\eta}{\eta+1}\right)^{\eta+1}.$$

It is easy to verify that for $p=2$, $\delta=1$, $\lambda=1$ and $\sigma(k)=k+7$, equation $(7.10.14)_1$ has an oscillatory solution $u(k)=(-1)^k k$.

## 7.11. Oscillation and Asymptotic Behavior for

$$\Delta^2(u(k)+p(k)u(k-\tau))+q(k)f(u(k+1-\sigma)) = 0, \quad k\in\mathbb{N} \tag{7.11.1}$$

where $p(k)$ and $q(k)$ are such that $q(k) \geq 0$ for all $k \in \mathbb{N}$, and are not zero for infinitely many values of $k$, $\tau \in \mathbb{N}(1)$, $\sigma \in \mathbb{N}$, and $f : \mathbb{R} \to \mathbb{R}$ is continuous with $uf(u) > 0$ for $u \neq 0$.

We shall need the following:

**Lemma 7.11.1.** Suppose that there exists a function $b(k)$, $k \in \mathbb{N}$ such that

$$b(k) = O(k) \quad \text{as} \quad k \to \infty \quad \text{and} \quad b(k) \leq p(k) \leq 0 \tag{7.11.2}$$

and let $u(k)$ be a nonoscillatory solution of (7.11.1).

(a) If (7.9.2), (7.9.4), (7.11.2) hold and $u(k)$ is eventually positive, then $z(k) = u(k) + p(k)u(k-\tau)$ and $\Delta z(k)$ are either both nonincreasing with

$$\lim_{k\to\infty} z(k) = \lim_{k\to\infty} \Delta z(k) = -\infty \tag{7.11.3}$$

or $\Delta z(k)$ is nonincreasing with

$$\lim_{k\to\infty} \Delta z(k) = 0, \quad \Delta z(k) > 0 \quad \text{and} \quad z(k) < 0 \quad \text{for all large } k. \tag{7.11.4}$$

(b) If (7.9.2), (7.9.4), (7.11.2) hold and $u(k)$ is eventually negative, then $z(k)$ and $\Delta z(k)$ are either both nondecreasing with

$$\lim_{k\to\infty} z(k) = \lim_{k\to\infty} \Delta z(k) = \infty \tag{7.11.5}$$

or $\Delta z(k)$ is nondecreasing with

$$\lim_{k\to\infty} \Delta z(k) = 0, \quad \Delta z(k) < 0 \quad \text{and} \quad z(k) > 0 \quad \text{for all large } k. \tag{7.11.6}$$

(c) If in addition to (7.9.2) and (7.9.4) there exists a $P_1 < 0$ such that

$$P_1 \leq p(k) \leq 0 \tag{7.11.7}$$

and $u(k)$ is eventually positive, then either (7.11.3) holds or $\Delta z(k)$ is nonincreasing with

$$\lim_{k\to\infty} z(k) = \lim_{k\to\infty} \Delta z(k) = 0, \quad \Delta z(k) > 0 \quad \text{and} \quad z(k) < 0 \quad \text{for all large } k. \tag{7.11.8}$$

(d) If (7.9.2), (7.9.4) and (7.11.7) hold and $u(k)$ is eventually negative, then either (7.11.5) holds or $\Delta z(k)$ is nondecreasing with

$$\lim_{k\to\infty} z(k) = \lim_{k\to\infty} \Delta z(k) = 0, \quad \Delta z(k) < 0 \quad \text{and} \quad z(k) > 0 \quad \text{for all large } k. \tag{7.11.9}$$

(c) If (7.9.2), (7.9.4), (7.11.7) hold and $P_1 \ge -1$, then (7.11.8) holds when $u(k)$ is eventually positive and (7.11.9) holds when $u(k)$ is eventually negative.

**Proof.** Let $u(k)$ be an eventually positive solution of (7.11.1). Then, there exists an integer $k_1 \ge k_0$ such that $u(k-\tau-\sigma) > 0$ for $k \in \mathbb{N}(k_1)$.

(a) From (7.11.1), we have $\Delta^2 z(k) = -q(k)f(u(k+1-\sigma)) \le 0$ for $k \in \mathbb{N}(k_1)$, which implies that $\Delta z(k)$ is nonincreasing for all $k \ge k_1$, so $z(k)$ is monotonic.

If there exists $k_2 \ge k_1$ such that $\Delta z(k_2) \le 0$, then since $q(k)$ is not identically zero for large $k$, there exists $k_3 \ge k_2$ such that $\Delta z(k) \le \Delta z(k_3) < 0$ for $k \ge k_3$, and a summation shows that $z(k) \to -\infty$ as $k \to \infty$. Moreover, $\Delta z(k) \to L$ where $0 > L \ge -\infty$. If $L = -\infty$, clearly (7.11.3) holds. If $L > -\infty$, summing (7.11.1) from $k_3$ to $k-1$ gives

$$\Delta z(k) = \Delta z(k_3) - \sum_{\ell=k_3}^{k-1} q(\ell)f(u(\ell+1-\sigma)).$$

Letting $k \to \infty$, we obtain

$$\sum_{\ell=k_3}^{\infty} q(\ell)f(u(\ell+1-\sigma)) = \Delta z(k_3) - L < \infty.$$

The last inequality together with (7.9.2) and (7.9.4) implies

$$\liminf_{k\to\infty} u(k) = 0. \tag{7.11.10}$$

Since $L < 0$ a summation shows that $z(k)$ is eventually negative, so we choose $k_4 \ge k_3$ such that $\Delta z(k) < L/2$ for $k > k_4$ and $z(k_4) < 0$. This implies that

$$z(k) \le z(k_4) + (k-k_4)L/2 < L(k-k_4)/2 \quad \text{for} \quad k \ge k_4$$

and hence $z(k) < Lk/4$ for $k > 2k_4$. From (7.11.2), we have

$$b(k)u(k-\tau) \le p(k)u(k-\tau) < z(k) < Lk/4.$$

Thus, $u(k-\tau) > Lk/(4b(k))$, which in view of (7.11.2) implies that $u(k)$ is bounded from below. But, this contradicts (7.11.10). Hence, we conclude that $\Delta z(k) > 0$ for $k \ge k_1$, and therefore $\Delta z(k) \to L \ge 0$ as $k \to \infty$ since $\Delta z(k)$ is nonincreasing for $k \ge k_1$. Summing (7.11.1), we obtain

$$\Delta z(k) = L + \sum_{\ell=k}^{\infty} q(\ell)f(u(\ell+1-\sigma)),$$

which again implies that $\liminf_{k\to\infty} u(k) = 0$. But, this is impossible if $L > 0$ since then $\Delta z(k) > L$ and we would have $u(k) \geq z(k) \to \infty$ as $k \to \infty$. Hence, we conclude that $L = 0$. Furthermore, if there exists a $k_5 \geq k_1$ such that $z(k_5) \geq 0$, then $\Delta z(k) > 0$ implies that $z(k) \geq z(k_6) > 0$ for $k \geq k_6 \geq k_5$, which again contradicts (7.11.10). Therefore, $z(k) < 0$ for $k \geq k_1$, which completes the proof of (a).

(b) The proof is similar to that of (a).

(c) First we note that (7.11.7) implies (7.11.2). Thus, from (a) either (7.11.3) or (7.11.4) holds. If (7.11.3) holds, we are done so suppose that (7.11.4) holds. Then, $\Delta z(k) > 0$ and $z(k) < 0$ and from the proof of (a) we obtain (7.11.10). Thus, $z(k) \to L_1 \leq 0$. If $L_1 < 0$, then $u(k) + P_1 u(k-\tau) \leq u(k) + p(k)u(k-\tau) = z(k) < L_1$ for $k \geq k_1$. But (7.11.10) implies that there exists an increasing sequence of integers $\{k_j\}$ and $u(k_j - \tau) \to 0$ as $j \to \infty$, which contradicts $u(k) > 0$. Thus, we conclude that $z(k) \to 0$ as $k \to \infty$ and the proof of part (c) is complete.

(d) The proof is similar to that of (c).

(e) Suppose that $u(k)$ is eventually positive and that (7.11.8) does not hold. Then, from part (c), (7.11.3) holds so $z(k) < 0$ for all large $k$. Since $P_1 \geq -1$,

$$u(k) \ < \ -p(k)u(k-\tau) \ \leq \ -P_1 u(k-\tau) \ \leq \ u(k-\tau)$$

for all large $k$. But the last inequality implies that $u(k)$ is bounded which contradicts (7.11.3). Therefore, (7.11.8) holds when $u(k)$ is eventually positive. The argument for the case when $u(k)$ is eventually negative is similar. ∎

**Theorem 7.11.2.** If (7.9.2), (7.9.4) and (7.9.5) hold, then every nonoscillatory solution $u(k)$ of (7.11.1) tends to zero as $k \to \infty$.

**Proof.** If $u(k)$ is eventually positive, then by part (e) of Lemma 7.11.1, (7.11.8) holds. Thus, $z(k) < 0$ for all sufficiently large $k$. Then, we have $u(k) < -p(k)u(k-\tau) \leq -P_2 u(k-\tau)$, and hence $u(k+\tau) \leq -P_2 u(k)$. By induction it follows that for sufficiently large $k$, $u(k+j\tau) \leq (-P_2)^j u(k)$ for every positive integer $j$. But, this implies that $u(k) \to 0$ as $k \to \infty$. The proof for the case $u(k)$ eventually negative is similar. ∎

**Example 7.11.1.** The difference equation

$$(7.11.11)\quad \Delta^2\left(u(k) - e^{-(\tau-1)}u(k-\tau)\right) + (e-1)^3 e^{2k-3\sigma+1} u^3(k+1-\sigma) = 0, \quad k \geq \tau + \sigma$$

satisfies all conditions of Theorem 7.11.2 for $\tau \geq 2$, $\sigma \geq 2$. In fact, $u(k) = e^{-k}$ is a nonoscillatory solution of (7.11.11) which tends to zero as $k \to \infty$.

**Theorem 7.11.3.** Suppose that $p(k) \geq 0$ and let $u(k)$ be a nonoscillatory solution of (7.11.1). Then,

(i) $|u(k)| \leq c_1 k$ for some constant $c_1 > 0$ and all $k \geq k_0 \in \mathbb{N}$

(ii) if $k/p(k)$ is bounded, then $u(k)$ is bounded

(iii) if $k/p(k) \to 0$ as $k \to \infty$, then $u(k) \to 0$ as $k \to \infty$.

**Proof.** Let $u(k)$ be an eventually positive solution of (7.11.1) and let $K \geq k_0$ be such that $u(k-\tau) > 0$ and $u(k-\sigma) > 0$ for $k \geq K$. From (7.11.1), we have $\Delta^2 z(k) \leq 0$ for $k \geq K$. Thus, $z(k) \leq z(K) + \Delta z(K)(k-K) \leq c_1 k$ for some constant $c_1 > 0$ and all sufficiently large $k$. Clearly, $u(k) \leq c_1 k$ if $p(k) \geq 0$, and hence (i) follows. Furthermore, since $p(k)u(k-\sigma) \leq c_1 k$, (ii) and (iii) are immediate. The proof when $u(k)$ is eventually negative is similar. ∎

**Example 7.11.2.** For the difference equations

$$\Delta^2\left(u(k) + \frac{1}{k}u(k-1)\right) + \frac{2}{(k-2)^3 k(k+1)(k+2)} u^3(k-2) = 0, \quad k \geq 3$$

$$\Delta^2(u(k) + ku(k-1)) + \frac{2}{(k+1)(k+2)(k+3)(k-3)^3} u^3(k-3) = 0, \quad k \geq 4$$

and

$$\Delta^2\left(u(k) + e^k(\ln k)u(k-1)\right) + \left(e^{3k-5}\ln\frac{(k+1)^2}{k(k+2)} - e^{2k-8}(e-1)^2\right) u^3(k-2) = 0, \quad k \geq 3$$

conclusions (i), (ii) and (iii) of Theorem 7.11.3 respectively follow. We also note that for these equations respectively $k$, $k/(k+1)$ and $e^{-k}$ are nonoscillatory solutions.

**Theorem 7.11.4.** If (7.9.2), (7.9.4) and (7.9.8) hold, then every bounded solution of (7.11.1) is either oscillatory or tends to zero as $k \to \infty$.

**Proof.** Assume that (7.11.1) has a bounded nonoscillatory solution $u(k)$ which is eventually positive, say for $k \geq k_1 \geq k_0$. By Lemma 7.11.1(c) either (7.11.3) or (7.11.8) holds. Condition (7.9.8) and the fact that $u(k)$ is bounded implies that (7.11.3) cannot hold. From (7.11.8) we have $z(k) < 0$ and $z(k) \to 0$ as $k \to \infty$. Hence, for any $\epsilon > 0$ there exists $K \geq k_1$

such that $-\epsilon < z(k) < u(k)+P_3u(k-\tau)$ or $u(k-\tau) < (-1/P_3)(u(k)+\epsilon)$ for $k \geq K$. Thus, we have

$$u(k) \;<\; -\frac{1}{P_3}u(k+\tau) - \frac{1}{P_3}\epsilon, \quad k \geq K$$

and hence

$$u(k) \;<\; \left(-\frac{1}{P_3}\right)^2 u(k+2\tau) + \left(-\frac{1}{P_3}\right)^2 \epsilon + \left(-\frac{1}{P_3}\right)\epsilon.$$

Continuing this, for all positive integers $m$, we obtain

$$u(k) \;<\; \left(-\frac{1}{P_3}\right)^m u(k+m\tau) + \sum_{i=1}^{m}\left(-\frac{1}{P_3}\right)^i \epsilon. \tag{7.11.12}$$

Let $\mu = 1 + 1/P_3 > 0$. Since $u(k)$ is bounded there exists a constant $L > 0$ such that $u(k) < L$ for $k \geq k_0$. Now choose $m_1$ large enough so that $(-1/P_3)^m < \epsilon/(\mu M)$ for $m > m_1$. Then, from (7.11.12), we have

$$u(k) \;<\; \frac{\epsilon}{\mu} + \epsilon\left[1-(-1/P_3)^m\right]/[1+1/P_3] \;<\; 2\epsilon/\mu.$$

Since $\epsilon$ is arbitrary, the above inequality implies that $u(k) \to 0$ as $k \to \infty$. The proof when $u(k)$ is eventually negative is similar. ∎

**Example 7.11.3.** The difference equation

$$\Delta^2(u(k) - 2u(k-2)) + \frac{7}{32}2^{2k}u^3(k-1) \;=\; 0, \quad k \geq 1$$

satisfies all conditions of Theorem 7.11.4 and has the solution $u(k) = 2^{-k}$ which tends to zero as $k \to \infty$.

**Theorem 7.11.5.** If (7.9.2), (7.9.4) and (7.11.7) with $P_1 \geq -1$ hold, then every unbounded solution of (7.11.1) is oscillatory.

**Proof.** Part (e) of Lemma 7.11.1 implies that all nonoscillatory solutions of (7.11.1) are bounded and hence the result follows. ∎

**Example 7.11.4.** The difference equation

$$\Delta^2\left(u(k) - \frac{1}{2}u(k-2)\right) + \frac{63}{8}2^{(2k+1)/3}u^{1/3}(k-1) \;=\; 0, \quad k \geq 2$$

satisfies all conditions of Theorem 7.11.5 and has an unbounded oscillatory solution $u(k) = (-1)^k 2^k$.

**Theorem 7.11.6.** Suppose in addition to (7.11.7) with $P_1 \geq -1$, the function $f$ is nondecreasing such that (7.9.19) holds, and

$$\sum_{k=k_0}^{\infty}\left(\sum_{\ell=k}^{\infty} q(\ell)\right) = \infty, \quad k_0 \in \mathbb{N}. \tag{7.11.13}$$

Then, all unbounded solutions of (7.11.1) are oscillatory.

**Proof.** Suppose (7.11.1) has an unbounded solution $u(k)$ such that $u(k-\tau-\sigma) > 0$ for $k \geq k_1 \geq k_0$. Then, for $z(k) = u(k) + p(k)u(k-\tau)$ it is clear that $\Delta^2 z(k) \leq 0$. Hence, $\Delta z(k)$ is nonincreasing and $z(k)$ is monotonic. Now, if $z(k)$ is eventually negative, then by (7.11.7) with $P_1 \geq -1$, we have

$$u(k) \leq -p(k)u(k-\tau) \leq u(k-\tau),$$

which contradicts the assumption that $u(k)$ is unbounded. Therefore, $z(k) \geq 0$ eventually. Furthermore, if $\Delta z(k)$ is eventually negative, then clearly $z(k)$ is eventually negative which, as above, is a contradiction. Thus, we have $z(k) > 0$ and $\Delta z(k) > 0$ for $k \geq k_1$. Since $0 < z(k) \leq u(k)$ and $f$ is nondecreasing, we have

$$\Delta^2 z(k) + q(k) f(z(k+1-\sigma)) \leq 0 \quad \text{for} \quad k \geq K = k_1 - 1 + \sigma.$$

For each $k \geq K$, a summation of the above inequality from $k$ to $k_2 - 1 > k$ leads to

$$\Delta z(k_2) - \Delta z(k) + f(z(k+1-\sigma)) \sum_{\ell=k}^{k_2-1} q(\ell) \leq 0$$

so that

$$f(z(k+1-\sigma)) \sum_{\ell=k}^{k_2-1} q(\ell) \leq \Delta z(k) - \Delta z(k_2).$$

Letting $k_2 \to \infty$, we obtain

$$f(z(k+1-\sigma)) \sum_{\ell=k}^{\infty} q(\ell) \leq \Delta z(k) \leq \Delta z(k-\sigma) \quad \text{for} \quad k \geq K.$$

Thus,

$$\sum_{\ell=k}^{\infty} q(\ell) \leq \frac{\Delta z(k-\sigma)}{f(z(k+1-\sigma))}.$$

Now since

$$\frac{z(k+1-\sigma) - z(k-\sigma)}{f(z(k+1-\sigma))} = \int_{z(k-\sigma)}^{z(k+1-\sigma)} \frac{dt}{f(z(k+1-\sigma))} \leq \int_{z(k-\sigma)}^{z(k+1-\sigma)} \frac{dt}{f(t)}$$

a summation from $k$ to $m$ yields

$$\sum_{\ell=k}^{m}\left(\sum_{j=\ell}^{\infty} q(j)\right) \leq \int_{z(k-\sigma)}^{z(m+1-\sigma)} \frac{dt}{f(t)},$$

which in view of (7.11.13) contradicts (7.9.19). ∎

**Example 7.11.5.** The difference equation

$$(7.11.14) \quad \Delta^2\left(u(k) - \frac{1}{2}u(k-2)\right) + \frac{2(k+3)}{(k-1)^3}u^3(k-1) = 0, \quad k \geq 2$$

satisfies all conditions of Theorem 7.11.6 and hence all unbounded solutions are oscillatory. One such solution is $u(k) = (-1)^k k$.

**Theorem 7.11.7.** Suppose that the function $f$ is nondecreasing, $0 \leq p(k) < 1$, $k \in \mathbb{N}$, and

$$(7.11.15) \qquad \sum_{\ell=k_0}^{\infty} q(\ell) f([1 - p(\ell+1-\sigma)]c) = \infty, \quad k_0 \in \mathbb{N}$$

for every constant $c > 0$. Then, the equation (7.11.1) is oscillatory.

**Proof.** As usual assume that (7.11.1) has a nonoscillatory solution $u(k)$, say $u(k) > 0$, $k \geq k_1 \geq k_0 + \tau + \sigma$. Since $z(k) = u(k) + p(k)u(k-\tau) > 0$ for $k \geq k_1$, as in Lemma 7.11.1(a) it follows that $\Delta z(k) > 0$ for $k \geq k_1$. Therefore, $z(k-\tau) \leq z(k) \leq u(k) + p(k)z(k-\tau)$, and hence $[1-p(k)]z(k-\tau) \leq u(k)$. Thus, we have

$$\begin{aligned} 0 &= \Delta^2 z(k) + q(k) f(u(k+1-\sigma)) \\ &\geq \Delta^2 z(k) + q(k) f([1-p(k+1-\sigma)]z(k+1-\tau-\sigma)) \end{aligned}$$

and since $z(k) > 0$ and $\Delta z(k) > 0$, there exists a constant $c > 0$ such that

$$\Delta^2 z(k) + q(k) f([1-p(k+1-\sigma)]c) \leq 0, \quad k \geq k_1.$$

Summing the above inequality, to obtain

$$\Delta z(k) - \Delta z(k_1) + \sum_{\ell=k_1}^{k-1} q(\ell) f([1-p(\ell+1-\sigma)]c) \leq 0,$$

which contradicts (7.11.15). The proof for the case when $u(k)$ is eventually negative is similar. ∎

**Example 7.11.6.** The difference equation

$$\Delta^2\left(u(k)+\frac{1}{k}u(k-1)\right)+\frac{2(2k^3+4k^2-1)}{k(k+1)(k+2)}u^3(k-1) \;=\; 0, \quad k\ge 1$$

satisfies all hypotheses of Theorem 7.11.7 and hence it is oscillatory. One such solution is $u(k)=(-1)^k$.

**Theorem 7.11.8.** If (7.9.2) and (7.9.4) hold and $p(k)$ is not eventually negative, then any solution $u(k)$ of (7.11.1) is either oscillatory or satisfies $\liminf_{k\to\infty}|u(k)|=0$.

**Proof.** Let $u(k)$ be a nonoscillatory solution of (7.11.1), say $u(k-\tau-\sigma)>0$, $k\ge k_1\ge k_0\in\mathbb{N}$. Then, (7.11.1) implies that $\Delta z(k)$ is nonoscillatory and $z(k)$ is monotonic. If there exists $k_2\ge k_1$ such that $\Delta z(k_2)\le 0$, then there exists $k_3>k_2$ such that $\Delta z(k)\le\Delta z(k_3)<0$ for $k\ge k_3$ and so $z(k)\to-\infty$ as $k\to\infty$. But, this contradicts the fact that $p(k)$ is not eventually negative. Hence, $\Delta z(k)>0$ for $k\ge k_1$ and therefore $\Delta z(k)\to L\ge 0$ as $k\to\infty$. Summing (7.11.1) from $k$ to $\infty$, we obtain

$$\Delta z(k) \;=\; L+\sum_{\ell=k}^{\infty}q(\ell)f(u(\ell+1-\sigma)),$$

which implies in view of (7.9.2) and (7.9.4) that $\liminf_{k\to\infty}u(k)=0$. The proof when $u(k-\tau-\sigma)<0$ for $k\ge k_1$ is similar. ∎

**Example 7.11.7.** The difference equation

$$(7.11.16)\quad \Delta^2\left(u(k)+(-1)^{k+1}2^{k-2}u(k-2)\right)+\frac{9}{32}2^{2k}u^3(k-1)=0,\ k\ge 2$$

satisfies all conditions of Theorem 7.11.8. Here $u(k)=(-1)^k/2^k$ is a solution of (7.11.16) satisfying the conclusion of Theorem 7.11.8. For the difference equation (7.11.16), Theorem 7.11.7 does not apply since (7.11.15) does not hold.

## 7.12. Classification of Solutions for

$$(7.12.1)\quad \Delta(a(k)\Delta(u(k)+p(k)u(k-\tau)))+q(k+1)f(u(k+1-\sigma)) \;=\; 0, \quad k\in\mathbb{N}$$

where $\tau\in\mathbb{N}(1)$, $\sigma\in\mathbb{N}$, functions $a(k)$, $p(k)$, $q(k)$ are defined on $\mathbb{N}$, $a(k)>0$ for all $k\in\mathbb{N}$ and $q(k)$ is not identically zero for large $k$, and the continuous function $f:\mathbb{R}\to\mathbb{R}$ is such that $uf(u)>0$ for all $u\ne 0$, and $f(u)-f(v)=g(u,v)(u-v)$ for $u,v\ne 0$, where $g(u,v)$ is a nonnegative function.

A nonoscillatory solution $u(k)$ of (7.12.1) is said to be *weakly oscillatory* if $\Delta u(k)$ changes sign for arbitrarily large values of $k$.

Let $S$ denote the set of all nontrivial solutions of (7.12.1). In view of their asymptotic behavior, all solutions of (7.12.1) may be a priori divided into the following classes:

$M^+ = \{u(k) \in S :$ there exists an integer $K \in \mathbb{N}$ such that $u(k)\Delta u(k) \geq 0$ for all $k \in \mathbb{N}(K)\}$;

$M^- = \{u(k) \in S : u(k)$ is nonoscillatory and there exists an integer $K \in \mathbb{N}$ such that $u(k)\Delta u(k) \leq 0$ for all $k \in \mathbb{N}(K)\}$;

$OS = \{u(k) \in S :$ for every integer $K \in \mathbb{N}$, there exists $k \geq K$ such that $u(k)u(k+1) \leq 0\}$;

$WOS = \{u(k) \in S : u(k)$ is nonoscillatory and for every $K \in \mathbb{N}$, there exists $k \geq K$ such that $\Delta u(k)\Delta u(k+1) \leq 0\}$.

In the following results we shall provide sufficient conditions which ensure the existence and nonexistence of the solutions of (7.12.1) in the above four classes when $q(k)$ is nonnegative, and changes sign for large $k \in \mathbb{N}$. We shall also study the asymptotic behavior of nonoscillatory solutions of (7.12.1).

**Theorem 7.12.1.** With respect to the difference equation (7.12.1) assume that the following hold

(i) $p(k)$ is nonnegative, and nondecreasing for all $k \in \mathbb{N}$

(ii) $\limsup_{k\to\infty} \sum_{\ell=k_0}^{k-1} q(\ell+1) = \infty, \quad k_0 \in \mathbb{N}.$

Then, $M^+ = \emptyset$.

**Proof.** Suppose that the equation (7.12.1) has a solution $u(k) \in M^+$. Since $u(k)\Delta u(k) \geq 0$ for all $k \in \mathbb{N}$ implies that $u(k)$ is nonoscillatory, without loss of generality we can assume that there exists an integer $k_1 \geq k_0 \in \mathbb{N}$ such that $u(k) > 0$, $\Delta u(k) \geq 0$, $u(k-\mu) > 0$, $\Delta u(k-\mu) \geq 0$ for all $k \geq k_1$. In fact, for $u(k) < 0$, $u(k-\mu) < 0$ for all large $k \in \mathbb{N}$ the proof is similar. Set $z(k) = u(k) + p(k)u(k-\tau)$, then in view of (i), $z(k) > 0$ and $\Delta z(k) \geq 0$ for all $k \geq k_1$. From the equation (7.12.1), we have

$$\text{(7.12.2)} \qquad \frac{a(k)\Delta z(k)}{f(u(k-\sigma))} - \frac{a(k_1)\Delta z(k_1)}{f(u(k_1-\sigma))} + \sum_{\ell=k_1}^{k-1} \frac{a(\ell)\Delta z(\ell) g(u(\ell+1-\sigma), u(\ell-\sigma))\Delta u(\ell-\sigma)}{f(u(\ell+1-\sigma))f(u(\ell-\sigma))} = -\sum_{\ell=k_1}^{k-1} q(\ell+1)$$

and hence

$$\frac{a(k)\Delta z(k)}{f(u(k-\sigma))} - \frac{a(k_1)\Delta z(k_1)}{f(u(k_1-\sigma))} \le -\sum_{\ell=k_1}^{k-1} q(\ell+1), \quad k \in \mathbb{N}(k_1).$$

Thus, from (ii) we find

$$\liminf_{k\to\infty} \frac{a(k)\Delta z(k)}{f(u(k-\sigma))} = -\infty,$$

which contradicts the assumption that $\Delta z(k) \ge 0$ for all large $k$. ∎

**Example 7.12.1.** In Theorem 7.12.1 the assumption (ii) cannot be dropped. For this, we consider the difference equation

$$(7.12.3)\quad \Delta\left(k\Delta\left(u(k) + \frac{k-1}{k}u(k-1)\right)\right) + \frac{3(k-2)^\alpha}{(k+1)(k+2)(k-3)^\alpha}(u(k-2))^\alpha = 0, \quad k \in \mathbb{N}(4)$$

where $\alpha > 0$ is the ratio of odd positive integers. For this difference equation assumption (i) holds, but (ii) is violated. The equation (7.12.3) has a solution $u(k) = (k-1)/k \in M^+$.

**Theorem 7.12.2.** With respect to the difference equation (7.12.1) assume that the following hold

(iii) $\tau \ge 1$ and $-1 \le p(k) \le 0$

(iv) $q(k) \ge 0$ for all $k \ge k_0 \in \mathbb{N}$

(v) $\lim_{k\to\infty} \sum_{\ell=k_0}^{k-1} q(\ell+1) = \infty$

(vi) $\sum_{\ell=k_0}^{\infty} \frac{1}{a(\ell)} = \infty.$

Then, $M^+ = \emptyset$.

**Proof.** As in Theorem 7.12.1, let (7.12.1) have a solution $u(k) \in M^+$ such that $u(k) > 0$, $\Delta u(k) \ge 0$, $u(k-\mu) > 0$, $\Delta u(k-\mu) \ge 0$ for all $k \ge k_1 \ge k_0$. Again, set $z(k) = u(k) + p(k)u(k-\tau)$. Then, in view of the assumption (iii) and the fact that $u(k) \in M^+$, we have $z(k) > u(k-\tau) + p(k)u(k-\tau) \ge 0$ for all $k \ge k_1$. Since equation (7.12.1) is the same as $\Delta(a(k)\Delta z(k)) = -q(k+1)f(u(k+1-\sigma))$, $k \in \mathbb{N}(k_1)$, from condition (iv) it follows that $a(k)\Delta z(k)$ is nonincreasing for all $k \in \mathbb{N}(k_1)$. Now suppose that $a(k)\Delta z(k) < 0$ for all $k \in \mathbb{N}(k_1)$, then there exists an integer $k_2 > k_1$ such that $a(k)\Delta z(k) \le a(k_2)\Delta z(k_2) < 0$ for all $k \in \mathbb{N}(k_2)$. Hence, we have

$$z(k) - z(k_2) \le \sum_{\ell=k_2}^{k-1} \frac{a(k_2)\Delta z(k_2)}{a(\ell)},$$

which in view of (vi) implies that $z(k) \to -\infty$ as $k \to \infty$, which is a contradiction. Thus, $a(k)\Delta z(k) \ge 0$. Now following as in Theorem 7.12.1, and using the condition (v), we obtain

$$\lim_{k\to\infty} \frac{a(k)\Delta z(k)}{f(u(k-\sigma))} = -\infty,$$

which is the required contradiction. ∎

**Example 7.12.2.** In Theorem 7.12.2 some of the assumptions cannot be dropped. For this, we consider the difference equation

$$\Delta\left(k^2\Delta(u(k) - 2u(k-1))\right) + \frac{2k+1}{k-2}u(k-2) = 0, \quad k \in \mathbb{N}(3) \tag{7.12.4}$$

for which conditions (iv) and (v) are satisfied, whereas (iii) and (vi) are violated. The equation (7.12.4) has a solution $u(k) = k \in M^+$.

**Theorem 7.12.3.** With respect to the difference equation (7.12.1) assume that either (a) or (b) holds, where

(a) $\tau \le \sigma$, and

(vii) condition (7.9.18) holds

(viii) $f$ is submultiplicative

(ix) $p(k)$ is nonnegative, and nonincreasing for all $k \in \mathbb{N}$

(x) $\limsup_{k\to\infty} \sum_{\ell=K}^{k} \frac{1}{f(1+p(\ell))a(\ell)} \sum_{\xi=K}^{\ell-1} q(\xi+1) = \infty$ for all $K \in \mathbb{N}$.

(b) conditions (vi) and (ix) hold, and

(xi) $q(k) > 0$ for all large $k \in \mathbb{N}$.

Then, $M^- = \emptyset$.

**Proof.** Let (a) hold, and that (7.12.1) has a solution $u(k) \in M^-$. Without loss of generality we can assume that there exists an integer $k_1 \ge k_0 \in \mathbb{N}$ such that $u(k) > 0$, $\Delta u(k) \le 0$, $u(k-\mu) > 0$, $\Delta u(k-\mu) \le 0$ for all $k \in \mathbb{N}(k_1)$. The proof is similar if $u(k) < 0$ and $\Delta u(k) \ge 0$ for all large $k$. Let $z(k) = u(k) + p(k)u(k-\tau)$, then in view of (ix), $z(k) > 0$ and $\Delta z(k) \le 0$ for all $k \in \mathbb{N}(k_1)$. Thus, on rewriting (7.12.2) as

$$\frac{a(k)\Delta z(k)}{f(u(k-\sigma))} = \frac{a(k_1)\Delta z(k_1)}{f(u(k_1-\sigma))}$$
$$- \sum_{\ell=k_1}^{k-1} \frac{a(\ell)\Delta z(\ell) g(u(\ell+1-\sigma), u(\ell-\sigma))\Delta u(\ell-\sigma)}{f(u(\ell+1-\sigma))f(u(\ell-\sigma))} - \sum_{\ell=k_1}^{k-1} q(\ell+1),$$

we get

$$\frac{\Delta z(k)}{f(u(k-\sigma))} \le -\frac{1}{a(k)}\sum_{\ell=k_1}^{k-1} q(\ell+1), \quad k \in \mathbb{N}(k_1). \tag{7.12.5}$$

Since $u(k)$ is nonincreasing and $\tau \le \sigma$, we have $z(k) \le (1+p(k))u(k-\sigma)$, and hence in view of (viii), we obtain

$$f(z(k)) \le f(1+p(k))f(u(k-\sigma)). \tag{7.12.6}$$

On combining (7.12.5) and (7.12.6), we find

$$-\frac{\Delta z(k)}{f(z(k))} \ge \frac{1}{f(1+p(k))a(k)}\sum_{\ell=k_1}^{k-1} q(\ell+1), \quad k \in \mathbb{N}(k_1). \tag{7.12.7}$$

Now, for $z(k+1) \le t \le z(k)$, we have $1/f(t) \ge 1/f(z(k))$, and hence it follows that

$$\int_{z(k+1)}^{z(k)} \frac{dt}{f(t)} \ge -\frac{\Delta z(k)}{f(z(k))}.$$

Using the above inequality in (7.12.7) and summing the resulting inequality from $k_1$ to $k$, we get

$$\sum_{\ell=k_1}^{k} \frac{1}{f(1+p(\ell))a(\ell)} \sum_{\xi=k_1}^{\ell-1} q(\xi+1) \le \int_{z(k+1)}^{z(k_1)} \frac{dt}{f(t)},$$

which in view of (x) implies that

$$\limsup_{k\to\infty} \int_{z(k+1)}^{z(k_1)} \frac{dt}{f(t)} = \infty. \tag{7.12.8}$$

But, this contradicts (vii), and hence (a) implies that $M^- = \emptyset$.

Now suppose that (b) holds. We proceed as earlier and define $w(k) = a(k)\Delta z(k)$, $k \in \mathbb{N}(k_1)$. Then, from (7.12.1) we have

$$w(k) = w(k_1) - \sum_{\ell=k_1}^{k-1} q(\ell+1)f(u(\ell+1-\sigma)). \tag{7.12.9}$$

Hence, it follows that

$$\Delta z(k) \le \frac{w(k_1)}{a(k)} < 0.$$

Summing the last inequality from $k_1$ to $k$ and using (vi), we get

$$z(k+1) \le z(k_1) + w(k_1)\sum_{\ell=k_1}^{k}\frac{1}{a(\ell)} \to -\infty \quad \text{as } k\to\infty,$$

which contradicts the assumption that $z(k)>0$ for all $k\in\mathbb{N}(k_1)$. Thus, (b) implies that $M^-=\emptyset$. ∎

**Example 7.12.3.** Some of the assumptions in Theorem 7.12.3 are essential. For this, we consider the neutral difference equations

$$\Delta\left(4^k\Delta\left(u(k)+\frac{1}{2}u(k-1)\right)\right)+4^{k-1}u(k-2) = 0, \quad k\in\mathbb{N} \tag{7.12.10}$$

and

$$\Delta\left(4^k\Delta\left(u(k)+\frac{1}{2}u(k-1)\right)\right)+4^{2(k-1)/3}(u(k-4))^{1/3}=0, \ k\in\mathbb{N}. \tag{7.12.11}$$

For both of these equations $u(k)=1/2^k$ is a solution, and hence $M^-\ne\emptyset$. For the equation (7.12.10) assumption (x) holds, while the condition (vii) is not satisfied. But, for the equation (7.12.11) assumption (x) does not hold, while condition (vii) is satisfied. Moreover, condition (vi) is not satisfied for either of the equations (7.12.10) and (7.12.11).

**Theorem 7.12.4.** With respect to the difference equation (7.12.1) assume that in addition to the conditions (v), (vi) and (xi) the following holds

(xii) $-1<p(1)\le p(k)\le 0$, for all $k\ge k_0\in\mathbb{N}$.

Then, $M^-=\emptyset$.

**Proof.** Suppose that (7.12.1) has a solution $u(k)\in M^-$. Once again without loss of generality we assume that there exists an integer $k_1\ge k_0\in\mathbb{N}$ such that $u(k)>0$, $\Delta u(k)\le 0$, $u(k-\mu)>0$, $\Delta u(k-\mu)\le 0$ for all $k\in\mathbb{N}(k_1)$. Let $z(k)=u(k)+p(k)u(k-\tau)$. We claim that $z(k)>0$ for all $k\in\mathbb{N}(k_1)$. Indeed, if $z(k)\le 0$, then in view of (xii) we have $u(k)<-p(1)u(k-\tau)$, which is the same as $u(k+\tau)<-p(1)u(k)$, and therefore $u(k+j\tau)<(-p(1))^ju(k)$, but this leads to $u(k)\to 0$ as $k\to\infty$, which is a contradiction. Since the equation (7.12.1) is the same as $\Delta(a(k)\Delta z(k))=-q(k+1)f(u(k+1-\sigma))$, $k\in\mathbb{N}(k_1)$ it follows that

$a(k)\Delta z(k)$ is decreasing for $k \in \mathbb{N}(k_1)$. Now, proceeding as in Theorem 7.12.2, in view of (vi) we find $a(k)\Delta z(k) > 0$, $k \in \mathbb{N}(k_1)$. We define $w(k) = (a(k)\Delta z(k))/f(z(k))$, $k \in \mathbb{N}(k_1)$ to obtain

$$(7.12.12)\quad \Delta w(k) = \frac{-q(k+1)f(u(k+1-\sigma))}{f(z(k+1))} - \frac{a(k)(\Delta z(k))^2 g(z(k), z(k+1))}{f(z(k))f(z(k+1))}$$
$$\leq \frac{-q(k+1)f(u(k+1-\sigma))}{f(z(k+1))}, \quad k \in \mathbb{N}(k_1).$$

From $z(k) = u(k) + p(k)u(k-\tau)$, we have in view of (xii) that $z(k+1) \leq u(k+1)$. Since $u(k) \in M^-$, we obtain $z(k+1) \leq u(k+1) \leq u(k+1-\sigma)$. Thus, $f(z(k+1)) \leq f(u(k+1-\sigma))$ for $k \in \mathbb{N}(k_1)$. Using this inequality in (7.12.12) and summing the resulting inequality, we get

$$w(k) \;\leq\; w(k_1) - \sum_{\ell=k_1}^{k-1} q(\ell+1),$$

which because of (v) implies that $w(k) \to -\infty$ as $k \to \infty$. This contradiction completes the proof. ∎

**Theorem 7.12.5.** With respect to the difference equation (7.12.1) assume that in addition to the condition (iv) the following hold

(xiii) $\tau$ is an odd positive integer

(xiv) $p(k) \equiv p \leq 0$ for all $k \in \mathbb{N}(k_0)$.

Then, $WOS = \emptyset$.

**Proof.** Let $u(k)$ be a weakly oscillatory solution of (7.12.1). Without loss of generality we assume that there exists an integer $k_1 \geq k_0 \in \mathbb{N}$ such that $u(k) > 0$, $u(k-\mu) > 0$ for all $k \in \mathbb{N}(k_1)$. The proof for the case $u(k) < 0$, $u(k-\mu) < 0$ for all large $k \in \mathbb{N}$ is similar. Let $z(k) = u(k) + pu(k-\tau)$. Then, we have $\Delta z(k) = \Delta u(k) + p\Delta u(k-\tau)$, $\Delta z(k+1) = \Delta u(k+1) + p\Delta u(k-\tau+1)$, and

$$(7.12.13)\quad \Delta z(k)\Delta z(k+1) = \Delta u(k)\Delta u(k+1) + p(\Delta u(k)\Delta u(k+1-\tau) + \Delta u(k+1)\Delta u(k-\tau)) + p^2 \Delta u(k-\tau)\Delta u(k+1-\tau).$$

Thus, in view of (xiii) and (xiv), we find $\Delta z(k)\Delta z(k+1) \leq 0$. Hence, $\Delta z(k)$ is oscillatory. Define $w(k) = a(k)\Delta z(k)$, so that $w(k)$ is also oscillatory. On the other hand, from (7.12.1) we have $\Delta w(k) = -q(k+1)f(u(k+1-\sigma))$, $k \in \mathbb{N}(k_1)$ and hence (iv) implies that $\Delta w(k) \leq 0$, and so $w(k)$ is nonincreasing. This contradiction completes the proof. ∎

**Remark 7.12.1.** From the above proof it is clear that Theorem 7.12.4 remains valid even if the condition (iv) is replaced by (iv)$'$ $q(k) \le 0$ for all $k \in \mathbb{N}(k_0)$.

**Theorem 7.12.6.** With respect to the difference equation (7.12.1) assume that in addition to the conditions (v) and (vi) the following holds

(xv) $p(k) \equiv p \ge 0$ for all $k \in \mathbb{N}(k_0)$.

Then, every solution of (7.12.1) is either oscillatory or weakly oscillatory.

**Proof.** From Theorem 7.12.1 it is clear that $M^+ = \emptyset$. Hence, to complete the proof it suffices to show that $M^- = \emptyset$. For this, let $u(k)$ be a solution of (7.12.1) which is in the class $M^-$. For this solution, as earlier we assume that $u(k) > 0$, $\Delta u(k) \le 0$, $u(k-\mu) > 0$, $\Delta u(k-\mu) \le 0$ for all $k \in \mathbb{N}(k_1)$, and define $z(k) = u(k) + pu(k-\tau)$. Then, in view of (xv), $z(k) > 0$ and $\Delta z(k) \le 0$ for all $k \in \mathbb{N}(k_1)$. Again, we define $w(k) = a(k)\Delta z(k)$ so that $w(k) \le 0$, $k \in \mathbb{N}(k_1)$, and from the equation (7.12.1), we have (7.12.9). Thus, from Abel's transformation Problem 1.9.7 it follows that

$$\begin{aligned} w(k) &= w(k_1) - f(u(k+1-\sigma)) \sum_{\ell=k_1}^{k-1} q(\ell+1) \\ &\quad + \sum_{\ell=k_1}^{k-1} \Delta f(u(\ell+1-\sigma)) \left( \sum_{\xi=k_1}^{\ell} q(\xi+1) \right). \end{aligned}$$

From the above relation, and the condition (v) with $k_1$ sufficiently large, we find that $w(k) \le w(k_1)$, i.e. $\Delta z(k) \le w(k_1)/a(k) < 0$. The rest of the proof of $M^- = \emptyset$ is the same as that of Theorem 7.12.3(b). ∎

**Theorem 7.12.7.** With respect to the difference equation (7.12.1) assume that in addition to the hypotheses of Theorem 7.12.4, $\tau \ge 1$. Then, every solution of (7.12.1) is either oscillatory or weakly oscillatory.

**Proof.** The proof follows from that of Theorems 7.12.2 and 7.12.4. ∎

**Theorem 7.12.8.** With respect to the difference equation (7.12.1) assume that in addition to the conditions (iv) (vi) and (xiii) the inequality $-1 \le p(k) \equiv p \le 0$ holds. Then, every solution of (7.12.1) belongs to the class $OS$.

**Proof.** The proof follows from that of Theorems 7.12.2, 7.12.4 and 7.12. 5. ∎

**Theorem 7.12.9.** If conditions (viii) (x) are satisfied, then for every solution $u(k) \in M^-$ of (7.12.1), $\lim_{k\to\infty} u(k) = 0$.

**Proof.** The proof follows from the same arguments as in the proof of Theorem 7.12.3(a), and the fact that (7.12.8) implies that $\lim_{k\to\infty} z(k) = 0$, and $z(k) \ge u(k)$ for all $k \in \mathbb{N}(k_1)$. ∎

**Theorem 7.12.10.** With respect to the difference equation (7.12.1) assume that in addition to the condition (i) the function $p(k)$ is bounded and the following holds

(xvi) $\displaystyle\limsup_{k\to\infty} \sum_{\ell=K}^{k-1} q(\ell+1) \sum_{\xi=K}^{\ell} \frac{1}{a(\xi)} = \infty$ for all $K \in \mathbb{N}(k_0)$.

Then, every solution of (7.12.1) in the class $M^+$ is unbounded.

**Proof.** Let $u(k)$ be a solution of (7.12.1) in $M^+$. As in Theorem 7.12.1 without loss of generality we assume that there exists an integer $k_1 \ge k_0 \in \mathbb{N}$ such that $u(k) > 0$, $\Delta u(k) \ge 0$, $u(k-\mu) > 0$, $\Delta u(k-\mu) \ge 0$ for all $k \in \mathbb{N}(k_1)$, and set $z(k) = u(k) + p(k)u(k-\tau)$, so that $z(k) > 0$ and $\Delta z(k) \ge 0$ for all $k \in \mathbb{N}(k_1)$. For the function

$$\phi(k) \;=\; -\frac{a(k)\Delta z(k)}{f(u(k-\sigma))} \sum_{\ell=k_1}^{k-1} \frac{1}{a(\ell)},$$

we have

$$\begin{aligned}\Delta\phi(k) \;=\;& -\frac{\Delta(a(k)\Delta z(k))}{f(u(k+1-\sigma))} \sum_{\ell=k_1}^{k} \frac{1}{a(\ell)} - \frac{\Delta z(k)}{f(u(k+1-\sigma))} \\ & + \frac{a(k)\Delta z(k) g(u(k-\sigma), u(k+1-\sigma))\Delta u(k-1)}{f(u(k-\sigma))f(u(k+1-\sigma))} \sum_{\ell=k_1}^{k-1} \frac{1}{a(\ell)} \\ \ge\;& q(k+1) \sum_{\ell=k_1}^{k} \frac{1}{a(\ell)} - \frac{\Delta z(k)}{f(u(k+1-\sigma))}.\end{aligned}$$

Summing the last inequality, we obtain

$$\phi(k) \ge \sum_{\ell=k_1}^{k-1} q(\ell+1) \sum_{\xi=k_1}^{\ell} \frac{1}{a(\xi)} - \sum_{\ell=k_1}^{k-1} \frac{\Delta z(\ell)}{f(u(\ell+1-\sigma))}. \tag{7.12.14}$$

Since $\Delta z(k)/f(u(k+1-\sigma))$ is positive for $k \in \mathbb{N}(k_1)$, the following limit exists

$$\lim_{k\to\infty} \sum_{\ell=k_1}^{k-1} \frac{\Delta z(\ell)}{f(u(\ell+1-\sigma))} \;=\; \beta \text{ (say)}.$$

We claim that $\beta = \infty$. Indeed, if $\beta < \infty$, then (xvi) combined with (7.12.14) leads to $\lim_{k\to\infty} w(k) = \infty$, which is a contradiction to the fact that $w(k)$ is negative for all $k \in \mathbb{N}(k_1)$. Now for all $k \geq k_1$, we have $f(u(k+1-\sigma)) \geq f(u(k_1+1-\sigma)) = c$, and consequently

$$\sum_{\ell=k_1}^{k-1} \frac{\Delta z(\ell)}{f(u(\ell+1-\sigma))} \leq \frac{1}{c}[z(k) - z(k_1)].$$

Thus, it follows that $\lim_{k\to\infty} z(k) = \infty$. Finally, since $z(k) = u(k) + p(k)u(k-\tau) \leq (1+p(k))u(k)$ the boundedness of $p(k)$ implies that $\lim_{k\to\infty} u(k) = \infty$. ∎

**Corollary 7.12.11.** Assume that $\tau \leq \sigma$ and the conditions (vii), (viii), (x), (xv) and (xvi) hold. Then, every bounded solution of (7.12.1) is either oscillatory, or weakly oscillatory.

**Proof.** The proof follows from Theorems 7.12.3(a) and 7.12.10. ∎

## 7.13. Existence of Solutions for (7.12.1)

In the previous section, to establish oscillatory behavior of solutions of (7.12.1), we have provided sufficient conditions which ensure that the classes $M^+$, $M^-$ and $WOS$ are empty. The purpose of this section is to prove the existence of solutions of (7.12.1) in these classes.

**Theorem 7.13.1.** With respect to the difference equation (7.12.1) assume that the following hold

(i) $\tau$ is an even positive integer

(ii) $p(k) \equiv p \geq 0$ and $\neq 1$ for all $k \in \mathbb{N}$

(iii) $q(k) > 0$ for all large values of $k$

(iv) $\displaystyle\sum_{k=k_0}^{\infty} \frac{1}{a(k)} = \infty$

(v) $\displaystyle\sum_{k=k_0}^{\infty} \frac{1}{a(k)} \sum_{\ell=k}^{\infty} q(\ell+1) < \infty.$

Then, $M^+ \neq \emptyset$.

**Proof.** We shall prove the existence of positive solutions of (7.12.1) in the class $M^+$, the existence of negative solutions in $M^+$ is similar.

First assume that $0 \leq p < 1$. Let $C = \max\{|f(u)| : 3(1-p) \leq u \leq 4\}$. Choose a positive integer $K \geq k_0$ sufficiently large, so that

$$C\sum_{k=K}^{\infty} \frac{1}{a(k)} \sum_{\ell=k}^{\infty} q(\ell+1) \leq \frac{1-p}{4}.$$

Consider the Banach space $B_K$ of all real functions $u(k)$, $k \in \mathbb{N}(K-\mu)$ with the sup norm $\|u\| = \sup_{k\in\mathbb{N}(K-\mu)} |u(k)|$. We define a subset $S$ of $B_K$ as

$$S = \{u \in B_K : 3(1-p) \le u(k) \le 4, \ k \in \mathbb{N}(K-\mu)\}.$$

Clearly, $S$ is a bounded, closed and convex subset of $B_K$. Now, we define an operator $T : S \to B_K$ as follows

$$Tu(k) = \begin{cases} 3 + p - pu(k-\tau) + \sum_{\ell=K}^{k-1} \frac{1}{a(\ell)} \sum_{\xi=\ell}^{\infty} q(\xi+1) f(u(\xi+1-\sigma)), & k \ge K \\ Tu(K) \ = \ 3 + p - pu(K-\tau), \quad K - \mu \le k \le K. \end{cases}$$

From the hypotheses this operator $T$ is continuous, and for $u \in S$, in view of $0 \le p < 1$, we have

$$Tu(k) \le \begin{cases} 3 + p + C \sum_{\ell=K}^{k-1} \frac{1}{a(\ell)} \sum_{\xi=\ell}^{\infty} q(\xi+1) \le 3 + p + \frac{1-p}{4} < 4, & k \ge K \\ 3 + p \ < \ 4, \quad K - \mu \le k \le K \end{cases}$$

and, similarly

$$Tu(k) \ \ge \ 3(1-p), \quad k \in \mathbb{N}(K-\mu).$$

Thus, $T(S) \subseteq S$. Therefore, by the Schauder fixed point theorem $T$ has a fixed point $u \in S$. It is clear that this $u(k)$ is a positive solution of (7.12.1).

Now assume that $p > 1$. Let $C = \max\{|f(u)| : 2(p-1) \le u \le 4p\}$. Choose a positive integer $K \ge k_0$ sufficiently large, so that

$$C \sum_{k=K}^{\infty} \frac{1}{a(k)} \sum_{\ell=k}^{\infty} q(\ell+1) \ \le \ \frac{p(p-1)}{4}.$$

Let $B_K$ be as above, and let

$$S = \{u \in B_K : 2(p-1) \le u(k) \le 4p, \ k \in \mathbb{N}(K-\mu)\}.$$

Again, $S$ is a bounded, closed and convex subset of $B_K$. Define an operator $T : S \to B_K$ as follows

$$Tu(k) = \begin{cases} 3p + 1 - \frac{1}{p} u(k+\tau) + \frac{1}{p} \sum_{\ell=K}^{k+\tau-1} \frac{1}{a(\ell)} \sum_{\xi=\ell}^{\infty} q(\xi+1) f(u(\xi+1-\sigma)), \\ \hfill k \ge K \\ Tu(K), \quad K - \mu \le k \le K. \end{cases}$$

For this continuous operator also it is easy to see that $T(S) \subseteq S$, and hence, by the Schauder fixed point theorem $T$ has a fixed point $u \in S$. Once again, it is clear that this $u(k)$ is a positive solution of (7.12.1).

Thus, we have established the existence of positive solutions for the equation (7.12.1), when $p \geq 0$, $p \neq 1$. Next, we shall show that $M^+ \neq \emptyset$. For this, suppose that $u(k) \in WOS$. Let $k_1 \geq k_0 \in \mathbb{N}$ be such that $u(k) > 0$, $u(k-\mu) > 0$ for all $k \in \mathbb{N}(k_1)$. Let $z(k) = u(k) + pu(k-\tau)$. Then, as in Theorem 7.12.5, we obtain (7.12.13), which in view of (i) and (ii) implies that $\Delta z(k)\Delta z(k+1) \leq 0$, and hence, $\Delta z(k)$ is oscillatory. Define $w(k) = a(k)\Delta z(k)$, so that $w(k)$ is also oscillatory. On the other hand, from (7.12.1), we have $\Delta w(k) = -q(k+1)f(u(k+1-\sigma))$, $k \in \mathbb{N}(k_1)$, and hence, (iii) implies that $\Delta w(k) \leq 0$, and so $w(k)$ is nonincreasing. This contradiction shows that $u(k) \notin WOS$. Also, from Theorem 7.12.3(b), we have $u(k) \notin M^-$. Thus, $u(k) \in M^+$. ∎

**Example 7.13.1.** Consider the difference equation

$$(7.13.1)\quad \Delta\left(\frac{1}{k}\Delta\left(u(k) + 2u(k-2)\right)\right) + \frac{3}{(k-2)^3 k(k+1)}(u(k-2))^3 = 0,$$

$$k \in \mathbb{N}(3)$$

for which all the conditions of Theorem 7.13.1 are satisfied. Thus, it follows that (7.13.1) has a solution $u(k)$ in the class $M^+$ such that $2 \leq u(k) \leq 8$ for all sufficiently large $k$. It also has an unbounded solution $u(k) = k$, which also belongs to the class $M^+$.

**Theorem 7.13.2.** With respect to the difference equation (7.12.1) assume that in addition to the condition (iii) the following hold

(vi) $\tau$ is an odd positive integer

(vii) $p(k) \equiv p \leq 0$ and $\neq -1$ for all $k \in \mathbb{N}$

(viii) $\displaystyle\sum_{k=k_0}^{\infty} \frac{1}{a(k)} \sum_{\ell=k_0}^{k-1} q(\ell+1) < \infty.$

Then, $M^+ \bigcup M^- \neq \emptyset$.

**Proof.** We shall prove the existence of positive solutions of (7.12.1) in the class $M^+ \bigcup M^-$, the existence of negative solutions in $M^+ \bigcup M^-$ is similar.

First assume that $-1 < p \leq 0$. Let $C = \max\{|f(u)| : 2(1+p)/3 \leq$

$u \leq 4/3\}$, and $K \geq k_0$ sufficiently large, so that

$$C \sum_{k=K}^{\infty} \frac{1}{a(k)} \sum_{\ell=K}^{k-1} q(\ell+1) \leq \frac{1+p}{3}.$$

Let the space $B_K$ be as in Theorem 7.13.1. We define a subset $S$ of $B_K$ as

$$S = \left\{ u \in B_K : \frac{2(1+p)}{3} \leq u(k) \leq \frac{4}{3}, \ k \in \mathbb{N}(K-\mu) \right\}.$$

Clearly, $S$ is a bounded, closed and convex subset of $B_K$. Now, we define an operator $T : S \to B_K$ as follows

$$Tu(k) = \begin{cases} 1 + p - pu(k-\tau) + \sum_{\ell=k}^{\infty} \frac{1}{a(\ell)} \sum_{\xi=K}^{\ell-1} q(\xi+1) f(u(\xi+1-\sigma)), \ k \geq K \\ Tu(K), \quad K-\mu \leq k \leq K. \end{cases}$$

This operator $T$ is continuous, and it is easy to see that $T(S) \subseteq S$, and hence, by the Schauder fixed point theorem $T$ has a fixed point $u \in S$. This fixed point is a positive solution of (7.12.1).

Now assume that $p < -1$. Let $C = \max\{|f(u)| : -p/2 \leq u \leq -2p\}$. Let $K \geq k_0$ be so large that

$$-\frac{C}{p} \sum_{k=K}^{\infty} \frac{1}{a(k)} \sum_{\ell=K}^{k-1} q(\ell+1) \leq -\frac{(1+p)}{4}.$$

Let $B_K$ be as above, and let

$$S = \left\{ u \in B_K : -\frac{p}{2} \leq u(k) \leq -2p, \ k \in \mathbb{N}(K-\mu) \right\}.$$

Again, $S$ is a bounded, closed and convex subset of $B_K$. Define an operator $T : S \to B_K$ as follows

$$Tu(k) = \begin{cases} -p - 1 - \frac{1}{p} u(k+\tau) + \frac{1}{p} \sum_{\ell=k+\tau}^{\infty} \frac{1}{a(\ell)} \sum_{\xi=K}^{\ell-1} q(\xi+1) f(u(\xi+1-\sigma)), \\ \hfill k \geq K \\ Tu(K), \quad K-\mu \leq k \leq K. \end{cases}$$

For this continuous operator also it is easy to see that $T(S) \subseteq S$, and hence, by the Schauder fixed point theorem $T$ has a fixed point $u \in S$. Once again, it is clear that this fixed point is a positive solution of (7.12.1).

This proves the existence of positive solutions $u(k)$ for the equation (7.12.1), when $p \le 0$, $p \ne -1$. Further, from Theorem 7.12.5, we have $u(k) \notin WOS$. Thus, in conclusion we find that $u(k) \in M^+ \bigcup M^-$. ∎

**Example 7.13.2.** For the difference equations

$$(7.13.2) \quad \Delta\left(4^k \Delta\left(u(k) - 4u(k-1)\right)\right) + 7\left(4^{5-k}\right)\left(u(k-2)\right)^5 = 0, \quad k \in \mathbb{N}(3)$$

and

$$(7.13.3) \quad \Delta\left(4^k \Delta\left(u(k) - \frac{1}{4}u(k-1)\right)\right) + 2^{4(k-2)/3}\left(u(k-2)\right)^{1/3} = 0, \quad k \in \mathbb{N}(3)$$

all the conditions of Theorem 7.13.2 are satisfied. In fact, $u(k) = 2^k \in M^+$ is a solution of (7.13.2), and $u(k) = 2^{-k} \in M^-$ is a solution of (7.13.3).

**Theorem 7.13.3.** With respect to the difference equation (7.12.1) assume that in addition to conditions (i), (ii), (iii) and (viii) the following holds

$$\text{(ix)} \quad \lim_{k \to \infty} \sum_{\ell=k_0}^{k-1} q(\ell+1) = \infty.$$

Then, $M^- \ne \emptyset$.

**Proof.** Once again we shall prove the existence of positive solutions of (7.12.1) in the class $M^-$, the existence of negative solutions in $M^-$ follows similarly.

First assume that $0 \le p < 1$. Let $C = \max\{|f(u)| : 5(1-p)/2 \le u \le 4\}$, and let $K \ge k_0$ be such that

$$C \sum_{k=K}^{\infty} \frac{1}{a(k)} \sum_{\ell=K}^{k-1} q(\ell+1) \le \frac{1-p}{2}.$$

Let $B_K$ be as before, and $S \subset B_K$ be defined as

$$S = \left\{ u \in B_K : \frac{5}{2}(1-p) \le u(k) \le 4, \ k \in \mathbb{N}(K-\mu) \right\}.$$

Clearly, $S$ is a bounded, closed and convex subset of $B_K$. Define an operator $T : S \to B_K$ as follows

$$Tu(k) = \begin{cases} 3 + p - pu(k-\tau) - \displaystyle\sum_{\ell=K}^{k-1} \frac{1}{a(\ell)} \sum_{\xi=K}^{\ell-1} q(\xi+1) f(u(\xi+1-\sigma)), & k \ge K \\ Tu(K) = 3 + p - pu(K-\tau), & K - \mu \le k \le K. \end{cases}$$

This operator $T$ is continuous, and as earlier it is easy to see that $T(S) \subseteq S$. Therefore, by the Schauder fixed point theorem $T$ has a fixed point $u \in S$. It is clear that this $u(k)$ is a positive solution of (7.12.1).

Now assume that $p > 1$. Let $C = \max\{|f(u)| : 5(p-1)/2 \le u \le 4p\}$, and let $K \ge k_0$ be such that

$$C \sum_{k=K}^{\infty} \frac{1}{a(k)} \sum_{\ell=K}^{k-1} q(\ell+1) \;\le\; \frac{p(p-1)}{2}.$$

Let $B_K$ be as above, and

$$S \;=\; \{u \in B_K : 5(p-1)/2 \le u(k) \le 4p,\ k \in \mathbb{N}(K-\mu)\}.$$

Again, $S$ is a bounded, closed and convex subset of $B_K$. Define an operator $T : S \to B_K$ as follows

$$Tu(k) = \begin{cases} 3p + 1 - \dfrac{1}{p}u(k+\tau) - \dfrac{1}{p}\displaystyle\sum_{\ell=K}^{k+\tau-1} \frac{1}{a(\ell)} \sum_{\xi=K}^{\ell-1} q(\xi+1) f(u(\xi+1-\sigma)), \\ \hfill k \ge K \\ Tu(K), \quad K - \mu \le k \le K. \end{cases}$$

For this continuous operator also it is easy to see that $T(S) \subseteq S$, and hence, by the Schauder fixed point theorem $T$ has a fixed point $u \in S$. Once again, it is clear that this $u(k)$ is a positive solution of (7.12.1).

Thus, we have obtained the existence of positive solutions $u(k)$ for the equation (7.12.1), when $p \ge 0$, $p \ne 1$. From the proof of Theorem 7.13.1, $u(k) \notin WOS$, and also from Theorem 7.12.1, $u(k) \notin M^+$. Hence, $u(k) \in M^-$. ∎

**Example 7.13.3.** Consider the difference equation

$$(7.13.4)\quad \Delta\left(4^k \Delta\left(u(k) + 4u(k-2)\right)\right) + 17\left(2^{8(k-1)/5}\right)\left(u(k-1)\right)^{3/5} \;=\; 0, \qquad k \in \mathbb{N}(2)$$

for which all the conditions of Theorem 7.13.3 are satisfied. In fact, $u(k) = 2^{-k}$ is a solution of (7.13.4) which belongs to the class $M^-$.

**Theorem 7.13.4.** With respect to the difference equation (7.12.1) assume that in addition to condition (iv) the following hold

(x) $p(k) \equiv p \ge 0$ for all $k \ge k_0 \in \mathbb{N}$

(xi) $\limsup_{k\to\infty} \sum_{\ell=K}^{k-1} q(\ell+1) = \infty$.

Then, $WOS \neq \emptyset$.

**Proof.** Suppose the result is not true, then there exists a solution $u(k)$ of (7.12.1) such that eventually, either

(7.13.5) $\qquad u(k) > 0 \quad \text{and} \quad \Delta u(k) > 0, \;$ or

(7.13.6) $\qquad u(k) > 0 \quad \text{and} \quad \Delta u(k) < 0 \quad \text{for all } \; k \in \mathbb{N}(K-\mu)$

holds. Assume that (7.13.5) holds. Let $z(k) = u(k) + pu(k-\tau)$, so that the equation (7.12.1) can be written as $\Delta(a(k)\Delta z(k)) = -q(k+1)f(u(k+1-\sigma))$. Let $w(k) = (a(k)\Delta z(k))/f(u(k+1-\sigma))$, $k \in \mathbb{N}(K)$. Then, $w(k) > 0$ so that $\Delta w(k) \leq -q(k+1)$, $k \in \mathbb{N}(K)$. Now, summing the last inequality from $K$ to $k-1$ with $K$ sufficiently large, we obtain

$$w(k) - w(K) \;\leq\; -\sum_{\ell=K}^{k-1} q(\ell+1).$$

In the above inequality the right side, in view of condition (xi), tends to $-\infty$. But, this contradicts the fact that $w(k) > 0$.

Next assume that (7.13.6) holds. From condition (xi) we find that there exists an integer $K_1 \geq K$ such that

$$\sum_{\ell=K_1}^{k-1} q(\ell+1) \;>\; 0. \tag{7.13.7}$$

For this, we set $F(k) = \sum_{\ell=K}^{k-1} q(\ell+1)$ and define $K_1 = \sup\{k \geq K : F(k) = 0\}$. Clearly, $F(K_1) = 0$ and $F(k) > 0$ for $k > K_1$. Then, $\sum_{\ell=K_1}^{k-1} q(\ell+1) = F(k) - F(K_1) = F(k) > 0$. Now from (7.13.6), (7.13.7) and Abel's transformation Problem 1.9.7 it follows for $k > K_1$ that

$$\sum_{\ell=K_1}^{k-1} q(\ell+1)f(u(\ell+1-\sigma))$$

$$= f(u(k+1-\sigma))\sum_{\ell=K_1}^{k-1} q(\ell+1) - \sum_{\ell=K_1}^{k-1} \Delta f(u(\ell+1-\sigma))\left(\sum_{\xi=K_1}^{\ell-1} q(\xi+1)\right) > 0.$$

Thus, on summing (7.12.1) from $K_1$ to $k-1$, we get

$$a(k)\Delta z(k) - a(K_1)\Delta z(K_1) \;=\; -\sum_{\ell=K_1}^{k-1} q(\ell+1)f(u(\ell+1-\sigma)) \;<\; 0.$$

Hence, $a(k)\Delta z(k) < a(K_1)\Delta z(K_1)$. It is clear that $\Delta z(K_1) < 0$. Therefore, $\Delta z(k) < a(K_1)\Delta z(K_1)/a(k)$. Again summing from $K_1$ to $k-1$, we obtain

$$z(k) - z(K_1) \ < \ a(K_1)\Delta z(K_1) \sum_{\ell=K_1}^{k-1} \frac{1}{a(\ell)},$$

which in view of (iv) implies that $z(k) \to -\infty$ as $k \to \infty$. But, by (7.13.6) we have $z(k) > 0$. This contradiction completes the proof. ∎

**Example 7.13.4.** Consider the difference equation

$$(7.13.8) \quad \Delta\left(k\Delta\left(u(k) + 2u(k-2)\right)\right) + \frac{(-1)^k(12k+6)}{\left(2-(-1)^k\right)^3}(u(k-1))^3 \ = \ 0, \qquad k \in \mathbb{N}(2)$$

for which all the conditions of Theorem 7.13.4 are satisfied. In fact, $u(k) = 2 + (-1)^k \in WOS$ is a solution of (7.13.8).

**Theorem 7.13.5.** With respect to the difference equation (7.12.1) assume that the conditions (i), (iii), (iv), (x) and (xi) are satisfied. Then, $OS \neq \emptyset$.

**Proof.** The result follows from the proof of Theorem 7.13.1, and Theorems 7.12.1 and 7.12.3(b). ∎

## 7.14. Oscillation of Mixed Difference Equations

Here we shall establish criteria for the oscillation of second order difference equations

$$(7.14.1)_\delta \quad \Delta^2(u(k)+au(k-\tau)-bu(k+\tau))+\delta[q(k)u(k-\sigma)+p(k)u(k+\eta))] = 0$$

$$(7.14.2)_\delta \quad \Delta^2(u(k)-au(k-\tau)+bu(k+\tau))+\delta[q(k)u(k-\sigma)+p(k)u(k+\eta))] = 0$$

$$(7.14.3)_\delta \quad \Delta^2(u(k)+au(k-\tau)+bu(k+\tau))+\delta[q(k)u(k-\sigma)+p(k)u(k+\eta))] = 0$$

and

$$(7.14.4)_\delta \quad \Delta^2(u(k)-au(k-\tau)-bu(k+\tau))+\delta[q(k)u(k-\sigma)+p(k)u(k+\eta))] = 0,$$

where $\delta = \pm 1$, $a$, $b$ are nonnegative real numbers, $\tau$, $\sigma$ and $\eta$ are positive integers, $p(k)$ and $q(k)$ are nonnegative functions on $\mathbb{N}$, and $p(k \pm \tau) = p(k)$, $q(k \pm \tau) = q(k)$, $k \in \mathbb{N}$.

We shall need the following:

**Lemma 7.14.1.** Assume that $p(k)$ and $q(k)$ are nonnegative functions, and $\sigma$ and $\eta$ are positive integers, $\eta > 2$. Then, the following hold:

(I) If

$$\limsup_{k\to\infty} \sum_{\ell=k}^{k+\eta-2} (k+\eta-\ell-1)p(\ell) > 1 \tag{7.14.5}$$

then the difference inequality

$$\Delta^2 v(k) \geq p(k)v(k+\eta) \tag{7.14.6}$$

has no eventually positive solution $v(k)$ which satisfies $\Delta^i v(k) > 0$ eventually, $i = 0, 1, 2$.

(II) If

$$\limsup_{k\to\infty} \sum_{\ell=k-\sigma}^{k} (k-\ell+1)q(\ell) > 1 \tag{7.14.7}$$

then the difference inequality

$$\Delta^2 z(k) \geq q(k)z(k-\sigma) \tag{7.14.8}$$

has no eventually positive solution $z(k)$ which satisfies $(-1)^i\Delta^i z(k) > 0$ eventually, $i = 0, 1, 2$.

**Proof.** (I) Let $v(k)$ be an eventually positive solution of (7.14.6) such that $\Delta^i v(k) > 0$ eventually, $i = 1, 2$. Summing (7.14.6) from $k$ to $s$, $s \geq k \geq k_1$ (say)

$$\Delta v(s+1) - \Delta v(k) \geq \sum_{\ell=k}^{s} p(\ell)v(\ell+\eta)$$

or

$$\Delta v(s+1) \geq \sum_{\ell=k}^{s} p(\ell)v(\ell+\eta). \tag{7.14.9}$$

Next, summing (7.14.9) from $k$ to $k+\eta-2$, $k \geq k_1$, we have

$$\begin{aligned} v(k+\eta) \geq v(k+\eta) - v(k+1) &\geq \sum_{s=k}^{k+\eta-2} \sum_{\ell=k}^{s} p(\ell)v(\ell+\eta) \\ &= \sum_{\ell=k}^{k+\eta-2} (k+\eta-\ell-1)p(\ell)v(\ell+\eta). \end{aligned}$$

Thus, we see that

$$\sum_{\ell=k}^{k+\eta-2} (k+\eta-\ell-1)p(\ell) \le 1,$$

which contradicts (7.14.5). This completes the proof of (I).

(II) Let $z(k)$ be an eventually positive solution of (7.14.8) such that $\Delta z(k) < 0$ and $\Delta^2 z(k) > 0$ eventually. In view of the monotonicity of $\Delta z(k)$, we obtain for every $k \ge i \ge k_1$

$$-z(i) \le (\Delta z(k))(k-i+1). \tag{7.14.10}$$

Thus, for every $\ell$, $k$ with $k-\sigma \le \ell \le k$ and $k \ge k_2 \ge k_1$

$$\ell - \sigma \ge k_1 \quad \text{and} \quad k - \sigma \ge k_1$$

and therefore, by (7.14.10) we have

$$z(\ell-\sigma) \ge (-\Delta z(k-\sigma))(k-\ell+1).$$

Now

$$\begin{aligned} 0 &\ge \Delta z(k-\sigma) + \sum_{\ell=k-\sigma}^{k} q(\ell) z(\ell-\sigma) \\ &\ge \Delta z(k-\sigma) + \sum_{\ell=k-\sigma}^{k} q(\ell)(k-\ell+1)(-\Delta z(k-\sigma)) \end{aligned}$$

or

$$1 \ge \sum_{\ell=k-\sigma}^{k} (k-\ell+1) q(\ell),$$

which contradicts (7.14.7). This completes the proof of (II). ∎

**Theorem 7.14.2.** Let $b > 0$, $\eta > 2$ and $\sigma > \tau$. If

$$\limsup_{k\to\infty} \sum_{\ell=k}^{k+\eta-2} (k+\eta-\ell-1)p(\ell) > 1 + a \tag{7.14.11}$$

and

$$\limsup_{k\to\infty} \sum_{\ell=k-(\sigma-\tau)}^{k} (k-\ell+1)q(\ell) > 1 + a \tag{7.14.12}$$

then $(7.14.1)_{-1}$ is oscillatory.

**Proof.** Let $u(k)$ be an eventually positive solution of $(7.14.1)_{-1}$, say $u(k) > 0$ for $k \geq k_0 \geq 0$. Set

$$z(k) = u(k) + au(k-\tau) - bu(k+\tau). \tag{7.14.13}$$

Then,

$$\Delta^2 z(k) = q(k)u(k-\sigma) + p(k)u(k+\eta) \geq 0 \quad \text{for} \quad k \geq k_1 \geq k_0, \tag{7.14.14}$$

which implies that $\Delta^i z(k)$, $i = 0, 1$ are eventually of one sign. Therefore, either

(I) $z(k) < 0$ eventually, or (II) $z(k) > 0$ eventually.

(I) Assume $z(k) < 0$ for $k \geq k_1$. Set

$$0 < y(k) = -z(k) = bu(k+\tau) - au(k-\tau) - u(k) \leq bu(k+\tau). \tag{7.14.15}$$

There exists $k_2 \geq k_1$ such that

$$u(k) \geq \frac{1}{b} y(k-\tau) \quad \text{for} \quad k \geq k_2. \tag{7.14.16}$$

Using (7.14.16) in (7.14.14), we have

$$\Delta^2 y(k) + \frac{q(k)}{b} y(k-(\sigma+\tau)) \leq 0 \quad \text{for} \quad k \geq k_2.$$

Since $\Delta^2 y(k) \leq 0$ and $y(k) > 0$ for $k \geq k_2$, we have $\Delta y(k) > 0$ for $k \geq k_3 \geq k_2$. There exist an integer $K \geq k_3$ and a constant $c > 0$ such that

$$y(k-(\sigma+\tau)) \geq c \quad \text{for} \quad k \geq K.$$

Thus

$$\Delta^2 y(k) + \frac{c}{b} q(k) \leq 0 \quad \text{for} \quad k \geq K$$

and hence

$$0 < \Delta y(k) \leq \Delta y(K) - \frac{c}{b} \sum_{\ell=K}^{k-1} q(\ell) \to -\infty \quad \text{as} \quad k \to \infty,$$

which is a contradiction.

(II) Assume $z(k) > 0$ for $k \geq k_1$. Set

$$w(k) = z(k) + az(k-\tau) - bz(k+\tau). \tag{7.14.17}$$

Then,

$$\Delta^2 w(k) = q(k)z(k-\sigma) + p(k)z(k+\eta) \tag{7.14.18}$$

and

$$\text{(7.14.19)}\quad \Delta^2(w(k)+aw(k-\tau)-bw(k+\tau)) \;=\; q(k)w(k-\sigma)+p(k)w(k+\eta).$$

Using the procedure of (I) we see that $w(k) > 0$ eventually. Now, there are two possibilities to consider:

(i) $\Delta z(k) > 0$ eventually, and (ii) $\Delta z(k) < 0$ eventually.

Suppose (i) holds. There exist positive constants $c_1$ and $c_2$ and an integer $K_1 \geq k_1$ such that

$$z(k-\sigma) \;\geq\; c_1 \text{ and } z(k+\eta) \;\geq\; c_2 \text{ for } k \geq K_1.$$

Thus

$$\Delta^2 w(k) \;\geq\; c_1 q(k) + c_2 p(k) \text{ for } k \geq K_1$$

and hence

$$\Delta w(k) \to \infty \text{ and } w(k) \to \infty \text{ as } k \to \infty.$$

Therefore, we conclude that $\Delta^i w(k) > 0,\ i = 0, 1, 2$ for $k \geq K_1$. From (7.14.18) and the facts that $z(k)$ is an increasing sequence, and $p(k)$ and $q(k)$ are periodic of period $\tau$, we obtain

$$\begin{aligned}\Delta^2 w(k-\tau) &= q(k-\tau)z(k-\tau-\sigma) + p(k-\tau)z(k-\tau+\eta)\\ &= q(k)z(k-\tau-\sigma) + p(k)z(k-\tau+\eta)\\ &\leq q(k)z(k-\sigma) + p(k)z(k+\eta)\\ &= \Delta^2 w(k) \quad \text{for } k \geq K_1.\end{aligned}$$

Using this fact in (7.14.19), we obtain

$$(1+a)\Delta^2 w(k) \;\geq\; p(k)w(k+\eta)$$

or

$$\text{(7.14.20)}\qquad \Delta^2 w(k) \;\geq\; \frac{p(k)}{1+a} w(k+\eta) \text{ for } k \geq K_1.$$

But in view of Lemma 7.14.1 (I) and condition (7.14.11), inequality (7.14.20) has no eventually positive solution $w(k)$ with $\Delta^i w(k) > 0,\ i = 0, 1, 2$ eventually, which is a contradiction. Next, suppose (ii) holds. First, we claim that $z(k) \to 0$ monotonically as $k \to \infty$. Otherwise, $z(k) \to c > 0$ as $k \to \infty$. There exists an integer $K_2 \geq k_1$ such that

$$z(k-\sigma) \;\geq\; \frac{c}{2} \text{ and } z(k+\eta) \;\geq\; \frac{c}{2} \text{ for } k \geq K_2.$$

Thus

$$\Delta^2 w(k) \;\geq\; \frac{c}{2}(q(k)+p(k)) \quad \text{for} \quad k \geq K_2$$

and hence

$$\Delta w(k) \to \infty \quad \text{and} \quad w(k) \to \infty \quad \text{as} \quad k \to \infty.$$

From (7.14.17), we see that $w(k) < \infty$ for all $k \geq K_2$, which is a contradiction. Therefore, we conclude that $z(k) \to 0$, $w(k) \to 0$ and $\Delta^2 w(k) \to 0$ monotonically as $k \to \infty$ and so, one can easily see that $\Delta w(k) < 0$ for $k \geq K_2$. From (7.14.18) and the facts that $z(k)$ is a decreasing sequence, and $p(k)$ and $q(k)$ are periodic of period $\tau$, we have

$$\begin{aligned}
\Delta^2 w(k-\tau) &= q(k-\tau)z(k-\tau-\sigma) + p(k-\tau)z(k-\tau+\eta) \\
&= q(k)z(k-\tau-\sigma) + p(k)z(k-\tau+\eta) \\
&\geq q(k)z(k-\sigma) + p(k)z(k+\eta) \\
&= \Delta^2 w(k) \quad \text{for} \quad k \geq K_2.
\end{aligned}$$

Using this fact in (7.14.19), we obtain

$$(1+a)\Delta^2 w(k-\tau) \;\geq\; q(k)w(k-\sigma)$$

or

$$\Delta^2 w(k) \;\geq\; \frac{q(k)}{1+a} w(k-(\sigma-\tau)) \quad \text{for} \quad k \geq K_2. \tag{7.14.21}$$

But in view of Lemma 7.14.1 (II) and condition (7.14.12), inequality (7.14.21) has no eventually positive solution $w(k)$ with $(-1)^i \Delta^i w(k) > 0$, $i = 0, 1, 2$ eventually, which is a contradiction. ∎

**Theorem 7.14.3.** Let $b > 0$ and $\eta > \tau + 2$. If

$$\limsup_{k\to\infty} \sum_{\ell=k}^{k+\eta-\tau-2} (k+\eta-\tau-\ell-1)p(\ell) \;>\; b \tag{7.14.22}$$

and

$$\limsup_{k\to\infty} \sum_{\ell=k-(\sigma+\tau)}^{k} (k-\ell+1)q(\ell) \;>\; b \tag{7.14.23}$$

then $(7.14.1)_1$ is oscillatory.

**Proof.** Let $u(k)$ be an eventually positive solution of $(7.14.1)_1$, say $u(k) > 0$ for $k \geq k_0 \geq 0$. Define $z(k)$ as in (7.14.13) and obtain

$$\Delta^2 z(k) \;=\; -q(k)u(k-\sigma) - p(k)u(k+\eta) \;\leq\; 0 \quad \text{for} \quad k \geq k_1 \geq k_0. \tag{7.14.24}$$

As in Theorem 7.14.2, we consider the two cases (I) and (II).

(I) Assume $z(k) < 0$ for $k \geq k_1$. Set $y(k)$ as in (7.14.15) and obtain (7.14.16). Using (7.14.16) in (7.14.24), we have

$$\Delta^2 y(k) \geq \frac{p(k)}{b} y(k + (\eta - \tau)) \quad \text{for} \quad k \geq k_2 \geq k_1$$

or

$$\Delta^2 y(k) \geq \frac{q(k)}{b} y(k - (\sigma + \tau)) \quad \text{for} \quad k \geq k_2.$$

The rest of the proof is similar to that of Theorem 7.14.2 cases (II) (i) and (II) (ii) respectively.

(II) Assume $z(k) > 0$ for $k \geq k_1$. Set $w(k)$ as in (7.14.17) and obtain

$$\Delta^2 w(k) + q(k) z(k - \sigma) + p(k) z(k + \eta) = 0$$

and

$$\Delta^2 \left( w(k) + a w(k - \tau) - b w(k + \tau) \right) + q(k) w(k - \sigma) + p(k) w(k + \eta) = 0.$$

It is easy to check that $w(k) > 0$ and $\Delta w(k) > 0$ and $\Delta z(k) > 0$ for $k \geq k_2 \geq k_1$. Thus, there exist positive constants $c_1$ and $c_2$ and an integer $K \geq k_2$ such that

$$z(k - \sigma) \geq c_1 \quad \text{and} \quad z(k + \eta) \geq c_2 \quad \text{for} \quad k \geq K$$

and hence

$$\Delta^2 w(k) + c_1 q(k) + c_2 p(k) \leq 0 \quad \text{for} \quad k \geq K.$$

Thus,

$$0 < \Delta w(k) \leq \Delta w(K) - \sum_{\ell=K}^{k-1} (c_1 q(k) + c_2 p(k)) \to -\infty \quad \text{as} \quad k \to \infty,$$

which is a contradiction. ∎

Next, we give the following two criteria for the oscillation of $(7.14.2)_\delta$, $\delta = \pm 1$.

**Theorem 7.14.4.** Let $a > 0$ and $\eta > \tau + 2$. If

$$\limsup_{k \to \infty} \sum_{\ell=k}^{k+(\eta-\tau)-2} (k + \eta - \tau - \ell - 1) p(\ell) > 1 + b \tag{7.14.25}$$

and

(7.14.26) $$\limsup_{k\to\infty} \sum_{\ell=k-\sigma}^{k} (k-\ell+1)q(\ell) > 1+b$$

then $(7.14.2)_{-1}$ is oscillatory.

**Theorem 7.14.5.** Let $a > 0$, $\sigma > \tau$ and $\eta + \tau > 2$. If

(7.14.27) $$\limsup_{k\to\infty} \sum_{\ell=k}^{k+\eta+\tau-2} (k+\eta+\tau-\ell-1)p(\ell) > a$$

and

(7.14.28) $$\limsup_{k\to\infty} \sum_{\ell=k-(\sigma-\tau)}^{k} (k-\ell+1)q(\ell) > a$$

then $(7.14.2)_1$ is oscillatory.

**Proof.** The proofs of Theorems 7.14.4 and 7.14.5 can be modeled on that of Theorems 7.14.2 and 7.14.3. ∎

The following two theorems are concerned with the oscillatory behavior of $(7.14.3)_\delta$, $\delta = \pm 1$.

**Theorem 7.14.6.** Let $\sigma > \tau$ and $\eta > \tau + 2$. If

(7.14.29) $$\limsup_{k\to\infty} \sum_{\ell=k}^{k+\eta-\tau-2} (k+\eta-\tau-\ell-1)p(\ell) > 1+a+b$$

and

(7.14.30) $$\limsup_{k\to\infty} \sum_{\ell=k-(\sigma-\tau)}^{k} (k-\ell+1)q(\ell) > (1+a+b)$$

then $(7.14.3)_{-1}$ is oscillatory.

**Proof.** Let $u(k)$ be an eventually positive solution of $(7.14.3)_{-1}$, say $u(k) > 0$ for $k \geq k_0 \geq 0$. Set

(7.14.31) $$z(k) = u(k) + au(k-\tau) + bu(k+\tau).$$

Then $z(k) > 0$ for $k \geq k_1 \geq k_0$ and

$$\Delta^2 z(k) = q(k)u(k-\sigma) + p(k)u(k+\eta) \geq 0 \quad \text{for } k \geq k_2 \geq k_1,$$

which implies that $\Delta^i z(k),\ i = 0,1$ are eventually of one sign. Next, we set

$$w(k) = z(k) + az(k-\tau) + bz(k+\tau). \tag{7.14.32}$$

Then,

$$\Delta^2 w(k) = q(k)z(k-\sigma) + p(k)z(k+\eta) \tag{7.14.33}$$

and

$$\Delta^2\left(w(k) + aw(k-\tau) + bw(k+\tau)\right) = q(k)w(k-\sigma) + p(k)z(k+\eta). \tag{7.14.34}$$

Now, we consider the two cases:

(i) $\Delta z(k) > 0$ for $k \ge k_2$, and (ii) $\Delta z(k) < 0$ for $k \ge k_2$.

(i) Assume $\Delta z(k) > 0,\ k \ge k_2$. Then $\Delta w(k) > 0$ for $k \ge k_3 \ge k_2$ and

$$\begin{aligned}\Delta^2 w(k-\tau) &= q(k-\tau)z(k-\tau-\sigma) + p(k-\tau)z(k-\tau+\eta)\\ &= q(k)z(k-\tau-\sigma) + p(k)z(k-\tau+\eta)\\ &\le q(k)z(k-\sigma) + p(k)z(k+\eta) = \Delta^2 w(k),\quad k \ge k_3 \ge k_4\end{aligned}$$

and

$$\begin{aligned}\Delta^2 w(k+\tau) &= q(k+\tau)z(k+\tau-\sigma) + p(k+\tau)z(k+\tau+\eta)\\ &= q(k)z(k+\tau-\sigma) + p(k)z(k+\tau+\eta)\\ &\ge q(k)z(k-\sigma) + p(k)z(k+\eta) = \Delta^2 w(k),\quad k \ge k_4.\end{aligned}$$

From (7.14.34), we see that

$$(1+a+b)\Delta^2 w(k+\tau) \ge p(k)w(k+\eta),\quad k \ge k_4$$

or

$$\Delta^2 w(k) \ge \frac{p(k)}{1+a+b} w(k+(\eta-\tau)),\quad k \ge k_4.$$

The rest of the proof is similar to that of Theorem 7.14.2 (II) (i).

(ii) Assume $\Delta z(k) < 0,\ k \ge k_2$. Then $\Delta w(k) < 0$ for $k \ge K_1 \ge k_2$ and for $k \ge K_2 \ge K_1$ we have

$$\Delta^2 w(k-\tau) \ge \Delta^2 w(k) \ge \Delta^2 w(k+\tau)$$

and from (7.14.34), we have

$$(1+a+b)\Delta^2 w(k-\tau) \ge q(k)w(k-\sigma)$$

or

$$\Delta^2 w(k) \geq \frac{q(k)}{1+a+b} w(k-(\sigma-\tau)) \quad \text{for} \quad k \geq K_3 \geq K_2.$$

The rest of the proof is similar to that of Theorem 7.14.2 (II) (ii). ∎

**Theorem 7.14.7.** If $\sum_{k=k_0}^{\infty} q(k) = \infty$, or $\sum_{k=k_0}^{\infty} p(k) = \infty$, then the equation $(7.14.3)_1$ is oscillatory.

**Proof.** The proof is similar to those of earlier results. ∎

Finally, we present the following two theorems for the oscillation of $(7.14.4)_\delta$, $\delta = \pm 1$.

**Theorem 7.14.8.** If conditions (7.14.5) and (7.14.7) hold, then $(7.14.4)_{-1}$ is oscillatory.

**Proof.** Let $u(k)$ be an eventually positive solution of $(7.14.4)_{-1}$, say $u(k) > 0$ for $k \geq k_0 \geq 0$. Set

$$z(k) = u(k) - au(k-\tau) - bu(k+\tau). \tag{7.14.35}$$

Then

$$\Delta^2 z(k) = q(k)u(k-\sigma) + p(k)u(k+\eta) \geq 0 \quad \text{for} \quad k \geq k_1 \geq k_0,$$

which implies that $\Delta^i z(k)$, $i = 0, 1$ are eventually of one sign. Next, we consider the two cases (I) and (II) as in Theorem 7.14.2.

(I) Suppose that $z(k) < 0$ for $k \geq k_1$. Set

$$0 < y(k) = -z(k) = au(k-\tau) + bu(k+\tau) - u(k) \tag{7.14.36}$$

and hence we see that

$$\Delta^2 y(k) + q(k)u(k-\sigma) + p(k)u(k+\eta) = 0$$

and conclude that $\Delta y(k) > 0$ for $k \geq k_2 \geq k_1$.

Next, we let

$$V(k) = ay(k-\tau) + by(k+\tau) - y(k). \tag{7.14.37}$$

Then

$$\Delta^2 V(k) = q(k)y(k-\sigma) + p(k)y(k+\eta) \geq 0 \quad \text{for} \quad k \geq k_1 \geq k_0,$$

which implies that $\Delta V(k)$ is eventually of one sign. Now, we consider the two cases:

(i) $\Delta V(k) > 0$ for $k \geq k_1$ and (ii) $\Delta V(k) < 0$ for $k \geq k_1$.

(i) Suppose that $\Delta V(k) > 0$ for $k \geq k_1$. Then, for the function $V(k)$ defined in (7.14.37), we obtain

$$\Delta^2 V(k) = q(k)y(k-\sigma) + p(k)y(k+\eta)$$

and

$$\Delta^2 \left(aV(k-\sigma) + bV(k+\eta) - V(k)\right) = q(k)V(k-\sigma) + p(k)V(k+\eta).$$

As in Theorem 7.14.2 (II) (i), we see that $V(k) > 0$ and $\Delta V(k) > 0$ eventually and

$$(a+b)\Delta^2 V(k+\tau) \geq p(k)V(k+\eta)$$

or

$$\Delta^2 V(k) \geq \frac{p(k)}{a+b} V(k + (\eta - \tau)) \quad \text{eventually.}$$

The rest of the proof is similar to that of Theorem 7.14.2 (II) (i).

(ii) $\Delta V(k) < 0$ for $k \geq k_1$. As in Theorem 7.14.2 (II) (ii), we observe that $V(k) > 0$ and $\Delta V(k) < 0$ eventually and hence, we obtain

$$(a+b)\Delta^2 V(k-\tau) \geq q(k)V(k-\sigma)$$

or

$$\Delta^2 V(k) \geq \frac{q(k)}{a+b} V(k - (\sigma - \tau)) \quad \text{eventually.}$$

Again, the rest of the proof is similar to that of Theorem 7.14.2 (II) (ii).

(II) Suppose that $z(k) > 0$ for $k \geq k_1$. The proof of this case is similar to that of Theorem 7.14.3 (II). ∎

**Theorem 7.14.9.** Let $a + b > 0$, $\sigma > \tau$ and $\eta > \tau + 2$. If

$$\limsup_{k\to\infty} \sum_{\ell=k}^{k+\eta-\tau-2} (k + \eta - \tau - \ell - 1)p(\ell) > a + b \tag{7.14.38}$$

and

$$\limsup_{k\to\infty} \sum_{\ell=k-(\sigma-\tau)}^{k} (k - \ell + 1)q(\ell) > a + b \tag{7.14.39}$$

then $(7.14.4)_1$ is oscillatory.

**Proof.** The proof is similar to those of earlier results. ∎

## 7.15. Oscillation and Nonoscillation for

$$\Delta^n(u(k)+p(k)u(k-\tau))+q(k)f(u(k-\sigma)) \;=\; 0, \quad k\in\mathbb{N} \tag{7.15.1}$$

where $\tau\in\mathbb{N}(1)$, $\sigma\in\mathbb{N}$, functions $p(k)$, $q(k)$ are defined on $\mathbb{N}$, $q(k)\geq 0$, $k\in\mathbb{N}$, and is not eventually identically zero, the continuous function $f:\mathbb{R}\to\mathbb{R}$ is such that $uf(u)>0$ for all $u\neq 0$. Further, with respect to (7.15.1), we shall assume that there exists a function $F:\mathbb{R}\to\mathbb{R}$ such that $F$ is continuous and nondecreasing and satisfies the inequality

$$-F(-uv)\;\geq\;F(uv)\;\geq\;CF(u)F(v) \quad \text{for } u,\ v>0,$$

where $C$ is a positive constant, and

$$|f(u)|\;\geq\;|F(u)|, \quad \frac{F(u)}{u}\;\geq\;\gamma\;>\;0, \text{ also } uF(u)>0 \text{ for } u\neq 0.$$

**Theorem 7.15.1.** (a) Let $n$ be even. If $0\leq p(k)<1$, and

$$\text{(i)}\quad \liminf_{k\to\infty}\left[\sum_{\ell=k-\sigma}^{k-1}q(\ell)F\left((1-p(\ell-\sigma))\left(\frac{\ell-\sigma}{2^{n-1}}\right)^{(n-1)}\right)\right]$$
$$>\left(\frac{\sigma}{\sigma+1}\right)^{\sigma+1}\frac{1}{C^2\gamma F\left(\frac{1}{(n-1)!}\right)},$$

then the equation (7.15.1) is oscillatory.

(b) Let $n$ be odd. If $0\leq p(k)\leq P_1<1$, where $P_1$ is a constant, and

$$\text{(ii)}\quad \liminf_{k\to\infty}\left[\sum_{\ell=k-\sigma}^{k-1}q(\ell)F\left(\left(\frac{\ell-\sigma}{2^{n-1}}\right)^{(n-1)}\right)\right]\;>\;\left(\frac{\sigma}{\sigma+1}\right)^{\sigma+1}\frac{1}{C^2\gamma F(M)}$$

for every $M\in(0,1)$, then every solution of (7.15.1) either oscillates or tends to zero as $k\to\infty$.

**Proof.** Let $u(k)$ be a nonoscillatory solution of (7.15.1) with $u(k)>0$, $u(k-\tau)>0$ and $u(k-\sigma)>0$ for all $k\geq k_0\geq K_0$. Setting $z(k)=u(k)+p(k)u(k-\tau)$, we get $z(k)\geq u(k)>0$ and

$$\Delta^n z(k)\;=\;-q(k)f(u(k-\sigma))\;<\;0, \quad k\in\mathbb{N}(k_0). \tag{7.15.2}$$

It follows from Theorem 1.8.11 that for $n\geq 2$,

$$\Delta^{n-1}z(k)\;>\;0, \quad k\in\mathbb{N}(k_0). \tag{7.15.3}$$

We claim that $\Delta z(k)\leq 0$ eventually. This is obvious from the equation (7.15.1) in the case $n=1$. For $n\geq 2$, we suppose on the contrary that $\Delta z(k)>0$ for $k\geq k_1>k_0$. Then,

(7.15.4)
$$(1-p(k))z(k) \le z(k)-p(k)z(k-\tau) = u(k)-p(k)p(k-\tau)u(k-2\tau) \le u(k)$$

for $k \ge k_2 > k_1$. Since $z(k)$ is positive and increasing, it follows from Corollary 1.8.12 and (7.15.4) that

$$u(k) \ge (1-p(k))z(k) \ge \frac{1-p(k)}{(n-1)!}\left(\frac{k}{2^{n-1}}\right)^{(n-1)} \Delta^{n-1}z(k), \quad k \ge 2^{n-1}k_2. \tag{7.15.5}$$

From (7.15.5) for $k \ge k_3 > k_2$, we find

$$\begin{aligned} & f(u(k-\sigma)) \\ & \ge F(u(k-\sigma)) \\ & \ge F\left(\frac{1-p(k-\sigma)}{(n-1)!}\left(\frac{k-\sigma}{2^{n-1}}\right)^{(n-1)} \Delta^{n-1}z(k-\sigma)\right) \\ & \ge C^2\gamma F\left(\frac{1}{(n-1)!}\right) F\left((1-p(k-\sigma))\left(\frac{k-\sigma}{2^{n-1}}\right)^{(n-1)}\right) \Delta^{n-1}z(k-\sigma). \end{aligned}$$

It follows from (7.15.2) and the above inequality that $\Delta^{n-1}z(k)$ is an eventually positive (cf. (7.15.3)) solution of

$$\Delta w(k) + q(k)C^2\gamma F\left(\frac{1}{(n-1)!}\right) F\left((1-p(k-\sigma))\left(\frac{k-\sigma}{2^{n-1}}\right)^{(n-1)}\right) \times w(k-\sigma) \le 0.$$

But, in view of condition (i), this is a contradiction to Theorem 6.20.5. Hence, $\Delta z(k) \le 0$ eventually.

Since $\Delta z(k) \le 0$ eventually, in Theorem 1.8.11 we must have $m = j = 0$, and

$$(-1)^i\Delta^i z(k) > 0, \quad 0 \le i \le n-1, \quad k \in \mathbb{N}(k_0). \tag{7.15.6}$$

If $n$ is even, (7.15.6) yields a contradiction to (7.15.3). This proves part (a) of the theorem.

Now, let $n$ be odd. Assume further that $u(k)$ does not tend to zero as $k \to \infty$. As $\Delta z(k) \le 0$ eventually, we have $z(k) \downarrow c$ as $k \to \infty$, where $0 < c < \infty$. Then, there exists $\epsilon > 0$ and an integer $k_4 > k_0$ such that

$$0 < \epsilon < c\,\frac{1-P_1}{1+P_1} < c$$

and

$$(7.15.7) \qquad c-\epsilon \;<\; z(k) \;\le\; z(k-\tau) \;<\; c+\epsilon, \quad k \in \mathbb{N}(k_4).$$

Thus, from (7.15.4) and (7.15.7), we find for $k \in \mathbb{N}(k_4)$ that

(7.15.8)

$$u(k) \ge z(k)-p(k)z(k-\tau) \ge z(k)-P_1 z(k-\tau) > (c-\epsilon)-P_1(c+\epsilon) > c_1 z(k),$$

where $c_1 = [(c-\epsilon)-P_1(c+\epsilon)]/(c+\epsilon) \in (0,1)$. Let $m = j$ be as in Corollary 1.8.12. We have for $k \ge k_5 > k_4$ that

$$(7.15.9) \qquad z(k) \;=\; \frac{z(k)}{z\left(2^{j+1-n}k\right)}\, z\left(2^{j+1-n}k\right) \;>\; c_2\, z\left(2^{j+1-n}k\right),$$

where $c_2 = (c-\epsilon)/(c+\epsilon) \in (0,1)$. Combining (7.15.8) and (7.15.9) and using Corollary 1.8.12, we get for $k \ge k_6 > k_5$ that

$$\begin{aligned} u(k) \;&>\; c_1c_2\, z\left(2^{j+1-n}k\right) \\ &\ge\; c_1c_2\, \frac{\left(2^{j+1-n}k-k_6\right)^{(n-1)}}{(n-1)!}\, \Delta^{n-1}z(k) \\ &\ge\; \frac{c_1c_2}{(n-1)!}\, 2^{(j+1-n)(n-1)}\left(k-2^n k_6\right)^{(n-1)}\, \Delta^{n-1}z(k). \end{aligned}$$

Thus, for $k \ge 2^{n+1}k_6 + n - 2$ it follows that

$$\begin{aligned} (7.15.10) \qquad u(k) \;&\ge\; \frac{c_1c_2}{(n-1)!}\, 2^{(j+1-n)(n-1)}\, \frac{1}{2^{n-1}}\, k^{(n-1)}\Delta^{n-1}z(k) \\ &\ge\; c_3\left(\frac{k}{2^{n-1}}\right)^{(n-1)} \Delta^{n-1}z(k), \end{aligned}$$

where $c_3 = c_1c_2\, 2^{(j-n)(n-1)}/(n-1)! \in (0,1)$. Now, by (7.15.10), for $k \ge k_7 > k_6$, we obtain

$$f(u(k-\sigma)) \;\ge\; F(u(k-\sigma)) \;\ge\; C^2\gamma F(c_3) F\left(\left(\frac{k-\sigma}{2^{n-1}}\right)^{(n-1)}\right) \Delta^{n-1}z(k-\sigma).$$

Using the above inequality in (7.15.2), we see that $\Delta^{n-1}z(k)$ is an eventually positive (cf. (7.15.3)) solution of

$$\Delta w(k) + q(k)C^2\gamma F(c_3)F\left(\left(\frac{k-\sigma}{2^{n-1}}\right)^{(n-1)}\right) w(k-\sigma) \;\le\; 0.$$

In view of condition (ii), this is a contradiction to Theorem 6.20.5. This completes the proof of part (b). ∎

**Theorem 7.15.2.** If $-1 < -P_2 \le p(k) \le 0$, where $P_2 > 0$ is a constant, and (ii) hold, then every solution of (7.15.1) either oscillates or tends to zero as $k \to \infty$.

**Proof.** Let $u(k)$ be a nonoscillatory solution of (7.15.1) with $u(k) > 0$, $u(k-\tau) > 0$ and $u(k-\sigma) > 0$ for all $k \ge k_0 \ge K_0$. Further, we assume that $u(k)$ does not tend to zero as $k \to \infty$. Setting $z(k) = u(k) + p(k)u(k-\tau)$, we get $z(k) \le u(k)$ and also inequality (7.15.2).

We claim that $\Delta u(k) \le 0$ eventually. Suppose on the contrary that $\Delta u(k) > 0$ for $k \ge k_1 > k_0$. Then, for $k \ge k_2 > k_1$, we have

$$z(k) \ \ge\ u(k) + p(k)u(k) \ \ge\ (1 - P_2)u(k) \ >\ 0. \tag{7.15.11}$$

Thus, inequality (7.15.3) follows from Theorem 1.8.11. Since $u(k)$ is unbounded, it follows from (7.15.11) that $z(k)$ is also unbounded, and hence $\Delta z(k) > 0$, $k \in \mathbb{N}(k_2)$. Applying Corollary 1.8.12, we find

$$u(k) \ \ge\ z(k) \ \ge\ \frac{1}{(n-1)!}\left(\frac{k}{2^{n-1}}\right)^{(n-1)} \Delta^{n-1} z(k), \quad k \ge 2^{n-1}k_2.$$

Therefore, in view of the above inequality, for $k \ge k_3 > k_2$ we obtain

$$\begin{aligned} f(u(k-\sigma)) \ &\ge\ F(u(k-\sigma)) \\ &\ge\ C^2\gamma F\left(\frac{1}{(n-1)!}\right) F\left(\left(\frac{k-\sigma}{2^{n-1}}\right)^{(n-1)}\right) \Delta^{n-1} z(k-\sigma). \end{aligned}$$

It follows from (7.15.2) and the above inequality that $\Delta^{n-1}z(k)$ is an eventually positive (cf. (7.15.3)) solution of

$$\Delta w(k) + q(k)C^2\gamma F\left(\frac{1}{(n-1)!}\right) F\left(\left(\frac{k-\sigma}{2^{n-1}}\right)^{(n-1)}\right) w(k-\sigma) \le 0. \tag{7.15.12}$$

In view of condition (ii), this contradicts Theorem 6.20.5. Hence, $\Delta u(k) \le 0$ eventually. Consequently, $u(k) \downarrow c$ as $k \to \infty$, where $0 < c < \infty$.

From the definition of $z(k)$, we find

$$\liminf_{k\to\infty} z(k) \ =\ \left(1 + \liminf_{k\to\infty} p(k)\right) c \ \ge\ (1 - P_2)c \ >\ 0.$$

Hence, $z(k)$ is eventually positive and (7.15.3) holds. Since $z(k) \le u(k)$ and $u(k)$ is nonincreasing eventually, $z(k)$ is also nonincreasing eventually. Thus, $z(k) \downarrow d$ as $k \to \infty$, where $0 < d < \infty$. Given $\epsilon \in (0, d)$, there exists an integer $k_4 > k_0$ such that

$$d - \epsilon \ <\ z(k) \ <\ d + \epsilon, \quad k \in \mathbb{N}(k_4). \tag{7.15.13}$$

Let $m = j$ be as in Corollary 1.8.12. For $k \geq k_5 > k_4$, using (7.15.13) and Corollary 1.8.12 successively, we find

$$\begin{aligned} z(k) &= \frac{z(k)}{z\left(2^{j+1-n}k\right)}\, z\left(2^{j+1-n}k\right) \\ &> \frac{d-\epsilon}{d+\epsilon}\, z\left(2^{j+1-n}k\right) \\ &\geq \frac{d-\epsilon}{d+\epsilon}\, \frac{\left(2^{j+1-n}k - k_5\right)^{(n-1)}}{(n-1)!}\, \Delta^{n-1}z(k) \\ &\geq \frac{d-\epsilon}{d+\epsilon}\, \frac{1}{(n-1)!}\, 2^{(j+1-n)(n-1)}(k - 2^k k_5)^{(n-1)}\, \Delta^{n-1}z(k). \end{aligned}$$

It follows that for $k \geq 2^{n+1}k_5 + n - 2$,

$$\begin{aligned} z(k) &\geq \frac{d-\epsilon}{d+\epsilon}\, \frac{1}{(n-1)!}\, 2^{(j+1-n)(n-1)}\, \frac{1}{2^{n-1}}\, k^{(n-1)}\Delta^{n-1}z(k) \\ &\geq d_1 \left(\frac{k}{2^{n-1}}\right)^{(n-1)} \Delta^{n-1}z(k), \end{aligned} \tag{7.15.14}$$

where $d_1 = 2^{(j-n)(n-1)}(d-\epsilon)/[(d+\epsilon)(n-1)!] \in (0,1)$. By (7.15.14), we obtain for $k \geq k_6 > k_5$,

$$\begin{aligned} f(u(k-\sigma)) &\geq F(u(k-\sigma)) \geq F(z(k-\sigma)) \\ &\geq C^2\gamma F(d_1) F\left(\left(\frac{k-\sigma}{2^{n-1}}\right)^{(n-1)}\right) \Delta^{n-1}z(k-\sigma). \end{aligned}$$

Using the above inequality in (7.15.2), we find that $\Delta^{n-1}z(k)$ is an eventually positive (cf. (7.15.3)) solution of

$$\Delta w(k) + q(k)C^2\gamma F(d_1) F\left(\left(\frac{k-\sigma}{2^{n-1}}\right)^{(n-1)}\right) w(k-\sigma) \leq 0.$$

In view of condition (ii), once again we get a contradiction to Theorem 6.20.5. ∎

**Theorem 7.15.3.** Let $p(k) \equiv -1$. If (ii) holds, then every solution of (7.15.1) either oscillates or tends to zero as $k \to \infty$.

**Proof.** Let $u(k)$ be a nonoscillatory solution of (7.15.1) with $u(k) > 0$, $u(k-\tau) > 0$ and $u(k-\sigma) > 0$ for all $k \geq k_0 \geq K_0$. Further, we assume that $u(k)$ does not tend to zero as $k \to \infty$. Setting $z(k) = u(k) - u(k-\tau)$, we get $z(k) < u(k)$ and also inequality (7.15.2).

If $z(k) < 0$ eventually, then $u(k) < u(k-\tau)$ eventually and therefore $u(k)$ is bounded. This in turn implies that $z(k)$ is bounded. If $z(k) >$

0 eventually, then $z(k)$ is also bounded. To show this, suppose that $z(k)$ is not bounded, i.e. $\Delta z(k) > 0$ for sufficiently large $k$. Using Corollary 1.8.12 and proceeding as in the proof of Theorem 7.15.2, we conclude that $\Delta^{n-1}z(k)$ is an eventually positive solution of (7.15.12), which is a contradiction due to condition (ii). Hence, $z(k)$ is bounded.

Next, let $\lim_{k\to\infty} u(k) = \mu > 0$. Given $\epsilon \in (0,\mu)$, there exists $k_1 > k_0$ such that $u(k-\sigma) > \mu - \epsilon$ for $k \ge k_1$. It follows that

$$f(u(k-\sigma)) \ge F(u(k-\sigma)) \ge F(\mu-\epsilon), \quad k \in \mathbb{N}(k_1). \tag{7.15.15}$$

Multiplying (7.15.1) by $(k-\sigma)^{(n-1)}$ and summing from $k_1$ to $k$, we get

(7.15.16)
$$\begin{aligned}\sum_{\ell=k_1}^{k}(\ell-\sigma)^{(n-1)}q(\ell)f(u(\ell-\sigma)) &= -\sum_{\ell=k_1}^{k}(\ell-\sigma)^{(n-1)}\Delta^n z(\ell)\\ &\le -\sum_{\ell=k_1}^{k}\ell^{(n-1)}\Delta^n z(\ell)\\ &= -\sum_{i=0}^{n-1}(-1)^i\Delta^i\ell^{(n-1)}\Delta^{n-i-1}z(\ell+i)\Bigg|_{\ell=k_1}^{k+1}\\ &< \infty, \quad k\in\mathbb{N}(k_1)\end{aligned}$$

where in the last inequality we have used the fact that $z(k)$ is bounded implies that $\Delta^{n-i-1}z(k)$, $0 \le i \le n-1$ is bounded. Coupling (7.15.15) and (7.15.16), we find

$$F(\mu-\epsilon)\sum_{\ell=k_1}^{k}(\ell-\sigma)^{(n-1)}q(\ell) < \infty, \quad k\in\mathbb{N}(k_1)$$

or equivalently,

$$\sum_{\ell=k_1}^{\infty}(\ell-\sigma)^{(n-1)}q(\ell) < \infty. \tag{7.15.17}$$

However, condition (ii) implies that

$$\sum_{\ell=k_1}^{\infty}(\ell-\sigma)^{(n-1)}q(\ell) = \infty,$$

which contradicts (7.15.17). ∎

**Theorem 7.15.4.** Let $p(k) \equiv 1$ and $\sum^{\infty} q(\ell) = \infty$.

(a) If $n$ is even, then the equation (7.15.1) is oscillatory.

(b) If $n$ is odd, then every solution of (7.15.1) either oscillates or tends to zero as $k \to \infty$.

**Proof.** Let $u(k)$ be a nonoscillatory solution of (7.15.1) with $u(k) > 0$, $u(k-\tau) > 0$ and $u(k-\sigma) > 0$ for all $k \ge k_0 \ge K_0$. Setting $z(k) = u(k) + u(k-\tau)$, we get $z(k) > 0$, $k \in K(k_0)$ and also inequalities (7.15.2) and (7.15.3). Summing (7.15.1) from $k_0$ to $(k-1)$, and using (7.15.3), we get

$$\Delta^{n-1} z(k_0) \;=\; \sum_{\ell=k_0}^{k-1} q(\ell) f(u(\ell-\sigma)) + \Delta^{n-1} z(k) \;>\; \sum_{\ell=k_0}^{k-1} q(\ell)\gamma u(\ell-\sigma),$$

which implies

$$\sum_{\ell=k_0}^{\infty} q(\ell) u(\ell-\sigma) \;<\; \infty. \tag{7.15.18}$$

We claim that if $\liminf_{k\to\infty} u(k) > 0$, then $\sum^{\infty} q(\ell) < \infty$. To show this, suppose on the contrary that $\sum^{\infty} q(\ell) = \infty$. Let $L = \inf_{\ell \ge k_0} u(\ell - \sigma)(>0)$. Then, we have

$$\sum_{\ell=k_0}^{\infty} q(\ell) u(\ell-\sigma) \;\ge\; L \sum_{\ell=k_0}^{\infty} q(\ell) \;=\; \infty,$$

which contradicts (7.15.18).

Case (a). $n$ is even. From Theorem 1.8.11, we see that $m = j$ is odd and hence $\Delta z(k) > 0$, $k \in \mathbb{N}(k_0)$. This means that for $k \ge k_1 > k_0$,

$$0 \;<\; z(k) - z(k-\tau) \;=\; u(k) - u(k-2\tau)$$

or, $u(k) > u(k-2\tau)$, $k \in \mathbb{N}(k_1)$. Therefore, $\liminf_{k\to\infty} u(k) > 0$. We have seen that this leads to $\sum^{\infty} q(\ell) < \infty$, which is a contradiction to $\sum^{\infty} q(k) = \infty$.

Case (b). $n$ is odd. We assume further that $u(k)$ does not tend to zero as $k \to \infty$. From Theorem 1.8.11, we see that $m = j$ is even. If $j \ge 2$, then again we have $\Delta z(k) > 0$, $k \in \mathbb{N}(k_0)$. Proceeding as in Case (a), we obtain a contradiction. If $j = 0$, then from Theorem 1.8.11 we have $\Delta z(k) < 0$, $k \in \mathbb{N}(k_0)$. Thus, $z(k) \downarrow \beta$ as $k \to \infty$, where $0 < \beta < \infty$. For $\epsilon \in (0,\beta)$, there exists an integer $k_1 > k_0$ such that

$$z(k) \;=\; u(k) + u(k-\tau) \;>\; \beta - \epsilon \;>\; 0, \quad k \in \mathbb{N}(k_1).$$

Hence, $\liminf_{k\to\infty} u(k) > 0$. We have noted earlier that this leads to $\sum^{\infty} q(\ell) < \infty$, which is a contradiction to $\sum^{\infty} q(k) = \infty$. ∎

**Remark 7.15.1.** Consider the following difference equation which is more general than (7.15.1)

$$\Delta^n(u(k)+p(k)u(\tau(k)))+q(k)f(u(k-\sigma)) = 0, \quad k\in\mathbb{N} \tag{7.15.19}$$

where $\tau:\mathbb{N}\to\mathbb{N}$ is nondecreasing and $\tau(k)\le k$, $\lim_{k\to\infty}\tau(k)=\infty$. We observe that with obvious slight modifications in the proofs, Theorems 7.15.1 7.15.4 also hold for the equation (7.15.19).

**Example 7.15.1.** Consider the difference equation

(7.15.20)
$$\Delta^n\left(u(k)-\frac{1}{2}u(k-\tau)\right)+(-1)^{n+\sigma+1}\frac{3}{e}2^{n-1}u(k-\sigma)\exp(|u(k-\sigma)|)=0,$$

where $\tau$ is any positive odd integer and $\sigma\in\mathbb{N}$ is such that $(n+\sigma)$ is odd. By taking $F(u)=u$, we note that all the conditions of Theorem 7.15.2 are satisfied. In fact, $u(k)=(-1)^k$ is an oscillatory solution of (7.15.20).

**Example 7.15.2.** Consider the difference equation

$$\Delta^n(u(k)-u(k-\tau))+(-1)^{n+\sigma+1}\,2^{n+1}\,u(k-\sigma)|u(k-\sigma)| = 0, \tag{7.15.21}$$

where $\tau$ is any positive odd integer and $\sigma\in\mathbb{N}$ is such that $(n+\sigma)$ is odd. By choosing $F(u)=f(u)=u|u|$, it is noted that all the hypotheses of Theorem 7.15.3 are fulfilled. In fact, the equation (7.15.21) has an oscillatory solution $u(k)=(-1)^k$.

**Example 7.15.3.** Consider the difference equation

$$\Delta^n\,(u(k)+u(k-\tau))+(-1)^{n+\sigma+1}2^n u(k-\sigma)(1+|u(k-\sigma)|)=0, \tag{7.15.22}$$

where $\tau\in\mathbb{N}(1)$, $\sigma\in\mathbb{N}$ is such that $(n+\sigma)$ is odd. By taking $F(u)=u$, we find that all the conditions of Theorem 7.15.4 are satisfied. It is noted that $u(k)=(-1)^k$ is an oscillatory solution of (7.15.22).

## 7.16. Oscillation and Nonoscillation for

$$\Delta\left(a(k)\Delta^{n-1}(u(k)-p(k)u(k-\tau))\right)+\delta q(k)f(u(\sigma(k)))=0,\ k\in\mathbb{N} \tag{7.16.1}$$

where $\delta=\pm1$, $\tau\in\mathbb{N}(1)$, $a(k)>0$ with $\Delta a(k)\ge0$, $k\in\mathbb{N}$ and $\sum^\infty 1/a(k)=\infty$, $1<c\le p(k)\le C<\infty$ for some real numbers $c$ and $C$, $q(k)\ge0$, $k\in\mathbb{N}$, $\sigma(k)\in\mathbb{N}$, $k\in\mathbb{N}$ and $\lim_{k\to\infty}\sigma(k)=\infty$, and $f:\mathbb{R}\to\mathbb{R}$ is continuous such that $uf(u)>0$ for $u\ne0$.

**Theorem 7.16.1.** Suppose that

$$\sum^{\infty} k^{n-1}\frac{q(k)}{a(k)} = \infty. \tag{7.16.2}$$

Then, the following hold

(a) every bounded solution $u(k)$ of (7.16.1) is oscillatory when $(-1)^n\delta = -1$

(b) every bounded solution $u(k)$ of (7.16.1) is either oscillatory or satisfies $\lim_{k\to\infty} u(k) = 0$ when $(-1)^n\delta = 1$.

**Proof.** Let $u(k)$ be an eventually positive solution of (7.16.1). Set $z(k) = u(k) - p(k)u(k-\tau)$. If $z(k)$ is eventually positive, then we have $u(k) > p(k)u(k-\tau) \geq cu(k-\tau)$, and therefore by induction $u(k) > c^j u(k-j\tau)$ or $u(k+j\tau) > c^j u(k)$ for every positive integer $j$. Letting $j\to\infty$, we find $\lim_{j\to\infty} u(k) = \infty$. Since this is a contradiction to $u(k)$ being bounded, we conclude that $z(k)$ is eventually negative. It follows from (7.16.1) that $\delta\Delta(a(k)\Delta^{n-1}z(k))$ is also eventually negative. Thus, eventually $\delta\Delta^{n-1}z(k)$ is either positive or negative. Suppose that it is eventually negative, then there is a $k_1 \in \mathbb{N}$ such that for $k \geq k_1$,

$$\delta a(k)\Delta^{n-1}z(k) \leq \delta a(k_1)\Delta^{n-1}z(k_1) < 0.$$

Dividing both sides of this inequality by $a(k)$ and summing from $k_1$ to $k$, we obtain

$$\delta\Delta^{n-1}z(k+1) \leq \delta a(k_1)\Delta^{n-1}z(k_1)\sum_{\ell=k_1}^{k}\frac{1}{a(\ell)}.$$

This inequality in view of $\sum^{\infty} 1/a(k) = \infty$ implies that $\delta\Delta^{n-1}z(k) \to -\infty$ as $k\to\infty$, which is a contradiction to $z(k)$ being bounded. Thus, $\delta\Delta^{n-1}z(k)$ is eventually positive. Now from (7.16.1), we have

$$\delta a(k)\Delta^n z(k) = -(\Delta a(k))\left(\delta\Delta^{n-1}z(k+1)\right) - q(k)f(u(\sigma(k))). \tag{7.16.3}$$

Since $\Delta a(k)$ and $q(k)$ are nonnegative, (7.16.3) implies that $\delta\Delta^n z(k)$ is eventually negative. In view of the fact that $z(k)$ is bounded, applying Theorem 1.8.11, we see that there are integers $k_2 \geq k_1$ and $m \in \{0,1\}$, $(-1)^{n-m}\delta = 1$, such that for $k \geq k_2$

$$\begin{aligned} &\Delta^j z(k) < 0, \quad j = 0,1,\cdots,m \\ &(-1)^{j-m}\Delta^j z(k) < 0, \quad j = m+1,\cdots,n. \end{aligned} \tag{7.16.4}$$

It is clear from (7.16.3) that

$$\delta\Delta^n z(k) + \frac{q(k)}{a(k)}f(u(\sigma(k))) \leq 0. \tag{7.16.5}$$

Multiplying (7.16.5) by $k^{n-1}$ and summing from $k_2$ to $k$ and then applying the summation by parts formula to the first term in the resulting inequality, we obtain

$$(7.16.6)\qquad \sum_{i=0}^{n-2}(-1)^{i+1}\delta\left(\Delta^i k_2^{n-1}\right)\left(\Delta^{n-i-1}z(k_2+i)\right)$$
$$+(-1)^{n-1}\delta\,(n-1)![z(k+n)-z(k_2+n-1)]$$
$$+\sum_{\ell=k_2}^{k}\ell^{n-1}\frac{q(\ell)}{a(\ell)}f(u(\sigma(\ell)))\ \le\ 0.$$

Since $z(k)$ is bounded, if we let $k\to\infty$ in (7.16.6) then we must have

$$(7.16.7)\qquad \sum_{\ell=k_2}^{\infty}\ell^{n-1}\frac{q(\ell)}{a(\ell)}f(u(\sigma(\ell)))\ <\ \infty.$$

From (7.16.2) and (7.16.7) it follows that $\liminf_{k\to\infty}u(k)=0$.

Now we shall show that $\lim_{k\to\infty}z(k)=0$. Clearly,

$$(7.16.8)\quad z(k+\tau)-z(k)\ =\ u(k+\tau)-(p(k+\tau)+1)u(k)+p(k)u(k-\tau).$$

Let $\{k_j\}$ be such that $k_j\to\infty$ as $j\to\infty$, and $u(k_j)\to 0$ as $j\to\infty$. Then, from (7.16.8) we get

$$\lim_{j\to\infty}\left[u(k_j+\tau)+p(k_j)u(k_j-\tau)\right]\ =\ 0.$$

As $u(k_j+\tau)>0$ and $p(k_j)u(k_j-\tau)>0$, we see that $p(k_j)u(k_j-\tau)\to 0$ as $j\to\infty$. If we now use the fact that $p(k)$ is bounded and $z(k_j)=u(k_j)-p(k_j)u(k_j-\tau)$, then we have $\lim_{k\to\infty}z(k)=0$.

If $(-1)^n\delta=-1$, then it follows from (7.16.4) that $m=1$ and consequently $z(k)$ is negative and decreasing. In this case, $\lim_{k\to\infty}z(k)=0$ is not possible, and therfore $u(k)$ must be oscillatory.

If $(-1)^n\delta=1$, then it follows from (7.16.4) that $m=0$ and so $z(k)$ increases to 0 as $k\to\infty$, i.e. given $\epsilon>0$ there exists a $k_3\ge k_2$ such that $z(k)>-\epsilon$ for all $k\ge k_3$. Thus, $u(k)-p(k)u(k-\tau)>-\epsilon$ for $k\ge k_3$, or $u(k)>-\epsilon+cu(k-\tau)$ for $k\ge k_3$, or $cu(k)<\epsilon+u(k+\tau)$ for $k\ge k_3$. By induction, we have

$$c^ju(k)\ <\ \epsilon+c\epsilon+\cdots+c^{j-1}\epsilon+u(k+j\tau)\quad \text{for}\quad k\ge k_3.$$

Let $M$ be a bound for $u(k)$, then it follows from the last inequality that

$$(7.16.9)\qquad u(k)\ <\ \frac{1-c^{-j}}{c-1}\epsilon+Mc^{-j}.$$

Since $\lim_{j\to\infty} c^{-j} = 0$ and $\epsilon > 0$ is arbitrary, (7.16.9) implies that $u(k) \to 0$ as $k \to \infty$. ∎

## 7.17. Problems

**7.17.1.** Consider the difference equation

$$\Delta(u(k) + pu(k-\tau)) + qu(k-\sigma) = 0, \quad k \in \mathbb{N} \tag{7.17.1}$$

where $p$, $q$ are constants, $\tau \in \mathbb{N}(1)$, and $\sigma \in \mathbb{N}$. Show that all solutions of (7.17.1) are oscillatory if and only if the characteristic equation

$$\lambda - 1 + \lambda^{-\tau}(\lambda - 1)p + q\lambda^{-\sigma} = 0$$

of (7.17.1) has no positive roots. In particular, deduce that all solutions of (7.17.1) are oscillatory

(i) for the case $\sigma = \tau = 1$ if and only if $q \ge p \ge 1$, or $p < 1$ and $q > (p+1)^2/4$

(ii) for the case $\sigma = 0$, $\tau = 1$ if and only if $p < 0$ and $q > 1 - p - 2\sqrt{-p}$, or $p = 0$ and $q = 1$.

**7.17.2.** Consider the difference equation (7.1.1) where $p = -1$, $\tau \in \mathbb{N}(1)$, $\sigma \in \mathbb{N}$, $0 \le kq(k) < \tau$ for $k \in \mathbb{N}(1)$, and

$$\sum_{k=1}^{\infty} q(k) \prod_{\ell=1}^{k} \left(1 - \frac{\ell q(\ell)}{\tau}\right)^{-1} = \infty.$$

Show that the equation (7.1.1) is oscillatory.

**7.17.3.** Consider the difference equation (7.1.1) where $p = -1$, $\tau \in \mathbb{N}(1)$, $\sigma \in \mathbb{N}$, $q(k) \ge 0$, $k \in \mathbb{N}$, and

$$\sum_{k=1}^{\infty} q(k) \exp\left(\frac{1}{\tau} \sum_{\ell=1}^{k} \ell q(\ell)\right) = \infty.$$

Show that the equation (7.1.1) is oscillatory.

**7.17.4.** Consider the difference equation (7.1.1) where $p = -1$, $\tau \in \mathbb{N}(1)$, $\sigma \in \mathbb{N}$, and $q(k) \ge 0$, $k \in \mathbb{N}$. Assume that there exists a function $0 \le \lambda(k) \le 1$, $k \in \mathbb{N}$ such that

$$\sum_{k=1}^{\infty} (1 - \lambda(k)) q(k) \exp\left[2\tau^{-1/2} \sum_{\ell=1}^{k} (\lambda(\ell) q(\ell))^{1/2}\right] = \infty.$$

Show that the equation (7.1.1) is oscillatory.

**7.17.5.** Consider the difference equation (7.1.1) where $p = -1$, $\tau \in \mathbb{N}(1)$, $\sigma \in \mathbb{N}$, and $q(k) \geq 0$, $k \in \mathbb{N}$. Show that

(i) if there exists a $\tau^* > \tau$ such that $4k^2 q(k) \geq \tau^*$ eventually, then the equation (7.1.1) is oscillatory

(ii) if $4k^2 q(k) \leq \tau$ eventually, then (7.1.1) has a nonoscillatory solution

(iii) if $\liminf_{k\to\infty} 4k^2 q(k) < \tau$, then (7.1.1) has a nonoscillatory solution

(iv) if $\limsup_{k\to\infty} 4k^2 q(k) > \tau$, then the equation (7.1.1) is oscillatory.

**7.17.6.** Consider the difference equation (7.1.1) where $p = -1$, $\tau \in \mathbb{N}(1)$, $\sigma \in \mathbb{N}$, and $q(k) = k^{-\alpha}$, $\alpha \in \mathbb{R}$. Show that the equation (7.1.1) is oscillatory if and only if $\alpha < 2$, or $\alpha = 2$ and $\tau < 4$. In particular, show that the *discrete Euler equation*

$$\Delta^2 u(k-1) + \gamma k^{-2} u(k) \;=\; 0, \qquad k \in \mathbb{N}(1)$$

is oscillatory if and only if $\gamma > 1/4$.

**7.17.7.** Consider the difference equation (7.1.1) where $p$ is a constant, $\tau \in \mathbb{N}(1)$, $\sigma \in \mathbb{N}$, $q(k) \geq 0$ eventually, and condition (7.1.3) holds. Let $u(k)$ be a nonoscillatory solution of (7.1.1). Show that

(i) if $p > -1$ and $p \neq 1$, then $\lim_{k\to\infty} u(k) = 0$

(ii) if $p = 1$, then construct an equation of the type (7.1.1) for which $\lim_{k\to\infty} u(k) \neq 0$

(iii) if $p = 1$, $q(k) > 0$ and $\displaystyle\limsup_{k\to\infty} \frac{q(k)}{q(k-\tau)} = \beta < \infty$, then $\lim_{k\to\infty} u(k) = 0$

(iv) if $p = 1$ and condition (7.1.3) is replaced by $\sum_{k=\tau}^{\infty} q^*(k) = \infty$ where $q^*(k) = \min\{q(k), q(k-\tau)\}$, then $\lim_{k\to\infty} u(k) = 0$.

**7.17.8.** Consider the difference equation (7.1.1) where $-1 < p < 0$, $\tau \in \mathbb{N}(1)$, $\sigma \in \mathbb{N}$, and $\liminf_{k\to\infty} q(k) = A > 0$ and $\limsup_{k\to\infty} q(k) = 1 - A$, or $\liminf_{k\to\infty} q(k) > \sigma^\sigma/(\sigma+1)^{\sigma+1}$. Show that the equation (7.1.1) is oscillatory.

**7.17.9.** Consider the difference equation (7.1.1) where $-1 \leq p \leq 0$, $q(k) \geq q > 0$, $\tau \in \mathbb{N}(1)$, $\sigma \in \mathbb{N}$, and

$$\inf_{k \geq K, \nu > 1} \left\{ \frac{1}{\nu} + q(k)\nu^\sigma - p\frac{q(k)}{q(k-\tau)}\nu^{\tau-1}(\nu - 1) \right\} > 1.$$

Show that the equation (7.1.1) is oscillatory.

**7.17.10.** Consider the difference equations (7.1.1) and (7.3.12) where $-1 < \overline{p} \le p < 0$, $\tau \in \mathbb{N}(1)$, $\sigma \in \mathbb{N}$, $h(k) \ge q(k) > 0$, $k \in \mathbb{N}$ and $\sum_{k=K}^{\infty} h(k) = \infty$. Show that the oscillation of (7.1.1) implies that of (7.3.12).

**7.17.11.** Consider the difference equation (7.3.18) where $\tau \in \mathbb{N}(1)$, $\sigma_i \in \mathbb{N}$, $1 \le i \le r$, $0 < \sigma_1 < \sigma_2 < \cdots < \sigma_r$, $q_i(k) \ge 0$, $k \in \mathbb{N}$, $1 \le i \le r$ and not identically zero. Let $u(k)$ be a solution of (7.3.18) and define $z(k) = u(k) + pu(k-\tau)$ and $w(k) = z(k) + pz(k-\tau)$. Show that

(i) if $p \in \mathbb{R}$ and the functions $q_i(k)$, $1 \le i \le r$ are periodic with period $\tau$, then $z(k)$ and $w(k)$ are solutions of (7.3.18) for all large $k$

(ii) if $-1 \le p < 0$ and $u(k)$ is eventually positive, then $z(k)$ is eventually positive and nonincreasing; further if $q_i(k)$, $1 \le i \le r$ are periodic with period $\tau$, then $w(k)$ is an eventually positive nonincreasing solution of (7.3.18)

(iii) if $p < -1$, there is an index $j \in \{1, \cdots, r\}$ such that $\sum^{\infty} q_j(k) = \infty$ and $u(k)$ is eventually positive, then $z(k)$ is eventually negative and nonincreasing; further if $q_i(k)$, $1 \le i \le r$ are periodic with period $\tau$, then $w(k)$ is an eventually positive nondecreasing solution of (7.3.18).

**7.17.12.** Consider the difference equation (7.3.18) where $\tau \in \mathbb{N}(1)$, $\sigma_i \in \mathbb{N}$, $1 \le i \le r$, $0 < \sigma_1 < \sigma_2 < \cdots < \sigma_r$, $q_i(k) \ge 0$, $k \in \mathbb{N}$, $1 \le i \le r$ and not identically zero. Further, assume that $q_i(k)$, $1 \le i \le r$ are periodic with periodic $\tau$ and denote $q(k) = \sum_{i=1}^{r} q_i(k)$, $k \in \mathbb{N}$. Show that

(i) if $-1 < p < 0$ and the inequality

$$\Delta v(k) + \frac{1}{1+p} q(k) v(k-\sigma_1) \le 0, \quad k \in \mathbb{N}$$

has no eventually positive solution, then the equation (7.3.18) is oscillatory

(ii) if $p = -1$ and the inequality

$$\Delta v(k) + q(k) v(k-\sigma_1) \le 0, \quad k \in \mathbb{N}$$

has no eventually positive solution, then the equation (7.3.18) is oscillatory

(iii) if $p > 0$, $\tau < \sigma_1$ and the inequality

$$\Delta v(k) + \frac{1}{1+p} q(k) v(k-(\sigma_1-\tau)) \le 0, \quad k \in \mathbb{N}$$

has no eventually positive solution, then the equation (7.3.18) is oscillatory

(iv) if $p < -1$, $\tau > \sigma_r$, there exists an index $j \in \{1, \cdots, r\}$ such that $\sum^{\infty} q_j(k) = \infty$, and the inequality

$$\Delta v(k) + \frac{1}{1+p} q(k) v(k + \tau - \sigma_r) \ \geq \ 0, \quad k \in \mathbb{N}$$

has no eventually positive solution, then the equation (7.3.18) is oscillatory

(v) if $p < -1$, $\tau > \sigma_r$, there exists an index $j \in \{1, \cdots, r\}$ such that $\sum^{\infty} q_j(k) = \infty$, and the inequality

$$\Delta v(k) + \frac{1}{p} q(k) v(k + \tau - \sigma_r) \ \leq \ 0, \quad k \in \mathbb{N}$$

has no eventually negative solution, then the equation (7.3.18) is oscillatory.

**7.17.13.** Consider the difference equation (7.3.18) where $-1 < p < 0$, $\tau \in \mathbb{N}(1)$, $\sigma_i$, $1 \leq i \leq r$ are integers with arbitrary sign, $q_i(k) \equiv q_i > 0$, $1 \leq i \leq r$, and

$$\sum_{i=1}^{r} q_i (-p)^{\mu_i} \left( \frac{\overline{\sigma}_i + 1}{\overline{\sigma}_i} \right)^{\overline{\sigma}_i} \left( \overline{\sigma}_i + 1 + \frac{p\tau}{1+p} \right) \ > \ 1 + p,$$

where $\mu_i$ is the smallest nonnegative integers such that $\overline{\sigma}_i = \mu_i \tau + \sigma_i > 0$, $1 \leq i \leq r$. Show that the equation (7.3.18) is oscillatory.

**7.17.14.** Consider the difference equation (7.1.1) where $|p| < 1$, $\tau \in \mathbb{N}(1)$, $\sigma \in \mathbb{N}$, and $q(k) \geq 0$, $k \in \mathbb{N}$. Assume that there exist a constant $\alpha > 0$ and an integer $K$ sufficiently large such that

$$(7.17.2) \quad 2|p|(2 - |p|) + \sum_{\ell=k}^{k+\sigma} q(\ell) \ \leq \ \alpha \ < \ \frac{3}{2} + \frac{(1 - 2|p|)^2}{2(\sigma + 1)}, \quad k \in \mathbb{N}(K).$$

Show that every oscillatory solution of (7.1.1) is bounded.

**7.17.15.** Consider the difference equation (7.1.1) where $|p| < 1$, $\tau \in \mathbb{N}(1)$, $\sigma \in \mathbb{N}$, $q(k) \geq 0$, $k \in \mathbb{N}$, and $\sum_{k=1}^{\infty} q(k) = \infty$. Assume that there exist a constant $\alpha > 0$ and an integer $K$ sufficiently large such that (7.17.2) holds. Show that every bounded oscillatory solution of (7.1.1) tends to zero as $k \to \infty$.

**7.17.16.** A combination of Theorem 7.2.6 and Problems 7.17.14 and 7.17.15 implies that in Theorem 7.4.1 condition (7.4.1) can be replaced by (7.17.2). Construct examples to show that conditions (7.4.1) and (7.17.2) in general cannot be compared.

**7.17.17.** Consider the difference equation (7.5.1) where $-1 < p \le 0$, $\tau \in \mathbb{N}(1)$, $\sigma_1$, $\sigma_2 \in \mathbb{N}$, $\sigma_1 > \sigma_2 + 1$, $q(k) \ge 0$, $h(k) \ge 0$, $\overline{q}(k) = q(k) - h(k - (\sigma_1 - \sigma_2)) \ge 0$, and

$$\sum_{\ell=k-(\sigma_1-\sigma_2)}^{k-1} h(\ell) \le 1 + p \quad \text{for all large } k, \qquad \liminf_{k\to\infty} \sum_{\ell=k-\sigma_1}^{k-1} \overline{q}(\ell) > 0$$

$$\inf_{k\ge K\in\mathbb{N},\lambda>0} \left\{ \frac{1}{\lambda} \prod_{\ell=k-\sigma_1+1}^{k} (1 - \lambda\overline{q}(\ell-1))^{-1} - p \prod_{\ell=k-\tau+1}^{k} (1 - \lambda\overline{q}(\ell-1))^{-1} \right.$$
$$\left. + \sum_{\ell=k-\sigma_1}^{k-\sigma_2-1} h(\ell - (\sigma_1 - \sigma_2)) \prod_{j=\ell+1}^{k} (1 - \lambda\overline{q}(j-1))^{-1} \right\} > 1.$$

Show that the equation (7.5.1) is oscillatory.

**7.17.18.** Consider the difference equation (7.5.1) where $p \le 0$, $\tau \in \mathbb{N}(1)$, $\sigma_1$, $\sigma_2 \in \mathbb{N}$, $\sigma_1 > \sigma_2 + 1$, $q(k) \ge 0$, $h(k) \ge 0$, $\overline{q}(k) = q(k) - h(k - (\sigma_1 - \sigma_2)) \ge 0$, $\sum_{k=1}^{\infty} \overline{q}(k) = \infty$, and there exist $K \in \mathbb{N}$ and $\lambda^* > 0$ such that $1 - \lambda^*\overline{q}(k-1) > 0$, $k = K, K+1, \cdots$, and

$$\sup_{k\ge K\in\mathbb{N}} \left\{ \frac{1}{\lambda^*} \prod_{\ell=k-\sigma_1+1}^{k} (1 - \lambda^*\overline{q}(\ell-1))^{-1} - p \prod_{\ell=k-\tau+1}^{k} (1 - \lambda^*\overline{q}(\ell-1))^{-1} \right.$$
$$\left. + \sum_{\ell=k-\sigma_1}^{k-\sigma_2-1} h(\ell - (\sigma_1 - \sigma_2)) \prod_{j=\ell+1}^{k} (1 - \lambda^*\overline{q}(j-1))^{-1} \right\} \le 1.$$

Show that (7.5.1) has a positive solution $u(k)$ with $\lim_{k\to\infty} u(k) = 0$.

**7.17.19.** Consider the difference equation (7.5.1) where $-1 < p < 0$, $\tau \in \mathbb{N}(1)$, $\sigma_1$, $\sigma_2 \in \mathbb{N}$, $\sigma_1 > \sigma_2 + 1$, $q(k) \ge 0$, $h(k) \ge 0$, $\overline{q}(k) = q(k) - h(k - (\sigma_1 - \sigma_2)) \ge 0$, $\sum_{k=1}^{\infty} \overline{q}(k) = \infty$, and $\limsup_{k\to\infty} \sum_{\ell=k-(\sigma_1-\sigma_2)}^{k-1} h(\ell) < 1 + p$. Show that

(i) every nonoscillatory solution of (7.5.1) tends to zero as $k \to \infty$

(ii) the equation (7.5.1) is oscillatory if and only if the difference inequality

$$\Delta(v(k) + pv(k-\tau)) + q(k)v(k-\sigma_1) - h(k)v(k-\sigma_2) \le 0, \qquad k \in \mathbb{N}$$

has no eventually positive solutions.

**7.17.20.** Show that in Theorem 7.7.3 the condition $p(k) \equiv 1$ can be replaced by $p(k) \ge 1$, $k \in \mathbb{N}$ and that $p(k-\sigma)q(k) \le \alpha q(k-\tau)$ for $k \ge \mu$ where $\alpha \ge 1$.

**7.17.21.** Consider the difference equation (7.7.1) where $\tau \in \mathbb{N}(1)$, $\sigma \in \mathbb{N}$, $q(k) \geq 0$, $p(k) \leq 0$ for all large $k$, and there exists a nonnegative number $\alpha$ such that $-p(k-\sigma)q(k) \leq \alpha q(k-\tau)$ for all large $k$. Show that the equation (7.7.1) is oscillatory provided the inequality

$$\Delta w(k) + \frac{q(k)}{1+\alpha} w(k-(\sigma-\tau)) \leq 0, \quad k \in \mathbb{N}$$

does not have an eventually positive solution.

**7.17.22.** Consider the difference equation (7.7.1) where $\tau \in \mathbb{N}(1)$, $\sigma \in \mathbb{N}$, $q(k) > 0$, $p(k) \leq 0$ for all large $k$, and there exists a nonnegative number $\alpha$ such that

$$\limsup_{k \to \infty} \frac{-p(k-\sigma)q(k)}{q(k-\tau)} \leq \alpha.$$

Show that the equation (7.7.1) is oscillatory provided the inequality

$$\Delta w(k) + \frac{q(k)}{1+\alpha+\epsilon} w(k-(\sigma-\tau)) < 0, \quad \epsilon > 0, \quad k \in \mathbb{N}$$

does not have an eventually positive solution.

**7.17.23.** Consider the difference equation (7.7.1) where $\tau \in \mathbb{N}(1)$, $\sigma \in \mathbb{N}$, $q(k) \geq 0$, $k \in \mathbb{N}$, there exists a positive integer $K$ such that (7.7.2) holds, and there exists a number $\alpha \in [0,1)$ such that $\alpha q(k-\tau) \leq p(k-\sigma)q(k)$ for all large $k$. Show that the equation (7.7.1) is oscillatory provided the inequality

$$\Delta w(k) + \frac{q(k)}{1-\alpha} w(k-(\sigma-\tau)) \leq 0, \quad k \in \mathbb{N}$$

does not have an eventually positive solution.

**7.17.24.** Show that the difference equation

$$\Delta\left(u(k) - \exp\left(k \sin\frac{k\pi}{2}\right) u(k-2)\right) + e^{-\beta k} u(k-\sigma) = 0, \quad k \in \mathbb{N}$$

where $\sigma \in \mathbb{N}$ and $\beta \in \mathbb{R}$ is oscillatory.

**7.17.25.** Show that the difference equation

$$\Delta\left(u(k) - \exp\left(\sin\frac{k\pi}{2\tau}\right) u(k-4\tau)\right) + ce^{-\beta k} u(k-\sigma) = 0, \quad k \in \mathbb{N}$$

where $\tau$, $\sigma \in \mathbb{N}(1)$, $c > 0$ and $\beta \leq 1/4\tau$ is oscillatory.

**7.17.26.** Consider the difference equation

$$(7.17.3)\quad \Delta(u(k) - p(k)u(k-\tau)) + q(k)u(k-\sigma_1) - h(k)u(k-\sigma_2) = 0, \quad k \in \mathbb{N}$$

where $\tau \in \mathbb{N}(1)$, $\sigma_1$, $\sigma_2 \in \mathbb{N}$, $\sigma_1 \geq \sigma_2$ and the functions $p(k)$, $q(k)$, $h(k)$ are defined on $\mathbb{N}$ with $q(k) \geq 0$, $h(k) \geq 0$, $k \in \mathbb{N}$. Assume that

(i) $p(k)$ and $q(k) - h(k-(\sigma_1-\sigma_2))$ are eventually nonnegative, and $q(k) - h(k-(\sigma_1-\sigma_2))$ is not identically zero for all large $k$

(ii) for all large $k$, $p(k) + \sum_{\ell=k-(\sigma_1-\sigma_2)}^{k-1} h(\ell) \leq 1$

(iii) $v(k)$ is an eventually positive solution of the inequality

$$\Delta(v(k) - p(k)v(k-\tau)) + q(k)v(k-\sigma_1) - h(k)v(k-\sigma_2) \leq 0, \quad k \in \mathbb{N}$$

and set

$$z(k) = v(k) - p(k)v(k-\tau) - \sum_{\ell=k-(\sigma_1-\sigma_2)}^{k-1} h(\ell)v(\ell-\sigma_2).$$

Show that eventually $z(k) > 0$ and $\Delta z(k) \leq 0$.

**7.17.27.** Assume that in Problem 7.17.26 condition (ii) is replaced by

(iv) for all large $k$, $p(k) + \sum_{\ell=k-(\sigma_1-\sigma_2)}^{k-1} h(\ell) \geq 1$, and

(v) there exists a nonnegative integer $s$ such that the functions

$$H_0(k) = \sum_{\ell=k}^{\infty}(q(\ell) - h(\ell-(\sigma_1-\sigma_2)))$$

$$H_j(k) = \sum_{\ell=k}^{\infty}\ell(q(\ell) - h(\ell-(\sigma_1-\sigma_2)))H_{j-1}(\ell), \quad j = 1, 2, \cdots, s$$

exist, and that

$$\sum_{\ell=1}^{\infty}\ell(q(\ell) - h(\ell-(\sigma_1-\sigma_2)))H_s(\ell) = \infty.$$

Show that eventually $z(k) < 0$ and $\Delta z(k) \leq 0$.

**7.17.28.** Consider the difference equation (7.17.3) where $\tau \in \mathbb{N}(1)$, $\sigma_1$, $\sigma_2 \in \mathbb{N}$, $\sigma_1 \geq \sigma_2$ and the functions $p(k)$, $q(k)$, $h(k)$ are defined on $\mathbb{N}$

with $q(k) \geq 0,\ h(k) \geq 0,\ k \in \mathbb{N}$. Assume that in addition to conditions (i) and (v) of Problems 7.17.26 and 7.17.27,

$$p(k) + \sum_{\ell=k-(\sigma_1-\sigma_2)}^{k-1} h(\ell) \equiv 1.$$

Show that the equation (7.17.3) is oscillatory. In particular, show that the difference equation

$$\Delta(u(k) - (1-\alpha)u(k-\tau)) + \left(\alpha + \frac{1}{(k+1)^{1.6}}\right) u(k-\sigma_1) - \alpha u(k-\sigma_2) = 0, \quad k \in \mathbb{N}$$

where $0 < \alpha < 1,\ \sigma_1 = \sigma_2 + 1,\ \tau = \sigma_2 > 0$ is oscillatory.

**7.17.29.** Consider the difference equation (7.17.3) where $\tau \in \mathbb{N}(1),\ \sigma_1,\ \sigma_2 \in \mathbb{N},\ \sigma_1 \geq \sigma_2$ and the functions $p(k),\ q(k),\ h(k)$ are defined on $\mathbb{N}$ with $q(k) \geq 0,\ h(k) \geq 0,\ k \in \mathbb{N}$. Assume that in addition to conditions (i), (ii) and (v) of Problems 7.17.26 and 7.17.27, $p(k-\sigma_1)[q(k) - h(k-(\sigma_1-\sigma_2))] \geq [q(k-\tau) - h(k-\sigma_1+\sigma_2-\tau)]$ for all large $k$. Show that the equation (7.17.3) is oscillatory.

**7.17.30.** Consider the difference equation (7.17.3) where $\tau \in \mathbb{N}(1),\ \sigma_1,\ \sigma_2 \in \mathbb{N},\ \sigma_1 \geq \sigma_2$ and the functions $p(k),\ q(k),\ h(k)$ are defined on $\mathbb{N}$ with $q(k) \geq 0,\ h(k) \geq 0,\ k \in \mathbb{N}$. Assume that in addition to conditions (i) and (iv), (v) of Problems 7.17.26 and 7.17.27, $p(k-\sigma_1)[q(k) - h(k-(\sigma_1-\sigma_2))] \leq c_1[q(k-\tau) - h(k-\sigma_1+\sigma_2-\tau)]$, $h(k)/[q(k) - h(k-(\sigma_1-\sigma_2))]$ is nondecreasing, and $[q(k) - h(k-(\sigma_1-\sigma_2))]h(k-\sigma_1) \leq c_2[q(k-\sigma_2) - h(k-\sigma_1)]$ for all large $k$, where $c_1, c_2$ are nonnegative constants and satisfy $c_1 + c_2(\sigma_1 - \sigma_2) = 1$. Show that the equation (7.17.3) is oscillatory. In particular, show that the difference equation

$$\Delta\left(u(k) - \frac{k+3}{2(k+2)}u(k-1)\right) + \left(\frac{1}{2} + \frac{1}{(k+1)^{1.6}}\right) u(k-2) - \frac{1}{2}u(k-1) = 0, \quad k \in \mathbb{N}$$

is oscillatory.

**7.17.31.** Show that in Problems 7.17.27 7.17.30 the condition (v) can be replaced by

$$\sum_{\ell=0}^{\infty} [q(\ell) - h(\ell - (\sigma_1 - \sigma_2))] \exp\left\{\frac{1}{\lambda}\sum_{j=1}^{k} j(q(j) - h(j - (\sigma_1 - \sigma_2)))\right\} = \infty,$$

where $\lambda = \max\{\tau, \sigma_1\} > 0$.

**7.17.32.** Consider the difference equation

$$(7.17.4) \qquad \Delta(u(k) + p(k)u(k-\tau)) = q(k)u(k-\sigma) + h(k)u(k), \quad k \in \mathbb{N}$$

where $\tau \in \mathbb{N}(1)$, $\sigma \in \mathbb{N}$ and the functions $p(k)$, $q(k)$, $h(k)$ are defined on $\mathbb{N}$. Assume that $p(k) \le p_0 < -1$, $q(k) < 0$ and $h(k) \le 0$ for all large $k$. Show that for an eventually positive solution $u(k)$ of (7.17.4) the function $z(k)$ defined by $z(k) = u(k) + p(k)u(k-\tau)$ satisfies $z(k) < 0$ and $\Delta z(k) < 0$ for all large $k$.

**7.17.33.** Consider the difference equation (7.17.4) where $\tau \in \mathbb{N}(1)$, $\sigma \in \mathbb{N}, \tau > \sigma$ and

(i) for all large $k$, $p(k) \le p_0 < -1$, $q(k) < 0, -1 < h(k) \le 0, h(k+\tau)q(k)p(k+\tau) \ge h(k)q(k+\tau)p(k+\tau-\sigma)$

(ii) $\liminf_{k\to\infty} \left[\dfrac{-\overline{h}(k+1)q(k)}{\overline{h}(k+\tau+1)p(k+\tau-\sigma)q(k+\tau)}\right] > 0$

(iii) $\liminf_{k\to\infty} \left[\dfrac{\overline{h}(k+1)q(k)}{\overline{h}(k+\tau-\sigma)p(k+\tau-\sigma)}\right] > 0$

and

(iv) $\liminf_{k\to\infty} \left[\dfrac{\overline{h}(k+1)q(k)(\tau-\sigma)}{\overline{h}(k+\tau-\sigma)p(k+\tau-\sigma)} - \dfrac{\overline{h}(k+1)q(k)}{\overline{h}(k+\tau+1)p(k+\tau-\sigma)q(k+\tau)}\right] > 1,$

where $\overline{h}(k) = \prod_{\ell=0}^{k-1}(1+h(\ell))^{-1}$, $k \in \mathbb{N}(1)$. Show that the difference equation (7.17.4) is oscillatory.

**7.17.34.** Consider the difference equation (7.17.4) where $\tau \in \mathbb{N}(1)$, $\sigma \in \mathbb{N}$, $\tau - \sigma = \omega > 1$ and in addition to conditions (i) and (iii) of Problem 7.17.33 assume that

$$\liminf_{k\to\infty} \left[\frac{\overline{h}(k+1)q(k)(\omega-1)^{1-\omega}\omega^{\omega}}{\overline{h}(k+\tau-\sigma)p(k+\tau-\sigma)} - \frac{\overline{h}(k+1)q(k)}{\overline{h}(k+\tau+1)p(k+\tau-\sigma)q(k+\tau)}\right] > 1.$$

Show that the difference equation (7.17.4) is oscillatory.

**7.17.35.** Consider the difference equation (7.17.4) where $\tau \in \mathbb{N}(1)$, $\sigma \in \mathbb{N}$ and for all large $k$, $p(k) \le p_0 < -1$, $q(k) < 0$, $-1 < h(k) \le 0$, $h(k-\tau)q(k)p(k-\sigma) \le h(k)q(k-\tau)p(k)$, and

$$\limsup_{k\to\infty} \left[\frac{-\overline{h}(k+1)q(k)p(k-\sigma)}{\overline{h}(k-\tau+1)q(k-\tau)}\right] < 1.$$

Show that the difference equation (7.17.4) is oscillatory.

**7.17.36.** Consider the difference equation (7.17.4) where $\tau \in \mathbb{N}(1)$, $\sigma \in \mathbb{N}$, $\tau \leq \sigma$ and for all large $k$, $-1 \leq p(k) \leq 0$, $q(k) < 0$, $-1 < h(k) \leq 0$, $h(k)q(k-\tau)p(k) \leq h(k-\tau)q(k)p(k-\tau)$,

$$\liminf_{k\to\infty}\left[\frac{\overline{h}(k+1)p(k)h(k)}{\overline{h}(k-\tau)} - \frac{\overline{h}(k+1)q(k)}{\overline{h}(k-\sigma)}\right] > 0$$

and

$$\liminf_{k\to\infty}\left[\frac{\overline{h}(k+1)p(k)h(k)}{\overline{h}(k-\tau)} - \frac{\overline{h}(k+1)q(k)}{\overline{h}(k-\sigma)} - \frac{\overline{h}(k+1)q(k)p(k-\sigma)}{\overline{h}(k-\tau+1)q(k-\tau)}\right] > 1.$$

Show that the difference equation (7.17.4) is oscillatory.

**7.17.37.** Consider the difference equation (7.17.4) where $\tau,\ \sigma \in \mathbb{N}(1)$, $\tau < \sigma$ and for all large $k$, $p(k) \geq 0$, $q(k) < 0$, $h(k) \leq 0$, $q(k)p(k-\sigma)/q(k-\tau) \leq \beta$, and

$$\liminf_{k\to\infty}\left\{\frac{1}{\sigma-\tau}\sum_{\ell=k-\sigma+\tau}^{k-1} -q(\ell)\right\} > \frac{(1+\beta)(\sigma-\tau)^{\sigma-\tau}}{(\sigma-\tau+1)^{\sigma-\tau+1}}.$$

Show that the difference equation (7.17.4) is oscillatory.

**7.17.38.** Consider the difference equation (7.8.1) where $\tau \in \mathbb{N}(1)$ and the functions $r(k) > 0$, $p(k) \geq 0$, $q(k) \geq 0$, $\sigma(k) \in \mathbb{N}$ for all $k \in \mathbb{N}$, and $\lim_{k\to\infty}(k-\sigma(k)) = 0$. Show that

(i) if $r(k) \equiv p(k) \equiv 1$ and $\sum_{\ell=0}^{\infty} q(\ell) = \infty$, then the equation (7.8.1) is oscillatory

(ii) if there exists a positive integer $K$ such that either (7.8.3) or (7.8.4) holds, and $p(k-\sigma(k))q(k) \geq r(k-\sigma(k))q(k-\tau)$, $k \in \mathbb{N}$ and

$$\sum_{\ell=0}^{\infty}\frac{q(\ell)}{r(\ell-\sigma(\ell))} = \infty \tag{7.17.5}$$

then the equation (7.8.1) is oscillatory

(iii) the difference equation

$$\Delta((k+1)u(k) - (k+1)u(k-2)) + u(k-1-\sin \pi k/2) = 0, \quad k \in \mathbb{N}$$

is oscillatory.

**7.17.39.** Consider the difference equation (7.8.1) where $\tau \in \mathbb{N}(1)$ and the functions $r(k) > 0$, $p(k) \geq 0$, $q(k) \geq 0$, $\sigma(k) \in \mathbb{N}$ for all $k \in \mathbb{N}$, and $\lim_{k\to\infty}(k-\sigma(k)) = 0$. Show that

(i) if $p(k) \geq r(k)$, $k \in \mathbb{N}$ and $u(k)$ is an eventually positive solution of (7.8.1), then there exists a $k_1 \in \mathbb{N}$ such that $z(k) = r(k)u(k) - p(k)u(k - \tau) < 0$ and $\Delta z(k) \leq 0$, $k \in \mathbb{N}(k_1)$
(ii) if $p(k) \geq r(k)$, $p(k - \sigma(k))q(k) \leq r(k - \sigma(k))q(k - \tau)$, $k \in \mathbb{N}$, $\sum_{\ell=0}^{\infty} q(\ell) = \infty$ and (7.17.5) hold, then the equation (7.8.1) is oscillatory.

**7.17.40.** Consider the difference equation

$$\Delta\left(u(k) - pu^{\alpha}(k - \tau)\right) + q(k)u^{\beta}(k - \sigma) = 0, \quad k \in \mathbb{N} \tag{7.17.6}$$

where $p$ is a constant, $\tau \in \mathbb{N}(1)$, $\sigma \in \mathbb{N}$, and $\alpha$ and $\beta$ are quotients of odd integers. Show that

(i) if $\alpha \in (0, 1)$, $\beta > 0$, $p > 0$ and $q(k) \geq 0$ eventually, then the equation (7.17.6) is oscillatory if and only if $\sum_{\ell=K}^{\infty} q(\ell) = \infty$, $K \in \mathbb{N}$
(ii) if $p \geq 0$ and there exists a $b \in (0, 1]$ such that

$$p\frac{k^b}{(k - \tau)^{\alpha b}} + k^b \sum_{\ell=k}^{\infty} \frac{q(\ell)}{(\ell - \sigma)^{\beta b}} \leq 1 \text{ for all large } k \in \mathbb{N},$$

then (7.17.6) has a positive solution $u(k)$ with $\lim_{k\to\infty} u(k) = 0$
(iii) condition $\sum_{\ell=K}^{\infty} q(\ell) = \infty$ is not sufficient for all solutions of (7.17.6) with $p < 0$ to be oscillatory
(iv) if $p < 0$ and $\sum_{k\in E} q(k) = \infty$ for every subset $E$ of $\mathbb{N}(1)$ such that $m(E \cap \{k, k + 1, \cdots, k + 2\tau - 1\}) \geq \tau$ for each $k \in \mathbb{N}(1)$, where $m(A)$ denotes the number of points in the set $A$, then every solution of (7.17.6) is either oscillatory or tends to zero as $k \to \infty$
(v) if $p < 0$, $Q(k) = -q(k) \geq 0$ and $\sum_{k\in E} Q(k) = \infty$, then every bounded solution of (7.17.6) is oscillatory.

**7.17.41.** Consider the difference equation

$$\Delta(u(k) + pu^{\alpha}(k - \tau)) + q(k)u^{\beta}(k - \sigma) = F(k) = \Delta f(k), \quad k \in \mathbb{N} \tag{7.17.7}$$

where $p$, $\tau$, $\sigma$, $\alpha$ and $\beta$ are as in Problem 7.17.40. Show that

(i) if $p \geq 0$, $q(k) \leq 0$ eventually, and $\limsup_{k\to\infty} f(k) = \infty$ and $\liminf_{k\to\infty} f(k) = -\infty$, then every bounded solution of (7.17.7) is oscillatory
(ii) if $p \geq 0$, $q(k) \geq 0$, and $\limsup_{k\to\infty} f(k) = \infty$ and $\liminf_{k\to\infty} f(k) = -\infty$, then the equation (7.17.7) is oscillatory
(iii) if $p > 0$, $q(k) \geq 0$ and $\sum_{k\in E} q(k)f_+^{\beta}(k) = \infty$, $\sum_{k\in E} q(k)f_-^{\beta}(k) = -\infty$ for every subset $E$ of $\mathbb{N}(1)$ such that $m(E \cap \{k, k+1, \cdots, k+2\tau -$

$1\}) \geq \tau$ for each $k \in \mathbb{N}(1)$, where $f_+(k) = \max\left\{\frac{f(k-\tau)}{2}, \left(\frac{f(k)}{2p}\right)^{1/\alpha}, 0\right\}$ and $f_-(k) = \min\left\{\frac{f(k-\tau)}{2}, \left(\frac{f(k)}{2p}\right)^{1/\alpha}, 0\right\}$, then the equation (7.17.7) is oscillatory.

**7.17.42.** Consider the difference equation

$$(7.17.8)\quad \Delta(u(k) - p(k)u(k-\tau)) + q(k)\prod_{i=1}^{r}|u(k-\sigma_i)|^{\alpha_i}\mathrm{sgn}\ u(k-\sigma_i) = 0, \quad k \in \mathbb{N}$$

where $\tau \in \mathbb{N}(1)$, $\sigma_i \in \mathbb{N}$, $1 \leq i \leq r$, $p(k) \geq 0$, $q(k) \geq 0$, $k \in \mathbb{N}$ and $q(k)$ is not identically zero for all large $k$, and $\alpha_i \geq 0$, $1 \leq i \leq r$ are such that $\sum_{i=1}^{r}\alpha_i = 1$. Suppose there exists a positive integer $K$ such that $p(K + \ell\tau) \leq 1$, $\ell \in \mathbb{N}$. Further, let $p(k) + q(k)\min\{\sigma_1, \cdots, \sigma_r\} > 0$, or $\min\{\sigma_1, \cdots, \sigma_r\} > 0$ and $q(k)$ does not vanish over sets of consecutive integers of the form $\{a, a+1, \cdots, a + \min\{\sigma_1, \cdots, \sigma_r\}\}$. Show that the equation (7.17.8) is oscillatory if and only if

$$(7.17.9)\quad \Delta(v(k) - p(k)v(k-\tau)) + q(k)\prod_{i=1}^{r}|v(k-\sigma_i)|^{\alpha_i}\mathrm{sgn}\ v(k-\sigma_i) \leq 0, \quad k \in \mathbb{N}$$

does not have an eventually positive solution.

(For $r = 1$ equation (7.17.8) is the same as (7.7.1).)

**7.17.43.** Let $\tau$, $\sigma_i$, $\alpha_i$, $1 \leq i \leq r$ be as in Problem 7.17.42. Further, let $p(k) \geq 1$, $q(k) \geq 0$, $k \in \mathbb{N}$ and $q(k)$ is not identically zero for all large $k$, and

$$(7.17.10)\qquad \sum_{k=0}^{\infty} q(k)\prod_{\ell=0}^{k-1}\left(1 + \frac{\ell q(\ell)}{\tau}\right) = \infty.$$

Show that for every eventually positive solution $v(k)$ of (7.17.9) the function $z(k)$ defined by $z(k) = v(k) - p(k)v(k-\tau)$ satisfies $z(k) > 0$ and $\Delta z(k) \leq 0$ for all large $k \in \mathbb{N}$.

**7.17.44.** Let $\tau$, $\sigma_i$, $\alpha_i$, $1 \leq i \leq r$ be as in Problem 7.17.42. Further, let $p(k) \equiv 1$, $q(k) \geq 0$, $k \in \mathbb{N}$ and $q(k)$ is not identically zero for all large $k$, and the condition (7.17.10) holds. Show that the equation (7.17.8) is oscillatory.

**7.17.45.** Let $\tau,\ \sigma_i,\ \alpha_i,\ 1 \le i \le r$ be as in Problem 7.17.42. Further, let $p(k) \ge 0,\ q(k) \ge 0,\ k \in \mathbb{N}$ and $q(k)$ is not identically zero for all large $k$, there exists a positive integer $K$ such that $p(K + \ell\tau) \le 1,\ \ell \in \mathbb{N}$, condition (7.17.10) holds, and for all large $k \in \mathbb{N}$

$$q(k)\prod_{i=1}^{r} p^{\alpha_i}(k-\sigma_i) \ \ge\ q(k-\tau). \tag{7.17.11}$$

Show that the equation (7.17.8) is oscillatory.

**7.17.46.** Show that in Problem 7.17.45 condition (7.17.11) can be replaced by that there exists a number $\alpha \in (0,1)$ such that for all large $k \in \mathbb{N}$

$$q(k)\prod_{i=1}^{r} p^{\alpha_i}(k-\sigma_i) \ \ge\ \alpha q(k-\tau)$$

and the inequality

$$\Delta w(k) + \frac{\alpha}{1-\alpha}q(k)w(k-\sigma-\tau) \ \le\ 0, \quad \sigma = \min\{\sigma_1,\cdots,\sigma_r\}, \quad k \in \mathbb{N}$$

does not have an eventually positive solution.

**7.17.47.** Consider the difference equation (7.9.1) where $p(k) \le 0,\ q(k) \ge 0,\ k \in \mathbb{N}$, and $p(k),\ q(k)$ are not zero for infinitely many values of $k,\ \tau \in \mathbb{N}(1),\ \sigma \in \mathbb{N}$, and $f : \mathbb{R} \to \mathbb{R}$ is a nondecreasing function with $uf(u) > 0$ for $u \ne 0$. Show that

(i) if there exists an integer $K \in \mathbb{N}$ such that $p(K + \ell\tau) \le 1,\ \ell \in \mathbb{N}$ then for any eventually positive solution $u(k)$ of (7.9.1), $z(k) = u(k) + p(k)u(k-\tau) > 0$ and $\Delta z(k) \le 0$ for all large $k \in \mathbb{N}$

(ii) if $p(k) < 0,\ k \in \mathbb{N}$ and there exists an integer $K \in \mathbb{N}$ such that $p(K+\ell\tau) \le 1$ for $\ell \in \mathbb{N}$, then the equation (7.9.1) has an eventually positive (bounded and eventually positive) solution if and only if the difference inequality

$$\Delta(v(k) + p(k)v(k-\tau)) + q(k)f(v(k-\sigma)) \ \le\ 0, \quad k \in \mathbb{N}$$

has an eventually positive (bounded and eventually positive) solution

(iii) the difference inequality

$$\Delta^2 z(k-1) + (1/\tau)q(k)f(z(k)) \ \le\ 0, \quad k \in \mathbb{N}$$

has an eventually positive (bounded and eventually positive) solution if and only if

$$\Delta(x(k) - x(k-\tau)) + q(k)f(x(k-\sigma)) \ \le\ 0, \quad k \in \mathbb{N}$$

has an eventually positive (bounded and eventually positive) solution

(iv) if $\gamma$ is a quotient of odd positive integers, then the generalized *Emden Fowler equation* (see (6.10.1))

$$\Delta(u(k)-u(k-\tau))+q(k)u^{\gamma}(k) \;=\; 0, \qquad k\in\mathbb{N} \tag{7.17.12}$$

has a bounded and eventually positive solution if and only if $\sum_{k=1}^{\infty} kq(k) < \infty$, in addition if $\gamma > 1$ or $0 < \gamma < 1$ then (7.17.12) has an eventually positive solution if and only if $\sum_{k=1}^{\infty} kq(k) < \infty$ or $\sum_{k=1}^{\infty} k^{\gamma}q(k) < \infty$ holds respectively.

**7.17.48.** Consider the difference equation (7.9.26), where $|p(k)| \le P < 1$, $\sum^{\infty}|q(k)| < \infty$, $|f(u)| \le A|u|^{\alpha}$, $0 < \alpha \le 1$, $u \in \mathbb{R}$, and $\tau \in \mathbb{N}(1)$, $\sigma \in \mathbb{N}$. Show that all solutions of (7.9.26) are bounded. In particular, deduce that all solutions of the difference equation

$$\Delta(u(k)+e^{-2}u(k-2)) = \frac{2(e-1)\sin u(k)}{e^{k+1}(1-e^{-k+1})^{1/3}\sin(1-e^{-k})}u^{1/3}(k-1) = 0, \quad k\in\mathbb{N}(1) \tag{7.17.13}$$

are bounded. In fact, $u(k) = 1-e^{-k}$ is a bounded solution of (7.17.13).

**7.17.49.** Consider the difference equation (7.9.26), where $p(k) = p > 0$, $q(k)$ is $\tau$ periodic, $f$ is nondecreasing, $\tau \in \mathbb{N}(1)$, $\sigma \in \mathbb{N}$, and

$$\begin{aligned} f(u+v) &\le f(u)+f(v), & u,v>0\\ f(u+v) &\ge f(u)+f(v), & u,v<0\\ f(\lambda u) &\le \lambda f(u), & \lambda\ge 0,\ u>0\\ f(\lambda u) &\ge \lambda f(u), & \lambda\ge 0,\ u<0. \end{aligned}$$

Show that every bounded solution of (7.9.26) is oscillatory.

**7.17.50.** Consider the difference equation

$$\Delta(u(k)-p(k)g(u(k-\tau)))+q(k)f(u(k-\sigma)) \;=\; 0, \quad k\in\mathbb{N} \tag{7.17.14}$$

where $p(k)$ and $q(k)$ are such that $q(k)\ge 0$ for all $k\in\mathbb{N}$, and not zero for infinitely many values of $k$, $\tau\in\mathbb{N}(1)$, $\sigma\in\mathbb{N}$ and $f,g:\mathbb{R}\to\mathbb{R}$ are continuous with $uf(u)>0$ and $ug(u)>0$ for all $u\ne 0$. Assume that the functions $f$ and $g$ are increasing, $0\le p(k)<1$, $k\in\mathbb{N}$, and there exists a positive function $h(k)$, $k\in\mathbb{N}$ such that for all sufficiently large $k$

$$p(k)g(1/h(k-\tau))h(k)+h(k)\sum_{\ell=k}^{\infty}q(\ell)f(1/h(\ell-\sigma)) \;\le\; 1.$$

Show that the equation (7.17.14) has a positive solution $u(k)$. Further, this solution $u(k) \to 0$ as $k \to \infty$ if $h(k) \to \infty$ as $k \to \infty$, and $u(k)$ is bounded if $h(k)$ is bounded away from zero.

**7.17.51.** Show that the Problem 7.17.50 is applicable to the following difference equations

(i) $\Delta(u(k) - 4^{2-k}u^3(k-1)) + 2^{3-2k}u^3(k-1) \; = \; 0, \quad k \in \mathbb{N}(3)$
with $h(k) = 1/k$, and one of its positive solution is $u(k) = 2^k$

(ii) $\Delta(u(k) - (1/8)u^3(k-1)) + q(k)u^{1/3}(k-3) \; = \; 0, \quad k \in \mathbb{N}(1)$
where $q(k) = 2^{-2k/3}\left(\frac{1}{4} - \frac{7}{16}4^{-k}\right) > 0$, with $h(k) = 2^k$, and one of its positive solution which tends to zero as $k \to \infty$ is $u(k) = 1/2^k$

(iii) $\Delta\left(u(k) - \frac{(k-1)^3}{2k^3}u^3(k-1)\right) + \frac{(k-1)^3}{k^4(k+1)}u^3(k-1) \; = \; 0, \quad k \in \mathbb{N}(1)$
with $h(k) = (k+2)/k$, and one of its positive bounded solution is $(k+1)/k$.

**7.17.52.** Consider the difference equation (7.17.14) where $p(k)$ and $q(k)$ are such that $q(k) \geq 0$ for all $k \in \mathbb{N}$, and not zero for infinitely many values of $k$, $\tau \in \mathbb{N}(1)$, $\sigma \in \mathbb{N}$ and $f, g : \mathbb{R} \to \mathbb{R}$ are continuous with $uf(u) > 0$ and $ug(u) > 0$ for all $u \neq 0$. Assume that the function $f$ is increasing and $\sum_{k=0}^{\infty} q(k) = \infty$. Show that the equation (7.17.14) is oscillatory if either of the following conditions hold

(i) in addition to (7.9.18) there exist constants $M > 0$ and $p \geq 0$ such that $|g(u)| \leq M|u|$ for all $u$ and $0 \leq p(k) \leq p$ with $pM < 1$; or

(ii) $\tau - \sigma \geq 1$, $g$ is increasing, there exist positive constants $M, A$ and $B$ such that $|g(u)| \geq M|u|$ for all $u$ and $A \leq p(k) \leq B$ with $MA > 1$, and $\int_c^{\infty} \frac{dt}{f(g^{-1}(t))} < \infty$ and $\int_{-\infty}^{-c} \frac{dt}{f(g^{-1}(t))} > -\infty$ for all $c > 0$; or

(iii) $\tau - \sigma \geq 1$, there exist positive constants $M_1, M_2, A$ and $B$ such that $M_1|u| \leq |g(u)| \leq M_2|u|$ for all $u$ and $A \leq p(k) \leq B$ with $M_1 A > 1$, and (7.9.19) is satisfied.

**7.17.53.** Consider the difference equation

$$(7.17.15) \quad \Delta(u(k-1) - pu(k-\tau-1)) \; = \; f(k, u(k), u(k-\tau)), \quad k \in \mathbb{N}$$

where $\tau \in \mathbb{N}(1)$, $0 \leq p < 1$, $f : \mathbb{N} \times \mathbb{R}^2 \to \mathbb{R}$ is continuous, and $f(k, u, v)$ is decreasing with respect to $u \in \mathbb{R}$. Assume that

(i) $u > v$ implies $f(k, u, v) < 0$, and

(ii) for $u \leq v$, $f(k, u, v) \leq q(k)g(u, v) + h(k)$,

where $g : \mathbb{R}^2 \to \mathbb{R}$ is continuous, $g(u,v)$ is decreasing with respect to $u$ and increasing with respect to $v$, and $g(u,u) \equiv 0$ for all $u \in \mathbb{R}$, and the functions $h(k)$ and $q(k)$ are such that $\sum_{k=1}^{\infty} |h(k)| < \infty$, $0 < q(k) < M < \infty$, $k \in \mathbb{N}$.

Show that every bounded solution of the equation (7.17.15) tends to a constant as $k \to \infty$.

**7.17.54.** Show that in Problem 7.17.53 the conditions (i) and (ii) can be replaced by

(i)′ $u < v$ implies $f(k,u,v) > 0$, and

(ii)′ for $u \ge v$, $f(k,u,v) \ge q(k)g(u,v) + h(k)$.

**7.17.55.** Assume that in Problem 7.17.53 the conditions (i) and (ii) are replaced by

(i)* for $u \ne v$, $(u-v)f(k,u,v) < 0$, and

(ii)* either (ii) or (ii)′ holds.

Show that every solution of the equation (7.17.15) tends to a constant as $k \to \infty$.

**7.17.56.** Consider the difference equation

$$(7.17.16)_\delta \quad \Delta(u(k) + p(k)u(\tau(k))) + \delta f(k, u(\sigma_1(k)), \cdots, u(\sigma_m(k))) = 0, \quad k \in \mathbb{N}$$

where the functions $p(k) \ge 0$, $k \in \mathbb{N}$, $\sigma_i(k) \in \mathbb{N}$, $\lim_{k\to\infty} \sigma_i(k) = \infty$, $1 \le i \le m$ and $f : \mathbb{N} \times \mathbb{R}^m \to \mathbb{R}$ is continuous, $f(k, u_1, \cdots, u_m)$ is nondecreasing in each $u_i$, $1 \le i \le m$ and $u_1 f(k, u_1, \cdots, u_m) > 0$ for $u_1 u_i > 0$, $1 \le i \le m$. Show that if there exist constants $\lambda, \mu, \nu$ and $\tau \in \mathbb{N}(1)$ such that

(i) $p(k) \le \lambda < 1$, $\tau(k) = k - \tau$, or $1 < \mu \le p(k) \le \nu$, $\tau(k) = k + \tau$ then the equation $(7.17.16)_{-1}$ has a bounded nonoscillatory solution if and only if

$$(7.17.17) \qquad \sum^{\infty} |f(k, b, \cdots, b)| < \infty \quad \text{for some constant } b \ne 0$$

(ii) $p(k) \le \lambda < 1$, $\tau(k) = k + \tau$, or $1 < \mu \le p(k) \le \nu$, $\tau(k) = k - \tau$ then the equation $(7.17.16)_1$ has a bounded nonoscillatory solution if and only if (7.17.17) holds.

**7.17.57.** Consider the difference equation

$$(7.17.18)_\delta \quad \Delta(u(k) + p(k)u(k-\tau)) + \delta q(k) f(u(k-\sigma)) = F(k), \quad k \in \mathbb{N}$$

where $q(k) \ge 0$, $k \in \mathbb{N}$, $f : \mathbb{R} \to \mathbb{R}$ is continuous with $uf(u) > 0$ for $u \ne 0$, $\sum^{\infty} F(k) < \infty$, and $\tau \in \mathbb{N}(1)$, $\sigma \in \mathbb{N}$ and $\delta = \pm 1$. Show that

(i) if (7.9.3) holds, then every nonoscillatory solution of $(7.17.18)_1$ is bounded

(ii) if (7.9.2), (7.9.4) and $p(k) \to 0$ as $k \to \infty$, then every nonoscillatory solution of $(7.17.18)_1$ tends to zero as $k \to \infty$

(iii) if (7.9.2) (7.9.4) and $p(k) \to 0$ as $k \to \infty$, then every bounded nonoscillatory solution of $(7.17.18)_{-1}$ tends to zero as $k \to \infty$.

**7.17.58.** Consider the difference equation $(7.17.18)_1$ where $f : \mathbb{R} \to \mathbb{R}$ is continuous with $uf(u) > 0$ for $u \ne 0$. Assume that there exist nonnegative constants $A, B$ and $\alpha$ with $0 < \alpha \le 1$ such that $|f(u)| \le A|u|^{\alpha} + B$, $\sum^{\infty} |q(k)| < \infty$ and $\sum^{\infty} F(k) < \infty$. Show that

(i) if there exists a constant $P$ such that $|p(k)| \le P < 1$, then all solutions of $(7.17.18)_1$ are bounded

(ii) if $p(k) \to 0$ as $k \to \infty$, then every oscillatory solution of $(7.17.18)_1$ tends to zero as $k \to \infty$.

**7.17.59.** Consider the difference equation

$$(7.17.19) \quad \Delta(a(k)\Delta(u(k) + p(k)u(k-\tau))) + q(k)f(u(k-\sigma)) = 0, \quad k \in \mathbb{N}$$

where $\tau \in \mathbb{N}(1)$, $\sigma \in \mathbb{N}$ and $a(k) > 0$, $k \in \mathbb{N}$ with $\sum_{k=0}^{\infty} 1/a(k) = \infty$, and the function $f : \mathbb{R} \to \mathbb{R}$ is such that $uf(u) > 0$ for $u \ne 0$. Show that

(i) if $0 \le p(k) \le 1$, $q(k) \ge 0$, $k \in \mathbb{N}(k_0)$, $f(u)/u \ge \gamma > 0$ for $u \ne 0$, and there exists a function $h(k)$ such that $h(k) > 0$ for $k \ge k_0$ and

$$\sum_{k=k_0+\sigma}^{\infty} \left[\gamma h(k)q(k)(1 - p(k-\sigma)) - \frac{a(k-\sigma)(\Delta h(k))^2}{4h(k)}\right] = \infty,$$

then the equation (7.17.19) is oscillatory

(ii) if $-1 \le p(k) \le 0$, $q(k) \ge 0$, $k \in \mathbb{N}(k_0)$, $f$ is a nondecreasing continuous function such that (7.9.19) holds, and

$$\sum_{k=k_0}^{\infty} \frac{1}{a(k)} \sum_{\ell=k+\sigma+1}^{\infty} q(\ell) = \infty,$$

then every unbounded solution of (7.17.19) is oscillatory

(iii) if $p(k) \equiv p \ge 0$, $f$ is nondecreasing, and $\sum^{\infty} q(k) = \infty$, then $\Delta u(k)$ of every solution $u(k)$ of (7.17.19) oscillates.

**7.17.60.** Consider the difference equation

$$\Delta^2(u(k) - pu(k-\tau)) = q(k)u(k-\sigma), \quad k \in \mathbb{N} \tag{7.17.20}$$

where $\tau \in \mathbb{N}(1)$. Show that

(i) if $p \geq 0$, $\sigma \in \mathbb{N}$, and $q(k) \geq 0$, $q(k) \not\equiv 0$, $k \in \mathbb{N}$, then (7.17.20) always has a positive solution which tends to infinity as $k \to \infty$

(ii) if $p > 1$, $\sigma \in \mathbb{N}(1)$, and $\sum_{k=K}^{\infty} k|q(k)| < \infty$, then (7.17.20) has a bounded positive solution

(iii) if $0 < p < 1$, $\sigma \in \mathbb{N}(1)$, $q(k) > 0$, $k \in \mathbb{N}$, and $\limsup_{k\to\infty} \sum_{\ell=k-\sigma}^{k-1} [\ell - (k-\sigma-1)] q(\ell) > 1$, then every bounded solution of (7.17.20) is oscillatory

(iv) if $p > 1$, $\sigma \in \mathbb{N}$, $q(k) \geq 0$, $k \in \mathbb{N}$, and $\sum_{k=K}^{\infty} kq(k) = \infty$, then every bounded solution of (7.17.20) is oscillatory.

**7.17.61.** Consider the difference equation

$$\Delta(a(k)\Delta(u(k) + p(k)u(k-\tau))) + q(k)u(k-\sigma(k)) = 0, \quad k \in \mathbb{N} \tag{7.17.21}$$

where $\tau \in \mathbb{N}(1)$, $0 \leq p(k) \leq 1$, $k \in \mathbb{N}$, $q(k) \geq 0$, $k \in \mathbb{N}$ with infinitely many nonzero terms, $\sigma(k) \in \mathbb{N}$, $k \in \mathbb{N}$, $k - \sigma(k)$ is nondecreasing and $\lim_{k\to\infty}(k - \sigma(k)) = \infty$, $a(k) > 0$, $k \in \mathbb{N}$ with $\sum_{k=0}^{\infty} 1/a(k) = \infty$. Define $Q(k) = q(k)(1 - p(k - \sigma(k))) \geq 0$ for all large $k$, and $r(k) = \sum_{\ell=0}^{k-1} 1/a(\ell)$, $k \in \mathbb{N}(1)$. Show that

(i) if $\sum^{\infty} Q(k) < \infty$, then the equation (7.17.21) is oscillatory

(ii) if $\sigma(k) \equiv \sigma$ and $r(k-\sigma-1)\sum_{\ell=k}^{\infty} Q(\ell) \geq \beta > 1/4$ for all large $k$, then the equation (7.17.21) is oscillatory.

**7.17.62.** Consider the difference equation (7.11.1) where $\tau \in \mathbb{N}(1)$, $\sigma \in \mathbb{N}$, $0 \leq p(k) < 1$, $k \in \mathbb{N}$, $q(k) \geq 0$ for all $k \in \mathbb{N}$ and is not zero for infinitely many values of $k$, $f : \mathbb{R} \to \mathbb{R}$ is continuous with $uf(u) > 0$ for $u \neq 0$ and is nondecreasing. Further, assume that $f(-uv) \geq f(uv) \geq Lf(u)f(v)$ for $u, v > 0$, and $f(u)/u^{\alpha} \geq d$ for $u \neq 0$, where $L$ and $d$ are positive constants and $\alpha$ is the ratio of odd positive integers. Show that the equation (7.11.1) is oscillatory if either of the following holds

(i) $\alpha \in (1, \infty)$ and $\sum_{k=1}^{\infty} kq(k)f(1 - p(k+1-\sigma)) = \infty$, or

(ii) $\alpha \in (0,1)$ and $\sum_{k=1}^{\infty} q(k)f((k-\tau-\sigma)/2)f(1-p(k+1-\sigma)) = \infty$.

**7.17.63.** Consider the difference equation (7.11.1) where $\tau \in \mathbb{N}(1)$, $\sigma \in \mathbb{N}$, $q(k) \geq 0$ for all $k \in \mathbb{N}$ and is not zero for infinitely many values of $k$, and $f : \mathbb{R} \to \mathbb{R}$ is continuous with $uf(u) > 0$ for $u \neq 0$. Further, assume that conditions (7.9.2) and (7.9.4) hold. Show that

(i) if there exists a constant $P$ such that $0 \le p(k) \le P < 1$, then equation (7.11.1) is oscillatory

(ii) if condition (7.9.8) holds, then every bounded solution $u(k)$ of (7.11.1) is either oscillatory or satisfies $u(k) \to 0$ as $k \to \infty$.

**7.17.64.** Consider the difference equation

$$(7.17.22) \quad \Delta^2(u(k) + p(k)u(k-\tau)) = q(k)f(u(k+1-\sigma)), \quad k \in \mathbb{N}$$

where $p(k)$ and $q(k)$ are such that $q(k) \ge 0$ for all $k \in \mathbb{N}$, and are not zero for infinitely many values of $k$, $\tau \in \mathbb{N}(1)$, $\sigma \in \mathbb{N}$, and $f : \mathbb{R} \to \mathbb{R}$ is continuous with $uf(u) > 0$ for $u \ne 0$. Assume that (7.9.2) and (7.9.4) hold, and there exists a constant $A < 0$ such that $A \le p(k) \le 0$. Show that

(i) if $u(k)$ is an eventually positive solution of (7.17.22), then for $z(k) = u(k) + p(k)u(k-\tau)$, $\Delta z(k)$ is nondecreasing, and either

$$(7.17.23) \quad \lim_{k\to\infty} z(k) = \lim_{k\to\infty} \Delta z(k) = \infty$$

or

$$(7.17.24) \quad \lim_{k\to\infty} z(k) = \lim_{k\to\infty} \Delta z(k) = 0, \quad \Delta z(k) < 0 \text{ and } z(k) > 0$$

(ii) if $u(k)$ is an eventually negative solution of (7.17.22), then $\Delta z(k)$ is nonincreasing, and either

$$(7.17.25) \quad \lim_{k\to\infty} z(k) = \lim_{k\to\infty} \Delta z(k) = -\infty$$

or

$$(7.17.26) \quad \lim_{k\to\infty} z(k) = \lim_{k\to\infty} \Delta z(k) = 0, \quad \Delta z(k) > 0 \text{ and } z(k) < 0.$$

**7.17.65.** The conclusion of Problem 7.17.64 is the best possible under its hypotheses. For this consider the difference equation

$$(7.17.27) \quad \Delta^2\left(u(k) - \frac{1}{2}e^{a\tau}u(k-\tau)\right) = \frac{(e^a-1)^2 e^{a\gamma\sigma} e^{(1-\gamma)ak}}{2^\gamma} u^\gamma(k-\sigma), \quad k \in \mathbb{N}$$

where $\tau \in \mathbb{N}(1)$, $\sigma \in \mathbb{N}$, $a$ is a constant, and $\gamma$ is the ratio of odd positive integers. Show that

(i) $u_1(k) = 2e^{ak}$ and $u_2(k) = -2e^{ak}$ are both nonoscillatory solutions of (7.17.27)

(ii) equation (7.17.27) satisfies all hypotheses of Problem 7.17.64 if either

(7.17.28) $$\gamma \leq 1 \quad \text{and} \quad a > 0, \quad \text{or}$$

(7.17.29) $$\gamma \geq 1 \quad \text{and} \quad a < 0$$

(iii) when (7.17.28) holds $z_1(k) = u_1(k) - (1/2)e^{a\tau}u_1(k-\tau) = e^{ak}$ satisfies (7.17.23), and $z_2(k) = u_2(k) - (1/2)e^{a\tau}u_2(k-\tau) = -e^{ak}$ satisfies (7.17.25)

(iv) when (7.17.29) holds $z_1(k)$ satisfies (7.17.24), and $z_2(k)$ satisfies (7.17.26).

**7.17.66.** Consider the difference equation (7.17.22) where $p(k)$ and $q(k)$ are such that $q(k) \geq 0$ for all $k \in \mathbb{N}$, and are not zero for infinitely many values of $k$, $\tau \in \mathbb{N}(1)$, $\sigma \in \mathbb{N}$, and $f : \mathbb{R} \to \mathbb{R}$ is continuous with $uf(u) > 0$ for $u \neq 0$. Show that

(i) if (7.9.2) and (7.9.4) hold, and there exists a constant $B$ such that $B \leq p(k) \leq -1$, then every nonoscillatory solution $u(k)$ of (7.17.22) satisfies $|u(k)| \to \infty$ as $k \to \infty$

(ii) if (7.9.5) holds, the function $f$ is nondecreasing such that (7.9.18) holds, and for $\sigma \geq 1$, $\sum_{k=k_0}^{\infty}\left(\sum_{\ell=k-\sigma}^{k} q(\ell)\right) = \infty$, $k_0 \in \mathbb{N}$, then every nonoscillatory solution $u(k)$ of (7.17.22) satisfies either $|u(k)| \to \infty$ or $u(k) \to 0$ as $k \to \infty$.

**7.17.67.** Consider the difference equation

$(7.17.30)_\delta$ $$\Delta^2(u(k) + pu(k+\delta\tau)) + q(k)f(u(\sigma(k))) = F(k), \quad k \in \mathbb{N}$$

where $\delta = \pm 1$, $\tau \in \mathbb{N}(1)$, $\sigma(k) \in \mathbb{N}$, $k \in \mathbb{N}$ with $\lim_{k\to\infty} \sigma(k) = \infty$, $p$ is a nonnegative constant, $q(k) \geq 0$ eventually, and $f : \mathbb{R} \to \mathbb{R}$ is continuous with $uf(u) > 0$ for $u \neq 0$. Assume that there exists an oscillatory function $h(k)$ such that $\Delta^2 h(k) = F(k)$. Show that

(i) if $\limsup_{k\to\infty}(h(k)/k) = \infty$ and $\liminf_{k\to\infty}(h(k)/k) = -\infty$, then equation $(7.17.30)_\delta$ is oscillatory

(ii) if in addition to condition (iii) of Theorem 7.10.3 and (7.10.9) there exists a function $r(k) \in \mathbb{N}(1)$, $k \in \mathbb{N}$ such that $(k - r(k))$ is increasing, $\lim_{k\to\infty}(k - r(k)) = \infty$, $\sigma(k) - \alpha \geq k - r(k)$, $k \in \mathbb{N}$ where $\alpha = 0$ if $\delta = -1$ and $\alpha = \tau$ if $\delta = 1$, and

$$\liminf_{k\to\infty} \frac{1}{r(k)} \sum_{\ell=k-r(k)}^{k-1} (\ell - r(\ell))q(\ell) > \gamma \limsup_{k\to\infty} \frac{r(k)^{r(k)}}{(r(k)+1)^{r(k)+1}},$$

then (a) the equation $(7.17.30)_{-1}$ is oscillatory provided $0 \le p < 1$, $\gamma = 2/M(1-p)$ and $\alpha = 0$, (b) the equation $(7.17.30)_1$ is oscillatory provided $p > 1$, $\gamma = 2p^2/M(p-1)$ and $\alpha = \tau$

(iii) if in addition to condition (iii) of Theorem 7.10.3 and (7.10.2) the function $\beta(k) = \min\{k, \sigma(k) - \alpha\}$ where $\alpha = 0$ if $\delta = -1$ and $\alpha = \tau$ if $\delta = 1$ is such that $\Delta\beta(k) > 0$, $\Delta^2\beta(k) \le 0$ for $k \in \mathbb{N}$, and $\sum^\infty \beta(k)q(k) = \infty$, then (a) the equation $(7.17.30)_{-1}$ is oscillatory provided $0 \le p < 1$, (b) the equation $(7.17.30)_1$ is oscillatory provided $p > 1$

(iv) if in addition to condition (iii) of Theorem 7.10.3 the function $f$ is nondecreasing, and $\sum^\infty q(k) = \infty$, then (a) the equation $(7.10.30)_{-1}$ is oscillatory provided $0 \le p < 1$, (b) the equation $(7.17.30)_1$ is oscillatory provided $p > 1$.

**7.17.68.** Consider the difference equation

$$\Delta(a(k-1)\Delta(u(k-1) + p(k-1)u(k-1-\tau))) + q(k)f(u(k-\sigma)) = 0, \quad k \in \mathbb{N}(1) \tag{7.17.31}$$

where $\tau$, $\sigma \in \mathbb{N}$, $0 \le p(k) \le 1$, $k \in \mathbb{N}$, $q(k) \ge 0$, $k \in \mathbb{N}$ with infinitely many nonzero terms, $a(k) > 0$, $k \in \mathbb{N}$ and $\sum^\infty 1/a(k) = \infty$, $f : \mathbb{R} \to \mathbb{R}$ is continuous, and there exists a constant $\gamma$ such that $f(u)/u \ge \gamma > 0$ for $u \ne 0$. Assume that there exists a function $\xi(k) > 0$, $k \in \mathbb{N}(1)$, and for $k \in \mathbb{N}(1)$ define

$$\begin{aligned}\psi(k) &= \xi(k)[\gamma q(k+1)(1 - p(k+1-\sigma)) + a(k-\sigma)\alpha^2(k) \\ &\qquad - \Delta(a(k-1-\sigma)\alpha(k-1))],\end{aligned}$$

where $\alpha(k) = -\Delta\xi(k)/2\xi(k)$. Let there exist a function $H(\ell, k)$, $\ell, k \in \mathbb{N}$ such that

(i) $H(\ell, \ell) = 0$, $\ell \in \mathbb{N}$ and $H(\ell, k) > 0$ for $\ell > k \ge 0$

(ii) $\Delta_k H(\ell, k) \le 0$ for $\ell > k \ge 0$, and there exists a nonnegative function $h(\ell, k)$, $\ell > k \ge 0$ with $\Delta_k H(\ell, k) \ge -h(\ell, k)\sqrt{H(\ell, k)}$ for $\ell > k \ge 0$.

Show that the equation (7.17.31) is oscillatory if

$$\limsup_{\ell\to\infty} \frac{1}{H(\ell, 0)} \sum_{k=0}^{\ell-1} \left( H(\ell, k)\psi(k) - \frac{a(k-\sigma)\xi^2(k+1)h^2(\ell, k)}{4\xi(k)} \right) = \infty.$$

In particular, with $H(\ell, k) = (\ell - k)^2$, $h(\ell, k) = 2$ and respectively the functions $\psi(k) = k+1$ and $(k+1)(k+2)$, $k \in \mathbb{N}$ show that the following equations are oscillatory

$$\Delta^2\left(u(k-1)+\frac{1}{\sqrt{k}}u(k-2)\right)+\frac{\lambda}{k^2}u(k-1)=0,\ k\in\mathbb{N}(1),\ \lambda>1/4$$

$$\Delta\left(\frac{1}{k}\Delta\left(u(k-1)+\frac{1}{k}u(k-2)\right)\right)+\frac{\lambda}{k(k+1)^2}u(k-1)=0,\ k\in\mathbb{N}(1),\ \lambda>1.$$

**7.17.69.** Consider the difference equation

$$\text{(7.17.32)}\quad \Delta(a(k-1)\Delta(u(k-1)+p(k-1)u(k-1-\tau)))+q(k)f(u(k-\sigma)) = F(k),\quad k\in\mathbb{N}(1)$$

where $\tau,\ \sigma\in\mathbb{N}$, $a(k)\neq 0,\ k\in\mathbb{N}$ and $f:\mathbb{R}\to\mathbb{R}$ is continuous with $uf(u)>0$ for $u\neq 0$, and there exist nonnegative constants $A, B$ and $\alpha$ with $0<\alpha\leq 1$ such that $|f(u)|\leq A|u|^{\alpha}+B$,

$$\sum_{k=K}^{\infty}\frac{1}{|a(k)|}\sum_{\ell=K}^{k}|q(\ell)|\ <\ \infty\quad\text{and}\quad\sum_{k=K}^{\infty}\frac{1}{|a(k)|}\sum_{\ell=K}^{k}|F(\ell)|\ <\ \infty.$$

Show that

(i) if there exists a constant $P$ such that $|p(k)|\leq P<1$, then all solutions of (7.17.32) are bounded

(ii) if $p(k)\to 0$ as $k\to\infty$, then every oscillatory solution of (7.17.32) tends to zero as $k\to\infty$.

In particular, deduce that all solutions of the difference equation

$$\text{(7.17.33)}\quad \Delta\left[(k+2)^2\Delta\left(u(k-1)+\frac{1}{(k+2)^4}u(k-2)\right)\right] -\frac{(k+1)^4+(k+3)^4}{(k+2)^2(k+3)^4}u(k-2) = \frac{(-1)^k\left[(k+2)^4+(k+3)^4\right]}{(k+2)^2(k+3)^4},\quad k\geq 2$$

are bounded, and every oscillatory solution tends to zero. In fact, $u(k)=(-1)^k/(k+3)^4$ is such a solution of (7.17.33).

**7.17.70.** Consider the difference equation

$$\text{(7.17.34)}\qquad \Delta^n(u(k)-pu(k-\tau))+q(k)u(k-\sigma)\ =\ 0,\qquad k\in\mathbb{N}$$

where $\tau\in\mathbb{N}(1)$, $\sigma\in\mathbb{N}$ and $n$ is odd. Show that

(i) if $p=-1$ and

$$\text{(7.17.35)}\qquad \sum^{\infty}k^{(n-1)}|q(k)|\ <\ \infty,$$

then (7.17.34) has an eventually positive solution

(ii) if $p = 1$ and $q(k) \geq 0$ for all large $k \in \mathbb{N}$, then a necessary and sufficient condition for (7.17.34) to have a bounded positive solution is

$$\sum^{\infty} k^{(n)} q(k) = \infty. \tag{7.17.36}$$

**7.17.71.** Consider the difference equation

$$\Delta^n (u(k) - p(k)u(k-\tau)) + q(k)u(k-\sigma) = 0, \quad k \in \mathbb{N} \tag{7.17.37}$$

where $\tau \in \mathbb{N}(1)$, $\sigma \in \mathbb{N}$, $n$ is odd, and $q(k) \geq 0$, $k \in \mathbb{N}$. Show that

(i) if $p(k) \equiv 1$ and

$$\sum_{k=0}^{\infty} k^n q(k) \sum_{\ell=k}^{\infty} (\ell - k + n)^{n-1} q(\ell) = \infty,$$

then the equation (7.17.37) is oscillatory

(ii) if $0 < p(k) \leq P < 1$ and

$$\sum^{\infty} k^{(n-1)} q(k) = \infty, \tag{7.17.38}$$

then every bounded nonoscillatory solution of (7.17.37) tends to zero as $k \to \infty$

(iii) if $1 \leq p(k) \leq P$ and (7.17.38) holds, then every bounded solution of (7.17.37) is oscillatory

(iv) if $0 \leq p(k) \leq P < 1$ and $\sum_{k=0}^{\infty} k^{(n-2)} q(k) = \infty$, then every unbounded solution of (7.17.37) is oscillatory.

**7.17.72.** Consider the difference equation

$$\Delta^n (u(k) - p(k)u(k-\tau)) + \delta q(k) u(\sigma(k+n-1)) = 0, \quad k \in \mathbb{N} \tag{7.17.39}$$

where $n \geq 2$, $\delta = \pm 1$, $|p(k)| \leq \lambda < 1$, $k \in \mathbb{N}$, $\sigma(k) \in \mathbb{N}$ and $\sigma(k) \leq k$ for $k \in \mathbb{N}$, $\lim_{k\to\infty} \sigma(k) = \infty$, and $q(k) > 0$, $k \in \mathbb{N}$. Show that

(a) if

(i) $p(k)p(k-\tau) \geq 0$ for all large $k \in \mathbb{N}$

(ii) the equation (7.17.39) has a nonoscillatory solution $u(k)$ satisfying $u(k)(u(k) - p(k)u(k-\tau)) > 0$ for all large $k \in \mathbb{N}$ and

$$\lim_{k\to\infty} \frac{u(k) - p(k)u(k-\tau)}{k^{(j)}} = \text{constant} \neq 0$$

for some $j \in \mathbb{N}(0, n-1)$,

then the following inequality holds

$$(7.17.40) \qquad \sum^{\infty} k^{(n-j-1)}(\sigma(k+n-1))^{(j)} q(k) \ < \ \infty$$

(b) if $(-1)^{n-j-1}\delta = 1$ and condition (i) holds, or $(-1)^{n-j-1}\delta = -1$ and $p(k) \geq 0$ for all large $k \in \mathbb{N}$, then (ii) holds provided the inequality (7.17.40) is satisfied

(c) if (i) holds, then for the equation (7.17.39) to have a nonoscillatory solution $u(k)$ satisfying $u(k)(u(k)-p(k)u(k-\tau)) > 0$ for all large $k \in \mathbb{N}$ and

$$(7.17.41) \quad \lim_{k\to\infty} \frac{u(k)-p(k)u(k-\tau)}{k^{(m)}} = 0, \quad \lim_{k\to\infty} \frac{u(k)-p(k)u(k-\tau)}{k^{(m-1)}} = \pm\infty$$

for some $m \in \mathbb{N}(1, n-1)$ such that $(-1)^{n-m-1}\delta = 1$, it is necessary that

$$\sum^{\infty} k^{(n-m-1)}(\sigma(k+n-1))^{(m-1)} q(k) \ < \ \infty$$

and

$$\sum^{\infty} k^{(n-m)}(\sigma(k+n-1))^{(m)} q(k) \ = \ \infty$$

(d) if $p(k) \geq 0$ and $\sigma(k) < k$ for all large $k$, and let $m \in \mathbb{N}(1, n-1)$ satisfy $(-1)^{n-m-1}\delta = 1$, then the equation (7.17.39) has a nonoscillatory solution $u(k)$ satisfying $u(k)(u(k)-p(k)u(k-\tau)) > 0$ for all large $k \in \mathbb{N}$ and (7.17.41), provided

$$\sum^{\infty} k^{(n-m-1)}(\sigma(k+n-1))^{(m)} q(k) \ < \ \infty$$

and

$$\sum^{\infty} k^{(n-m)}(\sigma(k+n-1))^{(m-1)} q(k) \ = \ \infty.$$

**7.17.73.** Consider the difference equations

$$(7.17.42) \qquad \Delta^n(u(k) - u(k-\tau)) + q(k)f(u(k-\sigma)) \ = \ 0, \qquad k \in \mathbb{N}$$

and

$$(7.17.43) \qquad \Delta^{n+1}u(k-1) + \frac{1}{\tau}q(k)f(u(k)) \ = \ 0, \qquad k \in \mathbb{N}$$

where $n$ is odd, $\tau \in \mathbb{N}(1)$, $\sigma \in \mathbb{Z}$, $q(k) \geq 0$ for all large $k \in \mathbb{N}$, and $f : \mathbb{R} \to \mathbb{R}$ is continuous, nondecreasing with $uf(u) > 0$ for all $u \neq 0$ and $f(-u) = -f(u)$. Show that the equation (7.17.42) is oscillatory

(i) if and only if the difference inequality

$$(7.17.44) \qquad \Delta^n(v(k) - v(k-\tau)) + q(k)f(v(k-\sigma)) \le 0, \quad k \in \mathbb{N}$$

has no eventually positive solution

(ii) if and only if (7.17.43) is oscillatory

(iii) if (7.9.19) and (7.17.36) hold

(iv) if $f(u) = u$, and

$$\sum_{k=m}^{\infty} \overline{q}(k) > \frac{\eta\tau}{m} \quad \text{for all large} \quad m \in \mathbb{N}$$

where $\eta > 1/4$, and

$$\overline{q}(k) = \begin{cases} q(k) & \text{if } n = 1 \\ \dfrac{1}{(n-2)!}\displaystyle\sum_{\ell=k}^{\infty}(\ell - k + n - 2)^{(n-2)}q(\ell) & \text{if } n \ge 3. \end{cases}$$

**7.17.74.** Consider the difference equation

$$(7.17.45) \qquad \Delta^n(u(k) - pu(k-\tau)) + q(k)f(u(k-\sigma)) = F(k), \quad k \in \mathbb{N}$$

where $\tau \in \mathbb{N}(1)$, $\sigma \in \mathbf{Z}$, and $f : \mathbb{R} \to \mathbb{R}$ is continuous. Assume that in addition to (7.17.35)

$$(7.17.46) \qquad \sum^{\infty} k^{(n-1)}|F(k)| < \infty.$$

Show that the equation (7.17.45) has a bounded positive solution

(i) if $p = 0$

(ii) if $p > 1$ and $f$ satisfies the Lipschitz condition

$$(7.17.47) \qquad |f(u) - f(v)| \le L|u - v|$$

for all $u, v \in [p, 2p]$

(iii) if $p < -1$ and $f$ satisfies (7.17.47) for all $u, v \in [-p, -2p]$

(iv) if $-1 < p < 0$ and $f$ satisfies (7.17.47) for all $u, v \in [-1/2p, -1/p]$

(v) if $0 < p < 1$ and $f$ satisfies (7.17.47) for all $u, v \in [1/2p, 1/p]$.

**7.17.75.** Consider the difference equation

$$(7.17.48)_\delta \qquad \Delta^n(u(k) - pu(k-\tau)) + \delta q(k)f(u(k-\sigma)) = F(k), \quad k \in \mathbb{N}$$

where $\delta = \pm 1$, $\tau \in \mathbb{N}(1)$, $\sigma \in \mathbb{Z}$, $q(k) \geq 0$, $k \in \mathbb{N}$ with infinitely many positive terms, and $f : \mathbb{R} \to \mathbb{R}$ is continuous with $uf(u) > 0$ for $u \neq 0$. Assume that there exists an oscillatory function $h(k)$ such that $\Delta^n h(k) = F(k)$, $k \in \mathbb{N}$ and $\lim_{k\to\infty} \Delta^j h(k) = 0$, $j = 1, \cdots, n-1$. Show that

(i) if $p = 0$, and (7.17.38) holds, then every bounded solution $u(k)$ of $(7.17.48)_\delta$ is oscillatory when $(-1)^n\delta = 1$, and is either oscillatory or such that $\lim_{k\to\infty} \Delta^j u(k) = 0$ for $j = 0, 1, \cdots, n-1$ when $(-1)^n\delta = -1$; further, if (7.17.46) holds then for this conclusion (7.17.38) is also a necessary condition

(ii) if $p > 1$ and (7.17.38) holds, then every bounded solution $u(k)$ of $(7.17.48)_\delta$ is oscillatory when $(-1)^n\delta = -1$, and is either oscillatory or such that $\lim_{k\to\infty} u(k) = 0$ when $(-1)^n\delta = 1$; further, if (7.17.46) holds and (7.17.47) is satisfied for all $u, v \in [p, 2p]$ then for this conclusion (7.17.38) is also a necessary condition.

**7.17.76.** State and prove results similar to those given in Problems 7.17.74 and 7.17.75 for the difference equation $(7.17.48)_\delta$ when $p$ is a function of $k$, and the function $p(k)$ is such that $1 < c \leq p(k) \leq C$, $k \in \mathbb{N}$ where $c$ and $C$ are numbers.

**7.17.77.** Consider the difference equation

$$(7.17.49) \qquad \Delta^n(u(k) - p(k)u(k-\tau)) + f(k, u(k-\sigma)) = 0, \qquad k \in \mathbb{N}$$

where $n$ is an odd integer, $\tau \in \mathbb{N}(1)$, $\sigma \in \mathbb{N}$ and $uf(k,u) > 0$ for all $u \neq 0$ and $k \in \mathbb{N}$. Assume that

(i) $0 \leq p(k) \leq p < 1$, $k \in \mathbb{N}$

(ii) for each $k \in \mathbb{N}$, $f(k,u)$ is nondecreasing in $u \in (0, \infty)$.

Show that

(a) if

$$\frac{pk}{k-\tau} + k\sum_{\ell=k}^{\infty} \frac{(\ell-k+1)\cdots(\ell-k+n-1)}{(n-1)!} f\left(\ell, \frac{1}{\ell-\sigma}\right) \leq 1$$

for $k \geq K \geq n+1$, then (7.17.49) has an eventually positive solution which converges to zero

(b) if there exists an integer $m \in \mathbb{N}(1, n-1)$ such that

$$\lim_{k\to\infty} p(k)\left(\frac{\tau}{k}\right)^m = c \in [0, 1)$$

and

$$\sum_{k=0}^{\infty} k^{n-m-1} f(k, \lambda(k-\sigma)^m) \ < \ \infty$$

for some $\lambda > 0$, then (7.17.49) has an eventually positive solution which diverges to infinity

(c) if $p(k) \to p^* \in [0,1)$ as $k \to \infty$, and

$$\sum_{k=0}^{\infty} k^{n-1} f(k, \lambda) \ < \ \infty$$

for some $\lambda > 0$, then (7.17.49) has an eventually positive solution which tends to a positive constant.

**7.17.78.** Consider the difference equation

(7.17.50) $\Delta\left(a(k)\left(\Delta^n(u(k) - p(k)u(k-\tau))\right)^{\gamma}\right) + f(k, u(k-\sigma)) = 0, \ \ k \in \mathbb{N}$

where $\tau \in \mathbb{N}(1)$, $\sigma \in \mathbb{N}$, $\gamma$ is a quotient of odd positive integers, $0 \le p(k) \le p < 1$, $k \in \mathbb{N}$, $a(k) > 0$, $k \in \mathbb{N}$, $f(k,u) : \mathbb{N} \times \mathbb{R} \to \mathbb{R}$ is continuous and $uf(k,u) > 0$ for all $u \neq 0$ and $k \in \mathbb{N}$. Assume that $f(k,u_1) \ge f(k,u_2)$ for all $u_1 \ge u_2 > 0$ or $u_1 \le u_2 < 0$, $k \in \mathbb{N}$, and $\lim_{k\to\infty} p(k)$ exists. Show that

(a) if $\sum^{\infty}(1/a(k))^{1/\gamma} = \infty$, then every bounded solution of (7.17.50) is oscillatory or tends to zero if and only if

$$\sum^{\infty} k^{n-1} \left( \frac{1}{a(k)} \sum_{\ell=k}^{\infty} |f(\ell, \lambda)| \right)^{1/\gamma} = \infty \qquad \text{for all } \ \lambda \neq 0$$

(b) if $\sum^{\infty}(1/a(k))^{1/\gamma} < \infty$, and $\sum^{\infty} |f(k,\lambda)| = \infty$ for all $\lambda \neq 0$, then a necessary and sufficient condition for (7.17.50) to have a nonoscillatory solution $u(k)$ which satisfies $\lim_{k\to\infty} u(k) = \alpha \neq 0$ is that

$$\sum^{\infty} k^{n-1} \left( \frac{1}{a(k)} \sum_{\ell=s}^{k-1} |f(\ell, \alpha)| \right)^{1/\gamma} \ < \ \infty \qquad \text{for some } \ \alpha \neq 0 \quad \text{and } \ s \in \mathbb{N}.$$

**7.17.79.** Consider the difference equation

(7.17.51) $\Delta\left(a(k)\Delta^{n-1}(u(k) + p(k)u(\tau(k)))\right) + f(k, u(\sigma(k))) = 0, \ \ k \in \mathbb{N}$

where $a(k) > 0$, $\Delta a(k) \ge 0$, $k \in \mathbb{N}$ and $\sum^{\infty} 1/a(k) = \infty$, $0 \le p(k) < 1$, $\tau(k) \le k$, $k \in \mathbb{N}$ and $\tau(k) \to \infty$, $\sigma(k) \to \infty$ as $k \to \infty$, and

$f(k,u): \mathbb{N}\times\mathbb{R}\to\mathbb{R}$ is continuous, nondecreasing in $u$ and $uf(k,u)>0$ for all $u\neq 0$ and $k\in\mathbb{N}$. Show that

(i) if for every $c>0$, $\sum^{\infty}(1/a(k))f(k,(1-p(\sigma(k))c))=\infty$, then

(a) the equation (7.17.51) is oscillatory when $n$ is even

(b) every solution $u(k)$ of (7.17.51) is either oscillatory or $\liminf_{k\to\infty}|u(k)|=\infty$ when $n$ is odd

(ii) if for every $c>0$, $\sum^{\infty}\left((k)^{(n-1)}/a(k)\right)f(k,(1-p(\sigma(k))c))=\infty$, then

(a) every bounded solution $u(k)$ of (7.17.51) is oscillatory when $n$ is even

(b) every bounded solution $u(k)$ of (7.17.51) is either oscillatory or $\liminf_{k\to\infty}|u(k)|=\infty$ when $n$ is odd.

**7.17.80.** Consider the difference equation

$$\Delta^n(u(k)+p(k)u(\tau(k)))+f(k,u(k),u(\sigma(k)))=F(k),\ k\in\mathbb{N} \tag{7.17.52}$$

where $n$ is even, and for all $k\in\mathbb{N}$, $0\le p(k)<1$, $\tau(k)\in\mathbb{N}$, $\sigma(k)\in\mathbb{N}$, $\sigma(k)\le k$, $\lim_{k\to\infty}\tau(k)=\infty$, $\lim_{k\to\infty}\sigma(k)=\infty$, $f(k,u,v)$ is continuous on $\mathbb{N}\times\mathbb{R}^2$ and not equivalently equal to zero for $uv>0$ and $k\in\mathbb{N}$, and $vf(k,u,v)>0$ for $uv>0$. Assume that

(i) $h(k)$ is an oscillatory function such that $\Delta^n h(k)=F(k)$ and $\lim_{k\to\infty}h(k)=0$

(ii) $|f(k,u,v)|\ge\phi(k)w\left(\frac{|v|}{[1-p(\sigma(k))][\sigma(k)]^{n-1}}\right)$, where $\phi(k)\ge 0$, $k\in\mathbb{N}$ and $w(u)>0$ for $u>0$ is continuous and nondecreasing on $\mathbb{R}_+$.

Show that if $\sum^{\infty}\phi(k)=\infty$ and $\int^{\pm a}\frac{du}{w(u)}<\infty$ for all $a>0$, then the equation (7.17.52) is oscillatory.

## 7.18. Notes

The qualitative study of solutions of neutral differential and difference equations is developing very rapidly. In fact, Bainov and Mishev [7] in their monograph cite over 150 recent publications on the oscillation of neutral differential equations alone. The equation (7.1.1) was first considered by Brayton and Willoughby [8] from the numerical point of view. All the results in Sections 7.1 7.3 are based on the work of Lalli et. al. [38 41], except Theorems 7.1.3, 7.2.6 and 7.2.11 which are respectively due to Yu and Wang [89], Zhou, Yu and Wong [108], and Shen, Wang and Qian [61].

These results improve/extend those of Georgiou, Grove and Ladas [21,22]. Theorem 7.4.1 is a contribution of Yu and Cheng [88]. Sections 7.5 and 7.6 once again contains the work of Lalli et. al. [38 41]. Results in Section 7.7 are borrowed from Chen, Lalli and Yu [12]. Lemma 7.8.1 and Theorem 7.8.2 are taken from Peil [60]. Results in Section 7.9 are adapted from Thandapani et. al. [76,82,85]. The nonhomogeneous difference equation $(7.10.1)_\delta$ with $p(k) \equiv p$ has been studied in Grace and Lalli [24,25], also see Lalli [41]. Section 7.11 contains the work of Thandapani et. al. [79]. All the results in Sections 7.12 and 7.13 are due to Agarwal, Manuel and Thandapani [1,2]. Sections 7.14 and 7.15, respectively, are based on the work of Agarwal and Grace [6], and Agarwal, Thandapani and Wong [3]. Theorem 7.16.1 is earlier proved in Zafer and Dahiya [90]. For several other related results see Agarwal and Wong [4], Agarwal and Grace [5], Budinčević [9,10], Chen and Zhang [11], Chen and Wang [13], Cheng et. al. [14,15], Dai et. al. [16 20], Grace and Lalli [23], Grace [26], Graef et. al. [27 30], Guan et. al. [31 33], Huang and Yu [34,35], Kordonis and Philos [36,37], Lalli and Grace [42,43], Li and Yeh [44], Li, Wang and Zhang [45], Li et al. [46 54], Lin and Cheng [55], Liu and Yan [56], Luo, He and Chen [57], Ou, Luo and Tang [59], Shengli and Cheng [62,63], Szmanda [64], Szafranski and Szmanda [65 70], Tang et. al. [71 74], Thandapani et. al. [75,77,78,80,81,83,84], Wang and Yu [86], Yang, Guan and Liu [87], Zafer [91 94], B.G. Zhang et. al. [95 100], G. Zhang and Cheng [101 105], Z. Zhang and Yu [106], Zhou et. al. [107,109].

## 7.19. References

[1]. R.P. Agarwal, M. Maria Susai Manuel and E. Thandapani, Oscillatory and nonoscillatory behavior of second order neutral delay difference equations, *Mathl. Comput. Modelling* **24**(1)(1996), 5 11.

[2]. R.P. Agarwal, M. Maria Susai Manuel and E. Thandapani, Oscillatory and nonoscillatory behavior of second order neutral delay difference equations II, *Appl. Math. Letters* **10**(2)(1997), 103 109.

[3]. R.P. Agarwal, E. Thandapani and P.J.Y. Wong, Oscillation of higher order neutral difference equations, *Appl. Math. Letters* **10**(1)(1997), 71 78.

[4]. R.P. Agarwal and P.J.Y. Wong, *Advanced Topics in Difference Equations*, *Kluwer*, Dordrecht, 1997.

[5]. R.P. Agarwal and S.R. Grace, On the oscillation of certain difference equations, *Mathl. Comput. Modelling*, to appear.

[6]. R.P. Agarwal and S.R. Grace, Oscillation theorems for certain difference equations, *Dynamic Sys. Appl.*, to appear.

[7]. D.D. Bainov and D.P. Mishev, *Oscillation Theory for Neutral Differential Equations with Delay*, *Adam Hilger*, New York, 1991.

[8]. R.K. Brayton and R.A. Willoughby, On the numerical integration of a symmetric system of difference differential equations of neutral type, *J. Math. Anal. Appl.* **18**(1967), 182 189.

[9]. M. Budinčević, Oscillations and the asymptotic behaviour of certain second order neutral difference equations, *Zb. Rad. Prirod.-Mat. Fak. Ser. Mat.* **21**(1991), 165 172.

[10]. M. Budinčević, Oscillation of a second order neutral difference equation, *Bull. Cl. Sci. Math. Nat. Sci. Math.* **22**(1997), 1 8.

[11]. M.P. Chen and B.G. Zhang, Oscillation and comparison theorems of difference equations with positive and negative coefficients, *Bull. Inst. Math. Acad. Sinica* **22**(1994), 295 306.

[12]. M.P. Chen, B.S. Lalli and J.S. Yu, Oscillation in neutral delay difference equations with variable coefficients, *Computers Math. Applic.* **29**(3) (1995), 5 11.

[13]. M.P. Chen and S.H. Wang, Oscillation in first order neutral delay difference equation with positive and negative coefficients, Preprint.

[14]. S.S. Cheng and G. Zhang, Monotone solutions of a higher order neutral difference equation, *Georgian Math. J.* **5**(1998), 49 54.

[15]. S.S. Cheng, G. Zhang and W.T. Li, On a higher order neutral difference equation, in *Recent Results in Mathematical Analysis*, to appear.

[16]. B.X. Dai, Oscillation and asymptotic behavior of a second order neutral difference equation. (Chinese), *Hunan Ann. Math.* **16**(3)(1996), 101 103.

[17]. B.X. Dai and D.Q. Lu, Oscillation in odd order neutral difference equations with variable coefficients. (Chinese), *J. Changsha Comm. Univ.* **13**(1997), 8 13.

[18]. B. X. Dai and X.J. Tu, Oscillation for a class of nonlinear neutral difference equations. (Chinese), *Hunan Daxue Xuebao* **24**(1997), 1 6.

[19]. B.X. Dai and L.H. Huang, Convergence of solutions for a class of neutral difference equation, *Computers Math. Applic.* **35**(10)(1998), 83 91.

[20]. B.X. Dai and L. Huang, Asymptotic behavior and existence of positive solutions for a neutral difference equation, *Ann. Diff. Eqns.* **14**(1998), 22 28.

[21]. D.A. Georgiou, E.A. Grove and G. Ladas, Oscillations of neutral difference equations, *Applicable Analysis* **33**(1989), 243 253.

[22]. D.A. Georgiou, E.A. Grove and G. Ladas, Oscillation of neutral difference equations with variable coefficients, in *Differential Equations: Stability and Control*, *Marcel Dekker*, New York, 1990, 165 173.

[23]. S.R. Grace and B.S. Lalli, Oscillation theorems for second order delay and neutral difference equations, *Utilitas Mathematica* **45**(1994), 197 211.

[24]. S.R. Grace and B.S. Lalli, Oscillation theorems for forced neutral difference equations, *J. Math. Anal. Appl.* **187**(1994), 91 106.

[25]. S.R. Grace and B.S. Lalli, Erratum: "Oscillation theorems for forced neutral difference equations", *J. Math. Anal. Appl.* **198**(1996), 922.

[26]. S.R. Grace, Oscillation of certain neutral difference equations of mixed type, *J. Math. Anal. Appl.* **224**(1998), 241 254.

[27]. J.R. Graef and P.W. Spikes, Asymptotic decay of oscillatory solutions of forced nonlinear difference equations, *Dynamic Sys. Appl.* **3**(1994), 95 102.

[28]. J.R. Graef and P.W. Spikes, Some results on the asymptotic behavior of solutions of a nonlinear difference equation with a forcing term, in *Proceedings of Dynamic Sys. Appl.* **1**(1994), 121 126.

[29]. J.R. Graef and P.W. Spikes, Boundedness and asymptotic behavior of solutions of a forced difference equation, *Internat. J. Math. & Math. Sci.* **17**(1994), 397 400.

[30]. J.R. Graef, A. Miciano, P.W. Spikes, P. Sundaram and E. Thandapani, Oscillatory and asymptotic behavior of solutions of nonlinear neutral type difference equations, *J. Austral. Math. Soc. (Series B)* **38**(1996), 163 171.

[31]. X.P. Guan, J. Yang and Q. Rong, Asymptotic and oscillatory behavior of $n$th order neutral difference equation with forced term, *Ann. Diff. Eqns.* **14**(1998), 143 149.

[32]. X.P. Guan, J.Yang, S.S. Cheng and W.T. Li, On a nonlinear neutral difference equation with positive and negative coefficients, Preprint.

[33]. X.P. Guan, J. Yang, S.T. Liu and S.S. Cheng, Oscillatory behavior of higher order nonlinear neutral difference equations, *Hokkaido Mathl. J.*, to appear.

[34]. L.H. Huang and J.S. Yu, Convergence in neutral delay difference equations, *Diff. Eqns. Dyn. Sys.*, to appear.

[35]. L.H. Huang and J.S. Yu, Asymptotic behavior of solutions for a class of difference equations, *J. Math. Anal. Appl.* **204**(1996), 830 839.

[36]. I.G. E. Kordonis and C.G. Philos, Oscillation of neutral difference

equations with periodic coefficients, *Computers Math. Applic.* **33**(7)(1997), 11 27.

[37]. I.G. E. Kordonis and C.G. Philos, On the behavior of the solutions for linear autonomous neutral delay difference equations, to appear.

[38]. B.S. Lalli, B.G. Zhang and J.Z. Li, On the oscillation of solutions and existence of positive solutions of neutral difference equations, *J. Math. Anal. Appl.* **158**(1991), 213 233.

[39]. B.S. Lalli and B.G. Zhang, On existence of positive solutions and bounded oscillations for neutral difference equations, *J. Math. Anal. Appl.* **166**(1992), 272 287.

[40]. B.S. Lalli and B.G. Zhang, Oscillation and comparison theorems for certain neutral difference equations, *J. Austral. Math. Soc. (Series B)* **34**(1992), 245 256.

[41]. B.S. Lalli, Oscillation theorems for neutral difference equations, *Computers Math. Applic.* **28**(1 3)(1994), 191 202.

[42]. B.S. Lalli and S.R. Grace, Oscillation theorems for second order delay and neutral difference equations, *Utilitas Mathematica* **45**(1994), 197 212.

[43]. B.S. Lalli and S.R. Grace, Oscillation theorems for second order neutral difference equations, *Appl. Math. Comput.* **62**(1994), 47 60.

[44]. H.J. Li and C.C. Yeh, Oscillation criteria for second order neutral delay difference equations, *Computers Math. Applic.* **36**(10-12)(1998),123 132.

[45]. J.W. Li, Z.C. Wang and H.Q. Zhang, Oscillation of neutral delay difference equations, *Diff. Eqns. Dyn. Sys.* **4**(1996), 113 121.

[46]. W.T. Li, Oscillation criteria for first order neutral nonlinear difference equations with variable coefficients, *Appl. Math. Letters* **10**(6)(1997), 101 106.

[47]. W.T. Li and Y. Chen, Oscillation and comparison of neutral difference equations, *J. Gansu Univ. Technol.* **1**(1997), 117 120.

[48]. W.T. Li and S.S. Cheng, Classification and existence of positive solutions of second order nonlinear neutral difference equations, *Funkcialaj Ekvacioj* **40**(1997), 371 393.

[49]. W.T. Li, Oscillation of higher order neutral nonlinear difference equations, *Appl. Math. Letters* **11**(4)(1998), 1 8.

[50]. W.T. Li and S.S. Cheng, Asymptotic trichotomy for positive solutions of a class of odd order nonlinear neutral difference equations, *Computers Math. Applic.* **35**(8)(1998), 101 108.

[51]. W.T. Li and S.S. Cheng, An oscillation theorem for a neutral difference

equation with positive and negative coefficients, *Tamk. J. Math.*, to appear.

[52]. W.T. Li and S.S. Cheng, On a neutral difference equation with positive and negative coefficients, *Southeast Asia Bull. Math.*, to appear.

[53]. W.T. Li, X.L. Fei and X.L. Fan, Existence and asymptotic behavior of positive solutions of higher order nonlinear difference equations with delay, *Mathl. Comput. Modeling*, to appear.

[54]. W.T. Li and B.G. Zhang, Existence and asymptotic behavior of nonoscillatory solutions of neutral nonlinear difference equations, to appear.

[55]. Y.Z. Lin and S.S. Cheng, Complete characterizations of an oscillatory neutral difference equation, *J. Math. Anal. Appl.* **221**(1998), 73 91.

[56]. B. Liu and J.R. Yan, Oscillation theorems for nonlinear neutral difference equations, J. Difference Eqns. Appl. **1**(1995), 307 315.

[57]. J. Luo, Z. He and H. Chen, Oscillation criteria for second order neutral difference equations, *Ann. Diff. Eqns.* **14**(1998), 262 266.

[58]. R.E. Moore, *Computational Functional Analysis, Ellis Harwood*, 1985.

[59]. C. Ou, W. Luo and H. Tang, On the global attractivity for a neutral difference equation with two delays, *Ann. Diff. Eqns.* **11**(1995), 420 427.

[60]. T. Peil, Oscillation criteria for a neutral delay difference equation, *PanAmer. Math. J.* **6**(1996), 25 39.

[61]. J.H. Shen, Z.C. Wang and X.Z. Qian, On existence of positive solutions of neutral difference equations, *Tamkang J. Math.* **25**(1994), 257 265.

[62]. X. Shengli and S.S. Cheng, Decaying solutions of neutral difference equations with delays, *Ann. Diff. Eqns.* **11**(1995), 331 345.

[63]. X. Shengli and S.S. Cheng, Oscillation criteria for a class of neutral difference equations, *Fasciculi Mathematici* **27**(1997), 101 121.

[64]. B. Szmanda, Properties of solutions of higher order difference equations, *Math. Comput. Modelling* **28**(10)(1998), 95 101.

[65]. Z. Szafranski and B. Szmanda, Oscillations of some linear difference equations, *Fasc. Math.* **25**(1995), 165 174.

[66]. Z. Szafranski and B. Szmanda, Oscillation and asymptotic behaviour of certain nonlinear difference equations, *Riv. Mat. Univ. Parma* **4**(1995), 231 240.

[67]. Z. Szafranski and B. Szmanda, Oscillations in certain difference equations, *Tamk. J. Math.* **27**(1996), 257 265.

[68]. Z. Szafranski and B. Szmanda, Oscillation of solutions of some nonlinear difference equations, *Publ. Mat.* **40**(1996), 127 133.

[69]. Z. Szafranski and B. Szmanda, Oscillation theorems for some nonlinear difference equations, *Appl. Math. Comput.* **83**(1997), 43–52.

[70]. Z. Szafranski and B. Szmanda, Oscillations of some difference equations, *Fasc. Math.* **28**(1998), 149–155.

[71]. X.H. Tang, Necessary and sufficient conditions for oscillation of solutions to a class of neutral difference equations with variable coefficients. (Chinese), *Hunan Daxue Xuebao* **23**(1996), 20–26.

[72]. X.H. Tang and J. Yan, Oscillation and nonoscillation of an odd-order nonlinear neutral difference equation, *Functional Differential Equations*, to appear.

[73]. X.H. Tang, Bounded positive solutions of neutral difference equations. (Chinese), *Hunan Jiaoyu Xueyuan Xuebao (Ziran Kexue)* **15**(5)(1997), 127–129.

[74]. X.H. Tang, J.S. Yu and D.H. Peng, Sharp conditions for the oscillation of neutral delay difference equations with summable coefficients, to appear.

[75]. E. Thandapani, Asymptotic and oscillatory behaviour of solutions of a second order nonlinear neutral delay difference equation, *Riv. Math. Univ. Parma* **1**(1992), 105–113.

[76]. E. Thandapani, Asymptotic and oscillatory behavior of solutions of nonlinear neutral delay difference equations, *Utilitas Mathematica* **45**(1994), 237–244.

[77]. E. Thandapani and P. Sundaram, Oscillation properties of first order nonlinear functional difference equations of neutral type, *Indian J. Math.* **36**(1994), 59–71.

[78]. E. Thandapani and P. Sundaram, On the asymptotic and oscillatory behaviour of solutions of second order nonlinear neutral difference equations, *Indian J. Pure Appl. Math.* **26**(1995), 1149–1160.

[79]. E. Thandapani, P. Sundaram, J.R. Graef and P.W. Spikes, Asymptotic properties of solutions of nonlinear second order neutral delay difference equations, *Dynamic Sys. Appl.* **4**(1995), 125–136.

[80]. E.Thandapani, P. Sundaram, J.R. Graef, A. Miciano and P.W. Spikes, Classification of nonoscillatory solutions of higher order neutral type difference equations, *Arch. Math. (Brnö)* **31**(1995), 263–277.

[81]. E. Thandapani, J.R. Graef and P.W. Spikes, On existence of positive solutions and oscillations of neutral difference equations of odd order, *J. Difference Eqns. Appl.* **2**(1996), 175–183.

[82]. E. Thandapani and P. Sundaram, Asymptotic and oscillatory behavior of solutions of first order nonlinear neutral difference equations, *Riv. Math. Pura Appl.* **18**(1996), 93–105.

[83]. E. Thandapani, P. Sundaram, J.R. Graef and P.W. Spikes, Asymptotic behavior and oscillation of solutions of neutral delay difference equations of arbitrary order, *Math. Slovaca* **47**(1997), 539 551.

[84]. E. Thandapani and L. Ramuppillai, Oscillation theorems for certain class of nonlinear difference equations, *Z. Anal. Anwendungen* **17**(1998), 513-519.

[85]. E. Thandapani, P. Sundaram and I. Györi, On the behavior of solutions of first order nonlinear neutral difference equations, to appear.

[86]. Z.C. Wang and J.S. Yu, Oscillation of neutral delay difference equations with summable coefficients, in *Proceedings of the 2nd International Conference on Difference Equations and Applications*, Vesprém, Hungary, August 7 11, 1995, eds. S. Elaydi, et. al. *Gordon & Breach*, New York, 1996, 629 639.

[87]. J. Yang, X. Guan and W. Liu, Oscillation and asymptotic behavior of second order nonlinear neutral difference equation, *Ann. Diff. Eqns.* **13**(1997), 94 106.

[88]. J.S. Yu and S.S. Cheng, A stability criterion for a neutral difference equation with delay, *Appl. Math. Letters* **7**(6)(1994), 75 80.

[89]. J.S. Yu and Z.C. Wang, Asymptotic behavior and oscillation in neutral delay difference equations, *Funkcialaj Ekvacioj* **37**(1994), 241 248.

[90]. A. Zafer and R.S. Dahiya, Oscillation of a neutral difference equation, *Appl. Math. Letters* **6**(2)(1993), 71 74.

[91]. A. Zafer, Oscillatory and asymptotic behavior of higher order difference equations, *Mathl. Comput. Modelling* **21**(4)(1995), 43 50.

[92]. A. Zafer, Oscillation of higher order neutral type difference equations, in *Proceedings of the 2nd International Conference on Difference Equations and Applications*, Vesprém, Hungary, August 7 11, 1995, eds. S. Elaydi, et. al. *Gordon & Breach*, New York, 1996, 641 647.

[93]. A. Zafer, Necessary and sufficient condition for oscillation of higher order nonlinear difference equations, *Computers Math. Applic.* **35**(10)(1998), 125 130.

[94]. A. Zafer, On the existence of positive solutions and oscillation of solutions of higher order difference equations with forcing terms, *Computers Math. Applic.* **36**(10 12)(1998), 27 35.

[95]. B.G. Zhang and P.X. Yan, Oscillation and comparison theorems for neutral difference equations, *Tamkang J. Math.* **25**(1994), 343 352.

[96]. B.G. Zhang and S.S. Cheng, Oscillation criteria and comparison theorems for delay difference equations, *Fasiculi Mathematici* **25**(1995), 13 32.

[97]. B.G. Zhang and H. Wang, The existence of oscillatory and nonoscillatory solutions of neutral difference equations, *Chinese J. Math.* **24**(1996), 377 393.

[98]. B.G. Zhang and H. Wang, The existence of nonoscillatory and oscillatory solutions of neutral difference equations, *Dynamic Sys. Appl.* **6**(1997), 411 428.

[99]. B.G. Zhang and G. Zhang, Oscillation of nonlinear difference equations of neutral type, *Dynamic Sys. Appl.* **7**(1998), 85 92.

[100]. B.G. Zhang and B. Yang, Oscillation in higher order nonlinear difference equations, Preprint.

[101]. G. Zhang and S.S. Cheng, Oscillation criteria for a neutral difference equation with delay, *Appl. Math. Letters* **8**(3)(1995), 13 17.

[102]. G. Zhang and S.S. Cheng, Elementary non existence criteria for a recurrence relation, *Chinese J. Math.* **24**(1996), 229 235.

[103]. G. Zhang and S.S. Cheng, Positive solutions of a nonlinear neutral difference equitation, *Nonlinear Analysis* **28**(1997), 729 738.

[104]. G. Zhang and S.S. Cheng, A necessary and sufficient oscillation condition for the discrete Euler equation, *PanAmer. Math. J.*, to appear.

[105]. G. Zhang and S.S. Cheng, Note on a discrete Emden Fowler equation, *PanAmer. Math. J.*, to appear.

[106]. Z. Zhang and J.S. Yu, Asymptotic behavior of solutions of neutral difference equations with positive and negative coefficients, *Math. Sci. Res. Hot Line* **2**(1998), 1 12.

[107]. Z. Zhou, L. Huang and B. Dai, Oscillation for nonlinear neutral difference equations, *Ann. Diff. Eqns.* **14**(1998), 468 473.

[108]. Z. Zhou, J.S. Yu and Z. Wang, Global attractivity of neutral difference equations, *Computers Math. Applic.* *36*(6)(1998), 1 10.

[109]. Z. Zhou and Q. Zhang, Linearized oscillations for even order neutral difference equations, *Math. Sci. Res. Hot Line* **2**(8)(1998), 11 17.

# Chapter 8
# Boundary Value Problems for Linear Systems

In general the theory and the construction of the solutions of boundary value problems is more difficult than those of initial value problems. Therefore, we begin this chapter by providing the necessary and sufficient conditions for the existence and uniqueness of the solutions of linear boundary value problems. For these problems explicit representations of the solutions are given in terms of Green's matrices. For the construction of the solutions we have included several algorithms which have been proposed recently. Although all these algorithms are the same in nature, namely convert the given boundary value problem to its equivalent initial value problem, in actual construction of the solutions one shows superiority over the others for which sometimes reasons can be explained. Most of these algorithms have been illustrated by solving discrete two point boundary value problems some of which are known to be unstable. The minimal solution of the difference equations which plays an important role in several branches of numerical analysis is introduced. For the construction of minimal solution classical algorithms of Miller and Olver are discussed.

## 8.1. Existence and Uniqueness

We begin with the observation that the existence and/or uniqueness of continuous boundary value problems do not imply the same for the corresponding discrete problems. For example, the continuous problem $u'' + \frac{\pi^2}{(K+1)^2}u = 0,\ u(0) = u(K+1) = 0,$ where $1 \le K \in \mathbb{N}$ has an infinite number of solutions $u(t) = c\sin\frac{\pi}{K+1}t$ ($c$ is arbitrary), whereas its discrete analog $u(k+1) - \left(2 - \frac{\pi^2}{(K+1)^2}\right)u(k) + u(k-1) = 0,\ k \in \mathbb{N}(1,K),\ u(0) = u(K+1) = 0$ has only one solution $u(k) \equiv 0$. The problem $u'' + \frac{\pi^2}{4(K+1)^2}u = 0,\ u(0) = 0,\ u(K+1) =$

1 has only one solution $u(t) = \sin\frac{\pi}{2(K+1)}t$ and its discrete analog $u(k+1) - \left(2 - \frac{\pi^2}{4(K+1)^2}\right)u(k) + u(k-1) = 0,\ k \in \mathbb{N}(1,K),\ u(0) = 0,\ u(K+1) = 1$ also has one solution. The continuous problem $u'' + 4\sin^2\frac{\pi}{2(K+1)}u = 0,\ u(0) = 0,\ u(K+1) = \epsilon\ (\neq 0)$ has only one solution $u(t) = \frac{\epsilon\sin\theta t}{\sin\theta(K+1)}$, where $\theta = 2\sin\frac{\pi}{2(K+1)}$, whereas its discrete analog $u(k+1) - \left(2 - 4\sin^2\frac{\pi}{2(K+1)}\right)u(k) + u(k-1) = 0,\ k \in \mathbb{N}(1,K),\ u(0) = 0,\ u(K+1) = \epsilon\ (\neq 0)$ has no solution.

**Theorem 8.1.1.** Let $\mathcal{U}(k,a),\ k \in \mathbb{N}(a,b)$ be the principal fundamental matrix solution of (1.2.12). Then, a necessary and sufficient condition for the existence of a unique solution of the boundary value problem (1.2.11), (1.5.3) is that the matrix

$$\mathcal{H} = \mathcal{L}[\mathcal{U}(k,a)] \tag{8.1.1}$$

be nonsingular. Further, this solution $\mathbf{u}(k)$ can be represented as

$$\mathbf{u}(k) = \mathcal{H}^1[\mathbf{b}(k)] + \mathcal{H}^2[\mathbf{l}], \tag{8.1.2}$$

where $\mathcal{H}^1$ is the linear operator mapping $B(a,b)$ into itself such that

$$\mathcal{H}^1[\mathbf{b}(k)] = \sum_{\ell=a+1}^{k}\mathcal{G}(k,\ell)\mathbf{b}(\ell-1) - \mathcal{U}(k,a)\mathcal{H}^{-1}\mathcal{L}\left[\sum_{\ell=a+1}^{k}\mathcal{G}(k,\ell)\mathbf{b}(\ell-1)\right],$$

$\mathcal{G}(k,\ell) = \mathcal{U}(k,a)\mathcal{U}^{-1}(\ell,a)$, and $\mathcal{H}^2$ is the linear operator mapping $\mathbb{R}^n$ into $B(a,b)$ such that

$$\mathcal{H}^2[\mathbf{l}] = \mathcal{U}(k,a)\mathcal{H}^{-1}\mathbf{l}.$$

**Proof.** From the considerations in Sections 2.5 and 2.6 any solution $\mathbf{u}(k),\ k \in \mathbb{N}(a,b)$ of (1.2.11) can be written as

$$\mathbf{u}(k) = \mathcal{U}(k,a)\mathbf{c} + \sum_{\ell=a+1}^{k}\mathcal{G}(k,\ell)\mathbf{b}(\ell-1), \tag{8.1.3}$$

where $\mathbf{c}$ is a constant vector.

The solution (8.1.3) satisfies (1.5.3) if and only if

$$\mathcal{L}[\mathcal{U}(k,a)]\mathbf{c} + \mathcal{L}\left[\sum_{\ell=a+1}^{k}\mathcal{G}(k,\ell)\mathbf{b}(\ell-1)\right] = \mathbf{l}. \tag{8.1.4}$$

Thus, from Lemma 2.2.1 the vector $\mathbf{c}$ can be determined uniquely if and only if $\det \mathcal{H} \neq 0$. Further, in such a case (8.1.4) gives

$$\mathbf{c} \;=\; \mathcal{H}^{-1}\mathbf{l} - \mathcal{H}^{-1}\mathcal{L}\left[\sum_{\ell=a+1}^{k} \mathcal{G}(k,\ell)\mathbf{b}(\ell-1)\right]. \tag{8.1.5}$$

Substituting (8.1.5) in (8.1.3), the result (8.1.2) follows. ∎

**Corollary 8.1.2.** A necessary and sufficient condition for the existence of a unique solution of the problem (1.2.11), (1.5.4) is that the matrix

$$\mathcal{H} \;=\; \sum_{i=1}^{r} \mathcal{L}^i\mathcal{U}(k_i,k_1) \tag{8.1.6}$$

be nonsingular. Further, this solution $\mathbf{u}(k)$, $k \in \mathbb{N}(k_1,k_r)$ can be written as

$$\mathbf{u}(k) \;=\; \mathcal{U}(k,k_1)\mathcal{H}^{-1}\mathbf{l} + \sum_{\ell=k_1+1}^{k_r} \mathcal{M}(k,\ell)\mathbf{b}(\ell-1), \tag{8.1.7}$$

where $\mathcal{M}(k,\ell)$ is the *Green's matrix* such that for $k_{i-1}+1 \le \ell \le k_i$, $2 \le i \le r$

$$\mathcal{M}(k,\ell) \;=\; \begin{cases} \mathcal{G}(k,\ell) - \mathcal{U}(k,k_1)\mathcal{H}^{-1}\sum_{j=i}^{r}\mathcal{L}^j\mathcal{G}(k_j,\ell), & k_{i-1}+1 \le \ell \le k \\ -\mathcal{U}(k,k_1)\mathcal{H}^{-1}\sum_{j=i}^{r}\mathcal{L}^j\mathcal{G}(k_j,\ell), & k+1 \le \ell \le k_i. \end{cases} \tag{8.1.8}$$

**Proof.** In this case (8.1.3) is written as

$$\mathbf{u}(k) \;=\; \mathcal{U}(k,k_1)\mathbf{c} + \sum_{\ell=k_1+1}^{k} \mathcal{G}(k,\ell)\mathbf{b}(\ell-1)$$

and (8.1.5) after arranging the terms becomes

$$\mathbf{c} \;=\; \mathcal{H}^{-1}\mathbf{l} - \mathcal{H}^{-1}\sum_{i=2}^{r}\;\sum_{\ell=k_{i-1}+1}^{k_i}\;\sum_{j=i}^{r}\mathcal{L}^j\mathcal{G}(k_j,\ell)\mathbf{b}(\ell-1). \qquad ∎$$

**Theorem 8.1.3.** Let the rank of the matrix $\mathcal{H}$ defined in (8.1.1) be $n-m$ $(1 \le m \le n)$. Then, the boundary value problem (1.2.11), (1.5.3) has a solution if and only if

$$(8.1.9) \qquad \mathcal{B}\mathbf{l} - \mathcal{B}\mathcal{L}\left[\sum_{\ell=a+1}^{k} \mathcal{G}(k,\ell)\mathbf{b}(\ell-1)\right] = 0,$$

where $\mathcal{B}$ is an $m \times n$ matrix whose row vectors are linearly independent vectors $\mathbf{d}^i$, $1 \le i \le m$ satisfying $\mathbf{d}^i\mathcal{H} = 0$.

In case (8.1.9) holds, any solution of (1.2.11), (1.5.3) can be given by

$$(8.1.10) \qquad \mathbf{u}(k) = \sum_{i=1}^{m} \alpha_i \mathbf{u}^i(k) + \mathcal{H}_1[\mathbf{b}(k)] + \mathcal{H}_2[\mathbf{l}],$$

where $\alpha_i$, $1 \le i \le m$ are arbitrary constants and $\mathbf{u}^i(k)$, $1 \le i \le m$ are $m$ linearly independent solutions of (1.2.12) satisfying $\mathcal{L}[\mathbf{u}^i(k)] = 0$, $\mathcal{H}_1$ is the linear operator mapping $B(a,b)$ into itself such that

$$\mathcal{H}_1[\mathbf{b}(k)] = \sum_{\ell=a+1}^{k} \mathcal{G}(k,\ell)\mathbf{b}(\ell-1) - \mathcal{U}(k,a)\mathcal{S}\mathcal{L}\left[\sum_{\ell=a+1}^{k} \mathcal{G}(k,\ell)\mathbf{b}(\ell-1)\right]$$

and $\mathcal{H}_2$ is the linear operator mapping $\mathbb{R}^n$ into $B(a,b)$ such that

$$\mathcal{H}_2[\mathbf{l}] = \mathcal{U}(k,a)\mathcal{S}\mathbf{l}.$$

The matrix $\mathcal{S}$ is an $n \times n$ matrix independent of $\mathbf{l} - \mathcal{L}\left[\sum_{\ell=a+1}^{k} \mathcal{G}(k,\ell)\mathbf{b}(\ell-1)\right]$ such that $\mathcal{H}\mathcal{S}\mathbf{p} = \mathbf{p}$ for any column vector $\mathbf{p}$ satisfying $\mathcal{B}\mathbf{p} = 0$.

**Proof.** From Lemma 2.2.1 the system (8.1.4) has a solution if and only if (8.1.9) holds. Further, in such a case the vector $\mathbf{c}$ can be given by

$$(8.1.11) \qquad \mathbf{c} = \sum_{i=1}^{m} \alpha_i \mathbf{c}^i + \mathcal{S}\mathbf{l} - \mathcal{S}\mathcal{L}\left[\sum_{\ell=a+1}^{k} \mathcal{G}(k,\ell)\mathbf{b}(\ell-1)\right],$$

where $\mathbf{c}^i$, $1 \le i \le m$ are $m$ linearly independent column vectors satisfying $\mathcal{H}\mathbf{c}^i = 0$. Let $\mathcal{U}(k,a)\mathbf{c}^i = \mathbf{u}^i(k)$, $1 \le i \le m$ then $\mathbf{u}^i(k)$ are linearly independent solutions of (1.2.12). Moreover,

$$\mathcal{L}[\mathbf{u}^i(k)] = \mathcal{L}[\mathcal{U}(k,a)\mathbf{c}^i] = \mathcal{L}[\mathcal{U}(k,a)]\mathbf{c}^i = \mathcal{H}\mathbf{c}^i = 0, \quad 1 \le i \le m.$$

Now substituting (8.1.11) in (8.1.3) we find (8.1.10). ∎

**Corollary 8.1.4.** Let the rank of the matrix $\mathcal{H}$ defined in (8.1.6) be $n-m$ $(1 \le m \le n)$. Then, (1.2.11), (1.5.4) has a solution if and only if

$$\mathcal{B}\mathbf{l} - \mathcal{B}\sum_{i=2}^{r}\mathcal{L}^i \sum_{\ell=k_1+1}^{k_i} \mathcal{G}(k_i,\ell)\mathbf{b}(\ell-1) \;=\; 0, \tag{8.1.12}$$

where $\mathcal{B}$ is an $m\times n$ matrix whose row vectors are linearly independent vectors $\mathbf{d}^i$, $1 \le i \le m$ satisfying $\mathbf{d}^i\mathcal{H} = 0$.

In case (8.1.12) holds , any solutions of (1.2.11), (1.5.4) can be given by

$$\mathbf{u}(k) \;=\; \sum_{i=1}^{m}\alpha_i\mathbf{u}^i(k) + \mathcal{U}(k,k_1)\mathcal{S}\mathbf{l} + \sum_{\ell=k_1+1}^{k_r}\mathcal{M}(k,\ell)\mathbf{b}(\ell-1), \tag{8.1.13}$$

where $\alpha_i$, $1 \le i \le m$ are arbitrary constants and $\mathbf{u}^i(k)$, $1\le i\le m$ are $m$ linearly independent solutions of (1.2.12) satisfying $\sum_{j=1}^{r}\mathcal{L}^j\mathbf{u}^i(k_j) = 0$, $\mathcal{S}$ is an $n\times n$ matrix independent of $\mathbf{l} - \sum_{i=2}^{r}\mathcal{L}^i\sum_{\ell=k_1+1}^{k_i}\mathcal{G}(k_i,\ell)\mathbf{b}(\ell-1)$ such that $\mathcal{H}\mathcal{S}\mathbf{p} = \mathbf{p}$ for any column vector $\mathbf{p}$ satisfying $\mathcal{B}\mathbf{p} = 0$, and $\mathcal{M}(k,\ell)$ is the *Green's matrix* such that for $k_{i-1}+1 \le \ell \le k_i$, $2 \le i \le r$

$$\mathcal{M}(k,\ell) \;=\; \begin{cases} \mathcal{G}(k,\ell) - \mathcal{U}(k,k_1)\mathcal{S}\displaystyle\sum_{j=i}^{r}\mathcal{L}^j\mathcal{G}(k_j,\ell), & k_{i-1}+1 \le \ell \le k \\ -\mathcal{U}(k,k_1)\mathcal{S}\displaystyle\sum_{j=i}^{r}\mathcal{L}^j\mathcal{G}(k_j,\ell), & k+1 \le \ell \le k_i. \end{cases} \tag{8.1.14}$$

**Example 8.1.1.** For the boundary value problem

$$\begin{bmatrix} u_1(k+1) \\ u_2(k+1) \end{bmatrix} = \begin{bmatrix} 0 & 1 \\ -1 & 2 \end{bmatrix}\begin{bmatrix} u_1(k) \\ u_2(k) \end{bmatrix} + \begin{bmatrix} b_1(k) \\ b_2(k) \end{bmatrix}, \tag{8.1.15}$$

$$k \in \mathbb{N}(0, K-1)$$

$$\begin{bmatrix} 1 & 0 \\ 0 & 0 \end{bmatrix}\begin{bmatrix} u_1(0) \\ u_2(0) \end{bmatrix} + \begin{bmatrix} 0 & 0 \\ 1 & 0 \end{bmatrix}\begin{bmatrix} u_1(K) \\ u_2(K) \end{bmatrix} = \begin{bmatrix} 0 \\ 0 \end{bmatrix} \tag{8.1.16}$$

it is easy to verify that

$$\mathcal{U}(k,0) \;=\; \begin{bmatrix} 1-k & k \\ -k & 1+k \end{bmatrix}, \quad \mathcal{H} \;=\; \begin{bmatrix} 1 & 0 \\ 1-K & K \end{bmatrix} \tag{8.1.17}$$

$$(8.1.18)\quad K\mathcal{M}(k,\ell) = \begin{cases} \begin{bmatrix} (K-k)(1+\ell) & -(K-k)\ell \\ (K-k-1)(1+\ell) & -(K-k-1)\ell \end{bmatrix}, \\ \qquad\qquad 1 \le \ell \le k \le K \\ \begin{bmatrix} (K-\ell-1)k & -(K-\ell)k \\ (K-\ell-1)(1+k) & -(K-\ell)(1+k) \end{bmatrix}, \\ \qquad\qquad 1 \le k+1 \le \ell \le K. \end{cases}$$

**Example 8.1.2.** For the system (8.1.15) together with the boundary conditions

$$(8.1.19)\quad \begin{bmatrix} 1 & 0 \\ 0 & 1 \end{bmatrix} \begin{bmatrix} u_1(0) \\ u_2(0) \end{bmatrix} + \begin{bmatrix} -1 & 1 \\ 0 & 0 \end{bmatrix} \begin{bmatrix} u_1(K) \\ u_2(K) \end{bmatrix} = \begin{bmatrix} 0 \\ 0 \end{bmatrix}$$

we find that $\mathcal{H} = \begin{bmatrix} 0 & 1 \\ 0 & 1 \end{bmatrix}$. Thus, $m = 1$ and we can take

$$(8.1.20)\quad \mathbf{u}^1(k) = \begin{bmatrix} 1-k \\ -k \end{bmatrix}, \quad \mathbf{d}^1 = [1\ \ -1], \quad \mathcal{S} = \begin{bmatrix} 0 & 0 \\ 1 & 0 \end{bmatrix}.$$

Further, the Green's matrix $\mathcal{M}(k,\ell)$ is given by

$$(8.1.21)\quad \mathcal{M}(k,\ell) = \begin{cases} \begin{bmatrix} 1+\ell & -\ell \\ 1+\ell & -\ell \end{bmatrix}, & 1 \le \ell \le k \le K \\ \begin{bmatrix} k & -k \\ 1+k & -1-k \end{bmatrix}, & 1 \le k+1 \le \ell \le K. \end{cases}$$

The condition (8.1.12) reduces to

$$(8.1.22)\quad \sum_{\ell=1}^{K} (b_1(\ell-1) - b_2(\ell-1)) = 0.$$

## 8.2. Method of Complementary Functions

We observe that any solution of the difference system (1.2.11) can be expressed as

$$(8.2.1)\quad \mathbf{u}(k) = \sum_{i=1}^{n} \mathbf{u}^i(k) u_i(a) + \mathbf{v}(k), \quad k \in \mathbb{N}(a,b)$$

where $\mathbf{u}^i(k)$, $1 \le i \le n$ are the solutions of the homogeneous system (1.2.12) satisfying

$$(8.2.2)\quad u_j^i(a) = \delta_{ij}, \quad 1 \le i,\ j \le n$$

and $\mathbf{v}(k)$ is the solution of (1.2.11) satisfying

$$(8.2.3)\quad v_j(a) = 0, \quad 1 \le j \le n.$$

The solution (8.2.1) satisfies (1.5.3) if and only if

$$(8.2.4)\qquad \sum_{i=1}^{n} \mathcal{L}\left[\mathbf{u}^i(k)\right] u_i(a) + \mathcal{L}[\mathbf{v}(k)] \;=\; \mathbf{l},$$

which is a system of $n$ linear algebraic equations in $n$ unknowns $u_i(a)$, $1 \le i \le n$. If the matrix $\mathcal{H}$ defined in (8.1.1) is nonsingular then the system (8.2.4) can be solved uniquely for $u_i(a)$, $1 \le i \le n$. Substituting these values in (8.2.1), we find the solution of (1.2.11), (1.5.3).

Thus, to obtain the solutions of (1.2.11), (1.5.3) we need $n$ solutions of (1.2.12) satisfying (8.2.2) and a particular solution $\mathbf{v}(k)$ of (1.2.11) satisfying (8.2.3), i.e. a total of $(n+1)$ solutions is necessary. Since all the solutions we compute are from the point $a$ up to $b$, this method is called *forward process*. Analogous to this method we have ***backward process*** in which all the necessary solutions are computed from the point $b$ up to $a$. For this, any solution of (1.2.11) can also be written as

$$(8.2.5)\qquad \mathbf{u}(k) \;=\; \sum_{i=1}^{n} \underline{\mathbf{u}}^i(k) u_i(b) + \underline{\mathbf{v}}(k), \quad k \in \mathbb{N}(a,b)$$

where $\underline{\mathbf{u}}^i(k)$, $1 \le i \le n$ are the solutions of the homogeneous system (1.2.12) satisfying

$$(8.2.6)\qquad \underline{u}^i_j(b) \;=\; \delta_{ij}, \quad 1 \le i,\ j \le n$$

and $\underline{\mathbf{v}}(k)$ is the solution of (1.2.11) satisfying

$$(8.2.7)\qquad \underline{v}_j(b) \;=\; 0, \quad 1 \le j \le n.$$

The solution (8.2.5) satisfies (1.5.3) if and only if

$$(8.2.8)\qquad \sum_{i=1}^{n} \mathcal{L}\left[\underline{\mathbf{u}}^i(k)\right] u_i(b) + \mathcal{L}[\underline{\mathbf{v}}(k)] \;=\; \mathbf{l}.$$

The system (8.2.8) provides the values of $u_i(b)$, $1 \le i \le n$ which we substitute in (8.2.5) to find the required solution.

In particular, for the boundary value problem (1.2.11), (1.5.4) in the above forward and backward processes we need to change the points $a$ to $k_1$ and $b$ to $k_r$, and (8.2.4) becomes

$$(8.2.9)\qquad \sum_{i=1}^{n}\left(\sum_{j=1}^{r} \mathcal{L}^j \mathbf{u}^i(k_j)\right) u_i(k_1) + \sum_{j=1}^{r} \mathcal{L}^j \mathbf{v}(k_j) \;=\; \mathbf{l}$$

whereas (8.2.8) reduces to

$$(8.2.10) \qquad \sum_{i=1}^{n}\left(\sum_{j=1}^{r}\mathcal{L}^j\underline{\mathbf{u}}^i(k_j)\right)u_i(k_r)+\sum_{j=1}^{r}\mathcal{L}^j\underline{\mathbf{v}}(k_j) = 1.$$

To construct the solution of (1.2.11), (1.5.3) by forward (backward) process we need to store $\mathbf{u}^i(k)$ $(\underline{\mathbf{u}}^i(k))$, $1 \le i \le n$ and $\mathbf{v}(k)$ $(\underline{\mathbf{v}}(k))$ at all the points of $\mathbb{N}(a,b)$ which may not be feasible. However, at least for the problem (1.2.11), (1.5.4) this difficulty may be simplified as follows: we store only $\mathbf{u}^i(k_j)$ $(\underline{\mathbf{u}}^i(k_j))$ and $\mathbf{v}(k_j)$ $(\underline{\mathbf{v}}(k_j))$ needed in (8.2.9) ((8.2.10)) and solve it for the $u_i(k_1)$ $(u_i(k_r))$, $1 \le i \le r$. The solution of (1.2.11), (1.5.4) is then obtained by computing the solution of (1.2.11) with these obtained values of $u_i(k_1)$ $(u_i(k_r))$, $1 \le i \le r$. This method of constructing the solution of (1.2.11), (1.5.4) is called *forward forward process* (*backward backward process*). Thus, for both of these methods we need to compute a total of $(n+2)$ appropriate solutions.

Next we shall show that the forward process for the problem (1.2.11), (1.5.8) requires only $(n-\beta_1)+1$ solutions instead of $(n+1)$. For this, we note that (8.2.1) in component form can be expressed as

$$(8.2.11) \qquad u_j(k) = \sum_{i=1, i\neq 1(s_1)}^{n} u_j^i(k)u_i(k_1)+w_j(k), \quad 1 \le j \le n,$$

$$k \in \mathbb{N}(k_1,k_r)$$

where

$$(8.2.12) \qquad w_j(k) = \sum_{i=1(s_1)} u_j^i(k)\ell_{1,i}+v_j(k), \quad 1 \le j \le n.$$

Obviously, $\mathbf{w}(k)$ defined in (8.2.12) is the solution of (1.2.11) satisfying

$$(8.2.13) \qquad w_j(k_1) = \begin{cases} \ell_{1,1(s_1)} & \text{if } j = 1(s_1) \\ 0 & \text{otherwise.} \end{cases}$$

Thus, to find (8.2.11) we need $(n-\beta_1)$ solutions of (1.2.12) satisfying

$$(8.2.14) \qquad u_j^i(k_1) = \begin{cases} 1 & \text{if } i = j \neq 1(s_1) \\ 0 & \text{otherwise} \end{cases}$$

and a particular solution of (1.2.11) satisfying (8.2.13), i.e. a total of $(n-\beta_1)+1$ solutions.

Now using the boundary conditions (1.5.8) other than at the point $k_1$, we find from (8.2.11) that

$$(8.2.15) \qquad \sum_{i=1,i\neq 1(s_1)}^{n} u^i_{j(s_j)}(k_j)u_i(k_1) = \ell_{j,j(s_j)} - w_{j(s_j)}(k_j), \quad 2 \le j \le r$$

which is a system of $(n-\beta_1)$ algebraic equations in $(n-\beta_1)$ unknowns $u_i(k_1),\ 1 \le i \le n,\ i \neq 1(s_1)$.

Similarly, the backward process for the problem (1.2.11), (1.5.8) requires only $(n-\beta_r)+1$ solutions. For this, in component form (8.2.5) can be written as

$$(8.2.16) \quad u_j(k) = \sum_{i=1,i\neq r(s_r)}^{n} \underline{u}^i_j(k)u_i(k_r) + \underline{w}_j(k), \ 1 \le j \le n, \ k \in \mathbb{N}(k_1,k_r)$$

where

$$(8.2.17) \qquad \underline{w}_j(k) = \sum_{i=r(s_r)} \underline{u}^i_j(k)\ell_{r,i} + \underline{v}_j(k), \quad 1 \le j \le n$$

and

$$(8.2.18) \qquad \underline{u}^i_j(k_r) = \begin{cases} 1 & \text{if } i = j \neq r(s_r) \\ 0 & \text{otherwise} \end{cases}$$

$$(8.2.19) \qquad \underline{w}_j(k_r) = \begin{cases} \ell_{1,r(s_r)} & \text{if } j = r(s_r) \\ 0 & \text{otherwise.} \end{cases}$$

The unknowns $u_i(k_r),\ 1 \le i \le n,\ i \neq r(s_r)$ are obtained from the system

$$(8.2.20) \quad \sum_{i=1,i\neq r(s_r)}^{n} \underline{u}^i_{j(s_j)}(k_j)u_i(k_r) = \ell_{j,j(s_j)} - \underline{w}_{j(s_j)}(k_j), \quad 1 \le j \le r-1.$$

In addition to the above observations we note that for the problem (1.2.11), (1.5.8) the forward forward (backward backward) process requires only $(n-\beta_1)+2((n-\beta_r)+2)$ solutions.

**Example 8.2.1.** From Theorem 1.6.1 the linear differential equation

$$(8.2.21) \qquad y'' = f(t)y + g(t)$$

together with the boundary conditions (1.6.11) has a unique solution provided $f(t) \geq 0$ for all $t \in [\alpha, \beta]$. For this problem (1.6.13) reduces to

$$(8.2.22)\quad \left(-1+\frac{h^2}{12}f_{k-1}\right)u(k-1)+\left(2+\frac{10h^2}{12}f_k\right)u(k)+\left(-1+\frac{h^2}{12}f_{k+1}\right)$$

$$\times\ u(k+1) \ = \ -\frac{h^2}{12}\left(g_{k-1}+10g_k+g_{k+1}\right), \quad k \in \mathbb{N}(1,K)$$

where $f_k = f(\alpha + kh)$ and $g_k = g(\alpha + kh)$.

If $\dfrac{h^2}{12}\max_{\alpha \leq t \leq \beta} f(t) < 1$, then in system form the boundary value problem (8.2.22), (1.6.14) can be written as

$$(8.2.23)\quad \begin{aligned} u_1(k+1) &= u_2(k) \\ u_2(k+1) &= \frac{1}{1-\frac{h^2}{12}f_{k+2}}\left\{\left(-1+\frac{h^2}{12}f_k\right)u_1(k)\right. \\ &\left. + \left(2+\frac{10h^2}{12}f_{k+1}\right)u_2(k) + \frac{h^2}{12}\left(g_k+10g_{k+1}+g_{k+2}\right)\right\}, \\ &\qquad k \in \mathbb{N}(0, K-1) \end{aligned}$$

$$(8.2.24)\quad u_1(0) \ = \ A, \quad u_2(K) \ = \ B.$$

For the problem (8.2.23), (8.2.24) we note that (8.2.11) reduces to

$$(8.2.25)\quad \begin{aligned} u_1(k) &= u_1^2(k)u_2(0) + w_1(k) \\ u_2(k) &= u_2^2(k)u_2(0) + w_2(k), \quad k \in \mathbb{N}(0,K) \end{aligned}$$

and (8.2.15) is simply

$$(8.2.26)\quad u_2^2(K)u_2(0) \ = \ B - w_2(K),$$

where $u^2(k)$ is the solution of the homogeneous system

$$(8.2.27)\quad \begin{aligned} u_1(k+1) &= u_2(k) \\ u_2(k+1) &= \frac{1}{1-\frac{h^2}{12}f_{k+2}}\left\{\left(-1+\frac{h^2}{12}f_k\right)u_1(k)\right. \\ &\left. + \left(2+\frac{10h^2}{12}f_{k+1}\right)u_2(k)\right\} \end{aligned}$$

$$(8.2.28)\quad u_1(0) \ = \ 0, \quad u_2(0) \ = \ 1$$

and $w(k)$ is the solution of the nonhomogeneous system (8.2.23) satisfying

$$w_1(0) = A, \quad w_2(0) = 0. \tag{8.2.29}$$

From (8.2.26) we find that the problem (8.2.23), (8.2.24) has a unique solution provided $u_2^2(K) \neq 0$, and in such a case

$$u_2(0) = \frac{B - w_2(K)}{u_2^2(K)}. \tag{8.2.30}$$

If $f(t) \geq 0$ in $[\alpha, \beta]$, then by induction we shall prove that $u_2^2(0) < u_2^2(1) < \cdots < u_2^2(K)$. For this, from (8.2.27) and (8.2.28) we have

$$u_2^2(1) = \frac{1}{1 - \frac{h^2}{12} f_2}\left(2 + \frac{10h^2}{12} f_1\right) > 1 = u_2^2(0).$$

Now we assume that $u_2^2(0) < u_2^2(1) < \cdots < u_2^2(m)$, $m \in \mathbb{N}(1, K-1)$. Then, since $u_1^2(m) = u_2^2(m-1) < u_2^2(m)$ from (8.2.27) it follows that

$$\begin{aligned} u_2^2(m+1) - u_2^2(m) &> \frac{1}{1 - \frac{h^2}{12} f_{m+2}}\left\{\left(-1 + \frac{h^2}{12} f_m\right) + \left(2 + \frac{10h^2}{12} f_{m+1}\right)\right. \\ &\quad \left. + \left(-1 + \frac{h^2}{12} f_{m+2}\right)\right\} u_2^2(m) \\ &= \frac{1}{1 - \frac{h^2}{12} f_{m+2}} \frac{h^2}{12} \left(f_m + 10 f_{m+1} + f_{m+2}\right) u_2^2(m) \geq 0. \end{aligned}$$

Similarly, for the problem (8.2.23), (8.2.24) equations (8.2.16) and (8.2.20) reduce to

$$\begin{aligned} u_1(k) &= \underline{u}_1^1(k) u_1(K) + \underline{w}_1(k) \\ u_2(k) &= \underline{u}_2^1(k) u_1(K) + \underline{w}_2(k), \quad k \in \mathbb{N}(0, K) \end{aligned} \tag{8.2.31}$$

and

$$u_1(K) = \frac{A - \underline{w}_1(0)}{\underline{u}_1^1(0)} \tag{8.2.32}$$

respectively, where $u^1(k)$ is the solution of the homogeneous system (8.2.27) satisfying

$$u_1(K) = 1, \quad u_2(K) = 0 \tag{8.2.33}$$

and $\underline{w}(k)$ is the solution of the nonhomogeneous system (8.2.23) satisfying

$$w_1(K) = 0, \quad w_2(K) = B. \tag{8.2.34}$$

**Remark 8.2.1.** In usual matrix form the boundary value problem (8.2.22), (1.6.14) is in fact the tridiagonal system of algebraic equations, which can be solved by, say, complete Gaussian elimination algorithm (CGEA, hereafter).

**Example 8.2.2.** Consider the boundary value problem

$$y'' = \frac{2}{t^2}y - \frac{1}{t}, \quad y(2) = y(3) = 0. \tag{8.2.35}$$

Since $f(t) = 2/t^2 > 0$ for all $t \in [2, 3]$, the problem has a unique solution $y(t) = \frac{1}{38}\left(19t - 5t^2 - \frac{36}{t}\right)$. We compute an approximate solution of (8.2.35) by its discrete analog (8.2.23), (8.2.24). For this discrete problem all the four methods discussed in this section work equally well. The errors obtained, as calculated from the exact solution $y(t)$ and approximate solution $u_1(k)$ with $h = 1/256$ are presented in Table 8.2.1.

**Example 8.2.3.** The boundary value problem

$$y'' = 400y, \quad y(0) = 1, \ y(5) = e^{-100} \tag{8.2.36}$$

has a unique solution $y(t) = e^{-20t}$. For the discrete analog (8.2.23), (8.2.24) of (8.2.36) the CGEA, the forward method and the forward forward method fail, whereas the backward as well as backward backward method works equally well. The errors obtained, as calculated from the exact solution $y(t)$ and approximate solution $u_1(k)$ with $h = 5/1024$ are presented in Table 8.2.2.

**Table 8.2.1.**

| $t$ | **Forw. Method** | **Forw.–Forw. Method** | **Back. Method** | **Back.–Back. Method** |
|---|---|---|---|---|
| 2.000 | $0.00000000D\ 00$ | $0.00000000D\ 00$ | $0.13877788D-16$ | $0.23399700D-13$ |
| 2.125 | $0.14066786D-12$ | $0.14115792D-12$ | $0.16245252D-12$ | $0.15509035D-12$ |
| 2.250 | $0.22427112D-12$ | $0.22550711D-12$ | $0.25087571D-12$ | $0.23015530D-12$ |
| 2.375 | $0.26694705D-12$ | $0.26395119D-12$ | $0.28074938D-12$ | $0.26009229D-12$ |
| 2.500 | $0.28290391D-12$ | $0.26557315D-12$ | $0.26370919D-12$ | $0.25433128D-12$ |
| 2.625 | $0.26833310D-12$ | $0.23850800D-12$ | $0.21972268D-12$ | $0.21974436D-12$ |
| 2.750 | $0.21402932D-12$ | $0.18873098D-12$ | $0.15880353D-12$ | $0.16098841D-12$ |
| 2.875 | $0.12252525D-12$ | $0.12078966D-12$ | $0.84389960D-13$ | $0.84855734D-13$ |
| 3.000 | $0.13877788D-16$ | $0.43965766D-13$ | $0.00000000D\ 00$ | $0.00000000D\ 00$ |

Table 8.2.2.

| $t$ | Backward Method | Back.-Back. Method |
|---|---|---|
| 0.0000 | $0.13877788D-16$ | $0.45796700D-14$ |
| 0.3125 | $0.22852521D-08$ | $0.22852521D-08$ |
| 0.6250 | $0.88231436D-11$ | $0.88231436D-11$ |
| 0.9375 | $0.25548996D-13$ | $0.25548996D-13$ |
| 1.2500 | $0.65761514D-16$ | $0.65761514D-16$ |
| 1.5625 | $0.15868689D-18$ | $0.15868689D-18$ |
| 1.8750 | $0.36760510D-21$ | $0.36760510D-21$ |
| 2.1875 | $0.82791842D-24$ | $0.82791842D-24$ |
| 2.5000 | $0.18265801D-26$ | $0.18265801D-26$ |
| 2.8125 | $0.39668929D-29$ | $0.39668929D-29$ |
| 3.1250 | $0.85087781D-32$ | $0.85087781D-32$ |
| 3.4375 | $0.18068376D-34$ | $0.18068376D-34$ |
| 3.7500 | $0.38051073D-37$ | $0.38051073D-37$ |
| 4.0625 | $0.79577125D-40$ | $0.79577125D-40$ |
| 4.3750 | $0.16543695D-42$ | $0.16543695D-42$ |
| 4.6875 | $0.41399291D-45$ | $0.41399291D-45$ |
| 5.0000 | $0.37200760D-43$ | $0.00000000D\ 00$ |

**Example 8.2.4.** The boundary value problem

$$y'' = (2m+1+t^2)y, \quad y(0) = \beta, \quad y(\infty) = 0 \tag{8.2.37}$$

where $m \geq 0$ and $\beta$ are known constants, is known as Holt's problem. This problem is a typical example where usual shooting methods fail [38]. Replacing the boundary condition $y(\infty) = 0$ by $y(T) = 0$ ($T$ finite) Holt [26] used finite difference methods (however, for $m = 0$, $\beta = 1$, $T = 12$; $m = 1$, $\beta = \pi^{-1/2}$, $T = 8$; $m = 2$, $\beta = 1/4$, $T = 8$ the results are unsatisfactory [26,38]), whereas Osborne [36] used a multiple shooting method and Roberts and Shipman [37] used a multipoint approach. In [3,4] we have formulated a new shooting method which gives accurate solutions of (8.2.37) for several different values of $m$ and $\beta$ up to $T = 18$ (this value of $T$ has been chosen in view of restricted computer capabilities). For the same and several other different values of $m$ and $\beta$ accurate solutions of (8.2.37) up to $T = 18$ have also been obtained in [13]. Here the error estimates in the solution of (8.2.37) when approximating $y(\infty) = 0$ by an appropriate boundary condition at $T$ are also available.

For the discrete analog (8.2.23), (8.2.24) of (8.2.37) (replacing $y(\infty) = 0$ by $y(18) = 0$) with $m = 2$, $\beta = 1/4$ and $h = 1/60$ the forward method and the forward forward method fail, whereas the backward as well as backward backward method works equally well. The numerical solution $u_1(k)$ is shown in Table 8.2.3.

**Example 8.2.5.** Consider the boundary value problem

$$y'' = (\sin 2t)y + \cos 2t, \quad y(-1) = y(1) = 0. \tag{8.2.38}$$

Although, the function $f(t) = \sin 2t$ changes sign in $[-1, 1]$, its discrete analog (8.2.23), (8.2.24) has a unique solution for all $K \geq 1$ (cf. Problem 8.10.2). For this discrete problem all the four methods work equally well. The numerical solution $u_1(k)$ for $h = 1/540$ is shown in Table 8.2.4.

**Table 8.2.3.**

| $t$ | Backward Method | Back.-Back. Method |
|---|---|---|
| 0.0 | $0.25000000D\ 00$ | $0.25000000D\ 00$ |
| 1.0 | $0.23407771D-01$ | $0.23407771D-01$ |
| 2.0 | $0.14143468D-02$ | $0.14143468D-02$ |
| 3.0 | $0.44114854D-04$ | $0.44114854D-04$ |
| 4.0 | $0.62609336D-06$ | $0.62609336D-06$ |
| 5.0 | $0.37645596D-08$ | $0.37645596D-08$ |
| 6.0 | $0.91929810D-11$ | $0.91929810D-11$ |
| 7.0 | $0.88796121D-14$ | $0.88796121D-14$ |
| 8.0 | $0.33341431D-17$ | $0.33341431D-17$ |
| 9.0 | $0.48088930D-21$ | $0.48088930D-21$ |
| 10.0 | $0.26416894D-25$ | $0.26416894D-25$ |
| 11.0 | $0.54925707D-30$ | $0.54925707D-30$ |
| 12.0 | $0.43020222D-35$ | $0.43020222D-35$ |
| 13.0 | $0.12646919D-40$ | $0.12646919D-40$ |
| 14.0 | $0.13914253D-46$ | $0.13914253D-46$ |
| 15.0 | $0.57160202D-53$ | $0.57160202D-53$ |
| 16.0 | $0.87512138D-60$ | $0.87512138D-60$ |
| 17.0 | $0.49854924D-67$ | $0.49854924D-67$ |
| 18.0 | $0.00000000D\ 00$ | $0.00000000D\ 00$ |

**Table 8.2.4.**

| $t$ | Forw. Method | Forw.–Forw. Method | Back. Method | Back.–Back. Method |
|---|---|---|---|---|
| −1.0 | $0.00000000D\ 00$ | $0.00000000D\ 00$ | $0.13877788D-16$ | $-0.52361250D-11$ |
| −0.9 | $-0.54219742D-01$ | $-0.54219742D-01$ | $-0.54219742D-01$ | $-0.54219742D-01$ |
| −0.8 | $-0.11017210D\ 00$ | $-0.11017210D\ 00$ | $-0.11017210D\ 00$ | $-0.11017210D\ 00$ |
| −0.7 | $-0.16531803D\ 00$ | $-0.16531803D\ 00$ | $-0.16531803D\ 00$ | $-0.16531803D\ 00$ |
| −0.6 | $-0.21715198D\ 00$ | $-0.21715198D\ 00$ | $-0.21715198D\ 00$ | $-0.21715198D\ 00$ |
| −0.5 | $-0.26336748D\ 00$ | $-0.26336748D\ 00$ | $-0.26336748D\ 00$ | $-0.26336748D\ 00$ |
| −0.4 | $-0.30200211D\ 00$ | $-0.30200211D\ 00$ | $-0.30200211D\ 00$ | $-0.30200211D\ 00$ |
| −0.3 | $-0.33154697D\ 00$ | $-0.33154697D\ 00$ | $-0.33154697D\ 00$ | $-0.33154697D\ 00$ |
| −0.2 | $-0.35101159D\ 00$ | $-0.35101159D\ 00$ | $-0.35101159D\ 00$ | $-0.35101159D\ 00$ |
| −0.1 | $-0.35994166D\ 00$ | $-0.35994166D\ 00$ | $-0.35994166D\ 00$ | $-0.35994166D\ 00$ |
| 0.0 | $-0.35839390D\ 00$ | $-0.35839390D\ 00$ | $-0.35839390D\ 00$ | $-0.35839390D\ 00$ |
| 0.1 | $-0.34687723D\ 00$ | $-0.34687723D\ 00$ | $-0.34687723D\ 00$ | $-0.34687723D\ 00$ |
| 0.2 | $-0.32627260D\ 00$ | $-0.32627260D\ 00$ | $-0.32627260D\ 00$ | $-0.32627260D\ 00$ |
| 0.3 | $-0.29774424D\ 00$ | $-0.29774424D\ 00$ | $-0.29774424D\ 00$ | $-0.29774424D\ 00$ |
| 0.4 | $-0.26265376D\ 00$ | $-0.26265376D\ 00$ | $-0.26265376D\ 00$ | $-0.26265376D\ 00$ |
| 0.5 | $-0.22248554D\ 00$ | $-0.22248554D\ 00$ | $-0.22248554D\ 00$ | $-0.22248554D\ 00$ |
| 0.6 | $-0.17878818D\ 00$ | $-0.17878818D\ 00$ | $-0.17878818D\ 00$ | $-0.17878818D\ 00$ |
| 0.7 | $-0.13313322D\ 00$ | $-0.13313322D\ 00$ | $-0.13313322D\ 00$ | $-0.13313322D\ 00$ |
| 0.8 | $-0.87088932D-01$ | $-0.87088932D-01$ | $-0.87088932D-01$ | $-0.87088932D-01$ |
| 0.9 | $-0.42204687D-01$ | $-0.42204687D-01$ | $-0.42204687D-01$ | $-0.42204687D-01$ |
| 1.0 | $0.22204460D-15$ | $-0.96123092D-11$ | $0.00000000D\ 00$ | $0.00000000D\ 00$ |

## 8.3. Method of Particular Solutions

We solve (1.2.11) with $(n+1)$ different sets of conditions

$$(8.3.1) \qquad \begin{aligned} u_j^i(a) &= \delta_{ij}, \quad 1 \le i, j \le n \\ u_j^{n+1}(a) &= 0, \quad 1 \le j \le n \end{aligned}$$

to obtain $\mathbf{u}^i(k)$, $1 \le i \le n+1$, i.e. $(n+1)$ particular solutions of (1.2.11). Next we introduce $(n+1)$ constants $c_i$, $1 \le i \le n+1$ and demand that the linear combination

$$(8.3.2) \qquad \mathbf{u}(k) = \sum_{i=1}^{n+1} c_i \mathbf{u}^i(k)$$

to be a solution of the problem (1.2.11), (1.5.3). For this, we must have

$$(8.3.3) \qquad \sum_{i=1}^{n} c_i = 1$$

and on substituting (8.3.2) in (1.5.3) we get $n$ more equations

$$(8.3.4) \qquad \sum_{i=1}^{n+1} \mathcal{L}[\mathbf{u}^i(k)] c_i = 1.$$

These $(n+1)$ equations (8.3.3), (8.3.4) are solved for the $(n+1)$ unknowns $c_i$, $1 \le i \le n+1$.

This method is theoretically the same as the forward process. For this, from (8.3.3) we have $c_{n+1} = 1 - \sum_{i=1}^{n} c_i$, and hence (8.3.2) can be written as

$$\mathbf{u}(k) = \sum_{i=1}^{n} c_i(\mathbf{u}^i(k) - \mathbf{u}^{n+1}(k)) + \mathbf{u}^{n+1}(k),$$

which is the same as (8.2.1). However, it uses only the nonhomogeneous system (1.2.11) in contrast with the forward process where (1.2.11) as well as the homogeneous system (1.2.12) is being used. But it leads to a system of $(n+1)$ equations instead of $n$ equations.

The method of particular solutions similar to backward, forward forward and backward backward processes can easily be formulated.

## 8.4. Method of Adjoints

As the name suggests we use the adjoint system (2.7.1) to obtain the solution of the problem (1.2.11), (1.5.5). We compute solutions of (2.7.1) backward once for each $u_q(k_i)$, $2 \le i \le r$ appearing in (1.5.5) with the conditions

$$(8.4.1) \qquad v_q^{p(i)}(k_i) = \alpha_{pq}^i, \quad 2 \le i \le r, \quad 1 \le p, q \le n$$

where $v_q^{p(i)}(k_i)$ is the $q$th component at $k_i$ for the $p$th backward solution.

Substituting (8.4.1) in the adjoint identity (2.7.9) with $k_0 = k_1$, we obtain

$$(8.4.2) \quad \sum_{q=1}^{n} \alpha_{pq}^i u_q(k_i) - \sum_{q=1}^{n} v_q^{p(i)}(k_1) u_q(k_1) = \sum_{\ell=k_1+1}^{k_i} \sum_{q=1}^{n} v_q^{p(i)}(\ell) b_q(\ell - 1),$$

$$2 \le i \le r.$$

Summing $(r-1)$ equations (8.4.2) and making use of (1.5.5), we get

(8.4.3)

$$\sum_{q=1}^{n} \left( \alpha_{pq}^1 + \sum_{i=2}^{r} v_q^{p(i)}(k_1) \right) u_q(k_1) = l_p - \sum_{i=2}^{r} \sum_{\ell=k_1+1}^{k_i} \sum_{q=1}^{n} v_q^{p(i)}(\ell) b_q(\ell - 1),$$

$$1 \le p \le n.$$

If the matrix $\left( \alpha_{pq}^1 + \sum_{i=2}^{r} v_q^{p(i)}(k_1) \right)$ is nonsingular, then the system (8.4.3) provides the unknowns $u_q(k_1)$, $1 \le q \le n$. The solution of the problem (1.2.11), (1.5.5) is obtained by computing the solution of (1.2.11) with these values of $u_q(k_1)$, $1 \le q \le n$. However, to evaluate the summation term in (8.4.3) we need to store the solutions of (2.7.1). This can be avoided at the cost of solving another $(r-1)$ systems. For this, we denote

$$w_{p(i)}(k) = - \sum_{\ell=k+1}^{k_i} \sum_{q=1}^{n} v_q^{p(i)}(\ell) b_q(\ell - 1), \quad 1 \le p \le n, \quad 2 \le i \le r$$

which is equivalent to solving

$$(8.4.4) \qquad w_{p(i)}(k) = - \sum_{q=1}^{n} v_q^{p(i)}(k+1) b_q(k) + w_{p(i)}(k+1)$$

$$(8.4.5) \qquad w_{p(i)}(k_i) = 0, \quad 1 \le p \le n, \quad 2 \le i \le r.$$

Thus, at the point $k_i$, $2 \le i \le r$ we solve a system of order $2n$ given by (2.7.1) and (8.4.4) subject to the conditions (8.4.1) and (8.4.5).

With this adjustment system (8.4.3) takes the form

$$\sum_{q=1}^{n}\left(\alpha^{1}_{pq}+\sum_{i=2}^{r} v_q^{p(i)}(k_1)\right)u_q(k_1) \;=\; l_p+\sum_{i=2}^{r} w_{p(i)}(k_1), \tag{8.4.6}$$

$$1\le p\le n.$$

This method of constructing the solution of (1.2.11), (1.5.5) is called the *backward forward process* and requires $(r-1)n$ backward solutions of the adjoint system (2.7.1) satisfying (8.4.1), $(r-1)$ backward solutions of (8.4.4) satisfying (8.4.5), and 1 forward solution of (1.2.11) with the obtained values of $u_q(k_1)$, $1\le q\le n$ from the system (8.4.6), i.e. a total of $(r-1)(n+1)+1$ solutions of $n$th order systems. In particular, if $r=2$ then once again we need $(n+2)$ solutions as in the forward forward or the backward backward process.

Similar to the backward forward process we have the *forward backward process*, and for this we solve (2.7.1) forward once for each $u_q(k_i)$, $1\le i\le r-1$ appearing in (1.5.5) with the conditions

$$\underline{v}_q^{\,p(i)}(k_i) \;=\; \alpha^{i}_{pq}, \quad 1\le i\le r-1, \quad 1\le p,q\le n \tag{8.4.7}$$

where $\underline{v}_q^{p(i)}(k_i)$ is the $q$th component at $k_i$ for the $p$th forward solution.

Substituting (8.4.7) in the adjoint identity (2.7.10) with $k_0=k_r$, we obtain

$$\sum_{q=1}^{n}\alpha^{i}_{pq}u_q(k_i)-\sum_{q=1}^{n}\underline{v}_q^{p(i)}(k_r)u_q(k_r) \;=\; -\sum_{\ell=k_i+1}^{k_r}\sum_{q=1}^{n}\underline{v}_q^{p(i)}(\ell)b_q(\ell-1), \tag{8.4.8}$$

$$1\le i\le r-1.$$

Summing $(r-1)$ equations (8.4.8) and making use of (1.5.5), we get

(8.4.9)

$$\sum_{q=1}^{n}\left(\alpha^{r}_{pq}+\sum_{i=1}^{r-1}\underline{v}_q^{p(i)}(k_r)\right)u_q(k_r) \;=\; l_p+\sum_{i=1}^{r-1}\sum_{\ell=k_i+1}^{k_r}\sum_{q=1}^{n}\underline{v}_q^{p(i)}(\ell)b_q(\ell-1),$$

$$1\le p\le n.$$

We introduce

$$\underline{w}_{p(i)}(k) \;=\; \sum_{\ell=k_i+1}^{k}\sum_{q=1}^{n}\underline{v}_q^{p(i)}(\ell)b_q(\ell-1), \quad 1\le p\le n, \quad 1\le i\le r-1$$

which is equivalent to solving

$$\underline{w}_{p(i)}(k) = -\sum_{q=1}^{n} \underline{v}_q^{p(i)}(k+1)b_q(k) + \underline{w}_{p(i)}(k+1) \tag{8.4.10}$$

$$\underline{w}_{p(i)}(k_i) = 0, \quad 1 \le p \le n, \quad 1 \le i \le r-1. \tag{8.4.11}$$

Thus, the system (8.4.9) is the same as

$$\sum_{q=1}^{n} \left( \alpha_{pq}^{r} + \sum_{i=1}^{r-1} \underline{v}_q^{p(i)}(k_r) \right) u_q(k_r) = l_p + \sum_{i=1}^{r-1} \underline{w}_{p(i)}(k_r), \quad 1 \le p \le n. \tag{8.4.12}$$

The solution of the problem (1.2.11), (1.5.5) is obtained by solving backward the system (1.2.11) with the obtained values of $u_q(k_r)$, $1 \le q \le n$ from the system (8.4.12).

Next we shall consider the system (1.2.11) together with the implicit separated conditions (1.5.7). We compute $(n - \beta_1)$ solutions of (2.7.1) backward with the conditions

$$v_q^{i(s_i)}(k_i) = \alpha_{i(s_i),q}, \quad 2 \le i \le r, \quad 1 \le s_i \le \beta_i, \quad 1 \le q \le n \tag{8.4.13}$$

where $v_q^{i(s_i)}(k_i)$ is the $q$th component at $k_i$ for the $s_i$th backward solution.

Substituting (8.4.13) in (2.7.9) with $k_0 = k_1$ and using (1.5.7), we obtain

$$\sum_{q=1}^{n} v_q^{i(s_i)}(k_1)u_q(k_1) = l_{i,i(s_i)} - \sum_{\ell=k_1+1}^{k_i} \sum_{q=1}^{n} v_q^{i(s_i)}(\ell)b_q(\ell-1), \quad 2 \le i \le r, \quad 1 \le s_i \le \beta_i. \tag{8.4.14}$$

We introduce

$$w_{i(s_i)}(k) = -\sum_{\ell=k+1}^{k_i} \sum_{q=1}^{n} v_q^{i(s_i)}(\ell)b_q(\ell-1), \quad 2 \le i \le r, \quad 1 \le s_i \le \beta_i$$

which is equivalent to solving

$$w_{i(s_i)}(k) = -\sum_{q=1}^{n} v_q^{i(s_i)}(k+1)b_q(k) + w_{i(s_i)}(k+1) \tag{8.4.15}$$

$$(8.4.16)\qquad w_{i(s_i)}(k_i) = 0, \quad 2 \le i \le r, \quad 1 \le s_i \le \beta_i.$$

Thus, the system (8.4.14) can be written as

$$(8.4.17)\qquad \sum_{q=1}^{n} v_q^{i(s_i)}(k_1)u_q(k_1) = l_{i,i(s_i)} + w_{i(s_i)}(k_1),$$

$$2 \le i \le r, \quad 1 \le s_i \le \beta_i.$$

System (8.4.17) together with (1.5.7) for $i = 1$, i.e.

$$(8.4.18)\qquad \sum_{q=1}^{n} \alpha_{1(s_1),q}u_q(k_1) = l_{1,1(s_1)}, \quad 1 \le s_1 \le \beta_1$$

form a system of $n$ equations in $n$ unknowns $u_q(k_1)$, $1 \le q \le n$. The solution of (1.2.11), (1.5.7) is obtained by solving forward the system (1.2.11) with these values of $u_q(k_1)$, $1 \le q \le n$.

In practice we couple the adjoint system (2.7.1) with the equation (8.4.15) and solve this system of $(n+1)$ equations from the points $k_i$, $2 \le i \le r$ to $k_1$ with the conditions (8.4.13) and (8.4.16).

Similarly, in the forward backward process for (1.2.11), (1.5.7) the unknowns $u_q(k_r)$, $1 \le q \le n$ are computed from the system

$$(8.4.19)\qquad \sum_{q=1}^{n} \underline{v}_q^{i(s_i)}(k_r)u_q(k_r) = l_{i,i(s_i)} + \underline{w}_{i(s_i)}(k_r),$$

$$1 \le i \le r-1, \ 1 \le s_i \le \beta_i$$

$$(8.4.20)\qquad \sum_{q=1}^{n} \alpha_{r(s_r),q}u_q(k_r) = l_{r,r(s_r)}, \quad 1 \le s_r \le \beta_r$$

where $\underline{v}_q^{i(s_i)}(k)$ is the $q$th component of the $s_i$th forward solution from the point $k_i$ of the adjoint system (2.7.1) satisfying

$$(8.4.21)\quad \underline{v}_q^{i(s_i)}(k_i) = \alpha_{i(s_i),q}, \quad 1 \le i \le r-1, \quad 1 \le s_i \le \beta_i, \quad 1 \le q \le n$$

and $\underline{w}_{i(s_i)}(k)$ is the forward solution of the initial value problem

$$(8.4.22)\qquad \underline{w}_{i(s_i)}(k) = -\sum_{q=1}^{n} \underline{v}_q^{i(s_i)}(k+1)b_q(k) + \underline{w}_{i(s_i)}(k+1)$$

$$(8.4.23)\qquad \underline{w}_{i(s_i)}(k_i) = 0, \quad 1 \le i \le r-1, \quad 1 \le s_i \le \beta_i.$$

The solution of (1.2.11), (1.5.7) is obtained by solving backward the system (1.2.11) with the obtained values of $u_q(k_r)$, $1 \le q \le n$.

**Example 8.4.1.** To apply backward forward process for the boundary value problem (8.2.23), (8.2.24) we note that (2.7.1), (8.4.15), (8.4.13) and (8.4.16) reduce to

$$\begin{aligned} v_1^2(k) &= -\frac{c_0(k)}{c_2(k)} v_2^2(k+1) \\ v_2^2(k) &= v_1^2(k+1) + \frac{c_1(k)}{c_2(k)} v_2^2(k+1) \\ w_2(k) &= -v_2^2(k+1) d(k) + w_2(k+1) \end{aligned} \tag{8.4.24}$$

$$v_1^2(K) = 0, \quad v_2^2(K) = 1, \quad w_2(K) = 0, \tag{8.4.25}$$

where $c_0(k) = 1 - \frac{h^2}{12} f_k$, $c_1(k) = 2 + \frac{10h^2}{12} f_{k+1}$, $c_2(k) = 1 - \frac{h^2}{12} f_{k+2}$ and $d(k) = \frac{1}{c_2(k)} \frac{h^2}{12} \left[ g_k + 10 g_{k+1} + g_{k+2} \right].$

Further, the system (8.4.17), (8.4.18) takes the form

$$\begin{aligned} v_1^2(0) u_1(0) + v_2^2(0) u_2(0) &= B + w_2(0) \\ u_1(0) &= A, \end{aligned}$$

which easily determines

$$\begin{aligned} u_1(0) &= A \\ u_2(0) &= \frac{B + w_2(0) - v_1^2(0) A}{v_2^2(0)}. \end{aligned} \tag{8.4.26}$$

The solution of (8.2.23), (8.2.24) is obtained by recursing forward the system (8.2.23) with the initial values (8.4.26).

Similarly, to apply forward backward process we find that (2.7.1), (8.4. 22), (8.4.21) and (8.4.23) reduce to

$$\begin{aligned} \underline{v}_1^1(k+1) &= \underline{v}_2^1(k) + \frac{c_1(k)}{c_0(k)} \underline{v}_1^1(k) \\ \underline{v}_2^1(k+1) &= -\frac{c_2(k)}{c_0(k)} \underline{v}_1^1(k) \\ \underline{w}_1(k+1) &= \underline{w}_1(k) + \underline{v}_2^1(k+1) d(k) \end{aligned} \tag{8.4.27}$$

$$\underline{v}_1^1(0) = 1, \quad \underline{v}_2^1(0) = 0, \quad \underline{w}_1(0) = 0. \tag{8.4.28}$$

Further, the system (8.4.19), (8.4.20) becomes

$$\begin{aligned} \underline{v}_1^1(K) u_1(K) + \underline{v}_2^1(K) u_2(K) &= A + \underline{w}_1(K) \\ u_2(K) &= B, \end{aligned}$$

which gives

$$(8.4.29) \qquad \begin{aligned} u_1(K) &= \frac{A + \underline{w}_1(K) - \underline{v}_2^1(K)B}{\underline{v}_1^1(K)} \\ u_2(K) &= B. \end{aligned}$$

The solution of (8.2.23), (8.2.24) is obtained by recursing backward the system (8.2.23) with the final values (8.4.29).

**Example 8.4.2.** For the discrete analog (8.2.23), (8.2.24) of the boundary value problem (8.2.35) both the methods discussed in this section work equally well. The errors obtained, as calculated from the exact solution $y(t)$ (see Example 8.2.2) and approximate solution $u_1(k)$ with $h = 1/256$ are presented in Table 8.4.1.

Table 8.4.1.

| $t$ | Back.-Ford. Method | Ford.-Back. Method |
|---|---|---|
| 2.000 | $0.00000000D\ 00$ | $0.64884418D-14$ |
| 2.125 | $0.10546078D-12$ | $0.10817389D-12$ |
| 2.250 | $0.15874801D-12$ | $0.15646512D-12$ |
| 2.375 | $0.17716384D-12$ | $0.16960044D-12$ |
| 2.500 | $0.16946167D-12$ | $0.15705492D-12$ |
| 2.625 | $0.14502982D-12$ | $0.12977119D-12$ |
| 2.750 | $0.10855900D-12$ | $0.93355879D-13$ |
| 2.875 | $0.65239480D-13$ | $0.49092674D-13$ |
| 3.000 | $0.16924973D-13$ | $0.00000000D\ 00$ |

**Example 8.4.3.** We apply the methods of this section to the discrete boundary value problem considered in Example 8.2.3. The results analogous to Table 8.2.2 are given in Table 8.4.2.

Table 8.4.2.

| $t$ | Back.-Ford. Method | Ford.-Back. Method |
|---|---|---|
| 0.0000 | | $0.16875390D-13$ |
| 0.3125 | | $0.22852524D-08$ |
| 0.6250 | | $0.88231446D-11$ |
| 0.9375 | | $0.25548999D-13$ |
| 1.2500 | | $0.65761522D-16$ |
| 1.5625 | | $0.15868691D-18$ |
| 1.8750 | | $0.36760514D-21$ |
| 2.1875 | Fails | $0.82791852D-24$ |
| 2.5000 | | $0.18265803D-26$ |
| 2.8125 | | $0.39668934D-29$ |
| 3.1250 | | $0.85087791D-32$ |
| 3.4375 | | $0.18068378D-34$ |
| 3.7500 | | $0.38051077D-37$ |
| 4.0625 | | $0.79577134D-40$ |
| 4.3750 | | $0.16543683D-42$ |
| 4.6875 | | $0.34217867D-45$ |
| 5.0000 | | $0.00000000D\ 00$ |

**Example 8.4.4.** For the discrete analog (8.2.23), (8.2.24) of (8.2.37) (replacing $y(\infty) = 0$ by $y(18) = 0$) with $m = 0$, $\beta = 1$ and $h = 1/60$ we apply both the methods of this section. The numerical solution $u_1(k)$ is shown in Table 8.4.3.

**Table** 8.4.3.

| $t$ | Back.-Ford. Method | Ford.-Back. Method |
|---|---|---|
| 0.0 | | $0.10000000D\ \ 01$ |
| 1.0 | | $0.25934255D\ \ 00$ |
| 2.0 | | $0.34564046D - 01$ |
| 3.0 | | $0.19885232D - 02$ |
| 4.0 | | $0.45958196D - 04$ |
| 5.0 | | $0.41255769D - 06$ |
| 6.0 | | $0.14129840D - 08$ |
| 7.0 | | $0.18272052D - 11$ |
| 8.0 | | $0.88629857D - 15$ |
| 9.0 | **Fails** | $0.16054990D - 18$ |
| 10.0 | | $0.10827935D - 22$ |
| 11.0 | | $0.27128207D - 27$ |
| 12.0 | | $0.25206584D - 32$ |
| 13.0 | | $0.86750117D - 38$ |
| 14.0 | | $0.11047255D - 43$ |
| 15.0 | | $0.52013857D - 50$ |
| 16.0 | | $0.90485908D - 57$ |
| 17.0 | | $0.58130757D - 64$ |
| 18.0 | | $0.00000000D\ \ 00$ |

**Example 8.4.5.** We apply the methods of this section to the discrete boundary value problem considered in Example 8.2.5. The results analogous to Table 8.2.4 are given in Table 8.4.4.

**Table** 8.4.4.

| $t$ | Back.-Ford. Method | Ford.-Back. Method |
|---|---|---|
| −1.0 | $0.00000000D\ \ 00$ | $0.22781405D - 13$ |
| −0.9 | $-0.54219742D - 01$ | $-0.54219742D - 01$ |
| −0.8 | $-0.11017210D\ \ 00$ | $-0.11017210D\ \ 00$ |
| −0.7 | $-0.16531803D\ \ 00$ | $-0.16531803D\ \ 00$ |
| −0.6 | $-0.21715198D\ \ 00$ | $-0.21715198D\ \ 00$ |
| −0.5 | $-0.26336748D\ \ 00$ | $-0.26336748D\ \ 00$ |
| −0.4 | $-0.30200211D\ \ 00$ | $-0.30200211D\ \ 00$ |
| −0.3 | $-0.33154697D\ \ 00$ | $-0.33154697D\ \ 00$ |
| −0.2 | $-0.35101159D\ \ 00$ | $-0.35101159D\ \ 00$ |
| −0.1 | $-0.35994166D\ \ 00$ | $-0.35994166D\ \ 00$ |
| 0.0 | $-0.35839390D\ \ 00$ | $-0.35839390D\ \ 00$ |
| 0.1 | $-0.34687723D\ \ 00$ | $-0.34687723D\ \ 00$ |
| 0.2 | $-0.32627260D\ \ 00$ | $-0.32627260D\ \ 00$ |
| 0.3 | $-0.29774424D\ \ 00$ | $-0.29774424D\ \ 00$ |
| 0.4 | $-0.26265376D\ \ 00$ | $-0.26265376D\ \ 00$ |
| 0.5 | $-0.22248554D\ \ 00$ | $-0.22248554D\ \ 00$ |
| 0.6 | $-0.17878818D\ \ 00$ | $-0.17878818D\ \ 00$ |
| 0.7 | $-0.13313322D\ \ 00$ | $-0.13313322D\ \ 00$ |
| 0.8 | $-0.87088932D - 01$ | $-0.87088932D - 01$ |
| 0.9 | $-0.42204687D - 01$ | $-0.42204687D - 01$ |
| 1.0 | $-0.58914799D - 12$ | $0.00000000D\ \ 00$ |

## 8.5. Method of Chasing

This method seems to be applicable only for some particular cases of (1.2.11) together with (1.5.7). For example, we shall formulate it for the system

$$(8.5.1)\qquad \begin{aligned} u_i(k+1) &= u_{i+1}(k), \quad 1 \le i \le n-1 \\ u_n(k+1) &= -\sum_{j=0}^{n-2} a_j(k) u_{j+1}(k) + b(k), \end{aligned}$$

which is equivalent to the difference equation (1.2.3) with $a_{n-1}(k) = 0$ and $a_n(k) = 1$.

For simplicity we shall rewrite the boundary conditions (1.5.7) as

$$(8.5.2)\qquad \sum_{j=1}^{n} \alpha_{ij} u_j(k_i) = l_i, \quad 1 \le i \le n$$

where $k_{\mu(i-1)+1} = k_{\mu(i-1)+2} = \cdots = k_{\mu(i)}$, $\mu(0) = 0$ and $\mu(i) = \sum_{j=1}^{i} \beta_j$, $1 \le i \le r$.

We observe that for a fixed $i$ in (8.5.2) at least one of $\alpha_{i,j}$, $1 \le j \le n$ is not zero, which we assume to be $\alpha_{i,i(i)}$. The subscript allows the possibility that $i(i)$ need not be the same as $i$. Thus, (8.5.2) can be written as

$$(8.5.3)\qquad u_{i(i)}(k_i) = \sum_{j=1, j\ne i(i)}^{n} \theta_{ij} u_j(k_i) + d_i,$$

where

$$\theta_{ij} = -\frac{\alpha_{ij}}{\alpha_{i,i(i)}}, \quad 1 \le j \le n, \ j \ne i(i), \quad \text{and} \quad d_i = \frac{l_i}{\alpha_{i,i(i)}}.$$

Based on the form of (8.5.3) we assume that the solution of (8.5.1) satisfies the relation

$$(8.5.4)\qquad u_{i(i)}(k) = \sum_{j=1, j\ne i(i)}^{n} \theta_{ij}(k) u_j(k) + d_i(k),$$

where the unknown functions $\theta_{ij}(k)$, $1 \le j \le n$, $j \ne i(i)$ and $d_i(k)$ are obtained as follows: Relation (8.5.4) is the same as

$$(8.5.5)\qquad u_{i(i)}(k+1) = \sum_{j=1, j\ne i(i)}^{n} \theta_{ij}(k+1) u_j(k+1) + d_i(k+1).$$

We shall use (8.5.1) and (8.5.4) to eliminate $u_n(k)$ from (8.5.5), however it depends on a particular value of $i(i)$ and we need to consider the following four cases:

**Case 1.** $i(i) = 1$ and $n \geq 3$

From (8.5.4), we have

$$u_n(k) = \frac{1}{\theta_{in}(k)}\left[u_1(k) - \sum_{j=2}^{n-1}\theta_{ij}(k)u_j(k) - d_i(k)\right]. \tag{8.5.6}$$

Using (8.5.1) and (8.5.6) in (8.5.5) and arranging the terms, we get

$$\begin{aligned}
&\left[\frac{\theta_{i,n-1}(k+1)}{\theta_{in}(k)} - \theta_{in}(k+1)a_0(k)\right]u_1(k) \\
&- \left[\frac{\theta_{i,n-1}(k+1)}{\theta_{in}(k)}\theta_{i2}(k) + \theta_{in}(k+1)a_1(k) + 1\right]u_2(k) \\
&+ \sum_{j=2}^{n-2}\left[\theta_{ij}(k+1) - \frac{\theta_{i,n-1}(k+1)}{\theta_{in}(k)}\theta_{i,j+1}(k) - \theta_{in}(k+1)a_j(k)\right]u_{j+1}(k) \\
&+ \left[d_i(k+1) + \theta_{in}(k+1)b(k) - \frac{\theta_{i,n-1}(k+1)}{\theta_{in}(k)}d_i(k)\right] = 0.
\end{aligned}$$

Thus, the following system must be satisfied

$$\begin{aligned}
\theta_{in}(k+1) &= -\left[a_0(k)\theta_{i2}(k) + a_1(k)\right]^{-1} \\
\theta_{i,n-1}(k+1) &= a_0(k)\theta_{in}(k)\theta_{in}(k+1) \\
\theta_{ij}(k+1) &= \left[a_0(k)\theta_{i,j+1}(k) + a_j(k)\right]\theta_{in}(k+1), \\
&\qquad j = n-2, \cdots, 2 \\
d_i(k+1) &= \left[a_0(k)d_i(k) - b(k)\right]\theta_{in}(k+1).
\end{aligned} \tag{8.5.7}$$

We also desire that the solution representation (8.5.4) satisfies the boundary condition (8.5.3). For this, we compare (8.5.3) and (8.5.4) at the point $k_i$, to obtain

$$\begin{aligned}
\theta_{ij}(k_i) &= \theta_{ij}, \quad j = n, \cdots, 2 \\
d_i(k_i) &= d_i.
\end{aligned} \tag{8.5.8}$$

In the remaining three cases we proceed as for the Case 1 and arrive at the following difference systems:

**Case 2.** $2 \le i(i) \le n-2$

(8.5.9)

$$\begin{aligned}
\theta_{in}(k+1) &= \theta_{i1}(k)\left[a_0(k)\theta_{i,i(i)+1}(k) - a_{i(i)}(k)\theta_{i_1}(k)\right]^{-1} \\
\theta_{i,n-1}(k+1) &= -a_0(k)\theta_{in}(k)\theta_{in}(k+1)/\theta_{i1}(k) \\
\theta_{ij}(k+1) &= \left[a_j(k)\theta_{in}(k)\theta_{in}(k+1) + \theta_{i,j+1}(k)\theta_{i,n-1}(k+1)\right]/\theta_{in}(k), \\
& \qquad j = n-2, \cdots, 1, \quad j \ne i(i),\ i(i)-1 \\
\theta_{i,i(i)-1}(k+1) &= \left[a_{i(i)-1}(k)\theta_{in}(k)\theta_{in}(k+1) - \theta_{i,n-1}(k+1)\right]/\theta_{in}(k) \\
d_i(k+1) &= \left[\theta_{i,n-1}(k+1)d_i(k) - b(k)\theta_{in}(k)\theta_{in}(k+1)\right]/\theta_{in}(k)
\end{aligned}$$

(8.5.10)
$$\begin{aligned}
\theta_{ij}(k_i) &= \theta_{ij}, \quad j = n, \cdots, 1, \quad j \ne i(i) \\
d\ (k_i) &= d_i.
\end{aligned}$$

**Case 3.** $i(i) = n-1$

(8.5.11)
$$\begin{aligned}
\theta_{in}(k+1) &= \theta_{i1}(k)/a_0(k)\theta_{in}(k) \\
\theta_{i,n-2}(k+1) &= \left[1 + a_{n-2}(k)\theta_{in}(k)\theta_{in}(k+1)\right]/\theta_{in}(k) \\
\theta_{ij}(k+1) &= \left[a_j(k)\theta_{in}(k)\theta_{in}(k+1) - \theta_{i,j+1}(k)\right]/\theta_{in}(k), \\
& \qquad j = n-3, \cdots, 1 \\
d_i(k+1) &= -\left[d_i(k) + b(k)\theta_{in}(k)\theta_{in}(k+1)\right]/\theta_{in}(k)
\end{aligned}$$

(8.5.12)
$$\begin{aligned}
\theta_{ij}(k_i) &= \theta_{ij}, \quad j = n, n-2, n-3, \cdots, 1 \\
d_i(k_i) &= d_i.
\end{aligned}$$

**Case 4.** $i(i) = n$

(8.5.13)
$$\begin{aligned}
&\theta_{i,n-1}(k+1) = -a_0(k)/\theta_{i1}(k) \\
&\theta_{ij}(k+1) = -a_j(k) - \theta_{i,j+1}(k)\theta_{i,n-1}(k+1), \quad j = n-2, \cdots, 1 \\
&d_i(k+1) = -\theta_{i,n-1}(k+1)d_i(k) + b(k)
\end{aligned}$$

(8.5.14)
$$\begin{aligned}
\theta_{ij}(k_i) &= \theta_{ij}, \quad j = n-1, \cdots, 1 \\
d_i(k_i) &= d_i.
\end{aligned}$$

For the particular value of $i(i)$ we solve the appropriate system from $k_i$ to $k_n$ and collect the values of $\theta_{ij}(k_n)$, $1 \le j \le n$, $j \ne i(i)$ and $d_i(k_n)$, thereby obtaining from (8.5.4) a new boundary relation at $k_n$

(8.5.15)
$$u_{i(i)}(k_n) = \sum_{j=1, j\ne i(i)}^{n} \theta_{ij}(k_n)u_j(k_n) + d_i(k_n).$$

Since in (8.5.2) we have $\beta_r$ relations at the point $k_n$ (which is in fact $k_r$) we can find $u_j(k_n)$, $1 \le j \le n$ if $n - \beta_r$ new relations of the type (8.5.15) are known. This in turn implies that we need to solve $n - \beta_r$ appropriate above difference systems. These systems are not necessarily different, especially because a difference system does not change as long as in (8.5.3) $i(i)$ is the same (we can have at most $n$ different difference systems). Finally, having obtained $u_j(k_n)$, $1 \le j \le n$ we solve the difference system (8.5.1) backward from $k_n$ to the point $k_1$.

**Example 8.5.1.** For the boundary value problem

$$\begin{aligned} u_1(k+1) &= u_2(k) \\ u_2(k+1) &= -a_0(k)u_1(k) + b(k), \quad 0 \le k \le K-1 \end{aligned} \tag{8.5.16}$$

$$\begin{aligned} u_2(0) &= l_1 + \alpha_0 u_1(0) \\ u_2(K) &= l_2 \end{aligned} \tag{8.5.17}$$

we assume that

$$u_2(k) = \theta_{21}(k)u_1(k) + d_2(k),$$

and find (Case 4) that the unknown functions $\theta_{21}(k)$ and $d_2(k)$ must satisfy

$$\begin{aligned} \theta_{21}(k+1) &= -a_0(k)/\theta_{21}(k) \\ d_2(k+1) &= -\theta_{21}(k+1)d_2(k) + b(k) \\ \theta_{21}(0) &= \alpha_0, \quad d_2(0) = l_1. \end{aligned} \tag{8.5.18}$$

To find the unknown $u_1(K)$ we use (8.5.15) which reduces to

$$u_2(K) = \theta_{21}(K)u_1(K) + d_2(K)$$

and hence

$$u_1(K) = \frac{l_2 - d_2(K)}{\theta_{21}(K)}. \tag{8.5.19}$$

The solution of (8.5.16), (8.5.17) is obtained by solving (8.5.16) backward with these values of $u_1(K)$ and $u_2(K)$.

**Example 8.5.2.** In Example 1.6.4 we have seen that the initial value problem (1.4.6), (1.4.7) can be approximated by the boundary value problem (1.6.7) (1.6.9). To solve this boundary value problem, since $a_0(k) = -(k+1)(k+2)$, $b(k) = -(k+1)$, $\alpha_0 = -1$, $l_1 = 1$ and $l_2 = 1/(K+3)$, equations (8.5.18) and (8.5.19) reduce to

$$\begin{aligned}\theta_{21}(k+1) &= (k+1)(k+2)/\theta_{21}(k)\\ d_2(k+1) &= -(k+1)\left[1+\frac{(k+2)}{\theta_{21}(k)}d_2(k)\right]\\ \theta_{21}(0) &= -1, \quad d_2(0) = 1\\ u_1(K) &= \frac{(K+3)^{-1}-d_2(K)}{\theta_{21}(K)}.\end{aligned}$$

We take $K = 10000$ and compute the approximate solution $u_1(k)$, and present it in Table 8.5.1.

**Table 8.5.1.**

| $k$ | $u_1(k)$ | $k$ | $u_1(k)$ | $k$ | $u_1(k)$ | $k$ | $u_1(k)$ |
|---|---|---|---|---|---|---|---|
| 1 | 0.36787944$D$ 00 | 10 | 0.83877070$D-01$ | 100 | 0.98048550$D-02$ | 1000 | 0.99800499$D-03$ |
| 2 | 0.26424112$D$ 00 | 20 | 0.45544884$D-01$ | 200 | 0.49506158$D-02$ | 2000 | 0.49950062$D-03$ |
| 3 | 0.20727665$D$ 00 | 30 | 0.31279674$D-01$ | 300 | 0.33112945$D-02$ | 3000 | 0.33311130$D-03$ |
| 4 | 0.17089341$D$ 00 | 40 | 0.23822729$D-01$ | 400 | 0.24875775$D-02$ | 4000 | 0.24987508$D-03$ |
| 5 | 0.14553294$D$ 00 | 50 | 0.19237754$D-01$ | 500 | 0.19920398$D-02$ | 5000 | 0.19992004$D-03$ |
| 6 | 0.12680236$D$ 00 | 60 | 0.16133165$D-01$ | 600 | 0.16611341$D-02$ | 6000 | 0.16661113$D-03$ |
| 7 | 0.11238350$D$ 00 | 70 | 0.13891533$D-01$ | 700 | 0.14245043$D-02$ | 7000 | 0.14281634$D-03$ |
| 8 | 0.10093197$D$ 00 | 80 | 0.12196915$D-01$ | 800 | 0.12468847$D-02$ | 8000 | 0.12496876$D-03$ |
| 9 | 0.91612293$D-01$ | 90 | 0.10870836$D-01$ | 900 | 0.11086488$D-02$ | 9000 | 0.11108643$D-03$ |

**Example 8.5.3.** Following Example 1.6.4 the initial value problem (1.9.4), (1.9.5) can be approximated by the boundary value problem

$$\begin{aligned}u(k+2) &= 25u(k) - \frac{4k+9}{(k+1)(k+2)}, \quad k \in \mathbb{N}(0, K-1)\\ u(1) &= 1 - 5u(0) \qquad (8.5.20)\\ u(K+1) &= \frac{1}{6(K+2)}.\end{aligned}$$

For this boundary value problem also we use the formulation of Example 8.5.1 for $K = 10000$ and present the numerical solution $u_1(k)$ in Table 8.5.2.

**Table 8.5.2.**

| $k$ | $u_1(k)$ | $k$ | $u_1(k)$ | $k$ | $u_1(k)$ | $k$ | $u_1(k)$ |
|---|---|---|---|---|---|---|---|
| 1 | 0.88392216$D-01$ | 10 | 0.15367550$D-01$ | 100 | 0.16528701$D-02$ | 1000 | 0.16652787$D-03$ |
| 2 | 0.58038920$D-01$ | 20 | 0.79975230$D-02$ | 200 | 0.82987267$D-03$ | 2000 | 0.83298623$D-04$ |
| 3 | 0.43138734$D-01$ | 30 | 0.54046330$D-02$ | 300 | 0.55401577$D-03$ | 3000 | 0.55540127$D-04$ |
| 4 | 0.34306330$D-01$ | 40 | 0.40812983$D-02$ | 400 | 0.41580006$D-03$ | 4000 | 0.41657988$D-04$ |
| 5 | 0.28468352$D-01$ | 50 | 0.32785146$D-02$ | 500 | 0.33277852$D-03$ | 5000 | 0.33327779$D-04$ |
| 6 | 0.24324906$D-01$ | 60 | 0.27396243$D-02$ | 600 | 0.27739240$D-03$ | 6000 | 0.27773920$D-04$ |
| 7 | 0.21232615$D-01$ | 70 | 0.23528767$D-02$ | 700 | 0.23781206$D-03$ | 7000 | 0.23806690$D-04$ |
| 8 | 0.18836924$D-01$ | 80 | 0.20618122$D-02$ | 800 | 0.20811650$D-03$ | 8000 | 0.20831163$D-04$ |
| 9 | 0.16926490$D-01$ | 90 | 0.18348317$D-02$ | 900 | 0.18501384$D-03$ | 9000 | 0.18516804$D-04$ |

**Example 8.5.4.** Following Example 1.6.4 the initial value problem (1.9.6), (1.9.7) can be approximated by the boundary value problem (since $u(k) \to \infty$, we can assume for large $K$, $u(K+2) = u(K+1)$)

$$
\begin{aligned}
u(k+2) &= 25u(k) - \frac{(3k+8)2^{k+1}}{(k+1)(k+2)}, \quad k \in \mathbb{N}(0, K-1) \\
u(1) &= 2 - 5u(0) \qquad\qquad (8.5.21) \\
u(K+1) &= \frac{2^{K+2}}{6(K+2)}.
\end{aligned}
$$

For this boundary value problem also we use the formulation of Example 8.5.1. In view of limited computer capabilities we choose $K = 250$ and present the numerical solution in Table 8.5.3.

**Table 8.5.3.**

| $k$ | $u_1(k)$ | $k$ | $u_1(k)$ | $k$ | $u_1(k)$ | $k$ | $u_1(k)$ |
|---|---|---|---|---|---|---|---|
| 0 | 0.33647224$D$ 00 | 10 | 0.27260357$D$ 02 | 110 | 0.33498134$D$ 31 | 210 | 0.22311824$D$ 61 |
| 1 | 0.31763882$D$ 00 | 20 | 0.14456393$D$ 05 | 120 | 0.31460579$D$ 34 | 220 | 0.21812163$D$ 64 |
| 2 | 0.41180592$D$ 00 | 30 | 0.99861672$D$ 07 | 130 | 0.29751110$D$ 37 | 230 | 0.21367554$D$ 67 |
| 3 | 0.60763709$D$ 00 | 40 | 0.77149316$D$ 10 | 140 | 0.28300147$D$ 40 | 240 | 0.20971754$D$ 70 |
| 4 | 0.96181456$D$ 00 | 50 | 0.63425992$D$ 13 | 150 | 0.27056586$D$ 43 | 250 | 0.24027412$D$ 73 |
| 5 | 0.15909272$D$ 01 | 60 | 0.54252148$D$ 16 | 160 | 0.25982047$D$ 46 | | |
| 6 | 0.27120305$D$ 01 | 70 | 0.47698737$D$ 19 | 170 | 0.25047150$D$ 49 | | |
| 7 | 0.47255616$D$ 01 | 80 | 0.42792501$D$ 22 | 180 | 0.24229027$D$ 52 | | |
| 8 | 0.83721921$D$ 01 | 90 | 0.38989275$D$ 25 | 190 | 0.23509613$D$ 55 | | |
| 9 | 0.15027929$D$ 02 | 100 | 0.35960999$D$ 28 | 200 | 0.22874447$D$ 58 | | |

## 8.6. Method of Imbedding: First Formulation

We partition the vector $\mathbf{u}(k)$ by setting $\mathbf{u}(k) = [\mathbf{v}(k), \mathbf{w}(k)]^T$, where $\mathbf{v}(k)$ is an $p \times 1$ vector, $\mathbf{w}(k)$ is an $q \times 1$ vector, and $p + q = n$. In general, the choice of the elements of $\mathbf{u}(k)$ which are to be $\mathbf{v}(k)$ and those which are to be $\mathbf{w}(k)$ is arbitrary, although for some problems one choice is more natural than any other. Once this setting is fixed, the difference system (1.2.11) can be written as

$$
\mathbf{v}(k+1) = \mathcal{A}^1(k)\mathbf{v}(k) + \mathcal{A}^2(k)\mathbf{w}(k) + \mathbf{b}^1(k) \qquad (8.6.1)
$$

$$
\mathbf{w}(k+1) = \mathcal{A}^3(k)\mathbf{v}(k) + \mathcal{A}^4(k)\mathbf{w}(k) + \mathbf{b}^2(k) \qquad (8.6.2)
$$

and the boundary conditions (1.5.4) take the form

$$
\sum_{i=1}^{r} {}_1\mathcal{M}^i\mathbf{v}(k_i) + \sum_{i=1}^{r} {}_2\mathcal{M}^i\mathbf{w}(k_i) = \mathbf{l}^1 \qquad (8.6.3)
$$

$$(8.6.4) \qquad \sum_{i=1}^{r} {}_3\mathcal{M}^i \mathbf{v}(k_i) + \sum_{i=1}^{r} {}_4\mathcal{M}^i \mathbf{w}(k_i) = \mathbf{l}^2,$$

where the matrices $\mathcal{A}^1(k)$, ${}_1\mathcal{M}^i$, $1 \le i \le r$ are of order $p \times p$, $\mathcal{A}^2(k)$, ${}_2\mathcal{M}^i$, $1 \le i \le r$ of order $p \times q$, $\mathcal{A}^3(k)$, ${}_3\mathcal{M}^i$, $1 \le i \le r$ of order $q \times p$, $\mathcal{A}^4(k)$, ${}_4\mathcal{M}^i$, $1 \le i \le r$ of order $q \times q$, and the vectors $\mathbf{b}^1(k)$, $\mathbf{l}^1$ are of order $p \times 1$, and $\mathbf{b}^2(k)$, $\mathbf{l}^2$ of order $q \times 1$.

There are two possible expressions for the development of the solution from (8.6.1), (8.6.2), a direct form and an inverse form. The direct form at a fixed point $k^* \in \mathbb{N}(k_1, k_r)$ is defined as

$$(8.6.5) \qquad \mathbf{v}(k) = \mathcal{R}^1(k,k^*)\mathbf{w}(k) + \mathcal{R}^2(k,k^*)\mathbf{v}(k^*) + \mathbf{c}^1(k,k^*)$$

$$(8.6.6) \qquad \mathbf{w}(k^*) = \mathcal{Q}^1(k,k^*)\mathbf{w}(k) + \mathcal{Q}^2(k,k^*)\mathbf{v}(k^*) + \mathbf{c}^2(k,k^*),$$

where the matrices $\mathcal{R}^1(k,k^*)$, $\mathcal{R}^2(k,k^*)$, $\mathcal{Q}^1(k,k^*)$, $\mathcal{Q}^2(k,k^*)$ are of orders $p\times q$, $p\times p$, $q\times q$ and $q\times p$ respectively, and the vectors $\mathbf{c}^1(k,k^*)$, $\mathbf{c}^2(k,k^*)$ are of orders $p \times 1$, $q \times 1$ respectively.

Relation (8.6.5) is the same as

$$(8.6.7) \quad \mathbf{v}(k{+}1) = \mathcal{R}^1(k{+}1,k^*)\mathbf{w}(k{+}1) + \mathcal{R}^2(k{+}1,k^*)\mathbf{v}(k^*) + \mathbf{c}^1(k{+}1,k^*).$$

Using (8.6.1), (8.6.2) and (8.6.5) in (8.6.7), we get

$$\begin{aligned}
&\left[\mathcal{A}^1(k) - \mathcal{R}^1(k+1,k^*)\mathcal{A}^3(k) - \mathcal{R}^2(k+1,k^*)\left(\mathcal{R}^2(k,k^*)\right)^{-1}\right]\mathbf{v}(k) + \\
&\left[\mathcal{A}^2(k) - \mathcal{R}^1(k+1,k^*)\mathcal{A}^4(k) + \mathcal{R}^2(k+1,k^*)\left(\mathcal{R}^2(k,k^*)\right)^{-1}\mathcal{R}^1(k,k^*)\right]\mathbf{w}(k) + \\
&\left[\mathbf{b}^1(k) - \mathcal{R}^1(k+1,k^*)\mathbf{b}^2(k) + \mathcal{R}^2(k+1,k^*)\left(\mathcal{R}^2(k,k^*)\right)^{-1}\mathbf{c}^1(k,k^*)\right. \\
&\qquad\qquad \left. - \mathbf{c}^1(k+1,k^*)\right] = 0.
\end{aligned}$$

Thus, for $k \ge k^*$ the following system must be satisfied

$$\begin{aligned}
&(8.6.8) \\
&\mathcal{R}^1(k+1,k^*) = \left[\mathcal{A}^2(k) + \mathcal{A}^1(k)\mathcal{R}^1(k,k^*)\right]\left[\mathcal{A}^4(k) + \mathcal{A}^3(k)\mathcal{R}^1(k,k^*)\right]^{-1} \\
&\mathcal{R}^2(k+1,k^*) = \left[\mathcal{A}^1(k) - \mathcal{R}^1(k+1,k^*)\mathcal{A}^3(k)\right]\mathcal{R}^2(k,k^*) \\
&\mathbf{c}^1(k+1,k^*) = \mathcal{A}^1(k)\mathbf{c}^1(k,k^*) - \mathcal{R}^1(k+1,k^*)\left[\mathbf{b}^2(k)\right. \\
&\qquad\qquad \left. + \mathcal{A}^3(k)\mathbf{c}^1(k,k^*)\right] + \mathbf{b}^1(k).
\end{aligned}$$

For $k \le k^*$, (8.6.8) can be conveniently written as

(8.6.9)

$$\mathcal{R}^1(k,k^*) = \left[\mathcal{R}^1(k+1,k^*)\mathcal{A}^3(k) - \mathcal{A}^1(k)\right]^{-1}\left[\mathcal{A}^2(k) - \mathcal{R}^1(k+1,k^*)\mathcal{A}^4(k)\right]$$
$$\mathcal{R}^2(k,k^*) = -\left[\mathcal{R}^1(k+1,k^*)\mathcal{A}^3(k) - \mathcal{A}^1(k)\right]^{-1}\mathcal{R}^2(k+1,k^*)$$
$$\mathbf{c}^1(k,k^*) = \left[\mathcal{R}^1(k+1,k^*)\mathcal{A}^3(k) - \mathcal{A}^1(k)\right]^{-1}\left[\mathbf{b}^1(k) - \mathcal{R}^1(k+1,k^*)\mathbf{b}^2(k) - \mathbf{c}^1(k+1,k^*)\right].$$

The initial conditions for the system (8.6.8) as well as (8.6.9) are obtained from the relation (8.6.5) and appear as

$$\mathcal{R}^1(k^*,k^*) = 0, \quad \mathcal{R}^2(k^*,k^*) = \mathcal{I}, \quad \mathbf{c}^1(k^*,k^*) = 0. \tag{8.6.10}$$

Similarly, from (8.6.6) and (8.6.2), (8.6.5) we find for $k \geq k^*$ that

(8.6.11)

$$\mathcal{Q}^1(k+1,k^*) = \mathcal{Q}^1(k,k^*)\left[\mathcal{A}^4(k) + \mathcal{A}^3(k)\mathcal{R}^1(k,k^*)\right]^{-1}$$
$$\mathcal{Q}^2(k+1,k^*) = \mathcal{Q}^2(k,k^*) - \mathcal{Q}^1(k+1,k^*)\mathcal{A}^3(k)\mathcal{R}^2(k,k^*)$$
$$\mathbf{c}^2(k+1,k^*) = \mathbf{c}^2(k,k^*) - \mathcal{Q}^1(k+1,k^*)\left[\mathbf{b}^2(k) + \mathcal{A}^3(k)\mathbf{c}^1(k,k^*)\right],$$

which is for $k \leq k^*$ better written as

(8.6.12)

$$\mathcal{Q}^1(k,k^*) = \mathcal{Q}^1(k+1,k^*)\left[\mathcal{A}^4(k) + \mathcal{A}^3(k)\mathcal{R}^1(k,k^*)\right]$$
$$\mathcal{Q}^2(k,k^*) = \mathcal{Q}^2(k+1,k^*) + \mathcal{Q}^1(k+1,k^*)\mathcal{A}^3(k)\mathcal{R}^2(k,k^*)$$
$$\mathbf{c}^2(k,k^*) = \mathbf{c}^2(k+1,k^*) + \mathcal{Q}^1(k+1,k^*)\left[\mathbf{b}^2(k) + \mathcal{A}^3(k)\mathbf{c}^1(k,k^*)\right].$$

The initial conditions for the system (8.6.11) as well as (8.6.12) are obtained from the relation (8.6.6) and appear as

$$\mathcal{Q}^1(k^*,k^*) = \mathcal{I}, \quad \mathcal{Q}^2(k^*,k^*) = 0, \quad \mathbf{c}^2(k^*,k^*) = 0. \tag{8.6.13}$$

Equations (8.6.8), (8.6.10), (8.6.11), (8.6.13) form a *RQ* forward system whereas (8.6.9), (8.6.10), (8.6.12), (8.6.13) form a *RQ* backward system.

The inverse form at a fixed point $k^* \in \mathbb{N}(k_1, k_r)$ is defined as

$$\mathbf{w}(k) = \mathcal{S}^1(k,k^*)\mathbf{v}(k) + \mathcal{S}^2(k,k^*)\mathbf{w}(k^*) + \mathbf{d}^1(k,k^*) \tag{8.6.14}$$

$$\mathbf{v}(k^*) = \mathcal{T}^1(k,k^*)\mathbf{v}(k) + \mathcal{T}^2(k,k^*)\mathbf{w}(k^*) + \mathbf{d}^2(k,k^*), \tag{8.6.15}$$

where the matrices $\mathcal{S}^1(k,k^*)$, $\mathcal{S}^2(k,k^*)$, $\mathcal{T}^1(k,k^*)$, $\mathcal{T}^2(k,k^*)$ are of orders $q\times p$, $q\times q$, $p\times p$ and $p\times q$ respectively, and the vectors $\mathbf{d}^1(k,k^*)$, $\mathbf{d}^2(k,k^*)$ are of orders $q\times 1$, $p\times 1$ respectively.

As above from (8.6.14) and (8.6.1), (8.6.2) for $k\geq k^*$, we obtain the system

(8.6.16)

$$\begin{aligned}\mathcal{S}^1(k+1,k^*) &= \left[\mathcal{A}^3(k)+\mathcal{A}^4(k)\mathcal{S}^1(k,k^*)\right]\left[\mathcal{A}^1(k)+\mathcal{A}^2(k)\mathcal{S}^1(k,k^*)\right]^{-1}\\ \mathcal{S}^2(k+1,k^*) &= \left[\mathcal{A}^4(k)-\mathcal{S}^1(k+1,k^*)\mathcal{A}^2(k)\right]\mathcal{S}^2(k,k^*)\\ \mathbf{d}^1(k+1,k^*) &= \mathcal{A}^4(k)\mathbf{d}^1(k,k^*)-\mathcal{S}^1(k+1,k^*)\left[\mathbf{b}^1(k)\right.\\ &\qquad\left.+\mathcal{A}^2(k)\mathbf{d}^1(k,k^*)\right]+\mathbf{b}^2(k),\end{aligned}$$

which is for $k\leq k^*$ written as

(8.6.17)

$$\begin{aligned}\mathcal{S}^1(k,k^*) &= \left[\mathcal{S}^1(k+1,k^*)\mathcal{A}^2(k)-\mathcal{A}^4(k)\right]^{-1}\times\\ &\qquad\left[\mathcal{A}^3(k)-\mathcal{S}^1(k+1,k^*)\mathcal{A}^1(k)\right]\\ \mathcal{S}^2(k,k^*) &= -\left[\mathcal{S}^1(k+1,k^*)\mathcal{A}^2(k)-\mathcal{A}^4(k)\right]^{-1}\mathcal{S}^2(k+1,k^*)\\ \mathbf{d}^1(k,k^*) &= \left[\mathcal{S}^1(k+1,k^*)\mathcal{A}^2(k)-\mathcal{A}^4(k)\right]^{-1}\times\\ &\qquad\left[\mathbf{b}^2(k)-\mathcal{S}^1(k+1,k^*)\mathbf{b}^1(k)-\mathbf{d}^1(k+1,k^*)\right].\end{aligned}$$

The initial conditions for the system (8.6.16) as well as (8.6.17) are obtained from the relation (8.6.14) and appear as

(8.6.18) $$\mathcal{S}^1(k^*,k^*)=0,\quad \mathcal{S}^2(k^*,k^*)=\mathcal{I},\quad \mathbf{d}^1(k^*,k^*)=0.$$

Finally, from (8.6.15) and (8.6.1), (8.6.14) for $k\geq k^*$, we find the system

(8.6.19)

$$\begin{aligned}\mathcal{T}^1(k+1,k^*) &= \mathcal{T}^1(k,k^*)\left[\mathcal{A}^1(k)+\mathcal{A}^2(k)\mathcal{S}^1(k,k^*)\right]^{-1}\\ \mathcal{T}^2(k+1,k^*) &= \mathcal{T}^2(k,k^*)-\mathcal{T}^1(k+1,k^*)\mathcal{A}^2(k)\mathcal{S}^2(k,k^*)\\ \mathbf{d}^2(k+1,k^*) &= \mathbf{d}^2(k,k^*)-\mathcal{T}^1(k+1,k^*)\left[\mathbf{b}^1(k)+\mathcal{A}^2(k)\mathbf{d}^1(k,k^*)\right],\end{aligned}$$

which is for $k\leq k^*$ written as

(8.6.20)
$$\begin{aligned}
\mathcal{T}^1(k,k^*) &= \mathcal{T}^1(k+1,k^*)\left[\mathcal{A}^1(k)+\mathcal{A}^2(k)\mathcal{S}^1(k,k^*)\right]\\
\mathcal{T}^2(k,k^*) &= \mathcal{T}^2(k+1,k^*)+\mathcal{T}^1(k+1,k^*)\mathcal{A}^2(k)\mathcal{S}^2(k,k^*)\\
\mathbf{d}^2(k,k^*) &= \mathbf{d}^2(k+1,k^*)+\mathcal{T}^1(k+1,k^*)\left[\mathbf{b}^1(k)+\mathcal{A}^2(k)\mathbf{d}^1(k,k^*)\right].
\end{aligned}$$

The initial conditions for the system (8.6.19) as well as (8.6.20) are obtained from the relation (8.6.15) and appear as

(8.6.21) $$\mathcal{T}^1(k^*,k^*) = \mathcal{I}, \quad \mathcal{T}^2(k^*,k^*) = 0, \quad \mathbf{d}^2(k^*,k^*) = 0.$$

Equations (8.6.16), (8.6.18), (8.6.19), (8.6.21) form a $ST$ forward system whereas (8.6.17), (8.6.18), (8.6.20), (8.6.21) form a $ST$ backward system.

The above formulation gives several methods to obtain the solution of the boundary value problem (1.2.11), (8.6.3), (8.6.4) which we list as follows:

**1.** ***RQ*** **Forward Process.** In $RQ$ forward system let $k^* = k_1$ and solve it for all $k \in \mathbb{N}(k_1,k_r)$. We store all the matrices $\mathcal{R}^1(k,k_1)$, $\mathcal{R}^2(k,k_1)$, $\mathcal{Q}^1(k,k_1)$, $\mathcal{Q}^2(k,k_1)$ and the vectors $\mathbf{c}^1(k,k_1)$, $\mathbf{c}^2(k,k_1)$ for all $k \in \mathbb{N}(k_1,k_r)$. At $k=k_i$, $2 \le i \le r$ the relations (8.6.5), (8.6.6) are

(8.6.22)$_i$ $$\mathbf{v}(k_i) = \mathcal{R}^1(k_i,k_1)\mathbf{w}(k_i)+\mathcal{R}^2(k_i,k_1)\mathbf{v}(k_1)+\mathbf{c}^1(k_i,k_1)$$

(8.6.23)$_i$ $$\mathbf{w}(k_1) = \mathcal{Q}^1(k_i,k_1)\mathbf{w}(k_i)+\mathcal{Q}^2(k_i,k_1)\mathbf{v}(k_1)+\mathbf{c}^2(k_i,k_1).$$

The systems (8.6.3), (8.6.4); (8.6.22)$_i$, (8.6.23)$_i$, $2 \le i \le r$ are solved for the unknowns $\mathbf{v}(k_i)$, $\mathbf{w}(k_i)$, $1 \le i \le r$. For $k \in \mathbb{N}(k_1,k_r)$, $k \ne k_i$, $1 \le i \le r$ the solution is then obtained by rearranging (8.6.5), (8.6.6) so that

(8.6.24) $$\mathbf{w}(k) = \left(\mathcal{Q}^1(k,k_1)\right)^{-1}\left[\mathbf{w}(k_1)-\mathcal{Q}^2(k,k_1)\mathbf{v}(k_1)-\mathbf{c}^2(k,k_1)\right]$$

(8.6.25) $$\mathbf{v}(k) = \mathcal{R}^1(k,k_1)\mathbf{w}(k)+\mathcal{R}^2(k,k_1)\mathbf{v}(k_1)+\mathbf{c}^1(k,k_1).$$

**2. Modified** ***RQ*** **Forward Process.** In $RQ$ forward process we store only the matrices and vectors required in (8.6.3), (8.6.4); (8.6.22)$_i$, (8.6.23)$_i$, $2 \le i \le r$ and that too until we solve it for $\mathbf{v}(k_i)$, $\mathbf{w}(k_i)$ for a fixed $i$, $1 \le i \le r$. This obtained vector $\mathbf{v}(k_i)$, $\mathbf{w}(k_i)$ is used to compute the solution $\mathbf{u}(k)$ of (1.2.11) for all $k \in \mathbb{N}(k_1,k_r)$. This vector $\mathbf{u}(k)$ is the required solution of the boundary value problem (1.2.11), (1.5.4).

**3. Repeated $RQ$ Forward Process.** We should expect in some problems that $RQ$ forward process will exhibit overflow. To cope with this situation, we can switch from the $RQ$ forward system to "a new" $RQ$ forward system prior to overflow, and to continue with the computation. In some problems multiple switching may be necessary. In general, one does not know where the $RQ$ forward system will overflow before actually attempting to solve the problem. In practice, one carries out the computation and if overflow occurs, one backs up and selects a switch point, say, $a_1 \in \mathbb{N}(k_1, k_r)$ where the solutions are still good, then one attempts to solve the problem by continuing the computation from that point.

To switch from the $RQ$ forward system to a new $RQ$ forward system at the switch point $a_1$, we consider $k^* = a_1$ in (8.6.5), (8.6.6) so that basically the $RQ$ forward system remains the same except $k^* = a_1$ instead of $k_1$.

We assume that to complete the forward computation only one switching at $a_1$ is needed, and $k_j < a_1 < k_{j+1}$ where $1 \le j \le r-1$, but fixed. The systems (8.6.3), (8.6.4); (8.6.5), (8.6.6) at $k^* = k_1$, $k = k_i$, $2 \le i \le j$ and $k = a_1$; (8.6.5), (8.6.6) at $k^* = a_1$, $k = k_i$, $j+1 \le i \le r$ are solved for the unknowns $\mathbf{v}(k_i)$, $\mathbf{w}(k_i)$, $1 \le i \le r$ and $\mathbf{v}(a_1)$, $\mathbf{w}(a_1)$. For $k \in \mathbb{N}(k_1, a_1 - 1)$, $k \ne k_i$, $1 \le i \le j$ the solution is obtained from (8.6.24), (8.6.25) whereas for $k \in \mathbb{N}(a_1 + 1, k_r)$, $k \ne k_i$, $j+1 \le i \le r$ it is obtained from

$$(8.6.26) \qquad \mathbf{w}(k) = \left(\mathcal{Q}^1(k, a_1)\right)^{-1} \left[\mathbf{w}(a_1) - \mathcal{Q}^2(k, a_1)\mathbf{v}(a_1) - \mathbf{c}^2(k, a_1)\right]$$

$$(8.6.27) \qquad \mathbf{v}(k) = \mathcal{R}^1(k, a_1)\mathbf{w}(k) + \mathcal{R}^2(k, a_1)\mathbf{v}(a_1) + \mathbf{c}^1(k, a_1).$$

The case where multiple switching is needed can be extended easily.

**4. $ST$ Forward Process.** In $ST$ forward system let $k^* = k_1$ and solve it for all $k \in \mathbb{N}(k_1, k_r)$. We store all the matrices $\mathcal{S}^1(k, k_1)$, $\mathcal{S}^2(k, k_1)$, $\mathcal{T}^1(k, k_1)$, $\mathcal{T}^2(k, k_1)$ and the vectors $\mathbf{d}^1(k, k_1)$, $\mathbf{d}^2(k, k_1)$ for all $k \in \mathbb{N}(k_1, k_r)$. The systems (8.6.3), (8.6.4); (8.6.14), (8.6.15) at $k^* = k_1$, $k = k_i$, $2 \le i \le r$ are solved for the unknowns $\mathbf{v}(k_i)$, $\mathbf{w}(k_i)$, $1 \le i \le r$. For $k \in \mathbb{N}(k_1, k_r)$, $k \ne k_i$, $1 \le i \le r$ the solution is then obtained by rearranging (8.6.14), (8.6.15) so that

$$(8.6.28) \qquad \mathbf{v}(k) = \left(\mathcal{T}^1(k, k_1)\right)^{-1} \left[\mathbf{v}(k_1) - \mathcal{T}^2(k, k_1)\mathbf{w}(k_1) - \mathbf{d}^2(k, k_1)\right]$$

$$(8.6.29) \qquad \mathbf{w}(k) = \mathcal{S}^1(k, k_1)\mathbf{v}(k_1) + \mathcal{S}^2(k, k_1)\mathbf{w}(k_1) + \mathbf{d}^1(k, k_1).$$

**5. Modified $ST$ Forward Process.** In $ST$ forward process we use the same technique as in the modified $RQ$ forward process.

**6. Repeated $ST$ Forward Process.** As in repeated $RQ$ forward process we switch from $ST$ forward system to $ST$ forward system as often as needed.

**7. Repeated $RQ-ST$ Forward Process.** We begin with $RQ(ST)$ forward system and whenever necessary switch to $ST(RQ)$ forward system.

$1'$. $RQ$ **Backward Process.** In $RQ$ backward system let $k^* = k_r$ and solve it backward for all $k \in \mathbb{N}(k_1, k_r)$. Rest of the technique is the same as in $RQ$ forward process.

Finally, we remark that corresponding to 2 7 we have $2' - 7'$ where forward is replaced by backward.

## 8.7. Method of Imbedding: Second Formulation

It seems that this method can be easily formulated only for the following boundary value problem

$$(8.7.1) \quad \mathbf{u}(k+1) = \mathcal{A}(k)\mathbf{u}(k) + \mathcal{B}(k)\mathbf{v}(k) + \mathbf{f}(k)$$

$$(8.7.2) \quad \mathbf{v}(k+1) = \mathcal{C}(k)\mathbf{u}(k) + \mathcal{D}(k)\mathbf{v}(k) + \mathbf{g}(k), \quad k \in \mathbb{N}(0, K-1)$$

$$(8.7.3) \quad \mathbf{u}(0) = \mathbf{c}, \quad \mathbf{v}(K) = \mathbf{d},$$

where the matrices $\mathcal{A}(k)$, $\mathcal{B}(k)$, $\mathcal{C}(k)$, $D(k)$ are of order $n \times n$, and the vectors $\mathbf{u}(k)$, $\mathbf{v}(k)$, $\mathbf{f}(k)$, $\mathbf{g}(k)$, $\mathbf{c}$, $\mathbf{d}$ are of order $n \times 1$.

We shall transform (8.7.1) (8.7.3) into an initial value problem by imbedding it with $K$ fixed, in a class of similar problems and relating the solutions of the problems for which the interval lengths are $K$ and $K+1$. For this, we shall denote the solution of (8.7.1) (8.7.3) by $\mathbf{u}(k, K)$, $\mathbf{v}(k, K)$ which emphasizes its dependence on $k$ as well as the length $K$. It is clear that we may represent this solution in the form

$$(8.7.4) \quad \mathbf{u}(k, K) = \mathbf{p}(k, K) + \mathcal{U}(k, K)\mathbf{d}$$

$$(8.7.5) \quad \mathbf{v}(k, K) = \mathcal{Q}(k, K) + \mathcal{V}(k, K)\mathbf{d}, \quad k \in \mathbb{N}(0, K)$$

where the $n \times n$ matrices $U(k, K)$ and $V(k, K)$ defined for all $k \in \mathbb{N}(0, K)$ and $K = 0, 1, \cdots$ are the solutions of the boundary value problem

$$(8.7.6) \quad \mathcal{U}(k+1, K) = \mathcal{A}(k)\mathcal{U}(k, K) + \mathcal{B}(k)\mathcal{V}(k, K)$$

$$(8.7.7) \quad \mathcal{V}(k+1, K) = \mathcal{C}(k)\mathcal{U}(k, K) + \mathcal{D}(k)\mathcal{V}(k, K), \quad k \in \mathbb{N}(0, K-1)$$

$$(8.7.8) \quad \mathcal{U}(0, K) = 0, \quad \mathcal{V}(K, K) = \mathcal{I}$$

and the $n \times 1$ vectors $\mathbf{p}(k,K)$ and $\mathbf{q}(k,K)$ defined for all $k \in \mathbb{N}(0,K)$ and $K = 0,1,\cdots$ satisfying the system

$$(8.7.9) \quad \mathbf{p}(k+1,K) = \mathcal{A}(k)\mathbf{p}(k,K) + \mathcal{B}(k)\mathbf{q}(k,K) + \mathbf{f}(k)$$

$$(8.7.10) \quad \mathbf{q}(k+1,K) = \mathcal{C}(k)\mathbf{p}(k,K) + \mathcal{D}(k)\mathbf{q}(k,K) + \mathbf{g}(k), \quad k \in \mathbb{N}(0,K-1)$$

$$(8.7.11) \quad \mathbf{p}(0,K) = \mathbf{c}, \quad \mathbf{q}(K,K) = 0.$$

For the process of length $K+1$, the equations corresponding to (8.7.6) (8.7.8) are

$$(8.7.12) \quad \mathcal{U}(k+1,K+1) = \mathcal{A}(k)\mathcal{U}(k,K+1) + \mathcal{B}(k)\mathcal{V}(k,K+1)$$

$$(8.7.13) \quad \mathcal{V}(k+1,K+1) = \mathcal{C}(k)\mathcal{U}(k,K+1) + \mathcal{D}(k)\mathcal{V}(k,K+1), \quad k \in \mathbb{N}(0,K)$$

$$(8.7.14) \quad \mathcal{U}(0,K+1) = 0, \quad \mathcal{V}(K+1,K+1) = \mathcal{I}.$$

On the interval of length $K$, the matrices $U(k,K+1)$ and $V(k,K+1)$, $k \in \mathbb{N}(0,K)$ satisfy the system

$$(8.7.15) \quad \mathcal{U}(k+1,K+1) = \mathcal{A}(k)\mathcal{U}(k,K+1) + \mathcal{B}(k)\mathcal{V}(k,K+1)$$

$$(8.7.16) \quad \mathcal{V}(k+1,K+1) = \mathcal{C}(k)\mathcal{U}(k,K+1) + \mathcal{D}(k)\mathcal{V}(k,K+1), \quad k \in \mathbb{N}(0,K-1)$$

$$(8.7.17) \quad \mathcal{U}(0,K+1) = 0, \quad \mathcal{V}(K,K+1) = \mathcal{V}(K,K+1).$$

Multiplying both sides of (8.7.6) (8.7.8) on the right with $V(K,K+1)$, to find

$$(8.7.18) \quad \mathcal{U}(k+1,K)\mathcal{V}(K,K+1) = \mathcal{A}(k)\mathcal{U}(k,K)\mathcal{V}(K,K+1) + \mathcal{B}(k)\mathcal{V}(k,K)\mathcal{V}(K,K+1)$$

$$(8.7.19) \quad \mathcal{V}(k+1,K)\mathcal{V}(K,K+1) = \mathcal{C}(k)\mathcal{U}(k,K)\mathcal{V}(K,K+1) + \mathcal{D}(k)\mathcal{V}(k,K)\mathcal{V}(K,K+1)$$

$$(8.7.20) \quad \mathcal{U}(0,K)\mathcal{V}(K,K+1) = 0, \quad \mathcal{V}(K,K)\mathcal{V}(K,K+1) = \mathcal{V}(K,K+1).$$

Assuming that (8.7.15) (8.7.17) has a unique solution, and comparing this system with (8.7.18) (8.7.20), we get

$$(8.7.21) \quad \mathcal{U}(k,K+1) = \mathcal{U}(k,K)\mathcal{V}(K,K+1)$$

$$(8.7.22) \quad \mathcal{V}(k,K+1) = \mathcal{V}(k,K)\mathcal{V}(K,K+1), \quad k \in \mathbb{N}(0,K).$$

Let $k = K$ in (8.7.21) so that it can be written as

$$(8.7.23) \quad \mathcal{U}(K,K+1) = \mathcal{R}(K)\mathcal{V}(K,K+1), \quad K = 0,1,\cdots$$

where $\mathcal{R}(K) = \mathcal{U}(K,K)$, $K = 0,1,\cdots$.

Using (8.7.14) and (8.7.23) in (8.7.13) at $k = K$ gives

$$\mathcal{I} = [\mathcal{C}(K)\mathcal{R}(K) + \mathcal{D}(K)]\,\mathcal{V}(K,K+1), \quad K = 0,1,\cdots$$

which is the same as

$$(8.7.24) \qquad \mathcal{V}(K, K+1) \;=\; [\mathcal{C}(K)\mathcal{R}(K) + \mathcal{D}(K)]^{-1}, \quad K = 0, 1, \cdots.$$

Using (8.7.23) and (8.7.24) in (8.7.12) at $k = K$ leads to

$$(8.7.25) \qquad \mathcal{R}(K+1) \;=\; [\mathcal{A}(K)\mathcal{R}(K) + \mathcal{B}(K)]\,[\mathcal{C}(K)\mathcal{R}(K) + \mathcal{D}(K)]^{-1}, \qquad K = 0, 1, \cdots.$$

The initial condition for (8.7.25) is obtained from (8.7.8) at $K = 0$ and appears as

$$(8.7.26) \qquad \mathcal{R}(0) \;=\; 0.$$

In particular (8.7.25), (8.7.26) determines $\mathcal{R}(k)$.

For $K \geq k$ the recurrence relations for the matrices $U(k, K)$ and $V(k, K)$ are obtained on using (8.7.24) in (8.7.21) and (8.7.22)

$$(8.7.27) \qquad \mathcal{U}(k, K+1) \;=\; \mathcal{U}(k, K)\,[\mathcal{C}(K)\mathcal{R}(K) + \mathcal{D}(K)]^{-1}$$

$$(8.7.28) \qquad \mathcal{V}(k, K+1) \;=\; \mathcal{V}(k, K)\,[\mathcal{C}(K)\mathcal{R}(K) + \mathcal{D}(K)]^{-1}.$$

Since $\mathcal{U}(K, K) = \mathcal{R}(K)$ and from (8.7.8), $\mathcal{V}(K, K) = \mathcal{I}$, $K = 0, 1, \cdots,$ the initial conditions at $K = k$ are

$$(8.7.29) \qquad \mathcal{U}(k, k) \;=\; \mathcal{R}(k), \quad \mathcal{V}(k, k) \;=\; \mathcal{I}.$$

Next we shall consider the vectors $\mathbf{p}(k, K)$ and $\mathbf{q}(k, K)$. For this, on the interval of length $K + 1$, equations (8.7.9) (8.7.11) become

$$(8.7.30) \qquad \mathbf{p}(k+1, K+1) \;=\; \mathcal{A}(k)\mathbf{p}(k, K+1) + \mathcal{B}(k)\mathbf{q}(k, K+1) + \mathbf{f}(k)$$

$$(8.7.31) \qquad \mathbf{q}(k+1, K+1) \;=\; \mathcal{C}(k)\mathbf{p}(k, K+1) + \mathcal{D}(k)\mathbf{q}(k, K+1) + \mathbf{g}(k), \qquad k \in \mathbb{N}(0, K)$$

$$(8.7.32) \qquad \mathbf{p}(0, K+1) \;=\; \mathbf{c}, \quad \mathbf{q}(K+1, K+1) \;=\; 0.$$

The difference vectors $\mathbf{z}(k, K) = \mathbf{p}(k, K+1) - \mathbf{p}(k, K)$ and $\mathbf{w}(k, K) = \mathbf{q}(k, K+1) - \mathbf{q}(k, K)$, $k \in \mathbb{N}(0, K)$ satisfy the system

$$(8.7.33) \qquad \mathbf{z}(k+1, K) \;=\; \mathcal{A}(k)\mathbf{z}(k, K) + \mathcal{B}(k)\mathbf{w}(k, K)$$

$$(8.7.34) \qquad \mathbf{w}(k+1, K) \;=\; \mathcal{C}(k)\mathbf{z}(k, K) + \mathcal{D}(k)\mathbf{w}(k, K), \quad k \in \mathbb{N}(0, K-1)$$

$$(8.7.35) \qquad \mathbf{z}(0, K) \;=\; 0, \quad \mathbf{w}(K, K) \;=\; \mathbf{q}(K, K+1).$$

The multiplication of equations (8.7.6) (8.7.8) by the vector $\mathbf{q}(K, K+1)$ yields the system

$$\begin{aligned}(8.7.36)\quad \mathcal{U}(k+1,K)\mathbf{q}(K,K+1) &= \mathcal{A}(k)\mathcal{U}(k,K)\mathbf{q}(K,K+1)\\ &\quad +\mathcal{B}(k)\mathcal{V}(k,K)\mathbf{q}(K,K+1)\\ (8.7.37)\quad \mathcal{V}(k+1,K)\mathbf{q}(K,K+1) &= \mathcal{C}(k)\mathcal{U}(k,K)\mathbf{q}(K,K+1)\\ &\quad +\mathcal{D}(k)\mathcal{V}(k,K)\mathbf{q}(K,K+1), \quad k\in\mathbb{N}(0,K-1)\\ (8.7.38)\quad \mathcal{U}(0,K)\mathbf{q}(K,K+1) = 0, &\quad \mathcal{V}(K,K)\mathbf{q}(K,K+1) = \mathbf{q}(K,K+1).\end{aligned}$$

Assuming that (8.7.33) (8.7.35) has a unique solution and comparing this system with (8.7.36) (8.7.38), we get

$$\begin{aligned}(8.7.39)\quad \mathbf{p}(k,K+1) &= \mathbf{p}(k,K)+\mathcal{U}(k,K)\mathbf{q}(K,K+1)\\ (8.7.40)\quad \mathbf{q}(k,K+1) &= \mathbf{q}(k,K)+\mathcal{V}(k,K)\mathbf{q}(K,K+1), \quad k\in\mathbb{N}(0,K).\end{aligned}$$

Let $k=K$ in (8.7.39) so that it can be written as

$$(8.7.41)\qquad \mathbf{p}(K,K+1) = \mathbf{s}(K)+\mathcal{R}(K)\mathbf{q}(K,K+1),$$

where $\mathbf{s}(K)=\mathbf{p}(K,K)$, $K=0,1,\cdots$.

Using (8.7.41) in (8.7.30) at $k=K$ gives

$$(8.7.42)\quad \mathbf{s}(K+1) = \mathcal{A}(K)\mathbf{s}(K)+[\mathcal{A}(K)\mathcal{R}(K)+\mathcal{B}(K)]\,\mathbf{q}(K,K+1)+\mathbf{f}(K).$$

Using (8.7.32) and (8.7.41) in (8.7.31) at $k=K$, to obtain

$$(8.7.43)\quad \mathbf{q}(K,K+1) = -\left[\mathcal{C}(K)\mathcal{R}(K)+\mathcal{D}(K)\right]^{-1}\left[\mathcal{C}(K)\mathbf{s}(K)+\mathbf{g}(K)\right].$$

Using (8.7.43) and (8.7.25) in (8.7.42), we find

$$(8.7.44)\quad \mathbf{s}(K+1) = \mathcal{A}(K)\mathbf{s}(K)-\mathcal{R}(K+1)\left[\mathcal{C}(K)\mathbf{s}(K)+\mathbf{g}(K)\right]+\mathbf{f}(K),$$
$$K=0,1,\cdots.$$

Since $\mathbf{s}(K)=\mathbf{p}(K,K)$ the initial condition for (8.7.44) is obtained from (8.7.11) and appears as

$$(8.7.45)\qquad \mathbf{s}(0) = \mathbf{c}.$$

In particular (8.7.44), (8.7.45) determines $\mathbf{s}(k)$.

For $K\geq k$ the recurrence relations for the vectors $\mathbf{p}(k,K)$ and $\mathbf{q}(k,K)$ are obtained on using (8.7.43) in (8.7.39) and (8.7.40)

$$\begin{aligned}(8.7.46)\quad \mathbf{p}(k,K+1) &= \mathbf{p}(k,K)-\mathcal{U}(k,K)[\mathcal{C}(K)\mathcal{R}(K)\\ &\quad +\mathcal{D}(K)]^{-1}\left[\mathcal{C}(K)\mathbf{s}(K)+\mathbf{g}(K)\right]\\ (8.7.47)\quad \mathbf{q}(k,K+1) &= \mathbf{q}(k,K)-\mathcal{V}(k,K)[\mathcal{C}(K)\mathcal{R}(K)\\ &\quad +\mathcal{D}(K)]^{-1}\left[\mathcal{C}(K)\mathbf{s}(K)+\mathbf{g}(K)\right].\end{aligned}$$

Since $\mathbf{p}(K,K) = \mathbf{s}(K)$ and from (8.7.11), $\mathbf{q}(K,K) = 0,\ K = 0,1,\cdots,$ the initial conditions at $K = k$ are

$$\text{(8.7.48)} \qquad \mathbf{p}(k,k) = \mathbf{s}(k), \quad \mathbf{q}(k,k) = 0.$$

All the necessary relations are now at hand, and we may summarize the method. For $K = 0,1,\cdots,k-1$ we employ the recurrence relations (8.7.25), (8.7.44) together with the initial conditions (8.7.26), (8.7.45) so that $\mathcal{R}(k)$ and $\mathbf{s}(k)$ are available. For $K \geq k$, we use the recurrence relations (8.7.25), (8.7.44), (8.7.27), (8.7.28), (8.7.46) and (8.7.47) together with the initial conditions $\mathcal{R}(k) = \mathcal{R}(k)$, $\mathbf{s}(k) = \mathbf{s}(k)$, (8.7.29) and (8.7.48). Finally, the required solution is obtained from (8.7.4), (8.7.5).

**Example 8.7.1.** Consider the boundary value problem (1.6.40), (1.6.41). On comparing it with (8.7.1) (8.7.3), we have $n = L$, $\mathcal{A}(k) = 0, \mathcal{B}(k) = \mathcal{I}$, $\mathbf{f}(k) = 0$, $\mathcal{C}(k) = -\mathcal{I}$, $\mathcal{D}(k) = \mathcal{Q}$, $\mathbf{g}(k) = -\mathbf{r}(k+1)$. Thus, to obtain the vector $\mathbf{u}(k)$, $k \leq K$ we need to solve

$$\begin{aligned} \mathcal{R}(k+1) &= [\mathcal{Q} - \mathcal{R}(k)]^{-1} \\ \mathbf{s}(k+1) &= \mathcal{R}(k+1)\,[\mathbf{s}(k) + \mathbf{r}(k+1)] \\ \mathcal{R}(0) &= 0, \quad \mathbf{s}(0) = \mathbf{c} \end{aligned}$$

and, at $k = m$ we adjoin

$$\begin{aligned} \mathcal{U}(m,k+1) &= \mathcal{U}(m,k)\mathcal{R}(k+1) \\ \mathbf{p}(m,k+1) &= \mathbf{p}(m,k) + \mathcal{U}(m,k)\mathbf{s}(k+1) \\ \mathcal{U}(m,m) &= \mathcal{R}(m), \quad \mathbf{p}(m,m) = \mathbf{s}(m) \end{aligned}$$

and finally

$$\mathbf{u}(m) = \mathbf{p}(m,K) + \mathcal{U}(m,K)\mathbf{d}.$$

## 8.8. Method of Sweep

We have seen that the second order recurrence relation

$$(8.8.1)_k \quad a_0(k)u(k-1) - a_1(k)u(k) + a_2(k)u(k+1) = -b(k), \quad k \in \mathbb{N}(1,K)$$

together with the boundary conditions

$$\text{(8.8.2)} \qquad u(0) = \alpha u(1) + \beta, \quad u(K+1) = \gamma u(K) + \delta$$

appears in several applications. Here, we shall formulate the known method of sweep which is very stable. For this, we shall assume that

$$\text{(8.8.3)} \quad a_0(k),\ a_1(k),\ a_2(k) > 0,\ a_1(k) \geq a_0(k) + a_2(k),\ 0 \leq \alpha < 1,\ 0 \leq \gamma < 1.$$

It is clear that the problem (8.2.22), (1.6.14) satisfies the above conditions provided $f(t) \geq 0,\ t \in [\alpha, \beta]$, and $\frac{h^2}{12} \max_{\alpha \leq t \leq \beta} f(t) < 1$.

From $(8.8.1)_K$ and $u(K+1) = \gamma u(K) + \delta$, we see that $u(K-1)$ is a linear function of $u(K)$. Also, on eliminating $u(K)$ from the equations $(8.8.1)_K$ and $(8.8.1)_{K-1}$ it is clear that $u(K-2)$ is a linear function of $u(K-1)$. Thus, we shall attempt to determine coefficients $\alpha(k), \beta(k), k = K+1, K, \cdots, 1$ such that the relation

$$(8.8.4) \qquad u(k-1) \;=\; \alpha(k)u(k) + \beta(k), \qquad k \in \mathbb{N}(1, K+1)$$

holds. For this, let $k = 1$ in (8.8.4), so that from $u(0) = \alpha u(1) + \beta$, we have

$$\alpha u(1) + \beta \;=\; u(0) \;=\; \alpha(1)u(1) + \beta(1),$$

which provides that

$$(8.8.5) \qquad \alpha(1) \;=\; \alpha, \qquad \beta(1) \;=\; \beta.$$

Next from (8.8.4), equation $(8.8.1)_k$ can be written as

$$(8.8.6) \quad a_0(k)(\alpha(k)u(k)+\beta(k))-a_1(k)u(k)+a_2(k)\frac{u(k)-\beta(k+1)}{\alpha(k+1)} \;=\; -b(k).$$

Since the relation (8.8.6) holds for all $k$, we can equate the coefficient of $u(k)$ to zero, to find the initial value problems involving first order nonlinear difference equations

$$(8.8.7) \quad \alpha(k+1) \;=\; \frac{a_2(k)}{a_1(k) - a_0(k)\alpha(k)}, \quad \alpha(1) \;=\; \alpha$$

$$(8.8.8) \quad \beta(k+1) \;=\; \frac{b(k) + a_0(k)\beta(k)}{a_1(k) - a_0(k)\alpha(k)}, \quad \beta(1) = \beta, \quad k \in \mathbb{N}(1, K).$$

It is easy to show that if conditions (8.8.3) are satisfied, then the denominator $a_1(k) - a_0(k)\alpha(k)$ in (8.8.7) as well as (8.8.8) does not vanish, and $0 \leq \alpha(k) < 1$. Indeed, if we rewrite the condition $a_1(k) \geq a_0(k) + a_2(k)$ as $a_1(k) = a_0(k) + a_2(k) + c(k),\ c(k) \geq 0$ then (8.8.7) can be written as

$$\alpha(k+1) \;=\; \frac{a_2(k)}{a_2(k) + a_0(k)(1-\alpha(k)) + c(k)}, \quad \alpha(1) \;=\; \alpha$$

from which the assertion is immediate.

Once the sequences $\alpha(k),\ \beta(k),\ k \in \mathbb{N}(1, K+1)$ are known, the solution of $(8.8.1)_k$, (8.8.2) can be obtained as follows: From (8.8.4), we have $u(K) = \alpha(K+1)u(K+1) + \beta(K+1)$, and hence from the boundary condition $u(K+1) = \gamma u(K) + \delta$ it follows that $u(K+1) = \gamma(\alpha(K+1)u(K+1) + \beta(K+1)) + \delta$, so that

$$u(K+1) = \frac{\gamma\beta(K+1)+\delta}{1-\gamma\alpha(K+1)}. \tag{8.8.9}$$

Since $0 \leq \alpha(K+1),\ \gamma < 1$ it is clear that $u(K+1)$ is well defined. Now from (8.8.4) the solution $u(k),\ k = K+1, K, \cdots, 1$ can be computed.

By the relation $0 \leq \alpha(k) < 1,$ it is clear that in the above process the errors will not be compiled. This method we shall call as the *forward sweep*.

As in the forward forward process we need not store $\alpha(k),\ \beta(k),\ k \in \mathbb{N}(1, K+1),$ rather only $\alpha(K+1),\ \beta(K+1)$ so that $u(K+1)$ from (8.8.9) and then $u(K)$ from (8.8.4) are known. The required solution $u(k)$ now can be computed by recursing backward $(8.8.1)_k$ with these known values of $u(K+1),\ u(K)$. However, often it gives unrealistic values. This process we shall call as the *forward sweep backward method.*

From $(8.8.1)_1$ and $u(0) = \alpha u(1) + \beta$ it is evident that $u(2)$ is a linear function of $u(1)$. Also, on eliminating $u(1)$ from $(8.8.1)_1$ and $(8.8.1)_2$ it is clear that $u(3)$ is a linear function of $u(2)$. Thus, in general $u(k+1)$ is a linear function of $u(k)$ alone. Therefore, we can write

$$u(k+1) = \gamma(k)u(k) + \delta(k), \quad k \in \mathbb{N}(0, K). \tag{8.8.10}$$

Now as in the forward sweep, we find the initial value problems

$$\gamma(k-1) = \frac{a_0(k)}{a_1(k) - a_2(k)\gamma(k)}, \quad \gamma(K) = \gamma \tag{8.8.11}$$

$$\delta(k-1) = \frac{b(k) + a_2(k)\delta(k)}{a_1(k) - a_2(k)\gamma(k)}, \quad \delta(K) = \delta, \quad k = K, K-1, \cdots, 1. \tag{8.8.12}$$

Once the sequences $\gamma(k),\ \delta(k),\ k \in \mathbb{N}(0, K)$ are known, the relation equivalent to (8.8.9) is immediately available and appears as

$$u(0) = \frac{\alpha\delta(0) + \beta}{1 - \alpha\gamma(0)}. \tag{8.8.13}$$

The required solution $u(k)$ is obtained by recursing forward the relation (8.8.10). This method we shall call as the *backward sweep.*

Instead of storing the sequences $\gamma(k),\ \delta(k),\ k \in \mathbb{N}(0, K),$ we can compute $u(0)$ from (8.8.13) and then $u(1)$ from (8.8.10), from the known values of $\gamma(0)$ and $\delta(0)$. Finally the required solution $u(k)$ is computed by recursing forward $(8.8.1)_k$ with these known values of $u(0)$ and $u(1)$. This process we shall call as *backward sweep forward method.*

**Example 8.8.1.** For the discrete analog (8.2.22), (1.6.14) of the boundary value problem (8.2.35) all the four methods discussed in this section work equally well. The errors obtained, as calculated from the exact solution $y(t)$ (see Example 8.2.2) and approximate solution $u(k)$ with $h = 1/256$ are presented in Table 8.8.1.

**Table** 8.8.1.

| $t$ | Forw. Sweep | Forw. Sweep–Back. Method | Back. Sweep– | Back. Sweep–Ford. Method |
|---|---|---|---|---|
| 2.000 | $0.00000000D\ \ 00$ | $0.22947398D-07$ | $0.00000000D\ \ 00$ | $0.00000000D\ \ 00$ |
| 2.125 | $0.12870868D-12$ | $0.25826906D-07$ | $0.60432142D-08$ | $0.94820384D-08$ |
| 2.250 | $0.19976382D-12$ | $0.23808074D-07$ | $0.70504433D-08$ | $0.19028611D-07$ |
| 2.375 | $0.22875278D-12$ | $0.19225328D-07$ | $0.53897599D-08$ | $0.28691833D-07$ |
| 2.500 | $0.22493812D-12$ | $0.13689967D-07$ | $0.26993845D-08$ | $0.38513395D-07$ |
| 2.625 | $0.19501588D-12$ | $0.83241260D-08$ | $0.12324682D-09$ | $0.48527049D-07$ |
| 2.750 | $0.14410608D-12$ | $0.39136916D-08$ | $0.15349350D-08$ | $0.58760408D-07$ |
| 2.875 | $0.77308819D-13$ | $0.10102117D-08$ | $0.17088811D-08$ | $0.69236284D-07$ |
| 3.000 | $0.00000000D\ \ 00$ | $0.00000000D\ \ 00$ | $0.00000000D\ \ 00$ | $0.79973681D-07$ |

**Example 8.8.2.** We apply the methods of this section to the discrete boundary value problem considered in Example 8.2.3. The results analogous to Table 8.2.2 are given in Table 8.8.2.

**Table** 8.8.2.

| $t$ | Forw. Sweep | Forw. Sweep–Back. Method | Back. Sweep– | Back. Sweep–Ford. Method |
|---|---|---|---|---|
| 0.0000 | $0.00000000D\ \ 00$ | $0.39154791D-12$ | $0.00000000D\ \ 00$ | |
| 0.3125 | $0.22852521D-08$ | $0.22852529D-08$ | $0.22852522D-08$ | |
| 0.6250 | $0.88231436D-11$ | $0.88231451D-11$ | $0.88231438D-11$ | |
| 0.9375 | $0.25548996D-13$ | $0.25548999D-13$ | $0.25548996D-13$ | |
| 1.2500 | $0.65761515D-16$ | $0.65761520D-16$ | $0.65761516D-16$ | |
| 1.5625 | $0.15868689D-18$ | $0.15868690D-18$ | $0.15868689D-18$ | |
| 1.8750 | $0.36760510D-21$ | $0.36760512D-21$ | $0.36760511D-21$ | |
| 2.1875 | $0.82791844D-24$ | $0.82791846D-24$ | $0.82791844D-24$ | |
| 2.5000 | $0.18265801D-26$ | $0.18265802D-26$ | $0.18265802D-26$ | **Fails** |
| 2.8125 | $0.39668930D-29$ | $0.39668931D-29$ | $0.39668930D-29$ | |
| 3.1250 | $0.85087783D-32$ | $0.85087784D-32$ | $0.85087783D-32$ | |
| 3.4375 | $0.18068376D-34$ | $0.18068376D-34$ | $0.18068376D-34$ | |
| 3.7500 | $0.38051074D-37$ | $0.38051074D-37$ | $0.38051074D-37$ | |
| 4.0625 | $0.79577127D-40$ | $0.79577127D-40$ | $0.79577127D-40$ | |
| 4.3750 | $0.16543696D-42$ | $0.16543696D-42$ | $0.16543696D-42$ | |
| 4.6875 | $0.41399292D-45$ | $0.41399292D-45$ | $0.41399292D-45$ | |
| 5.0000 | $0.00000000D\ \ 00$ | $0.00000000D\ \ 00$ | $0.37200760D-43$ | |

**Example 8.8.3.** We apply the methods of this section to the discrete boundary value problem considered in Example 8.2.4. The results analogous to Table 8.2.3 are presented in Table 8.8.3.

**Example 8.8.4.** We apply the methods of this section to the discrete boundary value problem considered in Example 8.2.5. The results analogous to Table 8.2.4 are presented in Table 8.8.4.

**Table 8.8.3.**

| $t$ | Forw. Sweep | Forw. Sweep–Back. Method | Back. Sweep– | Back. Sweep–Ford. Method |
|---|---|---|---|---|
| 0.0 | 0.25000000D 00 | 0.25381778D 00 | 0.25000000D 00 | |
| 1.0 | 0.23407771D−01 | 0.23763681D−01 | 0.23406242D−01 | |
| 2.0 | 0.14143468D−02 | 0.14356292D−02 | 0.14140353D−02 | |
| 3.0 | 0.44114854D−04 | 0.44767722D−04 | 0.44094352D−04 | |
| 4.0 | 0.62609336D−06 | 0.63514634D−06 | 0.62559284D−06 | |
| 5.0 | 0.37645596D−08 | 0.38173671D−08 | 0.37599485D−08 | |
| 6.0 | 0.91929810D−11 | 0.93171146D−11 | 0.91769720D−11 | |
| 7.0 | 0.88796121D−14 | 0.89940335D−14 | 0.88587505D−14 | |
| 8.0 | 0.33341431D−17 | 0.33747388D−17 | 0.33239780D−17 | |
| 9.0 | 0.48088930D−21 | 0.48635830D−21 | 0.47904278D−21 | |
| 10.0 | 0.26416894D−25 | 0.26693655D−25 | 0.26292144D−25 | **Fails** |
| 11.0 | 0.54925707D−30 | 0.55446825D−30 | 0.54612826D−30 | |
| 12.0 | 0.43020222D−35 | 0.43381844D−35 | 0.42729319D−35 | |
| 13.0 | 0.12646919D−40 | 0.12738372D−40 | 0.12548769D−40 | |
| 14.0 | 0.13914253D−46 | 0.13997238D−46 | 0.13788700D−46 | |
| 15.0 | 0.57160202D−53 | 0.57423386D−53 | 0.56559657D−53 | |
| 16.0 | 0.87512138D−60 | 0.87788000D−60 | 0.86467545D−60 | |
| 17.0 | 0.49854924D−67 | 0.49935094D−67 | 0.49183999D−67 | |
| 18.0 | 0.00000000D 00 | 0.00000000D 00 | 0.00000000D 00 | |

**Table 8.8.4.**

| $t$ | Forw. Sweep | Forw. Sweep–Back. Method | Back. Sweep– | Back. Sweep–Ford. Method |
|---|---|---|---|---|
| −1.0 | 0.00000000D 00 | −0.17399314D−06 | 0.00000000D 00 | 0.00000000D 00 |
| −0.9 | −0.54219742D−01 | −0.54219921D−01 | −0.54219748D−01 | −0.54219746D−01 |
| −0.8 | −0.11017210D 00 | −0.11017229D 00 | −0.11017212D 00 | −0.11017211D 01 |
| −0.7 | −0.16531803D 00 | −0.16531822D 00 | −0.16531805D 00 | −0.16531804D 00 |
| −0.6 | −0.21715198D 00 | −0.21715217D 00 | −0.21715201D 00 | −0.21715199D 00 |
| −0.5 | −0.26336748D 00 | −0.26336767D 00 | −0.26336751D 00 | −0.26336750D 00 |
| −0.4 | −0.30200211D 00 | −0.30200228D 00 | −0.30200214D 00 | −0.30200213D 00 |
| −0.3 | −0.33154697D 00 | −0.33154713D 00 | −0.33154699D 00 | −0.33154700D 00 |
| −0.2 | −0.35101159D 00 | −0.35101172D 00 | −0.35101160D 00 | −0.35101162D 00 |
| −0.1 | −0.35994166D 00 | −0.35994177D 00 | −0.35994166D 00 | −0.35994169D 00 |
| 0.0 | −0.35839390D 00 | −0.35839397D 00 | −0.35839387D 00 | −0.35839393D 00 |
| 0.1 | −0.34687723D 00 | −0.34687728D 00 | −0.34687719D 00 | −0.34687726D 00 |
| 0.2 | −0.32627260D 00 | −0.32627262D 00 | −0.32627255D 00 | −0.32627264D 00 |
| 0.3 | −0.29774424D 00 | −0.29774425D 00 | −0.29774418D 00 | −0.29774428D 00 |
| 0.4 | −0.26265376D 00 | −0.26265376D 00 | −0.26265370D 00 | −0.26265381D 00 |
| 0.5 | −0.22248554D 00 | −0.22248553D 00 | −0.22248548D 00 | −0.22248559D 00 |
| 0.6 | −0.17878818D 00 | −0.17878817D 00 | −0.17878813D 00 | −0.17878823D 00 |
| 0.7 | −0.13313322D 00 | −0.13313321D 00 | −0.13313319D 00 | −0.13313328D 00 |
| 0.8 | −0.87088932D−01 | −0.87088926D−01 | −0.87088909D−01 | −0.87088996D−01 |
| 0.9 | −0.42204687D−01 | −0.42204685D−01 | −0.42204676D−01 | −0.42204757D−01 |
| 1.0 | 0.00000000D 00 | 0.00000000D 00 | 0.00000000D 00 | −0.76962026D−07 |

## 8.9. Miller's and Olver's Algorithms

For the homogeneous difference equation (1.2.4) let $u_1(k), \cdots, u_n(k)$ be the linearly independent solutions, and let

$$\lim_{k\to\infty} \frac{u_1(k)}{u_i(k)} = 0, \quad i = 2, \cdots, n. \tag{8.9.1}$$

Then, the solution $u_1(k)$ is said to be *minimal* (Recessive). The importance of minimal solution in the study of special functions, orthogonal polynomials, quadrature formulas, and numerical methods for ordinary differential equations is well known, e.g. see Cash [14] and Wimp [54]. The difference equation $u(k+2) - u(k) = 0$ has no minimal solution, whereas the equation $u(k+2) - 3u(k+1) + 2u(k) = 0$ has a minimal solution. If a minimal solution exists then it is, up to a constant multiple, unique. For if $u_1(k)$ and $u_2(k)$ were two minimal solutions then simultaneously $\frac{u_1(k)}{u_2(k)} \to 0$ and $\frac{u_2(k)}{u_1(k)} \to 0$, which is impossible. Therefore, to compute the minimal solution of (1.2.4) only one appropriate initial condition is needed.

So far, necessary and sufficient conditions for the existence of minimal solution of (1.2.4) are not known. Further, from Problem 8.10.5 it follows that, even the exact initial conditions that guarantee the minimal solution cannot be used to generate it by recursing (1.2.4) in the forward direction. In fact, a small rounding error in the computation will lead to contain all the other solutions $u_i(k)$, $i = 2, \cdots, n$ which grow faster than $u_1(k)$, and consequently this will lead to overflow.

The general solution of the nonhomogeneous equation (1.2.3) can be written as $u(k) = \sum_{i=1}^{n} c_i u_i(k) + \overline{u}(k)$, where $\overline{u}(k)$ is a particular solution of (1.2.3), which is also assumed to be minimal, i.e.

$$\lim_{k\to\infty} \frac{\overline{u}(k)}{u_i(k)} = 0, \quad i = 2, \cdots, n. \tag{8.9.2}$$

Our interest is in the computation of the solution

$$u(k) = \left[\frac{u(0) - \overline{u}(0)}{u_1(0)}\right] u_1(k) + \overline{u}(k). \tag{8.9.3}$$

For this, Miller's and Olver's algorithms and their several refinements are well known. We shall discuss these algorithms only for the second order difference equations.

**Theorem 8.9.1.** For the second order difference equation

$$a_0(k)u(k) + a_1(k)u(k+1) + a_2(k)u(k+2) = b(k) \tag{8.9.4}$$

let the conditions (8.9.1) and (8.9.2) be satisfied. Then, for every large $K \in \mathbb{N}$ the boundary value problem

$$a_0(k)u^{(K)}(k) + a_1(k)u^{(K)}(k+1) + a_2(k)u^{(K)}(k+2) = b(k), \quad k \in \mathbb{N}(0, K-1) \tag{8.9.5}$$

$$u^{(K)}(0) = u(0), \quad U^{(K)}(K+1) = 0 \tag{8.9.6}$$

has a solution $u^{(K)}(k)$, and moreover, for fixed $k$, $\lim_{K\to\infty} u^{(K)}(k) \to u(k)$.

**Proof.** Since $u(k)$ defined in (8.9.3) is a particular solution of (8.9.5), any other solution of the same equation can be written as

$$u^{(K)}(k) = c_1^{(K)}u_1(k) + c_2^{(K)}u_2(k) + u(k). \tag{8.9.7}$$

This solution also satisfies the boundary conditions (8.9.6) if and only if

$$c_1^{(K)} = \frac{u(K+1)/u_2(K+1)}{u_1(0)/u_2(0) - u_1(K+1)/u_2(K+1)}, \quad c_2^{(K)} = -\frac{u_1(0)}{u_2(0)}c_1^{(K)}.$$

However, $c_1^{(K)}$ and $c_2^{(K)}$ tend to zero as $K \to \infty$ follows from (8.9.1) and (8.9.2). Therefore, from (8.9.7) it is clear that $\lim_{K\to\infty} u^{(K)}(k) = u(k)$.

In (8.9.6) the condition $u^{(K)}(K+1) = 0$ can be replaced by $u^{(K)}(K+1) = \overline{u}(K+1)$, where $\overline{u}(K+1)$ is an approximation of $u(K+1)$, if available.

**Miller's Algorithm:** An approximation to the minimal solution $u(k)$ of (2.16.8) can be obtained by recursing backward (8.9.5) $(b(k) = 0)$ with the conditions $u^{(K)}(K+1) = 0, u^{(K)}(K) = 1$ and then multiplying the computed solution $u^{(K)}(k)$ by $\dfrac{u(0)}{u^{(K)}(0)}$. It is interesting to note that this is precisely the backward process (cf. Section 8.2) to solve (8.9.5), (8.9.6) $(b(k) = 0)$. A disadvantage of this method is that one does not know a priori which value of $K$ must be used to obtain the required accuracy.

**Olver's Algorithm:** In (8.8.4) let $k$ to be $k+1$, $\alpha(k+1) = \dfrac{p(k)}{p(k+1)}$ and $\beta(k+1) = \dfrac{e(k)}{p(k+1)}$ so that for the equation (8.9.5) it takes the form

$$-p(k)u^{(K)}(k+1) + p(k+1)u^{(K)}(k) = e(k), \quad k \in \mathbb{N}(0, K) \tag{8.9.8}$$

and the system (8.8.7), (8.8.8) becomes

(8.9.9) $$p(k+1) = -\frac{a_1(k-1)}{a_2(k-1)}p(k) - \frac{a_0(k-1)}{a_2(k-1)}p(k-1),$$

$$p(0) = 0, \ p(1) = 1$$

(8.9.10) $$e(k) = \frac{a_0(k-1)}{a_2(k-1)}e(k-1) - \frac{b(k-1)}{a_2(k-1)}p(k), \quad e(0) = u(0),$$

$$k \in \mathbb{N}(1, K).$$

This system (8.9.9), (8.9.10) is solved in the forward direction to compute $p(k)$, $k \in \mathbb{N}(0, K+1)$ and $e(k)$, $k \in \mathbb{N}(0, K)$. The solution $u^{(K)}(k)$ of (8.9.5), (8.9.6) is then obtained by recursing backward (8.9.8) from the known $u^{(K)}(K+1) = 0$.

Thus, the method of forward sweep and Olver's algorithm are theoretically the same. However, (8.8.7) is a nonlinear first order difference equation while (8.9.9) is a second order linear equation.

In (8.9.8) let $K$ to be $K+1$, so that it takes the form

(8.9.11) $$-p(k)u^{(K+1)}(k+1)+p(k+1)u^{(K+1)}(k) = e(k), \quad k \in \mathbb{N}(0, K+1).$$

Subtracting (8.9.8) from (8.9.11) to obtain

$$\left[u^{(K+1)}(k) - u^{(K)}(k)\right] = \frac{p(k)}{p(k+1)}\left[u^{(K+1)}(k+1) - u^{(K)}(k+1)\right],$$

$$k \in \mathbb{N}(0, K)$$

which has the solution

$$\left[u^{(K+1)}(k) - u^{(K)}(k)\right] = \frac{p(k)}{p(K+1)}\left[u^{(K+1)}(K+1) - u^{(K)}(K+1)\right].$$

However, since $u^{(K)}(K+1) = 0$ and $u^{(K+1)}(K+1) = \dfrac{e(K+1)}{p(K+2)}$, we find

(8.9.12) $$\left[u^{(K+1)}(k) - u^{(K)}(k)\right] = \frac{p(k)}{p(K+1)}\frac{e(K+1)}{p(K+2)}.$$

Suppose that we wish to compute the minimal solution $u(k)$, $k \in \mathbb{N}(0, L)$ of (8.9.4) to $d$ decimal places for given values of the integers $L$ and $d$. The recurrence relations for $p(k)$ and $e(k)$, i.e. (8.9.9) and (8.9.10) are first applied for $k = 1, 2, \cdots, L, L+1, \cdots$ until a value of $k$ is reached for which

$$\left| \frac{p(L)e(k+1)}{p(k+1)p(k+2)} \right| < \frac{1}{2} \times 10^{-d}$$

and then $K$ is taken as $k-1$ and finally $u^{(K)}(K+1)$ is set equal to 0.

## 8.10. Problems

**8.10.1.** Consider the difference equation

$$u(k+1) - 2u(k) + u(k-1) = \phi_k, \quad k \in \mathbb{N}(1, K) \tag{8.10.1}$$

together with the boundary conditions (1.6.14). Show that

(i) the problem (8.10.1), (1.6.14) is equivalent to

$$u(k) = A + \frac{B-A}{K+1}k + \sum_{\ell=1}^{K} g(k,\ell)\phi_\ell, \tag{8.10.2}$$

where

$$g(k,\ell) = -\frac{1}{K+1}\begin{cases} (K-k+1)\ell, & 0 \le \ell \le k-1 \\ (K-\ell+1)k, & k \le \ell \le K+1 \end{cases} \tag{8.10.3}$$

(ii) the function $g(k,\ell) \le 0$ and

$$\sum_{\ell=1}^{K} -g(k,\ell) = \frac{k(K-k+1)}{2} \le \frac{(K+1)^2}{8} \tag{8.10.4}$$

$$\sum_{\ell=1}^{K} -g(k,\ell) \le \frac{(K+1)^2}{2\pi} \sin\frac{k\pi}{K+1} \tag{8.10.5}$$

$$\sum_{\ell=1}^{K} -g(k,\ell)\sin\frac{(\ell-1)\pi}{K+1} = \frac{1}{4\sin^2\frac{\pi}{2(K+1)}} \times \left[\sin\frac{(k-1)\pi}{K+1} + \sin\frac{\pi}{K+1} - \frac{2k}{K+1}\sin\frac{\pi}{K+1}\right] \tag{8.10.6}$$

$$\sum_{\ell=1}^{K} -g(k,\ell)\sin\frac{\ell\pi}{K+1} = \frac{1}{4\sin^2\frac{\pi}{2(K+1)}}\sin\frac{k\pi}{K+1} \tag{8.10.7}$$

$$(8.10.8)\qquad \sum_{\ell=1}^{K} -g(k,\ell)\sin\frac{(\ell+1)\pi}{K+1} = \frac{1}{4\sin^2\frac{\pi}{2(K+1)}} \times$$

$$\left[\sin\frac{(k+1)\pi}{K+1} - \sin\frac{\pi}{K+1} + \frac{2k}{K+1}\sin\frac{\pi}{K+1}\right].$$

**8.10.2.** Use Problem 8.10.1 to show that the discrete boundary value problem (8.2.22), (1.6.14) has a unique solution provided

$$(8.10.9)\qquad \theta = \frac{h^2}{12}\frac{10+2\cos\frac{\pi}{K+1}}{4\sin^2\frac{\pi}{2(K+1)}}\max_{a\le t\le b}|f(t)| < 1.$$

Further, show that the inequality (8.10.9) is best possible in the sense that if $\theta = 1$ then there are problems for which existence or uniqueness or both fail.

**8.10.3.** Show that the boundary value problem $(8.8.1)_k$, (1.6.14) where $a_0(k) > 0$, $a_2(k) > 0$, $a_1(k) > (a_0(k) + a_2(k)) + \delta$ $(\delta > 0)$, has a unique solution $u(k)$, which satisfies the inequality $u(k) \le \max\left\{|A|, |B|, \frac{1}{\delta}\max_{1\le k\le K}|b(k)|\right\}$.

**8.10.4.** Let in system (1.2.12) the matrix $\mathcal{A}(k)$ be periodic of period $K$. Show that for the boundary value problem (1.2.12), (1.5.6) the Green's matrix $\mathcal{M}(k,\ell)$ defined in (8.1.8) can be written as

$$(8.10.10)\quad \mathcal{M}(k,\ell) = \begin{cases} \mathcal{U}(k,0)\,(\mathcal{I}-\mathcal{U}(K,0))^{-1}\mathcal{U}^{-1}(\ell,0), & 1\le \ell\le k \\ \mathcal{U}(k+K,0)\,(\mathcal{I}-\mathcal{U}(K,0))^{-1}\mathcal{U}^{-1}(\ell,0), & k+1\le \ell\le K. \end{cases}$$

**8.10.5.** Let $u_1(k)$ be a minimal solution of the difference equation (1.2.4). Show that (1.2.4) is unstable for $u_1(k)$.

**8.10.6.** Consider the forward sweep for the boundary value problem

$$(8.10.11)\quad u(k-1) - 2\lambda u(k) + u(k+1) = -b(k), \quad \lambda > 1, \quad k \in \mathbb{N}(1,K)$$

$$(8.10.12)\qquad u(0) = \beta, \quad u(K+1) = \delta$$

to show that the solution $\alpha(k)$ of the resulting problem (8.8.7) can be generated on $\mathbb{N}$, and $0 \le \alpha(k) < 1$. Further, $\alpha(k)$ tends to the smaller root of the characteristic polynomial of (8.10.11).

**8.10.7.** Use Olver's algorithm to find the minimal solution of the problem

$$200u(k) - 102u(k+1) + u(k+2) = 0, \quad u(0) = \sqrt{3}$$

on the interval $\mathbb{N}(0,30)$ correct to 10 significant digits.

**8.10.8.** The Anger Weber functions satisfy the nonhomogeneous equation

$$u(k) - 2(k+1)u(k+1) + u(k+2) = -\frac{2}{\pi}\left[1 + (-1)^k\right] \tag{8.10.13}$$

with $u(0) = -0.568656627$. Show that the equation (8.10.13) has a minimal solution. Further, use Olver's algorithm to find this minimal solution on the interval $\mathbb{N}(0,20)$ correct to 10 decimal places.

**8.10.9.** Show that the following nonhomogeneous equations have minimal solutions and for their computation Olver's algorithm converges

(i) $u(k) - 2(k+1)u(k+1) + u(k+2) = \dfrac{1}{\sqrt{\pi}2^{k+1}\Gamma\left(k+\frac{5}{2}\right)}$

(ii) $(2k+3)u(k) + 2\left[1 + 2(2k+1)(2k+3)\right]u(k+1) - (2k+1)u(k+2)$ $= -8e^{-1/2}$.

**8.10.10.** Consider the singular perturbation boundary value problem (2.16.24), (1.6.14). Show that the point $k = K+1$ is a boundary layer point. Further, deduce that its solution $u(k)$ can be uniformly approximated in $\mathbb{N}(0, K+1)$ by

$$u(k) = v(k) + \epsilon^{K+1-k}w(k) + O(\epsilon) \quad \text{as} \quad \epsilon \to 0,$$

where $v(k)$ and $w(k)$ are the solutions of

$$av(k+1) + v(k) = 0, \quad v(0) = A$$

and

$$w(k+2) + aw(k+1) = 0, \quad w(K+1) = B - v(K+1).$$

## 8.11. Notes

The existence and uniqueness of solutions of linear boundary value problems has been a subject matter of numerous number of papers, e.g. Agarwal [1,5], Denkowski [18], Halanay [25], Rodriguez [40–42], Sugiyama [49] and Szafraniec [50]. The results of Section 8.1 are from Agarwal [1,5] and in particular include several known criterion. These results will be used in the next chapter to study boundary value problems for nonlinear systems. The method of complementary functions and the method of particular solutions are discussed in Agarwal [2], whereas the Examples 8.2.1 – 8.2.5 are from Usmani and Agarwal [53]. The application of adjoint equations (transpose equations) to compute certain sums was indicated by Clenshaw [15] (cf. Section 2.11). The formulation of the method of adjoints and the Examples 8.4.1 8.4.5 are due to Agarwal and Nanda [6]. The method of chasing for solving second order continuous boundary value problems is originally due to Gel'fand, see Agarwal [4, and references therein]. The method of chasing discussed in Section 8.5 as well as the Examples 8.5.1 8.5.4 are adapted from Gupta and Agarwal [24]. Invariant imbedding methods are well known for solving continuous two point boundary value problems, e.g. Roberts and Shipman [39] and Scott [44, and references therein]. The formulation of this powerful technique in Section 8.6 is based on Agarwal and Usmani [7], whereas Section 8.7 is due to Angel and Kalaba [8]. The method of sweep in Section 8.8 is based on Tikhonov et. al. [51], also see Godunov and Ryabenki [23] and Trigiante and Sivasundaram [52]. Miller's and Olver's algorithms and their several theoretical as well as computational refinements are available in Arscott et. al. [9 12], Cruyssen [16,17], Gautschi [19 21], Mattheij [27 31], Oliver [32,33], Olver [34,35], Sadowski and Lozier [43], Scraton [45], Shintani [46] and Zahar [55]. Cash [14] and Wimp [54] provide an up to date account for these algorithms. For other related results see Gautschi [22], and Spigler and Vianello [47,48].

## 8.12. References

[1]. R.P. Agarwal, On multipoint boundary value problems for discrete equations, *J. Math. Anal. Appl.* **96**(1983), 520 534.

[2]. R.P. Agarwal, Initial value methods for discrete boundary value problems, *J. Math. Anal. Appl.* **100**(1984), 513 529.

[3]. R.P. Agarwal and R.C. Gupta, On the solution of Holt's problem, *BIT* **24**(1984), 342 346.

[4]. R.P. Agarwal, On Gel'fand's method of chasing for solving multipoint boundary value problems, in *Equadiff 6: Proc. of the International Conference on Differential Equations and Their Applications* eds. J. Vosmansky and M. Zlámal, J.E. Purkyne University, Brnö, (1985), 267 274.

[5]. R.P. Agarwal, Computational methods for discrete boundary value problems, *Appl. Math. Comp.* **18**(1986), 15 41.

[6]. R.P. Agarwal and T.R. Nanda, Two new algorithms for discrete boundary value problems, *J. Appl. Math. Stoc. Anal.* **3**(1990), 1 13.

[7]. R.P. Agarwal and R.A. Usmani, On the formulation of invariant imbedding method to solve multipoint discrete boundary value problems, *Applied Mathematics Letters* **4**(1991), 17 22.

[8]. E. Angel and R. Kalaba, A one sweep numerical method for vector matrix difference equations with two point boundary conditions, *J. Optimization Theory and Appl.* **6**(1970), 345 355.

[9]. F.M. Arscott, The connection between some differential-equation eigenvalue problems and some related difference equation problems, in *Proc. Fifth Manitoba Conf. on Numerical Math. and Computing* (1975), 211 212.

[10]. F.M. Arscott, R. Lacroix and W.T. Shymanski, A three term recursion and the computing of Mathieu functions, in *Proc. Eighth Manitoba Conf. on Numerical Math. and Computing* (1978), 107 115.

[11]. F.M. Arscott, A Riccati type transformation of linear difference equations, *Congressus Numerantium,* **30**(1981), 197 202.

[12]. F.M. Arscott, P.J. Taylor and R.V.M. Zahar, On the numerical construction of ellipsoidal wave functions, *Mathematics of Computation* **40**(1983), 367 380.

[13]. K. Balla and M. Vicsek, On the reduction of Holt's problem to a finite interval, *Numer. Math.* **51**(1987), 291 302.

[14]. J.R. Cash, *Stable Recursions, Academic Press,* London, 1979.

[15]. C.W. Clenshaw, A note on the summation of Chebyshev series, *MTAC* **9**(1955), 118 120.

[16]. P. Van der Cruyssen, Linear difference equations and generalized continued fractions, *Computing* **22**(1979), 269 278.

[17]. P. Van der Cruyssen, A reformulation of Olver's algorithm for the numerical solution of second order linear difference equations, *Numer. Math.* **32**(1979), 159 166.

[18]. Z. Denkowski, Linear problems for systems of difference equations, *Annales Polonici Mathematici* **24**(1970), 77 86.

[19]. W. Gautschi, Recursive computation of certain integrals, *J. Assoc. Comp. Mach.* **8**(1961), 21 40.

[20]. W. Gautschi, Computational aspect of three term recurrence relations, *SIAM Rev.* **9**(1967), 24 82.

[21]. W. Gautschi, Recursive computation of the repeated integrals of the error function, *Math. Comp.* **15**(1967), 227 232.

[22]. W. Gautschi, The computation of special functions by linear difference equations, in *Proceedings of the 2nd International Conference on Difference Equations and Applications,* Veszprém, Hungary, August 7 11, 1995, eds. S. Elaydi, et. al., *Gordon & Breach,* New York, 1996, 213 243.

[23]. S.K. Godunov and V.S. Ryabenki, *Theory of Difference Schemes,*

*North Holland*, Amsterdam, 1964.

[24]. R.C. Gupta and R.P. Agarwal, A new shooting method for multi point discrete boundary value problems, *J. Math. Anal. Appl.* **112**(1985), 210 220.

[25]. A. Halanay, Solutions periodiques et presque periodiques des systems d'equations aux differences finies, *Arch. Rational Mech. Anal.* **12**(1963), 134 149.

[26]. J.F. Holt, Numerical solution of nonlinear two point boundary value problems by finite difference methods, *Commun. ACM* **7**(1964), 366 373.

[27]. R.M.M. Mattheij, Accurate estimates of solutions of second order recursions, *Linear Algebra Appl.* **12**(1975), 29 54.

[28]. R.M.M. Mattheij and A. van der Sluis, Error estimates for Miller's algorithm, *Numer. Math.* **26**(1976), 61 78.

[29]. R.M.M. Mattheij, Characterizations of dominant and dominated solutions of linear recursions, *Numer. Math.* **35**(1980), 421 442.

[30]. R.M.M. Mattheij, Stable computation of solutions of unstable linear initial value recursions, *BIT* **22**(1982), 79 93.

[31]. R.M.M. Mattheij, Accurate estimates for the fundamental solutions of discrete boundary value problems, *J. Math. Anal. Appl.* **101**(1984), 444 464.

[32]. J. Oliver, Relative error propagation in the recursive solution of linear recurrence relations, *Numer. Math.* **9**(1967), 323 340.

[33]. J. Oliver, The numerical solution of linear recurrence relations, *Numer. Math.* **11**(1968), 349 360.

[34]. F.W. Olver, Error analysis of Miller's recurrence algorithm, *Math. Comp.* **18**(1964), 65 74.

[35]. F.W. Olver, Numerical solution of second order linear difference equations, *J. Res. Nat. Bur. Standards* **71B**(1967), 111 129.

[36]. M.R. Osborne, On shooting methods for boundary value problems, *J. Math. Anal. Appl.* **27**(1969), 417 433.

[37]. S.M. Roberts and J.S. Shipman, Multipoint solution of two-point boundary-value problems, *J. Optimization Theory and Appl.* **7**(1971), 301 318.

[38]. S.M. Roberts and J.S. Shipman, *Two Point Boundary value Problems: Shooting Methods, Elsevier*, New York, 1972.

[39]. S.M. Roberts and J.S. Shipman, On the formulation of invariant imbedding problems, *J. Optimization Theory and Appl.* **28**(1979), 525 547.

[40]. J. Rodriguez, On resonant discrete boundary value problem, *Applicable Analysis* **19**(1985), 265-274.

[41]. J. Rodriguez, Resonance in nonlinear discrete systems with nonlinear constraints, *Proc. 24th Conf. Decision and Control* **IEEE**(1985), 1738 1743.

[42]. J. Rodriguez, On nonlinear discrete boundary value problems, *J. Math. Anal. Appl.* **114**(1986), 398 408.

[43]. W.L. Sadowski and D.W. Lozier, Use of Olver's algorithm to evaluate certain definite integrals of plasma physics involving Chebyshev polynomials, *J. Comput. Phys.* **10**(1972), 607 613.

[44]. M.R. Scott, *Invariant Imbedding and its Applications to Ordinary Differential Equations, Addison Wesley Pub. Comp.*, Reading, Massachusetts, 1973.

[45]. R.E. Scraton, A modification of Miller's recurrence algorithm, *BIT* **12**(1972), 242 251.

[46]. H. Shintani, Note on Miller's recurrence algorithm, *J. Sci. Hiroshima Univ.* **29**(1965), 121 133.

[47]. R. Spigler and M. Vianello, Liouville Green approximations for a class of linear oscillatory difference equations of the second order, *J. Comput. Appl. Math.* **41**(1992), 105 116.

[48]. R. Spigler and M. Vianello, Discrete and continuous Liouville Green Olever approximations: A unified treatment via Volterra Stieljes integral equations, *SIAM J. Math. Anal.* **25**(1994), 720 732.

[49]. S. Sugiyama, On periodic solutions of difference equations, *Bull. Sci. Engg. Resh. Lab. Waseda Univ.* **52**(1971), 89 94.

[50]. F.H. Szafraniec, Existence theorems for discrete boundary value problems, *Annales Polonici Mathematici* **21**(1968), 73 83.

[51]. A.N. Tikhonov, A.B. Vasil'eva and A.G. Sveshnikov, *Differential Equations, Springer Verlag*, Berlin, 1985.

[52]. D. Trigiante and S. Sivasundaram, A new algorithm for unstable three term recurrence relations, *Appl. Math. Comp.* **22**(1987), 277 289.

[53]. R.A. Usmani and R.P. Agarwal, On the numerical solution of two point discrete boundary value problems, *Appl. Math. Comp.* **25**(1988), 247 264.

[54]. J. Wimp, *Computation with Recurrence Relations, Pitman Advanced Publishing Program*, Boston, 1984.

[55]. R.V.M. Zahar, Mathematical analysis of Miller's algorithm, *Numer. Math.* **27**(1977), 165 170.

# Chapter 9
# Boundary Value Problems for Nonlinear Systems

It is well understood that working with generalized normed spaces for the systems, one achieves better qualitative as well as quantitative information about the solutions than what can be inferred by considering the usual norms. In particular, the component wise study enlarges the domain of existence and uniqueness of solutions, weakens the convergence conditions for the iterative methods, and provides the sharper error estimates. In Section 9.1, we define generalized normed spaces, state two fixed point theorems and collect some properties of square matrices which are needed throughout this chapter. In Section 9.2, we prove the existence and uniqueness of the solutions of the problem (1.2.8), (1.5.1). For this problem we also provide a priori sufficient conditions which ensure the convergence of Picard's iterative scheme to its unique solution. This is followed by the computational aspects of Picard's scheme on a floating point system. This includes the necessary and sufficient conditions for the convergence of the approximate Picard's iterative scheme, sufficient conditions for an oscillatory state, and the stopping criterion. An application of Picard's method to perturbed boundary value problems is discussed in Section 9.6. Next, for the problem (1.2.8), (1.5.4) we introduce various partial orderings in the space $B(k_1, k_r)$ and use them to prove the monotonic convergence of the Picard's scheme to its solutions. The monotone convergence of periodic boundary value problems is presented in Section 9.8. The convergence of the Newton's and approximate Newton's methods for the problem (1.2.8), (1.5.1) is discussed in Sections 9.9 and 9.10 respectively. In Section 9.11 we shall show that various initial value methods of Chapter 8 can be used in an iterative way to solve the nonlinear boundary value problems. This is followed by the invariant imbedding method which converts a given two point nonlinear boundary value problem to its equivalent initial value problems.

## 9.1. Preliminary Results from Analysis

We shall consider the inequalities between two vectors in $\mathbb{R}^n$ component wise, whereas between $n \times n$ matrices element wise.

**Definition 9.1.1.** Let $E$ be a real vector space. A *generalized norm* on $E$ is a mapping $\|\cdot\|_G : E \to \mathbb{R}^n_+$ denoted by $\|\mathbf{u}\|_G = (\alpha_1(\mathbf{u}), \cdots, \alpha_n(\mathbf{u}))$ such that

(i) $\|\mathbf{u}\|_G \geq 0$, i.e. $\alpha_i(\mathbf{u}) \geq 0$ for all $i$

(ii) $\|\mathbf{u}\|_G = 0$ if and only if $\mathbf{u} = 0$, i.e. $\alpha_i(\mathbf{u}) = 0$ for all $i$ if and only if $\mathbf{u} = 0$

(iii) $\|\lambda\mathbf{u}\|_G = |\lambda|\|\mathbf{u}\|_G$, i.e. $\alpha_i(\lambda\mathbf{u}) = |\lambda|\alpha_i(\mathbf{u})$ for all $i$

(iv) $\|\mathbf{u}+\mathbf{v}\|_G \leq \|\mathbf{u}\|_G + \|\mathbf{v}\|_G$, i.e. $\alpha_i(\mathbf{u}+\mathbf{v}) \leq \alpha_i(\mathbf{u}) + \alpha_i(\mathbf{v})$ for all $i$.

The space $(E, \|\cdot\|_G)$ is called a *generalized normed space*. The topology in this space is given in the following way: For each $\mathbf{u} \in E$, and $\epsilon > 0$, let $B_\epsilon(\mathbf{u}) = \{\mathbf{v} \in E : \|\mathbf{v} - \mathbf{u}\|_G < \epsilon\mathbf{w}\}$, where $\mathbf{w} = (1, \cdots, 1) \in \mathbb{R}^n$. Then, $\{B_\epsilon(\mathbf{u}) : \mathbf{u} \in E, \epsilon > 0\}$ forms a basis for a topology on $E$. The same topology can be induced by the usual norm $\|\cdot\|$ which is defined as follows: If $\|\mathbf{u}\|_G = (\alpha_1(\mathbf{u}), \cdots, \alpha_n(\mathbf{u}))$, then $\|\mathbf{u}\| = \max\{\alpha_1(\mathbf{u}), \cdots, \alpha_n(\mathbf{u})\}$. Since the topology of the normed space $(E, \|\cdot\|)$ is given by the basis of neighborhoods $V_\epsilon(\mathbf{u}) = \{\mathbf{v} \in E : \|\mathbf{v} - \mathbf{u}\| < \epsilon\}$, $\mathbf{u} \in E$, $\epsilon > 0$ and $V_\epsilon(\mathbf{u}) = B_\epsilon(\mathbf{u})$, both the above definitions of norm define the same topology on $E$ and are equivalent. Thus, from the topological point of view there is no need for introducing the generalized norm. However, we have more flexibility when working with generalized spaces.

Before we state fixed point theorems in generalized normed spaces we collect the following well known properties of matrices which will be used frequently without further mention.

**1.** For any square matrix $\mathcal{A}$, $\lim_{m\to\infty} \mathcal{A}^m = 0$ if and only if $\rho(\mathcal{A}) < 1$, where $\rho(\mathcal{A})$ denotes the spectral radius of $\mathcal{A}$.

**2.** For any square matrix $\mathcal{A}$, $(\mathcal{I}-\mathcal{A})^{-1}$ exists and $(\mathcal{I}-\mathcal{A})^{-1} = \sum_{m=0}^{\infty} \mathcal{A}^m$ if $\rho(\mathcal{A}) < 1$. Also, if $\mathcal{A} \geq 0$, then $(\mathcal{I} - \mathcal{A})^{-1}$ exists and is nonnegative if and only if $\rho(\mathcal{A}) < 1$.

**3.** If $0 \leq \mathcal{B} \leq \mathcal{A}$ and $\rho(\mathcal{A}) < 1$, then $\rho(\mathcal{B}) < 1$.

**4.** If $\mathcal{A} \geq 0$ then $\rho(3\mathcal{A}) = 3\rho(\mathcal{A}) < 1$ if and only if $\rho(2\mathcal{A}(\mathcal{I}-\mathcal{A})^{-1}) < 1$.

**5.** (Toeplitz Lemma). For a given square matrix $\mathcal{A} \geq 0$ with $\rho(\mathcal{A}) < 1$ we define the sequence $\{\mathbf{s}^m\}$, where $\mathbf{s}^m = \sum_{i=0}^{m} \mathcal{A}^{m-i}\mathbf{d}^i$, $m = 0, 1, \cdots$. Then, $\lim_{m\to\infty} \mathbf{s}^m = 0$ if and only if the sequence $\{\mathbf{d}^m\} \to 0$.

**6.** For any natural norm $\|\cdot\|$, $\rho(\mathcal{A}) \leq \|\mathcal{A}\|$. Also, if $\rho(\mathcal{A}) < 1$ then a natural norm can be found such that $\|\mathcal{A}\| < 1$.

**Theorem 9.1.1** (Schauder's Fixed Point Theorem). Let $E$ be a generalized Banach space (complete generalized normed linear space) and let $F \subset E$ be closed and convex. If $\mathcal{T} : F \to F$ is completely continuous, then $\mathcal{T}$ has a fixed point.

**Theorem 9.1.2** (Contraction Mapping Theorem). Let $E$ be a generalized Banach space, and let for $\mathbf{r} \in R^n_+$, $\mathbf{r} > 0$, $\overline{S}(\mathbf{u}^0, \mathbf{r}) = \{\mathbf{u} \in E : \|\mathbf{u} - \mathbf{u}^0\|_G \leq \mathbf{r}\}$. Let $\mathcal{T}$ map $\overline{S}(\mathbf{u}^0, \mathbf{r})$ into $E$, and

(i) for all $\mathbf{u}$, $\mathbf{v} \in \overline{S}(\mathbf{u}^0, \mathbf{r})$, $\|\mathcal{T}\mathbf{u} - \mathcal{T}\mathbf{v}\|_G \leq \mathcal{K}\|\mathbf{u} - \mathbf{v}\|_G$, , where $\mathcal{K} \geq 0$ is an $n \times n$ matrix with $\rho(\mathcal{K}) < 1$

(ii) $\mathbf{r}^0 = (\mathcal{I} - \mathcal{K})^{-1}\|\mathcal{T}\mathbf{u}^0 - \mathbf{u}^0\|_G \leq \mathbf{r}$.

Then, the following hold

1. $\mathcal{T}$ has a fixed point $\mathbf{u}^*$ in $\overline{S}(\mathbf{u}^0, \mathbf{r}^0)$
2. $\mathbf{u}^*$ is the unique fixed point of $\mathcal{T}$ in $\overline{S}(\mathbf{u}^0, \mathbf{r})$
3. the sequence $\{\mathbf{u}^m\}$ defined by $\mathbf{u}^{m+1} = \mathcal{T}\mathbf{u}^m$, $m = 0, 1, \cdots$ converges to $\mathbf{u}^*$ with $\|\mathbf{u}^* - \mathbf{u}^m\|_G \leq \mathcal{K}^m\mathbf{r}^0$
4. for any $\mathbf{u} \in \overline{S}(\mathbf{u}^0, \mathbf{r}^0)$, $\mathbf{u}^* = \lim_{m\to\infty} \mathcal{T}^m\mathbf{u}$
5. any sequence $\{\overline{\mathbf{u}}^m\}$ such that $\overline{\mathbf{u}}^m \in \overline{S}(\mathbf{u}^m, \mathcal{K}^m\mathbf{r}^0)$, $m = 0, 1, \cdots$ converges to $\mathbf{u}^*$.

For the nonlinear system of algebraic equations $\mathbf{g}(\mathbf{u}) = 0$ Newton's method (cf. Example 1.4.8) is

$$\mathbf{u}^{k+1} = \mathbf{u}^k - (\mathbf{g}_\mathbf{u}(\mathbf{u}^k))^{-1}\mathbf{g}(\mathbf{u}^k), \quad k = 0, 1, \cdots. \tag{9.1.1}$$

The following result provides sufficient conditions for its convergence in the maximum norm.

**Theorem 9.1.3** (Kantorovich's Theorem). Suppose that

(i) for the initial approximation $\mathbf{u}^0$ to the solution of the system $\mathbf{g}(\mathbf{u}) = 0$, $(\mathbf{g}_\mathbf{u}(\mathbf{u}^0))^{-1}$ exists and $\|(\mathbf{g}_\mathbf{u}(\mathbf{u}^0))^{-1}\| \leq \beta_0$

(ii) $\mathbf{u}^0$ satisfies $\mathbf{g}(\mathbf{u}) = 0$ approximately in the sense that $\|(\mathbf{g}_\mathbf{u}(\mathbf{u}^0))^{-1}$ $\mathbf{g}(\mathbf{u}^0)\| \leq \eta_0$

(iii) in the region defined by inequality (9.1.2), the components of the vector $\mathbf{g}(\mathbf{u})$ are twice continuously differentiable with respect to the components of $\mathbf{u}$ and satisfy $\sum_{j,\ell}^{n} \left|\frac{\partial^2 g_i}{\partial u_j \partial u_\ell}\right| \leq K$ for each $i$

(iv) the constants $\beta_0$, $\eta_0$ and $K$ satisfy $h_0 = \beta_0\eta_0 K \leq 1/2$.

Then, the system $\mathbf{g}(\mathbf{u}) = 0$ has a solution $\mathbf{u}$ in the cube

$$\|\mathbf{u} - \mathbf{u}^0\| \leq \frac{1 - \sqrt{1 - 2h_0}}{h_0}\eta_0. \tag{9.1.2}$$

Moreover, the successive approximations $\mathbf{u}^k$ defined by (9.1.1) exist and converge to $\mathbf{u}$ and the speed of convergence may be estimated by the inequality

$$\|\mathbf{u}^k - \mathbf{u}\| \leq \frac{1}{2^{k-1}}(2h_0)^{2^k-1}\eta_0,$$

which shows that the rate of convergence for the Newton method is quadratic.

## 9.2. Existence and Uniqueness

Throughout, we shall consider the linear space $E$ as $B(a,b)$, i.e. the space of all real $n$ vector functions defined on $\mathbb{N}(a,b)$. For $\mathbf{u}(k) = (u_1(k),\cdots,u_n(k)) \in B(a,b)$ we shall denote by $|\mathbf{u}(k)| = (|u_1(k)|,\cdots, |u_n(k)|)$, $\alpha_i(\mathbf{u}) = \sup_{k\in\mathbb{N}(a,b)}|u_i(k)|$ and $\|\mathbf{u}\|_G = \left(\sup_{k\in\mathbb{N}(a,b)}|u_1(k)|, \cdots, \sup_{k\in\mathbb{N}(a,b)}|u_n(k)|\right)$. The space $B(a,b)$ equipped with this $\|\cdot\|_G$ is a generalized normed space. If $\mathbf{u}\in\mathbb{R}^n$, then obviously $\mathbf{u}\in B(a,b)$, and hence $|\mathbf{u}| = \|\mathbf{u}\|_G = (|u_1|,\cdots,|u_n|)$. The same notations will be used for the $n\times n$ matrix valued functions also.

**Theorem 9.2.1.** With respect to the difference system (1.2.8) on $\mathbb{N}(a,b-1)$ and the boundary condition (1.5.1) we assume that the following conditions hold

(i) there exist an $n\times n$ nonsingular matrix $\mathcal{A}(k)$ defined on $\mathbb{N}(a,b-1)$, and a linear operator $\mathcal{L}$ mapping $B(a,b)$ into $\mathbb{R}^n$ such that if $\mathcal{U}(k,a)$ is the principal fundamental matrix of (1.2.12) then the matrix $\mathcal{H}$ defined in (8.1.1) is nonsingular

(ii) there exist nonnegative matrices $\mathcal{M}^1$ and $\mathcal{M}^2$ such that $\|H^1\|_G \leq \mathcal{M}^1$, $\|H^2\|_G \leq \mathcal{M}^2$, where the operators $\mathcal{H}^1$ and $\mathcal{H}^2$ are defined in Theorem 8.1.1

(iii) there exist nonnegative vectors $\mathbf{r}^1$ and $\mathbf{r}^2$ such that for all $k\in\mathbb{N}(a,b-1)$ and $\mathbf{u}\in B_1(a,b) = \{\mathbf{u}(k)\in B(a,b) : \|\mathbf{u}\|_G \leq 2\mathbf{r}\}$, $|\mathbf{f}(k,\mathbf{u}) - \mathcal{A}(k)\mathbf{u}| \leq \mathbf{r}^1$ and $\|\mathcal{L}[\mathbf{u}] \pm \mathcal{F}[\mathbf{u}]\|_G \leq \mathbf{r}^2$

(iv) $\mathcal{M}^1\mathbf{r}^1 + \mathcal{M}^2\mathbf{r}^2 \leq 2\mathbf{r}$.

Further, the function $\mathbf{f}(k,\mathbf{u})$ is continuous on $\mathbb{N}(a,b-1)\times\mathbb{R}^n$. Then, the boundary value problem (1.2.8), (1.5.1) has a solution in $B_1(a,b)$.

**Proof.** Boundary value problem (1.2.8), (1.5.1) is the same as

$$(9.2.1)\quad \mathbf{u}(k+1) = \mathcal{A}(k)\mathbf{u}(k) + \mathbf{f}(k,\mathbf{u}(k)) - \mathcal{A}(k)\mathbf{u}(k), \quad k\in\mathbb{N}(a,b-1)$$

$$(9.2.2)\quad \mathcal{L}[\mathbf{u}] = \mathcal{L}[\mathbf{u}] \pm \mathcal{F}[\mathbf{u}].$$

Hence from Theorem 8.1.1 it follows that

$$(9.2.3)\qquad \mathbf{u}(k) = \mathcal{H}^1[\mathbf{f}(k,\mathbf{u}(k)) - \mathcal{A}(k)\mathbf{u}(k)] + \mathcal{H}^2[\mathcal{L}[\mathbf{u}] \pm \mathcal{F}[\mathbf{u}]].$$

The mapping $\mathcal{T} : B(a,b) \to B(a,b)$ defined by

$$(9.2.4) \qquad \mathcal{T}\mathbf{u}(k) \;=\; \mathcal{H}^1[\mathbf{f}(k,\mathbf{u}(k)) - \mathcal{A}(k)\mathbf{u}(k)] + \mathcal{H}^2[\mathcal{L}[\mathbf{u}] \pm \mathcal{F}[\mathbf{u}]]$$

is completely continuous. Obviously, any fixed point of (9.2.4) is a solution of (1.2.8), (1.5.1).

The set $B_1(a,b) \subset B(a,b)$ is a closed convex subset of the Banach space $B(a,b)$. Further, for $\mathbf{u}(k) \in B_1(a,b)$ it is easy to see that

$$\|\mathcal{T}\mathbf{u}\|_G \;\le\; \mathcal{M}^1\mathbf{r}^1 + \mathcal{M}^2\mathbf{r}^2 \;\le\; 2\mathbf{r}.$$

Thus, $\mathcal{T}$ maps $B_1(a,b)$ into itself, and from Theorem 9.1.1 it follows that $\mathcal{T}$ has a fixed point in $B_1(a,b)$. ∎

**Definition 9.2.1.** A function $\overline{\mathbf{u}}(k) \in B(a,b)$ is called an *approximate solution* of (1.2.8), (1.5.1) if there exist $\mathbf{d}^1$ and $\mathbf{d}^2$ nonnegative vectors such that for all $k \in \mathbb{N}(a,b-1)$, $|\overline{\mathbf{u}}(k+1) - \mathbf{f}(k,\overline{\mathbf{u}}(k))| \le \mathbf{d}^1$ and $\|\mathcal{F}[\overline{\mathbf{u}}]\|_G \le \mathbf{d}^2$, i.e. there exist a function $\mathbf{q}(k)$ on $\mathbb{N}(a,b-1)$ and a constant vector $\mathbf{l}^1$ such that $\overline{\mathbf{u}}(k+1) = \mathbf{f}(k,\overline{\mathbf{u}}(k)) + \mathbf{q}(k)$, $k \in \mathbb{N}(a,b-1)$ and $\mathcal{F}[\overline{\mathbf{u}}] = \mathbf{l}^1$ with $|\mathbf{q}(k)| \le \mathbf{d}^1$ and $\|\mathbf{l}^1\|_G \le \mathbf{d}^2$.

**Theorem 9.2.2.** With respect to the boundary value problem (1.2.8), (1.5.1) we assume that there exists an approximate solution $\overline{\mathbf{u}}(k)$ and

(i) the function $\mathbf{f}(k,\mathbf{u})$ is continuously differentiable with respect to $\mathbf{u}$ in $\mathbb{N}(a,b-1) \times \mathbb{R}^n$ and $\mathbf{f}_{\mathbf{u}}(k,\mathbf{u})$ represents the Jacobian matrix of $\mathbf{f}(k,\mathbf{u})$ with respect to $\mathbf{u}$, $\mathcal{F}[\mathbf{u}]$ is continuously Fréchet differentiable in $B(a,b)$ and $\mathcal{F}_{\mathbf{u}}[\mathbf{u}]$ denotes the linear operator mapping $B(a,b)$ to $\mathbb{R}^n$

(ii) condition (i) of Theorem 9.2.1

(iii) condition (ii) of Theorem 9.2.1

(iv) there exist nonnegative matrices $\mathcal{M}^3$ and $\mathcal{M}^4$, and a positive vector $\mathbf{r}$ such that for all $k \in \mathbb{N}(a,b-1)$ and $\mathbf{u} \in \overline{S}(\overline{\mathbf{u}},\mathbf{r}) = \{\mathbf{u}(k) \in B(a,b) : \|\mathbf{u} - \overline{\mathbf{u}}\|_G \le \mathbf{r}\}$, $|\mathbf{f}_{\mathbf{u}}(k,\mathbf{u}) - \mathcal{A}(k)| \le \mathcal{M}^3$ and $\|\mathcal{F}_{\mathbf{u}}[\mathbf{u}] \pm \mathcal{L}\|_G \le \mathcal{M}^4$

(v) $\mathcal{K} = \mathcal{M}^1\mathcal{M}^3 + \mathcal{M}^2\mathcal{M}^4$, $\rho(\mathcal{K}) < 1$ and $(\mathcal{I} - \mathcal{K})^{-1}(\mathcal{M}^1\mathbf{d}^1 + \mathcal{M}^2\mathbf{d}^2) \le \mathbf{r}$.

Then, the following hold

1. there exists a solution $\mathbf{u}^*(k)$ of (1.2.8), (1.5.1) in $\overline{\mathbf{S}}(\overline{\mathbf{u}},\mathbf{r}^0)$, where $\mathbf{r}^0 = (\mathcal{I} - \mathcal{K})^{-1}\|\mathbf{u}^1 - \overline{\mathbf{u}}\|_G$
2. $\mathbf{u}^*(k)$ is the unique solution of (1.2.8), (1.5.1) in $\overline{S}(\overline{\mathbf{u}},\mathbf{r})$
3. the *Picard iterative sequence* $\{\mathbf{u}^m(k)\}$ defined by

$$(9.2.5) \qquad \begin{aligned} \mathbf{u}^{m+1}(k) &= \mathcal{H}^1[\mathbf{f}(k,\mathbf{u}^m(k)) - \mathcal{A}(k)\mathbf{u}^m(k)] + \mathcal{H}^2[\mathcal{L}[\mathbf{u}^m] \pm \mathcal{F}[\mathbf{u}^m]] \\ \mathbf{u}^0(k) &= \overline{\mathbf{u}}(k), \quad m = 0,1,\cdots \end{aligned}$$

converges to $\mathbf{u}^*(k)$ with $\|\mathbf{u}^* - \mathbf{u}^m\|_G \le \mathcal{K}^m \mathbf{r}^0$

4. for $\mathbf{u}^0(k) = \mathbf{u}(k) \in \overline{S}(\overline{\mathbf{u}}, \mathbf{r}^0)$ the iterative process (9.2.5) converges to $\mathbf{u}^*(k)$

5. any sequence $\{\overline{\mathbf{u}}^m(k)\}$ such that $\overline{\mathbf{u}}^m(k) \in \overline{S}(\mathbf{u}^m, \mathcal{K}^m\mathbf{r}^0)$, $m = 0, 1, \cdots$ converges to $\mathbf{u}^*(k)$.

**Proof.** From the Definition 9.2.1, the approximate solution $\overline{\mathbf{u}}(k)$ satisfies

$$\begin{aligned}\overline{\mathbf{u}}(k+1) &= \mathcal{A}(k)\overline{\mathbf{u}}(k) + \mathbf{f}(k, \overline{\mathbf{u}}(k)) + \mathbf{q}(k) - \mathcal{A}(k)\overline{\mathbf{u}}(k) \\ \mathcal{L}[\overline{\mathbf{u}}] &= \mathcal{L}[\overline{\mathbf{u}}] \pm \mathcal{F}[\overline{\mathbf{u}}] \mp \mathbf{l}^1.\end{aligned}$$

Thus, from Theorem 8.1.1 it follows that

$$\overline{\mathbf{u}}(k) = \mathcal{H}^1[\mathbf{f}(k, \overline{u}(k)) + \mathbf{q}(k) - \mathcal{A}(k)\overline{\mathbf{u}}(k)] + H^2[\mathcal{L}[\overline{\mathbf{u}}] \pm \mathcal{F}[\overline{\mathbf{u}}] \mp \mathbf{l}^1]. \tag{9.2.6}$$

We shall show that the operator $\mathcal{T} : \overline{S}(\overline{\mathbf{u}}, \mathbf{r}) \to B(a, b)$ defined in (9.2.4) satisfies the conditions of Theorem 9.1.2. For this, let $\mathbf{u}(k)$, $\mathbf{v}(k) \in \overline{S}(\overline{\mathbf{u}}, \mathbf{r})$, then from (9.2.4) we have

$$\begin{aligned}\mathcal{T}\mathbf{u}(k) - \mathcal{T}\mathbf{v}(k) &= \mathcal{H}^1[\mathbf{f}(k, \mathbf{u}(k)) - \mathbf{f}(k, \mathbf{v}(k)) - \mathcal{A}(k)(\mathbf{u}(k) - \mathbf{v}(k))] \\ &\quad + \mathcal{H}^2[\mathcal{L}[\mathbf{u} - \mathbf{v}] \pm (\mathcal{F}[\mathbf{u}] - \mathcal{F}[\mathbf{v}])] \\ &= \mathcal{H}^1\left[\int_0^1 [\mathbf{f}_{\mathbf{u}}(k, \mathbf{v}(k) + \theta_1(\mathbf{u}(k) - \mathbf{v}(k))) - \mathcal{A}(k)]\right. \\ &\quad \left. \times\, (\mathbf{u}(k) - \mathbf{v}(k))d\theta_1\right] \\ &\quad + H^2\left[\int_0^1 [\mathcal{L} \pm \mathcal{F}_{\mathbf{u}}[\mathbf{v} + \theta_2(\mathbf{u} - \mathbf{v})]]\,[\mathbf{u} - \mathbf{v}]d\theta_2\right]\end{aligned}$$

and hence from (iii) and (iv) and the fact that $\mathbf{v}(k) + \theta_i(\mathbf{u}(k) - \mathbf{v}(k)) \in \overline{S}(\overline{\mathbf{u}}, \mathbf{r})$, $i = 1, 2$ we obtain

$$\|\mathcal{T}\mathbf{u} - \mathcal{T}\mathbf{v}\|_G \le (\mathcal{M}^1\mathcal{M}^3 + \mathcal{M}^2\mathcal{M}^4)\|\mathbf{u} - \mathbf{v}\|_G = \mathcal{K}\|\mathbf{u} - \mathbf{v}\|_G.$$

Next from (9.2.6) and (9.2.4), we get

$$\mathcal{T}\overline{\mathbf{u}}(k) - \overline{\mathbf{u}}(k) = \mathcal{T}\mathbf{u}^0(k) - \mathbf{u}^0(k) = \mathcal{H}^1[-\mathbf{q}(k)] + \mathcal{H}^2[\pm\mathbf{l}^1]$$

and hence from the Definition 9.2.1 it follows that

$$\|\mathcal{T}\mathbf{u}^0 - \mathbf{u}^0\|_G \le \mathcal{M}^1\mathbf{d}^1 + \mathcal{M}^2\mathbf{d}^2. \tag{9.2.7}$$

Thus, from (v) we get $\mathbf{r}^0 = (\mathcal{I} - \mathcal{K})^{-1}\|\mathbf{u}^1 - \mathbf{u}^0\|_G \le (\mathcal{I} - \mathcal{K})^{-1}(\mathcal{M}^1\mathbf{d}^1 + \mathcal{M}^2\mathbf{d}^2) \le \mathbf{r}$.

Hence, the conditions of Theorem 9.1.2 are satisfied and the conclusions 1 5 follow. ∎

**Remark 9.2.1.** From the conclusion 3 and (9.2.7), we have

$$\|\mathbf{u}^* - \overline{\mathbf{u}}\|_G \ \le\ (\mathcal{I}-\mathcal{K})^{-1}\|\mathbf{u}^1 - \mathbf{u}^0\|_G \ \le\ (\mathcal{I}-\mathcal{K})^{-1}(\mathcal{M}^1\mathbf{d}^1 + \mathcal{M}^2\mathbf{d}^2).$$

**Definition 9.2.2.** Any solution $\hat{\mathbf{u}}(k) \in B(a,b)$ of (1.2.8), (1.5.1) is called *isolated* if $\mathcal{F}_{\mathbf{u}}[\hat{\mathbf{u}}][\mathcal{U}(k,a)]$ is nonsingular, where $\mathcal{U}(k,a)$ is the principal fundamental matrix solution of the variational system $\mathbf{u}(k+1) = \mathbf{f}_{\mathbf{u}}(k,\hat{\mathbf{u}}(k))\mathbf{u}(k)$, $k \in \mathbb{N}(a,b-1)$.

**Theorem 9.2.3.** Let $\hat{\mathbf{u}}(k)$ be an isolated solution of (1.2.8), (1.5.1). Then, there is no other solution of (1.2.8), (1.5.1) in a sufficiently small neighborhood of $\hat{\mathbf{u}}(k)$.

**Proof.** Let $\mathcal{U}(k,a)$ be as in the Definition 9.2.2. For this $\mathcal{U}(k,a)$ there exist nonnegative $n\times n$ matrices $\overline{\mathcal{M}}^1$ and $\overline{\mathcal{M}}^2$ such that $\|\mathcal{H}^1\|_G \le \overline{\mathcal{M}}^1$ and $\|\mathcal{H}^2\|_G \le \overline{\mathcal{M}}^2$, where $\mathcal{H}^1$ and $\mathcal{H}^2$ are defined in Theorem 8.1.1. Since $\mathbf{f}_{\mathbf{u}}(k,\mathbf{u})$ and $\mathcal{F}_{\mathbf{u}}[\mathbf{u}]$ are continuous, there exists a positive vector $\mathbf{r}^3$ such that for all $k \in \mathbb{N}(a,b-1)$ and $\mathbf{u} \in \overline{S}(\hat{\mathbf{u}},\mathbf{r}^3)$ we have $|\mathbf{f}_{\mathbf{u}}(k,\mathbf{u}) - \mathbf{f}_{\mathbf{u}}(k,\hat{\mathbf{u}})| \le \mathcal{M}^5$ and $\|\mathcal{F}_{\mathbf{u}}[\mathbf{u}] - \mathcal{F}_{\mathbf{u}}[\hat{\mathbf{u}}]\|_G \le \mathcal{M}^6$, where $\mathcal{M}^5$ and $\mathcal{M}^6$ are nonnegative $n\times n$ matrices such that $\rho\left(\overline{\mathcal{M}}^1\mathcal{M}^5 + \overline{\mathcal{M}}^2\mathcal{M}^6\right) < 1$.

Let $\hat{\mathbf{u}}^*(k)$ be any other solution of (1.2.8), (1.5.1). Then, for $\mathbf{u}(k) = \hat{\mathbf{u}}(k) - \hat{\mathbf{u}}^*(k)$, we find

$$\begin{aligned}(9.2.8)\qquad \mathbf{u}(k+1) &= \mathbf{f}(k,\hat{\mathbf{u}}(k)) - \mathbf{f}(k,\hat{\mathbf{u}}^*(k))\\ &= \int_0^1 [\mathbf{f}_{\mathbf{u}}(k,\hat{\mathbf{u}}^*(k) + \theta_3(\hat{\mathbf{u}}(k) - \hat{\mathbf{u}}^*(k)))]\,\mathbf{u}(k)d\theta_3\end{aligned}$$

and

$$(9.2.9)\qquad 0 \ = \ \mathcal{F}[\hat{\mathbf{u}}] - \mathcal{F}[\hat{\mathbf{u}}^*] \ = \ \int_0^1 \mathcal{F}_{\mathbf{u}}\left[\hat{\mathbf{u}}^* + \theta_4(\hat{\mathbf{u}} - \hat{\mathbf{u}}^*)\right][\mathbf{u}]d\theta_4.$$

From Theorem 8.1.1, the solution of (9.2.8), (9.2.9) can be written as

$$\begin{aligned}(9.2.10)\ \mathbf{u}(k) = \mathcal{H}^1&\left[\int_0^1 [\mathbf{f}_{\mathbf{u}}(k,\hat{\mathbf{u}}^*(k)+\theta_3(\hat{\mathbf{u}}(k)-\hat{\mathbf{u}}^*(k))) - \mathbf{f}_{\mathbf{u}}(k,\hat{\mathbf{u}}(k))]\mathbf{u}(k)d\theta_3\right]\\ &+\mathcal{H}^2\left[-\int_0^1 \left[\mathcal{F}_{\mathbf{u}}\left[\hat{\mathbf{u}}^* + \theta_4(\hat{\mathbf{u}} - \hat{\mathbf{u}}^*)\right] - \mathcal{F}_{\mathbf{u}}[\hat{\mathbf{u}}]\right][\mathbf{u}]d\theta_4\right].\end{aligned}$$

Since $\hat{\mathbf{u}}^*(k) + \theta_i(\hat{\mathbf{u}}(k) - \hat{\mathbf{u}}^*(k)) \in \overline{S}(\hat{\mathbf{u}},\mathbf{r}^3)$, $i = 3,4$ equation (9.2.10) provides

$$\|\mathbf{u}\|_G \ \le\ (\overline{\mathcal{M}}^1\mathcal{M}^5 + \overline{\mathcal{M}}^2\mathcal{M}^6)\|\mathbf{u}\|_G$$

and from $\rho\left(\overline{\mathcal{M}}^1\mathcal{M}^5+\overline{\mathcal{M}}^2\mathcal{M}^6\right)<1$, we get $\|\mathbf{u}\|_G\le 0$, which is not true, and hence $\hat{\mathbf{u}}(k)=\hat{\mathbf{u}}^*(k)$. ∎

**Theorem 9.2.4.** The solution $\mathbf{u}^*(k)$ of (1.2.8), (1.5.1) obtained in Theorem 9.2.2 is an isolated solution.

**Proof.** If not, then there exists a nonzero vector $\mathbf{p}$ such that $\mathcal{F}_{\mathbf{u}}[\mathbf{u}^*]$ $[\mathcal{U}(k,a)]\mathbf{p}=0$, where $\mathcal{U}(k,a)$ is the fundamental matrix solution of $\mathbf{u}(k+1)=\mathcal{F}_{\mathbf{u}}(k,\mathbf{u}^*(k))\mathbf{u}(k)$.

Let $\mathbf{z}(k)=\mathcal{U}(k,a)\mathbf{p}$, so that

$$\mathbf{z}(k+1)=\mathbf{f}_{\mathbf{u}}(k,\mathbf{u}^*(k))\mathbf{z}(k),\quad \mathcal{F}_{\mathbf{u}}[\mathbf{u}^*][\mathbf{z}] = 0. \tag{9.2.11}$$

From Theorem 8.1.1, the solution $\mathbf{z}(k)$ of the problem (9.2.11) can be written as

$$\mathbf{z}(k) = \mathcal{H}^1\left[\mathbf{f}_{\mathbf{u}}(k,\mathbf{u}^*(k))\mathbf{z}(k)-\mathcal{A}(k)\mathbf{z}(k)\right]+\mathcal{H}^2\left[\mathcal{L}[\mathbf{z}]\pm\mathcal{F}_{\mathbf{u}}[\mathbf{u}^*][\mathbf{z}]\right].$$

Thus, from (iii) (v) of Theorem 9.2.2 it follows that

$$\|\mathbf{z}\|_G \le \left(\mathcal{M}^1\mathcal{M}^3+\mathcal{M}^2\mathcal{M}^4\right)\|\mathbf{z}\|_G = \mathcal{K}\|\mathbf{z}\|_G$$

or $\|\mathbf{z}\|_G\le 0$, which implies that $\mathcal{U}(k,a)\mathbf{p}\equiv 0$. Since $\mathcal{U}(k,a)$ is nonsingular, we find that $\mathbf{p}=0$. This contradiction proves that $\mathbf{u}^*(k)$ is isolated. ∎

**Example 9.2.1.** The boundary value problem

$$y'' = \beta e^{\alpha y},\quad y(0) = y(1) = 0 \tag{9.2.12}$$

arises in applications involving the diffusion of heat generated by positive temperature dependent sources. For instance, if $\alpha=1$ it arises in the analysis of Joule losses in electrically conducting solids, with $\beta$ representing the square of the constant current and $e^y$ the temperature dependent resistance, or in frictional heating with $\beta$ representing the square of the constant shear stress and $e^y$ the temperature dependent fluidity.

If $\alpha\beta=0$, then the problem (9.2.12) has a unique solution,

(i) if $\beta=0$, then $y(t)\equiv 0$

(ii) if $\alpha=0$, then $y(t)=(\beta/2)t(t-1)$.

If $\alpha\beta<0$, then the problem (9.2.12) has as many solutions as the number of roots of the equation $c=\sqrt{2|\alpha\beta|}\cosh(c/4)$, also for each such $c_i$ the solution is

$$y_i(t) = -\frac{2}{\alpha}\left\{\ln\left(\cosh\left(\frac{1}{2}c_i\left(t-\frac{1}{2}\right)\right)\right)-\ln\left(\cosh\left(\frac{1}{4}c_i\right)\right)\right\}. \tag{9.2.13}$$

From the equation $c = \sqrt{2|\alpha\beta|}\cosh(c/4)$ it follows that if

$$\sqrt{\frac{|\alpha\beta|}{8}}\min_{c\geq 0}\frac{\cosh(c/4)}{(c/4)} \begin{array}{ll} <1, & (9.2.12) \text{ has two solutions} \\ =1, & (9.2.12) \text{ has one solution} \\ >1, & (9.2.12) \text{ has no solution.} \end{array}$$

If $\alpha\beta > 0$, then the problem (9.2.12) has a unique solution

$$(9.2.14) \qquad y_1(t) = \frac{2}{\alpha}\ln\left(c_1\Big/\cos\left(\frac{1}{2}c_1\left(t-\frac{1}{2}\right)\right)\right) - \frac{1}{\alpha}\ln(2\alpha\beta),$$

where $(1/4)c_1 \in (-\pi/2, \pi/2)$ is the root of the equation $c/4 = \sqrt{\alpha\beta/8}$ $\cos(c/4)$.

For the problem (9.2.12) we consider its discrete analog as

$$(9.2.15) \qquad \begin{array}{l} u(k+1) - 2u(k) + u(k-1) = \dfrac{\beta}{(K+1)^2}e^{\alpha u(k)}, \quad k \in \mathbb{N}(1, K) \\ u(0) = u(K+1) = 0, \end{array}$$

which is in system form can be written as

$$(9.2.16) \qquad \begin{bmatrix} u_1(k+1) \\ u_2(k+1) \end{bmatrix} = \begin{bmatrix} 0 & 1 \\ -1 & 2 \end{bmatrix}\begin{bmatrix} u_1(k) \\ u_2(k) \end{bmatrix} + \begin{bmatrix} 0 \\ \beta/(K+1)^2 e^{\alpha u_2(k)} \end{bmatrix},$$

$$k \in \mathbb{N}(0, K-1)$$

$$(9.2.17) \qquad \begin{bmatrix} 1 & 0 \\ 0 & 0 \end{bmatrix}\begin{bmatrix} u_1(0) \\ u_2(0) \end{bmatrix} + \begin{bmatrix} 0 & 0 \\ 0 & 1 \end{bmatrix}\begin{bmatrix} u_1(K) \\ u_2(K) \end{bmatrix} = \begin{bmatrix} 0 \\ 0 \end{bmatrix}.$$

In (9.2.16), (9.2.17) we shall assume that $|\alpha| \leq 1$ and $|\beta| \leq 1$.

Let $\overline{\mathbf{u}}(k) \equiv 0$ be an approximate solution of (9.2.16), (9.2.17) so that

$$\mathbf{d}^1 = \begin{bmatrix} 0 \\ 1/(K+1)^2 \end{bmatrix}, \quad \mathbf{d}^2 = \begin{bmatrix} 0 \\ 0 \end{bmatrix}.$$

For this approximate solution, we take $\mathcal{A}(k) = \begin{bmatrix} 0 & 1 \\ -1 & 2 \end{bmatrix}$, and $\mathcal{L}[\mathbf{u}] = \begin{bmatrix} 1 & 0 \\ 0 & 0 \end{bmatrix}\begin{bmatrix} u_1(0) \\ u_2(0) \end{bmatrix} + \begin{bmatrix} 0 & 0 \\ 0 & 1 \end{bmatrix}\begin{bmatrix} u_1(K) \\ u_2(K) \end{bmatrix}$, then the following are easy to compute

$$\mathcal{U}(k,0) = \begin{bmatrix} 1-k & k \\ -k & 1+k \end{bmatrix}, \quad \mathcal{H} = \mathcal{L}[\mathcal{U}(k,0)] = \begin{bmatrix} 1 & 0 \\ -K & 1+K \end{bmatrix},$$

$$\mathcal{H}^1[\mathbf{b}(k)] = \frac{1}{1+K}\sum_{\ell=1}^{k}\begin{bmatrix}(1+\ell)(1+K-k) & -\ell(1+K-k)\\(1+\ell)(K-k) & \ell(K-k)\end{bmatrix}\mathbf{b}(\ell-1)$$
$$+\frac{1}{1+K}\sum_{\ell=k+1}^{K}\begin{bmatrix}k(K-\ell) & -k(1+K-\ell)\\(1+k)(K-\ell) & -(1+k)(1+K-\ell)\end{bmatrix}\mathbf{b}(\ell-1),$$
$$\mathcal{H}^2[\mathbf{l}] = \frac{1}{1+K}\begin{bmatrix}1+K-k & k\\K-k & 1+k\end{bmatrix}\mathbf{l}$$

and hence

$$\|\mathcal{H}^1\|_G$$
$$\le \frac{1}{2(1+K)}\max_{0\le k\le K}\begin{bmatrix}k(1+K)(1+K-k)+2k & k(1+K)(1+K-k)\\(1+k)(1+K)(K-k)-2(K-k) & (1+k)(1+K)(K-k)\end{bmatrix}$$

$$\le \frac{1}{8}\begin{bmatrix}(2+K)^2 & (1+K)^2\\(1+K)^2 & (1+K)^2\end{bmatrix} = \mathcal{M}^1,$$

$$\|\mathcal{H}^2\|_G \le \begin{bmatrix}1 & 1\\1 & 1\end{bmatrix} = \mathcal{M}^2.$$

Also, we have

$$|\mathbf{f_u}(k,\mathbf{u}) - \mathcal{A}(k)| \le \left\|\begin{bmatrix}0 & 0\\0 & \alpha\beta/(1+K)^2e^{\alpha u_2(k)}\end{bmatrix}\right\|_G$$
$$\le \begin{bmatrix}0 & 0\\0 & 1/(1+K)^2e^r\end{bmatrix} = \mathcal{M}^3$$

for all $k \in \mathbb{N}(0, K-1)$ and $\mathbf{u} \in \overline{S}(0,\mathbf{r})$, where $\mathbf{r} = (r,r)$, $r > 0$. Further, since $\|\mathcal{F}_\mathbf{u}[\mathbf{u}] - \mathcal{L}\|_G = 0$, we can take $\mathcal{M}^4 = 0$.

Thus, we find

$$\mathcal{K} = \mathcal{M}^1\mathcal{M}^3 + \mathcal{M}^2\mathcal{M}^4 = \frac{1}{8}\begin{bmatrix}0 & e^r\\0 & e^r\end{bmatrix}$$

and $\rho(\mathcal{K}) < 1$ provided $(1/8)e^r < 1$, i.e. $r \le 2.07944\cdots$.

Hence, in view of (9.2.7) the assumptions of Theorem 9.2.2 are satisfied provided $(\mathcal{I} - \mathcal{K})^{-1}(\mathcal{M}^1\mathbf{d}^1 + \mathcal{M}^2\mathbf{d}^2) \le \mathbf{r}$, which implies that

$$\frac{1}{8}\left(1 - \frac{1}{8}e^r\right)^{-1} \le r. \tag{9.2.18}$$

Inequality (9.2.18) is satisfied if $0.14614\cdots \le r \le 2.0154\cdots$. Thus,

1. there exists a solution $\mathbf{u}^*(k)$ of (9.2.16), (9.2.17) in $\overline{S}(0,\mathbf{r}^0) = \{\mathbf{u}(k) \in B(0,K) : \|\mathbf{u}\|_G \le 0.14614\cdots(1,1)^T\}$
2. $\mathbf{u}^*(k)$ is the unique solution of (9.2.16), (9.2.17) in $\overline{S}(0,\mathbf{r}) = \{\mathbf{u}(k) \in B(0,K) : \|\mathbf{u}\|_G \le 2.0154\cdots(1,1)^T\}$
3. if $r = 2.0154$, then the following error estimate holds

$$\|\mathbf{u}^* - \mathbf{u}^m\|_G \le (0.937966\cdots)^m(0.14614\cdots)(1,1)^T.$$

## 9.3. Approximate Picard's Iterates

In Theorem 9.2.2 the conclusion 3 ensures that the Picard iterative sequence $\{\mathbf{u}^m(k)\}$ obtained from (9.2.5) converges to the solution $\mathbf{u}^*(k)$ of (1.2.8), (1.5.1). However, in practical evaluation this sequence is approximated by the computed sequence, say, $\{\mathbf{v}^m(k)\}$. To find $\mathbf{v}^{m+1}(k)$ the function $\mathbf{f}$ is approximated by $\mathbf{f}^m$, and the operator $\mathcal{F}$ by $\mathcal{F}^m$. Therefore, the computed sequence $\{\mathbf{v}^m(k)\}$ satisfies the recurrence relation

(9.3.1)
$$\begin{aligned} \mathbf{v}^{m+1}(k) &= \mathcal{H}^1\left[\mathbf{f}^m(k,\mathbf{v}^m(k)) - \mathcal{A}(k)\mathbf{v}^m(k)\right] + \mathcal{H}^2\left[\mathcal{L}[\mathbf{v}^m] \pm \mathcal{F}^m[\mathbf{v}^m]\right] \\ \mathbf{v}^0(k) &= \mathbf{u}^0(k) = \overline{\mathbf{u}}(k), \quad m = 0,1,\cdots. \end{aligned}$$

With respect to $\mathbf{f}^m$ and $\mathcal{F}^m$, we shall assume the following:

**Condition** $(c_1)$. For all $k \in \mathbb{N}(a,b-1)$ and $\mathbf{v}^m(k)$ obtained from (9.3.1) the following inequalities hold

$$|\mathbf{f}(k,\mathbf{v}^m(k)) - \mathbf{f}^m(k,\mathbf{v}^m(k))| \le \mathcal{M}^7|\mathbf{f}(k,\mathbf{v}^m(k))| \quad (9.3.2)$$

$$\|\mathcal{F}[\mathbf{v}^m] - \mathcal{F}^m[\mathbf{v}^m]\|_G \le \mathcal{M}^8\|\mathcal{F}[\mathbf{v}^m]\|_G, \quad (9.3.3)$$

where $\mathcal{M}^7$ and $\mathcal{M}^8$ are $n \times n$ nonnegative matrices with $\rho(\mathcal{M}^7)$, $\rho(\mathcal{M}^8) < 1$. Inequalities (9.3.2) and (9.3.3) correspond to the relative error in approximating $\mathbf{f}$ and $\mathcal{F}$ by $\mathbf{f}^m$ and $\mathcal{F}^m$. Further, since $\rho(\mathcal{M}^7)$, $\rho(\mathcal{M}^8) < 1$ these inequalities provide that

$$|\mathbf{f}(k,\mathbf{v}^m(k))| \le (\mathcal{I} - \mathcal{M}^7)^{-1}|\mathbf{f}^m(k,\mathbf{v}^m(k))| \quad (9.3.4)$$

and

$$\|\mathcal{F}[\mathbf{v}^m]\|_G \le (\mathcal{I} - \mathcal{M}^8)^{-1}\|\mathcal{F}^m[\mathbf{v}^m]\|_G. \quad (9.3.5)$$

**Theorem 9.3.1.** With respect to the boundary value problem (1.2.8), (1.5.1) we assume that there exists an approximate solution $\overline{\mathbf{u}}(k)$ and conditions (i) (iv) of Theorem 9.2.2 are satisfied. Further, let condition $(c_1)$ be satisfied, and $\rho(\mathcal{K}^1) < 1$, where

$$\mathcal{K}^1 = \mathcal{M}^1(\mathcal{I}+\mathcal{M}^7)\mathcal{M}^3+\mathcal{M}^2(\mathcal{I}+\mathcal{M}^8)\mathcal{M}^4+\mathcal{M}^1\mathcal{M}^7 \sup_{k\in\mathbb{N}(a,b-1)} |\mathcal{A}(k)|+\mathcal{M}^2\mathcal{M}^8\|\mathcal{L}\|_G$$

and

$$\mathbf{r}^4 = (\mathcal{I}-\mathcal{K}^1)^{-1}(\mathcal{M}^1\mathbf{d}^1+\mathcal{M}^2\mathbf{d}^2+\mathcal{M}^1\mathcal{M}^7(\mathcal{I}-\mathcal{M}^7)^{-1} \sup_{k\in\mathbb{N}(a,b-1)} |\mathbf{f}^0(k,\overline{\mathbf{u}}(k))|$$
$$+\mathcal{M}^2\mathcal{M}^8(\mathcal{I}-\mathcal{M}^8)^{-1}\|\mathcal{F}^0[\overline{\mathbf{u}}]\|_G) \leq \mathbf{r}.$$

Then, the following hold

1. all the conclusions 1 5 of Theorem 9.2.2 hold
2. the sequence $\{\mathbf{v}^m(k)\}$ obtained from (9.3.1) remains in $\overline{S}(\overline{\mathbf{u}},\mathbf{r}^4)$
3. the sequence $\{\mathbf{v}^m(k)\}$ converges to $\mathbf{u}^*(k)$ the solution of (1.2.8), (1.5.1) if and only if $\lim_{m\to\infty}\mathbf{a}^m = 0$, where

(9.3.6)
$$\mathbf{a}^m = \|\mathbf{v}^{m+1}(k) - \mathcal{H}^1[\mathbf{f}(k,\mathbf{v}^m(k)) - \mathcal{A}(k)\mathbf{v}^m(k)] - \mathcal{H}^2[\mathcal{L}[\mathbf{v}^m] \pm \mathcal{F}[\mathbf{v}^m]]\|_G$$

and, also

$$(9.3.7)\ \|\mathbf{u}^*-\mathbf{v}^{m+1}\|_G \leq (\mathcal{I}-\mathcal{K})^{-1}\Big[\mathcal{M}^1\mathcal{M}^7(\mathcal{I}-\mathcal{M}^7)^{-1} \sup_{k\in\mathbb{N}(a,b-1)} |\mathbf{f}^m(k,\mathbf{v}^m(k))|$$
$$+\mathcal{M}^2\mathcal{M}^8(\mathcal{I}-\mathcal{M}^8)^{-1}\|\mathcal{F}^m[\mathbf{v}^m]\|_G + \mathcal{K}\|\mathbf{v}^{m+1}-\mathbf{v}^m\|_G\Big].$$

**Proof.** Since $\mathcal{K}^1 \geq \mathcal{K}$, $\rho(\mathcal{K}^1) < 1$ implies that $\rho(\mathcal{K}) < 1$, and obviously $\mathbf{r}^4 \geq (\mathcal{I}-\mathcal{K})^{-1}(\mathcal{M}^1\mathbf{d}^1+\mathcal{M}^2\mathbf{d}^2)$, the conditions of Theorem 9.2.2 are satisfied and conclusion 1 follows.

To prove 2, we note that $\overline{\mathbf{u}}(k) \in \overline{S}(\overline{\mathbf{u}},\mathbf{r}^4)$, and from (9.2.6) and (9.3.1) we find

$$\begin{aligned}\mathbf{v}^1(k) - \overline{\mathbf{u}}(k) &= \mathcal{H}^1[\mathbf{f}^0(k,\overline{\mathbf{u}}(k)) - \mathbf{f}(k,\overline{\mathbf{u}}(k)) - \mathbf{q}(k)] \\ &\quad +\mathcal{H}^2[\pm(\mathcal{F}^0[\overline{\mathbf{u}}] - \mathcal{F}[\overline{\mathbf{u}}]) \pm \mathbf{l}^1]\end{aligned}$$

and hence

$$\begin{aligned}&\|\mathbf{v}^1 - \overline{\mathbf{u}}\|_G \\ &\leq \mathcal{M}^1\mathcal{M}^7 \sup_{k\in\mathbb{N}(a,b-1)} |\mathbf{f}(k,\overline{\mathbf{u}}(k))| + \mathcal{M}^1\mathbf{d}^1 + \mathcal{M}^2\mathcal{M}^8\|\mathcal{F}[\overline{\mathbf{u}}]\|_G + \mathcal{M}^2\mathbf{d}^2 \\ &\leq \mathcal{M}^1\mathcal{M}^7(\mathcal{I}-\mathcal{M}^7)^{-1} \sup_{k\in\mathbb{N}(a,b-1)} |\mathbf{f}^0(k,\overline{\mathbf{u}}(k))| \\ &\quad + \mathcal{M}^2\mathcal{M}^8(\mathcal{I}-\mathcal{M}^8)^{-1}\|\mathcal{F}^0[\overline{\mathbf{u}}]\|_G + \mathcal{M}^1\mathbf{d}^1 + \mathcal{M}^2\mathbf{d}^2 \ \leq \mathbf{r}^4.\end{aligned}$$

Now we assume that $\mathbf{v}^m(k) \in \overline{S}(\overline{\mathbf{u}}, \mathbf{r}^4)$ and will show that $\mathbf{v}^{m+1}(k) \in \overline{S}(\overline{\mathbf{u}}, \mathbf{r}^4)$. From (9.2.6) and (9.3.1), we have

$$\begin{aligned}
&\mathbf{v}^{m+1}(k) - \overline{\mathbf{u}}(k) \\
&= \mathcal{H}^1 \left[\mathbf{f}^m(k, \mathbf{v}^m(k)) - \mathbf{f}(k, \overline{\mathbf{u}}(k)) - \mathcal{A}(k)(\mathbf{v}^m(k) - \overline{\mathbf{u}}(k)) - \mathbf{q}(k)\right] \\
&\quad + \mathcal{H}^2 \left[\mathcal{L}[\mathbf{v}^m - \overline{\mathbf{u}}] \pm (\mathcal{F}^m[\mathbf{v}^m] - \mathcal{F}[\overline{\mathbf{u}}]) \pm \mathbf{l}^1\right] \\
&= \mathcal{H}^1 \Bigg[\mathbf{f}^m(k, \mathbf{v}^m(k)) - \mathbf{f}(k, \mathbf{v}^m(k)) - \mathbf{q}(k) + \int_0^1 [\mathbf{f}_{\mathbf{u}}(k, \overline{\mathbf{u}}(k) \\
&\quad + \theta_5(\mathbf{v}^m(k) - \overline{\mathbf{u}}(k))) - \mathcal{A}(k)] (\mathbf{v}^m(k) - \overline{\mathbf{u}}(k)) d\theta_5\Bigg] \\
&\quad + \mathcal{H}^2 \Bigg[\pm(\mathcal{F}^m[\mathbf{v}^m] - \mathcal{F}[\mathbf{v}^m]) \pm \mathbf{l}^1 \pm \int_0^1 [\mathcal{F}_{\mathbf{u}}[\overline{\mathbf{u}} + \theta_6(\mathbf{v}^m - \overline{\mathbf{u}})] \pm \mathcal{L}] \\
&\qquad\qquad \times [\mathbf{v}^m - \overline{\mathbf{u}}] d\theta_6\Bigg]
\end{aligned}$$

and hence

$$\begin{aligned}
\|\mathbf{v}^{m+1} - \overline{\mathbf{u}}\|_G \;\le\; & \mathcal{M}^1 \left[\mathcal{M}^7 \sup_{k \in \mathbb{N}(a,b-1)} |\mathbf{f}(k, \mathbf{v}^m(k))| + \mathcal{M}^3 \mathbf{r}^4 + \mathbf{d}^1\right] \\
& + \mathcal{M}^2 \left[\mathcal{M}^8 \|\mathcal{F}[\mathbf{v}^m]\|_G + \mathcal{M}^4 \mathbf{r}^4 + \mathbf{d}^2\right].
\end{aligned}$$

Next since

$$\begin{aligned}
(9.3.8)\quad |\mathbf{f}(k, \mathbf{v}^m(k))| \;\le\; & |\mathbf{f}(k, \mathbf{v}^m(k)) - \mathbf{f}(k, \overline{\mathbf{u}}(k)) - \mathcal{A}(k)(\mathbf{v}^m(k) - \overline{\mathbf{u}}(k))| \\
& + |\mathbf{f}(k, \overline{\mathbf{u}}(k))| + |\mathcal{A}(k)|\mathbf{r}^4 \\
\le\; & \mathcal{M}^3 \mathbf{r}^4 + (\mathcal{I} - \mathcal{M}^7)^{-1} |\mathbf{f}^0(k, \overline{\mathbf{u}}(k))| + |\mathcal{A}(k)|\mathbf{r}^4
\end{aligned}$$

and, similarly

$$(9.3.9)\qquad \|\mathcal{F}[\mathbf{v}^m]\|_G \le \mathcal{M}^4 \mathbf{r}^4 + (\mathcal{I} - \mathcal{M}^8)^{-1} \|\mathcal{F}^0[\overline{\mathbf{u}}]\|_G + \|\mathcal{L}\|_G \mathbf{r}^4$$

it follows that

$$\begin{aligned}
\|\mathbf{v}^{m+1} - \overline{\mathbf{u}}\|_G \le \mathcal{K}^1 \mathbf{r}^4 + \Bigg(& \mathcal{M}^1 \mathbf{d}^1 + \mathcal{M}^2 \mathbf{d}^2 + \mathcal{M}^1 \mathcal{M}^7 (\mathcal{I} - \mathcal{M}^7)^{-1} \times \\
& \sup_{k \in \mathbb{N}(a,b-1)} |\mathbf{f}^0(k, \overline{\mathbf{u}}(k))| + \mathcal{M}^2 \mathcal{M}^8 (\mathcal{I} - \mathcal{M}^8)^{-1} \|\mathcal{F}^0[\overline{\mathbf{u}}]\|_G\Bigg) \\
\le \mathcal{K}^1 \mathbf{r}^4 + (\mathcal{I} - \mathcal{K}^1)\mathbf{r}^4 \;=\; & \mathbf{r}^4.
\end{aligned}$$

This completes the proof of the conclusion 2.

Next we shall prove 3. From the definitions of $\mathbf{u}^{m+1}(k)$ and $\mathbf{v}^{m+1}(k)$, we have

$$\begin{aligned}
&\mathbf{u}^{m+1}(k) - \mathbf{v}^{m+1}(k) \\
&= -\mathbf{v}^{m+1}(k) + \mathcal{H}^1[\mathbf{f}(k,\mathbf{v}^m(k)) - \mathcal{A}(k)\mathbf{v}^m(k)] \\
&\quad + \mathcal{H}^2[\mathcal{L}[\mathbf{v}^m] \pm \mathcal{F}[\mathbf{v}^m]] + \mathcal{H}^1[\mathbf{f}(k,\mathbf{u}^m(k)) - \mathbf{f}(k,\mathbf{v}^m(k)) \\
&\quad - \mathcal{A}(k)(\mathbf{u}^m(k) - \mathbf{v}^m(k))] + \mathcal{H}^2[\mathcal{L}[\mathbf{u}^m - \mathbf{v}^m] \pm (\mathcal{F}[\mathbf{u}^m] - \mathcal{F}[\mathbf{v}^m])]
\end{aligned}$$

and hence as in part 2, we find

$$\|\mathbf{u}^{m+1} - \mathbf{v}^{m+1}\|_G \le \mathbf{a}^m + (\mathcal{M}^1\mathcal{M}^3 + \mathcal{M}^2\mathcal{M}^4)\,\|\mathbf{u}^m - \mathbf{v}^m\|_G.$$

Using the fact that $\|\mathbf{u}^0 - \mathbf{v}^0\|_G = 0$, the above inequality implies that

$$\|\mathbf{u}^{m+1} - \mathbf{v}^{m+1}\|_G \le \sum_{i=0}^{m} \mathcal{K}^{m-i}\mathbf{a}^i.$$

Thus, from the triangle inequality, we get

$$\|\mathbf{u}^* - \mathbf{v}^{m+1}\|_G \le \sum_{i=0}^{m} \mathcal{K}^{m-i}\mathbf{a}^i + \|\mathbf{u}^* - \mathbf{u}^{m+1}\|_G. \tag{9.3.10}$$

In (9.3.10) Theorem 9.2.2 ensures that $\lim_{m\to\infty}\|\mathbf{u}^* - \mathbf{u}^{m+1}\|_G = 0$. Thus, the condition $\lim_{m\to\infty}\mathbf{a}^m = 0$ is necessary and sufficient for the convergence of the sequence $\{\mathbf{v}^m(k)\}$ to $\mathbf{u}^*(k)$ follows from the Toeplitz lemma.

Finally, we shall prove (9.3.7). For this, we have

$$\begin{aligned}
\mathbf{u}^*(k) - \mathbf{v}^{m+1}(k) &= \mathcal{H}^1[\mathbf{f}(k,\mathbf{u}^*(k)) - \mathbf{f}^m(k,\mathbf{v}^m(k)) - \mathcal{A}(k)(\mathbf{u}^*(k) - \mathbf{v}^m(k))] \\
&\quad + \mathcal{H}^2\left[\mathcal{L}[\mathbf{u}^* - \mathbf{v}^m] \pm (\mathcal{F}[\mathbf{u}^*] - \mathcal{F}^m[\mathbf{v}^m])\right]
\end{aligned}$$

and as in part 2, we find

$$\begin{aligned}
&\|\mathbf{u}^* - \mathbf{v}^{m+1}\|_G \\
&\le \mathcal{M}^1\left[\mathcal{M}^3\|\mathbf{u}^* - \mathbf{v}^m\|_G + \mathcal{M}^7(\mathcal{I} - \mathcal{M}^7)^{-1}\sup_{k\in\mathbf{N}(a,b-1)}|\mathbf{f}^m(k,\mathbf{v}^m(k))|\right] \\
&\quad + \mathcal{M}^2\left[\mathcal{M}^4\|\mathbf{u}^* - \mathbf{v}^m\|_G + \mathcal{M}^8(\mathcal{I} - \mathcal{M}^8)^{-1}\|\mathcal{F}^m[\mathbf{v}^m]\|_G\right] \\
&\le \mathcal{K}\|\mathbf{u}^* - \mathbf{v}^{m+1}\|_G + \left[\mathcal{M}^1\mathcal{M}^7(\mathcal{I} - \mathcal{M}^7)^{-1}\sup_{k\in\mathbf{N}(a,b-1)}|\mathbf{f}^m(k,\mathbf{v}^m(k))|\right. \\
&\quad \left. + \mathcal{M}^2\mathcal{M}^8(\mathcal{I} - \mathcal{M}^8)^{-1}\|\mathcal{F}^m[\mathbf{v}^m]\|_G + \mathcal{K}\|\mathbf{v}^{m+1} - \mathbf{v}^m\|_G\right],
\end{aligned}$$

which is the same as (9.3.7). ∎

In our next result with respect to $\mathbf{f}^m$ and $\mathcal{F}^m$ we shall assume the following :

**Condition** $(c_2)$. For all $k \in \mathbb{N}(a, b-1)$ and $\mathbf{v}^m(k)$ obtained from (9.3.1) the following inequalities hold

$$|\mathbf{f}(k, \mathbf{v}^m(k)) - \mathbf{f}^m(k, \mathbf{v}^m(k))| \leq \mathbf{r}^5 \tag{9.3.11}$$

$$\|\mathcal{F}[\mathbf{v}^m] - \mathcal{F}^m[\mathbf{v}^m]\|_G \leq \mathbf{r}^6, \tag{9.3.12}$$

where $\mathbf{r}^5$ and $\mathbf{r}^6$ are $n \times 1$ nonnegative vectors. Inequalities (9.3.11), (9.3.12) correspond to the absolute error in approximating $\mathbf{f}$ and $\mathcal{F}$ by $\mathbf{f}^m$ and $\mathcal{F}^m$.

**Theorem 9.3.2.** With respect to the boundary value problem (1.2.8), (1.5.1) we assume that there exists an approximate solution $\overline{\mathbf{u}}(k)$ and conditions (i) (iv) of Theorem 9.2.2 are satisfied. Further, let condition $(c_2)$ be satisfied, and $\rho(\mathcal{K}) < 1$, also

$$\mathbf{r}^7 = (\mathcal{I} - \mathcal{K})^{-1}(\mathcal{M}^1(\mathbf{r}^5 + \mathbf{d}^1) + \mathcal{M}^2(\mathbf{r}^6 + \mathbf{d}^2)) \leq \mathbf{r}.$$

Then, the following hold

1. all the conclusions 1 5 of Theorem 9.2.2 hold
2. the sequence $\{\mathbf{v}^m(k)\}$ obtained from (9.3.1) remains in $\overline{S}(\overline{\mathbf{u}}, \mathbf{r}^7)$
3. the condition $\lim_{m\to\infty} \mathbf{a}^m = 0$ is necessary and sufficient for the convergence of $\{\mathbf{v}^m(k)\}$ to the solution $\mathbf{u}^*(k)$ of (1.2.8), (1.5.1) where $\mathbf{a}^m$ are defined in (9.3.6), and

$$\|\mathbf{u}^* - \mathbf{v}^{m+1}\|_G \leq (\mathcal{I} - \mathcal{K})^{-1}(\mathcal{M}^1\mathbf{r}^5 + \mathcal{M}^2\mathbf{r}^6 + \mathcal{K}\|\mathbf{v}^{m+1} - \mathbf{v}^m\|_G).$$

**Proof.** The proof is contained in Theorem 9.3.1. ∎

## 9.4. Oscillatory State

When the sequence $\{\mathbf{v}^m(k)\}$ from (9.3.1) is constructed on a floating point system, then the mutual distances of two distinct $\mathbf{v}^m(k)$ cannot be smaller than a certain fixed positive constant. If the conditions of Theorem 9.3.1 or Theorem 9.3.2 are satisfied, then from the conclusion 2, the number of distinct $\mathbf{v}^m(k)$ must be finite. Thus, it is necessary that

$$\mathbf{v}^{m+\mu}(k) = \mathbf{v}^m(k) \tag{9.4.1}$$

for a certain $m$ and a positive integer $\mu$. Also, once (9.4.1) has happened, then $\mathbf{v}^{m+\mu+\nu}(k)=\mathbf{v}^{m+\nu}(k),\ \nu=0,1,\cdots$.

Hence, the sequence $\{\mathbf{v}^m(k)\}$ oscillates, taking $\mu$ values

$$\mathbf{v}^m(k),\mathbf{v}^{m+1}(k),\cdots,\mathbf{v}^{m+\mu-1}(k). \tag{9.4.2}$$

**Theorem 9.4.1.** Let the conditions of Theorem 9.3.1 be satisfied, and let the sequence $\{\mathbf{v}^m(k)\}$ be obtained from (9.3.1) on a floating point system. Then, the sequence $\{\mathbf{v}^m(k)\}$ oscillates, taking a finite number of values after a certain $m$, and for $\mathbf{v}^m(k)$ in such an oscillatory state it holds that

(9.4.3)

$$\|\mathbf{v}^m-\mathbf{u}^*\|_G \le (\mathcal{I}-\mathcal{K}^1)^{-1}\left[\mathcal{M}^1\mathcal{M}^7\sup_{k\in\mathbb{N}(a,b-1)}|\mathbf{f}(k,\mathbf{u}^*(k))|+\mathcal{M}^2\mathcal{M}^8\|\mathcal{F}[\mathbf{u}^*]\|_G\right].$$

**Proof.** Since we have already observed that the sequence $\{\mathbf{v}^m(k)\}$ obtained on a floating point system oscillates, we need to show that the inequality (9.4.3) holds. For this, as earlier, successively we have

$$\begin{aligned}
&\|\mathbf{v}^{m+1}-\mathbf{u}^{m+1}\|_G\\
&\le \mathcal{M}^1\left[\mathcal{M}^3\|\mathbf{v}^m-\mathbf{u}^m\|_G+\mathcal{M}^7\sup_{k\in\mathbb{N}(a,b-1)}|\mathbf{f}(k,\mathbf{v}^m(k))|\right]\\
&+\mathcal{M}^2\left[\mathcal{M}^4\|\mathbf{v}^m-\mathbf{u}^m\|_G+\mathcal{M}^8\|\mathcal{F}[\mathbf{v}^m]\|_G\right]\\
&\le \mathcal{M}^1\left[\mathcal{M}^3\|\mathbf{v}^m-\mathbf{u}^m\|_G+\mathcal{M}^7\left(\mathcal{M}^3\|\mathbf{v}^m-\mathbf{u}^m\|_G+\sup_{k\in\mathbb{N}(a,b-1)}|\mathbf{f}(k,\mathbf{u}^m(k))|\right.\right.\\
&\left.\left.+\sup_{k\in\mathbb{N}(a,b-1)}|\mathcal{A}(k)|\|\mathbf{v}^m-\mathbf{u}^m\|_G\right)\right]+\mathcal{M}^2\left[\mathcal{M}^4\|\mathbf{v}^m-\mathbf{u}^m\|_G\right.\\
&\left.+\mathcal{M}^8\left(\mathcal{M}^4\|\mathbf{v}^m-\mathbf{u}^m\|_G+\|\mathcal{F}[\mathbf{u}^m]\|_G+\|\mathcal{L}\|_G\|\mathbf{v}^m-\mathbf{u}^m\|_G\right)\right]\\
&=\mathcal{K}^1\|\mathbf{v}^m-\mathbf{u}^m\|_G+\mathcal{M}^1\mathcal{M}^7\sup_{k\in\mathbb{N}(a,b-1)}|\mathbf{f}(k,\mathbf{u}^m(k))|+\mathcal{M}^2\mathcal{M}^8\|\mathcal{F}[\mathbf{u}^m]\|_G\\
&\le\mathcal{K}^1\|\mathbf{v}^m-\mathbf{u}^m\|_G+\mathcal{M}^1\mathcal{M}^7\left(\mathcal{M}^3\|\mathbf{u}^m-\mathbf{u}^*\|_G+\sup_{k\in\mathbb{N}(a,b-1)}|\mathbf{f}(k,\mathbf{u}^*(k))|\right.\\
&\left.+\sup_{k\in\mathbb{N}(a,b-1)}|\mathcal{A}(k)|\|\mathbf{u}^m-\mathbf{u}^*\|_G\right)+\mathcal{M}^2\mathcal{M}^8\left(\mathcal{M}^4\|\mathbf{u}^m-\mathbf{u}^*\|_G\right.\\
&\left.+\|\mathcal{F}[\mathbf{u}^*]\|_G+\|\mathcal{L}\|_G\|\mathbf{u}^m-\mathbf{u}^*\|_G\right)
\end{aligned}$$

$$\leq \mathcal{K}^1\|\mathbf{v}^m-\mathbf{u}^m\|_G+\mathcal{M}^1\mathcal{M}^7\sup_{k\in\mathbb{N}(a,b-1)}|\mathbf{f}(k,\mathbf{u}^*(k))|+\mathcal{M}^2\mathcal{M}^8\|\mathcal{F}[\mathbf{u}^*]\|_G$$

$$+\left(\mathcal{M}^1\mathcal{M}^7\mathcal{M}^3+\mathcal{M}^1\mathcal{M}^7\sup_{k\in\mathbb{N}(a,b-1)}|\mathcal{A}(k)|+\mathcal{M}^2\mathcal{M}^8\mathcal{M}^4+\mathcal{M}^2\mathcal{M}^8\|\mathcal{L}\|_G\right)\mathcal{K}^m\mathbf{r}^0$$

From the above inequality, we find

$$(9.4.4)\quad \|\mathbf{v}^m-\mathbf{u}^m\|_G$$
$$\leq (\mathcal{I}-\mathcal{K}^1)^{-1}\left[\mathcal{M}^1\mathcal{M}^7\sup_{k\in\mathbb{N}(a,b-1)}|\mathbf{f}(k,\mathbf{u}^*(k))|+\mathcal{M}^2\mathcal{M}^8\|\mathcal{F}[\mathbf{u}^*]\|_G\right]$$
$$+\sum_{i=0}^{m-1}(\mathcal{K}^1)^{m-i-1}\left(\mathcal{M}^1\mathcal{M}^7\mathcal{M}^3+\mathcal{M}^1\mathcal{M}^7\sup_{k\in\mathbb{N}(a,b-1)}|\mathcal{A}(k)|\right.$$
$$\left.+\mathcal{M}^2\mathcal{M}^8\mathcal{M}^4+\mathcal{M}^2\mathcal{M}^8\|\mathcal{L}\|_G\right)\mathcal{K}^i\mathbf{r}^0 \;=\; \mathbf{d}^m,\quad \text{say.}$$

Let $\mathbf{v}^\ell(k)$ be one of the given in (9.4.2), then since $\mathbf{v}^{\ell+q\mu}(k)=\mathbf{v}^\ell(k)$, $q=0,1,\cdots$ from (9.4.4), we find

$$(9.4.5)\qquad \|\mathbf{v}^\ell-\mathbf{u}^{\ell+q\mu}\|_G\;\leq\;\mathbf{d}^{\ell+q\mu}.$$

In inequality (9.4.5) as $q\to\infty$, $\mathbf{u}^{\ell+q\mu}(k)\to\mathbf{u}^*(k)$, and the second term of $\mathbf{d}^{\ell+q\mu}$ tends to zero by the Toeplitz lemma. This completes the proof of (9.4.3). ∎

**Remark 9.4.1.** Since $\mathbf{u}^*(k)\in\overline{S}(\overline{\mathbf{u}},\mathbf{r}^0)$ the right side of (9.4.3) can easily be estimated. Further, if for all $m$, $\mathbf{f}^m=\overline{\mathbf{f}}$ and $\mathcal{F}^m=\overline{\mathcal{F}}$, then inequalities (9.3.4) and (9.3.5) can be used in (9.4.3), to obtain

$$\|\mathbf{v}^m-\mathbf{u}^*\|_G\leq(\mathcal{I}-\mathcal{K}^1)^{-1}\left[\mathcal{M}^1\mathcal{M}^7(\mathcal{I}-\mathcal{M}^7)^{-1}\sup_{k\in\mathbb{N}(a,b-1)}|\overline{\mathbf{f}}(k,\mathbf{u}^*(k))|\right.$$
$$\left.+\mathcal{M}^2\mathcal{M}^8(\mathcal{I}-\mathcal{M}^8)^{-1}\|\overline{\mathcal{F}}[\mathbf{u}^*]\|_G\right].$$

**Theorem 9.4.2.** Let the conditions of Theorem 9.3.2 be satisfied, and let the sequence $\{\mathbf{v}^m(k)\}$ be obtained from (9.3.1) on a floating point system. Then, the conclusion of Theorem 9.4.1 holds with (9.4.3) replaced by

$$\|\mathbf{v}^m-\mathbf{u}^*\|_G\;\leq\;(\mathcal{I}-\mathcal{K})^{-1}\left(\mathcal{M}^1\mathbf{r}^5+\mathcal{M}^2\mathbf{r}^6\right).$$

**Proof.** The proof is contained in the proof of Theorem 9.4.1. ∎

## 9.5. Stopping Criterion

In order to detect whether the sequence $\{\mathbf{v}^m(k)\}$ obtained from (9.3.1) on a floating point system has attained an oscillatory state, it is necessary to store all $\mathbf{v}^m(k)$, $m = 0, 1, \cdots$ in the memory unit until the equality (9.4.1) is verified. However, in practical computations we stop the process by the inequality

$$\text{(9.5.1)} \qquad \|\mathbf{v}^m - \mathbf{v}^{m-1}\|_G \ \le\ \mathbf{r}^8,$$

where $\mathbf{r}^8$ is a nonnegative vector. Naturally, each component of $\mathbf{r}^8$ cannot be too small, since $\|\mathbf{v}^m - \mathbf{v}^{m-1}\|_G$ does not always tend to zero as $m \to \infty$. The following result provides a sufficient condition for the vector $\mathbf{r}^8$ so that the process can be stopped by the criterion (9.5.1).

**Theorem 9.5.1.** Let the conditions of Theorem 9.3.1 be satisfied, and let the sequence $\{\mathbf{v}^m(k)\}$ be obtained from (9.3.1) on a floating point system. Further, let $\mathcal{M}^7$ and $\mathcal{M}^8$ be such that $\rho\left[(\mathcal{I}-\mathcal{K}^1)^{-1}(\mathcal{K}^1-\mathcal{K})\right] < 1$. Then, the process (9.3.1) can be stopped by the criterion (9.5.1) if

$$\text{(9.5.2)} \quad \mathbf{r}^8 \ \ge\ 2(\mathcal{I}-2\mathcal{K}^1+\mathcal{K})^{-1}\left[\mathcal{M}^1\mathcal{M}^7 \sup_{k\in\mathbb{N}(a,b-1)} |\mathbf{f}(k,\mathbf{v}^{m-1}(k))| + \mathcal{M}^2\mathcal{M}^8\|\mathcal{F}[\mathbf{v}^{m-1}]\|_G\right].$$

**Proof.** Let $\mathbf{v}^{m-1}(k)$ be in an oscillatory state. Then, from the inequality (9.4.3), we have

$$\text{(9.5.3)} \quad \|\mathbf{v}^{m-1}-\mathbf{u}^*\|_G \le (\mathcal{I}-\mathcal{K}^1)^{-1}\left[\mathcal{M}^1\mathcal{M}^7 \sup_{k\in\mathbb{N}(a,b-1)} |\mathbf{f}(k,\mathbf{u}^*(k))| + \mathcal{M}^2\mathcal{M}^8\|\mathcal{F}[\mathbf{u}^*]\|_G\right].$$

Since $\mathbf{v}^{m-1}(k)$ is in an oscillatory state, $\mathbf{v}^m(k)$ is also in an oscillatory state, and hence from (9.4.3) and (9.5.3), we get

$$\text{(9.5.4)} \quad \|\mathbf{v}^m - \mathbf{v}^{m-1}\|_G \ \le\ 2(\mathcal{I}-\mathcal{K}^1)^{-1}\left[\mathcal{M}^1\mathcal{M}^7 \sup_{k\in\mathbb{N}(a,b-1)} |\mathbf{f}(k,\mathbf{u}^*(k))| + \mathcal{M}^2\mathcal{M}^8\|\mathcal{F}[\mathbf{u}^*]\|_G\right].$$

Next, as earlier we find

$$\text{(9.5.5)} \quad \left[\mathcal{M}^1\mathcal{M}^7 \sup_{k\in\mathbb{N}(a,b-1)} |\mathbf{f}(k,\mathbf{u}^*(k))| + \mathcal{M}^2\mathcal{M}^8\|\mathcal{F}[\mathbf{u}^*]\|_G\right]$$

$$\leq \left(\mathcal{M}^1\mathcal{M}^7\mathcal{M}^3+\mathcal{M}^1\mathcal{M}^7 \sup_{k\in\mathbb{N}(a,b-1)} |\mathcal{A}(k)|+\mathcal{M}^2\mathcal{M}^8\mathcal{M}^4+\mathcal{M}^2\mathcal{M}^8\|\mathcal{L}\|_G\right) \times$$

$$\|\mathbf{v}^{m-1}-\mathbf{u}^*\|_G+\left(\mathcal{M}^1\mathcal{M}^7 \sup_{k\in\mathbb{N}(a,b-1)} |\mathbf{f}(k,\mathbf{v}^{m-1}(k))|+\mathcal{M}^2\mathcal{M}^8\|\mathcal{F}[\mathbf{v}^{m-1}]\|_G\right)$$

and hence, from (9.5.3) it follows that

$$\|\mathbf{v}^{m-1}-\mathbf{u}^*\|_G \leq (\mathcal{I}-\mathcal{K}^1)^{-1}(\mathcal{K}^1-\mathcal{K})\|\mathbf{v}^{m-1}-\mathbf{u}^*\|_G+(\mathcal{I}-\mathcal{K}^1)^{-1}\times$$
$$\left(\mathcal{M}^1\mathcal{M}^7 \sup_{k\in\mathbb{N}(a,b-1)} |\mathbf{f}(k,\mathbf{v}^{m-1}(k))|+\mathcal{M}^2\mathcal{M}^8\|\mathcal{F}[\mathbf{v}^{m-1}]\|_G\right),$$

which provides that

$$\|\mathbf{v}^{m-1}-\mathbf{u}^*\|_G \leq (\mathcal{I}-2\mathcal{K}^1+\mathcal{K})^{-1}\left[\mathcal{M}^1\mathcal{M}^7 \sup_{k\in\mathbb{N}(a,b-1)} |\mathbf{f}(k,\mathbf{v}^{m-1}(k))|\right.$$
$$\left.+\mathcal{M}^2\mathcal{M}^8\|\mathcal{F}[\mathbf{v}^{m-1}]\|_G\right].$$

Substituting this estimate in (9.5.5), we obtain

$$\left[\mathcal{M}^1\mathcal{M}^7 \sup_{k\in\mathbb{N}(a,b-1)} |\mathbf{f}(k,\mathbf{u}^*(k))|+\mathcal{M}^2\mathcal{M}^8\|\mathbf{f}[\mathbf{u}^*]\|_G\right]$$
$$\leq \left[(\mathcal{K}^1-\mathcal{K})(\mathcal{I}-2\mathcal{K}^1+\mathcal{K})^{-1}+\mathcal{I}\right]\left[\mathcal{M}^1\mathcal{M}^7 \sup_{k\in\mathbb{N}(a,b-1)} |\mathbf{f}(k,\mathbf{v}^{m-1}(k))|\right.$$
$$\left.+\mathcal{M}^2\mathcal{M}^8\|\mathcal{F}[\mathbf{v}^{m-1}]\|_G\right]$$
$$= (\mathcal{I}-\mathcal{K}^1)(\mathcal{I}-2\mathcal{K}^1+\mathcal{K})^{-1}\left[\mathcal{M}^1\mathcal{M}^7 \sup_{k\in\mathbb{N}(a,b-1)} |\mathbf{f}(k,\mathbf{v}^{m-1}(k))|\right.$$
$$\left.+\mathcal{M}^2\mathcal{M}^8\|\mathcal{F}[\mathbf{v}^{m-1}]\|_G\right].$$

Substituting this estimate in (9.5.4), we finally get

$$\text{(9.5.6)} \quad \|\mathbf{v}^m-\mathbf{v}^{m-1}\|_G \leq 2\left(\mathcal{I}-2\mathcal{K}^1+\mathcal{K}\right)^{-1} \times$$
$$\left[\mathcal{M}^1\mathcal{M}^7 \sup_{k\in\mathbb{N}(a,b-1)} |\mathbf{f}(k,\mathbf{v}^{m-1}(k))|+\mathcal{M}^2\mathcal{M}^8\|\mathcal{F}[\mathbf{v}^{m-1}]\|_G\right].$$

Thus, (9.5.1) is satisfied if (9.5.2) holds. ∎

**Remark 9.5.1.** If for all $m$, $\mathbf{f}^m = \overline{\mathbf{f}}$ and $\mathcal{F}^m = \overline{\mathcal{F}}$, then inequalities (9.3.4) and (9.3.5) can be used in (9.5.6), and then (9.5.2) can be replaced by

$$\mathbf{r}^8 \geq 2(\mathcal{I}-2\mathcal{K}^1+\mathcal{K})^{-1}\left[\mathcal{M}^1\mathcal{M}^7(\mathcal{I}-\mathcal{M}^7)^{-1}\sup_{k\in\mathbb{N}(a,b-1)}|\overline{\mathbf{f}}(k,\mathbf{v}^{m-1}(k))| + \mathcal{M}^2\mathcal{M}^8(\mathcal{I}-\mathcal{M}^8)^{-1}\|\overline{\mathcal{F}}[\mathbf{v}^{m-1}]\|_G\right].$$

**Theorem 9.5.2.** Let the conditions of Theorem 9.3.2 be satisfied, and let the sequence $\{\mathbf{v}^m(k)\}$ be obtained from (9.3.1) on a floating point system. Then, the process (9.3.1) can be stopped by the criterion (9.5.1) if

$$\mathbf{r}^8 \;\geq\; 2(\mathcal{I}-\mathcal{K})^{-1}(\mathcal{M}^1\mathbf{r}^5+\mathcal{M}^2\mathbf{r}^6).$$

**Proof.** The proof is immediate. ∎

## 9.6. Application to the Perturbation Method

Here we shall consider the boundary value problem

$$(9.6.1)\qquad \mathbf{u}(k+1) \;=\; \mathbf{f}(k,\mathbf{u}(k)) + \lambda\mathbf{g}(k,\mathbf{u}(k),\lambda), \quad k\in\mathbb{N}(a,b-1)$$

$$(9.6.2)\qquad \mathcal{F}[\mathbf{u}] + \lambda\mathcal{G}[\mathbf{u},\lambda] \;=\; 0$$

as the perturbed problem of (1.2.8), (1.5.1). In (9.6.1), (9.6.2), $\lambda$ is a small parameter such that $\lambda\in\bigwedge = \{\lambda\in\mathbb{R} : |\lambda|\leq\rho\}$, $\rho>0$, $\mathbf{g}(k,\mathbf{u},\lambda)$ is continuously differentiable with respect to $\mathbf{u}$ in $\mathbb{N}(a,b-1)\times\mathcal{R}^n\times\bigwedge$, and $\mathbf{g}_{\mathbf{u}}(k,\mathbf{u},\lambda)$ represents the Jacobian matrix of $\mathbf{g}(k,\mathbf{u},\lambda)$ with respect to $\mathbf{u}$, $\mathcal{G}[\mathbf{u},\lambda]$ is continuously Fréchet differentiable in $B(a,b)\times\bigwedge$, and $\mathcal{G}_{\mathbf{u}}[\mathbf{u},\lambda]$ denotes the linear operator mapping $B(a,b)\times\bigwedge$ into $\mathbb{R}^n$.

Let $\hat{\mathbf{u}}(k)$ be an isolated solution of (1.2.8), (1.5.1). For $\lambda\neq 0$ we seek an approximate solution $\overline{\mathbf{u}}(k)$ of (9.6.1), (9.6.2) of the form $\overline{\mathbf{u}}(k) = \hat{\mathbf{u}}(k) - \lambda\mathbf{u}(k)$. We substitute this in (9.6.1), (9.6.2) and neglect the terms higher than order one in $\lambda$, to obtain

$$(9.6.3)\qquad \mathbf{u}(k+1) \;=\; \mathbf{f}_{\mathbf{u}}(k,\hat{\mathbf{u}}(k))\mathbf{u}(k) - \mathbf{g}(k,\hat{\mathbf{u}}(k),0), \quad k\in\mathbb{N}(a,b-1)$$

$$(9.6.4)\qquad F_{\mathbf{u}}[\hat{\mathbf{u}}][\mathbf{u}] \;=\; \mathcal{G}[\hat{\mathbf{u}},0].$$

Since $\hat{\mathbf{u}}(k)$ is isolated, by Definition 9.2.2 the matrix $\mathcal{F}_{\mathbf{u}}[\hat{\mathbf{u}}][\mathcal{U}(k,a)]$ is nonsingular, and from Theorem 8.1.1 the problem (9.6.3), (9.6.4) is equivalent to

$$(9.6.5)\qquad \mathbf{u}(k) \;=\; \mathcal{H}^1\left[-\mathbf{g}(k,\hat{\mathbf{u}}(k),0)\right] + \mathcal{H}^2\left[\mathcal{G}[\hat{\mathbf{u}},0]\right].$$

Next for this approximate solution $\overline{\mathbf{u}}(k)$ of (9.6.1), (9.6.2) we shall show that the conditions of Theorem 9.2.2 are satisfied. For this, we take $\mathcal{A}(k) = \mathbf{f_u}(k, \hat{\mathbf{u}}(k)), \mathcal{L} = \mathcal{F}_{\mathbf{u}}[\hat{\mathbf{u}}]$ so that condition (ii) is satisfied. As in the proof of Theorem 9.2.3, we have $\overline{\mathcal{M}}^1$ and $\overline{\mathcal{M}}^2$ such that $\|\mathcal{H}^1\|_G \leq \overline{\mathcal{M}}^1, \|\mathcal{H}^2\|_G \leq \overline{\mathcal{M}}^2$, and hence condition (iii) is also satisfied.

Let $\mathbf{d}^3$ and $\mathbf{d}^4$ be nonnegative constants such that $\sup_{k\in\mathbb{N}(a,b-1)} |\mathbf{g}(k, \hat{\mathbf{u}}(k), 0)| \leq \mathbf{d}^3$, $\|\mathcal{G}[\hat{\mathbf{u}}, 0]\|_G \leq \mathbf{d}^4$. Then, from (9.6.5) it follows that

$$\|\mathbf{u}(k)\|_G \;\leq\; \overline{\mathcal{M}}^1\mathbf{d}^3 + \overline{\mathcal{M}}^2\mathbf{d}^4 \;=\; \mathbf{d}^5, \text{ say.}$$

Let $\mathbf{r}^3$ be the positive vector as in Theorem 9.2.3. We choose a positive vector $\mathbf{r}^9$ and $\lambda$ so that

$$\mathbf{r}^9 + |\lambda|\mathbf{d}^5 \;\leq\; \mathbf{r}^3. \tag{9.6.6}$$

If $\mathbf{u}(k) \in \overline{S}(\overline{\mathbf{u}}, \mathbf{r}^9)$, then we find

$$\|\mathbf{u} - \hat{\mathbf{u}}\|_G \;\leq\; \|\mathbf{u} - \overline{\mathbf{u}}\|_G + \|\overline{\mathbf{u}} - \hat{\mathbf{u}}\|_G \;\leq\; \mathbf{r}^9 + |\lambda|\mathbf{d}^5 \;\leq\; \mathbf{r}^3$$

and hence $\overline{S}(\overline{\mathbf{u}}, \mathbf{r}^9) \subseteq \overline{S}(\hat{\mathbf{u}}, \mathbf{r}^3)$. As in the proof of Theorem 9.2.3 for all $\mathbf{u}(k) \in \overline{S}(\overline{\mathbf{u}}, \mathbf{r}^9)$, $|\mathbf{f_u}(k, \mathbf{u}) - \mathbf{f_u}(k, \hat{\mathbf{u}})| \leq \mathcal{M}^5, \|\mathcal{F}_{\mathbf{u}}[\mathbf{u}] - \mathcal{F}_{\mathbf{u}}[\hat{\mathbf{u}}]\|_G \leq \mathcal{M}^6$. Further, since $\mathbf{g_u}(k, \mathbf{u}, \lambda)$ and $\mathcal{G}_{\mathbf{u}}[\mathbf{u}, \lambda]$ are continuous, there exist $n \times n$ nonnegative matrices $\mathcal{M}^9$ and $\mathcal{M}^{10}$ such that for all $k \in \mathbb{N}(a, b-1)$, $\mathbf{u}(k) \in \overline{S}(\hat{\mathbf{u}}, \mathbf{r}^3)$, $\lambda \in \Lambda$, $|\mathbf{g_u}(k, \mathbf{u}, \lambda)| \leq \mathcal{M}^9$ and $\|\mathcal{G}_{\mathbf{u}}[\mathbf{u}, \lambda]\|_G \leq \mathcal{M}^{10}$. Thus, for all $k \in \mathbb{N}(a, b-1)$, $\mathbf{u}(k) \in \overline{S}(\overline{\mathbf{u}}, \mathbf{r}^9)$, $\lambda \in \Lambda$ we have

$$|\mathbf{f_u}(k, \mathbf{u}) + \lambda\mathbf{g_u}(k, \mathbf{u}, \lambda) - \mathbf{f_u}(k, \hat{\mathbf{u}})| \;\leq\; \mathcal{M}^5 + |\lambda|\mathcal{M}^9$$

and

$$\|\mathcal{F}_{\mathbf{u}}[\mathbf{u}] + \lambda\mathcal{G}_{\mathbf{u}}[\mathbf{u}, \lambda] - \mathcal{F}_{\mathbf{u}}[\hat{\mathbf{u}}]\|_G \;\leq\; \mathcal{M}^6 + |\lambda|\mathcal{M}^{10}.$$

Hence, the condition (iv) is also satisfied. To satisfy condition (v) we need $\rho(\mathcal{K}^\lambda) < 1$, where

$$\mathcal{K}^\lambda = \overline{\mathcal{M}}^1\mathcal{M}^5 + |\lambda|\overline{\mathcal{M}}^1\mathcal{M}^9 + \overline{\mathcal{M}}^2\mathcal{M}^6 + |\lambda|\overline{\mathcal{M}}^2\mathcal{M}^{10}.$$

However, in Theorem 9.2.3, $\rho(\overline{\mathcal{M}}^1\mathcal{M}^5 + \overline{\mathcal{M}}^2\mathcal{M}^6) < 1$, thus there exists a norm $\|\cdot\|$ such that $\|(\overline{\mathcal{M}}^1\mathcal{M}^5 + \overline{\mathcal{M}}^2\mathcal{M}^6)\| < 1$. Further, since $\rho(\mathcal{K}^\lambda) \leq \|\mathcal{K}^\lambda\|$, the inequality $\rho(\mathcal{K}^\lambda) < 1$ is satisfied provided

$$|\lambda| < \frac{1 - \|(\overline{\mathcal{M}}^1\mathcal{M}^5 + \overline{\mathcal{M}}^2\mathcal{M}^6)\|}{\|(\overline{\mathcal{M}}^1\mathcal{M}^9 + \overline{\mathcal{M}}^2\mathcal{M}^{10})\|}. \tag{9.6.7}$$

Next, we assume that for all $k \in \mathbb{N}(a, b-1)$, $\mathbf{u}(k) \in \overline{S}(\hat{\mathbf{u}}, \mathbf{r}^3)$ and $\lambda \in \bigwedge$, the following holds

$$|\mathbf{g}(k, \mathbf{u}, \lambda) - \mathbf{g}(k, \mathbf{u}, 0)| \leq |\lambda| \mathbf{d}^6$$

and

$$\|\mathcal{G}[\mathbf{u}, \lambda] - \mathcal{G}[\mathbf{u}, 0]\|_G \leq |\lambda| \mathbf{d}^7,$$

where $\mathbf{d}^6$ and $\mathbf{d}^7$ are nonnegative vectors.

An easy computation shows that

$$\begin{aligned}
&\overline{\mathbf{u}}(k+1) - \mathbf{f}(k, \overline{\mathbf{u}}(k)) - \lambda \mathbf{g}(k, \overline{\mathbf{u}}(k), \lambda) \\
&\quad = \lambda \int_0^1 [\mathbf{f_u}(k, \hat{\mathbf{u}}(k) - \theta_7 \lambda \mathbf{u}(k)) - \mathbf{f_u}(k, \hat{\mathbf{u}}(k))]\, \mathbf{u}(k) d\theta_7 \\
&\quad\quad - \lambda [\mathbf{g}(k, \hat{\mathbf{u}}(k) - \lambda \mathbf{u}(k), \lambda) - \mathbf{g}(k, \hat{\mathbf{u}}(k) - \lambda \mathbf{u}(k), 0)] \\
&\quad\quad + \lambda^2 \int_0^1 \mathbf{g_u}(k, \hat{\mathbf{u}}(k) - \theta_8 \lambda \mathbf{u}(k), 0) \mathbf{u}(k) d\theta_9.
\end{aligned}$$

Since $\hat{\mathbf{u}}(k) - \theta_i \lambda \mathbf{u}(k) \in \overline{S}(\hat{\mathbf{u}}, \mathbf{r}^3)$, $i = 7, 8$ we find

$$\text{(9.6.8)} \quad \begin{aligned} &|\overline{\mathbf{u}}(k+1) - \mathbf{f}(k, \overline{\mathbf{u}}(k)) - \lambda \mathbf{g}(k, \overline{\mathbf{u}}(k), \lambda)| \\ &\qquad \leq |\lambda| \mathcal{M}^5 \mathbf{d}^5 + |\lambda|^2 \mathbf{d}^6 + |\lambda|^2 \mathcal{M}^9 \mathbf{d}^5. \end{aligned}$$

Similarly, we obtain

$$\text{(9.6.9)} \qquad \|\mathcal{F}[\overline{\mathbf{u}}] + \lambda \mathcal{G}[\overline{\mathbf{u}}, \lambda]\|_G \leq |\lambda| \mathcal{M}^6 \mathbf{d}^5 + |\lambda|^2 \mathbf{d}^7 + |\lambda|^2 \mathcal{M}^{10} \mathbf{d}^5.$$

Thus, the second part of condition (v), i.e. $(\mathcal{I} - \mathcal{K})^{-1}(\mathcal{M}^1 \mathbf{d}^1 + \mathcal{M}^2 \mathbf{d}^2) \leq \mathbf{r}$ is satisfied provided

$$\text{(9.6.10)} \qquad |\lambda| (\mathcal{I} - \mathcal{K}^\lambda)^{-1} (\mathcal{K}^\lambda \mathbf{d}^5 + |\lambda| (\overline{\mathcal{M}}^1 \mathbf{d}^6 + \overline{\mathcal{M}}^2 \mathbf{d}^7)) \leq \mathbf{r}^8.$$

Therefore, if $|\lambda| < \rho$ and if (9.6.6), (9.6.7) and (9.6.10) are satisfied (which is always the case if $|\lambda|$ is sufficiently small), then the conditions of Theorem 9.2.2 for the system (9.6.1), (9.6.2) with this approximate solution $\overline{\mathbf{u}}(k)$ are satisfied, and hence all the corresponding conclusions 1 5 of Theorem 9.2.2 for this problem also follow. ∎

If we further assume that for all $k \in \mathbb{N}(a, b-1)$ and $\mathbf{u}(k) \in \overline{S}(\hat{\mathbf{u}}, \mathbf{r}^3)$, $|\mathbf{f_u}(k, \mathbf{u}) - \mathbf{f_u}(k, \hat{\mathbf{u}})| \leq \mathcal{P} \|\mathbf{u} - \hat{\mathbf{u}}\|_G$ and $\|\mathcal{F}_\mathbf{u}[\mathbf{u}] - \mathcal{F}_\mathbf{u}[\hat{\mathbf{u}}]\|_G \leq \mathcal{Q} \|\mathbf{u} - \hat{\mathbf{u}}\|_G$, where $\mathcal{P}$ and $\mathcal{Q}$ are symmetric tensors of the third order with nonnegative components, then the right side of (9.6.8) can be replaced by $|\lambda|^2 \left( \frac{1}{2} \mathcal{P} \mathbf{d}^5 \cdot \mathbf{d}^5 + \mathbf{d}^6 + \mathcal{M}^9 \mathbf{d}^5 \right)$ and that of (9.6.9) by $|\lambda|^2 \left( \frac{1}{2} \mathcal{Q} \mathbf{d}^5 \cdot \mathbf{d}^5 + \right.$

$\mathbf{d}^7 + \mathcal{M}^{10}\mathbf{d}^5\Big)$. With this replacement (9.6.10) takes the form

$$\begin{aligned}\mathbf{r}^{10} &= |\lambda|^2(\mathcal{I}-\mathcal{K}^\lambda)^{-1}\Big[\overline{\mathcal{M}}^1\left(\frac{1}{2}\mathcal{P}\mathbf{d}^5\cdot\mathbf{d}^5+\mathbf{d}^6+\mathcal{M}^9\mathbf{d}^5\right)\\ &\quad +\overline{\mathcal{M}}^2\left(\frac{1}{2}\mathcal{Q}\mathbf{d}^5\cdot\mathbf{d}^5+\mathbf{d}^7+\mathcal{M}^{10}\mathbf{d}^5\right)\Big] \;\le\; \mathbf{r}^9.\end{aligned}$$

Hence, if $\mathbf{u}^*(k)$ is the solution of (9.6.1), (9.6.2) then from Remark 9.2.1 it follows that

$$\|\mathbf{u}^* - \overline{\mathbf{u}}\|_G \;\le\; \mathbf{r}^{10},$$

i.e. the perturbation method produces an approximate solution within the error $0(\lambda^2)$.

## 9.7. Monotone Convergence

With respect to the difference equation (1.2.8) on $\mathbb{N}(k_1, k_r-1)$ and the boundary conditions (1.5.4) we assume that there exists an $n\times n$ nonsingular matrix $\mathcal{A}(k)$ defined on $\mathbb{N}(k_1,k_r-1)$ such that if $\mathcal{U}(k,k_1)$ is the principal fundamental matrix of (1.2.12) then the matrix $\mathcal{H}$ defined in (8.1.6) is nonsingular. Thus, from Corollary 8.1.2 any solution $\mathbf{u}(k)$ of (1.2.8), (1.5.4) also satisfies

$$\text{(9.7.1)}\qquad \mathbf{u}(k) = \mathcal{U}(k,k_1)\mathcal{H}^{-1}\mathbf{l} + \sum_{\ell=k_1}^{k_r-1}\overline{\mathcal{M}}(k,\ell)[\mathbf{f}(\ell,\mathbf{u}(\ell)) - \mathcal{A}(\ell)\mathbf{u}(\ell)],$$

where the Green's matrix $\overline{\mathcal{M}}(k,\ell) = \mathcal{M}(k,\ell+1)$ is rewritten as

$$\text{(9.7.2)}\qquad \overline{\mathcal{M}}(k,\ell) = \begin{cases}\mathcal{U}(k,k_1)(\mathcal{I}+\mathcal{A}^i)\mathcal{U}^{-1}(\ell,k_1)\mathcal{A}^{-1}(\ell), & k_{i-1}\le\ell\le k-1\\ \mathcal{U}(k,k_1)\mathcal{A}^i\mathcal{U}^{-1}(\ell,k_1)\mathcal{A}^{-1}(\ell), & k\le\ell\le k_i-1\end{cases}$$

and $\mathcal{A}^i = -\mathcal{H}^{-1}\sum_{j=i}^{r}\mathcal{L}^j\mathcal{U}(k_j,k_1),\;\; 2\le i\le r.$

Let $\mathcal{P}: B(k_1,k_r)\to B(k_1,k_r)$ and $\mathcal{Q}: B(k_1,k_r-1)\to B(k_1,k_r-1)$ be invertible linear operators. For $\mathbf{u},\mathbf{v}\in B(k_1,k_r)$ and $\overline{\mathbf{u}},\overline{\mathbf{v}}\in B(k_1,k_r-1)$ we define the relations $\le_{\mathcal{P}}$ and $\le_{\mathcal{Q}}$ by

$\mathbf{u} \le_{\mathcal{P}} \mathbf{v}$ if and only if $\mathcal{P}\mathbf{u}(k) \le \mathcal{P}\mathbf{v}(k)$, componentwise, for all $k\in\mathbb{N}(k_1,k_r)$

and

$$\overline{\mathbf{u}} \le_{\mathcal{Q}} \overline{\mathbf{v}} \text{ if and only if } \mathcal{Q}\overline{\mathbf{u}}(k) \le \mathcal{Q}\overline{\mathbf{v}}(k), \text{ componentwise, for all } k \in \mathbb{N}(k_1, k_r - 1).$$

The relation $\le_{\mathcal{P}}$ is a partial ordering in $B(k_1, k_r)$ and we say that $\le_{\mathcal{P}}$ is the partial ordering induced by $\mathcal{P}$; similarly, the relation $\le_{\mathcal{Q}}$ is a partial ordering in $B(k_1, k_r - 1)$ and we say that $\le_{\mathcal{Q}}$ is the partial ordering induced by $\mathcal{Q}$. If for some $k \in \mathbb{N}(k_1, k_r)$, $\mathcal{P}\mathbf{u}(k) \le \mathcal{P}\mathbf{v}(k)$, componentwise, then we shall say $\mathbf{u}(k) \le_{\mathcal{P}} \mathbf{v}(k)$.

Suppose there exist nonsingular $n \times n$ matrices $\hat{\mathcal{P}}, \hat{\mathcal{Q}}^2, \cdots, \hat{\mathcal{Q}}^r$ such that $\hat{\mathcal{P}}\mathcal{A}^i(\hat{\mathcal{Q}}^i)^{-1} \ge 0$ and $\hat{\mathcal{P}}(\mathcal{I} + \mathcal{A}^i)(\hat{\mathcal{Q}}^i)^{-1} \ge 0$, elementwise for $i = 2, \cdots, r$. Let $\le_{\mathcal{P}}$ be the partial ordering induced by $\mathcal{P} = \hat{\mathcal{P}}\mathcal{U}^{-1}(k, k_1)$ in $B(k_1, k_r)$ and $\le_{\mathcal{Q}}$ be the partial ordering induced by $\mathcal{Q}^2 = \hat{\mathcal{Q}}^2\mathcal{U}^{-1}(k, k_1)\mathcal{A}^{-1}(k), \cdots, \mathcal{Q}^r = \hat{\mathcal{Q}}^r\mathcal{U}^{-1}(k, k_1)\mathcal{A}^{-1}(k)$ in $B(k_1, k_r - 1)$.

**Theorem 9.7.1.** Assume that there exist functions $\mathbf{u}^0$ and $\mathbf{v}^0$ in $B(k_1, k_r)$ satisfying

(i) $\mathbf{u}^0 \le_{\mathcal{P}} \mathbf{v}^0$

(ii) $\sum_{i=1}^{r} \mathcal{L}^i\mathbf{u}^0(k_i) = \mathbf{l} = \sum_{i=1}^{r} \mathcal{L}^i\mathbf{v}^0(k_i)$

(iii) $\mathbf{u}^0(k+1) - \mathbf{f}(k, \mathbf{u}^0(k)) \le_{\mathcal{Q}} 0 \le_{\mathcal{Q}} \mathbf{v}^0(k+1) - \mathbf{f}(k, \mathbf{v}^0(k))$

(iv) if $\mathbf{u}, \mathbf{v} \in B(k_1, k_r)$ and $\mathbf{u}^0 \le_{\mathcal{P}} \mathbf{u} \le_{\mathcal{P}} \mathbf{v} \le_{\mathcal{P}} \mathbf{v}^0$, then $\mathbf{f}(k, \mathbf{u}) - \mathcal{A}(k)\mathbf{u} \le_{\mathcal{Q}} \mathbf{f}(k, \mathbf{v}) - \mathcal{A}(k)\mathbf{v}$.

Further, the function $\mathbf{f}(k, \mathbf{u})$ is continuous on $\mathbb{N}(k_1, k_r - 1) \times \mathcal{R}^n$. Then, the boundary value problem (1.2.8), (1.5.4) has at least one solution $\mathbf{u}^*(k)$ such that

$$\mathbf{u}^0 \le_{\mathcal{P}} \mathbf{u}^* \le_{\mathcal{P}} \mathbf{v}^0. \tag{9.7.3}$$

**Proof.** Let $B_2(k_1, k_r) = \{\mathbf{u} \in B(k_1, k_r) : \mathbf{u}^0 \le_{\mathcal{P}} \mathbf{u} \le_{\mathcal{P}} \mathbf{v}^0\}$. Obviously, $B_2(k_1, k_r)$ is a closed convex subset of the Banach space $B(k_1, k_r)$ with $\|\mathbf{u}\| = \max_{1 \le i \le n} \sup_{k \in \mathbb{N}(k_1, k_r)} |u_i(k)|$. We shall show that the continuous operator $\mathcal{T} : B(k_1, k_r) \to B(k_1, k_r)$ defined by

$$\mathcal{T}\mathbf{u}(k) = \mathcal{U}(k, k_1)\mathcal{H}^{-1}\mathbf{l} + \sum_{\ell=k_1}^{k_r-1} \overline{\mathcal{M}}(k, \ell)[\mathbf{f}(\ell, \mathbf{u}(\ell)) - \mathcal{A}(\ell)\mathbf{u}(\ell)] \tag{9.7.4}$$

maps $B_2(k_1, k_r)$ into itself.

Suppose $\mathbf{u},\ \mathbf{v} \in B_2(k_1, k_r)$ and $\mathbf{u} \leq_{\mathcal{P}} \mathbf{v}$. Then, we have

$$\begin{aligned}(\mathcal{T}\mathbf{v}-\mathcal{T}\mathbf{u})(k) &= \sum_{\ell=k_1}^{k_r-1}\overline{\mathcal{M}}(k,\ell)[\mathbf{f}(\ell,\mathbf{v}(\ell))-\mathcal{A}(\ell)\mathbf{v}(\ell)-\mathbf{f}(\ell,\mathbf{u}(\ell))+\mathcal{A}(\ell)\mathbf{u}(\ell)] \\ &= \sum_{i=2}^{r}\sum_{\ell=k_{i-1}}^{k-1} \mathcal{U}(k,k_1)(\mathcal{I}+\mathcal{A}^i)\mathcal{U}^{-1}(\ell,k_1)\mathcal{A}^{-1}(\ell)[\mathbf{f}(\ell,\mathbf{v}(\ell)) \\ &\quad -\mathcal{A}(\ell)\mathbf{v}(\ell)-\mathbf{f}(\ell,\mathbf{u}(\ell))+\mathcal{A}(\ell)\mathbf{u}(\ell)] \\ &\quad +\sum_{i=2}^{r}\sum_{\ell=k}^{k_i-1}\mathcal{U}(k,k_1)\mathcal{A}^i\mathcal{U}^{-1}(\ell,k_1)\mathcal{A}^{-1}(\ell)[\mathbf{f}(\ell,\mathbf{v}(\ell)) \\ &\quad -\mathcal{A}(\ell)\mathbf{v}(\ell)-\mathbf{f}(\ell,\mathbf{u}(\ell))+\mathcal{A}(\ell)\mathbf{u}(\ell)].\end{aligned}$$

Thus, it follows that

$$\begin{aligned}\mathcal{P}(\mathcal{T}\mathbf{v}-\mathcal{T}\mathbf{u})(k) &= \sum_{i=2}^{r}\sum_{\ell=k_{i-1}}^{k-1}\hat{\mathcal{P}}(\mathcal{I}+\mathcal{A}^i)(\hat{\mathcal{Q}}^i)^{-1}(\hat{\mathcal{Q}}^i\mathcal{U}^{-1}(\ell,k_1)\mathcal{A}^{-1}(\ell)) \\ &\quad \times[\mathbf{f}(\ell,\mathbf{v}(\ell))-\mathcal{A}(\ell)\mathbf{v}(\ell)-\mathbf{f}(\ell,\mathbf{u}(\ell))+\mathcal{A}(\ell)\mathbf{u}(\ell)] \\ &\quad +\sum_{i=2}^{r}\sum_{\ell=k}^{k_i-1}\hat{\mathcal{P}}\mathcal{A}^i(\hat{\mathcal{Q}}^i)^{-1}(\hat{\mathcal{Q}}^i\mathcal{U}^{-1}(\ell,k_1)\mathcal{A}^{-1}(\ell))[\mathbf{f}(\ell,\mathbf{v}(\ell)) \\ &\quad -\mathcal{A}(\ell)\mathbf{v}(\ell)-\mathbf{f}(\ell,\mathbf{u}(\ell))+\mathcal{A}(\ell)\mathbf{u}(\ell)].\end{aligned}$$

However, since $\hat{\mathcal{P}}(\mathcal{I}+\mathcal{A}^i)(\hat{\mathcal{Q}}^i)^{-1} \geq 0,\ \hat{\mathcal{P}}\mathcal{A}^i(\hat{\mathcal{Q}}^i)^{-1} \geq 0,\ 2 \leq i \leq r$ elementwise, and by (iv), $\mathbf{f}(\ell,\mathbf{u}(\ell)) - \mathcal{A}(\ell)\mathbf{u}(\ell) \leq_{\mathcal{Q}} \mathbf{f}(\ell,\mathbf{v}(\ell)) - \mathcal{A}(\ell)\mathbf{v}(\ell)$, it follows that $\mathcal{P}(\mathcal{T}\mathbf{v}-\mathcal{T}\mathbf{u})(k) \geq 0$, which is the same as $\mathcal{T}\mathbf{u} \leq_{\mathcal{P}} \mathcal{T}\mathbf{v}$, i.e. $\mathcal{T}$ is monotone in $B_2(k_1,k_r)$ with respect to $\leq_{\mathcal{P}}$.

We shall now show that $\mathbf{u}^0 \leq_{\mathcal{P}} \mathcal{T}\mathbf{u}^0$ and $\mathcal{T}\mathbf{v}^0 \leq_{\mathcal{P}} \mathbf{v}^0$, and then it will follow that $\mathcal{T}$ maps $B_2(k_1,k_r)$ into itself. For this, we note that the solution of the boundary value problem $\mathbf{u}(k+1) = \mathcal{A}(k)\mathbf{u}(k),\ \sum_{i=1}^{r}\mathcal{L}^i\mathbf{u}(k_i) = \sum_{i=1}^{r}\mathcal{L}^i\mathbf{v}(k_i) = \mathbf{l}$ is the same as $\mathcal{U}(k,k_1)\mathcal{H}^{-1}\mathbf{l}$. Therefore, it follows that

$$\mathbf{v}^0(k) \ = \ \mathcal{U}(k,k_1)\mathcal{H}^{-1}\mathbf{l} + \sum_{\ell=k_1}^{k_r-1}\overline{\mathcal{M}}(k,\ell)[\mathbf{v}^0(\ell+1)-\mathcal{A}(\ell)\mathbf{v}^0(\ell)]$$

and hence

$$(\mathbf{v}^0-\mathcal{T}\mathbf{v}^0)(k) \ = \ \sum_{\ell=k_1}^{k_r-1}\overline{\mathcal{M}}(k,\ell)[\mathbf{v}^0(\ell+1)-\mathbf{f}(\ell,\mathbf{v}^0(\ell))].$$

However, since by (iii), $\mathbf{v}^0(k+1)-\mathbf{f}(k,\mathbf{v}^0(k))\ {}_{\mathcal{Q}}\!\geq 0$, by the above argument we find that $\mathcal{P}(\mathbf{v}^0-\mathcal{T}\mathbf{v}^0) \geq 0$, i.e. $\mathcal{T}\mathbf{v}^0 \leq_{\mathcal{P}} \mathbf{v}^0$. The proof for $\mathbf{u}^0 \leq_{\mathcal{P}} \mathcal{T}\mathbf{u}^0$ is similar.

The existence of a fixed point $\mathbf{u}^*$ of $\mathcal{T}$ in $B_2(k_1, k_r)$ now follows as an application of Theorem 9.1.1. ∎

Let the sequences $\{\mathbf{u}^m\}$ and $\{\mathbf{v}^m\}$ inductively be defined as

$$(9.7.5) \qquad \mathbf{u}^{m+1}(k) = \mathcal{T}\mathbf{u}^m(k), \quad \mathbf{v}^{m+1}(k) = \mathcal{T}\mathbf{v}^m(k), \quad m = 0, 1, \cdots$$

where $\mathcal{T}$ is as in (9.7.4). We have shown in the proof of Theorem 9.7.1 that $\mathcal{T}$ is monotone in $B_2(k_1, k_r)$ with respect to $\leq_{\mathcal{P}}$, and so

$$\mathbf{u}^m \leq_{\mathcal{P}} \mathbf{u}^{m+1} \leq_{\mathcal{P}} \mathbf{v}^{m+1} \leq_{\mathcal{P}} \mathbf{v}^m, \quad m = 0, 1, \cdots.$$

Since $\mathcal{P}$ is invertible we obtain the following:

**Corollary 9.7.2.** Assume that the hypotheses of Theorem 9.7.1 be satisfied . Then, the sequences $\{\mathbf{u}^m\}$ and $\{\mathbf{v}^m\}$ defined in (9.7.5) converge in $B_2(k_1, k_r)$ to the solutions $\mathbf{u}$ and $\mathbf{v}$ of the boundary value problem (1.2.8), (1.5.4) and

$$\mathbf{u}^m \leq_{\mathcal{P}} \mathbf{u}^{m+1} \leq_{\mathcal{P}} \mathbf{u} \leq_{\mathcal{P}} \mathbf{v} \leq_{\mathcal{P}} \mathbf{v}^{m+1} \leq_{\mathcal{P}} \mathbf{v}^m, \quad m = 0, 1, \cdots.$$

Further, if $\mathbf{u}^*$ is any solution of (1.2.8), (1.5.4) satisfying (9.7.3), then $\mathbf{u} \leq_{\mathcal{P}} \mathbf{u}^* \leq_{\mathcal{P}} \mathbf{v}$.

**Remark 9.7.1.** In the hypothesis (ii) of Theorem 9.7.1 we have taken identity $=$ only for simplicity. The results on the existence of solutions and the monotone iterative convergence remain valid if $=$ is replaced by a suitable partial ordering $\leq_{\mathcal{S}}$ in $\mathbb{R}^n$. For this, once again for simplicity, we let $r = 2$. A suitable partial ordering $\leq_{\mathcal{S}}$ in $\mathbb{R}^n$ is then induced by $\hat{\mathcal{P}}[\mathcal{L}^1 + \mathcal{L}^2\mathcal{U}(k_2, k_1)]^{-1}$, and the hypothesis (ii) of Theorem 9.7.1 can be replaced by

$$(\text{ii})' \qquad [\mathcal{L}^1\mathbf{u}^0(k_1) + \mathcal{L}^2\mathbf{u}^0(k_2)] \leq_{\mathcal{S}} \mathbf{l} \leq_{\mathcal{S}} [\mathcal{L}^1\mathbf{v}^0(k_1) + \mathcal{L}^2\mathbf{v}^0(k_2)].$$

For this, in the above, the proof of $\mathbf{u}^0 \leq_{\mathcal{P}} \mathcal{T}\mathbf{u}^0$ and $\mathcal{T}\mathbf{v}^0 \leq_{\mathcal{P}} \mathbf{v}^0$ needs a slight modification which is based on the new representation

$$\begin{aligned} \mathbf{v}^0(k) &= \mathcal{U}(k, k_1)\left[\mathcal{L}^1 + \mathcal{L}^2\mathcal{U}(k_2, k_1)\right]^{-1}\left[\mathcal{L}^1\mathbf{v}^0(k_1) + \mathcal{L}^2\mathbf{v}^0(k_2)\right] \\ &\quad + \sum_{\ell=k_1}^{k_2-1} \overline{\mathcal{M}}(k, \ell)\left[\mathbf{v}^0(\ell+1) - \mathcal{A}(\ell)\mathbf{v}^0(\ell)\right]. \end{aligned}$$

To apply above results, in particular, we consider the periodic boundary value problem

$$(9.7.6) \qquad \begin{aligned} \Delta^2 u(k) &= f(k, u(k), \Delta u(k)), \quad k \in \mathbb{N}(0, K-1) \\ u(0) &= u(K), \quad u(1) = u(K+1). \end{aligned}$$

In system form the above problem is the same as

$$\begin{bmatrix} u_1(k+1) \\ u_2(k+1) \end{bmatrix} = \begin{bmatrix} u_1(k)+u_2(k) \\ u_2(k)+f(k,u_1(k),u_2(k)) \end{bmatrix} = \mathbf{f}(k,\mathbf{u}(k)), \quad k \in \mathbb{N}(0,K-1)$$

(9.7.7)

$$\begin{bmatrix} u_1(0) \\ u_2(0) \end{bmatrix} - \begin{bmatrix} u_1(K) \\ u_2(K) \end{bmatrix} = 0.$$

Let $\mathcal{E}$ be a real $2\times 2$ matrix with real eigenvalues $\lambda_1$ and $\lambda_2$ satisfying $-1<\lambda_2<0<\lambda_1$. Then, the problem (9.7.7) is equivalent to

$$\mathbf{u}(k+1) = (\mathcal{I}+\mathcal{E})\mathbf{u}(k)+\mathbf{f}(k,\mathbf{u}(k))-(\mathcal{I}+\mathcal{E})\mathbf{u}(k), \quad k\in\mathbb{N}(0,K-1)$$

(9.7.8)

$$\mathbf{u}(0)-\mathbf{u}(K) = 0.$$

**Theorem 9.7.3.** Let $\hat{\mathcal{P}}$ be such that $\hat{\mathcal{P}}\mathcal{E}\hat{\mathcal{P}}^{-1} = \mathrm{diag}\{\lambda_i\}$ and let $\hat{\mathcal{Q}}^2 = \hat{\mathcal{Q}} = \begin{bmatrix} -1 & 0 \\ 0 & 1 \end{bmatrix}\hat{\mathcal{P}}$. Let $\le_{\mathcal{P}}$ be the partial ordering induced by $\hat{\mathcal{P}}$ and $\le_{\mathcal{Q}}$, $\le_{\mathcal{S}}$ be partial orderings induced by $\hat{\mathcal{Q}}$. Further, let all conditions of Theorem 9.7.1, with (ii) replaced by (ii)$'$, be satisfied. Then, the boundary value problem (9.7.6) has a solution $u^*(k)$ such that

$$\begin{bmatrix} u^0 \\ \Delta u^0 \end{bmatrix} \le_{\mathcal{P}} \begin{bmatrix} u^* \\ \Delta u^* \end{bmatrix} \le_{\mathcal{P}} \begin{bmatrix} v^0 \\ \Delta v^0 \end{bmatrix}.$$

**Proof.** For (9.7.8), we have $\mathcal{U}(k,0) = (\mathcal{I}+\mathcal{E})^k$, $\mathcal{H} = \mathcal{I}-(\mathcal{I}+\mathcal{E})^K$, $\mathcal{A}^2 = \mathcal{A} = -[\mathcal{I}-(\mathcal{I}+\mathcal{E})^K]^{-1}[-\mathcal{I}](\mathcal{I}+\mathcal{E})^K = [\mathcal{I}-(\mathcal{I}+\mathcal{E})^K]^{-1}(\mathcal{I}+\mathcal{E})^K$. Since, $\hat{\mathcal{P}}\mathcal{E}\hat{\mathcal{P}}^{-1} = \mathrm{diag}\{\lambda_i\}$, $\hat{\mathcal{P}}\mathcal{A}\hat{\mathcal{P}}^{-1} = \mathrm{diag}\{[1-(1+\lambda_i)^K]^{-1}(1+\lambda_i)^K\}$, $-1<\lambda_2<0<\lambda_1$, and so, $\hat{\mathcal{P}}\mathcal{A}\hat{\mathcal{Q}}^{-1}\ge 0$ and $\hat{\mathcal{P}}(\mathcal{I}+\mathcal{A})\hat{\mathcal{Q}}^{-1}\ge 0$, elementwise. Let $\le_{\mathcal{P}}$ be the partial ordering induced by $\hat{\mathcal{P}}\mathcal{U}^{-1}(k,0)$, $\le_{\mathcal{Q}}$ be the partial ordering induced by $\hat{\mathcal{Q}}\mathcal{U}^{-1}(k,0)(\mathcal{I}+\mathcal{E})^{-1}$, and $\le_{\mathcal{S}}$ be the partial ordering induced by $\hat{\mathcal{P}}[\mathcal{I}-\mathcal{U}(K,0)]^{-1}$. Thus, in view of Theorem 9.7.1 and Remark 9.7.1 it suffices to show that $\le_{\mathcal{P}}$ is induced by $\hat{\mathcal{P}}$ and $\le_{\mathcal{Q}}$, $\le_{\mathcal{S}}$ are induced by $\hat{\mathcal{Q}}$. For this, since $\hat{\mathcal{P}}\mathcal{U}^{-1}(k,0) = \hat{\mathcal{P}}(\mathcal{I}+\mathcal{E})^{-k} = \mathrm{diag}\{(1+\lambda_i)^{-m}\}\hat{\mathcal{P}}$ and so, $\le_{\mathcal{P}}$ is induced by $\hat{\mathcal{P}}$. Further, $\hat{\mathcal{Q}}\mathcal{U}^{-1}(k,0)(\mathcal{I}+\mathcal{E})^{-1} = \mathrm{diag}\{(1+\lambda_i)^{-(m+1)}\}\hat{\mathcal{Q}}$ and $\hat{\mathcal{P}}[\mathcal{I}-\mathcal{U}(K,0)]^{-1} = \mathrm{diag}\{|1-(1+\lambda_i)^K|^{-1}\}\hat{\mathcal{Q}}$, $\le_{\mathcal{Q}}$ and $\le_{\mathcal{S}}$ are induced by $\hat{\mathcal{Q}}$. ∎

## 9.8. Periodic Boundary Value Problems

Let $T=\{t_0,t_1,\cdots,t_K\}$ denote a set of increasing time instances. For a function $\mathbf{u}: T\to\mathbb{R}^n$, we define the first order backward difference operator $\nabla \mathbf{u}(t_k) = \mathbf{u}(t_k) - \mathbf{u}(t_{k-1}),\ 1 \le k \le K$. In this section we shall develop a monotone iterative method for the following periodic boundary value problem

$$(9.8.1)\qquad \begin{aligned} \nabla \mathbf{u}(t_k) &= \mathbf{f}(t_k, \mathbf{u}(t_k)), \qquad 1 \le k \le K \\ \mathbf{u}(t_0) &= \mathbf{u}(t_K) \end{aligned}$$

where $\mathbf{f} : T \times \mathbb{R}^n \to \mathbb{R}^n$ is a continuous function.

Let $N_1$ and $N_2$ be two subsets of the index set $N = \{1, 2, \cdots, n\}$. We say that $D = \{N_1, N_2\}$ is a decomposition of $N$ if $N_1 \cap N_2 = \emptyset$ and $N_1 \cup N_2 = N$.

**Definition 9.8.1.** A function $\mathbf{f} = \mathbf{f}(t, \mathbf{u}) : T \times \mathbb{R}^n \to \mathbb{R}^n$ is said to possess the *D quasimonotonicity* if there exists a decomposition $D = \{N_1, N_2\}$ of $N$ such that for all $i \in N_1$, $f_i(t, \mathbf{u})$ is monotone nonincreasing in $u_j$, $j = 1, 2, \cdots, n$, $j \ne i$ and for all $i \in N_2$, $f_i(t, \mathbf{u})$ is monotone nondecreasing in $u_j$, $j = 1, 2, \cdots, n$, $j \ne i$.

Let $\le$ and $<$ denote the usual partial ordering in $\mathbb{R}^n$. Based on the decomposition $D = \{N_1, N_2\}$ of $N$, we introduce partial ordering $\le_D$ in $\mathbb{R}^n$ as follows: Let $\mathbf{u}, \mathbf{v} \in \mathbb{R}^n$ with $\mathbf{u} = (u_1, \cdots, u_n)^T$ and $\mathbf{v} = (v_1, \cdots, v_n)^T$, we say $\mathbf{u} \le_D \mathbf{v}$ if $u_i \le v_i$ for all $i \in N_1$ and $u_i \ge v_i$ for all $i \in N_2$. The ordering $\mathbf{u} <_D \mathbf{v}$ can be defined similarly with the strict inequalities.

**Definition 9.8.2.** Let $D = \{N_1, N_2\}$ be a decomposition of $N$. A function $\overline{\mathbf{u}} : T \to \mathbb{R}^n$ is called a *D upper periodic solution* of (9.8.1) if

$$\begin{aligned} \nabla \overline{\mathbf{u}}(t_k) &\le_D \mathbf{f}(t_k, \overline{\mathbf{u}}(t_k)), \qquad 1 \le k \le K \\ \overline{\mathbf{u}}(t_0) &= \overline{\mathbf{u}}(t_K) \end{aligned}$$

and a function $\underline{\mathbf{u}} : T \to \mathbb{R}^n$ is called a *D lower periodic solution* of (9.8.1) if

$$\begin{aligned} \nabla \underline{\mathbf{u}}(t_k) &\ge_D \mathbf{f}(t_k, \underline{\mathbf{u}}(t_k)), \qquad 1 \le k \le K \\ \underline{\mathbf{u}}(t_0) &= \underline{\mathbf{u}}(t_K). \end{aligned}$$

It is obvious that every solution of (9.8.1) is a $D$ upper periodic solution as well as a $D$ lower periodic solution.

**Lemma 9.8.1.** Let $u : T \to \mathbb{R}$ be a function such that

$$(9.8.2)\qquad \begin{aligned} \nabla u(t_k) - \sigma u(t_k) &\ge 0, \qquad 1 \le k \le K \\ u(t_0) &= u(t_K) \end{aligned}$$

where $0 \neq \sigma \leq 1$ is a given constant. Then,

$$u(t_k)\begin{cases} \geq 0, & \sigma < 0, \\ \leq 0, & 0 < \sigma \leq 1, \end{cases} \quad k = 0, 1, \cdots, K.$$

**Proof.** Let $\mathbf{v} = (u(t_1), \cdots, u(t_K))^T \in \mathbb{R}^K$. By (9.8.2),

$$\mathcal{A}\mathbf{v} \geq 0, \tag{9.8.3}$$

where

$$\mathcal{A} = \begin{bmatrix} 1-\sigma & & & & & -1 \\ -1 & 1-\sigma & & & & \\ & -1 & 1-\sigma & & & \\ & & \ddots & \ddots & & \\ & & & -1 & 1-\sigma & \\ & & & & -1 & 1-\sigma \end{bmatrix}.$$

Let $\mathcal{A}^{-1} = (a_{i,j})$. It can be easily checked that

$$a_{i,j} = \begin{cases} -\dfrac{(1-\sigma)^{j-i-1}}{1-(1-\sigma)^K}, & i = 1, 2, \cdots, K, \ j = i+1, \cdots, K \\ -\dfrac{(1-\sigma)^{K+j-i-1}}{1-(1-\sigma)^K}, & i = 1, 2, \cdots, K, \ j = 1, 2, \cdots, i. \end{cases} \tag{9.8.4}$$

If $\sigma < 0$, we find that $a_{i,j} > 0$, i.e. $\mathcal{A}^{-1} > 0$ and hence by (9.8.3), $\mathbf{v} \geq 0$. If $0 < \sigma \leq 1$, we have that $a_{i,j} \leq 0$, i.e. $\mathcal{A}^{-1} \leq 0$ and thus by (9.8.3), $\mathbf{v} \leq 0$. ∎

In $\mathbb{R}^n$, Lemma 9.8.1 can be stated as follows:

**Lemma 9.8.2.** Let $D = \{N_1, N_2\}$ be a decomposition of $N$ and $\mathbf{u} : T \to \mathbb{R}^n$ be a function such that

$$\begin{aligned} \nabla \mathbf{u}(t_k) - \mathcal{M}\mathbf{u}(t_k) &\geq_D 0, \qquad 1 \leq k \leq K \\ \mathbf{u}(t_0) &= \mathbf{u}(t_K) \end{aligned}$$

where $\mathcal{M} = \operatorname{diag}(M_1, \cdots, M_n)$ and $\mathbf{e} \geq (M_1, \cdots, M_n)^T >_D 0$ with $\mathbf{e} = (1, \cdots, 1)^T \in \mathbb{R}^n$. Then, $\mathbf{u}(t_k) \leq 0$, $k = 0, 1, \cdots, K$.

**Remark 9.8.1.** From (9.8.4) it is clear that for any function $\mathbf{g} : T \to \mathbb{R}^n$ and $\mathcal{M} = \operatorname{diag}(M_1, \cdots, M_n)$ with $M_i \neq 0$, $i = 1, \cdots, n$ the linear problem

$$\begin{aligned} \nabla \mathbf{u}(t_k) - \mathcal{M}\mathbf{u}(t_k) &= \mathbf{g}(t_k), \qquad 1 \leq k \leq K \\ \mathbf{u}(t_0) &= \mathbf{u}(t_K) \end{aligned}$$

is solvable uniquely.

For the periodic boundary value problem (9.8.1) we shall consider the following iterative process:

$$
\begin{aligned}
&\nabla \mathbf{u}^{(m)}(t_k) - \mathcal{M}^* \mathbf{u}^{(m)}(t_k) = \mathbf{f}(t_k, \mathbf{u}^{(m-1)}(t_k)) - \mathcal{M}^* \mathbf{u}^{(m-1)}(t_k), \\
&\qquad\qquad\qquad\qquad\qquad\qquad\qquad\qquad 1 \le k \le K \\
&\mathbf{u}^{(m)}(t_0) = \mathbf{u}^{(m)}(t_K)
\end{aligned} \tag{9.8.5}
$$

where $\mathcal{M}^* = \operatorname{diag}(M_1^*, \cdots, M_n^*)$ is specified later. From Remark 9.8.1 it is clear that the above iterative process is well defined provided $M_i^* \neq 0,\ i = 1, 2, \cdots, n$.

**Theorem 9.8.3.** Let $D = \{N_1, N_2\}$ be a decomposition of $N$, and $\overline{\mathbf{u}}$ and $\underline{\mathbf{u}}$ be respectively $D$ upper and $D$ lower periodic solutions of (9.8.1) such that $\overline{\mathbf{u}}(t) \ge \underline{\mathbf{u}}(t),\ t \in T$. Assume that $\mathbf{f}$ is $D$ quasimonotone in $W = \{\mathbf{u}(t) : \underline{\mathbf{u}}(t) \le \mathbf{u}(t) \le \overline{\mathbf{u}}(t),\ t \in T\}$ and

$$
\mathbf{f}(t, \mathbf{u}(t)) - \mathbf{f}(t, \tilde{\mathbf{u}}(t)) \le_D \mathcal{M}(\mathbf{u}(t) - \tilde{\mathbf{u}}(t)), \qquad t \in T \tag{9.8.6}
$$

whenever $\underline{\mathbf{u}}(t) \le \tilde{\mathbf{u}}(t) \le \mathbf{u}(t) \le \overline{\mathbf{u}}(t),\ t \in T$, where $\mathcal{M} = \operatorname{diag}(M_1, \cdots, M_n)$ with $\mathbf{e} \ge (M_1, \cdots, M_n)^T >_D 0$, $\mathbf{e} = (1, \cdots, 1)^T \in \mathbb{R}^n$. Then, two sequences $\{\overline{\mathbf{u}}^{(m)}(t_k)\}$ and $\{\underline{\mathbf{u}}^{(m)}(t_k)\}$ defined by (9.8.5) with $\mathcal{M}^* = \mathcal{M}$ and the initial functions $\overline{\mathbf{u}}^{(0)}(t_k) = \overline{\mathbf{u}}(t_k)$ and $\underline{\mathbf{u}}^{(0)}(t_k) = \underline{\mathbf{u}}(t_k)$ $(k = 0, 1, \cdots, K)$, converge monotonically to the solutions $\overline{\mathbf{u}}^*(t_k)$ and $\underline{\mathbf{u}}^*(t_k)$ of (9.8.1), respectively. Further, for all $m \ge 0$ and $1 \le k \le K$,

$$
\begin{aligned}
\underline{\mathbf{u}}(t_k) \le \underline{\mathbf{u}}^{(m)}(t_k) \le \underline{\mathbf{u}}^{(m+1)}(t_k) \le \underline{\mathbf{u}}^*(t_k) \le \overline{\mathbf{u}}^*(t_k) \le \overline{\mathbf{u}}^{(m+1)}(t_k) \\
\le \overline{\mathbf{u}}^{(m)}(t_k) \le \overline{\mathbf{u}}(t_k)
\end{aligned} \tag{9.8.7}
$$

and for any one solution $\mathbf{u}^*(t)$ of (9.8.1) in $W$, we have $\underline{\mathbf{u}}^*(t) \le \mathbf{u}^*(t) \le \overline{\mathbf{u}}^*(t),\ t \in T$.

**Proof.** We use induction to show that for all $k = 1, \cdots, K$

$$
\underline{\mathbf{u}}(t_k) \le \underline{\mathbf{u}}^{(m)}(t_k) \le \underline{\mathbf{u}}^{(m+1)}(t_k) \le \overline{\mathbf{u}}^{(m+1)}(t_k) \le \overline{\mathbf{u}}^{(m)}(t_k) \le \overline{\mathbf{u}}(t_k), \quad m = 0, 1, \cdots. \tag{9.8.8}
$$

From (9.8.5), Definition 9.8.2 and (9.8.6) for all $k = 1, \cdots, K$, we have

$$
\begin{aligned}
&\nabla(\overline{\mathbf{u}}^{(1)}(t_k) - \overline{\mathbf{u}}(t_k)) - \mathcal{M}(\overline{\mathbf{u}}^{(1)}(t_k) - \overline{\mathbf{u}}(t_k)) \ge_D 0 \\
&\nabla(\underline{\mathbf{u}}(t_k) - \underline{\mathbf{u}}^{(1)}(t_k)) - \mathcal{M}(\underline{\mathbf{u}}(t_k) - \underline{\mathbf{u}}^{(1)}(t_k)) \ge_D 0 \\
&\nabla(\underline{\mathbf{u}}^{(1)}(t_k) - \overline{\mathbf{u}}^{(1)}(t_k)) - \mathcal{M}(\underline{\mathbf{u}}^{(1)}(t_k) - \overline{\mathbf{u}}^{(1)}(t_k)) \\
&\qquad = \mathbf{f}(t_k, \underline{\mathbf{u}}(t_k)) - \mathbf{f}(t_k, \overline{\mathbf{u}}(t_k)) - \mathcal{M}(\underline{\mathbf{u}}(t_k) - \overline{\mathbf{u}}(t_k)) \ge_D 0.
\end{aligned}
$$

Thus, in view of Lemma 9.8.2 the above inequalities imply that (9.8.8) holds for $k = 0$. Now assume that (9.8.8) holds for some $k \geq 0$. Then, from (9.8.5) and (9.8.6) for all $k = 1, \cdots, K$, we have

$$\begin{aligned}&\nabla(\underline{\mathbf{u}}^{(m+1)}(t_k) - \underline{\mathbf{u}}^{(m+2)}(t_k)) - \mathcal{M}(\underline{\mathbf{u}}^{(m+1)}(t_k) - \underline{\mathbf{u}}^{(m+2)}(t_k))\\ &= \mathbf{f}(t_k, \underline{\mathbf{u}}^{(m)}(t_k)) - \mathbf{f}(t_k, \underline{\mathbf{u}}^{(m+1)}(t_k)) - \mathcal{M}(\underline{\mathbf{u}}^{(m)}(t_k) - \underline{\mathbf{u}}^{(m+1)}(t_k)) \geq_D 0.\end{aligned}$$

Again on applying Lemma 9.8.2 to the above inequality, we get $\underline{\mathbf{u}}^{(m+1)}(t_k) \leq \underline{\mathbf{u}}^{(m+2)}(t_k)$, $k = 1, 2, \cdots, K$. A similar argument gives that $\overline{\mathbf{u}}^{(m+2)}(t_k) \leq \overline{\mathbf{u}}^{(m+1)}(t_k)$ and $\underline{\mathbf{u}}^{(m+2)}(t_k) \leq \overline{\mathbf{u}}^{(m+2)}(t_k)$, $k = 1, 2, \cdots, K$. This completes the induction for (9.8.8).

In view of (9.8.8), there exist limits $\underline{\mathbf{u}}^*(t_k)$ and $\overline{\mathbf{u}}^*(t_k)$ such that

$$\lim_{m\to\infty} \underline{\mathbf{u}}^{(m)}(t_k) = \underline{\mathbf{u}}^*(t_k), \qquad \lim_{m\to\infty} \overline{\mathbf{u}}^{(m)}(t_k) = \overline{\mathbf{u}}^*(t_k), \qquad k = 0, 1, \cdots, K$$

and (9.8.7) holds. Letting $m \to \infty$ in (9.8.5) we see that $\underline{\mathbf{u}}^*(t_k)$ and $\overline{\mathbf{u}}^*(t_k)$ are solutions of (9.8.1).

Now if $\mathbf{u}^*(t)$ is any solution of (9.8.1) in $W$, then it is also a $D$ lower periodic solution of (9.8.1) in $W$. Using $\underline{\mathbf{u}}^{(0)}(t_k) = \mathbf{u}^*(t_k)$ and $\overline{\mathbf{u}}^{(0)}(t_k) = \overline{\mathbf{u}}(t_k)$ $(k = 1, 2, \cdots, K)$, the above arguments imply that $\overline{\mathbf{u}}^*(t_k) \geq \mathbf{u}^*(t_k)$ $(k = 1, 2, \cdots, K)$. Similarly, using $\mathbf{u}^*(t_k)$ as a $D$ upper periodic solution of (9.8.1) in $W$ we get $\underline{\mathbf{u}}^*(t_k) \leq \mathbf{u}^*(t_k)$ $(k = 1, 2, \cdots, K)$. This proves that $\underline{\mathbf{u}}^*(t_k) \leq \mathbf{u}^*(t_k) \leq \overline{\mathbf{u}}^*(t_k)$, $k = 1, 2, \cdots, K$. ∎

If in addition to the assumptions of Theorem 9.8.3, $\mathbf{f}$ satisfies some uniqueness condition in $W$, then $\mathbf{u}(t_k) = \overline{\mathbf{u}}^*(t_k) = \underline{\mathbf{u}}^*(t_k)$ $(k = 0, 1, \cdots, K)$ is the unique solution of (9.8.1) in $W$.

**Remark 9.8.2.** The condition (9.8.6) is equivalent to that for all $t \in T$,

$$\begin{aligned}f_j(t, \mathbf{u}(t)) - f_j(t, \tilde{\mathbf{u}}(t)) &\leq M_j(u_j(t) - \tilde{u}_j(t)), \qquad 1 \geq M_j > 0, \quad j \in N_1\\ f_j(t, \mathbf{u}(t)) - f_j(t, \tilde{\mathbf{u}}(t)) &\geq M_j(u_j(t) - \tilde{u}_j(t)), \qquad M_j < 0, \quad j \in N_2\end{aligned}$$

whenever $\underline{\mathbf{u}}(t) \leq \tilde{\mathbf{u}}(t) \leq \mathbf{u}(t) \leq \overline{\mathbf{u}}(t)$, $t \in T$.

**Remark 9.8.3.** It is clear that the monotone convergence of $\{\mathbf{u}^{(m)}(t_k)\}$ holds if we consider instead of (9.8.5) the following iterative scheme

$$\begin{aligned}&\nabla u_j^{(m)}(t_k) - M_j^* u_j^{(m)}(t_k)\\ &\quad = f_j(t_k, u_1^{(m)}(t_k), \cdots, u_{j-1}^{(m)}(t_k), u_j^{(m-1)}(t_k), \cdots, u_n^{(m-1)}(t_k))\\ &\qquad\qquad - M_j^* u_j^{(m-1)}(t_k), \qquad 1 \leq k \leq K,\ 1 \leq j \leq n\\ &u^{(m)}(t_0) = u^{(m)}(t_K).\end{aligned} \tag{9.8.9}$$

Moreover, the scheme (9.8.9) will give better results than (9.8.5).

**Theorem 9.8.4.** Assume that the conditions of Theorem 9.8.1 hold. Let $\{\overline{\mathbf{u}}^{(m)}(t_k)\}$ and $\{\underline{\mathbf{u}}^{(m)}(t_k)\}$ denote the sequences generated by (9.8.5) with $M^* = M$ and the initial functions $\overline{\mathbf{u}}^{(0)}(t_k) = \overline{\mathbf{u}}(t_k)$ and $\underline{\mathbf{u}}^{(0)}(t_k) = \underline{\mathbf{u}}(t_k)$, respectively. Also, let $\{\overline{\mathbf{u}}'^{(m)}(t_k)\}$ and $\{\underline{\mathbf{u}}'^{(m)}(t_k)\}$ denote sequences generated from (9.8.9) with $M_j^* = M_j$ $(j = 1, 2, \cdots, n)$ and the same initial functions $\overline{\mathbf{u}}'^{(0)}(t_k) = \overline{\mathbf{u}}(t_k)$ and $\underline{\mathbf{u}}'^{(0)}(t_k) = \underline{\mathbf{u}}(t_k)$, respectively. Then all these sequences have the same monotone convergence established in Theorem 9.8.3. Moreover,

$$\overline{\mathbf{u}}^{(m)}(t_k) \geq \overline{\mathbf{u}}'^{(m)}(t_k), \quad \underline{\mathbf{u}}^{(m)}(t_k) \leq \underline{\mathbf{u}}'^{(m)}(t_k), \qquad k = 0, 1, 2, \cdots, K, \quad m = 0, 1, \cdots. \tag{9.8.10}$$

**Proof.** From the monotone convergence of the sequences for all $k = 0, 1, 2, \cdots, K$ and $m = 0, 1, \cdots,$ we have

$$\begin{aligned} &\underline{\mathbf{u}}(t_k) \leq \underline{\mathbf{u}}^{(m)}(t_k) \leq \underline{\mathbf{u}}^{(m+1)}(t_k) \leq \overline{\mathbf{u}}^{(m+1)}(t_k) \leq \overline{\mathbf{u}}^{(m)}(t_k) \leq \overline{\mathbf{u}}(t_k) \\ &\underline{\mathbf{u}}(t_k) \leq \underline{\mathbf{u}}'^{(m)}(t_k) \leq \underline{\mathbf{u}}'^{(m+1)}(t_k) \leq \overline{\mathbf{u}}'^{(m+1)}(t_k) \leq \overline{\mathbf{u}}'^{(m)}(t_k) \leq \overline{\mathbf{u}}(t_k). \end{aligned} \tag{9.8.11}$$

Clearly (9.8.10) is true for $m = 0$. Suppose that (9.8.10) holds for some $m \geq 0$. Let $\overline{\mathbf{w}}^{(m+1)}(t_k) = \overline{\mathbf{u}}'^{(m+1)}(t_k) - \overline{\mathbf{u}}^{(m+1)}(t_k)$, $k = 0, 1, \cdots, K$. Then, $\overline{\mathbf{w}}^{(m+1)}(t_0) = \overline{\mathbf{w}}^{(m+1)}(t_K)$ and for $k = 1, 2, \cdots, K$ and $j = 1, 2, \cdots, n$,

$$\begin{aligned} &\nabla \overline{w}_j^{(m+1)}(t_k) - M_j \overline{w}_j^{(m+1)}(t_k) \\ &= f_j(t_k, \overline{u}_1'^{(m+1)}(t_k), \cdots, \overline{u}_{j-1}'^{(m+1)}(t_k), \overline{u}_j'^{(m)}(t_k), \cdots, \overline{u}_n'^{(m)}(t_k)) \\ &\quad - f_j(t_k, \overline{\mathbf{u}}^{(m)}(t_k)) - M_j \left(\overline{u}_j'^{(m)}(t_k) - \overline{u}_j^{(m)}(t_k)\right). \end{aligned} \tag{9.8.12}$$

If $1 \in N_1$, we have from (9.8.10), (9.8.12) and (9.8.6) that

$$\nabla \overline{w}_1^{(m+1)}(t_k) - M_1 \overline{w}_1^{(m+1)}(t_k) \geq 0$$

from which and Lemma 9.8.1 we find $\overline{w}_1^{(m+1)}(t_k) \leq 0$, $k = 1, 2, \cdots, K$. Similarly, we have $\overline{w}_1^{(m+1)}(t_k) \leq 0$, $k = 1, 2, \cdots, K$ if $1 \in N_2$. Assume that there exits some $n > \tilde{j} \geq 1$ such that $\overline{w}_j^{(m+1)}(t_k) \leq 0$ for all $j = 1, 2, \cdots, \tilde{j}$ and $k = 1, 2, \cdots, K$. If $\tilde{j} + 1 \in N_1$, we have from (9.8.10) (9.8.12) and (9.8.6) that

$$\nabla \overline{w}_{\tilde{j}+1}^{(m+1)}(t_k) - M_{\tilde{j}+1} \overline{w}_{\tilde{j}+1}^{(m+1)}(t_k) \geq 0$$

from which and Lemma 9.8.1 we get $\overline{w}_{\tilde{j}+1}^{(m+1)}(t_k) \leq 0$, $k = 1, 2, \cdots, K$. Similarly, we obtain $\overline{w}_{\tilde{j}+1}^{(m+1)}(t_k) \leq 0$, $k = 1, 2, \cdots, K$ if $\tilde{j} + 1 \in N_2$. By induction this proves $\overline{w}_j^{(m+1)}(t_k) \le 0$, i.e. $\overline{\mathbf{u}}'^{(m+1)}(t_k) \le \overline{\mathbf{u}}^{(m+1)}(t_k)$, $k = 1, 2, \cdots, K$. Analogously, we have $\underline{\mathbf{u}}^{(m+1)}(t_k) \le \underline{\mathbf{u}}'^{(m+1)}(t_k)$, $k = 1, 2, \cdots, K$. This shows that (9.8.10) holds for $k+1$. ∎

Now let $t_k = \pi k/K$, $k = 0, 1, \cdots, K$ and consider the periodic boundary value problem (9.8.1) where $\mathbf{u}(t) = (u_1(t), u_2(t))^T$ and $\mathbf{f} = (f_1, f_2)^T$ with

$$\begin{aligned} f_1(t_k, \mathbf{u}(t_k)) &= \frac{\pi}{K}(u_1(t_k) - u_2(t_k)), \qquad 1 \le k \le K \\ f_2(t_k, \mathbf{u}(t_k)) &= \frac{\pi}{K}\left(u_1(t_k) - 2u_2(t_k) + \frac{1}{2}\sin(t_k)\right), \qquad 1 \le k \le K. \end{aligned}$$

Here $N = \{1, 2\}$. We take $N_1 = \{1\}$ and $N_2 = \{2\}$ so that $D = \{N_1, N_2\}$ is a decomposition of $N$. Let $\underline{\mathbf{u}}(t) = (0, 0)^T$ and $\overline{\mathbf{u}}(t) = (1.5, 1)^T$. It is clear that $\overline{\mathbf{u}}(t) \ge \underline{\mathbf{u}}(t)$, $t \in T$ and $\mathbf{f}$ is $D$ quasimonotone in $W = \{\mathbf{u}(t) : \underline{\mathbf{u}}(t) \le \mathbf{u}(t) \le \overline{\mathbf{u}}(t),\ t \in T\}$. Moreover, we find easily that $\underline{\mathbf{u}}(t)$ and $\overline{\mathbf{u}}(t)$ are $D$ lower periodic solution and $D$ upper periodic solution, respectively. Set $\mathcal{M} = \mathrm{diag}(\pi/K, -2\pi/K)$ and we have

$$\mathbf{f}(t, \mathbf{u}(t)) - \mathbf{f}(t, \tilde{\mathbf{u}}(t)) \le_D \mathcal{M}(\mathbf{u}(t) - \tilde{\mathbf{u}}(t)), \qquad t \in T$$

whenever $\underline{\mathbf{u}}(t) \le \tilde{\mathbf{u}}(t) \le \mathbf{u}(t) \le \overline{\mathbf{u}}(t)$, $t \in T$. Let $K = 20$. We use iterative scheme (9.8.5) with $\mathcal{M}^* = \mathcal{M}$ to solve (9.8.1) and denote by $\mathbf{u}^{(m)}(t)$ the $m$th iteration. In practical computations, the monotone convergence of the sequence $\{\mathbf{u}^{(m)}(t)\}$ is observed. If $\mathbf{u}^{(0)}(t) = \overline{\mathbf{u}}(t)$, then $\{\mathbf{u}^{(m)}(t)\}$ is a monotone nonincreasing sequence (see Table 9.8.1). If $\mathbf{u}^{(0)}(t) = \underline{\mathbf{u}}(t)$, then $\{\mathbf{u}^{(m)}(t)\}$ is a monotone nondecreasing sequence (see Table 9.8.2). The monotonicity in Tables 9.8.1 and 9.8.2 agrees with the one described in Theorem 9.8.3.

**Table** 9.8.1.

| $m$ | $u_1^{(m)}(t_2)$ | $u_1^{(m)}(t_4)$ | $u_1^{(m)}(t_6)$ | $u_2^{(m)}(t_2)$ | $u_2^{(m)}(t_4)$ | $u_2^{(m)}(t_6)$ |
|---|---|---|---|---|---|---|
| 1 | 1.000000 | 1.000000 | 1.000000 | 0.830328 | 0.852288 | 0.889736 |
| 5 | 0.477053 | 0.501266 | 0.518673 | 0.375851 | 0.404161 | 0.449332 |
| 10 | 0.344072 | 0.368578 | 0.385917 | 0.249386 | 0.277880 | 0.323163 |
| 13 | 0.317788 | 0.342292 | 0.359629 | 0.237306 | 0.265802 | 0.311083 |
| 26 | 0.307271 | 0.331774 | 0.349111 | 0.228153 | 0.256648 | 0.301929 |

**Table** 9.8.2.

| $m$ | $u_1^{(m)}(t_2)$ | $u_1^{(m)}(t_4)$ | $u_1^{(m)}(t_6)$ | $u_2^{(m)}(t_2)$ | $u_2^{(m)}(t_4)$ | $u_2^{(m)}(t_6)$ |
|---|---|---|---|---|---|---|
| 1 | 0.000000 | 0.000000 | 0.000000 | 0.080328 | 0.102288 | 0.139736 |
| 5 | 0.227053 | 0.251266 | 0.268673 | 0.188351 | 0.216661 | 0.261832 |
| 10 | 0.297197 | 0.321703 | 0.339042 | 0.218136 | 0.246630 | 0.291913 |
| 13 | 0.302163 | 0.326667 | 0.344004 | 0.225588 | 0.254083 | 0.299364 |
| 23 | 0.306972 | 0.331475 | 0.348812 | 0.227992 | 0.256487 | 0.301768 |

Next, we construct the sequence $\{\mathbf{u}^{(m)}(t)\}$ by following the scheme (9.8.9) with $\mathcal{M}^* = \mathcal{M}$. Table 9.8.3 lists the values of $\mathbf{u}^{(m)}(t)$ when

$\mathbf{u}^{(0)}(t) = \overline{\mathbf{u}}(t)$, and Table 9.8.4 gives the corresponding values when $\mathbf{u}^{(0)}(t) = \underline{\mathbf{u}}(t)$. Tables 8.8.3 and 9.8.4 illustrate the monotonicity established in Theorem 9.8.3. Moreover, a comparison of the above Tables 9.8.1 9.8.4 shows that the scheme (9.8.9) indeed gives better results than obtained from (9.8.5).

**Table 9.8.3.**

| $m$ | $u_1^{(m)}(t_2)$ | $u_1^{(m)}(t_4)$ | $u_1^{(m)}(t_6)$ | $u_2^{(m)}(t_2)$ | $u_2^{(m)}(t_4)$ | $u_2^{(m)}(t_6)$ |
|---|---|---|---|---|---|---|
| 1 | 1.000000 | 1.000000 | 1.000000 | 0.580328 | 0.602288 | 0.639735 |
| 5 | 0.349749 | 0.374261 | 0.391611 | 0.249386 | 0.277880 | 0.323163 |
| 10 | 0.308459 | 0.332963 | 0.350299 | 0.228736 | 0.257231 | 0.302512 |
| 13 | 0.307293 | 0.331797 | 0 349133 | 0.228153 | 0.256648 | 0.301929 |

**Table 9.8.4.**

| $k$ | $u_1^{(m)}(t_2)$ | $u_1^{(m)}(t_4)$ | $u_1^{(m)}(t_6)$ | $u_2^{(m)}(t_2)$ | $u_2^{(m)}(t_4)$ | $u_2^{(m)}(t_6)$ |
|---|---|---|---|---|---|---|
| 1 | 0.000000 | 0.000000 | 0.000000 | 0.080328 | 0.102288 | 0.139736 |
| 5 | 0.287250 | 0.311761 | 0.329111 | 0.218136 | 0.246630 | 0.291913 |
| 10 | 0.306506 | 0.331010 | 0.348346 | 0.227759 | 0.256255 | 0.301535 |
| 13 | 0.307049 | 0.331553 | 0.348889 | 0.228031 | 0.256526 | 0.301807 |

## 9.9. Newton's Method

For the boundary value problem (1.2.8), (1.5.1) Newton's method leads to the construction of the sequence $\{\mathbf{u}^m(k)\}$ generated by the iterative scheme

$$(9.9.1)\quad \begin{aligned} \mathbf{u}^{m+1}(k+1) &= \mathbf{f}(k,\mathbf{u}^m(k)) + \mathbf{f}_{\mathbf{u}}(k,\mathbf{u}^m(k))(\mathbf{u}^{m+1}(k) - \mathbf{u}^m(k)) \\ &\mathcal{F}[\mathbf{u}^m] + \mathcal{F}_{\mathbf{u}}[\mathbf{u}^m][\mathbf{u}^{m+1} - \mathbf{u}^m] = 0, \quad m = 0, 1, \cdots \end{aligned}$$

where $\mathbf{u}^0(k) = \overline{\mathbf{u}}(k)$. In the following result we shall provide sufficient conditions so that this sequence $\{\mathbf{u}^m(k)\}$ indeed exists and converges to the unique solution of (1.2.8), (1.5.1).

**Theorem 9.9.1.** With respect to the boundary value problem (1.2.8), (1.5.1) we assume that there exists an approximate solution $\overline{\mathbf{u}}(k)$ and conditions (i) (iv) of Theorem 9.2.2 are satisfied. Further, let $3\rho(\mathcal{K}) < 1$, and $\mathbf{r}^{11} = (\mathcal{I} - 3\mathcal{K})^{-1}(\mathcal{M}^1\mathbf{d}^1 + \mathcal{M}^2\mathbf{d}^2) \le \mathbf{r}$. Then, the following hold

1. the sequence $\{\mathbf{u}^m(k)\}$ obtained by Newton's scheme (9.9.1) remains in $\overline{S}(\overline{\mathbf{u}}, \mathbf{r}^{11})$

2. the sequence $\{\mathbf{u}^m(k)\}$ converges to the unique solution $\mathbf{u}^*(k)$ of (1.2.8), (1.5.1)

3. a bound on the error involving the matrix $\mathcal{K}^* = 2\mathcal{K}(\mathcal{I} - \mathcal{K})^{-1}$ is given by

$$(9.9.2)\quad \|\mathbf{u}^m - \mathbf{u}^*\|_G \le (\mathcal{K}^*)^m(\mathcal{I} - \mathcal{K}^*)^{-1}\|\mathbf{u}^1 - \overline{\mathbf{u}}\|_G$$

$$(9.9.3)\quad \le (\mathcal{K}^*)^m(\mathcal{I} - 3\mathcal{K})^{-1}(\mathcal{M}^1\mathbf{d}^1 + \mathcal{M}^2\mathbf{d}^2).$$

**Proof.** First we shall show that $\{\mathbf{u}^m(k)\} \subseteq \overline{S}(\overline{\mathbf{u}}, \mathbf{r}^{11})$. For this, on $B(a,b)$ we define an implicit operator $\mathcal{T}$ as follows

$$(9.9.4)\quad \begin{aligned}\mathcal{T}\mathbf{u}(k) &= \mathcal{H}^1[\mathbf{f}(k,\mathbf{u}(k)) + \mathbf{f_u}(k,\mathbf{u}(k))(\mathcal{T}\mathbf{u}(k) - \mathbf{u}(k)) - \mathcal{A}(k)\mathcal{T}\mathbf{u}(k)] \\ &\quad + \mathcal{H}^2[\mathcal{L}[\mathcal{T}\mathbf{u}] - \mathcal{F}[\mathbf{u}] - \mathcal{F}_\mathbf{u}[\mathbf{u}][\mathcal{T}\mathbf{u} - \mathbf{u}]]\end{aligned}$$

whose form is patterned similar to that of (9.2.4) for (9.9.1).

Since $\overline{\mathbf{u}}(k) \in \overline{S}(\overline{\mathbf{u}}, \mathbf{r}^{11})$, it suffices to show that if $\mathbf{u}(k) \in \overline{S}(\overline{\mathbf{u}}, \mathbf{r}^{11})$, then $\mathcal{T}\mathbf{u}(k) \in \overline{S}(\overline{\mathbf{u}}, \mathbf{r}^{11})$. For this, let $\mathbf{u}(k) \in \overline{S}(\overline{\mathbf{u}}, \mathbf{r}^{11})$, then from (9.2.6) and (9.9.4), we have

$$\begin{aligned}\mathcal{T}\mathbf{u}(k) - \overline{\mathbf{u}}(k) &= \mathcal{H}^1[\mathbf{f}(k,\mathbf{u}(k)) + \mathbf{f_u}(k,\mathbf{u}(k))(\mathcal{T}\mathbf{u}(k) - \mathbf{u}(k)) \\ &\quad - \mathcal{A}(k)\mathcal{T}\mathbf{u}(k) - \mathbf{f}(k,\overline{\mathbf{u}}(k)) - \mathbf{q}(k) + \mathcal{A}(k)\overline{\mathbf{u}}(k)] \\ &\quad + \mathcal{H}^2[\mathcal{L}[\mathcal{T}\mathbf{u}] - \mathcal{F}[\mathbf{u}] - \mathcal{F}_\mathbf{u}[\mathbf{u}][\mathcal{T}\mathbf{u} - \mathbf{u}] - \mathcal{L}[\overline{\mathbf{u}}] + \mathcal{F}[\overline{\mathbf{u}}] - \mathbf{l}^1] \\ (9.9.5)\qquad &= \mathcal{H}^1[\mathbf{f}(k,\mathbf{u}(k)) - \mathbf{f}(k,\overline{\mathbf{u}}(k)) - \mathcal{A}(k)(\mathbf{u}(k) - \overline{\mathbf{u}}(k)) \\ &\quad + (\mathbf{f_u}(k,\mathbf{u}(k)) - \mathcal{A}(k))(\mathcal{T}\mathbf{u}(k) - \mathbf{u}(k)) - \mathbf{q}(k)] \\ &\quad + \mathcal{H}^2[-(\mathcal{F}[\mathbf{u}] - \mathcal{F}[\overline{\mathbf{u}}] - \mathcal{L}[\mathbf{u} - \overline{\mathbf{u}}]) - (\mathbf{f_u}[\mathbf{u}] - \mathcal{L})[\mathcal{T}\mathbf{u} - \mathbf{u}] - \mathbf{l}^1]\end{aligned}$$

and now as in Theorem 9.2.2 it follows that

$$\begin{aligned}\|\mathcal{T}\mathbf{u} - \overline{\mathbf{u}}\|_G &\le \mathcal{M}^1\left[\mathcal{M}^3\|\mathbf{u} - \overline{\mathbf{u}}\|_G + \mathcal{M}^3\|\mathcal{T}\mathbf{u} - \mathbf{u}\|_G + \mathbf{d}^1\right] \\ &\quad + \mathcal{M}^2\left[\mathcal{M}^4\|\mathbf{u} - \overline{\mathbf{u}}\|_G + \mathcal{M}^4\|\mathcal{T}\mathbf{u} - \mathbf{u}\|_G + \mathbf{d}^2\right] \\ &\le \mathcal{M}^1\left[2\mathcal{M}^3\|\mathbf{u} - \overline{\mathbf{u}}\|_G + \mathcal{M}^3\|\mathcal{T}\mathbf{u} - \overline{\mathbf{u}}\|_G + \mathbf{d}^1\right] \\ &\quad + \mathcal{M}^2\left[2\mathcal{M}^4\|\mathbf{u} - \overline{\mathbf{u}}\|_G + \mathcal{M}^4\|\mathcal{T}\mathbf{u} - \overline{\mathbf{u}}\|_G + \mathbf{d}^2\right] \\ &= 2\mathcal{K}\|\mathbf{u} - \overline{\mathbf{u}}\|_G + \mathcal{K}\|\mathcal{T}\mathbf{u} - \overline{\mathbf{u}}\|_G + \left(\mathcal{M}^1\mathbf{d}^1 + \mathcal{M}^2\mathbf{d}^2\right),\end{aligned}$$

which is the same as

$$\|\mathcal{T}\mathbf{u} - \overline{\mathbf{u}}\|_G \le 2\mathcal{K}(\mathcal{I} - \mathcal{K})^{-1}\|\mathbf{u} - \overline{\mathbf{u}}\|_G + (\mathcal{I} - \mathcal{K})^{-1}\left(\mathcal{M}^1\mathbf{d}^1 + \mathcal{M}^2\mathbf{d}^2\right).$$

Thus, we find that

$$\begin{aligned}\|\mathcal{T}\mathbf{u} - \overline{\mathbf{u}}\|_G &\le 2\mathcal{K}(\mathcal{I} - \mathcal{K})^{-1}(\mathcal{I} - 3\mathcal{K})^{-1}\left(\mathcal{M}^1\mathbf{d}^1 + \mathcal{M}^2\mathbf{d}^2\right) \\ &\quad + (\mathcal{I} - \mathcal{K})^{-1}\left(\mathcal{M}^1\mathbf{d}^1 + \mathcal{M}^2\mathbf{d}^2\right) \\ &= (\mathcal{I} - \mathcal{K})^{-1}\left[2\mathcal{K}(\mathcal{I} - 3\mathcal{K})^{-1} + \mathcal{I}\right]\left(\mathcal{M}^1\mathbf{d}^1 + \mathcal{M}^2\mathbf{d}^2\right) \\ &= (\mathcal{I} - \mathcal{K})^{-1}(\mathcal{I} - \mathcal{K})(\mathcal{I} - 3\mathcal{K})^{-1}\left(\mathcal{M}^1\mathbf{d}^1 + \mathcal{M}^2\mathbf{d}^2\right) \\ &= (\mathcal{I} - 3\mathcal{K})^{-1}\left(\mathcal{M}^1\mathbf{d}^1 + \mathcal{M}^2\mathbf{d}^2\right).\end{aligned}$$

Therefore, $\|\mathcal{T}\mathbf{u} - \overline{\mathbf{u}}\|_G \le \mathbf{r}^{11}$ follows from the definition of $\mathbf{r}^{11}$.

Next, we shall show the convergence of the sequence $\{\mathbf{u}^m(k)\}$. From (9.9.1), in view of (9.9.4), we have

$$
\begin{aligned}
(9.9.6)\quad \mathbf{u}^{m+1}(k)-\mathbf{u}^m(k) &= \mathcal{H}^1\big[\mathbf{f}(k,\mathbf{u}^m(k))+\mathbf{f}_{\mathbf{u}}(k,\mathbf{u}^m(k))\big(\mathbf{u}^{m+1}(k)\\
&\quad -\mathbf{u}^m(k)\big)-\mathcal{A}(k)\mathbf{u}^{m+1}(k)-\mathbf{f}(k,\mathbf{u}^{m-1}(k))\\
&\quad -\mathbf{f}_{\mathbf{u}}(k,\mathbf{u}^{m-1}(k))\big(\mathbf{u}^m(k)-\mathbf{u}^{m-1}(k)\big)+\mathcal{A}(k)\mathbf{u}^m(k)\big]\\
&\quad +\mathcal{H}^2\big[\mathcal{L}[\mathbf{u}^{m+1}]-\mathcal{F}[\mathbf{u}^m]-\mathcal{F}_{\mathbf{u}}[\mathbf{u}^m][\mathbf{u}^{m+1}-\mathbf{u}^m]-\mathcal{L}[\mathbf{u}^m]\\
&\quad +\mathcal{F}[\mathbf{u}^{m-1}]+\mathcal{F}_{\mathbf{u}}[\mathbf{u}^{m-1}][\mathbf{u}^m-\mathbf{u}^{m-1}]\big]\\
&= \mathcal{H}^1\big[\mathbf{f}(k,\mathbf{u}^m(k))-\mathbf{f}(k,\mathbf{u}^{m-1}(k))-\mathcal{A}(k)\big(\mathbf{u}^m(k)-\mathbf{u}^{m-1}(k)\big)\\
&\quad +\big(\mathbf{f}_{\mathbf{u}}(k,\mathbf{u}^m(k))-\mathcal{A}(k)\big)\big(\mathbf{u}^{m+1}(k)-\mathbf{u}^m(k)\big)\\
&\quad -\big(\mathbf{f}_{\mathbf{u}}(k,\mathbf{u}^{m-1}(k))-\mathcal{A}(k)\big)\big(\mathbf{u}^m(k)-\mathbf{u}^{m-1}(k)\big)\big]\\
&\quad +\mathcal{H}^2\big[-\big(\mathcal{F}[\mathbf{u}^m]-\mathcal{F}[\mathbf{u}^{m-1}]-\mathcal{L}[\mathbf{u}^m-\mathbf{u}^{m-1}]\big)\\
&\quad -(\mathcal{F}_{\mathbf{u}}[\mathbf{u}^m]-\mathcal{L})[\mathbf{u}^{m+1}-\mathbf{u}^m]+(\mathcal{F}_{\mathbf{u}}[\mathbf{u}^{m-1}]-\mathcal{L})[\mathbf{u}^m-\mathbf{u}^{m-1}]\big]
\end{aligned}
$$

and hence

$$
\begin{aligned}
\|\mathbf{u}^{m+1}-\mathbf{u}^m\|_G &\le \mathcal{M}^1\big[\mathcal{M}^3\|\mathbf{u}^m-\mathbf{u}^{m-1}\|_G+\mathcal{M}^3\|\mathbf{u}^{m+1}-\mathbf{u}^m\|_G\\
&\quad +\mathcal{M}^3\|\mathbf{u}^m-\mathbf{u}^{m-1}\|_G\big]+\mathcal{M}^2\big[\mathcal{M}^4\|\mathbf{u}^m-\mathbf{u}^{m-1}\|_G\\
&\quad +\mathcal{M}^4\|\mathbf{u}^{m+1}-\mathbf{u}^m\|_G+\mathcal{M}^4\|\mathbf{u}^m-\mathbf{u}^{m-1}\|_G\big]\\
&= 2\mathcal{K}\|\mathbf{u}^m-\mathbf{u}^{m-1}\|_G+\mathcal{K}\|\mathbf{u}^{m+1}-\mathbf{u}^m\|_G,
\end{aligned}
$$

which is the same as

$$\|\mathbf{u}^{m+1}-\mathbf{u}^m\|_G \le 2\mathcal{K}(\mathcal{I}-\mathcal{K})^{-1}\|\mathbf{u}^m-\mathbf{u}^{m-1}\|_G.$$

Thus, by an easy induction, we get

$$(9.9.7)\qquad \|\mathbf{u}^{m+1}-\mathbf{u}^m\|_G \le (2\mathcal{K}(\mathcal{I}-\mathcal{K})^{-1})^m\|\mathbf{u}^1-\mathbf{u}^0\|_G.$$

However, since $3\rho(\mathcal{K})<1$ implies that $\rho(2\mathcal{K}(\mathcal{I}-\mathcal{K})^{-1})<1$, from (9.9.7) it is clear that $\{\mathbf{u}^m(k)\}$ is a Cauchy sequence, and therefore converges to some $\mathbf{u}^*(k)\in\overline{S}(\overline{\mathbf{u}},\mathbf{r}^{11})$. This $\mathbf{u}^*(k)$ is the unique solution of (1.2.8), (1.5.1) can easily be verified.

The error bound (9.9.2) follows from (9.9.7) and the triangle inequality

$$
\begin{aligned}
\|\mathbf{u}^{m+p}-\mathbf{u}^m\| &\le \sum_{i=0}^{p-1}\|\mathbf{u}^{m+p-i}-\mathbf{u}^{m+p-i-1}\|_G\\
&\le \sum_{i=0}^{p-1}(\mathcal{K}^*)^{m+p-i-1}\|\mathbf{u}^1-\overline{\mathbf{u}}\|_G\\
&= (\mathcal{K}^*)^m\sum_{i=0}^{p-1}(\mathcal{K}^*)^i\|\mathbf{u}^1-\overline{\mathbf{u}}\|_G\\
&\le (\mathcal{K}^*)^m(\mathcal{I}-\mathcal{K}^*)^{-1}\|\mathbf{u}^1-\overline{\mathbf{u}}\|_G
\end{aligned}
$$

and now taking the limit as $p \to \infty$.

Finally, from (9.9.5) we have

$$\begin{aligned}\mathbf{u}^1(k) - \overline{\mathbf{u}}(k) &= \mathcal{H}^1[(\mathbf{f_u}(k, \overline{\mathbf{u}}(k)) - \mathcal{A}(k))(\mathbf{u}^1(k) - \overline{\mathbf{u}}(k)) - \mathbf{q}(k)] \\ &\quad + \mathcal{H}^2[-(\mathcal{F}_\mathbf{u}[\overline{\mathbf{u}}] - \mathcal{L})[\mathbf{u}^1 - \overline{\mathbf{u}}] - \mathbf{l}^1]\end{aligned}$$

and hence

$$\|\mathbf{u}^1 - \overline{\mathbf{u}}\|_G \le \mathcal{M}^1[\mathcal{M}^3\|\mathbf{u}^1 - \overline{\mathbf{u}}\|_G + \mathbf{d}^1] + \mathcal{M}^2[\mathcal{M}^4\|\mathbf{u}^1 - \overline{\mathbf{u}}\|_G + \mathbf{d}^2],$$

which is the same as

$$\|\mathbf{u}^1 - \overline{\mathbf{u}}\|_G \le (\mathcal{I} - \mathcal{K})^{-1}(\mathcal{M}^1\mathbf{d}^1 + \mathcal{M}^2\mathbf{d}^2). \tag{9.9.8}$$

Using this inequality in (9.9.2) and the fact that $(\mathcal{I} - \mathcal{K}^*)^{-1}(\mathcal{I} - \mathcal{K})^{-1} = (\mathcal{I} - 3\mathcal{K})^{-1}$, the required estimate (9.9.3) follows. ∎

In our next result in addition to the hypotheses of Theorem 9.9.1 we shall need the following conditions.

**Condition** $(d_1)$. For all $k \in \mathbb{N}(a, b-1)$ and $\mathbf{u}, \mathbf{v} \in \overline{S}(\overline{\mathbf{u}}, r^{11})$

$$|\mathbf{f}(k, \mathbf{u}) - \mathbf{f}(k, \mathbf{v}) - \mathbf{f_u}(k, \mathbf{v})(\mathbf{u} - \mathbf{v})| \le \mathcal{P} \cdot \|\mathbf{u} - \mathbf{v}\|_G \cdot \|\mathbf{u} - \mathbf{v}\|_G, \tag{9.9.9}$$

where $\mathcal{P} = (p_{ij\ell})$ is a symmetric tensor of the third order with nonnegative components. (Obviously, if $\mathbf{f}$ is twice continuously differentiable with respect to $\mathbf{u}$ for all $(k, \mathbf{u}) \in \mathbb{N}(a, b-1) \times \overline{S}(\overline{\mathbf{u}}, \mathbf{r}^{11})$ and all the second derivatives $\dfrac{\partial^2 f_i}{\partial u_j \partial u_\ell}$ are bounded there, then this condition is satisfied, with

$$p_{ij\ell} = \frac{1}{2} \sup_{(k,\mathbf{u}) \in \mathbb{N}(a,b-1) \times \overline{S}(\overline{\mathbf{u}}, \mathbf{r}^{11})} \left| \frac{\partial^2 f_i}{\partial u_j \partial u_\ell} \right|.\Bigg)$$

**Condition** $(d_2)$. For all $\mathbf{u}, \mathbf{v} \in \overline{S}(\overline{\mathbf{u}}, \mathbf{r}^{11})$

$$\|\mathcal{F}[\mathbf{u}] - \mathcal{F}[\mathbf{v}] - \mathcal{F}_\mathbf{u}[\mathbf{v}][\mathbf{u} - \mathbf{v}]\|_G \le \mathcal{Q} \cdot \|\mathbf{u} - \mathbf{v}\|_G \cdot \|\mathbf{u} - \mathbf{v}\|_G, \tag{9.9.10}$$

where $\mathcal{Q} = (q_{ij\ell})$ is a symmetric tensor of the third order with nonnegative components.

**Theorem 9.9.2.** Let in addition to the hypotheses of Theorem 9.9.1 conditions $(d_1)$ and $(d_2)$ be satisfied. Then, the following holds

$$\|\mathbf{u}^{m+1} - \mathbf{u}^m\|_G \le \mathcal{H} \cdot \|\mathbf{u}^m - \mathbf{u}^{m-1}\|_G \cdot \|\mathbf{u}^m - \mathbf{u}^{m-1}\|_G, \tag{9.9.11}$$

where $\mathcal{H} = (\mathcal{I} - \mathcal{K})^{-1}(\mathcal{M}^1\mathcal{P} + \mathcal{M}^2\mathcal{Q})$ is a tensor of the third order with nonnegative components.

**Proof.** From (9.9.6), we have

$$\begin{aligned}\mathbf{u}^{m+1}(k)-\mathbf{u}^m(k) &= \mathcal{H}^1\left[\mathbf{f}(k,\mathbf{u}^m(k)) - \mathbf{f}(k,\mathbf{u}^{m-1}(k))\right.\\ &\quad -\mathbf{f}_{\mathbf{u}}(k,\mathbf{u}^{m-1}(k))(\mathbf{u}^m(k) - \mathbf{u}^{m-1}(k))\\ &\quad \left.+(\mathbf{f}_{\mathbf{u}}(k,\mathbf{u}^m(k)) - \mathcal{A}(k))(\mathbf{u}^{m+1}(k) - \mathbf{u}^m(k))\right]\\ &\quad +\mathcal{H}^2\left[-\left(\mathcal{F}[\mathbf{u}^m]-\mathcal{F}[\mathbf{u}^{m-1}]-\mathcal{F}_{\mathbf{u}}[\mathbf{u}^{m-1}][\mathbf{u}^m-\mathbf{u}^{m-1}]\right)\right.\\ &\quad \left.-(\mathcal{F}_{\mathbf{u}}[\mathbf{u}^m] - \mathcal{L})[\mathbf{u}^{m+1}-\mathbf{u}^m]\right].\end{aligned}$$

Thus, on using the given hypotheses and the fact that $\{\mathbf{u}^m\} \subseteq \overline{S}(\overline{\mathbf{u}}, \mathbf{r}^{11})$, we obtain

$$\begin{aligned}&\|\mathbf{u}^{m+1}-\mathbf{u}^m\|_G\\ &\le \mathcal{M}^1\left[\mathcal{P}\cdot\|\mathbf{u}^m-\mathbf{u}^{m-1}\|_G\cdot\|\mathbf{u}^m-\mathbf{u}^{m-1}\|_G + \mathcal{M}^3\|\mathbf{u}^{m+1}-\mathbf{u}^m\|_G\right]\\ &\quad +\mathcal{M}^2\left[\mathcal{Q}\cdot\|\mathbf{u}^m-\mathbf{u}^{m-1}\|_G\cdot\|\mathbf{u}^m-\mathbf{u}^{m-1}\|_G + \mathcal{M}^4\|\mathbf{u}^{m+1}-\mathbf{u}^m\|_G\right]\\ &= (\mathcal{M}^1\mathcal{P}+\mathcal{M}^2\mathcal{Q})\cdot\|\mathbf{u}^m-\mathbf{u}^{m-1}\|_G\cdot\|\mathbf{u}^m-\mathbf{u}^{m-1}\|_G+\mathcal{K}\|\mathbf{u}^{m+1}-\mathbf{u}^m\|_G,\end{aligned}$$

which is the same as (9.9.11). ∎

**Remark 9.9.1.** In view of (9.9.7) and (9.9.8) the inequality (9.9.11) can be written as

$$\text{(9.9.12)}\quad \|\mathbf{u}^{m+1}-\mathbf{u}^m\|_G \le \mathcal{H}\cdot\left[(\mathcal{K}^*)^{m-1}(\mathcal{I}-\mathcal{K})^{-1}\left(\mathcal{M}^1\mathbf{d}^1+\mathcal{M}^2\mathbf{d}^2\right)\right]\cdot \left[(\mathcal{K}^*)^{m-1}(\mathcal{I}-\mathcal{K})^{-1}\left(\mathcal{M}^1\mathbf{d}^1+\mathcal{M}^2\mathbf{d}^2\right)\right].$$

**Example 9.9.1.** For the boundary value problem (9.2.16), (9.2.17) we follow as in Example 9.2.1 and note that $\rho(3\mathcal{K}) < 1$ if $(3/8)e^r < 1$, i.e. $r \le 0.980829\cdots$. Further, $(\mathcal{I} - 3\mathcal{K})^{-1}\left(\mathcal{M}^1\mathbf{d}^1 + \mathcal{M}^2\mathbf{d}^2\right) \le \mathbf{r}$ provided $\frac{1}{8}\left(1-\frac{3}{8}e^r\right)^{-1} \le r$, i.e. $0.238565607\cdots \le r \le 0.814141745\cdots$. Therefore, if $r = 0.238565607$ then all the conditions of Theorem 9.9.1 are satisfied and the Newton's scheme (9.9.1) for the problem (9.2.16), (9.2.17) converges to the unique solution $\mathbf{u}^*(k)$ in $\overline{S}(0,\mathbf{r}^{11}) = \{\mathbf{u}(k) \in B(0,K) : \|\mathbf{u}(k)\|_G \le 0.238565607(1,1)^T\}$. Moreover, since with this choice of $r$

$$\mathcal{K} = \begin{bmatrix} 0 & 0.158678373\\ 0 & 0.158678373\end{bmatrix}, \qquad (\mathcal{I}-\mathcal{K})^{-1} = \begin{bmatrix} 1 & 0.188606078\\ 0 & 1.188606079\end{bmatrix}$$

$$\mathcal{K}^* = \begin{bmatrix} 0 & 0.377212157\\ 0 & 0.377212157\end{bmatrix}, \qquad (\mathcal{I}-\mathcal{K}^*)^{-1} = \begin{bmatrix} 1 & 0.605683237\\ 0 & 1.605683237\end{bmatrix}$$

the error bound (9.9.3) reduces to

$$\|\mathbf{u}^* - \mathbf{u}^m\|_G \le (0.377212157)^m(0.238565607)(1,1)^T.$$

Further, we have

$$\mathcal{P} = \frac{1}{2}\left[\begin{array}{cc|cc} 0 & 0 & 0 & 0 \\ 0 & 0 & 0 & \frac{1}{(K+1)^2} \times 1.269426987 \end{array}\right], \quad \mathcal{Q} = 0.$$

Thus, it follows that

$$\mathcal{H} = \frac{1}{2}\left[\begin{array}{cc|cc} 0 & 0 & 0 & 0.188606078 \\ 0 & 0 & 0 & 0.188606078 \end{array}\right].$$

Hence, the error bound (9.9.11) gives

$$|u_1^{m+1} - u_1^m| \le (0.094303039)|u_2^m - u_2^{m-1}|^2 \tag{9.9.13}$$

$$|u_2^{m+1} - u_2^m| \le (0.094303039)|u_2^m - u_2^{m-1}|^2. \tag{9.9.14}$$

Since from (9.9.8), we have

$$\|\mathbf{u}^1 - \mathbf{u}^0\|_G \le 0.148575759(1,1)^T$$

inequality (9.9.14) easily determines

$$\begin{aligned} |u_2^{m+1} - u_2^m| &\le \left(0.094303039\ |u_2^1 - u_2^0|\right)^{2^m} (10.60411213) \\ &\le (0.014011145)^{2^m}(10.60411213). \end{aligned}$$

Finally, using this estimate in (9.9.13), we obtain the expected inequality

$$|u_1^{m+1} - u_1^m| \le (0.014011145)^{2^m}(10.60411213).$$

We also note that for this particular example the error bound (9.9.12) reduces to

$$\|\mathbf{u}^{m+1} - \mathbf{u}^m\|_G \le (0.377212157)^{2m-2}(0.0020817166)(1,1)^T.$$

## 9.10. Approximate Newton's Method

As in Section 9.3, we shall assume that the sequence $\{\mathbf{u}^m(k)\}$ generated by the iterative scheme (9.9.1) is approximated by the computed sequence $\{\mathbf{v}^m(k)\}$. To find $\mathbf{v}^{m+1}(k)$ the function $\mathbf{f}$ is approximated by $\mathbf{f}^m$, and the operator $\mathcal{F}$ by $\mathcal{F}^m$. Therefore, the computed sequence $\{\mathbf{v}^m(k)\}$ satisfies the recurrence relation

$$\begin{aligned} \mathbf{v}^{m+1}(k) = \mathcal{H}^1 & \left[\mathbf{f}^m(k, \mathbf{v}^m(k)) + \mathbf{f}_{\mathbf{u}}^m(k, \mathbf{v}^m(k))\left(\mathbf{v}^{m+1}(k)\right.\right. \\ & \left.\left. - \mathbf{v}^m(k)\right) - \mathcal{A}(k)\mathbf{v}^{m+1}(k)\right] \\ & + \mathcal{H}^2\left[-\mathcal{F}^m[\mathbf{v}^m] - \mathcal{F}_{\mathbf{u}}^m[\mathbf{v}^m][\mathbf{v}^{m+1} - \mathbf{v}^m] + \mathcal{L}[\mathbf{v}^{m+1}]\right] \\ \mathbf{v}^0(k) = \mathbf{u}^0(k) = & \ \overline{\mathbf{u}}(k), \quad m = 0, 1, \cdots. \end{aligned} \tag{9.10.1}$$

With respect to $\mathbf{f}^m$ and $\mathcal{F}^m$, we shall assume the following:

**Condition** $(p_1)$. For all $(k,\mathbf{u}) \in \mathbb{N}(a,b-1) \times \overline{S}(\overline{\mathbf{u}},\mathbf{r})$ the function $\mathbf{f}^m(k,\mathbf{u})$ is continuously differentiable with respect to $\mathbf{u}$, and $|\mathbf{f}^m_{\mathbf{u}}(k,\mathbf{u}) - \mathcal{A}(k)| \leq \mathcal{M}^3$. Also, for all $k \in \mathbb{N}(a,b-1)$ and $\mathbf{v}^m(k)$ obtained from (9.10.1) the inequality (9.3.2) holds.

**Condition** $(p_2)$. For all $\mathbf{u} \in \overline{S}(\overline{\mathbf{u}},\mathbf{r})$, $\mathcal{F}^m[\mathbf{u}]$ is continuously Fréchet differentiable and $\|\mathcal{F}^m_{\mathbf{u}}[\mathbf{u}] - \mathcal{L}\|_G \leq \mathcal{M}^4$. Also, for $\mathbf{v}^m(k)$ obtained from (9.10.1) the inequality (9.3.3) holds.

**Theorem 9.10.1.** With respect to the boundary value problem (1.2.8), (1.5.1) we assume that there exists an approximate solution $\overline{\mathbf{u}}(k)$ and conditions (i) (iv) of Theorem 9.2.2 are satisfied. Further, let conditions $(p_1)$ and $(p_2)$ be satisfied, and $\rho(\hat{\mathcal{K}}) < 1$, where

$$\hat{\mathcal{K}} = (\mathcal{I}-\mathcal{K})^{-1}\left[2\mathcal{K} + \mathcal{M}^1\mathcal{M}^7\left(\mathcal{M}^3 + \sup_{k\in\mathbb{N}(a,b-1)} |\mathcal{A}(k)|\right)\right.$$
$$\left. +\mathcal{M}^2\mathcal{M}^8(\mathcal{M}^4 + \|\mathcal{L}\|_G)\right]$$

and

$$\mathbf{r}^{12} = (\mathcal{I}-\hat{\mathcal{K}})^{-1}(\mathcal{I}-\mathcal{K})^{-1}\left[\mathcal{M}^1\mathbf{d}^1 + \mathcal{M}^2\mathbf{d}^2 + \mathcal{M}^1\mathcal{M}^7(\mathcal{I}-\mathcal{M}^7)^{-1}\times\right.$$
$$\left.\sup_{k\in\mathbb{N}(a,b-1)} |\mathbf{f}^0(k,\overline{\mathbf{u}}(k))| + \mathcal{M}^2\mathcal{M}^8(\mathcal{I}-\mathcal{M}^8)^{-1}\|\mathcal{F}^0[\overline{\mathbf{u}}]\|_G\right] \leq \mathbf{r}.$$

Then, the following hold

1. all the conclusions 1 3 of Theorem 9.9.1 hold
2. the sequence $\{\mathbf{v}^m(k)\}$ obtained from (9.10.1) remains in $\overline{S}(\overline{\mathbf{u}},\mathbf{r}^{12})$
3. the sequence $\{\mathbf{v}^m(k)\}$ converges to $\mathbf{u}^*(k)$ the solution of (1.2.8), (1.5.1) if and only if $\lim_{m\to\infty} \mathbf{b}^m = 0$, where

(9.10.2)
$$\mathbf{b}^m = \|\mathbf{v}^{m+1}(k) - \mathcal{H}^1[\mathbf{f}(k,\mathbf{v}^m(k)) + \mathbf{f}_{\mathbf{u}}(k,\mathbf{v}^m(k))(\mathbf{v}^{m+1}(k)-\mathbf{v}^m(k)) - \mathcal{A}(k)\mathbf{v}^{m+1}(k)] - \mathcal{H}^2[-\mathcal{F}[\mathbf{v}^m] - \mathcal{F}_{\mathbf{u}}[\mathbf{v}^m][\mathbf{v}^{m+1}-\mathbf{v}^m] + \mathcal{L}[\mathbf{v}^{m+1}]]\|_G$$

and, also

(9.10.3)
$$\|\mathbf{u}^* - \mathbf{v}^{m+1}\|_G$$
$$\leq (\mathcal{I}-\mathcal{K})^{-1}\left[\mathcal{M}^1\mathcal{M}^7(\mathcal{I}-\mathcal{M}^7)^{-1}\sup_{k\in\mathbb{N}(a,b-1)} |\mathbf{f}^m(k,\mathbf{v}^m(k))|\right.$$

$$+\mathcal{M}^2\mathcal{M}^8(\mathcal{I}-\mathcal{M}^8)^{-1}\|\mathcal{F}^m[\mathbf{v}^m]\|_G + 2\mathcal{K}\|\mathbf{v}^{m+1}-\mathbf{v}^m\|_G\Big].$$

**Proof.** Since $\hat{\mathcal{K}} \geq 2(\mathcal{I}-\mathcal{K})^{-1}\mathcal{K}$, $\rho(\hat{\mathcal{K}}) < 1$ implies that $\rho\left(2\mathcal{K}(\mathcal{I}-\mathcal{K})^{-1}\right) < 1$, which in turn implies $\rho(3\mathcal{K}) < 1$, and obviously $\mathbf{r}^{12} \geq \left[\mathcal{I}-2\mathcal{K}(\mathcal{I}-\mathcal{K})^{-1}\right]^{-1}(\mathcal{I}-\mathcal{K})^{-1}\left(\mathcal{M}^1\mathbf{d}^1+\mathcal{M}^2\mathbf{d}^2\right) = (\mathcal{I}-3\mathcal{K})^{-1}\left(\mathcal{M}^1\mathbf{d}^1+\mathcal{M}^2\mathbf{d}^2\right) = \mathbf{r}^{11}$, the conditions of Theorem 9.9.1 are satisfied and conclusion 1 follows.

To prove 2, we note that $\overline{\mathbf{u}}(k) \in \overline{S}(\overline{\mathbf{u}}, \mathbf{r}^{12})$, and from (9.2.6) and (9.10.1) we find

$$\begin{aligned}\mathbf{v}^1(k)-\overline{\mathbf{u}}(k) &= \mathcal{H}^1\left[\mathbf{f}^0(k,\overline{\mathbf{u}}(k)) - \mathbf{f}(k,\overline{\mathbf{u}}(k))\right.\\ &\quad \left.+(\mathbf{f}^0_{\mathbf{u}}(k,\overline{\mathbf{u}}(k)) - \mathcal{A}(k))(\mathbf{v}^1(k)-\overline{\mathbf{u}}(k)) - \mathbf{q}(k)\right]\\ &\quad +\mathcal{H}^2\left[-(\mathcal{F}^0[\overline{\mathbf{u}}]-\mathcal{F}[\overline{\mathbf{u}}]) - (\mathcal{F}^0_{\mathbf{u}}[\overline{\mathbf{u}}]-\mathcal{L})[\mathbf{v}^1-\overline{\mathbf{u}}] - \mathbf{l}^1\right]\end{aligned}$$

and hence

$$\begin{aligned}\|\mathbf{v}^1-\overline{\mathbf{u}}\|_G &\leq \mathcal{M}^1\left[\mathcal{M}^7 \sup_{k\in\mathbb{N}(a,b-1)} |\mathbf{f}(k,\overline{\mathbf{u}}(k))| + \mathcal{M}^3\|\mathbf{v}^1-\overline{\mathbf{u}}\|_G + \mathbf{d}^1\right]\\ &\quad +\mathcal{M}^2\left[\mathcal{M}^8\|\mathcal{F}[\overline{\mathbf{u}}]\|_G + \mathcal{M}^4\|\mathbf{v}^1-\overline{\mathbf{u}}\|_G + \mathbf{d}^2\right],\end{aligned}$$

which in view of (9.3.4) and (9.3.5) implies that

$$\begin{aligned}\|\mathbf{v}^1-\overline{\mathbf{u}}\|_G &\leq (\mathcal{I}-\mathcal{K})^{-1}\left[\mathcal{M}^1\mathbf{d}^1 + \mathcal{M}^2\mathbf{d}^2 + \mathcal{M}^1\mathcal{M}^7(\mathcal{I}-\mathcal{M}^7)^{-1}\times\right.\\ &\quad \left.\sup_{k\in\mathbb{N}(a,b-1)} |\mathbf{f}^0(k,\overline{\mathbf{u}}(k))| + \mathcal{M}^2\mathcal{M}^8(\mathcal{I}-\mathcal{M}^8)^{-1}\|\mathcal{F}^0[\overline{\mathbf{u}}]\|_G\right]\\ &\leq \mathbf{r}^{12}.\end{aligned}$$

Now we assume that $\mathbf{v}^m(k) \in \overline{S}(\overline{\mathbf{u}}, \mathbf{r}^{12})$ and will show that $\mathbf{v}^{m+1}(k) \in \overline{S}(\overline{\mathbf{u}}, \mathbf{r}^{12})$. From (9.2.6) and (9.10.1), we have

$$\begin{aligned}\mathbf{v}^{m+1}(k)-\overline{\mathbf{u}}(k) &= \mathcal{H}^1\left[\mathbf{f}^m(k,\mathbf{v}^m(k)) - \mathbf{f}(k,\mathbf{v}^m(k))\right.\\ &\quad +\mathbf{f}(k,\mathbf{v}^m(k)) - \mathbf{f}(k,\overline{\mathbf{u}}(k)) - \mathcal{A}(k)(\mathbf{v}^m(k)-\overline{\mathbf{u}}(k))\\ &\quad +(\mathbf{f}^m_{\mathbf{u}}(k,\mathbf{v}^m(k)) - \mathcal{A}(k))(\mathbf{v}^{m+1}(k)-\overline{\mathbf{u}}(k))\\ &\quad \left.-(\mathbf{f}^m_{\mathbf{u}}(k,\mathbf{v}^m(k)) - \mathcal{A}(k))(\mathbf{v}^m(k)-\overline{\mathbf{u}}(k)) - \mathbf{q}(k)\right]\\ &\quad +\mathcal{H}^2\left[-(\mathcal{F}^m[\mathbf{v}^m]-\mathcal{F}[\mathbf{v}^m]) - (\mathcal{F}[\mathbf{v}^m]-\mathcal{F}[\overline{\mathbf{u}}]-\mathcal{L}[\mathbf{v}^m-\overline{\mathbf{u}}])\right.\\ &\quad \left.-(\mathcal{F}^m_{\mathbf{u}}[\mathbf{v}^m]-\mathcal{L})[\mathbf{v}^{m+1}-\overline{\mathbf{u}}] + (\mathcal{F}^m_{\mathbf{u}}[\mathbf{v}^m]-\mathcal{L})[\mathbf{v}^m-\overline{\mathbf{u}}] - \mathbf{l}^1\right]\end{aligned}$$

and hence

$$\begin{aligned}\|\mathbf{v}^{m+1}-\overline{\mathbf{u}}\|_G \le \mathcal{M}^1 &\Bigg[\mathcal{M}^7 \sup_{k\in\mathbb{N}(a,b-1)} |\mathbf{f}(k,\mathbf{v}^m(k))| + \mathcal{M}^3\|\mathbf{v}^m-\overline{\mathbf{u}}\|_G \\ &+\mathcal{M}^3\|\mathbf{v}^{m+1}-\overline{\mathbf{u}}\|_G + \mathcal{M}^3\|\mathbf{v}^m-\overline{\mathbf{u}}\|_G + \mathbf{d}^1\Bigg] \\ &+\mathcal{M}^2\left[\mathcal{M}^8\|\mathcal{F}[\mathbf{v}^m]\|_G + \mathcal{M}^4\|\mathbf{v}^m-\overline{\mathbf{u}}\|_G + \mathcal{M}^4\|\mathbf{v}^{m+1}-\overline{\mathbf{u}}\|_G \right.\\ &\left.+\mathcal{M}^4\|\mathbf{v}^m-\overline{\mathbf{u}}\|_G + \mathbf{d}^2\right],\end{aligned}$$

which in view of (9.3.8) and (9.3.9) leads to

$$\begin{aligned}&(\mathcal{I}-\mathcal{K})\|\mathbf{v}^{m+1}-\overline{\mathbf{u}}\|_G \\ &\le \mathcal{M}^1\left[\mathcal{M}^7\left(\mathcal{M}^3 + \sup_{k\in\mathbb{N}(a,b-1)}|\mathcal{A}(k)|\right) + 2\mathcal{M}^3\right]\|\mathbf{v}^m-\overline{\mathbf{u}}\|_G \\ &\quad + \mathcal{M}^1\left[\mathcal{M}^7(\mathcal{I}-\mathcal{M}^7)^{-1}\sup_{k\in\mathbb{N}(a,b-1)}|\mathbf{f}^0(k,\overline{\mathbf{u}}(k))| + \mathbf{d}^1\right] \\ &\quad + \mathcal{M}^2\left[\mathcal{M}^8(\mathcal{M}^4 + \|\mathcal{L}\|_G) + 2\mathcal{M}^4\right]\|\mathbf{v}^m-\overline{\mathbf{u}}\|_G \\ &\quad + \mathcal{M}^2\left[\mathcal{M}^8(\mathcal{I}-\mathcal{M}^8)^{-1}\|\mathcal{F}^0[\overline{\mathbf{u}}]\|_G + \mathbf{d}^2\right] \\ &\le (\mathcal{I}-\mathcal{K})\hat{\mathcal{K}}\mathbf{r}^{12} + (\mathcal{I}-\mathcal{K})(\mathcal{I}-\hat{\mathcal{K}})\mathbf{r}^{12} = (\mathcal{I}-\mathcal{K})\mathbf{r}^{12}.\end{aligned}$$

Thus, $\|\mathbf{v}^{m+1}-\overline{\mathbf{u}}\|_G \le \mathbf{r}^{12}$ and this completes the proof of the conclusion 2.

Next we shall prove 3. From the definitions of $\mathbf{u}^{m+1}(k)$ and $\mathbf{v}^{m+1}(k)$, we have

$$\begin{aligned}&\mathbf{u}^{m+1}(k)-\mathbf{v}^{m+1}(k) \\ &= -\mathbf{v}^{m+1}(k) + \mathcal{H}^1\left[\mathbf{f}(k,\mathbf{v}^m(k)) + \mathbf{f_u}(k,\mathbf{v}^m(k))\times\right. \\ &\quad \left.(\mathbf{v}^{m+1}(k)-\mathbf{v}^m(k)) - \mathcal{A}(k)\mathbf{v}^{m+1}(k)\right] \\ &\quad + \mathcal{H}^2\left[-\mathcal{F}[\mathbf{v}^m] - \mathcal{F}_{\mathbf{u}}[\mathbf{v}^m][\mathbf{v}^{m+1}-\mathbf{v}^m] + \mathcal{L}[\mathbf{v}^{m+1}]\right] \\ &\quad + \mathcal{H}^1\left[\mathbf{f}(k,\mathbf{u}^m(k)) - \mathbf{f}(k,\mathbf{v}^m(k)) - \mathcal{A}(k)(\mathbf{u}^m(k)-\mathbf{v}^m(k))\right. \\ &\quad + (\mathbf{f_u}(k,\mathbf{u}^m(k)) - \mathcal{A}(k))(\mathbf{u}^{m+1}(k)-\mathbf{u}^m(k)) \\ &\quad \left.-(\mathbf{f_u}(k,\mathbf{v}^m(k)) - \mathcal{A}(k))(\mathbf{v}^{m+1}(k)-\mathbf{v}^m(k))\right] \\ &\quad + \mathcal{H}^2\left[-(\mathcal{F}[\mathbf{u}^m] - \mathcal{F}[\mathbf{v}^m] - \mathcal{L}[\mathbf{u}^m-\mathbf{v}^m])\right. \\ &\quad \left.-(\mathbf{f_u}[\mathbf{u}^m]-\mathcal{L})[\mathbf{u}^{m+1}-\mathbf{u}^m] + (\mathcal{F}_{\mathbf{u}}[\mathbf{v}^m]-\mathcal{L})[\mathbf{v}^{m+1}-\mathbf{v}^m]\right]\end{aligned}$$

and hence

$$\begin{aligned}\|\mathbf{u}^{m+1}-\mathbf{v}^{m+1}\|_G &\le \mathbf{b}^m + \mathcal{M}^1\left[\mathcal{M}^3\|\mathbf{u}^m-\mathbf{v}^m\|_G + \mathcal{M}^3\|\mathbf{u}^{m+1}-\mathbf{u}^m\|_G\right.\\ &\quad \left.+\mathcal{M}^3\|\mathbf{v}^{m+1}-\mathbf{v}^m\|_G\right] + \mathcal{M}^2\left[\mathcal{M}^4\|\mathbf{u}^m-\mathbf{v}^m\|_G\right.\\ &\quad \left.+\mathcal{M}^4\|\mathbf{u}^{m+1}-\mathbf{u}^m\|_G + \mathcal{M}^4\|\mathbf{v}^{m+1}-\mathbf{v}^m\|_G\right]\\ &\le \mathbf{b}^m + \mathcal{K}\|\mathbf{u}^m-\mathbf{v}^m\|_G + \mathcal{K}\|\mathbf{u}^{m+1}-\mathbf{u}^m\|_G\\ &\quad +\mathcal{K}\left(\|\mathbf{v}^{m+1}-\mathbf{u}^{m+1}\|_G + \|\mathbf{u}^{m+1}-\mathbf{u}^m\|_G + \|\mathbf{u}^m-\mathbf{v}^m\|_G\right),\end{aligned}$$

which implies that

$$\|\mathbf{u}^{m+1}-\mathbf{v}^{m+1}\|_G \le (\mathcal{I}-\mathcal{K})^{-1}\mathbf{b}^m + \mathcal{K}^*\|\mathbf{u}^m-\mathbf{v}^m\|_G + \mathcal{K}^*\|\mathbf{u}^{m+1}-\mathbf{u}^m\|_G.$$

Thus, from (9.9.7) we find that

$$\|\mathbf{u}^{m+1}-\mathbf{v}^{m+1}\|_G \le \left[(\mathcal{I}-\mathcal{K})^{-1}\mathbf{b}^m + (\mathcal{K}^*)^{m+1}\|\mathbf{u}^1-\mathbf{u}^0\|_G\right] + \mathcal{K}^*\|\mathbf{u}^m-\mathbf{v}^m\|_G.$$

Using the fact that $\|\mathbf{u}^0-\mathbf{v}^0\|_G = 0,$ the above inequality gives

$$\|\mathbf{u}^{m+1}-\mathbf{v}^{m+1}\|_G \le \sum_{i=0}^{m}(\mathcal{K}^*)^{m-i}\left[(\mathcal{I}-\mathcal{K})^{-1}\mathbf{b}^i + (\mathcal{K}^*)^{i+1}\|\mathbf{u}^1-\mathbf{u}^0\|_G\right].$$

Therefore, from the triangle inequality, we obtain

$$\begin{aligned}(9.10.4)\qquad \|\mathbf{u}^*-\mathbf{v}^{m+1}\|_G &\le \sum_{i=0}^{m}(\mathcal{K}^*)^{m-i}(\mathcal{I}-\mathcal{K})^{-1}\mathbf{b}^i\\ &\quad +(m+1)(\mathcal{K}^*)^{m+1}\|\mathbf{u}^1-\mathbf{u}^0\|_G + \|\mathbf{u}^*-\mathbf{u}^{m+1}\|_G.\end{aligned}$$

In (9.10.4) Theorem 9.9.1 ensures that $\lim_{m\to\infty}\|\mathbf{u}^*-\mathbf{u}^{m+1}\|_G = 0,$ and since $\rho(\mathcal{K}^*) < 1,$ $\lim_{m\to\infty}\left[(m+1)(\mathcal{K}^*)^{m+1}\|\mathbf{u}^1-\mathbf{u}^0\|_G\right] = 0.$ Thus, the condition $\lim_{m\to\infty}\mathbf{b}^m = 0$ is necessary and sufficient for the convergence of the sequence $\{\mathbf{v}^m(k)\}$ to $\mathbf{u}^*(k)$ follows from the Toeplitz lemma.

Finally, we shall prove (9.10.3). For this, we have

$$\begin{aligned}\mathbf{u}^*(k)-\mathbf{v}^{m+1}(k) &= \mathcal{H}^1\left[\mathbf{f}(k,\mathbf{u}^*(k)) - \mathbf{f}(k,\mathbf{v}^m(k)) - \mathcal{A}(k)(\mathbf{u}^*(k)\right.\\ &\quad -\mathbf{v}^m(k)) + \mathbf{f}(k,\mathbf{v}^m(k)) - \mathbf{f}^m(k,\mathbf{v}^m(k))\\ &\quad \left.-(\mathbf{f}^m_{\mathbf{u}}(k,\mathbf{v}^m(k)) - \mathcal{A}(k))(\mathbf{v}^{m+1}(k)-\mathbf{v}^m(k))\right]\\ &\quad +\mathcal{H}^2\left[-(\mathcal{F}[\mathbf{u}^*] - \mathcal{F}[\mathbf{v}^m] - \mathcal{L}[\mathbf{u}^*-\mathbf{v}^m])\right.\\ &\quad \left.-(\mathcal{F}[\mathbf{v}^m] - \mathcal{F}^m[\mathbf{v}^m]) + (\mathcal{F}^m_{\mathbf{u}}[\mathbf{v}^m] - \mathcal{L})[\mathbf{v}^{m+1}-\mathbf{v}^m]\right]\end{aligned}$$

and hence

$$\begin{aligned}\|\mathbf{u}^*-\mathbf{v}^{m+1}\|_G \le \mathcal{M}^1 & \left[\mathcal{M}^3\|\mathbf{u}^* - \mathbf{v}^m\|_G + \mathcal{M}^7 \sup_{k\in\mathbb{N}(a,b-1)} |\mathbf{f}(k,\mathbf{v}^m(k))| \right. \\ & \left. +\mathcal{M}^3\|\mathbf{v}^{m+1} - \mathbf{v}^m\|_G\right] + \mathcal{M}^2\left[\mathcal{M}^4\|\mathbf{u}^* - \mathbf{v}^m\|_G \right. \\ & \left. +\mathcal{M}^8\|\mathcal{F}[\mathbf{v}^m]\|_G + \mathcal{M}^4\|\mathbf{v}^{m+1} - \mathbf{v}^m\|_G\right] \\ \le & \ \mathcal{K}\|\mathbf{u}^*-\mathbf{v}^{m+1}\|_G + \mathcal{M}^1\mathcal{M}^7(\mathcal{I}-\mathcal{M}^7)^{-1} \sup_{k\in\mathbb{N}(a,b-1)} |\mathbf{f}^m(k,\mathbf{v}^m(k))| \\ & +\mathcal{M}^2\mathcal{M}^8(\mathcal{I}-\mathcal{M}^8)^{-1}\|\mathcal{F}^m[\mathbf{v}^m]\|_G + 2\mathcal{K}\|\mathbf{v}^{m+1}-\mathbf{v}^m\|_G,\end{aligned}$$

which is the same as (9.10.3). ∎

In our next result we shall need the following:

**Condition** $(p_3)$ $((p_4))$. In condition $(p_1)$ $((p_2))$ instead of (9.3.2) ((9.3.3)) the inequality (9.3.11) ((9.3.12)) holds.

**Theorem 9.10.2.** With respect to the boundary value problem (1.2.8), (1.5.1) we assume that there exists an approximate solution $\overline{\mathbf{u}}(k)$ and conditions (i) (iv) of Theorem 9.2.2 are satisfied. Further, let conditions $(p_3)$ and $(p_4)$ be satisfied, and $\rho(3\mathcal{K}) < 1$, also

$$\mathbf{r}^{13} = (\mathcal{I}-\mathcal{K}^*)^{-1}(\mathcal{I}-\mathcal{K})^{-1}\left(\mathcal{M}^1(\mathbf{r}^5+\mathbf{d}^1)+\mathcal{M}^2(\mathbf{r}^6+\mathbf{d}^2)\right) \le \mathbf{r}.$$

Then, the following hold

1. all the conclusions 1 3 of Theorem 9.9.1 hold
2. the sequence $\{\mathbf{v}^m(k)\}$ obtained from (9.10.1) remains in $\overline{S}(\overline{\mathbf{u}},\mathbf{r}^{13})$
3. the condition $\lim_{m\to\infty}\mathbf{b}^m = 0$ is necessary and sufficient for the convergence of $\{\mathbf{v}^m(k)\}$ to the solution $\mathbf{u}^*(k)$ of (1.2.8), (1.5.1) where $\mathbf{b}^m$ are defined in (9.10.2), and

$$\|\mathbf{u}^*-\mathbf{v}^{m+1}\|_G \le (\mathcal{I}-\mathcal{K})^{-1}\left(\mathcal{M}^1\mathbf{r}^5+\mathcal{M}^2\mathbf{r}^6+2\mathcal{K}\|\mathbf{v}^{m+1}-\mathbf{v}^m\|_G\right).$$

**Proof.** The proof is contained in Theorem 9.10.1. ∎

## 9.11. Initial–Value Methods

The method of complementary functions developed in Section 8.2 for linear problems can be used in an iterative way to solve the nonlinear boundary value problem (1.2.8), (1.5.1). For this, we assume the trial value of $\mathbf{u}(a)$ and find the solution $\mathbf{u}(k)$ of (1.2.8). Let us consider a nearby solution $\mathbf{u}(k)+\delta\mathbf{u}(k)$, where $\delta\mathbf{u}(k)$ is the first order correction to $\mathbf{u}(k)$ to produce the actual solution of (1.2.8), (1.5.1). The system of the nearby solution is

$$\text{(9.11.1)}\quad \mathbf{u}(k+1)+\delta\mathbf{u}(k+1) \;=\; \mathbf{f}(k,\mathbf{u}(k)+\delta\mathbf{u}(k)), \quad k\in\mathbb{N}(a,b-1).$$

Expanding the right side of (9.11.1) in a Taylor series up to and including first order terms, we obtain the variational system

$$\text{(9.11.2)}\qquad \delta\mathbf{u}(k+1) \;=\; \mathbf{f}_{\mathbf{u}}(k,\mathbf{u}(k))\delta\mathbf{u}(k).$$

In a similar way, the boundary conditions for the variational system are obtained and appear as

$$\text{(9.11.3)}\qquad \mathcal{F}_{\mathbf{u}}[\mathbf{u}][\delta\mathbf{u}(k)] \;=\; -\mathcal{F}[\mathbf{u}]_{cal},$$

where $\mathcal{F}[\mathbf{u}]_{cal}$ is the vector calculated from the solution $\mathbf{u}(k)$. Equations (9.11.2), (9.11.3) form a linear system and play the role of (1.2.11), (1.5.3).

Note that we have interpreted the variation $\delta\mathbf{u}(k)$ as the difference between the true (but unknown) and the calculated solution, i.e.

$$\text{(9.11.4)}\qquad \delta\mathbf{u}(k) \;=\; \mathbf{u}_{true}(k)-\mathbf{u}_{cal}(k).$$

Since equations (9.11.2) (9.11.4) are only approximate equations, the process of finding the true solution is iterative and terminates when $|\delta\mathbf{u}(k)|$, $k\in\mathbb{N}(a,b)$ is sufficiently small (less than a preassigned tolerance). Equations (9.11.2) (9.11.4) for the $m$th iteration are written as

$$\text{(9.11.5)}\qquad (\delta\mathbf{u}(k+1))^{(m)} \;=\; (\mathbf{f}_{\mathbf{u}}(k,\mathbf{u}(k)))^{(m)}(\delta\mathbf{u}(k))^{(m)}$$

$$\text{(9.11.6)}\qquad (\mathcal{F}_{\mathbf{u}}[\mathbf{u}])^{(m)}[(\delta\mathbf{u}(k))^{(m)}] \;=\; -(\mathcal{F}[\mathbf{u}]_{cal})^{(m)}$$

$$\text{(9.11.7)}\qquad (\delta\mathbf{u}(k))^{(m)} \;=\; \mathbf{u}_{true}(k)-(\mathbf{u}_{cal}(k))^{(m)}.$$

For the system (9.11.5), (9.11.6) we find that

$$\text{(9.11.8)}\quad (\delta\mathbf{u}(k))^{(m)} \;=\; \prod_{\ell=a}^{k-1}(\mathbf{f}_{\mathbf{u}}(a+k-1-\ell,\mathbf{u}(a+k-1-\ell)))^{(m)}\mathbf{c}^{(m)},$$

where

(9.11.9)

$$\mathbf{c}^{(m)} = -\left[(\mathcal{F}_{\mathbf{u}}[\mathbf{u}])^{(m)}\left[\prod_{\ell=a}^{k-1}(\mathbf{f}_{\mathbf{u}}(a+k-1-\ell,\mathbf{u}(a+k-1-\ell)))^{(m)}\right]\right]^{-1}(\mathbf{f}[\mathbf{u}]_{cal})^{(m)}.$$

From (9.11.7), we obtain new initial condition for the next iteration, which is

$$(9.11.10) \qquad (\mathbf{u}(a))^{(m+1)} = (\mathbf{u}(a))^{(m)} + \mathbf{c}^{(m)}, \quad m = 0, 1, \cdots.$$

If we denote the solution of (1.2.8) as $\mathbf{u}(k) = \mathbf{u}(k, \mathbf{u}(a))$, then the assumed conditions on the function $\mathbf{f}(k, \mathbf{u})$ imply that $\mathbf{u}(k, \mathbf{u}(a))$ continuously depends on the initial vector $\mathbf{u}(a)$. Thus, solving the boundary value problem (1.2.8), (1.5.1) is equivalent to finding $\mathbf{u}(a)$ for which $\mathcal{F}[\mathbf{u}(k, \mathbf{u}(a))] = \overline{\mathcal{F}}[\mathbf{u}(a)]$, say, is zero.

Assume that the $m$th approximation to $\mathbf{u}(a)$, which we denote as $(\mathbf{u}(a))^{(m)}$, has been found. Then, Newton's method provides the $(m+1)$ th approximation by the relation

$$\overline{\mathcal{F}}\left[(\mathbf{u}(a))^{(m)}\right] + \overline{\mathcal{F}}_{(\mathbf{u}(a))^{(m)}}((\mathbf{u}(a))^{(m+1)} - (\mathbf{u}(a))^{(m)}) = 0,$$

which in view of (9.11.8) and (9.11.10) is the same as

$$(9.11.11) \quad \mathcal{F}\left[\mathbf{u}(k, (\mathbf{u}(a))^{(m)})\right]_{cal} + \mathcal{F}_{\mathbf{u}}\left[\mathbf{u}(k, (\mathbf{u}(a))^{(m)})\right] \times \frac{\partial \mathbf{u}(k, (\mathbf{u}(a))^{(m)})}{\partial (\mathbf{u}(a))^{(m)}}(\delta \mathbf{u}(a))^{(m)} = 0.$$

The total variation in $\mathbf{u}(k, (\mathbf{u}(a))^{(m)})$ can be expressed as

$$(9.11.12) \qquad \delta \mathbf{u}(k, (\mathbf{u}(a))^{(m)}) = \frac{\partial \mathbf{u}(k, (\mathbf{u}(a))^{(m)})}{\partial (\mathbf{u}(a))^{(m)}} \cdot (\delta \mathbf{u}(a))^{(m)}.$$

Also, the solution of (9.11.5) is

$$(9.11.13)$$
$$\delta \mathbf{u}(k, (\mathbf{u}(a))^{(m)}) = \prod_{\ell=a}^{k-1} (\mathbf{f}_{\mathbf{u}}(a+k-1-\ell, \mathbf{u}(a+k-1-\ell, (\mathbf{u}(a))^{(m)})))^{(m)} (\delta \mathbf{u}(a))^{(m)}.$$

Therefore, from (9.11.12) and (9.11.13), we get

$$(9.11.14)$$
$$\frac{\partial \mathbf{u}(k, (\mathbf{u}(a))^{(m)})}{\partial (\mathbf{u}(a))^{(m)}} = \prod_{\ell=a}^{k-1} (\mathbf{f}_{\mathbf{u}}(a+k-1-\ell, \mathbf{u}(a+k-1-\ell, (\mathbf{u}(a))^{(m)})))^{(m)}.$$

Using (9.11.14) in (9.11.11) and rearranging the terms, we obtain $(\delta \mathbf{u}(a))^{(m)} = \mathbf{c}^{(m)}$. Thus, the method (9.11.10) used to find $\mathbf{u}(a)$ is equivalent to solving $\overline{\mathcal{F}}[\mathbf{u}(a)] = 0$ for $\mathbf{u}(a)$ by Newton's method. Therefore, a suitable application of the Kantorovich sufficiency Theorem 9.1.3 furnishes a theoretical basis for the convergence of the process and an estimate on the rate of convergence.

In particular, for the boundary conditions (1.5.2) equation (9.11.3) takes the form (1.5.4), where

$$\mathcal{L}^s = \left(\frac{\partial \phi_i}{\partial u_j(k_s)}\right), \quad 1 \le s \le r, \quad \mathbf{l} = \left(-\phi_{i(cal)}\right)$$

and $\mathbf{u}(k_i)$ is replaced by $\delta\mathbf{u}(k_i)$. Thus, as in Section 8.2, if we solve the linear system (9.11.5) with the initial conditions

$$\left(\delta u_j^i(k_1)\right)^{(m)} = \delta_{ij}, \quad 1 \le i, j \le n \tag{9.11.15}$$

then (9.11.6) in view of (8.2.9) reduces to

$$\sum_{i=1}^{n}\left(\sum_{j=1}^{r} \mathcal{L}^j \delta\mathbf{u}^i(k_j)\right)^{(m)} (\delta u_i(k_1))^{(m)} = (\mathbf{l})^{(m)}. \tag{9.11.16}$$

Thus, the *forward process* for the boundary value problem (1.2.8), (1.5.2) is obtained by the equations (9.11.5), (9.11.15), (9.11.16), (9.11.7) and (9.11.10) with $a$ replaced by $k_1$ and $\mathbf{c}^{(m)} = \left(\delta u_i(k_1)\right)^{(m)}$.

Similarly, the *backward process* for the problem (1.2.8), (1.5.2) consists of the equations (9.11.5),

$$\left(\delta \underline{u}_j^i(k_r)\right) = \delta_{ij}, \quad 1 \le i, j \le n \tag{9.11.17}$$

$$\sum_{i=1}^{n}\left(\sum_{j=1}^{r} \mathcal{L}^j \delta\underline{\mathbf{u}}^i(k_j)\right)^{(m)} (\delta \underline{u}_i(k_r)) = (\mathbf{l})^{(m)}, \tag{9.11.18}$$

(9.11.7) and (9.11.10) with a replaced by $k_r$ and $\mathbf{c}^{(m)} = \left(\delta \underline{u}_i(k_r)\right)^{(m)}$.

As we have indicated in Section 8.2, if the boundary conditions are such that $u_i(k_1)$ for some fixed indices $i$ are explicitly known, then in the initial vector $\mathbf{u}(k_1)$ these components are taken to be same known values. Obviously, then for these indices $i$, $\delta u_i(k_1) = 0$. Further, if the total number of such indices $i$ is $s$, then with this choice of $\mathbf{u}(k_1)$ in the forward process each iteration requires only $n - s$ solutions. A similar statement holds for the backward process.

We also note that for the nonlinear boundary value problem (1.2.8), (1.5.1), like the method of complementary functions, the method of adjoints discussed in Section 8.4 can also be used in an iterative way.

**Example 9.11.1.** Consider the boundary value problem

$$\begin{aligned} u(k+1) - 2u(k) + u(k-1) &= \frac{2}{(K+1)^2}u^3(k), \quad k \in \mathbb{N}(1, K) \\ u(0) = 1, \quad u(K+1) - u(K-1) &+ \frac{2}{K+1}u^2(K) = 0, \end{aligned} \tag{9.11.19}$$

which is the discrete analog of $y'' = 2y^3$, $y(1) = 1$, $y'(2) + y^2(2) = 0$ discussed previously by Fox [26]. The continuous problem has a unique solution $y(t) = 1/t$.

In system form (9.11.19) appears as

(9.11.20)

$$\begin{aligned} u_1(k+1) &= u_2(k) \\ u_2(k+1) &= 2u_2(k) - u_1(k) + \frac{2}{(K+1)^2}u_2^3(k), \quad k \in \mathbb{N}(0, K-1) \\ u_1(0) &= 1, \quad u_2(K) - u_1(K-1) + \frac{2}{K+1}u_1^2(K) = 0. \end{aligned}$$

For the above problem equation (9.11.5) becomes

(9.11.21)

$$\begin{bmatrix} \delta u_1(k+1) \\ \delta u_2(k+1) \end{bmatrix}^{(m)} = \begin{bmatrix} 0 & 1 \\ -1 & 2 + \dfrac{6}{(K+1)^2}u_2^2(K) \end{bmatrix}^{(m)} \begin{bmatrix} \delta u_1(k) \\ \delta u_2(k) \end{bmatrix}^{(m)},$$

$$k \in \mathbb{N}(0, K-1)$$

and the variational boundary conditions are

$$\begin{aligned} &\begin{bmatrix} 1 & 0 \\ 0 & 0 \end{bmatrix} \begin{bmatrix} \delta u_1(0) \\ \delta u_2(0) \end{bmatrix}^{(m)} + \begin{bmatrix} 0 & 0 \\ -1 & 0 \end{bmatrix} \begin{bmatrix} \delta u_1(K-1) \\ \delta u_2(K-1) \end{bmatrix}^{(m)} \\ &\quad + \begin{bmatrix} 0 & 0 \\ \dfrac{4}{K+1}u_1(K) & 1 \end{bmatrix}^{(m)} \begin{bmatrix} \delta u_1(K) \\ \delta u_2(K) \end{bmatrix}^{(m)} \\ &= \begin{bmatrix} 1 - u_1(0) \\ -u_2(K) + u_1(K-1) - \dfrac{2}{K+1}u_1^2(K) \end{bmatrix}_{cal}^{(m)}. \end{aligned} \tag{9.11.22}$$

For the system (9.11.21) the initial conditions (9.11.15) are

$$(9.11.23) \qquad \delta u_1^1(0) = 1, \quad \delta u_2^1(0) = 0$$

$$(9.11.24) \qquad \delta u_1^2(0) = 0, \quad \delta u_2^2(0) = 1.$$

The equation (9.11.16) reduces to

$$\left\{\begin{bmatrix} 1 & 0 \\ 0 & 0 \end{bmatrix}\begin{bmatrix} \delta u_1^1(0) \\ \delta u_2^1(0) \end{bmatrix}^{(m)} + \begin{bmatrix} 0 & 0 \\ -1 & 0 \end{bmatrix}\begin{bmatrix} \delta u_1^1(K-1) \\ \delta u_2^1(K-1) \end{bmatrix}^{(m)}\right.$$

$$\left. + \begin{bmatrix} 0 & 0 \\ \frac{4}{K+1}u_1(K) & 1 \end{bmatrix}^{(m)} \begin{bmatrix} \delta u_1^1(K) \\ \delta u_2^1(K) \end{bmatrix}^{(m)}\right\} (\delta u_1(0))^{(m)}$$

$$+ \left\{\begin{bmatrix} 1 & 0 \\ 0 & 0 \end{bmatrix}\begin{bmatrix} \delta u_1^2(0) \\ \delta u_2^2(0) \end{bmatrix}^{(m)} + \begin{bmatrix} 0 & 0 \\ -1 & 0 \end{bmatrix}\begin{bmatrix} \delta u_1^2(K-1) \\ \delta u_2^2(K-1) \end{bmatrix}^{(m)}\right.$$

$$\left. + \begin{bmatrix} 0 & 0 \\ \frac{4}{K+1}u_1(K) & 1 \end{bmatrix}^{(m)} \begin{bmatrix} \delta u_1^2(K) \\ \delta u_2^2(K) \end{bmatrix}^{(m)}\right\} (\delta u_2(0))^{(m)}$$

$$= \begin{bmatrix} 1 - u_1(0) \\ -u_2(K) + u_1(K-1) - \frac{2}{K+1}u_1^2(K) \end{bmatrix}^{(m)}_{(cal)},$$

which is the same as

$$(9.11.25) \quad (\delta u_1(0))^{(m)} = (1 - u_1(0))^{(m)}$$

(9.11.26)

$$\left[-(\delta u_1^1(K-1))^{(m)} + \left(\frac{4}{K+1}u_1(K)\delta u_1^1(K) + \delta u_2^1(K)\right)^{(m)}\right](\delta u_1(0))^{(m)}$$

$$+ \left[-(\delta u_1^2(K-1))^{(m)} + \left(\frac{4}{K+1}u_1(K)\delta u_1^2(K) + \delta u_2^2(K)\right)^{(m)}\right](\delta u_2(0))^{(m)}$$

$$= \left[-u_2(K) + u_1(K-1) - \frac{2}{K+1}u_1^2(K)\right]^{(m)}_{(cal)}.$$

The equation (9.11.10) reduces to

$$(9.11.27) \qquad (u_1(0))^{(m+1)} = (u_1(0))^{(m)} + (\delta u_1(0))^{(m)}$$

(9.11.28) $$(u_2(0))^{(m+1)} = (u_2(0))^{(m)} + (\delta u_2(0))^{(m)}.$$

Since $u_1(0) = 1$ is already known, the initial vector we choose is $\mathbf{u}(0) = (1, \alpha)^T$, where $\alpha$ is a known constant. With this choice of initial vector equations (9.11.25) and (9.11.27) immediately imply that $(\delta u_1(0))^{(m)} = 0$ for all $m$, and (9.11.26) becomes

(9.11.29)
$$\left[-(\delta u_1^2(K-1))^{(m)} + \left(\frac{4}{K+1}u_1(K)\delta u_1^2(K) + \delta u_2^2(K)\right)^{(m)}\right](\delta u_2(0))^{(m)}$$
$$= \left[-u_2(K) + u_1(K-1) - \frac{2}{K+1}u_1^2(K)\right]^{(m)}_{(cal)}$$

in which only the solution $\delta\mathbf{u}^2(k)$ is required. Thus, for each iteration we need to solve (9.11.21) only once with the conditions (9.11.24).

In Table 9.11.1 we present the converged numerical solution $u_1(k)$ obtained by taking $\alpha = 1$ for the several different values of $K$.

**Table** 9.11.1.

| $K$ / $t$ | 4 | 9 | 49 | 99 | Exact Solution |
|---|---|---|---|---|---|
| 1.2 | 0.834892169 | 0.833735968 | 0.833349665 | 0.833337485 | 0.833333333 |
| 1.4 | 0.716340926 | 0.714813225 | 0.714307103 | 0.714291215 | 0.714285714 |
| 1.6 | 0.627196585 | 0.625561175 | 0.625022777 | 0.625005951 | 0.625000000 |
| 1.8 | 0.557790149 | 0.556124286 | 0.555578689 | 0.555561713 | 0.555555555 |
| 2.0 | 0.502267329 | 0.500575173 | 0.500023466 | 0.500006385 | 0.500000000 |

**Example 9.11.2.** Consider the boundary value problem (9.2.16), (9.2.17) with $\alpha = \beta = 1$. For this problem $u_1(0) = 0$ is already known and we choose $u_2(0) = 0.001$, and apply the method of adjoints in an iterative way for the several different values of $K$. The numerical solution $u_1(k)$ for the fourth iteration is presented in Table 9.11.2.

**Table** 9.11.2.

| $K$ / $t$ | 4 | 9 | 49 | 99 |
|---|---|---|---|---|
| 0.2 | −0.0730539075 | −0.0732143313 | −0.0732662139 | −0.0732678385 |
| 0.4 | −0.108925786 | −0.109159129 | −0.109234569 | −0.109236932 |
| 0.6 | −0.108925786 | −0.109159129 | −0.10923457 | −0.109236932 |
| 0.8 | −0.0730539075 | −0.0732143314 | −0.0732662144 | −0.0732678385 |
| 1.0 | $6.18456397E-11$ | $-1.35514711E-10$ | $-7.43057172E-10$ | $5.90340221E-10$ |

| 199 | Exact Solution |
|---|---|
| −0.0732682461 | −0.0732683791 |
| −0.109237524 | −0.10923772 |
| −0.109237525 | −0.10923772 |
| −0.0732682478 | −0.0732683791 |
| $-1.48980917E-09$ | 0.0 |

## 9.12. Invariant Imbedding Method

Consider the difference system (1.2.8) on $\mathbb{N}(0, K-1)$ together with the boundary conditions

$$\mathbf{g}(\mathbf{u}(0)) + \mathbf{h}(\mathbf{u}(K)) = \mathbf{l}, \tag{9.12.1}$$

where $\mathbf{g}$ and $\mathbf{h}$ map $\mathbb{R}^n$ into $\mathbb{R}^n$ and $\mathbf{l}$ is the known vector. We shall vary $K$ between 1 and $K_1$ (for simplicity, we will assume that $K_1 = \infty$) and $\mathbf{l}$ in $\mathbb{R}^n$ to imbed the problem (1.2.8), (9.12.1) into a family of similar problems. Throughout, we shall assume that each such boundary value problem has a unique solution. Since the solution to the initial value problem for (1.2.8) exists and is unique for $k \in \mathbb{N}$, the solution of the boundary value problem can be continued to all of $\mathbb{N}$. This solution we shall denote by $\mathbf{u}(k, K, \mathbf{l})$ to emphasize its dependence on $K$ and $\mathbf{l}$. The method of invariant imbedding seeks to replace (1.2.8), (9.12.1) by initial value problems. For example, if we know the final value $\mathbf{u}(K, K, \mathbf{l})$ of the solution of (1.2.8), (9.12.1) and in addition if the backward Cauchy problem for (1.2.8) is uniquely solvable, then we can solve (1.2.8), (9.12.1) by backward recursion. The problem now is to find $\mathbf{u}(K, K, \mathbf{l})$. Using invariant imbedding, we set up a difference equation for $\mathbf{r}(K, \mathbf{l}) = \mathbf{u}(K, K, \mathbf{l})$. Under certain conditions on $\mathbf{g}$ and $\mathbf{h}$, we can determine a complete set of initial conditions for this equation. Once this is done, solving the original boundary value problem is reduced to solving two initial value problems.

**Theorem 9.12.1.** Let $\mathbf{u}(k, K, \mathbf{l})$ and $\mathbf{r}(K, \mathbf{l})$ be as above, and assume that the backward Cauchy problem for (1.2.8) is uniquely solvable. If the equation $\mathbf{u}(K+1, K+1, \mathbf{l}^1) = \mathbf{u}(K+1, K, \mathbf{l})$ can be solved uniquely for $\mathbf{l}^1$, then the functions $\mathbf{u}(k, K, \mathbf{l})$, $\mathbf{r}(K, \mathbf{l})$ satisfy the difference equations

$$\mathbf{u}(k, K+1, \mathbf{l} + \mathbf{h}(\mathbf{f}(K, \mathbf{r}(K, \mathbf{l}))) - \mathbf{h}(\mathbf{r}(K, \mathbf{l}))) = \mathbf{u}(k, K, \mathbf{l}), \quad k \in \mathbb{N}(0, K) \tag{9.12.2}$$

$$\mathbf{u}(k, k, \mathbf{l}) = \mathbf{r}(k, \mathbf{l}) \tag{9.12.3}$$

$$\mathbf{r}(K+1, \mathbf{l} + \mathbf{h}(\mathbf{f}(K, \mathbf{r}(K, \mathbf{l}))) - \mathbf{h}(\mathbf{r}(K, \mathbf{l}))) = \mathbf{f}(K, \mathbf{r}(K, \mathbf{l})). \tag{9.12.4}$$

In addition, if $\mathbf{g} + \mathbf{h}$ has an inverse, then $\mathbf{r}(K, \mathbf{l})$ has the initial value

$$\mathbf{r}(0, \mathbf{l}) = (\mathbf{g} + \mathbf{h})^{-1}(\mathbf{l}). \tag{9.12.5}$$

**Proof.** Consider $\mathbf{u}(k,K,\mathbf{l})$ and $\mathbf{u}(k,K+1,\mathbf{l}^1)$. In order that $\mathbf{u}(k,K+1,\mathbf{l}^1)$ constitutes the unique extension of $\mathbf{u}(k,K,\mathbf{l})$, it is necessary and sufficient that $\mathbf{u}(K+1,K,\mathbf{l}) = \mathbf{u}(K+1,K+1,\mathbf{l}^1)$. This follows from the existence and uniqueness of solutions to the backward Cauchy problem. By assumption, we can pick a unique $\mathbf{l}^1$ satisfying $\mathbf{u}(K+1,K,\mathbf{l}) = \mathbf{u}(K+1,K+1,\mathbf{l}^1)$. With this choice of $\mathbf{l}^1$, we arrive at the fundamental relation

$$\mathbf{u}(k,K+1,\mathbf{l}^1) = \mathbf{u}(k,K,\mathbf{l}), \quad k \in \mathbb{N}. \tag{9.12.6}$$

From (1.2.8), we get that

$$\mathbf{u}(K+1,K,\mathbf{l}) = \mathbf{f}(K,\mathbf{u}(K,K,\mathbf{l})). \tag{9.12.7}$$

Substituting this in (9.12.6) with $k = K+1$ gives

$$\mathbf{u}(K+1,K+1,\mathbf{l}^1) = \mathbf{f}(K,\mathbf{u}(K,K,\mathbf{l})). \tag{9.12.8}$$

Using the definition of $\mathbf{r}(K,\mathbf{l})$, (9.12.8) becomes

$$\mathbf{r}(K+1,\mathbf{l}^1) = \mathbf{f}(K,\mathbf{r}(K,\mathbf{l})). \tag{9.12.9}$$

We now eliminate $\mathbf{l}^1$ from (9.12.6), (9.12.9). From (9.12.1), we get

$$\mathbf{g}(\mathbf{u}(0,K+1,\mathbf{l}^1)) + \mathbf{h}(\mathbf{u}(K+1,K+1,\mathbf{l}^1)) = \mathbf{l}^1. \tag{9.12.10}$$

But $\mathbf{u}(K+1,K+1,\mathbf{l}^1) = \mathbf{f}(K,\mathbf{u}(K,K+1,\mathbf{l}^1))$, and putting $k = K$, $k = 0$ in (9.12.6) gives

$$\mathbf{r}(K,\mathbf{l}) = \mathbf{u}(K,K,\mathbf{l}) = \mathbf{u}(K,K+1,\mathbf{l}^1)$$

and

$$\mathbf{s}(K,\mathbf{l}) = \mathbf{u}(0,K,\mathbf{l}) = \mathbf{u}(0,K+1,\mathbf{l}^1).$$

Using these relations in (9.12.10), to obtain

$$\mathbf{g}(\mathbf{s}(K,\mathbf{l})) + \mathbf{h}(\mathbf{f}(K,\mathbf{r}(K,\mathbf{l}))) = \mathbf{l}^1. \tag{9.12.11}$$

Substituting this into (9.12.6), (9.12.9), we get

$$\mathbf{u}(k,K+1,\mathbf{g}(\mathbf{s}(K,\mathbf{l})) + \mathbf{h}(\mathbf{f}(K,\mathbf{r}(K,\mathbf{l})))) = \mathbf{u}(k,K,\mathbf{l}) \tag{9.12.12}$$

and

$$\mathbf{r}(K+1,\mathbf{g}(\mathbf{s}(K,\mathbf{l})) + \mathbf{h}(\mathbf{f}(K,\mathbf{r}(K,\mathbf{l})))) = \mathbf{f}(K,\mathbf{r}(K,\mathbf{l})). \tag{9.12.13}$$

From the definition of $\mathbf{s}(K,\mathbf{l})$ and (9.12.1), we get

$$\mathbf{g}(\mathbf{s}(K,\mathbf{l})) = \mathbf{l} - \mathbf{h}(\mathbf{r}(K,\mathbf{l})). \tag{9.12.14}$$

Putting this into (9.12.12), (9.12.13) gives (9.12.2), (9.12.4). To get (9.12.5), we put $K = 0$ into (9.12.1), giving $\mathbf{g}(\mathbf{r}(0, \mathbf{l})) + \mathbf{h}(\mathbf{r}(0, \mathbf{l})) = \mathbf{l}$. Using the fact that $\mathbf{g} + \mathbf{h}$ has an inverse immediately leads to (9.12.5). ∎

**Remark 9.12.1.** Equations (9.12.2) (9.12.5) can be viewed as the fundamental initial value formulation of (1.2.8), (9.12.1). It is a single sweep method in that recursion is carried out in the direction of increasing $k$ only.

**Theorem 9.12.2.** Let the function $\mathbf{u}(k, K, \mathbf{l})$ be defined by the initial value problems (9.12.2) (9.12.5). Assume also that the Cauchy problems for (9.12.2), (9.12.4) have unique solutions. Then, $\mathbf{u}(k, K, \mathbf{l})$ satisfies (1.2.8), (9.12.1).

**Proof.** Define

$$\mathbf{v}(k, K, \mathbf{l}) = \mathbf{u}(k+1, K, \mathbf{l}) - \mathbf{f}(k, \mathbf{u}(k, K, \mathbf{l})), \quad k \in \mathbb{N}(0, K) \tag{9.12.15}$$

and

$$\mathbf{l}^1 = \mathbf{l} + \mathbf{h}(\mathbf{f}(K, \mathbf{r}(K, \mathbf{l}))) - \mathbf{h}(\mathbf{r}(K, \mathbf{l})). \tag{9.12.16}$$

From (9.12.15), we get

$$\mathbf{v}(k, K+1, \mathbf{l}^1) = \mathbf{u}(k+1, K+1, \mathbf{l}^1) - \mathbf{f}(k, \mathbf{u}(k, K+1, \mathbf{l}^1)). \tag{9.12.17}$$

Because of (9.12.2), the above equation is the same as

$$\mathbf{v}(k, K+1, \mathbf{l}^1) = \mathbf{u}(k+1, K, \mathbf{l}) - \mathbf{f}(k, \mathbf{u}(k, K, \mathbf{l})) = \mathbf{v}(k, K, \mathbf{l}). \tag{9.12.18}$$

Similarly, we find

$$\begin{aligned} \mathbf{v}(k, k, \mathbf{l}) &= \mathbf{u}(k+1, k, \mathbf{l}) - \mathbf{f}(k, \mathbf{u}(k, k, \mathbf{l})) \\ &= \mathbf{u}(k+1, k+1, \mathbf{l}^1) - \mathbf{f}(k, \mathbf{u}(k, k, \mathbf{l})) \\ &= \mathbf{r}(k+1, \mathbf{l}^1) - \mathbf{f}(k, \mathbf{r}(k, \mathbf{l})) = 0, \end{aligned} \tag{9.12.19}$$

where we have used (9.12.4) in deriving (9.12.19). Therefore, $\mathbf{v}(k, K, \mathbf{l})$ satisfies (9.12.2) with the initial condition zero. By the assumed uniqueness of the initial value problem for (9.12.2), we see that $\mathbf{v}(k, K, \mathbf{l}) = 0$, and so $\mathbf{u}(k, K, \mathbf{l})$ satisfies (1.2.8).

To obtain (9.12.1), we define

$$\mathbf{w}(K, \mathbf{l}) = \mathbf{g}(\mathbf{u}(0, K, \mathbf{l})) + \mathbf{h}(\mathbf{r}(K, \mathbf{l})) - \mathbf{l}. \tag{9.12.20}$$

Now, using (9.12.2) and (9.12.4), we see that

$$\begin{aligned} \mathbf{w}(K+1, \mathbf{l}^1) &= \mathbf{g}(\mathbf{u}(0, K+1, \mathbf{l}^1)) + \mathbf{h}(\mathbf{r}(K+1, \mathbf{l}^1)) - \mathbf{l}^1 \\ &= \mathbf{g}(\mathbf{u}(0, K, \mathbf{l})) + \mathbf{h}(\mathbf{f}(K, \mathbf{r}(K, \mathbf{l}))) \\ &\quad -\mathbf{l} - \mathbf{h}(\mathbf{f}(K, \mathbf{r}(K, \mathbf{l}))) + \mathbf{h}(\mathbf{r}(K, \mathbf{l})) \\ &= \mathbf{g}(\mathbf{u}(0, K, \mathbf{l})) + \mathbf{h}(\mathbf{r}(K, \mathbf{l})) - \mathbf{l} = \mathbf{w}(K, \mathbf{l}). \end{aligned}$$

Also, using (9.12.3) and (9.12.5), we get

$$\mathbf{w}(0,l) = \mathbf{g}(\mathbf{u}(0,0,l)) + \mathbf{h}(\mathbf{r}(0,l)) - 1 = 1 - 1 = 0.$$

Therefore, we see that $\mathbf{w}(K,l)$ satisfies (9.12.2) with $k=0$ and initial condition zero. Again using the uniqueness assumptions, we get $\mathbf{w}(K,l) = 0$, and consequently $\mathbf{g}(\mathbf{u}(0,K,l)) + \mathbf{h}(\mathbf{u}(K,K,l)) = 1$. ∎

## 9.13. Problems

**9.13.1.** With respect to the boundary value problem (1.6.13), (1.6.14) let there exist a positive constant $r$ and $Q = \sup\{|f(t_k,u)| : k \in \mathbb{N}(0,K+1),\ |u| \le 2r\}$. Further, let $\max\{|A|,|B|\} \le r$ and $(\beta-\alpha) \le (8r/Q)^{1/2}$. Show that the problem (1.6.13), (1.6.14) has at least one solution.

**9.13.2.** With respect to the boundary value problem (1.6.13), (1.6.14) assume that for all $k \in \mathbb{N}(0,K+1)$ and $u \in \mathbb{R}$, $|f(t_k,u)| \le c_0 + c_1|u|^\alpha$, where $c_0$ and $c_1$ are nonnegative constants and $0 < \alpha < 1$. Show that the problem (1.6.13), (1.6.14) has at least one solution.

**9.13.3.** With respect to the boundary value problem (1.6.13), (1.6.14) assume that for all $k \in \mathbb{N}(0,K+1)$ and $u \in S$, $|f(t_k,u)| \le L\left|u - A - \frac{B-A}{K+1}k\right| + q$, where $L$ and $q$ are nonnegative constants such that

$$\theta = \frac{h^2}{12}\frac{\left(10 + 2\cos\frac{\pi}{K+1}\right)}{4\sin^2\frac{\pi}{2(K+1)}}L < 1, \tag{9.13.1}$$

and $S = \left\{u(k) \in B(0,K+1) : \left|u(k) - A - \frac{B-A}{K+1}k\right| \le \frac{1}{2\pi}(1-\theta)^{-1}(\beta-\alpha)^2 q\sin\frac{k\pi}{K+1}\right\}$. Show that the problem (1.6.13), (1.6.14) has at least one solution in $S$.

**9.13.4.** Let in Problem 9.13.3, $S = \mathbb{R}$, $q = 0$ and $A = B = 0$. Show that $u(k) = 0$, $k \in \mathbb{N}(0,K+1)$ is the only solution of (1.6.13), (1.6.14). Further, show that in this conclusion the inequality (9.13.1) is the best possible.

**9.13.5.** With respect to the boundary value problem (1.6.13), (1.6.14) assume that for all $k \in \mathbb{N}(0,K+1)$ and $u,v \in S$, $|f(t_k,u) - f(t_k,v)| \le L|u-v|$, where $L$ is a nonnegative constant satisfying (9.13.1), and $S = \Big\{u(k) \in B(0,K+1) : \left|u(k) - A - \frac{B-A}{K+1}k\right| \le \frac{1}{2\pi}(1-\theta)^{-1}(\beta-\alpha)^2 \sup_{\ell\in\mathbb{N}(0,K+1)}$

$\left|f\left(t_\ell, A+\dfrac{B-A}{K+1}\ell\right)\right| \sin\dfrac{k\pi}{K+1}\Big\}$. Show that (1.6.13), (1.6.14) has a unique solution $u(k)$ is $S$. Further, show that the iterative scheme

$$\begin{aligned} u^{m+1}(k+1)-2u^{m+1}(k)+u^{m+1}(k-1) &= \frac{1}{12}h^2(f(t_{k-1},u^m(k-1)) \\ &\quad + 10f(t_k,u^m(k)) + f(t_{k+1},u^m(k+1))) \\ u^{m+1}(0) = A, \quad u^{m+1}(K+1) &= B, \quad m=0,1,\cdots \end{aligned}$$

where $u^0(k) = A + \dfrac{B-A}{K+1}k$ converges to $u(k)$, and an estimate on the rate of convergence can be given by

$$\begin{aligned} |u(k)-u^m(k)| &\le \theta^m(1-\theta)^{-1}(\beta-\alpha)^2 \times \\ &\quad \frac{1}{2\pi}\sup_{\ell\in\mathbb{N}(0,K+1)}\left|f\left(t_\ell, A+\frac{B-A}{K+1}\ell\right)\right|\sin\frac{k\pi}{K+1}. \end{aligned}$$

**9.13.6.** With respect to the boundary value problem (1.6.13), (1.6.14) assume that the function $f(t_k,u)$ is nonincreasing in $u$, and let there exist functions $u^0(k)$ and $v^0(k)$ such that $u^0(k)\le v^0(k)$ for all $k\in\mathbb{N}(0,K+1)$ and

$$\begin{aligned} u^0(k+1)-2u^0(k)+u^0(k-1) &\ge \frac{1}{12}h^2(f(t_{k-1},u^0(k-1))+10f(t_k,u^0(k)) \\ &\quad +f(t_{k+1},u^0(k+1))) \\ v^0(k+1)-2v^0(k)+v^0(k-1) &\le \frac{1}{12}h^2(f(t_{k-1},v^0(k-1))+10f(t_k,v^0(k)) \\ &\quad +f(t_{k+1},v^0(k+1))), \quad k\in\mathbb{N}(1,K) \\ u^0(0) \le A \le v^0(0), \quad & u^0(K+1) \le B \le v^0(k+1). \end{aligned}$$

Show that the sequences $\{u^m(k)\},\{v^m(k)\}$ generated by the iterative schemes

$$\begin{aligned} u^{m+1}(k+1)-2u^{m+1}(k)+u^{m+1}(k-1) &= \frac{1}{12}h^2(f(t_{k-1},u^m(k-1)) \\ &\quad + 10f(t_k,u^m(k)) + f(t_{k+1},u^m(k+1))) \\ v^{m+1}(k+1)-2v^{m+1}(k)+v^{m+1}(k-1) &= \frac{1}{12}h^2(f(t_{k-1},v^m(k-1)) \\ &\quad + 10f(t_k,v^m(k)) + f(t_{k+1},v^m(k+1))), \quad k\in\mathbb{N}(1,K) \\ u^{m+1}(0) = A = v^{m+1}(0), \quad & u^{m+1}(K+1) = B = v^{m+1}(k+1) \end{aligned}$$

converge to the solutions $u(k)$, $v(k)$ of (1.6.13), (1.6.14). Further, show that

$$\begin{aligned} u^0(k) \le u^1(k) \le \cdots \le u^m(k) \le \cdots \le u(k) \le v(k) \le \cdots \\ \le v^m(k) \le \cdots \le v^1(k) \le v^0(k) \end{aligned}$$

and each solution $w(k)$ of this problem which is such that $u^0(k) \le w(k) \le v^0(k)$ satisfies $u(k) \le w(k) \le v(k)$.

**9.13.7.** For the continuous boundary value problem (9.2.12) instead of (9.2.15) consider the discrete analog (1.6.13), (1.6.14) and as in Example 9.2.1 use Theorem 9.2.2 to this discrete problem to discuss the existence, uniqueness and the convergence of the Picard iterative scheme.

**9.13.8.** For the continuous boundary value problem $y'' = e^y, y(0) = y(1) = 0$ consider the discrete analog (1.6.13), (1.6.14) and apply the method of complementary functions and the method of adjoints in an iterative way, and compare the obtained numerical results with those of presented in Table 9.11.2.

**9.13.9.** Consider the boundary value problem

(9.13.2)
$$\mathbf{u}(k+1) - 2\mathbf{u}(k) + \mathbf{u}(k-1) = \frac{1}{(K+1)^2}\mathbf{f}\left(\frac{k}{K+1}, \mathbf{u}(k)\right), \quad k \in \mathbb{N}(1,K)$$

(9.13.3)
$$\mathbf{u}(0) = \mathbf{c}, \quad \mathbf{u}(K+1) = \mathbf{d}$$

which has an immediate relation with the continuous boundary value problem $\mathbf{y}'' = \mathbf{f}(t, \mathbf{y}),\ \mathbf{y}(0) = \mathbf{c},\ \mathbf{y}(1) = \mathbf{d}$. Let $\mathbf{r} \in \mathbb{R}^n_+$ be a given positive vector and let there exist a $\mathbf{q} \in \mathbb{R}^n_+$ such that for all $k \in \mathbb{N}(1,K)$ and $\mathbf{u} \in B_1(0, K+1) = \{\mathbf{u}(k) \in B(0, K+1) : \|\mathbf{u}\|_G \le 2\mathbf{r}\}, \left|\mathbf{f}\left(\frac{k}{K+1}, \mathbf{u}\right)\right| \le \mathbf{q}$. Further, let $\max\{|\mathbf{c}|, |\mathbf{d}|\} \le \mathbf{r}$ and $\frac{1}{8}\mathbf{q} \le \mathbf{r}$. Show that (9.13.2), (9.13.3) has at least one solution in $B_1(0, K+1)$.

**9.13.10.** Consider the boundary value problem (9.13.2), (9.13.3) and assume that for all $k \in \mathbb{N}(1,K)$ and $\mathbf{u} \in \mathbb{R}^n$, $\left|f_i\left(\frac{k}{K+1}, \mathbf{u}\right)\right| \le p_i + \sum_{j=1}^{n} q_{ij}|u_j|^{\alpha(i,j)}$, $1 \le i \le n$ where $p_i$, $q_{ij}$ and $\alpha(i,j)$, $1 \le i, j \le n$ are nonnegative constants and $\alpha(i,j) < 1$. Show that (9.13.2), (9.13.3) has at least one solution.

**9.13.11.** Consider the boundary value problem (9.13.2), (9.13.3) and assume that for all $k \in \mathbb{N}(1,K)$ and $\mathbf{u} \in S = \Big\{ \mathbf{u}(k) \in B(0, K+1) : \Big|\mathbf{u}(k)$

$$-\mathbf{c}-\frac{\mathbf{d}-\mathbf{c}}{K+1}k\Bigg| \ \leq \ \frac{1}{2\pi}\left[\mathcal{I}-\frac{1}{4(K+1)^2\sin^2\frac{\pi}{2(K+1)}}\mathcal{M}\right]^{-1}\mathbf{q}\sin\frac{k\pi}{K+1}\Bigg\},$$

$\left|\mathbf{f}\left(\frac{k}{K+1},\mathbf{u}\right)\right| \leq M\left|\mathbf{u}-\mathbf{c}-\frac{\mathbf{d}-\mathbf{c}}{K+1}k\right|+\mathbf{q},$ where $M$ is an $n\times n$ nonnegative matrix with

$$\text{(9.13.4)}\qquad \frac{1}{4(K+1)^2\sin^2\frac{\pi}{2(K+1)}}\rho(M) \ < \ 1,$$

and $\mathbf{q}\in\mathbb{R}^n_+$. Show that (9.13.2), (9.13.3) has at least one solution in $S$.

**9.13.12.** Let in Problem 9.13.11, $S=\mathbb{R}^n$, $\mathbf{q}=0$ and $\mathbf{c}=\mathbf{d}=0$. Show that $\mathbf{u}(k)=0,\ k\in\mathbb{N}(0,K+1)$ is the only solution of (9.13.2), (9.13.3). Further, show that in this conclusion the inequality (9.13.4) is the best possible.

**9.13.13.** Consider the boundary value problem (9.13.2), (9.13.3) and assume that for all $k\in\mathbb{N}(1,K)$ and $\mathbf{u}^1,\mathbf{u}^2\in S=\Bigg\{\mathbf{u}(k)\in B(0,K+1):$

$$\left|\mathbf{u}(k)-\mathbf{c}-\frac{\mathbf{d}-\mathbf{c}}{K+1}k\right|\leq\frac{1}{2\pi}\left[\mathcal{I}-\frac{1}{4(K+1)^2\sin^2\frac{\pi}{2(K+1)}}\mathcal{M}\right]^{-1}\sup_{\ell\in\mathbb{N}(1,K)}$$

$$\left|\mathbf{f}\left(\frac{\ell}{K+1},\mathbf{c}+\frac{\mathbf{d}-\mathbf{c}}{K+1}\ell\right)\right|\sin\frac{k\pi}{K+1}\Bigg\},\quad \left|\mathbf{f}\left(\frac{k}{K+1},\mathbf{u}^1\right)-\mathbf{f}\left(\frac{k}{K+1},\mathbf{u}^2\right)\right|$$

$\leq\mathcal{M}|\mathbf{u}^1-\mathbf{u}^2|$, where $\mathcal{M}$ is an $n\times n$ nonnegative matrix satisfying (9.13.4). Show that (9.13.2), (9.13.3) has a unique solution $\mathbf{u}(k)$ in $S$. Further, show that the iterative scheme

$$\mathbf{u}^{m+1}(k+1)-2\mathbf{u}^{m+1}(k)+\mathbf{u}^{m+1}(k-1)=\frac{1}{(K+1)^2}\mathbf{f}\left(\frac{k}{K+1},\mathbf{u}^m(k)\right)$$
$$\mathbf{u}^{m+1}(0) \ = \ \mathbf{c},\quad \mathbf{u}^{m+1}(K+1) \ = \ \mathbf{d},\quad m=0,1,\cdots$$

where $\mathbf{u}^0(k)=\mathbf{c}+\frac{\mathbf{d}-\mathbf{c}}{K+1}k$ converges to $\mathbf{u}(k)$, and an estimate on the rate of convergence can be given by

$$|\mathbf{u}(k)-\mathbf{u}^m(k)|\leq\left(\frac{1}{4(K+1)^2\sin^2\frac{\pi}{2(K+1)}}\mathcal{M}\right)^k\left[\mathcal{I}-\frac{1}{4(K+1)^2\sin^2\frac{\pi}{2(K+1)}}\mathcal{M}\right]^{-1}$$
$$\times\frac{1}{2\pi}\sup_{\ell\in\mathbb{N}(1,K)}\left|\mathbf{f}\left(\frac{\ell}{K+1},\mathbf{u}^0(\ell)\right)\right|\sin\frac{k\pi}{K+1}.$$

**9.13.14.** Let $\mathbf{v}(k),\ \mathbf{w}(k) \in B(0, K+1)$ be two functions with $\mathbf{v}(k) \le \mathbf{w}(k)$ on $\mathbb{N}(0, K+1)$ and

$$\mathbf{v}(k+1) - 2\mathbf{v}(k) + \mathbf{v}(k-1) \ge \frac{1}{(K+1)^2}\mathbf{f}\left(\frac{k}{K+1}, \mathbf{v}(k)\right),$$

$$\mathbf{w}(k+1) - 2\mathbf{w}(k) + \mathbf{w}(k-1) \le \frac{1}{(K+1)^2}\mathbf{f}\left(\frac{k}{K+1}, \mathbf{w}(k)\right), \quad k \in \mathbb{N}(1, K)$$

also $\mathbf{v}(0) \le \mathbf{c} \le \mathbf{w}(0)$, $\mathbf{v}(K+1) \le \mathbf{d} \le \mathbf{w}(K+1)$. Further, let $\mathbf{f}\left(\frac{k}{K+1}, \mathbf{u}\right)$ be quasimonotone nonincreasing in $\mathbf{u}$, i.e. for fixed $k$, $f_i\left(\frac{k}{K+1}, \mathbf{u}\right)$ is nonincreasing in $u_j$ for $1 \le j \le n$, $j \ne i$. Show that the boundary value problem (9.13.2), (9.13.3) has at least one solution $\mathbf{u}(k)$ such that $\mathbf{v}(k) \le \mathbf{u}(k) \le \mathbf{w}(k)$.

**9.13.15.** The solution $\mathbf{v}(k)$ of the boundary value problem

$$\text{(9.13.5)} \quad \mathbf{v}(k+1) - 2\mathbf{v}(k) + \mathbf{v}(k-1) + \frac{1}{(K+1)^2}\mathbf{g}\left(\frac{k}{K+1}, \mathbf{v}(k)\right) = 0, \quad k \in \mathbb{N}(1, K)$$

$$\text{(9.13.6)} \quad \mathbf{v}(0) = \mathbf{c}^1, \quad \mathbf{v}(K+1) = \mathbf{d}^1$$

is said to be *maximal* if for any other solution $\mathbf{w}(k)$ of (9.13.5), (9.13.6) the inequality $\mathbf{w}(k) \le \mathbf{v}(k)$ holds for all $k \in \mathbb{N}(0, K+1)$. Let $\mathbf{g}\left(\frac{k}{K+1}, \mathbf{v}\right)$ be nonnegative and nondecreasing in $\mathbf{v}$ for all $(k, \mathbf{v}) \in \mathbb{N}(1, K) \times \mathbb{R}^n_+$. Further, let $\mathbf{c}^1$ and $\mathbf{d}^1$ be nonnegative and there exists a $\mathbf{r} \in \mathbb{R}^n_+$, $\mathbf{r} > 0$ such that for all $k \in \mathbb{N}(0, K+1)$

$$\mathbf{c}^1 + \frac{\mathbf{d}^1 - \mathbf{c}^1}{K+1}k - \frac{1}{(K+1)^2}\sum_{\ell=1}^{K} g(k,\ell)\mathbf{g}\left(\frac{\ell}{K+1}, \mathbf{r}\right) \le \mathbf{r}$$

where $g(k,\ell)$ is defined in (8.10.3). Show that the sequence $\{\mathbf{v}^m(k)\}$ generated by

$$\mathbf{v}^{m+1}(k) = \mathbf{c}^1 + \frac{\mathbf{d}^1 - \mathbf{c}^1}{K+1}k - \frac{1}{(K+1)^2}\sum_{\ell=1}^{K} g(k,\ell)\mathbf{g}\left(\frac{\ell}{K+1}, \mathbf{v}^m(\ell)\right)$$

$$\mathbf{v}^0(k) = \mathbf{c}^1 + \frac{\mathbf{d}^1 - \mathbf{c}^1}{K+1}k - \frac{1}{(K+1)^2}\sum_{\ell=1}^{K} g(k,\ell)\mathbf{g}\left(\frac{\ell}{K+1}, \mathbf{r}\right), \quad m = 0, 1, \cdots$$

converges to the maximal solution $\mathbf{v}(k)$ of (9.13.5), (9.13.6) and $\mathbf{v}(k) \in S_{\mathbf{r}} = \{\mathbf{v}(k) \in B(0, N+1) : \|\mathbf{v}(k)\|_G \le \mathbf{r}\}$.

**9.13.16.** Let the function $\mathbf{g}$ and the vectors $\mathbf{c}^1$ and $\mathbf{d}^1$ be as in Problem 9.13.15. Further, let $|\mathbf{c}| \le \mathbf{c}^1$, $|\mathbf{d}| \le \mathbf{d}^1$ and for all $(k, \mathbf{u}) \in \mathbb{N}(1, K) \times \mathbb{R}^n$, $\left|\mathbf{f}\left(\frac{k}{K+1}, \mathbf{u}\right)\right| \le \mathbf{g}\left(\frac{k}{K+1}, |\mathbf{u}|\right)$. Show that for any solution $\mathbf{u}(k) \in S_{\mathbf{r}}$ of (9.13.2), (9.13.3) the inequality $|\mathbf{u}(k)| \le \mathbf{v}(k)$ holds for all $k \in \mathbb{N}(0, K+1)$, where $\mathbf{v}(k)$ is the maximal solution of (9.13.5), (9.13.6).

## 9.14. Notes

The importance of generalized normed spaces in the study of systems of nonlinear equations has been recognized in numerous recent publications, e.g. Agarwal [4,7–10,13,14], Bernfeld and Lakshmikantham [22], Ortega and Rheinboldt [33], Perov and Kibenko [38], Schröder [43], Šeda [44], Shridharan and Agarwal [46,47], Urabe [50,51], Yamamoto [58–61]. The results collected in Section 9.1 are available at several places, e.g. Agarwal [4], Bernfeld and Lakshmikantham [22], Urabe [50]. All the results in Sections 9.2–9.6 are from Agarwal [5,6,12]. Eloe and Grimm [23] and Eloe [24] form the basis of Theorems 9.7.1 and 9.7.3. Monotone convergence of periodic boundary value problems presented in Section 9.8 is due to Wang and Agarwal [55]. The convergence of the Newton's method and the approximate Newton's method discussed in Sections 9.9 and 9.10 are taken from Agarwal [13]. Initial value methods for nonlinear boundary value problems presented in Section 9.11 are from Agarwal [6,12]. Theorems 9.12.1 and 9.12.2 are due to Golberg [28]. Discrete boundary value problems for nonlinear systems have also been discussed in Agarwal et. al. [16,18,20,34–36,56,57], Falb and DeJong [25], Pao [37], Rodriguez [40–42], Wang [53,54], and Zhuang, Cheng and Cheng [62]. The solutions of most of the problems given in Section 9.13 can be deduced from the results proved in Agarwal [9,11]. Results related to this chapter for the continuous boundary value problems are available in Agarwal et. al. [1–3,14,15,17,19], Falb and DeJong [25], Fujii and Hayashi [27], Hayashi [29], Mitsui [30], Ojika and Kasue [31], Ojika [32], Shintani and Hayashi [45], Urabe [48,49,52], and Yamamoto [60]. While attention for the continuous two point boundary value problems has been focused at several places, we refer to the monographs of Ascher, Mattheij and Russell [21], Roberts and Shipman [39].

## 9.15. References

[1]. R.P. Agarwal, The numerical solution of multipoint boundary value problems, *J. Comp. Appl. Math.* **5**(1979), 17–24.

[2]. R.P. Agarwal, On the periodic solutions of nonlinear second order

differential systems, *J. Comp. Appl. Math.* **5**(1979), 117 123.

[3]. R.P. Agarwal, On the method of complementary functions for nonlinear boundary value problems, *J. Optimization Theory and Appl.* **36**(1982), 139 144.

[4]. R.P. Agarwal, Contraction and approximate contraction with an application to multi point boundary value problems, *J. Comp. Appl. Math.* **9**(1983), 315 325.

[5]. R.P. Agarwal, On multipoint boundary value problems for discrete equations, *J. Math. Anal. Appl.* **96**(1983), 520 534.

[6]. R.P. Agarwal, Initial value methods for discrete boundary value problems, *J. Math. Anal. Appl.* **100**(1984), 513 529.

[7]. R.P. Agarwal, On Urabe's application of Newton's method to nonlinear boundary value problems, *Arch. Math. (Brnö)* **20**(1984), 113 124.

[8]. R.P. Agarwal, Component wise convergence of iterative methods for nonlinear Volterra integro differential systems with nonlinear boundary conditions, *Jour. Math. Phyl. Sci.* **18**(1984), 291 322.

[9]. R.P. Agarwal, On boundary value problems of second order discrete systems, *Applicable Analysis* **20**(1985), 1 17.

[10]. R.P. Agarwal and J. Vósmanský, Necessary and sufficient conditions for the convergence of approximate Picard's iterates for nonlinear boundary value problems, *Arch. Math. (Brnö)* **21**(1985), 171 176.

[11]. R.P. Agarwal, On Nomerov's method for solving two point boundary value problems, *Utilitas Mathematica* **28**(1985), 159 174.

[12]. R.P. Agarwal, Computational methods for discrete boundary value problems, *Appl. Math. Comp.* **18**(1986), 15 41.

[13]. R.P. Agarwal, Computational methods for discrete boundary value problems II, *J. Math. Anal. Appl.* **166**(1992), 540 562.

[14]. R.P. Agarwal, Component wise convergence of quasilinearization method for nonlinear boundary value problems, *Hiroshima Math. J.* **22**(1992), 525 541.

[15]. R.P. Agarwal, F.H. Wong and W.C. Lian, Existence of positive solutions of nonlinear second order differential systems, *Applicable Analysis* **63**(1996), 375 387.

[16]. R.P. Agarwal and P.J.Y. Wong, *Advanced Topics in Difference Equations*, *Kluwer*, Dordrecht, 1997.

[17]. R.P. Agarwal and D. O'Regan, A coupled system of boundary value problems, *Applicable Analysis* **69**(1998), 381 385.

[18]. R.P. Agarwal and D. O'Regan, A coupled system of difference equations, *Appl. Math. Comput.*, to appear.

[19]. R.P. Agarwal and D. O'Regan, Multiple solutions for a coupled system of boundary value problems, *Dynamics of Continuous, Discrete and Impulsive Systems*, to appear.

[20]. R.P. Agarwal and P.J.Y. Wong, Existence criteria for a system of two point boundary value problems, to appear.

[21]. U.M. Ascher, R.M.M. Mattheij and R.D. Russell, *Numerical Solution of Boundary Value Problems for Ordinary Differential Equations*, Prentice Hall Series in Comp. Math., Prentice Hall, New Jersey, 1988.

[22]. S.R. Bernfeld and V. Lakshmikantham, *An Introduction to Nonlinear Boundary Value Problems*, *Academic Press*, New York, 1974.

[23]. P.W. Eloe and L.J. Grimm, Differential systems and multipoint boundary value problems, *ZAMM* **62**(1982), 630 632.

[24]. P.W. Eloe, A boundary value problem for a system of difference equations, *Nonlinear Analysis* **7**(1983), 813 820.

[25]. P.L. Falb and J.L. DeJong, *Some Successive Approximation Methods in Control and Oscillation Theory*, *Academic Press*, New York, 1969.

[26]. L. Fox, *The Numerical Solution of Two Point Boundary Value Problems in Ordinary differential Equations*, *Oxford Univ. Press*, London, 1957.

[27]. M. Fujii and Y. Hayashi, Numerical solutions to problems of the least squares type for ordinary differential equations, *Hiroshima Math. J.* **13**(1983), 477 499.

[28]. M.A. Golberg, Derivation and validation of initial value methods for boundary value problems for difference equations, *J. Optimization Theory and Appl.* **7**(1971), 411 419.

[29]. Y. Hayashi, On a posteriori error estimation in the numerical solution of ordinary differential equations, *Hiroshima Math. J.* **9** (1979), 201 243.

[30]. T. Mitsui, The initial value adjusting method for problems of the least square type of ordinary differential equations, *Publ. RIMS Kyoto Univ.* **16**(1980), 785 810.

[31]. T. Ojika and Y. Kasue, Initial value adjusting method for the solution of nonlinear multipoint boundary value problems, *J. Math. Anal. Appl.* **69**(1979), 359 371.

[32]. T. Ojika, On quadratic convergence of the initial value adjusting method for nonlinear multipoint boundary value problems, *J. Math. Anal. Appl.* **73**(1980), 192 203.

[33]. J.M. Ortega and W. Rheinboldt, On a class of approximate iterative processes, *Arch. Rational Mech. Anal.* **23**(1967), 352 365.

[34]. P.Y.H. Pang and R.P. Agarwal, Periodic boundary value problems for first and second order discrete systems, *Mathl. Comput. Modelling* **16**(1992), 101 112.

[35]. P.Y.H. Pang and R.P. Agarwal, On discrete boundary value problems arising in transport phenomena, *Appl. Math. Comput.* **60**(1994), 193 214.

[36]. P.Y.H. Pang and R.P. Agarwal, Monotone iterative methods for a general class of discrete boundary value problems, *Computers Math. Applic.* **28**(1 3)(1994), 243 254.

[37]. C.V. Pao, Monotone methods for a finite difference system of reaction diffusion equation with time delay, *Computers Math. Applic.* **36**(10 12)(1998), 37 47.

[38]. A. Perov and A. Kibenko, On a certain general method for investigation of boundary value problems, *Izv. Akad. Nauk SSSR* **30** (1966), 249 264.

[39]. S.M. Roberts and J.S. Shipman, *Two Point Boundary Value Problems: Shooting Methods*, *Elsevier*, New York, 1972.

[40]. J. Rodriguez, On resonant discrete boundary value problem, *Applicable Analysis* **19**(1985), 265 274.

[41]. J. Rodriguez, Resonance in nonlinear discrete systems with nonlinear constraints, *Proc. 24th Conf. Decision and Control, IEEE* (1985), 1738 1743.

[42]. J. Rodriguez, On nonlinear discrete boundary value problems, *J. Math. Anal. Appl.* **114**(1986), 398 408.

[43]. J. Schröder, *Operator Inequalities*, *Academic Press*, New York, 1980.

[44]. V. Šeda, On a vector multipoint boundary value problem, *Arch. Math. (Brnö)* **22**(1986), 75 92.

[45]. H. Shintani and Y. Hayashi, A posteriori error estimates and iterative methods in the numerical solution of systems of ordinary differential equations, *Hiroshima Math. J.* **8**(1978), 101 121.

[46]. R. Shridharan and R.P. Agarwal, Stationary and nonstationary iterative methods for nonlinear boundary value problems, *Mathl. Comput. Modelling* **18**(1993), 43 62.

[47]. R. Shridharan and R.P. Agarwal, General iterative methods for nonlinear boundary value problems, *J. Austral. Math. Soc. Ser. B,* **37**(1995), 58 85.

[48]. M. Urabe, An existence theorem for multi point boundary value problems, *Funkcial. Ekvac.* **9**(1966), 43 60.

[49]. M. Urabe, The Newton method and its application to boundary value problems with nonlinear boundary conditions, *Proc. US Japan Seminar on Differential and Functional Equations*, Benjamin, New York, (1967), 383 410.

[50]. M. Urabe, Component wise error analysis of iterative methods practiced on a floating point system, *Mem. Fac. Sci., Kyushu Univ., Ser. A,*

*Math.* **27**(1973), 23 64.

[51]. M. Urabe, A posteriori component wise error estimation of approximate solutions to nonlinear equations, *Lecture Notes in Computer Sci.* **29**, Springer Verlag, Berlin (1975), 99 111.

[52]. M. Urabe, On the Newton method to solve problems of the least squares type for ordinary differential equations, *Mem. Facu. Sci., Kyushu Univ., Ser. A, Math.* **29**(1975), 173 183.

[53]. Y.M. Wang, Monotone enclosure for a class of discrete boundary value problems without monotone nonlinearities, *Computers Math. Applic.* **35**(6)(1998), 51 60.

[54]. Y.M. Wang, Monotone methods for a boundary value problem of second order discrete equation, *Computers Math. Applic.* **36**(6)(1998), 77 92.

[55]. Y.M. Wang and R.P. Agarwal, Remarks on periodic boundary value problems of first order discrete systems, *Intern. J. Computer Math.*, to appear.

[56]. Y.M. Wang and R.P. Agarwal, Monotone methods for solving a boundary value problem of second order discrete system, to appear.

[57]. P.J.Y. Wong and R.P. Agarwal, Fixed sign solutions of a system of higher order difference equations, *Jour. Comp. Appl. Math.*, to appear.

[58]. T. Yamamoto, Componentwise error estimates for approximate solutions of nonlinear equations, *JIP* **2**(1979), 121 126.

[59]. T. Yamamoto, Componentwise error estimates for approximate solutions of systems of equations, *Lecture Notes in Num. Appl. Anal.* **3**(1981), 1 22.

[60]. T. Yamamoto, An existence theorem of solution to boundary value problems and its application to error estimates, *Math. Japonica* **27** (1982), 301 318.

[61]. T. Yamamoto, A unified derivation of several error bounds for Newton's process, *J. Comp. Appl. Math.* **12**(1985), 179 191.

[62]. W. Zhuang, Y. Cheng and S.S. Cheng, Monotone methods for a discrete boundary value problem, *Computers Math. Applic.* **32**(1996), 41 49.

# Chapter 10
# Miscellaneous Properties of Solutions of Higher Order Linear Difference Equations

Disconjugacy property of a linear homogeneous differential equation allows the possibility of interpolation by its solutions. While this property has been investigated thoroughly for the differential equations, its discrete analogs are not fully developed. In this chapter we shall introduce disconjugacy, right disconjugacy, left disconjugacy, right disfocality, eventual disconjugacy and eventual right disfocality for the linear homogeneous difference equations, and for each such concept state several results which provide necessary and sufficient conditions. This includes Polya's factorization, and interrelationship between D Markov, D Fekete and D Descartes systems. This is followed by the statement of the discrete analog of a result due to Elias, which bounds the number of certain types of zeros of solutions of linear homogeneous difference equations on a discrete interval. A classification of solutions of these equations based on their behavior in a neighborhood of infinity is also included. Then, we provide explicit representations of polynomials passing through the given boundary conditions which also include (1.5.9) (1.5.14). Such polynomials are called discrete interpolating polynomials. This is followed by the explicit representations of Green's functions for several higher order boundary value problems. For these Green's functions we state several equalities and inequalities whose continuous analogs have proved to be very useful in providing disconjugacy tests and distance between consecutive zeros of the solutions of higher order differential equations. The explicit forms of interpolating polynomials and those of Green's functions help in establishing maximum principles for functions satisfying higher order inequalities. We state some such maximum principles. Finally, in this chapter we have included several results which provide error estimates in polynomial interpolation. Some of these results will be used in the next chapter to study higher order boundary value problems. To limit the size of this volume the proofs of the theorems in this chapter have not been given. However, we observe that almost all the proofs require special devices and no unified approach seems to be available.

## 10.1. Disconjugacy

Throughout, this chapter unless otherwise stated for the $n$th order linear difference equation (1.2.4) we shall assume that the coefficients $a_i(k)$ are defined on $\mathbb{N}(a, b-1)$, $a_n(k) \equiv 1$ and $a_0(k)$ satisfies

$$(10.1.1) \qquad (-1)^n a_0(k) \ > \ 0.$$

Further, whenever necessary we shall extend the domain of the coefficients $a_i(k)$ to $\mathbf{Z}$ by defining $a_i(k) = a_i(a)$ for $k < a$ and $a_i(k) = a_i(b-1)$ for $k \geq b$.

**Definition 10.1.1.** The difference equation (1.2.4) is called $r$ *disconjugate* on $\mathbb{N}(a, b-1+n)$ if no nontrivial solution has $n$ nodes on $\mathbb{N}(a, b-1+n)$.

**Definition 10.1.2.** The difference equation (1.2.4) is called *disconjugate* on $\mathbb{N}(a, b-1+n)$ if no nontrivial solution has $n$ generalized zeros on $\mathbb{N}(a, b-1+n)$.

**Theorem 10.1.1.** Condition (10.1.1) is necessary for (1.2.4) to be disconjugate on $\mathbb{N}(a, b-1+n)$.

**Theorem 10.1.2.** The difference equation (1.2.4) is $r$ disconjugate on $\mathbb{N}(a, b-1+n)$ if and only if it is disconjugate on $\mathbb{N}(a, b-1+n)$.

**Theorem 10.1.3.** The difference equation (1.2.4) is disconjugate on $\mathbb{N}(a, b-1+n)$ if and only if $u(k) \equiv 0$ is the only solution of (1.2.4) having $p\ (> 0)$ successive zeros at $k = a, a+1, \cdots, a+p-1$ and $n-p$ successive generalized zeros at $k = c, c+1, \cdots, c+n-p-1 \in \mathbb{N}(a, b-1+n)$ for some $c \geq a+p$.

**Theorem 10.1.4.** The difference equation (1.2.4) is disconjugate on $\mathbb{N}(a, b-1+n)$ if and only if there exists a fundamental set of solutions $u_1(k), \cdots, u_n(k)$ of (1.2.4) on $\mathbb{N}(a, b-1+n)$ such that $\det C(u_1, \cdots, u_i)(k) > 0$ on $\mathbb{N}(a, b+n-i)$, $1 \leq i \leq n$.

**Theorem 10.1.5** (Polya's Factorization). The difference equation (1.2.4) is disconjugate on $\mathbb{N}(a, b-1+n)$ if and only if there exist positive functions $h_i(k)$, $k \in \mathbb{N}(a, b-1+n-i)$, $0 \leq i \leq n$ such that

$$(10.1.2) \qquad L[u(k)] \ = \ \sum_{i=0}^{n} a_i(k)u(k+i) \ = \ h_n\Delta\{h_{n-1}\Delta[\cdots\Delta(h_0 u)]\}.$$

**Corollary 10.1.6.** The difference equation

$$\Delta^n u(k) = 0, \quad k \in \mathbb{N}(a, b-1) \tag{10.1.3}$$

is disconjugate on $\mathbb{N}(a, b-1+n)$.

**Theorem 10.1.7.** Let $0 \le h(k) \le g(k)$, $k \in \mathbb{N}(a, b-1)$ and assume that the equations $L[u(k)] = 0$ and $L[u(k)] + g(k)u(k) = 0$ are disconjugate on $\mathbb{N}(a, b-1+n)$. Then, the equation $L[u(k)] + h(k)u(k) = 0$ is disconjugate on $\mathbb{N}(a, b-1+n)$.

## 10.2. Right and Left Disconjugacy

**Definition 10.2.1.** Let $\mathbb{N}(c, d) \subseteq \mathbb{N}(a, b-1+n)$ with $d - c + 1 \ge n$, and let $1 \le j \le n-1$. Equation (1.2.4) is said to be *right* $(j, n-j)$ *disconjugate* on $\mathbb{N}(c, d)$ provided there is no nontrivial solution $u(k)$ of (1.2.4) and integers $\alpha$, $\beta \in \mathbb{N}(c, d)$ with $\alpha < \alpha + j \le \beta \le \beta + n - j - 1 \le d$ such that

$$\begin{aligned} u(\alpha + i) &= 0, \quad 0 \le i \le j-1 \\ u(\beta + i) &= 0, \quad 0 \le i \le n-j-2 \ \ (\text{if } n - j \ge 2) \end{aligned} \tag{10.2.1}$$

and $u(k)$ has a generalized zero at $\beta + n - j - 1$. Similarly, we say that (1.2.4) is *left* $(j, n-j)$ *disconjugate* on $\mathbb{N}(c, d)$ provided there is no nontrivial solution $u(k)$ of (1.2.4) and integers $\alpha$, $\beta \in \mathbb{N}(c, d)$ with $\alpha < \alpha + j \le \beta \le \beta + n - j - 1 \le d$ such that

$$\begin{aligned} u(\alpha + i) &= 0, \quad 0 \le i \le j-2 \ \ (\text{if } j \ge 2) \\ u(\beta + i) &= 0, \quad 0 \le i \le n-j-1 \end{aligned} \tag{10.2.2}$$

and $u(k)$ has a generalized zero at $\alpha + j - 1$.

**Remark 10.2.1.** If (1.2.4) is disconjugate on $\mathbb{N}(c, d)$ then it is right $(j, n-j)$ disconjugate for $1 \le j \le n-1$. However, right $(j, n-j)$ disconjugacy for some fixed $1 \le j \le n-1$ does not imply right $(n-j, j)$ disconjugacy or disconjugacy. For example, the difference equation $u(k+3) - u(k+2) - u(k+1) - u(k) = 0$ is right $(2, 1)$ disconjugate on $\mathbb{N}(0, 3)$, but there is a solution $u(k)$ with $u(0) = u(2) = 0$, $u(1) = u(3) = 1$ so that $u(k)$ has a generalized zero at $k = 3$. Hence, this difference equation is not right $(1, 2)$ disconjugate or disconjugate on $\mathbb{N}(0, 3)$.

**Theorem 10.2.1.** If the difference equation (1.2.4) is right $(\ell, n-\ell)$ disconjugate on $\mathbb{N}(a, b-1+n)$, where $\ell \in \{1, \cdots, n-1\}$ is fixed, then

(10.2.3)

$$(-1)^{j(n+\ell)}\begin{vmatrix} a_\ell(k) & a_{\ell+1}(k) & a_{\ell+j-1}(k) \\ a_{\ell-1}(k+1) & a_\ell(k+1) & a_{\ell+j-2}(k+1) \\ \cdots & \cdots & \cdots \\ a_{\ell-j+1}(k+j-1) & a_{\ell-j+2}(k+j-1) & a_\ell(k+j-1) \end{vmatrix} > 0$$

for $k \in \mathbb{N}(a, b-j)$ and $j = 1, 2, \cdots, b-a$. (Here, $a_j(k) \equiv 0$ for $j > n$ or $j < 0$.)

**Theorem 10.2.2.** Let $\ell \in \{1, \cdots, n-1\}$ be fixed, and let (10.2.3) hold for $k \in \mathbb{N}(a, b-j)$ and $j = 1, 2, \cdots, b-a$. Then, (1.2.4) is right $(\ell, n-\ell)$ disconjugate on $\mathbb{N}(a, b-1+n)$.

**Theorem 10.2.3.** A necessary and sufficient condition for (1.2.4) to be left $(\ell, n-\ell)$ disconjugate on $\mathbb{N}(a, b-1+n)$ for a fixed $\ell \in \{1, \cdots, n-1\}$ is (10.2.3) for $k \in \mathbb{N}(a, b-j)$ and $j = 1, 2, \cdots, b-a$.

**Theorem 10.2.4.** The following are equivalent:

(i) (1.2.4) is disconjugate on $\mathbb{N}(a, b-1+n)$

(ii) (1.2.4) is right $(\ell, n-\ell)$ disconjugate on $\mathbb{N}(a, b-1+n)$ for $\ell = 1, \cdots, n-1$

(iii) (1.2.4) is left $(\ell, n-\ell)$ disconjugate on $\mathbb{N}(a, b-1+n)$ for $\ell = 1, \cdots, n-1$

(iv) (10.2.3) hold for $\ell = 1, \cdots, n-1$ and $k \in \mathbb{N}(a, b-j)$ where $j = 1, 2, \cdots, b-a$.

**Theorem 10.2.5.** Let (10.2.3) hold for $k = a$, $\ell = 1, \cdots, n-1$ and $j = 1, 2, \cdots, b-a$. Then, (1.2.4) is disconjugate on $\mathbb{N}(a, b-1+n)$.

**Theorem 10.2.6.** Let $u_j(k, \ell)$ be a solution of (1.2.4) satisfying the partial set of initial conditions

(10.2.4) $$u_j(\ell + i, \ell) = \delta_{ij}, \quad 0 \le i \le j$$

for $1 \le j \le n-1$. Then,

(i) the difference equation (1.2.4) is right $(j, n-j)$ disconjugate on $\mathbb{N}(c, d)$ if and only if

(10.2.5) $$\det C(u_j(k, \ell), \cdots, u_{n-1}(k, \ell)) > 0$$

for $c \le \ell \le k - j \le d - n + 1$

(ii) the difference equation (1.2.4) is left $(j, n-j)$ disconjugate on $\mathbb{N}(c, d)$ if and only if

(10.2.6) $$(-1)^{j(n-j)} \det C\left(u_{n-j}(k, \ell), \cdots, u_{n-1}(k, \ell)\right) > 0$$

for $c \le k \le \ell - j \le d - n + 1$.

**Corollary 10.2.7.** If (1.2.4) is right $(j, n-j)$ disconjugate on $\mathbb{N}(c,d)$ then there exist solutions $u_j(k), \cdots, u_{n-1}(k)$ of (1.2.4) and a linear difference equation $L_1[u(k)] = 0$ of order $j$ such that

$$L[u(k)] = L_1 L_2[u(k)], \quad k \in \mathbb{N}(c, d-n) \tag{10.2.7}$$

where $L_2[u(k)] = \det C(u(k), u_j(k), \cdots, u_{n-1}(k))$.

**Definition 10.2.2.** Let $1 \le j \le n-1$. Equation (1.2.4) is said to be $\rho_j$ *disconjugate* on $\mathbb{N}(c,d)$ provided there does not exist a nontrivial solution $u(k)$ and an integer $\alpha \in \mathbb{N}(c, d+1-n)$ such that $u(\alpha+i) = 0$, $0 \le i \le j-1$ and when 'restricted' to $\mathbb{N}(c+j, d)$, $u(k)$ has $n-j$ generalized zeros.

**Remark 10.2.2.** To illustrate the meaning of the word 'restricted' in the above definition, consider the function $u(k)$ defined on $\mathbb{N}(0,2)$ by $u(0) = 1$, $u(1) = 0$, and $u(2) = 1$. This function has generalized zeros at $k = 1$ and $k = 2$, but when restricted to $\mathbb{N}(1,2)$, $u(k)$ has a generalized zero only at $k = 1$.

**Theorem 10.2.8.** Let $1 \le j \le n-1$. Then, (1.2.4) is right $(\ell, n-\ell)$ disconjugate on $\mathbb{N}(c,d)$ for $j \le \ell \le n-1$ if and only if (1.2.4) is $\rho_j$ disconjugate on $\mathbb{N}(c,d)$.

**Theorem 10.2.9.** Assume that $w(k)$ is a solution of

$$L[w(k)] \ge g(k)w(k), \quad k \in \mathbb{N}(a, b-1) \tag{10.2.8}$$

with $w(k) \ge 0$ on $\mathbb{N}(a, b-1)$, and $z(k)$ is a solution of

$$L[z(k)] \le h(k)z(k), \quad k \in \mathbb{N}(a, b-1) \tag{10.2.9}$$

such that $z(a+i) = w(a+i)$, $0 \le i \le n-1$. If

$$g(k) \ge h(k), \quad k \in \mathbb{N}(a, b-1) \tag{10.2.10}$$

and the equation

$$L[v(k)] = h(k)v(k), \quad k \in \mathbb{N}(a, b-1) \tag{10.2.11}$$

is right $(n-1, 1)$ disconjugate on $\mathbb{N}(a, b-1+n)$, then $w(k) \ge z(k)$ for all $k \in \mathbb{N}(a, b-1+n)$.

**Corollary 10.2.10.** If (10.2.11) is right $(n-1, 1)$ disconjugate on $\mathbb{N}(a, b-1+n)$ and (10.2.10) holds, then the equation

$$L[u(k)] = g(k)u(k), \quad k \in \mathbb{N}(a, b-1) \tag{10.2.12}$$

is right $(n-1, 1)$ disconjugate on $\mathbb{N}(a, b-1+n)$.

**Theorem 10.2.11.** Assume that $n$ is even (odd), $w(k)$ is a solution of (10.2.8) and $z(k)$ is a solution of (10.2.9) such that $z(a+i) = w(a+i)$, $0 \le i \le n-1$. If $z(k) \ge 0$ $(w(k) \ge 0)$, $k \in \mathbb{N}(a, b-1+n)$, (10.2.10) holds and (10.2.11) ((10.2.12)) is left $(1, n-1)$ disconjugate on $\mathbb{N}(a, b-1+n)$, then $(-1)^n w(k) \ge (-1)^n z(k)$, $k \in \mathbb{N}(a, b-1+n)$.

**Corollary 10.2.12.** If $n$ is even (odd), (10.2.10) holds, and (10.2.11) ((10.2.12)) is left $(1, n-1)$ disconjugate on $\mathbb{N}(a, b-1+n)$, then (10.2.12) ((10.2.11)) is left $(1, n-1)$ disconjugate on $\mathbb{N}(a, b-1+n)$.

**Definition 10.2.3.** Let $\mathbb{N}(c, d) \subseteq \mathbb{N}(a, b-1+n)$ with $d - c + 1 \ge n$, and let $1 \le j \le n-1$. Equation (1.2.4) is said to be $(j, n-j)$ *disconjugate* on $\mathbb{N}(c, d)$ provided it is both left and right $(j, n-j)$ disconjugate on $\mathbb{N}(c, d)$.

**Theorem 10.2.13.** Assume that $w(k)$ is a solution of (10.2.8) and $z(k)$ is a solution of (10.2.9) with $z(k) \ge 0$ on $\mathbb{N}(a, b-1)$, and

$$(10.2.13) \qquad \begin{aligned} z(a+i) &= w(a+i), \quad 0 \le i \le n-2 \\ z(b-1+n) &= w(b-1+n). \end{aligned}$$

If (10.2.11) is right $(n-1, 1)$ disconjugate on $\mathbb{N}(a, b-1+n)$, (10.2.12) is $(n-2, 2)$ disconjugate on $\mathbb{N}(a, b-1+n)$, and (10.2.10) holds, then $z(k) \ge w(k)$ on $\mathbb{N}(a, b-1+n)$.

**Corollary 10.2.14.** If (10.2.11) is right $(n-1, 1)$ disconjugate on $\mathbb{N}(a, b-1+n)$, (10.2.12) is $(n-2, 2)$ disconjugate on $\mathbb{N}(a, b-1+n)$, and (10.2.10) holds, then (10.2.11) is right $(n-2, 2)$ disconjugate on $\mathbb{N}(a, b-1+n)$.

**Theorem 10.2.15.** Let $1 \le j \le n-1$. Assume that $w(k)$ is a solution of (10.2.8) and $z(k)$ is a solution of (10.2.9) with

$$(10.2.14) \qquad \begin{aligned} z(a+i) &= w(a+i), \quad 0 \le i \le j-1 \\ z(b-1+n-i) &= w(b-1+n-i), \quad 0 \le i \le n-j-1. \end{aligned}$$

Further, in addition to (10.2.10) one of the following holds

1. $w(k) \ge 0$ on $\mathbb{N}(a, b-1)$ and (10.2.11) is either disconjugate on $\mathbb{N}(a, b-1)$ or $(i, n-i)$ disconjugate on $\mathbb{N}(a+j-i, b-1+n+j-i)$ for $j-1 \le i < n$, or

2. $z(k) \ge 0$ on $\mathbb{N}(a, b-1)$ and (10.2.12) is either disconjugate on $\mathbb{N}(a, b-1)$ or $(i, n-i)$ disconjugate on $\mathbb{N}(a+j-i, b-1+n+j-i)$ for $j-1 \le i < n$.

Then, $(-1)^{n-j} w(k) \ge (-1)^{n-j} z(k)$ on $\mathbb{N}(a, b-1+n)$.

## 10.3. Adjoint Equations

For the adjoint of (1.2.4) more than one formulations are possible. In fact, one form has already appeared in (2.11.14), another form which has proved to be more useful is defined as

$$(10.3.1) \qquad L^*[v(k)] = (-1)^n \sum_{i=0}^{n} a_i(k-i)v(k+n-i) = 0.$$

**Theorem 10.3.1.** Let $u(k)$ be a solution of (1.2.4) on $\mathbf{Z}$ and $p$ be an integer. If $a_i(k) = a_i(-k+p-n-i)$, $0 \le i \le n$ then $v(k) = u(-k+p)$ is a solution of the adjoint difference equation (10.3.1).

**Corollary 10.3.2.** If $L[u(k)] = 0$ has constant coefficients and $u(k)$ is a solution on $\mathbf{Z}$, then $v(k) = u(-k+p)$ is a solution of $L^*[v(k)] = 0$ for any integer $p$.

To obtain the Lagrange's identity for this new adjoint equation we define the quasi difference operators $\Delta_i$, $0 \le i \le n$ as follows:

$$\begin{aligned} \Delta_0 v(k) &= v(k) \\ \Delta_i v(k) &= \Delta(\Delta_{i-1}v(k)) + (-1)^i \alpha_{n-i}(k-n+i)v(k+i), \quad 1 \le i \le n \end{aligned}$$

where $\alpha_j(k) = \sum_{i=j}^{n} \binom{i}{j} a_i(k)$, $0 \le j \le n$. Obviously, $\Delta_n = L^*$.

**Theorem 10.3.3** (Lagrange's Identity). Let the functions $u(k)$ and $v(k)$ be defined on $\mathbb{N}(a, b-1+n)$. Then, for $k \in \mathbb{N}(a, b-1)$

$$(10.3.2) \qquad v(k+n)L[u(k)] - (-1)^n u(k)L^*[v(k)] = \Delta\left[\sum_{i=0}^{n-1}(-1)^i \Delta_i v(k+n-i-1)\Delta^{n-i-1}u(k)\right].$$

Let $u_j(k,\ell)$, $0 \le j \le n-1$ be the solution of (1.2.4) satisfying

$$(10.3.3) \qquad \Delta^i u_j(k,\ell)\Big|_{k=\ell} = \delta_{ij}, \quad 0 \le i \le n-1.$$

Then, $u_j(k,\ell)$ has $j$ zeros at $\ell, \ell+1, \cdots, \ell+j-1$ and $u_j(\ell+j,\ell) = 1$. Similarly, let $v_j(k,\ell)$, $0 \le j \le n-1$ be the solution of (10.3.1) satisfying

$$(10.3.4) \qquad \Delta_i v_j(k-i,\ell)\Big|_{k=\ell} = \delta_{ij}, \quad 0 \le i \le n-1.$$

Then, $v_j(k,\ell)$ has $j$ zeros at $\ell, \ell-1, \cdots, \ell-j+1$ and $v_j(\ell-j, j) = (-1)^j$.

**Theorem 10.3.4.** Let $p,\ q \in \mathbb{N}(0, n-1)$. Then, for $\ell \in \mathbb{N}(a, b-1+n)$ and $k \in \mathbb{N}(a, b-1)$

$$\Delta^p u_q(\ell, k) \ = \ (-1)^{p+q}\Delta_{n-q-1} v_{n-p-1}(k+q, \ell+n-1). \tag{10.3.5}$$

**Theorem 10.3.5.** let $0 \le j \le n-1$. Then, for $c \in \mathbb{N}(a, b-1+n)$ and $d \in \mathbb{N}(a, b-1)$

$$\begin{aligned} &\det C\left(u_j(k,c), \cdots, u_{n-1}(k,c)\right)\Big|_{k=d} \\ &= (-1)^{j(n-j)} \det C\left(v_j(k, d+n-1), \cdots, v_{n-1}(k, d+n-1)\right)\Big|_{k=c+j}. \end{aligned} \tag{10.3.6}$$

## 10.4. Right and Left Disconjugacy for the Adjoint Equation

**Definition 10.4.1.** Let $v(k)$ be a solution of the adjoint equation (10.3.1) which is restricted to $\mathbb{N}(c, d)$. We say that $v(k)$ has a *generalized zero* at $\alpha \in \mathbb{N}(c, d)$ provided either $v(\alpha) = 0$, or there exists an integer $m$, $1 \le m \le d - \alpha$ such that $(-1)^m v(\alpha)v(\alpha+m) > 0$ and $v(j) = 0$ for all $j \in \mathbb{N}(\alpha+1, \alpha+m-1)$.

**Definition 10.4.2.** Let $d - c + 1 \ge n$, and let $1 \le j \le n-1$. Equation (10.3.1) is said to be *left* $(n-j, j)$ *disconjugate* on $\mathbb{N}(c, d)$ provided there is no nontrivial solution $v(k)$ of (10.3.1) and integers $\alpha,\ \beta \in \mathbb{N}(c, d)$ with $\alpha < \alpha + n - j \le \beta \le d - j + 1$ such that

$$\begin{aligned} v(\alpha+i) &= 0, \quad 1 \le i \le n-j-1 \ \ (\text{if } n-j \ge 2) \\ v(\beta+i) &= 0, \quad 0 \le i \le j-1 \end{aligned} \tag{10.4.1}$$

and $v(k)$ has a generalized zero at $\alpha$. Similarly, we say that (10.3.1) is *right* $(n-j, j)$ *disconjugate* on $\mathbb{N}(c, d)$ provided there is no nontrivial solution $v(k)$ of (10.3.1) and integers $\alpha,\ \beta \in \mathbb{N}(c, d)$ with $\alpha < \alpha+n-j \le \beta \le d - j + 1$ such that

$$\begin{aligned} v(\alpha+i) &= 0, \quad 0 \le i \le n-j-1 \\ v(\beta+i) &= 0, \quad 1 \le i \le j-1 \ \ (\text{if } j \ge 2) \end{aligned} \tag{10.4.2}$$

and $v(k)$ has a generalized zero at $\beta$.

**Theorem 10.4.1.** Let $v_j(k, \ell)$, $0 \le j \le n-1$ be the solution of (10.3.1) satisfying (10.3.4). Then,

(i) the adjoint difference equation (10.3.1) is left $(n-j, j)$ disconjugate on $\mathbb{N}(c, d)$ if and only if

$$(-1)^{j(n-j)} C\left(v_j(k,\ell), \cdots, v_{n-1}(k,\ell)\right) \ > \ 0 \tag{10.4.3}$$

for $c \le k \le \ell - n + 1 \le d - n + 1$

(ii) the adjoint difference equation (10.3.1) is right $(n-j, j)$ disconjugate on $\mathbb{N}(c, d)$ if and only if

$$(10.4.4) \qquad C\left(v_{n-j}(k, \ell), \cdots, v_{n-1}(k, \ell)\right) \; > \; 0$$

for $c + n - j - 1 \le \ell < k \le d - j + 1$.

**Theorem 10.4.2.** The following hold

(i) the difference equation (1.2.4) is right $(j, n-j)$ disconjugate on $\mathbb{N}(c, d)$ if and only if the adjoint difference equation (10.3.1) is left $(n-j, j)$ disconjugate on $\mathbb{N}(c+j, d+j)$

(ii) the difference equation (1.2.4) is left $(j, n-j)$ disconjugate on $\mathbb{N}(c, d)$ if and only if the adjoint difference equation (10.3.1) is right $(n-j, j)$ disconjugate on $\mathbb{N}(c+j, d+j)$.

## 10.5. Right Disfocality

**Definition 10.5.1.** The linear difference equation (1.2.4) is said to be *right disfocal* on $\mathbb{N}(a, b-1+n)$ if and only if $u(k) \equiv 0$ is the only solution of (1.2.4) on $\mathbb{N}(a, b-1+n)$ such that $\Delta^{j-1}u(k)$ has a generalized zero at $s_j$, $1 \le j \le n$ where $a \le s_1 \le s_2 \le \cdots \le s_n \le b$.

**Definition 10.5.2.** Let $1 \le p \le n$ and $m_1, \cdots, m_p$ be positive integers such that $\sum_{i=1}^{p} m_i = n$. We say that (1.2.4) is $m_1, \cdots, m_p$ *right disfocal* on $\mathbb{N}(a, b-1+n)$ if and only if $u(k) \equiv 0$ is the only solution of (1.2.4) on $\mathbb{N}(a, b-1+n)$ such that, for each $1 \le i \le p$, $\Delta^{i-1}u(k)$ has $m_i$ generalized zeros at $s_{m_1+\cdots+m_{i-1}+1}, \cdots, s_{m_1+\cdots+m_i}$, where $a \le s_1 < \cdots < s_{m_1}$ in $\mathbb{N}(a, b+n-p)$, and $s_{m_1+\cdots+m_{i-1}} \le s_{m_1+\cdots+m_{i-1}+1} < \cdots < s_{m_1+\cdots+m_i}$ in $\mathbb{N}(a, b+n-p)$ for $2 \le i \le p$.

**Remark 10.5.1.** From the discrete Rolle's Theorem 1.8.1 with respect to the generalized zeros it immediately follows that if (1.2.4) is right disfocal on $\mathbb{N}(a, b-1+n)$, then it is $m_1, \cdots, m_p$ right disfocal on $\mathbb{N}(a, b-1+n)$ for all $m_1, \cdots, m_p$. In turn, if (1.2.4) is $m_1, \cdots, m_p$ right disfocal for some $m_1, \cdots, m_p$ then it is disconjugate on $\mathbb{N}(a, b-1+n)$.

Let the functions $u_1(k), \cdots, u_n(k)$ be defined on $\mathbb{N}(a, b-1+n)$. For $1 \le q \le n$ and indices $1 \le i_1 \le \cdots \le i_q \le n$, we define

$$D^q(i_1, \cdots, i_q)(k) \;=\; \det\left(\Delta^{i_j-1}u_\ell(k)\right), \quad 1 \le j, \; \ell \le q$$

where $k \in \mathbb{N}(a, b+n-i_q)$. Further, we define

$$D^q\left(i_1, \cdots, i_q; k_1, \cdots, k_q\right) \;=\; \det\left(\Delta^{i_j-1}u_\ell(k_j)\right), \quad 1 \le j, \; \ell \le q$$

where $a \le k_1 \le \cdots \le k_q$ in $\mathbb{N}(a, b+n-i_q)$.

**Definition 10.5.3.** Let the functions $u_1(k), \cdots, u_n(k)$ be defined on $\mathbb{N}(a, b-1+n)$. We shall say that $u_1(k), \cdots, u_n(k)$ forms a *D Markov system* on $\mathbb{N}(a, b-1+n)$ if $D^q(n-q+1, \cdots, n)(k) > 0$, $k \in \mathbb{N}(a, b)$, $1 \le q \le n$. We shall call $u_1(k), \cdots, u_n(k)$ a *D Fekete system* on $\mathbb{N}(a, b-1+n)$ if $D^q(i, \cdots, i+q-1)(k) > 0$, $k \in \mathbb{N}(a, b+n-i-q+1)$, $1 \le i \le n-q+1$, $1 \le q \le n$. We shall name $u_1(k), \cdots, u_n(k)$ a *D Descartes system* on $\mathbb{N}(a, b-1+n)$ if $D^q(i_1, \cdots, i_q)(k) > 0$, $k \in \mathbb{N}(a, b+n-i_q)$ for all sets of indices satisfying $1 \le i_1 < \cdots < i_q \le n$.

**Theorem 10.5.1.** The following are equivalent

(i) equation (1.2.4) is right disfocal on $\mathbb{N}(a, b-1+n)$

(ii) equation (1.2.4) has a D Markov system of solutions $u_1(k), \cdots, u_n(k)$ on $\mathbb{N}(a, b-1+n)$ satisfying the partial set of initial conditions

$$\begin{aligned} \Delta^{i-1} u_j(a) &= 0, \quad 1 \le i \le n-j \\ (-1)^{j-1} \Delta^{n-j} u_j(a) &> 0, \quad 1 \le j \le n \end{aligned}$$

(iii) equation (1.2.4) has a D Fekete system of solutions on $\mathbb{N}(a, b-1+n)$

(iv) equation (1.2.4) has a D Descartes system of solutions on $\mathbb{N}(a, b-1+n)$

(v) $u(k) \equiv 0$ is the only solution of the equation (1.2.4) such that for each $0 \le j \le n-1$, $u(a) = \cdots = \Delta^{n-j-1} u(a) = 0$, $\Delta^{n-j+1} u(k_1) = \cdots = \Delta^{n-1} u(k_1) = 0$, $a+1 \le k_1 \in \mathbb{N}(a, b)$, and $\Delta^{n-j} u(k)$ has a node at $k_2$ for some $k_2 \in \mathbb{N}(a, k_1)$.

**Remark 10.5.2.** If (1.2.4) has a D Markov system of solutions on $\mathbb{N}(a, b-1+n)$, then it does not follow that (1.2.4) is right disfocal on $\mathbb{N}(a, b-1+n)$. For example, consider the difference equation

$$u(k+2) - 2u(k+1) + 2u(k) = 0, \quad k = 0, 1, 2. \tag{10.5.1}$$

Let $u_1(k)$ and $u_2(k)$ be the solutions of (10.5.1) satisfying the initial conditions $u_1(0) = -2$, $\Delta u_1(0) = 1$ and $u_2(0) = 0$, $\Delta u_2(0) = -1$. For these solutions, we have

| $k$ | 0 | 1 | 2 | 3 | 4 |
|---|---|---|---|---|---|
| $u_1(k) = D^1(1)(k)$ | $-2$ | $-1$ | 2 | 6 | 8 |
| $\Delta u_1(k) = D^1(2)(k)$ | 1 | 3 | 4 | 2 | |
| $u_2(k)$ | 0 | $-1$ | $-2$ | $-2$ | 0 |
| $\Delta u_2(k)$ | $-1$ | $-1$ | 0 | 2 | |
| $D^2(1,2)(k)$ | 2 | 4 | 8 | 16. | |

Thus, $u_1(k)$, $u_2(k)$ forms a D Markov system of solutions of (10.5.1) on $\mathbb{N}(0,4)$. However, the solution $u(k) = \{0,1,2,2,0\}$ is a 'right focal solution' of (10.5.1) on $\mathbb{N}(0,4)$ since $u(0) = \Delta u(2) = 0$.

**Definition 10.5.4.** Let $r_1, \cdots, r_n$ be positive integers such that $n \geq r_1 \geq \cdots \geq r_j \geq r_{j+1} \geq \cdots \geq r_n = 1$, and $r_j \leq r_{j+1} + 1$, $1 \leq j \leq n-1$. Further, let the functions $u_1(k), \cdots, u_n(k)$ be defined on $\mathbb{N}(a, b-1+n)$. We shall say that $u_1(k), \cdots, u_n(k)$ forms a *D Fekete system with respect to* $\{r_q\}_{q=1}^n$ on $\mathbb{N}(a, b-1+n)$ if $D^q(i, \cdots, i+q-1)(k) > 0$, $k \in \mathbb{N}(a, b+n-i-q+1)$, $1 \leq i \leq r_q$, $1 \leq q \leq n$. We shall call $u_1(k), \cdots, u_n(k)$ a *D Descartes system with respect to* $\{r_q\}_{q=1}^n$ on $\mathbb{N}(a, b-1+n)$ if $D^{h+q}(i_1, \cdots, i_h, i, \cdots, i+q-1)(k) > 0$, $k \in \mathbb{N}(a, b+n-i-q+1)$, $1 \leq i_1 < \cdots < i_h < i \leq r_q$, $0 \leq h$, $1 \leq q \leq n$.

**Theorem 10.5.2.** The following are equivalent

(i) equation (1.2.4) is $m_1, \cdots, m_p$ right disfocal on $\mathbb{N}(a, b-1+n)$

(ii) equation (1.2.4) has a D Fekete system with respect to $\{r_q\}_{q=1}^n$ on $\mathbb{N}(a, b-1+n)$

(iii) equation (1.2.4) has a D Descartes system with respect to $\{r_q\}_{q=1}^n$ on $\mathbb{N}(a, b-1+n)$

(iv) there exists a system of solutions $u_1(k), \cdots, u_n(k)$ of (1.2.4) on $\mathbb{N}(a, b-1+n)$ such that $D^q(i_1, \cdots, i_q; k_1, \cdots, k_q) > 0$ for all sets of indices satisfying $1 \leq i_1 \leq \cdots \leq i_q \leq p$, and $i_j \leq r_{q-j+1}$, $1 \leq j \leq p$, and for all points $\{k_j\}_{j=1}^q$ satisfying $a \leq k_j < k_{j+1}$ in $\mathbb{N}(a, b+n-i_{j+1})$ if $i_j = i_{j+1}$, and $a \leq k_j \leq k_{j+1}$ in $\mathbb{N}(a, b+n-i_{j+1})$ if $i_j < i_{j+1}$, $1 \leq j \leq q-1$, $1 \leq q \leq n$.

## 10.6. Eventual Disconjugacy and Right Disfocality

We shall consider the difference equation (1.2.4) in its equivalent form

$$L[u(k)] = \sum_{i=0}^{n} b_i(k)\Delta^{n-i}u(k) = 0 \tag{10.6.1}$$

on $\mathbb{N}(a)$, where it is assumed that $b_0(k) = 1$.

**Definition 10.6.1.** The difference equation (10.6.1) is said to be *eventually disconjugate (eventually right disfocal)* if there exists $k_0 \geq a$, $k_0 \in \mathbb{N}(a)$ such that the equation (10.6.1) is disconjugate (right disconjugate) on $\mathbb{N}(k_0)$.

Let $\alpha(k)$ be a function defined on $\mathbb{N}(a)$, and let $p \geq 2$. If $\sum_{\ell=a}^{\infty}(\ell+1)\cdots(\ell+p-1)\alpha(\ell)$ converges, then we define $S_0(k,\alpha) = \alpha(k)$, $S_1(k,\alpha) =$

$\sum_{\ell=k}^{\infty}\alpha(\ell)$, and for $2 \le i \le p$

$$S_i(k,\alpha) = \sum_{\ell=k}^{\infty}((\ell+1-k)\cdots(\ell+(i-1)-k)/(i-1)!)\alpha(\ell).$$

**Theorem 10.6.1.** Suppose the sums $\sum^{\infty} k^{i-1}b_i(k)$, $1 \le i \le n$ are finite and $\sum^{\infty}|S_{i-1}(k,b_i)| < \infty$, $1 \le i \le n$. Then, the difference equation (10.6.1) is eventually disconjugate as well as eventually right disfocal.

## 10.7. A Classification of Solutions

Consider the linear $n$th order difference equation

$$L[u(k)] + p(k)u(k) = 0, \quad k \in \mathbb{N}(a) \tag{10.7.1}$$

where $p(k)$ is sign definite on $\mathbb{N}(a)$, and $L[u(k)] = h_n(k)\Delta\{\cdots\Delta[h_1(k)\Delta(h_0(k)u(k))]\}$ with $h_i(k) > 0$ on $\mathbb{N}(a)$, $0 \le i \le n$. Define quasi difference operators $\Delta_\nu$, $0 \le \nu \le n$, recursively by $\Delta_0 u(k) = h_0(k)u(k)$ and $\Delta_\nu u(k) = h_\nu(k)\Delta[\Delta_{\nu-1}u(k)]$, $1 \le \nu \le n$. We shall assume that $(-1)^n\left[(-1)^n\prod_{i=0}^{n}h_i(k) + p(k)\right] > 0$, $k \in \mathbb{N}(a)$. This condition corresponds to (10.1.1) for the difference equation (1.2.4).

Assume $c \in \mathbb{N}(a)$. Then, we define $S(u,c+)$ to be the maximum number of sign changes in the sequence $(\Delta_0 u(c),\cdots,(-1)^n\Delta_n u(c))$ where zeros are replaced by arbitrary nonzero numbers except if $\Delta_0 u(c) = 0$ (if and only if $\Delta_n u(c) = 0$) in which case $\Delta_0 u(c)$ and $(-1)^n\Delta_n u(c)$ can be replaced by nonzero real numbers $\alpha$ and $\beta$ respectively, where $\operatorname{sgn}\alpha\beta = \operatorname{sgn}\{(-1)^{n+1}p(k)\}$. Define $S(u,c-)$ to be the maximum number of sign changes in the sequence $(\Delta_0 u(c),\cdots,\Delta_n u(c))$ where zeros are replaced by arbitrary nonzero numbers except if $\Delta_0 u(c) = 0$ in which case $\Delta_0 u(c)$ and $\Delta_n u(c)$ can be replaced by nonzero real numbers $\alpha$ and $\beta$ respectively, where $\operatorname{sgn}\alpha\beta = -\operatorname{sgn} p(k)$.

**Theorem 10.7.1.** Assume $u(k)$ is a nontrivial solution of (10.7.1). Then,

(i) $S(u,c+)$ and $S(u,c-)$ are greater than or equal to the number of values of $i$ such that $\Delta_i u(c) = 0$, $0 \le i \le n-1$.

(ii) $S(u,c+) + S(u,c-) \ge n$. Further, if $\Delta_i u(c) \ne 0$, $0 \le i \le n-1$ then $S(u,c+) + S(u,c-) = n$.

(iii) $(-1)^{S(u,c-)}p(k) < 0$, $(-1)^{n-S(u,c+)}p(k) < 0$.

Let $u(k)$ be a solution of (10.7.1). We say $\Delta_\nu u(k)$ has a *zero point* at $k_0 \in \mathbb{N}(a)$ provided $\Delta_\nu u(k)$ has a generalized zero at $k_0$ and $\Delta_\nu u(k_0 - 1) \ne 0$ (if $k_0 > a$). Let $k_0$ be a zero point of $\Delta_\nu u(k)$.

Then we define the *multiplicity* $n_\nu(k_0)$ of the zero point $k_0$ as $n_\nu(k_0) = \max\{\ell : \Delta_\nu u(k)$ has $\ell - 1$ zeros starting at $k_0$ and a generalized zero at $k_0 + \ell - 1\}$. We say the zero point $k_0$ *extends* through $k_1$ $(\geq k_0)$ in case the multiplicity of the zero point of $\Delta_\nu u(k)$ at $k_0$ is at least $k_1 - k_0 + 1$ and $\Delta_\nu u(k_1) = 0$.

Let $\{k_{i\nu}\}$ be the zero point of $\Delta_\nu u(k)$ in $\mathbb{N}(a,b)$ which are not zeros of $\Delta_{\nu-1}u(k)$, and $\{k_{i0}\}$ be the zero points of $\Delta_0 u(k)$ in $\mathbb{N}(a,b)$ which are not zeros of $\Delta_{n-1}u(k)$. Further, let $< j >$ denote the greatest even integer less than or equal to $j$.

**Theorem 10.7.2.** Let $u(k)$ be a nontrivial solution of (10.7.1) and assume that $b \geq a$ is such that no quasi difference $\Delta_i u(k) \equiv 0$ on $\mathbb{N}(a,b)$ for $0 \leq i \leq n-1$ (if $b \geq a+n-1$ then this condition definitely holds). Then,

$$S(u,a+) + S(u,b-) + \sum_{\nu=0}^{n-1} \sum_{\mathbb{N}(a,b)} < n_\nu(k_{i\nu}) > \ \leq n, \tag{10.7.2}$$

where the sum over $\mathbb{N}(a,b)$ is understood not to contain any zero point which extends through $b$.

**Remark 10.7.1.** The inequality (10.7.2) in Theorem 10.7.2 is not true if we just assume $b \geq a$. To see this, let $u(k)$ be the solution of (10.7.1) satisfying $\Delta_i u(a) = 0$, $0 \leq i \leq n-2$, $\Delta_{n-1}u(a) = 1$. If $b = a+1$ then $S(u,a+) + S(u,b-) \geq 2n - 3$.

**Remark 10.7.2.** The inequality (10.7.2) in Theorem 10.7.2 is true with $a$, $b$ replaced by $c$, $d$ respectively, provided the solution is restricted to $\mathbb{N}(c)$.

**Corollary 10.7.3.** If $(-1)^{n-j}p(k) > 0$ on $\mathbb{N}(c,d)$ for some $j \in \{1, \cdots, n-1\}$, then (10.7.1) is right $(j, n-j)$ disconjugate on $\mathbb{N}(c, d+n)$.

**Corollary 10.7.4.** If $u(k)$ is a nontrivial solution of (10.7.1), then $S(u,k+) \geq S(u,c+)$ for all $k \in \mathbb{N}(c+n-1)$.

**Corollary 10.7.5.** If $u(k)$ is a nontrivial solution of (10.7.1), then $S(u) = \lim_{c\to\infty} S(u,c+)$ exists and $S(u)$ is an integer satisfying $(-1)^{n-S(u)}p(k) < 0$.

**Theorem 10.7.6.** If $u(k)$ is a nontrivial solution of (10.7.1), then $S(u,k+)$ is a nondecreasing function of $k$ for $k \in \mathbb{N}(a)$.

Assume that $0 \leq j \leq n$ and $(-1)^{n-j}p(k) < 0$. For a nontrivial solution $u(k)$ of (10.7.1), let $S(u) = \lim_{c\to\infty} S(u,c+)$ as in Corollary

10.7.5. We define $S_j = \{u(k) : u(k)$ is a nontrivial solution of (10.7.1) with $S(u) = j\}$.

**Example 10.7.1.** The difference equation $\Delta^4 u(k) - (1/16)u(k) = 0$ has solutions $u_1(k) = (1/2)^k$, $u_2(k) = (\sqrt{5}/2)^k \cos\theta k$, $u_3(k) = (\sqrt{5}/2)^k \sin\theta k$, $u_4(k) = (3/2)^k$, where $\theta = \tan^{-1} 1/2$. It is easy to see that $u_1(k) \in S_0$, $u_2(k), u_3(k) \in S_2$, and $u_4(k) \in S_4$.

**Example 10.7.2.** The difference equation $\Delta^4 u(k) + u(k) = 0$ has solutions $u_1(k) = \alpha^k \cos\beta k$, $u_2(k) = \alpha^k \sin\beta k$, $u_3(k) = \gamma^k \cos\delta k$, $u_4(k) = \gamma^k \sin\delta k$, where $\alpha = |1 - (1-i)/\sqrt{2}|$, $\beta = \tan^{-1} 1/(\sqrt{2} - 1)$, $\gamma = |1 + (1+i)/\sqrt{2}|$, and $\delta = \tan^{-1} 1/(\sqrt{2} + 1)$. It is easy to see that $u_1(k), u_2(k) \in S_1$ and $u_3(k), u_4(k) \in S_3$.

**Theorem 10.7.7.** The set of nontrivial solutions of (10.7.1) is the union of the sets $S_j$, where $0 \le j \le n$ satisfies $(-1)^{n-j}p(k) < 0$. Each of these sets $S_j$ is nonempty.

**Theorem 10.7.8.** If $\{u_i(k)\}$ is a sequence of solutions of (10.7.1) such that $S(u_i) = j$ for all $i \ge 1$ and $\{u_i(k)\}$ converges pointwise to a nontrivial solution $u(k)$ of (10.7.1) then $S(u) \le j$.

**Example 10.7.3.** Consider the difference equation $\Delta^4 u(k) - (1/16)u(k) = 0$ given in Example 10.7.1. Two solutions of this difference equation are $u(k) = (1/2)^k$, $v(k) = (3/2)^k$, where $u(k) \in S_0$ and $v(k) \in S_4$. Define $u_i(k) = (1/i)v(k) + u(k)$. It is easy to see that $u_i(k) \in S_4$ for all $i$ but the limit solution $u(k) \in S_0$.

## 10.8. Interpolating Polynomials

**Theorem 10.8.1.** The unique polynomial $P_{n-1}(k)$ of degree $n-1$ satisfying *conjugate boundary conditions*

$$P_{n-1}(k_i) = u(k_i) = A_i, \quad 1 \le i \le n \tag{10.8.1}$$

where $a = k_1 < k_2 < \cdots < k_n = b - 1 + n$ and each $k_i \in \mathbb{N}(a, b-1+n)$ can be written as

$$P_{n-1}(k) = \sum_{i=1}^{n} \prod_{j=1, j\ne i}^{n} \left(\frac{k - k_j}{k_i - k_j}\right) A_i. \tag{10.8.2}$$

**Remark 10.8.1.** Niccoletti boundary conditions (1.5.9) are obviously a particular case of (10.8.1). Interestingly, Hermite ($r$ point) boundary conditions (1.5.10) in view of (1.1.2) are also a special case of (10.8.1).

**Theorem 10.8.2.** The unique polynomial $P_{2m-1}(k)$ of degree $2m-1$ satisfying *osculatory boundary conditions*

(10.8.3) $P_{2m-1}(k_i) = u(k_i) = A_i,\ \Delta P_{2m-1}(k_i) = \Delta u(k_i) = B_i,\ 1 \le i \le m$

where $a = k_1 < k_1 + 2 < k_2 < k_2 + 2 < \cdots < k_{m-1} < k_{m-1} + 2 < k_m < k_m + 1 = b - 1 + 2m$ and each $k_i \in \mathbb{N}(a, b-1+2m)$ can be written as

$$P_{2m-1}(k) = \sum_{i=1}^{m} h_i(k) A_i + \sum_{i=1}^{m} \overline{h}_i(k) B_i, \tag{10.8.4}$$

where

(10.8.5)
$$h_i(k) = \left[1 - \left(1 + \prod_{j=1}^{m}\left(\frac{k_i - 1 - k_j}{k_i + 1 - k_j}\right)\right)(k - k_i)\right] \prod_{j=1, j\ne i}^{m} \frac{(k-k_j)^{(2)}}{(k_i - k_j)^{(2)}},$$

and

$$\overline{h}_i(k) = -\left(\prod_{j=1}^{m}\left(\frac{k_i - 1 - k_j}{k_i + 1 - k_j}\right)\right)(k - k_i) \prod_{j=1, j\ne i}^{m} \frac{(k-k_j)^{(2)}}{(k_i - k_j)^{(2)}}, \quad 1 \le i \le m. \tag{10.8.6}$$

**Theorem 10.8.3.** The unique polynomial $P_{2m-1}(k)$ of degree $2m-1$ satisfying *two point Taylor boundary conditions*

(10.8.7)
$$\Delta^i P_{2m-1}(a) = \Delta^i u(a) = A_i, \quad \Delta^i P_{2m-1}(b+m) = \Delta^i u(b+m) = B_i, \quad 0 \le i \le m-1$$

can be written as

(10.8.8)
$$P_{2m-1}(k) = (k-a)^{(m)} \sum_{i=0}^{m-1} \frac{(k-b-m)^{(i)}}{i!}\beta_i + (k-b-m)^{(m)} \sum_{i=0}^{m-1} \frac{(k-a)^{(i)}}{i!}\alpha_i,$$

where

$$\alpha_i = \Delta^i \left[\frac{P_{2m-1}(k)}{(k-b-m)^{(m)}}\right]\Bigg|_{k=a}, \quad \beta_i = \Delta^i \left[\frac{P_{2m-1}(k)}{(k-a)^{(m)}}\right]\Bigg|_{k=b+m}, \quad 0 \le i \le m-1.$$

Since in view of Problem 1.9.37 each $\alpha_i$ $(\beta_i)$ is explicitly known in terms of $A_j$ $(B_j)$, $0 \le j \le i)$, we have

(10.8.9)

$$P_{2m-1}(k) = (k-a)^{(m)}\sum_{i=0}^{m-1}\left(\sum_{j=i}^{m-1}\binom{j}{i}\frac{(k-b-m)^{(j)}(k-b-m-j-1)^{(m-j-1)}}{j!(-1)^{m-j-1}(m-j-1)!(b+m+j-a)^{(m)}}\right)B_i$$

$$+(k-b-m)^{(m)}\sum_{i=0}^{m-1}\left(\sum_{j=i}^{m-1}\binom{j}{i}\frac{(k-a)^{(j)}(k-a-j-1)^{(m-j-1)}}{j!(-1)^{m-j-1}(m-j-1)!(a+j-b-m)^{(m)}}\right)A_i.$$

We also have the representation

$$(10.8.10) \qquad P_{2m-1}(k) = \sum_{i=0}^{m-1} q_i(k)A_i + \sum_{i=0}^{m-1}\overline{q}_i(k)B_i,$$

where $q_i(k)$ and $\overline{q}_i(k)$, $0 \le i \le m-1$ are the polynomials of degree $2m-1$ satisfying $\Delta^r q_i(a) = \delta_{ir}$, $\Delta^r q_i(b+m) = 0$, $\Delta^r\overline{q}_i(a) = 0$, $\Delta^r\overline{q}_i(b+m) = \delta_{ir}$, $0 \le i,\ r \le m-1$ and appear as

(10.8.11)

$$q_i(k) = (b+2m-k-1)^{(m)}\sum_{j=0}^{m-i-1}\binom{m+j-1}{j}\frac{(k-a)^{(i+j)}}{i!(b+2m-i-1-a)^{(m+j)}},$$

and

(10.8.12)

$$\overline{q}_i(k) = (-1)^i(k-a)^{(m)}\sum_{j=0}^{m-i-1}\binom{m+j-1}{j}\frac{(b+m+i+j-k-1)^{(i+j)}}{i!(b+m+i+j-a)^{(m+j)}},$$

$$0 \le i \le m-1.$$

**Theorem 10.8.4.** The unique polynomial $P_{n-1}(k)$ of degree $n-1$ satisfying Hermite ($r$ point) boundary conditions (1.5.10) can be written as

$$(10.8.13) \qquad P_{n-1}(k) = \sum_{j=1}^{r}\sum_{\ell=0}^{p_j}\sum_{s=\ell}^{p_j}\binom{s}{\ell}\prod_{i=1,i\ne j}^{r}\frac{(k-k_i)^{(p_i+1)}}{(k_j+s-k_i)^{(p_i+1)}}\times$$
$$\frac{(k-k_j)^{(s)}(k-k_j-s-1)^{(p_j-s)}}{s!(-1)^{p_j-s}(p_j-s)!}A_{j,\ell}.$$

**Theorem 10.8.5.** The unique polynomial $P_{n-1}(k)$ of degree $n-1$ satisfying two point Hermite boundary conditions (also known as $(p, n-p)$ boundary conditions)

$$(10.8.14) \qquad \begin{aligned} \Delta^i P_{n-1}(a) &= A_i, \quad 0 \le i \le p-1 \ (p \ge 1) \\ \Delta^i P_{n-1}(b+p) &= B_i, \quad 0 \le i \le n-p-1 \end{aligned}$$

can be written as

$$(10.8.15) \qquad P_{n-1}(k) = \sum_{i=0}^{p-1} c_i(k)A_i + \sum_{j=0}^{n-p-1} d_j(k)B_j,$$

where

$$c_i(k) = (b+n-k-1)^{(n-p)} \sum_{\tau=0}^{p-i-1} \binom{n-p+\tau-1}{\tau} \times \frac{(k-a)^{(i+\tau)}}{i!(b+n-i-1-a)^{(n-p+\tau)}}, \quad i = 0, 1, \cdots, p-1$$

and

$$d_j(k) = (-1)^j (k-a)^{(p)} \sum_{\tau=0}^{n-p-j-1} \binom{p+\tau-1}{\tau} \times \frac{(b+p+j+\tau-k-1)^{(j+\tau)}}{j!(b+p+j+\tau-a)^{(p+\tau)}}, \quad j = 0, 1, \cdots, n-p-1.$$

**Remark 10.8.2.** It is clear that $c_i(k) \geq 0$, $0 \leq i \leq p-1$, $(-1)^j d_j(k) \geq 0$, $0 \leq j \leq n-p-1$, $k \in \mathbb{N}(a, b+p)$. Also,

$$(10.8.16) \qquad c_0(k) + d_0(k) = 1.$$

**Theorem 10.8.6.** The unique polynomial $P_{n-1}(k)$ of degree $n-1$ satisfying Abel Gontscharoff (*right focal point*) boundary conditions (1.5.11) can be written as

$$(10.8.17) \qquad P_{n-1}(k) = \sum_{i=0}^{n-1} T_i(k)A_i,$$

where

(10.8.18)

$$T_i(k) = \frac{1}{1!\cdots i!} \begin{vmatrix} 1 & (k_1)^{(1)} & (k_1)^{(2)} & & (k_1)^{(i-1)} & (k_1)^{(i)} \\ 0 & 1 & 2(k_2)^{(1)} & & (i-1)(k_2)^{(i-2)} & i(k_2)^{(i-1)} \\ \cdots & & & \cdots & & \cdots \\ 0 & 0 & 0 & & (i-1)! & i!(k_i)^{(1)} \\ 1 & (k)^{(1)} & (k)^{(2)} & & (k)^{(i-1)} & (k)^{(i)} \end{vmatrix}.$$

In particular

$$\begin{aligned} T_0(k) &= 1 \\ T_1(k) &= \left[(k)^{(1)} - (k_1)^{(1)}\right] \\ T_2(k) &= \frac{1}{2!}\left[\left((k)^{(2)} - (k_1)^{(2)}\right) - 2(k_2)^{(1)}\left((k)^{(1)} - (k_1)^{(1)}\right)\right]. \end{aligned}$$

**Remark 10.8.3.** An alternative representation of $T_i(k)$ is in terms of iterated summations

$$T_i(k) = \oint_{\ell_1=k_1}^{k-1} \oint_{\ell_2=k_2}^{\ell_1-1} \cdots \oint_{\ell_i=k_i}^{\ell_{i-1}-1} 1, \tag{10.8.19}$$

where for the integers $p$ and $q$ and any function $f(k)$

$$\oint_{\ell=p}^{q-1} f(\ell) = \begin{cases} \displaystyle\sum_{\ell=p}^{q-1} f(\ell) & \text{if } q \geq p \\ \displaystyle -\sum_{\ell=q}^{p-1} f(\ell) & \text{if } p \geq q. \end{cases}$$

**Theorem 10.8.7.** The unique polynomial $P_{n-1}(k)$ of degree $n-1$ satisfying *two point right focal boundary conditions*

$$\begin{aligned} \Delta^i P_{n-1}(a) &= \Delta^i u(a) = A_i, \quad 0 \leq i \leq p-1 \\ &\qquad\qquad (1 \leq p \leq n-1, \ \text{but fixed}) \\ \Delta^i P_{n-1}(b) &= \Delta^i u(b) = A_i, \quad p \leq i \leq n-1 \end{aligned} \tag{10.8.20}$$

can be written as

$$\begin{aligned} P_{n-1}(k) &= \sum_{i=0}^{p-1} \frac{(k-a)^{(i)}}{i!} A_i \\ &+ \sum_{i=0}^{n-p-1} \left( \sum_{j=0}^{i} \frac{(k-a)^{(p+j)}}{(p+j)!} \frac{(-1)^{i-j}}{(i-j)!} (b-a+i-j-1)^{(i-j)} \right) A_{p+i}. \end{aligned} \tag{10.8.21}$$

**Theorem 10.8.8.** The unique polynomial $P_{n-1}(k)$ of degree $n-1$ satisfying $(n,p)$ boundary conditions (1.5.12) can be written as

$$\begin{aligned} P_{n-1}(k) &= \sum_{i=0}^{n-2} \frac{(k-a)^{(i)}}{i!} A_i \\ &+ \left[ B - \sum_{i=0}^{n-p-2} \frac{(b+n-p-a-1)^{(i)}}{i!} A_{p+i} \right] \frac{(n-p-1)!}{(n-1)!} \frac{(k-a)^{(n-1)}}{(b+n-p-a-1)^{(n-p-1)}}. \end{aligned} \tag{10.8.22}$$

**Theorem 10.8.9.** The unique polynomial $P_{n-1}(k)$ of degree $n-1$ satisfying $(p,n)$ boundary conditions (1.5.13) can be written as

$$(10.8.23)\qquad P_{n-1}(k) \;=\; \sum_{i=0}^{n-2}\frac{(b+i-k)^{(i)}}{i!}(-1)^i A_i$$

$$+\left[B-\sum_{i=0}^{n-p-2}\frac{(b+i-a)^{(i)}}{i!}(-1)^i A_{p+i}\right]\frac{(n-p-1)!}{(n-1)!}(-1)^p\frac{(b+n-1-k)^{(n-1)}}{(b+n-p-a-1)^{(n-p-1)}}.$$

## 10.9. Green's Functions

Following Section 2.11 let $v_j(k),\ 1\le j\le n$ be a fixed set of linearly independent solutions of (1.2.4), and let $\phi(k)$ be any particular solution of (1.2.3). Then, any solution of (1.2.3) can be written as

$$(10.9.1)\qquad u(k) \;=\; \sum_{j=1}^{n} c_j v_j(k)+\phi(k), \quad k\in\mathbb{N}(a,b-1+n)$$

where $c_j,\ 1\le j\le n$ are fixed constants.

This solution satisfies the linearly independent boundary conditions

$$(10.9.2)\qquad l_i[u] \;=\; \sum_{\tau=0}^{n-1}\alpha_{i\tau}u(k_i+\tau) \;=\; A_i, \quad 1\le i\le n$$

where $a\le k_1\le\cdots\le k_n\le b$ and $\alpha_{i\tau},\ A_i,\ 1\le i\le n,\ 0\le\tau\le n-1$ are the known constants, if and only if the system

$$A_i \;=\; l_i\left[\sum_{j=1}^{n}c_jv_j+\phi\right] \;=\; \sum_{j=1}^{n}c_jl_i[v_j]+l_i[\phi], \quad 1\le i\le n$$

has a unique solution. Thus, by Lemma 2.2.1 the problem (1.2.3), (10.9.2) has a unique solution if and only if $\det(l_i[v_j])\neq 0$. Further, in such a case the existence of the fundamental system of solutions $\overline{v}_j(k),\ 1\le j\le n$ of (1.2.4) satisfying $l_i[\overline{v}_j]=\delta_{ij}$ is assured $(\det(l_i[\overline{v}_j])=1)$.

For convenience, we shall write $D_i(\ell)$ = cofactor of $\overline{v}_i(\ell+n-1)$ in the $\det\mathcal{V}(\ell)=\det(\overline{v}_i(\ell+j)),\ 1\le i\le n,\ 0\le j\le n-1$. Further, let $k_0=a,\ k_{n+1}=b,$ and $D_0(k)=D_{n+1}(k)=\overline{v}_0(k)=\overline{v}_{n+1}(k)=0$ on $\mathbb{N}(a,b-1+n)$. Then, in view of (2.11.8) the general solution of (1.2.3) can be written as

$$(10.9.3)\qquad u(k) \;=\; \sum_{j=1}^{n}c_j\overline{v}_j(k)+\sum_{\ell=a}^{k-n}\frac{1}{\det\mathcal{V}(\ell+1)}\sum_{j=0}^{n+1}D_j(\ell+1)\overline{v}_j(k)b(\ell),$$

$$k\in\mathbb{N}(a,b-1+n).$$

Since from the properties of $G(k,\ell+1)$ defined in (2.11.7)

$$u(k_i+\tau) = \sum_{j=1}^{n} c_j\overline{v}_j(k_i+\tau)+\sum_{\ell=a}^{k_i-1}\frac{1}{\det \mathcal{V}(\ell+1)}\sum_{j=0}^{n+1}D_j(\ell+1)\overline{v}_j(k_i+\tau)b(\ell),$$

$$0\le\tau\le n-1$$

boundary conditions (10.9.2) can be used to determine the constants $c_j$, $1\le j\le n$ which appear as

$$c_j = A_j - \sum_{\ell=a}^{k_j-1}\frac{1}{\det \mathcal{V}(\ell+1)}D_j(\ell+1)b(\ell), \quad 1\le j\le n.$$

Thus, the solution of (1.2.3), (10.9.2) can be written as

$$\begin{aligned}
u(k) &= \sum_{j=1}^{n}A_j\overline{v}_j(k) - \sum_{j=0}^{n+1}\sum_{\ell=k_0}^{k_j-1}\frac{1}{\det \mathcal{V}(\ell+1)}D_j(\ell+1)\overline{v}_j(k)b(\ell)\\
&\quad+\sum_{\ell=k_0}^{k-n}\frac{1}{\det \mathcal{V}(\ell+1)}\sum_{j=0}^{n+1}D_j(\ell+1)\overline{v}_j(k)b(\ell)\\
&= \sum_{j=1}^{n}A_j\overline{v}_j(k) - \sum_{i=0}^{n}\sum_{\ell=k_i}^{k_{i+1}-1}\frac{1}{\det \mathcal{V}(\ell+1)}\sum_{j=i+1}^{n+1}D_j(\ell+1)\overline{v}_j(k)b(\ell)\\
&\quad+\sum_{\ell=k_0}^{k-n}\frac{1}{\det \mathcal{V}(\ell+1)}\sum_{j=0}^{n+1}D_j(\ell+1)\overline{v}_j(k)b(\ell)\\
&= \sum_{j=1}^{n}A_j\overline{v}_j(k) + \sum_{\ell=a}^{b-1}g(k,\ell)b(\ell),
\end{aligned}$$

where for $k_{i+1}-k_i\ge 1$,

$$(10.9.4)\quad g(k,\ell) = \begin{cases} \dfrac{1}{\det \mathcal{V}(\ell+1)}\displaystyle\sum_{j=0}^{i}D_j(\ell+1)\overline{v}_j(k), & k_i\le\ell\le k-n\\[2ex] -\dfrac{1}{\det \mathcal{V}(\ell+1)}\displaystyle\sum_{j=i+1}^{n+1}D_j(\ell+1)\overline{v}_j(k), & \\ \qquad k-n+1\le\ell\le k_{i+1}-1, & 0\le i\le n\end{cases}$$

and for $k_{i+1}-k_i<1$, $g(k,\ell)=0$.

This function $g(k,\ell)$ is called the *Green's function* of the boundary value problem (1.2.4),

$$(10.9.5)\qquad l_i[u] = 0, \quad 1\le i\le n$$

and is uniquely determined on $\mathbb{N}(a, b-1+n) \times \mathbb{N}(a, b-1)$. The following properties of $g(k, \ell)$ are fundamental

(i) $\Delta^i g(k, \ell)$, $0 \le i \le n-1$ exists on $\mathbb{N}(a, b-1+n-i) \times \mathbb{N}(a, b-1)$

(ii) $g(k, \ell)$ as a function of $k$ satisfies

$$L[g(k,\ell)] = \sum_{i=0}^{n} a_i(k) g(k+i, \ell) = \delta_{k\ell}, \quad k \in \mathbb{N}(a, b-1)$$

(iii) $g(k, \ell)$ as a function of $k$ satisfies the homogeneous boundary conditions (10.9.5)

(iv) for any function $b(k)$ defined on $\mathbb{N}(a, b-1)$, the unique solution of the boundary value problem (1.2.3), (10.9.5) is given by

$$u(k) = \sum_{\ell=a}^{b-1} g(k, \ell) b(\ell).$$

**Theorem 10.9.1.** The Green's function $g(k, \ell)$ of the boundary value problem

(10.9.6) $$u(k+1) - 2u(k) + u(k-1) = 0, \quad k \in \mathbb{N}(1, K)$$

(10.9.7) $$\begin{aligned} a_0 u(0) - a_1 \Delta u(0)/h &= 0 \\ b_0 u(K+1) + b_1 \Delta u(K)/h &= 0 \quad (h > 0) \end{aligned}$$

exists if and only if $\lambda = a_0 b_0 (K+1) + a_0 b_1/h + b_0 a_1/h \neq 0$, and is given by

(10.9.8)
$$g(k, \ell) = -\frac{1}{\lambda} \begin{cases} (b_0(K+1-k)+b_1/h)(a_0\ell+a_1/h), & 1 \le \ell \le k-1 \\ (b_0(K+1-\ell)+b_1/h)(a_0 k+a_1/h), & k \le \ell \le K. \end{cases}$$

Further, if $a_0, a_1, b_0, b_1 \ge 0$ then

(10.9.9) $$-g(k, \ell) \ge 0, \quad (k, \ell) \in \mathbb{N}(0, K+1) \times \mathbb{N}(1, K).$$

**Theorem 10.9.2.** Let the difference equation (1.2.4) be disconjugate on $\mathbb{N}(a, b-1+n)$. Then, the Green's function $g(k, \ell)$ of the conjugate boundary value problem (1.2.4),

(10.9.10) $$u(k_i) = 0, \quad 1 \le i \le n$$

where $a = k_1 < k_2 < \cdots < k_n = b-1+n$ and each $k_i \in \mathbb{N}(a, b-1+n)$ exists on $\mathbb{N}(a, b-1+n) \times \mathbb{N}(a, b-1)$. Further, for all $(k, \ell) \in \mathbb{N}(a, b-1+n) \times \mathbb{N}(a, b-1)$

$$(10.9.11) \qquad (-1)^{n+\sigma(k)} g(k, \ell) \ \geq \ 0,$$

where $\sigma(k) = \text{card}\{i : k_i < k, \ 1 \leq i \leq n\}$.

**Theorem 10.9.3.** The Green's function $g(k, \ell)$ of the conjugate boundary value problem (10.1.3), (10.9.10) exists on $\mathbb{N}(a, b-1+n) \times \mathbb{N}(a, b-1)$, and is given by

$$(10.9.12) \quad g(k, \ell) \ = \ -\frac{1}{(n-1)!} \begin{cases} g_1(k, \ell) - (k-\ell-1)^{(n-1)}, \\ \qquad\qquad a \leq k_r - n + 1 \leq \ell \leq k - n \\ g_1(k, \ell), \quad k - n + 1 \leq \ell \leq k_{r+1} - n \\ \qquad\qquad 1 \leq r \leq n-1 \end{cases}$$

where

$$g_1(k, \ell) \ = \ \sum_{i=r+1}^{n} \prod_{j=1, j \neq i}^{n} \left( \frac{k - k_j}{k_i - k_j} \right) (k_i - \ell - 1)^{(n-1)}.$$

**Remark 10.9.1.** From Corollary 10.1.6 the difference equation (10.1.3) is disconjugate on $\mathbb{N}(a, b-1+n)$, therefore in particular the inequality (10.9.11) holds for the Green's function $g(k, \ell)$ defined in (10.9.12).

**Theorem 10.9.4.** The Green's function $g(k, \ell)$ of the osculatory boundary value problem (10.1.3) with $n = 2m$,

$$(10.9.13) \qquad u(k_i) \ = \ \Delta u(k_i) \ = \ 0, \quad 1 \leq i \leq m$$

where $a = k_1 < k_1 + 2 < k_2 < k_2 + 2 < \cdots < k_{m-1} < k_{m-1} + 2 < k_m < k_m + 1 = b - 1 + 2m$ and each $k_i \in \mathbb{N}(a, b-1+2m)$ exists on $\mathbb{N}(a, b-1+2m) \times \mathbb{N}(a, b-1)$ and is given by

(10.9.14)

$$g(k, \ell) \ = \ -\frac{1}{(2m-1)!} \begin{cases} g_1(k, \ell) - (k-\ell-1)^{(2m-1)}, \\ \qquad\qquad a \leq k_r - 2m + 2 \leq \ell \leq k - 2m \\ g_1(k, \ell), \quad k - 2m + 1 \leq \ell \leq k_{r+1} - 2m + 1 \\ \qquad\qquad 1 \leq r \leq m-1 \end{cases}$$

where

$$g_1(k, \ell) = \sum_{j=r+1}^{m} \left[ h_j(k)(k_j - \ell - 2m + 1) + \overline{h}_j(k)(2m-1) \right] (k_j - \ell - 1)^{(2m-2)}$$

and $h_j(k)$, $\overline{h}_j(k)$ are defined in (10.8.5) and (10.8.6). Further, for all $(k,\ell) \in \mathbb{N}(a, b-1+2m) \times \mathbb{N}(a, b-1)$

$$g(k,\ell) \geq 0. \tag{10.9.15}$$

**Theorem 10.9.5.** The Green's function $g(k,\ell)$ of the two point Taylor boundary value problem (10.1.3) with $n = 2m$,

$$\Delta^i u(a) = \Delta^i u(b+m) = 0, \quad 0 \le i \le m-1 \tag{10.9.16}$$

exists on $\mathbb{N}(a, b-1+2m) \times \mathbb{N}(a, b-1)$ and is given by

(10.9.17)

$$g(k,\ell) = -\frac{1}{(2m-1)!}\begin{cases} g_1(k,\ell) - (k-\ell-1)^{(2m-1)}, & a \le \ell \le k-2m \\ g_1(k,\ell), & k-2m+1 \le \ell \le b-1 \end{cases}$$

where

$$g_1(k,\ell) = \sum_{i=0}^{m-1} (2m-1)^{(i)}(b+m-\ell-1)^{(2m-i-1)}\overline{q}_i(k)$$

and $\overline{q}_i(k)$ are defined in (10.8.12). Further, for all $(k,\ell) \in \mathbb{N}(a, b-1+2m) \times \mathbb{N}(a, b-1)$

$$(-1)^m g(k,\ell) \geq 0. \tag{10.9.18}$$

**Theorem 10.9.6.** Let $u_j(k,\ell)$, $0 \le j \le n-1$ be the solution of (1.2.4) satisfying (10.3.3). Further, let (1.2.4) be $(p, n-p)$ disconjugate on $\mathbb{N}(a, b-1+n)$ for a fixed $p$, $1 \le p \le n-1$. Then, the Green's function $g(k,\ell)$ of the $(p, n-p)$ boundary value problem (1.2.4),

$$\begin{aligned} \Delta^i u(a) &= 0, \quad 0 \le i \le p-1 \\ \Delta^i u(b+p) &= 0, \quad 0 \le i \le n-p-1 \end{aligned} \tag{10.9.19}$$

exists. It is defined on $\mathbb{N}(a, b-1+n) \times \mathbb{N}(a, b-1)$ and can be expressed in the form

$$g(k,\ell) = \frac{1}{D}\begin{vmatrix} 0 & u_p(k,a) & u_{n-1}(k,a) \\ u_{n-1}(b+p,\ell+1) & u_p(b+p,a) & u_{n-1}(b+p,a) \\ \cdots & \cdots & \cdots \\ u_{n-1}(b-1+n,\ell+1) & u_p(b-1+n,a) & u_{n-1}(b-1+n,a) \end{vmatrix}$$

for $a \le k \le \ell+n-1$, and for $\ell+n \le k \le b-1+n$

$$g(k,\ell) = \frac{1}{D}\begin{vmatrix} u_{n-1}(k,\ell+1) & u_p(k,a) & u_{n-1}(k,a) \\ u_{n-1}(b+p,\ell+1) & u_p(b+p,a) & u_{n-1}(b+p,a) \\ \cdots & \cdots & \cdots \\ u_{n-1}(b-1+n,\ell+1) & u_p(b-1+n,a) & u_{n-1}(b-1+n,a) \end{vmatrix},$$

where

$$D = \begin{vmatrix} u_p(b+p,a) & \cdots & u_{n-1}(b+p,a) \\ \cdots & \cdots & \cdots \\ u_p(b-1+n,a) & & u_{n-1}(b-1+n,a) \end{vmatrix}.$$

**Theorem 10.9.7.** Assume that one of the following holds

1. equation (1.2.4) is disconjugate on $\mathbb{N}(a, b-1+n)$, or

2. $2 \le p \le n-1$ and equation (1.2.4) is $(j, n-j)$ disconjugate on $\mathbb{N}(a+p-j, b-1+n+p-j)$ for $j = p-1, \cdots, n-1$.

Then, the Green's function $g(k,\ell)$ for the $(p, n-p)$ boundary value problem (1.2.4), (10.9.19) satisfies

$$(10.9.20)\quad (-1)^{n-p} g(k,\ell) \ > \ 0, \quad k \in \mathbb{N}(a+p, b-1+p), \quad \ell \in \mathbb{N}(a, b-1).$$

**Theorem 10.9.8.** The Green's function $g(k,\ell)$ of the $(p, n-p)$ boundary value problem (10.1.3), (10.9.19) with $a = 0$ exists on $\mathbb{N}(0, b-1+n) \times \mathbb{N}(0, b-1)$, and is given by

(10.9.21)

$$g(k,\ell) = \begin{cases} \displaystyle\sum_{j=0}^{p-1}\left[\sum_{i=0}^{p-j-1}\binom{n-p+i-1}{i}\frac{k^{(j+i)}}{(b+n-j-1)^{(n-p+i)}}\right] \\ \quad \displaystyle\times\ \frac{(-\ell-1)^{(n-j-1)}}{j!(n-j-1)!}(b+n-k-1)^{(n-p)}, \quad 0 \le \ell \le k-1 \\ \displaystyle -\sum_{j=0}^{n-p-1}\left[\sum_{i=0}^{n-p-j-1}\binom{p+i-1}{i}\frac{(b+p+j+i-k-1)^{(j+i)}}{(b+p+j+i)^{(p+i)}}\right] \\ \quad \displaystyle\times\ (-1)^j\frac{(b+p-\ell-1)^{(n-j-1)}}{j!(n-j-1)!}k^{(p)}, \quad k \le \ell \le b-1. \end{cases}$$

**Theorem 10.9.9.** The Green's function $g(k,\ell)$ of the focal boundary value problem (10.1.3),

$$(10.9.22)\qquad \Delta^i u(a_{i+1}) \ = \ 0, \quad 0 \le i \le n-1$$

where $a \le a_1 \le \cdots \le a_n \le b$ exists on $\mathbb{N}(a, b-1+n) \times \mathbb{N}(a, b-1)$, and in terms of Abel Gontscharoff polynomials $T_i(k)$ (cf. (10.8.18) or

(10.8.19)) can be written as

$$(10.9.23)\quad g(k,\ell) = \begin{cases} \displaystyle\sum_{i=0}^{j-1} \frac{T_i(k)}{(n-i-1)!}(a_{i+1}-\ell-1)^{(n-i-1)}, \\ \qquad\qquad a_j \le \ell \le k-1 \\ \displaystyle -\sum_{i=j}^{n-1} \frac{T_i(k)}{(n-i-1)!}(a_{i+1}-\ell-1)^{(n-i-1)}, \\ \qquad\qquad k \le \ell \le a_{j+1}-1 \\ \qquad j = 0,1,\cdots,n \ (a_0 = a,\ a_{n+1} = b). \end{cases}$$

**Theorem 10.9.10.** The Green's function $g(k,\ell)$ of the two point focal boundary value problem (10.1.3),

$$(10.9.24)\quad \begin{aligned} \Delta^i u(a) &= 0, \quad 0 \le i \le p-1 \quad (1 \le p \le n-1, \text{ but fixed}) \\ \Delta^i u(b) &= 0, \quad p \le i \le n-1 \end{aligned}$$

exists on $\mathbb{N}(a, b-1+n) \times \mathbb{N}(a, b-1)$, and is given by

$$(10.9.25)\quad g(k,\ell) = (-1)^{n-p} \begin{cases} \displaystyle\sum_{\tau=a}^{\ell} g_0(k,\ell,\tau), & a \le \ell \le k-1 \\ \displaystyle\sum_{\tau=a}^{k-1} g_0(k,\ell,\tau), & k \le \ell \le b-1 \end{cases}$$

where

$$(10.9.26)\quad g_0(k,\ell,\tau) = \frac{(k-\tau-1)^{(p-1)}(\ell+n-p-1-\tau)^{(n-p-1)}}{(p-1)!(n-p-1)!}.$$

Further, for $0 \le i \le p-1$

$$(10.9.27)\quad \Delta^i g(k,\ell) = (-1)^{n-p} \begin{cases} \displaystyle\sum_{\tau=a}^{\ell} g_i(k,\ell,\tau), & a \le \ell \le k-1 \\ \displaystyle\sum_{\tau=a}^{k-1} g_i(k,\ell,\tau), & k \le \ell \le b-1 \end{cases}$$

where

$$(10.9.28)\quad g_i(k,\ell,\tau) = \frac{(k-\tau-1)^{(p-i-1)}(\ell+n-p-1-\tau)^{(n-p-1)}}{(p-i-1)!(n-p-1)!}$$

and for $0 \le i \le n-p-1$

(10.9.29)

$$\Delta^{i+p} g(k,\ell) = (-1)^{n-p+i} \begin{cases} 0, & a \le \ell \le k-1 \\ \dfrac{(\ell+n-p-1-i-k)^{(n-p-i-1)}}{(n-p-i-1)!}, & k \le \ell \le b-1. \end{cases}$$

From (10.9.25) (10.9.29) it is clear that

$$(10.9.30) \quad \begin{aligned} (-1)^{n-p}\Delta^i g(k,\ell) &> 0, \quad k \in \mathbb{N}(a+p-i, b-1+n-i), \\ & \qquad \ell \in \mathbb{N}(a, b-1), \quad 0 \le i \le p-1 \\ (-1)^{n-p+i}\Delta^{i+p} g(k,\ell) &> 0, \quad a \le k \le \ell \le b-1, \\ & \qquad 0 \le i \le n-p-1. \end{aligned}$$

**Theorem 10.9.11.** Let $1 \le p \le n$ but fixed, and let $\{r_1, \cdots, r_p\}$, $\{s_1, \cdots, s_{n-p}\}$ be a partition of $\{1, \cdots, n\}$ such that $r_1 < r_2 < \cdots < r_p$ and $s_1 < s_2 < \cdots < s_{n-p}$. Then, the Green's function $g(k,\ell)$ of the two point focal type boundary value problem (10.1.3),

$$(10.9.31) \quad \begin{aligned} \Delta^{r_i - 1} u(a) &= 0, \quad 1 \le i \le p \\ \Delta^{s_i - 1} u(b+n-s_i) &= 0, \quad 1 \le i \le n-p \end{aligned}$$

exists on $\mathbb{N}(a, b-1+n) \times \mathbb{N}(a, b-1)$, and

$$(10.9.32) \quad (-1)^{\sigma_i}\Delta^i g(k,\ell) \ge 0, \quad 0 \le i \le n-1$$

on $\mathbb{N}(a, b-1+n-i) \times \mathbb{N}(a, b-1)$, where $\sigma_i = \operatorname{card}\{j : s_j > i,\ 1 \le j \le n-p\}$.

**Theorem 10.9.12.** Let $1 \le p,\ q \le n$ but fixed. Then, the Green's function $g(k,\ell)$ of the two point boundary value problem (10.1.3),

$$(10.9.33) \quad \begin{aligned} \Delta^{i-1} u(a) &= 0, \quad 1 \le i \le p \\ \Delta^{i-1} u(b+n-i) &= 0, \quad q+1 \le i \le q+n-p \end{aligned}$$

exists on $\mathbb{N}(a, b-1+n) \times \mathbb{N}(a, b-1)$, and

$$(10.9.34) \quad (-1)^{n-p}\Delta^i g(k,\ell) \ge 0, \quad 0 \le i \le q$$

on $\mathbb{N}(a, b-1+n-i) \times \mathbb{N}(a, b-1)$.

**Remark 10.9.2.** If $q = 0$ then (10.9.33) represents conjugate type boundary conditions, and if $q = p$ then (10.9.33) reduces to two point focal boundary conditions.

**Theorem 10.9.13.** The Green's function $g(k,\ell)$ of the $(n,p)$ boundary value problem (10.1.3),

$$\begin{aligned} \Delta^i u(a) &= 0, \quad 0 \le i \le n-2 \\ \Delta^p u(b-1+n-p) &= 0, \quad (0 \le p \le n-1, \text{ but fixed}) \end{aligned} \tag{10.9.35}$$

exists on $\mathbb{N}(a, b-1+n) \times \mathbb{N}(a, b-1)$, and is given by

$$g(k,\ell) = -\frac{1}{(n-1)!}\begin{cases} g_1(k,\ell) - (k-\ell-1)^{(n-1)}, & a \le \ell \le k-n \\ g_1(k,\ell), & k-n+1 \le \ell \le b-1 \end{cases} \tag{10.9.36}$$

where

$$g_1(k,\ell) = \frac{(k-a)^{(n-1)}(b+n-p-\ell-2)^{(n-p-1)}}{(b+n-p-a-1)^{(n-p-1)}}.$$

Further,

$$-\Delta^i g(k,\ell) \ge 0, \quad 0 \le i \le p \tag{10.9.37}$$

on $\mathbb{N}(a, b-1+n-i) \times \mathbb{N}(a, b-1)$.

**Theorem 10.9.14.** The Green's function $g(k,\ell)$ of the $(p,n)$ boundary value problem (10.1.3),

$$\begin{aligned} \Delta^p u(a) &= 0, \quad (0 \le p \le n-1, \text{ but fixed}) \\ \Delta^i u(b+1) &= 0, \quad 0 \le i \le n-2 \end{aligned} \tag{10.9.38}$$

exists on $\mathbb{N}(a, b-1+n) \times \mathbb{N}(a, b-1)$, and is given by

$$g(k,\ell) = \frac{(-1)^{n+1}}{(n-1)!}\begin{cases} g_1(k,\ell), & a \le \ell \le k-1 \\ g_1(k,\ell) - (\ell+n-1-k)^{(n-1)}, & k \le \ell \le b-1 \end{cases} \tag{10.9.39}$$

where

$$g_1(k,\ell) = \frac{(b-1+n-k)^{(n-1)}(\ell+n-p-1-a)^{(n-p-1)}}{(b-1+n-p-a)^{(n-p-1)}}.$$

Further,

$$(-1)^{n+i+1}\Delta^i g(k,\ell) \ge 0, \quad 0 \le i \le p \tag{10.9.40}$$

on $\mathbb{N}(a, b-1+n-i) \times \mathbb{N}(a, b-1)$.

**Theorem 10.9.15.** The Green's function $g_m^1(k,\ell)$ of the Lidstone boundary value problem (10.1.3) with $n = 2m$,

$$\Delta^{2i}u(a) = \Delta^{2i}u(b-1+2m-2i) = 0, \quad 0 \le i \le m-1 \tag{10.9.41}$$

exists on $\mathbb{N}(a, b-1+2m) \times \mathbb{N}(a, b-1)$ and is recursively defined by

$$\begin{aligned} g_1^1(k,\ell) &= g_1(k,\ell), \quad \mathbb{N}(a,b+1) \times \mathbb{N}(a,b-1) \\ g_{i+1}^1(k,\ell) &= \sum_{k_1=a}^{b+2i-1} g_{i+1}(k,k_1) g_i^1(k_1,\ell), \\ &\quad \mathbb{N}(a,b+2i+1) \times \mathbb{N}(a,b-1), \quad i = 1,2,\cdots,m-1 \end{aligned} \tag{10.9.42}$$

where for each $1 \le i \le n$

(10.9.43)

$$g_i(k,\ell) = -\frac{1}{b-1+2i-a}\begin{cases} (b-1+2i-k)(\ell+1-a), & a \le \ell \le k-2 \\ (k-a)(b+2i-2-\ell), & k-1 \le \ell \le b+2i-3 \end{cases}$$

which itself is the Green's function of the boundary value problem

$$\begin{aligned} \Delta^2 u(k) &= 0, \quad k \in \mathbb{N}(a, b+2i-3) \\ u(a) &= u(b-1+2i) = 0 \end{aligned} \tag{10.9.44}$$

and is defined on $\mathbb{N}(a, b+2i-1) \times \mathbb{N}(a, b+2i-3)$.

Further, for all $(k,\ell) \in \mathbb{N}(a, b-1+2m) \times \mathbb{N}(a, b-1)$

$$(-1)^m g_m^1(k,\ell) \ge 0. \tag{10.9.45}$$

Our final result here shows that the unique polynomial $P_{2m-1}(k)$ of degree $2m-1$ satisfying Lidstone boundary conditions (1.5.14) can be represented in terms of $g_m^j(k,\ell)$, which are recursively defined as follows:

For a fixed $1 \le j \le m$

$$\begin{aligned} g_j^j(k,\ell) &= g_j(k,\ell), \quad \mathbb{N}(a,b+2j-1) \times \mathbb{N}(a,b+2j-3) \\ g_{i+1}^j(k,\ell) &= \sum_{k_1=a}^{b+2i-1} g_{i+1}(k,k_1) g_i^j(k_1,\ell), \\ &\quad \mathbb{N}(a,b+2i+1) \times \mathbb{N}(a,b+2j-3), \ i = j, j+1, \cdots, m-1. \end{aligned} \tag{10.9.46}$$

**Theorem 10.9.16.** The unique polynomial $P_{2m-1}(k)$ of degree $2m-1$ satisfying Lidstone boundary conditions (1.5.14) can be written as

$$\text{(10.9.47)} \quad P_{2m-1}(k) = \left(\frac{k-a}{b-1+2m-a}\right) B_0 + \left(1 - \frac{k-a}{b-1+2m-a}\right) A_0$$

$$+ \sum_{i=0}^{m-2} \sum_{\ell=a}^{b+2m-2i-3} g_m^{m-i}(k,\ell) \left[ \left( \frac{\ell-a}{b+2m-2i-3-a} \right) B_{2i+2} \right.$$

$$+\left(1-\frac{\ell-a}{b+2m-2i-3-a}\right)A_{2i+2}\Bigg].$$

## 10.10. Inequalities and Equalities for Green's Functions

**Theorem 10.10.1.** For the Green's function $g(k,\ell)$ of the conjugate boundary value problem (10.1.3), (10.9.10) defined in (10.9.12) the following hold

$$|g(k,\ell)| \le \left(\frac{n-1}{n}\right)^{n-1}\frac{(b-1+n-a)^{n-1}}{n!} \tag{10.10.1}$$

(10.10.2)

$$|\Delta^i g(k,\ell)| \le \frac{i}{n-1}\left(\frac{n-i-1}{n-1}\right)^{(n-i-1)/i}\frac{(b-1+n-a)^{n-i-1}}{(n-i-1)!},$$

$$1\le i\le n-1$$

$$\sum_{\ell=a}^{b-1}|g(k,\ell)| = \frac{1}{n!}\prod_{i=1}^{n}|k-k_i| \tag{10.10.3}$$

$$\le \frac{(n-1)^{n-1}}{n^n}\frac{(b-1+n-a)^n}{n!}. \tag{10.10.4}$$

**Corollary 10.10.2.** For the Green's function $g(k,\ell)$ of the osculatory boundary value problem (10.1.3) with $n=2m$, (10.9.13) defined in (10.9.14) the following hold

(10.10.5)

$$\sum_{\ell=a}^{b-1}|g(k,\ell)| = \frac{1}{(2m)!}\prod_{i=1}^{m}(k-k_i)^{(2)} \le \frac{(2m-1)^{2m-1}}{(2m)^{2m}}\frac{(b-1+2m-a)^{2m}}{(2m)!}.$$

**Corollary 10.10.3.** For the Green's function $g(k,\ell)$ of the two point Taylor boundary value problem (10.1.3) with $n=2m$, (10.9.16) defined in (10.9.17) the following hold

(10.10.6)

$$\sum_{\ell=a}^{b-1}|g(k,\ell)| = \frac{1}{(2m)!}(k-a)^{(m)}(b-1+2m-k)^{(m)} \le \left(\frac{1}{4}\right)^m\frac{(b+m-a)^{2m}}{(2m)!}.$$

**Corollary 10.10.4.** For the Green's function $g(k,\ell)$ of the $(p,n-p)$ boundary value problem (10.1.3), (10.9.19) the following hold

$$\sum_{\ell=a}^{b-1} |g(k,\ell)| = \frac{1}{n!}(k-a)^{(p)}(b-1+n-k)^{(n-p)} \tag{10.10.7}$$

$$\leq \frac{p^p(n-p)^{n-p}}{n^n}\frac{(b-a+n/2)^n}{n!}. \tag{10.10.8}$$

**Theorem 10.10.5.** For the Green's function $g(k,\ell)$ of the two point right focal boundary value problem (10.1.3), (10.9.24) defined in (10.9.25) the following hold

$$\sum_{\ell=a}^{b-1} |\Delta^i g(k,\ell)| = \left|\sum_{j=0}^{n-p}(-1)^j\binom{k-a}{n-i-j}\binom{b-a+j-1}{j}\right| \tag{10.10.9}$$

$$\leq \left|\sum_{j=0}^{n-p}(-1)^j\binom{b-a}{n-i-j}\binom{b-a+j-1}{j}\right| = C_{n,i}, \tag{10.10.10}$$

$$0 \leq i \leq p-1$$

$$\sum_{\ell=a}^{b-1} |\Delta^{i+p} g(k,\ell)| = \frac{(b+n-p-1-i-k)^{(n-p-i)}}{(n-p-i)!} \tag{10.10.11}$$

$$\leq \frac{(b+n-p-1-i-a)^{(n-p-i)}}{(n-p-i)!} = C_{n,i+p}, \tag{10.10.12}$$

$$0 \leq i \leq n-p-1.$$

**Theorem 10.10.6.** For the Green's function $g(k,\ell)$ of the $(n,p)$ boundary value problem (10.1.3), (10.9.35) defined in (10.9.36) the following hold

$$\sum_{\ell=a}^{b-1} |\Delta^i g(k,\ell)| = \frac{1}{(n-i-1)!}(k-a)^{(n-i-1)}\left[\frac{b-a}{n-p} - \frac{k-a-n+i+1}{n-i}\right] \tag{10.10.13}$$

$$\leq \left\{\begin{array}{ll} \dfrac{(p-i)}{(n-p)}\dfrac{(b-p+n-a)^{(n-i)}}{(n-i)!}, & 0 \leq i \leq p-1 \\ \dfrac{(n-p-1)^{n-p-1}}{(n-p)^{n-p}}\dfrac{(b-1+n-p-a)^{n-p}}{(n-p)!}, & i = p \end{array}\right\} \tag{10.10.14}$$

$$= D_{n,i}, \quad 0 \leq i \leq p.$$

**Theorem 10.10.7.** For the Green's function $g(k,\ell)$ of the $(p,n)$ boundary value problem (10.1.3), (10.9.38) defined in (10.9.39) the following hold

$$(10.10.15)\quad \sum_{\ell=a}^{b-1} |\Delta^i g(k,\ell)| = \frac{(b-1+n-i-k)^{(n-i-1)}}{(n-i-1)!}\left[\frac{b-a}{n-p} - \frac{b-k}{n-i}\right]$$

$$(10.10.16)\quad \le \begin{cases} \dfrac{(p-i)}{(n-p)}\dfrac{(b-1+n-i-a)^{(n-i)}}{(n-i)!}, & 0 \le i \le p-1 \\ \dfrac{(n-p-1)^{n-p-1}}{(n-p)^{n-p}}\dfrac{(b-1+n-p-a)^{n-p}}{(n-p)!}, & i = p \end{cases}$$

$$= E_{n,i}, \quad 0 \le i \le p.$$

**Theorem 10.10.8.** For the Green's function $g_m^1(k,\ell)$ of the Lidstone boundary value problem (10.1.3) with $n = 2m$, (10.9.41) the following holds

$$(10.10.17)\quad \sum_{\ell=a}^{b-1} |g_m^1(k,\ell)| \le \left(\frac{1}{8}\right)^m \prod_{i=1}^{m} (b+2i-1-a)^2.$$

## 10.11. Maximum Principles

**Theorem 10.11.1.** If $u(k)$ is defined on $\mathbb{N}(a,b+1)$, and $\Delta^2 u(k) \ge 0$, $k \in \mathbb{N}(a,b-1)$, and attains its maximum at some $k^* \in \mathbb{N}(a+1,b)$, then $u(k)$ is identically constant on $\mathbb{N}(a,b+1)$.

**Remark 10.11.1.** As a consequence of Theorem 10.11.1, $u(k) \le \max\{u(a), u(b+1)\}$, $k \in \mathbb{N}(a,b+1)$.

**Remark 10.11.2.** Theorem 10.11.1 holds if we reverse the inequality and replace "maximum" by "minimum".

The maximum principle stated in Theorem 10.11.1 does not necessarily hold for functions satisfying higher order inequalities. For example, let $u(k) = -(k/10-1)^2$, $k \in \mathbb{N}(0,20)$. For this function $\Delta^4 u(k) \ge 0$, $k \in \mathbb{N}(0,16)$, but $u(k)$ attains its maximum at $k = 10$ which is a point in $\mathbb{N}(1,19)$. Extensions of Theorem 10.11.1 are embodied in the following:

**Theorem 10.11.2.** Let $1 \le p \le n-1$ and $u(k)$ be defined on the discrete interval $\mathbb{N}(a,b+n-1)$, and

$$(10.11.1)\quad \Delta^n u(k) \ge 0, \quad k \in \mathbb{N}(a,b-1)$$

$$(10.11.2)\quad (-1)^{n-p}\Delta^i u(a) \ge 0, \quad i = 1,2,\cdots,p-1 \quad \text{(if such an } i \text{ exists)}$$

$$(10.11.3)\quad (-1)^{n-p+j}\Delta^j u(b+p) \ge 0, \quad j = 1,2,\cdots,n-p-1$$

(if such a $j$ exists).

Then, in the case $n-p$ even $u(k)$ attains its minimum and in the case $n-p$ odd $u(k)$ attains its maximum either at $a$ or $b+p$.

**Corollary 10.11.3.** Let $u(k)$ be defined on $\mathbb{N}(a, b-1+2m)$, and

$$\Delta^{2m}u(k) \geq 0, \quad k \in \mathbb{N}(a, b-1) \tag{10.11.4}$$

$$\begin{aligned} &(-1)^m \Delta^i u(a) \geq 0 \\ &(-1)^{m+i}\Delta^i u(b+m) \geq 0, \quad 1 \leq i \leq m-1 \end{aligned} \tag{10.11.5}$$

then in the case $m$ even ($m$ odd) $u(k)$, $k \in \mathbb{N}(a, b+m)$ attains its minimum (maximum) at either $a$ or $b+m$.

**Corollary 10.11.4.** Let $u(k)$ be defined on $\mathbb{N}(a, b-1+n)$, and satisfy the inequality (10.11.1). Further, let

$$\Delta^i u(a) \leq 0, \quad 1 \leq i \leq n-2 \tag{10.11.6}$$

then $u(k)$ attains its maximum at $a$ or $b-1+n$.

**Corollary 10.11.5.** Let $u(k)$ be defined on $\mathbb{N}(a, b-1+n)$, and satisfy the inequality (10.11.1). Further, let

$$(-1)^i \Delta^i u(b+1) \geq 0, \quad 1 \leq i \leq n-2 \tag{10.11.7}$$

then in the case $n$ odd ($n$ even) $u(k)$ attains its minimum (maximum) at $a$ or $b+1$.

**Remark 10.11.3.** When the inequalities in (10.11.1) (10.11.7) are reversed, the results remain true provided the word maximum (minimum) is replaced by minimum (maximum).

## 10.12. Error Estimates in Polynomial Interpolation

**Theorem 10.12.1.** Let $u(k)$ be a function defined on $\mathbb{N}(a, b-1+n)$, and satisfy the conjugate boundary conditions (10.8.1). Further, let $P_{n-1}(k)$ and $g(k, \ell)$ be as in (10.8.2) and (10.9.12) respectively. Then, for all $k \in \mathbb{N}(a, b-1+n)$ the following hold

$$u(k) = P_{n-1}(k) + \sum_{\ell=a}^{b-1} g(k, \ell)\Delta^n u(\ell) \tag{10.12.1}$$

(10.12.2)

$$|u(k) - P_{n-1}(k)| \leq \frac{(n-1)^{n-1}}{n^n} \frac{(b-1+n-a)^n}{n!} \max_{k \in \mathbb{N}(a, b-1)} |\Delta^n u(k)|.$$

**Theorem 10.12.2.** Let $u(k)$ be a function defined on $\mathbb{N}(a, b-1+2m)$, and satisfy the osculatory boundary conditions (10.8.3). Further, let $P_{2m-1}(k)$ and $g(k,\ell)$ be as in (10.8.4) and (10.9.14) respectively. Then, for all $k \in \mathbb{N}(a, b-1+2m)$ the following hold

$$u(k) = P_{2m-1}(k) + \sum_{\ell=a}^{b-1} g(k,\ell)\Delta^{2m}u(\ell) \tag{10.12.3}$$

(10.12.4)

$$|u(k) - P_{2m-1}(k)| \le \frac{(2m-1)^{2m-1}}{(2m)^{2m}} \frac{(b-1+2m-a)^{2m}}{(2m)!} \max_{k\in\mathbb{N}(a,b-1)} |\Delta^{2m}u(k)|.$$

**Theorem 10.12.3.** Let $u(k)$ be a function defined on $\mathbb{N}(a, b-1+2m)$, and satisfy the two point Taylor boundary conditions (10.8.7). Further, let $P_{2m-1}(k)$ and $g(k,\ell)$ be as in (10.8.10) and (10.9.17) respectively. Then, for all $k \in \mathbb{N}(a, b-1+2m)$, (10.12.3) holds and

$$|u(k) - P_{2m-1}(k)| \le \left(\frac{1}{4}\right)^m \frac{(b+m-a)^{2m}}{(2m)!} \max_{k\in\mathbb{N}(a,b-1)} |\Delta^{2m}u(k)|. \tag{10.12.5}$$

**Theorem 10.12.4.** Let $u(k)$ be a function defined on $\mathbb{N}(a, b-1+n)$, and satisfy the right focal boundary conditions (1.5.11). Further, let $P_{n-1}(k)$ and $g(k,\ell)$ be as in (10.8.17) and (10.9.23) respectively. Then, for all $k \in \mathbb{N}(a, b-1+n)$, (10.12.1) holds and

$$|\Delta^i(u(k) - P_{n-1}(k))| \le \frac{(b+n-1-a-i)^{(n-i)}}{(n-i)!} \binom{n-i-1}{\left[\frac{n-i-1}{2}\right]} \tag{10.12.6}$$

$$\times \max_{k\in\mathbb{N}(a,b-1)} |\Delta^n u(k)|, \quad k \in \mathbb{N}(a, b-1+n-i), \quad 0 \le i \le n-1.$$

Further, for each $0 \le i \le n-1$ the inequality (10.12.6) is the best possible.

**Theorem 10.12.5.** Let $\alpha \in \mathbb{N}(0, n-2)$, $\beta \in \mathbb{N}(1, n-1)$, and $a = a_1 = \cdots = a_{\alpha+1} < a_{\alpha+2} \le \cdots \le a_{n-\beta} \le a_{n-\beta+1} = \cdots = a_n = b$. Then, in Theorem 10.12.4 the inequalities (10.12.6) can be improved by

(10.12.7)

$$|\Delta^i(u(k) - P_{n-1}(k))| \le \frac{(b+n-1-a-i)^{(n-i)}}{(n-i)!} \begin{Bmatrix} \binom{n-i-1}{r^*}, \\ \quad 0 \le i \le n-\beta-1 \\ 1, \quad n-\beta \le i \le n-1 \end{Bmatrix}$$

$$\times \max_{k\in\mathbb{N}(a,b-1)} |\Delta^n u(k)|, \quad k \in \mathbb{N}(a, b-1+n-i), \quad 0 \le i \le n-1$$

where $r^* = \max\left\{\alpha - i, \beta, \left[\frac{n-i-1}{2}\right]\right\}$. Further, for each $0 \le i \le n-1$ the inequality (10.12.7) is the best possible.

**Theorem 10.12.6.** Let $u(k)$ be a function defined on $\mathbb{N}(a, b-1+n)$, and satisfy the two point right focal boundary conditions (10.8.20). Further, let $P_{n-1}(k)$ and $g(k,\ell)$ be as in (10.8.21) and (10.9.25) respectively. Then, for all $k \in \mathbb{N}(a, b-1+n)$, (10.12.1) holds and

$$|\Delta^i(u(k) - P_{n-1}(k))| \le C_{n,i} \max_{k \in \mathbb{N}(a,b-1)} |\Delta^n u(k)|, \tag{10.12.8}$$

$$k \in \mathbb{N}(a, b-1+n-i), \quad 0 \le i \le n-1$$

where $C_{n,i}$ are defined in (10.10.10) and (10.10.12).

**Theorem 10.12.7.** Let $u(k)$ be a function defined on $\mathbb{N}(a, b-1+n)$, and satisfy the $(n,p)$ boundary conditions (1.5.12). Further, let $P_{n-1}(k)$ and $g(k,\ell)$ be as in (10.8.22) and (10.9.36) respectively. Then, for all $k \in \mathbb{N}(a, b-1+n)$, (10.12.1) holds and

$$|\Delta^i(u(k) - P_{n-1}(k))| \le D_{n,i} \max_{k \in \mathbb{N}(a,b-1)} |\Delta^n u(k)|, \tag{10.12.9}$$

$$k \in \mathbb{N}(a, b-1+n-i), \quad 0 \le i \le p$$

where $D_{n,i}$ are defined in (10.10.14).

**Theorem 10.12.8.** Let $u(k)$ be a function defined on $\mathbb{N}(a, b-1+n)$, and satisfy the $(p,n)$ boundary conditions (1.5.13). Further, let $P_{n-1}(k)$ and $g(k,\ell)$ be as in (10.8.23) and (10.9.39) respectively. Then, for all $k \in \mathbb{N}(a, b-1+n)$, (10.12.1) holds and

$$|\Delta^i(u(k) - P_{n-1}(k))| \le E_{n,i} \max_{k \in \mathbb{N}(a,b-1)} |\Delta^n u(k)|, \tag{10.12.10}$$

$$k \in \mathbb{N}(a, b-1+n-i), \quad 0 \le i \le p$$

where $E_{n,i}$ are defined in (10.10.16).

**Theorem 10.12.9.** Let $u(k)$ be a function defined on $\mathbb{N}(a, b-1+2m)$, and satisfy the Lidstone boundary conditions (1.5.14). Further, let $P_{2m-1}(k)$ and $g^1_m(k,\ell)$ be as in (10.9.47) and (10.9.42) respectively. Then, for all $k \in \mathbb{N}(a, b-1+2m)$ the following hold

$$u(k) = P_{2m-1}(k) + \sum_{\ell=a}^{b-1} g^1_m(k,\ell)\Delta^{2m}u(\ell) \tag{10.12.11}$$

and

(10.12.12)

$$|u(k) - P_{2m-1}(k)| \leq \left(\frac{1}{8}\right)^m \prod_{i=1}^{m} (b + 2i - 1 - a)^2 \max_{k \in \mathbb{N}(a,b-1)} |\Delta^{2m} u(k)|.$$

## 10.13. Problems

**10.13.1.** Show that the difference equation

$$u(k+2) + pu(k+1) + qu(k) = 0, \quad k \in \mathbb{N}(a,b)$$

is disconjugate if $q > 0$, $p < 0$ and $p^2 - 4q \geq 0$.

**10.13.2.** Consider the second order linear difference equation

$$u(k+2) + a_1(k)u(k+1) + a_0(k)u(k) = 0, \quad k \in \mathbb{N}(a,b) \tag{10.13.1}$$

where $a_0(k) > 0$ on $\mathbb{N}(a,b)$. Show that (10.13.1) is disconjugate on $\mathbb{N}(a, b+2)$ if and only if $(-1)^m D_m(a) > 0$, $m \in \mathbb{N}(1, b-a+1)$ where $D_m(a)$ is the tridiagonal determinant defined by

$$D_m(a) = \begin{vmatrix} a_1(a) & 1 & 0 & \cdots & 0 \\ a_0(a+1) & a_1(a+1) & 1 & \cdots & 0 \\ 0 & a_0(a+2) & a_1(a+2) & \cdots & 0 \\ \vdots & \vdots & \vdots & & \vdots \\ 0 & 0 & 0 & \cdots & a_1(a+m-1) \end{vmatrix}.$$

**10.13.3.** Consider the difference equation

$$\Delta^2 u(k) + p(k)u(k+1) = 0, \quad k \in \mathbb{Z}. \tag{10.13.2}$$

Suppose that there exist $\epsilon_1 > 0$, $\epsilon_2 > 0$ such that

$$\limsup_{m \to \infty} \sum_{k=0}^{m} \arctan \frac{\epsilon_1}{2\alpha_k(p, \epsilon_1)} > \frac{\pi}{4}$$

and

$$\limsup_{m \to -\infty} \sum_{k=m}^{1} \arctan \frac{\epsilon_2}{2\beta_k(p, \epsilon_2)} > \frac{\pi}{4},$$

where

$$\alpha_0 = 1 + \epsilon_1, \quad \beta_1 = 1 + \epsilon_2$$

$$\alpha_k = \alpha_k(p,\epsilon_1) = \left(\epsilon_1 - \sum_{\ell=0}^{k-1} p(\ell) + 1\right) \prod_{\tau=0}^{k-1} \left(\epsilon_1 - \sum_{\ell=0}^{\tau-1} p(\ell) + 1\right)^2 \quad \text{for } k \geq 1$$

and

$$\beta_k = \beta_k(p,\epsilon_2) = \left(\epsilon_2 - \sum_{\ell=k-1}^{-1} p(\ell) + 1\right) \prod_{\tau=k+1}^{1} \left(\epsilon_2 - \sum_{\ell=\tau-1}^{-1} p(\ell) + 1\right)^2 \quad \text{for } k \leq 0.$$

Show that (10.13.2) is conjugate in $\mathbf{Z}$.

**10.13.4.** With respect to the difference equation (10.13.2) suppose that

$$\liminf_{m\to\infty} \frac{1}{m+1} \sum_{\tau=0}^{m} \sum_{\ell=0}^{\tau-1} p(\ell) \ = \ c_1 > 0$$

and

$$\liminf_{m\to-\infty} \frac{1}{-m+2} \sum_{\tau=m}^{1} \sum_{\ell=\tau-1}^{-1} p(\ell) \ = \ c_2 > 0.$$

Show that (10.13.2) is conjugate in $\mathbf{Z}$.

**10.13.5.** In the difference equation (10.13.2) suppose that $p(k) \geq 0$, $k \in \mathbb{N}(1)$. Show that (10.13.2) is conjugate in $\mathbb{N}(K)$, $K \in \mathbb{N}(1)$ if there exist integers $L$, $M$ with $K < L < M$ such that

$$\frac{1}{L-K} \ < \ \sum_{k=L}^{M} p(k).$$

**10.13.6.** Consider the third order linear difference equation

$$\text{(10.13.3)} \quad \Delta^3 u(k-1) + p(k)\Delta u(k) + q(k)u(k) \ = \ 0, \quad k \in \mathbb{N}(a+1, b+1)$$

whose solutions are defined on $\mathbb{N}(a, b+3)$. We say that $u(k)$ has a *generalized zero* of order $m$ at $a$ provided $u(a+i) = 0$, $0 \leq i \leq m-1$; and at $k_0 > a$ provided $u(k_0-1) \neq 0$, $u(k_0+i) = 0$, $1 \leq i \leq m-2$ and $(-1)^m u(k_0-1)u(k_0+m-1) \geq 0$. Show that

(i) no nontrivial solution of (10.13.3) can have a generalized zero of order three or more

(ii) if $p(k) \leq 0$, $k \in \mathbb{N}(a+1, b+2)$ and $q(k) \geq 0$, $k \in \mathbb{N}(a+1, b+1)$, then (10.13.3) is $(1,2)$ disconjugate on $\mathbb{N}(a, b+3)$, i.e. no nontrivial solution of (10.13.3) has a generalized zero at some $k_1 \in \mathbb{N}(a, b+1)$ followed by a double generalized zero in $\mathbb{N}(k_1+1, b+3)$

(iii) if $p(k) \le 0,\ q(k) \ge 0,\ k \in \mathbb{N}(a+1,b+1)$ and for all $k_0 \in \mathbb{N}(a+1,b+1)$

$$|p(k)| \ \ge\ 2\sum_{\ell=k_0}^{k-1} q(\ell), \qquad k_0 \le \ell \le b+1$$

then (10.13.3) is disconjugate on $\mathbb{N}(a, b+3)$.

**10.13.7.** Consider the difference equation

$$\Delta^4 u(k-2) + q(k)u(k) \ =\ 0, \qquad k \in \mathbb{N}(a+2, b+2) \tag{10.13.4}$$

whose solutions are defined on $\mathbb{N}(a, b+4)$. We say $u(k)$ has a *generalized zero* of order $m$ at $a$ if $u(a+\ell) = 0,\ 0 \le \ell \le m-1$, and at $k_0 \in \mathbb{N}(a+1, b+n-m+1)$ if $u(k_0-1) \ne 0$ and $u(k_0+\ell) = 0,\ 0 \le \ell \le m-1$ or $u(k_0) \ne 0,\ u(k_0+\ell) = 0,\ 1 \le \ell \le m-1$ and $(-1)^m u(k_0)u(k_0+m) > 0$. Show that

(i) no nontrivial solution of (10.13.4) has a generalized zero of order four or more

(ii) equation (10.13.4) is $(2,2)$ disconjugate on $\mathbb{N}(a, b+4)$, i.e. no nontrivial solution of (10.13.4) has two distinct generalized zeros, each of order two or more, in $\mathbb{N}(a, b+4)$ provided

$$\left(\frac{b-a+5}{2}\right)^{(3)} \sum_{\ell=a+2}^{b+2} q_-(\ell) \ <\ 24, \tag{10.13.5}$$

where $q_-(\ell) = \max\{0, -q(\ell)\}$. Further, construct an example to show that inequality (10.13.5) is sharp.

**10.13.8.** Consider the $n$th order linear difference equation

$$\Delta^n u(k-p) + q(k)u(k) \ =\ 0, \qquad k \in \mathbb{N}(a+p, b+p) \tag{10.13.6}$$

where $a \le b$ and $1 \le p \le n-1$, whose solutions are defined on $\mathbb{N}(a, b+n)$. Equation (10.13.6) is called $(n-1, 1)$ *disfocal* on $\mathbb{N}(a, b+n)$ if $u(k)$ is a nontrivial solution of (10.13.6) with a generalized zero (see Problem 10.13.7) of order $n-1$ at $k_0 \in \mathbb{N}(a, b)$, then $\Delta^{n-1}u(k)$ is of one sign on $\mathbb{N}(k_0, b+1)$. Show that (10.13.6) is $(n-1, 1)$ disfocal on $\mathbb{N}(a, b+n)$ provided

$$\binom{b+n-1-a}{n-1} \sum_{\ell=a+p}^{b+p} q_+(k) \ <\ 1. \tag{10.13.7}$$

Further, construct an example to show that the inequality (10.13.7) is sharp.

**10.13.9.** Consider the $2m$th order linear difference equation

$$\sum_{i=0}^{m} \Delta^i \left[r_i(k)\Delta^i u(k-i)\right] = 0, \quad k \in \mathbb{N}(a+m, b+m) \tag{10.13.8}$$

where $r_i(k)$ are defined on $\mathbb{N}(a+m, b+m+i)$, $0 \le i \le m$ and $r_m(k) > 0$ on $\mathbb{N}(a+m, b+2m)$, whose solutions are defined on $\mathbb{N}(a, b+2m)$. Show that

(i) no nontrivial solution of (10.13.8) can have a generalized zero (see Problem 10.13.6) of order $2m$

(ii) if $(-1)^{m+i} r_i(k) \ge 0$, $k \in \mathbb{N}(a+m, b+m+i)$, $0 \le i \le n-1$ then (10.13.8) is $(m, m)$ disconjugate on $\mathbb{N}(a, b+2m)$, i.e. no nontrivial solution of (10.13.8) has a generalized zero of order $m$ followed by a generalized zero of order $m$ in $\mathbb{N}(a, b+2m)$.

**10.13.10.** Let $D$ be a subset of any normed linear space $S$ of functions which map $\mathbb{N}(a, b+2)$ to $\mathbb{R}$, and let $f(k, x, y) : \mathbb{N}(a, b+2) \times \mathbb{R}^2 \to \mathbb{R}$. Assume that $f \in C^{(2)}$ with respect to $x$ and $y$ for each $k \in \mathbb{N}(a, b+2)$. Consider the discrete functional

$$J[u] = \sum_{k=a+1}^{b+2} f(k, u(k), \Delta u(k-1)), \tag{10.13.9}$$

subject to $u(k) \in D$. We call $D$ a set of *admissible functions*. Let $u(k) \in D$, then $\eta(k) \in S$ is called an *admissible variation* for $u(k)$ if $u(k) + \epsilon\eta(k) \in D$ for sufficiently small $\epsilon$. Let in $D$, $\|u\| = \max_{k \in \mathbb{N}(a,b+2)} |u(k)|$, and let $u_0(k) \in D$. Then, $J[u]$ has a *local maximum (minimum)* at $u_0(k)$ if there is a $\delta > 0$ such that $J[u] \le (\ge)\ J[u_0]$ for all $u(k) \in D$ with $\|u - u_0\| < \delta$. $J[u]$ is said to have a *local extremum* at $u_0(k)$ if it has either a local maximum or a local minimum. If in addition $J[u] > J[u_0]$ for $u(k) \not\equiv u_0(k)$ in $D$ with $\|u - u_0\| < \delta$, then $J[u]$ has a *proper local minimum* at $u_0(k)$. Show that

(i) if $J[u]$, $u(k) \in D$ has a local extremum at $u_0(k) \in D$, and $\eta(k)$ is an admissible variation for $u_0(k)$, then the *first variation* of $J[u]$ along $u_0(k)$

$$J_1[\eta] = J_1[\eta, u_0] = \sum_{k=a+1}^{b+2} \left[f_x(k, u_0(k), \Delta u_0(k-1))\eta(k) + f_y(k, u_0(k), \Delta u_0(k-1))\Delta\eta(k-1)\right] = 0 \tag{10.13.10}$$

(ii) if $J[u]$, $u(k) \in D$ has a local minimum (maximum) at $u_0(k) \in D$, then the *second variation* of $J[u]$ along $u_0(k)$, $J_2[\eta, u_0] \ge (\le)\ 0$ for all $\eta(k)$ which are admissible variations for $u_0(k)$, where

$$(10.13.11)\quad J_2[\eta] = J_2[\eta, u_0] = \sum_{k=a+1}^{b+2} \left[P(k)\eta^2(k) + 2Q(k)\eta(k)\Delta\eta(k-1) + R(k)[\Delta\eta(k-1)]^2\right]$$

and

$$\begin{aligned} P(k) &= P(k,u) = f_{xx}(k, u(k), \Delta u(k-1)) \\ Q(k) &= Q(k,u) = f_{xy}(k, u(k), \Delta u(k-1)) \\ R(k) &= R(k,u) = f_{yy}(k, u(k), \Delta u(k-1)). \end{aligned}$$

**10.13.11.** In Problem 10.13.10 let $D = \{u(k) : \mathbb{N}(a,b+2) \to \mathbb{R} \mid \alpha u(a) + \beta\Delta u(a) = A,$ where $\alpha \neq \beta,\ \alpha^2+\beta^2 > 0$ and $\gamma u(b+1) + \delta\Delta u(b+1) = B,$ where $\delta \neq 0,\ \gamma^2+\delta^2 > 0\}$, and define the set of admissible variations $D_0 = \{\eta(k) : \mathbb{N}(a,b+2) \to \mathbb{R} \mid \alpha\eta(a) + \beta\Delta\eta(a) = 0,$ where $\alpha \neq \beta,\ \alpha^2+\beta^2 > 0$ and $\gamma\eta(b+1) + \delta\Delta\eta(b+1) = 0,$ where $\delta \neq 0,\ \gamma^2+\delta^2 > 0\}$. Show that

(i) if $J[u],\ u(k) \in D$ has a local extremum at $u_0(k) \in D$, then $u_0(k)$ satisfies

$$(10.13.12)\quad Lu(k) = \begin{cases} \left(\dfrac{\beta}{\beta-\alpha}\right) f_y(a+1, u(a+1), \Delta u(a)) & \text{for } k = a+1 \\ 0 \quad \text{for } k \in \mathbb{N}(a+2, b) \\ -f_x(b+2, u(b+2), \Delta u(b+1)) \\ \quad -\left(\dfrac{\delta-\gamma}{\delta}\right) f_y(b+2, u(b+2), \Delta u(b+1)) & \text{for } k = b+1 \end{cases}$$

where $L$ is the Euler Langrange operator defined by

$$(10.13.13)\quad Lu(k) = f_x(k, u(k), \Delta u(k-1)) - \Delta f_y(k, u(k), \Delta u(k-1))$$

(ii) if $J[u]$ subject to $u(k) \in D_1 = \{u(k) : \mathbb{N}(a,b+2) \to \mathbb{R} \mid u(a) = A\}$ has a local extremum at $u_0(k)$, then $u_0(k)$ satisfies $Lu(k) = 0,\ k \in \mathbb{N}(a+1, b+1),\ u_0(a) = A$ and $u_0(k)$ satisfies the transversality condition

$$(10.13.14)\quad f_x(b+2, u(b+2), \Delta u(b+1)) + f_y(b+2, u(b+2), \Delta u(b+1)) = 0$$

(iii) if $J[u]$ subject to $u(k) \in D_2 = \{u(k) : \mathbb{N}(a,b+2) \to \mathbb{R} \mid u(b+2) = B\}$ has a local extremum at $u_0(k)$, then $u_0(k)$ satisfies $Lu(k) = 0,\ k \in \mathbb{N}(a+1, b+1),\ u_0(b+2) = B$ and $u_0(k)$ satisfies the transversality condition

$$(10.13.15)\quad f_y(a+1, u(a+1), \Delta u(a)) = 0$$

(iv) if $J[u]$ subject to $u(k) \in D_3 = \{u(k) : \mathbb{N}(a, b+2) \to \mathbb{R}\}$ has a local extremum at $u_0(k)$, then $u_0(k)$ satisfies $Lu(k) = 0$, $k \in \mathbb{N}(a+1, b+1)$ and $u_0(k)$ satisfies the transversality conditions (10.13.14) and (10.13.15).

**10.13.12.** Let $D$ be as in Problem 10.13.11, and define

$$\begin{aligned}
p(k-1) &= p(k-1, u) = Q(k, u) + R(k, u), \quad k \in \mathbb{N}(a+1, b+2) \\
q(k) &= q(k, u) = \Delta Q(k, u) - P(k, u), \quad k \in \mathbb{N}(a+1, b+1) \\
c(k) &= c(k, u) = q(k, u) - p(k, u) - p(k-1, u), \quad k \in \mathbb{N}(a+1, b+1) \\
C &= P(b+2) + 2Q(b+2) + R(b+2) \\
K_a &= -c(a+1) + 2p(a)\left(\frac{\beta}{\alpha-\beta}\right) + R(a+1)\left(\frac{\alpha}{\alpha-\beta}\right)^2 \\
K_b &= -c(b+1) + 2p(b+1)\left(\frac{\gamma-\delta}{\delta}\right) + C\left(\frac{\gamma-\delta}{\delta}\right)^2.
\end{aligned}$$

Show that if $J[u]$, $u \in D$ has a local minimum (maximum) at $u_0(k) \in D$, then the following Legendre's necessary condition holds

$$c(k, u_0) \le (\ge) \begin{cases} K_a & \text{if } k = a+1 \\ 0 & \text{if } k \in \mathbb{N}(a+2, b) \\ K_b & \text{if } k = b+1. \end{cases}$$

**10.13.13.** Let $D$ and $D_0$ be as in Problem 10.13.11. Suppose that $u_0(k) \in D$ satisfies the condition (10.13.12). Show that if $J_2[\eta, u_0]$ is positive (negative) definite on $D_0$, then $J[u]$ has a proper local minimum (maximum) at $u_0(k)$.

**10.13.14.** Show that $J_2[\eta]$, $\eta \in D_0^c = \{\eta(k) : \mathbb{N}(a, b+2) \to \mathbb{R} \mid \eta(a) = \eta(b+2) = 0\}$ can be written as

$$J_2[\eta] = \sum_{k=a+1}^{b+2} \left[p(k-1)[\Delta\eta(k-1)]^2 - q(k)\eta^2(k)\right]$$

and hence the Euler Langrange equation for $J_2[\eta]$ is the self adjoint difference equation

$$\text{(10.13.16)} \qquad \mathcal{L}u(k) = \Delta[p(k-1)\Delta u(k-1)] + q(k)u(k) = 0, \qquad k \in \mathbb{N}(a+1, b+1).$$

This equation is called the *Jacobi equation* for $J[u]$, $u \in D^c = \{u(k) : \mathbb{N}(a, b+2) \to \mathbb{R} \mid u(a) = A,\ u(b+2) = B\}$.

**10.13.15.** Let $D^c$ and $D_0^c$ be as in Problem 10.13.14. Consider the quadratic functional $\mathcal{Q}$ on $D_0^c$ as

$$(10.13.17) \qquad \mathcal{Q}[\eta] = \sum_{k=a+1}^{b+2} \left[p(k-1)[\Delta\eta(k-1)]^2 - q(k)\eta^2(k)\right]$$

where $p(k) > 0$, $k \in \mathbb{N}(a, b+1)$. Show that the self adjoint equation (10.13.16) is disconjugate on $\mathbb{N}(a, b+2)$ if and only if $\mathcal{Q}$ is positive definite on $D_0^c$. Hence, if $u_0(k) \in D^c$ satisfies the Euler Lagrange equation $Lu(k) = 0$, $k \in \mathbb{N}(a+1, b+1)$ and the corresponding Jacobi equation $\mathcal{L}u(k) = 0$ is disconjugate on $\mathbb{N}(a, b+2)$, then $J[u]$, $u \in D^c$ has a local minimum at $u_0(k)$.

**10.13.16.** Show that if the self adjoint equation (10.13.16) is disconjugate on $\mathbb{N}(a, b+2)$, then

$$\sum_{k=a+1}^{b+1} q(k) < p(a) + p(b+1).$$

**10.13.17.** (Sturm's Comparison Theorem). Consider the self adjoint equations

$$\mathcal{L}_i u(k) = \Delta[p_i(k-1)\Delta u(k-1)] + q_i(k)u(k) = 0, \quad k \in \mathbb{N}(a+1, b+1), \ i = 1, 2$$

where $p_2(k) \geq p_1(k) > 0$, $k \in \mathbb{N}(a, b+1)$ and $q_1(k) \geq q_2(k)$, $k \in \mathbb{N}(a+1, b+1)$. Show that if $\mathcal{L}_1 u(k) = 0$ is disconjugate on $\mathbb{N}(a, b+2)$, then $\mathcal{L}_2 u(k) = 0$ is disconjugate on $\mathbb{N}(a, b+2)$.

**10.13.18.** (Reid's Roundabout Theorem). Show that the following are equivalent:

(i) the equation (10.13.16) is disconjugate on $\mathbb{N}(a, b+2)$

(ii) the equation (10.13.16) has a solution $u(k)$ such that $p(k-1)u(k-1)u(k) > 0$ on $\mathbb{N}(a+1, b+2)$

(iii) the Riccati equation

$$(10.13.18) \qquad \Delta z(k) + q(k) + \frac{z^2(k)}{z(k) + p(k-1)} = 0$$

has a solution $z(k)$ on $\mathbb{N}(a+1, b+2)$ satisfying $z(k) + p(k-1) > 0$ on $\mathbb{N}(a+1, b+2)$

(iv) the quadratic functional $\mathcal{Q}$ defined in (10.13.17) is positive definite on $D_0^c$.

**10.13.19.** Let the operator $\mathcal{L}$ be as in (10.13.16), where $p(k)$ is defined and nonzero on $\mathbb{N}(a, b+1)$, and $q(k)$ is defined on $\mathbb{N}(a+1, b+1)$. Let $H$ be a constant such that $p(b+1) - H \neq 0$. Define the operator $\mathcal{M} = \mathcal{M}(p, q, H)$ on the set of functions with domain $\mathbb{N}(a, b+2)$ by

$$\mathcal{M}u(k) = \begin{cases} \mathcal{L}u(k), & k \in \mathbb{N}(a+1, b) \\ \mathcal{L}u(b+1) - Hu(b+2), & k = b+1. \end{cases}$$

The second order difference equation $\mathcal{M}u(k) = 0$, $k \in \mathbb{N}(a+1, b+1)$ is called *C disfocal* (C stands for Coppel) on $\mathbb{N}(a, b+2)$ if whenever $u(k)$ is its nontrivial solution satisfying $\Delta u(b+1) = 0$, then $p(k-1)u(k)u(k-1) > 0$ for all $k \in \mathbb{N}(a+1, b+1)$. Show that

(i) $\mathcal{M}u(k) = 0$, $k \in \mathbb{N}(a+1, b+1)$ is C disfocal on $\mathbb{N}(a, b+2)$ if and only if the quadratic functional $\mathcal{Q}$ defined by

$$\mathcal{Q}[\eta] = \sum_{k=a+1}^{b+1} \left[p(k-1)[\Delta\eta(k-1)]^2 - q(k)\eta^2(k)\right] + H\eta^2(b+2)$$

is positive definite on $D_0^f = \{\eta(k) : \mathbb{N}(a, b+2) \to \mathbb{R} \mid \eta(a) = \Delta\eta(b+1) = 0\}$
(ii) if $u_0(k) \in D^f = \{u(k) : \mathbb{N}(a, b+2) \to \mathbb{R} \mid u(a) = A,\ \Delta u(b+1) = B\}$ satisfies the Euler Lagrange equation $Lu(k) = 0$, $k \in \mathbb{N}(a+1, b+1)$, and the corresponding equation $\mathcal{M}u(k) = 0$, $k \in \mathbb{N}(a+1, b+1)$ with $H = P(b+2) + Q(b+2)$ is C disfocal on $\mathbb{N}(a, b+2)$, then $J[u]$, $u \in D^f$ has a local minimum at $u_0(k)$.

**10.13.20.** (Sturm's Comparison Theorem). Consider the second order difference equations $\mathcal{M}_i(p_i, q_i, H_i) = 0$, $k \in \mathbb{N}(a+1, b+1)$, $i = 1, 2$ where $p_2(k) \geq p_1(k) > 0$, $k \in \mathbb{N}(a+1, b+1)$, $q_1(k) \geq q_2(k)$ on $\mathbb{N}(a, b+1)$ and $H_2 \geq H_1$. Show that if $\mathcal{M}_1 u(k) = 0$ is C disfocal on $\mathbb{N}(a, b+2)$, then $\mathcal{M}_2 u(k) = 0$ is C disfocal on $\mathbb{N}(a, b+2)$.

**10.13.21.** Find the minimum for the functional

$$J[u] = \sum_{k=a+1}^{b+1} \{(1/8)^{k-1}[\Delta u(k-1)]^2 - 3(1/8)^k u^2(k)\}$$

subject to $u \in D^c$ and $u \in D^f$.

**10.13.22.** Let $f(k, x, y) : \mathbb{N}(a, b+4) \times \mathbb{R}^2 \to \mathbb{R}$. Assume that $f \in C^{(2)}$ with respect to $x$ and $y$ for each $k \in \mathbb{N}(a, b+4)$. Consider the functional

$$J[u] = \sum_{k=a+2}^{b+4} f(k, u(k), \Delta^2 u(k-2))$$

subject to $u(k) \in \mathcal{D} = \{u(k) : \mathbb{N}(a, b+4) \to \mathbb{R} \mid u(a) = A,\ \Delta u(a) = B,\ \Delta^2 u(b+2) = C,\ \Delta^3 u(b+1) = D\}$. Show that

(i) if $J[u],\ u(k) \in D$ has a local extremum at $u_0(k) \in D$, and $\eta(k) \in D_0 = \{\eta(k) : \mathbb{N}(a, b+4) \to \mathbb{R} \mid \eta(a) = \Delta\eta(a) = \Delta^2\eta(b+2) = \Delta^3\eta(b+1) = 0\}$, then the first variation of $J[u]$ along $u_0(k)$

$$\begin{aligned} J_1[\eta] = J_1[\eta, u_0] = & \sum_{k=a+2}^{b} Eu_0(k)\eta(k) + \big[Eu_0(b+1) - f_x(b+3) - f_y(b+3) \\ & -2f_x(b+4)\big]\eta(b+1) + \big[Eu_0(b+2) + 2f_x(b+3) \\ & +2f_y(b+3) + 3f_x(b+4) - f_y(b+4)\big]\eta(b+2) \ = \ 0, \end{aligned}$$

where $Eu_0(k) = f_x(k) + \Delta^2 f_y(k)$, and

$$\begin{aligned} f_x(k) &= f_x(k, u_0(k), \Delta^2 u_0(k-2)) \\ f_y(k) &= f_y(k, u_0(k), \Delta^2 u_0(k-2)) \end{aligned}$$

and hence

$$\begin{aligned} & Eu_0(k) = 0, \quad k \in \mathbb{N}(a+2, b) \\ & Eu_0(b+1) = f_x(b+3) + f_y(b+3) + 2f_x(b+4) \\ & Eu_0(b+2) = -2f_x(b+3) - 2f_y(b+3) - 3f_x(b+4) + f_y(b+4) \end{aligned}$$

(ii) if $J[u],\ u(k) \in D$ has a local minimum (maximum) at $u_0(k) \in D$, then the second variation of $J[u]$ along $u_0(k)$, $J_2[\eta, u_0] \geq\ (\leq 0)$ for all $\eta(k) \in D_0$, where

$$\begin{aligned} J_2[\eta] = J_2[\eta, u_0] = & \sum_{k=a+2}^{b+2} \big[P(k)\eta^2(k) + 2Q(k)\eta^2(k) \\ & -4Q(k)\eta(k)\eta(k-1) + 2Q(k)\eta(k)\eta(k-2) + R(k)\eta^2(k) \\ & +4R(k)\eta^2(k-1) + R(k)\eta^2(k-2) - 4R(k)\eta(k)\eta(k-1) \\ & +2R(k)\eta(k)\eta(k-2) - 4R(k)\eta(k-1)\eta(k-2)\big] \\ & + \sum_{k=b+3}^{b+4} P(k)\eta^2(k) \end{aligned}$$

and

$$\begin{aligned} P(k) &= f_{xx}(k, u_0(k), \Delta^2 u_0(k-2)) \\ Q(k) &= f_{xy}(k, u_0(k), \Delta^2 u_0(k-2)) \\ R(k) &= f_{yy}(k, u_0(k), \Delta^2 u_0(k-2)). \end{aligned}$$

**10.13.23.** Consider the $2m$th order linear difference equation

$$(10.13.19) \qquad \sum_{i=0}^{m} \Delta^i \left(r_i(k-i)\Delta^i u(k-i)\right) \ = \ 0, \quad k \in \mathbb{N}(a)$$

where $r_i(k)$ are defined on $\mathbb{N}(a+m-i)$, $0 \le i \le m$ and $r_m(k) > 0$ on $\mathbb{N}(a)$. Whenever necessary we extend the domain of the functions $r_i$ to $\mathbb{Z}$ by defining $r_i(k) = r_i(a+m-i)$ for $k < a+m-i$, $0 \le i \le m$.

For a given function $u(k)$, $k \in \mathbb{N}(a)$ on $\mathbb{N}(a+m)$ we define the nonlinear operator $F$ by

$$\begin{aligned} Fu(k) &= (-1)^m \sum_{i=1}^{m} (-1)^i \Delta^{i-1} u(k-1) \left( r_i(k-1) \Delta^i u(k-1) \right) \\ &\quad -(-1)^m \sum_{j=2}^{m} \sum_{i=j}^{m} (-1)^{i-j} \Delta^{i-j} u(k) \Delta^{j-1} \left( r_i(k-j) \Delta^i u(k-j) \right). \end{aligned}$$

Show that

(i) If $u(k)$ is a solution of (10.13.19), then

$$\Delta F u(k) = (-1)^m r_0(k) [u(k)]^2 + \sum_{i=1}^{m} (-1)^{m+i} r_i(k-1) \left[ \Delta^i u(k-1) \right]^2.$$

In particular, if

$$(10.13.20) \qquad (-1)^{m+i} r_i(k) \ge 0, \quad k \in \mathbb{N}(a+m-i), \quad 0 \le i \le m-1$$

then $F$ is nondecreasing along solutions $u(k)$ of (10.13.19) on $\mathbb{N}(a+m)$.
(ii) If (10.13.20) holds and $u(k)$ is a nontrivial solution of (10.13.19) with a generalized zero (see Problem 10.13.6) of order at least $m$ at $k_0 \ge a$, then

(a) $Fu(k_0) < 0$ provided $k_0 > a$, and either

(I) $Fu(k_0+1) > 0$, if $(-1)^m u(k_0-1) u(k_0+m-1) > 0$, or

(II) $u(k)$ has exactly $m+\ell$ consecutive zeros starting at $k_0$, where $0 \le \ell \le m-1$, and

$$(10.13.21) \qquad Fu(k_0+s) = 0, \quad 1 \le s \le \ell+1$$

with

$$(10.13.22) \qquad Fu(k_0+\ell+2) > 0.$$

(b) Case (II) holds if $k_0 = a$, and (10.13.21), (10.13.22) both follow.
(iii) If (10.13.20) holds, then (10.13.19) is $(m, m)$ disconjugate on $\mathbb{N}(a)$.
(iv) If (10.13.20) holds, then any nontrivial solution $u(k)$ of (10.13.19) with a generalized zero of order at least $m$ is a type P solution, i.e. $Fu(k) > 0$ near $\infty$. In particular, (10.13.19) has $2m$ linearly independent type P solutions.

(v) If (10.13.20) holds, then (10.13.19) has $m$ linearly independent type Q solutions, i.e. $Fu(k) \le 0$ in a neighborhood of $\infty$.

(vi) If (10.13.20) holds, then all solutions of (10.13.19) are type P or type Q solutions.

(vii) If (10.13.20) holds and $u(k)$ is a type Q solution of (10.13.19), then

$$\sum_{k=a+m-i}^{\infty} (-1)^{m-i} r_i(k) \left[\Delta^i u(k)\right]^2 < \infty, \quad 0 \le i \le m.$$

Further, if $r_0(k) \ne 0$ in a neighborhood of $\infty$, then every nontrivial type Q solution of (10.13.19) is a strict type Q solution.

(viii) If (10.13.20) holds and $\liminf_{k\to\infty}(-1)^m r_0(k) > 0$, then (10.13.19) has $m$ linearly independent type Q solutions $v_\ell(k)$ satisfying $\lim_{k\to\infty} v_\ell(k) = 0,\ 1 \le \ell \le m$.

**10.13.24.** Show that

(i) the Green's function $g(k,\ell)$ of the boundary value problem

$$\begin{aligned} -\Delta^2 u(k) &= 0, \quad k \in \mathbb{N}(0,K) \\ \alpha u(0) - \beta \Delta u(0) &= 0 \\ \gamma u(K+1) + \delta \Delta u(K+1) &= 0 \end{aligned}$$

exists on $\mathbb{N}(0,K+2)\times\mathbb{N}(0,K)$ if and only if $\rho = \alpha\gamma(K+1)+\alpha\delta+\beta\gamma \ne 0$, and is given by

$$g(k,\ell) = \frac{1}{\rho}\begin{cases} [\beta+\alpha(\ell+1)][\delta+\gamma(K+1-k)], & 0 \le \ell \le k-1 \\ (\beta+\alpha k)[\delta+\gamma(K-\ell)], & k \le \ell \le K \end{cases}$$

(ii) if $\alpha > 0,\ \gamma > 0,\ \beta \ge 0,\ \delta \ge \gamma$ so that $\rho > 0$, then

(a) for $(k,\ell) \in \mathbb{N}(0,K+2) \times \mathbb{N}(0,K)$

$$0 \le g(k,\ell) \le Lg(\ell,\ell)$$

where $L = \begin{cases} (\beta+\alpha)/\beta, & \beta > 0 \\ 2, & \beta = 0 \end{cases}$

(b) for $(k,\ell) \in \mathbb{N}(m,K) \times \mathbb{N}(0,K)$ where $m \in \mathbb{N}(1,K)$

$$g(k,\ell) \ge \frac{(\beta+\alpha)(\delta+\gamma)}{(\beta+\alpha K)(\delta+\gamma K)} g(\ell,\ell).$$

**10.13.25.** Show that for the Green's function $g(k,\ell)$ of the $(n,p)$ boundary value problem (10.1.3), (10.9.35) defined in (10.9.36) with $a = 0$ the following hold

(i) for $(k,\ell) \in \mathbb{N}(0, b-1+n) \times \mathbb{N}(0, b-1)$

$$-g(k,\ell) \le \frac{(b+n-1)^{(n-1)}}{(n-1)!(b+n-p-1)^{(n-p-1)}}(b+n-p-\ell-2)^{(n-p-1)}$$

(ii) for $(k,\ell) \in \mathbb{N}(n-1, b+n-p-1) \times \mathbb{N}(0, b-1)$

$$-g(k,\ell) \ge \frac{(b+n-p-\ell-2)^{(n-p-1)}}{(b+n-p-1)^{(n-p-1)}}\left[1 - \frac{(b-\ell-1)^{(p)}}{(b)^{(p)}}\right].$$

**10.13.26.** Show that the Green's function $g(k,\ell)$ of the boundary value problem

$$\begin{aligned} (-1)^m \Delta^m \left[p(k-m)\Delta^m u(k-m)\right] &= 0, \quad k \in \mathbb{N}(m, b+m) \\ \Delta^i u(0) &= 0, \quad 0 \le i \le m-1 \\ \Delta^{m+i} u(b+1) &= 0, \quad 0 \le i \le m-1 \end{aligned}$$

where $p(k) > 0$, $k \in \mathbb{N}(0, b+m)$, exists and is given by

$$g(k,\ell) = \begin{cases} \displaystyle\sum_{\tau=0}^{k-1} \frac{(k-\tau-1)^{(m-1)}}{(m-1)!} \frac{(\ell-\tau-1)^{(m-1)}}{(m-1)!p(\tau)}, & k-1 \le \ell - m \\ \displaystyle\sum_{\tau=0}^{\ell-m} \frac{(k-\tau-1)^{(m-1)}}{(m-1)!} \frac{(\ell-\tau-1)^{(m-1)}}{(m-1)!p(\tau)}, & \ell - m < k-1. \end{cases}$$

Further, show that

$\Delta^i g(k,\ell) \ge 0$ for $k \in \mathbb{N}(0, b+2m-i)$, $\ell \in \mathbb{N}(m, b+m)$, $0 \le i \le m-1$.

## 10.14. Notes

The landmark paper of Hartman [37] has resulted in the tremendous interest in establishing discrete analogs of the known results for the ordinary differential equations. Theorem 10.1.1 is due to Hankerson [30], Theorems 10.1.2 10.1.5 and Corollary 10.1.6 are from Hartman [37], and Theorem 10.1.7 is proved by Eloe [24]. Theorem 10.2.1 is due to Peterson [45], whereas Theorems 10.2.2 10.2.5 are proved by Peil [43,44], however some parts of these results have appeared earlier in Hankerson [30] and Peterson [46]. Theorem 10.2.6 is a contribution of Peterson [47], whereas rest of the results in Section 10.2 are from Hankerson [34]. Some of these results are modifications of the theorems of Peterson [49]. The adjoint difference equation (10.3.1) has appeared in Peterson [46 48], whereas all the results related to this equation in Section 10.3 are from Hankerson [30]. Theorems 10.4.1 and 10.4.2 are proved in Hankerson [30], however a slightly weaker form of Theorem 10.4.1 is available in Peterson [47]. Theorem 10.5.1 is from Eloe [20], whereas Theorem 10.5.2 is due to Eloe and Henderson [25]. Theorem 10.6.1 has been proved by Eloe [23]. All the results in Section 10.7 are proved in Hankerson and Peterson [31,32]. Discrete interpolating polynomials given in Section 10.8 are constructed in Agarwal and Lalli [5], and Agarwal, O'Regan and Wong [8]. Green's function for an $n$th order linear difference equation together with two point boundary conditions first appeared in the work of Böcher [13]. Theorem 10.9.1 is due to Gaines [29], whereas Theorem 10.9.2 is from Hartman [37]. Theorems 10.9.3 10.9.5 are proved in Agarwal and Lalli [5]. Theorems 10.9.6 and 10.9.7 are from Peterson [45,48], also see Hankerson and Peterson [33]. Theorem 10.9.8 is established in Agarwal, O'Regan and Wong [8], whereas Theorem 10.9.9 is taken from Agarwal [7]. Theorem 10.9.10 has appeared in Hankerson and Peterson [35]. Eloe [21] has proved Theorems 10.9.11 and 10.9.12. All the remaining results in Section 10.9 are taken from Agarwal and Lalli [5]. Several other results for the discrete Green's functions are available in Teptin [56 58]. Inequalities (10.10.1) and (10.10.2) are due to Teptin [55], whereas all the remaining results in Section 10.10 are proved in Agarwal and Lalli [5]. Results in Sections 10.11 and 10.12 are also from Agarwal and Lalli [5], except Theorems 10.12.4 and 10.12.5 which are recent contributions of Wong [62]. In recent years discrete calculus of variations has attracted several researchers, our Problems 10.13.10 10.13.22 are based on Harmsen [36], whereas Problem 10.13.23 is due to Anderson [10,11]. For several other similar results see Ahlbrandt and Peterson [9], Atici and Peterson [12], Došlý and Řehák [15], Eloe and Peil [26], Henderson and Peterson [38], Peil [41,42], Peterson and Ridenhour [50,51]. Continuous analogs of most of the results presented in this chapter are available in Agarwal [1,3], Agarwal and Usmani [2], Agarwal and Wong [4,6], Coppel [14], Dunninger [16], Elias [17 19], Eloe and Henderson [22], Etgen, Jones and Taylor [27,28], Muldowney [39,40], Protter and Weinberger [52], Šeda [53], Shui Nee Chow, Dunninger and Lasota [54], Trench [59,60], Wong and Agarwal [61].

## 10.15. References

[1]. R.P. Agarwal, *Boundary Value Problems for Higher Order Differential Equations*, *World Scientific*, Singapore, 1986.

[2]. R.P. Agarwal and R.A. Usmani, Iterative methods for solving right focal point boundary value problems, *J. Comp. Appl. Math.* **14**(1986), 371 390.

[3]. R.P. Agarwal, Some new results for two point problems for higher order differential equations, *Funkcialaj Ekvacioj* **29**(1986), 197 212.

[4]. R.P. Agarwal and P.J.Y. Wong, Lidstone polynomials and boundary value problems, *Computers Math. Applic.* **17**(1989), 1397 1421.

[5]. R.P. Agarwal and B.S. Lalli, Discrete: Polynomial interpolation, Green's functions, maximum principles, error bounds and boundary value problems, *Computers Math. Applic.* **25**(1993), 3 39.

[6]. R.P. Agarwal and P.J.Y. Wong, *Error Inequalities in Polynomial Interpolation and Their Applications*, Kluwer, Dordrecht, 1993.

[7]. R.P. Agarwal, *Focal Boundary Value Problems for Differential and Difference Equations*, Kluwer, Dordrecht, 1998.

[8]. R.P. Agarwal, D. O'Regan and P.J.Y. Wong, *Positive Solutions of Differential, Difference and Integral Equations*, Kluwer, Dordrecht, 1999.

[9]. C. Ahlbrandt and A. Peterson, The $(n,n)$ disconjugacy of a $2n$th order linear difference equation, *Computers Math. Applic.* **28**(1 3)(1994), 1 9.

[10]. D.R. Anderson, *Discrete Hamiltonian Systems*, Ph.D. Dissertation, *University of Nebraska*, Lincoln, 1997.

[11]. D.R. Anderson, A $2n$th order linear difference equation, *Communications in Applied Analysis* **2**(1998), 521 529.

[12]. F. Atici and A. Peterson, Bounds for positive solutions for a focal boundary value problem, *Computers Math. Applic.* **36**(10 12)(1998), 99 107.

[13]. M. Bôcher, Boundary value problems and Green's functions for linear differential and difference equations, *Ann. Math.* **13**(1911/12), 71 88.

[14]. W. Coppel, *Disconjugacy, Lecture Notes in Math.* **220**, *Springer Verlag*, Berlin, 1971.

[15]. O. Došlý and P. Řehák, Conjugacy criteria for second order linear difference equations, *Arch. Math. (Brnö)* **34**(1998), 301 310.

[16]. D.R. Dunninger, Maximum principles for fourth order ordinary differential inequalities, *J. Math. Anal. Appl.* **82**(1981), 399 405.

[17]. U. Elias, Oscillatory solutions and extremal points for a linear differential equation, *Arch. Rat. Mech. Anal.* **71**(1979), 177 198.

[18]. U. Elias, A classification of the solutions of a differential equation according to their asymptotic behaviour, *Proc. Roy. Soc. Edinburgh Sect. A* **83**(1979), 25 38.

[19]. U. Elias, A classification of the solutions of a differential equation according to their behaviour at infinity, II, *Proc. Roy. Soc. Edinburgh Sect. A* **100**(1985), 53 66.

[20]. P.W. Eloe, Criteria for right disfocality of linear difference equations, *J. Math. Anal. Appl.* **120**(1986), 610 621.

[21]. P.W. Eloe, Sign properties of Green's functions for two classes of boundary value problems, *Canad. Math. Bull.* **30**(1987), 28 35.

[22]. P.W. Eloe and J. Henderson, Some analogues of Markov and Descartes systems for right disfocality, *Proc. Amer. Math. Soc.* **99**(1987), 543 548.

[23]. P.W. Eloe, Eventual disconjugacy and right disfocality of linear difference equations, *Canad. Math. Bull.* **31**(1988), 362 373.

[24]. P.W. Eloe, A comparison theorem for linear difference equations, *Proc. Amer. Math. Soc.* **103**(1988), 451 457.

[25]. P.W. Eloe and J. Henderson, Analogues of Fekete and Descartes systems of solutions for difference equations, *J. Appr. Theory* **59**(1989), 38 52.

[26]. P.W. Eloe and T. Peil, Sign properties of Green's functions for difference equations, in *Proceedings of the World Congress of Nonlinear Analysts*, Tampa, Florida, August 19 26, 1992, ed. V. Lakshmikantham, *Walter de Gruyter*, Berlin, 1996, 1121 1130.

[27]. G.J. Etgen, G.D. Jones and W.E. Taylor, Jr., Structure of the solution space of certain linear equations, *J. Diff. Eqns.* **59**(1985), 229 242.

[28]. G. J. Etgen, G.D. Jones and W.E. Taylor Jr., On the factorization of ordinary linear differential operators, *Trans. Amer. Math. Soc.* **297**(1986), 717 728.

[29]. R. Gaines, Difference equations associated with boundary value problems for second order nonlinear ordinary differential equations, *SIAM J. Numer. Anal.* **11**(1974), 411 434.

[30]. D. Hankerson, *Boundary Value Problems for $n$th Order Difference Equations*, Ph. D. dissertation, *University of Nebraska*, Lincoln, 1986.

[31]. D. Hankerson and A. Peterson, On a theorem of Elias for difference equations, in *Proc. of the Seventh International Conference on Nonlinear Analysis and Applications*, ed. V. Lakshmikantham, *Marcel Dekker*, New York, (1987), 229 234.

[32]. D. Hankerson and A. Peterson, A classification of the solutions of a difference equation according to their behavior at infinity, *J. Math. Anal. Appl.* **136**(1988), 249 266.

[33]. D. Hankerson and A. Peterson, A positivity result applied to difference equations, *J. Appr. Theory* **59**(1989), 76 86.

[34]. D. Hankerson, An existence and uniqueness theorem for difference equations, *SIAM J. Math. Anal.* **20**(1989), 1208 1217.

[35]. D. Hankerson and A. Peterson, Comparison of eigenvalues for focal point problems for $n$th order difference equations, *Differential and Integral*

*Equations* **3**(1990), 363 380.

[36]. B.J. Harmsen, *The Discrete Calculus of Variations*, Ph.D. Thesis, *University of Nebraska*, Lincoln, 1995.

[37]. P. Hartman, Difference equations: Disconjugacy, principal solutions, Green's functions, complete monotonicity, *Trans. Amer. Math. Soc.* **246**(1978), 1 30.

[38]. J. Henderson and A. Peterson, Disconjugacy for a third order linear difference equation, *Computers Math. Applic.* **28**(1 3)(1994), 131 139.

[39]. J. S. Muldowney, A necessary and sufficient condition for disfocality, *Proc. Amer. Math. Soc.* **74**(1979), 49 55.

[40]. J. S. Muldowney, On invertibility of linear ordinary differential boundary value problems, *SIAM J. Math. Anal.* **12**(1981), 368 384.

[41]. T. Peil, *Criteria for Disconjugacy and Disfocality for an $n$th Order Linear Difference Equation,* Ph.D. Dissertation, *University of Nebraska*, Lincoln, 1990.

[42]. T. Peil, Disconjugacy of second order linear difference equations, in *Differential Equations: Stability and Control*, ed. S. Elaydi, *Marcel Dekker*, 1991, 409 414.

[43]. T. Peil, Disconjugacy for $n$th order linear difference equations, *J. Math. Anal. Appl.* **160**(1991), 132 148.

[44]. T. Peil, Criteria for right disfocality of an $n$th order linear difference equation, *Rocky Mount. J. Math.* **22**(1992), 1523 1543.

[45]. A. Peterson, Boundary value problems for $n$th order linear difference equations, *SIAM J. Math. Anal.* **15**(1984), 124 132.

[46]. A. Peterson, Boundary value problems and Green's functions for linear difference equations, in *Proc. of the Twelfth Midwest Differential Equations Conference*, ed. J. L. Henderson, *University of Missouri Rolla*, Rolla, MO, (1985), 79 100.

[47]. A. Peterson, On $(k, n-k)$ disconjugacy for linear difference equations, in *Proc. of the International Conference on Qualitative Properties of Differential Equations*, eds. W. Allegretto and G.J. Butler, *University of Alberta*, 1984, Edmonton, (1986), 329 337.

[48]. A. Peterson, Green's function for $(k, n-k)$ boundary value problems for linear difference equations, *J. Math. Anal. Appl.* **124**(1987), 127 138.

[49]. A. Peterson, Existence and uniqueness theorems for nonlinear difference equations, *J. Math. Anal. Appl.* **125**(1987), 185 191.

[50]. A. Peterson and J. Ridenhour, A disfocality criterion for an $n$th order difference equation, in *Proceedings of the 1st International Conference on Difference Equations*, San Antonio, Texas, U.S.A. May 25 28, 1994, eds. S. Elaydi, et. al., *Gordon & Breach*, Luxembourg, 1995, 411 418.

[51]. A. Peterson and J. Ridenhour, The (2,2) disconjugacy of a fourth order difference equation, *J. Difference Eqns. Appl.* **1**(1995), 87 93.

[52]. M.H. Protter and H.F. Weinberger, *Maximum Principles in Differential Equations, Prentice Hall, Englewood Cliffs*, 1967.

[53]. V. Šeda, Two remarks on boundary value problems for ordinary differential equations, *J. Diff. Eqns.* **26**(1977), 278 290.

[54]. Shui Nee Chow, D.R. Dunninger and A. Lasota, A maximum principle for fourth order differential equations, *J. Diff. Eqns.* **14**(1973), 101 105.

[55]. A.L. Teptin, Estimates of Green's function for a many point boundary value problem, *Differential Equations* **17**(1981), 641 647.

[56]. A.L. Teptin, On the sign of the Green's function of a certain difference boundary value problem, *Differ. Uravn.* **17**(1981), 2283 2286 (Russian).

[57]. A.L. Teptin, On the difference Green's function changing sign only on the lines $x = const$, $s = const$, *Sov. Math.* **25** No.8(1981), 55 61.

[58]. A.L. Teptin, On the sign of the Green's function of a multipoint difference boundary value problem, *Sov. Math.* **26** No.1(1982), 107 109.

[59]. W.F. Trench, A sufficient condition for eventual disconjugacy, *Proc. Amer. Math. Soc.* **52**(1975), 139 146.

[60]. W.F. Trench, Eventual disconjugacy of a linear differential equation, *Proc. Amer. Math. Soc.* **89**(1983), 461 466.

[61]. P.J.Y. Wong and R.P. Agarwal, Abel Gontscharoff interpolation error bounds for derivatives, *Proc. Royal Soc. Edinburgh* **119A**(1991), 367 372.

[62]. P.J.Y. Wong, Best error estimates for discrete Abel Gontscharoff interpolation, *J. Approximation Theory* **97**(1999), 65 81.

# Chapter 11

# Boundary Value Problems for Higher Order Difference Equations

Results stated in Chapter 10 play a fundamental role in the study of various higher order boundary value problems including those discussed in Section 1.5. Using these results we provide easily verifiable sets of necessary and sufficient conditions so that each of these boundary value problems has at least one solution. Sufficient conditions ensuring the uniqueness of these solutions are also included. This is followed by the convergence of the constructive methods: Picard's method, the approximate Picard's method, quasilinearization, and the approximate quasilinearization. The results obtained herein are more explicit than those discussed in Chapter 9 for the systems of difference equations. The monotonic convergence of the Picard's iterative method is analyzed in Section 11.4. Next, we shall show that the initial value methods discussed in Chapters 8 and 9 for constructing the solutions of boundary value problems can also be used to prove the existence and uniqueness theorems for the higher order discrete boundary value problems. In Section 10.9, we have noticed that the uniqueness of the solutions of the linear boundary value problems implies the existence of the solutions. The argument employed in proving this assertion is algebraic and is based on the linear structure of the fundamental system of solutions of the difference equations and the linearity of the boundary conditions. In Section 11.6 sufficient conditions which guarantee this property for the nonlinear boundary value problems are provided.

## 11.1. Existence and Uniqueness

Inequalities obtained in Section 10.12 will be used here to provide easier tests for the local existence and uniqueness of the solutions of higher order boundary value problems.

**Theorem 11.1.1.** With respect to the conjugate boundary value problem (1.2.7), (10.8.1) assume that

(i) $M > 0$ is a given real number and the function $f(k, u_0, u_1, \cdots, u_{n-1})$ is continuous on the compact set: $\mathbb{N}(a, b-1) \times D_0$, where

$$D_0 = \{(u_0, u_1, \cdots, u_{n-1}) : |u_i| \le 2M, \quad 0 \le i \le n-1\}$$

and $\max_{\mathbb{N}(a,b-1)\times D_0} |f(k, u_0, u_1, \cdots, u_{n-1})| \le Q$

(ii) $\max_{\mathbb{N}(a,b-1+n)} |P_{n-1}(k)| \le M$, where $P_{n-1}(k)$ is the conjugate interpolating polynomial defined in (10.8.2)

(iii) $\dfrac{(n-1)^{n-1}}{n^n} \dfrac{(b-1+n-a)^n}{n!} Q \le M$.

Then, (1.2.7), (10.8.1) has a solution in $D_0$.

**Proof.** In view of (10.12.1) the problem (1.2.7), (10.8.1) is equivalent to the equation

$$(11.1.1)\quad u(k) = P_{n-1}(k) + \sum_{\ell=a}^{b-1} g(k,\ell) f(\ell, u(\ell), u(\ell+1), \cdots, u(\ell+n-1)),$$

where $g(k,\ell)$ is the Green's function of the conjugate boundary value problem (10.1.3), (10.9.10) defined in (10.9.12). Let $S(a, b-1+n)$ be the space of all real functions defined on $\mathbb{N}(a, b-1+n)$. We shall equip the space $S(a, b-1+n)$ with the norm $\|u\| = \max_{\mathbb{N}(a,b-1+n)} |u(k)|$, so that it becomes a Banach space. Now define an operator $T : S(a, b-1+n) \to S(a, b-1+n)$ as follows

$$(11.1.2)\quad Tu(k) = P_{n-1}(k) + \sum_{\ell=a}^{b-1} g(k,\ell) f(\ell, u(\ell), u(\ell+1), \cdots, u(\ell+n-1)).$$

Obviously, $u(k)$ is a solution of (1.2.7), (10.8.1) if and only if $u(k)$ is a fixed point of $T$. The set $S_1 = \{u(k) \in S(a, b-1+n) : \|u\| \le 2M\}$ is a closed convex subset of the Banach space $S(a, b-1+n)$. Since

$$\Delta^n \left[\sum_{\ell=a}^{b-1} g(k,\ell) f(\ell, u(\ell), u(\ell+1), \cdots, u(\ell+n-1))\right]$$
$$= f(k, u(k), u(k+1), \cdots, u(k+n-1))$$

for any $u(k) \in S_1$, in view of (11.1.2) and (10.12.2), it follows that

$$|Tu(k) - P_{n-1}(k)| \le \frac{(n-1)^{n-1}}{n^n} \frac{(b-1+n-a)^n}{n!} Q$$

and therefore

$$\begin{aligned} \|Tu\| &\le \max_{\mathbb{N}(a,b-1+n)} |P_{n-1}(k)| + \frac{(n-1)^{n-1}}{n^n}\frac{(b-1+n-a)^n}{n!}Q \\ &\le M + M = 2M. \end{aligned}$$

Thus, $T$ maps $S_1$ into itself and that $\overline{T(S_1)}$ is compact. By the Schauder fixed point theorem, the operator $T$ has a fixed point in $S_1$. Thus, the boundary value problem (1.2.7), (10.8.1) has a solution in $D_0$. ∎

**Theorem 11.1.2.** With respect to the osculatory boundary value problem (1.2.7) with $n = 2m$, (10.8.3) assume that

(i) $M > 0$ is a given real number and the function $f(k, u_0, u_1, \cdots, u_{2m-1})$ is continuous on the compact set: $\mathbb{N}(a, b-1) \times D_0$, where

$$D_0 = \{(u_0, u_1, \cdots, u_{2m-1}) : |u_i| \le 2M, \quad 0 \le i \le 2m-1\}$$

and $\max_{\mathbb{N}(a,b-1)\times D_0} |f(k, u_0, u_1, \cdots, u_{2m-1})| \le Q$

(ii) $\max_{\mathbb{N}(a,b-1+2m)} |P_{2m-1}(k)| \le M$, where $P_{2m-1}(k)$ is the osculatory interpolating polynomial defined in (10.8.4)

(iii) $\dfrac{(2m-1)^{2m-1}}{(2m)^{2m}} \dfrac{(b-1+2m-a)^{2m}}{(2m)!}Q \le M.$

Then, (1.2.7) with $n = 2m$, (10.8.3) has a solution in $D_0$.

**Theorem 11.1.3.** With respect to the two point Taylor boundary value problem (1.2.7) with $n = 2m$, (10.8.7) assume that

(i) condition (i) of Theorem 11.1.2

(ii) condition (ii) of Theorem 11.1.2 with $P_{2m-1}(k)$ as the two point Taylor interpolating polynomial defined in (10.8.10)

(iii) $\left(\dfrac{1}{4}\right)^m \dfrac{(b+m-a)^{2m}}{(2m)!}Q \le M.$

Then, (1.2.7) with $n = 2m$, (10.8.7) has a solution in $D_0$.

**Theorem 11.1.4.** With respect to the two point right focal boundary value problem (1.2.6), (10.8.20) assume that

(i) $M_i > 0$, $0 \le i \le n-1$ are given real numbers and the function $f(k, u_0, u_1, \cdots, u_{n-1})$ is continuous on the compact set: $\mathbb{N}(a, b-1) \times D_0$, where

$$D_0 = \{(u_0, u_1, \cdots, u_{n-1}) : |u_i| \le 2M_i, \quad 0 \le i \le n-1\}$$

and $\max\limits_{\mathbb{N}(a,b-1)\times D_0} |f(k,u_0,u_1,\cdots,u_{n-1})| \le Q$

(ii) $\max\limits_{\mathbb{N}(a,b-1+n-i)} |\Delta^i P_{n-1}(k)| \le M_i,\ 0 \le i \le n-1$ where $P_{n-1}(k)$ is the two point right focal interpolating polynomial defined in (10.8.21)

(iii) $C_{n,i}Q \le M_i,\ 0 \le i \le n-1$ where $C_{n,i}$ are defined in (10.10.10) and (10.10.12).

Then, (1.2.6), (10.8.20) has a solution in $D_0$.

**Proof.** For the problem (1.2.6), (10.8.20) equations corresponding to (11.1.1) and (11.1.2) are

$$(11.1.3)\quad u(k) = P_{n-1}(k) + \sum_{\ell=a}^{b-1} g(k,\ell) f\left(\ell, u(\ell), \Delta u(\ell), \cdots, \Delta^{n-1}u(\ell)\right)$$

and

$$(11.1.4)\quad Tu(k) = P_{n-1}(k) + \sum_{\ell=a}^{b-1} g(k,\ell) f\left(\ell, u(\ell), \Delta u(\ell), \cdots, \Delta^{n-1}u(\ell)\right),$$

where $g(k,\ell)$ is the Green's function of the two point right focal boundary value problem (10.1.3), (10.9.24) defined in (10.9.25). The space $S(a,b-1+n)$ we shall equip with the norm $\|u\| = \max\left\{\|\Delta^i u(k)\|,\ 0 \le i \le n-1\right\}$ where $\|\Delta^i u(k)\| = \max\limits_{\mathbb{N}(a,b-1+n-i)} |\Delta^i u(k)|$. The set $S_1 = \{u(k) \in S(a,b-1+n) : \|\Delta^i u(k)\| \le 2M_i,\ 0 \le i \le n-1\}$ is a closed convex subset of the Banach space $S(a,b-1+n)$, and as in Theorem 11.1.1 in view of (11.1.4) and (10.12.8) for any $u(k) \in S_1$ it follows that

$$\begin{aligned}\|\Delta^i Tu(k)\| &\le \max_{\mathbb{N}(a,b-1+n-i)} |\Delta^i P_{n-1}(k)| + C_{n,i}Q \\ &\le 2M_i,\quad 0 \le i \le n-1\end{aligned}$$

from which the conclusion is immediate. ∎

**Theorem 11.1.5.** With respect to the $(n,p)$ boundary value problem (1.2.7), (1.5.12) assume that

(i) condition (i) of Theorem 11.1.1

(ii) condition (ii) of Theorem 11.1.1 with $P_{n-1}(k)$ as the $(n,p)$ interpolating polynomial defined in (10.8.22)

(iii) $D_{n,0}Q \le M$, where $D_{n,0}$ is defined in (10.10.14).

Then, (1.2.7), (1.5.12) has a solution in $D_0$.

**Theorem 11.1.6.** With respect to the $(p,n)$ boundary value problem (1.2.7), (1.5.13) assume that

(i) condition (i) of Theorem 11.1.1

(ii) condition (ii) of Theorem 11.1.1 with $P_{n-1}(k)$ as the $(p,n)$ interpolating polynomial defined in (10.8.23)

(iii) $E_{n,0}Q \le M$, where $E_{n,0}$ is defined in (10.10.16).

Then, (1.2.7), (1.5.13) has a solution in $D_0$.

**Theorem 11.1.7.** With respect to the Lidstone boundary value problem (1.2.7) with $n = 2m$, (1.5.14) assume that

(i) condition (i) of Theorem 11.1.2

(ii) condition (ii) of Theorem 11.1.2 with $P_{2m-1}(k)$ as the Lidstone interpolating polynomial defined in (10.9.47)

(iii) $\left(\frac{1}{8}\right)^m \prod_{i=1}^{m}(b+2i-1-a)^2 Q \le M.$

Then, (1.2.7) with $n = 2m$, (1.5.14) has a solution in $D_0$.

Hereafter, we shall prove results only for the two point right focal boundary value problem (1.2.6), (10.8.20) whereas, for the other problems analogous results can easily be stated.

**Theorem 11.1.8.** Suppose that the function $f(k,u_0,u_1,\cdots,u_{n-1})$ is continuous and on $\mathbb{N}(a,b-1)\times\mathbb{R}^n$

$$|f(k,u_0,u_1,\cdots,u_{n-1})| \le \lambda + \sum_{i=0}^{n-1}\lambda_i|u_i|^{\alpha(i)}, \tag{11.1.5}$$

where $0 \le \alpha(i) < 1$, $\lambda$ and $\lambda_i$, $0 \le i \le n-1$ are nonnegative constants. Then, (1.2.6), (10.8.20) has a solution.

**Proof.** We shall show that the conditions of Theorem 11.1.4 are satisfied. For this, the inequality (11.1.5) implies that on $\mathbb{N}(a,b-1)\times D_0$

$$|f(k,u_0,u_1,\cdots,u_{n-1})| \le \lambda + \sum_{i=0}^{n-1}\lambda_i(2M_i)^{\alpha(i)} = Q_1 \text{ (say)}.$$

Thus, it suffices to choose $M_i$, $0 \le i \le n-1$ so large that condition (ii) of Theorem 11.1.4 holds and $C_{n,i}Q_1 \le M_i$, $0 \le i \le n-1$. ∎

Theorem 11.1.4 is a local existence result whereas Theorem 11.1.8 does not require any condition on the constants $C_{n,i}$ or the boundary conditions.

The question: what happens if $\alpha(i) = 1,\ 0 \le i \le n-1$ in (11.1.5) is considered in the next result.

**Theorem 11.1.9.** Suppose that the function $f(k, u_0, u_1, \cdots, u_{n-1})$ is continuous and on $\mathbb{N}(a, b-1) \times D_1$

$$|f(k, u_0, u_1, \cdots, u_{n-1})| \le \lambda + \sum_{i=0}^{n-1} \lambda_i |u_i|, \tag{11.1.6}$$

where

$$D_1 = \left\{ (u_0, u_1, \cdots, u_{n-1}) : |u_i| \le \max_{\mathbb{N}(a,b-1+n-i)} \left|\Delta^i P_{n-1}(k)\right| + C_{n,i} \frac{\lambda + c}{1-\theta},\ 0 \le i \le n-1 \right\}$$

and

$$c = \sum_{i=0}^{n-1} \lambda_i \max_{\mathbb{N}(a,b-1+n-i)} \left|\Delta^i P_{n-1}(k)\right|,$$

$$\theta = \sum_{i=0}^{n-1} C_{n,i} \lambda_i < 1.$$

Then, (1.2.6), (10.8.20) has a solution in $D_1$.

**Proof.** The boundary value problem (1.2.6), (10.8.20) can be written as

$$\Delta^n v(k) = f\left(k, v(k) + P_{n-1}(k), \Delta v(k) + \Delta P_{n-1}(k), \cdots, \Delta^{n-1} v(k) + \Delta^{n-1} P_{n-1}(k)\right) \tag{11.1.7}$$

$$\begin{aligned} \Delta^i v(a) &= 0, \quad 0 \le i \le p-1 \\ \Delta^i v(b) &= 0, \quad p \le i \le n-1. \end{aligned} \tag{11.1.8}$$

We define $S_2(a, b-1+n)$ as the space of all real functions defined on $\mathbb{N}(a, b-1+n)$ satisfying the boundary conditions (11.1.8). If we introduce in $S_2(a, b-1+n)$ the norm $\|v\| = \max_{\mathbb{N}(a,b-1)} |\Delta^n v(k)|$, then it becomes a Banach space. We shall show that the mapping $T : S_2(a, b-1+n) \to S_2(a, b-1+n)$ defined by

$$Tv(k) = \sum_{\ell=a}^{b-1} g(k,\ell) f(\ell, v(\ell) + P_{n-1}(\ell), \cdots) \tag{11.1.9}$$

maps the ball $S_3 = \left\{ v(k) \in S_2(a, b-1+n) : \|v\| \le \frac{\lambda + c}{1-\theta} \right\}$ into itself.

For this, let $v(k) \in S_3$ then from Theorem 10.12.6 on $\mathbb{N}(a, b-1+n-i)$, we have

$$|\Delta^i v(k)| \ \le \ C_{n,i}\frac{\lambda + c}{1-\theta}, \quad 0 \le i \le n-1$$

and hence on $\mathbb{N}(a, b-1+n-i)$

$$|\Delta^i v(k) + \Delta^i P_{n-1}(k)| \le \max_{\mathbb{N}(a,b-1+n-i)} |\Delta^i P_{n-1}(k)| + C_{n,i}\frac{\lambda+c}{1-\theta}, \ 0 \le i \le n-1$$

which implies that $(k, v(k) + P_{n-1}(k), \Delta v(k) + \Delta P_{n-1}(k), \cdots, \Delta^{n-1}v(k) + \Delta^{n-1}P_{n-1}(k)) \in \mathbb{N}(a, b-1) \times D_1$.

Further, from (11.1.9) we have

$$\|Tv\| \ = \ \max_{\mathbb{N}(a,b-1)} |f(k, v(k) + P_{n-1}(k), \cdots)|$$

and hence in view of (11.1.6) it follows that

$$\begin{aligned} \|Tv\| \ &\le \ \lambda + \sum_{i=0}^{n-1} \lambda_i \max_{\mathbb{N}(a,b-1+n-i)} |\Delta^i v(k) + \Delta^i P_{n-1}(k)| \\ &\le \ \lambda + c + \sum_{i=0}^{n-1} \lambda_i C_{n,i}\frac{\lambda+c}{1-\theta} \\ &= \ \lambda + c + \theta\frac{\lambda+c}{1-\theta} \ = \ \frac{\lambda+c}{1-\theta}. \end{aligned}$$

Thus, the operator $T$ has a fixed point in $S_3$. This fixed point $v(k)$ is a solution of (11.1.7), (11.1.8) and hence the problem (1.2.6), (10.8.20) has a solution $u(k) = v(k) + P_{n-1}(k)$. ∎

**Theorem 11.1.10.** Suppose that the boundary value problem (1.2.6), (10.9.24) has a nontrivial solution $u(k)$ and the condition (11.1.6) with $\lambda = 0$ is satisfied on $\mathbb{N}(a, b-1) \times D_2$, where

$$D_2 \ = \ \{(u_0, u_1, \cdots, u_{n-1}) : \ |u_i| \le C_{n,i}M, \quad 0 \le i \le n-1\}$$

and $M = \max_{\mathbb{N}(a,b-1)} |\Delta^n u(k)|$. Then, it is necessary that $\theta \ge 1$.

**Proof.** Since $u(k)$ is a nontrivial solution of (1.2.6), (10.9.24) it is necessary that $M \ne 0$, and Theorem 10.12.6 implies that $(k, u(k), \Delta u(k), \cdots,$

$\Delta^{n-1}u(k)) \in \mathbb{N}(a,b-1)\times D_2$. Thus, we have

$$\begin{aligned} M = \max_{\mathbb{N}(a,b-1)} |\Delta^n u(k)| &= \max_{\mathbb{N}(a,b-1)} \left| f\left(k, u(k), \Delta u(k), \cdots, \Delta^{n-1}u(k)\right) \right| \\ &\le \sum_{i=0}^{n-1} \lambda_i \max_{\mathbb{N}(a,b-1+n-i)} |\Delta^i u(k)| \\ &\le \sum_{i=0}^{n-1} \lambda_i C_{n,i} M \;=\; \theta M \end{aligned}$$

and hence $\theta \ge 1$. ∎

Conditions of Theorem 11.1.10 ensure that in (11.1.6) at least one of the $\lambda_i$, $0 \le i \le n-1$ will not be zero, otherwise on $\mathbb{N}(a, b-1+n)$ the solution $u(k)$ will coincide with a polynomial of degree at most $n-1$ and will not be a nontrivial solution of (1.2.6), (10.9.24). Further, $u(k) \equiv 0$ is obviously a solution of (1.2.6), (10.9.24), and if $\theta < 1$ then it is also unique.

**Theorem 11.1.11.** Suppose that for all $(k, u_0, u_1, \cdots, u_{n-1})$, $(k, v_0, v_1, \cdots, v_{n-1}) \in \mathbb{N}(a, b-1)\times D_1$ the function $f$ satisfies the Lipschitz condition

(11.1.10) $$|f(k, u_0, u_1, \cdots, u_{n-1}) - f(k, v_0, v_1, \cdots, v_{n-1})| \le \sum_{i=0}^{n-1} \lambda_i |u_i - v_i|,$$

where $\lambda = \max_{\mathbb{N}(a,b-1)} |f(k, 0, 0, \cdots, 0)|$. Then, the boundary value problem (1.2.6), (10.8.20) has a unique solution in $D_1$.

**Proof.** Lipschitz condition (11.1.10) in particular implies (11.1.6) and the continuity of $f$ on $\mathbb{N}(a, b-1) \times D_1$, therefore the existence of a solution of (1.2.6), (10.8.20) follows from Theorem 11.1.9. To show the uniqueness let $u(k)$ and $v(k)$ be two solutions of (1.2.6), (10.8.20) in $D_1$. Then, in view of (11.1.3) and Theorem 10.12.6 it follows that

$$\begin{aligned} |\Delta^n(u(k) - v(k))| &\le \max_{\mathbb{N}(a,b-1)} \sum_{i=0}^{n-1} \lambda_i \left|\Delta^i(u(k) - v(k))\right| \\ &\le \sum_{i=0}^{n-1} \lambda_i C_{n,i} \left|\Delta^n(u(k) - v(k))\right| \\ &= \theta \left|\Delta^n(u(k) - v(k))\right|. \end{aligned}$$

Since $\theta < 1$, we find that $\Delta^n(u(k) - v(k)) = 0$, $k \in \mathbb{N}(a, b-1)$. But, then $u(k) = v(k)$, $k \in \mathbb{N}(a, b-1+n)$ follows from the boundary conditions (10.8.20). ∎

## 11.2. Picard's and Approximate Picard's Methods

In Sections 9.2 and 9.3 Picard's and Approximate Picard's methods have been successfully used to construct the solutions of the boundary value problems for the nonlinear systems. These methods have an important characteristic, that bounds of the difference between iterates and the solution are easily available. In this section we shall discuss these methods only for the boundary value problem (1.2.6), (10.8.20). For other problems analogous results can be stated without much difficulty. For this, we need

**Definition 11.2.1.** A function $\overline{u}(k)$ defined on $\mathbb{N}(a, b-1+n)$ is called an *approximate solution* of (1.2.6), (10.8.20) if there exist $\delta$ and $\epsilon$ nonnegative constants such that

$$\max_{\mathbb{N}(a,b-1)} \left|\Delta^n \overline{u}(k) - f\left(k, \overline{u}(k), \Delta\overline{u}(k), \cdots, \Delta^{n-1}\overline{u}(k)\right)\right| \leq \delta \tag{11.2.1}$$

and

$$\max_{\mathbb{N}(a,b-1+n-i)} \left|\Delta^i P_{n-1}(k) - \Delta^i \overline{P}_{n-1}(k)\right| \leq \epsilon C_{n,i}, \quad 0 \leq i \leq n-1 \tag{11.2.2}$$

where $P_{n-1}(k)$ and $\overline{P}_{n-1}(k)$ are the two point right focal interpolating polynomials satisfying (10.8.20) and

$$\begin{aligned} \Delta^i \overline{P}_{n-1}(a) &= \Delta^i \overline{u}(a), \quad 0 \leq i \leq p-1 \\ \Delta^i \overline{P}_{n-1}(b) &= \Delta^i \overline{u}(b), \quad p \leq i \leq n-1 \end{aligned} \tag{11.2.3}$$

respectively, and the constants $C_{n,i}$ are defined in (10.10.10) and (10.10.12).

Inequality (11.2.1) means that there exists a function $\eta(k)$, $k \in \mathbb{N}(a, b-1)$ such that

$$\Delta^n \overline{u}(k) = f\left(k, \overline{u}(k), \Delta\overline{u}(k), \cdots, \Delta^{n-1}\overline{u}(k)\right) + \eta(k), \quad k \in \mathbb{N}(a, b-1)$$

where $\max_{\mathbb{N}(a,b-1)} |\eta(k)| \leq \delta$. Thus, the approximate solution $\overline{u}(k)$ can be expressed as

$$\overline{u}(k) = \overline{P}_{n-1}(k) + \sum_{\ell=a}^{b-1} g(k,\ell)\left[f\left(\ell, \overline{u}(\ell), \Delta\overline{u}(\ell), \cdots, \Delta^{n-1}\overline{u}(\ell)\right) + \eta(\ell)\right]. \tag{11.2.4}$$

In what follows, we shall consider the Banach space $S(a, b-1+n)$ and for $u(k) \in S(a, b-1+n)$ the norm is $\|u\| = \max\{\|\Delta^i u(k)\|/C_{n,i},\ 0 \leq i \leq n-1\}$.

**Theorem 11.2.1.** With respect to the boundary value problem (1.2.6), (10.8.20) assume that there exists an approximate solution $\overline{u}(k)$ and

(i) the function $f$ satisfies the Lipschitz condition (11.1.10) on $\mathbb{N}(a,b-1)\times D_3$, where

$$D_3=\Big\{(u_0,u_1,\cdots,u_{n-1}):\ \big|u_i-\Delta^i\overline{u}(k)\big|\le\mu C_{n,i},\ \ k\in\mathbb{N}(a,b-1+n-i),\ \ 0\le i\le n-1\Big\}$$

(ii) $\theta<1$

(iii) $(1-\theta)^{-1}(\epsilon+\delta)\le\mu$.

Then, the following hold

(1) there exists a solution $u^*(k)$ of (1.2.6), (10.8.20) in $\overline{S}(\overline{u},\mu_0)$

(2) $u^*(k)$ is the unique solution of (1.2.6), (10.8.20) in $\overline{S}(\overline{u},\mu)$

(3) the *Picard iterative sequence* $\{u_m(k)\}$ defined by

$$\begin{aligned}u_{m+1}(k)&=P_{n-1}(k)+\sum_{\ell=a}^{b-1}g(k,\ell)f\left(\ell,u_m(\ell),\cdots,\Delta^{n-1}u_m(\ell)\right)\\ u_0(k)&=\overline{u}(k),\quad m=0,1,\cdots\end{aligned}\tag{11.2.5}$$

converges to $u^*(k)$ with $\|u^*-u_m\|\le\theta^m\mu_0$

(4) for $u_0(k)=u(k)\in\overline{S}(\overline{u},\mu_0)$ the iterative process (11.2.5) converges to $u^*(k)$

(5) any sequence $\{\overline{u}_m(k)\}$ such that $\overline{u}_m(k)\in\overline{S}(u_m,\theta^m\mu_0)$, $m=0,1,\cdots$ converges to $u^*(k)$ where $\mu_0=(1-\theta)^{-1}\|u_1-\overline{u}\|$.

**Proof.** We shall show that the operator $T:\overline{S}(\overline{u},\mu)\to S(a,b-1+n)$ defined in (11.1.4) satisfies the conditions of Theorem 9.1.2. Let $u(k)\in\overline{S}(\overline{u},\mu)$, then from the definition of norm, we have

$$\|u-\overline{u}\|=\max\left\{\max_{\mathbb{N}(a,b-1+n-i)}\left|\Delta^i u(k)-\Delta^i\overline{u}(k)\right|/C_{n,i},\ 0\le i\le n-1\right\}\le\mu,$$

which implies that $\left|\Delta^i u(k)-\Delta^i\overline{u}(k)\right|\le\mu C_{n,i}$. $k\in\mathbb{N}(a,b-1+n-i)$, $0\le i\le n-1$. Thus, $\left(u(k),\Delta u(k),\cdots,\Delta^{n-1}u(k)\right)\in D_3$. Further, if $u(k),\ v(k)\in\overline{S}(\overline{u},\mu)$, then $Tu(k)-Tv(k)$ satisfies the conditions of

Theorem 10.12.6 with $P_{n-1}(k) \equiv 0$, and we get

$$\begin{aligned}\left|\Delta^j Tu(k) - \Delta^j Tv(k)\right| &\le C_{n,j} \max_{\mathbb{N}(a,b-1)} |f(k,u(k),\cdots) - f(k,v(k),\cdots)| \\ &\le C_{n,j} \sum_{i=0}^{n-1} \lambda_i \max_{\mathbb{N}(a,b-1+n-i)} \left|\Delta^i u(k) - \Delta^i v(k)\right| \\ &\le C_{n,j} \sum_{i=0}^{n-1} \lambda_i C_{n,i} \|u - v\|, \quad 0 \le j \le n-1\end{aligned}$$

and hence

$$\left|\Delta^j Tu(k) - \Delta^j Tv(k)\right| / C_{n,j} \le \theta \|u - v\|, \quad 0 \le j \le n-1$$

from which it follows that $\|Tu - Tv\| \le \theta \|u - v\|$.

Next, from (11.1.4) and (11.2.4), we have

$$\begin{aligned}T\overline{u}(k) - \overline{u}(k) &= Tu_0(k) - u_0(k) \\ &= P_{n-1}(k) - \overline{P}_{n-1}(k) - \sum_{\ell=a}^{b-1} g(k,\ell)\eta(\ell).\end{aligned} \tag{11.2.6}$$

The function $w(k) = -\sum_{\ell=a}^{b-1} g(k,\ell)\eta(\ell)$ satisfies the conditions of Theorem 10.12.6 with $P_{n-1}(k) \equiv 0$, and $\Delta^n w(k) = -\eta(k)$, thus

$$\max_{\mathbb{N}(a,b-1)} |\Delta^n w(k)| = \max_{\mathbb{N}(a,b-1)} |\eta(k)| \le \delta$$

and hence

$$\left|\Delta^j w(k)\right| \le C_{n,j}\delta, \quad 0 \le j \le n-1.$$

Using these inequalities and (11.2.2) in (11.2.6), we obtain

$$\left|\Delta^j Tu_0(k) - \Delta^j u_0(k)\right| \le (\epsilon + \delta) C_{n,j}, \quad 0 \le j \le n-1$$

which is the same as

$$\left|\Delta^j Tu_0(k) - \Delta^j u_0(k)\right| / C_{n,j} \le (\epsilon + \delta), \quad 0 \le j \le n-1$$

and hence $\|Tu_0 - u_0\| \le (\epsilon + \delta)$. Thus, from the hypothesis (iii) it follows that $(1-\theta)^{-1}\|Tu_0 - u_0\| \le (1-\theta)^{-1}(\epsilon+\delta) \le \mu$.

Hence, the conditions of Theorem 9.1.2 are satisfied and conclusions (1) (5) follow. ∎

In Theorem 11.2.1 the conclusion (3) ensures that the sequence $\{u_m(k)\}$ obtained from (11.2.5) converges to the solution $u^*(k)$ of (1.2.6), (10.8.20).

However, in practical evaluation this sequence is approximated by the computed sequence, say, $\{v_m(k)\}$. To find $v_{m+1}(k)$ the function $f$ is approximated by $f_m$. Therefore, the computed sequence $\{v_m(k)\}$ satisfies the recurrence relation

$$
\begin{aligned}
v_{m+1}(k) &= P_{n-1}(k) + \sum_{\ell=a}^{b-1} g(k,\ell) f_m\left(\ell, v_m(\ell), \cdots, \Delta^{n-1} v_m(\ell)\right) \\
v_0(k) = u_0(k) &= \overline{u}(k), \quad m = 0, 1, \cdots.
\end{aligned}
\tag{11.2.7}
$$

With respect to $f_m$, we shall assume the following:

**Condition** $(c_1)$. For all $k \in \mathbb{N}(a, b-1)$ and $\Delta^i v_m(k)$, $0 \le i \le n-1$ obtained from (11.2.7) the following inequality is satisfied

$$
|f(k, v_m(k), \cdots) - f_m(k, v_m(k), \cdots)| \le \nu |f(k, v_m(k), \cdots)|, \tag{11.2.8}
$$

where $\nu$ is a nonnegative constant.

Inequality (11.2.8) corresponds to the relative error in approximating $f$ by $f_m$ for the $(m+1)$th iteration.

**Theorem 11.2.2.** With respect to the boundary value problem (1.2.6), (10.8.20) assume that there exists an approximate solution $\overline{u}(k)$ and the condition $(c_1)$ is satisfied. Further, assume that

(i) condition (i) of Theorem 11.2.1

(ii) $\theta_1 = (1+\nu)\theta < 1$

(iii) $\mu_1 = (1-\theta_1)^{-1}(\epsilon + \delta + \nu F) \le \mu$, where

$$
F = \max_{\mathbb{N}(a,b-1)} \left| f(k, \overline{u}(k), \Delta \overline{u}(k), \cdots, \Delta^{n-1} \overline{u}(k)) \right|.
$$

Then, the following hold

(1) all the conclusions (1) (5) of Theorem 11.2.1 are valid

(2) the sequence $\{v_m(k)\}$ obtained from (11.2.7) remains in $\overline{S}(\overline{u}, \mu_1)$

(3) the sequence $\{v_m(k)\}$ converges to $u^*(k)$, the solution of (1.2.6), (10.8.20) if and only if $\lim_{m\to\infty} w_m = 0$, where

(11.2.9)

$$
w_m = \left\| v_{m+1}(k) - P_{n-1}(k) - \sum_{\ell=a}^{b-1} g(k,\ell) f\left(\ell, v_m(\ell), \cdots, \Delta^{n-1} v_m(\ell)\right) \right\|
$$

and

(11.2.10)

$$\|u^* - v_{m+1}\| \leq (1-\theta)^{-1}\left[\theta\|v_{m+1} - v_m\| + \nu \max_{\mathbb{N}(a,b-1)} |f(k, v_m(k), \cdots)|\right].$$

**Proof.** Since $\theta_1 < 1$ implies $\theta < 1$ and obviously $\mu_0 \leq \mu_1$, conditions of Theorem 11.2.1 are satisfied and conclusion (1) follows.

To prove (2), we note that $\overline{u}(k) \in \overline{S}(\overline{u}, \mu_1)$ and from (11.2.4) and (11.2.7), we find

$$\begin{aligned} v_1(k) - \overline{u}(k) &= P_{n-1}(k) - \overline{P}_{n-1}(k) + \sum_{\ell=a}^{b-1} g(k,\ell)\,[f_0(\ell, \overline{u}(\ell), \cdots) \\ &\qquad - f(\ell, \overline{u}(\ell), \cdots) - \eta(\ell)]. \end{aligned}$$

Thus, from Theorem 10.12.6, we get

$$\left|\Delta^j v_1(k) - \Delta^j \overline{u}(k)\right| \leq (\epsilon+\delta)C_{n,j} + C_{n,j}\nu F, \quad 0 \leq j \leq n-1$$

and hence

$$\|v_1 - \overline{u}\| \leq (\epsilon + \delta + \nu F) \leq \mu_1.$$

Now we assume that $v_m(k) \in \overline{S}(\overline{u}, \mu_1)$ and will show that $v_{m+1}(k) \in \overline{S}(\overline{u}, \mu_1)$. From (11.2.4) and (11.2.7), we have

$$\begin{aligned} v_{m+1}(k) - \overline{u}(k) &= P_{n-1}(k) - \overline{P}_{n-1}(k) + \sum_{\ell=a}^{b-1} g(k,\ell)[f_m(\ell, v_m(\ell), \cdots) \\ &\qquad - f(\ell, \overline{u}(\ell), \cdots) - \eta(\ell)] \end{aligned}$$

and Theorem 10.12.6 provides

$$\begin{aligned} &\left|\Delta^j v_{m+1}(k) - \Delta^j \overline{u}(k)\right| \\ &\quad \leq (\epsilon+\delta)C_{n,j} + C_{n,j} \max_{\mathbb{N}(a,b-1)} [|f_m(k, v_m(k), \cdots) - f(k, v_m(k), \cdots)| \\ &\qquad + |f(k, v_m(k), \cdots) - f(k, \overline{u}(k), \cdots)|] \\ &\quad \leq C_{n,j}\left[\epsilon+\delta+\nu F+(1+\nu) \max_{\mathbb{N}(a,b-1)} |f(k, v_m(k), \cdots) - f(k, \overline{u}(k), \cdots)|\right] \\ &\quad \leq C_{n,j}\left[\epsilon+\delta+\nu F+(1+\nu)\sum_{i=0}^{n-1}\lambda_i \max_{\mathbb{N}(a,b-1+n-i)} \left|\Delta^i v_m(k) - \Delta^i \overline{u}(k)\right|\right] \\ &\quad \leq C_{n,j}\,[\epsilon+\delta+\nu F+(1+\nu)\theta\|v_m - \overline{u}\|], \quad 0 \leq j \leq n-1. \end{aligned}$$

Hence, we get

$$\left|\Delta^j v_{m+1}(k) - \Delta^j \overline{u}(k)\right|/C_{n,j} \leq (\epsilon+\delta+\nu F) + \theta_1\|v_m - \overline{u}\|, \quad 0 \leq j \leq n-1$$

which gives

$$\|v_{m+1} - \overline{u}\| \le (1-\theta_1)\mu_1 + \theta_1\mu_1 = \mu_1.$$

This completes the proof of (2).

From the definitions of $u_{m+1}(k)$ and $v_{m+1}(k)$, we have

$$\begin{aligned} u_{m+1}(k) - v_{m+1}(k) &= P_{n-1}(k) + \sum_{\ell=a}^{b-1} g(k,\ell) f(\ell, v_m(\ell), \cdots) - v_{m+1}(k) \\ &\quad + \sum_{\ell=a}^{b-1} g(k,\ell)[f(\ell, u_m(\ell), \cdots) - f(\ell, v_m(\ell), \cdots)] \end{aligned}$$

and hence, as earlier we find

$$\|u_{m+1} - v_{m+1}\| \le w_m + \theta\|u_m - v_m\|.$$

Since $u_0(k) = v_0(k)$, the above inequality provides

$$\|u_{m+1} - v_{m+1}\| \le \sum_{i=0}^{m} \theta^{m-i} w_i.$$

Thus, from the triangle inequality, we get

$$\|u^* - v_{m+1}\| \le \sum_{i=0}^{m} \theta^{m-i} w_i + \|u^* - u_{m+1}\|. \tag{11.2.11}$$

In (11.2.11), Theorem 11.2.1 ensures that $\lim_{m\to\infty} \|u^* - u_{m+1}\| = 0$. Thus, the condition $\lim_{m\to\infty} w_m = 0$ is necessary and sufficient for the convergence of the sequence $\{v_m(k)\}$ to $u^*(k)$ follows from the Toeplitz lemma.

Finally, to prove (11.2.10), we note that

$$\begin{aligned} u^*(k) - v_{m+1}(k) &= \sum_{\ell=a}^{b-1} g(k,\ell)[f(\ell, u^*(\ell), \cdots) - f(\ell, v_m(\ell), \cdots) \\ &\quad + f(\ell, v_m(\ell), \cdots) - f_m(\ell, v_m(\ell), \cdots)] \end{aligned}$$

and as earlier, we find

$$\begin{aligned} \|u^* - v_{m+1}\| &\le \theta\|u^* - v_m\| + \nu \max_{\mathbb{N}(a,b-1)} |f(k, v_m(k), \cdots)| \\ &\le \theta\|u^* - v_{m+1}\| + \theta\|v_{m+1} - v_m\| + \nu \max_{\mathbb{N}(a,b-1)} |f(k, v_m(k), \cdots)|, \end{aligned}$$

which is the same as (11.2.10). ∎

In our next result, we shall assume

**Condition** $(c_2)$. For all $k \in \mathbb{N}(a, b-1)$ and $\Delta^i v_m(k)$, $0 \le i \le n-1$

obtained from (11.2.7) the following inequality is satisfied

(11.2.12) $$|f(k, v_m(k), \cdots) - f_m(k, v_m(k), \cdots)| \leq \nu_1,$$

where $\nu_1$ is a nonnegative constant.

Inequality (11.2.12) corresponds to the absolute error in approximating $f$ by $f_m$ for the $(m+1)$th iteration.

**Theorem 11.2.3.** With respect to the boundary value problem (1.2.6), (10.8.20) assume that there exists an approximate solution $\overline{u}(k)$ and the condition $(c_2)$ is satisfied. Further, assume that

(i) condition (i) of Theorem 11.2.1

(ii) condition (ii) of Theorem 11.2.1

(iii) $\mu_2 = (1-\theta)^{-1}(\epsilon + \delta + \nu_1) \leq \mu$.

Then, the following hold

(1) all the conclusions (1) (5) of Theorem 11.2.1 are valid

(2) the sequence $\{v_m(k)\}$ obtained from (11.2.7) remains in $\overline{S}(\overline{u}, \mu_2)$

(3) the condition $\lim_{m\to\infty} w_m = 0$ is necessary and sufficient for the convergence of $\{v_m(k)\}$ to the solution $u^*(k)$ of (1.2.6), (10.8.20) where $w_m$ are defined in (11.2.9), and

$$\|u^* - v_{m+1}\| \leq (1-\theta)^{-1}\left[\theta\|v_{m+1} - v_m\| + \nu_1\right].$$

**Proof.** The proof is contained in Theorem 11.2.2. ■

## 11.3. Quasilinearization and Approximate Quasilinearization

Newton's method which has been used in Section 9.8 to solve boundary value problems for the nonlinear systems when applied to higher order differential equations has been labeled as quasilinearization. Here, once again we shall discuss this method only for the discrete boundary value problem (1.2.6), (10.8.20), whereas analogous results for the other problems can be stated easily. For this, following the notations and definitions of the previous section we shall provide sufficient conditions so that the sequence $\{u_m(k)\}$ generated by the quasilinear iterative scheme

(11.3.1) $$\Delta^n u_{m+1}(k) = f\left(k, u_m(k), \Delta u_m(k), \cdots, \Delta^{n-1} u_m(k)\right) + \sum_{i=0}^{n-1}\left(\Delta^i u_{m+1}(k) - \Delta^i u_m(k)\right)\frac{\partial}{\partial \Delta^i u_m(k)} f(k, u_m(k), \cdots)$$

$$\begin{aligned} \Delta^i u_{m+1}(a) &= A_i, \quad 0 \le i \le p-1 \\ \Delta^i u_{m+1}(b) &= A_i, \quad p \le i \le n-1, \quad m = 0, 1, \cdots \end{aligned} \tag{11.3.2}$$

with $u_0(k) = \overline{u}(k)$, converges to the unique solution $u^*(k)$ of the boundary value problem (1.2.6), (10.8.20).

**Theorem 11.3.1.** With respect to the boundary value problem (1.2.6), (10.8.20) assume that there exists an approximate solution $\overline{u}(k)$ and

(i) the function $f(k, u_0, u_1, \cdots, u_{n-1})$ is continuously differentiable with respect to all $u_i$, $0 \le i \le n-1$ on $\mathbb{N}(a, b-1) \times D_3$

(ii) there exist $\lambda_i$, $0 \le i \le n-1$ nonnegative constants such that for all $(k, u_0, u_1, \cdots, u_{n-1}) \in \mathbb{N}(a, b-1) \times D_3$

$$\left| \frac{\partial}{\partial u_i} f(k, u_0, u_1, \cdots, u_{n-1}) \right| \le \lambda_i, \quad 0 \le i \le n-1$$

(iii) $3\theta < 1$

(iv) $\mu_3 = (1-3\theta)^{-1}(\epsilon + \delta) \le \mu$.

Then, the following hold

(1) the sequence $\{u_m(k)\}$ generated by the process (11.3.1), (11.3.2) remains in $\overline{S}(\overline{u}, \mu_3)$

(2) the sequence $\{u_m(k)\}$ converges to the unique solution $u^*(k)$ of (1.2.6), (10.8.20)

(3) a bound on the error is given by

$$\|u_m - u^*\| \le \left(\frac{2\theta}{1-\theta}\right)^m \left(1 - \frac{2\theta}{1-\theta}\right)^{-1} \|u_1 - \overline{u}\| \tag{11.3.3}$$

$$\le \left(\frac{2\theta}{1-\theta}\right)^m \left(1 - \frac{2\theta}{1-\theta}\right)^{-1} (1-\theta)^{-1}(\epsilon + \delta). \tag{11.3.4}$$

**Proof.** First, we shall show that the sequence $\{u_m(k)\}$ remains in $\overline{S}(\overline{u}, \mu_3)$. We define an implicit operator $T$ as follows

$$\begin{aligned} Tu(k) = {} & P_{n-1}(k) + \sum_{\ell=a}^{b-1} g(k, \ell) \Big[ f(\ell, u(\ell), \cdots) \\ & + \sum_{i=0}^{n-1} (\Delta^i Tu(\ell) - \Delta^i u(\ell)) \frac{\partial}{\partial \Delta^i u(\ell)} f(\ell, u(\ell), \cdots) \Big] \end{aligned} \tag{11.3.5}$$

whose form is patterned on the summation equation representation of (11. 3.1), (11.3.2).

Since $\overline{u}(k) \in \overline{S}(\overline{u}, \mu_3)$, it is sufficient to show that if $u(k) \in \overline{S}(\overline{u}, \mu_3)$, then $Tu(k) \in \overline{S}(\overline{u}, \mu_3)$. For this, if $u(k) \in \overline{S}(\overline{u}, \mu_3)$ then $(u(k), \Delta u(k), \cdots, \Delta^{n-1}u(k)) \in D_3$ and from (11.2.4) and (11.3.5), we have

$$Tu(k) - \overline{u}(k) = P_{n-1}(k) - \overline{P}_{n-1}(k) + \sum_{\ell=a}^{b-1} g(k,\ell)\Bigg[f(\ell, u(\ell), \cdots)$$
$$+ \sum_{i=0}^{n-1} (\Delta^i Tu(\ell) - \Delta^i u(\ell)) \frac{\partial}{\partial \Delta^i u(\ell)} f(\ell, u(\ell), \cdots) - f(\ell, \overline{u}(\ell), \cdots) - \eta(\ell)\Bigg].$$

Thus, an application of Theorem 10.12.6 provides

$$|\Delta^j Tu(k) - \Delta^j \overline{u}(k)| \le \epsilon C_{n,j} + C_{n,j} \max_{\mathbb{N}(a,b-1)} \Bigg[|f(k, u(k), \cdots) -$$
$$f(k, \overline{u}(k), \cdots)| + \sum_{i=0}^{n-1} \lambda_i \left\{|\Delta^i Tu(k) - \Delta^i \overline{u}(k)| + |\Delta^i u(k) - \Delta^i \overline{u}(k)|\right\} + \delta\Bigg]$$

and hence, we get

$$|\Delta^j Tu(k) - \Delta^j \overline{u}(k)| / C_{n,j} \le (\epsilon + \delta) + \sum_{i=0}^{n-1} C_{n,i}\lambda_i \left[\|Tu - \overline{u}\| + 2\|u - \overline{u}\|\right],$$
$$0 \le j \le n-1.$$

From the above inequality, we find

$$\|Tu - \overline{u}\| \le (\epsilon + \delta) + \theta\|Tu - \overline{u}\| + 2\theta\|u - \overline{u}\|,$$

which gives

$$\|Tu - \overline{u}\| \le (1-\theta)^{-1}\left[(\epsilon + \delta) + 2\theta\mu_3\right].$$

Thus, $\|Tu - \overline{u}\| \le \mu_3$ follows from the definition of $\mu_3$.

Next, we shall show the convergence of the sequence $\{u_m(k)\}$. From (11.3.1), (11.3.2) we have

$$(11.3.6)\quad u_{m+1}(k) - u_m(k) = \sum_{\ell=a}^{b-1} g(k,\ell)\Bigg[f(\ell, u_m(\ell), \cdots) - f(\ell, u_{m-1}(\ell), \cdots)$$
$$+ \sum_{i=0}^{n-1} \Bigg\{(\Delta^i u_{m+1}(\ell) - \Delta^i u_m(\ell)) \frac{\partial}{\partial \Delta^i u_m(\ell)} f(\ell, u_m(\ell), \cdots)$$
$$- (\Delta^i u_m(\ell) - \Delta^i u_{m-1}(\ell)) \frac{\partial}{\partial \Delta^i u_{m-1}(\ell)} f(\ell, u_{m-1}(\ell), \cdots)\Bigg\}\Bigg].$$

Thus, from Theorem 10.12.6 and the fact that $\{u_m(k)\} \subseteq \overline{S}(\overline{u}, \mu_3)$, we get

$$\left|\Delta^j u_{m+1}(k) - \Delta^j u_m(k)\right| \le C_{n,j} \max_{\mathbb{N}(a,b-1)} \left[2\sum_{i=0}^{n-1} \lambda_i \left|\Delta^i u_m(k) - \Delta^i u_{m-1}(k)\right| + \sum_{i=0}^{n-1} \lambda_i \left|\Delta^i u_{m+1}(k) - \Delta^i u_m(k)\right|\right]$$

and hence

$$\left|\Delta^j u_{m+1}(k) - \Delta^j u_m(k)\right| / C_{n,j} \ \le\ 2\theta\|u_m - u_{m-1}\| + \theta\|u_{m+1} - u_m\|, \quad 0 \le j \le n-1$$

which provides

$$\|u_{m+1} - u_m\| \ \le\ 2\theta\|u_m - u_{m-1}\| + \theta\|u_{m+1} - u_m\|$$

or

$$\|u_{m+1} - u_m\| \ \le\ \frac{2\theta}{1-\theta}\|u_m - u_{m-1}\|$$

and by an easy induction, we get

$$\|u_{m+1} - u_m\| \ \le\ \left(\frac{2\theta}{1-\theta}\right)^m \|u_1 - \overline{u}\|. \tag{11.3.7}$$

Since $3\theta < 1$, inequality (11.3.7) implies that $\{u_m(k)\}$ is a Cauchy sequence and hence converges to some $u^*(k) \in \overline{S}(\overline{u}, \mu_3)$. This $u^*(k)$ is the unique solution of (1.2.6), (10.8.20) and can easily be verified.

The error bound (11.3.3) follows from (11.3.7) and the triangle inequality

$$\begin{aligned}
\|u_{m+p} - u_m\| &\le \|u_{m+p} - u_{m+p-1}\| + \cdots + \|u_{m+1} - u_m\| \\
&\le \left[\left(\frac{2\theta}{1-\theta}\right)^{m+p-1} + \cdots + \left(\frac{2\theta}{1-\theta}\right)^m\right] \|u_1 - \overline{u}\| \\
&\le \left(\frac{2\theta}{1-\theta}\right)^m \left(1 - \frac{2\theta}{1-\theta}\right)^{-1} \|u_1 - \overline{u}\|
\end{aligned}$$

and now taking $p \to \infty$.

Next, from (11.2.4), (11.3.1), (11.3.2) we have

$$u_1(k) - u_0(k) \ =\ P_{n-1}(k) - \overline{P}_{n-1}(k) - \sum_{\ell=a}^{b-1} g(k,\ell) \ \times \left[\sum_{i=0}^{n-1} \left(\Delta^i u_1(\ell) - \Delta^i u_0(\ell)\right) \frac{\partial}{\partial \Delta^i u_0(\ell)} f(\ell, u_0(\ell), \cdots) - \eta(\ell)\right]$$

and as earlier, we find

$$\|u_1 - u_0\| \;\le\; (1-\theta)^{-1}(\epsilon + \delta). \tag{11.3.8}$$

Using (11.3.8) in (11.3.3) the inequality (11.3.4) follows. ∎

**Theorem 11.3.2.** Let the conditions of Theorem 11.3.1 be satisfied. Further, let $f(k, u_0, u_1, \cdots, u_{n-1})$ be continuously twice differentiable with respect to all $u_i$, $0 \le i \le n-1$ on $\mathbb{N}(a, b-1) \times D_3$ and

$$\left|\frac{\partial^2}{\partial u_i \partial u_j} f(k, u_0, u_1, \cdots, u_{n-1})\right| \;\le\; \lambda_i \lambda_j \zeta, \quad 0 \le i,\ j \le n-1.$$

Then, the following hold

$$\|u_{m+1} - u_m\| \;\le\; \alpha\|u_m - u_{m-1}\|^2 \;\le\; \frac{1}{\alpha}\left(\alpha\|u_1 - u_0\|\right)^{2^m} \tag{11.3.9}$$

$$\le\; \frac{1}{\alpha}\left[\frac{1}{2}\zeta(\epsilon+\delta)\left(\frac{\theta}{1-\theta}\right)^2\right]^{2^m},$$

where $\alpha = \left(\zeta\theta^2/2(1-\theta)\right)$. Thus, the convergence is quadratic if $\frac{1}{2}\zeta(\epsilon + \delta)\left(\frac{\theta}{1-\theta}\right)^2 < 1.$

**Proof.** From $\{u_m(k)\} \subseteq \overline{S}(\overline{u}, \mu_3)$ it follows that for all $m$, $(u_m(k), \Delta u_m(k), \cdots, \Delta^{n-1}u_m(k)) \in D_3$. Further, since $f$ is twice continuously differentiable, we have

$$\begin{aligned} f(k, u_m(k), \cdots) \;=&\; f(k, u_{m-1}(k), \cdots) \\ &+ \sum_{i=0}^{n-1}\left(\Delta^i u_m(k) - \Delta^i u_{m-1}(k)\right)\frac{\partial}{\partial \Delta^i u_{m-1}(k)} f(k, u_{m-1}(k), \cdots) \\ &+ \frac{1}{2}\left[\sum_{i=0}^{n-1}\left(\Delta^i u_m(k) - \Delta^i u_{m-1}(k)\right)\frac{\partial}{\partial p_i(k)}\right]^2 f(k, p_0(k), \cdots, p_{n-1}(k)), \end{aligned} \tag{11.3.10}$$

where $p_i(k)$ lies between $\Delta^i u_{m-1}(k)$ and $\Delta^i u_m(k)$, $0 \le i \le n-1$.

Using (11.3.10) in (11.3.6), we get

$$
\begin{aligned}
&u_{m+1}(k) - u_m(k) \\
&= \sum_{\ell=a}^{b-1} g(k,\ell)\left\{ \sum_{i=0}^{n-1} \left(\Delta^i u_{m+1}(\ell) - \Delta^i u_m(\ell)\right) \frac{\partial}{\partial \Delta^i u_m(\ell)} f(\ell, u_m(\ell), \cdots) \right. \\
&\left. + \frac{1}{2}\left[\sum_{i=0}^{n-1} \left(\Delta^i u_m(\ell) - \Delta^i u_{m-1}(\ell)\right) \frac{\partial}{\partial p_i(\ell)}\right]^2 f(\ell, p_0(\ell), \cdots, p_{n-1}(\ell)) \right\}.
\end{aligned}
$$

Thus, Theorem 10.12.6 provides

$$
\begin{aligned}
\left|\Delta^j u_{m+1}(k) - \Delta^j u_m(k)\right| \;\leq\; & C_{n,j}\left[\sum_{i=0}^{n-1} \lambda_i C_{n,i} \|u_{m+1} - u_m\| \right. \\
& \left. + \frac{1}{2}\left(\sum_{i=0}^{n-1} \lambda_i C_{n,i}\right)^2 \zeta \|u_m - u_{m-1}\|^2 \right]
\end{aligned}
$$

and hence

$$
\|u_{m+1} - u_m\| \;\leq\; \theta\|u_{m+1} - u_m\| + \frac{1}{2}\zeta\theta^2\|u_m - u_{m-1}\|^2,
$$

which is the same as the first part of the inequality (11.3.9). The second part of (11.3.9) follows by an easy induction. Finally, the last part is an application of (11.3.8). ∎

In Theorem 11.3.1 the conclusion (3) ensures that the sequence $\{u_m(k)\}$ generated from (11.3.1), (11.3.2) converges linearly to the unique solution $u^*(k)$ of the boundary value problem (1.2.6), (10.8.20). Theorem 11.3.2 provides sufficient conditions for its quadratic convergence. However, in practical evaluation this sequence is approximated by the computed sequence, say, $\{v_m(k)\}$ which satisfies the recurrence relation

$$
\begin{aligned}
\text{(11.3.11)}\quad \Delta^n v_{m+1}(k) \;=\; & f_m\left(k, v_m(k), \Delta v_m(k), \cdots, \Delta^{n-1} v_m(k)\right) \\
& + \sum_{i=0}^{n-1} \left(\Delta^i v_{m+1}(k) - \Delta^i v_m(k)\right) \frac{\partial}{\partial \Delta^i v_m(k)} f_m(k, v_m(k), \cdots)
\end{aligned}
$$

$$
\text{(11.3.12)}\quad
\begin{aligned}
\Delta^i v_{m+1}(a) &= A_i, \quad 0 \leq i \leq p-1 \\
\Delta^i v_{m+1}(b) &= B_i, \quad p \leq i \leq n-1, \quad m = 0, 1, \cdots
\end{aligned}
$$

where $v_0(k) = u_0(k) = \overline{u}(k)$.

With respect to $f_m$, we shall assume the following:

**Condition** $(d_1)$. (i) The function $f_m(k, u_0, u_1, \cdots, u_{n-1})$ is continuously differentiable with respect to all $u_i$, $0 \le i \le n-1$ on $\mathbb{N}(a, b-1) \times D_3$ and

$$\left| \frac{\partial}{\partial u_i} f_m(k, u_0, u_1, \cdots, u_{n-1}) \right| \le \lambda_i, \quad 0 \le i \le n-1$$

(ii) condition $(c_1)$ is satisfied.

**Theorem 11.3.3.** With respect to the boundary value problem (1.2.6), (10.8.20) assume that there exists an approximate solution $\overline{u}(k)$ and the condition $(d_1)$ is satisfied. Further, we assume

(i) conditions (i) and (ii) of Theorem 11.3.1

(ii) $\theta_2 = (3+\nu)\theta < 1$

(iii) $\mu_4 = (1-\theta_2)^{-1}(\epsilon + \delta + \nu F) \le \mu$, where

$$F = \max_{\mathbb{N}(a,b-1)} \left| f\left(k, \overline{u}(k), \Delta\overline{u}(k), \cdots, \Delta^{n-1}\overline{u}(k)\right) \right|.$$

Then, the following hold

(1) all the conclusions (1) (3) of Theorem 11.3.1 are valid

(2) the sequence $\{v_m(k)\}$ obtained from (11.3.11), (11.3.12) remains in $\overline{S}(\overline{u}, \mu_4)$

(3) the sequence $\{v_m(k)\}$ converges to $u^*(k)$ the solution of (1.2.6), (10.8.20) if and only if $\lim_{m\to\infty} w_m = 0$, where $w_m$ are defined in (11.2.9), and

$$(11.3.13) \quad \|u^* - v_{m+1}\| \le (1-\theta)^{-1} \left[ 2\theta \|v_{m+1} - v_m\| + \nu \max_{\mathbb{N}(a,b-1)} |f(k, v_m(k), \cdots)| \right].$$

**Proof.** Since $\theta_2 < 1$ implies $3\theta < 1$ and obviously $\mu_3 \le \mu_4$, the conditions of Theorem 11.3.1 are satisfied and part (1) follows.

To prove (2), we note that $\overline{u}(k) \in \overline{S}(\overline{u}, \mu_4)$ and from (11.2.4), (11.3.11),

(11.3.12) we have

$$\begin{aligned} v_1(k) - \overline{u}(k) &= P_{n-1}(k) - \overline{P}_{n-1}(k) + \sum_{\ell=a}^{b-1} g(k,\ell) \Bigg[ f_0(\ell, v_0(\ell), \cdots) \\ &\quad + \sum_{i=0}^{n-1} \left(\Delta^i v_1(\ell) - \Delta^i v_0(\ell)\right) \frac{\partial}{\partial \Delta^i v_0(\ell)} f_0(\ell, v_0(\ell), \cdots) \\ &\quad - f(\ell, v_0(\ell), \cdots) - \eta(\ell) \Bigg] \end{aligned}$$

and Theorem 10.12.6 provides

$$\|v_1 - \overline{u}\| \ \le \ (\epsilon + \delta + \nu F) + \theta \|v_1 - v_0\|$$

and hence

$$\|v_1 - \overline{u}\| \ \le \ (1-\theta)^{-1}(\epsilon + \delta + \nu F) \ \le \ \mu_4. \tag{11.3.14}$$

Thus, $v_1(k) \in \overline{S}(\overline{u}, \mu_4)$. Next, we assume that $v_m(k) \in \overline{S}(\overline{u}, \mu_4)$ and will show that $v_{m+1}(k) \in \overline{S}(\overline{u}, \mu_4)$. From (11.2.4), (11.3.11), (11.3.12) we have

$$\begin{aligned} v_{m+1}(k) - \overline{u}(k) &= P_{n-1}(k) - \overline{P}_{n-1}(k) + \sum_{\ell=a}^{b-1} g(k,\ell) \Bigg[ f_m(\ell, v_m(\ell), \cdots) \\ &\quad + \sum_{i=0}^{n-1} \left(\Delta^i v_{m+1}(\ell) - \Delta^i v_m(\ell)\right) \frac{\partial}{\partial \Delta^i v_m(\ell)} f_m(\ell, v_m(\ell), \cdots) \\ &\quad - f(\ell, v_0(\ell), \cdots) - \eta(\ell) \Bigg] \end{aligned}$$

and from Theorem 10.12.6, we get

$$\begin{aligned} &\left|\Delta^j v_{m+1}(k) - \Delta^j \overline{u}(k)\right| \\ &\quad \le (\epsilon + \delta) C_{n,j} + C_{n,j} \max_{\mathbf{N}(a,b-1)} \Bigg[ \sum_{i=0}^{n-1} \lambda_i \left|\Delta^i v_{m+1}(k) - \Delta^i v_m(k)\right| \\ &\qquad + (1+\nu)|f(k, v_m(k), \cdots) - f(k, v_0(k), \cdots)| + \nu |f(k, v_0(k), \cdots)| \Bigg] \end{aligned}$$

and hence, we find

$$\begin{aligned} \|v_{m+1} - \overline{u}\| &\le (\epsilon + \delta + \nu F) + \theta \|v_{m+1} - v_m\| + (1+\nu)\theta \|v_m - v_0\| \\ &\le (\epsilon + \delta + \nu F) + (2+\nu)\theta \|v_m - v_0\| + \theta \|v_{m+1} - v_0\|. \end{aligned}$$

From the last inequality, we obtain

$$\|v_{m+1} - \overline{u}\| \leq (1-\theta)^{-1}\left[(\epsilon + \delta + \nu F) + (2+\nu)\theta\mu_4\right] = \mu_4.$$

This completes the proof of part (2).

Next, from the definitions of $u_{m+1}(k)$ and $v_{m+1}(k)$, we have

$$\begin{aligned} u_{m+1}(k) - v_{m+1}(k) &= P_{n-1}(k) + \sum_{\ell=a}^{b-1} g(k,\ell) f(\ell, v_m(\ell), \cdots) - v_{m+1}(k) \\ &+ \sum_{\ell=a}^{b-1} g(k,\ell) \Bigg[ f(\ell, u_m(\ell), \cdots) - f(\ell, v_m(\ell), \cdots) \\ &+ \sum_{i=0}^{n-1} \left(\Delta^i u_{m+1}(\ell) - \Delta^i u_m(\ell)\right) \frac{\partial}{\partial \Delta^i u_m(\ell)} f(\ell, u_m(\ell), \cdots) \Bigg] \end{aligned}$$

and hence as earlier, we find

$$\|u_{m+1} - v_{m+1}\| \leq w_m + \theta\|u_m - v_m\| + \theta\|u_{m+1} - u_m\|. \tag{11.3.15}$$

Using (11.3.7) in (11.3.15), we get

$$\|u_{m+1} - v_{m+1}\| \leq w_m + \theta\|u_m - v_m\| + \theta\left(\frac{2\theta}{1-\theta}\right)^m \|u_1 - \overline{u}\|.$$

Since $u_0(k) = v_0(k) = \overline{u}(k)$, the above inequality provides

$$\|u_{m+1} - v_{m+1}\| \leq \sum_{i=0}^{m} \theta^{m-i}\left[w_i + \theta\left(\frac{2\theta}{1-\theta}\right)^i \|u_1 - \overline{u}\|\right]. \tag{11.3.16}$$

Using (11.3.16) in the triangle inequality, we obtain

$$\|v_{m+1} - u^*\| \leq \|u_{m+1} - u^*\| + \sum_{i=0}^{m} \theta^{m-i}\left[w_i + \theta\left(\frac{2\theta}{1-\theta}\right)^i \|u_1 - \overline{u}\|\right]. \tag{11.3.17}$$

In (11.3.17), Theorem 11.3.1 ensures that $\lim_{m\to\infty}\|u_{m+1} - u^*\| = 0$. Thus, from the Toeplitz lemma $\lim_{m\to\infty}\|v_{m+1} - u^*\| = 0$ if and only if $\lim_{m\to\infty}\left[w_m + \theta\left(\frac{2\theta}{1-\theta}\right)^m \|u_1 - \overline{u}\|\right] = 0$. However, $\lim_{m\to\infty}\left(\frac{2\theta}{1-\theta}\right)^m = 0$, and hence if and only if $\lim_{m\to\infty} w_m = 0$.

Finally, to prove (11.3.13) we note that

$$\begin{aligned} u^*(k) - v_{m+1}(k) &= \sum_{\ell=a}^{b-1} g(k,\ell) \Bigg[ f(\ell, u^*(\ell), \cdots) - f(\ell, v_m(\ell), \cdots) \\ &\quad + f(\ell, v_m(\ell), \cdots) - f_m(\ell, v_m(\ell), \cdots) \\ &\quad - \sum_{i=0}^{n-1} \left(\Delta^i v_{m+1}(\ell) - \Delta^i v_m(\ell)\right) \frac{\partial}{\partial \Delta^i v_m(\ell)} f_m(\ell, v_m(\ell), \cdots) \Bigg] \end{aligned}$$

and hence

$$\begin{aligned} \|u^* - v_{m+1}\| &\le \theta\|u^* - v_m\| + \theta\|v_{m+1} - v_m\| + \nu \max_{\mathbb{N}(a,b-1)} |f(k, v_m(k), \cdots)| \\ &\le 2\theta\|v_{m+1} - v_m\| + \nu \max_{\mathbb{N}(a,b-1)} |f(k, v_m(k), \cdots)| + \theta\|u^* - v_{m+1}\|, \end{aligned}$$

which is the same as (11.3.13). ∎

**Theorem 11.3.4.** Let the conditions of Theorem 11.3.3 be satisfied. Further, let $f_m = f_0$ for all $m = 1, 2, \cdots$ and $f_0(k, u_0, u_1, \cdots, u_{n-1})$ be continuously twice differentiable with respect to all $u_i$, $0 \le i \le n-1$ on $\mathbb{N}(a, b-1) \times D_3$ and

$$\left| \frac{\partial^2}{\partial u_i \partial u_j} f_0(k, u_0, u_1, \cdots, u_{n-1}) \right| \le \lambda_i \lambda_j \zeta, \quad 0 \le i,\ j \le n-1.$$

Then, the following hold

$$\begin{aligned} \|v_{m+1} - v_m\| &\le \alpha\|v_m - v_{m-1}\|^2 \le \frac{1}{\alpha}\left(\alpha\|v_1 - v_0\|\right)^{2^m} \\ &\le \frac{1}{\alpha}\left[\frac{1}{2}\zeta(\epsilon + \delta + \nu F)\left(\frac{\theta}{1-\theta}\right)^2\right]^{2^m}, \end{aligned} \tag{11.3.18}$$

where $\alpha$ is the same as in Theorem 11.3.2.

**Proof.** As in the proof of Theorem 11.3.2, we have

$$\begin{aligned} &v_{m+1}(k) - v_m(k) \\ &= \sum_{\ell=a}^{b-1} g(k,\ell) \Bigg\{ \sum_{i=0}^{n-1} \left(\Delta^i v_{m+1}(\ell) - \Delta^i v_m(\ell)\right) \frac{\partial}{\partial \Delta^i v_m(\ell)} f_0(\ell, v_m(\ell), \cdots) \\ &\quad + \frac{1}{2}\left[\sum_{i=0}^{n-1} \left(\Delta^i v_m(\ell) - \Delta^i v_{m-1}(\ell)\right) \frac{\partial}{\partial p_i(\ell)}\right]^2 f_0(\ell, p_0(\ell), \cdots, p_{n-1}(\ell)) \Bigg\}, \end{aligned}$$

where $p_i(k)$ lies between $\Delta^i v_{m-1}(k)$ and $\Delta^i v_m(k)$, $0 \le i \le n-1$.

Thus, as earlier we get

$$\|v_{m+1} - v_m\| \;\le\; \theta\|v_{m+1} - v_m\| + \frac{1}{2}\,\zeta\theta^2\|v_m - v_{m-1}\|^2,$$

which is the same as the first part of (11.3.18). The last part of (11.3.18) follows from (11.3.14). ∎

## 11.4. Monotone Convergence

Consider the boundary value problem

$$(11.4.1)\qquad L[u(k)] \;=\; f(k, u(k), u(k+1), \cdots, u(k+n-1)) \;=\; f[k,u], \qquad k \in \mathbb{N}(a, b-1)$$

$$(11.4.2)\qquad \mathcal{P}[u] \;=\; \mathbf{l},$$

where $L[u(k)] = \sum_{i=0}^{n} a_i(k)u(k+i)$, $a_n(k) = 1$, $a_0(k) \neq 0$ and $a_i(k)$, $0 \le i \le n-1$ are defined on $\mathbb{N}(a, b-1)$, $f : \mathbb{N}(a, b-1) \times \mathbb{R}^n \to \mathbb{R}$, $\mathcal{P} : S(a, b-1+n) \to \mathbb{R}^n$ is linear and continuous, where as earlier $S(a, b-1+n)$ is the space of all real functions defined on $\mathbb{N}(a, b-1+n)$, and $\mathbf{l} \in \mathbb{R}^n$ is a given vector.

With respect to $L$ and $\mathcal{P}$, we shall assume the following:

**Condition** $(P_1)$. $u(k) \equiv 0$ is the only solution of the homogeneous boundary value problem $L[u(k)] = 0$, $\mathcal{P}[u] = 0$.

Thus, in view of Section 10.9 for this homogeneous problem the Green's function $g(k, \ell)$ exists on $\mathbb{N}(a, b-1+n) \times \mathbb{N}(a, b-1)$, the problem $L[u(k)] = 0$, (11.4.2) has a unique solution $\phi_{\mathbf{l}}(k)$, and the problem (11.4.1), (11.4.2) is equivalent to

$$(11.4.3)\qquad u(k) \;=\; \phi_{\mathbf{l}}(k) + \sum_{\ell=a}^{b-1} g(k,\ell) f[\ell, u].$$

As in Section 11.1 we shall equip the space $S(a, b-1+n)$ with the norm $\|u\| = \max_{\mathbb{N}(a,b-1+n)} |u(k)|$, so that it becomes a Banach space.

**Theorem 11.4.1.** Suppose that condition $P_1$ holds and $f$ is continuous and bounded. Then, for any $\mathbf{l} \in \mathbb{R}^n$ the problem (11.4.1), (11.4.2) has a solution.

**Proof.** Define an operator $T : S(a, b-1+n) \to S(a, b-1+n)$ as follows

$$(11.4.4)\qquad Tu(k) \;=\; \phi_{\mathbf{l}}(k) + \sum_{\ell=a}^{b-1} g(k,\ell) f[\ell, u].$$

Obviously, $u(k)$ is a solution of (11.4.1), (11.4.2) if and only if $u(k)$ is a fixed point of $T$. Let $Q = \sup\{|f(k,u_1,\cdots,u_n)| : (k,u_1,\cdots,u_n) \in \mathbb{N}(a,b-1)\times\mathbb{R}^n\}$, $\phi = \|\phi_1(k)\|$, and $G = \max_{k\in\mathbb{N}(a,b-1+n)} \sum_{\ell=a}^{b-1} |g(k,\ell)|$. Let $S_1 = \{u(k) \in S(a,b-1+n) : \|u\| \leq \phi + QG\}$, and note that the continuous operator $T$ defined in (11.4.4) maps the closed convex set $S_1$ into itself and that $\overline{T(S_1)}$ is compact. By the Schauder fixed point theorem, the operator $T$ has a fixed point in $S_1$. Thus, the problem (11.4.1), (11.4.2) has a solution in $S_1$. ∎

**Condition** $(P_2)$. The sign of the Green's function $g(k,\ell)$ of the problem $L[u(k)] = 0$, $\mathcal{P}[u] = 0$ is independent of $\ell$.

The motivation of this condition comes from the sign properties of the Green's functions stated in Section 10.9, e.g. the inequality (10.9.11).

**Theorem 11.4.2.** Suppose that

(i) conditions $P_1$ and $P_2$ hold, and let $\{I_1, I_2\}$ be a partition of $\mathbb{N}(a,b-1+n)$ such that

$$(11.4.5) \qquad \begin{aligned} g(k,\ell) &\leq 0 \quad \text{for } (k,\ell) \in I_1 \times \mathbb{N}(a,b-1), \\ g(k,\ell) &\geq 0 \quad \text{for } (k,\ell) \in I_2 \times \mathbb{N}(a,b-1) \end{aligned}$$

(ii) $f$ satisfies the Lipschitz condition (11.1.10) with $\lambda_i = \beta$, $0 \leq i \leq n-1$ on $\mathbb{N}(a,b-1)\times\mathbb{R}^n$

(iii) there exist functions $v_0(k)$ and $w_0(k)$ in the Banach space $S(a,b-1+n)$ satisfying

$$(11.4.6) \qquad \mathcal{P}[v_0] = 1 = \mathcal{P}[w_0]$$

and such that for $k \in \mathbb{N}(a,b-1)$,

$$(11.4.7) \quad L[v_0(k)] - f[k,v_0] + A_0(k) \leq 0 \leq L[w_0(k)] - f[k,w_0] - A_0(k),$$

where

$$(11.4.8) \qquad A_0(k) = \beta\sum_{i=0}^{n-1} |v_0(k+i) - w_0(k+i)|.$$

Then, there exists a solution $u(k)$ of the problem (11.4.1), (11.4.2) such that

$$(11.4.9) \quad v_0(k) \geq u(k) \geq w_0(k) \text{ for } k \in I_1, \quad v_0(k) \leq u(k) \leq w_0(k) \text{ for } k \in I_2.$$

**Proof.** We shall first show that

(11.4.10) $\quad v_0(k) \geq w_0(k)$ for $k \in I_1$ and $v_0(k) \leq w_0(k)$ for $k \in I_2$.

By (11.4.6), $v_0(k) - w_0(k)$ satisfies $\mathcal{P}[v_0 - w_0] = 0$, and hence

$$v_0(k) - w_0(k) = \sum_{\ell=a}^{b-1} g(k,\ell)L[v_0(\ell) - w_0(\ell)].$$

However, in view of (11.4.7), (11.4.8) and the Lipschitz condition (11.1.10) with $\lambda_i = \beta, \ 0 \leq i \leq n-1$ it follows that $L[v_0(k) - w_0(k)] \leq 0$ on $\mathbb{N}(a, b-1)$. Inequalities (11.4.10) now directly follow from (11.4.5).

For each $u \in S(a, b-1+n)$ and $k \in \mathbb{N}(a, b-1)$, we define $\overline{u}(k+j), \ 0 \leq j \leq n-1$ as follows

$$\overline{u}(k+j) = \begin{cases} \left.\begin{array}{l} v_0(k+j) \text{ if } u(k+j) > v_0(k+j) \\ u(k+j) \text{ if } w_0(k+j) \leq u(k+j) \leq v_0(k+j) \\ w_0(k+j) \text{ if } u(k+j) < w_0(k+j) \end{array}\right\}, & k+j \in I_1 \\ \left.\begin{array}{l} v_0(k+j) \text{ if } u(k+j) < v_0(k+j) \\ u(k+j) \text{ if } w_0(k+j) \geq u(k+j) \geq v_0(k+j) \\ w_0(k+j) \text{ if } u(k+j) > w_0(k+j) \end{array}\right\}, & k+j \in I_2. \end{cases}$$

For $k \in \mathbb{N}(a, b-1)$, we define $\overline{f}[k,u] \equiv f(k, \overline{u}(k), \overline{u}(k+1), \cdots, \overline{u}(k+1-n))$. The function $\overline{f}$ is continuous and bounded on $\mathbb{N}(a, b-1) \times \mathbb{R}^n$ and so, by Theorem 11.4.1, the boundary value problem $L[u(k)] = \overline{f}[k,u], \ \mathcal{P}[u] = 1$ has a solution $u(k)$. We shall show that this solution $u(k)$ satisfies (11.4.9), which in turn implies that $u(k)$ is a solution of (11.4.1), (11.4.2). For this, we note that $v_0(k) - u(k)$ satisfies $\mathcal{P}[v_0 - u] = 0$, and hence for all $k \in \mathbb{N}(a, b-1+n)$

$$v_0(k) - u(k) = \sum_{\ell=a}^{b-1} g(k,\ell)L[v_0(\ell) - u(\ell)].$$

For $k \in \mathbb{N}(a, b-1)$, in view of Lipschitz condition (11.1.10) with $\lambda_i = \beta, \ 0 \leq i \leq n-1$, we have

$$L[v_0(k) - u(k)] \leq f[k, v_0] - \overline{f}[k,u] - A_0(k) \leq 0.$$

Thus, from (11.4.5) it follows that $v_0(k) \geq u(k), \ k \in I_1$ and $v_0(k) \leq u(k), \ k \in I_2$. The proof for $u(k) \geq w_0(k), \ k \in I_1$ and $u(k) \leq w_0(k), \ k \in I_2$ is similar. ∎

**Corollary 11.4.3.** Assume that all the hypotheses of Theorem 11.4.2 are satisfied, and define the sequences $\{v_m(k)\}$ and $\{w_m(k)\}$ as follows

$$v_{m+1}(k) = \phi_1(k) + \sum_{\ell=a}^{b-1} g(k,\ell)(f[\ell, v_m] - A_m(\ell)),$$

$$w_{m+1}(k) = \phi_1(k) + \sum_{\ell=a}^{b-1} g(k,\ell)(f[\ell, w_m] + A_m(\ell)),$$

$$A_m(k) = \beta \sum_{i=0}^{n-1} |v_m(k+i) - w_m(k+i)|, \quad m = 0, 1, \cdots.$$

If $u(k)$ is any solution of (11.4.1), (11.4.2) satisfying (11.4.9), then for each $m \geq 0$

$$v_m(k) \geq v_{m+1}(k) \geq u(k) \geq w_{m+1}(k) \geq w_m(k), \quad k \in I_1$$
$$v_m(k) \leq v_{m+1}(k) \leq u(k) \leq w_{m+1}(k) \leq w_m(k), \quad k \in I_2.$$

**Remark 11.4.1.** Due to the generality of the boundary conditions (11.4.2) we require that $v_0(k)$ and $w_0(k)$ satisfy (11.4.6). However, for some particular boundary value problems this condition can be weakened. For this, let $\phi_{v_0}(k)$ and $\phi_{w_0}(k)$ be the unique solutions of the problems $L[u(k)] = 0$, $\mathcal{P}[u] = \mathcal{P}[v_0]$ and $L[u(k)] = 0$, $\mathcal{P}[u] = \mathcal{P}[w_0]$ respectively. If

$$\text{(11.4.11)} \quad \begin{array}{l} \phi_{v_0}(k) \geq \phi_1(k) \geq \phi_{w_0}(k), \quad k \in I_1 \\ \phi_{v_0}(k) \leq \phi_1(k) \leq \phi_{w_0}(k), \quad k \in I_2 \end{array}$$

then the hypothesis that $v_0(k)$ and $w_0(k)$ satisfy (11.4.11) can replace the condition (11.4.6).

**Theorem 11.4.4.** Suppose that

(i) condition (i) of Theorem 11.4.2

(ii) $f : \mathbb{N}(a, b-1) \times \mathbb{R} \to \mathbb{R}$ is continuous and for $u_1, u_2 \in \mathbb{R}$, $u_1 \geq u_2$

$$\text{(11.4.12)} \quad f(k, u_1) \leq f(k, u_2), \quad k \in I_1 \quad \text{and} \quad f(k, u_1) \geq f(k, u_2), \quad k \in I_2$$

(iii) there exist functions $v_0(k)$ and $w_0(k)$ in the Banach space $S(a, b-1+n)$ satisfying (11.4.6), and

$$\text{(11.4.13)} \quad v_0(k) \geq w_0(k), \quad k \in I_1 \quad \text{and} \quad v_0(k) \leq w_0(k), \quad k \in I_2,$$

$$\text{(11.4.14)} \quad L[v_0(k)] - f(k, v_0(k)) \leq 0 \leq L[w_0(k)] - f(k, w_0(k)),$$
$$k \in \mathbb{N}(a, b-1).$$

Then, the sequences $\{v_m(k)\}$ and $\{w_m(k)\}$ defined by

$$v_{m+1}(k) = \phi_1(k) + \sum_{\ell=a}^{b-1} g(k,\ell) f(\ell, v_m(\ell)),$$

$$w_{m+1}(k) = \phi_1(k) + \sum_{\ell=a}^{b-1} g(k,\ell) f(\ell, w_m(\ell)), \quad m = 0, 1, \cdots$$

converge in $S(a, b-1+n)$ to $v(k)$ and $w(k)$ respectively, where $v(k)$ and $w(k)$ are solutions of the boundary value problem

$$L[u(k)] = f(k, u(k)), \quad k \in \mathbb{N}(a, b-1), \quad \mathcal{P}[u] = 1. \tag{11.4.15}$$

Further, for each $m \geq 0$

$$v_m(k) \geq v_{m+1}(k) \geq v(k) \geq w(k) \geq w_{m+1}(k) \geq w_m(k), \quad k \in I_1$$
$$v_m(k) \leq v_{m+1}(k) \leq v(k) \leq w(k) \leq w_{m+1}(k) \leq w_m(k), \quad k \in I_2.$$

**Proof.** The proof is similar to that of Theorem 11.4.2. ∎

Now we shall consider the boundary value problem (1.2.6), (10.8.20). For this, four cases arise: (i) $n$ is even, $p$ is odd, (ii) $n$ is even, $p$ is even, (iii) $n$ is odd, $p$ is odd, (iv) $n$ is odd, $p$ is even. We shall consider only the case (i), whereas results for the other three cases can be stated analogously. For $u, v \in S(a, b-1+n)$ we say that $u \leq_S v$ if and only if $\Delta^i u(k) \leq \Delta^i v(k)$, $k \in \mathbb{N}(a, b-1+n-i)$, $i \in J_1 = \{j : 0 \leq j \leq p\} \bigcup \{j : p < j \text{ (odd)} \leq n-1\}$, and $\Delta^i u(k) \geq \Delta^i v(k)$, $k \in \mathbb{N}(a, b-1+n-i)$, $i \in J_2 = \{j : p < j \text{ (even)} \leq n-1\}$. Thus, from Theorem 10.9.10, $\Delta^i g(k, \ell) \leq 0$, $(k, \ell) \in \mathbb{N}(a, b-1+n-i) \times \mathbb{N}(a, b-1)$, $i \in J_1$, and $\Delta^i g(k, \ell) \geq 0$, $(k, \ell) \in \mathbb{N}(a, b-1+n-i) \times N(a, b-1)$, $i \in J_2$.

**Theorem 11.4.5.** With respect to the boundary value problem (1.2.6), (10.8.20) we assume that $n$ is even, $p$ is odd, and

(i) $f(k, u_0, u_1, \cdots, u_{n-1})$ is continuous on $\mathbb{N}(a, b-1) \times \mathbb{R}^n$, and nonincreasing in $u_i$ for all $i \in J_1$ and nondecreasing in $u_i$ for all $i \in J_2$

(ii) there exist functions $v_0(k)$ and $w_0(k)$ in the Banach space $S(a, b-1+n)$ (with the norm $\|u\| = \max\{\|\Delta^i u(k)\| = \max_{\mathbb{N}(a,b-1+n-i)} |\Delta^i u(k)|, 0 \leq i \leq n-1\}$) such that

$$v_0 \leq_S w_0, \tag{11.4.16}$$

$$\Delta^n w_0(k) - f\left(k, w_0(k), \Delta w_0(k), \cdots, \Delta^{n-1} w_0(k)\right) \leq 0$$
$$\leq \Delta^n v_0(k) - f\left(k, v_0(k), \Delta v_0(k), \cdots, \Delta^{n-1} v_0(k)\right),$$
$$k \in \mathbb{N}(a, b-1), \tag{11.4.17}$$

$$P_{n-1, v_0} \leq_S P_{n-1} \leq_S P_{n-1, w_0}, \tag{11.4.18}$$

where $P_{n-1}(k)$ is defined in (10.8.21), and $P_{n-1,v_0}(k)$ and $P_{n-1,w_0}(k)$ are the polynomials of degree $n-1$ satisfying

$$\Delta^i P_{n-1,v_0}(a) = \Delta^i v_0(a), \quad 0 \leq i \leq p-1$$
$$\Delta^i P_{n-1,v_0}(b) = \Delta^i v_0(b), \quad p \leq i \leq n-1$$

and

$$\Delta^i P_{n-1,w_0}(a) = \Delta^i w_0(a), \quad 0 \le i \le p-1$$
$$\Delta^i P_{n-1,w_0}(b) = \Delta^i w_0(b), \quad p \le i \le n-1$$

respectively.

Then, the sequences $\{v_m\}$, $\{w_m\}$ where $v_m(k)$ and $w_m(k)$ are defined by the iterative schemes

$$v_{m+1}(k) = P_{n-1}(k) + \sum_{\ell=a}^{b-1} g(k,\ell) f\left(\ell, v_m(\ell), \Delta v_m(\ell), \cdots, \Delta^{n-1} v_m(\ell)\right)$$

$$w_{m+1}(k) = P_{n-1}(k) + \sum_{\ell=a}^{b-1} g(k,\ell) f\left(\ell, w_m(\ell), \Delta w_m(\ell), \cdots, \Delta^{n-1} w_m(\ell)\right), \quad m = 0, 1, \cdots$$

converge in $S(a, b-1+n)$ to the solutions $v(k)$ and $w(k)$ of (1.2.6), (10.8.20). Further,

$$v_0 \le_S v_1 \le_S \cdots \le_S v_m \le_S \cdots \le_S v \le_S w \le_S \cdots \le_S w_m \le_S \cdots \le_S w_1 \le_S w_0.$$

Also, each solution $z(k)$ of this problem which is such that $v_0 \le_S z \le_S w_0$ satisfies $v \le_S z \le_S w$.

**Proof.** The proof is similar to that of earlier results. ∎

## 11.5. Initial–Value Methods

In Chapters 8 and 9 initial value methods have been used to construct the solutions of linear and nonlinear boundary value problems. The purpose of this section is to use these methods to prove some existence and uniqueness results for higher order boundary value problems. First, for a given $1 \le p \le n-1$ we shall consider the $(p, n-p)$ boundary value problem

$$(11.5.1) \quad L[u(k)] = \sum_{i=0}^{n} a_i(k) u(k+i) = f(k, u(k)), \quad k \in \mathbb{N}(a, b-1)$$

$$(11.5.2) \quad \begin{aligned} u(a+i) &= A_i, \quad 0 \le i \le p-1 \\ u(b-1+n-i) &= B_i, \quad 0 \le i \le n-p-1. \end{aligned}$$

In (11.5.1) the functions $a_i(k)$ are defined on $\mathbb{N}(a, b-1)$, $a_n(k) \equiv 1$ and $a_0(k)$ satisfies (10.1.1), and the function $f(k,u)$ is defined on $\mathbb{N}(a, b-1) \times \mathbb{R}$.

**Theorem 11.5.1.** Assume that $f(k,u)$ is continuous on $\mathbb{N}(a,b-1)\times\mathbb{R}$, and there is a function $h(k)$ defined on $\mathbb{N}(a,b-1)$ such that

$$(11.5.3) \qquad f(k,u)-f(k,v) \ \geq\ h(k)(u-v), \quad k\in\mathbb{N}(a,b-1)$$

holds whenever $u \geq v$. If $L[u(k)] = h(k)u(k)$ is right $(n-1,1)$ disconjugate on $\mathbb{N}(a,b-1+n)$, then (11.5.1), (11.5.2) with $p=n-1$ has a unique solution.

**Proof.** Let $u(k,m)$ be the unique solution of (11.5.1) satisfying the initial conditions $u(a+i)=A_i$, $0\leq i\leq n-2$, $u(a+n-1)=m$. Let $S=\{u(b-1+n): m\in\mathbb{R}\}$. By the continuous dependence of solutions on initial conditions $S$ is an interval. To prove the existence of a solution it suffices to show that $S$ is not bounded above as well as below.

Define the sequence of integral means $\{f_r(k,u)\}$ of $f(k,u)$ by

$$f_r(k,u) \ =\ \frac{r}{2}\int_{u-1/r}^{u+1/r} f(k,v)dv, \quad r \ =\ 1,2,\cdots$$

for $k\in\mathbb{N}(a,b-1)$, $u\in\mathbb{R}$. It is clear that $f_r(k,u)\to f(k,u)$ uniformly on compact subsets of $\mathbb{N}(a,b-1)\times\mathbb{R}$, the functions $f_r(k,u)$, $\partial f_r(k,u)/\partial u$ are continuous on $\mathbb{N}(a,b-1)\times\mathbb{R}$, and $\partial f_r(k,u)/\partial u \geq h(k)$, $k\in\mathbb{N}(a,b-1)$.

Let $u_r(k,m)$ be the solution of the initial value problem $L[u(k)] = f_r(k,u(k))$, $u(a+i)=A_i$, $0\leq i\leq n-2$, $u(a+n-1)=m$. For $m_1>m_2$, we have

$$(11.5.4) \quad u_r(k,m_1)-u_r(k,m_2) \ =\ \frac{\partial u_r(k,\overline{m})}{\partial m}(m_1-m_2), \quad \overline{m}\in(m_2,m_1)$$

where $\partial u_r(k,\overline{m})/\partial m$ is the solution of the initial value problem

$$L[u(k)] = \frac{\partial f_r(k,u_r(k,\overline{m}))}{\partial u}u(k),\ u(a+i)=0,\ 0\leq i\leq n-2,\ u(a+n-1)=1.$$

Since the equation $L[u(k)] = h(k)u(k)$ is right $(n-1,1)$ disconjugate on $\mathbb{N}(a,b-1+n)$ and $\partial f_r(k,u_r(k,\overline{m}))/\partial u \geq h(k)$, $k\in\mathbb{N}(a,b-1)$, from Corollary 10.2.10 it follows that the equation $L[u(k)] = \dfrac{\partial f_r(k,u_r(k,\overline{m}))}{\partial u}u(k)$ is also right $(n-1,1)$ disconjugate on $\mathbb{N}(a,b-1+n)$. Thus, $\partial u_r(k,\overline{m})/\partial m$ as well as the solution $v(k)$ of the initial value problem $L[v(k)] = h(k)v(k)$, $v(a+i)=0$, $0\leq i\leq n-2$, $v(a+n-1)=1$ is positive on $\mathbb{N}(a+n-1,b-1+n)$. Further, from Theorem 10.2.9 and (11.5.4) it follows that

$$u_r(k,m_1)-u_r(k,m_2) \ \geq\ v(k)(m_1-m_2), \quad k\in\mathbb{N}(a,b-1+n).$$

Let $k = b-1+n$ and use the continuous dependence of solutions on initial conditions to get that

$$(11.5.5)\quad u(b-1+n,m_1) - u(b-1+n,m_2) \ \geq\ v(b-1+n)(m_1 - m_2).$$

Since $v(b-1+n) > 0$, it follows that $\lim_{m\to\infty} u(b-1+n,m) = \infty$, and $\lim_{m\to-\infty} u(b-1+n,m) = -\infty$. Hence the existence part of the proof is complete. The uniqueness part of the proof is also immediate from (11.5.5). ∎

**Theorem 11.5.2.** Assume that $f(k,u)$ is continuous on $\mathbb{N}(a,b-1)\times\mathbb{R}$, and $n$ is even (odd) and there is a function $h(k)$ defined on $\mathbb{N}(a,b-1)$ such that (11.5.3) $(f(k,u) - f(k,v) \leq h(k)(u-v),\ k \in \mathbb{N}(a,b-1))$ holds whenever $u \geq v$. If $L[u(k)] = h(k)u(k)$ is left $(1,n-1)$ disconjugate on $\mathbb{N}(a,b-1+n)$, then (11.5.1), (11.5.2) with $p=1$ has a unique solution.

**Proof.** The proof is similar to that of Theorem 11.5.1. ∎

**Theorem 11.5.3.** Assume that $f(k,u)$ is continuous on $\mathbb{N}(a,b-1)\times\mathbb{R}$, and there are functions $g(k)$ and $h(k)$ defined on $\mathbb{N}(a,b-1)$ such that

$$(11.5.6)\quad h(k)(u-v) \ \leq\ f(k,u) - f(k,v) \ \leq\ g(k)(u-v), \quad k \in \mathbb{N}(a,b-1)$$

holds whenever $u \geq v$. If $L[u(k)] = h(k)u(k)$ is right $(n-1,1)$ disconjugate on $\mathbb{N}(a,b-1+n)$ and $L[u(k)] = g(k)u(k)$ is $(n-2,2)$ disconjugate on $\mathbb{N}(a,b-1+n)$, then (11.5.1), (11.5.2) with $p=n-2$ has a unique solution.

**Proof.** By Theorem 11.5.1 there exists a unique solution $u(k,m)$ of the boundary value problem (11.5.1), $u(a+i) = A_i$, $0 \leq i \leq n-3$, $u(a+n-2) = m$, $u(b-1+n) = B_0$. Let $S = \{u(b-2+n,m) : m \in \mathbb{R}\}$. From Problem 11.7.2, $S$ is an interval and to prove the existence part it suffices to show that $S$ is neither bounded below nor above. For $m_1 > m_2$ let $z(k) = (u(k,m_1) - u(k,m_2))/(m_1-m_2)$, so that $L[z(k)] = (L[u(k,m_1)] - L[u(k,m_2)])/(m_1-m_2) = (f(k,u(k,m_1)) - f(k,u(k,m_2)))/(m_1-m_2)$. We define

$$r(k) = \begin{cases} (f(k,u(k,m_1)) - f(k,u(k,m_2)))/(u(k,m_1) - u(k,m_2)), \\ \qquad\qquad\qquad\qquad\qquad\qquad u(k,m_1) \neq u(k,m_2) \\ h(k), \quad u(k,m_1) = u(k,m_2). \end{cases}$$

Then, $z(k)$ is the solution of the boundary value problem $L[z(k)] = r(k)z(k)$, $z(a+i) = 0$, $0 \leq i \leq n-3$, $z(a+n-2) = 1$, $z(b-1+n) = 0$. Since $h(k) \leq r(k) \leq g(k)$, $k \in \mathbb{N}(a,b-1)$ and $L[u(k)] = h(k)u(k)$ is right $(n-1,1)$ disconjugate, we have that both $L[z(k)] = r(k)z(k)$ and

$L[v(k)] = g(k)v(k)$ are right disconjugate by Corollary 10.2.10. Now using the fact that $L[v(k)] = g(k)v(k)$ is also right $(n-2,2)$ disconjugate, from Corollary 10.2.14 it follows that $L[z(k)] = r(k)z(k)$ is right $(n-2,2)$ disconjugate. Thus, we find that $z(k) \geq 0$ on $\mathbb{N}(a, b-1+n)$. Hence, in view of Theorem 10.2.13 we obtain that $z(k) \geq v(k)$, where $v(k)$ is the solution of $L[v(k)] = g(k)v(k)$, $v(a+i) = z(a+i)$, $0 \leq i \leq n-2$, $v(b-1+n) = z(b-1+n)$. Therefore, $u(k, m_1) - u(k, m_2) \geq v(k)(m_1 - m_2)$, $k \in \mathbb{N}(a, b-1+n)$. Letting $k = b-2+n$, we find that

$$(11.5.7) \quad u(b-2+n, m_1) - u(b-2+n, m_2) \geq v(b-2+n)(m_1 - m_2).$$

Since $L[v(k)] = g(k)v(k)$ is both $(n-1,1)$ and $(n-2,2)$ disconjugate on $\mathbb{N}(a, b-1+n)$ it follows that $v(k) > 0$ on $\mathbb{N}(a+n-2, b-2+n)$. Thus, from (11.5.7) it is clear that $\lim_{m\to\infty} u(b-2+n, m) = \infty$, and $\lim_{m\to-\infty} u(b-2+n, m) = -\infty$.

For the uniqueness of solutions, suppose on the contrary that $u_1(k)$ and $u_2(k)$ are distinct solutions of the boundary value problem (11.5.1), (11.5.2) with $p = n-2$. Since solutions of (11.5.1), (11.5.2) with $p = n-1$ are unique, we can write $u_1(k) = u(k, m_1)$, $u_2(k) = u(k, m_2)$, for some $m_1 \neq m_2$. Without loss of generality we can assume that $m_1 > m_2$. But then (11.5.7) shows that $u(b-2+n, m_1) \neq u(b-2+n, m_2)$, which contradicts the assumption that both $u_1(k)$ and $u_2(k)$ were solutions of the same problem. Hence the uniqueness condition is satisfied. ∎

**Theorem 11.5.4.** Let the function $f(k, u)$ be as in Theorem 11.5.3. If $L[u(k)] = h(k)u(k)$ and $L[u(k)] = g(k)u(k)$ are disconjugate on $\mathbb{N}(a, b-1+n)$, then the boundary value problem (11.5.1), (11.5.2) has a unique solution.

**Proof.** The proof is by induction on decreasing values of $p$. The cases $p = n-1$ and $p = n-2$ are contained in Theorems 11.5.1 and 10.5.3 respectively. Assume $p \leq n-3$ and that the theorem holds if $p$ is replaced by $p+1$. Then, there exists a unique solution $u(k, m)$ of the boundary value problem (11.5.1), $u(a+i) = A_i$, $0 \leq i \leq p-1$, $u(a+p) = m$, $u(b-1+n-i) = B_i$, $0 \leq i \leq n-p-2$. For $m_1 > m_2$ let $z(k)$ and $r(k)$ be as in the proof of Theorem 11.5.3. Then, $z(k)$ is the solution of the boundary value problem $L[z(k)] = r(k)z(k)$, $z(a+i) = 0$, $0 \leq i \leq p-1$, $z(a+p) = 1$, $z(b-1+n-i) = 0$, $0 \leq i \leq n-p-2$. Since $h(k) \leq r(k) \leq g(k)$, $k \in \mathbb{N}(a, b-1)$ and both $L[u(k)] = h(k)u(k)$ and $L[u(k)] = g(k)u(k)$ are disconjugate on $\mathbb{N}(a, b-1+n)$, from a slight modification of Theorem 10.1.7 it follows that $L[z(k)] = r(k)z(k)$ is also disconjugate on $\mathbb{N}(a, b-1+n)$.

Now consider the case that $n-p$ is odd, the case for $n-p$ even is similar. Let $v(k)$ be the solution of $L[v(k)] = h(k)v(k)$, $v(a+i) = z(a+i)$, $0 \le i \le p$, $v(b-1+n-i) = z(b-1+n-i)$, $0 \le i \le n-p-2$. By the disconjugacy assumptions, we have $v(k) \ge 0$ on $\mathbb{N}(a, b-1+n)$. Thus, from Theorem 10.2.15 it follows that $z(k) \ge v(k)$, $k \in \mathbb{N}(a, b-1+n)$. The remainder of the proof is similar to that of Theorem 11.5.3. ∎

Next, we shall consider the difference equation

$$\Delta\left(\rho(k)\Delta^{n-1}u(k)\right) = f\left(k, u(k), \Delta u(k), \cdots, \Delta^{n-1}u(k)\right), \quad k \in \mathbb{N}(0, b-1) \tag{11.5.8}$$

together with the $(n,p)$ boundary conditions (1.5.12) with $a = 0$. In (11.5.8) the function $\rho(k)$ is defined and positive on $\mathbb{N}(0,b)$, and the function $f(k, u_0, u_1, \cdots, u_{n-1})$ is defined and continuous on $\mathbb{N}(0, b-1) \times R^n$.

**Lemma 11.5.5.** Let $q \in \mathbb{N}$, and $u(k)$ be a function defined on $\mathbb{N}(0, n+q)$ such that $\Delta^i u(0) = \epsilon_i$, $0 \le i \le n-1$.

(i) If $\epsilon_i > 0$, $0 \le i \le n-1$ and $\Delta^{n-1}u(k) > 0$ on $\mathbb{N}(0, q+1)$, then $\Delta^j u(k) > 0$ on $\mathbb{N}(0, n+q-j)$, and hence $\Delta^j u(k)$ is strictly increasing on $\mathbb{N}(0, n+q-j)$, $0 \le j \le n-2$.

(ii) If $\epsilon_i = 0$, $0 \le i \le n-2$, then

$$\begin{aligned} \Delta^j u(k) &= 0, \quad k \in \mathbb{N}(0, n-j-2) \\ \Delta^j u(n-j-1) &= \epsilon_{n-1}, \quad 0 \le j \le n-1 \end{aligned} \tag{11.5.9}$$

also, if $\Delta^{n-1}u(k) > 0$, $k \in \mathbb{N}(0, q+1)$ then $\Delta^j u(k) > 0$, $k \in \mathbb{N}(n-j-1, n+q-j)$, $0 \le j \le n-2$ and for such $k$

$$u(k) \le \frac{1}{j!}(k-n+j+1)^{(j)}\Delta^j u(k), \quad 1 \le j \le n-2. \tag{11.5.10}$$

**Proof.** From (1.8.7) it follows that

$$\Delta^j u(k) = \sum_{i=j}^{n-2} \frac{(k)^{(i-j)}}{(i-j)!}\epsilon_i + \frac{1}{(n-j-2)!} \sum_{\ell=0}^{k-n+j+1} (k-\ell-1)^{(n-j-2)}\Delta^{n-1}u(\ell), \quad 0 \le j \le n-2 \tag{11.5.11}$$

from which part (i) is immediate.

For part (ii) the equality (11.5.11) reduces to

(11.5.12)

$$\Delta^j u(k) = \frac{1}{(n-j-2)!}\sum_{\ell=0}^{k-n+j+1}(k-\ell-1)^{(n-j-2)}\Delta^{n-1}u(\ell), \quad 0 \le j \le n-2$$

and from this (11.5.9) is clear. Further, if $\Delta^{n-1}u(k) > 0,\ k \in \mathbb{N}(0, q+1)$ then from (11.5.12), $\Delta^j u(k) > 0,\ k \in \mathbb{N}(n-j-1, n+q-j)$ is also immediate. Now, in view of (1.8.6) and (11.5.9), we have

$$u(k) = \frac{1}{(j-1)!}\sum_{\ell=n-j-1}^{k-j}(k-\ell-1)^{(j-1)}\Delta^j u(\ell), \quad 1 \le j \le n-2$$

and hence on using the increasing nature of $\Delta^j u(k),\ k \in \mathbb{N}(n-j-1, n+q-j)$, we find

$$\begin{aligned} u(k) &\le \frac{1}{(j-1)!}\sum_{\ell=n-j-1}^{k-j}(k-\ell-1)^{(j-1)}\Delta^j u(k) \\ &= \frac{1}{j!}(k-n+j+1)^{(j)}\Delta^j u(k). \qquad \blacksquare \end{aligned}$$

**Remark 11.5.1.** Throughout, in Lemma 11.5.5 the strict inequalities can be replaced by with equalities.

**Lemma 11.5.6.** Assume that $a_i(k),\ 0 \le i \le n-1$ are defined and nonnegative on $\mathbb{N}(0, b-1)$. Then, for each $\alpha > 0$ the solution of the initial value problem

$$\Delta\left(\rho(k)\Delta^{n-1}v(k)\right) = \sum_{i=0}^{n-1}a_i(k)\Delta^i v(k) \tag{11.5.13}$$

$$\Delta^i v(0) = 0, \quad 0 \le i \le n-2, \quad \Delta^{n-1}v(0) = \alpha > 0 \tag{11.5.14}$$

has the property that $\Delta^j v(k) \ge 0$ for all $k \in \mathbb{N}(0, b-1+n-j)$ and in particular for all $k \in \mathbb{N}(n-j-1,\ b-1+n-j)$ the strict inequality $\Delta^j u(k) > 0,\ 0 \le j \le n-1$ holds.

**Proof.** Let $q \in \mathbb{N}(1, b)$ be the first point where $\Delta^{n-1}v(k) \le 0$, then from Lemma 11.5.5, $\Delta^j v(k) \ge 0$ for all $k \in \mathbb{N}(0, n+q-j-2)$, and in particular $\Delta^j v(q-1) \ge 0,\ 0 \le j \le n-2$. However, from the difference equation (11.5.13), we have

$$\rho(q)\Delta^{n-1}v(q) = \rho(q-1)\Delta^{n-1}v(q-1) + \sum_{i=0}^{n-1}a_i(q-1)\Delta^i v(q-1) > 0.$$

This contradiction completes the proof. $\blacksquare$

**Lemma 11.5.7.** Assume that

(i) $g(k, u_0, u_1, \cdots, u_{n-1})$ is defined on $\mathbb{N}(0, b-1) \times \mathbb{R}^n$ and nondecreasing in $u_0, u_1, \cdots, u_{n-1}$ for a fixed $k \in \mathbb{N}(0, b-1)$, also for $\lambda > 1$

$$\lambda g(k, u_0, u_1, \cdots, u_{n-1}) \leq g(k, \lambda u_0, \lambda u_1, \cdots, \lambda u_{n-1})$$

(ii) for a fixed $k \in \mathbb{N}(0, b-1)$ and $u_i \in \mathbb{R}^+$, $0 \leq i \leq n-1$

$$f(k, u_0, u_1, \cdots, u_{n-1}) \geq g(k, u_0, u_1, \cdots, u_{n-1}) + l(k)u_0 + \sum_{i=1}^{n-2} a_i(k)u_i,$$

where $a_i(k) \geq 0$, $1 \leq i \leq n-2$ and $l(k)$ are defined on $\mathbb{N}(0, b-1)$ and

$$l(k) + \sum_{i=1}^{n-2} a_i(k) \frac{i!}{(k-n+i+1)^{(i)}} \geq 0 \tag{11.5.15}$$

(iii) $u(k, 0, \beta)$ is the solution of (11.5.8) satisfying the initial conditions

$$\Delta^i u(0) = 0, \quad 0 \leq i \leq n-2, \quad \Delta^{n-1} u(0) = \beta \tag{11.5.16}$$

(iv) there exists a solution $v(k, 0, \alpha)$ of the difference equation

$$\begin{aligned} \Delta\left(\rho(k)\Delta^{n-1}v(k)\right) &= g\left(k, v(k), \Delta v(k), \cdots, \Delta^{n-1}v(k)\right) \\ &\quad + l(k)v(k) + \sum_{i=1}^{n-2} a_i(k)\Delta^i v(k) \end{aligned} \tag{11.5.17}$$

satisfying the initial conditions (11.5.14) such that $\Delta^{n-1}v(k, 0, \alpha) > 0$ for all $k \in \mathbb{N}(0, b)$.

Then, for all $k \in \mathbb{N}(0, b-1+n-i)$

$$0 \leq \frac{\beta - \epsilon}{\alpha} \Delta^i v(k, 0, \alpha) \leq \Delta^i u(k, 0, \beta), \quad 0 \leq i \leq n-1 \tag{11.5.18}$$

where $\epsilon > 0$ and $\beta - \epsilon > \alpha$. In particular $\Delta^i u(k, 0, \beta) > 0$ for all $k \in \mathbb{N}(n-i-1, b-1+n-i)$, $0 \leq i \leq n-1$.

**Proof.** Since $\Delta^{n-1}v(k, 0, \alpha) > 0$ for all $k \in \mathbb{N}(0, b)$ and $\Delta^i v(0, 0, \alpha) = 0$, $0 \leq i \leq n-2$, Lemma 11.5.5 ensures that $\Delta^i v(k, 0, \alpha) \geq 0$, $k \in \mathbb{N}(0, b-1+n-i)$, and in particular strict inequality holds for all $k \in \mathbb{N}(n-i-1, b-1+n-i)$, $0 \leq i \leq n-1$. Thus, it suffices to show that $\frac{\beta-\epsilon}{\alpha}\Delta^i v(k, 0, \alpha) \leq \Delta^i u(k, 0, \beta)$, $0 \leq i \leq n-1$ holds on $\mathbb{N}(0, b-1+n-i)$. For this, we define a function $\phi(k)$, $k \in \mathbb{N}(0, b-1+n)$ as follows $\phi(k) = u(k, 0, \beta) - \frac{\beta-\epsilon}{\alpha} v(k, 0, \alpha)$. Then, $\Delta^i \phi(0) = 0$, $0 \leq i \leq n-2$ and

$\Delta^{n-1}\phi(0) = \epsilon > 0$, and from Lemma 11.5.5 and Remark 11.5.1 note that we need to prove $\Delta^{n-1}\phi(k) \geq 0$, $k \in \mathbb{N}(0,b)$. Let $q \in \mathbb{N}(1,b)$ be the first point where $\Delta^{n-1}\phi(q) < 0$. Then, from Lemma 11.5.5, $\Delta^j\phi(k) \geq 0$, $k \in \mathbb{N}(0, n+q-j-2)$, $0 \leq j \leq n-1$. Hence, in particular $\Delta^j\phi(q-1) \geq 0$, $0 \leq j \leq n-1$. Since $\rho(k) > 0$, $k \in \mathbb{N}(0,b)$, we have

(11.5.19)
$$\Delta\left(\rho(q-1)\Delta^{n-1}\phi(q-1)\right) = \rho(q)\Delta^{n-1}\phi(q) - \rho(q-1)\Delta^{n-1}\phi(q-1) < 0.$$

Next, using the conditions on the functions and the inequality (11.5.19), we successively obtain

$$\begin{aligned}
& f\left(q-1, u(q-1), \Delta u(q-1), \cdots, \Delta^{n-1}u(q-1)\right) \\
&\quad = \Delta\left(\rho(q-1)\Delta^{n-1}u(q-1)\right) \\
&\quad = \Delta\left(\rho(q-1)\Delta^{n-1}\phi(q-1)\right) + \frac{\beta-\epsilon}{\alpha}\Delta\left(\rho(q-1)\Delta^{n-1}v(q-1)\right) \\
&\quad < \frac{\beta-\epsilon}{\alpha}\Delta\left(\rho(q-1)\Delta^{n-1}v(q-1)\right) \\
&\quad = \frac{\beta-\epsilon}{\alpha}\Bigg[g\left(q-1, v(q-1), \Delta v(q-1), \cdots, \Delta^{n-1}v(q-1)\right) \\
&\qquad + l(q-1)v(q-1) + \sum_{i=1}^{n-2} a_i(q-1)\Delta^i v(q-1)\Bigg] \\
&\quad \leq g\left(q-1, \frac{\beta-\epsilon}{\alpha}v(q-1), \frac{\beta-\epsilon}{\alpha}\Delta v(q-1), \cdots, \frac{\beta-\epsilon}{\alpha}\Delta^{n-1}v(q-1)\right) \\
&\qquad + \frac{\beta-\epsilon}{\alpha}\left[l(q-1)v(q-1) + \sum_{i=1}^{n-2} a_i(q-1)\Delta^i v(q-1)\right] \\
&\quad \leq f\left(q-1, u(q-1), \Delta u(q-1), \cdots, \Delta^{n-1}u(q-1)\right) \\
&\qquad - l(q-1)u(q-1) - \sum_{i=1}^{n-2} a_i(q-1)\Delta^i u(q-1) \\
&\qquad + \frac{\beta-\epsilon}{\alpha}\left[l(q-1)v(q-1) + \sum_{i=1}^{n-2} a_i(q-1)\Delta^i v(q-1)\right] \\
&\quad = f\left(q-1, u(q-1), \Delta u(q-1), \cdots, \Delta^{n-1}u(q-1)\right) \\
&\qquad - \left[l(q-1)\phi(q-1) + \sum_{i=1}^{n-2} a_i(q-1)\Delta^i \phi(q-1)\right] \\
&\quad \leq f\left(q-1, u(q-1), \Delta u(q-1), \cdots, \Delta^{n-1}u(q-1)\right) \\
&\qquad - \left[l(q-1) + \sum_{i=1}^{n-2} \frac{a_i(q-1)(i)!}{(q-n+i)^{(i)}}\right]\phi(q-1),
\end{aligned}$$

which is not true from (11.5.15) and the fact that $\phi(q-1) \geq 0$. This contradiction completes the proof. ∎

**Corollary 11.5.8.** Assume that $u(k,0,\beta)$ be as in Lemma 11.5.7, and let for a fixed $k \in \mathbb{N}(0,b-1)$ and $u_i \in \mathbb{R}_+$, $0 \leq i \leq n-1$

$$f(k,u_0,u_1,\cdots,u_{n-1}) \geq \sum_{i=0}^{n-1} a_i(k)u_i, \tag{11.5.20}$$

where $a_i(k) \geq 0$, $0 \leq i \leq n-1$ are defined on $\mathbb{N}(0,b-1)$. Further, let $v(k,0,\alpha)$ be the solution of (11.5.13), (11.5.14). Then, the conclusion of Lemma 11.5.7 follows.

**Proof.** In view of Lemma 11.5.6 we see that all the conditions of Lemma 11.5.7 are satisfied. ∎

**Theorem 11.5.9.** In addition to the assumption (i) of Lemma 11.5.7, we assume that

(i) for a fixed $k \in \mathbb{N}(0,b-1)$ and $u_i \geq \overline{u}_i$, $0 \leq i \leq n-1$

$$\begin{aligned} & f(k,u_0,u_1,\cdots,u_{n-1}) - f(k,\overline{u}_0,\overline{u}_1,\cdots,\overline{u}_{n-1}) \\ & \geq g(k,u_0-\overline{u}_0,\cdots,u_{n-1}-\overline{u}_{n-1}) + l(k)(u_0-\overline{u}_0) + \sum_{i=1}^{n-2} a_i(k)(u_i-\overline{u}_i), \end{aligned} \tag{11.5.21}$$

where $a_i(k) \geq 0$, $1 \leq i \leq n-2$ and $l(k)$ are defined on $\mathbb{N}(0,b-1)$ and (11.5.15) holds

(ii) for each $\alpha > 0$ condition (iv) of Lemma 11.5.7 holds.

Then, the boundary value problem (11.5.8), (1.5.12) with $a = 0$ has a unique solution.

**Proof.** Let $\overline{A}$ denote the vector $(A_0, A_1, \cdots, A_{n-2})$ and $u(k,\overline{A},\gamma_i)$, $i = 1,2$ be the solutions of (11.5.8), $\Delta^i u(0,\overline{A},\gamma_i) = A_i$, $0 \leq i \leq n-2$, $\Delta^{n-1}u(0,\overline{A},\gamma_i) = \gamma_i$. For $\gamma_1 > \gamma_2$, we define $w(k,\overline{A},\gamma_1,\gamma_2) = u(k,\overline{A},\gamma_1) - u(k,\overline{A},\gamma_2)$, then $w(k,\overline{A},\gamma_1,\gamma_2)$ is the solution of the initial value problem

$$\begin{aligned} \Delta\left(\rho(k)\Delta^{n-1}w(k,\overline{A},\gamma_1,\gamma_2)\right) &= F\left(k, w(k,\overline{A},\gamma_1,\gamma_2),\cdots,\right. \\ & \qquad \left.\Delta^{n-1}w(k,\overline{A},\gamma_1,\gamma_2)\right), \quad k \in \mathbb{N}(0,b-1) \\ \Delta^i w(0,\overline{A},\gamma_1,\gamma_2) &= 0, \quad 0 \leq i \leq n-2 \\ \Delta^{n-1}w(0,\overline{A},\gamma_1,\gamma_2) &= \gamma_1 - \gamma_2 > 0, \end{aligned} \tag{11.5.22}$$

where

$$F(k,\cdots) = f(k, w(k,\overline{A},\gamma_1,\gamma_2) + u(k,\overline{A},\gamma_2),\cdots) - f(k,u(k,\overline{A},\gamma_2),\cdots).$$

Thus, in view of condition (i), in Lemma 11.5.7 the function $f$ can be replaced by $F$, and in conclusion the solutions $w(k,\overline{A},\gamma_1,\gamma_2)$ of (11.5.22) and $v(k,0,\alpha)$ of (11.5.17), (11.5.14) with $\gamma_1-\gamma_2>\alpha>0$ satisfy

$$0 \le \frac{\gamma_1-\gamma_2}{\alpha}\Delta^i v(k,0,\alpha) \le \Delta^i w(k,\overline{A},\gamma_1,\gamma_2),\ k\in\mathbb{N}(0,b-1+n-i),\ 0\le i\le n-1$$

and $\Delta^i w(k,\overline{A},\gamma_1,\gamma_2)>0$ for all $k\in\mathbb{N}(n-i-1,b-1+n-i)$.

The above inequality in particular implies that

$$0 \le \frac{\gamma_1-\gamma_2}{\alpha}\Delta^p v(b-1+n-p,0,\alpha) \le \Delta^p w(b-1+n-p,\overline{A},\gamma_1,\gamma_2).$$

The rest of the proof is similar to that of Theorem 11.5.3. ∎

**Corollary 11.5.10.** Let for a fixed $k\in\mathbb{N}(0,b-1)$ and $u_i\ge\overline{u}_i$, $0\le i\le n-1$

$$f(k,u_0,u_1,\cdots,u_{n-1})-f(k,\overline{u}_0,\overline{u}_1,\cdots,\overline{u}_{n-1}) \ge \sum_{i=0}^{n-1}a_i(k)(u_i-\overline{u}_i),$$

where $a_i(k)\ge 0$, $0\le i\le n-1$ are defined on $\mathbb{N}(a,b-1)$, (in particular $f$ is nondecreasing in all $u_i$, $0\le i\le n-1$). Then, the boundary value problem (11.5.8), (1.5.12) has a unique solution.

## 11.6. Uniqueness Implies Existence

Here we shall consider the difference equation (1.2.5) together with the boundary conditions (10.8.1), where $k_1<k_2<\cdots<k_n$ and each $k_i\in\mathbb{N}(a)$. For convenience, we shall assume that $k_1=m_1$, $k_i-k_{i-1}=m_i$ $(\ge 1)$, $2\le i\le n$ and call (10.8.1) as $(m_1,m_2,\cdots,m_n)$ conjugate boundary conditions. Throughout, for the $(m_1,m_2,\cdots,m_n)$ conjugate boundary value problem (1.2.5), (10.8.1) we shall assume that the following conditions are satisfied.

**Condition 11.6.1.** The function $f:\mathbb{N}(a)\times\mathbb{R}^n\to\mathbb{R}$ is continuous and the equation $u_n=f(k,u_0,\cdots,u_{n-1})$ can be solved for $u_0$ as a continuous function of $u_1,\cdots,u_n$ for each $k\in\mathbb{N}(a)$.

**Condition 11.6.2.** Given $m_1\in\mathbb{N}(a)$ and $m_2,\cdots,m_n\in\mathbb{N}(1)$, if $k_1=m_1$ and $k_i=k_{i-1}+m_i$, $2\le i\le n$ and if $u(k)$ and $v(k)$ are solutions of (1.2.5) such that $u(k_1)=v(k_1)$ and $u(k)-v(k)$ has a generalized zero at $k_i$, $2\le i\le n$, then it follows that $u(k)=v(k)$ on $\mathbb{N}(k_1,k_n)$.

As a consequence of condition 11.6.1 it follows that (1.2.5) is an $n$th order difference equation on any subinterval of $\mathbb{N}(a)$, that solutions of initial value problems for (1.2.5) are unique and exist on $\mathbb{N}(a)$, and that solutions of (1.2.5) depend continuously on initial conditions. Further, condition 11.6.2 in fact implies that $u(k) = v(k)$ on $\mathbb{N}(a)$.

**Theorem 11.6.1.** For the difference equation (1.2.5) let the conditions 11.6.1 and 11.6.2 be satisfied. Then, each $(m_1, m_2, \cdots, m_n)$ conjugate boundary value problem (1.2.5), (10.8.1) has a unique solution on $\mathbb{N}(a)$.

**Proof.** We note that the condition 11.6.2 implies the uniqueness of all such solutions. The proof of the existence of solutions is by induction on $m_2, \cdots, m_n$. To begin, let $m_i = 1,\ 2 \le i \le n$ so that (1.2.5), (10.8.1) becomes an initial value problem for which a unique solution on $\mathbb{N}(a)$ exists. Assume now that $m_i = 1,\ 2 \le i \le n-1,\ m_n > 1$ and each $(m_1, 1, \cdots, 1, h)$ conjugate boundary value problem, where $1 \le h < m_n$, for (1.2.5) has a unique solution on $\mathbb{N}(a)$. Let $v_1(k)$ be the solution of the $(m_1, 1, \cdots, 1, m_n - 1)$ conjugate boundary value problem for (1.2.5) satisfying $v_1(k_i) = A_i,\ 1 \le i \le n-1,\ v_1(k_n - 1) = 0$ (see the definition of $k_i,\ 1 \le i \le n$). Now define $S_1 = \{r \in \mathbb{R} :$ there is a solution $u(k)$ of (1.2.5) satisfying $u(k_i) = v_1(k_i),\ 1 \le i \le n-1$ and $u(k_n) = r\}$. Since $v_1(k_n) \in S_1$, $S_1$ is nonempty. Moreover, from Problem 11.7.5 it follows that $S_1$ is an open subset of $\mathbb{R}$. We claim that $S_1$ is also a closed subset of $\mathbb{R}$. If not, then there exist $r_0 \in \overline{S}_1 \backslash S_1$ and a strictly monotone sequence $\{r_m\} \subset S_1$ such that $\lim_{m \to \infty} r_m = r_0$. We may assume without loss of generality that $r_m \uparrow r_0$. For each $m \in \mathbb{N}(1)$, let $u_m(k)$ denote the corresponding solution of (1.2.5) satisfying $u_m(k_i) = v_1(k_i),\ 1 \le i \le n-1,\ u_m(k_n) = r_m$. From the condition 11.6.2 it follows that $u_m(k) < u_{m+1}(k)$ on $\mathbb{N}(k_{n-1}+1)$, for all $m \in \mathbb{N}(1)$. Furthermore, the induction hypothesis implies the existence of unique solutions of $(m_1, 1, \cdots, 1, m_n - 1)$ conjugate boundary value problems for (1.2.5), which when coupled with Problem 11.7.6 along with $r_0 \notin S_1$ implies that $u_m(k_n - 1) \uparrow \infty$ as $m \to \infty$. Moreover, by Problem 11.7.4 there exists $k_0 \in \mathbb{N}(k_n + 1, k_n + n - 1)$ such that $u_m(k_0) \uparrow \infty$ as $m \to \infty$.

Now let $z(k)$ denote the solution of the $(m_1 + 1, 1, \cdots, 1, m_n - 1)$ conjugate boundary value problem for (1.2.5) satisfying $z(k_i) = v_1(k_i),\ 2 \le i \le n-1,\ z(k_{n-1}+1) = 0,\ z(k_n) = r_0$. Since $u_m(k_n - 1) \uparrow \infty$ and $u_m(k_0) \uparrow \infty$, whereas $u_m(k_n) = r_m < r_0 = z(k_n)$ for all $m \in \mathbb{N}(1)$, it follows that for some $M \in \mathbb{N}(1)$, $z(k) - u_M(k)$ has a generalized zero at $k_n$, and also a generalized zero (or zero) at some $\ell_0 \in \mathbb{N}(k_n + 1, k_0)$. Furthermore, $z(k_i) - u_M(k_i) = 0,\ 2 \le i \le n-1$ and hence from condition 11.6.2, $z(k) = u_M(k)$ on $\mathbb{N}(a)$, which is a contradiction. Hence, $S_1$ is also closed and consequently $S_1 = \mathbb{R}$. Choosing $A_n \in S_1$ it follows that there exists a solution $u(k)$ of (1.2.5) satisfying $u(k_i) = A_i$, $1 \le i \le n$. In particular, given $m_1 \in \mathbb{N}(a)$, $m_i = 1$, $2 \le i \le n-1$ and $m_n \ge 1$ each $(m_1, 1, \cdots, 1, m_n)$ conjugate boundary value problem for (1.2.5) has a unique solution on $\mathbb{N}(a)$.

For the next part of the proof we induct on $m_{n-1}$. For this, we now assume that $m_i = 1$, $2 \le i \le n-2$, $m_{n-1} > 1$, $m_n \ge 1$ and that there exists a unique solution of each $(m_1, 1, \cdots, 1, p, m_n)$ conjugate boundary value problem for (1.2.5) on $\mathbb{N}(a)$, where $1 \le p < m_{n-1}$. Let $m_n = 1$ and $v_2(k)$ be the solution of the $(m_1, 1, \cdots, 1, m_{n-1}-1, 1)$ conjugate boundary value problem for (1.2.5) satisfying $v_2(k_i) = A_i$, $1 \le i \le n-2$, $v_2(k_{n-1} - 1) = 0$, $v_2(k_{n-1}) = A_{n-1}$. This time, we define $S_2 = \{r \in \mathbb{R} :$ there is a solution $u(k)$ of (1.2.5) satisfying $u(k_i) = v_2(k_i)$, $1 \le i \le n-1$ and $u(k_n) = r\}$. Again, since $v_2(k_n) \in S_2$, $S_2$ is nonempty, also in view of Problem 11.7.5, $S_2$ is an open subset of $\mathbb{R}$. We claim that $S_2$ is also closed. If not, then there exist $r_0 \in \overline{S}_2 \backslash S_2$ and a strictly monotone sequence $\{r_m\} \subset S_2$ such that $\lim_{m\to\infty} r_m = r_0$. We may assume again that $r_m \uparrow r_0$. Let $u_m(k)$ denote the corresponding solution of (1.2.5) satisfying $u_m(k_i) = v_2(k_i)$, $1 \le i \le n-1$, $u_m(k_n) = r_m$. From the condition 11.6.2 it follows that $u_m(k) > u_{m+1}(k)$ on $\mathbb{N}(k_{n-2}+1,\ k_{n-1}-1)$ and $u_m(k) < u_{m+1}(k)$ on $\mathbb{N}(k_n)$, for all $m \in \mathbb{N}(1)$. Since $r_0 \notin S_2$ and since there exist unique solutions of $(m_1, 1, \cdots, 1, m_{n-1} - 1, 1)$ problems, Problem 11.7.6 implies that $u_m(k_{n-1} - 1) \downarrow -\infty$ as $m \to \infty$ and Problem 11.7.4 implies that there exists $k_0 \in \mathbb{N}(k_n + 1, k_n + n - 1)$ such that $u_m(k_0) \uparrow \infty$ as $m \to \infty$.

Now let $z(k)$ denote the solution of the $(m_1 + 1, 1, \cdots, 1, m_{n-1} - 1, 1)$ conjugate boundary value problem for (1.2.5) satisfying $z(k_i) = v_2(k_i)$, $2 \le i \le n-2$, $z(k_{n-2} + 1) = 0$, $z(k_{n-1}) = v_2(k_{n-1})$, $z(k_n) = r_0$. Since $u_m(k_{n-1} - 1) \downarrow -\infty$, whereas $z(k_{n-1}) - u_m(k_{n-1}) = 0$ and $z(k_n) - u_m(k_n) > 0$ for all $m \in \mathbb{N}(1)$, it follows that for all $m$ sufficiently large, $z(k) - u_m(k)$ has a generalized zero at $k_n$. Since $u_m(k_0) \uparrow \infty$, there exists $M \in \mathbb{N}(1)$ such that $z(k) - u_M(k)$ has a generalized zero at $k_n$ and a generalized zero (or zero) at some $\ell_0 \in \mathbb{N}(k_n + 1, k_0)$. We also have that $z(k_i) - u_M(k_i) = 0$, $2 \le i \le n-1$, and condition 11.6.2 implies that $z(k) = u_M(k)$ on $\mathbb{N}(a)$, which is again a contradiction. Thus, $S_2$ is closed and $S_2 = \mathbb{R}$. Choosing $A_n \in S_2$ it follows that there exists a solution $u(k)$ of (1.2.5) satisfying $u(k_i) = A_i$, $1 \le i \le n$. In summary, given $m_1 \in \mathbb{N}(a)$, $m_i = 1$, $2 \le i \le n-2$, $m_{n-1} > 1$, $m_n = 1$ each $(m_1, 1, \cdots, 1, m_{n-1}, 1)$ conjugate boundary value problem for (1.2.5) has a unique solution on $\mathbb{N}(a)$.

Still assuming the inductive hypotheses associated with $m_{n-1} > 1$, we assume in addition that $m_n > 1$ and that given $m_1 \in \mathbb{N}(a)$ and $m_i = 1$, $2 \le i \le n-2$ there exists a unique solution of each $(m_1, 1, \cdots, 1, m_{n-1}, h)$ conjugate boundary value problem, where $1 \le h < m_n$ for (1.2.5) on $\mathbb{N}(a)$. Let $z_3(k)$ be the solution of the $(m_1, 1, \cdots, 1, m_{n-1}, m_n - 1)$ conjugate problem for (1.2.5) satisfying $z_3(k_i) = A_i$, $1 \le i \le n-1$, $z_3(k_n - 1) = 0$. We define $S_3 = \{r \in \mathbb{R} :$ there is a solution $u(k)$ of (1.2.5) satisfying $u(k_i) = v_3(k_i)$, $1 \le i \le n-1$ and $u(k_n) = r\}$. As before $S_3$ is nonempty open subset of $\mathbb{R}$ and we claim that $S_3$ is also closed. Assuming again that the claim is false, let $r_0 \in \overline{S}_3 \backslash S_3$ and $\{r_m\} \subset S_3$ with $r_m \uparrow r_0$ be as in the previous considerations, and let $u_m(k)$ denote the solutions of (1.2.5) satisfying $u_m(k_i) = v_3(k_i)$, $1 \le i \le n-1$, $u_m(k_n) = r_m$. Condition 11.6.2 implies that $u_m(k) < u_{m+1}(k)$ on $\mathbb{N}(k_{n-1}+1)$, for all $m \in \mathbb{N}(1)$, and because of the existence of unique solutions of $(m_1, 1, \cdots, m_{n-1}, m_n - 1)$ problems for (1.2.5) along with $r_0 \notin S_3$, Problem 11.7.6 implies that $u_m(k_n - 1) \uparrow \infty$, as $m \to \infty$, and Problem 11.7.4 implies that for some $k_0 \in \mathbb{N}(k_n + 1, k_n + n - 1)$, $u_m(k_0) \uparrow \infty$ as $m \to \infty$.

Now let $z(k)$ be the solution of the $(m_1 + 1, 1, \cdots, 1, m_n - 1, m_n)$ boundary value problem for (1.2.5) satisfying $z(k_i) = v_3(k_i)$, $2 \le i \le n-2$, $z(k_{n-2}+1) = 0$, $z(k_{n-1}) = v_3(k_{n-1})$, $z(k_n) = r_0$. Such a solution $z(k)$ exists by the primary induction hypotheses on $m_{n-1}$. Because of the unbounded conditions on $\{u_m(k_n - 1)\}$ and $\{u_m(k_0)\}$, while $z(k_n) > u_m(k_n)$, for all $m \in \mathbb{N}(1)$, there exists $M \in \mathbb{N}(1)$ such that $z(k) - u_M(k)$ has a generalized zero at $k_n$ and a generalized zero at some $\ell_0 \in \mathbb{N}(k_n + 1, k_0)$. Moreover, $z(k_i) - u_M(k_i) = 0$, $2 \le i \le n-1$ from which it follows that $z(k) = u_M(k)$ on $\mathbb{N}(a)$. This contradiction completes the proof of $S_3$ being closed. Thus, $S_3 = \mathbb{R}$ and choosing $A_n \in S_3$ the corresponding solution $u(k)$ of (1.2.5) satisfying $u(k_n) = A_n$ is the desired solution. In particular, given $m_1 \in \mathbb{N}(a)$, $m_i = 1$, $2 \le i \le m-2$ and $m_n \ge 1$ each $(m_1, 1, \cdots, 1, m_{n-1}, m_n)$ conjugate boundary value problem for (1.2.5) has a unique solution on $\mathbb{N}(a)$. This completes the induction on $m_{n-1}$.

Now we shall induct on $m_{n-2}$. For this, our assumption is that $m_{n-2} > 1$ and that given $m_1 \in \mathbb{N}(a)$, $m_i = 1$, $2 \le i \le n-3$ and $m_{n-1}$, $m_n \ge 1$ there exists a unique solution of each $(m_1, 1, \cdots, 1, q, m_{n-1}, m_n)$ conjugate boundary value problem, where $1 \le q < m_{n-2}$ for (1.2.5) on $\mathbb{N}(a)$. Under this assumption, we will be concerned with the solutions of $(m_1, 1, \cdots, 1, m_{n-2}, 1, 1)$ followed by $(m_1, 1, \cdots, 1, m_{n-2}, 1, m_n)$, $m_n > 1$ followed by $(m_1, 1, \cdots, 1, m_{n-2}, m_{n-1}, 1)$, $m_{n-1} > 1$ followed by $(m_1, 1, \cdots, 1, m_{n-2}, m_{n-1}, m_n)$, $m_{n-1}$, $m_n > 1$ boundary value problems for (1.2.5).

Let $m_1 \in \mathbb{N}(a)$, $m_2 = \cdots = m_{n-3} = m_{n-1} = m_n = 1$ and $v_4(k)$ be the solution of the $(m_1, 1, \cdots, 1, m_{n-2} - 1, 1, 1)$ conjugate boundary value problem for (1.2.5) satisfying $v_4(k_i) = A_i$, $1 \leq i \leq n-3$, $v_4(k_{n-2} - 1) = 0$, $v_4(k_i) = A_i$, $i = n-2, n-1$. Defining $S_4 = \{r \in \mathbb{R} :$ there is a solution $u(k)$ of (1.2.5) satisfying $u(k_i) = v_4(k_i)$, $1 \leq i \leq n-1$ and $u(k_n) = r\}$, $S_4$ is nonempty and open. If we assume $S_4$ is not closed, then let $r_0$ and $\{r_m\}$, with $r_m \uparrow r_0$ be as earlier and $u_m(k)$ denote the corresponding solution of (1.2.5). It follows in this case that $u_m(k_{n-2} - 1) \uparrow \infty$ as $m \to \infty$ and for some $k_0 \in \mathbb{N}(k_n+1, k_n+n-1)$, $u_m(k_0) \uparrow \infty$ as $m \to \infty$. Denoting by $z(k)$ the solution of the $(m_1 + 1, 1, \cdots, 1, m_{n-2} - 1, 1, 1)$ problem for (1.2.5) satisfying $z(k_i) = v_4(k_i)$, $2 \leq i \leq n-3$, $z(k_{n-3} + 1) = 0$, $z(k_i) = v_4(k_i)$, $i = n-2, n-1$, $z(k_n) = r_0$ it follows that for some $M \in \mathbb{N}(1)$, $z(k) - u_M(k)$ has a generalized zero at $k_n$, a generalized zero at some $\ell_0 \in \mathbb{N}(k_n + 1, k_0)$, and zeros at $k_i$, $2 \leq i \leq n-1$. Again, it contradicts condition 11.6.2, and hence $S_4$ is closed. Thus, we can select $A_n \in S_4$ and the corresponding solution is the desired solution of the $(m_1, 1, \cdots, 1, m_{n-2}, 1, 1)$ problem for (1.2.5).

In addition to our assumptions on $m_{n-2} > 1$ we assume that $m_n > 1$ and that given $m_1 \in \mathbb{N}(a)$, $m_2 = \cdots = m_{n-3} = m_{n-1} = 1$, each $(m_1, 1, \cdots, 1, m_{n-2}, 1, h)$ conjugate boundary value problem, where $1 \leq h < m_n$, for (1.2.5) has a unique solution on $\mathbb{N}(a)$. Let $v_5(k)$ be the solution of the $(m_1, 1, \cdots, 1, m_{n-2}, 1, m_n - 1)$ problem for (1.2.5) satisfying $v_5(k_i) = A_i$, $1 \leq i \leq n-1$, $v_5(k_n - 1) = 0$. Defining $S_5$ in the standard way, $S_5$ is nonempty and open. If we assume $S_5$ is not closed, then let $r_0$ and $\{r_m\}$ with $r_m \uparrow r_0$ be as usual, and let $u_m(k)$ be the appropriate solution of (1.2.5). By the existence of unique solutions of $(m_1, 1, \cdots, 1, m_{n-2}, 1, m_n - 1)$ problems for (1.2.5), we have that $u_m(k_n - 1) \uparrow \infty$. Also, $u_m(k_0) \uparrow \infty$, where $k_0$ is as usual. In this case, now let $z(k)$ be the solution of the $(m_1+1, 1, \cdots, 1, m_{n-2}-1, 1, m_n)$ problem for (1.2.5) satisfying $z(k_i) = v_5(k_i)$, $2 \leq i \leq n-3$, $z(k_{n-3}+1) = 0$, $z(k_i) = v_5(k_i)$, $i = n-2$, $n-1$, $z(k_n) = r_0$. Then, there exists $M \in \mathbb{N}(1)$ such that $z(k) - u_M(k)$ has a generalized zero at $k_n$, a generalized zero at some $\ell_0 \in \mathbb{N}(k_n+1, k_0)$, and zeros at $k_i$, $2 \leq i \leq n-1$ which is the usual contradiction. Thus, $S_5$ is closed, and we conclude the existence of unique solutions of $(m_1, 1, \cdots, 1, m_{n-2}, 1, m_n)$ conjugate boundary value problems for (1.2.5) on $\mathbb{N}(a)$.

In addition to the primary inductive hypotheses on $m_{n-2}$, we assume now that $m_{n-1} > 1$ and that given $m_1 \in \mathbb{N}(a)$, $m_i = 1$, $2 \leq i \leq n-3$, $m_n \geq 1$ there exists a unique solution of each $(m_1, 1, \cdots, 1, m_{n-2}, p, m_n)$ conjugate boundary value problem, where $1 \leq p < m_{n-1}$, for (1.2.5) on

$\mathbb{N}(a)$. Let $v_6(k)$ be the solution of the $(m_1, 1, \cdots, 1, m_{n-2}, m_{n-1} - 1, 1)$ problem for (1.2.5) satisfying $v_6(k_i) = A_i,\ 1 \le i \le n-2,\ v_6(k_{n-1} - 1) = 0,\ v_6(k_{n-1}) = A_{n-1}$. The corresponding set $S_6$ will be nonempty and open. Repeating the pattern, we assume $S_6$ is not closed and make the usual arguments using $r_0$, $\{r_m\}$ and the corresponding solutions of (1.2.5). In this case $u_m(k_{n-1} - 1) \downarrow -\infty$ as $k \to \infty$, and for some $k_0 \in \mathbb{N}(k_n + 1, k_n + n - 1)$, $u_m(k_0) \uparrow \infty$ as $k \to \infty$. With $z(k)$ the solution of the $(m_1 + 1, 1, \cdots, 1, m_{n-2} - 1, m_{n-1}, 1)$ boundary value problem for (1.2.5) satisfying $z(k_i) = v_6(k_i),\ 2 \le i \le n-3,\ z(k_{n-3} + 1) = 0,\ z(k_i) = v_6(k_i),\ i = n-2,\ n-1,\ z(k_n) = r_0$ it follows that for some $M \in \mathbb{N}(1)$, $z(k) - u_M(k)$ has a generalized zero at $k_n$, a generalized zero at some $\ell_0 \in \mathbb{N}(k_n + 1, k_0)$, and zeros at $k_i$, $2 \le i \le n-1$. This is a contradiction to condition 11.6.2, and hence $S_6$ is closed, consequently each $(m_1, 1, \cdots, 1, m_{n-2}, m_{n-1}, 1)$ problem for (1.2.5) has a unique solution on $\mathbb{N}(a)$.

For the final step under the primary induction hypotheses on $m_{n-2} > 1$ and the induction hypotheses on $m_{n-1} > 1$, we assume in addition that $m_n > 1$ and given $m_1 \in \mathbb{N}(a)$ and $m_i = 1,\ 2 \le i \le n-3$ there exists a unique solution of each $(m_1, 1, \cdots, 1, m_{n-2}, m_{n-1}, h)$ conjugate boundary value problem, where $1 \le h < m_n$, for (1.2.5) on $\mathbb{N}(a)$. Let $v_7(k)$ be the solution of the $(m_1, 1, \cdots, 1, m_{n-2}, m_{n-1}, m_n - 1)$ boundary value problem for (1.2.5) satisfying $v_7(k_i) = A_i,\ 1 \le i \le n-1,\ v_7(k_n - 1) = 0$. Defining the nonempty open set $S_7$ as earlier, and making the usual assumption that $S_7$ is not closed, let $r_0, \{r_m\}$ and $u_m(k)$ be the appropriate values and solutions. We can argue that $u_m(k_n - 1) \uparrow \infty$ and $u_m(k_0) \uparrow \infty$ for some $k_0 \in \mathbb{N}(k_n + 1, k_n + n - 1)$. If $z(k)$ is the solution of the $(m_1 + 1, 1, \cdots, 1, m_{n-2} - 1, m_{n-1}, m_n)$ conjugate boundary value problem for (1.2.5) satisfying $z(k_i) = v_7(k_i),\ 2 \le i \le n-3,\ z(k_{n-3}+1) = 0,\ z(k_i) = v_7(k_i),\ i = n-2, n-1,\ z(k_n) = r_0$, then there exists $M \in \mathbb{N}(1)$ such that $z(k) - u_M(k)$ has a generalized zero at $k_n$, a generalized zero at some $\ell_0 \in \mathbb{N}(k_n + 1, k_0)$ and zeros at $k_i$, $2 \le i \le n-1$. This contradicts condition 11.6.2, hence $S_7$ is closed, and as in each of the above cases, the $(m_1, 1, \cdots, m_{n-2}, m_{n-1}, m_n)$ problem for (1.2.5) has a unique solution on $\mathbb{N}(a)$.

The above arguments exhibit the entire pattern for the induction scheme in obtaining solutions of the boundary value problems. For the general step, if $2 \le s \le n-3$ and $m_s > 1$ then we need to proceed through $2^{n-s}$ inductive steps, wherein we induct on $m_n, m_{n-1}, \cdots, m_{s+1}$, following the pattern in the above parts of the proof. ∎

**Definition 11.6.1.** Let $2 \le p \le n$ and let $m_1, \cdots, m_p$ be positive integers such that $\sum_{i=1}^{p} m_i = n$. Let $s_0 = 0$ and for $1 \le j \le p$, $s_j = \sum_{i=1}^{j} m_i$. For points $a \le k_p < k_{p-1} < \cdots < k_1$, where each $k_i \in \mathbb{N}(a)$ and $k_j + m_j + 1 \le k_{j-1}$, $2 \le j \le p$ conditions

$$\Delta^i u(k_j) = A_{i+1}, \quad s_{j-1} \le i \le s_j - 1, \quad 1 \le j \le p \tag{11.6.1}$$

are called $(m_p, \cdots, m_1)$ *left focal boundary conditions.*

**Condition 11.6.3.** Given $2 \le p \le n$, positive integers $m_1, \cdots, m_p$ such that $\sum_{i=1}^{p} m_i = n$, and points $a \le k_p < k_{p-1} < \cdots < k_1$ where each $k_i \in \mathbb{N}(a)$ and $k_j + m_j + 1 \le k_{j-1}$, $2 \le j \le p$ if $u(k)$ and $v(k)$ are solutions of (1.2.5) such that $\Delta^i(u(k) - v(k))$, $s_{j-1} \le i \le s_j - 1$, ($s_0 = 0$ and $s_j = \sum_{i=1}^{j} m_i$, $1 \le j \le p$), has a generalized zero at $k_j$, $1 \le j \le p$ then it follows that $u(k) = v(k)$ on $\mathbb{N}(k_p, k_1 + m_1 - 1)$.

As a consequence of condition 11.6.3 it follows that each $(m_p, \cdots, m_1)$ left focal boundary value problem (1.2.5), (11.6.1) has at most one solution on $\mathbb{N}(a)$. Further, if conditions 11.6.1 and 11.6.3 are satisfied then the conclusion of Theorem 11.6.1 holds.

**Theorem 11.6.2.** For the difference equation (1.2.5) let the conditions 11.6.1 and 11.6.3 be satisfied. Then, for each $1 \le p \le n-1$ and $0 \le q \le n-p$, and the points $a \le k_2 < k_1$, where $k_2, k_1 \in \mathbb{N}(a)$ and $k_2 + p + 1 \le k_1$, there exists a unique solution of (1.2.5) satisfying

$$\begin{aligned} \Delta^i u(k_2) &= A_{i+(n-p)-q+1}, \quad q \le i \le p+q-1 \\ \Delta^i u(k_1) &= A_{i+1}, \quad 0 \le i \le n-p-1 \end{aligned}$$

on $\mathbb{N}(a)$, for every choice of $A_i \in \mathbb{R}$, $1 \le i \le n$.

**Theorem 11.6.3.** For the difference equation (1.2.5) let the conditions 11.6.1 and 11.6.3 be satisfied. Then, for $2 \le p \le n$ each $(m_p, \cdots, m_1)$ left focal boundary value problem (1.2.5), (11.6.1) has a unique solution on $\mathbb{N}(a)$.

## 11.7. Problems

**11.7.1.** Prove Theorem 11.4.5.

**11.7.2.** Assume that $c, d \in \mathbb{N}(a, b+p)$, $c + p \le d$ and that solutions of the $(p, n-p)$ boundary value problem (11.5.1),

$$\begin{aligned} u(c+i) &= C_i, \quad 0 \le i \le p-1 \\ u(d+i) &= D_i, \quad 0 \le i \le n-p-1 \end{aligned} \tag{11.7.1}$$

are unique on $\mathbb{N}(c, d+n-p-1)$. By using Brouwer theorem on the invariance of domain show that for a given solution $u(k)$ of (11.5.1) there exists an $\epsilon > 0$ such that if $\overline{\gamma} = (\gamma_0, \cdots, \gamma_{p-1})$ and $\overline{\delta} = (\delta_0, \cdots, \delta_{n-p-1})$ satisfy $|\gamma_i| < \epsilon$, $0 \le i \le p-1$, $|\delta_i| < \epsilon$, $0 \le i \le n-p-1$ then the boundary value problem

$$(11.7.2)\qquad \begin{aligned} L[v(k)] &= f(k, v(k)), \quad k \in \mathbb{N}(a, b-1) \\ v(c+i) &= u(c+i) + \gamma_i, \quad 0 \le i \le p-1 \\ v(d+i) &= u(d+i) + \delta_i, \quad 0 \le i \le n-p-1 \end{aligned}$$

has a unique solution $v(k, \overline{\gamma}, \overline{\delta})$. Furthermore, as $\epsilon \to 0$ the solutions $v(k, \overline{\gamma}, \overline{\delta})$ converge to $u(k)$.

**11.7.3.** Let in addition to the uniqueness assumption of $(p, n-p)$ boundary value problem (11.5.1), (11.7.1) in Problem 11.7.2, $\partial f/\partial u$ exists and is continuous. Further, the variational equation $L[z(k)] = \dfrac{\partial f}{\partial u}(k, u(k))z(k)$ is right $(p, n-p)$ disconjugate along all solutions $u(k)$ of (11.5.1). Show that for a given solution $u(k)$ of (11.5.1) there exists an $\epsilon > 0$ such that the boundary value problem (11.5.1),

$$\begin{aligned} v(c+i) &= u(c+i), \quad 0 \le i \le p-2 \\ v(c+p-1) &= u(c+p-1) + \lambda \\ v(d+i) &= u(d+i), \quad 0 \le i \le n-p-1 \end{aligned}$$

has a unique solution $v(k, \lambda)$ for $|\lambda| < \epsilon$. Furthermore, $w(k) = \dfrac{\partial v(k, \lambda)}{\partial \lambda}$ exists for $|\lambda| < \epsilon$ and is the solution of the variational equation with $u(k) = v(k, \lambda)$ satisfying the boundary conditions

$$\begin{aligned} w(c+i) &= 0, \quad 0 \le i \le p-2 \\ w(c+p-1) &= 1 \\ w(d+i) &= 0, \quad 0 \le i \le n-p-1. \end{aligned}$$

**11.7.4.** For the difference equation (1.2.5) let the condition 11.6.1 be satisfied. Further, let there exist a sequence $\{u_m(k)\}$ of solutions of (1.2.5), an interval $\mathbb{N}(k_0, k_0+n-1) \subset \mathbb{N}(a)$, and an $M > 0$ such that $|u_m(k)| \le M$, for all $k \in \mathbb{N}(k_0, k_0+n-1)$ and $m \in \mathbb{N}(1)$. Show that there exists a subsequence $\{u_{m,j}(k)\}$ of $\{u_m(k)\}$ that converges pointwise on $\mathbb{N}(a)$ to a solution of (1.2.5).

**11.7.5.** For the difference equation (1.2.5) let the conditions 10.6.1 and 11.6.2 be satisfied. Show that for a given solution $u(k)$ of (1.2.5) on $\mathbb{N}(a)$, points $k_1 < k_2 < \cdots < k_n$ belonging to $\mathbb{N}(a)$, an interval $\mathbb{N}(k_1, b) \subset$

$\mathbb{N}(a)$ where $b \geq k_n$, and $\epsilon > 0$, there exists a $\delta(\epsilon, \mathbb{N}(k_1, b)) > 0$ such that, if $|u(k_i) - A_i| < \delta$, $1 \leq i \leq n$ then there exists a solution $v(k)$ of (1.2.5) satisfying $v(k_i) = A_i$, $1 \leq i \leq n$ and $|v(k) - u(k)| < \epsilon$ for all $k \in \mathbb{N}(k_1, b)$.

**11.7.6.** For the difference equation (1.2.5) let the conditions 10.6.1 and 11.6.2 be satisfied. Further, let there exist a sequence $\{u_m(k)\}$ of solutions of (1.2.5) and an $M > 0$ such that $|u_m(k_i)| \leq M$, $1 \leq i \leq n$ and $m \in \mathbb{N}(1)$. Show that there exists a subsequence $\{u_{m,j}(k)\}$ that converges pointwise on $\mathbb{N}(a)$. In particular, for this subsequence, if $\lim_{j\to\infty} u_{m,j}(k_i) = A_i$, $1 \leq i \leq n$ then show that $\{u_{m,j}(k)\}$ converges pointwise on $\mathbb{N}(a)$ to the solution of the $(m_1, m_2, \cdots, m_n)$ conjugate boundary value problem (1.2.5), (10.8.1).

**11.7.7.** For the difference equation (1.2.5) let the conditions 10.6.1 and 11.6.3 be satisfied. Let $2 \leq p \leq n$ and positive integers $m_1, \cdots, m_p$ such that $\sum_{i=1}^{p} m_i = n$ be given and let $s_j$, $0 \leq j \leq p$ be the corresponding partial sums. Show that for a given solution $u(k)$ of (1.2.5) on $\mathbb{N}(a)$, points $a \leq k_p < k_{p-1} < \cdots < k_1$, where each $k_i \in \mathbb{N}(a)$ and $k_j + m_j + 1 \leq k_{j-1}$, $2 \leq j \leq p$ an interval $\mathbb{N}(a, b)$, $b \geq k_1 + m_1 - 1$ and an $\epsilon > 0$, there exists a $\delta(\epsilon, \mathbb{N}(a, b)) > 0$ such that, if $\left|\Delta^i u(k_j) - A_{i+1}\right| < \delta$, $s_{j-1} \leq i \leq s_j - 1$, $1 \leq j \leq p$ then there exists a solution $v(k)$ of (1.2.5) satisfying $\Delta^i v(k_j) = A_{i+1}$, $s_{j-1} \leq i \leq s_j - 1$, $1 \leq j \leq p$ and $\left|\Delta^i v(k) - \Delta^i u(k)\right| < \epsilon$, $0 \leq i \leq n-1$ for all $k \in \mathbb{N}(a, b)$.

**11.7.8.** Prove Theorem 11.6.2.

**11.7.9.** Prove Theorem 11.6.3.

**11.7.10.** Let $(-1)^{n-r} p(k) > 0$ on $\mathbb{N}(c, d)$ for some $0 \leq r \leq n$ and $c,\ d \in \mathbb{N}(a)$ with $c + n - 1 \leq d$. Show that the boundary value problem

$$\begin{aligned} L[u(k)] + p(k)u(k) &= f(k), \quad k \in \mathbb{N}(c, d) \\ \Delta_i u(c) - \alpha_i \Delta_{i+1} u(c) &= A_i, \quad i \in \{i_1, \cdots, i_r\} \\ \Delta_j u(d) - \beta_j \Delta_{j+1} u(d) &= B_j, \quad j \in \{j_1, \cdots, j_{n-r}\} \end{aligned}$$

where the operator $L$ and the quasi-differences $\Delta_i$ are the same as in Section 10.7, $\alpha_i$, $\beta_j \geq 0$ and $0 \leq i_1 < \cdots < i_r \leq n-1$, $0 \leq j_1 < \cdots < j_{n-r} \leq n-1$, has a unique solution.

## 11.8. Notes.

In recent years the theory of boundary value problems for higher order differential equations has advanced profoundly, e.g. Agarwal [1] contains an in-depth and up-to-date coverage of more than 250 research publications. All the results in Sections 11.1 – 11.3 have been taken from Agarwal and Lalli [6]. Continuous analogs of these results are available in Agarwal [1], Agarwal and Wong [5]. An important feature of Theorems 11.2.2 and 11.2.3 is that these results reduce to Theorem 11.2.1 when $\nu = 0$ and $\nu_1 = 0$ respectively. It will be of interest to obtain similar results when the approximating function $f_m$ satisfies other error criteria. While in Section 11.3 we have succeeded in establishing the convergence of the quasilinear methods for the higher order equations, the known monotonic convergence property shared by second order continuous problems needs investigations. Theorems 11.4.1 11.4.4 are due to Eloe [8], whereas Theorem 10.4.5 is from Agarwal and Lalli [6]. Similar results for the continuous boundary value problems are available in Agarwal [1,3], Agarwal and Usmani [4], Šeda [20]. Theorems 11.5.1 and 11.5.2 are from Peterson [19], whereas Theorems 11.5.3 and 11.5.4 are borrowed from Hankerson [9,10]. Rest of the results in Section 11.5 are proved in Agarwal [2]. Theorem 11.6.1 is due to Henderson [16]. Its continuous analog has been proved independently by Hartman [11] and Klassen [17]. Theorems 11.6.2 and 11.6.3 have been proved in Henderson [14,15], whereas Henderson [12, 13] contains their continuous analogs. Existence and uniqueness of second and fourth order discrete boundary value problems have also been discussed in Lasota [18] and Denkowski [7] respectively.

## 11.9. References

[1]. R.P. Agarwal, *Boundary Value Problems for Higher Order Differential Equations, World Scientific,* Singapore, 1986.

[2]. R.P. Agarwal, Initial and boundary value problems for $n$th order difference equations, *Math. Slovaca* **36**(1986), 39 47.

[3]. R.P. Agarwal, Monotone convergence of iterative methods for $(n,p)$ and $(p,n)$ boundary value problems, *J. Comp. Appl. Math.* **21**(1988), 223 230.

[4]. R.P. Agarwal and R.A. Usmani, Monotone convergence of iterative methods for right focal point boundary value problems, *J. Math. Anal. Appl.* **130**(1988), 451 459.

[5]. R.P. Agarwal and P.J.Y. Wong, Lidstone polynomials and boundary value problems, *Computers Math. Applic.* **17**(1989), 1397 1421.

[6]. R.P. Agarwal and B.S. Lalli, Discrete: polynomial interpolation, Green's functions, maximum principles, error bounds and boundary value problems, *Computers Math. Applic.* **25**(1993), 3 39.

[7]. Z. Denkowski, The boundary value problems for ordinary nonlinear

differential and difference equations of the fourth order, *Annales Polonici Mathematici* **24**(1970), 87 102.

[8]. P.W. Eloe, Difference equations and multipoint boundary value problems, *Proc. Amer. Math. Soc.* **86**(1982), 253 259.

[9]. D. Hankerson, *Boundary Value Problems for nth Order Difference Equations*, Ph.D. dissertation, *University of Nebraska, Lincoln,* 1986.

[10]. D. Hankerson, An existence and uniqueness theorem for difference equations, *SIAM J. Math. Anal.* **20**(1989), 1208 1217.

[11]. P. Hartman, On $n-$parameter families and interpolation problems for nonlinear ordinary differential equations, *Trans. Amer. Math. Soc.* **154**(1971), 201 226.

[12]. J. Henderson, Existence of solutions of right focal point boundary value problems for ordinary differential equations, *Nonlinear Analysis* **5**(1981), 989 1002.

[13]. J. Henderson, Uniqueness of solutions of right focal point boundary value problems for ordinary differential equations, *J. Diff. Eqns.* **41**(1981), 218 227.

[14]. J. Henderson, Focal boundary value problems for nonlinear difference equations, I, *J. Math. Anal. Appl.* **141**(1989), 559 567.

[15]. J. Henderson, Focal boundary value problems for nonlinear difference equations, II, *J. Math. Anal. Appl.* **141**(1989), 568 579.

[16]. J. Henderson, Existence theorems for boundary value problems for $n$th order nonlinear difference equations, *SIAM J. Math. Anal.* **20**(1989), 468 478.

[17]. G. Klaasen, Existence theorems for boundary value problems for $n$th order ordinary differential equations, *Rocky Mountain J. Math.* **3**(1973), 457 472.

[18]. A. Lasota, A discrete boundary value problem, *Annales Polonici Mathematici* **20**(1968), 183 190.

[19]. A. Peterson, Existence and uniqueness theorems for nonlinear difference equations, *J. Math. Anal. Appl.* **125**(1987), 185 191.

[20]. V. Šeda, Two remarks on boundary value problems for ordinary differential equations, *J. Diff. Eqns.* **26**(1977), 278 290.

# Chapter 12

# Sturm–Liouville Problems and Related Inequalities

This chapter is devoted to special type of boundary value problems which lead to the concepts of eigenvalues and eigenfunctions, orthogonality, and finite Fourier series. While in relation to differential equations these notions play a fundamental role in the study of mathematical physics and engineering, and have resulted in a vast amount of advanced mathematics, in the discrete case their importance is not fully explored, except that most of these problems are equivalent to some special matrix eigenvalue problems. We shall exploit this equivalence to derive Wirtinger and Opial type inequalities. Next, in this chapter we shall touch upon cone theory and use it to prove the existence and the comparison theorems for the least positive eigenvalues of the $(p, n-p)$ discrete boundary value problems. Finally, as a further application to cone theory we shall discuss positive solutions and nonlinear eigenvalue problems for third order difference equations.

For convenience, throughout this chapter a row as well as column vector $\mathbf{u}$ in $\mathbb{R}^n$ is denoted as $\mathbf{u} = (u_1, \cdots, u_n)$.

## 12.1. Sturm–Liouville Problems

Obviously, the homogeneous linear boundary value problems may have nontrivial solutions. If the coefficients of the difference equation and/or of the boundary conditions depend upon a parameter, then one of the pioneer problems of mathematical physics is to determine the value(s) of the parameter for which such nontrivial solutions exist. These special values of the parameter are called ***eigenvalues*** and the corresponding nontrivial solutions are called ***eigenfunctions***. Boundary value problem which consists of the difference equation

$$\Delta(p(k-1)\Delta u(k-1)) + q(k)u(k) + \lambda r(k)u(k) = 0, \quad k \in \mathbb{N}(1, K) \tag{12.1.1}$$

and the boundary conditions

$$u(0) = \alpha u(1), \quad u(K+1) = \beta u(K) \tag{12.1.2}$$

is called *Sturm Liouville problem.* In the difference equation (12.1.1), $\lambda$ is a parameter, and the functions $p, q$ and $r$ are defined on $\mathbb{N}(0,K)$, $\mathbb{N}(1,K)$ and $\mathbb{N}(1,K)$ respectively, and $p(k) > 0,\ k \in \mathbb{N}(0,K)$, $r(k) > 0,\ k \in \mathbb{N}(1,K)$. In the boundary conditions (12.1.2), $\alpha$ and $\beta$ are known constants.

The following results in which the existence of the eigenvalues of (12.1.1), (12.1.2) is tacitly assumed are fundamental.

**Theorem 12.1.1.** The eigenvalues of the Sturm Liouville problem (12.1. 1), (12.1.2) are simple, i.e. if $\lambda$ is an eigenvalue of (12.1.1), (12.1.2) and $\phi_1(k)$ and $\phi_2(k)$ are the corresponding eigenfunctions, then $\phi_1(k)$ and $\phi_2(k)$ are linearly dependent on $\mathbb{N}(0, K+1)$.

**Proof.** In the expanded form the difference equation (12.1.1) is the same as

$$(12.1.3)\quad \begin{aligned} p(k)u(k+1) - (p(k)+p(k-1))u(k) + (q(k)+\lambda r(k))u(k) \\ +p(k-1)u(k-1) \;=\; 0, \quad k \in \mathbb{N}(1,K). \end{aligned}$$

Therefore, if $\phi_1(k)$ and $\phi_2(k)$ both are solutions of (12.1.3), then from Problem 2.16.21 if follows that

$$det\ C(\phi_1,\phi_2)(k) \;=\; det\ C(0)\prod_{\ell=0}^{k-1}\frac{p(\ell)}{p(\ell+1)}, \quad k \in \mathbb{N}(0,K)$$

and hence $p(k)\, det\ C(\phi_1,\phi_2)(k) = c$ (constant). To find the value of $c$, we note that $\phi_1(0) = \alpha\phi_1(1)$ and $\phi_2(0) = \alpha\phi_2(1)$. This implies that $det\, C(\phi_1,\phi_2)(0) = 0$, and hence $c$ is zero. Thus, $p(k)\, det\, C(\phi_1,\phi_2)(k) = 0$, $k \in \mathbb{N}(0,K)$, i.e. $\phi_1(k)$ and $\phi_2(k)$ are linearly dependent on $\mathbb{N}(0, K+1)$.

**Definition 12.1.1.** The set of functions $\{\phi_m(k),\ m = 1, 2, \cdots\}$ each of which is defined on $\overline{\mathbb{N}}$ is said to be *orthogonal* on $\overline{\mathbb{N}}$ with respect to the nonnegative function $r(k),\ k \in \overline{\mathbb{N}}$ if

$$\sum_{\ell\in\overline{\mathbb{N}}} r(\ell)\phi_\mu(\ell)\phi_\nu(\ell) \;=\; 0 \quad \text{for all } \mu \neq \nu.$$

The function $r(k)$ is called the *weight function.*

**Theorem 12.1.2.** Let $\lambda_m,\ m = 1, 2, \cdots$ be the eigenvalues of the Sturm Liouville problem (12.1.1), (12.1.2) and $\phi_m(k),\ m = 1, 2, \cdots$ be the corresponding eigenfunctions. Then, the set $\{\phi_m(k),\ m = 1, 2, \cdots\}$ is orthogonal on $\mathbb{N}(1,K)$ with respect to the weight function $r(k)$.

**Proof.** Let $\lambda_\mu$ and $\lambda_\nu$ be two distinct eigenvalues of (12.1.1), (12.1.2) and $\phi_\mu(k)$ and $\phi_\nu(k)$ be the corresponding eigenfunctions. Then, the two equations

$$\Delta(p(k-1)\Delta\phi_\mu(k-1)) + q(k)\phi_\mu(k) + \lambda_\mu r(k)\phi_\mu(k) = 0 \tag{12.1.4}$$

and

$$\Delta(p(k-1)\Delta\phi_\nu(k-1)) + q(k)\phi_\nu(k) + \lambda_\nu r(k)\phi_\nu(k) = 0, \ k \in \mathbb{N}(1, K) \tag{12.1.5}$$

are satisfied. We multiply (12.1.4) by $\phi_\nu(k)$ and (12.1.5) by $\phi_\mu(k)$ and subtract one resulting equation from the other, to obtain

$$\begin{aligned}
&(\lambda_\mu - \lambda_\nu)r(k)\phi_\mu(k)\phi_\nu(k) \\
&\quad = \phi_\mu(k)\Delta(p(k-1)\Delta\phi_\nu(k-1)) - \phi_\nu(k)\Delta(p(k-1)\Delta\phi_\mu(k-1)).
\end{aligned}$$

Summing this relation from $k = 1$ to $k = K$ and using (1.8.5), to get

$$\begin{aligned}
&(\lambda_\mu - \lambda_\nu)\sum_{\ell=1}^{K} r(\ell)\phi_\mu(\ell)\phi_\nu(\ell) \\
&\quad = \phi_\mu(\ell)p(\ell-1)\Delta\phi_\nu(\ell-1)\Big|_{\ell=1}^{K+1} - \sum_{\ell=1}^{K}\Delta\phi_\mu(\ell)p(\ell)\Delta\phi_\nu(\ell) \\
&\qquad - \phi_\nu(\ell)p(\ell-1)\Delta\phi_\mu(\ell-1)\Big|_{\ell=1}^{K+1} + \sum_{\ell=1}^{K}\Delta\phi_\nu(\ell)p(\ell)\Delta\phi_\mu(\ell)
\end{aligned}$$

$$= p(\ell-1)(\phi_\nu(\ell)\phi_\mu(\ell-1) - \phi_\mu(\ell)\phi_\nu(\ell-1))\Big|_{\ell=1}^{K+1}. \tag{12.1.6}$$

In view of the boundary conditions (12.1.2) the right side of (12.1.6) clearly vanishes. Hence, we have

$$(\lambda_\mu - \lambda_\nu)\sum_{\ell=1}^{K} r(\ell)\phi_\mu(\ell)\phi_\nu(\ell) = 0. \tag{12.1.7}$$

However, since $\lambda_\mu \neq \lambda_\nu$ the result follows. ∎

**Theorem 12.1.3.** Let $\lambda_1$ and $\lambda_2$ be two eigenvalues of the Sturm Liouville problem (12.1.1), (12.1.2) and $\phi_1(k)$ and $\phi_2(k)$ be the corresponding eigenfunctions. Then, $\phi_1(k)$ and $\phi_2(k)$ are linearly dependent on $\mathbb{N}(0, K+1)$ only if $\lambda_1 = \lambda_2$.

**Proof.** The proof is a direct consequence of the equality (12.1.7). ∎

**Theorem 12.1.4.** For the Sturm Liouville problem (12.1.1), (12.1.2) eigenvalues are real.

**Proof.** Let $\lambda = \xi + i\zeta$ be a complex eigenvalue and $\phi(k) = v(k) + iw(k)$ be the corresponding eigenfunction of (12.1.1), (12.1.2). Then, it is easily seen that the equations

$$\Delta(p(k-1)\Delta v(k-1)) + q(k)v(k) + (\xi v(k) - \zeta w(k))r(k) = 0$$

and

$$\Delta(p(k-1)\Delta w(k-1)) + q(k)w(k) + (\zeta v(k) + \xi w(k))r(k) = 0, \quad k \in \mathbb{N}(1,K)$$

are satisfied. Further

$$v(0) = \alpha v(1), \quad w(0) = \alpha w(1), \quad v(K+1) = \beta v(K), \quad w(K+1) = \beta w(K).$$

Thus, as in Theorem 12.1.2, we find

$$\sum_{\ell=1}^{K}[-(\xi v(\ell) - \zeta w(\ell))w(\ell) + (\zeta v(\ell) + \xi w(\ell))v(\ell)]r(\ell) = 0,$$

which is the same as

$$\zeta \sum_{\ell=1}^{K}(w^2(\ell) + v^2(\ell))r(\ell) = 0.$$

Hence, it is necessary that $\zeta = 0$, i.e. $\lambda$ is real. ∎

**Example 12.1.1.** For the Sturm Liouville problem

$$\begin{aligned} \Delta^2 u(k-1) + (2-s)u(k) + \lambda u(k) &= 0, \quad k \in \mathbb{N}(1,K), \\ &\qquad s > 0 \text{ is a number} \\ u(0) = u(K+1) &= 0 \end{aligned} \tag{12.1.8}$$

the eigenvalues are $\lambda_m = s - 2\cos\left(\frac{m\pi}{K+1}\right)$, $1 \le m \le K$ and the corresponding eigenfunctions are $\phi_m(k) = \sin\left(\frac{mk\pi}{K+1}\right)$, $1 \le m \le K$. In particular, for $s = 2$ the eigenvalues simplify to $\lambda_m = 4\sin^2\frac{m\pi}{2(K+1)}$, $1 \le m \le K$.

**Example 12.1.2.** For the Sturm Liouville problem

$$\begin{aligned} \Delta^2 u(k-1) + 2\lambda u(k) &= 0, \quad k \in \mathbb{N}(1,K) \\ u(0) = u(K+1) &= 0 \end{aligned} \tag{12.1.9}$$

the eigenvalues are $\lambda_m = 2\sin^2\dfrac{m\pi}{2(K+1)}, \ 1 \le m \le K$ and the corresponding eigenfunctions are $\phi_m(k) = \sin\left(\dfrac{mk\pi}{K+1}\right), \ 1 \le m \le K.$

**Example 12.1.3.** For the Sturm Liouville problem

$$\Delta^2 u(k-1) + \lambda u(k) = 0, \quad k \in \mathbb{N}(1,K) \tag{12.1.10}$$

$$u(0) = u(1), \quad u(K+1) = u(K) \tag{12.1.11}$$

the eigenvalues are $\lambda_m = 4\sin^2\dfrac{m\pi}{2K}, \ 0 \le m \le K-1$ and the corresponding eigenfunctions are $\phi_m(k) = \cos\dfrac{m\pi(2k-1)}{2K}, \ 0 \le m \le K-1.$

**Example 12.1.4.** For the eigenvalue problem

$$\begin{aligned} cu(k+1) - au(k) + bu(k-1) + \lambda u(k) &= 0, \\ & k \in \mathbb{N}(1,K), \ bc > 0 \\ u(0) = u(K+1) &= 0 \end{aligned} \tag{12.1.12}$$

the eigenvalues are $\lambda_m = a - 2\sqrt{bc}\cos\left(\dfrac{m\pi}{K+1}\right), \ 1 \le m \le K$ and the corresponding eigenfunctions are $\phi_m(k) = (b/c)^{k/2}\sin\left(\dfrac{mk\pi}{K+1}\right), \ 1 \le m \le K.$

## 12.2. Eigenvalue Problems for Symmetric Matrices

Let $\mathcal{A}$ be a real symmetric $n \times n$ matrix, and $\mathcal{R}$ be an $n \times n$ diagonal matrix with positive diagonal elements. For the matrix eigenvalue problem

$$\mathcal{A}\mathbf{u} = \lambda\mathcal{R}\mathbf{u} \tag{12.2.1}$$

the following results are well known.

1. There exist exactly $n$ real eigenvalues $\lambda_m, \ 1 \le m \le n$ which need not be distinct.

2. Corresponding to each eigenvalue $\lambda_m$ there exists an eigenvector $\mathbf{u}^m$ which can be so chosen that $n$ vectors $\mathbf{u}^1, \cdots, \mathbf{u}^n$ are mutually orthogonal with respect to the matrix $\mathcal{R} = diag\ (r_{11}, \cdots, r_{nn})$, i.e. $(\mathbf{u}^\mu)\mathcal{R}\mathbf{u}^\nu = \sum_{i=1}^n r_{ii}u_i^\mu u_i^\nu = 0$ if $\mu \ne \nu$. In particular, these vectors are linearly independent.

3. If the real symmetric matrix $\mathcal{A}$ is tridiagonal of the form

$$(12.2.2)\qquad \mathcal{H}_n(\mathbf{g},\mathbf{h}) = \begin{bmatrix} g_1 & h_1 & & & & \\ h_1 & g_2 & h_2 & & & \\ & h_2 & \cdot & \cdot & & \\ & & \cdot & & & \\ & & & h_{n-2} & g_{n-1} & h_{n-1} \\ & & & & h_{n-1} & g_n \end{bmatrix}$$

where $\mathbf{g} = (g_1, \cdots, g_n)$, $\mathbf{h} = (h_1, \cdots, h_{n-1})$ and $h_i^2 > 0$, $1 \le i \le n-1$ then the eigenvalues $\lambda_m$ of (12.2.1) are real and distinct.

4. If $\mathcal{R} = \mathcal{I}$ and the eigenvalues $\lambda_m$, $1 \le m \le n$ (of $\mathcal{A}$ ) are arranged in an increasing order, i.e. $\lambda_1 \le \cdots \le \lambda_n$ then for any vector $\mathbf{u} \in \mathbb{R}^n$,

$$(12.2.3)\qquad \lambda_1(\mathbf{u},\mathbf{u}) \le (\mathcal{A}\mathbf{u},\mathbf{u}) \le \lambda_n(\mathbf{u},\mathbf{u}),$$

where $(\mathbf{u},\mathbf{v}) = \sum_{i=1}^n u_i v_i$ is the usual scalar product. In case $\lambda_1 < \lambda_2$ the equality $\lambda_1(\mathbf{u},\mathbf{u}) = (\mathcal{A}\mathbf{u},\mathbf{u})$ holds if and only if $\mathbf{u}$ is a scalar multiple of $\mathbf{u}^1$. Similarly, if $\lambda_{n-1} < \lambda_n$ the equality $(\mathcal{A}\mathbf{u},\mathbf{u}) = \lambda_n(\mathbf{u},\mathbf{u})$ holds if and only if $\mathbf{u}$ is a scalar multiple of $\mathbf{u}^n$. Further, for any vector $\mathbf{u}$ orthogonal to $\mathbf{u}^1$,

$$(12.2.4)\qquad \lambda_2(\mathbf{u},\mathbf{u}) \le (\mathcal{A}\mathbf{u},\mathbf{u}).$$

In case $\lambda_4 > \lambda_3 = \lambda_2 > \lambda_1$, then a vector $\mathbf{u}$ orthogonal to $\mathbf{u}^1$ satisfies the equality $\lambda_2(\mathbf{u},\mathbf{u}) = (\mathcal{A}\mathbf{u},\mathbf{u})$ if and only if $\mathbf{u}$ is a linear combination of $\mathbf{u}^2$ and $\mathbf{u}^3$.

5. If the real symmetric matrix $\mathcal{A}$ is positive definite also, i.e. for every $\mathbf{u} \in \mathbb{R}^n$, $(\mathcal{A}\mathbf{u},\mathbf{u}) > 0$ then the eigenvalues $\lambda_m$, $1 \le m \le n$ are positive. Thus, in particular, if $\mathcal{R} = \mathcal{I}$ and $\mathcal{A} = \mathcal{H}_n(\mathbf{g},\mathbf{h})$ is positive definite, then the eigenvalues $\lambda_m$, $1 \le m \le n$ (of $\mathcal{A}$ ) can be arranged in an strictly increasing order, i.e. $0 < \lambda_1 < \cdots < \lambda_n$.

**Example 12.2.1.** For the positive functions $r(k)$, $k \in \mathbb{N}(0,n)$ and $p(k)$, $k \in \mathbb{N}(1,n)$ the matrix $\mathcal{H}_n(\mathbf{g},\mathbf{h})$, where

$$\mathbf{g} = \left(\frac{r(0)+r(1)}{p(1)}, \cdots, \frac{r(n-1)+r(n)}{p(n)}\right)$$

and $\mathbf{h} = \left(-\frac{r(1)}{\sqrt{p(1)p(2)}}, \cdots, -\frac{r(n-1)}{\sqrt{p(n-1)p(n)}}\right)$ is positive definite. This is clear from the equality

$$(12.2.5)\qquad (\mathcal{H}_n(\mathbf{g},\mathbf{h})\mathbf{u},\mathbf{u}) = \sum_{\ell=1}^{n} \frac{r(\ell-1)+r(\ell)}{p(\ell)} u^2(\ell)$$

$$
\begin{aligned}
&-2\sum_{\ell=1}^{n-1}\frac{r(\ell)}{\sqrt{p(\ell)p(\ell+1)}}u(\ell)u(\ell+1)\\
&= \frac{r(0)}{p(1)}u^2(1)+\sum_{\ell=1}^{n-1}\frac{r(\ell)}{p(\ell)p(\ell+1)}(\sqrt{p(\ell+1)}\,u(\ell)\\
&\quad -\sqrt{p(\ell)}u(\ell+1))^2+\frac{r(n)}{p(n)}u^2(n).
\end{aligned}
$$

**Example 12.2.2.** For the positive functions $r(k),\ k\in\mathbb{N}(0,n-1)$ and $p(k),\ k\in\mathbb{N}(1,n)$ the matrix $\mathcal{H}_n(\mathbf{g},\mathbf{h})$ where

$$
\mathbf{g} \;=\; \left(\frac{r(0)+r(1)}{p(1)},\cdots,\frac{r(n-2)+r(n-1)}{p(n-1)},\frac{r(n-1)}{p(n)}\right)
$$

and $\mathbf{h} \;=\; \left(-\frac{r(1)}{\sqrt{p(1)p(2)}},\cdots,-\frac{r(n-1)}{\sqrt{p(n-1)p(n)}}\right)$ is positive definite.

**Example 12.2.3.** For the positive functions $r(k),\ k\in\mathbb{N}(1,n)$ and $p(k),\ k\in\mathbb{N}(1,n)$ the matrix $\mathcal{H}_n(\mathbf{g},\mathbf{h})$ where

$$
\mathbf{g} \;=\; \left(\frac{r(1)}{p(1)},\frac{r(1)+r(2)}{p(2)},\cdots,\frac{r(n-1)+r(n)}{p(n)}\right)
$$

and $\mathbf{h} \;=\; \left(-\frac{r(1)}{\sqrt{p(1)p(2)}},\cdots,-\frac{r(n-1)}{\sqrt{p(n-1)p(n)}}\right)$ is positive definite.

## 12.3. Matrix Formulation of Sturm–Liouville Problems

Let in the equation (12.1.3), $s(k)=p(k)+p(k-1)-q(k),\ k\in\mathbb{N}(1,K)$ so that it can be written as

$$
\begin{aligned}
(12.3.1)\qquad -p(k-1)u(k-1)+s(k)u(k)-p(k)u(k+1)&\\
=\lambda r(k)u(k),\qquad k\in\mathbb{N}(1,K).&
\end{aligned}
$$

Thus, for $k=1,\ K$ we have the equations

$$
(12.3.2)\qquad -p(0)u(0)+s(1)u(1)-p(1)u(2) \;=\; \lambda r(1)u(1)
$$

and

$$
(12.3.3)\quad -p(K-1)u(K-1)+s(K)u(K)-p(K)u(K+1) \;=\; \lambda r(K)u(K),
$$

which in view of the boundary conditions (12.1.2) take the form

$$
(12.3.4)\qquad \overline{s}(1)u(1)-p(1)u(2) \;=\; \lambda r(1)u(1)
$$

and

$$(12.3.5) \qquad -p(K-1)u(K-1)+\overline{s}(K)u(K) \;=\; \lambda r(K)u(K),$$

where $\overline{s}(1)=s(1)-\alpha p(0)$ and $\overline{s}(K)=s(K)-\beta p(K)$.

The $K$ equations (12.3.4), (12.3.1) for $k\in\mathbb{N}(2,K-1)$ and (12.3.5) can be written in the system form (12.2.1), where the $K\times K$ matrix $\mathcal{A}$ is real, symmetric and tridiagonal of the form $\mathcal{H}_K(\mathbf{s},\mathbf{p})$, with $\mathbf{s}=(\overline{s}(1),s(2),\cdots,s(K-1),\overline{s}(K))$ and $\mathbf{p}=(-p(1),\cdots,-p(K-1))$, $\mathcal{R}$ is a $K\times K$ diagonal matrix defined as $\mathcal{R}=diag\,(r(1),\cdots,r(K))$, and $\mathbf{u}=(u(1),\cdots,u(K))$.

Since $p(k)>0$, $k\in\mathbb{N}(0,K)$ and $r(k)>0$, $k\in\mathbb{N}(1,K)$ it follows that (i) the problem (12.1.1), (12.1.2) has exactly $K$ real eigenvalues $\lambda_m$, $1\le m\le K$ which are distinct, and (ii) corresponding to each eigenvalue $\lambda_m$ there exists an eigenfunction $\phi_m(k)$, $k\in\mathbb{N}(1,K)$. These eigenfunctions $\phi_m(k)$, $1\le m\le K$ are mutually orthogonal with respect to the function $r(k)$, i.e. $\sum_{\ell=1}^{K} r(\ell)\phi_\mu(\ell)\phi_\nu(\ell)=0$, if $\mu\ne\nu$. In particular, these eigenfunctions are linearly independent on $\mathbb{N}(1,K)$.

Thus, the matrix formulation (12.2.1) of the Sturm Liouville problem (12.1.1), (12.1.2) is more informative than the conclusions of Theorems 12.1.1 12.1.4. We further note that if the condition $p(k)>0$, $k\in\mathbb{N}(0,K)$ does not hold then also the problem (12.1.1), (12.1.2) has exactly $K$ real eigenvalues $\lambda_m$, $1\le m\le K$ but may not be distinct, and with respect to $r(k)$ mutually orthogonal eigenfunctions $\phi_m(k)$, $1\le m\le K$ can be chosen.

If $p(0)=0$, then in (12.3.2) the quantity $u(0)$ is not involved, so that the boundary condition $u(0)=\alpha u(1)$ is not needed. A similar remark holds in the case when $p(K)=0$.

**Example 12.3.1.** For the eigenvalue problem

$$(12.3.6) \qquad \Delta(p(k-1)\Delta u(k-1))+\lambda u(k) \;=\; 0, \quad k\in\mathbb{N}(1,K)$$

$$(12.3.7) \qquad u(0) \;=\; 0,$$

where $p(k)=1$, $k\in\mathbb{N}(0,K-1)$, $p(K)=0$ the eigenvalues are $\lambda_m=4\sin^2\left(\dfrac{2m-1}{2(2K+1)}\right)\pi$, $1\le m\le K$ and the corresponding eigenfunctions are $\phi_m(k)=\sin\left(\dfrac{2m-1}{2K+1}\right)k\pi$, $1\le m\le K$.

**Example 12.3.2.** For the difference equation (12.3.6) where $p(k)=1$, $k\in\mathbb{N}(1,K-1)$, $p(0)=p(K)=0$ the eigenvalues are $\lambda_m=$

$4\sin^2\dfrac{(m-1)\pi}{2K}$, $1 \le m \le K$ and the corresponding eigenfunctions are $\phi_m(k) = \cos\left(\dfrac{(m-1)(2k-1)\pi}{2K}\right)$, $1 \le m \le K$.

## 12.4. Symmetric, Antisymmetric and Periodic Boundary Conditions

Consider the difference equation (12.1.1) on $\mathbb{N}(0, K+1)$ together with the *symmetric boundary conditions*

$$(12.4.1)\qquad u(-k) = u(k), \quad u(K+1+k) = u(K+1-k), \quad k \in \mathbb{Z}.$$

It is clear that the symmetric boundary value problem (12.1.1), (12.4.1) extends the definition of $u(k)$ to all integers $k$.

Once again in (12.1.3) we let $s(k) = p(k) + p(k-1) - q(k)$, $k \in \mathbb{N}(0, K+1)$ so that besides (12.3.1) the equations corresponding to $k = 0$ and $k = K+1$ are

$$(12.4.2)\qquad -p(-1)u(-1) + s(0)u(0) - p(0)u(1) = \lambda r(0)u(0)$$

and

$$(12.4.3)\quad -p(K)u(K)+s(K+1)u(K+1)-p(K+1)u(K+2) = \lambda r(K+1)u(K+1).$$

In these equations we use (12.4.1) for $k = 1$ to eliminate $u(-1)$ and $u(K+2)$. Thus, the resulting equations can be written as

$$(12.4.4)\qquad s(0)u(0) - (p(-1) + p(0))u(1) = \lambda r(0)u(0)$$

and

$$(12.4.5)\quad -(p(K) + p(K+1))u(K) + s(K+1)u(K+1) = \lambda r(K+1)u(K+1).$$

If $p(-1) = p(K+1) = 0$, then $(K+2)$ equations (12.4.4), (12.3.1) and (12.4.5) lead to a system of the form (12.2.1), where $\mathcal{A}$ is a $(K+2)\times(K+2)$ real symmetric tridiagonal matrix $\mathcal{H}_{K+2}(\mathbf{s}, \mathbf{p})$, with $\mathbf{s} = (s(0), \cdots, s(K+1))$ and $\mathbf{p} = (-p(0), \cdots, -p(K))$, $\mathcal{R}$ is a $(K+2)\times(K+2)$ diagonal matrix defined as $\mathcal{R} = diag\ (r(0), ..., r(K+1))$, and $\mathbf{u} = (u(0), \cdots, u(K+1))$.

If $p(-1)p(K+1) \neq 0$, then the $(K+2)$ equations (12.4.4), (12.3.1) and (12.4.5) lead to a system of the form (12.2.1) with a nonsymmetric matrix $\mathcal{A}$. However, in this case equations (12.4.4) and (12.4.5) can be written as

$$(12.4.6)\qquad \frac{p(0)s(0)}{p(-1)+p(0)}u(0) - p(0)u(1) = \lambda\frac{p(0)r(0)}{p(-1)+p(0)}u(0)$$

and

$$(12.4.7)\quad -p(K)u(K)+\frac{p(K)s(K+1)}{p(K)+p(K+1)}u(K+1) = \lambda\frac{p(K)r(K+1)}{p(K)+p(K+1)}u(K+1).$$

The $(K+2)$ equations (12.4.6), (12.3.1) and (12.4.7) does lead to a system of the form (12.2.1) with a symmetric matrix $\mathcal{A}$. Further, for this system if $\lambda_\mu$ and $\lambda_\nu$ are the two distinct eigenvalues and $\phi_\mu(k)$ and $\phi_\nu(k)$, $k \in \mathbb{N}(0, K+1)$ are the corresponding eigenfunctions, then it follows that

$$\frac{p(0)r(0)}{p(-1)+p(0)}\phi_\mu(0)\phi_\nu(0) + \sum_{\ell=1}^{K} r(\ell)\phi_\mu(\ell)\phi_\nu(\ell) + \frac{p(K)r(K+1)}{p(K)+p(K+1)}\phi_\mu(K+1)\phi_\nu(K+1) = 0.$$

Thus, the eigenfunctions are orthogonal on $\mathbb{N}(0, K+1)$ with respect to the weight function

$$\overline{r}(k) = \begin{cases} p(0)r(0)/(p(-1)+p(0)), & k=0 \\ r(k), \quad k \in \mathbb{N}(1,K) \\ p(K)r(K+1)/(p(K)+p(K+1)), & k = K+1. \end{cases}$$

Now we shall consider the difference equation (12.1.1) on $\mathbb{N}(0, K+1)$ together with the *antisymmetric boundary conditions*

$$(12.4.8)\quad u(-k) = -u(k), \quad u(K+1+k) = -u(K+1-k), \quad k \in \mathbb{Z}.$$

In particular, these conditions imply that $u(0) = u(K+1) = 0$. Thus, the equations (12.4.2) and (12.4.3) reduce to

$$(12.4.9)\qquad -p(-1)u(-1) - p(0)u(1) = 0$$

and

$$(12.4.10)\qquad -p(K)u(K) - p(K+1)u(K+2) = 0.$$

Therefore, in view of (12.4.8) for $k=1$ these equations are

$$(12.4.11)\qquad (p(-1) - p(0))u(1) = 0$$

and

$$(12.4.12)\qquad (p(K+1) - p(K))u(K) = 0.$$

Since for a nontrivial solution $u(1)u(K) \neq 0$, it is necessary that $p(-1) = p(0)$ and $p(K) = p(K+1)$.

Finally, we note that the boundary conditions $u(0) = u(K+1) = 0$ is a particular case of (12.1.2), and therefore all the results of Section 12.1.1 hold for the antisymmetric boundary value problem (12.1.1) on $\mathbb{N}(0, K+1)$, (12.4.8) also. Further, since

$$\sum_{\ell=0}^{K+1} r(\ell)\phi_\mu(\ell)\phi_\nu(\ell) = \sum_{\ell=1}^{K} r(\ell)\phi_\mu(\ell)\phi_\nu(\ell) = 0 \quad (\lambda_\mu \neq \lambda_\nu)$$

the eigenfunctions are orthogonal over $\mathbb{N}(0, K+1)$ as well as $\mathbb{N}(1, K)$ with respect to the function $r(k)$.

In view of the above considerations it is clear that the eigenfunctions of the symmetric (antisymmetric) boundary value problem are even (odd) periodic functions of period $2K+2$. Further, if the functions $p(k)$, $q(k)$ and $r(k)$ are periodic of period $2K+2$, and if $p(-k) = p(k-1)$, $q(-k) = q(k)$ and $r(-k) = r(k)$ on $\mathbb{Z}$, then the eigenfunctions of these problems satisfy the difference equation (12.1.1) on $\mathbb{Z}$.

In the rest of this section we shall consider the difference equation (12.1.1) on $\mathbb{N}(0, K+1)$ together with the periodic boundary conditions

$$u(K+1+k) = u(k), \quad k \in \mathbb{Z}. \tag{12.4.13}$$

Since $u(K) = u(-1)$, $u(K+1) = u(0)$ and $u(K+2) = u(1)$, equations (12.4.2) and (12.4.3) can be written as

$$s(0)u(0) - p(0)u(1) - p(-1)u(K) = \lambda r(0)u(0) \tag{12.4.14}$$

and

$$s(K+1)u(0) - p(K+1)u(1) - p(K)u(K) = \lambda r(K+1)u(0). \tag{12.4.15}$$

Thus, this problem is meaningful only if (12.4.14) is the same as (12.4.15). For this, either $p(k)$, $q(k)$ and $r(k)$ must be periodic of period $K+1$, or more generally, are such that

$$\frac{p(K+1)}{p(0)} = \frac{p(K)}{p(-1)} = \frac{q(K+1)}{q(0)} = \frac{r(K+1)}{r(0)}. \tag{12.4.16}$$

For $k = K$ equation (12.3.1) can be written as

$$-p(K)u(0) - p(K-1)u(K-1) + s(K)u(K) = \lambda r(K)u(K). \tag{12.4.17}$$

The $(K+1)$ equations (12.4.14), (12.3.1) for $k \in \mathbb{N}(1, K-1)$ and (12.4.17) lead to a system of the form (12.2.1), where $\mathcal{A}$ is a $(K+1) \times$

$(K+1)$ real symmetric matrix of the form

$$\mathcal{A} = \begin{bmatrix} s(0) & -p(0) & & & & & -p(-1) \\ -p(0) & s(1) & -p(1) & & & & \\ & -p(1) & \cdot & \cdot & & & \\ & & & \cdot & & & \\ & & & & -p(K-2) & s(K-1) & -p(K-1) \\ -p(K) & & & & & -p(K-1) & s(K) \end{bmatrix},$$

$\mathcal{R}$ is a $(K+1)\times(K+1)$ diagonal matrix defined as $\mathcal{R} = diag\ (r(0), \ldots, r(K))$, and $\mathbf{u} = (u(0), \cdots, u(K))$. Further, for this system if $\lambda_\mu$ and $\lambda_\nu$ are two distinct eigenvalues and $\phi_\mu(k)$ and $\phi_\nu(k)$, $k \in \mathbb{N}(0, K)$ are the corresponding eigenfunctions, then it follows that

$$\sum_{\ell=0}^{K} r(\ell)\phi_\mu(\ell)\phi_\nu(\ell) \ = \ 0. \tag{12.4.18}$$

However, since $\phi_\mu(0)\phi_\nu(0) = \phi_\mu(K+1)\phi_\nu(K+1)$ the orthogonality relation (12.4.18) can be written as

$$\frac{1}{2}r(0)\phi_\mu(0)\phi_\nu(0) + \sum_{\ell=1}^{K} r(\ell)\phi_\mu(\ell)\phi_\nu(\ell) + \frac{1}{2}r(K+1)\phi_\mu(K+1)\phi_\nu(K+1) = 0.$$

Finally, we note that if $p(k)$, $q(k)$ and $r(k)$ are periodic of period $K+1$ then the eigenfunctions of this problem satisfy the difference equation on $\mathbb{Z}$.

**Example 12.4.1.** For the difference equation (12.1.10) on $\mathbb{N}(0, K+1)$ together with the symmetric boundary conditions (12.4.1) the eigenvalues are $\lambda_m = 4\sin^2 \dfrac{m\pi}{2(K+1)}$, $0 \le m \le K+1$ and the corresponding eigenfunctions are $\phi_m(k) = \cos\left(\dfrac{mk\pi}{K+1}\right)$, $0 \le m \le K+1$. Further, since $p = 1$, $q = 0$ and $r = 1$ for $\mu \ne \nu$ it follows that

$$\frac{1}{2} + \sum_{\ell=1}^{K} \cos\left(\frac{\mu\ell\pi}{K+1}\right)\cos\left(\frac{\nu\ell\pi}{K+1}\right) + \frac{1}{2}(-1)^{\mu+\nu} \ = \ 0.$$

**Example 12.4.2.** For the difference equation (12.1.10) on $\mathbb{Z}$ together with the periodic boundary conditions

$$u(2K+2+k) \ = \ u(k), \quad k \in \mathbb{Z} \tag{12.4.19}$$

the $K+2$ distinct eigenvalues are $\lambda_m = 4\sin^2\frac{m\pi}{2(K+1)}$, $0 \le m \le K+1$ and corresponding to $\lambda_0 = 0$ the eigenfunction is $\phi_0(k) = 1$; corresponding to $\lambda_{K+1} = 4$ the eigenfunction is $\phi_{K+1}(k) = \cos\pi k$; whereas corresponding to the remaining eigenvalues each correspond to the two linearly independent eigenfunctions $\phi_m^{(1)}(k) = \cos\left(\frac{m\pi k}{K+1}\right)$, $\phi_m^{(2)}(k) = \sin\left(\frac{m\pi k}{K+1}\right)$, $1 \le m \le K$. Further, for $0 \le \mu, \nu \le K+1$ it follows that

$$\sum_{\ell=-K-1}^{K+1} \sin\left(\frac{\mu\pi\ell}{K+1}\right)\sin\left(\frac{\nu\pi\ell}{K+1}\right) = 0, \quad \mu \ne \nu$$

$$\sum_{\ell=-K-1}^{K+1} \sin\left(\frac{\mu\pi\ell}{K+1}\right)\cos\left(\frac{\nu\pi\ell}{K+1}\right) = 0,$$

and

$$\frac{1}{2}(-1)^{\mu+\nu} + \sum_{\ell=-K}^{K} \cos\left(\frac{\mu\pi\ell}{K+1}\right)\cos\left(\frac{\nu\pi\ell}{K+1}\right) + \frac{1}{2}(-1)^{\mu+\nu} = 0, \quad \mu \ne \nu.$$

## 12.5. Discrete Fourier Series

Let $a, b \in \mathbb{Z}$ and $\{\phi_m(k),\ a \le m \le b\}$ be an orthogonal set of functions on $\mathbb{N}(a,b)$ with respect to the positive weight function $r(k)$, $k \in \mathbb{N}(a,b)$. Since, orthogonality of these functions $\phi_m(k)$, $a \le m \le b$, in particular, implies their linear independence on $\mathbb{N}(a,b)$, any function $u(k)$, $k \in \mathbb{N}(a,b)$ can be expressed as a linear combination of $\phi_m(k)$, $a \le m \le b$, i.e.

$$u(k) = \sum_{m=a}^{b} c_m\phi_m(k), \quad k \in \mathbb{N}(a,b) \tag{12.5.1}$$

where the constants $c_m$, $a \le m \le b$ can be determined as follows: We multiply both sides of (12.5.1) by $r(k)\phi_n(k)$, $a \le n \le b$, sum the results from $k = a$ to $k = b$, and use the orthogonality of the functions $\phi_m(k)$, $a \le m \le b$ on $\mathbb{N}(a,b)$, to obtain

$$\sum_{k=a}^{b} r(k)\phi_n(k)u(k) = \sum_{m=a}^{b} c_m\left(\sum_{k=a}^{b} r(k)\phi_n(k)\phi_m(k)\right) = c_n\sum_{k=a}^{b} r(k)\phi_n^2(k)$$

and hence

$$c_m = \frac{\sum_{k=a}^{b} r(k)\phi_m(k)u(k)}{\sum_{k=a}^{b} r(k)\phi_m^2(k)}, \quad a \le m \le b. \tag{12.5.2}$$

In particular, if the functions $\phi_m(k),\ a \le m \le b$ are orthonormal, i.e. for each $m$, $\sum_{k=a}^{b} r(k)\phi_m^2(k) = 1$ then the constants $c_m$ simplify to

$$(12.5.3) \qquad c_m = \sum_{k=a}^{b} r(k)\phi_m(k)u(k), \quad a \le m \le b.$$

The relation (12.5.1) is called the *discrete Fourier series,* and the constants $c_m$ in (12.5.2) are the corresponding *discrete Fourier coefficients.*

**Example 12.5.1.** From Example 12.1.1 it is clear that the functions $\phi_m(k) = \sin\left(\frac{mk\pi}{K+1}\right),\ 1 \le m \le K$ are orthogonal on $\mathbb{N}(1,K)$ with respect to the weight function $r(k) = 1,\ k \in \mathbb{N}(1,K)$. Further, since for each $1 \le m \le K$, $\sum_{k=1}^{K} \sin^2\left(\frac{mk\pi}{K+1}\right) = \frac{K+1}{2}$, for any function $u(k)$ defined on $\mathbb{N}(1,K)$ it follows from (12.5.1) and (12.5.2) that

$$(12.5.4) \qquad u(k) = \sum_{m=1}^{K} c_m \sin\left(\frac{mk\pi}{K+1}\right), \quad k \in \mathbb{N}(1,K)$$

where

$$c_m = \frac{2}{K+1}\sum_{k=1}^{K} u(k)\sin\left(\frac{mk\pi}{K+1}\right), \quad 1 \le m \le K.$$

The relation (12.5.4) is called the *discrete Fourier sine series.*

**Example 12.5.2.** From Example 12.4.1 it is clear that the functions $\phi_m(k) = \cos\left(\frac{mk\pi}{K+1}\right),\ 0 \le m \le K+1$ are orthogonal on $\mathbb{N}(0,K+1)$ with respect to the weight function $r(k) = \begin{cases} 1/2, & k = 0, K+1 \\ 1, & k \in \mathbb{N}(1,K) \end{cases}$. Further, since $\sum_{k=0}^{K+1} r(k)\cos^2\left(\frac{mk\pi}{K+1}\right) = \begin{cases} K+1, & m = 0, K+1 \\ (K+1)/2, & 1 \le m \le K \end{cases}$, for any function $u(k)$ defined on $\mathbb{N}(0,K+1)$ it follows from (12.5.1) and (12.5.2) that

$$(12.5.5) \qquad u(k) = \sum_{m=0}^{K+1} c_m \cos\left(\frac{mk\pi}{K+1}\right), \quad k \in \mathbb{N}(0,K+1)$$

where

$$\begin{aligned} c_0 &= \frac{1}{K+1}\sum_{k=0}^{K+1} r(k)u(k), \\ c_m &= \frac{2}{K+1}\sum_{k=0}^{K+1} r(k)u(k)\cos\left(\frac{mk\pi}{K+1}\right), \quad 1 \le m \le K \\ c_{K+1} &= \frac{1}{K+1}\sum_{k=0}^{K+1} r(k)(-1)^k u(k). \end{aligned}$$

The relation (12.5.5) is called the *discrete Fourier cosine series.*

If $k \in \mathbb{Z}$, then the representation (12.5.4) defines an odd periodic function of periodic $2(K+1)$, which agrees with $u(k)$, $k \in \mathbb{N}(1, K)$ and is zero when $k = 0$ and $K+1$. Similarly, the representation (12.5.5) defines an even periodic function of period $2(K+1)$, which agrees with $u(k)$, $k \in \mathbb{N}(0, K+1)$.

**Example 12.5.3.** From Example 12.4.2 it is clear that $(2K+2)$ functions $\phi_m^{(1)}(k) = \cos\left(\frac{mk\pi}{K+1}\right)$, $0 \le m \le K+1$, $\phi_m^{(2)}(k) = \sin\left(\frac{mk\pi}{K+1}\right)$, $1 \le m \le K$ are orthogonal on $\mathbb{N}(-K-1, K+1)$ with respect to the weight function $r(k) = \begin{cases} 1/2, & k = \pm(K+1) \\ 1, & k \in \mathbb{N}(-K, K) \end{cases}$. Further, since $\sum_{k=-K-1}^{K+1} r(k) \times \sin^2\left(\frac{mk\pi}{K+1}\right) = K+1$, $1 \le m \le K$ and $\sum_{k=-K-1}^{K+1} r(k)\cos^2\left(\frac{mk\pi}{K+1}\right) = \begin{cases} 2(K+1), & m = 0, K+1 \\ K+1, & 1 \le m \le K \end{cases}$, for any function $u(k)$ defined on $\mathbb{N}(-K-1, K+1)$ it follows from (12.5.1) and (12.5.2) that

$$\overline{u}(k) = c_0 + \sum_{m=1}^{K}\left(c_m\cos\left(\frac{mk\pi}{K+1}\right) + d_m\sin\left(\frac{mk\pi}{K+1}\right)\right) + c_{K+1}\cos \pi k, \quad k \in \mathbb{N}(-K-1, K+1) \tag{12.5.6}$$

where

$$\begin{aligned} c_0 &= \frac{1}{2(K+1)}\sum_{k=-K-1}^{K+1} r(k)u(k), \\ c_m &= \frac{1}{K+1}\sum_{k=-K-1}^{K+1} r(k)u(k)\cos\left(\frac{mk\pi}{K+1}\right), \quad 1 \le m \le K \end{aligned}$$

$$c_{K+1} = \frac{1}{2(K+1)} \sum_{k=-K-1}^{K+1} r(k)u(k)\cos \pi k,$$

$$d_m = \frac{1}{K+1} \sum_{k=-K-1}^{K+1} r(k)u(k)\sin\left(\frac{mk\pi}{K+1}\right), \quad 1 \le m \le K$$

and $\overline{u}(k) = \begin{cases} u(k), & k \in \mathbb{N}(-K, K) \\ (u(-K-1) + u(K+1))/2, & k = \pm(K+1) \end{cases}$.

## 12.6. Wirtinger Type Inequalities

**Theorem 12.6.1.** For any function $u(k)$, $k \in \mathbb{N}(0, K+1)$ satisfying $u(0) = u(K+1) = 0$ the following inequalities hold

$$(12.6.1)\quad 4\sin^2 \frac{\pi}{2(K+1)} \sum_{\ell=1}^{K} u^2(\ell) \le \sum_{\ell=0}^{K} (\Delta u(\ell))^2 \le 4\cos^2 \frac{\pi}{2(K+1)} \sum_{\ell=1}^{K} u^2(\ell).$$

In the left (right) of (12.6.1) equality holds if and only if $u(k) = c\sin\left(\frac{k\pi}{K+1}\right)$ $\left(u(k) = c(-1)^{k-1}\sin\left(\frac{k\pi}{K+1}\right)\right)$, where $c$ is an arbitrary constant.

**Proof.** We note that

$$(12.6.2)\quad \sum_{\ell=0}^{K} (\Delta u(\ell))^2 = \sum_{\ell=0}^{K} (u(\ell+1) - u(\ell))^2 = 2\sum_{\ell=1}^{K} u^2(\ell) - 2\sum_{\ell=1}^{K-1} u(\ell)u(\ell+1)$$
$$= (\mathcal{A}\mathbf{u}, \mathbf{u}),$$

where $\mathcal{A}$ is a $K \times K$ real symmetric tridiagonal matrix $\mathcal{H}_K(\mathbf{g}, \mathbf{h})$, with $\mathbf{g} = (2, \cdots, 2)$, $\mathbf{h} = (-1, \cdots, -1)$, and $\mathbf{u} = (u(1), \cdots, u(K))$.

For $s = 2$ writing the problem (12.1.8) in the system form (12.2.1), we find from Example 12.1.1 that for this matrix $\mathcal{A}$ the $K$ distinct eigenvalues are $\lambda_m = 4\sin^2 \frac{m\pi}{2(K+1)}$, $1 \le m \le K$ and the corresponding eigenvectors are $\mathbf{u}^m = \left\{\phi_m(k) = \sin\left(\frac{mk\pi}{K+1}\right),\ 1 \le k \le K\right\}$, $1 \le m \le K$. Therefore, $\lambda_1 = 4\sin^2 \frac{\pi}{2(K+1)}$, $\phi_1(k) = \sin\left(\frac{k\pi}{K+1}\right)$; and $\lambda_K = 4\sin^2 \frac{K\pi}{2(K+1)} = 4\cos^2 \frac{\pi}{2(K+1)}$, $\phi_K(k) = \sin\left(\frac{Kk\pi}{K+1}\right) = (-1)^{k-1}\sin\left(\frac{k\pi}{K+1}\right)$. The inequalities (12.6.1) now follow from (12.2.3) and (12.6.2). ∎

**Theorem 12.6.2.** For any function $u(k)$, $k \in \mathbb{N}(0,K)$ satisfying $u(0) = 0$ the following inequalities hold

$$(12.6.3)\quad 4\sin^2\frac{\pi}{2(2K+1)}\sum_{\ell=1}^{K}u^2(\ell) \le \sum_{\ell=0}^{K-1}(\Delta u(\ell))^2 \le 4\cos^2\frac{\pi}{2K+1}\sum_{\ell=1}^{K}u^2(\ell).$$

In the left (right) of (12.6.3) equality holds if and only if $u(k) = c\sin\left(\frac{k\pi}{2K+1}\right)$ $\left(u(k) = c(-1)^{k-1}\sin\left(\frac{2k\pi}{2K+1}\right)\right)$, where $c$ is an arbitrary constant.

**Proof.** As in Theorem 12.6.1, we have

$$\begin{aligned}(12.6.4)\quad \sum_{\ell=0}^{K-1}(\Delta u(\ell))^2 &= 2\sum_{\ell=1}^{K-1}u^2(\ell) + u^2(K) - 2\sum_{\ell=1}^{K-1}u(\ell)u(\ell+1)\\ &= (\mathcal{A}\mathbf{u},\mathbf{u}),\end{aligned}$$

where $\mathcal{A}$ is a $K\times K$ real symmetric tridiagonal matrix $\mathcal{H}_K(\mathbf{g},\mathbf{h})$, with $\mathbf{g} = (2,\cdots,2,1)$, $\mathbf{h} = (-1,\cdots,-1)$, and $\mathbf{u} = (u(1),\cdots,u(K))$.

Writing the problem (12.3.6), (12.3.7) in the system form (12.2.1), we find from Example 12.3.1 that for this matrix $\mathcal{A}$ the $K$ distinct eigenvalues are $\lambda_m = 4\sin^2\left(\frac{2m-1}{2(2K+1)}\right)\pi$, $1 \le m \le K$ and the corresponding eigenfunctions are $\mathbf{u}^m = \left\{\phi_m(k) = \sin\left(\frac{2m-1}{2K+1}\right)k\pi,\ 1\le k\le K\right\}$, $1 \le m \le K$. Therefore, $\lambda_1 = 4\sin^2\frac{\pi}{2(2K+1)}$, $\phi_1(k) = \sin\left(\frac{k\pi}{2K+1}\right)$; and $\lambda_K = 4\sin^2\left(\frac{2K-1}{2(2K+1)}\right)\pi = 4\cos^2\frac{\pi}{2K+1}$, $\phi_K(k) = \sin\left(\frac{2K-1}{2K+1}\right)k\pi = (-1)^{k-1}\sin\left(\frac{2k\pi}{2K+1}\right)$. The inequalities (12.6.3) now follow from (12.2.3) and (12.6.4). ∎

**Theorem 12.6.3.** For any function $u(k)$, $k \in \mathbb{N}(0,K+1)$ satisfying $u(0) = u(K+1)$, $\sum_{\ell=0}^{K}u(\ell) = 0$ the following inequalities hold

$$(12.6.5)\quad 4\sin^2\frac{\pi}{K+1}\sum_{\ell=0}^{K}u^2(\ell) \le \sum_{\ell=0}^{K}(\Delta u(\ell))^2 \le 4\sin^2\frac{\left[\frac{K+1}{2}\right]\pi}{K+1}\sum_{\ell=0}^{K}u^2(\ell).$$

In the left of (12.6.5) equality holds if any only if $u(k) = c_1\cos\left(\frac{2k\pi}{K+1}\right) + c_2\sin\left(\frac{2k\pi}{K+1}\right)$, where $c_1$ and $c_2$ are arbitrary constants.

**Proof.** Since $u(0) = u(K+1)$, we find that

$$\sum_{\ell=0}^{K}(\Delta u(\ell))^2 = 2\sum_{\ell=1}^{K} u^2(\ell) - 2u(0)u(1) - 2\sum_{\ell=1}^{K-1} u(\ell)u(\ell+1) - 2u(K)u(0) = (\mathcal{A}\mathbf{u}, \mathbf{u}), \tag{12.6.6}$$

where $\mathcal{A}$ is the $(K+1)\times(K+1)$ real symmetric matrix of the form

$$\mathcal{A} = \begin{bmatrix} 2 & -1 & & & & -1 \\ -1 & 2 & -1 & & & \\ & -1 & \cdot & \cdot & & \\ & & \cdot & & & \\ & & & -1 & 2 & -1 \\ -1 & & & & -1 & 2 \end{bmatrix}$$

and $\mathbf{u} = (u(0), \cdots, u(K))$.

Writing the problem (12.1.10), $u(-1) = u(K)$, $u(0) = u(K+1)$ in the system form (12.2.1), we find from Problem 12.11.5 that for this matrix $\mathcal{A}$ the least eigenvalue is $\lambda_0 = 0$, and corresponding to this eigenvalue the eigenvector is $\mathbf{u}^0 = (1, \cdots, 1)$. A vector $\mathbf{u} = (u(0), \cdots, u(K))$ is orthogonal to $\mathbf{u}^0$ if and only if $\sum_{\ell=0}^{K} u(\ell) = 0$. Further, from Problem 12.11.5, $\lambda_1 = 4\sin^2\dfrac{\pi}{K+1}$ with multiplicity two, i.e. it corresponds to two linearly independent eigenvectors $\mathbf{u}^1 = \left\{\phi_1^{(1)}(k) = \cos\left(\dfrac{2k\pi}{K+1}\right),\ 0 \le k \le K\right\}$, and $\mathbf{v}^1 = \left\{\phi_1^{(2)}(k) = \sin\left(\dfrac{2k\pi}{K+1}\right),\ 0 \le k \le K\right\}$. Thus, from (12.2.3) and (12.6.6) the left inequality in (12.6.5) follows. To show the right part of (12.6.5) it suffices to note from Problem 12.11.5 that the greatest eigenvalue of this matrix $A$ is $4\sin^2\dfrac{[(K+1)/2]\pi}{K+1}$. ∎

A generalization of Theorem 12.6.1 is embodied in the following:

**Theorem 12.6.4.** Let $r(k)$, $k \in \mathbb{N}(0, K)$ and $p(k)$, $k \in \mathbb{N}(1, K+1)$ be positive functions and let $Q_k(t)$, $k \in \mathbb{N}(0, K)$ be polynomials defined by

$$\frac{r(k+1)}{\sqrt{p(k+1)p(k+2)}}Q_{k+1}(t) = \left(\frac{r(k)+r(k+1)}{p(k+1)} - t\right)Q_k(t) - \frac{r(k)}{\sqrt{p(k)p(k+1)}}Q_{k-1}(t), \quad Q_0(t) = 1,\ Q_{-1}(t) = 0. \tag{12.6.7}$$

Then, for any function $u(k)$, $k \in \mathbb{N}(0, K+1)$ satisfying $u(0) = u(K+1) = 0$ the following inequalities hold

$$\lambda_1 \sum_{\ell=1}^{K} p(\ell)u^2(\ell) \le \sum_{\ell=0}^{K} r(\ell)(\Delta u(\ell))^2 \le \lambda_K \sum_{\ell=1}^{K} p(\ell)u^2(\ell), \tag{12.6.8}$$

where $\lambda_1$ and $\lambda_K$ are the minimal and maximal zeros of the polynomial $Q_K(t)$. In the left (right) of (12.6.8) equality holds if and only if $u(k) = \frac{c}{\sqrt{p(k)}}Q_{k-1}(\lambda)$, $k \in \mathbb{N}(1, K)$ where $\lambda = \lambda_1(\lambda_K)$ and $c$ is an arbitrary constant.

**Proof.** In (12.6.8) we substitute $u(k) = v(k)/\sqrt{p(k)}$, $k \in \mathbb{N}(0, K+1)$ so that it takes the form

$$\begin{aligned}\lambda_1 \sum_{\ell=1}^{K} v^2(\ell) &\le \sum_{\ell=0}^{K} \frac{r(\ell)}{p(\ell)p(\ell+1)}(\sqrt{p(\ell+1)}v(\ell) - \sqrt{p(\ell)}v(\ell+1))^2 \\ &\le \lambda_K \sum_{\ell=1}^{K} v^2(\ell).\end{aligned} \tag{12.6.9}$$

Thus, in view of Example 12.2.1 inequalities (12.6.9) are equivalent to

$$\lambda_1(\mathbf{v}, \mathbf{v}) \le (\mathcal{H}_K(\mathbf{g}, \mathbf{h})\mathbf{v}, \mathbf{v}) \le \lambda_K(\mathbf{v}, \mathbf{v}), \tag{12.6.10}$$

where the tridiagonal matrix $\mathcal{H}_K(\mathbf{g}, \mathbf{h})$ is with

$$\mathbf{g} = \left(\frac{r(0) + r(1)}{p(1)}, \cdots, \frac{r(K-1) + r(K)}{p(K)}\right)$$

and $\mathbf{h} = \left(-\frac{r(1)}{\sqrt{p(1)p(2)}}, \cdots, -\frac{r(K-1)}{\sqrt{p(K-1)p(K)}}\right)$, and $\mathbf{v} = (v(1), \cdots, v(K))$.

Now let $\mathbf{v} = (Q_0(t), \cdots, Q_{K-1}(t))$, then from (12.6.7) it follows that

$$\mathcal{H}_K(\mathbf{g}, \mathbf{h})\mathbf{v} = t\mathbf{v} + \frac{r(K)}{\sqrt{p(K)p(K+1)}}Q_K(t)\mathbf{e}^K, \tag{12.6.11}$$

where $\mathbf{e}^K = (0, \cdots, 0, 1)$.

Thus, if $t$ is such that $Q_K(t) = 0$, then $t$ is an eigenvalue of the matrix $\mathcal{H}_K(\mathbf{g}, \mathbf{h})$ and $\mathbf{v}$ is an eigenvector. Conversely, if $t$ is an eigenvalue and $\mathbf{v}$ is an eigenvector of the matrix $\mathcal{H}_K(\mathbf{g}, \mathbf{h})$, then $Q_K(t) = 0$, i.e. $t$ is a zero of the polynomial $Q_K(t)$. Hence, if for the real symmetric tridiagonal and positive definite matrix $\mathcal{H}_K(\mathbf{g},\mathbf{h})$, $0 < \lambda_1 < \cdots < \lambda_K$ are the eigenvalues then these are also the zeros of $Q_K(t)$. The desired inequalities (12.6.10), or equivalently (12.6.8), now follow from (12.2.3). ■

**Remark 12.6.1.** If in Theorem 12.6.4, $p(k) = r(k) = 1$ then (12.6.7) reduces to

$$Q_{k+1}(t) = (2-t)Q_k(t) - Q_{k-1}(t), \quad Q_0(t) = 1, \; Q_1(t) = 2-t \tag{12.6.12}$$

which can be solved to obtain $Q_k(t) = \dfrac{\sin(k+1)\theta}{\sin\theta}$, where $2\cos\theta = 2-t$. Thus, $Q_K(t) = \dfrac{\sin(K+1)\theta}{\sin\theta}$ and hence $\lambda_m = 4\sin^2\dfrac{m\pi}{2(K+1)}$, $1 \le m \le K$. For $\lambda_1 = 4\sin^2\dfrac{\pi}{2(K+1)}$ the corresponding $\theta = \dfrac{\pi}{K+1}$, and therefore $(Q_0(\lambda_1),\cdots,Q_{K-1}(\lambda_1)) = \left(c_1\sin\dfrac{\pi}{K+1},\cdots,c_1\sin\dfrac{K\pi}{K+1}\right)$. Similarly, for $\lambda_K = 4\sin^2\dfrac{K\pi}{2(K+1)}$ the corresponding $\theta = \dfrac{K\pi}{K+1}$, and $(Q_0(\lambda_K),\cdots,Q_{K-1}(\lambda_K)) = \left(c_1\sin\dfrac{\pi}{K+1}, -c_1\sin\dfrac{2\pi}{K+1},\cdots,c_1(-1)^{K-1}\right.$ $\left.\times\ \sin\dfrac{K\pi}{K+1}\right)$. Thus, in this case Theorem 12.6.4 gives the conclusions of Theorem 12.6.1.

**Corollary 12.6.5.** For any function $u(k)$, $k \in \mathbb{N}(0, K+1)$ satisfying $u(0) = u(K+1) = 0$ the following inequalities hold

$$\begin{aligned} 4\sin^2\frac{\pi}{2(K+1)}\sum_{\ell=1}^{K}\ell^2u^2(\ell) \;&\le\; \sum_{\ell=0}^{K}\ell(\ell+1)(\Delta u(\ell))^2 \\ &\le\; 4\cos^2\frac{\pi}{2(K+1)}\sum_{\ell=1}^{K}\ell^2u^2(\ell). \end{aligned} \tag{12.6.13}$$

In the left (right) of (12.6.13) equality holds if and only if $u(k) = c\sin\left(\dfrac{k\pi}{K+1}\right)$ $\left(u(k) = c(-1)^{k-1}\sin\left(\dfrac{k\pi}{K+1}\right)\right)$, where $c$ is an arbitrary constant.

**Proof.** It suffices to note that for $p(k) = k^2$ and $r(k) = k(k+1)$, (12.6.7) reduces to (12.6.12). ■

**Corollary 12.6.6.** For any function $u(k)$, $k \in \mathbb{N}(0, K+1)$ satisfying

$u(0) = u(K+1) = 0$ the following inequality holds

$$\sum_{\ell=0}^{K} \ell(\Delta u(\ell))^2 \le \lambda_K \sum_{\ell=1}^{K} u^2(\ell), \tag{12.6.14}$$

where $\lambda_K$ is the maximal zero of the Laguerre polynomial $L_K(t)$. In (12.6.14) equality holds if any only if $u(k) = cL_{k-1}(\lambda_K)$, $k \in \mathbb{N}(1,K)$, where $c$ is an arbitrary constant and $L_{k-1}(t)$, $k \in \mathbb{N}(1,K)$ are Laguerre polynomials.

**Proof.** It suffices to note that for $p(k) = 1$ and $r(k) = k$, (12.6.7) reduces to

$$(k+1)Q_{k+1}(t) = (2k+1-t)Q_k(t) - kQ_{k-1}(t), \quad Q_0(t) = 1, \ Q_{-1}(t) = 0$$

whose solution is $Q_k(t) = L_k(t)$. ∎

The proof of the following result is similar to that of Theorem 12.6.4.

**Theorem 12.6.7.** Let $r(k)$, $k \in \mathbb{N}(0, K-1)$ and $p(k)$, $k \in \mathbb{N}(1,K)$ be positive functions and let $Q_k(t)$, $k \in \mathbb{N}(0, K-1)$ be polynomials defined by (12.6.7). Then, for any function $u(k)$, $k \in \mathbb{N}(0,K)$ satisfying $u(0) = 0$ the following inequalities hold

$$\lambda_1 \sum_{\ell=1}^{K} p(\ell)u^2(\ell) \le \sum_{\ell=0}^{K-1} r(\ell)(\Delta u(\ell))^2 \le \lambda_K \sum_{\ell=1}^{K} p(\ell)u^2(\ell), \tag{12.6.15}$$

where $\lambda_1$ and $\lambda_K$ are the minimal and maximal zeros of the polynomial $R_K(t)$ which is defined as

$$R_K(t) = \left(\frac{r(K-1)}{p(K)} - t\right) Q_{K-1}(t) - \frac{r(K-1)}{\sqrt{p(K-1)p(K)}} Q_{K-2}(t). \tag{12.6.16}$$

In the left (right) of (12.6.15) equality holds if and only if $u(k) = \frac{c}{\sqrt{p(k)}} \times Q_{k-1}(\lambda)$, $k \in \mathbb{N}(1,K)$ where $\lambda = \lambda_1$ $(\lambda_K)$ and $c$ is an arbitrary constant.

**Remark 12.6.2.** As in Remark 12.6.1 for $p(k) = r(k) = 1$ we have $Q_k(t) = \dfrac{\sin(k+1)\theta}{\sin\theta}$, where $2\cos\theta = 2 - t$. Therefore, from (12.6.16) we find that $R_K(t) = (1-t)Q_{K-1}(t) - Q_{K-2}(t) = \cos\left(\dfrac{2K+1}{2}\right)\theta \Big/ \cos\dfrac{\theta}{2}$. Thus, $\lambda_m = 4\sin^2\left(\dfrac{2m-1}{2(2K+1)}\right)\pi$, $1 \le m \le K$ and hence (12.6.15) reduces to (12.6.3).

**Corollary 12.6.8.** For any function $u(k)$, $k \in \mathbb{N}(0.K)$ satisfying $u(0) = 0$ the following inequality holds

(12.6.17) $$\sum_{\ell=0}^{K-1} \ell(\Delta u(\ell))^2 \le \lambda_K \sum_{\ell=1}^{K} u^2(\ell),$$

where $\lambda_K$ is the maximal zero of the generalized Laguerre polynomial $L_K^{(-1)}(t) = \sum_{i=1}^{K} \binom{K-1}{K-i} \frac{(-t)^i}{i!}$. In (12.6.17) the equality holds if and only if $u(k) = cL_{k-1}(\lambda_K)$, $k \in \mathbb{N}(1, K)$, where $c$ is an arbitrary constant and $L_{k-1}(t)$, $k \in \mathbb{N}(1, K)$ are Laguerre polynomials.

**Proof.** Following the proof of Corollary 12.6.6 it suffices to note that $R_K(t) = KL_K^{(-1)}(t)$. ∎

## 12.7. Generalized Wirtinger Type Inequalities

For a given function $u(k)$, $k \in \mathbb{N}(0, K)$ with the related vector $\mathbf{u} = (u(0), \cdots, u(K))$ we define the operators $\overline{\Delta}^m : \mathbb{R}^{K+1} \to \mathbb{R}^{K+1}$ and $\overline{\nabla}^m : \mathbb{R}^{K+1} \to \mathbb{R}^{K+1}$, $m \in \mathbb{N}(0, K)$ by the formulae $\overline{\Delta}^m \mathbf{u} = (\Delta_0^m u(0), \cdots, \Delta_0^m u(K))$ and $\overline{\nabla}^m \mathbf{u} = (\nabla_0^m u(0), \cdots, \nabla_0^m u(K))$, where

$$\Delta_0^m u(k) = \begin{cases} \Delta^m u(k), & k \in \mathbb{N}(0, K-m) \\ 0, & k \in \mathbb{N}(K-m+1, K) \end{cases}$$

and

$$\nabla_0^m u(k) = \begin{cases} 0, & k \in \mathbb{N}(0, m-1) \\ \nabla^m u(k), & k \in \mathbb{N}(m, K). \end{cases}$$

The set of all vectors $\mathbf{u} = (u(0), \cdots, u(K))$ satisfying

(12.7.1) $$u(0) = \cdots = u(m-1) = u(K-m+1) = \cdots = u(K) = 0$$

is denoted by $L_{2m}^{K+1}$, where the integer $m$ $(2m \le K)$ is fixed. The operator $\mathcal{T} : L_{2m}^{K+1} \to \mathbb{R}^{K+1}$ is defined by the formula

(12.7.2) $$\mathcal{T}\mathbf{u} = (-1)^m \overline{\nabla}^m \overline{\Delta}^m \mathbf{u}.$$

**Theorem 12.7.1.** For any function $u(k)$, $k \in \mathbb{N}(0, K)$ satisfying (12.7.1) the following inequality holds

(12.7.3) $$\sum_{\ell=0}^{K} u^2(\ell) \le \frac{1}{\lambda_{1,m}} \sum_{\ell=0}^{K-m} (\Delta^m u(\ell))^2,$$

where $\lambda_{1,m}$ is the smallest positive number such that the equation

$$\mathcal{T}\mathbf{u} = \lambda \mathbf{u} \tag{12.7.4}$$

has nontrivial solution in $L_{2m}^{K+1}$.

**Proof.** $\mathbb{R}^{K+1}$ with the usual scalar product $(\mathbf{u},\mathbf{v}) = \sum_{i=0}^{K} u_i v_i$ and the Euclidean norm $\|\mathbf{u}\| = (\mathbf{u},\mathbf{u})^{1/2}$ is obviously a Hilbert space. Thus, $L_{2m}^{K+1}$ as a subset of $\mathbb{R}^{K+1}$ is also a Hilbert space. Since for $\mathbf{u} = (u(0),\cdots,u(K))$, $\mathbf{v} = (v(0),\cdots,v(K)) \in L_{2m}^{K+1}$, we have

$$(\mathcal{T}\mathbf{u},\mathbf{v}) = \sum_{\ell=0}^{K} v(\ell)(-1)^m \nabla_0^m \Delta_0^m u(\ell) \tag{12.7.5}$$

from the repeated use of

$$\sum_{\ell=0}^{K} u(\ell)\Delta_0 v(\ell) + \sum_{\ell=0}^{K} \nabla_0 u(\ell) v(\ell) = u(\ell)v(\ell)\Big|_{\ell=0}^{K}, \tag{12.7.6}$$

which is the same as (1.8.5), it follows that

$$\begin{aligned}
(\mathcal{T}\mathbf{u},\mathbf{v}) &= (-1)^m \left[ v(\ell)\nabla_0^{m-1}\Delta_0^m u(\ell)\Big|_{\ell=0}^{K} - \sum_{\ell=0}^{K} \Delta_0 v(\ell)\nabla_0^{m-1}\Delta_0^m u(\ell) \right] \\
&= (-1)^{m+1} \sum_{\ell=0}^{K} \Delta_0 v(\ell)\nabla_0^{m-1}\Delta_0^m u(\ell) \\
&= (-1)^{m+2} \sum_{\ell=0}^{K} \Delta_0^2 v(\ell)\nabla_0^{m-2}\Delta_0^m u(\ell) \\
&\cdots \\
&= (-1)^{m+m} \sum_{\ell=0}^{K} \Delta_0^m v(\ell)\Delta_0^m u(\ell) \qquad (12.7.7) \\
&= (-1)^{2m+1} \sum_{\ell=0}^{K} \nabla_0 \Delta_0^m v(\ell)\Delta_0^{m-1} u(\ell) \\
&= (-1)^{2} \sum_{\ell=0}^{K} \nabla_0^2 \Delta_0^m v(\ell)\Delta_0^{m-2} u(\ell) \\
&\cdots \\
&= (-1)^{m} \sum_{\ell=0}^{K} \nabla_0^m \Delta_0^m v(\ell) u(\ell) \\
&= (\mathbf{u},\mathcal{T}\mathbf{v}). \qquad (12.7.8)
\end{aligned}$$

Thus, the operator $\mathcal{T}$ on $L_{2m}^{K+1}$ is self adjoint. Further, from (12.7.7), we have

$$(12.7.9) \qquad (\mathcal{T}\mathbf{u}, \mathbf{u}) \;=\; G(\mathbf{u}) \;=\; \|\Delta_0^m \mathbf{u}\|^2.$$

Since $\mathcal{T}$ is self adjoint it has real eigenvalues. Hence, by (12.7.4) and (12.7.9) for the vectors $\mathbf{u} \in L_{2m}^{K+1}$ satisfying the condition $\|\mathbf{u}\| = 1$ it follows that

$$(12.7.10) \qquad G(\mathbf{u}) \;=\; \lambda,$$

where $\lambda$ is an eigenvalue of $\mathcal{T}$. Therefore, the minimum of the nonnegative function $G : \mathbb{R}^{K+1} \to \mathbb{R}$ over the set $L_{2m}^{K+1} \cap S$, where $S = \{\mathbf{u} \in \mathbb{R}^{K+1} : \|\mathbf{u}\| = 1\}$ is equal to the smallest eigenvalue of the operator $\mathcal{T}$. From the continuity of $G$, and the compactness of the set $L_{2m}^{K+1} \cap S$ it follows that the minimum is attained, and therefore, is positive. We denote it by $\lambda_{1,m}$. Thus, for any $\mathbf{u} \in L_{2m}^{K+1}$, $\|\mathbf{u}\| > 0$ we have $G(\mathbf{u}/\|\mathbf{u}\|) \ge \lambda_{1,m}$, which is the same as (12.7.3). Since the case $\|\mathbf{u}\| = 0$ is trivial, the proof is completed. ∎

It is clear that (12.7.3) for $m = 1$ and $K = K + 1$ reduces to the left inequality (12.6.1).

## 12.8. Generalized Opial Type Inequalities

**Theorem 12.8.1.** Let $r(k)$ and $p(k)$, $k \in \mathbb{N}(1, K)$ be positive functions and let $Q_k(t)$, $k \in \mathbb{N}(0, K)$ be polynomials defined by

$$(12.8.1) \quad \frac{r(k+1)}{2\sqrt{p(k)p(k+1)}} Q_k(t) = \left(\frac{r(k)}{p(k)} - t\right) Q_{k-1}(t) - \frac{r(k)}{2\sqrt{p(k-1)p(k)}} Q_{k-2}(t),$$
$$Q_0(t) \;=\; Q_0 \neq 0, \quad Q_{-1}(t) \;=\; 0.$$

Then, for any function $u(k)$, $k \in \mathbb{N}(0, K)$ satisfying $u(0) = 0$ the following inequalities hold

$$(12.8.2) \quad \lambda_1 \sum_{\ell=1}^{K} p(\ell) u^2(\ell) \;\le\; \sum_{\ell=1}^{K} r(\ell) u(\ell) \nabla u(\ell) \;\le\; \lambda_K \sum_{\ell=1}^{K} p(\ell) u^2(\ell),$$

where $\lambda_1$ and $\lambda_K$ are the minimal and maximal zeros of the polynomial $Q_K(t)$. In the left (right) of (12.8.2) equality holds if and only if $u(k) = \dfrac{c}{\sqrt{p(k)}} Q_{k-1}(\lambda)$, $k \in \mathbb{N}(1, K)$ where $\lambda = \lambda_1(\lambda_K)$ and $c$ is an arbitrary constant.

**Proof.** In (12.8.1) we substitute $u(k) = v(k)/\sqrt{p(k)}$, $k \in \mathbb{N}(1, K)$ so that it takes the form

$$\begin{aligned}\lambda_1 \sum_{\ell=1}^{K} v^2(\ell) &\le \sum_{\ell=1}^{K} \frac{r(\ell)v(\ell)}{p(\ell)\sqrt{p(\ell-1)}}\left(\sqrt{p(\ell-1)}v(\ell) - \sqrt{p(\ell)}v(\ell-1)\right) \\ &\le \lambda_K \sum_{\ell=1}^{K} v^2(\ell).\end{aligned}$$

Thus, as in Theorem 12.6.4 the above inequality is equivalent to (12.6.10), where the tridiagonal matrix $\mathcal{H}_K(\mathbf{g},\mathbf{h})$ is with $\mathbf{g} = \left(\frac{r(1)}{p(1)}, \cdots, \frac{r(K)}{p(K)}\right)$ and $\mathbf{h} = \left(-\frac{r(2)}{2\sqrt{p(1)p(2)}}, \cdots, -\frac{r(K)}{2\sqrt{p(K-1)p(K)}}\right)$, and $\mathbf{v} = (v(1), \cdots, v(K))$. The rest of the proof is similar to that of Theorem 12.6.4. ∎

**Corollary 12.8.2.** For any function $u(k)$, $k \in \mathbb{N}(0,K)$ satisfying $u(0) = 0$ the following inequalities hold

$$(12.8.3)\quad 2\sin^2\frac{\pi}{2(K+1)} \sum_{\ell=1}^{K} u^2(\ell) \le \sum_{\ell=1}^{K} u(\ell)\nabla u(\ell) \le 2\cos^2\frac{\pi}{2(K+1)} \sum_{\ell=1}^{K} u^2(\ell).$$

In the left (right) of (12.8.3) equality holds if and only if $u(k) = c\sin\left(\frac{k\pi}{K+1}\right)$ $\left(u(k) = c(-1)^{k-1}\sin\left(\frac{k\pi}{K+1}\right)\right)$, where $c$ is an arbitrary constant.

**Proof.** It suffices to note that for $p(k) = r(k) = 1$, (12.8.1) reduces to

$$Q_k(t) = 2(1-t)Q_{k-1}(t) - Q_{k-2}(t), \quad Q_0(t) = 1, \; Q_1(t) = 2(1-t)$$

which can be solved to obtain $Q_k(t) = \sin(k+1)\theta/\sin\theta$, where $t = 2\sin^2\theta/2$. ∎

**Remark 12.8.1.** Inequalities (12.8.3) can be written as

$$-\cos\frac{\pi}{K+1} \sum_{\ell=1}^{K} u^2(\ell) \le \sum_{\ell=2}^{K} u(\ell)u(\ell-1) \le \cos\frac{\pi}{K+1} \sum_{\ell=1}^{K} u^2(\ell),$$

which is the same as

$$(12.8.4)\qquad \left|\sum_{\ell=2}^{K} u(\ell)u(\ell-1)\right| \le \cos\frac{\pi}{K+1} \sum_{\ell=1}^{K} u^2(\ell).$$

**Corollary 12.8.3.** Let the functions $r(k)$ and $p(k)$ be recursively given by

$$r(k) = (2k+s-1)p(k), \quad k \in \mathbb{N}(1,K)$$
$$p(k+1) = \frac{4k(k+s)}{(2k+s+1)^2}p(k), \quad k \in \mathbb{N}(1,K-1)$$

with $r(1) = 1$ and $s > -1$. Then, for any function $u(k)$, $k \in \mathbb{N}(0,K)$ satisfying $u(0) = 0$ the inequalities (12.8.2) hold, where $\lambda_1$ and $\lambda_K$ are the minimal and maximal zeros of the normalized generalized Laguerre polynomial $\overline{L}_K^{(s)}(t) = L_K^{(s)}(t)/\|L_K^{(s)}\|$ with $L_K^{(s)}(t) = \sum_{i=0}^{K}\binom{K+s}{K-i}\frac{(-t)^i}{i!}$ and $\|L_K^{(s)}\| = \sqrt{\Gamma(K+s+1)/K!}$. In the left (right) of (12.8.2) equality holds if and only if $u(k) = \frac{c}{\sqrt{p(k)}}\overline{L}_{k-1}^{(s)}(\lambda)$, where $\lambda = \lambda_1(\lambda_K)$ and $c$ is an arbitrary constant.

**Proof.** For this choice of $p(k)$ and $r(k)$ it suffices to note that (12.8.1) reduces to

$$\sqrt{k(k+s)}Q_k(t) = (2k+s-1-t)Q_{k-1}(t) - \sqrt{(k-1)(k+s-1)}Q_{k-2}(t),$$

which is the recurrence relation for the normalized Laguerre polynomials $\overline{L}_k^{(s)}(t)$. ∎

## 12.9. Comparison Theorems for Eigenvalues

Consider the system of difference equations

$$(12.9.1) \qquad (-1)^{m-p}L[\mathbf{u}(k)] = \lambda\mathcal{P}(k)\mathbf{u}(k+p), \quad k \in \mathbb{N}(0,K-1)$$

and

$$(12.9.2) \qquad (-1)^{m-p}L[\mathbf{u}(k)] = \Lambda\mathcal{Q}(k)\mathbf{u}(k+p), \quad k \in \mathbb{N}(0,K-1)$$

together with the boundary conditions

$$(12.9.3) \qquad \begin{aligned} \Delta^i\mathbf{u}(0) &= 0, \quad 0 \le i \le p-1 \\ \Delta^i\mathbf{u}(K+p) &= 0, \quad 0 \le i \le m-p-1 \end{aligned}$$

where $L[\mathbf{u}(k)] = \sum_{i=0}^{m} a_i(k)\mathbf{u}(k+i)$, $a_i(k)$ are defined on $\mathbb{N}(0,K-1)$, $a_m(k) \equiv 1$ and $a_0(k)$ satisfies (10.1.1); $1 \le p \le m-1$ is a fixed integer; $\lambda$ and $\Lambda$ are parameters, and the $n \times n$ matrices $\mathcal{P}(k)$ and $\mathcal{Q}(k)$ are defined on $\mathbb{N}(0,K-1)$.

Our aim here is to prove the existence and the comparison theorems for the least positive eigenvalues of (12.9.1), (12.9.3) and (12.9.2), (12.9.3) respectively. For this, we shall need Theorem 10.9.7 which is restated as follows:

**Theorem 12.9.1.** Assume that one of the following holds

1. $L[u(k)] = 0$ is disconjugate on $\mathbb{N}(0, K-1+m)$, or
2. $2 \le p \le m-1$ and $L[u(k)] = 0$ is $(j, m-j)$ disconjugate on $\mathbb{N}(p-j, K-1+m+p-j)$ for $j = p-1, \cdots, m-1$.

Then, the Green's function $g(k,\ell)$ of the boundary value problem

$$\begin{aligned}(-1)^{m-p}L[u(k)] &= 0\\ \Delta^i u(0) &= 0, \quad 0 \le i \le p-1, \quad \Delta^i u(K+p) = 0, \quad 0 \le i \le m-p-1\end{aligned}$$

satisfies

$$g(k,\ell) > 0, \quad k \in \mathbb{N}(p, K-1+p), \quad \ell \in \mathbb{N}(0, K-1). \tag{12.9.4}$$

We shall also need some results from the cone theory. For this, let $B$ be a Banach space. A closed nonempty subset $P$ of $B$ is called a *cone* provided that whenever $\mathbf{u}, \mathbf{v} \in P$ it follows that $\alpha\mathbf{u} + \beta\mathbf{v} \in P$ for all $\alpha \ge 0, \beta \ge 0$ and whenever $\mathbf{u}, -\mathbf{u} \in P$, then $\mathbf{u} = 0$. We say that a cone P is *reproducing* provided $B = P - P = \{\mathbf{u} - \mathbf{v} : \mathbf{u}, \mathbf{v} \in P\}$. We write $\mathbf{u} \le \mathbf{v}$ provided $\mathbf{v} - \mathbf{u} \in P$. If $\mathcal{M}$ and $\mathcal{N}$ are operators on $B$, then we write $\mathcal{M} \le \mathcal{N}$ (with respect to $P$) provided $\mathcal{M}\mathbf{u} \le \mathcal{N}\mathbf{u}$ for all $\mathbf{u} \in P$. A bounded linear operator $\mathcal{M}$ is $\mathbf{u}^0$ *positive* provided $\mathbf{u}^0 \in P$ and for each nonzero $\mathbf{u} \in P$, there are positive numbers $c_1, c_2$ (which in general depend on $\mathbf{u}$) such that $c_1\mathbf{u}^0 \le \mathcal{M}\mathbf{u} \le c_2\mathbf{u}^0$.

**Theorem 12.9.2.** Assume that $P$ is a reproducing cone and $\mathcal{M}$ is a linear compact operator which leaves the cone $P$ invariant. Further, assume that there is a nontrivial $\mathbf{u}^0 \in B$ and an $\epsilon_0 > 0$ such that $\mathcal{M}\mathbf{u}^0 \ge \epsilon_0\mathbf{u}^0$. Then, $\mathcal{M}$ has at least one eigenvector $\mathbf{z}^0 \in P$ with corresponding eigenvalue $\lambda_0 \ge \epsilon_0$ such that $\lambda_0$ is an upper bound for the moduli of the eigenvalues of $\mathcal{M}$.

**Theorem 12.9.3.** Assume that $P$ is a reproducing cone and $\mathcal{M}$ is a compact $\mathbf{u}^0$ positive linear operator. Then, $\mathcal{M}$ has an essentially unique eigenvector in $P$ and the corresponding eigenvalue is simple, positive, and larger than the modulus of any other eigenvalue of $\mathcal{M}$.

**Theorem 12.9.4.** Assume that $\mathcal{M}$ and $\mathcal{N}$ are linear operators and that at least one of them is $\mathbf{u}^0$ positive. If $\mathcal{M} \le \mathcal{N}$ and there exist nontrivial $\mathbf{u}^1, \mathbf{u}^2 \in P$, $\lambda_1, \lambda_2 > 0$ such that $\mathcal{M}\mathbf{u}^1 \ge \lambda_1\mathbf{u}^1$ and $\mathcal{N}\mathbf{u}^2 \le \lambda_2\mathbf{u}^2$, then $\lambda_1 \le \lambda_2$ and if $\lambda_1 = \lambda_2$ then $\mathbf{u}^1$ is a scalar multiple of $\mathbf{u}^2$.

Let $B(0, K-1+m)$ be the space of all $n$ vector functions defined on $\mathbb{N}(0, K-1+m)$. The Banach space that we are interested in here is $B = \{\mathbf{u}(k) \in B(0, K-1+m) : \Delta^i \mathbf{u}(0) = 0,\ 0 \le i \le p-1,\ \Delta^i \mathbf{u}(K+p) = 0,\ 0 \le i \le m-p-1\}$, where the norm on $B$ is defined by $\|\mathbf{u}\| = \max_{\mathbb{N}(p,K-1+p)} |\mathbf{u}(k)|$, and $|\cdot|$ is the Euclidean norm. Let $\rho$ be a reproducing cone in $\mathbb{R}^n$ and define the cone $P$ by $P = \{\mathbf{u} \in B : \mathbf{u}(k) \in \rho,\ k \in \mathbb{N}(p, K-1+p)\}$. It is clear that $P$ is a reproducing cone. Define operators $\mathcal{M}$ and $\mathcal{N}$ on $B$ by

$$\mathcal{M}\mathbf{u}(k) = \sum_{\ell=0}^{K-1} g(k,\ell)\mathcal{P}(\ell)\mathbf{u}(\ell+p) \tag{12.9.5}$$

and

$$\mathcal{N}\mathbf{u}(k) = \sum_{\ell=0}^{K-1} g(k,\ell)\mathcal{Q}(\ell)\mathbf{u}(\ell+p) \tag{12.9.6}$$

for $k \in \mathbb{N}(0, K-1+m)$. It is easy to verify that the operators $\mathcal{M}$ and $\mathcal{N}$ are compact linear operators. Further, if $\lambda_0 \neq 0$ is an eigenvalue of $\mathcal{M}$ and $\mathbf{w}^0(k)$ is the corresponding eigenvector, then $\mathcal{M}\mathbf{w}^0(k) = \lambda_0 \mathbf{w}^0(k)$, and hence

$$(-1)^{m-p} L[\mathbf{w}^0(k)] = \frac{1}{\lambda_0}\mathcal{P}(k)\mathbf{w}^0(k+p)$$

and $\mathbf{w}^0(k)$ satisfies the boundary conditions (12.9.3). This is summarized in the following:

**Remark 12.9.1.** $\lambda_0 \neq 0$ is an eigenvalue of $\mathcal{M}$ with the corresponding eigenfunction $\mathbf{w}^0(k)$ if and only if $1/\lambda_0$ is an eigenvalue of (12.9.1), (12.9.3) with the corresponding eigenfunction $\mathbf{w}^0(k)$. Similar statement holds for the operator $\mathcal{N}$ and the eigenvalue problem (12.9.2), (12.9.3).

**Theorem 12.9.5.** In addition to the conditions of Theorem 12.9.1 assume that $\mathcal{Q}(k)\rho \subseteq \rho$ for $k \in \mathbb{N}(0, K-1)$, and for each nontrivial $\mathbf{u} \in P$ there is a $k_{\mathbf{u}} \in \mathbb{N}(0, K-1)$ such that $\mathcal{Q}(k_{\mathbf{u}})\mathbf{u}(k_{\mathbf{u}}+p) \in \rho^0$ (interior of $\rho$). Then, the boundary value problem (12.9.2), (12.9.3) has a smallest positive eigenvalue $\Lambda_0$ and $\Lambda_0$ is smaller than the modulus of any other eigenvalue of (12.9.2), (12.9.3). Furthermore, there is an essentially unique eigenfunction $\mathbf{w}^0(k)$ corresponding to $\Lambda_0$ and either $\mathbf{w}^0 \in P^0$ or $-\mathbf{w}^0 \in P^0$.

**Proof.** First we shall show that $\mathcal{N} : P\backslash\{0\} \to P^0$. For this, let $0 \neq \mathbf{u} \in P$ and $\mathbf{v}(k) = \mathcal{N}\mathbf{u}(k)$. Obviously, $\mathbf{v}(k)$ satisfies the boundary conditions (12.9.3), and $\mathbf{v}(k) \in \rho$ for all $k \in \mathbb{N}(p.K-1+p)$. By hypothesis, there is a $k_{\mathbf{u}} \in \mathbb{N}(0, K-1)$ such that $\mathcal{Q}(k_{\mathbf{u}})\mathbf{u}(k_{\mathbf{u}}+p) \in \rho^0$. Thus, in view of Theorem 12.9.1 it follows that $g(k, k_{\mathbf{u}})\mathcal{Q}(k_{\mathbf{u}})\mathbf{u}(k_{\mathbf{u}}+p) \in \rho^0$. Hence, $\mathbf{v}(k) \in \rho^0$, $k \in \mathbb{N}(p, K-1+p)$, and from this it is clear that $\mathbf{v} \in P^0$.

Next, we shall prove that $\mathcal{N}$ is $\mathbf{u}^0$ positive. For this, since $\mathcal{N} : P\backslash\{0\} \to P^0$, $P^0 \neq \emptyset$. Let $\mathbf{u}^0 \in P^0$ and $0 \neq \mathbf{u} \in P$. Since $\mathbf{u}^0 \in P^0$ and $\mathcal{N}\mathbf{u} \in P^0$, we can choose numbers $c_2$ sufficiently large and $c_1 > 0$ sufficiently small so that $\mathbf{u}^0 - \frac{1}{c_2}\mathcal{N}\mathbf{u} \in P$ and $\mathcal{N}\mathbf{u} - c_1\mathbf{u}^0 \in P$. Thus, it follows that $c_1\mathbf{u}^0 \leq \mathcal{N}\mathbf{u} \leq c_2\mathbf{u}^0$ with respect to $P$ and so $N$ is $\mathbf{u}^0$ positive. The conclusion of the theorem now follows from Theorem 12.9.3 and Remark 12.9.1. ∎

**Theorem 12.9.6.** In addition to the conditions of Theorem 12.9.1 assume that $\mathcal{P}(k)$ and $\mathcal{Q}(k)$ satisfy the assumptions concerning $\mathcal{Q}(k)$ in Theorem 12.9.5. If $\mathcal{P}(k) \leq \mathcal{Q}(k)$ with respect to $\rho$, $k \in \mathbb{N}(0, K-1)$, then the smallest positive eigenvalues $\lambda_0$ and $\Lambda_0$ of (12.9.1), (12.9.3) and (12.9.2), (12.9.3) respectively, satisfy $\Lambda_0 \leq \lambda_0$. Furthermore, if $\Lambda_0 = \lambda_0$ then $\mathcal{P}(k)\mathbf{w}^0(k+p) = \mathcal{Q}(k)\mathbf{w}^0(k+p)$, $k \in \mathbb{N}(0, K-1)$ where $\mathbf{w}^0(k)$ is as in Theorem 12.9.5.

**Proof.** By Theorem 12.9.5, $\lambda_0 > 0$ and $\Lambda_0 > 0$ exist. We will now show that $\mathcal{M} \leq \mathcal{N}$ with respect to $P$. For $\mathbf{u} \in P$, we have

$$\begin{aligned} \mathcal{M}\mathbf{u}(k) &= \sum_{\ell=0}^{K-1} g(k,\ell)\mathcal{P}(\ell)\mathbf{u}(\ell+p) \\ &\leq \sum_{\ell=0}^{K-1} g(k,\ell)\mathcal{Q}(\ell)\mathbf{u}(\ell+p) = \mathcal{N}\mathbf{u}(k), \quad k \in \mathbb{N}(0, K-1+m). \end{aligned}$$

Further, $\Delta^i\mathcal{M}\mathbf{u}(0) = \Delta^i\mathcal{N}\mathbf{u}(0) = 0$, $0 \leq i \leq p-1$, and $\Delta^i\mathcal{M}\mathbf{u}(K+p) = \Delta^i\mathcal{N}\mathbf{u}(K+p) = 0$, $0 \leq i \leq m-p-1$. Thus, Theorem 12.9.4 implies that $\Lambda_0 \leq \lambda_0$.

If $\Lambda_0 = \lambda_0$, then by Theorem 12.9.4 the eigenfunctions $\mathbf{v}(k)$ and $\mathbf{w}(k)$ of (12.9.1), (12.9.3) and (12.9.2), (12.9.3) respectively are scalar multiples of each other, say $\mathbf{w}(k) = c\mathbf{v}(k)$. Thus, it follows that

$$(-1)^{m-p}L[\mathbf{w}(k)] = \lambda_0\mathcal{Q}(k)\mathbf{w}(k+p) = \lambda_0\mathcal{P}(k)\mathbf{w}(k+p), \quad k \in \mathbb{N}(0, K-1).$$

Hence, $P(k)\mathbf{w}^0(k+p) = \mathcal{Q}(k)\mathbf{w}^0(k+p)$, $k \in \mathbb{N}(0, K-1)$ where $\mathbf{w}^0(k) = \mathbf{w}(k)$. ∎

In out next result we shall use the cone $\rho_1$, which is a quadrant in $\mathbb{R}^n$, and in terms of $\delta_i \in \{-1, 1\}$, $1 \leq i \leq n$ is defined as $\rho_1 =$

$\{\mathbf{u} \in \mathbb{R}^n : \delta_i u_i \geq 0,\ 1 \leq i \leq n\}$. The related cone $P_1$ in $B$ is then $P_1 = \{\mathbf{u} \in B : \mathbf{u}(k) \in \rho_1,\ k \in \mathbb{N}(p, K-1+p)\}$.

**Theorem 12.9.7.** In addition to the conditions of Theorem 12.9.1 assume that $\delta_i \delta_j p_{ij}(k) \geq 0$ on $\mathbb{N}(0, K-1)$ for $1 \leq i,\ j \leq n$ and that there is a $k_0 \in \mathbb{N}(0, K-1)$ and an $i_0 \in \mathbb{N}(1, n)$ such that $p_{i_0 i_0}(k_0) > 0$. Then, the eigenvalue problem (12.9.1), (12.9.3) has a least positive eigenvalue $\lambda_0$ which is a lower bound on the modulus of the eigenvalues of (12.9.1), (12.9.3) and satisfies

$$\lambda_0^{-1} \ \geq\ g(k_0 + p, k_0) p_{i_0 i_0}(k_0).$$

Furthermore, there is an eigenfunction $\mathbf{z}^0(k)$ corresponding to $\lambda_0$ satisfying $\delta_i(\mathbf{z}^0(k))_i \geq 0,\ k \in \mathbb{N}(0, K-1+m),\ 1 \leq i \leq n$.

**Proof.** First we shall show that $\mathcal{M} : P_1 \to P_1$. For this, let $\mathbf{u} \in P_1$ and consider

$$\delta_i(\mathcal{M}\mathbf{u})_i(k) \ = \ \sum_{\ell=0}^{K-1} g(k,\ell) \sum_{j=1}^{n} \delta_i \delta_j p_{ij}(\ell) \delta_j u_j(\ell + p) \ \geq\ 0,$$

$$1 \leq i \leq n, \quad k \in \mathbb{N}(0, K-1+m).$$

Further, $\mathcal{M}\mathbf{u}(k)$ satisfies the boundary conditions (12.9.3). Hence, $\mathcal{M} : P_1 \to P_1$.

Define $\mathbf{w} \in P_1$ by setting $w_i(k) = 0$ on $\mathbb{N}(0, K-1+m)$ for $i \neq i_0$, and set $w_{i_0}(k) = \begin{cases} 0, & k \neq k_0 + p \\ \delta_{i_0}, & k = k_0 + p \end{cases}$ where $i_0$ and $k_0$ are as in the statement of the theorem. Since $\epsilon_0 = g(k_0 + p, k_0) p_{i_0 i_0}(k_0) > 0$, for $i \neq i_0$ we have $\delta_i(\mathcal{M}\mathbf{w})_i(k) \geq 0 = \epsilon_0 \delta_i w_i(k),\ k \in \mathbb{N}(0, K-1+m)$. Further, for $k \neq k_0 + p,\ \delta_{i_0}(\mathcal{M}\mathbf{w})_{i_0}(k) \geq 0 = \epsilon_0 \delta_{i_0} w_{i_0}(k)$. We also have that

$$\begin{aligned} \delta_{i_0}(\mathcal{M}\mathbf{w})_{i_0}(k_0 + p) \ &= \ \sum_{\ell=0}^{K-1} g(k_0 + p, \ell) \sum_{j=1}^{n} \delta_{i_0} \delta_j p_{i_0 j}(\ell) \delta_j \mathbf{w}_j(\ell + p) \\ &= \ g(k_0 + p, k_0) p_{i_0 i_0}(k_0) \delta_{i_0} w_{i_0}(k_0 + p) \\ &= \ \epsilon_0 \delta_{i_0} w_{i_0}(k_0 + p). \end{aligned}$$

Thus, it follows that $\mathcal{M}\mathbf{w} \geq \epsilon_0 \mathbf{w}$ with respect to $P_1$. The conclusion now follows from Theorem 12.9.2. ∎

## 12.10. Positive Solutions of

(12.10.1) $\quad \Delta^3 u(k) + \lambda a(k) f(u(k)) \ = \ 0, \quad k \in \mathbb{N}(2, K+2)$

(12.10.2) $\quad u(0) \ = \ u(1) \ = \ u(K+3) \ = \ 0$

where $f: R_+ \to R_+$ is continuous, and $a(k)$ is a positive function defined on $\mathbb{N}(0, K+2)$.

To establish the positive solutions of (12.10.1), (12.10.2) we shall need the following Krasnosel'skii fixed point theorem.

**Theorem 12.10.1.** [28] Let $B$ be a Banach space, and let $P \subset B$ be a cone in $B$. Assume $\Omega_1$, $\Omega_2$ are open subsets of $B$ with $0 \in \Omega_1 \subset \overline{\Omega}_1 \subset \Omega_2$, and let

$$A : P \cap (\overline{\Omega}_2 \backslash \Omega_1) \to P$$

be a completely continuous operator such that, either

(i) $\|Au\| \le \|u\|$, $u \in P \cap \partial\Omega_1$ and $\|Au\| \ge \|u\|$, $u \in P \cap \partial\Omega_2$, or

(ii) $\|Au\| \ge \|u\|$, $u \in P \cap \partial\Omega_1$ and $\|Au\| \le \|u\|$, $u \in P \cap \partial\Omega_2$.

Then, $A$ has a fixed point in $P \cap (\overline{\Omega}_2 \backslash \Omega_1)$.

We shall also require certain upper and lower bounds for the Green's function $g(k, \ell)$ of the boundary value problem $-\Delta^3 u(k) = 0$, $k \in \mathbb{N}(2, K+2)$, (12.10.2). For this, we recall from Theorem 10.9.2 that

$$g(k, \ell) > 0 \quad \text{on} \quad \mathbb{N}(2, K+2) \times \mathbb{N}(0, K) \tag{12.10.3}$$

so that from the boundary conditions (12.10.2)

$$g(k, \ell) \ge 0 \quad \text{on} \quad \mathbb{N}(0, K+3) \times \mathbb{N}(0, K). \tag{12.10.4}$$

Now for each $\ell \in \mathbb{N}(0, K)$, let $\tau(\ell) \in \mathbb{N}(0, K+3)$ be defined by

$$g(\tau(\ell), \ell) = \max_{k \in \mathbb{N}(0, K+3)} g(k, \ell). \tag{12.10.5}$$

(Clearly, $\tau(\ell) \in \mathbb{N}(2, K+2)$.) So, we can state

$$g(k, \ell) \le g(\tau(\ell), \ell), \quad k \in \mathbb{N}(0, K+3), \quad \ell \in \mathbb{N}(0, K). \tag{12.10.6}$$

As a lower bound, we have

$$g(k, \ell) \ge \frac{2}{(K+1)(K+2)} g(\tau(\ell), \ell), \ k \in \mathbb{N}(2, K+2), \ \ell \in \mathbb{N}(0, K) \tag{12.10.7}$$

which is a special case of a very general result proved in [3].

For our construction, let

$$B = \{u \mid \mathbb{N}(0, K+3) \to \mathbb{R} : u(0) = u(1) = u(K+3) = 0\}$$

with the norm $\|u\| = \max_{k\in\mathbb{N}(0,K+3)} |u(k)|$. Clearly, $(B, \|\cdot\|)$ is a Banach space. We define a cone $P$, by

$$P = \left\{ u \in B \ : \ u(k) \ge 0 \quad \text{on} \quad \mathbb{N}(0, K+3), \quad \text{and} \quad \min_{k\in\mathbb{N}(2,K+2)} u(k) \ge \frac{2}{(K+1)(K+2)}\|u\| \right\}.$$

Also, we define the number $\sigma \in \mathbb{N}(0, K+3)$ by

$$\sum_{\ell=2}^{K} g(\sigma,\ell)a(\ell) = \max_{k\in\mathbb{N}(0,K+3)} \sum_{\ell=2}^{K} g(k,\ell)a(\ell). \tag{12.10.8}$$

(Clearly, $\sigma \in \mathbb{N}(2, K+2)$.)

In what follows, we shall also denote by

$$f_0 = \lim_{u\to 0+} \frac{f(u)}{u} \quad \text{and} \quad f_\infty = \lim_{u\to\infty} \frac{f(u)}{u}. \tag{12.10.9}$$

**Theorem 12.10.2.** With respect to the boundary value problem (12.10.1), (12.10.2) assume that $\lambda = 1$. Further, let either

(i) $f_0 = 0$ and $f_\infty = \infty$, (i.e. $f$ is *superlinear*), or

(ii) $f_0 = \infty$ and $f_\infty = 0$, (i.e. $f$ is *sublinear*),

then (12.10.1), (12.10.2) has at least one solution in $P$.

**Proof.** We define an operator $A : P \to B$ by

$$Au(k) = \sum_{\ell=0}^{K} g(k,\ell)a(\ell)f(u(\ell)), \quad u \in P. \tag{12.10.10}$$

Clearly, it suffices to exhibit a fixed point of $A$ that lies in the cone $P$.

We first observe from (12.10.4) that if $u \in P$, then $Au(k) \ge 0$ on $\mathbb{N}(0, K+3)$. Further, from the properties of the Green's function $g(k,\ell)$, $Au$ satisfies the boundary conditions (12.10.2). Moreover, for $u \in P$, we have from (12.10.6)

$$Au(k) \le \sum_{\ell=0}^{K} g(\tau(\ell),\ell)a(\ell)f(u(\ell)), \quad k \in \mathbb{N}(0, K+3)$$

and so

$$\|Au\| \le \sum_{\ell=0}^{K} g(\tau(\ell),\ell)a(\ell)f(u(\ell)). \tag{12.10.11}$$

Thus, if $u \in P$, then (12.10.7) and (12.10.11) imply

$$\begin{aligned}\min_{k\in\mathbb{N}(2,K+2)} Au(k) &= \min_{k\in\mathbb{N}(2,K+2)} \sum_{\ell=0}^{K} g(k,\ell)a(\ell)f(u(\ell)) \\ &\geq \sum_{\ell=0}^{K} \frac{2}{(K+1)(K+2)} g(\tau(\ell),\ell)a(\ell)f(u(\ell)) \\ &\geq \frac{2}{(K+1)(K+2)}\|Au\|.\end{aligned}$$

As a consequence $A: P \to P$. In addtion, it is immediate that $A$ is completely continuous.

Let us now consider the cases of the theorem.

Case (i). Assume $f_0 = 0$ and $f_\infty = \infty$. From $f_0 = 0$, there exist $\eta > 0$ and $H_1 > 0$ such that $f(u) \leq \eta u$ for $0 < u \leq H_1$, and

$$\eta \sum_{\ell=0}^{K} g(\tau(\ell),\ell)a(\ell) \leq 1.$$

Define the open subset $\Omega_1 = \{u \in B : \|u\| < H_1\}$. Then, for $u \in P \cap \partial\Omega_1$ we have $\|u\| = H_1$, and from (12.10.6),

$$\begin{aligned}Au(k) &\leq \sum_{\ell=0}^{K} g(\tau(\ell),\ell)a(\ell)f(u(\ell)) \\ &\leq \sum_{\ell=0}^{K} g(\tau(\ell),\ell)a(\ell)\eta u(\ell) \\ &\leq \eta\sum_{\ell=0}^{K} g(\tau(\ell),\ell)a(\ell)\|u\| \leq \|u\|, \quad k \in \mathbb{N}(0,K+3)\end{aligned}$$

and we conclude

(12.10.12) $$\|Au\| \leq \|u\| \quad \text{for } u \in P \cap \partial\Omega_1.$$

Next from $f_\infty = \infty$ there exist $\lambda > 0$ and $\overline{H}_2 > 0$ such that $f(u) \geq \lambda u$ for $u \geq \overline{H}_2$, and

$$\lambda\left(\frac{2}{(K+1)(K+2)}\right)\sum_{\ell=2}^{K} g(\sigma,\ell)a(\ell) \geq 1.$$

Let $H_2 = \max\{2H_1, ((K+1)(K+2)/2)\overline{H}_2\}$, and define $\Omega_2 = \{u \in B : \|u\| < H_2\}$. If $u \in P$ with $\|u\| = H_2$, then $\min_{k \in \mathbb{N}(2,K+2)} u(k) \geq (2/(K+1)(K+2))\|u\| \geq \overline{H}_2$, and

$$\begin{aligned} Au(\sigma) &= \sum_{\ell=0}^{K} g(\sigma,\ell)a(\ell)f(u(\ell)) \\ &\geq \sum_{\ell=2}^{K} g(\sigma,\ell)a(\ell)f(u(\ell)) \\ &\geq \sum_{\ell=2}^{K} g(\sigma,\ell)a(\ell)\lambda u(\ell) \\ &\geq \lambda\left(\frac{2}{(K+1)(K+2)}\right)\sum_{\ell=2}^{K} g(\sigma,\ell)a(\ell)\|u\| \ \geq \ \|u\|. \end{aligned}$$

Consequently, $\|Au\| \geq \|u\|$, and so

(12.10.13) $$\|Au\| \ \geq \ \|u\| \quad \text{for } u \in P \cap \partial\Omega_2.$$

Now an application of Theorem 12.10.1 to (12.10.12) and (12.10.13) yields that $A$ has a fixed point $u \in P \cap (\overline{\Omega}_2 \backslash \Omega_1)$. As such, the fixed point $u$ is a desired solution of (12.10.1), (12.10.2) for the case of $f$ superlinear.

Case (ii). Assume $f_0 = \infty$ and $f_\infty = 0$. From $f_0 = \infty$ there exists $\overline{\eta} > 0$ and $J_1 > 0$ such that $f(u) \geq \overline{\eta} u$ for $0 < u \leq J_1$, and

$$\overline{\eta}\left(\frac{2}{(K+1)(K+2)}\right)\sum_{\ell=2}^{K} g(\sigma,\ell)a(\ell) \ \geq \ 1.$$

In this case, we define $\Omega_1 = \{u \in B : \|u\| < J_1\}$. Then, for $u \in P \cap \partial\Omega_1$ we have $f(u(\ell)) \geq \overline{\eta} u(\ell)$, $\ell \in \mathbb{N}(0,K)$, and moreover, $u(\ell) \geq (2/(K+1)(K+2))\|u\|$, $\ell \in \mathbb{N}(2,K+2)$. Thus,

$$\begin{aligned} Au(\sigma) &\geq \sum_{\ell=2}^{K} g(\sigma,\ell)a(\ell)f(u(\ell)) \\ &\geq \sum_{\ell=2}^{K} g(\sigma,\ell)a(\ell)\overline{\eta} u(\ell) \\ &\geq \overline{\eta}\left(\frac{2}{(K+1)(K+2)}\right)\sum_{\ell=2}^{K} g(\sigma,\ell)a(\ell)\|u\| \ \geq \ \|u\|. \end{aligned}$$

From which we have

(12.10.14) $$\|Au\| \ \geq \ \|u\| \quad \text{for } u \in P \cap \partial\Omega_1.$$

For the final part of the proof we deal with $f_\infty = 0$. In this case there exist $\overline{\lambda} > 0$ and $\overline{J}_2 > 0$ such that $f(u) \le \overline{\lambda} u$ for $u \ge \overline{J}_2$, and

$$\overline{\lambda} \sum_{\ell=0}^{K} g(\tau(\ell), \ell) a(\ell) \ \le \ 1.$$

There are two subcases: (I) $f$ is bounded, and (II) $f$ is unbounded.

Subcase (I). Suppose $M > 0$ is such that $f(u) \le M$ for all $0 < u < \infty$. Let $J_2 = \max\{2J_1, \ M\sum_{\ell=0}^{K} g(\tau(\ell), \ell)a(\ell)\}$. Then, for $u \in P$ with $\|u\| = J_2$, we have

$$Au(k) \ \le \ M \sum_{\ell=0}^{K} g(\tau(\ell), \ell) u(\ell) \ \le \ \|u\|, \quad k \in \mathbb{N}(0, K+3)$$

so that $\|Au\| \le \|u\|$. So, if $\Omega_2 = \{u \in B : \|u\| < J_2\}$, then

(12.10.15) $$\|Au\| \ \le \ \|u\| \quad \text{for } u \in P \cap \partial\Omega_2.$$

Subcase (II). For this, let $J_2 > \max\{2J_1, \overline{J}_2\}$ be such that $f(u) \le f(J_2)$ for $0 < u \le J_2$. If we choose $u \in P$ with $\|u\| = J_2$, then

$$\begin{aligned} Au(k) \ &\le \ \sum_{\ell=0}^{K} g(\tau(\ell), \ell) a(\ell) f(u(\ell)) \\ &\le \ \sum_{\ell=0}^{K} g(\tau(\ell), \ell) a(\ell) f(J_2) \\ &\le \ \sum_{\ell=0}^{K} g(\tau(\ell), \ell) a(\ell) \overline{\lambda} J_2 \\ &= \ \overline{\lambda} \sum_{\ell=0}^{K} g(\tau(\ell), \ell) a(\ell) \|u\| \ \le \ \|u\|, \quad k \in \mathbb{N}(0, K+3). \end{aligned}$$

In this case, if $\Omega_2 = \{u \in B : \|u\| < J_2\}$, then

(12.10.16) $$\|Au\| \ \le \ \|u\| \quad \text{for } u \in P \cap \partial\Omega_2.$$

Thus, from the two subcases, an applcation of Theorem 12.10.1 to (12.10.14), (12.10.15) or (12.10.14), (12.10.16) yields a fixed point of $A$ which belongs to $P \cap (\overline{\Omega}_2 \backslash \Omega_1)$. This fixed point is a solution of (12.10.1), (12.10.2) for the case of $f$ sublinear. ∎

**Theorem 12.10.3.** With respect to the boundary value problem (12.10.1), (12.10.2) assume that $f_0$ and $f_\infty$ both exist as positive real numbers. Then, for each $\lambda$ satisfying

$$\frac{(K+1)(K+2)}{2\left(\sum_{\ell=2}^{K} g(\sigma,\ell)a(\ell)\right) f_\infty} < \lambda < \frac{1}{\left(\sum_{\ell=0}^{K} g(\tau(\ell),\ell)a(\ell)\right) f_0} \tag{12.10.17}$$

there exists at least one solution of (12.10.1), (12.10.2) that lies in $P$.

**Proof.** Let $\lambda$ be given as in (12.10.17). Then, let $\epsilon > 0$ be such that

$$\frac{(K+1)(K+2)}{2\left(\sum_{\ell=2}^{K} g(\sigma,\ell)a(\ell)\right)(f_\infty - \epsilon)} \le \lambda \le \frac{1}{\left(\sum_{\ell=0}^{K} g(\tau(\ell),\ell)a(\ell)\right)(f_0 + \epsilon)}.$$

We define an operator $A : P \to B$ by

$$Au(k) = \lambda \sum_{\ell=0}^{K} g(k,\ell)a(\ell)f(u(\ell)), \quad u \in P \tag{12.10.18}$$

and seek a fixed point of $A$ that lies in the cone $P$.

The arguments identical to those in the first part of the proof of Theorem 12.10.2 yield that $A : P \to P$ and that $A$ is completely continuous.

The existence of $f_0$ as a positive real number implies that there exists $H_1 > 0$ such that $f(u) \le (f_0 + \epsilon)u$ for $0 < u \le H_1$. So, for $u \in P$ with $\|u\| = H_1$, from (12.10.6) we have

$$\begin{aligned} Au(k) &\le \lambda \sum_{\ell=0}^{K} g(\tau(\ell),\ell)a(\ell)f(u(\ell)) \\ &\le \lambda \sum_{\ell=0}^{K} g(\tau(\ell),\ell)a(\ell)(f_0+\epsilon)u(\ell) \\ &\le \lambda \sum_{\ell=0}^{K} g(\tau(\ell),\ell)a(\ell)(f_0+\epsilon)\|u\| \le \|u\|, \quad k \in \mathbb{N}(0, K+3). \end{aligned}$$

Therefore, $\|Au\| \le \|u\|$. Hence, if we set $\Omega_1 = \{u \in B : \|u\| < H_1\}$, then

$$\|Au\| \le \|u\| \quad \text{for } u \in P \cap \partial\Omega_1. \tag{12.10.19}$$

Next, since $f_\infty$ is a positive real number, there exists an $\overline{H}_2 > 0$ such that $f(u) \ge (f_\infty - \epsilon)u$ for all $u \ge \overline{H}_2$. Let $H_2 = \max\{2H_1, ((K+1)(K+2)/2)\overline{H}_2\}$, and define $\Omega_2 = \{u \in B : \|u\| < H_2\}$. If $u \in P$ with

$\|u\| = H_2$, then $\min_{k\in\mathbf{N}(2,K+2)} u(k)(2/(K+1)(K+2))\|u\| \geq \overline{H}_2$, and

$$\begin{aligned} Au(\sigma) &\geq \lambda \sum_{\ell=2}^{K} g(\sigma,\ell)a(\ell)f(u(\ell)) \\ &\geq \lambda \sum_{\ell=2}^{K} g(\sigma,\ell)a(\ell)(f_\infty - \epsilon)u(\ell) \\ &\geq \lambda\left(\frac{2}{(K+1)(K+2)}\right)\sum_{\ell=2}^{K} g(\sigma,\ell)a(\ell)(f_\infty-\epsilon)\|u\| \;\geq\; \|u\|. \end{aligned}$$

Thus, $\|Au\| \geq \|u\|$, and so

$$\|Au\| \;\geq\; \|u\| \quad \text{for } u \in P \cap \partial\Omega_2. \tag{12.10.20}$$

Applying Theorem 12.10.1 to (12.10.19) and (12.10.20) we conclude that $A$ has a fixed point $u(k) \in P \cap (\overline{\Omega}_2 \backslash \Omega_1)$. This fixed point $u(k)$ is a solution of (12.10.1), (12.10.2) corresponding to the given value of $\lambda$. ∎

**Theorem 12.10.4.** With respect to the boundary value problem (12.10.1), (12.10.2) assume that $f_0$ and $f_\infty$ both exist as positive real numbers. Then, for each $\lambda$ satisfying

$$\frac{(K+1)(K+2)}{2\left(\sum_{\ell=2}^{K} g(\sigma,\ell)a(\ell)\right) f_0} \;<\; \lambda \;<\; \frac{1}{\left(\sum_{\ell=0}^{K} g(\tau(\ell),\ell)a(\ell)\right) f_\infty} \tag{12.10.21}$$

there exists at least one solution of (12.10.1), (12.10.2) that lies in $P$.

**Proof.** Let $\lambda$ be given as in (12.10.21), and choose $\epsilon > 0$ such that

$$\frac{(K+1)(K+2)}{2\left(\sum_{\ell=2}^{K} g(\sigma,\ell)a(\ell)\right)(f_0-\epsilon)} \;\leq\; \lambda \;\leq\; \frac{1}{\left(\sum_{\ell=0}^{K} g(\tau(\ell),\ell)a(\ell)\right)(f_\infty+\epsilon)}.$$

Let $A$ be the cone preserving, completely continuous operator that is defined in (12.10.18).

Beginning with $f_0$ there exists an $H_1 > 0$ such that $f(u) \geq (f_0 - \epsilon)u$ for $0 < u \leq H_1$. So for $u \in P$ and $\|u\| = H_1$, we have

$$\begin{aligned} Au(\sigma) &\geq \lambda \sum_{\ell=2}^{K} g(\sigma,\ell)a(\ell)f(u(\ell)) \\ &\geq \lambda \sum_{\ell=2}^{K} g(\sigma,\ell)a(\ell)(f_0 - \epsilon)u(\ell) \\ &\geq \lambda\left(\frac{2}{(K+1)(K+2)}\right)\sum_{\ell=2}^{K} g(\sigma,\ell)a(\ell)(f_0-\epsilon)\|u\| \;\geq\; \|u\|. \end{aligned}$$

Therefore, if we let $\Omega_1 = \{u \in B : \|u\| < H_1\}$, then

$$\|Au\| \geq \|u\| \quad \text{for} \quad u \in P \cap \partial\Omega_1. \tag{12.10.22}$$

There remains to consider $f_\infty$. There exists $\overline{H}_2 > 0$ such that $f(u) \leq (f_\infty + \epsilon)u$ for all $u \geq \overline{H}_2$. There are two subcases: (a) $f$ is bounded, and (b) $f$ is unbounded.

For Case (a), suppose $M > 0$ is such that $f(u) \leq M$ for all $0 < u < \infty$. Let $H_2 = \max\{2H_1,\ M\lambda\sum_{\ell=0}^{K} g(\tau(\ell),\ell)a(\ell)\}$. Also, let $\Omega_2 = \{u \in B : \|u\| < H_2\}$. Then, for $u \in P \cap \partial\Omega_2$, we have

$$Au(k) \leq \lambda M \sum_{\ell=0}^{K} g(\tau(\ell),\ell)a(\ell) \leq \|u\|, \quad k \in \mathbb{N}(0, K+3)$$

and so

$$\|Au\| \leq \|u\| \quad \text{for} \quad u \in P \cap \partial\Omega_2. \tag{12.10.23}$$

For Case (b), let $H_2 > \max\{2H_1, \overline{H}_2\}$ be such that $f(u) \leq f(H_2)$ for $0 < u \leq H_2$. Let $\Omega_2 = \{u \in B : \|u\| < H_2\}$. Choosing $u \in P \cap \partial\Omega_2$, we have

$$\begin{aligned} Au(k) &\leq \lambda \sum_{\ell=0}^{K} g(\tau(\ell),\ell)a(\ell)f(u(\ell)) \\ &\leq \lambda \sum_{\ell=0}^{K} g(\tau(\ell),\ell)a(\ell)f(H_2) \\ &\leq \lambda \sum_{\ell=0}^{K} g(\tau(\ell),\ell)a(\ell)(f_\infty + \epsilon)H_2 \\ &= \lambda \sum_{\ell=0}^{K} g(\tau(\ell),\ell)a(\ell)(f_\infty + \epsilon)\|u\| \leq \|u\|, \quad k \in \mathbb{N}(0, K+3), \end{aligned}$$

and so

$$\|Au\| \leq \|u\| \quad \text{for} \quad u \in P \cap \partial\Omega_2. \tag{12.10.24}$$

In each of the subcases we apply Theorem 12.10.1 to (12.10.22), (12.10. 23) or (12.10.22), (12.10.24) to obtain a fixed point of $A$ that lies in $P \cap (\overline{\Omega}_2 \backslash \Omega_1)$. This fixed point is a solution of (12.10.1), (12.10.2) corresponding to the given $\lambda$. ■

## 12.11. Problems

**12.11.1.** Let $\lambda$ be an eigenvalue of the Sturm Liouville problem (12.1.1),

$$\alpha u(0) + \beta \Delta u(0) = 0 \tag{12.11.1}$$

$$\gamma u(K) + \delta \Delta u(K) = 0, \tag{12.11.2}$$

where $\alpha^2 + \beta^2 \neq 0$, $\gamma^2 + \delta^2 \neq 0$. Show that if $q(k) \leq 0$ on $\mathbb{N}(1, K)$, $\alpha\beta \leq 0$ and $\gamma\delta \geq 0$, then $\lambda \geq 0$, and if in addition $q(k) > 0$ at two consecutive integers in $\mathbb{N}(1, K)$, then $\lambda > 0$.

**12.11.2.** Let $u_1(k, \lambda)$, $u_2(k, \lambda)$ be the linearly independent solutions of (12.1.1) satisfying the initial conditions

$$u_1(0, \lambda) = 1, \quad \Delta u_1(0, \lambda) = 0$$
$$u_2(0, \lambda) = 0, \quad \Delta u_2(0, \lambda) = 1$$

and let $u(k, \lambda) = \beta u_1(k, \lambda) - \alpha u_2(k, \lambda)$, which is a nontrivial solution of (12.1.1) and satisfies the boundary condition (12.11.1). Show that the eigenvalues of the Sturm Liouville problem (12.1.1), (12.11.1), (12.11.2) are the zeros of $g(\lambda) = \gamma u(K, \lambda) + \delta \Delta u(K, \lambda)$, $\lambda \in \mathbb{R}$ and the zeros of $g(\lambda)$ are simple.

**12.11.3.** Show that the Sturm Liouville problem (12.1.1), $u(0) = u(K+1) = 0$ has $K$ eigenvalues $\lambda_1 < \cdots < \lambda_K$. Further, if $\lambda_i$ are the eigenvalues and $\phi_i(k)$, $1 \leq i \leq K$ are the corresponding eigenfunctions, then $\phi_i(k)$ has exactly $i - 1$ generalized zeros in $\mathbb{N}(1, K)$.

**12.11.4.** Let $\lambda$ be an eigenvalue and $\phi(k)$ be the corresponding eigenfunction of (12.1.1), (12.11.1), (12.11.2). Show that the nonhomogeneous problem

$$\Delta\left(p(k-1)\Delta u(k-1)\right) + q(k)u(k) + \lambda r(k)u(k) = f(k), \quad k \in \mathbb{N}(1, K)$$

(12.11.1), (12.11.2) has a solution if and only if $\phi(k)$ and $f(k)$ are orthogonal on $\mathbb{N}(1, K)$, i.e. $\sum_{k=1}^{K} \phi(k)f(k) = 0$.

**12.11.5.** For the difference equation (12.1.10) together with the periodic boundary conditions $u(-1) = u(K)$, $u(0) = u(K+1)$ show that the distinct eigenvalues are $\lambda_m = 4\sin^2 \dfrac{m\pi}{K+1}$, $0 \leq m \leq \left[\dfrac{K+1}{2}\right]$ and corresponding to $\lambda_0 = 0$ the eigenfunction is $\phi_0(k) = 1$; corresponding to $\lambda_{(K+1)/2}$ (which is possible only when $K$ is odd) the eigenfunction is $\phi_{(K+1)/2}(k) = \cos 2\pi k$; whereas corresponding to the remaining eigenvalues each corresponds to the two linearly independent eigenfunctions $\phi_m^{(1)}(k) = \cos\left(\dfrac{2m\pi k}{K+1}\right)$, $\phi_m^{(2)}(k) = \sin\left(\dfrac{2m\pi k}{K+1}\right)$, $1 \leq m \leq \left[\dfrac{K}{2}\right]$.

**12.11.6.** Show that for any function $u(k)$, $k \in \mathbb{N}(0, K)$ satisfying $u(K) = 0$, $u(K+1) = u(0)$ the following inequality holds

$$\sum_{\ell=0}^{K} (\Delta u(\ell))^2 \ \geq \ 4 \ \sin \frac{\pi}{2(K+1)} \sum_{\ell=0}^{K} u^2(\ell). \tag{12.11.3}$$

In (12.11.3) equality holds if and only if $u(k) = c \sin \dfrac{(k+1)\pi}{(K+1)}$, where $c$ is an arbitrary constant.

**12.11.7.** Show that for any function $u(k)$, $k \in \mathbb{N}(1, K)$ satisfying $\sum_{\ell=1}^{K} u(\ell) = 0$ the following inequality holds

$$\sum_{\ell=1}^{K-1} (\Delta u(\ell))^2 \ \geq \ 4 \ \sin^2 \frac{\pi}{2K} \sum_{\ell=1}^{K} u^2(\ell) + 2K \ \sin \frac{\pi}{2K} \times \left(\sin \frac{\pi}{K} - \sin \frac{\pi}{2K}\right) (u(1) + u(K))^2. \tag{12.11.4}$$

In (12.11.4) equality holds if and only if $u(k) = c \sin \dfrac{(2k-1)\pi}{2K}$, where $c$ is an arbitrary constant.

**12.11.8.** Show that for any function $u(k)$, $k \in \mathbb{N}(1, K)$ satisfying $u(K+1) = u(1)$, $\sum_{\ell=1}^{K} u(\ell) = 0$, and $K = 2m$ the following inequality holds

$$\sum_{\ell=1}^{K} (\Delta u(\ell))^2 \ \geq \ 4 \ \sin^2 \frac{\pi}{K} \sum_{\ell=1}^{K} u^2(\ell) + K \sin \frac{\pi}{K} \times \left(\sin \frac{2\pi}{K} - \sin \frac{\pi}{K}\right) (u(m) + u(2m))^2. \tag{12.11.5}$$

In (12.11.5) equality holds if and only if $u(k) = c_1 \cos \dfrac{2k\pi}{K} + c_2 \sin \dfrac{2k\pi}{K}$, where $c_1$ and $c_2$ are arbitrary constants.

**12.11.9.** Show that for any complex valued function $u(k)$, $k \in \mathbb{N}(0, K)$ and $1 \leq m \leq K$ the following best possible inequalities hold

$$-\gamma \sum_{\ell=0}^{K} |u(\ell)|^2 \ \leq \ \sum_{\ell=0}^{K-m} (u(\ell)\overline{u}(\ell+m) + \overline{u}(\ell)u(\ell+m)) \ \leq \ \gamma \sum_{\ell=0}^{K} |u(\ell)|^2,$$

where $\gamma = 2\cos \dfrac{\pi}{r+2}$ and $r = [K/m]$.

**12.11.10.** Show that for any real or complex valued function $u(k)$, $k \in \mathbb{N}(0, K)$ and $1 \le m \le K$, $i = 1,2,3,4$ the following best possible inequalities hold

$$\alpha_i^{\pm} \sum_{\ell=0}^{K} |u(\ell)|^2 \le \sum^{i} |u(\ell) \pm u(\ell + m)|^2 \le \beta_i^{\pm} \sum_{\ell=0}^{K} |u(\ell)|^2, \tag{12.11.6}$$

where the summation symbols are defined by

$$\sum^{1} = \sum_{\ell=0}^{K-m}$$

$$\sum^{2} = \sum_{\ell=0}^{K} \text{ with } u(K+1) = \cdots = u(K+m) = 0$$

$$\sum^{3} = \sum_{\ell=-m}^{K-m} \text{ with } u(-m) = \cdots = u(-1) = 0$$

$$\sum^{4} = \sum_{\ell=-m}^{K} \text{ with } u(-m) = \cdots = u(-1) = 0 = u(K+1) = \cdots = u(K+m),$$

(It is clear that the cases $i = 2$ and $3$ are the same apart from the notation of the variables $u(k)$. Hence, there are 6 different cases in (12.11.6) corresponding to $i = 1, 2$ or $3, 4$ and the $+$ and $-$ sign.) and

$$\alpha_1^+ = \alpha_1^- = 0, \quad \beta_1^+ = \beta_1^- = 4\cos^2 \frac{\pi}{2(r+1)}, \quad r = [K/m]$$

$$\alpha_2^+ = \alpha_2^- = \alpha_3^+ = \alpha_3^- = 4\sin^2 \frac{\pi}{2(2r+3)}$$

$$\beta_2^+ = \beta_2^- = \beta_3^+ = \beta_3^- = 4\cos^2 \frac{\pi}{2r+3}$$

$$\alpha_4^+ = \alpha_4^- = 4\sin^2 \frac{\pi}{2(r+2)}, \quad \beta_4^+ = \beta_4^- = 4\cos^2 \frac{\pi}{2(r+2)}.$$

**12.11.11.** Use the relation

$$\mathcal{H}_K^2(\mathbf{g}, \mathbf{h}) = \begin{bmatrix} 5 & -4 & 1 & & & & & & \\ -4 & 6 & -4 & 1 & & & & & \\ 1 & -4 & 6 & -4 & 1 & & & & \\ & 1 & -4 & 6 & -4 & 1 & & & \\ & & & & & \cdot \; \cdot & & & \\ & & & & & \cdot & & & \\ & & & & & 1 & -4 & 6 & -4 \\ & & & & & & 1 & -4 & 5 \end{bmatrix},$$

where $\mathbf{g} = (2, \cdots, 2)$ and $\mathbf{h} = (-1, \cdots, -1)$ to show that for any function $u(k)$, $k \in \mathbb{N}(0, K+1)$ satisfying $u(0) = u(K+1) = 0$ the following inequalities hold

$$
\begin{aligned}
(12.11.7) \qquad 16\sin^4\frac{\pi}{2(K+1)}\sum_{\ell=1}^{K}u^2(\ell) \ &\le\ \sum_{\ell=0}^{K-1}(\Delta^2 u(\ell))^2 \\
&\le\ 16\cos^4\frac{\pi}{2(K+1)}\sum_{\ell=1}^{K}u^2(\ell).
\end{aligned}
$$

In the left (right) of (12.11.7) equality holds if and only if $u(k) = c\sin\left(\frac{k\pi}{K+1}\right)$ $\left(u(k) = c(-1)^{k-1}\sin\left(\frac{k\pi}{K+1}\right)\right)$, where $c$ is an arbitrary constant.

**12.11.12.** Use the relation

$$
\mathcal{H}_K^2(\mathbf{g}, \mathbf{h}) = \begin{bmatrix}
2 & -3 & 1 & & & & & & \\
-3 & 6 & -4 & 1 & & & & & \\
1 & -4 & 6 & -4 & 1 & & & & \\
 & 1 & -4 & 6 & -4 & 1 & & & \\
 & & & & & \cdot\ \cdot & & & \\
 & & & & & \cdot & & & \\
 & & & & & 1 & -4 & 6 & -3 \\
 & & & & & & 1 & -3 & 2
\end{bmatrix},
$$

where $\mathbf{g} = (1, 2, \cdots, 2, 1)$ and $\mathbf{h} = (-1, \cdots, -1)$ to show that for any function $u(k)$, $k \in \mathbb{N}(0, K+1)$ satisfying $u(0) = u(1)$, $u(K+1) = u(K)$, $\sum_{\ell=1}^{K} u(\ell) = 0$ the following inequalities hold

$$
\begin{aligned}
(12.11.8) \qquad 16\sin^4\frac{\pi}{2K}\sum_{\ell=1}^{K}u^2(\ell) \ &\le\ \sum_{\ell=0}^{K-1}(\Delta^2 u(\ell))^2 \\
&\le\ 16\cos^4\frac{\pi}{2K}\sum_{\ell=1}^{K}u^2(\ell).
\end{aligned}
$$

In the left (right) of (12.11.8) equality holds if and only if $u(k) = c\cos\left(\frac{(2k-1)\pi}{2K}\right)$ $\left(u(k) = c\cos\left(\frac{(K-1)(2k-1)\pi}{2K}\right)\right)$, where $c$ is an arbitrary constant.

**12.11.13.** Show that for any function $u(k)$, $k \in \mathbb{N}(0, K)$ satisfying $u(0) = u(1)$, $u(K+1) = u(K)$ the following inequality holds

$$
(12.11.9) \qquad \sum_{\ell=0}^{K-1}(\Delta^2 u(\ell))^2 \ \le\ 16\ \cos^4\frac{\pi}{2K}\sum_{\ell=1}^{K}u^2(\ell).
$$

In (12.11.9) equality holds if and only if $u(k) = c(-1)^k \sin \frac{(2k-1)\pi}{K}$, where $c$ is an arbitrary constant.

**12.11.14.** For any $K$ dimensional vector $\mathbf{u} = (u(1), \cdots, u(K))^T$ we introduce a periodically extended $K$ vector by setting $u(k+\tau K) = u(k)$ for $k = 1, \cdots, K$ and $\tau \in \mathbb{N}$, and denote the $m$th difference of $\mathbf{u}$ by $\mathbf{u}^{(m)} = (\Delta^{(m)}u(1), \cdots, \Delta^{(m)}u(k))^T$, where

$$\Delta^{(m)}u(k) = \sum_{\ell=0}^{m}(-1)^{m-\ell}\binom{m}{\ell}u(k-[m/2]+\ell), \quad 1 \le k \le K.$$

Show that

$$\left(\mathbf{u}^{(m)},\ \mathbf{u}^{(m)}\right) \ \ge\ \left(4\sin^2\frac{\pi}{K}\right)^m (\mathbf{u},\ \mathbf{u}). \tag{12.11.10}$$

In (12.11.10) equality holds if and only if $\mathbf{u}$ is the periodic extension of a vector of the form $c_1\mathbf{u} + c_2\mathbf{v}$, where $\mathbf{u} = (u(1), \cdots, u(K))^T$ and $\mathbf{v} = (v(1), \cdots, v(K))^T$ have the components $u(k) = \cos\frac{2k\pi}{K}$, $v(k) = \sin\frac{2k\pi}{K}$ and $c_1$, $c_2$ are arbitrary constants.

**12.11.15.** Let $m, K \in \mathbb{N}(1)$, and define $c(m) = 1 - [m/2]$ and $d(m) = K - [(m+1)/2]$. Show that for any function $u(k)$, $k \in \mathbb{N}(1, K)$ satisfying $u(\tau) = u(1-\tau)$, $u(K+1-\tau) = u(K+\tau)$, $c(m) \le \tau \le 0$ the following inequality holds

$$\sum_{\ell=c(m)}^{d(m)} (\Delta^m u(\ell))^2 \ \le\ \left(4\cos^2\frac{\pi}{2K}\right)^m \sum_{\ell=1}^{K} u^2(\ell). \tag{12.11.11}$$

In (12.11.11) equality holds if and only if $u(k) = c(-1)^k \sin \frac{(2k-1)\pi}{K}$, where $c$ is an arbitrary constant.

**12.11.16.** Show that if in addition to conditions of Problem 12.11.15, $\sum_{\ell=1}^{K} u(\ell) = 0$, then the following inequality holds

$$\sum_{\ell=c(m)}^{d(m)} (\Delta^m u(\ell))^2 \ \ge\ \left(4\sin^2\frac{\pi}{2K}\right)^m \sum_{\ell=1}^{K} u^2(\ell). \tag{12.11.12}$$

In (12.11.12) equality holds if and only if $u(k) = c\cos\frac{(2k-1)\pi}{2K}$, where $c$ is an arbitrary constant.

**12.11.17.** Suppose that the function $u(k)$, $k \in \mathbb{N}(0, K+1)$ satisfies the conditions $u(0) = u(K+1) = 0$ and the inequality

$$|\Delta^2 u(k-1)| \le L + L_0|u(k)| + L_1|\Delta u(k)|, \quad k \in \mathbb{N}(1, K).$$

If the constants $L$, $L_0$, $L_1$ are nonnegative and if

$$\rho_{K+1} = \frac{1}{4\sin^2 \frac{\pi}{2(K+1)}} L_0 + \frac{1}{2}\left[\frac{K+2}{2}\right] L_1 < 1, \tag{12.11.13}$$

then show that

$$\left(\sum_{\ell=1}^{K} u^2(\ell)\right)^{1/2} \le \frac{\sqrt{K}L}{4\sin^2 \frac{\pi}{2(K+1)}(1 - \rho_{K+1})}$$

and

$$\left(\sum_{\ell=0}^{K} (\Delta u(\ell))^2\right)^{1/2} \le \frac{\sqrt{K}L}{2\sin \frac{\pi}{2(K+1)}(1 - \rho_{K+1})}.$$

**12.11.18.** Consider the difference equation

$$\Delta^2 u(k-1) = f(k, u(k), \Delta u(k)), \quad k \in \mathbb{N}(1, K) \tag{12.11.14}$$

together with the boundary conditions (1.6.14).

(i) If $f(k, u, v)$ is continuous on $\mathbb{N}(1, K) \times \mathbb{R}^2$ and satisfies

$$|f(k, u, v)| \le L + L_0|u| + L_1|v|,$$

where the constants $L$, $L_0$, $L_1$ are nonnegative and satisfy (12.11.13), then show that the problem (12.11.14), (1.6.14) has at least one solution.

(ii) If $f(k, u, v)$ satisfies the Lipschitz condition

$$|f(k, u, v) - f(k, \overline{u}, \overline{v})| \le L_0|u - \overline{u}| + L_1|v - \overline{v}|$$

on $\mathbb{N}(1, K) \times \mathbb{R}^2$ and (12.11.13) holds, then show that the problem (12.11.14), (1.6.14) has a unique solution.

**12.11.19.** Suppose that the function $u(k)$, $k \in \mathbb{N}(0, K)$ satisfies the conditions $u(0) = u(1) = u(K-1) = u(K) = 0$ and the inequality

$$|\nabla_0^2 \Delta_0^2 u(k)| \le L + L_0|u(k)| + L_1|\Delta_0 u(k)| + L_2|\nabla_0 \Delta_0 u(k)|, \quad k \in \mathbb{N}(0, K).$$

If the constants $L$, $L_0$, $L_1$, $L_2$ are nonnegative and if

$$\rho_K = \frac{1}{\lambda_{1,2}}L_0 + \frac{1}{2\sqrt{\lambda_{1,2}}\sin\frac{\pi}{2K}}L_1 + \frac{1}{\sqrt{\lambda_{1,2}}}L_2 < 1, \tag{12.11.15}$$

where $\lambda_{1,2}$ is defined in Theorem 12.7.1, then show that

$$\left(\sum_{\ell=0}^{K} u^2(\ell)\right)^{1/2} \leq \frac{L\sqrt{K-3}}{\lambda_{1,2}(1-\rho_K)},$$

$$\left(\sum_{\ell=0}^{K} (\Delta_0 u(\ell))^2\right)^{1/2} \leq \frac{L\sqrt{K-3}}{2\sqrt{\lambda_{1,2}}\sin\frac{\pi}{2K}(1-\rho_K)}$$

and

$$\left(\sum_{\ell=0}^{K} (\Delta_0^2 u(\ell))^2\right)^{1/2} \leq \frac{L\sqrt{K-3}}{\sqrt{\lambda_{1,2}}(1-\rho_K)}.$$

**12.11.20.** Consider the boundary value problem

$$\begin{aligned} \nabla^2\Delta^2 u(k) &= f(k, u(k), \Delta u(k), \nabla\Delta u(k)), \quad k \in \mathbb{N}(2, K-2) \\ u(0) = A_0, \quad \Delta u(0) &= A_1, \quad u(K) = B_0, \quad \nabla u(K) = B_1. \end{aligned} \tag{12.11.16}$$

(i) If $f(k, u, v, w)$ is continuous on $\mathbb{N}(2, K-2) \times \mathbb{R}^3$ and satisfies

$$|f(k, u, v, w)| \leq L + L_0|u| + L_1|v| + L_2|w|,$$

where the constants $L$, $L_0$, $L_1$, $L_2$ are nonnegative and satisfy (12.11.15), then show that the problem (12.11.16) has at least one solution.

(ii) If $f(k, u, v, w)$ satisfies the Lipschitz condition

$$|f(k, u, v, w) - f(k, \overline{u}, \overline{v}, \overline{w})| \leq L_0|u - \overline{u}| + L_1|v - \overline{v}| + L_2|w - \overline{w}|$$

on $\mathbb{N}(2, K-2) \times \mathbb{R}^3$ and (12.11.15) holds, then show that the problem (12.11.16) has a unique solution.

**12.11.21.** Let $u(k)$ be a periodic function of period $K$, and let the function $v(k)$ be defined on $\mathbb{N}(1, K)$. For the function $w(k) = \sum_{\ell=1}^{K} u(k+\ell)v(\ell)$ show that

$$\sum_{\ell=1}^{K} |w(\ell)|^r \leq \left(\sum_{\ell=1}^{K} |u(\ell)|^r\right)\left(\sum_{\ell=1}^{K} |v(\ell)|\right)^r,$$

where $r$ is a positive integer.

**12.11.22.** For any periodic function $u(k)$ of period $K$ satisfying $\sum_{\ell=1}^{K} u(\ell) = 0$, and $n$ any positive integer, show that

(i) $u(k) = -\frac{1}{K}\sum_{\ell=1}^{K}(K-\ell)\Delta u(k+\ell-1)$

(ii) $u(k) = \frac{1}{2K}\sum_{\ell=1}^{K}\ell(K-\ell)\Delta^2 u(k+\ell-1)$

(iii) $\sum_{\ell=1}^{K}|u(\ell)|^n \le ((K-1)/2)^n \sum_{\ell=1}^{K}|\Delta u(\ell)|^n$

(iv) $\sum_{\ell=1}^{K}|u(\ell)|^n \le (K^2-1)/12)^n \sum_{\ell=1}^{K}|\Delta^2 u(\ell-1)|^n$.

**12.11.23.** For any periodic function $u(k)$ of period $K$ satisfying $\sum_{\ell=1}^{K} u(\ell) = 0$, and $n$ any positive integer, show that

$$\max_{\mathbb{N}(1,K)} |u(k)| \le \frac{1}{2^{n-1}K} M_{n,K} \max_{\mathbb{N}(1,K)} |\Delta^n u(k)|, \tag{12.11.17}$$

where

for $K$ even, $n$ odd

$$M_{n,K} = \sum_{\ell=0}^{(K-2)/2}\left[\sin(2\ell+1)\frac{\pi}{K}\right]^{-n-1}$$

for $K$ even, $n$ even, $K/2$ odd

$$M_{n,K} = \sum_{\ell=0}^{(K-2)/2}(-1)^{\ell}\left[\sin(2\ell+1)\frac{\pi}{K}\right]^{-n-1}$$

for $K$ even, $n$ even, $K/2$ even

$$M_{n,K} = \sum_{\ell=0}^{(K-2)/2}(-1)^{\ell}\cos(2\ell+1)\frac{\pi}{K}\left[\sin(2\ell+1)\frac{\pi}{K}\right]^{-n-1}$$

for $K$ odd, $n$ odd

$$M_{n,K} = \sum_{\ell=0}^{(K-3)/2}\left[1+\cos(2\ell+1)\frac{\pi}{K}\right]\left[\sin(2\ell+1)\frac{\pi}{K}\right]^{-n-1}$$

for $K$ odd, $n$ even

$$M_{n,K} = \sum_{\ell=0}^{(K-3)/2} (-1)^{\ell} \cos(2\ell+1)\frac{\pi}{K}\left[1+\cos(2\ell+1)\frac{\pi}{K}\right]\left[\sin(2\ell+1)\frac{\pi}{K}\right]^{-n-1}.$$

In (12.11.17) equality holds for the periodic function $u(k)$ of period $K$ satisfying $\sum_{\ell=1}^{K} u(\ell) = 0$ and defined by $\Delta^K u(k) = v(k)$, where

$$\text{for } K \text{ even,} \quad v(k) = \begin{cases} 1, & k \in \mathbb{N}(0,(K-2)/2) \\ -1, & k \in \mathbb{N}(K/2, K-1) \end{cases}$$

and

$$\text{for } K \text{ odd,} \quad v(k) = \begin{cases} 1. & k \in \mathbb{N}(0,(K-3)/2) \\ 0, & k = (K-1)/2 \\ -1, & k \in \mathbb{N}((K+1)/2, K-1). \end{cases}$$

**12.11.24.** (Rayleigh's Inequality). Let $\lambda_1$ be the smallest eigenvalue of (12.1.1), $u(0) = u(K+1) = 0$. Show that

$$\lambda_1 \le \frac{\sum_{k=1}^{K+1} p(k-1)\left[\Delta v(k-1)\right]^2 - \sum_{k=1}^{K} q(k)v^2(k)}{\sum_{k=1}^{K} r(k)v^2(k)}, \tag{12.11.18}$$

where $v(k)$ is any nontrivial function defined on $\mathbb{N}(0, K+1)$ with $v(0) = v(K+1) = 0$. Further, show that in (12.11.18) equality holds if and only if $v(k)$ is an eigenfunction corresponding to $\lambda_1$.

**12.11.25.** (Lyapunov's Inequality). If the boundary value problem $\Delta^2 u(k-1) + q(k)u(k) = 0$, $k \in \mathbb{N}(1,K)$, $u(0) = u(K+1) = 0$ where $q(k) \ge 0$, $k \in \mathbb{N}(1,K)$, has a nontrivial solution $u(k)$, then show that

$$\sum_{\ell=1}^{K} q(\ell) \ge \begin{cases} \dfrac{(2m+1)}{m(m+1)} & \text{if } K = 2m \\ \dfrac{2}{(m+1)} & \text{if } K = 2m+1 \end{cases}$$

and this inequality is best possible in the sense that if for any $K$ equality holds then there exist functions $q(k) \ge 0$, $k \in \mathbb{N}(1,K)$ and $u(k) \ne 0$, $k \in \mathbb{N}(0, K+1)$ such that $q(k) = -\Delta^2 u(k-1)/u(k)$, $k \in \mathbb{N}(1,K)$, $u(0) = u(K+1) = 0$.

**12.11.26.** Let the function $u(k)$ be defined on $\mathbb{N}(0,K)$, and $q^{-1} + (q')^{-1} = 1$, where $1 \le q \le \infty$. Show that

(i) if $u(0) = 0$, then

$$\sum_{\ell=1}^{K} |u(\ell)| \le \left(\sum_{\ell=0}^{K-1} (K-\ell)^{q'}\right)^{1/q'} \left(\sum_{\ell=0}^{K-1} |\Delta u(\ell)|^q\right)^{1/q}$$

(ii) if $u(0) = u(1) = 0$, then

$$\sum_{\ell=2}^{K} |u(\ell)| \le \left(\sum_{\ell=1}^{K-1}\left[\frac{\ell(\ell+1)}{2}\right]^{q'}\right)^{1/q'} \left(\sum_{\ell=0}^{K-2} |\Delta^2 u(\ell)|^q\right)^{1/q}.$$

**12.11.27.** For the computation of the eigenvalues of the boundary value problem

$$y'' + (\lambda r(t) - q(t))y = 0, \quad y(a) = y(b) = 0 \tag{12.11.19}$$

where $r,\ q \in C[a,b]$ and $r(t) > 0,\ q(t) \ge 0$ for all $t \in [a,b]$ show that the generalized matrix eigenvalue problem

$$(\mathcal{A} + 180h^2\mathcal{Q})\mathbf{u} = 180\Lambda h^2\mathcal{R}\mathbf{u} \tag{12.11.20}$$

provides a fourth order approximation, i.e. if $\lambda$ is a fixed eigenvalue of (12.11.19) and $\Lambda$ is the corresponding approximation obtained from (12.11.20), then $\left|1 - \dfrac{\lambda}{\Lambda}\right| = O(h^4)$. In (12.11.20), $h = (b-a)/(K+1)$, $\mathbf{u} = (u(1),\cdots,u(K))$, $u(k) = y(t_k)$, $t_k = a + kh$, $k \in \mathbb{N}(0, K+1)$, $\mathcal{A} = 180\mathcal{H}_K(\mathbf{g},\mathbf{h}) + 15\mathcal{H}_K^2(\mathbf{g},\mathbf{h}) + 2\mathcal{H}_K^3(\mathbf{g},\mathbf{h})$, $\mathbf{g} = (2,\cdots,2)$, $\mathbf{h} = (-1,\cdots,-1)$, $\mathcal{Q} = diag\ (q(t_1),\cdots,q(t_K))$, and $\mathcal{R} = diag\ (r(t_1),\cdots,r(t_K))$.

**12.11.28.** Use (12.11.20) to compute the approximation of the first eigenvalue of the following boundary value problems

$$y'' + \left(\frac{\lambda}{1+t^2} - (1+t)^4\right)y = 0, \quad y(0) = y(1) = 0$$

and

$$y'' + (\lambda\cos t - \sin t)y = 0, \quad y(0) = y(1) = 0$$

with $h = 2^{-p}$, $3 \le p \le 8$. Does this computation justify the order of convergence to be four?

**12.11.29.** For the computation of the eigenvalues of the boundary value problem

$$y'''' - (\lambda r(t) - q(t))y = 0 \tag{12.11.21}$$

$$y''(a) = y'''(a) = y''(b) = y'''(b) = 0, \tag{12.11.22}$$

where $r,\ q \in C[a,b]$ and $r(t) > 0,\ q(t) > 0$ for all $t \in [a,b]$ show that the generalized matrix eigenvalue problem

$$(\mathcal{A} + h^4\mathcal{Q})\mathbf{u} = \Lambda h^4\mathcal{R}\mathbf{u} \tag{12.11.23}$$

provides a second order approximation. In (12.11.23), $h = (b-a)/(K-1)$, $\mathbf{u} = (u(1), \cdots, u(K))$, $u(k) = y(t_k)$, $t_k = a+(k-1)h$, $k \in \mathbb{N}(1,K)$, $K \geq 5$, $\mathcal{A}$ is a singular five band symmetric matrix

$$\mathcal{A} = \begin{bmatrix} 1 & -2 & 1 & & & & & & \\ -2 & 5 & -4 & 1 & & & & & \\ 1 & -4 & 6 & -4 & 1 & & & & \\ & 1 & -4 & 6 & -4 & 1 & & & \\ & & & & & \cdot & \cdot & & \\ & & & & & \cdot & & & \\ & & & & & 1 & -4 & 5 & -2 \\ & & & & & & 1 & -2 & 1 \end{bmatrix},$$

$$\mathcal{Q} = diag\left(\frac{7}{12}q(t_1), \frac{11}{12}q(t_2), q(t_3), \cdots, q(t_{K-2}), \frac{11}{12}q(t_{K-1}), \frac{7}{12}q(t_K)\right)$$

and

$$\mathcal{R} = diag\left(\frac{7}{12}r(t_1), \frac{11}{12}r(t_2), r(t_3), \cdots, r(t_{K-2}), \frac{11}{12}r(t_{K-1}), \frac{7}{12}r(t_K)\right).$$

**12.11.30.** Use (12.11.23) to compute the approximation of the first eigenvalue of the following boundary value problems

(12.11.24)
$$\begin{aligned} y'''' - \left(\frac{\lambda}{1+t^4} - 1 - t^2\right)y &= 0 \\ y''(1) = y'''(1) = y''(2) = y'''(2) &= 0 \end{aligned}$$

and

(12.11.25)
$$\begin{aligned} y'''' - (\lambda(2+\sin t) - \cosh t)y &= 0 \\ y''(0) = y'''(0) = y''(1) = y'''(1) &= 0 \end{aligned}$$

with $h = 2^{-p}$, $2 \leq p \leq 8$. Does this computation justify the order of convergence to be two?

**12.11.31.** In Problem 12.11.29 let the matrix $\mathcal{A}$ be a singular seven band symmetric matrix

$$\mathcal{A} = \begin{bmatrix} 6 & -13 & 8 & -1 & & & & \\ -13 & 36 & -34 & 12 & -1 & & & \\ 8 & -34 & 54 & -39 & 12 & -1 & & \\ -1 & 12 & -39 & 56 & -39 & 12 & -1 & \\ & \cdots & \cdots & & \cdots & & \cdots & \\ & & -1 & 12 & -39 & 54 & -34 & 8 \\ & & & -1 & 12 & -34 & 36 & -13 \\ & & & & -1 & 8 & -13 & 6 \end{bmatrix},$$

$$\mathcal{Q} = diag\left(\frac{17}{12}q(t_1), \frac{26}{3}q(t_2), \frac{59}{12}q(t_3), 6q(t_4), \cdots, 6q(t_{K-3}), \right.$$
$$\left. \frac{59}{12}q(t_{K-2}), \frac{26}{3}q(t_{K-1}), \frac{17}{12}q(t_K)\right)$$

and

$$\mathcal{R} = diag\left(\frac{17}{12}r(t_1), \frac{26}{3}r(t_2), \frac{59}{12}r(t_3), 6r(t_4), \cdots, 6r(t_{K-3}), \right.$$
$$\left. \frac{59}{12}r(t_{K-2}), \frac{26}{3}r(t_{K-1}), \frac{17}{12}r(t_K)\right).$$

Show that with this replacement the generalized matrix eigenvalue problem (12.11.23) provides a third order approximation for the computation of the eigenvalues of the boundary value problem (12.11.21), (12.11.22). Further, use this method to compute the approximation of the first eigenvalue of the problems (12.11.24) and (12.11.25) with $h = 2^{-p}$, $3 \leq p \leq 8$ and justify the order of convergence of the method to be three.

**12.11.32.** For the computation of the eigenvalues of the boundary value problem (12.11.21) with $q(t) \geq 0$,

$$y(a) = y'(a) = y(b) = y'(b) = 0 \tag{12.11.26}$$

show that the generalized matrix eigenvalue problem (12.11.23) provides a second order approximation, where $h = (b-a)/(K+1)$, $\mathbf{u} = (u(1), \cdots, u(K))$, $u(k) = y(t_k)$, $t_k = a + kh$, $k \in \mathbb{N}(0, K+1)$, $\mathcal{A}$ is a five band symmetric matrix

$$\mathcal{A} = \begin{bmatrix} 7 & -4 & 1 & & & & & & \\ -4 & 6 & -4 & 1 & & & & & \\ 1 & -4 & 6 & -4 & 1 & & & & \\ & 1 & -4 & 6 & -4 & 1 & & & \\ & & & & & \ddots & & & \\ & & & & & 1 & -4 & 6 & -4 \\ & & & & & & 1 & -4 & 7 \end{bmatrix},$$

$\mathcal{Q} = diag\,(q(t_1), \cdots, q(t_K))$, and $\mathcal{R} = diag\,(r(t_1), \cdots, r(t_K))$. Further, use this method to compute the approximation of the first eigenvalue of the problem

$$y'''' - \frac{\lambda}{t^4}y = 0, \quad y(1) = y'(1) = y(e) = y'(e) = 0 \tag{12.11.27}$$

with $h = 2^{-p}$, $3 \le p \le 8$ and justify the order of convergence of the method to be two.

**12.11.33.** If in Problem 12.11.32 the matrix $\mathcal{A}$ is replaced by the seven band symmetric matrix

$$\mathcal{A} = \begin{bmatrix} \frac{38}{3} & -7 & 2 & -\frac{1}{6} & & & & & \\ -7 & \frac{113}{12} & -\frac{13}{2} & 2 & -\frac{1}{6} & & & & \\ 2 & -\frac{13}{2} & \frac{28}{3} & -\frac{13}{2} & 2 & -\frac{1}{6} & & & \\ -\frac{1}{6} & 2 & -\frac{13}{2} & \frac{28}{3} & -\frac{13}{2} & 2 & -\frac{1}{6} & & \\ & \cdots & \cdots & & \cdots & & \cdots & & \\ & & -\frac{1}{6} & 2 & -\frac{13}{2} & \frac{28}{3} & -\frac{13}{2} & 2 \\ & & & -\frac{1}{6} & 2 & -\frac{13}{2} & \frac{113}{12} & -7 \\ & & & & -\frac{1}{6} & 2 & -7 & \frac{38}{3} \end{bmatrix},$$

then show that the resulting generalized matrix eigenvalue problem (12.11.23) provides a fourth order approximation for the computation of the eigenvalues of the boundary value problem (12.11.21) with $q(t) \ge 0$, (12.11.26). Further, use this method to compute the approximation of the first eigenvalue of the problem (12.11.27) with $h = 2^{-p}$, $3 \le p \le 8$ and justify the order of convergence of the method to be four.

**12.11.34.** Let $\delta_i$, $1 \le i \le n$ and $P_1^0$ be as in Section 12.9. If in addition to the conditions of Theorem 12.9.1, $\delta_i\delta_j q_{ij}(k) > 0$, $k \in \mathbb{N}(0, K-1)$, $1 \le i, j \le n$ then show that the boundary value problem (12.9.2), (12.9.3) has a smallest positive eigenvalue $\Lambda_0$ which is smaller than the modulus of any other eigenvalue of (12.9.2), (12.9.3). Furthermore, there is an essentially unique eigenfunction $\mathbf{w}^0 \in P_1^0$ or $-\mathbf{w}^0 \in P_1^0$.

**12.11.35.** If $\mathcal{P}(k)$ satisfies the hypothesis of Theorem 12.9.7, then show that the eigenvalue problem

$$-\Delta^2\mathbf{u}(k) = \lambda\mathcal{P}(k)\mathbf{u}(k+1), \quad \mathbf{u}(0) = \mathbf{u}(K+1) = 0$$

has a smallest positive eigenvalue $\lambda_0$ which satisfies

$$\lambda_0^{-1} \ge \frac{(k_0+1)(K-k_0)}{K+1} P_{i_0 i_0}(k_0).$$

**12.11.36.** If the conditions of Theorem 12.9.7 are satisfied, then show that the least positive eigenvalue $\lambda_0$ of (12.9.1), (12.9.3) satisfies

$$g(k_0+p,k_0)P_{i_0i_0}(k_0) \ \le\ \lambda_0^{-1} \ \le\ G\sum_{\ell=0}^{K-1}\|\mathcal{P}(\ell)\|,$$

where $G = \max\{g(k,\ell) : k \in \mathbb{N}(p, K-1+p),\ \ell \in \mathbb{N}(0,K-1)\}$ and $\|\mathcal{P}(\ell)\| = \max_{1\le i\le n}\sum_{j=1}^{n}\delta_i\delta_j p_{ij}(\ell)$.

**12.11.37.** Let $\delta_i$, $1 \le i \le n$ be as in Section 12.9. If in addition to the conditions of Theorem 12.9.1 there is an $i_0 \in \mathbb{N}(1,n)$ and a $k_0 \in \mathbb{N}(0,K-1)$ such that $p_{i_0i_0}(k_0) > 0$, and $0 \le \delta_i\delta_j p_{ij}(k) \le \delta_i\delta_j q_{ij}(k)$ and $q_{ij}(k) \not\equiv 0$ on $\mathbb{N}(0,K-1)$ for $1 \le i,\ j \le n$ then show that the eigenvalue problems (12.9.1), (12.9.3) and (12.9.2), (12.9.3) have smallest positive eigenvalues $\lambda_0$ and $\Lambda_0$ respectively. Furthermore, $\Lambda_0 \le \lambda_0$ and $\Lambda_0 = \lambda_0$ if and only if $\mathcal{P}(k) = \mathcal{Q}(k)$ on $\mathbb{N}(0,K-1)$.

**12.11.38.** Consider the two point focal boundary value problem

$$\Delta^n u(k) \ =\ (-1)^{n-p}\lambda\sum_{i=0}^{n-1} q_i(k)\Delta^i u(k), \quad k \in \mathbb{N}(0,K-1) \tag{12.11.28}$$

$$\begin{aligned}&\Delta^i u(0) \ =\ 0, \quad 0 \le i \le p-1 \ \ (1 \le p \le n-1, \ \text{but fixed})\\ &\Delta^i u(K) = 0, \quad p \le i \le n-1\end{aligned} \tag{12.11.29}$$

where $\lambda$ is a parameter, and the functions $q_i(k)$, $0 \le i \le n-1$ are defined on $\mathbb{N}(0,K-1)$. Further,

$q_i(k) \ge 0,\ k \in \mathbb{N}(p-i, K-1),\ \ 0 \le i \le p-1$
$(-1)^i q_{p+i}(k) \ge 0,\ k \in \mathbb{N}(0,K-1),\ 0 \le i \le n-p-1$
if $1 \le p \le n-2$, then $\sum_{i=0}^{p-1} q_i(K-1) > 0$, and
if $p = n-1$, then $q_{n-2}(K-1) > 0$ and $q_{n-1}(K-1) > 0$.

Show that the problem (12.11.28), (12.11.29) has a smallest positive eigenvalue $\lambda_0$ and $\lambda_0$ is smaller than the modulus of any other eigenvalue of (12.11.28), (12.11.29). Furthermore, there is an essentially unique eigenfunction $u_0(k)$ corresponding to $\lambda_0$, and either $u_0(k)$ or $-u_0(k)$ satisfies

$$\begin{aligned}&\Delta^i u(k) > 0, \quad k \in \mathbb{N}(p-i, K-1+n-i), \quad 0 \le i \le p-1\\ &(-1)^i\Delta^{p+i}u(k) > 0, \quad k \in \mathbb{N}(0,K-1), \quad 0 \le i \le n-p-1.\end{aligned} \tag{12.11.30}$$

**12.11.39.** Consider the difference equations

$$\Delta^n u(k) \ =\ (-1)^{n-p}\lambda\sum_{i=0}^{p-1} q_i(k)\Delta^i u(k), \quad k \in \mathbb{N}(0,K-1) \tag{12.11.31}$$

and

$$(12.11.32) \quad \Delta^n u(k) = (-1)^{n-p}\Lambda \sum_{i=0}^{p-1} Q_i(k)\Delta^i u(k), \quad k \in \mathbb{N}(0, K-1)$$

together with the boundary conditions (12.11.29), where $\lambda$ and $\Lambda$ are parameters, $1 \le p \le n-2$, and the functions $q_i(k)$, $Q_i(k)$, $0 \le i \le p-1$ are defined on $\mathbb{N}(0, K-1)$. Further,

$$p > 1 \text{ and } \sum_{i=0}^{p-2} q_i(K-1) > 0, \quad \text{or} \quad \sum_{\ell=k}^{K-1} q_{p-1}(\ell) > 0, \quad k \in \mathbb{N}(1, K-1)$$

and

$$0 \le \sum_{\ell=k}^{K-1} q_i(\ell) \le \sum_{\ell=k}^{K-1} Q_i(\ell), \quad k \in \mathbb{N}(p-i, K-1), \quad 0 \le i \le p-1.$$

Show that the problem (12.11.31), (12.11.29) ((12.11.32), (12.11.29)) has a smallest positive eigenvalue $\lambda_0(\Lambda_0)$ and $\lambda_0(\Lambda_0)$ is smaller than the modulus of any other eigenvalue of (12.11.31), (12.11.29) ((12.11.32), (12.11.29)). Furthermore, there is an essentially unique eigenfunction $u_0(k)(v_0(k))$ corresponding to $\lambda_0(\Lambda_0)$, and either $u_0(k)(v_0(k))$ or $-u_0(k)(-v_0(k))$ satisfies (12.11.30). Also, show that $\Lambda_0 \le \lambda_0$ and $\Lambda_0 = \lambda_0$ if and only if $q_i(k) = Q_i(k)$, $k \in \mathbb{N}(p-i, K-1)$, $0 \le i \le p-1$.

## 12.12. Notes.

An elementary discussion of discrete Sturm–Liouville problems is available at several places, e.g. see the books by Fort [18], Hildebrand [25], Levy and Lessman [30]. Our treatment in Section 12.1 is similar to the theory of continuous Sturm–Liouville problems presented in Agarwal and Gupta [1]. Results of Section 12.2 on eigenvalue problems for symmetric matrices can be found in Parlett [40], and Usmani [47]. Theorems 12.6.1 - 12.6.3 have been adapted from the landmark paper of Fan, Taussky and Todd [15]. An alternative proof of Theorem 12.6.1 based on the discrete Fourier series representations has been given in Lasota [29]. Redheffer [42] unifies Theorems 12.6.1 and 12.6.2 and claims to provide an 'easier' proof. Losonczi [31] contains some generalizations of these results. Theorems 12.6.4 and 12.6.7 are due to Milovanović and Milovanović [33]. Theorem 12.7.1 is from Denkowski [12], whereas Denkowski [13] contains some of its applications to fourth order discrete boundary value problems. Several other related inequalities have been established by Alzer [5], Block [6], Chen [9], Cheng [10, 11], Egerváry and Szász [14], Fejér [16], Fink [17], Goodman and Lee [20], Lunter [32], Milovanović and Milovanović [35, 36], Novotna

[37 39], Pfeffer [41], Shisha [43], Szegö [44], and Yin [49]. Some of these interesting inequalities are included as problems. Theorem 12.8.1 is taken from Milovanović and Milovanović [34]. While the existence and the comparison theorems for the least positive eigenvalues of higher order continuous boundary value problems have been studied extensively in Ahmad and Lazer [4], Gentry and Travis [19], Hankerson and Peterson [22], Keener and Travis [26,27], Travis [45] and several others, whereas for the discrete case very few such results are known. For a systematic treatment of cone theory see Guo and Lakshmikantham [21], and Krasnosel'skii [28]. Theorems 12.9.2 and 12.9.3 are proved in Krasnosel'skii [28], whereas Theorem 12.9.4 is due to Travis [45]. Rest of the results in Section 12.9 are from Hankerson and Peterson [23]. Similar results for the focal point discrete boundary value problems are available in Hankerson and Peterson [24]. Section 12.10 contains the work of Agarwal and Henderson [2]. Recent work on the positive solutions of differential, difference and integral equations has been documented in Agarwal, O'Regan and Wong [3]. Several discretizations which provide approximations to the eigenvalues of the continuous Sturm Liouville type boundary value problems based on the work of Chawla and Katti [7], Chawla [8], Usmani and Agarwal [46,48] have been included as problems.

## 12.13. References


[1]. R.P. Agarwal and R.C. Gupta, *Essentials of Ordinary Differential Equations*, *McGraw Hill*, Singapore, New York, 1991.

[2]. R.P. Agarwal and J. Henderson, Positive solutions and nonlinear eigenvalue problems for third order difference equations, *Computers Math. Applic.* **36**(10 12)(1998), 347 355.

[3]. R.P. Agarwal, D. O'Regan and P.J.Y. Wong, *Positive Solutions of Differential, Difference and Integral Equations*, *Kluwer*, Dordrecht, 1999.

[4]. S. Ahmad and A. Lazer, An N dimensional extension of the Sturm separation and comparison theory to a class of nonselfadjoint systems, *SIAM J. Math. Anal.* **8**(1978), 1137 1150.

[5]. H. Alzer, Converses of two inequalities by Ky Fan, O. Taussky and J. Todd, *J. Math. Anal. Appl.* **161**(1991), 142 147.

[6]. H.D. Block, Discrete analogues of certain integral inequalities, *Proc. Amer. Math. Soc.* **8**(1957), 852 859.

[7]. M.M. Chawla and C.P. Katti, A new symmetric five diagonal finite difference method for computing eigenvalues of fourth order two point bound-

ary value problems, *J. Comp. Appl. Math.* **8**(1982), 135 136.

[8]. M.M. Chawla, A new fourth order finite difference method for computing eigenvalues of fourth order two point boundary value problems, *IMA J. Numer. Anal.* **3**(1983), 291 293.

[9]. W. Chen, On a question of H. Alzer, *Arch. Math.* **62**(1994), 315 320.

[10]. S.S. Cheng, A discrete analogue of the inequality of Lyapunov, *Hokkaido Math. J.* **12**(1983), 105 112.

[11]. S.S. Cheng, Discrete quadratic Wirtinger's inequalities, *Linear Algebra Appl.* **85**(1987), 57 73.

[12]. Z. Denkowski, Inequalities of Wirtinger's type and their discrete analogues, *Zeszyty Naukowe U.J., Prace Math.* **15**(1971), 27 37.

[13]. Z. Denkowski, The boundary value problems for ordinary non linear differential and difference equations of the fourth order, *Annales Polonici Mathematici* **24**(1970), 87 102.

[14]. E. Egerváry and O. Szász, Einige Extremalprobleme im Bereiche der trigonometrischen Polynome, *Math. Z.* **27**(1928), 641 692.

[15]. K. Fan, O. Taussky and J. Todd, Discrete analogs of inequalities of Wirtinger, *Monatsh. Math.* **59**(1955), 73 90.

[16]. L. Fejér, Über trigonometrische Polynome, *J. Reine Angew. Math.* **146**(1915), 53 82.

[17]. A.M. Fink, Discrete inequalities of generalized Wirtinger type, *Aequationes Mathematicae* **11**(1974), 31 39.

[18]. T. Fort, *Finite Differences and Difference Equations in the Real Domain, The Clarendon Press*, Oxford, 1948.

[19]. R.D. Gentry and C.C. Travis, Comparison of eigenvalues associated with linear differential equations of arbitrary order, *Trans. Amer. Math. Soc.* **223**(1976), 167 179.

[20]. T.N.T. Goodman and S.L. Lee, Inequalities involving periodic sequences, *Bull. Malasian Math. Soc.* **2**(1979), 1 11.

[21]. D. Guo and V. Lakshmikantham, *Nonlinear Problems in Abstract Cones, Academic Press*, New York, 1988.

[22]. D. Hankerson and A. Peterson, Comparison theorems for eigenvalue problems for nth order differential equations, *Proc. Amer. Math. Soc.* **104**(1988), 1204 1211.

[23]. D. Hankerson and A. Peterson, A positivity result applied to difference equations, *J. Appr. Theory* **59**(1989), 76 86.

[24]. D. Hankerson and A. Peterson, Comparison of eigenvalues for focal point problems for $n$ th order difference equations, *Differential and Integral*

*Equations* **3**(1990), 363 380.

[25]. F.B. Hildebrand, *Finite Difference Equations and Simulations*, *Prentice Hall*, Englewood Cliffs, N.J., 1968.

[26]. M.S. Keener and C.C. Travis, Positive cones and focal points for a class of $n$th order differential equations, *Trans. Amer. Math. Soc.* **237**(1978), 331 351.

[27]. M.S. Keener and C.C. Travis, Sturmian theory for a class of nonselfadjoint differential systems, *Ann. Mat. Pura Appl.* **123**(1980), 247 266.

[28]. M.A. Krasnosel'skii, *Positive Solutions of Operator Equations*, *Fizmatgiz*, Moscow, 1962: English Translation Noordhoff, Groningen, The Netherlands, 1964.

[29]. A. Lasota, A discrete boundary value problem, *Annales Polonici Mathematici* **20**(1968), 183 190.

[30]. H. Levy and F. Lessman, *Finite Difference Equations*, *Sir Issac Pitman and Sons*, Ltd., London, 1959.

[31]. L. Losonczi, On some discrete quadratic inequalities, in General Inequalities 5, ed. W. Walter, ISNM 80, Birkhäuser Verlag, Basel, (1987), 73 85.

[32]. G. Lunter, New proofs and a generalization of inequalities of Fan, Taussky and Todd, *J. Math. Anal. Appl.* **185**(1994), 464 476.

[33]. G.V. Milovanović and I.Ž. Milovanović, On discrete inequalities of Wirtinger's type, *J. Math. Anal. Appl.* **88**(1982), 378 387.

[34]. G.V. Milovanović and I.Ž. Milovanović, Some discrete inequalities of Opial's type, *Acta Sci. Math.* **47**(1984), 413 417.

[35]. G.V. Milovanović and I.Ž. Milovanović, Discrete inequalities of Wirtinger's type for higher differences, *J. Inequ. Appl.* **1**(1997), 301 310.

[36]. G.V. Milovanović and I.Ž. Milovanović, Discrete inequalities of Wirtinger's type, in *Recent Progress in Inequalities*, ed. G.V. Milovanović, *Kluwer*, Dordrecht, (1998), 289 308.

[37]. J. Novotna, Variations of discrete analogues of Wirtinger's inequality, *Časopis Pěst. Mat.* **105**(1980), 278 285.

[38]. J. Novotna, Discrete analogues of Wirtinger's inequality for a two dimensional array, *Časopis Pěst. Mat.* **105**(1980), 354 362.

[39]. J. Novotna, A sharpening of discrete analogues of Wirtinger's inequality, *Časopis Pěst. Mat.* **108**(1983), 70 77.

[40]. B.N. Parlett, *The Symmetric Eigenvalue Problem,* Prentice Hall Series in Comp. Math., *Prentice Hall*, New Jersey, 1980.

[41]. A.M. Pfeffer, On certain discrete inequalities and their continuous analogs, *J. Res. Nat. Bur. Standards* **B70**(1966), 221 231.

[42]. R.M. Redheffer, Easy proofs of hard inequalities, in *General Inequalities 3*, eds. E.F. Beckenbach and W. Walter, ISNM 64, *Birkhäuser Verlag*, Basel, (1983), 123 140.

[43]. O. Shisha, On the discrete version of Wirtinger's inequality, *Amer. Math. Monthly* **80**(1973), 755 760.

[44]. G. Szegö, Koeffizientenabschätzungen bei ebenen und räumlichen harmonischen Entwicklungen, *Math. Ann.* **96**(1926/27), 601 632.

[45]. C.C. Travis, Comparison of eigenvalues for linear differential equations of order $2n$, *Trans. Amer. Math. Soc.* **177**(1973), 363 374.

[46]. R.A. Usmani and R.P. Agarwal, New symmetric finite difference methods for computing eigenvalues of a boundary value problem, *Comm. Appl. Numer. Methods* **1**(1985), 305 309.

[47]. R.A. Usmani, *Applied Linear Algebra*, *Marcel Dekker*, New York, 1987.

[48]. R.A. Usmani and R.P. Agarwal, Some higher order methods for computing eigenvalues of two point boundary value problems, *Comm. Appl. Numer. Methods* **3**(1987), 5 9.

[49]. X.R. Yin, A converse inequality of Fan, Taussky and Todd, *J. Math. Anal. Appl.* **182**(1994), 654 657.

# Chapter 13
# Difference Inequalities in Several Independent Variables

Inequalities developed in Chapter 4 have natural extensions for functions of $m$ independent variables. These inequalities are used as a fundamental tool in the study of related partial difference equations. We begin this chapter with the recently established discrete analog of Riemann's function. This function is repeatedly used to study linear Gronwall type inequalities. Next we shall provide an upper estimate on the Riemann's function which is quite adequate in practical applications and provides Wendroff's type estimates rather easily. This is followed by several nonlinear inequalities. Inequalities involving higher order differences in two independent variables are also directly considered. For this the relevant Taylor's formula in two independent variables is included. Next we move to multidimensional linear as well as nonlinear discrete inequalities, and wherever possible provide upper bounds in terms of known functions. This is followed by convolution type inequalities. Here the upper estimate appears in terms of discrete resolvent function. Finally, we shall develop Opial's and Wirtinger's type inequalities in two independent variables.

## 13.1. Discrete Riemann's Function

Once again, let $\mathbb{N} = \{0, 1, \cdots\}$ be the set of natural numbers including zero, and the product $\mathbb{N} \times \cdots \times \mathbb{N}$ ($m$ times) be denotes by $\mathbb{N}^m$. A point $(x_1, \cdots, x_m)$ in $\mathbb{N}^m$ is denoted by $x$, whereas $\overline{x}_i$ represents $(x_1, \cdots, x_{i-1}, x_{i+1}, \cdots, x_m)$, and $(\overline{x}_i, \cdot)$ stands for $(x_1, \cdots, x_{i-1}, \cdot, x_{i+1}, \cdots, x_m)$, also for all $s, x \in \mathbb{N}^m$, $0 \le s \le x$ represents $0 \le s_i \le x_i$, $1 \le i \le m$. For a given function $u(x)$ on $\mathbb{N}^m$, the first order difference with respect to the variable $x_i$ is defined as $\Delta_{x_i} u(x) = u(\overline{x}_i, x_i + 1) - u(k)$, and the second order difference with respect to the variables $x_i$ and $x_j$ is defined as $\Delta_{x_i}\Delta_{x_j} u(x) = \Delta_{x_i} u(\overline{x}_j, x_j + 1) - \Delta_{x_i} u(x) = u(x_1, \cdots, x_{i-1}, x_i + 1, x_{i+1}, \cdots, x_{j-1}, x_j + 1, x_{j+1}, \cdots, x_m) - u(\overline{x}_i, x_i + 1) - u(\overline{x}_j, x_j + 1) + u(x)$. The higher order differences are defined analogously. The $\mathbf{S}_{\ell=s}^{x-1} u(\ell)$ represents the $m$ fold sum $\sum_{\ell_1=s_1}^{x_1-1} \cdots \sum_{\ell_m=s_m}^{x_m-1} u(\ell_1, \cdots, \ell_m)$, and $\Delta_x^m u(x)$

denotes $\Delta_{x_1}\cdots\Delta_{x_m}u(x_1,\cdots,x_m)$. The empty sums and products are taken to be 0 and 1, respectively.

**Lemma 13.1.1.** Let $g(x)$ be defined on $\mathbb{N}^m$, then the function $V(s;x)$, $s \le x-1$, $(s;x) \in \mathbb{N}^m \times \mathbb{N}^m$ is a solution of

$$(13.1.1) \qquad (-1)^m \Delta_s^m V(s;x) = g(s)V(s+1;x)$$

$$(13.1.2) \qquad V(\overline{s}_i, x_i; x) = 1, \quad 1 \le i \le m$$

if and only if

$$(13.1.3) \qquad V(s;x) = 1 + \mathbf{S}_{\ell=s}^{x-1} g(\ell)V(\ell+1;x).$$

**Proof.** From (13.1.1), we have

$$(-1)^m \Delta_{\overline{\ell}_m}^{m-1}[V(\overline{\ell}_m, \ell_m+1;x) - V(\ell;x)] = g(\ell)V(\ell+1;x),$$

and hence on summing it from $\ell_m = s_m$ to $\ell_m = x_m - 1$, we get

$$(-1)^m \Delta_{\overline{\ell}_m}^{m-1}\left(V(\overline{\ell}_m, \ell_m;x)\Big|_{\ell_m=s_m}^{x_m}\right) = \sum_{\ell_m=s_m}^{x_m-1} g(\ell)V(\ell+1;x),$$

which is from (13.1.2) is the same as

$$(-1)^{m+1}\Delta_{\overline{\ell}_m}^{m-1}V(\overline{\ell}_m, s_m;x) = \sum_{\ell_m=s_m}^{x_m-1} g(\ell)V(\ell+1;x).$$

Continuing in this way, we obtain

$$(-1)^{m+m-1}\Delta_{\ell_1}V(\overline{s}_1,\ell_1;x) = \mathbf{S}_{\overline{\ell}_1=\overline{s}_1}^{\overline{x}_1-1} g(\ell)V(\ell+1;x)$$

and hence on summing it from $\ell_1 = s_1$ to $\ell_1 = x_1 - 1$, we have

$$(-1)^{2m-1}\left(V(\overline{s}_1,\ell_1;x)\Big|_{\ell_1=s_1}^{x_1}\right) = \mathbf{S}_{\ell=s}^{x-1} g(\ell)V(\ell+1;x),$$

which is from (13.1.2) is the same as

$$-1 + V(s;x) = \mathbf{S}_{\ell=s}^{x-1} g(\ell)V(\ell+1;x). \qquad \blacksquare$$

**Lemma 13.1.2.** The problem (13.1.1), (13.1.2) or equivalently (13.1.3), has a unique solution $V(s;x)$. Further, if $g(x) \ge 0$ on $\mathbb{N}^m$, then $V(s;x) \ge 1$ on $\mathbb{N}^m \times \mathbb{N}^m$.

**Proof.** For the iterates

$$(13.1.4) \qquad \begin{aligned} V_0(s;x) &= 1 \\ V_{n+1}(s;x) &= 1 + \mathbf{S}_{\ell=s}^{x-1}\, g(\ell)V_n(\ell+1;x), \quad n = 0, 1, \cdots \end{aligned}$$

an easy induction gives

$$|V_n(s;x) - V_{n-1}(s;x)| \le G^n \frac{1}{(n!)^m} \prod_{i=1}^{m} (x_i - s_i)^n,$$

where $G = \max_{0 \le \ell \le x-1} |g(\ell)|$.

Therefore, for $(s;x) \in \mathbb{N}^m \times \mathbb{N}^m$ it follows that

$$\begin{aligned} &|V_0(s;x)| + \sum_{k=1}^{n} |V_k(s;x) - V_{k-1}(s;x)| \\ &\quad \le 1 + \sum_{k=1}^{n} G^k \frac{1}{k!} \left[ \prod_{i=1}^{m} (x_i - s_i) \right]^k \le \exp\left[ G \prod_{i=1}^{m} (x_i - s_i) \right] \end{aligned}$$

and hence the sequence $\{V_n(s;x)\}$ converges to a solution $V(s;x)$ of (13.1.3). The uniqueness of $V(s;x)$ and the inequality $V(s;x) \ge 1$ on $\mathbb{N}^m \times \mathbb{N}^m$ (when $g(x) \ge 0$ on $\mathbb{N}^m$) are obvious from (13.1.4). ∎

**Lemma 13.1.3.** Let $g(x) \ge 0$ and $h(x)$ be defined on $\mathbb{N}^m$ and the following inequality holds

$$(13.1.5) \qquad \Delta_x^m u(x) \le g(x)u(x) + h(x),$$

where

$$(13.1.6) \qquad u(\overline{x}_i, 0) = 0, \quad 1 \le i \le m.$$

Then, for all $x \in \mathbb{N}^m$

$$(13.1.7) \qquad u(x) \le \mathbf{S}_{s=0}^{x-1}\, h(s)V(s+1;x),$$

where $V(s;x)$ is the solution of (13.1.1), (13.1.2).

**Proof.** From (13.1.1) and (13.1.5), we have

$$(13.1.8) \qquad \mathbf{S}_{s=0}^{x-1}\, V(s+1;x)\Delta_s^m u(s) - \mathbf{S}_{s=0}^{x-1}\, (-1)^m \Delta_s^m V(s;x)u(s) \le \mathbf{S}_{s=0}^{x-1}\, h(s)V(s+1;x).$$

An application of (1.8.5) provides

$$(13.1.9)\ \mathbf{S}_{s=0}^{x-1} (-1)^m u(s)\Delta_s^m V(s;x) = (-1)^m \mathbf{S}_{\bar{s}_m=0}^{\bar{x}_m-1} \left[ u(s)\Delta_{\bar{s}_m}^{m-1} V(s;x)\Big|_{s_m=0}^{x_m} - \sum_{s_m=0}^{x_m-1} \Delta_{s_m} u(s)\Delta_{\bar{s}_m}^{m-1} V(\bar{s}_m, s_m+1;x)\right].$$

Using (13.1.2) and (13.1.6), the right side of (13.1.9) reduces to

$$(-1)^{m+1} \sum_{s_m=0}^{x_m-1} \mathbf{S}_{\bar{s}_m=0}^{\bar{x}_m-1} \Delta_{s_m} u(s)\Delta_{\bar{s}_m}^{m-1} V(\bar{s}_m, s_m+1;x).$$

Repeating the above arguments successively, we obtain

$$(-1)^{2m-1} \sum_{s_m=0}^{x_m-1} \cdots \sum_{s_2=0}^{x_2-1} \left[ \Delta_{s_m} \cdots \Delta_{s_2} u(s) V(s_1, s_2+1, \cdots, s_m+1;x)\Big|_{s_1=0}^{x_1} - \sum_{s_1=0}^{x_1-1} \Delta_s^m u(s) V(s+1;x)\right],$$

which is the same as

$$(-1)^{2m-1} \sum_{s_m=0}^{x_m-1} \cdots \sum_{s_2=0}^{x_2-1} \Delta_{s_m} \cdots \Delta_{s_2} u(\bar{s}_1, x_1) + \mathbf{S}_{s=0}^{x-1} \Delta_s^m u(s) V(s+1;x)$$

or

$$-u(x) + \mathbf{S}_{s=0}^{x-1} \Delta_s^m u(s) V(s+1;x).$$

Substituting this in (13.1.8), the result (13.1.7) follows. ∎

**Remark 13.1.1** For all $g(x)$ and $h(x)$, equality in (13.1.5) implies equality in (13.1.7), and hence $V(s;x)$ the solution of (13.1.1), (13.1.2) is the discrete analog of *Riemann's function.*

**Corollary 13.1.4.** Let $g(x)$ and $h(x)$ be as in Lemma 13.1.3, and $\phi(x)$, $\psi(x)$ be defined on $\mathbb{N}^m$ and satisfy

$$\begin{aligned} \Delta_x^m \phi(x) &\le g(x)\phi(x) + h(x) \\ \Delta_x^m \psi(x) &\ge g(x)\psi(x) + h(x) \\ \phi(\bar{x}_i, 0) &= \psi(\bar{x}_i, 0), \quad 1 \le i \le m. \end{aligned}$$

Then, for all $x \in \mathbb{N}^m$

$$\phi(x) \le \psi(x).$$

**Lemma 13.1.5.** Let $g(x)$ be as in Lemma 13.1.3, and $V(s;x)$ be the solution of (13.1.1), (13.1.2). Let $W(s;x)$ be defined for all $s \le x-1$, $(s;x) \in \mathbb{N}^m \times \mathbb{N}^m$ and

$$(-1)^m \Delta_s^m W(s;x) \;\geq\; g(s)W(s+1;x) \tag{13.1.10}$$

$$W(\overline{s}_i, x_i; x) \;=\; 1, \quad 1 \leq i \leq m. \tag{13.1.11}$$

Then, for all $s \leq x-1$, $(s;x) \in \mathbb{N}^m \times \mathbb{N}^m$

$$V(s;x) \;\leq\; W(s;x).$$

**Proof.** Let $\phi(s;x)$ be defined and nonnegative for all $s \leq x-1$, $(s;x) \in \mathbb{N}^m \times \mathbb{N}^m$ so that

$$(-1)^m \Delta_s^m W(s;x) \;=\; g(s)W(s+1;x) + \phi(s;x). \tag{13.1.12}$$

Next we define the iterates as follows

$$\begin{aligned} W_0(s;x) &= V(s;x) \\ W_{n+1}(s;x) &= 1 + \mathbf{S}_{\ell=s}^{x-1} g(\ell)W_n(\ell+1;x) + \mathbf{S}_{\ell=s}^{x-1} \phi(\ell;x), \; n = 0, 1, \cdots. \end{aligned}$$

Obviously, $W_n(s;x) \geq V(s;x)$ for all $n \geq 1$, and as in Lemma 13.1.2 the sequence $\{W_n(s;x)\}$ converges to $W(s;x)$ which is the solution of (13.1.12), (13.1.11). ∎

## 13.2. Linear Inequalities

In what follows we shall assume that the functions which appear in the inequalities are real valued, nonnegative and defined on $\mathbb{N}^m$.

**Theorem 13.2.1.** Let for all $x \in \mathbb{N}^m$ the following inequality be satisfied

$$u(x) \;\leq\; p(x) + q(x)\mathbf{S}_{s=0}^{x-1} f(s)u(s). \tag{13.2.1}$$

Then, for all $x \in \mathbb{N}^m$

$$u(x) \;\leq\; p(x) + q(x)\mathbf{S}_{s=0}^{x-1} f(s)p(s)V(s+1;x), \tag{13.2.2}$$

where $V(s;x)$ is the solution of

$$\begin{aligned} (-1)^m \Delta_s^m V(s;x) &= f(s)q(s)V(s+1;x), \quad s \leq x-1 \\ V(\overline{s}_i, x_i; x) &= 1, \quad 1 \leq i \leq m. \end{aligned}$$

**Proof.** Define a function $v(x)$ on $\mathbb{N}^m$ as follows

$$v(x) \;=\; \mathbf{S}_{s=0}^{x-1} f(s)u(s).$$

For this function, we have

$$\Delta_x^m v(x) \;=\; f(x)u(x), \quad v(\overline{x}_i, 0) \;=\; 0, \quad 1 \leq i \leq m. \tag{13.2.3}$$

Since $u(x) \le p(x) + q(x)v(x)$, and $f(x) \ge 0$, from (13.2.3) we get

$$\Delta_x^m v(x) \le f(x)p(x) + f(x)q(x)v(x), \quad v(\overline{x}_i, 0) = 0, \quad 1 \le i \le m.$$

Now an application of Lemma 13.1.3 provides

$$v(x) \le \mathbf{S}_{s=0}^{x-1} f(s)p(s)V(s+1;x). \tag{13.2.4}$$

The result (13.2.2) follows from (13.2.4) and the inequality $u(x) \le p(x) + q(x)v(x)$. ∎

**Remark 13.2.1.** The inequality (13.2.2) is the best possible in the sense that equality in (13.2.1) implies equality in (13.2.2).

**Theorem 13.2.2.** Let for all $x \in \mathbb{N}^m$ the following inequality be satisfied

$$u(x) \le p(x) + q(x) \sum_{i=1}^{r} E_i(x,u), \tag{13.2.5}$$

where

$$E_i(x,u) = \mathbf{S}_{x^1=0}^{x-1} f_{i1}(x^1) \mathbf{S}_{x^2=0}^{x^1-1} f_{i2}(x^2) \cdots \mathbf{S}_{x^i=0}^{x^{i-1}-1} f_{ii}(x^i)u(x^i). \tag{13.2.6}$$

Then, for all $x \in \mathbb{N}^m$

$$u(x) \le p(x) + q(x) \mathbf{S}_{s=0}^{x-1} \left( \sum_{i=1}^{r} \Delta_s^m E_i(s,p) \right) V(s+1;x), \tag{13.2.7}$$

where $V(s;x)$ is the solution of

$$(-1)^m \Delta_s^m V(s;x) = \left( \sum_{i=1}^{r} \Delta_s^m E_i(s,q) \right) V(s+1;x), \quad s \le x-1$$
$$V(\overline{s}_i, x_i; x) = 1, \quad 1 \le i \le m.$$

**Proof.** The proof uses the arguments of Theorem 4.1.4 and Theorem 13.2.1. ∎

**Condition (c).** We say that condition (c) is satisfied if for all $x \in \mathbb{N}^m$ the inequality (13.2.5) holds, where

$$f_{ii}(x) = f_i(x), \quad 1 \le i \le r$$
$$f_{i+1,i}(x) = f_{i+2,i}(x) = \cdots = f_{r,i}(x) = g_i(x), \quad 1 \le i \le r-1.$$

In our next result for all $x \in \mathbb{N}^m$ we shall denote

$$\phi_j(x) = \max\left\{0,\ \sum_{i=1}^{r-j+1} q(x)f_i(x) - g_{r-j+1}(x),\ g_i(x) - g_{r-j+1}(x),\right.$$

$$\left. 1 \le i \le r-j\right\},\quad 1 \le j \le r$$

where $g_r(x) = 0$ for all $x \in \mathbb{N}^m$.

**Theorem 13.2.3.** Let the condition (c) be satisfied. Then, for all $x \in \mathbb{N}^m$

$$(13.2.8)_j \qquad u(x) \le p(x) + q(x)\psi_j(x), \quad 1 \le j \le r$$

where

$$\psi_j(x) = \mathbf{S}_{s=0}^{x-1}\left[p(s)\sum_{i=1}^{r-j+1} f_i(s) + g_{r-j+1}(s)\psi_{j-1}(s)\right] V_j(s+1;x),\ 1 \le j \le r$$

and $V_j(s;x),\ 1 \le j \le r$ are the solutions of

$$\begin{aligned}(-1)^m\Delta_s^m V_j(s;x) &= \phi_j(s)V_j(s+1;x), \quad s \le x-1\\ V_j(\overline{s}_i, x_i; x) &= 1, \quad 1 \le i \le m.\end{aligned}$$

**Proof.** The proof is similar to that of Theorem 4.1.5 and Theorem 13.2.1. ∎

**Theorem 13.2.4.** Let for all $x \in \mathbb{N}^m$ the following inequality be satisfied

$$(13.2.9) \qquad u(x) \le p_0(x) + \sum_{i=1}^{r} p_i(x)\mathbf{S}_{s=0}^{x-1} q_i(s)u(s).$$

Then, for all $x \in \mathbb{N}^m$

$$(13.2.10) \qquad u(x) \le F_r[p_0(x)],$$

where

$$\begin{aligned}F_i &= D_iD_{i-1}\cdots D_0\\ D_0[w] &= w\\ D_j[w] &= w + (F_{j-1}[p_j])\,\mathbf{S}_{s=0}^{x-1} q_j(s)w(s)V_j(s+1;x)\end{aligned}$$

and $V_j(s;x),\ 1 \le j \le r$ are the solutions of

$$\begin{aligned}(-1)^m\Delta_s^m V_j(s;x) &= q_j(s)F_{j-1}[p_j(s)]V_j(s+1;x), \quad s \le x-1\\ V_j(\overline{s}_i, x_i, ; x) &= 1, \quad 1 \le j \le r.\end{aligned}$$

**Proof.** The proof is similar to that of Theorem 4.1.6 and Theorem 13.2.1. ∎

## 13.3. Wendroff Type Inequalities

Let $W(s;x)$ be any function defined for all $s \le x-1$, $(s;x) \in \mathbb{N}^m \times \mathbb{N}^m$ and

$$(13.3.1) \qquad \begin{aligned} (-1)^m \Delta_s^m W(s;x) &\ge f(s)q(s)W(s+1;x), \quad s \le x-1 \\ W(\overline{s}_i, x_i; x) &= 1, \quad 1 \le i \le m. \end{aligned}$$

Then, from Lemma 13.1.5 it follows that in (13.2.2), $V(s+1;x)$ can be replaced by $W(s+1;x)$. However, finding a suitable $W(s;x)$ in advance which satisfies (13.3.1) seems to be quite difficult. Therefore, for the function $V(s;x)$ we shall provide an upper estimate which is quite adequate in practical applications.

**Lemma 13.3.1.** Let $V(s;x)$ be as in Theorem 13.2.1. Then, for all $s \le x-1$, $(s;x) \in \mathbb{N}^m \times \mathbb{N}^m$

$$(13.3.2) \qquad V(s;x) \le \prod_{\ell_1=s_1}^{x_1-1} \left(1 + \mathbf{S}_{\overline{\ell}_1=\overline{s}_1}^{\overline{x}_1-1} f(\ell)q(\ell)\right).$$

**Proof.** Since $f(x)q(x) \ge 0$ for all $x \in \mathbb{N}^m$, Lemma 13.1.2 implies that $V(s;x) \ge 1$. Therefore, $(-1)^m \Delta_s^m V(s;x) \ge 0$, which on following the proof of Lemma 13.1.1 gives that $(-1)^i \Delta_{s_1} \cdots \Delta_{s_i} V(s;x) \ge 0$, $1 \le i \le m$. Now since

$$(-1)^m \Delta_{s_m} \left[\frac{\Delta_{\overline{s}_m}^{m-1} V(s;x)}{V(\overline{s}_m+1, s_m; x)}\right] + (-1)^m \Delta_{\overline{s}_m}^{m-1} V(s;x) \times \left[\frac{1}{V(\overline{s}_m+1, s_m; x)} - \frac{1}{V(s+1;x)}\right] = f(s)q(s)$$

it follows that

$$(13.3.3) \qquad (-1)^m \Delta_{s_m} \left[\frac{\Delta_{\overline{s}_m}^{m-1} V(s;x)}{V(\overline{s}_m+1, s_m; x)}\right] \le f(s)q(s).$$

In (13.3.3) keeping $\overline{s}_m$ fixed and setting $s_m = \ell_m$ and summing over $\ell_m = s_m$ to $\ell_m = x_m - 1$, to obtain

$$(-1)^{m+1} \left[\frac{\Delta_{\overline{s}_m}^{m-1} V(s;x)}{V(\overline{s}_m+1, s_m; x)}\right] \le \sum_{\ell_m=s_m}^{x_m-1} f(\overline{s}_m, \ell_m) q(\overline{s}_m, \ell_m).$$

Repeating the above arguments successively with respect to $s_{m-1}, \cdots, s_2$, we find

$$(-1)^{2m-1}\left[\frac{\Delta_{s_1} V(s;x)}{V(\overline{s}_1, s_1+1;x)}\right] \ \le\ \mathbf{S}_{\overline{\ell}_1=\overline{s}_1}^{\overline{x}_1-1} f(\overline{\ell}_1, s_1)q(\overline{\ell}_1, s_1),$$

which is the same as

$$V(s;x) \ \le\ \left[1+\mathbf{S}_{\overline{\ell}_1=\overline{s}_1}^{\overline{x}_1-1} f(\overline{\ell}_1, s_1)q(\overline{\ell}_1, s_1)\right] V(\overline{s}_1, s_1+1;x).$$

The above inequality easily provides (13.3.2). ∎

**Corollary 13.3.2.** Let $V(s;x)$ be as in Theorem 13.2.1. Then, for all $s \le x-1$, $(s;x) \in \mathbb{N}^m \times \mathbb{N}^m$

$$V(s;x) \ \le\ \min_{1\le i\le m}\left\{\prod_{\ell_i=s_i}^{x_i-1}\left(1+\mathbf{S}_{\overline{\ell}_i=\overline{s}_i}^{\overline{x}_i-1} f(\ell)q(\ell)\right)\right\}.$$

**Theorem 13.3.3.** Let for all $x \in \mathbb{N}^m$ the inequality (13.2.1) be satisfied. Then, for all $x \in \mathbb{N}^m$

$$u(x) \ \le p(x) + q(x)\mathbf{S}_{s=0}^{x-1} f(s)p(s) \ \times \min_{1\le i\le m}\left\{\prod_{\ell_i=s_i+1}^{x_i-1}\left(1+\mathbf{S}_{\overline{\ell}_i=\overline{s}_i+1}^{\overline{x}_i-1} f(\ell)q(\ell)\right)\right\}. \tag{13.3.4}$$

**Remark 13.3.1.** For $m=1$, (13.3.4) is the same as (4.1.2) with $a=0$.

**Corollary 13.3.4.** Let in Theorem 13.2.1, $p(x)$ be nondecreasing and $q(x) \ge 1$. Then, for all $x \in \mathbb{N}^m$

$$u(x) \ \le\ p(x)q(x)\min_{1\le i\le m}\left\{\prod_{\ell_i=0}^{x_i-1}\left(1+\mathbf{S}_{\overline{\ell}_i=0}^{\overline{x}_i-1} f(\ell)q(\ell)\right)\right\}. \tag{13.3.5}$$

**Proof.** For such $p(x)$ and $q(x)$, inequality (13.2.2) gives

$$\begin{aligned} u(x) \ &\le\ p(x)q(x)\left[1+\mathbf{S}_{s=0}^{x-1} f(s)q(s)V(s+1;x)\right] \\ &=\ p(x)q(x)\left[1+\mathbf{S}_{s=0}^{x-1}(-1)^m\Delta_s^m V(s;x)\right]. \end{aligned} \tag{13.3.6}$$

Now using $V(\overline{s}_i, x_i; x) = 1$, $1 \le i \le m$ it follows that

$$\begin{aligned} u(x) \ &\le\ p(x)q(x)\left[1+(-1)^{2m-1}\sum_{s_1=0}^{x_1-1}\Delta_{s_1} V(s_1, 0, \cdots, 0; x)\right] \\ &=\ p(x)q(x)\left[1+(-1)^{2m-1}(V(x_1, 0, \cdots, 0; x) - V(0;x))\right] \end{aligned}$$

$$(13.3.7) \qquad = p(x)q(x)V(0;x).$$

The inequality (13.3.5) is now immediate from Corollary 13.3.2. ∎

**Theorem 13.3.5.** Let for all $x \in \mathbb{N}^m$ the inequality (13.2.5) be satisfied. Then, for all $x \in \mathbb{N}^m$

$$(13.3.8) \quad u(x) \le p(x) + q(x)\mathbf{S}_{s=0}^{x-1}\left(\sum_{i=0}^{r}\Delta_s^m E_i(s,p)\right) \times \min_{1\le j\le m}\left\{\prod_{\ell_j=s_j+1}^{x_j-1}\left(1+\mathbf{S}_{\overline{\ell}_j=\overline{s}_j+1}^{\overline{x}_j-1}\sum_{i=1}^{r}\Delta_s^m E_i(\ell,q)\right)\right\}.$$

Further, if $p(x)$ is nondecreasing and $q(x) \ge 1$, then

$$(13.3.9) \quad u(x) \le p(x)q(x)\min_{1\le j\le m}\left\{\prod_{\ell_j=0}^{x_j-1}\left(1+\sum_{i=1}^{r}\Delta_{\ell_j}E_i(\overline{x}_j,\ell_j,q)\right)\right\}.$$

**Remark 13.3.2.** Results which use the estimates on the corresponding functions $V_j(s;x)$, $1 \le j \le r$ in Theorems 13.2.3 and 13.2.4 can be stated analogously.

**Theorem 13.3.6.** Let for all $x$, $X \in \mathbb{N}^m$ such that $x \le X$ the following inequality be satisfied

$$(13.3.10) \qquad u(X) \ge u(x) - q(X)\mathbf{S}_{\ell=x+1}^{X} f(\ell)u(\ell).$$

Then, for all $x$, $X \in \mathbb{N}^m$, $x \le X$

$$(13.3.11) \quad u(X) \ge u(x)\left[\min_{1\le j\le m}\left\{\prod_{\ell_j=x_j+1}^{X_j}\left(1+q(X)\mathbf{S}_{\overline{\ell}_j=\overline{x}_j+1}^{\overline{X}_j}f(\ell)\right)\right\}\right]^{-1}.$$

**Proof.** With the transformation $x = X-\alpha$, $\ell = X-\beta$ where $0 \le \alpha,\beta \le X$, $\alpha,\beta \in \mathbb{N}^m$, inequality (13.3.10) can be written as

$$u(X) \ge u(X-\alpha) - q(X)\mathbf{S}_{\beta=0}^{\alpha-1} f(X-\beta)u(X-\beta).$$

Therefore, if $u(X-\alpha_1) = \overline{u}(\alpha_1)$, $f(X-\beta_1) = \overline{f}(\beta_1)$ where $0 \le \alpha_1,\beta_1 \le X$, then it follows that

$$(13.3.12) \qquad \overline{u}(\alpha) \le u(X) + \mathbf{S}_{\beta=0}^{\alpha-1} q(X)\overline{f}(\beta)\overline{u}(\beta).$$

Since the inequality (13.3.12) satisfies the hypotheses of Corollary 13.3.4, from (13.3.7) it follows that

$$\overline{u}(\alpha) \le u(X)V(0;\alpha), \tag{13.3.13}$$

where $V(\beta;\alpha)$ is the solution of the equation

$$V(\beta;\alpha) = 1 + \mathbf{S}_{\tau=\beta}^{\alpha-1} q(X)\overline{f}(\tau)V(\tau+1;\alpha). \tag{13.3.14}$$

However, from Corollary 13.3.2 we have

$$V(\beta;\alpha) \le \min_{1\le j\le m}\left\{\prod_{\tau_j=\beta_j}^{\alpha_j-1}\left(1+q(X)\mathbf{S}_{\overline{\tau}_j=\overline{\beta}_j}^{\overline{\alpha}_j-1}\overline{f}(\tau)\right)\right\}.$$

Using the above estimate in (13.3.13), to obtain

$$\overline{u}(\alpha) \le u(X)\min_{1\le j\le m}\left\{\prod_{\tau_j=0}^{\alpha_j-1}\left(1+q(X)\mathbf{S}_{\overline{\tau}_j=0}^{\overline{\alpha}_j-1}\overline{f}(\tau)\right)\right\},$$

which is the same as

$$\begin{aligned} u(x) &\le u(X)\min_{1\le j\le m}\left\{\prod_{\tau_j=0}^{X_j-x_j-1}\left(1+q(X)\mathbf{S}_{\overline{\tau}_j=0}^{\overline{X}_j-\overline{x}_j-1}f(X-\tau)\right)\right\} \\ &= u(X)\min_{1\le j\le m}\left\{\prod_{\ell_j=x_j+1}^{X_j}\left(1+q(X)\mathbf{S}_{\overline{\ell}_j=\overline{x}_j+1}^{\overline{X}_j}f(\ell)\right)\right\}. \qquad \blacksquare \end{aligned}$$

## 13.4. Nonlinear Inequalities

Our first result for the nonlinear case is connected with the following inequality

$$u(x) \le p(x)\left[q+\sum_{i=1}^{r}H_i(x,u)\right], \tag{13.4.1}$$

where

$$H_i(x,u) = \mathbf{S}_{x^1=0}^{x-1} f_{i1}(x^1)u^{\alpha_{i1}}(x^1)\cdots\mathbf{S}_{x^i=0}^{x^{i-1}-1} f_{ii}(x^i)u^{\alpha_{ii}}(x^i) \tag{13.4.2}$$

and $\alpha_{ij}$, $1\le j\le i$, $1\le i\le r$ are nonnegative constants and the constant $q>0$.

In the following result we shall denote $\alpha_i = \sum_{j=1}^{i} \alpha_{ij}$ and $\alpha = \max_{1 \le i \le r} \alpha_i$.

**Theorem 13.4.1.** Let for all $x \in \mathbb{N}^m$ the inequality (13.4.1) be satisfied. Then, for all $x \in \mathbb{N}^m$

$$u(x) \le qp(x) \min_{1 \le i \le m} \left\{ \prod_{\ell_i=0}^{x_i-1} (1 + \Delta_{\ell_i} Q(\overline{x}_i, \ell_i)) \right\}, \text{ if } \alpha = 1 \tag{13.4.3}$$

$$\le p(x) \left[ q^{1-\alpha} + (1-\alpha) Q(x) \right]^{1/(1-\alpha)}, \text{ if } \alpha \ne 1 \tag{13.4.4}$$

where $Q(x) = \sum_{i=1}^{r} H_i(x,p) q^{\alpha_i - \alpha}$ and when $\alpha > 1$, we assume that $q^{1-\alpha} + (1-\alpha)Q(x) > 0$ for all $x \in \mathbb{N}^m$.

**Proof.** The inequality (13.4.1) can be written as $u(x) \le p(x)v(x)$, where $v(x) = q + \sum_{i=1}^{r} H_i(x,u)$. Thus, on using the nondecreasing nature of $v(x)$, we find

$$\Delta_x^m v(x) \le \sum_{i=1}^{r} \Delta_x^m H_i(x,p) v^{\alpha_i}(x).$$

Since $v(x) \ge q$, we get

$$\Delta_x^m v(x) \le \sum_{i=1}^{r} \Delta_x^m H_i(x,p) q^{\alpha_i - \alpha} v^{\alpha}(x) = \Delta_x^m Q(x) v^{\alpha}(x).$$

Once again on using the nondecreasing nature of $v(x)$, the above inequality gives

$$\Delta_{x_m} \left[ \frac{\Delta_{\overline{x}_m}^{m-1} v(x)}{v^{\alpha}(x)} \right] \le \Delta_x^m Q(x). \tag{13.4.5}$$

In (13.4.5) setting $x_m = \ell_m$ and summing over $\ell_m = 0$ to $\ell_m = x_m - 1$, we find on using $\Delta_{\overline{x}_m}^{m-1} v(\overline{x}_m, 0) = \Delta_{\overline{x}_m}^{m-1} Q(\overline{x}_m, 0) = 0$ that

$$\frac{\Delta_{\overline{x}_m}^{m-1} v(x)}{v^{\alpha}(x)} \le \Delta_{\overline{x}_n}^{m-1} Q(x).$$

Repeating the above arguments successively with respect to $x_{m-1}, \cdots, x_{i+1}, x_{i-1}, \cdots, x_1$ we get

$$\frac{\Delta_{x_i} v(x)}{v^{\alpha}(x)} \le \Delta_{x_i} Q(x). \tag{13.4.6}$$

If $\alpha = 1$, the result (13.4.3) immediately follows from (13.4.6) and the fact that $v(\overline{x}_i, 0) = q$.

If $\alpha \neq 1$, we have

$$\frac{\Delta_{x_i} v^{1-\alpha}(x)}{1-\alpha} = \int_{x_i}^{x_i+1} \frac{dv(\overline{x}_i, t)}{v^{\alpha}(\overline{x}_i, t)} \leq \frac{\Delta_{x_i} v(x)}{v^{\alpha}(x)}$$

and from (13.4.6), we obtain

$$\text{(13.4.7)} \qquad \frac{\Delta_{x_i} v^{1-\alpha}(x)}{1-\alpha} \leq \Delta_{x_i} Q(x).$$

In (13.4.7) setting $x_i = \ell_i$ and summing over $\ell_i = 0$ to $\ell_i = x_i - 1$, we get the required inequality (13.4.4). ∎

For the next result we shall need the class $T$ (see Definition 4.2.1).

**Theorem 13.4.2.** Let for all $x \in \mathbb{N}^m$ the following inequality be satisfied

$$\text{(13.4.8)} \qquad u(x) \leq p(x) + \sum_{i=1}^{r_1} E_i(x, u) + \sum_{i=1}^{r_2} p_i(x) \mathbf{S}_{s=0}^{x-1} q_i(s) W_i(u(s)),$$

where (i) $p(x) \geq 1$ and nondecreasing, (ii) $p_i(x) \geq 1$, $1 \leq i \leq r_2$, (iii) $W_i \in T$, $1 \leq i \leq r_2$. Then, for all $x \in \mathbb{N}^m$

$$\text{(13.4.9)} \qquad u(x) \leq p(x) v(x) e(x) \prod_{i=1}^{r_2} J_i(x),$$

where

$$e(x) = \prod_{i=1}^{r_2} p_i(x), \quad v(x) = \min_{1 \leq j \leq m} \left\{ \prod_{\ell_j=0}^{x_j-1} \left( 1 + \sum_{i=1}^{r_1} \Delta_{\ell_j} E_i(\overline{x}_j, \ell_j, e) \right) \right\}$$

$$J_0(x) = 1, \quad J_j(x) = G_j^{-1} \left[ G_j(1) + \mathbf{S}_{s=0}^{x-1} q_j(s) v(s) e(s) \prod_{i=1}^{j-1} J_i(s) \right], \quad 1 \leq j \leq r_2$$

and

$$G_j(w) = \int_{w_0}^{w} \frac{dt}{W_j(t)}, \quad w \geq w_0 \geq 1$$

as long as

$$G_j(1) + \mathbf{S}_{s=0}^{x-1} q_j(s) v(s) e(s) \prod_{i=1}^{j-1} J_i(s) \in Dom\left(G_j^{-1}\right), \quad 1 \leq j \leq r_2.$$

**Proof.** The proof is similar to that of Theorem 4.2.3 and Theorem 13.4.1. ■

**Theorem 13.4.3.** In addition to the hypotheses of Theorem 13.4.2 let $p_i(x)$, $1 \le i \le r_2$ be nondecreasing. Then, for all $x \in \mathbb{N}^m$

$$u(x) \le p(x)v^*(x)\prod_{i=1}^{r_2} J_i^*(x), \tag{13.4.10}$$

where $v^*(x)$ is the same as $v(x)$ in Theorem 13.4.2 with $e(x) = 1$;

$$J_0^*(x) = 1, \ J_j^*(x) = p_j(x)G_j^{-1}\left[G_j(1) + \mathbf{S}_{s=0}^{x-1} q_j(s)v^*(s)p_j(s)\prod_{i=1}^{j-1} J_i^*(s)\right], \quad 1 \le j \le r_2$$

as long as

$$G_j(1) + \mathbf{S}_{s=0}^{x-1} q_j(s)v^*(s)p_j(s)\prod_{i=1}^{j-1} J_i^*(s) \in Dom\left(G_j^{-1}\right), \quad 1 \le j \le r_2$$

and $G_j$, $1 \le j \le r_2$ are the same as in Theorem 13.4.2.

**Remark 13.4.1.** Results analogous to Theorems 4.2.5 and 4.2.6 can be stated similarly.

## 13.5. Inequalities Involving Partial Differences

The following result is two dimensional discrete *Taylor's formula.*

**Lemma 13.5.1.** Let the function $u(k,\ell)$ be defined on $\mathbb{N}\times\mathbb{N}$. Then, for $0 \le i \le r_1 - 1$, $0 \le j \le r_2 - 1$ ($r_1$, $r_2$ positive integers) and $(k,\ell) \in \mathbb{N}\times\mathbb{N}$

$$\Delta_k^i\Delta_\ell^j u(k,\ell) = \phi_{ij}(k,\ell) + \frac{1}{(r_1-i-1)!(r_2-j-1)!} \times \sum_{\tau=0}^{k-r_1+i}\sum_{\eta=0}^{\ell-r_2+j}(k-\tau-1)^{(r_1-i-1)}(\ell-\eta-1)^{(r_2-j-1)}\Delta_\tau^{r_1}\Delta_\eta^{r_2}u(\tau,\eta), \tag{13.5.1}$$

where

(13.5.2)

$$\phi_{ij}(k,\ell) = \sum_{\alpha=i}^{r_1-1}\frac{(k)^{(\alpha-i)}}{(\alpha-i)!}\Delta_k^\alpha\Delta_\ell^j u(0,\ell) + \sum_{\beta=j}^{r_2-1}\frac{(\ell)^{(\beta-j)}}{(\beta-j)!}\Delta_k^i\Delta_\ell^\beta u(k,0)$$

$$-\sum_{\alpha=i}^{r_1-1}\sum_{\beta=j}^{r_2-1}\frac{(k)^{(\alpha-i)}}{(\alpha-i)!}\frac{(\ell)^{(\beta-j)}}{(\beta-j)!}\Delta_k^\alpha\Delta_\ell^\beta u(0,0).$$

**Proof.** From (1.8.7) it follows that

$$\begin{aligned}\phi_{ij}(k,\ell) &= \sum_{\beta=j}^{r_2-1}\frac{(\ell)^{(\beta-j)}}{(\beta-j)!}\Delta_k^i\Delta_\ell^\beta u(k,0)+\sum_{\alpha=i}^{r_1-1}\frac{(k)^{(\alpha-i)}}{(\alpha-i)!}\times\\ &\quad \left[\Delta_k^\alpha\Delta_\ell^j u(0,\ell)-\sum_{\beta=j}^{r_2-1}\frac{(\ell)^{(\beta-j)}}{(\beta-j)!}\Delta_k^\alpha\Delta_\ell^\beta u(0,0)\right]\\ &= \sum_{\beta=j}^{r_2-1}\frac{(\ell)^{(\beta-j)}}{(\beta-j)!}\Delta_k^i\Delta_\ell^\beta u(k,0)+\sum_{\alpha=i}^{r_1-1}\frac{(k)^{(\alpha-i)}}{(\alpha-i)!}\times\\ &\quad \frac{1}{(r_2-j-1)!}\sum_{\eta=0}^{\ell-r_2+j}(\ell-\eta-1)^{(r_2-j-1)}\Delta_k^\alpha\Delta_\eta^{r_2}u(0,\eta).\end{aligned}$$

Thus, the right side of (13.5.1) is the same as

$$\sum_{\beta=j}^{r_2-1}\frac{(\ell)^{(\beta-j)}}{(\beta-j)!}\Delta_k^i\Delta_\ell^\beta u(k,0)+\frac{1}{(r_2-j-1)!}\sum_{\eta=0}^{\ell-r_2+j}(\ell-\eta-1)^{(r_2-j-1)}\times$$

$$\left[\sum_{\alpha=i}^{r_1-1}\frac{(k)^{(\alpha-i)}}{(\alpha-i)!}\Delta_k^\alpha\Delta_\eta^{r_2}u(0,\eta)+\frac{1}{(r_1-i-1)!}\sum_{\tau=0}^{k-r_1+i}(k-\tau-1)^{(r_1-i-1)}\Delta_\tau^{r_1}\Delta_\eta^{r_2}u(\tau,\eta)\right],$$

which is on applying (1.8.7) successively leads to

$$\begin{aligned}&\sum_{\beta=j}^{r_2-1}\frac{(\ell)^{(\beta-j)}}{(\beta-j)!}\Delta_k^i\Delta_\ell^\beta u(k,0)+\frac{1}{(r_2-j-1)!}\sum_{\eta=0}^{\ell-r_2+j}(\ell-\eta-1)^{(r_2-j-1)}\Delta_k^i\Delta_\eta^{r_2}u(k,\eta)\\ &\qquad = \Delta_k^i\Delta_\ell^j u(k,\ell). \qquad \blacksquare\end{aligned}$$

**Theorem 13.5.2.** Let for all $k,\ell\in\mathbb{N}\times\mathbb{N}$ the following inequality be satisfied

$$(13.5.3)\quad \Delta_k^{r_1}\Delta_\ell^{r_2}u(k,\ell)\le p(k,\ell)+q(k,\ell)\sum_{i=0}^{r_1}\sum_{j=0}^{r_2}\sum_{\tau=0}^{k-1}\sum_{\eta=0}^{\ell-1}h_{ij}(\tau,\eta)\Delta_\tau^i\Delta_\eta^j u(\tau,\eta).$$

Then, for all $(k,\ell)\in\mathbb{N}\times\mathbb{N}$

$$(13.5.4)\quad \Delta_k^{r_1}\Delta_\ell^{r_2}u(k,\ell)\le p(k,\ell)+q(k,\ell)\sum_{\tau=0}^{k-1}\sum_{\eta=0}^{\ell-1}A_1(\tau,\eta)V(\tau+1,\eta+1;k,\ell),$$

where

$$(13.5.5)\quad A_1(k,\ell) = h_{r_1r_2}(k,\ell)p(k,\ell) + \sum_{j=0}^{r_2-1} h_{r_1j}(k,\ell) \times$$

$$\left[\sum_{\beta=j}^{r_2-1} \frac{(\ell)^{(\beta-j)}}{(\beta-j)!}\Delta_k^{r_1}\Delta_\ell^{\beta}u(k,0) + \frac{1}{(r_2-j-1)!}\sum_{\eta=0}^{\ell-r_2+j}(\ell-\eta-1)^{(r_2-j-1)}p(k,\eta)\right]$$

$$+\sum_{i=0}^{r_1-1} h_{ir_2}(k,\ell)\left[\sum_{\alpha=i}^{r_1-1}\frac{(k)^{(\alpha-i)}}{(\alpha-i)!}\Delta_k^{\alpha}\Delta_\ell^{r_2}u(0,\ell) + \frac{1}{(r_1-i-1)!}\times\right.$$

$$\left.\sum_{\tau=0}^{k-r_1+i}(k-\tau-1)^{(r_1-i-1)}p(\tau,\ell)\right] + \sum_{i=0}^{r_1-1}\sum_{j=0}^{r_2-1} h_{ij}(k,\ell)\left[\phi_{ij}(k,\ell)\right.$$

$$\left.+\frac{1}{(r_1-i-1)!(r_2-j-1)!}\sum_{\tau=0}^{k-r_1+i}\sum_{\eta=0}^{\ell-r_2+j}(k-\tau-1)^{(r_1-i-1)}(\ell-\eta-1)^{(r_2-j-1)}p(\tau,\eta)\right]$$

and $V(\tau,\eta;k,\ell)$, $\tau \le k-1$, $\eta \le \ell-1$ is the solution of

$$(13.5.6)\qquad \Delta_\tau\Delta_\eta V(\tau,\eta;k,\ell) = B_1(\tau,\eta)V(\tau+1,\eta+1;k,\ell)$$

$$(13.5.7)\qquad V(k,\eta;k,\ell) = V(\tau,\ell;k,\ell) = 1,$$

where

$$(13.5.8)\quad B_1(k,\ell) = h_{r_1r_2}(k,\ell)q(k,\ell)$$

$$+\sum_{j=0}^{r_2-1} h_{r_1j}(k,\ell)\frac{1}{(r_2-j-1)!}\sum_{\eta=0}^{\ell-r_2+j}(\ell-\eta-1)^{(r_2-j-1)}q(k,\eta)$$

$$+\sum_{i=0}^{r_1-1} h_{ir_2}(k,\ell)\frac{1}{(r_1-i-1)!}\sum_{\tau=0}^{k-r_1+i}(k-\tau-1)^{(r_1-i-1)}q(\tau,\ell)$$

$$+\sum_{i=0}^{r_1-1}\sum_{j=0}^{r_2-1} h_{ij}(k,\ell)\frac{1}{(r_1-i-1)!(r_2-j-1)!}\times$$

$$\sum_{\tau=0}^{k-r_1+i}\sum_{\eta=0}^{\ell-r_2+j}(k-\tau-1)^{(r_1-i-1)}(\ell-\eta-1)^{(r_2-j-1)}q(\tau,\eta).$$

**Proof.** Define a function $v(k,\ell)$ on $\mathbb{N}\times\mathbb{N}$ as follows

$$v(k,\ell) = \sum_{i=0}^{r_1}\sum_{j=0}^{r_2}\sum_{\tau=0}^{k-1}\sum_{\eta=0}^{\ell-1} h_{ij}(\tau,\eta)\Delta_\tau^i\Delta_\eta^j u(\tau,\eta),$$

then (13.5.3) can be written as

$$\Delta_k^{r_1}\Delta_\ell^{r_2}u(k,\ell) \;\le\; p(k,\ell)+q(k,\ell)v(k,\ell). \tag{13.5.9}$$

From the definition of $v(k,\ell)$, we have

$$\begin{aligned}\Delta_k\Delta_\ell v(k,\ell) &= \sum_{i=0}^{r_1}\sum_{j=0}^{r_2}h_{ij}(k,\ell)\Delta_k^i\Delta_\ell^j u(k,\ell)\\ v(0,\ell) &= v(k,0) \;=\; 0.\end{aligned} \tag{13.5.10}$$

Using Lemma 13.5.1 and (1.8.7) in (13.5.10), we get

$$\begin{aligned}\Delta_k\Delta_\ell v(k,\ell) &= h_{r_1r_2}(k,\ell)\Delta_k^{r_1}\Delta_\ell^{r_2}u(k,\ell)\\ &+\sum_{j=0}^{r_2-1}h_{r_1j}(k,\ell)\left[\sum_{\beta=j}^{r_2-1}\frac{(\ell)^{(\beta-j)}}{(\beta-j)!}\Delta_k^{r_1}\Delta_\ell^{\beta}u(k,0)\right.\\ &\left.+\frac{1}{(r_2-j-1)!}\sum_{\eta=0}^{\ell-r_2+j}(\ell-\eta-1)^{(r_2-j-1)}\Delta_k^{r_1}\Delta_\eta^{r_2}u(k,\eta)\right]\\ &+\sum_{i=0}^{r_1-1}h_{ir_2}(k,\ell)\left[\sum_{\alpha=i}^{r_1-1}\frac{(k)^{(\alpha-i)}}{(\alpha-i)!}\Delta_k^{\alpha}\Delta_\ell^{r_2}u(0,\ell)\right.\\ &\left.+\frac{1}{(r_1-i-1)!}\sum_{\tau=0}^{k-r_1+i}(k-\tau-1)^{(r_1-i-1)}\Delta_\tau^{r_1}\Delta_\ell^{r_2}u(\tau,\ell)\right]\\ &+\sum_{i=0}^{r_1-1}\sum_{j=0}^{r_2-1}h_{ij}(k,\ell)\left[\phi_{ij}(k,\ell)+\frac{1}{(r_1-i-1)!(r_2-j-1)!}\times\right.\\ &\left.\sum_{\tau=0}^{k-r_1+i}\sum_{\eta=0}^{\ell-r_2+j}(k-\tau-1)^{(r_1-i-1)}(\ell-\eta-1)^{(r_2-j-1)}\Delta_\tau^{r_1}\Delta_\eta^{r_2}u(\tau,\eta)\right].\end{aligned} \tag{13.5.11}$$

Using (13.5.9) in (13.5.11) and the nondecreasing nature of $v(k,\ell)$, we obtain

$$\begin{aligned}\Delta_k\Delta_\ell v(k,\ell) &\le A_1(k,\ell)+B_1(k,\ell)v(k,\ell)\\ v(0,\ell) &= v(k,0) \;=\; 0.\end{aligned} \tag{13.5.12}$$

Thus, as an application of Lemma 13.1.3 it follows that

$$v(k,\ell) \;=\; \sum_{\tau=0}^{k-1}\sum_{\eta=0}^{\ell-1}A_1(\tau,\eta)V(\tau+1,\eta+1;k,\ell). \tag{13.5.13}$$

Substituting (13.5.13) in (13.5.9), the result (13.5.4) follows. ∎

**Remark 13.5.1.** From (13.5.4) and Lemma 13.5.1 an upper estimate for $\Delta_k^i \Delta_\ell^j u(k,\ell)$ , $0 \le i \le r_1 - 1,\ 0 \le j \le r_2 - 1$ is readily available. Indeed, we have

$$\Delta_k^i \Delta_\ell^j u(k,\ell) \le \phi_{ij}(k,\ell) + \frac{1}{(r_1-i-1)!(r_2-j-1)!} \times \sum_{\tau=0}^{k-r_1+i} \sum_{\eta=0}^{\ell-r_2+j} (k-\tau-1)^{(r_1-i-1)} (\ell-\eta-1)^{(r_2-j-1)} \times \left[ p(\tau,\eta) + q(\tau,\eta) \sum_{\tau_1=0}^{\tau-1} \sum_{\eta_1=0}^{\eta-1} A_1(\tau_1,\eta_1) V(\tau_1+1,\eta_1+1;\tau,\eta) \right]. \tag{13.5.14}$$

**Corollary 13.5.3.** In Theorem 13.5.2 the inequality (13.5.4) can be replaced by

$$\Delta_k^{r_1} \Delta_\ell^{r_2} u(k,\ell) \le p(k,\ell) + q(k,\ell) \left[ \sum_{\tau=0}^{k-1} \sum_{\eta=0}^{\ell-1} A_1(\tau,\eta) \right] \times \prod_{\tau=0}^{k-1} \left( 1 + \sum_{\eta=0}^{\ell-1} B_1(\tau,\eta) \right). \tag{13.5.15}$$

**Proof.** From (13.5.12), we have

$$v(k,\ell) \le \sum_{\tau=0}^{k-1} \sum_{\eta=0}^{\ell-1} A_1(\tau,\eta) + \sum_{\tau=0}^{k-1} \sum_{\eta=0}^{\ell-1} B_1(\tau,\eta) v(\tau,\eta).$$

Therefore, from Corollary 13.3.4, we obtain

$$v(k,\ell) \le \left[ \sum_{\tau=0}^{k-1} \sum_{\eta=0}^{\ell-1} A_1(\tau,\eta) \right] \prod_{\tau=0}^{k-1} \left( 1 + \sum_{\eta=0}^{\ell-1} B_1(\tau,\eta) \right). \tag{13.5.16}$$

Substituting (13.5.16) in (13.5.9) the inequality (13.5.15) follows. ∎

**Theorem 13.5.4.** Let for all $k,\ell \in \mathbb{N} \times \mathbb{N}$ the following inequality be satisfied

$$\Delta_k^r \Delta_\ell^r u(k,\ell) \le p(k,\ell) + q(k,\ell) \sum_{i=0}^{r} \sum_{\tau=0}^{k-1} \sum_{\eta=0}^{\ell-1} h(\tau,\eta) \Delta_\tau^i \Delta_\eta^i u(\tau,\eta), \tag{13.5.17}$$

where $q(k,\ell) \ge 1$. Then, for all $(k,\ell) \in \mathbb{N} \times \mathbb{N}$

(13.5.18) $$\Delta_k^r\Delta_\ell^r u(k,\ell) \le p(k,\ell) + q(k,\ell)\sum_{\tau=0}^{k-1}\sum_{\eta=0}^{\ell-1}[H(\tau,\eta) + h(\tau,\eta)B_r(\tau,\eta)]V_{r+1}(\tau+1,\eta+1;k,\ell),$$

where

$$H(k,\ell) = h(k,\ell)\left\{p(k,\ell) + \sum_{i=0}^{r-1}\left[\phi_{ii}(k,\ell) + \frac{1}{((r-i-1)!)^2}\times \sum_{\tau=0}^{k-r+i}\sum_{\eta=0}^{\ell-r+i}(k-\tau-1)^{(r-i-1)}(\ell-\eta-1)^{(r-i-1)}p(\tau,\eta)\right]\right\},$$

$B_0(k,\ell) \equiv 0$, $B_i(k,\ell) = \sum_{\tau=0}^{k-1}\sum_{\eta=0}^{\ell-1}[H(\tau,\eta) + B_{i-1}(\tau,\eta)]V_i(\tau+1,\eta+1;k,\ell)$, $1 \le i \le r$ and $V_i(\tau,\eta;k,\ell)$, $\tau \le k-1$, $\eta \le \ell-1$, $1 \le i \le r+1$ are the solutions of

$$\begin{aligned}
\Delta_\tau\Delta_\eta V_1(\tau,\eta;k,\ell) &= [h(\tau,\eta)q(\tau,\eta) + h(\tau,\eta) + rq(\tau,\eta) \\
&\quad +(r-1)]V_1(\tau+1,\eta+1;k,\ell) \\
\Delta_\tau\Delta\eta V_i(\tau,\eta;k,\ell) &= [h(\tau,\eta)q(\tau,\eta) + h(\tau,\eta) + (r-i+1)q(\tau,\eta) \\
&\quad +(r-i-1)]V_i(\tau+1,\eta+1;k,\ell), \quad 2 \le i \le r \\
\Delta_\tau\Delta_\eta V_{r+1}(\tau,\eta;k,\ell) &= [h(\tau,\eta)q(\tau,\eta) - h(\tau,\eta)]V_{r+1}(\tau+1,\eta+1;k,\ell) \\
V_j(k,\eta;k,\ell) &= V_j(\tau,\ell;k,\ell) = 1, \quad 1 \le j \le r+1.
\end{aligned}$$

**Proof.** Define a function $v_1(k,\ell)$ on $\mathbb{N}\times\mathbb{N}$ as follows

(13.5.19) $$v_1(k,\ell) = \sum_{i=0}^{r}\sum_{\tau=0}^{k-1}\sum_{\eta=0}^{\ell-1}h(\tau,\eta)\Delta_\tau^i\Delta_\eta^i u(\tau,\eta),$$

then (13.5.17) can be written as

(13.5.20) $$\Delta_k^r\Delta_\ell^r u(k,\ell) \le p(k,\ell) + q(k,\ell)v_1(k,\ell).$$

From the definition of $v_1(k,\ell)$ we have

$$\Delta_k\Delta_\ell v_1(k,\ell) = h(k,\ell)\left[\Delta_k^r\Delta_\ell^r u(k,\ell) + \sum_{i=0}^{r-1}\Delta_k^i\Delta_\ell^i u(k,\ell)\right].$$

Therefore, from (13.5.20) and Lemma 13.5.1, we obtain

(13.5.21) $$\Delta_k\Delta_\ell v_1(k,\ell) + h(k,\ell)v_1(k,\ell) \le H(k,\ell) + h(k,\ell)q(k,\ell)v_1(k,\ell) + h(k,\ell)v_2(k,\ell),$$

where

$$(13.5.22)\quad v_2(k,\ell) = v_1(k,\ell) + \sum_{i=0}^{r-1} \frac{1}{((r-i-1)!)^2} \times \sum_{\tau=0}^{k-r+i}\sum_{\eta=0}^{\ell-r+i} (k-\tau-1)^{(r-i-1)}(\ell-\eta-1)^{(r-i-1)} q(\tau,\eta)v_1(\tau,\eta).$$

Since $v_1(k,\ell) \le v_2(k,\ell)$, it follows from (13.5.22) and (13.5.21) that

$$\Delta_k\Delta_\ell v_2(k,\ell) + v_2(k,\ell) \le H(k,\ell) + [h(k,\ell)q(k,\ell) + h(k,\ell) + q(k,\ell)]v_2(k,\ell) + v_3(k,\ell),$$

where

$$v_3(k,\ell) = v_2(k,\ell) + \sum_{i=0}^{r-2} \frac{1}{((r-i-2)!)^2} \sum_{\tau=0}^{k-r+i+1}\sum_{\eta=0}^{\ell-r+i+1} (k-\tau-1)^{(r-i-2)} \times (\ell-\eta-1)^{(r-i-2)} q(\tau,\eta)v_2(\tau,\eta).$$

Once again, on using $v_2(k,\ell) \le v_3(k,\ell)$, we find

$$\Delta_k\Delta_\ell v_3(k,\ell) + v_3(k,\ell) \le H(k,\ell) + [h(k,\ell)q(k,\ell) + h(k,\ell) + 2q(k,\ell) + 1]v_3(k,\ell) + v_4(k,\ell),$$

where

$$v_4(k,\ell) = v_3(k,\ell) + \sum_{i=0}^{r-3} \frac{1}{((r-i-3)!)^2} \sum_{\tau=0}^{k-r+i+2}\sum_{\eta=0}^{\ell-r+i+2} (k-\tau-1)^{(r-i-3)} \times (\ell-\eta-1)^{(r-i-3)} q(\tau,\eta)v_3(\tau,\eta).$$

Continuing in this way, we get

$$(13.5.23)\quad \Delta_k\Delta_\ell v_r(k,\ell) + v_r(k,\ell) \le H(k,\ell) + [h(k,\ell)q(k,\ell) + h(k,\ell) + (r-1)q(k,\ell) + (r-2)]v_r(k,\ell) + v_{r+1}(k,\ell),$$

where $v_{r+1}(k,\ell) = v_r(k,\ell) + \sum_{\tau=0}^{k-1}\sum_{\eta=0}^{\ell-1} q(\tau,\eta)v_r(\tau,\eta)$, and hence on using $v_r(k,\ell) \le v_{r+1}(k,\ell)$ in (13.5.23), we obtain

$$\Delta_k\Delta_\ell v_{r+1}(k,\ell) \le H(k,\ell) + [h(k,\ell)q(k,\ell) + h(k,\ell) + rq(k,\ell) + (r-1)]v_{r+1}(k,\ell).$$

Since $v_j(k,0) = v_j(0,\ell) = 0,\ 1 \le j \le r+1$, an application of Lemma 13.1.3 provides

$$(13.5.24)\quad v_{r+1}(k,\ell) \le \sum_{\tau=0}^{k-1}\sum_{\eta=0}^{\ell-1} H(\tau,\eta)V_1(\tau+1,\eta+1;k,\ell) = B_1(k,\ell).$$

Using (13.5.24) in (13.5.23) and applying Lemma 13.1.3, we get $v_r(k,\ell) \leq B_2(k,\ell)$. Continuing in this way, we find $v_2(k,\ell) \leq B_r(k,\ell)$. Thus, from (13.5.21), we obtain

$$\Delta_k \Delta_\ell v_1(k,\ell) \;\leq\; H(k,\ell) + h(k,\ell)B_r(k,\ell) + h(k,\ell)(q(k,\ell) - 1)v_1(k,\ell)$$

and Lemma 13.1 3 finally gives

$$(13.5.25)\quad v_1(k,\ell) \leq \sum_{\tau=0}^{k-1}\sum_{\eta=0}^{\ell-1}[H(\tau,\eta)+h(\tau,\eta)B_r(\tau,\eta)]V_{r+1}(\tau+1,\eta+1;k,\ell).$$

The result (13.5.18) now follows from (13.5.20) and (13.5.25). ∎

**Remark 13.5.2.** For $q(k,\ell)$ not necessarily greater than 1, the conclusion of Theorem 13.5.3 remains valid if $V_r$, $V_{r+1}$ and $B_r$ are replaced by $V_r^*$, $V_{r+1}^*$ and $B_r^*$ defined by

$$\Delta_\tau \Delta_\eta V_r^*(\tau,\eta;k,\ell) = [h(\tau,\eta)q(\tau,\eta) + h(\tau,\eta) + q(\tau,\eta)]V_r^*(\tau+1,\eta+1;k,\ell)$$
$$\Delta_\tau \Delta_\eta V_{r+1}^*(\tau,\eta;k,\ell) = h(\tau,\eta)q(\tau,\eta)V_{r+1}^*(\tau+1,\eta+1;k,\ell)$$
$$V_r^*(k,\eta;k,\ell) = V_r^*(\tau,\ell;k,\ell) = V_{r+1}^*(k,\eta;k,\ell) = V_{r+1}^*(\tau,\ell;k,\ell) = 1,$$

and $B_r^*(k,\ell) = \sum_{\tau=0}^{k-1}\sum_{\eta=0}^{\ell-1}[H(\tau,\eta) + B_{r-1}(\tau,\eta)]V_r^*(\tau+1,\eta+1;k,\ell)$.

**Remark 13.5.3.** The result which can be deduced from Theorem 13.5.2 for the inequality (13.5.17) does not seem to be comparable with the one obtained in Theorem 13.5.4.

## 13.6. Multidimensional Linear Inequalities

The multidimensional version of Lemma 13.1.3 is stated in the following:

**Lemma 13.6.1.** Let the $n \times n$ matrix $\mathcal{A}(x)$ be defined and nonnegative on $\mathbb{N}^m$. Let $n$ vector functions $\mathbf{h}(k)$ and $\mathbf{u}(k)$ be defined on $\mathbb{N}^m$. Further, let for all $x \in \mathbb{N}^m$ the following inequality be satisfied

$$\Delta_x^m \mathbf{u}(x) \;\leq\; \mathcal{A}(x)\mathbf{u}(x) + \mathbf{h}(x),$$

where

$$\mathbf{u}(\overline{x}_i, 0) \;=\; 0, \quad 1 \leq i \leq m.$$

Then, for all $x \in \mathbb{N}^m$

$$\mathbf{u}(x) \;\leq\; \mathbf{S}_{s=0}^{x-1}\mathcal{V}(s+1;x)\mathbf{h}(s),$$

where the $n \times n$ matrix $\mathcal{V}(s;x)$, $s \le x-1$, $(s;x) \in \mathbb{N}^m \times \mathbb{N}^m$ is a solution of

$$\begin{aligned}(-1)^m \Delta_s^m \mathcal{V}(s;x) &= \mathcal{V}(s+1;x)\mathcal{A}(s)\\ \mathcal{V}(\overline{s}_i, x_i; x) &= \mathcal{I}, \quad 1 \le i \le m\end{aligned}$$

or equivalently,

$$\mathcal{V}(s;x) = \mathcal{I} + \mathbf{S}_{\ell=s}^{x-1} \mathcal{V}(\ell+1;x)\mathcal{A}(\ell).$$

**Theorem 13.6.2.** Let the $n \times n$ matrices $\mathcal{G}(x)$ and $\mathcal{H}(x)$ be defined and nonnegative on $\mathbb{N}^m$, and the $n$ vector functions $\mathbf{p}(x)$ and $\mathbf{u}(x)$ be defined on $\mathbb{N}^m$. Further, let for all $x \in \mathbb{N}^m$ the following inequality be satisfied

$$\mathbf{u}(x) \le \mathbf{p}(x) + \mathcal{G}(x)\mathbf{S}_{s=0}^{x-1}\mathcal{H}(s)\mathbf{u}(s). \tag{13.6.1}$$

Then, for all $x \in \mathbb{N}^m$

$$\mathbf{u}(x) \le \mathbf{p}(x) + \mathcal{G}(x)\mathbf{S}_{s=0}^{x-1}\mathcal{V}(s+1;x)\mathcal{H}(s)\mathbf{p}(s), \tag{13.6.2}$$

where $\mathcal{V}(s;x)$ satisfies

$$\mathcal{V}(s;x) = \mathcal{I} + \mathbf{S}_{\ell=s}^{x-1}\mathcal{V}(\ell+1;x)\mathcal{H}(\ell)\mathcal{G}(\ell). \tag{13.6.3}$$

**Proof.** The proof is similar to that of Theorem 13.2.1. ∎

**Theorem 13.6.3.** Let in addition to hypotheses of Theorem 13.6.2, $\mathbf{u}(x) \ge 0$ for all $x \in \mathbb{N}^m$. Then, for all $x \in \mathbb{N}^m$

$$u_i(x) \le p_i(x) + \max_{1 \le j \le n} g_{ij}(x) q(x), \tag{13.6.4}$$

where

$$q(x) = \sum_{s_1=0}^{x_1-1} \alpha(\overline{x}_1, s_1) \prod_{\ell_1=s_1+1}^{x_1-1} (1 + \beta(\overline{x}_1, \ell_1))$$

and

$$\alpha(\overline{x}_1, x_1) = \sum_{j,r=1}^{n} \mathbf{S}_{\overline{s}_1=0}^{\overline{x}_1-1} h_{jr}(\overline{s}_1, x_1) p_r(\overline{s}_1, x_1)$$

$$\beta(\overline{x}_1, x_1) = \max_{1 \le \tau \le n} \sum_{j,r=1}^{n} \mathbf{S}_{\overline{s}_1=0}^{\overline{x}_1-1} h_{jr}(\overline{s}_1, x_1) g_{r\tau}(\overline{s}_1, x_1).$$

**Proof.** The proof is similar to that of Theorem 4.4.5. ∎

**Theorem 13.6.4.** Let in addition to hypotheses of Theorem 13.6.3, $\mathbf{p}(x) \geq 0$ for all $x \in \mathbb{N}^m$. Then, for all $x \in \mathbb{N}^m$

$$(13.6.5) \quad u^*(x) \leq p^*(x) + g^*(x)\mathbf{S}_{s=0}^{x-1} h^*(s)p^*(s) \times \min_{1\leq j\leq m}\left\{\prod_{\ell_j=s_j+1}^{x_j-1}\left(1+\mathbf{S}_{\bar{\ell}_j=\bar{s}_j+1}^{\bar{x}_j-1} h^*(\ell)g^*(\ell)\right)\right\},$$

where

$$u^*(x) = \max_{1\leq i\leq n} u_i(x), \quad p^*(x) = \max_{1\leq i\leq n} p_i(x),$$

$$g^*(x) = \sum_{j=1}^{n}\left(\max_{1\leq i\leq n} g_{ij}(x)\right), \quad \text{and} \quad h^*(x) = \max_{1\leq j\leq n}\left(\sum_{r=1}^{n} h_{jr}(x)\right).$$

**Proof.** The proof is similar to that of Theorem 4.4.6. ∎

**Theorem 13.6.5.** Let the hypotheses of Theorem 13.6.4 be satisfied. Then, for all $x \in \mathbb{N}^m$

$$(13.6.6) \quad u_j(x) \leq p_j(x) + \sum_{i=1}^{n} g_{ji}(x)\mathbf{S}_{s=0}^{x-1}\left(\sum_{\tau=1}^{n}\sum_{\eta=1}^{n} h_{\tau\eta}(s)p_\eta(s)\right) \times \min_{1\leq k\leq m}\left\{\prod_{\ell_k=s_k+1}^{x_k-1}\left(1+\mathbf{S}_{\bar{\ell}_k=\bar{s}_k+1}^{\bar{x}_k-1}\|\mathcal{H}(\ell)\|\|\mathcal{G}(\ell)\|\right)\right\},$$

where $\|\mathcal{G}\|$ is any $n\times n$ matrix norm such that $|g_{ij}| \leq \|\mathcal{G}\|$.

**Proof.** In component form the inequality (13.6.2) is the same as

$$u_j(x) \leq p_j(x) + \sum_{i=1}^{n}\sum_{\tau=1}^{n}\sum_{\eta=1}^{n} g_{ji}(x)\mathbf{S}_{s=0}^{x-1} v_{i\tau}(s+1;x)h_{\tau\eta}(s)p_\eta(s).$$

Hence, it follows that

$$u_j(x) \leq p_j(x) + \sum_{i=1}^{n}\sum_{\tau=1}^{n}\sum_{\eta=1}^{n} g_{ji}(x)\mathbf{S}_{s=0}^{x-1}\|\mathcal{V}(s+1;x)\|h_{\tau\eta}(s)p_\eta(s)$$

$$(13.6.7) \quad = p_j(x) + \sum_{i=1}^{n} g_{ji}(x)\mathbf{S}_{s=0}^{x-1}\left(\sum_{\tau=1}^{n}\sum_{\eta=1}^{n} h_{\tau\eta}(s)p_\eta(s)\right)\|\mathcal{V}(s+1;x)\|.$$

Next from (13.6.3), we have

$$\|\mathcal{V}(s;x)\| \leq 1 + \mathbf{S}_{\ell=s}^{x-1}\|\mathcal{V}(\ell+1;x)\|\|\mathcal{H}(\ell)\|\|\mathcal{G}(\ell)\|,$$

which is a 1 dimensional inequality. Hence Corollary 13.3.2 gives that

$$\|\mathcal{V}(s;x)\| \;\le\; \min_{1\le k\le m}\left\{\prod_{\ell_k=s_k}^{x_k-1}\left(1+\mathbf{S}_{\bar{\ell}_k=\bar{s}_k}^{\bar{x}_k-1}\|\mathcal{H}(\ell)\|\|\mathcal{G}(\ell)\|\right)\right\}. \tag{13.6.8}$$

Using (13.6.8) in (13.6.7) the resulting inequality (13.6.6) follows. ∎

**Remark 13.6.1.** Let $\|\mathbf{u}\|$ be any vector norm and $\|\mathcal{G}\|$ be the matrix compatible norm, and the conditions of Theorem 13.6.4 are satisfied, then it is easy to get

$$\begin{aligned}\|\mathbf{u}(x)\| \;\le\; & \|\mathbf{p}(x)\| + \|\mathcal{G}(x)\|\mathbf{S}_{s=0}^{x-1}\|\mathcal{H}(s)\|\|\mathbf{p}(s)\| \times \\ & \min_{1\le j\le m}\left\{\prod_{\ell_j=s_j+1}^{x_j-1}\left(1+\mathbf{S}_{\bar{\ell}_j=\bar{s}_j+1}^{\bar{x}_j-1}\|\mathcal{H}(\ell)\|\|\mathcal{G}(\ell)\|\right)\right\}.\end{aligned} \tag{13.6.9}$$

**Theorem 13.6.6.** Let the $n\times n$ matrix $\mathcal{K}(x,s)$ be defined and nonnegative on $\mathbb{N}^m\times\mathbb{N}^m$. Let $n$ vector functions $\mathbf{p}(x)$ and $\mathbf{u}(x)$ be defined and nonnegative on $\mathbb{N}^m$. Further, let for all $x\in\mathbb{N}^m$ the following inequality be satisfied

$$\mathbf{u}(x) \;\le\; \mathbf{p}(x) + \mathbf{S}_{s=0}^{x-1}\mathcal{K}(x,s)\mathbf{u}(s). \tag{13.6.10}$$

Then, for all $x\in\mathbb{N}^m$

$$\mathbf{u}(x) \;\le\; \left[\mathcal{I} + \mathbf{S}_{s=0}^{x-1}\mathcal{V}(s+1;x)\mathcal{K}^*(x,s)\right]\mathbf{p}^*(x), \tag{13.6.11}$$

where $\mathbf{p}^*(x) = \sup\{\mathbf{p}(\ell) : 0\le\ell\le x\}$, $\mathcal{K}^*(x,s) = \sup\{\mathcal{K}(\ell,s) : 0\le\ell\le x\}$, and $\mathcal{V}(s;x)$ satisfies

$$\mathcal{V}(s;x) \;=\; \mathcal{I} + \mathbf{S}_{\ell=s}^{x-1}\mathcal{V}(\ell+1;x)\mathcal{K}^*(x,\ell). \tag{13.6.12}$$

**Proof.** For any fixed point $X$ in $\mathbb{N}^m$, it follows that

$$\mathbf{u}(x) \;\le\; \mathbf{p}^*(X) + \mathbf{S}_{s=0}^{x-1}\mathcal{K}^*(X,s)\mathbf{u}(s), \quad \text{for all} \quad 0\le x\le X.$$

Thus, Theorem 13.6.2 implies that

$$\mathbf{u}(x) \le \left[\mathcal{I} + \mathbf{S}_{s=0}^{x-1}\mathcal{V}(s+1;x)\mathcal{K}^*(X,s)\right]\mathbf{p}^*(X), \text{ for all } 0\le x\le X \tag{13.6.13}$$

where

$$\mathcal{V}(s;x) \;=\; \mathcal{I} + \mathbf{S}_{\ell=s}^{x-1}\mathcal{V}(\ell+1;x)\mathcal{K}^*(X,\ell). \tag{13.6.14}$$

In particular (13.6.13) and (13.6.14) hold for $x = X$. Thus, replacing $X$ by $x$ in the resulting equations (13.6.13) and (13.6.14), we get the desired inequality (13.6.11). ■

## 13.7. Multidimensional Nonlinear Inequalities

In this section we are concerned with comparing the solutions $\mathbf{u}(x)$, $x \in \mathbf{N}^m$ of the nonlinear difference equation

$$\Delta_x^m \mathbf{u}(x) = \mathbf{f}(x, \mathbf{u}(x)) \tag{13.7.1}$$

with solutions $\mathbf{v}(x)$ and $\mathbf{w}(x)$ of the corresponding nonlinear difference inequalities

$$\Delta_x^m \mathbf{v}(x) \leq \mathbf{f}(x, \mathbf{v}(x)) \tag{13.7.2}$$

and

$$\Delta_x^m \mathbf{w}(x) \geq \mathbf{f}(x, \mathbf{w}(x)), \tag{13.7.3}$$

respectively.

In what follows $(i)x$ denotes a point $(x_1, \cdots, x_m)$ in which $i$ variables are zero. There are $\binom{m}{i}$ total such possibilities. Thus, if at the $m$ hyperplanes $x = (1)x$ the function $\mathbf{u}(x)$ is known, then a recursive argument can be used to ensure the existence and uniqueness of the solutions of (13.7.1). This is apparent from the summation representation

$$\mathbf{u}(x) = \sum_{i=1}^{m} (-1)^{i+1} \sum_{i} \mathbf{u}((i)x) + \mathbf{S}_{s=0}^{x-1} \mathbf{f}(s, \mathbf{u}(s)), \tag{13.7.4}$$

where $\sum_i$ represents the summation over all the possibilities $(i)x$. From these notations it is also clear that the solutions $\mathbf{v}(x)$ and $\mathbf{w}(x)$ of the inequalities (13.7.2) and (13.7.3) have the summation representation

$$\mathbf{v}(x) \leq \sum_{i=1}^{m} (-1)^{i+1} \sum_{i} \mathbf{v}((i)x) + \mathbf{S}_{s=0}^{x-1} \mathbf{f}(s, \mathbf{v}(s)) \tag{13.7.5}$$

and

$$\mathbf{w}(x) \geq \sum_{i=1}^{m} (-1)^{i+1} \sum_{i} \mathbf{w}((i)x) + \mathbf{S}_{s=0}^{x-1} \mathbf{f}(s, \mathbf{w}(s)). \tag{13.7.6}$$

**Theorem 13.7.1.** Let $\mathbf{u}(x)$, $\mathbf{v}(x)$ and $\mathbf{w}(x)$ be the solutions of (13.7.1), (13.7.2) and (13.7.3) respectively, and

$$(13.7.7)\quad \sum_{i=1}^{m}(-1)^{i+1}\sum_{i}\mathbf{v}((i)x) \le \sum_{i=1}^{m}(-1)^{i+1}\sum_{i}\mathbf{u}((i)x) \le \sum_{i=1}^{m}(-1)^{i+1}\sum_{i}\mathbf{w}((i)x).$$

Further, let for all fixed $x \in \mathbb{N}^m$, and $1 \le i \le n$ the function $f_i(x, u_1, \cdots, u_n)$ is nondecreasing with respect to all $u_1, \cdots, u_n$.

Then, for all $x \in \mathbb{N}^m$

$$(13.7.8)\qquad \mathbf{v}(x) \le \mathbf{u}(x) \le \mathbf{w}(x).$$

**Proof.** As we have noted $\mathbf{u}(x)$, $\mathbf{v}(x)$ and $\mathbf{w}(x)$ have the representations (13.7.4), (13.7.5) and (13.7.6) respectively. Thus, for all $x = (j)x \in \mathbb{N}^m$, $1 \le j \le m$, (13.7.8) follows from (13.7.7) and the fact that $\mathbf{S}_{s=0}^{j(x)-1}\mathbf{f}(s, \mathbf{u}(s)) = 0$.

If $\mathbf{u}(x) \le \mathbf{w}(x)$ is not true for all $x \in \mathbb{N}^m$, then there is some $1 \le j \le n$ and an $x^*$, $0 < x^* \in \mathbb{N}^m$ such that $u_j(x^*) > w_j(x^*)$ and $\mathbf{u}(x) \le \mathbf{w}(x)$ for all $0 \le x < x^*$. However, since $f_j$ is nondecreasing in $u_1, \cdots, u_n$, from (13.7.6) it follows that

$$\begin{aligned} w_j(x^*) &\ge \sum_{i=1}^{m}(-1)^{i+1}\sum_{i} w_j((i)x^*) + \mathbf{S}_{s=0}^{x^*-1} f_j(s, \mathbf{w}(s)) \\ &\ge \sum_{i=1}^{m}(-1)^{i+1}\sum_{i} u_j((i)x^*) + \mathbf{S}_{s=0}^{x^*-1} f_j(s, \mathbf{u}(s)) = u_j(x^*). \end{aligned}$$

This contradiction completes the proof of $\mathbf{u}(x) \le \mathbf{w}(x)$ for all $x \in \mathbb{N}^m$. The inequality $\mathbf{v}(x) \le \mathbf{u}(x)$ can be proved analogously. ■

**Remark 13.7.1.** It is easy to verify that

$$(13.7.9)\quad \sum_{i=1}^{m}(-1)^{i+1}\sum_{i}\mathbf{u}((i)x) = \sum_{j=0}^{m-1}\left(\sum_{s_1=0}^{x_1-1}\cdots\sum_{s_j=0}^{x_j-1}\Delta_{s_1}\cdots\Delta_{s_j}\mathbf{u}(s_1, \cdots, s_j, 0, x_{j+2}, \cdots, x_m)\right)$$

and hence inequality (13.7.7) certainly holds if for all $0 \le j \le m-1$

$$\begin{aligned} &\Delta_{x_1}\cdots\Delta_{x_j}\mathbf{v}(x_1, \cdots, x_j, 0, x_{j+2}, \cdots, x_m) \\ &\quad \le \Delta_{x_1}\cdots\Delta_{x_j}\mathbf{u}(x_1, \cdots, x_j, 0, x_{j+2}, \cdots, x_m) \\ &\quad \le \Delta_{x_1}\cdots\Delta_{x_j}\mathbf{w}(x_1, \cdots, x_j, 0, x_{j+2}, \cdots, x_m). \end{aligned}$$

**Remark 13.7.2.** If the strict inequality holds in (13.7.7), then strict inequality holds in (13.7.8).

**Theorem 13.7.2.** Assume that the following conditions hold

(i) $\mathbf{u}(x,\mu)$ is the solution of the problem

$$\Delta_x^m \mathbf{u}(x) = \mathbf{f}(x, \mathbf{u}(x), \mu) \tag{13.7.10}$$

$$\mathbf{u}((i)x) = \mathbf{a}([\overline{x}_i], \mu), \tag{13.7.11}$$

where $\mu$ is an $r$ dimensional vector, and $[\overline{x}_i]$ represents the points in $\mathbf{N}^{(m-i)}$ of nonzero variables in $(i)x$

(ii) for all fixed $x$, $0 \le x \le X$, $X \in \mathbf{N}^m$ and $1 \le j \le n$ the function $f_j(x, u_1, \cdots, u_n, \mu_1, \cdots, \mu_r)$ is nondecreasing with respect to $u_1, \cdots, u_n$ and $\mu_1, \cdots, \mu_r$

(iii) for all fixed $[\overline{x}_i]$, $0 \le [\overline{x}_i] \le [\overline{X}_i]$, and $1 \le j \le n$ the function $\sum_{i=1}^m (-1)^{i+1} \sum_i a_j([\overline{x}_i], \mu_1, \cdots, \mu_r)$ is strictly increasing in $\mu_1, \cdots, \mu_r$.

Then, for all $x$, $0 \le x \le X$ the solution $\mathbf{u}(x,\mu)$ of (13.7.10), (13.7.11) is a strictly increasing function of $\mu$, i.e. if $\mu^1 < \mu^2$ then $\mathbf{u}(x,\mu^1) < \mathbf{u}(x,\mu^2)$.

Furthermore, if (a) for all fixed $x, 0 \le x \le X$ the function $\mathbf{f}(x, \mathbf{u}, \mu)$ is continuous with respect to $\mathbf{u}$ and $\mu$, and (b) for all fixed $[\overline{x}_i]$, $0 \le [\overline{x}_i] \le [\overline{X}_i]$ the function $\mathbf{a}([\overline{x}_i], \mu)$ is continuous with respect to $\mu$, then for all $0 \le x \le X$, $\lim_{\mu \to 0} \mathbf{u}(x,\mu) = \mathbf{u}(x)$, where $\mathbf{u}(x)$ is the solution of (13.7.1) satisfying

$$\mathbf{u}((i)x) = \mathbf{a}([\overline{x}_i]). \tag{13.7.12}$$

Moreover, if $X < \infty$, then $\lim_{\mu \to 0} \mathbf{u}(x,\mu) = \mathbf{u}(x)$ is uniform.

**Proof.** Let $\mu^1 < \mu^2$, then since

$$\mathbf{u}(x,\mu^k) = \sum_{i=1}^m (-1)^{i+1} \sum_i \mathbf{a}([\overline{x}_i], \mu^k) + \mathbf{S}_{s=0}^{x-1} \mathbf{f}(s, \mathbf{u}(s,\mu^k), \mu^k), \quad k = 1, 2$$

conditions (ii) and (iii) imply that

$$\mathbf{u}(x,\mu^2) > \sum_{i=1}^m (-1)^{i+1} \sum_i \mathbf{a}([\overline{x}_i], \mu^1) + \mathbf{S}_{s=0}^{x-1} \mathbf{f}(s, \mathbf{u}(s,\mu^2), \mu^1)$$

and now for all $x$, $0 \le x \le X$ the inequality $\mathbf{u}(x,\mu^1) < \mathbf{u}(x,\mu^2)$ follows as in the proof of Theorem 13.7.1.

The rest of the conclusion is a consequence of the continuity assumptions. ∎

**Theorem 13.7.3.** Let for all fixed $x$, $0 \le x \le X$, and $1 \le i \le n$ the function $f_i(x, u_1, \cdots, u_n)$ is nondecreasing with respect to all $u_1, \cdots, u_n$. Let there exist a function $\mathbf{v}(x, \mathbf{u})$ defined for all $0 \le x \le X$, $\mathbf{u} \in \mathbb{R}^n$ which is such that for any function $\mathbf{w}(x)$ defined for all $x$, $0 \le x \le X$

$$\Delta_x^m \mathbf{v}(x, \mathbf{w}(x)) \le \mathbf{f}(x, \mathbf{v}(x, \mathbf{w}(x))). \tag{13.7.13}$$

Further, let the solution $\mathbf{u}(x)$ of (13.7.1) be such that

$$\sum_{i=1}^{m}(-1)^{i+1}\sum_{i}\mathbf{v}((i)x, \mathbf{w}((i)x)) \le \sum_{i=1}^{n}(-1)^{i+1}\sum_{i}\mathbf{u}((i)x). \tag{13.7.14}$$

Then, for all $x$, $0 \le x \le X$

$$\mathbf{v}(x, \mathbf{w}(x)) \le \mathbf{u}(x). \tag{13.7.15}$$

**Proof.** Let $\mathbf{z}(x) = \mathbf{v}(x, \mathbf{w}(x))$, then from (13.7.13) it follows that

$$\Delta_x^m \mathbf{z}(x) = \Delta_x^m \mathbf{v}(x, \mathbf{w}(x)) \le \mathbf{f}(x, \mathbf{v}(x, \mathbf{w}(x)) = \mathbf{f}(x, \mathbf{z}(x)).$$

Also, (13.7.14) is the same as

$$\sum_{i=1}^{m}(-1)^{i+1}\sum_{i}\mathbf{z}((i)x) \le \sum_{i=1}^{m}(-1)^{i+1}\sum_{i}\mathbf{u}((i)x).$$

Thus, for all $x$, $0 \le x \le X$ Theorem 13.7.1 gives that $\mathbf{z}(x) = \mathbf{v}(x, \mathbf{w}(x)) \le \mathbf{u}(x)$. ∎

**Theorem 13.7.4.** Assume that the following conditions hold

(i) for all $x$, $0 \le x \le X$ and $\mathbf{u}$, $\mathbf{v} \in \mathbb{R}^n$

$$|\mathbf{f}(x, \mathbf{u}) - \mathbf{f}(x, \mathbf{v})| \le \mathbf{g}(x, |\mathbf{u} - \mathbf{v}|), \tag{13.7.16}$$

where the function $\mathbf{g}(x, \mathbf{w})$ is defined for all $x$, $0 \le x \le X$, $\mathbf{w} \in \mathbb{R}_+^n$; and for all fixed $x$, $0 \le x \le X$, and $1 \le i \le n$ the function $g_i(x, w_1, \cdots, w_n)$ is nondecreasing with respect to $w_1, \cdots, w_n$

(ii) there exist functions $\mathbf{u}^1(x)$, $\mathbf{u}^2(x)$, $\mathbf{z}^1(x)$ and $\mathbf{z}^2(x)$ which are defined for all $x$, $0 \le x \le X$ and satisfy the inequalities

$$|\Delta_x^m \mathbf{u}^1(x) - \mathbf{f}(x, \mathbf{u}^1(x))| \le \mathbf{z}^1(x) \tag{13.7.17}$$

and

$$|\Delta_x^m \mathbf{u}^2(x) - \mathbf{f}(x, \mathbf{u}^2(x))| \le \mathbf{z}^2(x) \tag{13.7.18}$$

(iii) $\mathbf{u}(x)$ is a solution of the difference equation

$$\Delta_x^m \mathbf{u}(x) = \mathbf{g}(x, \mathbf{u}(x)) + \mathbf{z}^1(x) + \mathbf{z}^2(x), \tag{13.7.19}$$

which satisfies the inequality

$$\left| \sum_{i=1}^{m} (-1)^{i+1} \sum_{i} (\mathbf{u}^1((i)x) - \mathbf{u}^2((i)x)) \right| \le \sum_{i=1}^{m} (-1)^{i+1} \sum_{i} \mathbf{u}((i)x). \tag{13.7.20}$$

Then, for all $x$, $0 \le x \le X$

$$|\mathbf{u}^1(x) - \mathbf{u}^2(x)| \le \mathbf{u}(x). \tag{13.7.21}$$

**Proof.** Inequalities (13.7.17) and (13.7.18) give

$$|\Delta_x^m (\mathbf{u}^1(x) - \mathbf{u}^2(x)) - (\mathbf{f}(x, \mathbf{u}^1(x)) - \mathbf{f}(x, \mathbf{u}^2(x)))| \le \mathbf{z}^1(x) + \mathbf{z}^2(x)$$

and hence, we have

$$\left| \mathbf{S}_{s=0}^{x-1} \Delta_s^m (\mathbf{u}^1(s) - \mathbf{u}^2(s)) - \mathbf{S}_{s=0}^{x-1} (\mathbf{f}(s, \mathbf{u}^1(s)) - \mathbf{f}(s, \mathbf{u}^2(s))) \right| \le \mathbf{S}_{s=0}^{x-1} (\mathbf{z}^1(s) + \mathbf{z}^2(s)),$$

which implies that

$$|\mathbf{u}^1(x) - \mathbf{u}^2(x)| \le \left| \sum_{i=1}^{m} (-1)^{i+1} \sum_{i} (\mathbf{u}^1((i)x) - \mathbf{u}^2((i)x)) \right| + \mathbf{S}_{s=0}^{x-1} |\mathbf{f}(s, \mathbf{u}^1(s)) - \mathbf{f}(s, \mathbf{u}^2(s))| + \mathbf{S}_{s=0}^{x-1} (\mathbf{z}^1(s) + \mathbf{z}^2(s)).$$

Using (13.7.16) and (13.7.20) in the above inequality, to obtain

$$\mathbf{w}(x) \le \sum_{i=1}^{m} (-1)^{i+1} \sum_{i} \mathbf{u}((i)x) + \mathbf{S}_{s=0}^{x-1} (\mathbf{g}(s, \mathbf{w}(s)) + \mathbf{z}^1(s) + \mathbf{z}^2(s)), \tag{13.7.22}$$

where $\mathbf{w}(x) = |\mathbf{u}^1(x) - \mathbf{u}^2(x)|$.

Since $\mathbf{u}(x)$, the solution of (13.7.19) has the summation representation

$$\mathbf{u}(x) = \sum_{i=1}^{m} (-1)^{i+1} \sum_{i} \mathbf{u}((i)x) + \mathbf{S}_{s=0}^{x-1} (\mathbf{g}(s, \mathbf{u}(s)) + \mathbf{z}^1(s) + \mathbf{z}^2(s)) \tag{13.7.23}$$

the inequality $\mathbf{w}(x) \le \mathbf{u}(x)$ follows on comparing (13.7.22) and (13.7.23) as in Theorem 13.7.1. ∎

**Theorem 13.7.5.** Assume that $\mathbf{u}(x,\mu)$ is the solution of (13.7.10), (13.7.12) and the following conditions hold

(i) $\lim_{\mu\to\mu^0} \mathbf{f}(x,\mathbf{u},\mu) = \mathbf{f}(x,\mathbf{u},\mu^0)$ uniformly for all $x$, $0 \le x \le X$ and $\mathbf{u} \in \mathbb{R}^n$

(ii) for all $0 \le x \le X$, $\mathbf{u}^1$, $\mathbf{u}^2 \in \mathbb{R}^n$ and $\mu \in \mathbb{R}^r$

$$|\mathbf{f}(x,\mathbf{u}^1,\mu) - \mathbf{f}(x,\mathbf{u}^2,\mu)| \le \mathbf{g}(x, |\mathbf{u}^1 - \mathbf{u}^2|),$$

where $\mathbf{g}(x,\mathbf{v})$ is defined for all $x$, $0 \le x \le X$, $\mathbf{v} \in \mathbb{R}^n_+$, $\mathbf{g}(x,0) = 0$, and for all fixed $x$ and $1 \le i \le n$ the function $g_i(x, v_1, \cdots, v_n)$ is nondecreasing with respect to $v_1, \cdots, v_n$.

Then, for any given $n$ dimensional vector $\epsilon > 0$ there exists an $r$ dimensional vector $\delta(\epsilon) > 0$ such that for all $x$, $0 \le x \le X < \infty$

$$|\mathbf{u}(x,\mu) - \mathbf{u}(x,\mu^0)| \le \epsilon$$

provided $|\mu - \mu^0| \le \delta(\epsilon)$.

**Proof.** Since $\mathbf{g}(x,0) = 0$ for all $x$, $0 \le x \le X$, the solution $\mathbf{v}(x,0)$ of $\Delta^m_x \mathbf{v}(x) = \mathbf{g}(x,\mathbf{v}(x))$ satisfying $\mathbf{v}((i)x) = 0$ is identically zero. Hence, for any $\epsilon > 0$ there exists an $n$ dimensional vector $\eta = \eta(\epsilon)$ such that the solution $\mathbf{v}(x,0,\eta)$ of the difference system $\Delta^m_x \mathbf{v}(x) = \mathbf{g}(x,\mathbf{v}(x)) + \eta$, satisfying $\mathbf{v}((i)x) = 0$ has the property that $\mathbf{v}(x,0,\eta) \le \epsilon$. Furthermore, because of (i) given $\eta > 0$ there exists a $\delta = \delta(\eta) > 0$ such that $|\mathbf{f}(x,\mathbf{u},\mu) - \mathbf{f}(x,\mathbf{u},\mu^0)| \le \eta$, provided $|\mu - \mu^0| \le \delta(\eta)$.

Now let $\epsilon > 0$ be given, then since

$$\begin{aligned} |\mathbf{u}(x,\mu) - \mathbf{u}(x,\mu^0)| &\le \mathbf{S}^{x-1}_{s=0} |\mathbf{f}(s,\mathbf{u}(s,\mu),\mu) - \mathbf{f}(x,\mathbf{u}(s,\mu^0),\mu^0)| \\ &\le \mathbf{S}^{x-1}_{s=0} [\mathbf{g}(s, |\mathbf{u}(s,\mu) - \mathbf{u}(s,\mu^0)|) + \eta] \end{aligned}$$

as in Theorem 13.7.1 it follows that $|\mathbf{u}(x,\mu) - \mathbf{u}(x,\mu^0)| \le \mathbf{v}(x,0,\eta) \le \epsilon$. Clearly, $\delta$ depends on $\epsilon$ since $\eta$ does. ∎

## 13.8. Convolution Type Inequalities

The Laurent transform ($Z$ transform) introduced in Chapter 2 for functions of one independent variable can be naturally generalized for functions of $m$ independent variables.

**Definition 13.8.1.** We say that a given function $u(x)$, $x \in \mathbb{N}^m$ has the property $(L)$, if there exist $\rho_1 > 0, \cdots, \rho_m > 0$ such that

$$\mathbf{S}^{\infty}_{x=0} |u(x)| \rho_1^{-x_1} \times \cdots \times \rho_m^{-x_m} < \infty.$$

Of course, $\rho_1, \cdots, \rho_m$ depend on the function $u(x)$.

**Definition 13.8.2.** For the function $u(x)$ having the property $(L)$, the Laurent transform is the function $U(z) = U(z_1, \cdots, z_m)$, given by

$$U(z) \; = \; \mathbf{S}_{x=0}^{\infty} \, u(x) z_1^{-x_1} \times \cdots \times z_m^{-x_m}, \quad |z_1| > \rho_1, \cdots, |z_m| > \rho_m.$$

In our first result for all $x \in \mathbf{N}^m$ we shall consider the inequality

$$u(x) \; \le \; p(x) + \mathbf{S}_{s=0}^{x-1} \, f(x-1-s) u(s), \tag{13.8.1}$$

where $p(x)$ and $u(x)$ are not necessarily nonnegative, and $p(x)$ and $f(x)$ have the property $(L)$. For this, it is clear that $u(x) \le v(x)$ for all $x \in \mathbf{N}^m$, where $v(x)$ is the solution of the equation

$$v(x) \; = \; p(x) + \mathbf{S}_{s=0}^{x-1} \, f(x-1-s) v(s). \tag{13.8.2}$$

Let $P(z)$ and $F(z)$ be the Laurent transforms of $p(x)$ and $f(x)$, then from (13.8.2) it follows that

$$V(z) \; = \; P(z) + \frac{1}{(z_1 \times \cdots \times z_m)} F(z) V(z), \tag{13.8.3}$$

where $V(z)$ is the Laurent transform of $v(x)$, defined for $|z_1|, \cdots, |z_m|$ sufficiently large. From (13.8.3), we have

$$\begin{cases} V(z) \; = \; P(z) + \dfrac{1}{(z_1 \times \cdots \times z_m)} R(z) P(z) \\[2ex] R(z) \; = \; \dfrac{(z_1 \times \cdots \times z_m) F(z)}{(z_1 \times \cdots \times z_m) - F(z)}. \end{cases} \tag{13.8.4}$$

The function $R(z)$ may be written as

$$R(z) \; = \; \mathbf{S}_{x=0}^{\infty} \, r(x) z_1^{-x_1} \times \cdots \times z_m^{-x_m}, \tag{13.8.5}$$

where the coefficient $r(x)$ is given by the formula

$$r(x) \; = \; (2\pi i)^{-m} \int \cdots \int R(z_1, \cdots, z_m) z_1^{x_1 - 1} \times \cdots \times z_m^{x_m - 1} \, dz_1 \times \cdots \times dz_m, \tag{13.8.6}$$

the integral being calculated on the product of the circumferences $|z_1| = \delta_1, \cdots, |z_m| = \delta_m$ with $\delta_1, \cdots, \delta_m$ sufficiently large such that this product lies in the domain $|z_1| > \rho_1, \cdots, |z_m| > \rho_m$ in which $R(z)$ is holomorphic.

Now since $F(z) = \mathbf{S}_{x=0}^{\infty} f(x) z_1^{-x_1} \times \cdots \times z_m^{-x_m}$ and $f(x) \geq 0$ for all $x \in \mathbb{N}^m$, from the expansion $R(z) = F(z) \sum_{j=0}^{\infty} \left( \frac{F(z)}{z_1 \times \cdots \times z_m} \right)^j$ it follows that $r(x) \geq 0$. Finally, from (13.8.4) we obtain

$$v(x) = p(x) + \mathbf{S}_{s=0}^{x-1} r(x-1-s) p(s), \tag{13.8.7}$$

and hence we can state

**Theorem 13.8.1.** Let for all $x \in \mathbb{N}^m$ the inequality (13.8.1) be satisfied, where $p(x)$ and $u(x)$ are not necessarily nonnegative, and $p(x)$ and $f(x)$ have the property $(L)$. Then, for all $x \in \mathbb{N}^m$

$$u(x) \leq p(x) + \mathbf{S}_{s=0}^{x-1} r(x-1-s) p(s), \tag{13.8.8}$$

where $r(x)$ is given by (13.8.6). Further, if $p(x)$ is nondecreasing, then

$$u(x) \leq p(x) \left( 1 + \mathbf{S}_{s=0}^{x-1} r(s) \right). \tag{13.8.9}$$

**Corollary 13.8.2.** Let for all $x \in \mathbb{N}^m$ the inequality (13.8.1) be satisfied, where $p(x)$ is nondecreasing for all $x \in \mathbb{N}^m$, and $f(x)$ has the property $(L)$. Then, for all $x \in \mathbb{N}^m$ bound (13.8.9) holds.

**Proof.** In this case inequality (13.8.1) can be written as

$$w(x) \leq 1 + \mathbf{S}_{s=0}^{x-1} f(x-1-s) w(s),$$

where $w(x) = u(x)/(p(x) + \epsilon)$ $(\epsilon > 0)$. Thus, from Theorem 13.8.1 it follows that

$$u(x) \leq (p(x) + \epsilon) \left( 1 + \mathbf{S}_{s=0}^{x-1} r(s) \right).$$

The result (13.8.9) now follows by taking $\epsilon \to 0$ in the above inequality. ∎

**Corollary 13.8.3.** Let the conditions of Theorem 13.8.1 be satisfied. Further, let $p(x)$ be nondecreasing for all $x \in \mathbb{N}^m$, and $F(z)$ is defined and $F(z) - z_1 \times \cdots \times z_m \neq 0$ for $|z_1| > \rho_1, \cdots, |z_m| > \rho_m$ with $\rho_1, \cdots, \rho_m \in (0,1)$. Then, for all $x \in \mathbb{N}^m$

$$u(x) \leq p(x) \left( 1 - \mathbf{S}_{s=0}^{\infty} f(s) \right)^{-1}. \tag{13.8.10}$$

**Proof.** These conditions ensure that the function $R(z)$ defined in (13.8.4) is holomorphic for $|z_1| > \rho_1, \cdots, |z_m| > \rho_m$ and we can put $z_1 = \cdots =$

$z_m = 1$ in (13.8.5) to conclude that the series $\mathbf{S}_{x=0}^{\infty} r(x)$ converges and has the sum $R(1,\cdots,1)$. Now since

$$1+\mathbf{S}_{x=0}^{\infty} r(x) = 1+R(1,\cdots,1) = (1-F(1,\cdots,1))^{-1} = \left(1-\mathbf{S}_{x=0}^{\infty} f(x)\right)^{-1},$$

from (13.8.9) the inequality (13.8.10) follows. ∎

**Remark 13.8.1.** From the hypotheses of Corollary 13.8.3 it is clear that $\mathbf{S}_{x=0}^{\infty} f(x) < 1$. An example of the function $f(x)$ which satisfies the conditions of Corollary 13.8.3 is given by $f(x) = \prod_{i=1}^{m} x_i a_i^{x_i}$, where each $a_i \in \left(0, (3-\sqrt{5})/2\right)$.

**Corollary 13.8.4.** Let the conditions of Corollary 13.8.2 be satisfied. Further, let $F(z)$ be defined and $F(z) - z_1 \times \cdots \times z_m \neq 0$ for $|z_1| > \rho_1, \cdots, |z_m| > \rho_m$ with $\rho_1, \cdots, \rho_m \in (0,1)$. Then, for all $x \in \mathbb{N}^m$ inequality (13.8.10) holds.

## 13.9. Opial and Wirtinger Type Inequalities in Two Variables

**Theorem 13.9.1.** Let $r_1$ and $r_2$ be fixed positive integers and $u(k,\ell)$ be a function defined on $\mathbb{N}\times\mathbb{N}$ such that $u(k,\ell) = 0$ for all $0 \le k \le r_1 - 1$, $\ell \in \mathbb{N}$ and $k \in \mathbb{N}$, $0 \le \ell \le r_2 - 1$. Then, for $0 \le i \le r_1 - 1$, $0 \le j \le r_2 - 1$ and $(k,\ell) \in \mathbb{N}\times\mathbb{N}$

$$\sum_{\tau=1}^{k-r_1+i}\sum_{\eta=1}^{\ell-r_2+j} \left|\Delta_\tau^i \Delta_\eta^j u(\tau+r_1-i-1, \eta+r_2-j-1)\right| \left|\Delta_\tau^{r_1}\Delta_\eta^{r_2} u(\tau,\eta)\right| \tag{13.9.1}$$

$$\le \frac{1}{2\sqrt{2}(r_1-i)!(r_2-j)!}\left(\frac{r_1-i}{2r_1-2i-1}\right)^{1/2}\left(\frac{r_2-j}{2r_2-2j-1}\right)^{1/2} \times (k)^{(r_1-i)}(\ell)^{(r_2-j)} \sum_{\tau=0}^{k-r_1+i}\sum_{\eta=0}^{\ell-r_2+j} \left|\Delta_\tau^{r_1}\Delta_\eta^{r_2} u(\tau,\eta)\right|^2.$$

**Proof.** From Lemma 13.5.1, we have

$$\Delta_k^i \Delta_\ell^j u(k,\ell) = \frac{1}{(r_1-i-1)!(r_2-j-1)!}\sum_{\tau=0}^{k-r_1+i}\sum_{\eta=0}^{\ell-r_2+j}(k-\tau-1)^{(r_1-i-1)} \times (\ell-\eta-1)^{(r_2-j-1)}\Delta_\tau^{r_1}\Delta_\eta^{r_2} u(\tau,\eta).$$

Therefore, by the Schwarz inequality it follows that

$$(13.9.2)\quad |\Delta_k^i \Delta_\ell^j u(k,\ell)| \;\le\; \frac{1}{(r_1-i-1)!(r_2-j-1)!}\times$$

$$\left[\sum_{\tau=0}^{k-r_1+i}\sum_{\eta=0}^{\ell-r_2+j}\left[(k-\tau-1)^{(r_1-i-1)}\right]^2\left[(\ell-\eta-1)^{(r_2-j-1)}\right]^2\right]^{1/2}\times$$

$$\left[\sum_{\tau=0}^{k-r_1+i}\sum_{\eta=0}^{\ell-r_2+j}\left|\Delta_\tau^{r_1}\Delta_\eta^{r_2}u(\tau,\eta)\right|^2\right]^{1/2}.$$

Let

$$\phi(\eta) \;=\; \frac{1}{(\ell-\eta-r_2+j+1)}\left[(\ell-\eta)^{(r_2-j)}\right]^2.$$

Then, since

$$\begin{aligned}
\Delta\phi(\eta) &= \frac{1}{(\ell-\eta-r_2+j)}(\ell-\eta-1)^2\cdots(\ell-\eta-r_2+j)^2\\
&\quad -\frac{1}{(\ell-\eta-r_2+j+1)}(\ell-\eta)^2\cdots(\ell-\eta-r_2+j+1)^2\\
&= \left[(\ell-\eta-1)^{(r_2-j-1)}\right]^2\left[(\ell-\eta-r_2+j)-\frac{(\ell-\eta)^2}{(\ell-\eta-r_2+j+1)}\right]\\
&= -\left[(\ell-\eta-1)^{(r_2-j-1)}\right]^2\left[(2r_2-2j-1)+\frac{(r_2-j-1)^2}{(\ell-\eta-r_2+j+1)}\right],
\end{aligned}$$

we find that

$$\begin{aligned}
\sum_{\eta=0}^{\ell-r_2+j}\left[(\ell-\eta-1)^{(r_2-j-1)}\right]^2 &= \sum_{\eta=0}^{\ell-r_2+j}\frac{-\Delta\phi(\eta)}{(2r_2-2j-1)+\dfrac{(r_2-j-1)^2}{(\ell-\eta-r_2+j+1)}}\\
&\le \frac{1}{(2r_2-2j-1)}\sum_{\eta=0}^{\ell-r_2+j}[-\Delta\phi(\eta)]\\
&= \frac{1}{(2r_2-2j-1)}[\phi(0)-\phi(\ell-r_2+j+1)]\\
&= \frac{1}{(2r_2-2j-1)}\left[(\ell)^{(r_2-j-1)}\right]^2(\ell-r_2+j+1).
\end{aligned}$$

Similarly, we also have

$$\sum_{\tau=0}^{k-r_1+i}\left[(k-\tau-1)^{(r_1-i-1)}\right]^2 \le \frac{1}{(2r_1-2i-1)}\left[(k)^{(r_1-i-1)}\right]^2(k-r_1+i+1).$$

using these estimates in (13.9.2), we get

$$(13.9.3)\quad |\Delta_k^i \Delta_\ell^j u(k,\ell)| \;\le\; \frac{1}{(r_1-i-1)!(r_2-j-1)!}\times$$

$$\frac{1}{(2r_1-2i-1)^{1/2}(2r_2-2j-1)^{1/2}}(k)^{(r_1-i-1)}(k-r_1+i+1)^{1/2}\times$$

$$(\ell)^{(r_2-j-1)}(\ell-r_2+j+1)^{1/2}\left[\sum_{\tau=0}^{k-r_1+i}\sum_{\eta=0}^{\ell-r_2+j}\left|\Delta_\tau^{r_1}\Delta_\eta^{r_2}u(\tau,\eta)\right|^2\right]^{1/2}.$$

From this inequality, and by the Schwarz inequality, we obtain

$$\sum_{\tau=0}^{k-r_1+i}\sum_{\eta=0}^{\ell-r_2+j}\left|\Delta_\tau^i\Delta_\eta^j u(\tau+r_1-i-1,\eta+r_2-j-1)\right|\left|\Delta_\tau^{r_1}\Delta_\eta^{r_2}u(\tau,\eta)\right|$$

$$\le \frac{1}{(r_1-i-1)!(r_2-j-1)!(2r_1-2i-1)^{1/2}(2r_2-2j-1)^{1/2}}\times$$

$$\sum_{\tau=0}^{k-r_1+i}\sum_{\eta=0}^{\ell-r_2+j}(\tau+r_1-i-1)^{(r_1-i-1)}\tau^{1/2}(\eta+r_2-j-1)^{(r_2-j-1)}\eta^{1/2}\times$$

$$\left|\Delta_\tau^{r_1}\Delta_\eta^{r_2}u(\tau,\eta)\right|\left[\sum_{\tau_1=0}^{\tau-1}\sum_{\eta_1=0}^{\eta-1}\left|\Delta_{\tau_1}^{r_1}\Delta_{\eta_1}^{r_2}u(\tau_1,\eta_1)\right|^2\right]^{1/2}$$

$$\le \frac{1}{(r_1-i-1)!(r_2-j-1)!(2r_1-2i-1)^{1/2}(2r_2-2j-1)^{1/2}}\times \tag{13.9.4}$$

$$\left[\sum_{\tau=0}^{k-r_1+i}\sum_{\eta=0}^{\ell-r_2+j}\left[(\tau+r_1-i-1)^{(r_1-i-1)}\right]^2\tau\left[(\eta+r_2-j-1)^{(r_2-j-1)}\right]^2\eta\right]^{1/2}\times$$

$$\left[\sum_{\tau=0}^{k-r_1+i}\sum_{\eta=0}^{\ell-r_2+j}\left|\Delta_\tau^{r_1}\Delta_\eta^{r_2}u(\tau,\eta)\right|^2\sum_{\tau_1=0}^{\tau-1}\sum_{\eta_1=0}^{\eta-1}\left|\Delta_{\tau_1}^{r_1}\Delta_{\eta_1}^{r_2}u(\tau_1,\eta_1)\right|^2\right]^{1/2}.$$

Let $\psi(\eta)=\left[(\eta+r_2-j-1)^{(r_2-j)}\right]^2$. Then, since

$$\begin{aligned}\Delta\psi(\eta) &= (r_2-j)(2\eta+r_2-j)\left[(\eta+r_2-j-1)^{(r_2-j-1)}\right]^2\\ &\ge 2(r_2-j)\eta\left[(\eta+r_2-j-1)^{(r_2-j-1)}\right]^2,\end{aligned}$$

we find that

$$\sum_{\eta=0}^{\ell-r_2+j}\eta\left[(\eta+r_2-j-1)^{(r_2-j-1)}\right]^2$$

$$=\frac{1}{2(r_2-j)}\left[\psi(\ell-r_2+j+1)-\psi(0)\right] = \frac{1}{2(r_2-j)}\left[(\ell)^{(r_2-j)}\right]^2.$$

Similarly, we also have

$$\sum_{\tau=0}^{k-r_1+i} \tau\left[(\tau+r_1-i-1)^{(r_1-i-1)}\right]^2 \;\le\; \frac{1}{2(r_1-i)}\left[(k)^{(r_1-i)}\right]^2.$$

Using these estimates in (13.9.4), we find

$$(13.9.5)\quad \sum_{\tau=0}^{k-r_1+i}\sum_{\eta=0}^{\ell-r_2+j} \left|\Delta_\tau^i\Delta_\eta^j u(\tau+r_1-i-1,\eta+r_2-j-1)\right|\left|\Delta_\tau^{r_1}\Delta_\eta^{r_2}u(\tau,\eta)\right|$$

$$\le \frac{1}{2(r_1-i)!(r_2-j)!}\left(\frac{r_1-i}{2r_1-2i-1}\right)^{1/2}\left(\frac{r_2-j}{2r_2-2j-1}\right)^{1/2}\times$$

$$(k)^{(r_1-i)}(\ell)^{(r_2-j)}\left[\sum_{\tau=0}^{k-r_1+i}\sum_{\eta=0}^{\ell-r_2+j} v(\tau,\eta)\Delta_\tau\Delta_\eta v(\tau,\eta)\right]^{1/2},$$

where

$$v(\tau,\eta) \;=\; \sum_{\tau_1=0}^{\tau-1}\sum_{\eta_1=0}^{\eta-1}\left|\Delta_{\tau_1}^{r_1}\Delta_{\tau_2}^{r_2}u(\tau_1,\eta_1)\right|^2.$$

Now since

$$\begin{aligned} v(\tau,\eta)\Delta_\tau\Delta_\eta v(\tau,\eta) &= \Delta_\tau\Delta_\eta v^2(\tau,\eta) - v(\tau+1,\eta+1)[v(\tau+1,\eta+1) \\ &\quad -v(\tau,\eta)] + v(\tau,\eta+1)[v(\tau,\eta+1)-v(\tau,\eta)] \\ &\quad +v(\tau+1,\eta)[v(\tau+1,\eta)-v(\tau,\eta)] \\ &\le \Delta_\tau\Delta_\eta v^2(\tau,\eta) - v(\tau+1,\eta)[v(\tau+1,\eta+1) \\ &\quad -v(\tau,\eta)] + v(\tau,\eta)[v(\tau,\eta+1)-v(\tau,\eta)] \\ &\quad +v(\tau+1,\eta)[v(\tau+1,\eta)-v(\tau,\eta)] \\ &= \Delta_\tau\Delta_\eta v^2(\tau,\eta) - v(\tau+1,\eta)[v(\tau+1,\eta+1) \\ &\quad -v(\tau+1,\eta)] + v(\tau,\eta)[v(\tau,\eta+1)-v(\tau,\eta)] \\ &\le \Delta_\tau\Delta_\eta v^2(\tau,\eta) - v(\tau,\eta)\Delta_\tau\Delta_\eta v(\tau,\eta) \end{aligned}$$

and hence

$$v(\tau,\eta)\Delta_\tau\Delta_\eta v(\tau,\eta) \;\le\; \frac{1}{2}\Delta_\tau\Delta_\eta v^2(\tau,\eta).$$

Thus, we have

$$\sum_{\tau=0}^{k-r_1+i}\sum_{\eta=0}^{\ell-r_2+j} v(\tau,\eta)\Delta_\tau\Delta_\eta v(\tau,\eta) \;\le\; \frac{1}{2}v^2(\tau,\eta)\Big|_{(0,0)}^{(k-r_1+i+1,\ell-r_2+j+1)}$$

$$= \frac{1}{2}\left[\sum_{\tau_1=0}^{k-r_1+i}\sum_{\eta_1=0}^{\ell-r_2+j}\left|\Delta_{\tau_1}^{r_1}\Delta_{\eta_1}^{r_2}u(\tau_1,\eta_1)\right|\right]^2.$$

Using the above estimate in (13.9.5) the resulting inequality (13.9.1) follows. ∎

**Theorem 13.9.2.** Let the functions $u_i(k,\ell)$, $i = 1,2$ be defined on $\mathbb{N}(0,K)\times\mathbb{N}(0,L)$, and $u_i(k,0) = u_i(0,\ell) = u_i(k,L) = u_i(K,\ell) = 0$ for all $k\in\mathbb{N}(0,K)$ and $\ell\in\mathbb{N}(0,L)$. Then, the following inequality holds

$$(13.9.6)\qquad \sum_{\tau=1}^{K-1}\sum_{\eta=1}^{L-1}|u_1(\tau,\eta)||u_2(\tau,\eta)| \le \sum_{i=1}^{2}\frac{1}{p_i}\left[\frac{K}{2}\right]^{p_i}\left[\frac{L}{2}\right]^{p_i}\sum_{\tau=0}^{K-1}{}'\sum_{\eta=0}^{L-1}{}''|\Delta_\tau\Delta_\eta u_i(\tau,\eta)|^{p_i},$$

where $p_1,\ p_2 > 1$ are such that $\dfrac{1}{p_1}+\dfrac{1}{p_2} = 1$, and $'$ and $''$ delete $\tau = \left[\dfrac{L+1}{2}\right]-1$, and $\eta = \left[\dfrac{K+1}{2}\right]-1$ in their respective summations.

**Proof.** From the assumptions the following identities are obvious

$$(13.9.7)\qquad u_i(k,\ell) = \sum_{\tau=0}^{k-1}\sum_{\eta=0}^{\ell-1}\Delta_\tau\Delta_\eta u_i(\tau,\eta)$$

$$(13.9.8)\qquad u_i(k,\ell) = -\sum_{\tau=0}^{k-1}\sum_{\eta=\ell}^{L-1}\Delta_\tau\Delta_\eta u_i(\tau,\eta)$$

$$(13.9.9)\qquad u_i(k,\ell) = -\sum_{\tau=k}^{K-1}\sum_{\eta=0}^{\ell-1}\Delta_\tau\Delta_\eta u_i(\tau,\eta)$$

$$(13.9.10)\qquad u_i(k,\ell) = \sum_{\tau=k}^{K-1}\sum_{\eta=\ell}^{L-1}\Delta_\tau\Delta_\eta u_i(\tau,\eta).$$

Identity (13.9.7) with an application of Hölder's inequality with indices $\dfrac{p_i}{p_i-1}$ and $p_i$ gives

$$|u_i(k,\ell)| \le (k\ell)^{(p_i-1)/p_i}\left(\sum_{\tau=0}^{k-1}\sum_{\eta=0}^{\ell-1}|\Delta_\tau\Delta_\eta u_i(\tau,\eta)|^{p_i}\right)^{1/p_i}.$$

Therefore, from Young's inequality $ab \le \dfrac{a^{p_1}}{p_1} + \dfrac{b^{p_2}}{p_2}$ it follows that

$$|u_1(k,\ell)||u_2(k,\ell)| \ \le\ \sum_{i=1}^{2} \frac{1}{p_i}(k\ell)^{p_i-1}\left(\sum_{\tau=0}^{k-1}\sum_{\eta=0}^{\ell-1} |\Delta_\tau\Delta_\eta u_i(\tau,\eta)|^{p_i}\right).$$

Hence, we find that

$$\sum_{\tau=1}^{k-1}\sum_{\eta=1}^{\ell-1} |u_1(\tau,\eta)||u_2(\tau,\eta)|$$

$$\le\ \sum_{i=1}^{2}\frac{1}{p_i}\sum_{\tau=1}^{k-1}\sum_{\eta=1}^{\ell-1}(\tau\eta)^{p_i-1}\left(\sum_{\tau_1=0}^{\tau-1}\sum_{\eta_1=0}^{\eta-1}|\Delta_{\tau_1}\Delta_{\eta_1}u_i(\tau_1,\eta_1)|^{p_i}\right)$$

$$=\ \sum_{i=1}^{2}\frac{1}{p_i}\sum_{\tau=1}^{k-1}\tau^{p_i-1}\sum_{\tau_1=0}^{\tau-1}\left\{\sum_{\eta=0}^{\ell-2}\left(\sum_{\eta_1=\eta+1}^{\ell-1}\eta_1^{p_i-1}\right)|\Delta_{\tau_1}\Delta_\eta u_i(\tau_1,\eta)|^{p_i}\right\}$$

$$\le\ \sum_{i=1}^{2}\frac{1}{p_i}(\ell-1)^{p_i}\sum_{\eta=0}^{\ell-2}\sum_{\tau=0}^{k-2}\left(\sum_{\tau_1=\tau+1}^{k-1}\tau_1^{p_i-1}\right)|\Delta_\tau\Delta_\eta u_i(\tau,\eta)|^{p_i}$$

$$\le\ \sum_{i=1}^{2}\frac{1}{p_i}(k-1)^{p_i}(\ell-1)^{p_i}\sum_{\tau=0}^{k-2}\sum_{\eta=0}^{\ell-2}|\Delta_\tau\Delta_\eta u_i(\tau,\eta)|^{p_i}. \tag{13.9.11}$$

Similarly, from (13.9.8) it follows that

$$\sum_{\tau=1}^{k-1}\sum_{\eta=\ell}^{L-1} |u_1(\tau,\eta)||u_2(\tau,\eta)|$$

$$\le\ \sum_{i=1}^{2}\frac{1}{p_i}\sum_{\tau=1}^{k-1}\sum_{\eta=\ell}^{L-1}(\tau(L-\eta))^{p_i-1}\left(\sum_{\tau_1=0}^{\tau-1}\sum_{\eta_1=\eta}^{L-1}|\Delta_{\tau_1}\Delta_{\eta_1}u_i(\tau_1,\eta_1)|^{p_i}\right)$$

$$\le\ \sum_{i=1}^{2}\frac{1}{p_i}(k-1)^{p_i}\sum_{\tau=0}^{k-2}\sum_{\eta=\ell}^{L-1}\left(\sum_{\eta_1=\ell}^{\eta}(L-\eta_1)^{p_i-1}\right)|\Delta_\tau\Delta_\eta u_i(\tau,\eta)|^{p_i}$$

$$\le\ \sum_{i=1}^{2}\frac{1}{p_i}(k-1)^{p_i}(L-\ell)^{p_i}\sum_{\tau=0}^{k-2}\sum_{\eta=\ell}^{L-1}|\Delta_\tau\Delta_\eta u_i(\tau,\eta)|^{p_i}. \tag{13.9.12}$$

Let $\ell = \left[\dfrac{L+1}{2}\right]$ in (13.9.11) and (13.9.12), and summing the resulting inequalities, we find

$$\sum_{\tau=1}^{k-1}\sum_{\eta=1}^{L-1}|u_1(\tau,\eta)||u_2(\tau,\eta)| \tag{13.9.13}$$

$$\leq \sum_{i=1}^{2} \frac{1}{p_i}(k-1)^{p_i}\left[\frac{L}{2}\right]^{p_i} \sum_{\tau=0}^{k-2}\sum_{\eta=0}^{L-1}{}'' |\Delta_\tau \Delta_\eta u_i(\tau,\eta)|^{p_i}.$$

Following as above for the equalities (13.9.9) and (13.9.10) to obtain

$$\sum_{\tau=k}^{K-1}\sum_{\eta=1}^{L-1} |u_1(\tau,\eta)||u_2(\tau,\eta)| \tag{13.9.14}$$

$$\leq \sum_{i=1}^{2} \frac{1}{p_i}(K-k)^{p_i}\left[\frac{L}{2}\right]^{p_i} \sum_{\tau=k}^{K-1}\sum_{\eta=0}^{L-1}{}'' |\Delta_\tau \Delta_\eta u_i(\tau,\eta)|^{p_i}.$$

Let $k = \left[\frac{K+1}{2}\right]$ in (13.9.13) and (13.9.14), and summing the resulting inequalities, we get the required inequality (13.9.6). ∎

**Corollary 13.9.3.** Let in Theorem 13.9.2, $u_1(k,\ell) = u_2(k,\ell)$ and $p_1 = p_2 = 2$. Then, the following inequality holds

$$\sum_{\tau=1}^{K-1}\sum_{\eta=1}^{L-1} |u_1(\tau,\eta)|^2 \leq \left[\frac{K}{2}\right]^2\left[\frac{L}{2}\right]^2 \sum_{\tau=0}^{K-1}{}'\sum_{\eta=0}^{L-1}{}'' |\Delta_\tau \Delta_\eta u_1(\tau,\eta)|^2. \tag{13.9.15}$$

**Theorem 13.9.4.** Let $u_i(k,\ell)$, $i = 1,2$ be as in Theorem 13.9.2. Then, the following inequality holds

$$\sum_{\tau=1}^{K-1}\sum_{\eta=1}^{L-1} |u_1(\tau,\eta)|^{p_1}|u_2(\tau,\eta)|^{p_2} \tag{13.9.16}$$

$$\leq \frac{1}{p_1+p_2}\left[\frac{K}{2}\right]^{p_1+p_2}\left[\frac{L}{2}\right]^{p_1+p_2} \sum_{i=1}^{2} p_i \sum_{\tau=0}^{K-1}{}'\sum_{\eta=0}^{L-1}{}'' |\Delta_\tau \Delta_\eta u_i(\tau,\eta)|^{p_1+p_2},$$

where $p_1,\ p_2 \geq 1$.

**Proof.** The proof is similar to that of Theorem 13.9.2 except we use Hölder's inequality with indices $\dfrac{p_1+p_2}{p_1+p_2-1}$ and $p_1+p_2$, and the inequality $pa^{p+q}+qb^{p+q}-(p+q)a^pb^q \geq 0$, where $a,b \geq 0$ and $p,q > 0$. ∎

**Theorem 13.9.5.** Let $u_i(k,\ell)$, $i = 1,2$ be as in Theorem 13.9.2. Then, the following inequality holds

$$\sum_{\tau=1}^{K-1}\sum_{\eta=1}^{L-1} |u_1(\tau,\eta)||u_2(\tau,\eta)| \leq \frac{1}{2}\sum_{i=1}^{2}\frac{1}{p_i}\max\left\{\left[\frac{K}{2}\right]^{p_i}, \left[\frac{L}{2}\right]^{p_i}\right\} \tag{13.9.17}$$

$$\times \sum_{\tau=0}^{K-1}\sum_{\eta=0}^{L-1}\left(|\Delta_\tau u_i(\tau,\eta)|^{p_i}+|\Delta_\eta u_i(\tau,\eta)|^{p_i}\right),$$

where $p_1$, $p_2 > 1$ are such that $\dfrac{1}{p_1}+\dfrac{1}{p_2}=1$.

**Proof.** From the given conditions the following identities hold

$$(13.9.18)\quad \begin{aligned} u_i(k,\ell) &= \sum_{\tau=0}^{k-1}\Delta_\tau u_i(\tau,\ell), & u_i(k,\ell) &= -\sum_{\tau=k}^{K-1}\Delta_\tau u_i(\tau,\ell)\\ u_i(k,\ell) &= \sum_{\eta=0}^{\ell-1}\Delta_\eta u_i(k,\eta), & u_i(k,\ell) &= -\sum_{\eta=\ell}^{L-1}\Delta_\eta u_i(k,\eta). \end{aligned}$$

Thus, as in Theorem 13.9.2 it follows that

$$\sum_{\tau=1}^{K-1}|u_1(\tau,\ell)||u_2(\tau,\ell)| \le \sum_{i=1}^{2}\frac{1}{p_i}\left[\frac{K}{2}\right]^{p_i}\sum_{\tau=0}^{K-1}{}'|\Delta_\tau u_i(\tau,\ell)|^{p_i},$$

which also gives

$$(13.9.19)\quad \sum_{\tau=1}^{K-1}\sum_{\eta=1}^{L-1}|u_1(\tau,\eta)||u_2(\tau,\eta)| \le \sum_{i=1}^{2}\frac{1}{p_i}\left[\frac{K}{2}\right]^{p_i}\sum_{\tau=0}^{K-1}\sum_{\eta=0}^{L-1}|\Delta_\tau u_i(\tau,\eta)|^{p_i},$$

and similarly, we find

$$(13.9.20)\quad \sum_{\tau=1}^{K-1}\sum_{\eta=1}^{L-1}|u_1(\tau,\eta)||u_2(\tau,\eta)| \le \sum_{i=1}^{2}\frac{1}{p_i}\left[\frac{L}{2}\right]^{p_i}\sum_{\tau=0}^{K-1}\sum_{\eta=0}^{L-1}|\Delta_\eta u_i(\tau,\eta)|^{p_i}.$$

Now addition of (13.9.19) and (13.9.20) gives the required inequality (13.9.17). ∎

**Theorem 13.9.6.** Let $u_i(k,\ell)$, $i=1,2$ be as in Theorem 13.9.2. Then, the following inequality holds

$$(13.9.21)\quad \sum_{\tau=1}^{K-1}\sum_{\eta=1}^{L-1}|u_1(\tau,\eta)|^{p_1}|u_2(\tau,\eta)|^{p_2} \le \max\left\{\left[\frac{K}{2}\right]^{p_1+p_2},\left[\frac{L}{2}\right]^{p_1+p_2}\right\}$$

$$\times\frac{1}{2(p_1+p_2)}\sum_{i=1}^{2}p_i\sum_{\tau=0}^{K-1}\sum_{\eta=0}^{L-1}\left(|\Delta_\tau u_i(\tau,\eta)|^{p_1+p_2}+|\Delta_\eta u_i(\tau,\eta)|^{p_1+p_2}\right),$$

where $p_1$, $p_2 \ge 1$.

**Proof.** The proof is similar to that of Theorems 13.9.5 and 13.9.4. ∎

**Corollary 13.9.7.** Let in Theorem 13.9.6, $u_1(k,\ell) = u_2(k,\ell)$ and $p_1 = p_2$. Then, the following inequality holds

$$\sum_{\tau=1}^{K-1}\sum_{\eta=1}^{L-1} |u_1(\tau,\eta)|^{2p_1} \le \frac{1}{2}\max\left\{\left[\frac{K}{2}\right]^{2p_1}, \left[\frac{L}{2}\right]^{2p_1}\right\} \times \sum_{\tau=0}^{K-1}\sum_{\eta=0}^{L-1} \left(|\Delta_\tau u_1(\tau,\eta)|^{2p_1} + |\Delta_\eta u_1(\tau,\eta)|^{2p_1}\right). \tag{13.9.22}$$

## 13.10. Problems

**13.10.1.** Let for all $x \in \mathbb{N}^m$ the inequality (13.2.1) be satisfied, and $p(x)$ and $q(x)$ be nondecreasing for all $x \in \mathbb{N}^m$. Show that for all $x \in \mathbb{N}^m$

$$u(x) \le p(x) \min_{1\le j\le m} \left\{ \prod_{\ell_j=0}^{x_j-1} \left(1 + q(x)\mathbf{S}_{\overline{\ell}_j=0}^{\overline{x}_j-1} f(\ell)\right)\right\}.$$

**13.10.2.** Prove Theorem 13.2.2.

**13.10.3.** Prove Theorem 13.2.3.

**13.10.4.** Prove Theorem 13.2.4.

**13.10.5.** Let the function $f(k,\ell,u,v)$ be defined for all $k,\ell \in \mathbb{N}$ and $u,v \in \mathbb{R}$, and nondecreasing in $u,v$. Further, let the functions $\phi(k,\ell)$ and $\psi(k,\ell)$ be defined for all $k,\ell \in \mathbb{N}$, and satisfy the inequalities

$$\begin{aligned}
\phi(k+1,\ell) &\le f(k,\ell,\phi(k,\ell),\phi(\ell,k))\\
\psi(k+1,\ell) &\ge f(k,\ell,\psi(k,\ell),\psi(\ell,k))\\
\phi(0,\ell) &\le \psi(0,\ell).
\end{aligned}$$

Show that for all $k,\ell \in \mathbb{N}$, $\phi(k,\ell) \le \psi(k,\ell)$.

**13.10.6.** Let for all $x$, $X-1 \in \mathbb{N}^m$ such that $x \le X-1$ the following inequality be satisfied

$$u(X) \ge u(x) - q(X)\mathbf{S}_{\ell=x}^{X-1} f(\ell)u(\ell).$$

Show that for all $x$, $X-1 \in \mathbb{N}^m$, $x \le X-1$

$$u(X) \ge u(x) \prod_{\ell_1=x_1}^{X_1-1} \left(1 - q(X)\mathbf{S}_{\overline{\ell}_1=\overline{x}_1}^{\overline{X}_1-1} f(\ell)\right)$$

as long as $1 - q(X)\mathbf{S}_{\bar{\ell}_1=\bar{x}_1}^{\bar{X}_1-1} f(\ell) > 0$.

**13.10.7.** Prove Theorem 13.4.2.

**13.10.8.** Prove Theorem 13.4.3.

**13.10.9.** Find results analogous to Theorems 4.2.5 and 4.2.6, for functions of $m$ independent variables.

**13.10.10.** Let for all $x, X \in \mathbb{N}^m$ such that $x \le X$ the following inequality be satisfied

$$u(X) \ \ge\ u(x) - q(X)W^{-1}\left(\mathbf{S}_{\ell=x+1}^{X} f(\ell)W(u(\ell))\right),$$

where the function $W$ is positive, increasing, convex and submultiplicative on $(0,\infty)$ and $\lim_{u\to\infty} W(u) = \infty$. Show that for all $x$, $X \in \mathbb{N}^m$, $x \le X$

$$u(X) \ \ge\ \alpha(X)W^{-1}\left[\alpha^{-1}(X)W(u(x)) \prod_{\ell_1=x_1+1}^{X_1} \left[1 + \beta(X)W(q(X)\beta^{-1}(X)) \times \ \mathbf{S}_{\bar{\ell}_1=\bar{x}_1+1}^{\bar{X}_1} f(\ell)\right]^{-1}\right],$$

where the functions $\alpha(x)$ and $\beta(x)$ are positive and $\alpha(x) + \beta(x) = 1$ for all $x \in \mathbb{N}^m$.

**13.10.11.** Let for all $x$, $X \in \mathbb{N}^m$ such that $x \le X$ the following inequality be satisfied

$$u(X) \ \ge\ u(x) - q(X)\mathbf{S}_{\ell=x+1}^{X}\ f(\ell)W(u(\ell)),$$

where the function $W$ is continuous, positive and nondecreasing on $[0,\infty)$. Show that for all $x$, $X \in \mathbb{N}^m$, $x \le X$

$$u(X) \ \ge\ G^{-1}\left[G(u(x)) - q(X)\mathbf{S}_{\ell=x+1}^{X} f(\ell)\right],$$

where

$$G(w) \ =\ \int_{w_0}^{w} \frac{dt}{W(t)}, \quad w > 0 \quad \text{and arbitrary} \quad w_0 \ge 0$$

as long as $G(u(x)) - q(X)\mathbf{S}_{\ell=x+1}^{X} f(\ell) \in Dom\,(G^{-1})$.

**13.10.12.** Let for all $(k,\ell) \in \mathbb{N} \times \mathbb{N}$ the following inequality be satisfied

$$\Delta_k^r \Delta_\ell^r u(k,\ell) \ \le\ p(k) + q(\ell) + \sum_{i=0}^{r} \sum_{\tau=0}^{k-1} \sum_{\eta=0}^{\ell-1} h(\tau,\eta)\Delta_\tau^i \Delta_\eta^i u(\tau,\eta),$$

where $\Delta_k p(k) \geq 0$, $\Delta_\ell q(\ell) \geq 0$, $p(0) = q(0)$. Show that for all $(k, \ell) \in \mathbb{N} \times \mathbb{N}$

$$\Delta_k^r \Delta_\ell^r u(k, \ell) \leq B_i(k, \ell), \quad 1 \leq i \leq r+1$$

where

$$B_1(k, \ell) = [p(k) + q(\ell)] \prod_{\tau=0}^{k-1} \left(1 + \sum_{\eta=0}^{\ell-1} [h(\tau, \eta) + r]\right)$$

and

$$B_i(k, \ell) = [p(k) + q(\ell)] + \sum_{\tau=0}^{k-1} \sum_{\eta=0}^{\ell-1} (h(\tau, \eta) + (r - i + 1)) B_{i-1}(\tau, \eta),$$

$$2 \leq i \leq r+1.$$

**13.10.13.** Prove Theorem 13.6.2.

**13.10.14.** Prove Theorem 13.6.3.

**13.10.15.** Prove Theorem 13.6.4.

**13.10.16.** Assume that $\mathbf{u}(x, \mathbf{T}(\mathbf{a}))$ is the solution of (13.7.1), (13.7.12) where $\mathbf{T}(\mathbf{a})$ denotes the term $\sum_{i=1}^{m} (-1)^{i+1} \sum_i \mathbf{a}([\overline{x}_i])$. Further, let $\mathbf{v}(x, 0)$ be the solution of the problem $\Delta_x^m \mathbf{v}(x) = \mathbf{F}(x, \mathbf{v}(x))$, $\mathbf{v}((i)x) = 0$, where the function $\mathbf{F}(x, \mathbf{v})$ for $0 \leq x \leq X$, $\mathbf{v} \in \mathbb{R}_+^n$ is defined as $\mathbf{F}(x, \mathbf{v}) = \sup_{|\mathbf{u} - \mathbf{T}(\mathbf{a})| \leq \mathbf{v}} |\mathbf{f}(x, \mathbf{u})|$. Show that for all $x$, $0 \leq x \leq X$, $|\mathbf{u}(x, \mathbf{T}(\mathbf{a})) - \mathbf{T}(\mathbf{a})| \leq \mathbf{v}(x, 0)$.

**13.10.17.** Assume that condition (i) of Theorem 13.7.4 is satisfied, and $\mathbf{u}(x, \mathbf{T}(\mathbf{a}))$ is as in Problem 13.10.16. Further, assume that $\mathbf{u}(x, \mathbf{T}(\mathbf{b}))$ is the solution of (13.7.1) satisfying $\mathbf{u}((i)x) = \mathbf{b}([\overline{x}_i])$. Show that for all $x$, $0 \leq x \leq X$, $|\mathbf{u}(x, \mathbf{T}(\mathbf{a})) - \mathbf{u}(x, \mathbf{T}(\mathbf{b}))| \leq \mathbf{v}(x)$, where $\mathbf{v}(x)$ is a solution of $\Delta_x^m \mathbf{v}(x) = \mathbf{g}(x, \mathbf{v}(x))$, satisfying $|\mathbf{T}(\mathbf{a}) - \mathbf{T}(\mathbf{b})| \leq \mathbf{T}(\mathbf{v})$.

**13.10.18.** Assume that for all $x$, $0 \leq x \leq X$, $\mathbf{u} \in \mathbb{R}^n$, $|\mathbf{f}(x, \mathbf{u})| \leq \mathbf{g}(x, |\mathbf{u}|)$ where the function $\mathbf{g}(x, \mathbf{v})$ is defined for all $x$, $0 \leq x \leq X$, $\mathbf{v} \in \mathbb{R}_+^n$; and for all fixed $x$, $0 \leq x \leq X$, and $1 \leq i \leq n$ the function $g_i(x, v_1, \cdots, v_n)$ is nondecreasing with respect to $v_1, \cdots, v_n$. Further, let $\mathbf{u}(x)$ be any solution of (13.7.1), and $\mathbf{v}(x)$ be a solution of $\Delta_x^m \mathbf{v}(x) = \mathbf{g}(x, \mathbf{v}(x))$ such that $|\mathbf{T}(\mathbf{u})| \leq \mathbf{T}(\mathbf{v})$. Show that (i) if $\mathbf{v}(x)$ is bounded, so is $\mathbf{u}(x)$, and (ii) if $\mathbf{v}(x) \to 0$ as $\|x\| = (x_1^2 + \cdots + x_m^2)^{1/2} \to \infty$, so is $\mathbf{u}(x)$.

**13.10.19.** Let for all $x \in \mathbb{N}^m$ the following inequality be satisfied

$$u(x) \leq p(x) + q(x) \mathbf{S}_{s=0}^{x-1} g(s) u(s) + \mathbf{S}_{s=0}^{x-1} f(x - s - 1) u(s),$$

volving higher order partial differences, *Nonlinear Analysis* **27**(1996), 429 454.

[16]. R.P. Agarwal and Q. Sheng, Sharp integral inequalities in $n$ independent variables, *Nonlinear Analysis* **26**(1996), 179 210.

[17]. R.P. Agarwal and P.Y.H. Pang, Opial-type inequalities involving higher order partial derivatives of two functions, In *General Inequalities 7*, ed. C. Bandle, ISNM 123, *Birkhäuser Verlag*, Basel (1997), 157 178.

[18]. R.P. Agarwal, J. Pečarić and I. Brnetic, Improved discrete inequalities in $n$ independent variables, *Applied Math. Letters* **11**(2)(1998), 91 97.

[19]. D. Bainov and P. Simeonov, *Integral Inequalities and Applications*, *Kluwer*, Dordrecht, 1992.

[20]. P.R. Beesack, Lower bounds for discrete inequalities of Gollwitzer Langenhop type, *An. st. Univ. Iasi* **30**(1984), 25 30.

[21]. A. Corduneanu, A discrete integral inequality of convolution type in two independent variables, *An. st. Univ. Iasi* **32**(1986), 51 56.

[22]. E. Magnucka Blandi, J. Popenda and R.P. Agarwal, Best possible Gronwall inequalities, *Mathl. Comput. Modelling* **26**(3)(1997), 1 8.

[23]. D.S. Mitrinović, J. Pečarić and A.M. Fink, *Inequalities Involving Functions and their Integrals and Derivatives*, *Kluwer*, Dordrecht, 1991.

[24]. B.G. Pachpatte and S.M. Singare, Discrete generalised Gronwall inequalities in three independent variables, *Pacific J. Math.* **82**(1979), 197 210.

[25]. B.G. Pachpatte, On some fundamental discrete inequalities in two independent variables, *Tamk. Jour. Math.* **12**(1981), 21 33.

[26]. B.G. Pachpatte, On certain discrete inequalities in two independent variables, *Soochow Jour. Math.* **11**(1985), 37 41.

[27]. B.G. Pachpatte, On some new multidimensional discrete inequalities, *Tamk. Jour. Math.* **17**(1986), 21 29.

[28]. B.G. Pachpatte, On certain multidimensional discrete inequalities, *Chinese J. Math.* **14**(1986), 185-195.

[29]. P.Y.H. Pang and R.P. Agarwal, On an integral inequality and its discrete analogue, *J. Math. Anal. Appl.* **194**(1995), 569 577.

[30]. P.Y.H. Pang and R.P. Agarwal, Discrete polar coordinates and a new discrete inequality, *Dynamics of Continuous, Discrete and Impulsive Systems* **2**(1996), 181 191.

[31]. J. Popenda, On the discrete analogy of Gronwall Wendroff inequality, *Demonstratio Mathematica* **18**(1985), 1083 1103.

[32]. J. Popenda, On the discrete inequalities of Gronwall Bellman type, *An. st. Univ. Iasi* **33**(1987), 47 52.

[33]. J. Popenda and R.P. Agarwal, On discrete Gronwall inequalities in many variables, *Computers Math. Applic.* **38**(1)(1999), 63 70.

[34]. S.M. Singare and B.G. Pachpatte, Wendroff type discrete inequalities and their applications, *Jour. Math. Phy. Sci.* **13**(1979), 149 167.

[35]. S.M. Singare and B.G. Pachpatte, On some fundamental discrete inequalities of the Wendroff type, *An. st. Univ. Iasi* **26**(1980), 85 94.

[36]. S.M. Singare and B.G. Pachpatte, On certain discrete inequalities of the Wendroff type, *Indian J. Pure Appl. Math.* **11**(1980), 727 736.

[37]. E. Thandapani and R.P. Agarwal, Some new discrete inequalities in two independent variables, *An. st. Univ. Iasi* **27**(1981), 269 278.

[38]. E. Thandapani and R.P. Agarwal, On some new inequalities in $n$ independent variables, *J. Math. Anal. Appl.* **86**(1982), 542 561.

[39]. E. Thandapani, On some new discrete inequalities in two independent variables involving higher order differences, *Jour. Math. Phy. Sci.* **21**(1987), 377 389.

[40]. E. Thandapani, Discrete inequalities in $n$ independent variables of Gronwall Bellman type, *Applicable Analysis* **30**(1988), 189 199.

[41]. E.H. Yang, On some new discrete generalizations of Gronwall's inequality, *J. Math. Anal. Appl.* **129**(1988), 505 516.

[42]. C.C. Yeh, Discrete inequalities of the Gronwall Bellman type in $n$ independent variables, *J. Math. Anal. Appl.* **105**(1985), 322 332.

[43]. C.C. Yeh, Discrete inequalities of the Gronwall Bellman type in $n$ independent variables, II, *J. Math. Anal. Appl.* **106**(1985), 282 285.

# Author Index

# Subject Index